This book studies existence and necessary conditions, such as Pontryagin's maximum principle for optimal control problems described by ordinary and partial differential equations. These necessary conditions are obtained from Kuhn–Tucker theorems for nonlinear programming problems in infinite dimensional spaces.

The optimal control problems include control constraints, state constraints, and target conditions. Evolution partial differential equations are studied using semigroup theory, abstract differential equations in linear spaces, integral equations, and interpolation theory. Existence of optimal controls is established for arbitrary control sets by means of a general theory of relaxed controls.

Applications include nonlinear systems described by partial differential equations of hyperbolic and parabolic type; the latter case deals with pointwise constraints on the solution and the gradient. The book also includes results on convergence of suboptimal controls.

H. O. Fattorini is Professor of Mathematics at the University of California, Los Angeles.

ENCYCLOPEDIA OF MATHEMATICS AND ITS APPLICATIONS

EDITED BY G.-C. ROTA

Volume 62

Infinite Dimensional Optimization and Control Theory

ENCYCLOPEDIA OF MATHEMATICS AND ITS APPLICATIONS

Infinite Dimensional Optimization and Control Theory

H. O. FATTORINI

CAMBRIDGE
UNIVERSITY PRESS

PUBLISHED BY THE PRESS SYNDICATE OF THE UNIVERSITY OF CAMBRIDGE
The Pitt Building, Trumpington Street, Cambridge, United Kingdom

CAMBRIDGE UNIVERSITY PRESS
The Edinburgh Building, Cambridge CB2 2RU, UK   http://www.cup.cam.ac.uk
40 West 20th Street, New York, NY 10011-4211, USA   http://www.cup.org
10 Stamford Road, Oakleigh, Melbourne 3166, Australia

First published 1999

Printed in the United States of America

Typeset in 10/12 Times Roman in LaTeX 2_ε [TB]

A catalog record for this book is available from the British Library

Library of Congress Cataloging-in-Publication Data
Fattorini, H. O. (Hector O.), 1938–
Infinite dimensional optimization and control theory / H. O. Fattorini.
p.   cm. – (Cambridge studies in advanced mathematics: 54)
ISBN (invalid) 0-521-42125-6 (ho)
1. Mathematical optimization.   2. Calculus of variations.
3. Control theory.   I. Title.   II. Series.
QA402.5.F365 1996
003'.5 – dc20   95-47377
                                                                    CIP

ISBN 0 521 45125 6 hardback

\# 33439458

To Natalia

CONTENTS

FOREWORD

An initial value or initial-boundary value problem for an evolution partial differential equation (an equation whose solutions depend on time) can usually be written as an abstract differential equation

$$y'(t) = f(t, y(t)) \tag{1}$$

in a suitable function space, the function f describing the action of the equation on the space variables, with boundary conditions (if any) included in the definition of the space or of the domain of f. The similarity of (1) with a true ordinary differential equation is only formal (f may not be everywhere defined, bounded or continuous) but gives heuristic insight into the problem, suggests ways to extend results from ordinary to partial differential equations and stresses unification, discovery of common threads and economy of thought. The "abstract" approach is not the best in all situations (for instance, many controllability results depend on properties of partial differential equations lost, or difficult to reformulate, in the translation to (1)), but it applies very well to optimization problems, where it is expedient to obtain general statements such as Pontryagin's maximum principle and then elucidate what the principle says for equations in various classes. Many of the techniques are (modulo some fine tuning) oblivious to the type of equation and are at least formally similar to classical procedures for systems of ordinary differential equations.

This work is on the Pontryagin's maximum principle for equations of the form (1), on its applications to diverse control systems described by partial differential equations, including control and state constraints and target conditions, and on other related questions such as existence and relaxation of controls. It is understood for use by nonspecialists, and with this in view incorporates blocks of auxiliary material (Sections **2.0**, **5.0** and **12.0** and portions of other chapters).

Those familiar with Pontryagin's maximum principle know that the key to its meaning lies not so much in its proof but in understanding what it says or does not say as applied to a particular optimal control problem, and finite dimensional systems are perhaps the best area where one can gain insight without much overhead. This motivates Part I (the first four chapters), which give a cursory introduction to some control problems for ordinary differential equations and a large number of examples, all classical in the literature. Part of the material (such as the nonlinear programming theory in Chapters 2 and 3) is also used later.

Part II (Chapters 5 to 11) is on infinite dimensional control systems, with Chapter 9 on linear systems. Linear problems are amenable to separation techniques more elementary than nonlinear programming, and there are linear theorems that do not yet have full nonlinear counterparts.

Part III (Chapters 12 to 14) is on relaxed controls; these appear when one tries to insure existence of optimal controls, something not always attainable with the "original" controls with which the problem is outfitted.

There are many obvious shortcuts through this book. To mention one, the fastest way to the infinite dimensional maximum principle begins with **7.1** and **7.2** on the general nonlinear programming problem in Banach spaces and then proceeds to the maximum principle with state constraints in Chapter 10, with assistance of various sections in Chapters 5 and 6; for the parabolic problems in Chapter 11, some of the material in Chapters 7 and 8 is needed.

All through, "Examples" are results either informally proved or left to the reader as exercises.

The references have no pretension of completeness. They only include works that deal with control problems through the abstract evolution equation (1) *and* are directly related with the results in this book, in particular with the maximum principle. When appropriate, we include papers that arrive at similar results by other methods.

We have also attempted to include a modicum of references to subjects not treated in any detail in this work (for instance, controllability, stabilization and the Hamilton-Jacobi approach to optimality); here, the words "...and other papers" invite the reader to perform further search. When possible, we have deferred to the extensive references in several recent books in control theory.

Acknowledgments

In 1986 I was invited by the Ministry of Education and Justice of Argentina to participate in the 1st. National School of Applied Mathematics, held that year in Potrero de Los Funes. I taught a course on finite dimensional control theory to an audience of graduate students from various Argentine universities. In 1987, a second National School of Applied Mathematics was held in Santa Fe de la Vera Cruz. The audience consisted of some of my 1986 students and some new ones, and the lectures were on infinite dimensional control theory, corresponding to various subjects in Parts II and III of this book. Notes were written by students on both occasions and were the initial impulse for the project. I am happy to acknowledge here my debt to the many individuals in the Secretary of Science and Technology and the National Research Council for this opportunity, as well as to the local organizers.

The second, and most important lucky break came with Halina Frankowska's visit to UCLA in 1987, during which the nonlinear programming theory underlying most of this work came into being, as well as the strategy to approach problems with state constraints.

Later, S. S. Sritharan provided the impetus, the motivation and much of the technical means for application of the nonlinear programming theory to problems in fluid mechanics. Gieri Simonett read large parts of the manuscript and is responsible for many improvements. Finally, Tomáš Roubíček contributed substantially to Part III and Wolfgang Rueß clarified various functional analytic constructs. It goes without saying that none of the above authors is in any way responsible for errors, or misinterpretations of their results or suggestions.

During my scientific career, I had the good fortune to be able to attend many workshops on state-of-the art control theory. These included several Oberwolfach meetings and most of the Vorau conferences. The influence of the latter in the development of infinite dimensional control theory can hardly be overestimated; various concepts fundamental in this work were presented in Vorau conferences and published in their Proceedings.

Part I

Finite Dimensional Control Problems

1

Calculus of Variations and Control Theory

1.1. Calculus of Variations: Surface of Revolution of Minimum Area

Consider the surfaces Σ of revolution about the x-axis whose boundary consists of the circles

$$\Gamma_a = \{(x, y, z); x = a, y^2 + z^2 = r_a\}$$
$$\Gamma_b = \{(x, y, z); x = b, y^2 + z^2 = r_b\}$$

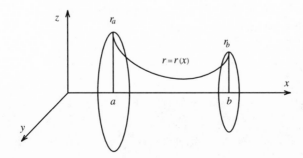

Figure 1.1.

$(a < b, r_a, r_b \geq 0)$. Among these surfaces, we look for one having minimum area. If $r = r(x) \geq 0$ $(a \leq x \leq b)$ is the equation of the surface, the area is[1]

$$A(r) = 2\pi \int_a^b r(x)\sqrt{1 + r'(x)^2}dx.$$

---

[1] If $r(x)$ is negative for some x the expression for the area is incorrect unless one replaces $r(x)$ by $|r(x)|$ in the integrand. This is ignored in some textbook treatments.

Since Γ_a and Γ_b are the boundary of Σ,

$$r(a) = r_a, \qquad r(b) = r_b. \tag{1.1.1}$$

More generally, we may minimize

$$J(y) = \int_a^b F(x, y(x), y'(x))dx. \tag{1.1.2}$$

The "variable" in J is itself a function $y(x)$ defined in the interval $a \leq x \leq b$. An expression like this is called a **functional**; the **admissible** functions $y(x)$ in the integrand belong to the space $C^{(1)}[a, b]$ of continuously differentiable functions in $a \leq x \leq b$ and satisfy boundary conditions, in this case (1.1.1). Admissible functions are an *affine* subspace of $C^{(1)}[a, b]$: if $y(x)$ satisfies (1.1.1) and $v(x)$ belongs to the subspace $C_0^{(1)}[a, b]$ of $C^{(1)}[a, b]$ defined by $v(a) = v(b) = 0$ then

$$y(x) + hv(x) \tag{1.1.3}$$

is admissible for any h. This makes viable the argument below, where we assume that $F(x, y, y')$ is everywhere defined and continuously differentiable with respect to y and y', with $F, \partial F/\partial y, \partial F/\partial y'$ continuous in all variables.

Assume that $\bar{y}(\cdot) \in C^{(1)}[a, b]$ is a **minimizing element** or a **minimum** of $J(y)$ (that is, $J(\bar{y}) \leq J(y)$ for all admissible $y(\cdot)$). Let $v(\cdot) \in C_0^{(1)}[a, b]$. Then

$$J(\bar{y} + hv) \geq J(\bar{y})$$

for all real h, hence $\phi(h) = J(\bar{y} + hv)$ has a minimum at $h = 0$. This implies $\phi'(0) = 0$. This condition can be written

$$\phi'(0) = \frac{d}{dh}\bigg|_{h=0} \phi(h) = \frac{d}{dh}\bigg|_{h=0} J(\bar{y} + hv)$$

$$= \frac{d}{dh}\bigg|_{h=0} \int_a^b F(x, \bar{y}(x) + hv(x), \bar{y}'(x) + hv'(x))dx$$

$$= \int_a^b \left\{ \frac{\partial F}{\partial y}(x, \bar{y}(x), \bar{y}'(x))v(x)dx + \frac{\partial F}{\partial y'}(x, \bar{y}(x), \bar{y}'(x))v'(x) \right\} dx = 0 \tag{1.1.4}$$

for any $v(\cdot) \in C_0^{(1)}[a, b]$.

Lemma 1.1.1. *Let $f(x), g(x)$ be continuous in $a \leq x \leq b$. Assume that for every $v \in C_0^{(1)}[a, b]$ we have*

$$\int_a^b \{f(x)v(x) + g(x)v'(x)\}dx = 0. \tag{1.1.5}$$

Then (after possible modification in a null set) $g(\cdot) \in C^{(1)}[a, b]$ and $g'(x) \equiv f(x)$.

Proof. Assume first that $f(x) \equiv 0$; (1.1.5) and the boundary conditions imply

$$\int_a^b (g(x) - c)v'(x)dx = 0 \qquad (1.1.6)$$

for any c. Define

$$v(x) = \int_a^x (g(\xi) - c_0)d\xi$$

where c_0 is such that $v(b) = 0$; then $v(\cdot) \in C_0^{(1)}[a, b]$. Replacing this particular $v(\cdot)$ and c_0 in (1.1.6) we obtain $g(x) \equiv c_0$. In the general case, define

$$F(x) = \int_0^x f(\xi)\,d\xi$$

and integrate (1.1.5) by parts, obtaining

$$\int_a^b \{g(x) - F(x)\}v'(x)dx = 0,$$

for all $v(\cdot) \in C_0^{(1)}[a, b]$ so that $g(x)$ and $F(x)$ differ by a constant. ∎

Using this result in (1.1.4) we deduce that $\partial F(x, \bar{y}(x), \bar{y}'(x))/\partial y'$ is a continuously differentiable function of x and that $y(x) = \bar{y}(x)$ satisfies the **Euler equation**

$$\frac{\partial F}{\partial y}(x, y(x), y'(x)) - \frac{d}{dx}\frac{\partial F}{\partial y'}(x, y(x), y'(x)) = 0. \qquad (1.1.7)$$

Hence (assuming J has a minimum in $C^{(1)}[a, b]$), the minimization problem reduces to solving (1.1.7) with boundary conditions (1.1.1). However, the theory of boundary value problems for differential equations is not as simple as that of initial value problems (where the value of a solution and its derivative are given at a single point). For a glimpse on boundary value problems see Elsgolts [1970, p. 165] or Gelfand–Fomin [1963, p. 16]; for a more complete treatment, see Keller [1968].

If $y(\cdot)$ is twice continuously differentiable we may apply the chain rule in the right side of (1.1.7) and obtain a *bona fide* second order differential equation for $\bar{y}(\cdot)$ (however, $y(x)$ may not be so smooth; see Example 1.1.3). If $y(\cdot) \in C^{(2)}[a, b]$ and, in addition, $F(x, y, y') = F(y, y')$ is independent of x, we multiply (1.1.7) by $y'(x)$ and integrate, obtaining

$$F(y(x), y'(x)) - y'(x)\frac{\partial F}{\partial y'}(y(x), y'(x)) = \beta, \qquad (1.1.8)$$

where β is a constant. Conversely, any solution of (1.1.8) with $y'(x) \neq 0$ necessarily satisfies the Euler equation.

The Euler equation for the minimal area problem is

$$\sqrt{1+r'^2} - \frac{d}{dx}\frac{rr'}{\sqrt{1+r'^2}} = 0.^{(2)} \tag{1.1.9}$$

Equation (1.1.8) is

$$r(x) = \beta\sqrt{1+r'(x)^2} \tag{1.1.10}$$

with solution $r(x) \equiv 0$ for $\beta = 0$, and

$$r(x) = \beta \cosh\left(\frac{x-\alpha}{\beta}\right) \tag{1.1.11}$$

for $\beta \neq 0$, where α is arbitrary. (Gelfand–Fomin [1963, p. 20]).

Example 1.1.2. There exist either two, one, or no $r(x)$ of the form (1.1.11) satisfying the boundary conditions (1.1.1). For a proof, see Bliss [1925, p. 90] or Cesari [1983, p. 143]. We just take a look at the case $a = 0, b = 1, r_a = r_b = r$. To satisfy the boundary conditions we must make $\alpha = 1/2$ and solve the transcendental equation

$$\phi(\beta) = \beta \cosh\left(\frac{1}{2\beta}\right) = r. \tag{1.1.12}$$

If m is the minimum of the function of $\beta > 0$ on the left-hand side, then the equation has no solution if $r < m$, one solution if $r = m$, and two solutions if $r > m$. We have $m = 0.7544\ldots$, attained at $\beta = 0.4167\ldots$.

For $r_a = r_b = 1$ (1.1.12) has two solutions, $\beta = 0.2350\ldots$ and $\beta = 0.8483\ldots$.

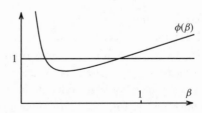

Figure 1.2.

Using formula (1.1.11) the area integral reduces to

$$2\pi\beta \int_0^1 \cosh^2((2x-1)/2\beta)dx = \pi\beta + \pi\beta^2 \sinh(1/\beta).$$

The surface corresponding to $\beta = 0.2350\ldots$ (resp. to $\beta = 0.8483\ldots$) has area $6.8456\ldots$ (resp. $5.9918\ldots$).

[(2)] A minimizing element $\bar{r}(x)$ of $A(r)$ satisfies the Euler equation (1.1.9) if $\bar{r}(x) > 0$; otherwise, $\bar{r}(x) + hv(x)$ may be zero for h arbitrarily small invalidating (1.1.4). Note also that no solution of (1.1.9) may be zero anywhere.

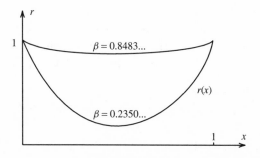

Figure 1.3.

Obviously, the first surface cannot be a minimum. The second is, although this is far from obvious; in fact, it is not even clear whether a minimal surface exists. On the other hand, if $r_a = r_b = m$, the only solution of the form (1.1.11) satisfying the boundary conditions (1.1.1), whose area is 4.2903... is not a minimum of the functional. In fact, the "surface" consisting of the two disks spanned by Γ_0 and Γ_1 connected by the segment of the x-axis between them has area $2\pi\beta^2 \cosh^2(1/2\beta) = 3.5762... < 4.2903...$. Obviously, this is not one of the surfaces allowed to compete for the minimum, but it can be approximated by smooth surfaces of revolution having almost the same area, for instance, $r_n(x) = (x^n + (1 - x)^n)\beta \cosh(1/2\beta)$ for large n (see Figure 1.4).

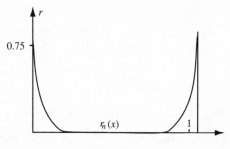

Figure 1.4.

Taking $r_a = r_b = m' > m$ with m' sufficiently close to m we obtain two functions of the form (1.1.11) satisfying the boundary conditions, none of which is a minimum.

For a complete solution of the minimal surface problem in the spirit of control theory, see **10.5**; classical treatments are given in Bliss [1925, Ch. IV] or Cesari [1983, p.143].

The results on the functional (1.1.2) extend easily to functionals depending on n functions $y_1(x), \ldots, y_n(x)$ and their derivatives,

$$J(y_1, \ldots, y_n) = \int_a^b F(x, y_1(x), \ldots, y_n(x), y_1'(x), \ldots, y_n'(x))dx. \quad (1.1.13)$$

Dealing with each $y_j(x)$ separately, we deduce that if F is everywhere defined and continuously differentiable with respect to each y_j and y_j' with F and partial derivatives continuous in all arguments, then a minimum $\bar{y}_1(x), \dots, \bar{y}_n(x)$ of (1.1.11) where each $\bar{y}_j(x)$ belongs to $C^{(1)}[a, b]$ must satisfy the **Euler equations**

$$\frac{\partial F}{\partial y_j} - \frac{d}{dx}\left(\frac{\partial F}{\partial y_j'}\right) = 0 \quad (j = 1, 2, \dots, n). \tag{1.1.14}$$

Example 1.1.3. (Gelfand–Fomin [1963, p. 16]) The functional

$$J(y) = \int_{-1}^{1} y^2(x)\big(2x - y'(x)\big)^2 dx$$

with boundary conditions $y(-1) = 0$, $y(1) = 1$ attains its minimum (zero) for $\bar{y}(x) = 0$ $(x \le 0)$, $\bar{y}(x) = x^2$ $(x \ge 0)$. The minimum $\bar{y}(\cdot)$ does not belong to $C^{(2)}[a, b]$.

Example 1.1.4. (Gelfand–Fomin [1963, p. 17]) Assume $F(x, y, y')$ has continuous partials up to order two in all variables. Let $\bar{y}(\cdot) \in C^{(1)}[a, b]$ be a solution of Euler's equation (1.7) with

$$\frac{\partial^2 F}{\partial y'^2}(x, \bar{y}(x), \bar{y}'(x)) \ne 0 \quad (a \le x \le b). \tag{1.1.15}$$

Then $\bar{y}(\cdot) \in C^{(2)}[a, b]$. This applies to the minimal area problem (where $\partial^2 F(r, r')/\partial r'^2 = r(1 + r'^2)^{-3/2}$) as follows: if $r = r(x)$ is the equation of a minimal surface in $C^{(1)}[a, b]$ with $r(x) > 0$, then $r(\cdot) \in C^{(2)}[a, b]$.

1.2. Interpretation of the Results

All we have shown on the problem of minimizing (1.1.2) is that if $\bar{y}(\cdot) \in C^{(1)}[a, b]$ is a minimum, then $\bar{y}(\cdot)$ satisfies the Euler equation (1.1.7). Thus, we only have *necessary* conditions for a minimum. They may not be sufficient: a solution of (1.1.7) satisfying the boundary conditions may not be a minimum of $J(y)$, as we have seen in Example 1.1.2. We meet the same problem in calculus trying to find the minima of a function $f(x) = f(x_1, x_2, \dots, x_m)$ in m-dimensional Euclidean space $\mathbb{R}^m$: at a minimum $\bar{x} = (\bar{x}_1, \bar{x}_2, \dots, \bar{x}_m)$ of f we have

$$\frac{\partial f}{\partial x_1} = \frac{\partial f}{\partial x_2} = \dots = \frac{\partial f}{\partial x_m} = 0 \tag{1.2.1}$$

but these conditions are not sufficient. Points $\bar{x} \in \mathbb{R}^m$ where (1.2.1) holds are called **extremals** of the function f, and we use the same terminology for the functional (1.1.2): a function $y \in C^{(1)}[a, b]$ satisfying (1.1.7) and the boundary conditions is an **extremal** of J. For instance, the inner surface in Figure 1.3 for $r_a = r_b = 1$ is an extremal but not a minimum.

In some cases, necessary conditions in combination with existence theorems give the actual minima of a functional. For instance, if a minimizing element $\bar{y}(\cdot) \in C^1[a, b]$ exists and solutions of the boundary value problem are unique, then the solution of the boundary value problem must be the minimum. However, this may fail as seen in Example 1.1.2 for the minimal area problem; solutions in $C^{(1)}[a, b]$ may not exist or the boundary value problem may have multiple solutions. Another problem without smooth solutions is

Example 1.2.1. (Gelfand–Fomin [1963, p. 61]) The functional

$$J(y) = \int_{-1}^{1} y^2(x)(1 - y'(x))^2 dx$$

with boundary conditions $y(-1) = 0$, $y(1) = 1$ attains its minimum (zero) for $\bar{y}(x) = 0$ $(x \leq 0)$, $\bar{y}(x) = x$ $(x \geq 0)$. The minimum $\bar{y}(\cdot)$ does not belong to $C^{(1)}[a, b]$.

Proper treatment of variational problems (and of control problems) needs a less demanding definition of solution; see **10.5** for more on this.

1.3. Mechanics and Calculus of Variations

Consider a mechanical system with a finite number of degrees of freedom. We denote by $q_1, q_2, \ldots, q_n$ the **generalized coordinates** of the system, in terms of which the Cartesian coordinates can be determined in 3-space. The n-dimensional point $q = (q_1, q_2, \ldots, q_n)$ moves arbitrarily in a region of Euclidean space $\mathbb{R}^n$ or, more generally, in an n-dimensional differential manifold. Assume for simplicity that the system consists of a finite number of particles with Cartesian coordinates $\mathbf{r}_1, \mathbf{r}_2, \ldots, \mathbf{r}_p \in \mathbb{R}^3$:

$$\mathbf{r}_j = \mathbf{r}_j(q_1, \ldots, q_n) \quad (1 \leq j \leq p), \tag{1.3.1}$$

and that the forces acting on the system are due to a potential,

$$F_j = -\nabla_{\mathbf{r}_j} U(\mathbf{r}_1, \ldots, \mathbf{r}_j, \ldots, \mathbf{r}_p) \quad (j = 1, 2, \ldots, p).$$

The Lagrangian of the system is

$$L = \sum_{j=1}^{p} \frac{m_j}{2} \|\mathbf{r}'_j\|^2 - U(\mathbf{r}_1, \ldots, \mathbf{r}_p) = T - U \tag{1.3.2}$$

where the **kinetic energy** T and the **potential energy** U are expressed in terms of the generalized coordinates. The motion of the system is described by **Hamilton's**

principle (Kompaneyets [1978, p. 17]): the possible motions $q_1(t), q_2(t), \ldots, q_n(t)$ of the system in a time interval $t_0 \le t \le t_1$ are extremals of the **action integral**

$$S = \int_{t_0}^{t_1} L(q_1, q_2, \ldots, q_n, q_1', q_2', \ldots, q_n') dt, \qquad (1.3.3)$$

that is, they solve the Euler equations

$$\frac{d}{dt}\left(\frac{\partial L}{\partial q_j'}\right) - \frac{\partial L}{\partial q_j} = 0, \qquad (1.3.4)$$

((see (1.1.14)). These are the **Euler-Lagrange equations** of mechanics and are combined with initial or boundary conditions: usually, the initial position and velocity of the system (that is, the position and velocity at $t = t_0$) are given. It also makes sense to specify the position of the system at different times t_0 and t_1, which produces a boundary value problem in the sense of **1.1.**

Example 1.3.1. A **simple pendulum** is a particle of mass m connected to the origin by a rigid rod of length l and allowed to move on the (x, y)-plane.

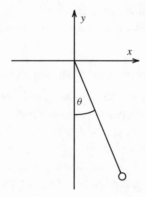

Figure 1.5.

Only one generalized coordinate, the angle θ, is necessary. It moves in the circle obtained identifying points modulo 2π on the line. The force of gravity comes from the potential energy $U = mgy$ (g the acceleration of gravity). Since $\mathbf{r} = l(\sin\theta, -\cos\theta)$, $\mathbf{r}' = l\theta'(\cos\theta, \sin\theta)$ and

$$L = T - U = \frac{m}{2}l^2\theta'^2 - mgl(1 - \cos\theta), \qquad (1.3.5)$$

(where we have taken arbitrarily the stable equilibrium position as having potential energy zero). The Euler-Lagrange equation is the nonlinear pendulum equation

$$\theta'' + \frac{g}{l}\sin\theta = 0. \qquad (1.3.6)$$

1.4. Optimal Control: Fuel Optimal Landing of a Space Vehicle

A space vehicle on a vertical trajectory tries to land smoothly (that is, with velocity zero) on the surface of a planet (see Figure 1.6). Denote by $h(t)$ the height at time t (so that $v(t) = h'(t)$ is the instantaneous velocity). Since combustible is being consumed, the mass $m(t)$ of the vehicle is a nonincreasing function of t. If we call $u(t)$ the instantaneous upwards thrust, Newton's law gives $m(t)h''(t) = -gm(t) + u(t)$, where g is the acceleration of gravity. Assuming that the thrust is proportional to the rate of decrease of mass (that is, proportional to the rate at which combustible is used up) we introduce $v(t) = h'(t)$ as a variable and obtain the following first-order system of differential equations:

$$h'(t) = v(t), \qquad v'(t) = -g + \frac{u(t)}{m(t)}, \qquad m'(t) = -Ku(t),$$

where $K > 0$. At the initial time $t_0 = 0$ we have initial conditions

$$h(0) = h_0, \qquad v(0) = v_0, \qquad m(0) = m_0.$$

The vehicle will land softly at time $\bar{t} \geq 0$ if

$$h(\bar{t}) = 0, \qquad v(\bar{t}) = 0.$$

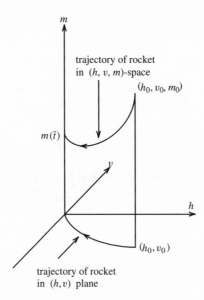

Figure 1.6.

The thrust cannot be negative or arbitrarily large:

$$0 \leq u(t) \leq R$$

for some $R > 0$. We have an optimization problem if we try to land minimizing the amount of combustible

$$m(0) - m(\bar{t}) = K \int_0^{\bar{t}} u(\tau)d\tau = J(u)$$

consumed from $t = 0$ until the landing time $t = \bar{t}$.

A complete solution of the landing problem is given in **4.7** and **4.8**.

1.5. Optimal Control Problems Described by Ordinary Differential Equations

The rocket landing problem is a particular case of a general **optimal control problem**: minimize a **cost functional** or **performance index** of the form

$$y_0(\bar{t}, u) = \int_0^{\bar{t}} f_0(\tau, y(\tau), u(\tau))d\tau + g_0(\bar{t}, y(\bar{t})) \qquad (1.5.1)$$

among all the solutions (or **trajectories**) of the vector differential equation

$$y'(t) = f(t, y(t), u(t)), \qquad y(0) = \zeta \qquad (1.5.2)$$

with $y(t)$ and $f(t, y, u)$ m-vector functions and ζ a m-vector. The system (1.5.2) is called the **state equation** of the system. The **control** $u(t)$ is a k-vector function satisfying a **control constraint**

$$u(t) \in U \qquad (1.5.3)$$

where $U \subseteq \mathbb{R}^k$ is the **control set** (more generally, we may use control sets $U(t)$ depending on time). Controls satisfying (1.5.3) are called **admissible**. In general, the problem includes a **target condition**

$$y(\bar{t}) \in Y \qquad (1.5.4)$$

where $Y \subseteq \mathbb{R}^m$ is the **target set**. The **terminal time** $\bar{t}$ at which the target condition (1.5.4) is to be satisfied may be fixed or free. The problem may also include **state constraints**

$$y(t) \in M(t) \qquad (1.5.5)$$

to be satisfied in $0 \le t \le \bar{t}$.

To fit the rocket landing problem in this scheme, we take $m = 3, k = 1$, $y(t) = (h(t), v(t), m(t))$, $f(t, y, u) = (v, -g + u/m, -Ku)$, $\zeta = (h_0, v_0, m_0)$, and $U = [0, R]$. The target set is the half line $Y = \{0\} \times \{0\} \times [m_e, \infty)$, and in the cost functional, we have $f_0(t, y, u) = Ku$, $g_0 = 0$. The problem actually includes two state constraints, namely

$$h(t) \ge 0, \qquad m(t) \ge m_e > 0. \qquad (1.5.6)$$

The second says that the mass at time t cannot be less than the mass m_e of the rocket with empty fuel tanks, and is automatically satisfied since $u(t) \geq 0$; the first warns that we must not drive the rocket into the ground. These state constraints are of the form (1.5.5) with $M(t) = [0, \infty) \times (-\infty, \infty) \times [m_e, \infty)$ for all t.

A prerequisite to the solution of the general optimal control problem is the **controllability problem**: can we find an admissible control $u(t)$ such that the corresponding trajectory $y(t)$ with $y(0) = \zeta$ satisfies the target condition (1.5.4) and the state constraint (1.5.5)? This controllability problem may not have a solution. For instance, in the landing problem, Y cannot be hit at all (that is, soft landing is impossible) if the initial amount of combustible m_0 is insufficient.

1.6. Calculus of Variations and Optimal Control. Spike Perturbations

Optimal control problems are similar to problems of calculus of variations; both deal with minimizing functionals. One may try to apply to control problems the arguments in **1.1**, based on affine perturbations $\bar{u}(t) + hv(t)$ of an optimal control $\bar{u}(t)$. However, it is not clear how to take h and $v(t)$ in order that the target condition (1.5.4) be satisfied by the trajectory corresponding to $u(t) + hv(t)$. Even if we ignore the target condition, we must be sure that the control $\bar{u}(t) + hv(t)$ is admissible, that is,

$$\bar{u}(t) + hv(t) \in U.$$

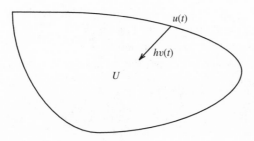

Figure 1.7.

For $h > 0$, this requires $v(t)$ to "point into U" at $\bar{u}(t)$ (see Figure 1.7). **Spike** or **needle** perturbations are better suited to control constraints and are defined as follows. Let $\bar{t} > 0, 0 < s \leq \bar{t}, 0 \leq h \leq s$ and v an element of the control set U. Given an admissible control $u(t)$ we define a new control $u_{s,h,v}(t)$ by

$$u_{s,h,v}(t) = \begin{cases} v & (s - h < t \leq s) \\ u(t) & \text{elsewhere} \end{cases} \tag{1.6.1}$$

(see Figure 1.8).

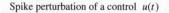

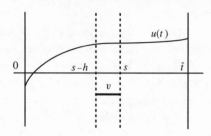

Figure 1.8.

Obviously, $u_{s,h,v}(t)$ is an admissible control. If $\bar{u}(t)$ is optimal in an interval $0 \le t \le \bar{t}$,

$$y_0(\bar{t}, \bar{u}_{s,h,v}) \ge y_0(\bar{t}, \bar{u}) \tag{1.6.2}$$

for s, h, v arbitrary, thus if

$$\xi_0(t) = \frac{d}{dh}\bigg|_{h=0+} y_0(t, \bar{u}_{s,h,v}) = \lim_{h \to 0+} \frac{1}{h}\big(y_0(t, \bar{u}_{s,h,v}) - y_0(t, \bar{u})\big) \tag{1.6.3}$$

exists, we have

$$\xi_0(\bar{t}) \ge 0 \tag{1.6.4}$$

for s, v arbitrary.

Spike perturbations can also be defined for $h < 0$ (the spike stands to the right of s). Since $y_0(\bar{t}, \bar{u}_{s,h,v}) \ge y_0(\bar{t}, \bar{u})$ for h of any sign, this should improve (1.6.4) to $\xi_0(\bar{t}) = 0$ for all s, v. However, the function $h \to y_0(\bar{t}, \bar{u}_{s,h,v})$ may not have a two-sided derivative at $h = 0$ (see Example 1.6.1).

We compute *formally* $\xi_0(t)$ for arbitrary t; justification is postponed to **2.3**. The first step is to calculate

$$\xi(t) = \frac{d}{dh}\bigg|_{h=0+} y(t, u_{s,h,v}) = \lim_{h \to 0+} \frac{1}{h}\big(y(t, u_{s,h,v}) - y(t, u)\big), \tag{1.6.5}$$

where $y(t, u)$ denotes the solution of (1.5.2) corresponding to the control $u(t)$. We have

$$\xi'(t) = \lim_{h \to 0+} \frac{1}{h}\big(y'(t, \bar{u}_{s,h,v}) - \bar{y}'(t, \bar{u})\big)$$

$$= \lim_{h \to 0+} \frac{1}{h}\big\{f(t, y(t, \bar{u}_{s,h,v}), \bar{u}_{s,h,v}(t)) - f(t, y(t, \bar{u}_{s,h,v}), \bar{u}(t))\big\}$$

$$+ \lim_{h \to 0+} \frac{1}{h}\big\{f(t, y(t, \bar{u}_{s,h,v}), \bar{u}(t)) - f(t, y(t, \bar{u}), \bar{u}(t))\big\}. \tag{1.6.6}$$

The function $f(t, y(t, \bar{u}_{s,h,v}), \bar{u}_{s,h,v}(t)) - f(t, y(t, \bar{u}_{s,h,v}), \bar{u}(t))$ is zero except in $s - h < t \leq s$, where $\bar{u}_{s,h,v}(t) = v$. Assuming that $y(t, \bar{u}_{s,h,v}) \approx y(s, \bar{u})$ and $\bar{u}(t) \approx \bar{u}(s)$ in the interval $s - h \leq t \leq s$ for h small enough, the first limit on the right should be the same as

$$\lim_{h \to 0+} \frac{1}{h} \chi_h(t) \{ f(s, y(s, \bar{u}), v) - f(s, y(s, \bar{u}), \bar{u}(s)) \}$$

($\chi_h(t)$ the characteristic function of $s - h < t \leq s$), which equals

$$\{ f(s, y(s, \bar{u}), v) - f(s, y(s, \bar{u}), u(s)) \} \delta(t - s)$$

(δ the Dirac delta). The limit in the second term is computed by the chain rule. We obtain in this way the **variational equation** for $\xi(t)$,

$$\xi'(t) = \partial_y f(t, y(t, \bar{u}), \bar{u}(t)) \xi(t) + \{ f(s, y(s, \bar{u}), v) - f(s, y(s, \bar{u}), \bar{u}(s)) \} \delta(t - s)$$
$$(0 \leq t \leq \bar{t}), \quad \xi(0) = 0. \quad (1.6.7)$$

Equivalently, $\xi(t) = 0$ for $t < s$ and

$$\xi'(t) = \partial_y f(t, y(t, \bar{u}), \bar{u}(t)) \xi(t) \quad (s \leq t \leq \bar{t}),$$
$$\xi(s) = \{ f(s, y(s, \bar{u}), v) - f(s, y(s, \bar{u}), \bar{u}(s)) \}. \quad (1.6.8)$$

In both equations, $\partial_y f$ denotes the Jacobian matrix of f with respect to the y variables. To figure out the limit (1.6.3) for a cost functional of the form (1.5.1), we write the integrand in the form

$$f_0(\tau, y(\tau, \bar{u}_{s,h,v}), \bar{u}_{s,h,v}(\tau)) - f_0(\tau, y(\tau, \bar{u}), \bar{u}(\tau))$$
$$= \{ f_0(\tau, y(\tau, \bar{u}_{s,h,v}), \bar{u}_{s,h,v}(\tau)) - f_0(\tau, y(\tau, \bar{u}_{s,h,v}), \bar{u}(\tau)) \}$$
$$+ \{ f_0(\tau, y(\tau, \bar{u}_{s,h,v}), \bar{u}(\tau)) - f_0(\tau, y(\tau, \bar{u}), \bar{u}(\tau)) \}$$

and argue as in the computation of (1.6.5). The final result is

$$\xi_0(t) = \{ f_0(s, y(s, \bar{u}), v) - f_0(s, y(s, \bar{u}), \bar{u}(s)) \}$$
$$+ \int_s^t \langle \nabla_y f_0(\tau, y(\tau, \bar{u}), \bar{u}(\tau)), \xi(\tau) \rangle d\tau + \langle \nabla_y g_0(t, y(t, \bar{u})), \xi(t) \rangle. \quad (1.6.9)$$

where ∇ denotes gradient and $\langle \cdot, \cdot \rangle$ inner product in $\mathbb{R}^m$. From all the transgressions in the argument, the worst is perhaps the continuity assumption that $\bar{u}(t) \approx \bar{u}(s)$ near s; optimal controls are often discontinuous. A correct proof needs some measure theory.

Replacing in (1.6.4), a necessary condition for $\bar{u}(\cdot)$ to be optimal is obtained. Some work will be needed in Chapter 2 to put this result in a usable form; we only show here how it works in a particular problem.

Example 1.6.1. Consider the control system

$$y'(t) = -u(t), \qquad y(0) = 1 \tag{1.6.10}$$

in the interval $0 \le t \le \bar{t}$, with control constraint

$$0 \le u(t) \le 1 \tag{1.6.11}$$

and cost functional

$$y_0(\bar{t}, u) = \int_0^{\bar{t}} u(\tau)^2 d\tau + y(\bar{t})^2. \tag{1.6.12}$$

Applications buffs may imagine a reservoir being pumped out at rate $u(t)$. The first term in the functional reflects the cost of pumping, thus minimization of $y_0(t, u)$ means draining the reservoir as much as possible while minimizing cost.

We have $f(t, y, u) \equiv -u$, so that the variational equation is

$$\xi'(t) = -\delta(t - s)\{v - u(s)\}, \qquad \xi(0) = 0$$

with solution $\xi(t) = -v(t - s)\{v - u(s)\}$, $v(\cdot)$ the Heaviside function $v(t) = 1 (t \ge 0), u(t) = 0 (t < 0)$. Assuming an optimal control $\bar{u}(t)$ exists and taking into account that $f_0(t, y, u) = u^2, g(t, y) = y^2$, (1.6.4) and (1.6.9) give $\xi_0(\bar{t}) = \{v^2 - \bar{u}(s)^2\} - 2y(\bar{t}, \bar{u})\{v - \bar{u}(s)\} \ge 0$, or $v^2 - 2y(\bar{t}, \bar{u})v \ge \bar{u}(s)^2 - 2y(\bar{t}, \bar{u})\bar{u}(s)$ for $0 \le s \le \bar{t}$ and $0 \le v \le 1$, so that

$$\bar{u}(s)^2 - 2y(\bar{t}, \bar{u})\bar{u}(s) = \min_{0 \le v \le 1}\{v^2 - 2y(\bar{t}, \bar{u})v\}. \tag{1.6.13}$$

This is a protoexample of Pontryagin's maximum (minimum) principle and shows one of its features: it gives the optimal control $\bar{u}(s)$, but only in terms of the unknown optimal trajectory $y(\bar{t}, \bar{u})$. In some cases (here for instance), it is possible to compute $\bar{u}(s)$ anyway. In fact, the initial value problem (1.6.10) and the fact that $0 \le \bar{u}(t) \le 1$ imply that $0 \le y(\bar{t}, \bar{u}) \le 1$. Accordingly, the minimum of $v^2 - 2y(\bar{t}, \bar{u})v$ in $0 \le v \le 1$ is $\bar{u}(s) \equiv y(\bar{t}, \bar{u})$. Replacing in the equation and making $t = \bar{t}$, we get $y(\bar{t}, \bar{u}) = 1 - \bar{t}y(\bar{t}, \bar{u})$, so that

$$\bar{u}(s) \equiv \frac{1}{1 + \bar{t}} \quad (0 \le s \le \bar{t}).$$

Note that if h is of arbitrary sign and $t > s + h$ we have

$$y(t, \bar{u}_{s,h,v}) = (t - |h|)\bar{u}^2 + |h|v^2 + (1 - (t - |h|)\bar{u} - |h|v)^2$$

for $t > s$; thus $h \to y(t, \bar{u}_{s,h,v})$ is not (two-sided) differentiable at $h = 0$. Computing the limit (1.6.3) for $h < 0$ just produces (1.6.4) again.

1.7. Optimal Control: Minimum Drag Nose Shape in Hypersonic Flow

A vehicle moves through a fluid in the direction of the positive y-axis with uniform speed V. Its nose is a body of revolution whose projection on the x, y plane is a curve described by parametric equations $x = x(t), y = y(t)$ $(0 \leq t \leq T)$ with $(x(0), y(0)) = (0, h), (y(0), y(T)) = (r, 0)$; h is the **height** of the nose and r its **maximum radius**. The quotient r/h is the **fineness ratio**.

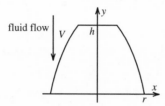

Figure 1.9.

A model due to Newton (see Goursat [1942, p. 658], McShane [1978/1989]) proposes that the drag normal to each surface element is proportional to the square $V^2 x'(t)^2/(x'(t)^2 + y'(t)^2)$ of the normal component of the velocity vector. The resultant of all these forces (obviously in the y-direction) is then the integral of $V^2 x'(t)^3/(x'(t)^2 + y'(t)^2)^{3/2}$ over the surface with respect to the area element $2\pi x(t)(x'(t)^2 + y'(t)^2)^{1/2} dt$, thus is proportional to the integral

$$\int_0^T \frac{x(t)x'(t)^3}{x'(t)^2 + y'(t)^2} dt.^{(1)} \tag{1.7.1}$$

It is easy to see that, without further conditions, the minimum of the integral is $-\infty$, but there are physical reasons to consider only nondecreasing $x(t)$ and nonincreasing $y(t)$; if this is not the case (Figure 1.10) there may be parts of the surface of the body isolated from the flow by stagnant fluid or by other parts of the body.[2]

This problem is treated in textbooks such as Goursat [1942] using classical calculus of variations, but it admits a more natural formulation as a control problem. We set

$$x'(t) = u(t), \qquad y'(t) = -v(t), \qquad x(0) = 0, \qquad y(0) = h \tag{1.7.2}$$

in a variable interval $0 \leq t \leq \bar{t}$ with control set $U = [0, \infty) \times [0, \infty)$, target condition

$$(x(\bar{t}), y(\bar{t})) = (r, 0), \tag{1.7.3}$$

[1] For air flow, this model is said to be "...very good at hypersonic speeds but *not* very good at subsonic speeds" in Bryson–Ho [1969, p. 52]. Hence the title of this section.

[2] Ignoring these conditions has led some authors to brand Newton's drag model as "absurd." See McShane [1978/1989] for a refutation; careful reading of Newton's original formulation of the problem reveals that monotonicity of $x(t)$ and $y(t)$ is actually required. See also Goldstine [1980, p. 7].

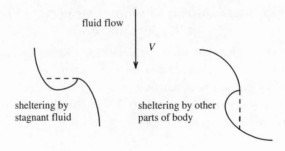

fluid flow

v

sheltering by
stagnant fluid

sheltering by other
parts of body

Figure 1.10.

and cost functional

$$y_0(t, u, v) = \int_0^t \frac{x(\tau)u(\tau)^3}{u(\tau)^2 + v(\tau)^2} \, d\tau. \tag{1.7.4}$$

A difference with the soft landing problem is that the control set is unbounded; thus, conditions on the controls $u(\cdot)$, $v(\cdot)$ are needed in order that (1.7.4) be finite. Since $xu^3/(u^2 + v^2) \leq xu$, it is enough to require the controls to be integrable. Another difference is that the parameter t in the minimum drag nose shape problem has no physical meaning so that we are free to reparametrize the curve at our pleasure. This will be used in the solution of this problem, presented in **4.9**, **13.7** and **13.8**.

1.8. Control of Functional Differential Equations: Optimal Forest Growth

Let $N(t)$ represent a population (bacteria in a test tube, people in a city, trees in a forest). The Malthusian model assumes a rate of growth proportional to the population: $N'(t) = aN(t)$. This gives the exponential growth law $N(t) = e^{at}N(0)$, which is only accurate for relatively small values of $N(t)$; overcrowding and competition for resources lower the rate of growth. A more realistic model assumes a steadily decreasing, eventually negative growth coefficient $a(N)$. Assuming $a(N)$ linear, Verhulst's **logistic equation**

$$N'(t) = (a - bN(t))N(t) \tag{1.8.1}$$

results, where $a, b > 0$. This model gives good results for bacteria populations, but does not describe accurately phenomena such as forest growth. In fact, the inhibiting effects of new trees on the growth rate are negligible until these have reached a certain "adult" size. Thus, the growth rate should be a function not of $N(t)$ but of $N(t - h)$ for a suitable time delay $h > 0$, leading to the **delayed logistic equation**

$$N'(t) = (a - bN(t - h))N(t). \tag{1.8.2}$$

Similar delay effects are observed in the influence of overcrowding in human populations (for an elementary exposition of logistic equations with and without delays see Haberman [1977, p. 119ff]). Equation (1.8.2) (the same as (1.8.1)) has two equilibrium solutions: one is $N(t) \equiv 0$, the other

$$N(t) \equiv N_e = a/b. \tag{1.8.3}$$

Assume tree seeds are planted, and trees are logged with seeding and logging rates $u_0(t)$ and $u_1(t)$ respectively. Let k be the time it takes a seed to become a baby tree. Then the equation becomes

$$N'(t) = (a - bN(t - h))N(t) + cu_0(t - k) - u_1(t) \tag{1.8.4}$$

where the coefficient c ($0 \leq c \leq 1$) accounts for the fraction of seeds that actually result in a tree. To start the equation we need to know the forest population in an interval of length h,

$$N(t) = N_0(t) \quad (t_0 - h \leq t \leq t_0). \tag{1.8.5}$$

To "attain the equilibrium population N_e" has at least two meanings. One is

$$N(\bar{t}) = N_e \tag{1.8.6}$$

and says the population is at equilibrium at $t = \bar{t}$ but not necessarily afterwards. If the population is to stay at equilibrium, the target condition must be

$$N(t) = N_e \quad (\bar{t} - h \leq t \leq \bar{t}), \tag{1.8.7}$$

which guarantees that $N(t) = N_e$ for all $t \geq \bar{t}$ if $cu_0(t - k) - u_1(t) = 0$ for $t \geq \bar{t}$. Target conditions of the form (1.8.6) are called **Euclidean**; those of the form (1.8.7) are called **functional**. To see the reason for this name, consider for instance the space $C[-h, 0]$ of continuous functions defined in the interval $-h \leq t \leq 0$. Given a function $y(t)$, denote by $y_t(\cdot)$ the **section** of $y(\cdot)$ defined by

$$y_t(\tau) = y(t + \tau) \quad (-h \leq \tau \leq 0). \tag{1.8.8}$$

Then the target condition (1.8.7) can be written as an ordinary target condition in the space $C[-h, 0]$:

$$N_{\bar{t}}(\cdot) = N_e \tag{1.8.9}$$

where N_e denotes the constant function. An "optimal net profit" problem is, for instance, to maximize the functional

$$J(t, u_1, u_2) = \alpha \int_0^t u_1(\tau)d\tau - \beta \int_0^t u_0(\tau)d\tau$$

with $\alpha, \beta > 0$ at some fixed time $\bar{t} > 0$; the first term represents the profit from logging and the second, the cost of seeding. Clearly, $u_0(t), u_1(t)$ are nonnegative and it is reasonable to include upper bounds on both rates:

$$0 \le u_0(t) \le R, \qquad 0 \le u_1(t) \le S. \tag{1.8.10}$$

Straight maximization of the profit may result in destruction of the forest at time $\bar{t}$, thus we supplement the problem with a (functional) target condition, say

$$|N(t) - N_e| \le \varepsilon \quad (\bar{t} - h \le t \le \bar{t}) \tag{1.8.11}$$

(N_e the equilibrium solution (1.8.3)), and terminate seeding at time $\bar{t} - k$. If the equilibrium position is stable, this means the forest population will stay near equilibrium after $\bar{t}$. Admissible controls for this problem are pairs $(u_0(t), u_1(t))$, u_0 defined in $-k \le t \le \bar{t}$, u_1 defined in $0 \le t \le \bar{t}$, and satisfying (1.8.10) in their respective intervals of definition.

Another growth model is described by the integrodifferential equation

$$N'(t) = \left(a - \left(\int_{-h}^{0} b(\tau)N(t + \tau)d\tau \right) \right) N(t) + cu_0(t - k) - u_1(t) \tag{1.8.12}$$

which takes into account the inhibiting effects of new trees of all sizes on the growth rate.

1.9. Control of Partial Differential Equations: Optimal Cooling of a Plate and Optimal Stabilization of a Vibrating Membrane

Consider a plate occupying a domain Ω with boundary Γ in 2-dimensional Euclidean space $\mathbb{R}^2$. In suitable units, the nonlinear partial differential equation

$$\frac{\partial y(t, x)}{\partial t} = \Delta y(t, x) - f(y(t, x)) + u(t, x) \quad (x \in \Omega), \tag{1.9.1}$$

$$\frac{\partial y(t, x)}{\partial \nu} = 0 \quad (x \in \Gamma), \tag{1.9.2}$$

($x = (x_1, x_2)$, Δ the Laplacian, $\partial/\partial \nu$ the outer normal derivative on Γ) describes the temperature $y(t, x)$ in Ω. The sum $-f(y(t, x)) + u(t, x)$ means applied heat, the first term through feedback, the second as a control, subject to the constraint

$$0 \le u(t, x) \le R. \tag{1.9.3}$$

Condition (1.9.2) indicates the boundary is insulated (there is no heat flow through the boundary). A conceivable problem is to drive the temperature from an initial value

$$y(0, x) = \zeta(x) \tag{1.9.4}$$

to a final value

$$y(\bar{t}, x) = \bar{y}(x) \tag{1.9.5}$$

in time $\bar{t}$, minimizing the cost of heating/cooling, which might be measured in the form

$$y_0(t, u) = \int_{(0,t)\times\Omega} u(\tau, x)^2 dx \, d\tau.$$

A number of variants are possible. For instance, control may be a k-dimensional function $(u_1(t), u_2(t), \ldots, u_k(t))$ of t entering the equation as $u(t, x) = \Sigma g_j(x) u_j(x)$, or the **exact** or **point** target condition (1.9.5) may be weakened to approximate conditions such as $|y(\bar{t}, x) - \bar{y}(x)| \le \varepsilon$ $(x \in \Omega)$ or

$$\int_\Omega (y(\bar{t}, x) - \bar{y}(x))^2 dx \le \varepsilon.$$

Control may also be applied on the boundary,

$$\frac{\partial y(t, x)}{\partial \nu} = g(y(t, x), u(t, x)) \quad (x \in \Gamma). \tag{1.9.6}$$

Engineers call systems such as (1.9.1)-(1.9.2) **distributed parameter systems**; when control appears in the boundary condition as in (1.9.6), we have a **boundary control system**.

For a peek into the results, we consider the distributed parameter system (1.9.1) with boundary condition (1.9.2), control constraint (1.9.3), fixed terminal time $\bar{t}$, and no target condition. The cost functional measures the final deviation from the target:

$$y_0(t, u) = \int_\Omega (y(\bar{t}, x) - \bar{y}(x))^2 dx. \tag{1.9.7}$$

We do spike perturbations in time,

$$u_{s,h,v}(t, x) = \begin{cases} v(x) & (s - h < t \le s) \\ u(t, x) & \text{elsewhere} \end{cases} \tag{1.9.8}$$

where $v(\cdot)$ is an element of the (functional) control set U defined by $0 \le v(x) \le R$. If $\bar{u}(t, x)$ is an optimal control,

$$y_0(\bar{t}, \bar{u}_{s,h,v}) \ge y_0(\bar{t}, \bar{u})$$

for $s, h, v(\cdot)$ arbitrary; thus, if

$$\xi_0(t) = \frac{d}{dh}\bigg|_{h=0+} y_0(t, \bar{u}_{s,h,v}) = \lim_{h \to 0+} \frac{1}{h}\big(y_0(t, \bar{u}_{s,h,v}) - y_0(t, \bar{u})\big), \tag{1.9.9}$$

we have

$$\xi_0(\bar{t}) \geq 0 \tag{1.9.10}$$

for s, $v(\cdot)$ arbitrary. As in **1.6**, the first step in the computation of $\xi_0(t)$ is to figure out

$$\xi(t, x) = \frac{d}{dh}\bigg|_{h=0+} y(t, x, u_{s,h,v})$$

$$= \lim_{h \to 0+} \frac{1}{h}\big(y(t, x, u_{s,h,v}) - y(t, x, u)\big), \tag{1.9.11}$$

where $y(t, x, u)$ indicates the solution of (1.9.1)-(1.9.2) corresponding to $u = u(t, x)$. A formal computation similar to that in the lines following (1.6.5) reveals that $\xi(t, x)$ is the solution of the linear initial value problem

$$\frac{\partial \xi(t, x)}{\partial t} = \Delta\xi(t, x) - \frac{\partial f(y(t, \bar{u}, x))}{\partial y}\xi(t, x)$$

$$+ (v(x) - \bar{u}(s, x))\delta(t - s) \quad (0 \leq t \leq \bar{t}, x \in \Omega) \tag{1.9.12}$$

$$\frac{\partial \xi(t, x)}{\partial v} = 0 \quad (0 \leq t \leq \bar{t}, x \in \Gamma). \tag{1.9.13}$$

$$\xi(0, x) = 0 \quad (x \in \Omega) \tag{1.9.14}$$

(δ the Dirac delta) or, equivalently,

$$\frac{\partial \xi(t, x)}{\partial t} = \Delta\xi(t, x) - \frac{\partial f(y(t, \bar{u}, x))}{\partial y}\xi(t, x) \quad (s \leq t \leq \bar{t}, x \in \Omega), \tag{1.9.15}$$

$$\frac{\partial \xi(t, x)}{\partial v} = 0 \quad (s \leq t \leq \bar{t}, x \in \Gamma), \tag{1.9.16}$$

$$\xi(s, x) = v(x) - \bar{u}(s, x) \quad (x \in \Omega). \tag{1.9.17}$$

We then compute $\xi_0(t)$ and use (1.9.10), obtaining

$$\int_\Omega (y(\bar{t}, \bar{u}, x) - \bar{y}(x))\xi(\bar{t}, x)dx \geq 0, \tag{1.9.18}$$

where $0 < s \leq t$ and $v(\cdot) \in U$. Now, let $z(t, x)$ be the solution of the backwards equation

$$\frac{\partial z(t, x)}{\partial t} = -\Delta z(t, x) + \frac{\partial f(y(t, \bar{u}, x))}{\partial y}z(t, x) \quad (0 \leq t \leq \bar{t}, x \in \Omega) \tag{1.9.19}$$

$$\frac{\partial z(t, x)}{\partial v} = 0 \quad (0 \leq t \leq \bar{t}, x \in \Gamma), \tag{1.9.20}$$

$$z(\bar{t}, x) = y(\bar{t}, \bar{u}, x) - \bar{y}(x) \quad (x \in \Omega). \tag{1.9.21}$$

Then

$$\int_\Omega z(\bar{t}, x)\xi(\bar{t}, x)dx - \int_\Omega z(s, x)\xi(s, x)dx$$

$$= \int_{(s,\bar{t})\times\Omega} \frac{\partial}{\partial t}\big(z(t, x)\xi(t, x)\big)dxdt$$

$$= \int_{(s,\bar{t})\times\Omega} \left(\frac{\partial z(t, x)}{\partial t}\xi(t, x) + z(t, x)\frac{\partial \xi(t, x)}{\partial t} \right)dx\,dt = 0$$

in view of the divergence theorem. Accordingly, (1.9.18) is

$$\int_\Omega z(s, x)(v(x) - \bar{u}(s, x))dx \geq 0$$

for all s, $v(\cdot)$, or

$$\int_\Omega z(s, x)\bar{u}(s, x) = \min_{v(\cdot)\in U} \int_\Omega z(s, x)v(x)dx \qquad (1.9.22)$$

for $0 < s \leq 1$, another protoexample of Pontryagin's minimum principle. Due to the control constraint, (1.9.22) implies

$$\bar{u}(s, x) = \begin{cases} R & \text{where } z(s, x) < 0, \\ 0 & \text{where } z(s, x) > 0 \end{cases} \qquad (1.9.23)$$

but gives no information on $u(s, x)$ in the set $e \subseteq (0, \bar{t}) \times \Omega$ where $z(t, x) = 0$. If $y(\bar{t}, \bar{u}, x) \neq y(\bar{t})$ then $z(s, x)$ is nontrivial (that is, not identically zero), but there is some distance from this property to the statement that e has measure zero; when it does, (1.9.23) gives information on the optimal control $\bar{u}(t, x)$ almost everywhere and deserves to be called a **bang-bang principle**. For more on this, see Chapter 11, in particular Problem 11.6.7 and the Miscellaneous Comments to Part II.

Typically for the maximum principle, (1.9.22) does not determine $\bar{u}$ directly; in fact, both the equation (1.9.19) for $z(s, x)$ and the final condition (1.9.21) presuppose knowledge of the unknown optimal solution $y(s, x, \bar{u})$. Even in the linear case, where (1.9.19) does not depend on the optimal solution, we must know $y(\bar{t}, \bar{u}, x)$ in (1.9.21). We shall take up the study of optimal control problems described by parabolic equations in Part II of this work.

Another problem with some claims to realism is that of bringing to equilibrium a vibrating membrane occupying the domain Ω and glued to the boundary; in suitable units, the corresponding system is

$$\frac{\partial^2 y(t, x)}{\partial t^2} = \Delta y(t, x) - f(y(t, x)) + u(t, x) \quad (x \in \Omega), \qquad (1.9.24)$$

$$y(t, x) = 0 \quad (x \in \Gamma), \qquad (1.9.25)$$

where the sum $-f(y(t, x)) + u(t, x)$ represents an applied force, the first term through feedback, the second as a control subject to a constraint, for instance

$$|u(x, t)| \leq R \tag{1.9.26}$$

or

$$\int_\Omega u(x, t)^2 dx \leq R. \tag{1.9.27}$$

Bringing the membrane to equilibrium from the initial conditions

$$y(0, x) = \zeta_0(x), \qquad \frac{\partial y(0, x)}{\partial t} = \zeta_1(x) \tag{1.9.28}$$

in time $\bar{t}$ amounts to the target condition(s)

$$y(\bar{t}, x) = 0, \qquad \frac{\partial y(\bar{t}, x)}{\partial t} = 0, \tag{1.9.29}$$

and the membrane will stay at equilibrium if the restoring force $f(y)$ satisfies $f(0) = 0$ and application of the control force terminates at $t = \bar{t}$. As in the cooling problem, control could be finite dimensional or applied through the boundary condition; the *distributed parameter* or *boundary* labels apply.

The two control problems above do not include state constraints for reasons of simplicity, but more realistic modeling must take them into account. For instance, equation (1.9.1) can only be expected to describe the evolution of the temperature $y(t, x)$ in an actual heating/cooling process within a certain range, which justifies the first of the bounds

$$|y(t, x)| \leq K, \qquad |\nabla y(t, x)| \leq L. \tag{1.9.30}$$

The second restriction reflects the fact that, in cooling a material such as glass, large temperature gradients may produce cracks and should be avoided. Of course, constraints such as (1.9.30) make the problem much harder to handle. However, we shall see in Chapter 11 that the minimum principle (1.9.22) still holds with a different definition of $z(t, x)$.

In the vibration model (1.9.24), the integral state constraint

$$\frac{1}{2} \int_\Omega \left\{ \left(\frac{\partial y}{\partial t} \right)^2 + \left(\frac{\partial y}{\partial x} \right)^2 \right\} dx \leq E^2 \tag{1.9.31}$$

puts a bound on the energy and reminds us of the fact that the wave equation (1.9.24) is just an approximation to the "true" nonlinear equation describing the vibration of the membrane, and that this approximation is only valid at low energy levels.

1.10. Finite Dimensional and Infinite Dimensional Control Problems

There is an important difference between the landing problem in **1.4** and the optimal control problems in **1.8** and **1.9**. In the former, as well as in any other optimal control problem that fits into the ordinary differential model in **1.5**, the **state** of the system is a finite dimensional vector; in the soft landing problem, this vector is the 3-dimensional vector $(h(t), v(t), m(t))$.

The state of the system described by the delay differential equation (1.8.4) (or the functional differential equation (1.8.12)) is a finite dimensional vector as well, namely the 1-dimensional vector $N(t)$. This way to look at the equation is adequate if one deals with the Euclidean target condition (1.8.6). However if the target condition is functional like (1.8.7) it is natural to consider as state of the system at time t the section y_t, which belongs to the infinite dimensional space $C[-h, 0]$.

The state of the system (1.9.1)-(1.9.2) at time t is a function $y(t, \cdot)$, thus an element of an infinite dimensional function space, for example $C(\overline{\Omega})$ or $L^2(\Omega)$. Same for the system (1.9.24)-(1.9.25): the state of the system at time t is the vector $(y(t, \cdot), y_t(t, \cdot))$ in a suitable energy space such as $H_0^1(\Omega) \times L^2(\Omega)$.

Ordinary differential equations like (1.5.2) and partial differential equations like (1.9.10) or (1.9.24) are specimens with something in common: all are *evolution equations*, that is, they describe a system's evolution in time. It comes as no surprise that their control theory contains many common elements, a thread running through this work. However, the treatment of systems whose states lie in an infinite dimensional space is much more involved, and complete generalizations of finite dimensional results are often unavailable.

2

Optimal Control Problems
Without Target Conditions

2.0. Elements of Measure and Integration Theory. The Lebesgue Integral. Metric, Banach, and Hilbert Spaces

Let S be an arbitrary set. A family Φ of subsets $\{a, b, c, d, e, \ldots\}$ of S is called a **field** if the family contains the null set $\emptyset$, the complement e^c of each member e, and the union $e_1 \cup e_2 \cup \ldots$ of each finite collection of its members. A field also contains the intersection of each finite collection of its members: this follows from one of De Morgan's laws. A field Φ is a σ-**field** if it contains the union of each countable collection of its members; De Morgan's law shows that a σ-field contains as well the intersection of each countable collection of its members.

A real or complex valued function μ defined in a field Φ is called a **set function**. A set function is **real** if its values are real; it is **positive** if its values are nonnegative. We call μ a **measure** if it satisfies $\mu(\emptyset) = 0$ and

$$\mu\left(\bigcup e_j\right) = \sum \mu(e_j) \tag{2.0.1}$$

for every finite collection $e_1, e_2, \ldots, e_n$ of pairwise disjoint sets in Φ. A measure μ is **countably additive** or σ-**additive** in a field Φ if (2.0.1) holds as well for countable families $\{e_j\}$ of pairwise disjoint sets in Φ whose union belongs to Φ. Since countably additive measures are more the rule than the exception, everybody uses the word "measure" for "countably additive measure"; if a measure is not countably additive, it is called a **finitely additive measure**. All definitions above make sense if μ takes values in a linear space. Such measures make an appearance in Chapter 10.

If μ is a real set function, the values $-\infty$ and $+\infty$ may be admitted; if they are not, μ is called **finite**. If μ is a measure, $+\infty$ does not interfere with (2.0.1), however, the values $-\infty$ and $+\infty$ should not be both present to avoid senseless expressions such as $-\infty + (+\infty)$.

We outline a classical method, due to Carathéodory, to construct measures from arbitrary set functions, and use it below to construct the Lebesgue measure. The first step (Theorem 2.0.1) is on finitely additive measures and will come in handy in Part III.

Let λ be an arbitrary set function λ defined in a field Φ (the value $+\infty$ admitted). A set $d \in \Phi$ is called a λ-**set** if and only if

$$\lambda(e) = \lambda(e \cap d) + \lambda(e \cap d^c) \tag{2.0.2}$$

for all $e \in \Phi$, where d^c is the complement of d; the possibility $+\infty = +\infty$ is admitted. In words, (2.0.2) means that d "divides any set $e \in \Phi$ additively with respect to λ." Magically, this condition produces a subfield Φ_λ of Φ over which λ graduates to a finitely additive measure. We call e in (2.0.2) the **test set**.

Theorem 2.0.1 (Carathéodory). *Let the set function λ satisfy $\lambda(\emptyset) = 0$. Then the family $\Phi_\lambda \subseteq \Phi$ of all λ-sets in Φ is a field on which λ is a finitely additive measure.*

Proof. For $d = \emptyset$ and $d = S$ (2.0.2) reduces to $\lambda(e) = \lambda(e) + \lambda(\emptyset)$. Moreover, if (2.0.2) holds for d, it holds as well for d^c. Thus, to show that Φ_λ is a field and that λ is a finitely additive measure in Φ_λ, it suffices to show that if $\{c_1, c_2, \ldots\}$ is a finite collection of sets in Φ_λ then $\cup c_j \in \Phi_\lambda$ and that $\lambda(\cup c_j) = \Sigma \lambda(c_j)$ if the sets are pairwise disjoint. We do a little better on the last equality showing that

$$\lambda \left(e \cap \left(\bigcup c_j \right) \right) = \sum \lambda(e \cap c_j) \tag{2.0.3}$$

for every $e \in \Phi$ if the sets $c_j \in \Phi_\lambda$ are pairwise disjoint (to deduce additivity we take $e = S$). It is obviously sufficient to work with two sets $c, d \in \Phi_\lambda$. Pick $e \in \Phi$. Since c is a λ-set,

$$\lambda(e \cap d) = \lambda(e \cap d \cap c) + \lambda(e \cap d \cap c^c) \quad \text{(test set } e \cap d\text{)}, \tag{2.0.4}$$

and, since d is a λ-set,

$$\lambda(e) = \lambda(e \cap d) + \lambda(e \cap d^c) \quad \text{(test set } e\text{)},$$

$$\lambda(e \cap (c \cap d)^c) = \lambda(e \cap (c \cap d)^c \cap d) + \lambda(e \cap (c \cap d)^c \cap d^c)$$
$$\text{(test set } e \cap (c \cap d)^c\text{)}. \tag{2.0.5}$$

We have $(c \cap d)^c \cap d = (c^c \cup d^c) \cap d = (c^c \cap d) \cup (d^c \cap d) = c^c \cap d$ and $(c \cap d)^c \cap d^c = (c^c \cup d^c) \cap d^c = (c^c \cap d^c) \cup (d^c \cap d^c) = d^c$, thus, the last equality can be written

$$\lambda(e \cap (c \cap d)^c) = \lambda(e \cap c^c \cap d) + \lambda(e \cap d^c). \tag{2.0.6}$$

Combining (2.0.4) and the first inequality (2.0.5), we obtain

$$\lambda(e) = \lambda(e \cap c \cap d) + \lambda(e \cap c^c \cap d) + \lambda(e \cap d^c) \qquad (2.0.7)$$

and then replace $\lambda(e \cap c^c \cap d) = \lambda(e \cap (c \cap d)^c) - \lambda(e \cap d^c)$ as given by (2.0.6) on the right-hand side of (2.0.7). The result is

$$\lambda(e) = \lambda(e \cap c \cap d) + \lambda(e \cap (c \cap d)^c). \qquad (2.0.8)$$

We have subtracted $\lambda(e \cap d^c)$, which is unjustified if $\lambda(e \cap d^c) = +\infty$. However, if $\lambda(e \cap d^c) = +\infty$, then by (2.0.6), $\lambda(e \cap (c \cap d)^c) = +\infty$, and by (2.0.5), $\lambda(e) = +\infty$, thus (2.0.8) holds anyway.

It follows from (2.0.8) that $c \cap d$ is a λ-set for arbitrary λ-sets c, d; since $(c \cup d)^c = c^c \cap d^c$, $c \cup d$ is a λ-set as well. If c and d are disjoint, we use the fact that d is a λ-set:

$$\lambda(e \cap (c \cup d)) = \lambda(e \cap (c \cup d) \cap d) + \lambda(e \cap (c \cup d) \cap d^c)$$

$$\text{(test set } e \cap (c \cup d)). \qquad (2.0.9)$$

The right side of (2.0.9) equals $\lambda(e \cap c) + \lambda(e \cap d)$, which ends the proof of (2.0.3) and thus of Theorem 2.0.1. ∎

A function λ defined in a σ-field Φ is an **outer measure** if $\lambda(\emptyset) = 0$, $0 \leq \lambda(e) \leq +\infty$ $(e \in \Phi)$, $\lambda(d) \leq \lambda(e)$ for $d, e \in \Phi, d \subseteq e$, and

$$\lambda\left(\bigcup e_j\right) \leq \sum \lambda(e_j) \qquad (2.0.10)$$

for every finite or countable collection $e_1, e_2, \ldots$ in Φ.

Theorem 2.0.2 (Carathéodory). *Assume λ is an outer measure in a σ-field Φ. Then Φ_λ is a σ-field and λ is a σ-additive measure in Φ_λ.*

Proof. Let $\{c_j\}$ be a finite or countable collection in Φ_λ. We may write $c = c_1 \cup c_2 \cup c_3 \cup \cdots = c_1 \cup (c_2 \backslash c_1) \cup (c_3 \backslash (c_1 \cup c_2)) \cup \ldots$, where $b \backslash d = b \cap d^c$ is the set theoretical difference, thus in showing that $c \in \Phi_\lambda$, we may assume that the c_j are pairwise disjoint. The set $c_1 \cup c_2 \cup \cdots \cup c_n$ is a λ-set, so that taking $e \in \Phi$ and using (2.0.3) in the second equality,

$$\lambda(e) = \lambda(e \cap (c_1 \cup \cdots \cup c_n)) + \lambda(e \cap (c \cup \cdots \cup c_n)^c)$$

$$= \lambda(e \cap c_1) + \cdots + \lambda(e \cap c_n) + \lambda(e \cap (c_1 \cup \cdots \cup c_n)^c)$$

$$\geq \lambda(e \cap c_1) + \cdots + \lambda(e \cap c_n) + \lambda(e \cap c^c). \qquad (2.0.11)$$

Since $e = (e \cap c) \cup (e \cap c^c)$, we obtain, combining with (2.0.11),

$$\lambda(e \cap c) + \lambda(e \cap c^c) \geq \lambda(e) \geq \sum_{j=1}^{n} \lambda(e \cap c_j) + \lambda(e \cap c^c)$$

for all n. We then let $n \to \infty$ and use the equality $e \cap c = \cup(e \cap c_j)$ and (2.0.10), obtaining

$$\lambda(e \cap c) + \lambda(e \cap c^c) \geq \lambda(e) \geq \sum_{j=1}^{\infty} \lambda(e \cap c_j) + \lambda(e \cap c^c)$$

$$\geq \lambda(e \cap c) + \lambda(e \cap c^c).$$

This shows that $c \in \Phi_\lambda$ and that

$$\lambda(e \cap c) = \sum_{j=1}^{\infty} \lambda(e \cap c_j), \tag{2.0.12}$$

ending the proof of Theorem 2.0.2. ∎

Remark 2.0.3. If $d \in \Phi$ and $\lambda(d) = 0$, then $d \in \Phi_\lambda$. In fact, we have $\lambda(e) \leq \lambda(e \cap d) + \lambda(e \cap d^c)$; on the other hand, $\lambda(e \cap d^c) \leq \lambda(e)$ and $\lambda(e \cap d) \leq \lambda(d) = 0$, so that $\lambda(e) \geq 0 + \lambda(e \cap d^c) = \lambda(e \cap d) + \lambda(e \cap d^c)$.

Theorems 2.0.1 and 2.0.2 are used in the construction of the Lebesgue measure in $\mathbb{R}^m$. We call $P \subseteq \mathbb{R}^m$ a **parallelepipedon** if

$$P = \{(x_1, x_2, \ldots, x_m); a_j < x_j < b_j, j = 1, 2, \ldots, m\}$$

with the a_j and b_j finite and $a_1 \leq b_1, \ldots, a_m \leq b_m$. The **(hyper)volume** of a parallelepipedon is $v(P) = (b_1 - a_1) \cdots (b_m - a_m)$. We define a set function λ in all subsets of $\mathbb{R}^m$ by the formula

$$\lambda(e) = \inf \sum_{n=1}^{\infty} v(P_n) \tag{2.0.13}$$

the infimum taken over all finite or countable families $\{P_n\}$ of parallelepipeda with

$$e \subseteq \bigcup_{n=1}^{\infty} P_n; \tag{2.0.14}$$

we let $\lambda(e) = +\infty$ if no sum on the right side of (2.0.13) is finite. We prove easily that λ is an outer measure in the field of all subsets of $\mathbb{R}^m$ (called **Lebesgue outer measure**) and that $\lambda(P) = v(P)$ for a parallelepipedon.

Carathéodory's Theorems 2.0.1 and 2.0.2 applied to λ produce a σ-additive measure μ in a σ-field Φ. This measure is the celebrated **Lebesgue measure**, and Φ_λ is the σ-field of all **Lebesgue measurable** (or simply **measurable**) subsets of $\mathbb{R}^m$. We show easily using the definition that any half-space $x_k > a$ or $x_k < b$ is Lebesgue measurable; since any parallelepipedon is a finite intersection of these half spaces, it is Lebesgue measurable. Obviously, any open set in $\mathbb{R}^m$ can be written as a union of parallelepipeda, and (by slightly inflating each one) we may assume that the a_j, b_j in the definition of P are rational. Since the family of all parellelepipeda with rational a_j, b_j is countable, it follows that every open set is a countable union of parallelepipeda, thus is Lebesgue measurable. Hence, their complements, closed sets, are measurable, and so are G_δ and F_σ sets (respectively, countable intersections of open sets and countable unions of closed sets).

Let $\Phi_b(C)$ be the σ-field generated by the closed subsets of $\mathbb{R}^m$ (that is, the intersection of all σ-fields that contain all closed subsets of $\mathbb{R}^m$). Since closed sets are Lebesgue measurable, Φ_λ is among the σ-fields in the intersection; thus we have $\Phi_b(C) \subseteq \Phi_\lambda$. The sets in $\Phi_b(C)$ are called **Borel sets** in $\mathbb{R}^m$; plainly, G_δ and F_σ sets are Borel sets.

By Remark 2.0.3, every $d \subseteq \mathbb{R}^m$ with $\lambda(d) = 0$ belongs to Φ_λ (i.e., is Lebesgue measurable with $\mu(e) = 0$). These sets are called **null sets**. Obviously, countable unions of null sets are null sets, and sets contained in null sets are null sets. Lebesgue measurable sets and Borel sets are "the same modulo null sets" in the following sense: given a Lebesgue measurable set d there exist Borel sets c, e with $c \subseteq d \subseteq e$ and $\mu(e \backslash d) = 0$. More precisely, we may assume that e (resp. c) is a G_δ set (resp. a F_σ set). Accordingly, d is the union of a Borel set and a null set.

In Part I and most of Part II, only Lebesgue measure will be used. To simplify, the Lebesgue measure μ of a measurable set e will be denoted by $|e|$, and we indicate in the same way the outer Lebesgue measure of an arbitrary set $e \subseteq \mathbb{R}^m$.

Let $\Omega \subseteq \mathbb{R}^m$ be Lebesgue measurable. A function $f : \Omega \to \mathbb{R}$ is **countably valued** if it can be written in the form

$$f(x) = \sum f_n \chi_n(x) \tag{2.0.15}$$

(finite or countable sum), where each $\chi_n(\cdot)$ is the characteristic function of a measurable set e_n and the sets e_n are pairwise disjoint with union Ω. The function $f : \Omega \to \mathbb{R}$ is **measurable** if and only if one of the two equivalent properties holds: (a) $f(\cdot)$ is the uniform limit in Ω of a sequence $f_n(\cdot)$ of countably valued functions, (b) $f^{-1}((-\infty, a))$ is Lebesgue measurable for every a. It is clear that the definition doesn't mind modification of f in a null set, thus we may allow a measurable function to be undefined (or infinite) in a null set.

Something happens **almost everywhere** (in short a.e.) in Ω or **for almost all** $x \in \Omega$ if it holds for every $x \in \Omega$ except for x in a null set. For instance, we say that two functions f, g coincide a.e. in Ω if and only if $f(x) = g(x)$ except in a null set, or that a sequence $\{f_n(x)\}$ converges a.e. to $f(x)$ if $\lim_{n \to \infty} f_n(x) = f(x)$ for almost all x. If this is the case and the $f_n(\cdot)$ are measurable, so is $f(x)$. Countably valued functions are measurable. Sums and products of measurable functions are measurable. A continuous function is measurable; this follows immediately from (b). See Natanson [1955] for many other properties of measurable functions and also the Examples at the end of the section.

A countably valued function $f(x)$ is (Lebesgue) **integrable** if $\Sigma |f_n| \mu(e_n) < \infty$; its **integral** is

$$\int_\Omega f(x)dx = \sum f_n \mu(e_n), \qquad (2.0.16)$$

where, if $f_n = 0$ we take $f_n \mu(e_n) = |f_n| \mu(e_n) = 0$ even if $\mu(e_n) = +\infty$. Although a given countably valued function can be written in the form (2.0.15) in many ways, $\int_\Omega f dx$ does not depend on the particular representation.

A measurable function $f : \Omega \to \mathbb{R}$ is (Lebesgue) **integrable** if there exists a sequence $\{f_n\}$ of integrable countably valued functions such that $f_n(\cdot) \to f(\cdot)$ uniformly in Ω and

$$\int_\Omega |f_n(x) - f_m(x)|dx \to 0 \quad \text{as } m, n \to \infty. \qquad (2.0.17)$$

By definition,

$$\int_\Omega f(x)dx = \lim_{n\to\infty} \int_\Omega f_n(x)dx. \qquad (2.0.18)$$

The definition of integral (or of integrability) does not depend on the particular sequence $\{f_n\}$ chosen. Two functions such that $f(x) = g(x)$ a.e. are integrable or nonintegrable together and, if integrable, have the same integral; a measurable function $f(\cdot)$ is **integrable** if and only if $|f(\cdot)|$ is integrable and

$$\left| \int_\Omega f(x)dx \right| \le \int_\Omega |f(x)|dx. \qquad (2.0.19)$$

The Lebesgue integral has all the expected properties: the class of Lebesgue integrable functions is a subspace over which $\int_\Omega$ is linear, and the integral of a function f such that $f(x) \ge 0$ a.e. is nonnegative. Additivity of domains holds as well.

The integral of $f(\cdot)$ over a set e is **equicontinuous** with respect to e in the sense that, given $\varepsilon > 0$, there exists $\delta > 0$ with

$$\int_e |f(x)|dx \le \varepsilon \quad \text{if } e \text{ is measurable and } |e| \le \delta. \qquad (2.0.20)$$

In applications, one has to pass to the limit under the integral sign, which means justifying

$$\int_\Omega f(x)dx = \lim_{n\to\infty} \int_\Omega f_n(x)dx \qquad (2.0.21)$$

for a sequence $\{f_n(\cdot)\}$ such that $\lim_{n\to\infty} f_n(x) = f(x)$ a.e.

Theorem 2.0.4 (Vitali). *Let $\{f_n(\cdot)\}$ be a sequence of integrable functions converging to $f(x)$ almost everywhere. Assume that (a) The integrals of the $\{f_n(\cdot)\}$ are equicontinuous in the sense that, given $\varepsilon > 0$ there exists $\delta > 0$ with*

$$\int_e |f_n(x)|dx \leq \varepsilon \quad \text{if } e \text{ is measurable and } |e| < \delta \qquad (2.0.22)$$

uniformly with respect to n, (b) for each $\varepsilon > 0$ there exists a measurable set $\Omega_0 \subseteq \Omega$ of finite measure such that

$$\int_{\Omega \setminus \Omega_0} |f_n(x)|dx \leq \varepsilon \qquad (2.0.23)$$

for all n. Then $f(\cdot)$ is integrable and (2.0.21) holds.

For a proof, see Dunford–Schwartz [1958, p. 150]. Of course (2.0.23) is unnecessary when $\mu(\Omega) < +\infty$.

As a consequence of Vitali's theorem we obtain

Theorem 2.0.5 (Lebesgue's dominated convergence theorem). *Let $\{f_n(\cdot)\}$ be a sequence of integrable functions converging to $f(\cdot)$ almost everywhere. Assume that*

$$|f_n(x)| \leq g(x) \quad (x \in \Omega) \qquad (2.0.24)$$

where $g(x)$ is integrable. Then (2.0.21) holds.

Theorem 2.0.5 follows immediately from Theorem 2.0.4 using equicontinuity of the integral of $g(\cdot)$ (see (2.0.20)).

Below, $m = 1$ and $\Omega = [0, T]$; $f(\cdot)$ is a function integrable in $[0, T]$. A point $t \in (0, T)$ is said to be a **Lebesgue point** of $f(t)$ if

$$\lim_{h \to 0} \frac{1}{h} \int_t^{t+h} |f(\tau) - f(t)|d\tau = 0 \qquad (2.0.25)$$

(the limit is two-sided).

Theorem 2.0.6. *Almost every point of $[0, T]$ is a Lebesgue point of f.*

For the proof (of a much more general result) see Dunford–Schwartz [1958, p. 217]. At a Lebesgue point we have

$$\frac{1}{h} \int_t^{t+h} f(\tau)d\tau \to f(t). \qquad (2.0.26)$$

Theorem 2.0.6 implies that, if $g(\cdot)$ is integrable in $0 \leq t \leq T$ and

$$f(t) = C + \int_0^t g(\tau)d\tau, \qquad (2.0.27)$$

then $g'(t)$ exists almost everywhere and

$$f'(t) = g(t). \tag{2.0.28}$$

It follows directly from equicontinuity of the integral that $f(\cdot)$ is **absolutely continuous** in the sense that, for every $\varepsilon > 0$ there exists $\delta > 0$ such that

$$\sum |f(b_j) - f(a_j)| \leq \varepsilon$$

for every finite collection $\{(a_j, b_j)\}$ of pairwise disjoint subintervals of the interval $[0, T]$ such that

$$\sum (b_j - a_j) \leq \delta.$$

Conversely, we have

Theorem 2.0.7. *Let $f(\cdot)$ be absolutely continuous in $0 \leq t \leq T$. Then the derivative $g(t) = f'(t)$ exists almost everywhere in $0 \leq t \leq T$. Moreover, $g(\cdot)$ is integrable in $0 \leq t \leq T$ and (2.0.27) holds for some C.*

For a proof, see Natanson [1955, p. 266] or Kolmogorov–Fomin [1970, p. 314 ff].

A **metric space** is a set V endowed with a real valued **metric** or **distance** $d(u, v)$, defined for pairs of elements of V and satisfying

$$d(u, v) \geq 0$$
$$d(u, v) = 0 \quad \text{if and only if } u = v$$
$$d(u, v) = d(v, u)$$
$$d(u, w) \leq d(u, v) + d(v, w)$$

for $u, v, w \in V$. The last property is called the **triangle inequality**. A sequence $\{u_n\} \subseteq V$ is **convergent** to $u \in V$ if $d(u_n, u) \to 0$; u is called the **limit** of $\{u_n\}$. A set $A \subseteq V$ is **closed** if it contains the limit of every convergent sequence contained in A. A **Cauchy sequence** in V is a sequence $\{u_n\}$ such that $d(u_n, u_m) \to 0$ as $m, n \to \infty$. A **complete** metric space is one where every Cauchy sequence is convergent.

Let E be a linear space over the real or the complex field. A real valued function $\| \cdot \|$ defined in E is a **norm** if it satisfies

$$\|y\| \geq 0$$
$$\|y\| = 0 \quad \text{if and only if } y = 0$$
$$\|\lambda y\| = |\lambda| \, \|y\|$$
$$\|y + z\| \leq \|y\| + \|z\|$$

where $y, z \in E$ and λ is a scalar. A space E so equipped is called a **normed** space, and it is a metric space with $d(y, z) = \|y - z\|$. If this metric space is complete we call E a **Banach space**; when necessary, we specify a **real** or a **complex** Banach space. The simplest example of real Banach space is Euclidean space $E = \mathbb{R}^m$ endowed with its usual norm

$$\|y\| = \|(y_1, y_2, \ldots, y_m)\| = (\Sigma|y_j|^2)^{1/2}. \tag{2.0.29}$$

Other norms may be used, for instance $\|y\|_\infty = \|(y_1, y_2, \ldots, y_m)\|_\infty = \max |y_j|$ or the p-analogue of the Euclidean norm $\|y\|_p = (\Sigma|y_j|^p)^{1/p}$, $1 \leq p < \infty$. All these norms (and all others) are **equivalent** in the sense that, if $\| \cdot \|, \| \cdot \|'$ are two arbitrary norms in $\mathbb{R}^m$, then there exist nonzero constants c, C such that $c\| \cdot \| \leq \| \cdot \|' \leq C\| \cdot \|$; obviously, equivalent norms in any normed space produce the same convergent (and Cauchy) sequences. In the space of $m \times m$ matrices (which is of course "the same as" $\mathbb{R}^{m \times m}$), other norms are useful, for instance, the **operator norm**

$$\|A\| = \max_{\|y\| \leq 1} \|Ay\| \tag{2.0.30}$$

associated with a given norm $\| \cdot \|$ in $\mathbb{R}^m$. Operator norms corresponding to different norms are equivalent (in fact, equivalent to any other norm in $\mathbb{R}^{m \times m}$). If the Euclidean norm (2.0.29) is used and A is a symmetric matrix $((Ax, y) = (x, Ay)$ for all $x, y \in \mathbb{R}^m)$, the operator norm can also be calculated in the form

$$\|A\| = \max_{\|y\| \leq 1} |(Ay, y)|. \tag{2.0.31}$$

To see this, let $\{y_k\}$ be an orthonormal basis of $\mathbb{R}^m$ consisting of eigenvectors of A. Then $\|Ay\|^2 = (Ay, Ay) = (\Sigma\lambda_j\alpha_j y_j, \Sigma\lambda_j\alpha_j y_j) = \Sigma\lambda_j^2\alpha_j^2$, and $(Ay, y) = (\Sigma\lambda_j\alpha_j y_j, \Sigma\alpha_j y_j) = \Sigma\lambda_j\alpha_j^2, \{\lambda_j\}$ the corresponding eigenvalues; thus, both (2.0.30) and (2.0.31) give $\max |\lambda_j|$. For a proof that does not use eigenvectors (and thus applies to operators), see Riesz–Nagy [1955, p. 229]. Obvious modifications apply to complex matrices.

A **scalar** or **inner** product in a complex linear space E is a function $(\cdot, \cdot)$ defined for pairs of elements of E, taking complex values and satisfying

$$(x, x) \geq 0$$
$$(x, x) = 0 \quad \text{if and only if } x = 0$$
$$(x, y) = \overline{(y, x)}$$
$$(x, y + z) = (x, y) + (x, z)$$
$$(x, \lambda y) = \lambda(x, y).$$

For a real linear space, $(\cdot, \cdot)$ takes real values, and complex conjugation is unnecessary in the third law. A (real or complex) space so equipped is called an

inner product space. An inner product gives rise to a norm through the formula $|x| = +\sqrt{(x,x)}$, thus an inner product space is a normed space. The inner product and its associated norm are related by the **Cauchy-Schwarz inequality**

$$|(x,y)| \leq \|x\|\|y\| \tag{2.0.32}$$

(see Goffman–Pedrick [1983, p. 165]). When complete, H is called a **Hilbert space**. The simplest complex (resp. real) Hilbert space is $\mathbb{C}^m$ (resp. $\mathbb{R}^m$) endowed with the inner product

$$(x,y) = \sum_{j=1}^{m} \bar{x}_j y_j \tag{2.0.33}$$

(drop conjugates for $\mathbb{R}^m$).

Projection Theorem 2.0.8. *Let N be a convex closed set in a Hilbert space H and let $z \in H$. Then there exists a unique element $y_0 \in N$ such that*

$$\|z - y_0\| = \min_{y \in N} \|z - y\|. \tag{2.0.34}$$

We have

$$(z - y_0, y - y_0) \leq 0 \quad (y \in N). \tag{2.0.35}$$

The proof is in almost every functional analysis textbook; see, for instance Goffman–Pedrick [1983, p. 169]. We note that if N is a subspace, we have equality in (2.0.35); in fact, $y - y_0$ is an arbitrary element of the subspace, thus (2.0.35) will hold as well for $y_0 - y$, and we have

$$(z - y_0, y) = 0 \quad (y \in N). \tag{2.0.36}$$

Given a set $N \subseteq H$, we define

$$N^\perp = \{x \in H; (x,y) = 0 \ (y \in N)\}. \tag{2.0.37}$$

It is obvious that $N^\perp$ is a closed subspace of H.

Corollary 2.0.9. *Let N be a closed subspace of H. Then*

$$N = N \oplus N^\perp$$

in the sense that every $z \in H$ can be uniquely written in the form $z = y + x$ with $y \in N, x \in N^\perp$ and $\|z\|^2 = \|y\|^2 + \|x\|^2$.

Proof. Applying the Projection Theorem 2.0.8 we find $y \in N$ such that $z - y \in N^\perp$; we set $x = z - y$. The norm relation is immediate from (2.0.36), and uniqueness follows taking the difference of two alleged decompositions of z. ∎

If E is a complex Banach, a **linear functional** on E is a function $\phi : E \to \mathbb{C}$ such that

$$\phi(y + z) = \phi(y) + \phi(z), \qquad \phi(\alpha y) = \alpha\phi(y) \quad (y, z \in E, \alpha \in \mathbb{C}).$$

In a real Banach space the definition is the same but $\phi : E \to \mathbb{R}$ and $\alpha \in \mathbb{R}$. A linear functional in E is continuous if and only if it is **bounded**:

$$|\phi(y)| \leq C\|y\| \quad (y \in E).$$

The **norm** $\|\phi\|$ of a bounded linear functional is the least C such that the inequality holds; in other words,

$$\|\phi\| = \sup_{\|y\| \leq 1} |\phi(y)|. \tag{2.0.38}$$

The bounded linear functionals in E are a linear space E^* under the natural definitions of addition and multiplication by a scalar. Endowed with the norm (2.0.38), E^* becomes a normed space (in fact, a Banach space; see Goffman–Pedrick [1983, p. 73]). We call E^* the **dual** of E. Elements of E will also be called x^*, y^*, ... the action of $y^* \in E^*$ on $y \in E$ indicated by

$$y^*(y) = \langle y^*, y \rangle \quad \text{or} \quad y^*(y) = \langle y, y^* \rangle. \tag{2.0.39}$$

The functional $\langle \cdot, \cdot \rangle$ in $E^* \times E$ is the **duality pairing** of E and E^*.

Characterization of the dual is simple for Hilbert spaces.

Theorem 2.0.10. *Let ϕ be a bounded linear functional in a real or complex Hilbert space H. Then there exists a unique $y_\phi \in H$ with*

$$\phi(y) = (y_\phi, y) \quad (y \in H), \qquad \|y_\phi\|_H = \|\phi\|_{H^*}. \tag{2.0.40}$$

Proof. The functional ϕ is linear, so that $N = \{z; \phi(z) = 0\}$ is a subspace of H. Since ϕ is continuous, N is closed. If $\phi \equiv 0$ there is nothing to prove; take $y_\phi = 0$. If $\phi \not\equiv 0$ then $N \neq H$, hence $N^\perp \neq 0$. If $y, z \in N^\perp$, we can find α, β such that $\phi(\alpha y + \beta z) = \alpha\phi(y) + \beta\phi(z) = 0$, thus, $\alpha y + \beta z \in N$; since $\alpha y + \beta z \in N^\perp$ as well and $N \cap N^\perp = \{0\}$, $\alpha y + \beta z = 0$. Accordingly, $N^\perp$ is one-dimensional, that is, $N^\perp = \{\alpha y_0; \alpha \in \mathbb{C}\}$; we may assume that $\|y_0\| = 1$. Take z arbitrary and, using Corollary 2.0.9 write $z = \alpha y_0 + x, x \in N$. Then $\phi(z) = \alpha\phi(y_0)$, so that the first relation (2.0.40) will be satisfied for $y_\phi = \phi(y_0)y_0$. The second relation is obvious. ∎

Theorem 2.0.10 shows that the dual H^* can be identified algebraically and metrically with H; in other words, H^* "is" H, or $H^* = H$ under the identification (2.0.40).

A metric space V is called **separable** if there exists a sequence $\{u_n\}$ **dense** in V in the sense that, for each $u \in V$ and each $\varepsilon > 0$ there exists an element u_k with $d(u, u_k) \leq \varepsilon$; the definition applies in particular in normed and inner product spaces.

A sequence $\{y_n\}$ in a Hilbert space H is **weakly convergent** to $y \in H$ if and only if

$$(y_n, z) \to (y, z)$$

for every $z \in E$.

Theorem 2.0.11. *Let H be a separable Hilbert space, $\{y_n\}$ a sequence in H with $\|y_n\| < C$. Then there exists a subsequence of $\{y_n\}$ (denoted with the same symbol) and an element $\bar{y} \in H$, $\|\bar{y}\| \leq C$ such that*

$$y_n \to \bar{y} \quad weakly.$$

The proof is an application of Cantor's **diagonal sequence** method. Since it will be used several times later, it is worth learning. Let $\{z_n\}$ be a sequence dense in H. Then, using the Bolzano-Weierstrass theorem we can select a subsequence $\{y_{1n}\}$ of $\{y_n\}$ such that (y_{1n}, z_1) is convergent, then a subsequence $\{y_{2n}\}$ of this subsequence such that (y_{2n}, z_2) is convergent, ... and so on. Arrange all these subsequences in an infinite table (or matrix):

$$
\begin{array}{cccc}
y_{11} & y_{12} & y_{13} & \cdots \\
y_{21} & y_{22} & y_{23} & \cdots \\
y_{31} & y_{32} & y_{33} & \cdots \\
\multicolumn{4}{c}{\cdots\cdots\cdots\cdots\cdots}
\end{array}
$$

This table has the following properties: (a) each row $\{y_{kn}\}$ is a subsequence of each of the rows above, (b) (y_{kn}, z_j) is convergent for $1 \leq j \leq k$. The **diagonal sequence** $\{y_{11}, y_{22}, y_{33}, \ldots\}$ is (except for a finite number of initial elements) a subsequence of each $\{y_{kn}\}$, thus (y_{nn}, z_j) is convergent for $j = 1, 2, \ldots$. Rename $\{y_{nn}\}$ as y_n. For $z \in H$ arbitrary write $(y_n, z) = (y_n, z - z_j) + (y_n, z_j)$ and use boundedness of $\|y_n\|$ and convergence of (y_n, z_j). We obtain convergence of (y_n, z) for all $z \in H$. Define a linear functional in H by

$$\phi(y) = \lim_{n \to \infty} (y_n, y).$$

Since $\|y_n\| \leq C$ we have $\|\phi\| \leq C$ and, by Theorem 2.0.10, there exists $y_\phi \in H$ such that $\phi(y) = (y_\phi, y)$. This element fulfills the claims in Theorem 2.0.11. ∎

A set $D \subseteq H$ (H a separable Hilbert space) is **weakly closed** if and only if it contains the limit of all weakly convergent sequences contained in D. Every weakly closed set is closed, but the converse is not true if H is infinite dimensional. However, we have

Theorem 2.0.12. *Let D be convex and closed. Then D is weakly closed.*

Although a more general result will be proved in Chapter 5, this is a quick Hilbert space proof. Assume there is a sequence $\{y_n\} \subseteq D$ such that $y_n \to z \notin D$ weakly. By the Projection Theorem there exists $x \in D$ with $(z - x, y - x) \leq 0$ $(y \in D)$. Putting $y = y_n$ and taking weak limits we get $(z - x, z - x) \leq 0$, so that $z = x$, a contradiction. ∎

We introduce two (classes of) Banach spaces. The first is $C(K; \mathbb{R}^m)$ (K a closed bounded set in $\mathbb{R}^k$), the space of all $\mathbb{R}^m$-valued continuous functions $f(\cdot) = (f_1(\cdot), f_2(\cdot), \ldots, f_m(\cdot))$ defined in K and equipped with the norm

$$\|f\| = \max_{x \in K} \|f(x)\|. \tag{2.0.41}$$

We check easily that $C(K; \mathbb{R}^m)$ is a Banach space. The second is $L^p(\Omega; \mathbb{R}^m)$ (Ω a measurable subset of $\mathbb{R}^k$), consisting of all vectors $f(\cdot) = (f_1(\cdot), f_2(\cdot), \ldots, f_m(\cdot))$ of measurable functions defined in Ω and such that

$$\|f\|_p = \left(\int_\Omega \|f(x)\|^p dx \right)^{1/p} < \infty. \tag{2.0.42}$$

Strictly speaking, $\| \cdot \|_p$ cannot be a norm since $\|f\|_p = 0$ does not imply $f(x) \equiv 0$ but only $f(x) = 0$ a.e. Hence, the correct definition of $L^p(\Omega; \mathbb{R}^m)$ is by equivalence classes with respect to the equivalence relation: $f(\cdot) \approx g(\cdot)$ if and only if $f(x) = g(x)$ a.e. Otherwise, all norm properties for $\| \cdot \|_p$ are obvious except the triangle inequality,

$$\left(\int_\Omega \|f(x) + g(x)\|^p dx \right)^{1/p} \leq \left(\int_\Omega \|f(x)\|^p dx \right)^{1/p} + \left(\int_\Omega \|g(x)\|^p dx \right)^{1/p} \tag{2.0.43}$$

which is known as **Minkowski's inequality**. It is plain for $p = 1$; for $p > 1$ it is a consequence of **Hölder's inequality**

$$\int_\Omega |(f(x), g(x))| dx \leq \left(\int_\Omega \|f(x)\|^p dx \right)^{1/p} \left(\int_\Omega \|g(x)\|^q dx \right)^{1/q} \tag{2.0.44}$$

where $1/q + 1/p = 1$ and $(\cdot, \cdot)$ is the scalar product in $\mathbb{R}^m$. For $p = 2$, Hölder's inequality is simply the **Schwarz inequality** corresponding to the scalar product

$$(f, g) = \int_\Omega (f(x), g(x)) \, dx \tag{2.0.45}$$

which makes $L^2(\Omega, \mathbb{R}^m)$ an inner product space. We complete the definition of the spaces $L^p(\Omega; \mathbb{R}^m)$ for $p = \infty$; $L^\infty(\Omega; \mathbb{R}^m)$ is the space of all vectors

$f(\cdot) = (f_1(\cdot), f_2(\cdot), \ldots, f_m(\cdot))$ of measurable functions defined in Ω and such that

$$\|f\|_\infty = \text{ess. sup}_{x \in \Omega} \|f(x)\| < \infty \qquad (2.0.46)$$

equipped with $\| \cdot \|_\infty$. Here, the ess. sup (essential supremum) of a real valued measurable function $\eta(\cdot)$ is the least C such that $\eta(x) \leq C$ a.e. The spaces $L^p(\Omega; \mathbb{R}^m)$ are complete, thus Banach spaces (Goffman–Pedrick [1983, p. 135]); $L^2(\Omega; \mathbb{R}^m)$ is a Hilbert space. L^p convergence does not imply pointwise convergence even almost everywhere; however, if the sequence $\{f_n(\cdot)\}$ is convergent in L^p to $f(\cdot)$ there exists a subsequence a.e. convergent to $f(\cdot)$ (Goffman–Pedrick [1983, p. 119]).

Let U be an arbitrary set, $G : U \to U$. A point $u \in U$ is a **fixed point** of G if $G(u) = u$.

Theorem 2.0.13 (Cacciopoli–Banach). *Let U be a complete metric space, and let $G : U \to U$ satisfy*

$$d(G(u), G(v)) \leq \alpha d(u, v) \quad (u, v \in U) \qquad (2.0.47)$$

with $\alpha < 1$. Then G has a unique fixed point in U.

Proof. Select $u_0 \in U$ arbitrary and define inductively a sequence $\{u_n\} \subseteq U$ by $u_{n+1} = G(u_n), n = 1, 2, \ldots$. We have $d(u_n, u_{n-1}) \leq \alpha d(u_{n-1}, u_{n-2}) \leq \cdots \leq \alpha^{n-1} d(u_1, u_0)$. It follows that if $m < n$ then

$$d(u_n, u_m) = \sum_{k=m+1}^n d(u_k, u_{k-1}) \leq d(u_1, u_0) \sum_{k=m+1}^n \alpha^{k-1} \leq \alpha^m/(1 - \alpha),$$

so that $\{u_n\}$ is a Cauchy sequence, hence convergent to $u \in U$. Since (2.0.47) implies continuity of G, u is a fixed point of G. Uniqueness follows noting that if u, u' are fixed points, then $d(u, u') = d(G(u), G(u')) \leq \alpha d(u, u')$, so that $d(u', u) = 0$. ∎

A subset K of a metric space U is **precompact** if every sequence $\{u_k\} \subseteq K$ has a convergent subsequence; if, in addition, the limit belongs to K then K is **compact**. A bounded subset of $\mathbb{R}^m$ is precompact; if the set is bounded and closed it is compact (consequence of the Bolzano–Weierstrass theorem). The converse is true as well; compact sets in $\mathbb{R}^m$ are bounded and closed.

The following fixed point result is related to compactness.

Theorem 2.0.14 (Schauder). *Let X be a closed convex subset of a Banach space E, G a continuous map from X into a precompact subset $Y \subseteq X$. Then G has a fixed point.*

For a proof, see Kantorovich–Akilov [1964, p. 645] or Schwartz [1969, p. 96].

We call $C(0, T; E)$ the space of all E-valued continuous functions defined in $[0, T]$, where E is a Banach space. Any $f(\cdot) \in C(0, T; E)$ is bounded; if not, there exists a sequence $\{t_n\} \subset [0, T]$ with $\{f(t_n)\}$ unbounded. By compactness of $[0, T]$ we may assume that $t_n \to \bar{t} \in [0, T]$, which contradicts continuity at $\bar{t}$. We prove similarly that $f(\cdot)$ assumes its maximum, and define

$$\|f\| = \max_{0 \le t \le T} \|f(t)\|_E. \tag{2.0.48}$$

We show easily that $C(0, T; E)$ is a Banach space. A set $\mathcal{F} \subseteq C(0, T; E)$ is **equicontinuous** if for every $\varepsilon > 0$ there exists $\delta > 0$ such that $\|f(t') - f(t)\| \le \varepsilon$ for $|t' - t| \le \delta$ for every $f(\cdot) \in \mathcal{F}$.

Theorem 2.0.15 (Arzelà–Ascoli). *Assume the sequence* $\{f_n(\cdot)\} \subseteq C(0, T; E)$ *is equicontinuous and that* $\{f_n(t); n = 1, 2, \ldots\} \subseteq E$ *is precompact for each* $t \in [0, T]$. *Then* $\{f_n(\cdot)\}$ *has a subsequence convergent in* $C(0, T; E)$.

Proof. Let $\{t_k\}$ be a countable dense set (say, the rationals) in $[0, T]$. Using precompactness select a subsequence $\{f_{1n}(\cdot)\}$ such that $\{f_{1n}(t_1)\}$ is convergent, a subsequence $\{f_{2n}(\cdot)\}$ of $\{f_{1n}(\cdot)\}$ such that $\{f_{2n}(t_2)\}$ is convergent, ... and apply the diagonal sequence trick to obtain a subsequence $\{f_{nn}(\cdot)\}$ such that $\{f_{nn}(t_k)\}$ is convergent for all k. Given $\varepsilon > 0$, select $\delta > 0$ such that $\|f_n(t) - f_n(t')\| \le \varepsilon/3$ if $\|t - t'\| \le \delta$ for all n and a finite subset $\{t_1, t_2, \ldots, t_n\}$ of $\{t_k\}$ such that for every $t > 0$ there exists t_j, $1 \le j \le n$ with $\|t - t_j\| \le \delta$. Finally, take n_0 so large that $\|f_{nn}(t_j) - f_{mm}(t_j)\| \le \varepsilon/3$ for $n, m > n_0, j = 1, 2, \ldots, n$. We then have

$\|f_{nn}(t) - f_{mm}(t)\|$
$$\le \|f_{nn}(t) - f_{nn}(t_j)\| + \|f_{nn}(t_j) - f_{mm}(t_j)\| + \|f_{mm}(t_j) - f_{mm}(t)\| \le \varepsilon$$

for every $t \in [0, T]$. This ends the proof. ∎

There are obvious generalizations; for instance, $[0, T]$ can be replaced by any compact set $K \subseteq \mathbb{R}^m$, or even by any compact metric space, and E can be any metric space (without essential changes in the proof). For an even more general version, see Kelley [1955, Theorem 17, p. 233].

The following two results deal with interchanging the order of integration in a double integral.

Theorem 2.0.16 (Fubini). *Let* $\Omega \subseteq \mathbb{R}^m$, $\Delta \subseteq \mathbb{R}^k$ *be measurable sets and let* $f(x, y)$ *be integrable in* $\Omega \times \Delta$ *with respect to the Lebesgue measure* $dxdy$ *in* $\mathbb{R}^m \times \mathbb{R}^k$. *Then* $f(x, y)$ *is integrable in* Δ *for almost every* $x \in \Omega$ *and integrable in* Ω

for almost every $y \in \Delta$. *The function* $x \to \int_{\Delta} f(x, y) dy$ *is integrable in* Ω, *the function* $y \to \int_{\Omega} f(x, y) dx$ *is integrable in* Δ *and*

$$\int_{\Omega} \left(\int_{\Delta} f(x, y) dy \right) dx = \int_{\Omega \times \Delta} f(x, y) dx dy$$

$$= \int_{\Delta} \left(\int_{\Omega} f(x, y) dx \right) dy. \qquad (2.0.49)$$

Theorem 2.0.17 (Tonelli). *Let* Ω, Δ *be as in Theorem* 2.0.16, $f(x, y)$ *nonnegative and measurable in* $\Omega \times \Delta$. *Assume that* $f(x, \cdot)$ *is integrable in* Δ *for almost all* $x \in \Omega$, *and that* $x \to \int_{\Delta} f(x, y) dy$ *is integrable in* Ω. *Then* $f(x, y)$ *is integrable in* $\Omega \times \Delta$ *(so that Fubini's Theorem applies and the double integral can be computed as in* (2.0.49)).

For proofs (of much more general statements), see Dunford–Schwartz [1958, pp. 190–194].

Example 2.0.18. A function f defined in a linear space E is **homogeneous** if $f(\lambda y) = \lambda f(y)$ $(\lambda \geq 0, y \in E)$. Let f, g be two homogeneous continuous functions in $\mathbb{R}^m$ such that $f(x) = 0$ only if $x = 0$; same for g. Show that there exist nonzero constants c, C such that $c\|f(y)\| \leq \|g(y)\| \leq C\|f(y)\|$ $(y \in \mathbb{R}^m)$. (*Hint:* consider max $\|f(y)\|$ and max $\|g(y)\|$ in $\|x\| = 1$ and the corresponding minima.) As a consequence, any two norms in $\mathbb{R}^m$ are equivalent.

Example 2.0.19. Show that the half-spaces defined by $x_k < a$ or by $x_k > a$ are Lebesgue measurable.

Example 2.0.20. If $d \subseteq \mathbb{R}^m$ is Lebesgue measurable there exists $e \supseteq d$ open such that $|e \backslash d| \leq \varepsilon$. (*Hint:* Since you can write d as a countable union of disjoint bounded measurable sets, you may assume $|d| < \infty$.) Remember that the measure of d coincides with its outer measure.

Example 2.0.21. If $d \subseteq \mathbb{R}^m$ is Lebesgue measurable, there exists a closed set $c \subseteq d$ such that $|d/c| \leq \varepsilon$. (Use Example 2.0.20 and take complements.)

Example 2.0.22. If $d \subseteq \mathbb{R}^m$ is Lebesgue measurable there exist a G_δ set e and a F_σ set c with $c \subseteq d \subseteq e$ and $|e/c| = 0$.

Example 2.0.23. The set $(-\infty, a)$ in the definition (b) of a measurable function may be replaced by $(-\infty, a]$, (b, ∞), $[b, \infty)$; if the inverse image of one of these intervals is Lebesgue measurable the same is true of the others.

Example 2.0.24. Show that definitions (a) and (b) of a measurable function are equivalent.

Example 2.0.25. If $f_n(x)$ is a sequence of measurable functions and $\lim_{n \to \infty} f_n(x)$ $= f(x)$ a.e. then $f(x)$ is measurable.

Example 2.0.26. Show that the integral of a simple function is well defined (that is, does not depend on the particular expression (2.0.15)).

Example 2.0.27. Let $\{f_n(\cdot)\}$ be a sequence of integrable simple functions satisfying (2.0.17). Then there exists $\Omega_0 \subseteq \Omega$ of finite measure such that $\int_{\Omega \setminus \Omega_0} |f_n(x)| dx$ $\leq \varepsilon$ $(n = 1, 2, \ldots)$. This is obviously true for a single f_n or for a finite number $f_1, \ldots, f_m$. Take m so large that $\int_\Omega |f_n(x) - f_m(x)| dx \leq \varepsilon/2$ for $n \geq m$.

Example 2.0.28. If it exists, the integral (2.0.18) does not depend on the sequence $\{f_n\}$. If $\{g_n\}$ is another sequence of simple functions satisfying the assumptions, use Example 2.0.27 for $f_n - g_n$ to show that $\int_\Omega |f_n(x) - g_n(x)| dx \to 0$.

Example 2.0.29. If $\int_\Omega |f(x)| dx = 0$ then $f(x) = 0$ a.e.

Example 2.0.30. Show that $\int_\Omega f(x) \geq 0$ if $f(x) \geq 0$ a.e.

Example 2.0.31. Sometimes, the Lebesgue integral is defined with (2.0.17) and (2.0.18) but only requiring that the sequence $\{f_n(x)\}$ be convergent a.e. to $f(x)$. This seemingly weaker definition of integral (called $w - \int_\Omega f(x) dx$ in the lines below) is equivalent. It is enough to show this for nonnegative functions. We prove first that $w - \int$ is well defined and is nonnegative for a.e. nonnegative functions. This implies $w - \int_\Omega f(x) dx \leq w - \int_\Omega g(x) dx$ if $f(x) \leq g(x)$ a.e. so that the integral is monotonic. If $f(\cdot)$ is a nonnegative w-integrable function, it is measurable (Example 2.0.25); moreover, using Example 2.0.27 we show that there exists $\Omega_0 \subseteq \Omega$ of finite measure such that $w - \int_{\Omega \setminus \Omega_0} f(x) dx \leq \varepsilon$. Let $n \geq 1$, $e_{mn} = \{x \in \Omega; m/n < f(x) \leq (m+1)/n\}$, $m = 0, 1, \ldots, g_n(x) =$ $\Sigma_m (m/n) \chi_{mn}(x)$, χ_{mn} the characteristic function of e_{mn}. Then, since $g_n(x) \leq$ $f(x)$, each g_n is integrable (we argue with partial sums) and $\int_{\Omega \setminus \Omega_0} g_n(x) dx \leq \varepsilon$ for all n. We have $|g_n(x) - g_m(x)| \leq |g_n(x) - f(x)| + |f(x) - g_m(x)| \leq 1/m + 1/n$ so that

$$\int_\Omega |g_n(x) - g_m(x)| dx \leq \varepsilon + |\Omega_0| \left(\frac{1}{m} + \frac{1}{n} \right).$$

2.1. Control Systems Described by Ordinary Differential Equations

Fixing $u(\cdot)$ in the control system

$$y'(t) = f(t, y(t), u(t)), \qquad y(s) = \zeta, \qquad (2.1.1)$$

we may limit ourselves to examine the "uncontrolled" equation

$$y'(t) = f(t, y(t)), \qquad y(s) = \zeta. \tag{2.1.2}$$

The function $f(t, y)$ is defined in $[0, T] \times E$, where $E = \mathbb{R}^m$ and $0 \le s < T$. We study (2.1.2) under either Hypotheses I or II below.

Hypothesis I. $f(t, y)$ is measurable in t for y fixed, and continuous in y for t fixed. For every $c > 0$ there exists $K(\cdot, c) \in L^1(0, T)$ such that

$$\|f(t, y)\| \le K(t, c) \quad (0 \le t \le T, \|y\| \le c). \tag{2.1.3}$$

Hypothesis II. $f(t, y)$ is measurable in t for y fixed. For every $c > 0$ there exist $K(\cdot, c), L(\cdot, c) \in L^1(0, T)$ such that (2.1.3) holds and

$$\|f(t, y') - f(t, y)\| \le L(t, c)\|y' - y\| \quad (0 \le t \le T, \|y\|, \|y'\| \le c). \tag{2.1.4}$$

Hypothesis II is obviously stronger than Hypothesis I.

By definition, solutions $y(\cdot)$ of (2.1.2) are absolutely continuous in $s \le t \le T$ and satisfy (2.1.2) a.e. Under Hypothesis I, the function $t \to f(t, y(t))$ is measurable for any $y(\cdot) \in C(0, T; E)$ (approximate $y(\cdot)$ by a piecewise constant function and use y-continuity of $f(t, y)$), hence it belongs to $L^1(0, T; E)$ by (2.1.3). It follows that the initial value problem (2.1.2) and the integral equation

$$y(t) = \zeta + \int_s^t f(\tau, y(\tau))d\tau. \tag{2.1.5}$$

are equivalent (see **2.0** for properties of absolutely continuous functions). We examine (2.1.5) under Hypothesis II and later (Theorem 2.1.12) under Hypothesis I.

Theorem 2.1.1. *The initial value problem* (2.1.2) *(or, equivalently, the integral equation* (2.1.5)*) has a unique solution in some interval* $s \le t \le T'$, *where* $s < T' \le T$.

Proof. Consider the operator

$$(\mathcal{G}y)(t) = \zeta + \int_s^t f(\tau, y(\tau))d\tau \tag{2.1.6}$$

in the closed ball $B(\zeta, 1) \subseteq C(s, T'; E)$ of center $\zeta(t) \equiv \zeta$ and radius 1. We have

$$\|(\mathcal{G}y)(t) - \zeta\| \le \int_s^{T'} K(\tau, \|\zeta\| + 1)d\tau,$$

thus $\mathcal{G}$ maps $B(\zeta, 1)$ into $B(\zeta, 1)$ for $T' - s \leq \delta_1$ sufficiently small. On the other hand,

$$\|(\mathcal{G}z)(t) - (\mathcal{G}y)(t)\| \leq \max_{s \leq t \leq T'} \|z(t) - y(t)\| \int_s^{T'} L(\tau, \|\zeta\| + 1)d\tau$$

so that again by absolute continuity of the integral, $\mathcal{G}$ is a contraction for $T' - s \leq a$ sufficiently small $\delta_2 \leq \delta_1$. Theorem 2.1.1 then follows from the Cacciopoli–Banach Theorem 2.0.13. ∎

The uniqueness statement in Theorem 2.1.1 is limited to intervals $s \leq t \leq T'$ where $\mathcal{G}$ is a contraction. A more general uniqueness (as well as continuous dependence) result follows from

Lemma 2.1.2. *Let* $\alpha(\cdot), \eta(\cdot) \in L^\infty(0, T)$, $\beta(\cdot) \in L^1(0, T)$, $\alpha(t), \eta(t), \beta(t) \geq 0$ *a.e. Assume that*

$$\eta(t) \leq \alpha(t) + \int_s^T \beta(\tau)\eta(\tau)d\tau \quad (s \leq t \leq T). \tag{2.1.7}$$

Then

$$\eta(t) \leq \alpha(t) + \frac{1}{B(t)} \int_s^t \alpha(\tau)\beta(\tau)B(\tau)d\tau \quad (s \leq t \leq T) \tag{2.1.8}$$

where

$$B(t) = \exp\left(-\int_s^t \beta(\tau)d\tau\right).$$

Proof. Setting

$$N(t) = \int_s^T \beta(\tau)\eta(\tau)d\tau$$

(2.1.7) is $N'(\tau) \leq \alpha(\tau)\beta(\tau) + \beta(\tau)N(\tau)$, hence $B(\tau)N'(\tau) - \beta(\tau)B(\tau)N(\tau) \leq \alpha(\tau)\beta(\tau)B(\tau)$, or $(B(\tau)N(\tau))' \leq \alpha(\tau)\beta(\tau)B(\tau)$. Integrating in $s \leq \tau \leq t$,

$$N(t) \leq \frac{1}{B(t)} \int_s^t \alpha(\tau)\beta(\tau)B(\tau)d\tau.$$

Combining with (2.1.7), (2.1.8) follows. ∎

Corollary 2.1.3 (Gronwall). *Assume* $\alpha(t) = \alpha$ *in (2.1.7). Then*

$$\eta(t) \leq \alpha \exp\left(\int_s^t \beta(\tau)d\tau\right) \quad (s \leq t \leq T). \tag{2.1.9}$$

Proof. Use the equality $B'(\tau) = -\beta(\tau)B(\tau)$ in the integral on the right-hand side of (2.1.8). ∎

Corollary 2.1.4. *Let $y_1(\cdot)$ and $y_2(\cdot)$ be solutions of (2.1.2) in the interval $s \leq t \leq T$ with initial conditions ζ_1 and ζ_2, respectively. Then, if c is a bound for $\|y_1(t)\|$, $\|y_2(t)\|$ in $s \leq t \leq T$, we have*

$$\|y_1(t) - y_2(t)\| \leq \|\zeta_1 - \zeta_2\| \exp\left(\int_s^t L(\tau, c) d\tau \right) \quad (s \leq t \leq T), \qquad (2.1.10)$$

where $L(\cdot, c)$ is the function in (2.1.4). In particular, if $\zeta_1 = \zeta_2 = \zeta$ we have $y_1(t) = y_2(t)$ in $s \leq t \leq T$.

Proof. By (2.1.4) we have

$$\|y_1(t) - y_2(t)\| \leq \|\zeta_1 - \zeta_2\| + \int_s^t L(\tau, c) \|y_1(\tau) - y_2(\tau)\| d\tau,$$

thus (2.1.10) follows from Corollary 2.1.3. ∎

Lemma 2.1.5. *Let $y(t)$ be a solution of (2.1.2) in an interval $[s, T')$. Assume that*

$$\|y(t)\| \leq c \quad (s \leq t < T'). \qquad (2.1.11)$$

Then $y(\cdot)$ can be extended to an interval $[s, T'']$ with $T' > T''$. (that is, a solution of (2.1.2) coinciding with $y(\cdot)$ in $s \leq t < T'$ exists in $[0, T''])$.

Proof. The integral equation (2.1.5) implies

$$\|y(t') - y(t)\| \leq \int_t^{t'} K(\tau, c) d\tau \to 0$$

as $t' - t \to 0$, for $t < t' < T'$. Completeness insures that $\lim_{t \to T'-} y(t)$ exists, thus $y(\cdot)$ can be extended as a continuous function to $[s, T']$ and is a solution of (2.1.5) there. Solving then the integral equation

$$y(t) = y(T') + \int_{T'}^t f(\tau, y(\tau)) d\tau$$

in some interval $[T', T'']$, we obtain the extension. ∎

Corollary 2.1.6. *The solution $y(\cdot)$ of (2.1.2) exists in $s \leq t \leq T$ or in an interval $[s, T_m)$, $T_m \leq T$ and*

$$\limsup_{t \to T_m^-} \|y(t)\| = \infty. \qquad (2.1.12)$$

Proof. Assuming that $y(\cdot)$ does not exist in $s \leq t \leq T$, let $[0, T_m)$ be the union of all intervals $[s, T')$ where $y(\cdot)$ exists. If (2.1.12) is false, then $\|y(t)\|$ is bounded in $0 \leq t < T_m$ and Lemma 2.1.5 applies. ∎

In certain cases, the bound (2.1.11) can be established *a priori*.

Lemma 2.1.7. *Assume that there exists a function* $c(\cdot) \in L^1(0, T)$, $c(t) \geq 0$ *a.e.*
such that

$$(y, f(t, y)) \leq c(t)(1 + \|y\|^2) \quad (0 \leq t \leq T, y \in E). \tag{2.1.13}$$

Then (2.1.11) *holds for any solution* $y(\cdot)$ *in any interval* $[0, T')$.

Proof. We have $(\|y(t)\|^2)' = (y(t), y(t))' = 2(y(t), y'(t)) = 2(y(t), f(t, y(t))) \leq$
$c(t)(1 + \|y(t)\|^2)$. Integrating and applying Corollary 2.1.3, (2.1.11) results. *A pri-*
ori bounds can be obtained by more refined methods, for instance, using Lyapunov
functions (see Elsgolts [1970] or Jordan–Smith [1987]).   ∎

Remark 2.1.8. If $s > 0$, solutions of (2.1.5) can be constructed in intervals $(T', s]$,
$T' < s$. All results above translate to "left-of-s" solutions.

When $f(t, y)$ is defined for all $t \geq 0$, we require Hypothesis II in each finite
interval $[0, T]$ ($K(\cdot, c)$, $L(\cdot, c)$ may depend on T). Of the results above, only
Corollary 2.1.6 needs some reformulation.

Corollary 2.1.9. *Under Assumption II for every* $T > 0$, *either the solution* $y(\cdot)$
exists in $s \leq t < \infty$ *or in an interval* $[s, T_m)$, $T_m < \infty$, *and* (2.1.12) *holds*.

The same applies to solutions for $t < s$ if $f(t, y)$ is defined in $-\infty < t < \infty$.
The linear initial value problem

$$y'(t) = A(t)y(t) + g(t), \qquad y(s) = \zeta \tag{2.1.14}$$

enjoys special privileges. Here, $A(\cdot)$ is a $m \times m$ matrix function (equivalently, a
$\mathbb{R}^{m \times m}$-valued function), and $g(\cdot)$ is a $\mathbb{R}^m$-valued function, both defined in $0 \leq t \leq T$.
If we assume that

$$A(\cdot) \in L^1(0, T; \mathbb{R}^{m \times m}), \quad g(\cdot) \in L^1(0, T; \mathbb{R}^m), \tag{2.1.15}$$

it is obvious that Hypothesis II and (2.1.13) hold, so that there is global existence
and uniqueness.

Solutions of a linear equation can be expressed by means of a useful explicit
formula. For this, we solve the matrix initial value problem

$$\partial_t S(t, s) = A(t)S(t, s) \quad (0 \leq t \leq T), \qquad S(s, s) = I, \tag{2.1.16}$$

($\partial_t = \partial/\partial_t$) or, equivalently, the associated matrix integral equation

$$S(t, s) = I + \int_s^t A(\tau)S(\tau, s)d\tau \tag{2.1.17}$$

in the interval $0 \le t \le T$. Our observations on linear equations apply to (2.1.17); its unique matrix valued solution $S(t, s)$ is called the **solution matrix** of the homogeneous equation

$$y'(t) = A(t)y(t). \tag{2.1.18}$$

In the following result, $\|B\|$ denotes the operator norm of a matrix B corresponding to the usual Euclidean norm $\| \cdot \|$, that is, $\|B\| = \sup_{\|x\| \le 1} \|Bx\|$. (This matrix norm can be substituted by any other matrix norm such as $\|A\| = (\Sigma |a_{jk}|^2)^{1/2}$, possibly with modification of constants, since all matrix norms are equivalent; see **2.0**, especially Example 2.0.18.)

Lemma 2.1.10. (a) The solution matrix $S(t, s)$ is continuous in $0 \le s, t \le T$; precisely, if $\alpha(t) = \|A(t)\|$ then

$$\|S(t', s') - S(t, s)\| \le C \left| \int_t^{t'} \alpha(\tau)d\tau \right| + C \left| \int_s^{s'} \alpha(\sigma)d\sigma \right|. \tag{2.1.19}$$

(b) For each t the (absolutely continuous) function $s \to S(t, s)$ satisfies the matrix differential equation

$$\partial_s S(t, s) = -S(t, s)A(s) \quad (0 \le s \le T), \qquad S(t, t) = I, \tag{2.1.20}$$

almost everywhere or, equivalently, the integral equation

$$S(t, s) = I + \int_s^t S(t, \sigma)A(\sigma)d\sigma. \tag{2.1.21}$$

(c) Let $y(t)$ be the (only) solution of the inhomogeneous initial value problem (2.1.14). Then $y(t)$ is given by the variation-of-constants formula

$$y(t) = S(t, s)\zeta + \int_s^t S(t, \tau)g(\tau)d\tau. \tag{2.1.22}$$

Proof. Note that the absolute value bars outside the integrals in (2.1.19) (and other formulas below) take care of the possibility that $t > t'$ or $s > s'$. To show (a), we note first that, by (2.1.17),

$$\|S(t, s)\| \le 1 + \left| \int_s^t \alpha(\tau)\|S(\tau, s)\|d\tau \right| \tag{2.1.23}$$

so that, by Corollary 2.1.3, $\|S(t, s)\|$ is uniformly bounded in $0 \le s, t \le T$. Write $S(t', s') - S(t, s) = (S(t', s') - S(t, s')) + (S(t, s') - S(t, s))$. The first term is estimated with (2.1.17):

$$\|S(t', s') - S(t, s')\| \le C \left| \int_t^{t'} \alpha(\tau)d\tau \right|,$$

where C is a bound for $\|S(t, s)\|$. The second term is

$$
\begin{aligned}
S(t, s') - S(t, s) = &-\int_s^{s'} A(\tau)S(\tau, s)d\tau \\
&+ \int_{s'}^t A(\tau)(S(\tau, s') - S(\tau, s))d\tau
\end{aligned}
$$

which gives

$$
\begin{aligned}
\|S(t, s') - S(t, s)\| \le &\left| \int_s^{s'} \alpha(\tau)\|S(\tau, s)\|d\tau \right| \\
&+ \left| \int_{s'}^t \alpha(\tau)\|S(\tau, s') - S(\tau, s)\|d\tau \right|.
\end{aligned}
$$

By Corollary 2.1.3,

$$
\|S(t, s') - S(t, s)\| \le C' \left| \int_s^{s'} \alpha(\tau)d\tau \right|.
$$

Putting together both estimations, (a) follows.

There are at least two ways to show (2.1.22). One is to calculate $y'(t)$ directly. The other is to replace the right-hand side in the integral equation corresponding to the initial value problem (2.1.14); everything reduces to

$$
\begin{aligned}
y(t) &= S(t, s)\zeta + \int_s^t S(t, \tau)g(\tau)d\tau \\
&= \zeta + \int_s^t \left(A(\tau) \left(S(\tau, s)\zeta + \int_s^\tau S(\tau, \sigma)g(\sigma)d\sigma \right) + g(\tau) \right) d\tau,
\end{aligned}
$$

which is proved using (2.1.17) and interchanging σ-integration and τ-integration by Fubini's theorem.

To establish (b), we apply the freshly obtained formula (2.1.22) to the function $y(t) \equiv \zeta \in \mathbb{R}^m$ arbitrary. Since $y'(t) = 0 = A(t)y(t) + g(t)$ with $g(t) = -A(t)\zeta$ and $y(t) = \zeta$, we have

$$
\zeta = S(t, s)\zeta - \int_s^t S(s, \sigma)A(\sigma)\zeta d\sigma,
$$

which is (2.1.21). This ends the proof of Lemma 2.1.10. ■

We note a consequence of (2.1.21). Taking adjoints and interchanging s and t we obtain

$$
S(s, t)^* = I - \int_s^t A(\tau)^* S(s, \tau)^* d\tau.
$$

Comparing with the equation

$$S^*(t, s) = I - \int_s^t A(\tau)^* S^*(\tau, s)d\tau,$$

defining the solution matrix of the backwards adjoint equation

$$y(t) = -A(t)^* y(t), \tag{2.1.24}$$

we obtain

Corollary 2.1.11. *Let $S(t, s)$ be the solution matrix of (2.1.18), $S^*(t, s)$ the solution matrix of (2.1.24). Then*

$$S^*(t, s) = S(s, t)^*. \tag{2.1.25}$$

We note the autonomous (that is, time independent) case of (2.1.18)

$$y'(t) = Ay(t), \tag{2.1.26}$$

where the solution operator $S(t, s)$ admits the explicit expression $S(t, s) = e^{A(t-s)}$, the matrix on the right-hand side computed by the exponential series $e^B = \Sigma t^n B^n / n!$ or by the functional calculus (Dunford–Schwartz [1958, p. 556]). In this setting, formula (2.1.25) results from the relation $(e^B)^* = e^{B^*}$.

We close this section with a result on (2.1.2) under Hypothesis I; it guarantees local existence but not uniqueness.

Theorem 2.1.12. *Under Hypothesis I, the initial value problem (2.1.2) (or, equivalently, the integral equation (2.1.5)) has a solution in some interval $s \le t \le T'$, where $s < T' \le T$.*

To prove Theorem 2.1.12 we choose T' (as in the first part of the proof of Theorem 2.1.1) in such a way that the operator $\mathcal{G}$ in (2.1.6) maps the closed ball $B(\zeta, 1) \subseteq C(s, T'; E)$ into itself. Then it is enough to show that $\mathcal{G}$ satisfies the assumptions of Schauder's Theorem 2.0.14, that is, that $\mathcal{G}$ is continuous and that for every sequence $\{y_n(\cdot)\} \subseteq B(\zeta, 1)$ the sequence $\{\mathcal{G}y_n(\cdot)\}$ has a convergent subsequence. The latter follows from the Arzelà–Ascoli Theorem; in fact, $\{(\mathcal{G}y_n)(\cdot)\}$ is uniformly bounded and

$$\|(\mathcal{G}y_n)(t') - (\mathcal{G}y_n)(t)\| \le \int_t^{t'} K(\tau, \|\zeta\| + 1)d\tau$$

so that $\{(\mathcal{G}y_n)(\cdot)\}$ is equicontinuous. To show continuity of $\mathcal{G}$, consider a sequence $\{y_n(\cdot)\}$ convergent in $B(\zeta, 1)$ to $y(\cdot)$. Assume $\{\mathcal{G}y_n\}$ is not uniformly convergent in $s \leq t \leq T'$ to $\mathcal{G}y$. Then there exist $\varepsilon > 0$ and a sequence $\{t_n\} \subset [s, T']$ so that $\|(\mathcal{G}y_n)(t_n) - (\mathcal{G}y)(t_n)\| \geq \varepsilon$; we may assume that $t_n \to \bar{t} \in [s, T']$. Write

$$(\mathcal{G}y_n)(t_n) - (\mathcal{G}y)(t_n) = \int_s^{T'} \chi_n(\tau)\{f(\tau, y_n(\tau))d\tau - f(\tau, y(\tau))\}d\tau,$$

where χ_n is the characteristic function of $[0, t_n]$. On account of y-continuity of $f(t, y)$, the integrand tends to zero (except perhaps at $\bar{t}$) and is bounded; thus, we obtain a contradiction from the dominated convergence Theorem 2.0.5.

Remark 2.1.13. Extension Lemma 2.1.5 generalizes (in a way) to the present setting; in fact, extension of $y(t)$ to the closed interval $s \leq t \leq T'$ only depends on Hypothesis I, and we can solve to the right of T' using Theorem 2.1.12. Accordingly, if $y(t)$ cannot be extended to the right of T', $\|y(t)\|$ must become unbounded as $t \to T'$. The maximal interval of existence $[s, T_m)$ in Corollary 2.1.6 can still be defined, but it does not depend uniquely on the initial condition (two different solutions may even coincide in an initial interval). However, an *a priori* bound still implies existence in the whole interval $s \leq t \leq T$.

Remark 2.1.14. The equivalence between the differential equation (2.1.2) and the integral equation (2.1.5) holds under the weaker assumption that, for every $y(\cdot) \in C(0, T; E)$, the function $t \to f(t, y(t))$ is Lebesgue integrable. However, without y-continuity, even local existence is in question. Assumptions of this type will be used in Part III of this book.

The control system (2.1.1) is handled by reduction to (2.1.2). The **control set** U is an arbitrary set. We denote by $F(0, T; U)$ the set of all functions $u(\cdot)$ defined in $0 \leq t \leq T$ with values in U. The function $f(t, y, u)$ is defined in $[0, T] \times E \times U$ and takes values in E. We postulate a subspace $C_{\text{ad}}(0, T; U) \subseteq F(0, T; U)$ (the **admissible control space**) such that Hypothesis I or II holds for $f(t, y) = f(t, y, u(t))$ for every $u(\cdot) \in C_{\text{ad}}(0, T; U)$.[1]

We call $y(t, u)$ the **solution** or **trajectory** of (2.2.1) corresponding to the control $u(\cdot)$. Under Hypothesis II, $y(t, u)$ is uniquely defined in a maximal interval of existence (depending on $u(\cdot)$). Under Hypothesis I, $y(t, u)$ denotes one of the trajectories corresponding to u (there may be many).

Example 2.1.15. Show the statements on the constant coefficient equation (2.1.26). In particular, show that the series for e^{Az} converges for all complex z uniformly on compact subsets of the complex plane and that e^{Az} can also be expressed by

---

[1] Throughout Parts I and II of this book, we take the point of view of not requiring any particular properties of admissible controls $u(\cdot)$ as elements of $C_{\text{ad}}(0, T; U)$ (this is inevitable if we give no structure to the control set U). The only requirements are on $f(t, y, u(t))$.

the contour integral

$$\frac{1}{2\pi i} \int_C f(\lambda)(\lambda I - A)^{-1} d\lambda \qquad (2.1.27)$$

where $f(z) = e^{\lambda z}$ and C is a simple closed curve enclosing the eigenvalues of A in its interior. See Dunford–Schwartz [1958, Chapter VII].

Example 2.1.16. Let $\lambda_1, \ldots, \lambda_p$ $(p \le m)$ be the eigenvalues of A, m_j the multiplicity of λ_j, so that $m_1 + \cdots + m_p = m$. Let P_j be the matrix obtained from (2.1.27) with $f(z) \equiv 1$, and C a simple closed curve enclosing λ_j in its interior but no other eigenvalue. Then we have

$$e^{Az} = \sum_{j=1}^{p} \sum_{k=0}^{m_j - 1} \frac{z^k e^{\lambda_j z}}{k!} (A - \lambda_j I)^k P_j.$$

(Use the Jordan canonical form or the functional calculus.)

2.2. Existence Theory for Optimal Control Problems

For existence purposes, we consider the control system

$$y'(t) = f(t, y(t), u(t)), \qquad y(0) = \zeta \qquad (2.2.1)$$

in an interval $0 \le t \le T$ with an admissible control space $C_{ad}(0, T; U)$ (see 2.1), state constraint and target condition

$$y(t, u) \in M(t), \qquad y(t, u) \in Y, \qquad (2.2.2)$$

and a cost functional $y_0(t, u)$.[1] A sequence of admissible controls $\{u^n(\cdot)\}$, $u^n(\cdot) \in C_{ad}(0, t_n; U)$ is called a **minimizing sequence** if $y(t, u^n)$ exists in $0 \le t \le t_n$ and

$$\lim_{n \to \infty} \text{dist}(y(t, u^n), M(t)) = 0 \quad (0 \le t \le t_n) \qquad (2.2.3)$$

$$\lim_{n \to \infty} \text{dist}(y(t_n, u^n), Y) = 0, \qquad (2.2.4)$$

$$\limsup_{n \to \infty} y_0(t_n, u^n) \le m, \qquad (2.2.5)$$

where

$$m = \inf y_0(\bar{t}, u), \qquad (2.2.6)$$

the infimum taken in the class of all admissible controls $u(\cdot)$ such that the trajectory $y(t, u)$ satisfies the target condition and the state constraints. If the terminal time $\bar{t}$ is fixed, then $t_n \equiv \bar{t}$.

---

[1] In general, the cost functional $y_0(t, u)$ depends on the solution $y(\tau, u)$ for $0 \le \tau \le t$; thus, it may not be defined in the whole interval.

If there are no admissible controls satisfying the target condition, we say that $m = +\infty$; on the other hand, if $m = -\infty$ the problem obviously has no solution. It is plain that if $m < \infty$ a minimizing sequence must exist; in fact, we can even make sure that the exact target condition and state constraint (2.2.2) hold for each term of the sequence.

The pattern of existence proofs is: given a minimizing sequence, select a subsequence such that $\{u^n(\cdot)\}$ and $\{y(\cdot, u^n)\}$ are convergent; then take limits. However, the following situation is typical: uniform convergence of (a subsequence of) the sequence of trajectories $\{y(t, u^n)\}$ can be achieved (using the Arzelà-Ascoli theorem), but the sequence $\{u^n(\cdot)\}$ is at most weakly convergent in some L^p space. Nonlinearities wreak havoc in this situation, since weak convergence of a sequence $u^n(\cdot)$ does not justify taking limits in $f(t, y(t, u^n), u^n(t))$ or in a cost functional.

In all the examples below, $U \subseteq \mathbb{R}^k$ for some k. We denote by $F(0, T; U)$ the space of all U-valued functions defined in $0 \leq t \leq T$, and by $F_m(0, T; U)$ the subspace of $F(0, T; U)$ consisting of measurable functions with the usual equivalence relation. Unless otherwise stated, $C_{ad}(0, T; U) = F_m(0, T; U)$, and there are no state constraints.

Example 2.2.1.

$$y'(t) = u(t), \qquad y(0) = 0$$
$$U = [0, 1], \qquad Y = \{-1\}, \qquad \bar{t} = 1,$$
$$y_0(t, u): \text{ unspecified.}$$

We have $y(t, u) \geq 0$ for any trajectory; thus, the target cannot be reached. There are no optimal controls or minimizing sequences.

Example 2.2.2.

$$y'(t) = -y(t)^2 u(t), \qquad y(0) = 1$$
$$U = [-1, 1], \qquad Y = \mathbb{R}, \qquad \bar{t} = 1,$$
$$y_0(t, u) = \int_0^t u(\tau)d\tau.$$

The solution is

$$y(t, u) = \left(\int_0^t u(\tau)d\tau + 1 \right)^{-1},$$

thus, we exclude the rogue control $u(t) \equiv -1$ from $C_{ad}(0, \bar{t}; U)$ (it causes the solution to blow up at terminal time). Obviously, $m = -1$. However, this value of $y_0(t, u)$ can only be attained by the rejected control; thus, there are no optimal controls. There are minimizing sequences, for instance, $u^n(t) \equiv 1/n - 1$.

Example 2.2.3.

$$y'(t) = u(t), \qquad y(0) = 0$$

$$U = [0, 1], \qquad Y = \{1\}, \qquad \bar{t}: \text{unspecified},$$

$$y_0(t, u) = \int_0^t u(\tau)^2 d\tau.$$

Minimizing sequences exist: take $u^n(t) \equiv 1/n$ $(0 \le t \le t_n = n)$. We have $y_0(t_n, u^n) = 1/n \to m = 0$. The only aspirant to optimal control is $\bar{u}(t) \equiv 0$, but it does not produce the target condition.

Example 2.2.4.

$$y'(t) = u(t), \qquad y(0) = 0$$

$$U = [0, \infty), \qquad Y = \mathbb{R}, \qquad \bar{t} = 1,$$

$$y_0(t, u) = \int_0^t \frac{1}{1 + u(\tau)} d\tau.$$

Obviously $m = 0$, but no optimal control exists. Minimizing sequences exist, for instance, $u^n(t) \equiv n$.

A proper treatment of the existence problem requires more measure theory and will be postponed until Part III of this book. We limit ourselves here to the system

$$y'(t) = f(t, y(t)) + F(t, y(t))u(t), \qquad y(0) = \zeta. \qquad (2.2.7)$$

The control set U is a subset of $\mathbb{R}^k$, $f(t, y)$ is a $\mathbb{R}^m$-valued function and $F(t, y)$ is a $k \times m$-valued matrix function, both defined in $[0, T] \times \mathbb{R}^m$. The admissible control space is

$$C_{ad}(0, T; U) = F_m(0, T; U) \cap L^2(0, T; \mathbb{R}^k). \qquad (2.2.8)$$

Intersection with L^2 is redundant if U is bounded.

Lemma 2.2.5. *Let U be convex and closed in $\mathbb{R}^k$. Then $C_{ad}(0, T; U)$ is weakly closed in $L^2(0, T; \mathbb{R}^k)$.*

Proof. Clearly $C_{ad}(0, T; U)$ is convex in $L^2(0, T; \mathbb{R}^k)$; accordingly, it is enough to show that it is closed and to apply Theorem 2.0.12. Let $\{u^n(\cdot)\}$ be a sequence in $C_{ad}(0, T; U)$ converging to $\bar{u}(\cdot) \in L^2(0, T; \mathbb{R}^k)$. Then we may select a subsequence convergent a.e. so that the condition $u(t) \in U$ a. e is preserved in the limit. ∎

The cost functional $y_0(t, u)$ is **weakly lower semicontinuous** if, for every sequence $\{u^n(\cdot)\} \subseteq C_{ad}(0, \bar{t}; U)$ such that (a) $u^n(\cdot) \to \bar{u}(\cdot) \in C_{ad}(0, \bar{t}; U)$ weakly in $L^2(0, \bar{t}; \mathbb{R}^k)$, (b) $y(t, u^n)$ exists in $0 \le t \le \bar{t}$ and $y(\cdot, u^n) \to y(\cdot)$ in $C(0, \bar{t}; E)$ we have

$$y_0(\bar{t}, \bar{u}) \le \limsup_{n \to \infty} y_0(\bar{t}, u^n). \tag{2.2.9}$$

(this definition is equivalent to the one with lim inf; it suffices to take subsequences). Let $u^n(\cdot) \in C_{ad}(0, t_n; U), t_n \le T$. Extend $u^n(\cdot)$ to $[0, T]$ by

$$u^n(t) = u = \text{ fixed element of } U \quad (t \ge t_n). \tag{2.2.10}$$

Likewise, extend $y(t, u^n)$ by

$$y(t, u^n) = y(t_n, u^n) \quad (t > t_n), \tag{2.2.11}$$

the extended sequences named in the same way. The cost functional $y_0(t, u)$ is **equicontinuous with respect to** t if for every sequence $\{u^n(\cdot)\}$ as above such that, (a) $t_n \to \bar{t}$, (b) $u^n(\cdot) \to \bar{u}(\cdot) \in C_{ad}(0, T; U)$ weakly in $L^2(0, T; \mathbb{R}^k)$, (c) $y(t, u^n)$ exists in $0 \le t \le t_n$ and $y(\cdot, u^n) \to y(\cdot)$ in $C(0, T; E)$, we have

$$|y_0(t_n, u^n) \to y_0(\bar{t}, u^n)| \to 0 \quad \text{as } n \to \infty. \tag{2.2.12}$$

Theorem 2.2.6. *Assume that $f(t, y)$ and $F(t, y)$ satisfy Hypothesis II in $0 \le t \le T$,[2] $f(t, y)$ with $K_1(\cdot, c), K(\cdot, c) \in L^1(0, \bar{t})$, $F(t, y)$ with $K_2(\cdot, c)$, $L_2(\cdot, c) \in L^2(0, T)$, and that the cost functional $y_0(t, u)$ is weakly lower semicontinuous and equicontinuous with respect to t. Further, assume (a) $-\infty < m < \infty$ (b) U is nonempty, convex, and closed, (c) the state constraint sets $M(t)$ and the target set Y are closed. Then, if there exists a minimizing sequence $\{u^n(\cdot)\}$, $u^n(\cdot) \in C_{ad}(0, t_n; U)$ with $\{t_n\}$ bounded, $\{u^n(\cdot)\}$ bounded in L^2 and $\{y(t, u^n)\}$ uniformly bounded, there exists an optimal control $\bar{u}(\cdot)$, which is the weak L^2 limit of a subsequence of $\{u^n(\cdot)\}$.*

Proof. We extend each $u^n(\cdot)$ and $y(\cdot, u^n)$ to $0 \le t \le T$ as in (2.2.10) and (2.2.11). By boundedness of t_n and of $\{u^n(\cdot)\}$ in $L^2(0, T; \mathbb{R}^m)$, we may assume, passing if necessary to a subsequence, that

$$t_n \to \bar{t}, \quad u^n(\cdot) \to \bar{u}(\cdot) \in C_{ad}(0, T; U) \text{ weakly.}$$

Write the integral equation (2.1.5) for each u^n:

$$y(t, u^n) = \zeta + \int_0^t f(\tau, y(\tau, u^n))d\tau$$

$$+ \int_0^t F(\tau, y(\tau, u^n))u^n(\tau)d\tau \quad (0 \le t \le t_n). \tag{2.2.13}$$

---

[2] In the result below, as in others in the book, statements such as "$f(t, y)$ satisfies Hypothesis II in $0 \le t \le T$" carry the implicit qualifier "for T large enough"; for instance, in Theorem 2.2.6, $T > \bar{t} = \lim t_n$.

Let c be a bound for $\|y(t, u^n)\|$, C a bound for $\|u^n\|_{L^2}$. Using the Schwarz inequality in the second integral,

$$\|y(t', u^n) - y(t, u^n)\|$$

$$\leq \int_t^{t'} K_1(\tau, c)d\tau + C\left(\int_t^{t'} K_2(\tau, c)^2 d\tau\right)^{1/2} \quad (0 \leq t \leq t' \leq t_n) \quad (2.2.14)$$

and we deduce that $\{y(\cdot, u^n)\}$ is equicontinuous in $0 \leq t \leq T$. Then the Arzelà-Ascoli Theorem 2.0.15 can be used to deduce that, if necessary passing to a subsequence, $\{y(\cdot, u^n)\}$ is uniformly convergent to a continuous function $y(\cdot)$ in $0 \leq t \leq T$.

We take limits in (2.2.13). In the first integral we use y-continuity of $f(t, y)$, the bound (2.1.3) in Hypothesis II, and the dominated convergence theorem. In the second, we note that y-continuity of $F(t, y)$, the corresponding bound (2.1.3) with $K_2(\cdot, c) \in L^2$, and again the dominated convergence theorem imply that $\{F(\tau, y(\tau, u^n))\}$ is convergent to $F(\tau, y(\tau))$ in the norm of $L^2(0, \bar{t}; \mathbb{R}^{k \times m})$; equivalently, each entry is convergent in $L^2(0, \bar{t})$. Since $\{u^n(\cdot)\}$ is weakly convergent, we can take limits under the integral sign and obtain

$$y(t) = \zeta + \int_0^t f(\tau, y(\tau))d\tau + \int_0^t F(\tau, y(\tau))\bar{u}(\tau)d\tau \quad (2.2.15)$$

so that $y(t) = y(t, \bar{u})$. The fact that $y(t, \bar{u}) \in M(t)$ for $0 \leq t \leq \bar{t}$ and that $y(\bar{t}, \bar{u}) \in Y$ follows from (2.2.3), (2.2.4) and the uniform convergence of the sequence $\{y(\cdot, u^n)\}$. That

$$y_0(\bar{t}, \bar{u}) = m \quad (2.2.16)$$

results directly from (2.2.5), uniform convergence of $\{y(t, u^n)\}$ and weak lower semicontinuity and t-equicontinuity of $y_0(t, u)$. This ends the proof. $\blacksquare$

Using the duality theory of L^p spaces (see **5.0**), the hypothesis that $K_2(\cdot, c)$, $L_2(\cdot, c) \in L^2(0, T)$ can be weakened to $K_2(\cdot, c)$, $L_2(\cdot, c) \in L^q(0, T)$ with $q > 1$; the control space is $F_m(0, T; U) \cap L^p(0, T; \mathbb{R}^k)$ with $1/p + 1/q = 1$.

This is why Theorem 2.2.6 does not apply to the systems in Examples 2.2.1 to 2.2.4:

Example 2.2.1: There are no minimizing sequences.

Example 2.2.2: There are no minimizing sequences with $\{y(\cdot, u^n)\}$ bounded.

Example 2.2.3: There are no minimizing sequences with $\{t_n\}$ bounded.

Example 2.2.4: There are no minimizing sequences $\{u^n\}$ bounded in L^2 (or any L^p) space.

The following example shows that boundedness in L^1 is not enough.

Example 2.2.7.

$$y'(t) = u(t), \qquad y(0) = 0,$$
$$U = [0, \infty), \qquad Y = \{1\}, \qquad \bar{t} = 1$$
$$y_0(t, u) = \int_0^t \tau u(\tau) \, d\tau$$

$C_{\mathrm{ad}}(0, 1; U) = F(0, 1; U) \cap L^1(0, 1)$. Minimizing sequences bounded in $L^1(0, \bar{t})$ exist, for instance, $u^n(t) = n$ $(0 \le t < 1/n)$, $u^n(t) = 0$ $(1/n \le t \le 1)$, and $m = 0$. However, there is no admissible control with $y(\bar{t}, u) = 1$, $y_0(\bar{t}, u) = 0$; the "optimal control" is $u(t) = \delta(t)$, the Dirac delta.

In many optimal control problems, the control set U is bounded, thus the L^2 boundedness assumption on $\{u^n(\cdot)\}$ is automatically satisfied if $\{t_n\}$ is bounded. On the other hand, *a priori* boundedness of $\{y(\cdot, u^n)\}$ results if additional hypotheses are imposed on $f(t, y)$ and $F(t, y)$. For instance,

Lemma 2.2.8. *Assume there exist functions* $c_1(\cdot) \in L^1(0, T)$, $c_2(\cdot) \in L^2(0, T)$ *with*

$$(y, f(t, y) + F(t, y)u)$$
$$\le c_1(t)(1 + \|y\|^2) + c_2(t)(1 + \|y\|^2)\|u\| \quad (y \in \mathbb{R}^m). \quad (2.2.17)$$

Then, $y(\cdot, u^n)$ *is uniformly bounded if* $\{u^n(\cdot)\}$ *is bounded in* L^2.

The proof results directly from Lemma 2.1.7. There is an L^q version where one takes $c_2(\cdot) \in L^q(0, T)$, $q > 1$.

The examples below illustrate the role of the other hypotheses (linearity in u, convexity of the control set U, weak lower semicontinuity of the cost functional) in Theorem 2.2.6.

Example 2.2.9.

$$y'(t) = u(t), \qquad y(0) = 0,$$
$$U = \{-1\} \cup \{1\}, \qquad Y = \mathbb{R}, \qquad \bar{t} = 1$$
$$y_0(t, u) = \int_0^t y(\tau)^2 d\tau.$$

If $u^n(t)$ is defined by

$$u^n(t) = \begin{cases} 1 & (2k/n \le t < (2k+1)/2n) \\ -1 & (2k+1)/2n \le t < (2k+2)/2n \end{cases} \qquad (2.2.18)$$

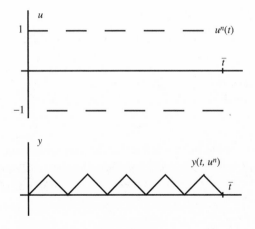

Figure 2.1.

$(k = 0, 1, \ldots, n - 1)$ then $u^n(\cdot) \to 0$ weakly in $L^2(0, 1)$. The sequence $\{u^n(\cdot)\}$ is a minimizing sequence with $y_0(t, u^n) \to 0$ uniformly in $0 \le t \le \bar{t}$. However, there is no optimal control, that is, no admissible control $\bar{u}(\cdot)$ with $y_0(\bar{t}, u) = 0$; if there were one, $y(t, \bar{u}) \equiv 0$, thus, $\bar{u} \equiv 0$. This control does not take values in U.

Example 2.2.10.

$$y'(t) = u(t), \qquad y(0) = 0,$$
$$U = [-1, 1], \qquad Y = \mathbb{R}, \qquad \bar{t} = 1,$$
$$y_0(t, u) = \int_0^t \{y(\tau)^2 + (u(\tau)^2 - 1)^2\} d\tau.$$

The sequence (2.2.18) is a minimizing sequence with $y_0(t, u^n) \to 0$ uniformly. Again, there is no optimal control; if $\bar{u}(\cdot)$ were optimal, $y(t, \bar{u}) \equiv 0$, thus $\bar{u} \equiv 0$ and $y_0(\bar{t}, \bar{u}) = 1$, a contradiction.

Example 2.2.11.

$$y_1'(t) = \cos(\pi u(t)/2), \qquad y_2'(t) = \sin(\pi u(t)/2),$$
$$y_1(0) = y_2(0) = 0$$
$$U = [-1, 1], \qquad\qquad Y = \{(0, 0)\}, \qquad \bar{t} = 1,$$
$$y_0(t, u) = \int_0^t \{y_1(\tau)^2 + y_2(\tau)^2\} d\tau.$$

If u^n is defined by (2.2.18), then $\{u^n\}$ is a minimizing sequence with $y_0(t, u^n) \to m = 0$ uniformly. If $\bar{u}(\cdot)$ is an admissible control attaining the minimum, we must have $y_1(t, \bar{u}) \equiv y_2(t, \bar{u}) \equiv 0$ so that $\cos(\pi \bar{u}(t)/2) \equiv \sin(\pi \bar{u}(t)/2) \equiv 0$ a. e; the second identity implies that $\bar{u}(t) \equiv 0$, a contradiction.

Existence fails in these examples because:

Example 2.2.9. The control set U is not convex (so that weak limits of U-valued controls may not be U-valued).

Example 2.2.10. The cost functional $y_0(t, u)$ is not weakly lower semicontinuous; $u^n(\cdot) \to 0$ weakly in $L^2(0, \bar{t})$ and $y(\bar{t}, u^n) \to 0$, but $y_0(\bar{t}, \bar{u}) = 1$.

Example 2.2.11. The right-hand side does not depend linearly on u.

Explanation of failure in Example 2.2.11 is too simplistic. Nonlinearity of a system $y'(t) = f(t, y(t), u(t))$ may not allow taking weak limits of minimizing sequences "inside $f(t, y(t), u(t))$," but this may not prevent existence of optimal controls.

Example 2.2.12.

$$y'(t) = 1 + u(t) - u(t)^2, \qquad y(0) = 0,$$
$$U = [-1, 1], \qquad Y = \{0\}, \qquad \bar{t} = 1,$$
$$y_0(t, u) = \int_0^t y(\tau)^2 d\tau.$$

The sequence (2.2.18) is a minimizing sequence with $y_0(\bar{t}, u^n) \to 0$ uniformly. The control $\bar{u}(t) \equiv 0$ (weak limit of u^n) is not optimal. However, we obtain an optimal control setting $1 + \bar{u}(t) - \bar{u}(t)^2 = 0$:

$$\bar{u}(t) \equiv (1 - \sqrt{5})/2,$$

the other root discarded in view of the control constraint $|u| \leq 1$. Comparing Example 2.2.12 with Example 2.2.11, we note that in the latter

$$f(t, y, U) = \{1 + u - u^2; -1 \leq u \leq 1\} = [-1, 5/4]$$

is convex; on the other hand, in the former, $f(t, y, U)$ is the nonconvex half circle $\{(\cos(\pi u/2), \sin(\pi u/2)); -1 \leq u \leq 1\}$. We may suspect that what is essential for existence is not linearity in u or convexity of the control set but convexity of $f(t, y, U)$. This suspicion is on target (see Remark 13.6.4).

Remark 2.2.13. Theorem 2.2.6 becomes especially simple for the linear system

$$y'(t) = A(t)y(t) + B(t)u(t), \qquad y(0) = \zeta, \qquad (2.2.19)$$

where $A(t)$ is a $m \times m$ matrix, and $B(t)$ is a $k \times m$ matrix. If we assume that

$$A(\cdot) \in L^1(0, T; \mathbb{R}^{m \times m}), \qquad B(\cdot) \in L^2(0, T; \mathbb{R}^{k \times m}) \qquad (2.2.20)$$

and use the admissible control space (2.2.8), all assumptions in Theorem 2.2.6 are satisfied. Solutions exist globally. Using Lemma 2.2.8 or the variation-of-constants formula (2.1.22), boundedness of $\{y(\cdot, u^n)\}$ for a minimizing sequence is automatic. Invoking the Arzelà-Ascoli theorem for convergence of $\{y(\cdot, u^n)\}$ is unnecessary; in fact, let $\{u^n(\cdot)\}$ be a sequence in $L^2(0, T; \mathbb{R}^k)$ such that $u^n(\cdot) \to \bar{u}(\cdot)$ weakly. If the sequence of trajectories $\{y(\cdot, u^n)\}$ does not converge uniformly to $\{y(\cdot, \bar{u})\}$, we can find a sequence $\{t_n\} \subset [0, \bar{t}]$ with $\|y(t_n, u^n) - y(t_n, \bar{u})\| \geq \varepsilon > 0$. Using (2.1.22), we obtain a contradiction with weak convergence of $\{u^n(\cdot)\}$ to $\bar{u}(\cdot)$.

Example 2.2.14. (The Banach-Saks Theorem). Let E be a Hilbert space, $\{u^n\}$ a sequence such that $u^n \to \bar{u}$ weakly. Then there exists a double sequence $\{\alpha_{nj}\}$ ($\alpha_{nj} \geq 0, \alpha_{nj} = 0$ if $j \geq n$, $\alpha_{nj} = 0$ for $n \geq n_0(j)$, $\Sigma_j \alpha_{nj} = 1$) such that $v^n = \Sigma_j \alpha_{nj} u^j \to \bar{u}$ strongly. (the choice of the sequence $\{\alpha_{nj}\}$ in the original theorem is different). As shown by Mazur, this result remains true if E is a Banach space and $u^n \to \bar{u}$ in the E^*-weak topology. See **5.0** for the definition of this topology; for a proof of Mazur's theorem see Yosida [1978, p. 120].

Example 2.2.15. A (real-valued) function ϕ defined in a convex set $U \subseteq \mathbb{R}^k$ is **convex** if

$$\phi(\alpha_1 u_1 + \cdots + \alpha_n u_n) \leq \alpha_1 \phi(u_1) + \cdots + \alpha_n \phi(u_n)$$

for every finite collection $\{\alpha_j\}$ with $\alpha_j \geq 0$, $\Sigma \alpha_j = 1$.

When ϕ is twice continuously differentiable in a convex set U with nonempty interior,

(a) For $k = 1$, $\phi(u)$ is convex in U if and only if $\phi''(u) \geq 0$ there.

(b) For $k > 1$, $\phi(u)$ is convex in U if and only if the Hessian matrix $H(u) = \{\partial_j \partial_j \phi(u)\}$ is positive definite for each $u \in U$ (that is, $(H(u)y, y) \geq 0$ for all $y \in \mathbb{R}^m$). See Rockafellar [1970, p. 27].

Example 2.2.16. Let $y_0(t, u)$ be a cost functional of the form

$$y_0(t, u) = \int_0^t f_0(\tau, y(\tau, u), u(\tau)) d\tau \qquad (2.2.21)$$

where $f_0(t, y, u) : [0, T] \times \mathbb{R}^m \times U \to \mathbb{R}$ (U a convex set in $\mathbb{R}^k$); we assume $f_0(t, y, u)$ measurable in t for y, u fixed and continuous in y, u for t fixed, continuity in y uniform with respect to u. Assume $f_0(t, y, u)$ is convex in u for each t, y, and that for each $c > 0$ there exists $K_0(\cdot, c) \in L^2(0, T)$ such that

$$|f_0(t, y, u)| \leq K_0(t, c)\|u\| \quad (0 \leq t \leq T, \|y\| \leq c, u \in U). \qquad (2.2.22)$$

Then $y_0(t, y, u)$ is weakly lower semicontinuous and equicontinuous with respect to t.

To show this we apply the Banach-Saks Theorem to the weakly convergent sequence $\{u^n(\cdot)\}$ in the definition of weak lower semicontinuity; then we select a subsequence of $\{v^n(\cdot)\}$ (equally named) such that $v^n(\tau) \to \bar{u}(\tau)$ a.e. We have

$$
\begin{aligned}
f_0(\tau, y(\tau, \bar{u}), \bar{u}(\tau)) = {} & f_0(\tau, y(\tau, \bar{u}), \Sigma\alpha_{nj}u^j(\tau)) \\
& + \{f_0(\tau, y(\tau, \bar{u}), \bar{u}(\tau)) - f_0(\tau, y(\tau, \bar{u}), \Sigma\alpha_{nj}u^j(\tau))\}
\end{aligned}
$$

where the integral of the second term converges to zero in view of a.e. convergence of $\Sigma\alpha_{nj}u^j(\cdot)$ to $\bar{u}(\cdot)$, (2.2.22), the Schwarz inequality, and uniform L^2 boundedness of $\Sigma\alpha_{nj}u^j(\tau)$. On the other hand,

$$
\begin{aligned}
f_0(\tau, & y(\tau, \bar{u}), \Sigma\alpha_{nj}u^j(\tau)) \\
& \leq \Sigma\alpha_{nj} f_0(\tau, y(\tau, \bar{u}), u^j(\tau)) \\
& = \Sigma\alpha_{nj} f_0(\tau, y(\tau, u^j), u^j(\tau)) \\
& + \Sigma\alpha_{nj}\{f_0(\tau, y(\tau, \bar{u}), u^j(\tau)) - f_0(\tau, y(\tau, u^j), u^j(\tau))\};
\end{aligned}
$$

hence, using the continuity assumptions and the dominated convergence theorem, it follows that if $y_0(\bar{t}, u^n) \leq C$ for all n then $y_0(\bar{t}, \bar{u}) \leq C$.

Equicontinuity results from the bound (2.2.22); by the Schwarz inequality,

$$
\int_t^{t'} |f_0(\tau, y(\tau, u), u(\tau))| dt \leq C \left(\int_t^{t'} |K_0(\tau, c)|^2 d\tau \right)^{1/2}.
$$

Remark 2.2.17. Theorem 2.2.6 uses only continuity, not Lipschitz continuity of $f(t, y)$ and $F(t, y)$, and thus remains true under Hypothesis I. Since in this setting trajectories may not be unique, $y_0(t, u)$ depends not only on u but also on the trajectory.

2.3. Trajectories and Spike Perturbations

We consider the control system

$$
y'(t) = f(t, y(t), u(t)), \qquad y(0) = \zeta \tag{2.3.1}
$$

in $E = \mathbb{R}^m$. The admissible control space $C_{ad}(0, T; U)$ is called **spike complete** if given $s \in (0, \bar{t}]$, $h < s$, and $v \in U$, the **spike perturbation**

$$
u_{s,h,v}(t) = \begin{cases} v & (s - h < t \leq s) \\ u(t) & \text{elsewhere} \end{cases} \tag{2.3.2}
$$

belongs to $C_{ad}(0, T; U)$ for every $u(\cdot) \in C_{ad}(0, T; U)$.

Assume $\bar{u}(\cdot)$ is a solution of an optimal control problem for (2.3.1) with cost functional $y_0(t, u)$, no state constraints or target condition, and fixed terminal time

$\bar{t}$. Then we must have

$$\xi_0(\bar{t}, s, \bar{u}, v) = \lim_{h \to 0+} \frac{1}{h}(y_0(\bar{t}, \bar{u}_{s,h,v}) - y_0(\bar{t}, \bar{u})) \geq 0 \qquad (2.3.3)$$

for every $s \in (0, \bar{t}]$ and every $v \in U$, assuming that the limit exists (see **1.6**). We justify here the formal computations in **1.6**, beginning with the calculation of the limit

$$\xi(t, s, \bar{u}, v) = \lim_{h \to 0+} \frac{1}{h}(y(t, \bar{u}_{s,h,v}) - y(t, \bar{u})). \qquad (2.3.4)$$

A prerequisite is Lemma 2.3.1 below, which shows that if $y(t, \bar{u})$ exists in $0 \leq t \leq \bar{t}$ then the same is true for $y(t, \bar{u}_{s,h,v})$ for h small enough. Of course, $y(t, \bar{u}_{s,h,v}) = y(t, \bar{u})$ for $0 \leq t \leq s - h$.

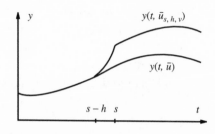

Figure 2.2.

Lemma 2.3.1. *Assume that* $f(t, y) = f(t, y, u(t))$ *satisfies Hypothesis II for every* $u(\cdot) \in C_{ad}(0, \bar{t}; U)$ *and that* $C_{ad}(0, \bar{t}; U)$ *is spike complete. Let* $\bar{u}(\cdot) \in C_{ad}(0, \bar{t}; U)$ *be such that* $y(t, \bar{u})$ *exists in* $0 \leq t \leq \bar{t}$, *and let* $s \in (0, \bar{t}], v \in U$. *Then there exists* $\delta > 0$ *such that if* $0 \leq h \leq \delta$ *then* $y(t, \bar{u}_{s,h,v})$ *exists in* $0 \leq t \leq \bar{t}$. *Moreover, there exists* $K(\cdot) \in L^1(0, \bar{t})$ *depending on* $\bar{u}(\cdot)$ *and* v *but not on* h *with*

$$\|y(t, \bar{u}_{s,h,v}) - y(t, \bar{u})\| \leq \int_{s-h}^{\min(s,t)} K(\tau)d\tau \quad (t \geq s - h). \qquad (2.3.5)$$

Proof. For each $h > 0$ we denote by $[0, t_h]$ the largest interval where $y(t, \bar{u}_{s,h,v})$ exists and satisfies

$$\|y(t, \bar{u}_{s,h,v}) - y(t, \bar{u})\| \leq 1. \qquad (2.3.6)$$

Since $y(t, \bar{u}_{s,h,v}) = y(t, \bar{u})$ in $0 \leq t \leq s - h$, Lemma 2.1.5 guarantees that $t_h > s - h$. We have

$$y(t, \bar{u}_{s,h,v}) - y(t, \bar{u})$$

$$= \int_0^t \{f(\tau, y(\tau, \bar{u}_{s,h,v}), \bar{u}_{s,h,v}(\tau)) - f(\tau, y(\tau, \bar{u}), \bar{u}_{s,h,v}(\tau))\}d\tau$$

$$+ \int_0^t \{f(\tau, y(\tau, \bar{u}), \bar{u}_{s,h,v}(\tau)) - f(\tau, y(\tau, \bar{u}), \bar{u}(\tau))\}d\tau. \qquad (2.3.7)$$

Call $K(t, c)$, $L(t, c)$ (resp. $K_v(t, c)$, $L_v(t, c)$) the functions in (2.1.3), (2.1.4) corresponding to $f(t, y, u(t))$ (resp. to $f(t, y, u_{s,h,v}(t))$ with $h = s$), and let c be the maximum of $\|y(t, \bar{u})\|$ in $0 \le t \le \bar{t}$. The integrand of the first integral is bounded by $L_v(\tau, c + 1)\|y(\tau, \bar{u}_{s,h,v}) - y(\tau, \bar{u})\|$ in $s - h < \tau \le \min(s, t_h)$ and by $L(\tau, c + 1)\|y(\tau, \bar{u}_{s,h,v}) - y(\tau, \bar{u})\|$ elsewhere. On the other hand, the integrand of the second integral is bounded by $K_v(\tau, c) + K(\tau, c)$ in $s - h < \tau \le \min(s, t_h)$; thus if $K_0(\tau, c) = K(\tau, c) + K_v(\tau, c)$ and $L_0(\tau, c) = L(\tau, c) + L_v(\tau, c)$ we have

$$\|y(t, \bar{u}_{s,h,v}) - y(t, \bar{u})\|$$
$$\le \int_{s-h}^{\min(s,t)} K(\tau, c)d\tau + \int_{s-h}^{t} L(\tau, c)\|y(\tau, \bar{u}_{s,h,v}) - y(\tau, \bar{u})\|d\tau.$$

By Gronwall's Lemma 2.1.2 inequality (2.3.5) with $K(\tau) = CK_0(\tau, c)$ follows in the interval $[0, t_h]$, where C is independent of h. For h sufficiently small, (2.3.5) shows that $\|y(t, \bar{u}_{s,h,v}) - y(t, \bar{u})\| < 1$ in $[0, t_h]$, which contradicts the maximality of $[0, t_h]$ with respect to (2.3.6) unless $t_h = \bar{t}$ (by Lemma 2.1.5 we can extend $y(t, \bar{u}_{s,h,v})$ to a larger interval where $\|y(t, \bar{u}_{s,h,v}) - y(t, \bar{u})\| < 1$ as well). This ends the proof.   ∎

Remark 2.3.2. Estimating (2.3.7) in the same way but simply taking the norm of the integrand in the second integral, we obtain

$$\|y(t, \bar{u}_{s,h,v}) - y(t, \bar{u})\|$$
$$\le C \int_{s-h}^{\min(s,t)} \|f(\tau, y(\tau, \bar{u}), v) - f(\tau, y(\tau, \bar{u}), \bar{u}(\tau))\|d\tau. \quad (2.3.8)$$

Generalizing the definition in **2.0** in an obvious way, a **Lebesgue point** of an integrable E-valued function $g(t)$ defined in $0 \le t \le \bar{t}$ is any point s where

$$\lim_{h \to 0} \frac{1}{h} \int_{s-h}^{s} \|g(\tau) - g(s)\|d\tau = 0 \qquad (2.3.9)$$

(two sided limit). Almost every $t \in [0, \bar{t}]$ is a Lebesgue point of each component of $g(\cdot)$ (see **2.0**), and a Lebesgue point of all components is a Lebesgue point of $g(\cdot)$: thus, almost every $t \in [0, \bar{t}]$ is a Lebesgue point of $g(\cdot)$. Half of the definition suffices below since we only take limits as $h \to 0+$; points where $\lim_{h \to 0+}$ exists are **left Lebesgue points**.

Inequality (2.3.8) implies that if s is a left Lebesgue point of both $f(\tau, y(\tau, \bar{u}), v)$ and $f(\tau, y(\tau, \bar{u}), \bar{u}(\tau))$, then

$$\|y(t, \bar{u}_{s,h,v}) - y(t, \bar{u})\| \le Ch \quad (0 \le t \le \bar{t}, h > 0). \qquad (2.3.10)$$

The computation of (2.3.4) requires a new assumption for the uncontrolled equation $y'(t) = f(t, y(t))$.

Hypothesis III. The partial derivatives $\partial_k f_j(t, y) = \partial f_j(t, y)/\partial y_k$ exist and $f(t, y), \partial_k f_j(t, y)$ are measurable in t for y fixed and continuous in y for t fixed.[1] For every $c > 0$ there exist $K(\cdot, c), L(\cdot, c) \in L^1(0, T)$ such that

$$\|f(t, y)\| \le K(t, c) \quad (0 \le t \le T, \|y\| \le c), \tag{2.3.11}$$

$$\|\partial_y f(t, y)\| \le L(t, c) \quad (0 \le t \le T, \|y\| < c), \tag{2.3.12}$$

where $\partial_y f$ is the Jacobian matrix

$$\partial_y f = \begin{bmatrix} \partial_1 f_1 & \cdots & \partial_m f_1 \\ \cdots\cdots\cdots\cdots\cdots \\ \partial_1 f_m & \cdots & \partial_m f_m \end{bmatrix}.$$

Hypothesis III is stronger than Hypothesis II; in fact, (2.1.4) follows from (2.3.12) and the mean value theorem.

Theorem 2.3.3. *Assume that $f(t, y) = f(t, y, u(t))$ satisfies Hypothesis III for every $u(\cdot) \in C_{ad}(0, \bar{t}; U)$ and that $C_{ad}(0, \bar{t}; U)$ is spike complete. Let $\bar{u}(\cdot) \in C_{ad}(0, \bar{t}; U)$ be such that $y(t, \bar{u})$ exists in $0 \le t \le \bar{t}$, and let e be the set of all left Lebesgue points of both functions*

$$f(\cdot, y(\cdot, \bar{u}), \bar{u}(\cdot)), \qquad f(\cdot, y(\cdot, \bar{u}), v) \tag{2.3.13}$$

in $(0, \bar{t}]$. Then, if $s \in e$, the limit (2.3.4) exists uniformly in $s \le t \le \bar{t}$ and

$$\xi(t, s, \bar{u}, v) = \begin{cases} 0 & (0 \le t < s) \\ S(t, s; \bar{u})\{f(s, y(s, \bar{u}), v) - f(s, y(s, \bar{u}), \bar{u}(s))\} \\ & (s \le t \le \bar{t}) \end{cases} \tag{2.3.14}$$

where $S(t, s; \bar{u})$ is the solution matrix of the **variational equation**

$$\xi'(t) = \partial_y f(t, y(t, \bar{u}), \bar{u}(t))\xi(t). \tag{2.3.15}$$

Proof. In view of Lemma 2.3.1, $y(t, \bar{u}_{s,h,v})$ exists in $0 \le t \le \bar{t}$ for h sufficiently small. By definition of the solution matrix $S(t, s; \bar{u})$, the function $\xi(t) = \xi(t, s, \bar{u}, v)$ in (2.3.14) satisfies the variational equation (2.3.15) in $s \le t \le \bar{t}$ with initial condition $\xi(s) = f(s, y(s, \bar{u}), v) - f(s, y(s, \bar{u}), \bar{u}(s))$ or, equivalently,

---

[1] There are several redundancies in this definition. Existence and continuity of the partial derivatives imply continuity of $f(t, y)$ for t fixed. The pointwise limit of measurable functions is measurable; thus, measurability of $\partial_k f_j(t, y)$ for y fixed is automatic. The same observation applies to various forthcoming definitions.

$$\xi(t, s, \bar{u}, v) = f(s, y(s, \bar{u}), v) - f(s, y(s, \bar{u}), \bar{u}(s))$$

$$+ \int_s^t \partial_y f(\tau, y(\tau, \bar{u}), \bar{u}(\tau)) \xi(\tau, s, \bar{u}, v) d\tau \quad (s \le t \le \bar{t}).$$

Since $\xi(\tau, s, \bar{u}, v) = 0$ for $\tau < s$, this equation can be written

$$\xi(t, s, \bar{u}, v) = f(s, y(s, \bar{u}), v) - f(s, y(s, \bar{u}), \bar{u}(s)) v(t - s)$$

$$+ \int_0^t \partial_y f(\tau, y(\tau, \bar{u}), \bar{u}(\tau)) \xi(\tau, s, \bar{u}, v) d\tau \quad (0 \le t \le \bar{t}), \quad (2.3.16)$$

where $v(t)$ is the Heaviside function (the characteristic function of $t \ge 0$). Define

$$\eta(t, h) = \frac{1}{h}(y(t, \bar{u}_{s,h,v}) - y(t, \bar{u})) - \xi(t, s, \bar{u}, v) \quad (0 \le t \le \bar{t})$$

and combine (2.3.16) with the integral equations for $y(t, \bar{u}_{s,h,v})$ and $y(t, \bar{u})$. The result is

$$\eta(t, h) = \int_0^t \left\{ \frac{1}{h} \{ f(\tau, y(\tau, \bar{u}_{s,h,v}), \bar{u}_{s,h,v}(\tau)) - f(\tau, y(\tau, \bar{u}), \bar{u}(\tau)) \} \right.$$

$$\left. - \partial_y f(\tau, y(\tau, \bar{u}), \bar{u}(\tau)) \xi(\tau, s, \bar{u}, v) \right\} d\tau$$

$$- \{ f(s, y(s, \bar{u}), v) - f(s, y(s, \bar{u}), \bar{u}(s)) \} v(t - s)$$

$$= \int_0^{\bar{t}} \partial_y f(\tau, y(\tau, \bar{u}), \bar{u}(\tau)) \eta(\tau, h) d\tau$$

$$+ \int_0^t \frac{1}{h} \{ f(\tau, y(\tau, \bar{u}_{s,h,v}), \bar{u}_{s,h,v}(\tau)) - f(\tau, y(\tau, \bar{u}), \bar{u}(\tau))$$

$$- \partial_y f(\tau, y(\tau, \bar{u}), \bar{u}(\tau))(y(\tau, \bar{u}_{s,h,v}) - y(\tau, \bar{u})) \} d\tau$$

$$- \{ f(s, y(s, \bar{u}), v) - f(s, y(s, \bar{u}), \bar{u}(s)) \} v(t - s)$$

$$= \int_0^{\bar{t}} \partial_y f(\tau, y(\tau, \bar{u}), \bar{u}(\tau)) \eta(\tau, h) d\tau$$

$$+ \int_0^t \frac{1}{h} \{ f(\tau, y(\tau, \bar{u}_{s,h,v}), \bar{u}(\tau)) - f(\tau, y(\tau, \bar{u}), \bar{u}(\tau))$$

$$- \partial_y f(\tau, y(\tau, \bar{u}), \bar{u}(\tau))(y(\tau, \bar{u}_{s,h,v}) - y(\tau, \bar{u})) \} d\tau$$

$$+ \frac{1}{h} \int_0^t \{ f(\tau, y(\tau, \bar{u}_{s,h,v}), \bar{u}_{s,h,v}(\tau)) - f(\tau, y(\tau, \bar{u}_{s,h,v}), \bar{u}(\tau)) \} d\tau$$

$$- \{ f(s, y(s, \bar{u}), v) - f(s, y(s, \bar{u}), \bar{u}(s)) \} v(t - s)$$

$$= \int_0^t \partial_y f(\tau, y(\tau, \bar{u}), \bar{u}(\tau)) \eta(\tau, h) d\tau + \kappa_1(t, h) + \kappa_2(t, h), \quad (2.3.17)$$

$\kappa_1(t, h)$ the second integral on the right side, $\kappa_2(t, h)$ the third integral and the last term. Using (2.3.12) for $f(\tau, y(\tau, \bar{u}), \bar{u}(\tau))$, we deduce the existence of $L(\cdot) \in L^1(0, \bar{t})$ such that

$$\|\eta(t, h)\| \le \|\kappa_1(t, h)\| + \|\kappa_2(t, h)\| + \int_0^t L(\tau)\|\eta(\tau, h)\|d\tau; \qquad (2.3.18)$$

thus, Theorem 2.3.3 will follow (via Lemma 2.1.2) if we show

$$\kappa_1(t, h) \to 0 \quad \text{uniformly in } 0 \le t \le \bar{t} \qquad (2.3.19)$$

$$\kappa_2(t, h) \to 0 \quad \text{uniformly in } s \le t \le \bar{t}. \qquad (2.3.20)$$

We do (2.3.19) first. Note that, by virtue of (2.3.10) and the mean value theorem, the integrand $\rho(\tau, h)$ of $\kappa_1(t, h)$ is uniformly bounded; moreover, it tends to zero pointwise by well known properties of the Jacobian matrix. If (2.3.19) does not hold, there exists a sequence $\{h_n\} \subseteq \mathbb{R}_+, h_n \to 0$ and a sequence $\{t_n\} \subseteq [0, \bar{t}]$ such that $\|\kappa_1(t_n, h_n)\| \ge \delta > 0$; passing if necessary to a subsequence, we may assume that $t_n \to t \in [0, \bar{t}]$. We then obtain a contradiction to the dominated convergence theorem writing

$$\kappa_1(t_n, h) = \int_0^{\bar{t}} \chi_n(\tau)\rho(\tau, h_n)d\tau,$$

where $\chi_n(\cdot)$ is the characteristic function of $0 \le t \le t_n$.

To show (2.3.20), note that, since $\bar{u}_{s,h,v}(\tau) = \bar{u}(\tau)$ for $t \ge s$, $\kappa_2(t, h)$ is constant in $t \ge s$. Hence, we only need to consider $t = s$, where

$$\begin{aligned}
\kappa_2(s, h) = &\frac{1}{h} \int_{s-h}^s \{f(\tau, y(\tau, \bar{u}_{s,h,v}), v) - f(\tau, y(\tau, \bar{u}), v)\}d\tau \\
&- \frac{1}{h} \int_{s-h}^s \{f(\tau, y(\tau, \bar{u}_{s,h,v}), \bar{u}(\tau)) - f(\tau, y(\tau, \bar{u}), \bar{u}(\tau))\}d\tau \\
&+ \frac{1}{h} \int_{s-h}^s \{f(\tau, y(\tau, \bar{u}), v) - f(\tau, y(\tau, \bar{u}), \bar{u}(\tau))\}d\tau \\
&- \{f(s, y(s, \bar{u}), v) - f(s, y(s, \bar{u}), \bar{u}(s))\}v(t - s).
\end{aligned}$$

The first two terms tend to zero on account of the local Lipschitz condition resulting from (2.3.12) and the estimate (2.3.10); the rest tends to zero by the left Lebesgue point condition for both functions (2.3.13). The proof of Theorem 2.3.3 is then complete. ∎

Example 2.3.4. Let $f(t, y, u)$ be continuous in all three variables and continuously differentiable with respect to y in $[0, T] \times \mathbb{R}^m \times U$, where $U \subseteq \mathbb{R}^k$ is closed and bounded, and let $C_{\text{ad}}(0, T; U)$ be the set of all measurable $\mathbb{R}^k$-valued functions taking values in U almost everywhere. Then $C_{ad}(0, T; U)$ is spike complete and $f(t, y, u(t))$ satisfies Hypothesis III for each $u(\cdot) \in C_{ad}(0, T; U)$ with $K(\cdot, c) \equiv K(c)$ and $L(\cdot, c) \equiv L(c)$ independent of $u(\cdot)$.

2.4. Cost Functionals and Spike Perturbations

We compute $\xi_0(\bar{t}, s, u, v)$ in (2.3.3) for the cost functional

$$y_0(t, u) = \int_0^t f_0(\tau, y(\tau, u), u(\tau))d\tau + g_0(t, y(t, u)) \qquad (2.4.1)$$

under the following companion of Hypothesis III:

Hypothesis III^0. The partial derivatives $\partial_k f_0(t, y) = \partial f_0(t, y)/\partial y_k$ exist, and $f(t, y), \partial_k f_0(t, y)$ are measurable in t for y fixed and continuous in y for t fixed. For every $c > 0$ there exist $K_0(\cdot, c), L_0(\cdot, c) \in L^1(0, T)$ such that

$$|f_0(t, y)| \le K_0(t, c) \quad (0 \le t \le T, \|y\| \le c), \qquad (2.4.2)$$

$$\|\partial_y f_0(t, y)\| \le L_0(t, c) \quad (0 \le t \le T, \|y\| \le c), \qquad (2.4.3)$$

where $\partial_y f$ is the gradient vector $\partial_y f = (\partial_1 f_0, \ldots, \partial_m f_0)$.

Theorem 2.4.1. *Assume that* $f(t, y) = f(t, y, u(t))$ *(resp.* $f_0(t, y) = f_0(t, y, u(t))$*) satisfies Hypothesis III (resp. Hypothesis III^0) for every* $u(\cdot) \in C_{ad}(0, \bar{t}; U)$*, that* $C_{ad}(0, \bar{t}; U)$ *is spike complete, and that the derivatives* $\partial_k g_0(t, y)$ *exist. Let* $\bar{u} \in C_{ad}(0, \bar{t}; U)$ *be such that* $y(t, \bar{u})$ *exists in* $0 \le t \le \bar{t}$*, and let* e *be the set of left Lebesgue points of all four functions*

$$f(\cdot, y(\cdot, \bar{u}), \bar{u}(\cdot)), \quad f(\cdot, y(\cdot, \bar{u}), v),$$
$$f_0(\cdot, y(\cdot, \bar{u}), \bar{u}(\cdot)), \quad f_0(\cdot, y(\cdot, \bar{u}), v) \qquad (2.4.4)$$

in $(0, \bar{t}]$*. Then, if* $s \in e$*, the limit*

$$\xi_0(t, s, \bar{u}, v) = \lim_{h \to 0+} \frac{1}{h}(y_0(t, \bar{u}_{s,h,v}) - y_0(t, \bar{u})) \qquad (2.4.5)$$

exists uniformly in $s \le t \le \bar{t}$ *and is given by*

$$\xi_0(t, s, \bar{u}, v) = \{f_0(s, y(s, \bar{u}), v) - f_0(s, y(s, \bar{u}), u(s))\}$$

$$+ \int_s^t \langle \partial_y f_0(\tau, y(\tau, \bar{u}), \bar{u}(\tau)), \xi(\tau, s, \bar{u}, v)\rangle d\tau$$

$$+ \langle \partial_y g_0(t, y(t, \bar{u})), \xi(t, s, \bar{u}, v)\rangle, \qquad (2.4.6)$$

$\xi(t, s, \bar{u}, v)$ *given by* (2.3.14).

Proof. For $t \geq s$ we have

$$\frac{1}{h}(y_0(t, \bar{u}_{s,h,v}) - y_0(t, \bar{u}))$$

$$= \int_0^t \frac{1}{h}\big\{ f_0(\tau, y(\tau, \bar{u}_{s,h,v}), \bar{u}_{s,h,v}(\tau)) - f_0(\tau, y(\tau, \bar{u}), \bar{u}_{s,h,v}(\tau)) \big\} d\tau$$

$$+ \frac{1}{h} \int_{s-h}^s \big\{ f_0(\tau, y(\tau, \bar{u}), v) - f_0(\tau, y(\tau, \bar{u}), \bar{u}(\tau)) \big\} d\tau$$

$$+ \frac{1}{h}\big(g_0(t, y(t, \bar{u}_{s,h,v})) - g_0(t, y(t, \bar{u})) \big).$$

Limits as $h \to 0$ are taken as follows.

First integral: By (2.3.10) and the local Lipschitz condition emanating from (2.4.3), the integrand is uniformly bounded. By Theorem 2.3.3, the chain rule and the fact that $\bar{u}_{s,h,v}(\tau)$ is eventually constant for each τ for h sufficiently small, the integrand converges to $\langle \partial_y f_0(\tau, y(\tau, \bar{u}), \bar{u}(\tau)), \xi(\tau, s, \bar{u}, v) \rangle$. We use then the dominated convergence theorem.

Second integral: We use the left Lebesgue point condition for $f_0(\cdot, y(\cdot, \bar{u}), \bar{u}(\cdot))$, $f_0(\cdot, y(\cdot, \bar{u}), v)$.

Third term: We use Theorem 2.3.3 and the chain rule. ∎

Example 2.4.2. Let $f_0(t, y, u)$ be continuous in all three variables and continuously differentiable with respect to y in $[0, T] \times \mathbb{R}^m \times U$, where $U \subseteq \mathbb{R}^k$ is closed and bounded, and let $C_{ad}(0, T; U)$ be the set of all measurable $\mathbb{R}^k$-valued functions taking values in U almost everywhere. Then $f_0(t, y, u(t))$ satisfies Hypothesis III^0 for each $u(\cdot) \in C_{ad}(0, T; U)$ with $K_0(\cdot, c) \equiv K_0(c)$ and $L_0(\cdot, c) \equiv L_0(c)$ independent of $u(\cdot)$.

2.5. Optimal Control Problems without Target Condition: The Hamiltonian Formalism

Let $\phi(t, v)$ be a function defined in $[0, T] \times U$ (U an arbitrary set) such that $\phi(t, v)$ is integrable in t for v fixed. The **left Lebesgue set** of $\phi(t, U)$ is the set d of all $t \in [0, T]$ such that t is a left Lebesgue point of $\phi(t, v)$ for every $v \in U$. We say that $\phi(t, v)$ is **regular** in U if its left Lebesgue set has full measure in $[0, T]$.

Consider the optimal control problem of minimizing the cost functional

$$y_0(\bar{t}, u) = \int_0^{\bar{t}} f_0(\tau, y(\tau, u), u(\tau)) d\tau + g_0(\bar{t}, y(\bar{t}, u)) \qquad (2.5.1)$$

among all trajectories of

$$y'(t) = f(t, y(t), u(t)) \qquad (2.5.2)$$

in a fixed interval $0 \le t \le \bar{t}$, with no state constraints or target condition. Then, if $\bar{u}(\cdot)$ is an optimal control, (2.3.3) must be satisfied,

$$\xi_0(\bar{t}, s, \bar{u}, v) = \lim_{h \to 0+} \frac{1}{h} \big(y_0(\bar{t}, \bar{u}_{s,h,v}) - y_0(\bar{t}, \bar{u}) \big) \ge 0 \qquad (2.5.3)$$

wherever the limit exists. Accordingly, if the assumptions in Theorem 2.4.1 for the computation of this limit are in place and, in addition, the functions $f(\cdot, y(\cdot, \bar{u}), v)$ and $f_0(\cdot, y(\cdot, \bar{u}), v)$ are regular in the control set U, (2.3.3) will be satisfied for arbitrary s in the set $e = d \cap d_0 \cap c \cap c_0$, where d (resp. d_0) is the left Lebesgue set of $f(\cdot, y(\cdot, \bar{u}), U)$ (resp. of $f_0(\cdot, y(\cdot, \bar{u}), U)$) and c is the set of all left Lebesgue points of $f(\cdot, y(\cdot, \bar{u}), \bar{u}(\cdot))$ and $f_0(\cdot, y(\cdot, \bar{u}), \bar{u}(\cdot))$. Now, we have $\xi(t, s, \bar{u}, \bar{u}(s)) = 0$, thus $\xi_0(t, s, \bar{u}, \bar{u}(s)) = 0$ and we can write (2.5.3) as

$$\xi_0(\bar{t}, s, \bar{u}, \bar{u}(s)) = \min_{v \in U} \xi_0(\bar{t}, s, \bar{u}, v) \qquad (2.5.4)$$

which deserves to be called a **minimum principle**. We can cast it in a pleasingly symmetric way using Hamiltonians. To this end, we consider the ordinary differential system

$$\mathbf{y}'(t) = \mathbf{f}(t, \mathbf{y}(t), u(t)), \qquad \mathbf{y}(0) = (0, \zeta) \qquad (2.5.5)$$

in the interval $0 \le t \le \bar{t}$, where $\mathbf{y}(t) = (y_0(t), y(t))$ and

$$\mathbf{f}(t, \mathbf{y}, u) = (f_0(t, y, u), f(t, y, u)).$$

This system is called the **augmented system** and lives in $\mathbf{E} = \mathbb{R} \times E = \mathbb{R}^{m+1}$; it includes in the same equation the control system (2.5.2) and the cost functional (2.5.1).

Let $\mathbf{y} = (y_0, y_1, \ldots, y_m) = (y_0, y)$ and $\mathbf{z} = (z_0, z_1, \ldots, z_m) = (z_0, z)$ be generic points in $\mathbb{R} \times E$. Define the **Hamiltonian** $H(t, \mathbf{y}, \mathbf{z}, u)$ by

$$H(t, \mathbf{y}, \mathbf{z}, u) = H(t, y, \mathbf{z}, u) = \langle \mathbf{z}, \mathbf{f}(t, y, u) \rangle$$
$$= z_0 f_0(t, y, u) + z_1 f_1(t, y, u) + \cdots + z_m f_m(t, y, u). \qquad (2.5.6)$$

We can write (2.5.5) in the form

$$\mathbf{y}'(t) = \nabla_{\mathbf{z}} H(t, \mathbf{y}(t), \mathbf{z}, u(t))^{(1)} \qquad \mathbf{y}(0) = (0, \zeta), \qquad (2.5.7)$$

where $\nabla_{\mathbf{z}}$ denotes gradient with respect to the z_j (note that, although H depends on $\mathbf{z}$, $\nabla_{\mathbf{z}} H$ does not). Consider next the backwards initial value (or final value) problem

$$\mathbf{z}'(s) = -\nabla_{\mathbf{y}} H(s, \mathbf{y}(s), \mathbf{z}(s), u(s)), \qquad \mathbf{z}(\bar{t}) = (1, \partial_y g_0(\bar{t}, y(\bar{t}, u))) \qquad (2.5.8)$$

---

[1] Throughout Part I, ∂ is a generic symbol for derivatives, gradients, and Jacobian matrices; in Parts II and III, it indicates Fréchet derivatives. For reasons of tradition, we use ∇ for gradients in the canonical equations.

in the same interval $0 \leq s \leq \bar{t}$, where $\mathbf{y}(t)$ is the solution of (2.5.7) (or, equivalently, of (2.5.5)). This system is linear (so that existence of a unique solution in $0 \leq t \leq \bar{t}$ is insured as long as $y(t)$ exists in the whole interval). Coordinate by coordinate, (2.5.8) is

$$z_0' = 0$$
$$z_1' = -(\partial_1 f_0)z_0 - (\partial_1 f_1)z_1 - \cdots - (\partial_1 f_m)z_m$$
$$\cdots\cdots\cdots\cdots\cdots\cdots\cdots\cdots\cdots\cdots\cdots\cdots\cdots\cdots \qquad (2.5.9)$$
$$z_m' = -(\partial_m f_0)z_0 - (\partial_m f_1)z_1 - \cdots - (\partial_m f_m)z_m.$$

The first equation results from y_0-independence of the Hamiltonian. In view of this equation and of the "final condition" (2.5.8),

$$z_0(s) = 1 \quad (0 \leq s \leq \bar{t}). \qquad (2.5.10)$$

Hence, the rest of the system (2.5.9) can be written

$$z'(s) = -\partial_y f(s, y(s), u(s))^* z(s) - z_0 \partial_y f_0(s, y(s), u(s)), \qquad (2.5.11)$$

where $z(s) = (z_1(s), z_2(s), \ldots, z_m(s))$, $\partial_y f$ is the Jacobian matrix and $\partial_y f_0$ is the gradient; the final condition is $z(\bar{t}) = \partial_y g_0(\bar{t}, y(\bar{t}, u))$.

Let $S(t, s; u)$ be the solution operator of the variational equation

$$\xi'(t) = \partial_y f(t, y(t, u), u(t))\xi(t).$$

Then the solution operator $S^*(t, s; u)$ of the adjoint variational equation

$$z'(s) = -\partial_y f(s, y(s, u), u(s))^* z(s)$$

is given by $S^*(s, t; u) = S(t, s; u)^*$ (Corollary 2.1.11). Using the variation-of-constants formula (2.1.22) and minding the final condition, we can write the solution of (2.5.11) in the form

$$z(s) = S(\bar{t}, s, u)^* \partial_y g_0(\bar{t}, y(\bar{t}, u)) + \int_s^{\bar{t}} S(\sigma, s; u)^* \partial_y f_0(\sigma, y(\sigma, u), u(\sigma))d\sigma. \qquad (2.5.12)$$

Using then (2.3.14) and (2.4.6), we obtain

$$\xi_0(\bar{t}, s, \bar{u}, v) = \{f_0(s, y(s, \bar{u}), v) - f_0(s, y(s, \bar{u}), \bar{u}(s))\}$$
$$+ \int_s^{\bar{t}} \langle S(\sigma, s; \bar{u})^* \partial_y f_0(\sigma, y(\sigma, \bar{u}), \bar{u}(\sigma)), \{f(s, y(s, \bar{u}), v) - f(s, y(s, \bar{u}), \bar{u}(s))\}\rangle d\sigma$$
$$+ \langle S(\bar{t}, s; \bar{u})^* \partial_y g_0(\bar{t}, y(\bar{t}, \bar{u}), \{f(s, y(s, \bar{u}), v) - f(s, y(s, \bar{u}), \bar{u}(s))\}\rangle$$
$$= \{f_0(s, y(s, \bar{u}), v) - f_0(s, y(s, \bar{u}), \bar{u}(s))\}$$
$$+ \langle \bar{z}(s), f(s, y(s, \bar{u}), v) - f(s, y(s, \bar{u}), \bar{u}(s))\rangle. \qquad (2.5.13)$$

Theorem 2.5.1. *Assume* $f(t, y) = f(t, y, u(t))$ *(resp.* $f_0(t, y) = f_0(t, y, u(t))$*)* *satisfies Hypothesis III (resp. Hypothesis III^0) in the interval* $0 \leq t \leq \bar{t}$ *for every* $u(\cdot) \in C_{ad}(0, \bar{t}; U)$, *that* $C_{ad}(0, \bar{t}; U)$ *is spike complete, and that the derivatives* $\partial_k g_0(t, y)$ *exist. Let* $\bar{u}(\cdot)$ *be an optimal control such that* $f(t, y(t, \bar{u}), v)$, $f_0(t, y(t, \bar{u}), v)$ *are regular in* U *(*$y(t, \bar{u})$ *the optimal trajectory). Finally, let* $\bar{\mathbf{z}}(s) = (1, \bar{z}(s))^{(2)}$ *be the solution of the system (2.5.8) with* $\mathbf{y}(s) = \mathbf{y}(s, \bar{u}) = (y_0(s, \bar{u}), y(s, \bar{u}))$ *and final condition* $\bar{z}(\bar{t}) = (1, \nabla_y g_0(\bar{t}, y(\bar{t}, \bar{u})))$. *Then*

$$H(s, \mathbf{y}(s, \bar{u}), \bar{\mathbf{z}}(s), \bar{u}(s)) = \min_{v \in U} H(s, \mathbf{y}(s, \bar{u}), \bar{\mathbf{z}}(s), v) \qquad (2.5.14)$$

almost everywhere in $0 \leq s \leq \bar{t}$.

The minimum principle (2.5.14) is an obvious restatement of (2.5.3). It is traditional to use *maximum* rather than minimum principles; we can turn (2.5.14) into a minimum principle by replacing the final condition in (2.5.8) by $(-1, -\nabla_y g_0(\bar{t}, y(\bar{t}, \bar{u})))$, which amounts to changing the sign of $z_0, \bar{z}(s)$.

The function $\bar{\mathbf{z}}(\cdot) = (1, \bar{z}(\cdot))$ in the minimum principle is called the **adjoint vector** or the **costate**; sometimes these names are used for $\bar{z}(\cdot)$ alone. Likewise, the word **state** is used to indicate $\mathbf{y}(t, \bar{u}) = (y_0(t, \bar{u}), y(t, \bar{u}))$ or just $y(t, \bar{u})$. We shall use the names for $y(t, \bar{u})$ and $\bar{z}(t)$; when the first coordinate is adjoined, we use the names **augmented state** and **augmented costate** (or **augmented adjoint vector**). The original equation (2.5.2) is the **state equation**. Equations (2.5.7) and (2.5.8) are called the **canonical equations**.

Remark 2.5.2. The minimum principle (2.5.14) is in general far from an "explicit solution" of the optimal control problem, since the Hamiltonian contains the unknown optimal augmented trajectory $(y_0(t, \bar{u}), y(t, \bar{u}))$ and the adjoint vector $\bar{z}(t)$, that depends itself on the unknown optimal control $\bar{u}$ and on the unknown final condition $\bar{\mathbf{z}}(\bar{t}) = (1, \nabla_y g_0(\bar{t}, y(\bar{t}, \bar{u})))$. When $g_0 = 0$, this final condition is explicitly known: $\mathbf{z}(\bar{t}) = (1, 0)$.

Example 2.5.3. We redo Example 1.6.1 (now rigorously and completely) using Theorem 2.5.1. The problem is to minimize

$$y_0(\bar{t}, u) = \int_0^{\bar{t}} u(\tau)^2 d\tau + y(\bar{t})^2$$

subject to the state equation

$$y'(t) = -u(t), \qquad y(0) = 1$$

---

(2) The notation $\bar{z}(s, \bar{u})$ would be more symmetric, but we avoid it since $\bar{z}(s)$ depends also on an unknown final condition.

with control set $U = [0, 1]$. Theorem 2.2.6 implies that an optimal control exists. Call $(1, p(t))$ the adjoint vector. The Hamiltonian is $H = u^2 - pu$, so that the second of the two canonical equations is $p' = 0$, and, in view of the final condition $p(\bar{t}) = \nabla_y g(\bar{t}, y(\bar{t}, \bar{u})) = 2y(\bar{t}, \bar{u})$, we have $p(t) \equiv 2y(\bar{t}, \bar{u})$. Accordingly, the minimum principle (2.5.14) is

$$\bar{u}(s)^2 - 2y(\bar{t}, \bar{u})\bar{u}(s) = \min_{0 \leq v \leq 1} \{v^2 - 2y(\bar{t}, \bar{u})v\}$$

which is exactly (1.6.13); from then on, we argue just as in Example 1.6.1.

2.6. Invariance of the Hamiltonian

Let $y(t)$, $z(t)$ be absolutely continuous functions satisfying the canonical equations

$$y'(t) = \nabla_z H(t, y(t), z(t)), \qquad z'(t) = -\nabla_y H(t, y(t), z(t)) \qquad (2.6.1)$$

a.e. in $0 \leq t \leq \bar{t}$, where $H : [0, \bar{t}] \times \mathbb{R}^m \times \mathbb{R}^m \to \mathbb{R}$ is continuously differentiable in all variables.

Theorem 2.6.1.

$$H(t, y(t), z(t)) = \int \partial_t H(t, y(t), z(t))dt \quad (0 \leq t \leq \bar{t}), \qquad (2.6.2)$$

∂_t *the partial derivative of H with respect to t,* $\int$ *an indefinite integral. In particular, if H is independent of t,*

$$H(y(t), z(t)) = constant \quad (0 \leq t \leq \bar{t}). \qquad (2.6.3)$$

The result follows showing that $H(t, y(t), y(t))$ is absolutely continuous and then differentiating with respect to t.   ■

There are applications to mechanics. We recall the Euler-Lagrange equations in **1.3**,

$$\frac{d}{dt}\left(\frac{\partial L}{\partial q_j'}\right) - \frac{\partial L}{\partial q_j} = 0 \qquad (2.6.4)$$

with $L = L(q_1, \ldots, q_n, q_1', \ldots, q_n')$. The Euler-Lagrange equations are a system of n second order equations. Mathematicians might reduce it to a first order system by taking $q_1', \ldots, q_n'$ as new unknowns, but physicists prefer to do the reduction using the **generalized momenta**

$$p_j = \frac{\partial L}{\partial q_j'}. \qquad (2.6.5)$$

To do this, one needs to know that the q_j' can be obtained back from the p_j.

Let $F(x) = F(x_1, x_2, \ldots, x_n)$ be a function defined in a neighborhood of a point $\bar{x} = (\bar{x}_1, \bar{x}_2, \ldots, \bar{x}_n)$ in $\mathbb{R}^n$. The **Legendre transformation** generated by F is

$$y_j = \frac{\partial F}{\partial x_j}(x_1, x_2, \ldots, x_n) \quad (j = 1, 2, \ldots, n). \tag{2.6.6}$$

Theorem 2.6.2. *Assume F is twice continuously differentiable in a neighborhood of $\bar{x}$ and that the Hessian matrix $\{\partial^2 F/\partial x_j \partial x_k\}$ is nonsingular at $x = \bar{x}$. Then the transformation (2.6.6) generated by F has a continuously differentiable inverse in a neighborhood of $\bar{y} = \nabla F(\bar{x})$; moreover, the inverse is the Legendre transformation generated by the conjugate function*

$$G(y) = \sum_{j=1}^{n} x_j(y) y_j - F(x(y)). \tag{2.6.7}$$

Proof. The Jacobian matrix of the transformation (2.6.6) is the Hessian matrix of F, thus, the existence and continuous differentiability of the inverse transformation are a consequence of the inverse function theorem. To show that the inverse transformation is generated by (2.6.7) we differentiate, obtaining

$$\frac{\partial G(y)}{\partial y_k} = \sum_{j=1}^{n} \frac{\partial x_j(y)}{\partial y_k} y_j + x_k(y) - \sum_{j=1}^{n} \frac{\partial F(x(y))}{\partial x_j} \frac{\partial x_j(y)}{\partial y_k}.$$

By (2.6.6) the first and third terms cancel out.   ∎

Assume that the Lagrangian is continuously differentiable in all variables, and consider equations (2.6.5) as defining a transformation from the space of the $q'_1, \ldots, q'_n$ into the space of the $p_1, \ldots, p_n$ (in this transformation the q_j are parameters). Theorem 2.6.2 certifies that this transformation will have (locally) a continuously differentiable inverse assuming that the Hessian matrix of L with respect to the variables $q'_1, \ldots, q'_n$ is not zero (for arguments on the meaning of this condition on the Hessian see Gantmajer [1970, p. 46]. The inverse transformation is the Legendre transformation generated by

$$H(p_1, \ldots, p_n, q_1, \ldots, q_n) = \sum_{j=1}^{n} p_j q'_j - L(q_1, \ldots, q_n, q'_1, \ldots, q'_n) \tag{2.6.8}$$

with the q'_j as functions of the p_j. The function H is called the **Hamiltonian** of the system, and the **canonical equations** are

$$q'_j = \frac{\partial H}{\partial p_j}, \qquad p'_j = -\frac{\partial H}{\partial q_j}. \tag{2.6.9}$$

By Theorem 2.6.2 the first group of canonical equations expresses the inverse of the transformation (2.6.5) (the Hamiltonian is the conjugate of the Lagrangian), while the second group of canonical equations duplicates the Euler–Lagrange equations (1.3.4) (note that $\partial H/\partial q_j = \partial L/\partial q_j$). Theorem 2.6.1 then applies to show that, in case the Hamiltonian is independent of time, it is constant along trajectories $(q_1(t), \ldots, q_n(t))$ of the mechanical system:

$$H(p_1(t), \ldots, p_n(t), q_1(t), \ldots, q_n(t)) = \text{constant}. \tag{2.6.10}$$

This is physically interesting since $H(p_1, \ldots, p_n, q_1, \ldots, q_n)$ is the total energy of the system; this follows from the definition (2.6.8) of H and the expressions for the kinetic and potential energy in **1.3**. For more on the physics, see Gantmajer [1970, p. 71 ff].

We prove below a "control version" of Theorem 2.6.1, where the function H in the canonical equations depends on a parameter $u \in U$ (U is an arbitrary set) and $H : [0, \bar{t}] \times \mathbb{R}^m \times \mathbb{R}^m \times U \to \mathbb{R}$.

Theorem 2.6.3. *Assume $H(t, y, z, u)$ is continuously differentiable with respect to t, y, z for u fixed. Let $\bar{u}(\cdot)$ be a U-valued function*[1] *and $y(\cdot), z(\cdot)$ absolutely continuous $\mathbb{R}^m$-valued functions, all defined in $0 \le t \le \bar{t}$ and such that the canonical equations*

$$y'(t) = \nabla_z H(t, y(t), z(t), \bar{u}(t)), \qquad z'(t) = -\nabla_y H(t, y(t), z(t), \bar{u}(t)) \tag{2.6.11}$$

are satisfied. Finally, assume that there exists a function $L(\cdot) \in L^1(0, \bar{t})$ such that

$$\|\nabla_{(s,y,z)} H(s, y(s), z(s), \bar{u}(t))\| \le L(s) \quad (0 \le s, t \le \bar{t}) \tag{2.6.12}$$

and that

$$H(s, y(s), z(s), \bar{u}(s)) = \min_{v \in U} H(s, y(s), z(s), v) \tag{2.6.13}$$

a.e. in $0 \le t \le \bar{t}$. Then the function $s \to \partial_s H(s, y(s), z(s), \bar{u}(s))$ belongs to $L^1(0, \bar{t})$ and

$$H(s, y(s), z(s), \bar{u}(s)) = \int \partial_s H(s, y(s), z(s), \bar{u}(s)) ds \tag{2.6.14}$$

a.e. in $0 \le s \le \bar{t}$, so that $s \to H(s, y(s), z(s), \bar{u}(s))$ is absolutely continuous; if H is independent of t,

$$H(s, y(s), z(s), \bar{u}(s)) = constant \quad a.e. \ in \ 0 \le t \le \bar{t}. \tag{2.6.15}$$

---

[1] There are no assumptions (on measurability or anything else) on $H(t, y, z, u)$ as a function of u or on $\bar{u}(t)$ as a function of t. The canonical equations (2.6.11) imply that the right-hand sides belong to $L^1(0, \bar{t})$, in particular, that they are measurable.

The proof requires various auxiliary steps. A function $\phi(s, t)$ defined in $a \leq s, t \leq b$ is **absolutely continuous in s uniformly with respect to** t if, given $\varepsilon > 0$ there exists $\delta > 0$ independent of t such that

$$\sum |\phi(b_j, t) - \phi(a_j, t)| \leq \varepsilon \quad (a \leq t \leq b) \tag{2.6.16}$$

for every finite collection of pairwise disjoint intervals $(a_j, b_j) \subseteq [a, b]$ with $\Sigma(b_j - a_j) \leq \delta$ (see **2.0** for the definition of plain absolute continuity). The estimate (2.6.12) is easily seen to imply that

$$\phi(s, t) = H(s, y(s), z(s), \bar{u}(t)) \tag{2.6.17}$$

is absolutely continuous in s uniformly with respect to t in $[0, \bar{t}]$.

Lemma 2.6.4. *Let $\phi(s, t)$ be defined in $0 \leq s, t \leq \bar{t}$, and such that*

$$\phi(s, s) = \min_{0 \leq t \leq \bar{t}} \phi(s, t) \tag{2.6.18}$$

in a set d dense in $[0, \bar{t}]$. Assume that $\phi(s, t)$ is absolutely continuous with respect to s uniformly in $t \in [0, \bar{t}]$. Then the function $\phi(s, s)$ can be uniquely extended to a function $\psi(s)$ absolutely continuous in $0 \leq s \leq \bar{t}$.

Proof. It follows from (2.6.18) that, if $\psi(s) = \phi(s, s)$ then $\psi(b) - \psi(a) \leq \phi(b, a) - \phi(a, a), \psi(a) - \psi(b) \leq \phi(a, b) - \phi(b, b)$, so that

$$|\psi(b) - \psi(a)| \leq |\phi(b, a) - \phi(a, a)| + |\phi(b, b) - \phi(a, b)|. \tag{2.6.19}$$

Hence, it follows from (2.6.16) that, given $\varepsilon > 0$, there exists $\delta > 0$ such that

$$\sum |\psi(b_j) - \psi(a_j)| \leq \varepsilon \tag{2.6.20}$$

for every finite collection of pairwise disjoint intervals $(a_j, b_j) \subseteq [a, b]$ with $\Sigma(b_j - a_j) \leq \delta$, as long as $a_j, b_j \in d$. This, in particular, implies that the function $\psi(s)$ is uniformly continuous in d, thus, it admits a unique (uniformly) continuous extension to the whole interval (Example 2.6.6 below). An obvious approximation argument based on continuity of the extension shows that (2.6.20) will hold even if $a_j, b_j \notin d$. ∎

Proof of Theorem 2.6.3. If $\phi(s, t)$ is the function in (2.6.17) we obtain from (2.6.13) (setting $v = \bar{u}(t)$) that $\phi(s, s) \leq \phi(s, t) \, (0 \leq t \leq \bar{t})$ for s in a set e of full measure in $[0, \bar{t}]$, so that (2.6.18) holds in e. Lemma 2.6.4 then implies that $\psi(s) = \phi(s, s) = H(s, y(s), z(s), \bar{u}(s))$ can be extended to an absolutely continuous function in $0 \leq s \leq \bar{t}$. It only remains to show that

$$\psi'(s) = \partial_s H(s, y(s), z(s), \bar{u}(s)) \quad \text{a.e. in } 0 \leq s \leq \bar{t}. \tag{2.6.21}$$

This we do as follows. Note first that if c is the set (of full measure in $[0, \bar{t}]$) where $y(s), z(s)$ are differentiable, then (by the chain rule) $\phi(s, t)$ is differentiable with respect to s for $s \in c$; also, since $\psi(s)$ is absolutely continuous, it is differentiable in a set d of full measure in $[0, \bar{t}]$. If $s \in e \cap c \cap d$, we have

$$\frac{\psi(s+h) - \psi(s)}{h} = \frac{\phi(s+h, s+h) - \phi(s, s)}{h}$$

$$= \frac{\phi(s+h, s+h) - \phi(s+h, s)}{h} + \frac{\phi(s+h, s) - \phi(s, s)}{h}.$$

The left-hand side and the second term on the right-hand side have a limit as $h \to 0$, thus, the first term on the right must have a limit as well. Since, by (2.6.18) $\phi(s+h, s+h) \le \phi(s+h, s)$, this term is nonnegative when $h < 0$, nonpositive, when $h > 0$; thus, its limit must be zero. We obtain in this way that

$$\psi'(s) = \frac{\partial}{\partial s} H(s, y(s), z(s), \bar{u}(t)) \Big|_{s=t}$$

$$= \partial_s H(s, y(s), z(s), \bar{u}(s)) + \langle y'(s), \nabla_y H(s, y(s), z(s), \bar{u}(s)) \rangle$$

$$+ \langle z'(s), \nabla_z H(t, y(t), z(t), \bar{u}(t)) \rangle = \partial_s H(s, y(s), z(s), \bar{u}(s))$$

so that (2.6.21) holds. This completes the proof.   ∎

The specific control application is a complement to Theorem 2.5.1. We limit ourselves to the autonomous case.

Theorem 2.6.5. *Assume that $f(y, u), f_0(y, u), C_{ad}(0, \bar{t}; U)$ and $\bar{u}(\cdot)$ satisfy the assumptions of Theorem 2.5.1. Then*

$$H(\mathbf{y}(t, \bar{u}), \bar{\mathbf{z}}(t), \bar{u}(t)) = constant \quad a.e. \ in \ 0 \le t \le \bar{t}. \tag{2.6.22}$$

Proof. Hypotheses III and III^0 for $f(y(t), \bar{u}(t))$ and $f_0(y(t), \bar{u}(t))$ imply the assumptions of Theorem 2.6.3 for $\mathbf{y}(t, \bar{u}), \bar{\mathbf{z}}(t), \bar{u}(t)$, and $H(\mathbf{y}, \mathbf{z}, t)$ given by (2.5.6); (2.6.13) is the conclusion of Theorem 2.5.1.   ∎

Since (2.6.15) is a consequence of (2.5.14), Theorem 2.6.5 contributes nothing towards the identification of $\bar{u}(\cdot)$; it just gives additional information.

Example 2.6.6. Let X be a metric space, $D \subseteq X$ a dense subset, $f : X \to \mathbb{R}$ a uniformly continuous function defined in D. Then f admits a (unique) uniformly continuous extension to all of X.

The extension can be defined by $f(x) = \lim f(x_n)$, where $\{x_n\}$ is a sequence in D such that $x_n \to x \in X$; uniform continuity of f guarantees that $\{f(x_n)\}$ is Cauchy (hence convergent) and that $f(x)$ does not depend on $\{x_n\}$.

Example 2.6.7. Let $f(y, u)$, $f_0(y, u)$ be continuous in y, u and continuously differentiable with respect to y in $\mathbb{R}^m \times U$, where $U \subseteq \mathbb{R}^k$ is closed and bounded, and let $C_{\mathrm{ad}}(0, T; U)$ be the set of all measurable $\mathbb{R}^k$-valued functions taking values in U almost everywhere. Then all the assumptions in Theorem 2.6.5 are satisfied.

2.7. The Linear-Quadratic Problem: Existence and Uniqueness of Optimal Controls

Consider a linear frictionless oscillator (a particle of mass m moving on a line under Hooke's law, that is, under the action of a linear restoring force $f(x) = -kx$) subject to an exterior force $u(t)$. The equation is

$$mx''(t) = -kx(t) + u(t) \tag{2.7.1}$$

where $x(t)$ is the position of the particle at time t, measured from the equilibrium position $x = 0$. The **energy** of the unexcited oscillator ($u(t) \equiv 0$) is

$$E(t) = \tfrac{1}{2}\big(mx'(t)^2 + kx(t)^2\big).$$

The exterior force may or may not be subject to a constraint. Given an initial position and velocity

$$x(0) = x_0, \qquad x'(0) = x_1$$

our aim is to **stabilize** the oscillating system in a fixed time interval $0 \le t \le \bar{t}$. This may be understood in many ways; here, we try to keep the average energy on the interval $0 \le t \le \bar{t}$ as small as possible, that is, we minimize the integral

$$\int_0^{\bar{t}} \big(mx'(\tau)^2 + kx(\tau)^2\big)d\tau.$$

We write this problem as one of the control problems in **2.2**. Setting $y(t) = (y_1(t), y_2(t)) = (x(t), x'(t))$, the forced equation (2.7.1) becomes the linear system

$$y'(t) = Ay(t) + Bu(t), \qquad y(0) = \zeta = (x_0, x_1) \tag{2.7.2}$$

with

$$A = \begin{bmatrix} 0 & 1 \\ -k/m & 0 \end{bmatrix} \qquad B = \begin{bmatrix} 0 \\ 1/m \end{bmatrix}.$$

The energy can be expressed in the form

$$E(t) = \tfrac{1}{2}(Ry(t), y(t)) \tag{2.7.3}$$

where R is the symmetric positive definite matrix

$$R = \begin{bmatrix} k & 0 \\ 0 & m \end{bmatrix}.$$

Linear oscillations of systems comprising many coupled elements (such as space structures) can also be written in the form (2.7.2), where A is a $m \times m$ matrix; the external force $u(t)$ is a k-dimensional vector and B is a $k \times m$ matrix. The energy of these systems is expressed in the form (2.7.3), R a positive definite matrix. The constraints on the vector u, if any, are of the form

$$u(t) \in U \tag{2.7.4}$$

where U is a control set in $\mathbb{R}^k$.

We study in this section and the next the **unconstrained** problem $U = \mathbb{R}^k$. The spectacular success of the treatment (down to the implementation of optimal control by feedback) is due to the fact that, in the absence of constraints, the problem is actually a classical variational problem like those in **1.1**; also, it can be treated as a quadratic minimization problem in Hilbert space. Our approach here is based on the minimum principle (2.5.14) and is capable of dealing with control constraints (see **2.9**).

Direct minimization of the average energy is not reasonable since very large controls may have to be used: we minimize instead the average of $(Ry(t), y(t)) + c\|u(t)\|^2$ in the control interval, where the term $c\|u(t)\|^2$, $c > 0$ limits the size of u.

To gain generality without additional work, we consider the nonautonomous equation

$$y'(t) = A(t)y(t) + B(t)u(t), \quad y(0) = \zeta \tag{2.7.5}$$

and the cost functional

$$y_0(\bar{t}, u) = \int_0^{\bar{t}} \{(R(\tau)y(\tau), y(\tau)) + c(\tau)\|u(\tau)\|^2\} d\tau \tag{2.7.6}$$

The assumptions on (2.7.5) are those in Remark 2.2.13,

$$A(\cdot) \in L^1(0, T; \mathbb{R}^{m \times m}), \qquad B(\cdot) \in L^2(0, T; \mathbb{R}^{k \times m}), \tag{2.7.7}$$

and the assumptions on the cost functional (2.7.6) are: each $R(t)$ is positive definite $((R(t)y, y) \geq 0$ for $y \in \mathbb{R}^m)$, $c(\cdot)$ is measurable, $c(t) \geq c > 0$, and

$$R(\cdot) \in L^1(0, T; \mathbb{R}^{m \times m}), \qquad c(\cdot) \in L^\infty(0, T). \tag{2.7.8}$$

The control space (see (2.2.8)) is

$$C_{\text{ad}}(0, \bar{t}; U) = F_m(0, T; U) \cap L^2(0, T; \mathbb{R}^k),$$

where we assume U convex and closed. For existence purposes, we may include a state constraint $y(t, u) \in M(t)$ $(0 \leq t \leq \bar{t})$ and a target condition $y(\bar{t}, u) \in Y$ with $M(t)$ and Y closed.

Theorem 2.7.1. *The linear-quadratic problem has a solution.*

Proof. To apply Theorem 2.2.6 we only have to show that the cost functional (2.7.6) is weakly lower semicontinuous. This follows from Example 2.2.15 but can also be proved directly. Let $\{u^n(\cdot)\}$ be a weakly convergent sequence in $L^2(0, \bar{t}; \mathbb{R}^k)$. Then, by the Arzelà-Ascoli theorem, $\{y(t, u^n)\}$ is uniformly convergent. We take limits in the integral involving $(R(\tau)y(\tau, u^n), y(\tau, u^n))$ using the dominated convergence theorem. In the second part of the integrand, we observe that in any Hilbert space (in this case $L^2(0, \bar{t}; \mathbb{R}^k)$), the unit ball is weakly closed (Theorem 2.0.12). Hence, we have

$$\|v\| \leq \liminf_{n \to \infty} \|v^n\|$$

for every sequence $\{v^n\}$ weakly convergent to v, and we can apply this observation to the sequence $\{c(\cdot)^{1/2} u^n(\cdot)\}$. The existence of the minimizing sequence $\{u^n(\cdot)\}$ required in Theorem 2.2.6 is obvious.   ∎

Theorem 2.7.2. *If the state constraint sets $M(t)$ and the target set Y are convex, the optimal control is unique.*

Proof. It follows from properties of symmetric positive definite matrices that $H = C(0, \bar{t}; \mathbb{R}^m) \times L^2(0, \bar{t}; \mathbb{R}^k)$ is an inner product space equipped with

$$((y, u), (z, v)) = \int_0^{\bar{t}} \{(R(\tau)y(\tau), z(\tau)) + c(\tau)(u(\tau), v(\tau))\} d\tau, \qquad (2.7.9)$$

and with the equivalence relation $(y, u) \approx 0$ if $((y, u), (y, u)) = 0$, which is equivalent to $u(t) \equiv 0$ a.e., $\int (R(\tau)y(\tau), y(\tau)) d\tau \equiv 0$. Since the equation is linear, we have $y(t, \alpha u_1 + (1 - \alpha)u_2) = \alpha y(t, u_1) + (1 - \alpha)y(t, u_2)$. If $0 \leq \alpha \leq 1$,

$$\|(y(\cdot, \alpha u_1 + (1 - \alpha)u_2), \alpha u_1 + (1 - \alpha)u_2)\|$$

$$\leq \alpha \|(y(\cdot, u_1), u_1(\cdot))\| + (1 - \alpha) \|(y(\cdot, u_2), u_2(\cdot))\|$$

where $\| \cdot \|$ is the norm associated with the inner product (2.7.9), with strict inequality if $0 < \alpha < 1$ and $u_1 \neq u_2$. If $m = \min y_0(\bar{t}, u)$ and u_1, u_2 are optimal controls, we have $\|(y(\cdot, u_1), u_1(\cdot))\| = y_0(\bar{t}, u_1)^{1/2} = \sqrt{m}$, $\|(y(\cdot, u_2), u_2(\cdot))\| = y_0(\bar{t}, u_2)^{1/2} = \sqrt{m}$. Hence, the control $u(\cdot) = \alpha u_1(\cdot) + (1 - \alpha)u_2(\cdot)$ satisfies $y_0(\bar{t}, u)^{1/2} < \alpha y_0(\bar{t}, u_1)^{1/2} + (1 - \alpha)y_0(\bar{t}, u_2)^{1/2} < \sqrt{m}$, which contradicts optimality.   ∎

2.8. The Unconstrained Linear-Quadratic Problem: Feedback, the Riccati Equation

We apply the minimum principle (Theorem 2.5.1) to the linear-quadratic problem (with no control constraints or target condition) under the hypotheses in **2.7**. We may and will assume $c(t) \equiv 1$, replacing $B(t)$ by $B(t)/c(t)$ in (2.7.5).

The Hamiltonian is

$$H(s, \mathbf{y}, \mathbf{z}, v) = \langle R(s)y, y\rangle + \|v\|^2 + \langle z, A(s)y + B(s)v\rangle$$

taking into account that $z_0 = 1$. To use the minimum principle (2.5.14), we eliminate from both sides terms not containing v and $\bar{u}(s)$; we obtain

$$\|\bar{u}(s)\|^2 + \langle B(s)^*\bar{z}(s), \bar{u}(s)\rangle = \min_{v \in U}\{\|v\|^2 + \langle B(s)^*\bar{z}(s), v\rangle\} \qquad (2.8.1)$$

a.e. in the control interval $0 \le s \le \bar{t}$, where $\bar{u}(\cdot)$ is the (unique) optimal control uncovered in 2.7, and $\bar{z}(\cdot)$ is the solution of the adjoint variational equation

$$\bar{z}'(s) = -A(s)^*\bar{z}(s) - 2R(s)y(s, \bar{u}), \qquad \bar{z}(\bar{t}) = 0 \qquad (2.8.2)$$

in $0 \le s \le \bar{t}$. Assuming that $U = \mathbb{R}^m$, if $B^*(s)\bar{z}(s) \ne 0$ then the minimum of $\|v\|^2 + \langle B^*(s)\bar{z}(s), v\rangle$ in $\|v\| = c$ is attained when v is a multiple of $-B^*(s)\bar{z}(s)$:

$$v = -c\frac{B(s)^*\bar{z}(s)}{\|B(s)^*\bar{z}(s)\|}. \qquad (2.8.3)$$

The minimum is $m(c) = c^2 - c\|B(s)^*\bar{z}(s)\|$ which, as function of c, reaches its own minimum at $c = \|B(s)^*\bar{z}(s)\|/2$. Accordingly, (2.8.1) identifies $\bar{u}(s)$ as

$$\bar{u}(s) = -\tfrac{1}{2}B(s)^*\bar{z}(s). \qquad (2.8.4)$$

If $B^*\bar{z}(s) = 0$ then (2.8.1) implies $\bar{u}(s) = 0$, hence, (2.8.4) is valid in any case.

We solve (2.8.2) by the variation-of-constants formula, keeping in mind that if $S(t, s)$ is the solution matrix of $y'(t) = A(t)y(t)$ then the solution matrix of $z'(s) = -A(s)^*z(s)$ is $S^*(s, t) = S(t, s)^*$. Rewrite (2.8.4) (with t as variable) in the form

$$\bar{u}(t) = -B(t)^* \int_t^{\bar{t}} S(\sigma, t)^* R(\sigma)y(\sigma, \bar{u})d\sigma. \qquad (2.8.5)$$

We now *assume* a feedback law of the form

$$\bar{u}(t) = -B(t)^* P(t)y(t, \bar{u}) \quad (0 \le t \le \bar{t}) \qquad (2.8.6)$$

where $P(t)$ is a $m \times m$ matrix and, operating formally, we obtain a differential equation for $P(t)$. The rigorous argument goes in reverse: we construct $P(t)$ using the equation and show that (2.8.6) reproduces the (only) optimal control $\bar{u}(t)$ given by (2.8.4).

This is the formal work. Replace (2.8.6) in the state equation (2.7.5):

$$y'(t, \bar{u}) = (A(t) - B(t)B(t)^* P(t))y(t, \bar{u}). \qquad (2.8.7)$$

If $V(t, s)$ is the solution matrix of this equation, we have $y(\sigma, \bar{u}) = V(\sigma, t)y(t, \bar{u})$; thus, (2.8.6) will hold if

$$P(t) = \int_t^{\bar{t}} S(\sigma, t)^* R(\sigma) V(\sigma, t) d\sigma. \tag{2.8.8}$$

This is not an explicit expression for $P(t)$ but an integral equation, since $V(\sigma, t)$ depends on $P(t)$. Differentiating with respect to t we obtain

$$P'(t) = -S(t, t)^* R(t) V(t, t) + \int_t^{\bar{t}} \partial_t S(\sigma, t)^* R(\sigma) V(\sigma, t) d\sigma$$

$$+ \int_t^{\bar{t}} S(\sigma, t)^* R(\sigma) \partial_t V(\sigma, t) d\sigma$$

$$= -R(t) - \int_t^{\bar{t}} A(t)^* S(\sigma, t)^* R(\sigma) V(\sigma, t) d\sigma$$

$$- \int_t^{\bar{t}} S(\sigma, t)^* R(\sigma) V(\sigma, t)(A(t) - B(t) B(t)^* P(t)) d\sigma,$$

where we have used $\partial_t S(\sigma, t)^* = \partial_t S^*(t, \sigma) = -A(t)^* S^*(t, \sigma) = A(t)^* S(\sigma, t)^*$ and $\partial_t V(\sigma, t) = -V(\sigma, t)(A(t) - B(t) B^*(t) P(t))$ (see (2.1.20)). We obtain the equation

$$P'(t) = -R(t) - A(t)^* P(t) - P(t) A(t)$$

$$+ P(t) B(t) B(t)^* P(t), \quad P(\bar{t}) = 0 \quad (0 \le t \le \bar{t}), \tag{2.8.9}$$

known as the (matrix) **Riccati equation**, since for $m = 1$ it reduces to Count Riccati's equation $p'(t) = -r(t) - 2a(t) p(t) + b(t)^2 p(t)^2$.

We check below that the function $f(t, P) = -R(t) - A(t)^* P - PA(t) + PB(t) B(t)^* P$ satisfies Hypothesis II in **2.1** in the space $\mathbb{R}^{m \times m}$. Measurability of $t \to f(t, P)$ and continuity of $P \to f(t, P)$ are obvious. If $\| \cdot \|$ is the operator norm of matrices,

$$\|f(t, P)\| \le \|R(t)\| + 2\|A(t)\| \|P\| + \|B(t)\|^2 \|P\|^2;$$

since $R(\cdot), A(\cdot) \in L^1(0, T; \mathbb{R}^{m \times m})$ and $B(\cdot) \in L^2(0, T; \mathbb{R}^{k \times m})$, (2.1.3) holds. As for (2.1.4), writing $P'B(t) B(t)^* P' - PB(t) B(t)^* P = (P' - P) B(t) B(t)^* P' + PB(t) B(t)^* (P' - P)$,

$$\|f(t, P') - f(t, P)\| \le 2\|A(t)\| \|P' - P\| + \|B(t)\|^2 (\|P'\| + \|P\|) \|P' - P\|.$$

It then follows from Theorem 2.1.1 that (2.8.9) can be solved (backwards) locally. Let $(t_0, \bar{t}]$ be the maximal backwards interval of existence. Integrating (2.8.9) we obtain the integral equation (2.8.8), which combined with (2.8.5) gives (2.8.6) in $(t_0, \bar{t}]$, *now as a theorem.* However, we still have to show that $P(t)$ exists in $0 \le t \le \bar{t}$. We do this establishing an a priori bound

$$\|P(t)\| \le C \tag{2.8.10}$$

and applying Lemma 2.1.5. To obtain this bound we take adjoints in the Riccati equation (2.8.9):

$$(P(t)^*)' = -R(t) - P(t)^*A(t) - A(t)^*P(t)^* + P(t)^*B(t)B(t)^*P(t).$$

This is (2.8.9) again, this time with $P(t)^*$ as variable; by uniqueness, $P(t)^* = P(t)$ in $(t_0, \bar{t}]$, that is, each $P(t)$ is symmetric there. If we can show that

$$0 \le \langle P(t)\zeta, \zeta \rangle \le C\|\zeta\| \tag{2.8.11}$$

then, since for any symmetric positive definite matrix P we have $\|P\| = \max_{\|\zeta\| \le 1} \langle P\zeta, \zeta \rangle$ (see (2.0.31)), (2.8.10) will result.

Let $t_0 < t_1 < \bar{t}$. We consider the linear-quadratic problem in the interval $t_1 \le t \le \bar{t}$, initial condition given at $t = t_1$,

$$y(t_1) = \zeta \tag{2.8.12}$$

with cost functional

$$y_0(t_1, \bar{t}, u) = \int_{t_1}^{\bar{t}} \{(R(\tau)y(\tau), y(\tau)) + \|u(\tau)\|^2\}d\tau. \tag{2.8.13}$$

Plainly, the feedback law (2.8.6) holds for this problem. Let $\bar{u}_1(\cdot)$ be the optimal control and $y(\cdot, \bar{u}_1)$ the corresponding optimal trajectory. Using the differential equation (2.8.7) satisfied by $y(t, \bar{u}_1)$ and the equation for $P(t)$ we obtain

$$\begin{aligned}
(P(t)&y(t, \bar{u}_1), y(t, \bar{u}_1))' \\
&= (P'(t)y(t, \bar{u}_1), y(t, \bar{u}_1)) + 2(P(t)y'(t, \bar{u}_1), y(t, \bar{u}_1)) \\
&= ((-R(t) - A(t)^*P(t) - P(t)A(t) + P(t)B(t)B(t)^*P(t))y(t, \bar{u}_1), y(t, \bar{u}_1)) \\
&\quad + 2(P(t)(A(t) - B(t)B(t)^*P(t))y(t, \bar{u}_1), y(t, \bar{u}_1)) \\
&= -(R(t)y(t, \bar{u}_1), y(t, \bar{u}_1)) - (B(t)^*P(t)y(t, \bar{u}_1), B(t)^*P(t)y(t, \bar{u}_1)) \\
&= -(R(t)y(t, \bar{u}_1), y(t, \bar{u}_1)) - (\bar{u}_1(t), \bar{u}_1(t)). \tag{2.8.14}
\end{aligned}$$

We integrate (2.8.14) in $t_1 \le t \le \bar{t}$ keeping in mind that $P(\bar{t}) = 0$: the result is

$$(P(t_1)\zeta, \zeta) = y_0(t_1, \bar{t}, \bar{u}_1). \tag{2.8.15}$$

The initial condition ζ is arbitrary and the right hand side of (2.8.15) is nonnegative, so that $(P(t_1)\zeta, \zeta) \ge 0$ for $\zeta \in \mathbb{R}^m$. Note that for the control $u(t) \equiv 0$ we have

$$y_0(t_1, \bar{t}, u) = \int_{t_1}^{\bar{t}} (R(\tau)y(\tau), y(\tau))d\tau \tag{2.8.16}$$

where $y(t) = S(t, t_1)\zeta$. Accordingly, $y_0(t_1, \bar{t}, 0) \le C\|\zeta\|^2$, where C does not depend on t_1. By optimality of $\bar{u}_1$ we must have $y_0(t_1, \bar{t}, \bar{u}_1) \le y_0(t_1, \bar{t}, 0)$. Hence, $(P(t_1)\zeta, \zeta) \le C\|\zeta\|^2$ with a constant that does not depend on t_1, thus, the inequality extends to the entire maximal interval of existence $(t_0, \bar{t}]$. We have thus shown existence of $P(t)$ in the whole control interval $0 \le t \le \bar{t}$, and established (2.8.6) globally.

Formula (2.8.6) is an example of a **feedback** or **closed loop** solution of an optimal control problem (that is, a formula that expresses the optimal control $\bar{u}$ in terms of the optimal trajectory). Feedback implemented by (2.8.6) is **instantaneous** in that $\bar{u}(t)$ only depends on $\bar{y}(t)$. This is the nicest kind. In other problems, other feedback laws may be acceptable, such as $\bar{u}(t) = \mathcal{F}[\bar{y}(\cdot)]$, where $\mathcal{F}$ is a functional that depends only on *present* or *past* values of the trajectory $\bar{y}(\cdot)$. A feedback solution eliminates the control parameter $u(t)$ and puts a control problem in "automatic pilot". In contrast, an **open loop** solution presupposes a "pilot" steering with an optimal control $\bar{u}(t)$. A feedback solution is more useful than an open loop solution, but it may be difficult or impossible to construct or to implement (see **4.3**, in particular Remark 4.3.6).

Feedback in mechanical systems is implemented by **sensors** that observe the trajectory and **actuators** that do the controlling; between them, processors compute with the observations and send the results to the actuators. The only computation needed in the linear-quadratic problem is matrix/vector multiplication for each t.

2.9. The Constrained Linear-Quadratic Problem

We take a look at the linear-quadratic problem with control constraints. The assumptions are the same as in **2.8** and, to fix ideas, we assume the control set is the unit ball $U = \{u; \|u\| \le 1\}$ of $\mathbb{R}^k$. Since U is convex and closed the existence-uniqueness arguments in **2.7** work. Using the minimum principle, the initial steps of the argument are the same as in **2.8**; in particular, (2.8.1) holds. If, for a given s we have $\|B(s)^*\bar{z}(s)\| \le 2$, then $\bar{u}(s)$ is given by (2.8.4):

$$\bar{u}(s) = -\tfrac{1}{2}B(s)^*\bar{z}(s).$$

If, on the other hand, $\|B(s)^*\bar{z}(s)\| > 2$,

$$\bar{u}(s) = -\frac{B(s)^*\bar{z}(s)}{\|B(s)^*\bar{z}(s)\|}.$$

These two formulas can be unified as

$$\bar{u}(s) = -\Phi(B^*(s)\bar{z}(s)) \qquad (2.9.1)$$

where Φ is the (Lipschitz) continuous function

$$\Phi(v) = \frac{v}{\max(2, \|v\|)}.$$

Accordingly, we can write the equation (2.7.5) satisfied by $\bar{y}(t) = y(t, \bar{u})$ and the equation (2.8.2) satisfied by $\bar{z}(t)$ as the system

$$\bar{y}'(t) = A(t)\bar{y}(t) - B(t)\Phi(B(t)^*\bar{z}(t)) \qquad (2.9.2)$$

$$\bar{z}'(t) = -A(t)\bar{z}(t) - 2R(t)\bar{y}(t). \qquad (2.9.3)$$

in the interval $0 \le t \le \bar{t}$. We may call this system an "explicit solution" of the constrained linear-quadratic problem, but note that initial conditions for $\bar{y}(\cdot)$ and $\bar{z}(\cdot)$ are given at opposite extremes of the interval $0 \le t \le \bar{t}$,

$$\bar{y}(0) = \zeta, \qquad \bar{z}(\bar{t}) = 0, \qquad (2.9.4)$$

so that we have a *boundary value* problem rather than an initial value problem. Short of explicit solutions, we can use formula (2.9.1) and the system (2.9.2)-(2.9.3) to deduce smoothness properties of the optimal control $\bar{u}(\cdot)$. For instance, if $B(s)$ is continuous, it follows from (2.9.1) that the optimal control $\bar{u}(\cdot)$ will be continuous as well. This is news, since we only start with the assumption $\bar{u}(\cdot) \in L^2(0, \bar{t}; \mathbb{R}^k)$. In the same vein, increasing smoothness of the coefficients results in increasing (piecewise) smoothness of the control. To realize this, we note that the function $\Phi(v)$, although only Lipschitz continuous in $\mathbb{R}^k$ is infinitely differentiable in $\|u\| < 2$ and $\|u\| > 2$. Hence,

Theorem 2.9.1. (*a*) *Let $B(\cdot)$ be continuous. Then $\bar{u}(\cdot)$ is continuous in $0 \le t \le \bar{t}$.* (*b*) *Let $A(t), B(t), R(t)$ be infinitely differentiable. Then $\bar{u}(\cdot)$ is infinitely differentiable in the subsets of $0 \le t \le \bar{t}$ defined by $\|B(s)^*z(s, \bar{u})\| > 2$ and by $\|B(s)^*z(s, \bar{u})\| < 2$.*

Theorem 2.9.1 results from the fact that a solution of the system,

$$y'(t) = f(t, y(t))$$

in an interval (a, b) is infinitely differentiable if $f(t, y)$ is infinitely differentiable in a neighborhood of $\{(t, y(t)); a < t < b\} \subset \mathbb{R}^{m+1}$; we just take derivatives repeatedly.   ∎

3

Abstract Minimization Problems: The Minimum Principle for the Time Optimal Problem

3.1. Abstract Minimization Problems

Optimal control problems include different sorts of equations, target conditions and constraints. We try a unified formulation.

Abstract minimization problem. Let V be a metric space, E a normed space, Y a subset of E. Given functions $f : V \to E$, $f_0 : V \to \mathbb{R}$, characterize the elements of V satisfying

$$\text{minimize} \quad f_0(u) \tag{3.1.1}$$

$$\text{subject to} \quad f(u) \in Y. \tag{3.1.2}$$

The constraint $f(u) \in Y$ is called the **target condition**; Y is the **target set**. State constraints can also be included (see Example 3.1.2 below). We do not assume that either f or f_0 are defined in all of V.

For obvious reasons, (3.1.1)-(3.1.2) is called an **abstract nonlinear programming problem** (see **4.10**). We call E the **state space**.

The optimal control problems in Chapter 1 can be put in this form, as seen below.

Example 3.1.1. A control problem described by an ordinary differential system

$$y'(t) = f(t, y(t), u(t)), \qquad y(0) = \zeta \tag{3.1.3}$$

with fixed terminal time $\bar{t}$, a cost functional $y_0(t, u)$, a control constraint

$$u(t) \in U \tag{3.1.4}$$

and a target condition

$$y(\bar{t}, u) \in Y \tag{3.1.5}$$

becomes an abstract minimization problem taking

$$f(u) = y(\bar{t}, u), \qquad f_0(u) = y_0(\bar{t}, u). \tag{3.1.6}$$

The state space is $E = \mathbb{R}^m$, and the metric space V is the space $C_{ad}(0, \bar{t}; U)$ of admissible controls (see **2.1**) equipped with a suitable distance; one that works well is the **Ekeland distance**

$$d(u(\cdot), v(\cdot)) = |\{t; u(t) \neq v(t)\}|, \qquad (3.1.7)$$

where $|\cdot|$ indicates Lebesgue measure. This description is oversimplified; in general, V will be a subspace of $C_{ad}(0, \bar{t}; U)$.

Example 3.1.2. Consider a control problem described by a functional differential equation

$$y'(t) = \left(a - \left(\int_{-h}^{0} b(\tau)y(t + \tau)d\tau\right)\right)y(t) + u(t) \qquad (3.1.8)$$

with fixed terminal time $\bar{t} > 0$ (see **1.8**). There are (at least) two natural types of target condition. One is the Euclidean target condition $y(\bar{t}, u) \in Y \subseteq \mathbb{R}^m$, where we take $E = \mathbb{R}^m$ as state space. The functions in (3.1.1)-(3.1.2) are again given by $f(u) = y(\bar{t}, u)$, $f_0(\bar{t}, u) = y_0(\bar{t}, u)$, and the control space $C_{ad}(0, \bar{t}; U)$ is the same in Example 3.1.1. If we use a functional target condition, for instance

$$y(\bar{t} + \tau) = \bar{y}(\tau), \quad (-h \leq \tau \leq 0) \qquad (3.1.9)$$

with $\bar{y}(\cdot)$ continuous, we may take $E = C(-h, 0; \mathbb{R}^m)$ equipped, for instance, with the supremum norm; the target condition (3.1.9) can then be written as an ordinary target condition $y_{\bar{t}} = \bar{y}$ where $y_t(\cdot)$ is the **section** of $y(\cdot)$ defined by $y_t(\tau) = y(t + \tau) \ (-h \leq \tau \leq 0)$. The functions are

$$f(u) = y_{\bar{t}}(\cdot) \qquad f_0(u) = y_0(\bar{t}, u). \qquad (3.1.10)$$

The control space and distance are the same. There are other natural choices of state spaces, for instance, $E = L^2(-h, 0; \mathbb{R}^m)$.

Example 3.1.3. Consider the distributed parameter system in **1.9**,

$$\frac{\partial y(t, x)}{\partial t} = \Delta y(t, x) - f(y(t, x)) + u(t, x) \quad (x \in \Omega), \qquad (3.1.11)$$

$$\frac{\partial y(t, x)}{\partial \nu} = 0 \quad (x \in \Gamma), \qquad (3.1.12)$$

with control constraint

$$0 \leq u(x, t) \leq R. \qquad (3.1.13)$$

We take as $C_{ad}(0, \bar{t}; U)$ the subset of $L^\infty((0, \bar{t}) \times \Omega)$ defined by (3.1.13) almost everywhere; the distance is

$$d(u(\cdot, \cdot), v(\cdot, \cdot)) = |\{t; u(t, \cdot) \neq v(t, \cdot)\}|. \qquad (3.1.14)$$

We have a wide choice of state spaces; for instance, $E = C(\overline{\Omega})$ or $E = L^2(\Omega)$ (see **1.9**). The functions in the abstract minimization problem are

$$f(u) = y(\overline{t}, \cdot, u), \qquad f_0(\overline{t}, u) = y_0(\overline{t}, u), \qquad (3.1.15)$$

where $y_0(t, u)$ is the cost functional and $y(t, x, u)$ is the solution of (3.1.11)-(3.1.12) corresponding to $u = u(\cdot, \cdot)$. We may also include a target condition $y(\overline{t}, \cdot) \in Y \subseteq$ state space.

Example 3.1.4. State constraints are included in the abstract formulation. For instance, add to the system in Example 3.1.1 the constraint

$$y(t, u) \in M(t) \subseteq \mathbb{R}^m \quad (0 \le t \le \overline{t}). \qquad (3.1.16)$$

We take now $E = C(0, \overline{t}; \mathbb{R}^m) \times \mathbb{R}^m$ as state space. The control space is the same in Example 3.1.1 and

$$f(u) = (y(\cdot, u), y(\overline{t}, u)), \qquad f_0(u) = y_0(\overline{t}, u).$$

Let $\mathbf{M}(\overline{t}) = \{y(\cdot) \in C(0, \overline{t}; \mathbb{R}^m); y(t) \in M(t) \ (0 \le t \le \overline{t})\}$. Then the state constraint (3.1.16) and the target condition (3.1.5) can be serviced at the same time as a target condition (3.1.2), writing

$$f(u) \in \mathbf{Y} = \mathbf{M}(\overline{t}) \times Y.$$

We may of course simplify by including the target condition (3.1.5) into the state constraint (3.1.16) (replace $M(\overline{t})$ by $M(\overline{t}) \cap Y$), but in practice this may not be convenient. Note that in including state constraints we have gone from a finite dimensional to an infinite dimensional state space.

Example 3.1.5. So far, we have had fixed arrival time. Consider again the control problem (3.1.3)-(3.1.4)-(3.1.5), this time with free arrival time. Let $T >$ optimal arrival time $\overline{t}$. The metric space V is the space of all pairs (t, u) where $0 \le t \le T$ and $u(\cdot) \in C_{\mathrm{ad}}(0, t; U)$. The distance between pairs $(s, u(\cdot))$, $(t, v(\cdot))$ is

$$d((s, u(\cdot)), (t, v(\cdot))) = t - s + d_s(u(\cdot), v(\cdot)), \qquad (3.1.17)$$

if $t \ge s$, where $d_s(u(\cdot), v(\cdot))$ is the Ekeland distance in $0 \le \tau \le s$; if $t \le s$ the roles of s and t are reversed. The problem becomes an abstract minimization problem taking $E = \mathbb{R}^m$ and

$$f((t, u)) = y(t, u), \qquad f_0((t, u)) = y_0(t, u). \qquad (3.1.18)$$

The simplest variable terminal time problem is the **time optimal problem**, where we minimize $\overline{t}$. Here a different abstract formulation is more convenient. To

fix ideas, consider the time optimal problem for the ordinary differential system
(3.1.3) with control constraint (3.1.4) and target condition (3.1.5). If $\bar{t}$ is the optimal
time and $y(t, \bar{u})$ is the trajectory corresponding to the optimal control $\bar{u}$, we have

$$y(\bar{t}, \bar{u}) \in Y.$$

On the other hand, by time optimality, if u is an arbitrary admissible control and
$\{t_n\}$ is a sequence with $t_n < \bar{t}$,

$$y(t_n, u) \notin Y \quad \text{for all } n.$$

Finally, because of continuity of trajectories (in particular of the optimal trajectory)
we have $y(t_n, \bar{u}) \to y(\bar{t}, \bar{u}) \in Y$ if $t_n \to \bar{t}$, so that

$$\text{dist}(y(t_n, \bar{u}), Y) \to 0.$$

This leads to the

Abstract time optimal problem. Let $\{V_n\}$ be a sequence of metric spaces, E a
normed space, Y a subset of E. Given a sequence of functions $f_n : V_n \to E$ such
that

$$f_n(V_n) \cap Y = \emptyset \quad (n = 1, 2, \ldots), \tag{3.1.19}$$

characterize the sequences $\{\bar{u}^n\}, \bar{u}^n \in V_n$ that satisfy

$$\text{dist}(f_n(\bar{u}^n), Y) \to 0 \quad \text{as } n \to \infty. \tag{3.1.20}$$

Any such sequence is called a **solution** of the abstract time optimal problem.
When applied to the system defined by (3.1.3), the space V_n is the control space
$C_{\text{ad}}(0, t_n; U)$ in the interval $0 \le t \le t_n$ and $\bar{u}^n$ is the restriction of the optimal
control $\bar{u}$ to $0 \le t \le t_n$.

3.2. Ekeland's Variational Principle

To try to understand the abstract minimization problem (3.1.1)-(3.1.2) we do the
baby problem where $E = \mathbb{R}$, $V = \mathbb{R}^2$; f and f_0 are continuously differentiable
scalar-valued functions. The target set is a point $Y = \{\bar{y}\}$. Of course, we can use
Lagrange multipliers: if the minimum m in (3.1.1) is finite and $\bar{u}$ is a solution of
the problem, there exist multipliers z_0, z such that

$$z_0 \partial f_0(\bar{u}) + z \partial f(\bar{u}) = 0. \tag{3.2.1}$$

where ∂ is an arbitrary directional derivative. A naive attempt to obtain the multipliers is: take a continuously differentiable function $\phi(x)$ having a unique minimum
at $x = 0$ (for instance, $\phi(x) = x^2$) and define

$$F(u) = \phi(f_0(u) - m) + \phi(f(u) - \bar{y}).$$

If $\bar{u}$ is a solution of (3.1.1)-(3.1.2) then $\bar{u}$ is a minimum of $F(u)$, thus

$$\partial F(\bar{u}) = 0.$$

However,

$$\partial F(\bar{u}) = \phi'(f_0(\bar{u}) - m)\partial f_0(\bar{u}) + \phi'(f(\bar{u}) - \bar{y})\partial f(\bar{u})$$

so that the (obviously counterfeit) aspirants to Lagrange multipliers are $z_0 = \phi'(f_0(u) - m)$ and $z = \phi'(f(\bar{u}) - \bar{y})$, both zero for $u = \bar{u}$. A conceivable way out is to use the function

$$F(u) = \{(f_0(u) - m)^2 + (f(u) - \bar{y})^2\}^{1/2}.$$

We have

$$\partial F(u) = \frac{(f_0(u) - m)\partial f_0(u) + (f(u) - \bar{y})\partial f(u)}{\{(f_0(u) - m)^2 + (f(u) - \bar{y})^2\}^{1/2}}$$

$$= z_0\partial f_0(u) + z\partial f(u)$$

where the multipliers z_0, z satisfy

$$z_0^2 + z^2 = 1 \tag{3.2.2}$$

and are thus nontrivial. However, since $x^{1/2}$ is not differentiable for $x = 0$, the calculation is invalid precisely at the point $u = \bar{u}$ where we need it.

Another try: we take $\varepsilon > 0$ and replace $F(u)$ with

$$F_\varepsilon(u) = \{(f_0(u) - (m - \varepsilon))^2 + (f(u) - \bar{y})^2\}^{1/2}. \tag{3.2.3}$$

The inside of the curly brackets is never zero; if it were there would be some u satisfying the target condition (3.1.2) with $f_0(u) = m - \varepsilon$, contradicting the definition of m. Hence $F_\varepsilon(u)$ is continuously differentiable in $\mathbb{R}^2$ with

$$\partial F_\varepsilon(u) = \frac{(f_0(u) - (m - \varepsilon))\partial f_0(u) + (f(u) - \bar{y})\partial f(u)}{\{(f_0(u) - (m - \varepsilon))^2 + (f(u) - \bar{y})^2\}^{1/2}}$$

$$= z_0\partial f_0(u) + z\partial f(u) \tag{3.2.4}$$

where z_0, z satisfy (3.2.2). However, $\bar{u}$ may no longer be a minimum of F_ε; we can only say that

$$F_\varepsilon(\bar{u}) = \varepsilon \le \inf F_\varepsilon(u) + \varepsilon \tag{3.2.5}$$

so that $\bar{u}$ is just an **approximate minimum** or ε-**minimum** of F_ε. Regrettably, ε-minima have no rights; directional derivatives at $u = \bar{u}$ have no special properties. However, we have the following related result.

Lemma 3.2.1. *Let $G(u)$ be continuously differentiable and bounded below in $\mathbb{R}^n$. Assume that $\bar{u}$ is an ε-minimum of $G(u)$, that is, that (3.2.5) holds for $G(u)$. Then there exists $\tilde{u} \in \mathbb{R}^n$ such that*

$$\|\tilde{u} - \bar{u}\| \le \sqrt{\varepsilon}, \qquad \|\partial G(\tilde{u})\| \le \sqrt{\varepsilon} \qquad (3.2.6)$$

for every directional derivative ∂.

We shall not prove Lemma 3.2.1 since it is a particular case of Ekeland's variational principle (Theorem 3.2.2 below). We note, however, that the one-dimensional version is intuitive; if the conclusion is false, we must have either

$$G'(u) \ge \sqrt{\varepsilon} \quad (\bar{u} - \sqrt{\varepsilon} \le u \le \bar{u} + \sqrt{\varepsilon})$$

or

$$G'(u) \le -\sqrt{\varepsilon} \quad (\bar{u} - \sqrt{\varepsilon} \le u \le \bar{u} + \sqrt{\varepsilon}).$$

Assuming the second inequality holds, we have

$$G(\bar{u} + \sqrt{\varepsilon}) = G(\bar{u}) + \int_{\bar{u}}^{\bar{u}+\sqrt{\varepsilon}} G'(u)du < G(\bar{u}) - \varepsilon < \min G$$

which is a contradiction. The second alternative is discarded in a similar way integrating $G'(u)$ from $\bar{u} - \sqrt{\varepsilon}$ to $\bar{u}$.

This is how we obtain the Lagrange multipliers from Lemma 3.2.1. Let $\{\varepsilon_n\}$ be a sequence of positive numbers tending to zero. Call $F_n(u)$ the function (3.2.3) corresponding to $\varepsilon = \varepsilon_n$. The point $\bar{u}$ is an ε_n-minimum of F_n; thus, if $\tilde{u} = \tilde{u}^n$ is the point provided by Lemma 3.2.1 we have, using (3.2.6),

$$|z_{0n}\partial f_0(\tilde{u}^n) + z_n\partial f(\tilde{u}^n)| \le \sqrt{\varepsilon_n}$$

where for each n, the multipliers z_{0n}, z_n satisfy (3.2.2). Since f_0, f are continuously differentiable, $\partial f_0(\tilde{u}^n) \to \partial f_0(\bar{u})$ and $\partial f(\tilde{u}^n) \to \partial f(\bar{u})$ as $n \to \infty$; on the other hand, by the Bolzano-Weierstrass theorem, the sequence $\{(z_{0n}, z_n)\} \subset \mathbb{R}^2$ has a subsequence convergent to $(z_0, z) \in \mathbb{R}^2$ satisfying (3.2.2). Taking limits we obtain (3.2.1) with nontrivial Lagrange multipliers.

Ekeland's variational principle (Theorem 3.2.2 below) is an enormously far-reaching version of Lemma 3.2.1; it applies in situations where not only f_0, f are not differentiable (or continuous or everywhere defined) but where it does not even make sense to talk about derivatives. Even in dimension 1 it produces a much more general result.

Let V be a metric space with distance $d(u, v)$. A function $F : V \to \mathbb{R} \cup \{+\infty\}$ is **lower semicontinuous** at $u \in V$ if and only if for every sequence $\{u_n\} \subset V$ such that $u_n \to u$ we have

$$F(u) \le \liminf_{n\to\infty} F(u_n). \qquad (3.2.7)$$

Note that if $F(u) < \infty$, (3.2.7) implies that for every $\varepsilon > 0$ there exists $\delta > 0$ such that

$$F(v) \geq F(u) - \varepsilon \quad \text{if } d(v, u) \leq \delta;$$

if not, there would exist $u_n \in V$ with $d(u, u_n) \leq 1/n$ and $F(u_n) \leq F(u) - \varepsilon$, which contradicts (3.2.7). The inequality does not follow if $F(u) = \infty$; we just take with $V = [-1, 1], u = 0$,

$$F(u) = \begin{cases} 1/|u| & (u \neq 0) \\ +\infty & (u = 0). \end{cases}$$

Note also that lim inf in (3.2.7) can be replaced by lim sup or even by lim, the latter assuming that the limit exists; we just take adequate subsequences.

Theorem 3.2.2 (Ekeland's variational principle). *Let V be a complete metric space, $F : V \to \mathbb{R} \cup \{+\infty\}$ a lower semicontinuous function not identically $+\infty$. Assume that F is bounded below and that $\bar{u} \in V$ is an ε-minimum of F. Then there exists $\tilde{u} \in V$ such that*

$$F(\tilde{u}) \leq F(\bar{u}) \tag{3.2.8}$$

$$d(\tilde{u}, \bar{u}) \leq \sqrt{\varepsilon} \tag{3.2.9}$$

$$F(u) \geq F(\tilde{u}) - \sqrt{\varepsilon} d(\tilde{u}, u) \quad (u \in V). \tag{3.2.10}$$

Proof (Crandall). We construct inductively a sequence $\{u_n\}$ as follows. The first term is $u_0 = \bar{u}$ ($\bar{u}$ the ε-minimum). Assuming that $u_0, u_1, \ldots, u_n$ have been chosen, we select u_{n+1} on the basis of the following alternative: (a) If

$$F(v) \geq F(u_n) - \sqrt{\varepsilon} d(u_n, v) \quad (\text{all } v \in V) \tag{3.2.11}$$

we take $u_{n+1} = u_n$, (note that if (a) occurs at the first step, Theorem 3.2.2 holds with $u = \bar{u}$). (b) If (3.2.11) is false, there exists $v \in V$ such that

$$F(v) < F(u_n) - \sqrt{\varepsilon} d(u_n, v). \tag{3.2.12}$$

Let S_n be the set of all such v. Then, if $u \in S_n$ we have

$$\inf_{v \in S_n} F(v) \leq F(u) < F(u_n).$$

Hence, we can choose u_{n+1} in S_n such that

$$F(u_{n+1}) - \inf_{v \in S_n} F(v) \leq \frac{1}{2} \left(F(u_n) - \inf_{v \in S_n} F(v) \right)$$

or, equivalently,

$$2F(u_{n+1}) - F(u_n) \leq \inf_{v \in S_n} F(v). \tag{3.2.13}$$

We show that $\{u_n\}$ is a Cauchy sequence. If alternative (a) occurs at some step it will occur always from then on, so that $u_n = u_{n+1} = u_{n+2} = \cdots$ and there is nothing to prove. We may then assume that the sequence $\{u_n\}$ is constructed according to alternative (b) at every step. In view of (3.2.12) and the fact that $u_{n+1} \in S_n$,

$$F(u_{n+1}) < F(u_n) - \sqrt{\varepsilon}d(u_n, u_{n+1}) \quad (n = 1, 2, \ldots). \tag{3.2.14}$$

Let $n < m$. Adding up copies of (3.2.14)

$$\sqrt{\varepsilon}d(u_n, u_{n+1}) < F(u_n) - F(u_{n+1}),$$
$$\sqrt{\varepsilon}d(u_{n+1}, u_{n+2}) < F(u_{n+1}) - F(u_{n+2}),$$
$$\cdots\cdots\cdots\cdots\cdots\cdots\cdots\cdots\cdots\cdots\cdots\cdots\cdots\cdots\cdots$$
$$\sqrt{\varepsilon}d(u_{m-1}, u_m) < F(u_{m-1}) - F(u_m)$$

and using the triangle inequality, we obtain

$$\sqrt{\varepsilon}d(u_n, u_m) \le \sqrt{\varepsilon}\sum_{j=n}^{m-1} d(u_j, u_{j+1})$$

$$< \sum_{j=n}^{m-1}(F(u_j) - F(u_{j+1})) = F(u_n) - F(u_m). \tag{3.2.15}$$

On the other hand, it results from (3.2.14) that

$$F(u_0) > F(u_1) > \ldots > F(u_n) > \ldots.$$

The function F is bounded below; thus, the sequence $\{F(u_n)\}$ is bounded below and $\lim_{n\to\infty} F(u_n)$ exists. Accordingly, $F(u_n) - F(u_m) \to 0$ as $m, n \to \infty$, and (3.2.15) implies that $\{u_n\}$ is a Cauchy sequence as claimed; since V is complete,

$$\tilde{u} = \lim_{n\to\infty} u_n$$

exists. We show that $\tilde{u}$ satisfies all the statements in Theorem 3.2.2. Note first that, taking into account that (a) may occur at some step, we have

$$F(u_0) \ge F(u_1) \ge \ldots \ge F(u_n) \ge \ldots, \tag{3.2.16}$$

strict inequality between $F(u_n)$ and $F(u_{n+1})$ corresponding to alternative (b). Likewise, we have

$$\sqrt{\varepsilon}d(u_n, u_m) \le F(u_n) - F(u_m) \tag{3.2.17}$$

for $n < m$. This has already been proved in (3.2.15) (with strict inequality) assuming that alternative (b) occurs at each step; if this is not the case, $\{u_n\}$ becomes stationary at some step, and (3.2.17) holds anyway.

With (3.2.16) and (3.2.17) in hand, we proceed to show (3.2.8) and (3.2.9). For this, we only use the fact that $u_{n+1} \in S_n$. Since F is lower semicontinuous, it follows from (3.2.16) that

$$F(\tilde{u}) \leq \lim_{n \to \infty} F(u_n) \leq F(u_0) = F(\bar{u}) \qquad (3.2.18)$$

which is (3.2.8). To show (3.2.9), we use (3.2.17) for $n = 0$ and take limits as $m \to \infty$. Since $u_0 = \bar{u}$, the result is

$$\sqrt{\varepsilon} d(\bar{u}, \tilde{u}) \leq F(\bar{u}) - \lim_{m \to \infty} F(u_m) \leq \left(\inf_{v \in V} F(v) + \varepsilon \right) - \inf_{v \in V} F(v) = \varepsilon.$$

Finally, assume (3.2.10) does not hold. Then there exists $v \in V$ such that

$$F(v) < F(\tilde{u}) - \sqrt{\varepsilon} d(\tilde{u}, v). \qquad (3.2.19)$$

Taking limits in (3.2.17) as $m \to \infty$,

$$\sqrt{\varepsilon} d(u_n, \tilde{u}) \leq F(u_n) - \lim_{m \to \infty} F(u_m) \leq F(u_n) - F(\tilde{u}) \qquad (3.2.20)$$

where we have used lower semicontinuity of F at $\tilde{u}$. We rewrite inequality (3.2.19) as $\sqrt{\varepsilon} d(\tilde{u}, v) < F(\tilde{u}) - F(v)$ and add it to (3.2.20), using the triangle inequality $\sqrt{\varepsilon} d(u_n, v) \leq \sqrt{\varepsilon} d(u_n, \tilde{u}) + \sqrt{\varepsilon} d(v, \tilde{u})$ on the left side. We obtain $\sqrt{\varepsilon} d(u_n, v) \leq F(u_n) - F(v)$, or

$$F(v) < F(u_n) - \sqrt{\varepsilon} d(u_n, v). \qquad (3.2.21)$$

This holds for all $n \geq 0$, so that $v \in \bigcap_{n \geq 1} S_n$. By virtue of (3.2.13),

$$2F(u_{n+1}) - F(u_n) \leq F(v).$$

Letting $n \to \infty$, $\lim_{n \to \infty} F(u_n) \leq F(v)$; thus, again by lower semicontinuity

$$F(\tilde{u}) < F(v),$$

which contradicts (3.2.19). This ends the proof.   ■

3.3. The Abstract Time Optimal Problem

Let E be a Banach space. A subset $W \subseteq E$ is a **cone** if

$$w \in W, \lambda \geq 0 \quad \text{imply} \quad \lambda w \in W.$$

A convex cone W can be characterized by the property

$$v, w \in W, \ \lambda, \mu \geq 0 \quad \text{imply} \quad \lambda v + \mu w \in W.$$

Let Z be an arbitrary subset of E. The (negative) **polar cone** of Z is the subset Z^- of the dual space E^* defined by

$$Z^- = \{y^* \in E^*; \ \langle y^*, y \rangle \leq 0, y \in Z\},$$

$\langle \cdot, \cdot \rangle$ the duality pairing of E and E^*. Obviously, Z^- is a closed convex cone.

Let Y be a closed set in a Banach space E. Given $\bar{y} \in Y$, the (Bouligand) **contingent cone** to Y at $\bar{y}$ is the set $K_Y(\bar{y})$ of all $w \in E$ such that there exists a sequence $\{h_k\} \subset \mathbb{R}_+ = \{$positive real numbers$\}$ with $h_k \to 0$ and a sequence $\{y_k\} \subseteq Y$ with $y_k \to \bar{y}$ and

$$\frac{y_k - \bar{y}}{h_k} \to w \quad \text{as } k \to \infty, \tag{3.3.1}$$

or, equivalently,

$$y_k = \bar{y} + h_k w + o(h_k) \quad \text{as } k \to \infty. \tag{3.3.2}$$

The **intermediate cone** $I_Y(\bar{y})$ consists of all w such that, for every sequence $\{h_k\} \subset \mathbb{R}_+$ with $h_k \to 0$, there exists a sequence $\{y_k\} \subset Y$ with $y_k \to \bar{y}$ such that (3.3.1) (or, equivalently, (3.3.2)) holds. $K_Y(\bar{y})$ and $I_Y(\bar{y})$ are closed cones, in general nonconvex (see Examples 3.3.5 and 3.3.6 below). Finally, the (Clarke) **tangent cone** $T_Y(\bar{y})$ to Y at $\bar{y}$ consists of all w such that, for every sequence $\{h_k\} \subset \mathbb{R}_+$ with $h_k \to 0$ and for every sequence $\{\bar{y}_k\} \subseteq Y$ with $\bar{y}_k \to \bar{y}$ there exists a sequence $\{y_k\} \subset Y$ with $y_k \to \bar{y}$ and

$$\frac{y_k - \bar{y}_k}{h_k} \to w \quad \text{as } k \to \infty, \tag{3.3.3}$$

or, equivalently,

$$y_k = \bar{y}_k + h_k w + o(h_k). \tag{3.3.4}$$

Membership of an element $w \in E$ in the cones $K_Y(\bar{y})$, $I_Y(\bar{y})$, and $T_Y(\bar{y})$ is checked in all three cases by showing that (3.3.3) or, equivalently, (3.3.4) hold under the conditions shown in the "membership table," where $\{y_k\}, \{\bar{y}_k\} \subset Y$ satisfy $y_k \to \bar{y}, \bar{y}_k \to \bar{y}$ and $\{h_k\} \subset \mathbb{R}_+$ satisfies $h_k \to 0$.

	$K_Y(\bar{y})$	$I_Y(\bar{y})$	$T_Y(\bar{y})$
$\{\bar{y}_k\}$	$\bar{y}_k = \bar{y}$	$\bar{y}_k = \bar{y}$	Arbitrarily given
$\{h_k\}$	Must be found such that (3.3.3) holds	Arbitrarily given	Arbitrarily given
$\{y_k\}$	Must be found such that (3.3.3) holds	Must be found such that (3.3.3) holds	Must be found so that (3.3.3) holds

The definitions of $K_Y(\bar{y})$, $I_Y(\bar{y})$, $T_Y(\bar{y})$ are increasingly demanding, so that

$$T_Y(\bar{y}) \subseteq I_Y(\bar{y}) \subseteq K_Y(\bar{y}).$$

The three cones coincide for convex sets and differential varieties (Examples 3.3.4 and 3.3.14 below), but are in general different (Example 3.3.9).

The **normal cone** $N_Y(\bar{y}) \subseteq E^*$ to Y at $\bar{y}$ is the (closed, convex) cone defined by

$$N_Y(\bar{y}) = T_Y(\bar{y})^-.$$

Let V be a metric space, E a Banach space, g a function defined in a subset $D(g) \subseteq V$. We call $\xi \in E$ a **variation** of g at u if and only if there exists a sequence $\{h_k\} \subset \mathbb{R}_+$ with $h_k \to 0$ and a sequence $\{u_k\} \subset D(g)$ such that

$$d(u_k, u) \le h_k, \qquad \lim_{k \to \infty} \frac{g(u_k) - g(u)}{h_k} = \xi, \tag{3.3.5}$$

or, equivalently,

$$d(u_k, u) \le h_k, \qquad g(u_k) = g(u) + h_k \xi + o(h_k) \quad \text{as } k \to \infty. \tag{3.3.6}$$

The set of all variations of g at u is called Var $g(u)$.

The vector $\xi \in E$ is a (one-sided) **directional derivative** of g at u if and only if there exists a $D(g)$-valued function $h \to u(h)$, defined in $0 \le h \le \delta > 0$ such that

$$d(u(h), u) \le h, \qquad \lim_{h \to 0+} \frac{g(u(h)) - g(u)}{h} = \xi \tag{3.3.7}$$

or, equivalently,

$$d(u(h), u) \le h, \quad g(u(h)) = g(u) + h \xi + o(h) \quad \text{as } h \to 0+. \tag{3.3.8}$$

We shall denote by Der $g(u)$, the set of all directional derivatives of g at u; obviously, Der $g(u) \subseteq$ Var $g(u)$. The two sets coincide, for instance, when g is a smooth function in $\mathbb{R}^n$ (Example 3.3.12 below) but they are in general different. Both sets are closed and star-shaped with respect to zero (Example 3.3.11).

In this chapter we only deal with the abstract time optimal problem for a Hilbert space E; if only interested in finite dimensional problems, the reader may assume that $E = \mathbb{R}^m$. The abstract time optimal problem for a general Banach space will be seen in Chapter 7.

Lemma 3.3.1. *Let V be a complete metric space, E a real Hilbert space, and $f : D(f) \to E$, where $D(f) \subseteq V$, $D(f) \neq \emptyset$. Assume that $Y \subseteq E$ is closed, that $f(D(f)) \cap Y = \emptyset$ and that for each $y \in Y$ the real valued function*

$$\Phi(u, y) = \begin{cases} \|f(u) - y\| & (u \in D(f)) \\ +\infty & (u \notin D(f)) \end{cases} \tag{3.3.9}$$

is lower semicontinuous with respect to u. Finally, let $\bar{u} \in D(f)$, $\bar{y} \in Y$,

$$\varepsilon = \|f(\bar{u}) - \bar{y}\|. \tag{3.3.10}$$

Then there exist $\tilde{u} \in D(f)$, $\tilde{y} \in Y$ with

$$\|f(\tilde{u}) - \tilde{y}\| \le \|f(\bar{u}) - \bar{y}\|, \tag{3.3.11}$$

$$d(\tilde{u}, \bar{u}) + \|\tilde{y} - \bar{y}\| \le \sqrt{\varepsilon}, \tag{3.3.12}$$

and $z \in E$ such that[1]

$$\|z\| = 1, \qquad \langle z, \xi - w \rangle \ge -\sqrt{\varepsilon}(1 + \|w\|) \tag{3.3.13}$$

for either $\xi \in \mathrm{Var}\, f(\tilde{u})$ and $w \in I_Y(\tilde{y})$ or $\xi \in \mathrm{Der}\, f(\tilde{u})$ and $w \in K_Y(\tilde{u})$.

Proof. Consider the function Φ defined by (3.3.9) in the space $V \times Y$ equipped with the distance $d((u, y), (u', y')) = d(u, u') + \|y - y'\|$ (which distance makes $V \times Y$ complete). The postulated lower semicontinuity of $\Phi(u, y)$ for y fixed implies joint lower semicontinuity of $\Phi(u, y)$ in $V \times Y$ (Example 3.3.13 below). We have

$$\Phi(\bar{u}, \bar{y}) = \|f(\bar{u}) - \bar{y}\| = \varepsilon \le 0 + \varepsilon \le \inf_{u \in V, y \in Y} \Phi(u, y) + \varepsilon.$$

Applying Ekeland's variational principle (Theorem 3.2.2) to Φ in $V \times Y$, we obtain an element $(\tilde{u}, \tilde{y}) \in V \times Y$ satisfying (3.3.11) and (3.3.12) and such that

$$\Phi(u, y) \ge \Phi(\tilde{u}, \tilde{y}) - \sqrt{\varepsilon}(d(u, \tilde{u}) + \|y - \tilde{y}\|) \quad ((u, y) \in V \times Y). \tag{3.3.14}$$

(Note that (3.3.11) implies in particular $\|f(\tilde{u}) - \tilde{y}\| < \infty$, thus $\tilde{u} \in D(f)$). Let $\xi \in \mathrm{Var}\, f(\tilde{u})$ and let $\{h_k\}$, $\{u_k\}$ be the sequences employed in (3.3.6) for the construction of ξ; we have $d(u_k, \tilde{u}) \le h_k$ and

$$f(u_k) = f(\tilde{u}) + h_k \xi + o(h_k) \quad \text{as } k \to \infty. \tag{3.3.15}$$

On the other hand, if $w \in I_Y(\tilde{y})$ there exists a sequence $\{y_k\} \subseteq Y$ such that

$$y_k = \tilde{y} + h_k w + o(h_k) \quad \text{as } k \to \infty. \tag{3.3.16}$$

Write (3.3.14) for $u = u_k$ and $y = y_k$, remembering that $d(u_k, \tilde{u}) \le h_k$:

$$\|f(\tilde{u}) + h_k \xi - \tilde{y} - h_k w + o(h_k)\| - \|f(\tilde{u}) - \tilde{y}\|$$
$$\ge -\sqrt{\varepsilon} h_k (1 + \|w + o(1)\|). \tag{3.3.17}$$

In a Hilbert space we have $\|y + h\xi\|^2 = \|y\|^2 + 2h(y, \xi) + h^2\|\xi\|^2$, which implies $(d/dh)|_{h=0}\|y + h\xi\|^2 = 2(y, \xi)$. Accordingly, if $y \ne 0$ we obtain, writing $\|y + h\xi\| = (\|y + h\xi\|^2)^{1/2}$,

$$\frac{d}{dh}\bigg|_{h=0} \|y + h\xi\| = \frac{(y, \xi)}{\|y\|}. \tag{3.3.18}$$

---

[1] We use the fact that in a real Hilbert space E the scalar product $(\cdot, \cdot)$ coincides with the duality pairing $\langle \cdot, \cdot \rangle$ of E and its dual E^*.

Accordingly, dividing by h_k in (3.3.17), letting $k \to \infty$ and keeping in mind that $f(D(f)) \cap Y = \emptyset$ (so that $f(\tilde{u}) - \tilde{y} \neq 0$), we obtain (3.3.13) with

$$z = \frac{f(\tilde{u}) - \tilde{y}}{\|f(\tilde{u}) - \tilde{y}\|}. \tag{3.3.19}$$

This ends the proof for the case $\xi \in \operatorname{Var} f(\tilde{u})$, $w \in I_Y(\tilde{y})$. On the other hand, assume $w \in K_Y(\tilde{y})$, and let $\{y_k\}, \{h_k\}$ be the sequences in (3.3.2). If $\xi \in \operatorname{Der} f(\tilde{u})$, then (3.3.8) implies $g(u(h_k)) = g(\tilde{u}) + h_k \xi + o(h_k)$. We write (3.3.14) for $u_k = u(h_k)$ and $y = y_k$ and argue in the same way, completing the proof. ∎

The key to the argument is *to be able to use the same sequence* $\{h_k\}$ in (3.3.15) and (3.3.16). This requires either the combination $\xi \in \operatorname{Var} f(\tilde{u})$ and $w \in I_Y(\tilde{y})$, where $\{h_k\}$ is determined by ξ, or $\xi \in \operatorname{Der} f(\tilde{u})$ and $w \in K_Y(\tilde{u})$, where $\{h_k\}$ is determined by w.

We bring back the abstract time optimal problem in **3.1**. We are given a sequence $\{V_n\}$ of metric spaces, a normed space E, a subset $Y \subseteq E$, and a sequence $\{f_n\}$ of functions $f_n : D(f_n) \to E$, $D(f_n) \subseteq V_n$, $f_n(D(f_n)) \cap Y = \emptyset$ ($n = 1, 2, \ldots$). A *solution* of the abstract time optimal problem is a sequence $\{\bar{u}^n\}$, $\bar{u}^n \in D(f_n)$ such that $\operatorname{dist}(f_n(\bar{u}^n), Y) \to 0$.

A sequence $\{\bar{y}^n\} \subseteq Y$ is **associated with** $\{\bar{u}^n\}$ if $\|f(\bar{u}^n) - \bar{y}^n\| \to 0$.

The precise assumptions on the abstract time optimal problem are:

(*a*) V_n is complete for each n and E is a real Hilbert space.

(*b*) For each $n = 1, 2, \ldots$ and each $y \in Y$ the real valued function

$$\Phi_n(u, y) = \begin{cases} \|f_n(u) - y\| & (u \in D(f_n)) \\ +\infty & (u \notin D(f_n)) \end{cases} \tag{3.3.20}$$

is lower semicontinuous with respect to u.

(*c*) The target set Y is closed.

We use the following notation below: given a sequence of sets $\{D_n\}$ in E, $\liminf_{n \to \infty} D_n$ is the set of all ξ such that there exists a sequence $\{\xi_n\}, \xi_n \in D_n$ with $\xi_n \to \xi$.

Theorem 3.3.2. *Let* $\{\bar{u}^n\}$ *be a solution of the abstract time optimal problem,* $\{\bar{y}^n\} \subset Y$ *an associated sequence,*

$$\varepsilon_n = \|f(\bar{u}^n) - \bar{y}^n\|. \tag{3.3.21}$$

Then there exist sequences $\{\tilde{u}^n\}, \tilde{u}^n \in D(f_n), \{\tilde{y}^n\} \subseteq Y$ *such that*

$$\|f_n(\tilde{u}^n) - \tilde{y}^n\| \leq \|f_n(\bar{u}^n) - \bar{y}^n\|, \tag{3.3.22}$$

$$d_n(\tilde{u}^n, \bar{u}^n) + \|\tilde{y}^n - \bar{y}^n\| \leq \sqrt{\varepsilon_n} \tag{3.3.23}$$

$(d_n$ the distance in $V_n)$, and a sequence $\{z_n\} \subseteq E$ such that

$$\|z_n\| = 1, \qquad \langle z_n, \xi^n - w^n \rangle \geq -\sqrt{\varepsilon_n}(1 + \|w^n\|) \qquad (3.3.24)$$

for either $\xi^n \in \text{Var} f_n(\tilde{u}^n)$ and $w^n \in I_Y(\tilde{y}^n)$ or for $\xi^n \in \text{Der} f_n(\tilde{u}^n)$ and $w^n \in K_Y(\tilde{y}^n)$. If z is a limit point of $\{z_n\}$ then we have

$$\|z\| = 1, \qquad \langle z, \xi \rangle \geq 0 \qquad (3.3.25)$$

for $\xi \in \liminf_{n\to\infty} \text{Var} f_n(\tilde{u}^n)$. If $\tilde{y}^n \to \tilde{y}$, then

$$z \in N_Y(\tilde{y}). \qquad (3.3.26)$$

Proof. Applying Lemma 3.3.1 to each term $\tilde{u}^n$ of a solution of the abstract time optimal problem and each term $\tilde{y}^n$ of an associated sequence, we obtain an element $(\tilde{u}, \tilde{y}) = (\tilde{u}^n, \tilde{y}^n)$ of $D(f_n) \times Y$ satisfying (3.3.22), (3.3.23) and

$$\Phi_n(u, y) \geq \Phi_n(\tilde{u}^n, \tilde{y}^n) - \sqrt{\varepsilon_n}(d_n(u, \tilde{u}^n) + \|y - \tilde{y}^n\|) \quad ((u, y) \in V_n \times Y) \qquad (3.3.27)$$

from which (3.3.24) follows as in Lemma 3.3.1: z_n is given by

$$z_n = \frac{f_n(\tilde{u}^n) - \tilde{y}^n}{\|f_n(\tilde{u}^n) - \tilde{y}^n\|}. \qquad (3.3.28)$$

The inequality in (3.3.25) is obtained taking $w^n = 0$ in (3.3.24) and passing to the limit. It only remains to prove (3.3.26). By (3.3.18), if $y, \eta \in E$, $y \neq 0$ we have

$$\|y + h\eta\| = \|y\| + h(y, \eta)/\|y\| + r(y, \eta, h) \qquad (3.3.29)$$

where, for y, η fixed, $\|r(y, \eta, h)\| = o(h)$ as $h \to 0$ (convergence is uniform for η bounded but we we don't need this). Let $w \in T_Y(\tilde{y})$. Using (3.3.29) choose a sequence $\{h_n\} \in \mathbb{R}_+$ with $h_n \to 0$ and such that

$$\frac{1}{h_n}\|r(f_n(\tilde{u}^n) - \tilde{y}^n, w, -h_n)\| \to 0. \qquad (3.3.30)$$

In view of (3.3.23), if $\tilde{y}_n \to \tilde{y}$ then $\tilde{y}^n \to \tilde{y}$ as well; thus, by definition of the tangent cone $T_Y(\tilde{y})$ there exists a sequence $\{y^n\} \in Y$ such that

$$y^n = \tilde{y}^n + h_n w + o(h_n). \qquad (3.3.31)$$

We use (3.3.27) for $u = \tilde{u}^n$ and $y = y^n$. Taking into account that the distance from $(\tilde{u}^n, y^n)$ to $(\tilde{u}^n, \tilde{y}^n)$ in $V_n \times Y$ is $d_n((\tilde{u}^n, y^n), (\tilde{u}^n, \tilde{y}^n)) = \|y^n - \tilde{y}^n\|$, we obtain

$$\|f(\tilde{u}^n) - y^n\| \geq \|f(\tilde{u}^n) - \tilde{y}^n\| - \sqrt{\varepsilon_n}\|y^n - \tilde{y}^n\|. \qquad (3.3.32)$$

We have $\|x^n\| - \|q_n\| \le \|x^n + q_n\| \le \|x^n\| + \|q_n\|$ for arbitrary sequences $\{x^n\}$, $\{q^n\} \subseteq E$, hence

$$\|x^n + o(h_n)\| = \|x^n\| + o(h_n). \tag{3.3.33}$$

Using (3.3.31), rewrite (3.3.32) in the form

$$\|f(\tilde{u}^n) - \tilde{y}^n - h_n w + o(h_n)\| \ge \|f(\tilde{u}^n) - \tilde{y}^n\| - \sqrt{\varepsilon_n}\|h_n w + o(h_n)\|. \tag{3.3.34}$$

In the left side we use (3.3.33) in the form $\|f(\tilde{u}^n) - \tilde{y}^n - h_n w + o(h_n)\| = \|f(\tilde{u}^n) - \tilde{y}^n - h_n w\| + o(h_n)$, and then (3.3.29); on the right, note that, by (3.3.33), $\|h_n w + o(h_n)\| = h_n(\|w\| + o(1))$. We obtain

$$\|f_n(\tilde{u}^n) - \tilde{y}^n\| - h_n \langle z_n, w \rangle + r(f_n(\tilde{u}^n) - \tilde{y}^n, w, -h_n) + o(h_n)$$
$$\ge \|f_n(\tilde{u}^n) - \tilde{y}^n\| - \sqrt{\varepsilon_n} h_n(\|w\| + o(1)).$$

Crossing out $\|f_n(\tilde{u}^n) - \tilde{y}^n\|$ from both sides, dividing by h_n and letting $n \to \infty$, we obtain $\langle z, w \rangle \le 0$ as claimed. This ends the proof. ∎

Example 3.3.3. Directional derivatives can be defined in the following alternate way: Der $g(u)$ consists of all $\xi \in E$ such that for every sequence $\{h_k\} \subset \mathbb{R}_+$ with $h_k \to 0$ there exists a sequence $\{u_k\} \subseteq D(g)$ such that (3.3.5) (equivalently, (3.3.6)) holds.

In fact, let $\xi \in \text{Der } g(u)$. Then, given $\{h_k\} \subset \mathbb{R}_+$ with $h_k \to 0$, (3.3.5) will hold for $u_k = u(h_k)$. Conversely, assume the definition above holds. We take $h_k = 1/k$. If $\{u_k\}$ is a sequence satisfying (3.3.5), define $u(h) = u_k$ ($h_k \le h < h_{k-1}$). We have $d(u(h), u) = d(u, u_k) \le h_k \le h$ in each interval; on the other hand,

$$\frac{g(u(h)) - g(u)}{h} = \frac{h_k}{h} \frac{g(u_k) - g(u)}{h_k}$$

also in each interval, where $1 \le h/h_k \le h_{k-1}/h_k = k/(k-1) \to 1$ as $h \to 0$. Hence, $\xi \in \text{Der } g(u)$.

Example 3.3.4. If Y is a closed convex set,

$$K_Y(\bar{y}) = I_Y(\bar{y}) = T_Y(\bar{y}) = \overline{\bigcup_{h>0} h(Y - \bar{y})}$$

for any $\bar{y} \in Y$ (see Aubin–Ekeland [1984, p. 407]).

Example 3.3.5. $K_Y(\bar{y})$, $I_Y(\bar{y})$ and $T_Y(\bar{y})$ are closed cones. (For the first and third cone see Aubin–Ekeland [1984, pp. 405–407]; the proof is similar for the intermediate cone.)

Example 3.3.6. Let $E = \mathbb{R}^2$, $\overline{\mathbb{R}}_+ = \{$nonnegative real numbers$\}$, and let $Y = (\{0\} \times \overline{\mathbb{R}}_+) \cup (\overline{\mathbb{R}}_+ \times \{0\})$ the union of the two positive semiaxes. We have $K_Y(0) = I_Y(0) = Y$.

Example 3.3.7. Let Y be as in Example 3.3.6. Then $T_Y(0) = \{0\}$. In fact, if $T_Y(0) \neq \{0\}$, then $T_Y(0)$ contains an element w with $\|w\| = 1$. Obtaining it as a limit (3.3.3) with $\bar{y}_k = (2^{-k}, 0)$, $h_k = 2^{-k}$, we deduce that $w = (1, 0)$; using then $\bar{y}_k = (0, 2^{-k})$, $h_k = 2^{-k}$, $w = (0, 1)$.

Example 3.3.8. Let $E = \mathbb{R}$ and $Z = \{0\} \cup \{2^{-n}; n \geq 1\}$. Then $K_Z(0) = \overline{\mathbb{R}}_+$, $I_Z(0) = \{0\}$.

Example 3.3.9. There exists a closed set $Y \subset \mathbb{R}^2$ and a point $\bar{y} \in Y$ such that the two inclusions $K_Y(\bar{y}) \subset I_Y(\bar{y}) \subset T_Y(\bar{y})$ are strict. For instance, we may take $Y = (\{0\} \times Z) \cup (\overline{\mathbb{R}}_+ \times \{0\})$, where Z is the set in Example 3.3.8.

Example 3.3.10. The tangent cone $T_Y(\bar{y})$ is convex. To see this, since $T_Y(\bar{y})$ is a cone, it is only necessary to show that if $w_1, w_2 \in T_Y(\bar{y})$ then $w_1 + w_2 \in T_Y(\bar{y})$. Let $\{\bar{y}_k\}$, $\{h_k\}$, and $\{y_k\}$ be the sequences in (3.3.4) for w_1. Since $y_k \to \bar{y}$, there exists a sequence $\{z_k\} \subseteq Y$ such that $z_k = y_k + h_k w_2 + o(h_k) = \bar{y}_k + h_k w_1 + h_k w_2 + o(h_k)$.

Example 3.3.11. The set Var $g(u)$ of all variations of g at u and the set Der $g(u)$ of all directional derivatives of g at u are star-shaped with respect to the origin and closed. To show that Der $g(u)$ is closed, let $\{\xi_n\} \subseteq \text{Der} \, g(u)$ be such that $\xi_n \to \xi$, and let $u_n(\cdot) : [0, \delta_n] \to V$ be the function used in the definition (3.3.7) of ξ_n; we may assume all $\delta_n = \delta > 0$. Pick a sequence $\{\varepsilon_n\} \subset \mathbb{R}_+$ with $\varepsilon_n \to 0$ and a decreasing sequence $\{\rho_n\}$ with $\rho_n \to 0$, $\|h^{-1}(g(u_n(h)) - g(u)) - \xi_n\| \leq \varepsilon_n$ for $0 \leq h \leq \rho_n$. Define $u(h) = u_n(u)$ for $\rho_{n+1} \leq h < \rho_n$. Then, for h in this interval we have

$$\|h^{-1}(g(u(h)) - g(u)) - \xi\| \leq \|h^{-1}(g(u_n(h)) - g(u)) - \xi_n\| + \|\xi - \xi_n\|.$$

Example 3.3.12. Let $g(u)$ be a function of a real variable differentiable at u. Then Var $g(u) = \text{Der} \, g(u) = \{tg'(u); -1 \leq t \leq 1\}$. Similarly, if $g(u) = g(u_1, \ldots, u_n)$ is a function of n variables having continuous partials in a neighborhood of $u = (u_1, \ldots, u_n)$ then Var $g(u) = \text{Der} \, g(u) = \{t\partial_\alpha g(u); -1 \leq t \leq 1, \|\alpha\| = 1\}$, where α ranges over all directions (that is, vectors of norm 1) in $\mathbb{R}^n$, and ∂_α indicates derivative in the direction α.

Example 3.3.13. Under the assumptions on $\Phi(u, y)$ in Lemma 3.3.1, $\Phi(u, y)$ is jointly lower semicontinuous in $V \times Y$.

Example 3.3.14. A **differential variety** S in $\mathbb{R}^m$ is any set locally defined by a set of equations $f_1(x) = f_2(x) = \cdots f_k(x) = 0$, where the functions $f_1, f_2, \ldots, f_k$ $(k < m)$ are continuously differentiable and the Jacobian matrix $\partial_y f = [\partial_j f_k]$ has constant rank k. A vector $w \in \mathbb{R}^m$ is a **tangent vector** at $\bar{x} = (x_1, x_2, \ldots, x_m) \in S$ if $w = \phi'(0)$, where $\phi(h)$ is a continuously differentiable function defined in an interval $|h| \leq \delta$, taking values in S and such that $\phi(0) = \bar{x}$. The set $T(\bar{x}, S)$ of all vectors w tangent to S at $\bar{x}$ is a k-dimensional linear subspace of $\mathbb{R}^m$, and $K_S(\bar{x}) = I_S(\bar{x}) = T_S(\bar{x}) = T(\bar{x}, S)$.

3.4. The Control Spaces

Let $|\cdot|$ be Lebesgue outer measure in $0 \leq t \leq T$ (see **2.0**). We identify elements of the admissible control space $C_{\text{ad}}(0, T; U)$ that coincide except in sets of (outer) measure zero and introduce a distance in $C_{\text{ad}}(0, T; U)$ by

$$d(u(\cdot), v(\cdot)) = |\{t \in [0, T]; u(t) \neq v(t)\}|.^{(1)} \tag{3.4.1}$$

It is obvious that $d(u, v) \geq 0$, $d(u, v) = d(v, u)$, and $d(u, v) = 0$ if and only if $u(\cdot) = v(\cdot)$ (equivalence taken into account). Moreover,

$$\{t; u(t) \neq w(t)\} \subseteq \{t; u(t) \neq v(t)\} \cup \{t; v(t) \neq w(t)\},$$

so that, taking outer measure, the triangle inequality $d(u, w) \leq d(u, v) + d(v, w)$ results.

We say that a sequence $\{u_n(\cdot)\} \subset C_{\text{ad}}(0, \bar{t}; U)$ is **stationary** if there exists a set e with $|e| = 0$ such that for every $t \in [0, T] \backslash e$ there exists n (depending on t) such that

$$u_n(t) = u_{n+1}(t) = u_{n+2}(t) = \cdots. \tag{3.4.2}$$

Obviously, a stationary sequence is pointwise convergent outside of e. The space $C_{\text{ad}}(0, \bar{t}; U)$ is called **saturated** if the limit of every stationary sequence belongs to $C_{\text{ad}}(0, \bar{t}; U)$.

Lemma 3.4.1. *Assume* $C_{\text{ad}}(0, \bar{t}; U)$ *is saturated. Then, equipped with* d, *it is a complete metric space.*

Proof. Let $\{u_n(\cdot)\}$ be a Cauchy sequence in $C_{\text{ad}}(0, \bar{t}; U)$. A Cauchy sequence is convergent if any of its subsequences is convergent (Example 3.4.2 below), so we begin by selecting a subsequence (denoted in the same way) such that

$$d(u_n, u_{n+1}) \leq 2^{-n}. \tag{3.4.3}$$

---

(1) Since we do not put any requirements on the admissible controls except on the behavior of $f(t, y, u(t))$, the set in (3.4.1) may not be Lebesgue measurable.

Let $d_n = \{t; u_n(t) \neq u_{n+1}(t)\}$, $e_n = \bigcup_{k \geq n} d_k$. Then

$$|e_n| \leq \sum_{k=n}^{\infty} |d_k| \leq \sum_{k=n}^{\infty} 2^{-k} = 2^{1-n}. \tag{3.4.4}$$

Since $e_1 \supseteq e_2 \supseteq \ldots$, if $e = \bigcap_{n \geq 1} e_n$ then $|e| \leq |e_n|$, so that $|e| = 0$. Assume $t \notin e$. Then, for some $n = n(t)$ we have $t \notin e_n$, so that $t \notin d_m$ for $m \geq n(t)$. It follows that (3.4.2) holds, thus the sequence $\{u_n(\cdot)\}$ is stationary. Since $C_{\text{ad}}(0, \bar{t}; U)$ is saturated, the function

$$u(t) = u_{n(t)}(t) \quad (t \notin e)$$

belongs to $C_{\text{ad}}(0, \bar{t}; U)$. It only remains to be shown that $d(u_n, u) \to 0$, that is, that

$$|\{t; u_n(t) \neq u(t)\}| \to 0 \tag{3.4.5}$$

as $n \to \infty$. This is seen as follows. If $u(t) \neq u_n(t)$ then there exists $k \geq n$ such that $t \in d_k = \{t; u_k(t) \neq u_{k+1}(t)\} \subseteq e_n$. Then, making use of (3.4.4) we deduce that $|\{t; u_n(t) \neq u(t)\}| \leq |e_n| \leq 2^{1-n}$. This ends the proof. ∎

Example 3.4.2. Let U be a metric space, $\{u_n\}$ a Cauchy sequence in U. Then (a) a subsequence satisfying (3.4.3) exists, (b) if $\{u_n\}$ has a convergent subsequence, it is itself convergent.

Example 3.4.3. Two admissible controls $u(\cdot)$, $v(\cdot)$ equivalent according to the definition above produce the same trajectory: $y(t, u) \equiv y(t, v)$.

3.5. Continuity of the Solution Map

We study the map $u \to y(t, u)$, where $y(t, u)$ is the solution of

$$y'(t) = f(t, y(t), u(t)), \qquad y(0) = \zeta \tag{3.5.1}$$

corresponding to $u \in C_{\text{ad}}(0, \bar{t}; U)$.

Let $e \subseteq [0, \bar{t}]$ be an arbitrary (possibly nonmeasurable) set. There exist measurable sets (denoted $[e]$ and called **measurable envelopes** of e) satisfying

$$e \subseteq [e], \qquad |e| = |[e]|, \tag{3.5.2}$$

where $|\cdot|$ denotes outer Lebesgue measure (Lebesgue measure for measurable sets). To construct a measurable envelope it suffices to select a decreasing sequence $\{d_n\}$ of measurable sets with $d_n \supseteq e$, $|d_n| \to |e|$ and then take $[e] = \cap d_n$.

Lemma 3.5.1. *Assume* (3.5.1) *satisfies Hypothesis II in* $0 \leq t \leq \bar{t}$ *for each* $u(\cdot) \in C_{\text{ad}}(0, \bar{t}; U)$, *with* $K(\cdot, c)$ *and* $L(c, \cdot)$ *independent of* $u(\cdot)$. *Let* $y(t, \bar{u})$

exist in $0 \leq t \leq \bar{t}$. Then there exists $\delta > 0$ such that, if $u(\cdot), v(\cdot) \in B(\bar{u}, \delta) =$
$\{u \in C_{ad}(0, \bar{t}; U); d(u, \bar{u}) \leq \delta\}$, d the distance (3.4.1), $y(t, u)$ and $y(t, v)$ exist
as well in $0 \leq t \leq \bar{t}$ and

$$\|y(t, v) - y(t, u)\| \leq C \int_{[\{\tau \in [0, \bar{t}]; u(\tau) \neq v(\tau)\}]} K(\tau, c + 1) d\tau \qquad (3.5.3)$$

in $0 \leq t \leq \bar{t}$, where c is a bound for $\|y(t, \bar{u})\|$ in $0 \leq t \leq \bar{t}$ and C does not depend
on u, v.

Lemma 3.5.1 is a close relative of Lemma 2.3.1 and is proved in the same way. If $[0, t_u]$ is the largest interval where $y(t, u)$ exists and satisfies $\|y(t, u) - y(t, \bar{u})\| \leq 1$, we have

$$y(t, u) - y(t, \bar{u}) = \int_0^t \{f(\tau, y(\tau, u), u(\tau)) - f(\tau, y(\tau, u), \bar{u}(\tau))\} d\tau$$

$$+ \int_0^t \{f(\tau, y(\tau, u), \bar{u}(\tau)) - f(\tau, y(\tau, \bar{u}), \bar{u}(\tau))\} d\tau, \quad (3.5.4)$$

thus

$$\|y(t, u) - y(t, \bar{u})\| \leq 2 \int_{[\{\tau \in [0, \bar{t}]; u(\tau) \neq v(\tau)\}]} K(\tau, c + 1) d\tau$$

$$+ \int_0^t L(\tau, c + 1) \|y(\tau, u) - y(\tau, \bar{u})\| d\tau. \quad (3.5.5)$$

We apply Corollary 2.1.3 and obtain the estimate (3.5.3) for $u, \bar{u}$ in $0 \leq t \leq t_u$; taking $d(u, \bar{u}) \leq \delta$ sufficiently small, the maximality of $[0, t_u]$ implies $t_u = \bar{t}$. We argue in the same way for $v(\cdot)$ and deduce that $y(t, v)$ exists as well in $0 \leq t \leq \bar{t}$. Finally, we write (3.5.4) for u and v and estimate in the same way. ∎

3.6. Continuity of the Solution Operator
of the Variational Equation

Lemma 3.6.1. *Let the control system (3.5.1) satisfy Hypothesis III in $0 \leq t \leq \bar{t}$
for each $u(\cdot) \in C_{ad}(0, \bar{t}; U)$ with $K(\cdot, c)$ and $L(c, \cdot)$ independent of $u(\cdot)$, and let
$y(t, \bar{u})$ exist in $0 \leq t \leq \bar{t}$. Let $S(t, s; u)$ be the solution matrix of the variational
equation*

$$\xi'(t) = \partial_y f(t, y(t, u), u(t)) \xi(t). \qquad (3.6.1)$$

*Then $S(t, s; u)$ is uniformly continuous in $[0, \bar{t}] \times [0, \bar{t}] \times B(\bar{u}, \delta)$, δ the constant
in Lemma 3.5.1, $B(\bar{u}, \delta)$ equipped with the distance (3.4.1).*

Proof. The assumptions and Theorem 3.5.1 imply that

$$\|\partial_y f(t, y(t, u), u(t))\| \le L(t, c+1) \quad (0 \le t \le \bar{t}) \tag{3.6.2}$$

(c a bound for $\|y(t, \bar{u})\|$ in $0 \le t \le \bar{t}$) for all $u(\cdot) \in B(\bar{u}, \delta)$. It follows from Lemma 2.1.10 that

$$\|S(t, s, u)\| \le C \tag{3.6.3}$$

$$\|S(t', s'; u) - S(t, s; u)\|$$

$$\le C \left| \int_t^{t'} L(\tau, c+1)d\tau \right| + C \left| \int_s^{s'} L(\sigma, c+1)d\sigma \right| \tag{3.6.4}$$

both inequalities holding uniformly in $0 \le s, s', t, t' \le \bar{t}$, $u(\cdot) \in B(\bar{u}, \delta)$. It is then enough to show that

$$\|S(t, s; v) - S(t, s; u)\| \to 0 \tag{3.6.5}$$

as $d(v, u) \to 0$, uniformly with respect to $s, t \in [0, \bar{t}]$. To this end, we use the integral equation defining $S(t, s; u)$ for $u, v \in B(\bar{u}, \delta)$,

$$S(t, s; v) - S(t, s; u)$$

$$= \int_s^t \left\{ \partial_y f(\tau, y(\tau, v), v(\tau)) - \partial_y f(\tau, y(\tau, v), u(\tau)) \right\} S(\tau, s; v)d\tau$$

$$+ \int_s^t \left\{ \partial_y f(\tau, y(\tau, v), u(\tau)) - \partial_y f(\tau, y(\tau, u), u(\tau)) \right\} S(\tau, s; v)d\tau$$

$$+ \int_s^t \partial_y f(\tau, y(\tau, u), u(\tau)) \left\{ S(\tau, s; v) - S(\tau, s; u) \right\} d\tau,$$

and estimate:

$$\|S(t, s; v) - S(t, s, u)\|$$

$$\le 2C \int_{[\{\tau \in [0, \bar{t}];\, u(\tau) \neq v(\tau)\}]} L(\tau, c+1)d\tau$$

$$+ C \int_s^t \|\partial_y f(\tau, y(\tau, v), u(\tau)) - \partial_y f(\tau, y(\tau, u), u(\tau))\| d\tau$$

$$+ \left| \int_s^t L(\tau, c+1) \|S(\tau, s; v) - S(\tau, s; u)\| d\tau \right|, \tag{3.6.6}$$

so that the result will follow from Corollary 2.1.3 if we show that the first two terms on the right side of (3.6.6) converge to zero uniformly with respect to s, t as $d(v, u) \to 0$. This is obvious for the first; the second term is handled using Lemma 3.5.1, continuity of $\partial_y f(\tau, y, u)$ with respect to y, the bound (3.6.2), and the dominated convergence theorem. ■

3.7. The Minimum Principle for the Time Optimal Problem

We apply the results in **3.3** to the time optimal problem for

$$y'(t) = f(t, y(t), u(t)), \qquad y(0) = \zeta \tag{3.7.1}$$

with target condition $y(\bar{t}) \in Y$. The control constraint $u(t) \in U$ is included in the definition of the space $C_{ad}(0, \bar{t}; U)$.

Theorem 3.7.1. *Let* (3.7.1) *satisfy Hypothesis III with* $K(\cdot, c)$, $L(\cdot, c)$ *independent of* $u(\cdot) \in C_{ad}(0, \bar{t}; U)$, *and assume the target set* Y *is closed. Further, let* $C_{ad}(0, T; U)$ *be spike complete and saturated. Then, if* $f(t, y(t, u), v)$ *is regular in* U [1] *for every* $u(\cdot) \in C_{ad}(0, T; U)$ *and* $\bar{u}(\cdot)$ *is a time optimal control, there exists* $z \in E$, $\|z\| = 1$, $z \in N_Y(\bar{y}) = $ *normal cone to* Y *at* $\bar{y} = y(\bar{t}, \bar{u})$ *such that, if* $\bar{z}(s)$ *is the solution of*

$$\bar{z}'(s) = -\partial_y f(s, y(s, \bar{u}), \bar{u}(s))^* \bar{z}(s), \qquad \bar{z}(\bar{t}) = z \tag{3.7.2}$$

in $0 \le s \le t$, *then*

$$\langle \bar{z}(s), f(s, y(s, \bar{u}), \bar{u}(s)) \rangle = \min_{v \in U} \langle \bar{z}(s), f(s, y(s, \bar{u}), v) \rangle \tag{3.7.3}$$

a.e. in $0 \le s \le \bar{t}$.

For the proof, we apply Theorem 3.3.2 with the following cast. We select a sequence $\{t_n\} \subset [0, \bar{t})$ such that $t_n \to \bar{t}$. For each n, the space $C_{ad}(0, t_n; U)$ is equipped with the distance $d_n(u, v) = |\{t \in [0, t_n]; u(t) \ne v(t)\}|$ (see (3.4.1)). Since $C_{ad}(0, T; U)$ is saturated and spike complete, so is $C_{ad}(0, t_n; U)$ for each n, and Lemma 3.4.1 insures completeness. The functions f_n in the abstract time optimal problem are

$$f_n(u) = y(t_n, u). \tag{3.7.4}$$

defined in $V_n = D(f_n) = B(\bar{u}, \delta)$, δ the constant in Lemma 3.5.1 in the interval $0 \le t \le t_n$. By Lemma 3.5.1 the f_n are continuous, which is in excess of the requirements in Theorem 3.3.2. Obviously, $f_n(V_n) \cap Y = \emptyset$ (the target cannot be hit in time $t_n < \bar{t}$). The solution $\{\bar{u}^n\}$ in Theorem 3.3.2 consists of the restrictions of the optimal control $\bar{u}(\cdot)$ to $0 \le t \le t_n$. Since all trajectories (in particular the optimal trajectory) are continuous, we have $f_n(\bar{u}^n) = y(t_n, \bar{u}) \to y(\bar{t}, \bar{u})$, so that the definition of solution applies. Theorem 3.3.2 produces a $z \in E$ satisfying (3.3.25) and (3.3.26). To interpret (3.3.25) we must construct elements of the set $\liminf_{n \to \infty} \text{Var } f_n(\bar{u}^n)$, $\{\bar{u}^n\}$ the sequence in Theorem 3.3.2. Since this sequence is not explicitly known, we do the construction for arbitrary sequences $\{u_n\}$, $u^n \in C_{ad}(0, t_n; U)$, $d_n(u^n, \bar{u}^n) \to 0$; d_n is the distance in $C_{ad}(0, t_n; U)$).

---

[1] See definition in **2.5**.

It has been shown in Theorem 2.3.3 (under assumptions much more general than those in Theorem 3.7.1) that if $s \in (0, t]$ and $u(\cdot) \in C_{ad}(0, \bar{t}; U)$ then if $u_{s,h,v}$ is the spike perturbation in **2.3**, the limit

$$\xi(t, s, u, v) = \lim_{h \to 0+} \frac{1}{h}(y(t, u_{s,h,v}) - y(t, u)) \qquad (3.7.5)$$

exists almost everywhere (precisely in the intersection of the left Lebesgue sets of the functions $f(s, y(s, u), u(s))$, $f(s, y(s, u), v)$) and equals

$$\xi(t, s, u, v) = \begin{cases} 0 & (0 \le t \le s) \\ S(t, s; u)\{f(s, y(s, u), v) - f(s, y(s, u), u(s))\} & (3.7.6) \\ & (s \le t \le \bar{t}) \end{cases}$$

where $S(t, s; u)$ is the solution operator of the variational equation

$$\xi'(t) = \partial_y f(t, y(t, u), u(t))\xi(t). \qquad (3.7.7)$$

Since $d(u, u_{s,h,v}) \le h$, $\xi(t, s, u, v)$ is a directional derivative of $f(u) = y(t, u)$ at u (depending on the parameters s and v), that is, $\xi(t, s, u, v) \in \text{Der } f(u)$.

Lemma 3.7.2. *Let $\bar{u} \in C_{ad}(0, \bar{t}; U)$ be such that $y(t, \bar{u})$ exists in $0 \le t \le \bar{t}$, and let $u^n \in C_{ad}(0, t_n; U)$ with $t_n \to \bar{t}$. Assume that*

$$\sum_{n=1}^{\infty} d_n(u^n, \bar{u}) < \infty. \qquad (3.7.8)$$

Then there exists a set e of full measure in $0 \le t \le \bar{t}$ such that, if $s \in e$ and $v \in U$, then the directional derivative $\xi(t_n, s, u^n, v)$ exists if $t_n > s$ and

$$\xi(t_n, s, u^n, v) \to \xi(\bar{t}, s, \bar{u}, v).$$

Proof. In view of (3.7.8), if $d_n = \{t; 0 \le t \le t_n, \bar{u}(t) \ne u^n(t)\}$ then $\sum |d_n| < \infty$. Accordingly, the set

$$d = \bigcap_{m=1}^{\infty} \bigcup_{n=m}^{\infty} d_n$$

has outer measure (thus measure) zero. Let $e(u^n)$ be the set of all left Lebesgue points of $f(\tau, y(\tau, u^n), u^n(\tau))$ and $f(\tau, y(\tau, u^n), v)$ in $[0, t_n]$, and let

$e_n = e(u^n) \cup [t_n, \bar{t}]$. Define

$$e = \left(\bigcap_{n=1}^{\infty} e_n \right) \setminus d. \qquad (3.7.9)$$

Obviously, e is total in $[0, \bar{t}]$. Let $s \in e$. Then $s \in e(u^n)$ for all n with $t_n > s$. Accordingly, $\xi(t_n, s, u^n, v)$ exists and equals (3.7.6) with $t = t_n, u = u^n$. Moreover, $s \notin d$, thus $s \notin d_n$ for $n \geq n_0$, and it follows that $u^n(s) = \bar{u}(s)$ for $n \geq n_0$. To show that (3.7.6) holds for $t = \bar{t}, u = \bar{u}$, it is enough to use Lemma 3.6.1 on continuity of the solution operator $S(t, s; u)$.   ∎

Proof of Theorem 3.7.1. We use the elements $\xi(\bar{t}, s, \bar{u}, v) \in \liminf_{n \to \infty} \operatorname{Der} f_n(\tilde{u}_n)$ in Lemma 3.7.2; note that the "fast convergence" condition (3.7.8) is a consequence of (3.3.23) if the sequence $\{\varepsilon_n\}$ is suitably chosen (for instance $\varepsilon_n = 1/n^3$). Inequality (3.3.25) implies

$$\langle z, \xi(\bar{t}, s, \bar{u}, v) \rangle \geq 0 \qquad (3.7.10)$$

for $v \in U$ and s in the set e in Lemma 3.7.1; using (3.7.6) and taking adjoints,

$$\langle S(\bar{t}, s; \bar{u})^* z, f(s, y(s, \bar{u}), v) - f(s, y(s, \bar{u}), \bar{u}(s)) \rangle \geq 0 \qquad (3.7.11)$$

for $v \in U$ and $s \in e$. Since $S(t, s; \bar{u})^* = S^*(s, t, \bar{u})$, $S^*(s, t, \bar{u})$ the solution operator of the adjoint variational equation, $S(\bar{t}, s, \bar{u})^* z$ is the solution $\bar{z}(s)$ of (3.7.2) in $0 \leq s \leq \bar{t}$. That $z \in N_Y(y(\bar{t}, \bar{u}))$ is (3.3.26), guaranteed by the fact that $\bar{y}^n = y(t_n, \bar{u}) \to \bar{y} = y(\bar{t}, u)$. This ends the proof.   ∎

The minimum principle can be reformulated using the Hamiltonian formalism. We use the **reduced Hamiltonian**

$$H(s, y, z, u) = \langle z, f(s, y, u) \rangle \quad (y, z \in E). \qquad (3.7.12)$$

The canonical equations for the optimal trajectory and the adjoint vector or costate $\bar{z}(t)$ (see **2.5**) are

$$y'(t, \bar{u}) = \nabla_z H(t, y(t, \bar{u}), \bar{z}(s), \bar{u}(t)), \qquad y(0, \bar{u}) = \zeta, \qquad (3.7.13)$$

$$\bar{z}'(s) = -\nabla_y H(s, y(s, \bar{u}), \bar{z}(s), \bar{u}(s)), \qquad z(\bar{t}) = z. \qquad (3.7.14)$$

The minimum principle (3.7.3) becomes

$$H(s, y(s, \bar{u}), \bar{z}(s), \bar{u}(s)) = \min_{v \in U} H(s, y(s, \bar{u}), \bar{z}(s), v) \qquad (3.7.15)$$

almost everywhere in $0 \leq s \leq \bar{t}$.

Remark 3.7.3. The condition that $z \in N_Y(\bar{y}(\bar{t}, \bar{u}))$ is called the **transversality condition**. The information that it gives about z is limited; it hinges on knowing the point $y(\bar{t}, \bar{u})$ where the optimal trajectory hits the target set. Transversality gives no information for a point target set $Y = \{\bar{y}\}$ since $T_Y(\bar{y}) = \{0\}$, $N_Y(\bar{y}) = E$. See Remark 4.2.4 on transversality for general optimal problems.

Example 3.7.4. If $C_{ad}(0, T; U)$ is not saturated, it is not complete. As any metric space, it can be **completed** (see Goffman–Pedrick [1983, p. 20] for precise definitions and proofs). Completion takes a very simple form in this case; it suffices to adjoin to $C_{ad}(0, T; U)$ all limits of stationary sequences (see **3.4**) and use the same definition of distance in the enlarged space $\overline{C}_{ad}(0, T; U)$. To show that $\overline{C}_{ad}(0, T; U)$ is complete, we approximate a Cauchy sequence $\{u_n(\cdot)\}$ by a sequence $\{v_n(\cdot)\}$ in $C_{ad}(0, T; U)$ and argue as in **3.4**. If $f(t, y, u(t))$ satisfies Hypothesis I for a control space $C_{ad}(0, T; U)$ with $K(\cdot, c)$ independent of $u(\cdot)$ then the same is true replacing $C_{ad}(0, T; U)$ by $\overline{C}_{ad}(0, T; U)$; similar statements hold for Hypotheses II and III. This suggests that we can eliminate the assumption that $C_{ad}(0, T; U)$ is saturated in Theorem 3.7.1; we just replace the system by the "completed" system. However, this is a different system and optimality may not be preserved. A more reasonable inference is that one should work from the beginning with a complete space $C_{ad}(0, T; U)$, since incompleteness may preclude existence.

Example 3.7.5. Let $f(t, y, u)$ be continuous in all three variables and continuously differentiable with respect to y in $[0, T] \times \mathbb{R}^m \times U$, where $U \subseteq \mathbb{R}^k$ is closed and bounded, and let $C_{ad}(0, T; U)$ be the set of all measurable $\mathbb{R}^k$-valued functions taking values in U almost everywhere. Then $C_{ad}(0, T; U)$ and $f(t, y, u)$ satisfy all the assumptions in Theorem 3.7.1.

3.8. Time Optimal Capture of a Wandering Particle

A particle of mass m moves on the x axis. At time $t = 0$, a force $u(t)$ is applied:

$$mx''(t) = u(t). \tag{3.8.1}$$

Choosing units, we may assume that $m = 1$. The force $u(t)$ obeys the constraint

$$|u(t)| \leq 1, \tag{3.8.2}$$

and the admissible control space $C_{ad}(0, T; U)$ consists of all measurable functions $u(\cdot)$ satisfying (3.8.2). The objective is to drive the particle to equilibrium in minimum time from the initial position-velocity $(x(0), x'(0)) = (\xi, \eta)$ in the phase plane. We use the phase plane notation (x, y) instead of (y_1, y_2); also, the coordinates of the adjoint vector will be named p, q instead of z_1, z_2.

The first order system is

$$x'(t) = y(t), \qquad y'(t) = u(t), \tag{3.8.3}$$

with $(x(0), y(0)) = (\xi, \eta)$; the target condition is

$$x(\bar{t}) = y(\bar{t}) = 0. \tag{3.8.4}$$

Existence of a time optimal control $\bar{u}(\cdot)$ is guaranteed by Theorem 2.2.6 modulo construction of the required minimizing sequence. It is enough to show that we can drive $(x(0), x'(0))$ to equilibrium in finite time with a control satisfying (3.8.2). This is easy to show, but we skip the proof since we prove later that the drive can be done with controls of a special form.

The Hamiltonian is $H(x, y, p, q, u) = py + qu$, hence the canonical equations for (p, q) are

$$p' = 0, \qquad q' = -p \tag{3.8.5}$$

with solutions

$$p(t) = a, \qquad q(t) = -at + b. \tag{3.8.6}$$

The fact that the final condition $(p(\bar{t}), q(\bar{t})) = (a, -a\bar{t} + b)$ is not zero insures that a and b cannot be both zero. In particular, $q(t)$ is not identically zero. Eliminating terms in the Hamiltonian that do not depend on $\bar{u}(s)$, the maximum principle[1] reduces to

$$q(s)\bar{u}(s) = \max_{|v| \le 1} q(s)v$$

so that an optimal control $\bar{u}(\cdot)$ must satisfy $\bar{u}(s) = \operatorname{sign} q(s)$ and must therefore be of one of the four following forms:

$$\bar{u}(t) = -1 \quad (0 \le t \le \bar{t}) \qquad\qquad \bar{u}(t) = 1 \quad (0 \le t \le \bar{t})$$

$$\bar{u}(t) = \begin{cases} -1 & (0 \le t \le t_1) \\ 1 & (t_1 < t \le \bar{t}) \end{cases} \qquad \bar{u}(t) = \begin{cases} 1 & (0 \le t \le t_1) \\ -1 & (t_1 < t \le \bar{t}), \end{cases}$$

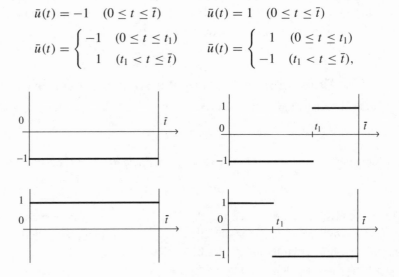

Figure 3.1.

where, in the last two cases, the switching point t_1 is unknown. A control like this is called **bang-bang** (it switches between two values). Bang-bang controls that switch more than once appear in next section.

---

[1] To argue in accordance with other books, we use the maximum rather than the minimum principle (see the end of the previous section). This amounts to changing the sign of the final condition $z = (p(\bar{t}), q(\bar{t}))$, so that z belongs to $-N_Y(\bar{y}) = -N_Y((x(\bar{t}), y(\bar{t}))$ rather than to $N_Y(\bar{y})$. This makes no difference for a point target problem since $N_Y(\bar{y})$ is the whole space; the only significant information is that $z \ne 0$.

In this problem (and in others) we use the following approach to the identification of optimal controls:

(a) Show that an optimal control exists.

(b) Using the maximum principle, select all postulants to optimal controls (those that satisfy the maximum principle).

(c) From all these, select the optimal one(s) (possibly using something else than the maximum principle).

In this section and the next, step (b) produces a unique control; since (a) guarantees existence, the lone postulant must be optimal.

We figure out the trajectories of (3.8.3) corresponding to the possible values $-1, 1$ of the optimal control. For $\bar{u}(t) = -1$,

$$y = -t + K_-, \qquad x = -\frac{t^2}{2} + K_- t + L_- = -\frac{y^2}{2} + C_-, \qquad (3.8.7)$$

K_-, L_-, C_- constants. The trajectories are parabolas over which we move clockwise.

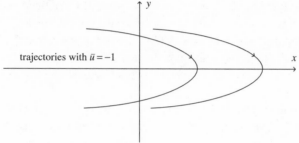

Figure 3.2.

For $\bar{u}(t) = 1$,

$$y = t + K_+, \qquad x = \frac{t^2}{2} + K_+ t + L_+ = \frac{y^2}{2} + C_+, \qquad (3.8.8)$$

K_+, L_+, C_+ constants. The trajectories are parabolas over which we also move clockwise.

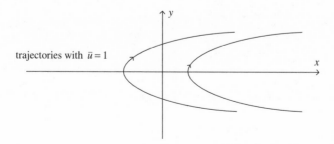

Figure 3.3.

Final approach to the origin must be along one of the two half-parabolas

$$\Pi^+ = \{(x, y); x = y^2/2 \ (y \le 0)\}, \qquad \Pi^- = \{(x, y); x = -y^2/2 \ (y \ge 0)\},$$

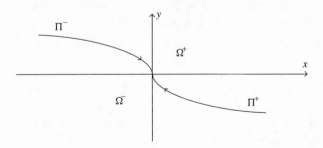

Figure 3.4.

thus any trajectory ending up at the origin must either be a part of Π^+ or Π^- or must link up with Π^+ or Π^-: only one "bus transfer" (a jump from the family (3.8.7) to the family (3.8.8) or vice versa) is permitted. We figure out how (ξ, η) can be driven to the origin in this way and also the driving time.

Due to the parametric equations (3.8.7)-(3.8.8) of the parabolas, the time elapsed over the part of a trajectory corresponding to $\bar{u} = -1$ is the distance swept by the projection on the y-axis, and the same is true for trajectories corresponding to $\bar{u} = 1$.

Let Ω^+ be the region of the (x, y) plane above $\Pi^+ \cup \Pi^-$, and let $(\xi, \eta) \in \Omega^+$. To drive to the origin using controls of the prescribed form we first catch a parabola of the family (3.8.7),

$$x = -\frac{y^2}{2} + \left(\xi + \frac{\eta^2}{2}\right)$$

until we hit Π^+; this happens at the switching point (ζ, v), where

$$\zeta = -\frac{v^2}{2} + \left(\xi + \frac{\eta^2}{2}\right) = \frac{v^2}{2}$$

so that

$$v = -\sqrt{\xi + \frac{\eta^2}{2}}.$$

The time elapsed on this part of the drive is $t_1 = \eta - v$; hence the switching time is

$$t_1 = \eta + \sqrt{\xi + \frac{\eta^2}{2}}.$$

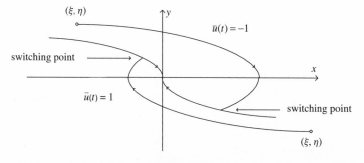

Figure 3.5.

We then switch to $\bar{u} = 1$ and drive along Π^+ to the origin. The time elapsed on the second part of the drive is $-\nu$, thus the total driving time is

$$\bar{t} = \eta + 2\sqrt{\xi + \frac{\eta^2}{2}}.$$

It is plain that the control $\bar{u}(\cdot)$ used is the only one of regulation form driving (ξ, η) to the origin, so that it must be the optimal control. We have also identified the optimal time.

A completely symmetric argument applies to the region Ω_- below $\Pi^+ \cup \Pi^-$.

The system can be given a feedback law that drives it optimally to the origin from any initial position (ξ, η); it suffices to define

$$P(x, y) = \begin{cases} -1 & (x, y) \in \Omega^+ \\ 1 & (x, y) \in \Omega^- \end{cases}$$

and rewrite (3.8.3) in the form

$$x'(t) = y(t), \qquad y'(t) = P(x(t), y(t)). \tag{3.8.9}$$

Example 3.8.1. Do the capture problem replacing the target set $Y = \{(0, 0)\}$ with $Y = \{(x, y); x^2 + y^2 \le 1\}$ and $Y = \{(x, y); |x|, |y| \le 1\}$. Check the transversality condition; if the maximum rather than the minimum principle is used, the final condition $z = (p(\bar{t}), q(\bar{t}))$ belongs to $-N_Y((x(\bar{t}), y(\bar{t}))$ rather than to $N_Y((x(\bar{t}), y(\bar{t}))$.

3.9. Time Optimal Stopping of an Oscillator

We reconsider the controlled linear oscillator in **2.7**,

$$mx''(t) = -kx(t) + u(t). \tag{3.9.1}$$

Choosing units, we may assume that $m = 1, k = 1$. The control constraint is

$$|u(t)| \leq 1. \tag{3.9.2}$$

The objective is to bring the oscillator to equilibrium

$$x(\bar{t}) = 0, \qquad x'(\bar{t}) = 0 \tag{3.9.3}$$

from an arbitrary initial position-velocity $(x(0), x'(0)) = (\xi, \eta)$ in optimal time $\bar{t}$. The first order system for $(x(t), y(t)) = (x(t), x'(t))$ is

$$x'(t) = y(t), \qquad y'(t) = -x(t) + u(t) \tag{3.9.4}$$

with initial condition $(x(0), y(0)) = (\xi, \eta)$ and target condition

$$x(\bar{t}) = y(\bar{t}) = 0. \tag{3.9.5}$$

The remarks on existence in **3.8** apply here. Using the same notation in last section, we call (p, q) the adjoint vector. The Hamiltonian is $H(x, y, p, q, u) = py - qx + qu$, thus the canonical equations for p, q are

$$p'(t) = q(t), \qquad q'(t) = -p(t) \tag{3.9.6}$$

with solutions

$$p(t) = c \sin(t + a), \qquad q(t) = c \cos(t + a). \tag{3.9.7}$$

The maximum principle is

$$q(t)\bar{u}(t) = \max_{|v| \leq 1} q(t)v \tag{3.9.8}$$

with $(p(\bar{t}), q(\bar{t})) \neq 0$, thus

$$\bar{u}(t) = \pm \mathrm{sign}\,(\cos(t + a)). \tag{3.9.9}$$

This can be reformulated thus: there exist points $t_1, \ldots, t_n$ with

$$0 \leq t_1 \leq t_2 \leq \cdots \leq t_n \leq \bar{t}, \qquad t \leq \pi, \qquad \bar{t} - t_n \leq \pi,$$
$$t_2 - t_1 = t_3 - t_2 = \cdots = t_n - t_{n+1} = \pi \tag{3.9.10}$$

such that $\bar{u}(t)$ changes sign (from 1 to -1 or from -1 to 1) when t crosses each t_k.

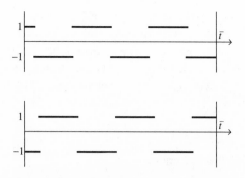

Figure 3.6.

As in **3.8**, we show that each point (ξ, η) in the phase plane can be driven to the origin by means of only one control $\bar{u}(t)$ of regulation form. This control must then be the optimal control.

For $\bar{u} = 1$, (3.9.4) is

$$(x - 1)' = y, \qquad y' = -(x - 1)$$

with solutions

$$x(t) = c\sin(t + a) + 1, \qquad y(t) = c\cos(t + a). \tag{3.9.11}$$

Hence, these trajectories are circles with center at $(1, 0)$ and arbitrary radius (see Figure 3.7). On the other hand, for $\bar{u} = -1$,

$$(x + 1)' = y, \qquad y' = -(x + 1)$$

with solutions

$$x(t) = c\sin(t + a) - 1, \qquad y(t) = c\cos(t + a). \tag{3.9.12}$$

Trajectories are circles with center at $(-1, 0)$ and arbitrary radius. Over both

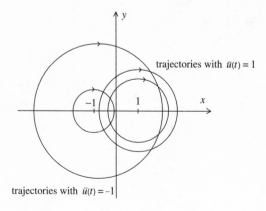

Figure 3.7.

families of circles we move clockwise with constant angular speed 1. The form (3.9.9) of the aspirants to optimal control determines that we must reach the origin jumping back and forth from trajectories of one of the two families (3.9.11) or (3.9.12) to trajectories of the other. In particular, final approach to the origin (corresponding to the interval $t_n \le t \le \bar{t} = $ optimal time in (3.9.10)) must be along one of the half-circles

$$\Gamma^+ = \{(x, y); (x - 1)^2 + y^2 = 1, y \le 0\},$$
$$\Gamma^- = \{(x, y); (x + 1)^2 + y^2 = 1, y \ge 0\}.$$

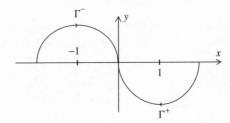

Figure 3.8.

(Figure 3.8). We figure out which points (ξ, η) in the plane can be driven to the origin using a trajectory that enters the origin along Γ^-. In the region Ω_1^+ (Figure 3.9), this is only possible driving with $u = 1$ until hitting Γ^-, switching

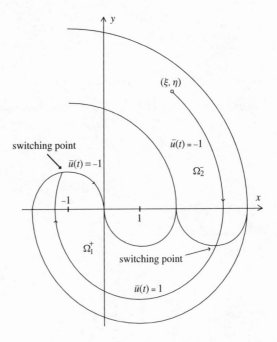

Figure 3.9.

to $u = -1$, and driving to the origin along Γ^-. In the region Ω_2^-, we drive first with $u = -1$ until the boundary of Ω_1^+ and then follow the driving directions for Ω_1^+. A similar argument applies to trajectories entering through Γ^+. We may then divide the plane in regions as shown in Figure 3.10. In regions with a superindex

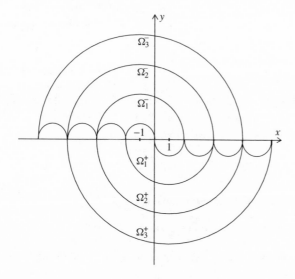

Figure 3.10.

"+" (resp. with a superindex "$-$") the control is $u = 1$ (resp. $u = -1$).

The feedback solution of the problem is obvious: if

$$\Omega^+ = \bigcup_{n=1}^{\infty} \Omega_n^+, \qquad \Omega^- = \bigcup_{n=1}^{\infty} \Omega_n^-,$$

(Figure 3.11), the system will drive itself optimally to the origin provided with the feedback law

$$P(x, y) = \begin{cases} 1 & ((x, y) \in \Omega^+) \\ -1 & ((x, y) \in \Omega^-) \end{cases}$$

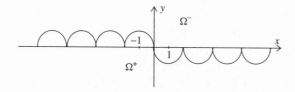

Figure 3.11.

Some of this reasoning applies to **controlled nonlinear oscillators**

$$x''(t) = -\phi(x(t)) + u(t) \tag{3.9.13}$$

such as the nonlinear pendulum (**1.3**, equation (1.3.6)). The continuously differentiable function $\phi(x)$ satisfies

$$x\phi(x) > 0 \quad (x \neq 0) \tag{3.9.14}$$

(that is, $-\phi(x)$ is a **restoring** force). The control constraint is again (9.2), and the oscillator is to be brought to equilibrium from an initial position (ξ, η) in minimum time $\bar{t}$. The first order system is

$$x'(t) = y(t) \qquad y'(t) = -\phi(x(t)) + u(t). \tag{3.9.15}$$

It satisfies Hypothesis III, so that by Theorem 2.1.1 solutions exist locally and are unique. We show that an *a priori* bound (2.1.11) can be found, so that by Lemma 2.1.5 the solutions exist for all t. Define

$$\Phi(x) = \int_0^x \phi(\xi)d\xi > 0 \quad (-\infty < x < \infty) \tag{3.9.16}$$

(the **potential energy** of the oscillator). By (3.9.14),

$$\Phi(x) > 0 \quad \text{for } x \neq 0. \tag{3.9.17}$$

Let $x(t)$ be a solution of (3.9.13) in an arbitrary interval. Multiplying by $x'(t)$ and integrating,

$$\frac{1}{2}x'(t)^2 + \Phi(x(t)) - \int_0^t x'(\tau)u(\tau)d\tau = E. \tag{3.9.18}$$

Let $E(t)$ be the total energy (the sum of the first two terms). Then

$$E(t) = E(0) + \int_0^t x'(\tau)u(\tau)d\tau$$

$$\leq E(0) + \frac{1}{2}\int_0^t (x'(\tau)^2 + 1)d\tau < E(0) + \frac{T}{2} + \int_0^t E(\tau)d\tau,$$

so that by Lemma 2.1.3 we deduce that $E(t)$ (thus $x(t)^2 + x'(t)^2$) is bounded in any interval $[0, T)$ where the solution exists; by Lemma 2.1.5, solutions exist for all t. The argument shows that if $\{u^n(\cdot)\}$ is a minimizing sequence, the trajectories $(x(t, u^n), y(t, u^n))$ must be bounded; that $\{t_n\}$ must be bounded is obvious (we are minimizing time). Hence, to apply Theorem 2.2.6 we only have to show that the oscillator can be driven to equilibrium in finite time. This will follow from the discussion below, where we use the fact that the hypotheses for the maximum principle (Theorem 3.7.1) are satisfied.

The Hamiltonian is $H(x, y, p, q, u) = py - q\phi(x) + qu$, and the canonical equations for (p, q) are

$$p'(t) = \phi'(x(t))q(t), \qquad q'(t) = -p(t) \tag{3.9.19}$$

hence

$$q''(t) = -\phi'(x(t))q(t) \quad (0 \leq t \leq \bar{t}). \tag{3.9.20}$$

Assume that $q(t)$ has an infinite number of zeros in the control interval $0 \leq t \leq \bar{t}$. Then these zeros must have an accumulation point t, at which point we must have $q'(t) = 0$. By uniqueness of solutions of the linear differential equation (3.9.20) (equivalently, of the system (3.9.19)), $q(\cdot)$ must be identically zero. Hence $p(\cdot)$ is identically zero as well, which contradicts the fact, guaranteed by the maximum principle, that $(p(\bar{t}), q(\bar{t})) \neq 0$. Thus, $q(\cdot)$ may only have a finite number of isolated zeros, and the maximum principle

$$q(t)u(t) = \max_{|v| \leq 1} q(t)v$$

implies that the optimal control

$$\bar{u}(t) = \operatorname{sign} q(t) \tag{3.9.21}$$

is bang-bang with a finite number of switchings. However, we know nothing about the length of the intervals where $\bar{u}(t)$ is constant. To obtain a better picture of the situation we figure out the trajectories corresponding to values $\bar{u} = \pm 1$ of the control. In general, if $\bar{u}(t) = u_0$ we obtain multiplying (3.9.13) by $x'(t)$ and integrating that

$$\tfrac{1}{2}x'^2 + \Phi(x) - u_0 x = C.$$

Accordingly, any solution $x(t)$ is contained in one of the curves

$$x' = \pm\sqrt{2}\sqrt{C - \Phi(x) + u_0 x} \tag{3.9.22}$$

in the phase plane, defined where the expression under the radical is nonnegative; these curves are symmetric with respect to the x-axis. To simplify, we assume now that $\phi(x)$ is strictly increasing in $-\infty < x < \infty$ and that $\Phi(x) \pm x \to +\infty$ as $|x| \to +\infty$. It follows that

$$\Psi_+(x) = \Phi(x) - x$$

has a unique minimum $x_+ > 0$ in $-\infty < x < \infty$, since $\Psi'_+(x) = \phi(x) - 1$ can only vanish once; moreover, $\Psi_+(x)$ is strictly increasing in $x_+ \leq x < \infty$, strictly decreasing in $-\infty < x \leq x_+$. We show in the same way that

$$\Psi_-(x) = \Phi(x) + x$$

has a unique minimum $x_- < 0$ and is strictly decreasing on the left of x_-, strictly increasing on the right. We conclude that all trajectories with $\bar{u} = 1$ are contained in simple closed curves containing x_+, and that all trajectories with $\bar{u} = -1$ are contained in simple closed curves containing x_-. A simple argument (see

Haberman [1977, p. 72]) shows that the solutions are periodic and the phase plane trajectories move clockwise. At this point the picture is a distorted version of Fig. 3.7 for the linear oscillator, the circles replaced by closed curves. We move optimally to the origin by jumping a finite number of times from one family of curves to the other. Final approach to the origin will be made along the two curves Γ^+ and Γ^- of equation (3.9.22) with $\bar{u} = 1$ (resp. $\bar{u} = -1$) that pass through the origin. Without information on the number of switchings, the aspirants to optimal control are now too many: we can drive (ξ, η) to the origin with infinitely many different bang-bang controls.

Example 3.9.1. Do the stopping problem for (3.9.1) replacing the target set $Y = \{(0, 0)\}$ with $Y = \{(x, y); x^2 + y^2 \leq 1\}$. Check the transversality condition.

3.10. Higher Dimensional Problems

To identify optimal controls from possible candidates may be very difficult in simple two-dimensional examples, and usually impossible in dimension 3 or higher. However, the maximum principle may yield nontrivial information. To see this, we look at the time optimal problem for a linear system

$$y'(t) = Ay(t) + Bu(t), \qquad y(0) = \zeta \qquad (3.10.1)$$

with $E = \mathbb{R}^m$ and control set $U \subseteq \mathbb{R}^k$ bounded, closed and convex. The admissible control space $C_{\text{ad}}(0, T; U)$ is the set of all $\mathbb{R}^k$-valued measurable vector functions such that $u \in U$ a.e. Existence of a time optimal control follows from Theorem 2.2.6 (see Remark 2.2.13) if ζ can be driven to zero in finite time. Given any optimal control $\bar{u}(t)$ the maximum principle produces a nonzero vector $z \in E$ such that

$$\langle B^* S(\bar{t} - s)^* z, \bar{u}(s) \rangle = \max_{v \in U} \langle B^* S(\bar{t} - s)^* z, v \rangle \qquad (3.10.2)$$

a.e. in $0 \leq s \leq \bar{t}$. This relinquishes no news on $\bar{u}(t)$ in the set

$$e = e(\bar{t}, z) = \{s; 0 \leq s \leq \bar{t}, B^* S(\bar{t} - s)^* z = 0\}. \qquad (3.10.3)$$

Now, $s \to B^* S(\bar{t} - s)^* z$ is analytic, thus, if not identically zero, its zeros cannot have an accumulation point; either $e = [0, \bar{t}]$ or

$$e = \{t_1, t_2, \ldots, t_k\}, \quad 0 \leq t_1 < t_2 < \cdots < t_k < \bar{t}.$$

If the latter holds, $k > 1$ and U is the unit ball in $\mathbb{R}^k$ then

$$\bar{u}(s) = \frac{B^* S(\bar{t} - s)^* z}{\| B^* S(\bar{t} - s)^* z \|} \qquad (t \notin e) \qquad (3.10.4)$$

which implies that $\|\bar{u}(t)\| = 1$ and that $\bar{u}(\cdot)$ is piecewise analytic. For $k = 1$,

(3.10.4) shows that $\bar{u}(\cdot)$ is bang-bang with a finite number of switchings. This can be generalized for k arbitrary. Assume that U is a polyhedron. It satisfies the **general position condition** if for every nonzero vector v along the direction of one of the edges, the vectors

$$Bv, ABv, \ldots, A^{m-1}Bv \tag{3.10.5}$$

are linearly independent.

Theorem 3.10.1. *Let the general position condition be satisfied. Then any optimal control $\bar{u}(t)$ is bang-bang (with a finite number of switchings) and takes values on the vertices of U.*

Proof. For any fixed t, the function $u \to \langle S(\bar{t}-t)^*z, Bu \rangle$ from U into $\mathbb{R}$, being linear, must attain its maximum either (a) on a unique vertex of U or (b) on an entire edge of U. Hence, it is enough to show that (b) may happen only in a finite set. If this is not true, there exists a sequence $\{t_n\} \subset [0, \bar{t}]$ such that $\langle S(\bar{t}-t_n)^*z, Bu \rangle$ attains its maximum on an entire edge (depending on n). Since there are only a finite number of edges we may assume, passing to a subsequence, that all the functions $\langle S(\bar{t}-t_n)^*z, Bu \rangle$ attain their maximum on the same edge J; thus, they are constant there. If $v \neq 0$ is a vector along J we can write $v = u_1 - u_0$ with $u_0, u_1 \in J$, so that $\langle S(\bar{t}-t_n)^*z, Bv \rangle = 0$; by analyticity,

$$\langle S(\bar{t}-t)^*z, Bv \rangle = 0 \quad (0 \le t \le \bar{t}). \tag{3.10.6}$$

Differentiating (3.10.6) repeatedly and setting $t = \bar{t}$,

$$\langle z, A^j Bv \rangle = 0 \quad (j = 1, 2, \ldots, m-1).$$

In view of the general position condition $z = 0$, a contradiction. ∎

Theorem 3.10.2. *Let the assumptions of Theorem 3.10.1 be satisfied. Then the optimal control $\bar{u}(\cdot)$ is unique.*

In fact, let $\bar{u}_1(\cdot), \bar{u}_2(\cdot)$ be two time optimal controls in $0 \le t \le \bar{t} =$ optimal time. By linearity and convexity,

$$\bar{u}(t) = \tfrac{1}{2}\bar{u}_1(t) + \tfrac{1}{2}\bar{u}_2(t)$$

will also be an optimal control. If $\bar{u}_1$ and $\bar{u}_2$ do not coincide almost everywhere, since both are bang-bang, there exists an interval where they take values in different vertices. Then the optimal control $\bar{u}(t)$ does not take values in a vertex in that interval, which contradicts Theorem 3.10.1. Obviously, the same uniqueness argument works for convex sets such that a nontrivial convex combination of boundary points belongs to the interior. ∎

When the set (3.10.3) is the whole interval $0 \leq t \leq \bar{t}$, the maximum principle collapses. The **rank condition** below prevents this. It refers to the matrix

$$M = [B, AB, \ldots, A^{m-1}B] \tag{3.10.7}$$

of m rows and mk columns.

Theorem 3.10.3. *The function*

$$t \to B^*S(\bar{t} - t)^*z \tag{3.10.8}$$

is not identically zero for any $z \neq 0$ if and only if

$$\operatorname{rank} M = m. \tag{3.10.9}$$

Proof. Let z be such that $B^*S(\bar{t} - t)^*z$ is identically zero. Differentiate repeatedly and take $t = \bar{t}$:

$$B^*(A^*)^j z = 0 \quad (j = 0, 1, \ldots, m - 1).$$

This means the m-dimensional vector z is orthogonal to all vectors of the form $y = A^j Bv$, $v \in \mathbb{R}^k$, in particular to all columns of the matrix M; it follows from the rank condition that $z = 0$. Conversely, if rank $M < m$ there exists $z \in \mathbb{R}^m$ such that z is orthogonal to all the columns of M, so that $\langle z, A^j Bv \rangle = 0$ for all $v \in \mathbb{R}^k$, $j = 1, 2, \ldots, m - 1$. By the Cayley–Hamilton theorem (Efimov–Rozendorf [1975, p. 252]), the same holds for all $j \geq m$ so that $B^*(A^*)^j z = 0$ for all j. This implies that (3.10.8) is identically zero.   ∎

Novanishing of $B^*S(\bar{t} - t)^*z$ for $z \neq 0$ is related to **controllability**. In this context, we consider unconstrained controls; we may take $C_{\mathrm{ad}}(0, \bar{t}; U) = C(0, \bar{t}; \mathbb{R}^k)$. We say that (3.10.1) is **controllable in time** $\bar{t}$ if, given $\zeta, \bar{y} \in E$ there exists $u \in C(0, \bar{t}; \mathbb{R}^k)$ such that

$$y(\bar{t}, \zeta, u) = \bar{y},$$

where $y(t, \zeta, u)$ is the solution of (3.10.1). In other words, a system is controllable if we can drive from any $\zeta \in E$ to any $\bar{y} \in E$ by means of some control u.

Theorem 3.10.4. *The system* (3.10.1) *is controllable in time* $\bar{t}$ *if and only if the function* (3.10.8) *is not identically zero for any $z \neq 0$, or, equivalently, if the rank condition* (3.10.9) *is satisfied.*

Proof. Let $R(\bar{t})$ be the subspace of E consisting of all elements of the form

$$y = \int_0^{\bar{t}} S(\bar{t} - \tau)Bu(\tau)d\tau$$

with $u \in C(0, \bar{t}; E)$. If (3.10.1) is controllable in time $\bar{t}$ then $R(\bar{t}) = E$ (we can drive from the origin to any point $\bar{y}$). Conversely, if $R(\bar{t}) = E$ the element

$y = \bar{y} - S(\bar{t})\zeta$ belongs to $R(\bar{t})$, thus (3.10.1) is controllable. Now, $R(\bar{t}) \neq E$ if and only if there exists $z \neq 0$ with $\langle z, y \rangle = 0$ for all $y \in R(\bar{t})$, which is

$$\int_0^{\bar{t}} \langle B^* S(\bar{t} - \tau)^* z, u(\tau) \rangle d\tau = \int_0^{\bar{t}} \langle z, S(\bar{t} - \tau) Bu(\tau) \rangle d\tau = 0$$

for all $u(\cdot) \in C(0, \bar{t}; \mathbb{R}^k)$. This implies $B^* S(\bar{t} - t)^* z = 0$. Conversely, if $B^* S(\bar{t} - t)^* z = 0$, $\langle z, y \rangle = 0$ for $y \in R(\bar{t})$. This ends the proof. ∎

We note that the result holds as well for other control spaces. For instance, we may take as $C_{\mathrm{ad}}(0, \bar{t}; U)$ the space of all polynomials with coefficients in $\mathbb{R}^k$ or (going in the other direction), $L^1(0, \bar{t}; \mathbb{R}^k)$.

Example 3.10.5. The particular case (3.10.1) of the maximum principle used here can be proved directly. Let

$$R(\bar{t}, \zeta) = \{y \in E; \, y = y(\bar{t}, \zeta, u), u \in C_{\mathrm{ad}}(0, \bar{t}; U)\}$$

($\bar{t} = $ optimal time). Then $R(\bar{t}, \zeta)$ is convex and closed. If $\{t_n\} \subseteq [0, \bar{t})$ is a sequence with $t_n \to \bar{t}$, $y(t_n, \zeta, u) \notin R(\bar{t}, \zeta)$. Apply the separation theorem for a point and a convex set in $\mathbb{R}^m$; see Bellman et al. [1956]. For infinite dimensional generalizations of this argument, see Chapter 9.

Example 3.10.6. Let $\lambda_1, \ldots, \lambda_p$ ($p \leq m$) be the eigenvalues of A and m_j the multiplicity of λ_j, $r = \max m_j$. Show that the rank condition (3.10.9) can be satisfied with an $r \times m$ matrix B but with no $k \times m$ matrix for $k < r$. In other words, r is the "minimum number of parameters needed to achieve controllability." It is easier to check nonvanishing of (3.10.8).

Example 3.10.7. If the control set U is a polyhedron satisfying the general position condition (3.10.5) and the eigenvalues of A are real (arbitrary multiplicities allowed), the number of switching points of an optimal control cannot exceed $m - 1$. See Pontryagin et al. [1962, p. 122].

4

Abstract Minimization Problems: The Minimum Principle for General Optimal Control Problems

4.1. The Abstract Minimization Problem

The problem, introduced in **3.1**, is: given a metric space V, a normed space E, a subset $Y \subseteq E$ and functions $f : D(f) \to E$, $f_0 : D(f_0) \to \mathbb{R}$ $(D(f), D(f_0) \subseteq V)$, characterize the solutions $\bar{u} \in D(f) \cap D(f_0)$ of

$$\text{minimize } f_0(u) \tag{4.1.1}$$

$$\text{subject to } f(u) \in Y. \tag{4.1.2}$$

Let m be the infimum. We require

$$-\infty < m < \infty. \tag{4.1.3}$$

In fact, if $m = \infty$, there is no u satisfying (4.1.2); if $m = -\infty$, the problem has no solution. We also assume that

(a) V is complete and E is a real Hilbert space,

(b) $D(f) \cap D(f_0) \neq \emptyset$ and the real valued functions

$$\Phi(u, y) = \begin{cases} \|f(u) - y\| & (u \in D(f)) \\ +\infty & (u \notin D(f)) \end{cases} \tag{4.1.4}$$

$$\Phi_0(u, y) = \begin{cases} \max(f_0(u), m - \varepsilon) & (u \in D(f_0)) \\ +\infty & (u \notin D(f_0)) \end{cases} \tag{4.1.5}$$

are lower semicontinuous with respect to u, the first for any $y \in Y$, the second for ε sufficiently small,

(c) the target set Y is closed.

We can say something not only about a solution $\bar{u}$ of the abstract minimization problem but about an **approximate** or **suboptimal solution**, which is a sequence

$\{\bar{u}^n\} \subseteq D(f) \cap D(f_0)$ with

$$\limsup_{n \to \infty} f_0(\bar{u}^n) \leq m, \qquad \lim_{n \to \infty} \text{dist}(f(\bar{u}^n), Y) = 0. \qquad (4.1.6)$$

A sequence $\{\bar{y}^n\} \subseteq Y$ is **associated** with $\{\bar{u}^n\}$ if there exists $\{\varepsilon_n\} \subset \mathbb{R}_+, \varepsilon_n \to 0$ with

$$f(\bar{u}_n) \leq m + \varepsilon_n, \qquad \|f(\bar{u}^n) - \bar{y}^n\| \leq \varepsilon_n. \qquad (4.1.7)$$

We write $\mathbf{f}(u) = (f_0, f)(u) = (f_0(u), f(u)) \in \mathbf{E} = \mathbb{R} \times E$.

Theorem 4.1.1. *Let* $\{\bar{u}^n\} \subseteq D(f) \cap D(f_0)$ *be a suboptimal solution of* (1.1)-(1.2) *and* $\{\bar{y}^n\} \subseteq Y$ *a sequence satisfying* (4.1.7). *Then there exists sequences* $\{\tilde{u}^n\} \subseteq D(f) \cap D(f_0), \{\tilde{y}^n\} \subseteq Y$ *with*

$$d(\tilde{u}^n, \bar{u}^n) + \|\tilde{y}^n - \bar{y}^n\| \leq \sqrt[4]{5}\sqrt{\varepsilon_n} \qquad (4.1.8)$$

and a sequence $(z_{0n}, z_n) \subseteq \mathbb{R} \times E$ *such that*

$$z_{0n}^2 + \|z_n\|^2 = 1, \qquad z_{0n} \geq 0,$$
$$z_{0n}\xi_0^n + \langle z_n, \xi^n - w^n \rangle \geq -\sqrt[4]{5}\sqrt{\varepsilon_n}(1 + \|w^n\|) \qquad (4.1.9)$$

for $(\xi_0^n, \xi^n) \in \text{Var}\,(f_0, f)(\tilde{u}^n)$ *and* $w^n \in I_Y(\tilde{y}^n)$ *or for* $(\xi_0^n, \xi^n) \in \text{Der}\,(f_0, f)(\tilde{u}^n)$ *and* $w^n \in K_Y(\tilde{y}^n)$. *If* (z_0, z) *is a limit point of* $\{(z_{0n}, z_n)\}$, *we have*

$$z_0^2 + \|z\|^2 = 1, \qquad z_0 \geq 0, \qquad z_0\xi_0 + \langle z, \xi \rangle \geq 0 \qquad (4.1.10)$$

for $(\xi_0, \xi) \in \liminf_{n \to \infty} \text{Var}\,(f_0, f)(\tilde{u}^n)$. *If* $\bar{y}^n \to \bar{y}$ *then*

$$z \in N_Y(\bar{y}). \qquad (4.1.11)$$

Theorem 4.1.1 follows straight from Theorem 3.3.2 on the abstract time optimal problem. The sequence of metric spaces is $V_n = V$ and the Hilbert space is $\mathbf{E} = \mathbb{R} \times E$ with its Hilbert product norm $\|(z_0, z)\| = (z_0^2 + \|z\|^2)^{1/2}$. We define functions $\mathbf{f}_n : D(\mathbf{f}_n) = D(f_0) \cap D(f) \to \mathbf{E} = \mathbb{R} \times E$ by

$$\mathbf{f}_n(u) = (\max(f_0(u) + \varepsilon_n, m), f(u)),$$

and a target set $\mathbf{Y}$ in $\mathbb{R} \times E$ by

$$\mathbf{Y} = \{m\} \times Y,$$

which is closed if Y is closed. We have $\mathbf{f}_n(D(\mathbf{f}_n)) \cap \mathbf{Y} = \emptyset$ (otherwise we would have $u \in V$ with $f(u) \in Y$ and $f_0(u) \leq m - \varepsilon_n < m$). The sequence $\{\bar{u}^n\}$ is a solution of the abstract time optimal problem; a sequence associated to $\{\bar{u}^n\}$ is $\{\bar{\mathbf{y}}^n\} = \{(m, \bar{y}^n)\}$, since

$$\|\mathbf{f}_n(\bar{u}^n) - \bar{\mathbf{y}}^n\| = \{(\max(f_0(\bar{u}^n) + \varepsilon_n, m) - m)^2 + \|f(\bar{u}^n) - \bar{y}^n\|^2\}^{1/2} \leq \sqrt{5}\varepsilon_n.$$

Finally, if $\Phi_n(u, y)$ is the function (3.3.20) corresponding to $\mathbf{f}_n(u)$, we have

$$
\begin{aligned}
\Phi_n(u, \mathbf{y}) &= \Phi_n(u, (m, y)) \\
&= \|\mathbf{f}_n(u) - (m, y)\| \\
&= \left\{ (\max(f_0(u) + \varepsilon_n, m) - m)^2 + \|f(u) - y\|^2 \right\}^{1/2}
\end{aligned}
$$

for $(m, y) \in \{m\} \times Y$, thus lower semicontinuity of $\Phi_n(u, \mathbf{y})$ follows from hypothesis (b) (Example 3.3.13 and Example 4.1.2 below). All assumptions verified, we apply Theorem 3.3.2 obtaining sequences $\{\tilde{u}^n\} \subset V$ and $\{\tilde{\mathbf{y}}^n\} = \{(m, \tilde{y}^n)\} \in \{m\} \times Y = \mathbf{Y}$ such that (3.3.22), (3.3.23) and (3.3.27) hold, and a sequence $\{\mathbf{z}_n\} = \{(z_{0n}, z_n)\}$ satisfying $\|\mathbf{z}_n\|^2 = z_{0n}^2 + \|z_n\|^2 = 1$ and the other conclusions of Theorem 3.3.2. Inequality (3.3.22) guarantees that $\tilde{u}^n \in D(\mathbf{f}_n) = D(f_0) \cap D(f)$ (see the comments after (3.3.14)). Since we have $\|\tilde{\mathbf{y}}^n - \bar{\mathbf{y}}^n\| = \|(m, \tilde{y}^n) - (m, \bar{y}^n)\| = \|\tilde{y}^n - \bar{y}^n\|$, (3.3.23) yields (4.1.8). In view of formula (3.3.28), the sequence $\{(z_{0n}, z_n)\}$ is given by

$$
\begin{aligned}
(z_{0n}, z_n) &= \frac{\mathbf{f}(\tilde{u}^n) - \tilde{\mathbf{y}}^n}{\|\mathbf{f}(\tilde{u}^n) - \tilde{\mathbf{y}}^n\|} \\
&= \frac{(\max(f_0(\tilde{u}^n) + \varepsilon_n, m) - m, \, f(\tilde{u}^n) - \tilde{y}^n)}{\left\{ (\max(f_0(\tilde{u}^n) + \varepsilon_n, m) - m)^2 + \|f(\tilde{u}^n) - \tilde{y}^n\|^2 \right\}^{1/2}}
\end{aligned}
\tag{4.1.12}
$$

so that $z_{0n} \geq 0$. To show the last inequality (4.1.9) note that

$$
I_{\mathbf{Y}}((m, \tilde{y}^n)) = \{0\} \times I_Y(\tilde{y}^n), \qquad K_{\mathbf{Y}}((m, \tilde{y}^n)) = \{0\} \times K_Y(\tilde{y}^n). \tag{4.1.13}
$$

Using (3.3.24),

$$
\begin{aligned}
z_{0n}\xi_0^n + \langle z_n, \xi^n - w^n \rangle &= \langle (z_{0n}, z_n), (\xi_0^n, \xi^n) - (0, w^n) \rangle \\
&\geq -\sqrt[4]{5}\sqrt{\varepsilon_n}(1 + \|w^n\|)
\end{aligned}
\tag{4.1.14}
$$

for $(\xi_0^n, \xi^n) \in \mathrm{Var}\,\mathbf{f}_n(\tilde{u}^n)$ and $w^n \in I_Y(\tilde{y}^n)$ or for $(\xi_0^n, \xi^n) \in \mathrm{Der}\,\mathbf{f}_n(\tilde{u}^n)$ and $w^n \in K_Y(\tilde{y}^n)$. This is not quite the right inequality, since (ξ_0^n, ξ^n) is in $\mathrm{Var}\,\mathbf{f}_n(\tilde{u}^n)$ ($\mathrm{Der}\,\mathbf{f}_n(\tilde{u}^n)$) rather than in $\mathrm{Var}\,(f_0, f)(\tilde{u}^n) = \mathrm{Var}\,\mathbf{f}(\tilde{u}^n)$ ($\mathrm{Der}\,\mathbf{f}(\tilde{u}^n)$). We clarify this first for Der. Note that there are two possibilities:

(i) $f_0(\tilde{u}^n) + \varepsilon_n > m$. In this case, every directional derivative (ξ_0^n, ξ^n) of $\mathbf{f}$ at $\tilde{u}^n$ is a directional derivative of $\mathbf{f}_n$ at $\tilde{u}^n$.

(ii) $f_0(\tilde{u}^n) + \varepsilon_n \leq m$. In this case, the first coordinate of $\mathbf{f}_n(\tilde{u}^n) - \mathbf{y}^n$ is zero, so that $z_{0n} = 0$ (formula (4.1.12)). Hence ξ_0^n in (4.1.14) is irrelevant, and we only need to prove that if $(\xi_0^n, \xi^n) \in \mathrm{Der}\,\mathbf{f}(\tilde{u}^n)$ then *there exists an element of* $\mathrm{Der}\,\mathbf{f}_n(\tilde{u}^n)$ *with the same second coordinate.* To see this let $\tilde{u}^n(h)$ be the function employed in (3.3.7) for the construction of the derivative (ξ_0^n, ξ^n). Then we have

$$\xi_0^n = \lim_{h \to 0+} \frac{f_0(\tilde{u}^n(h)) - f_0(\tilde{u}^n(h))}{h}$$

$$= \lim_{h \to 0+} \frac{(f_0(\tilde{u}^n(h)) + \varepsilon_n) - (f_0(\tilde{u}^n(h)) + \varepsilon_n)}{h},$$

and the fact that ξ^n is the second coordinate of an element of Der $\mathbf{f}(\tilde{u}^n)$ follows from Example 4.1.3 (a). The argument for Var is similar, this time using (b). This completes the proof of the second inequality (4.1.9). All relations (4.1.10) for a limit point are obvious. It follows from Theorem 3.3.2 that if $\tilde{y}^n \to \bar{y}$ then $(z_0, z) \in N_{\mathbf{Y}}((m, \bar{y})) = T_{\mathbf{Y}}((m, \bar{y}))^- = (\{0\} \times T_Y(\bar{y}))^- = \{0\}^- \times T_Y(\bar{y})^- = \mathbb{R} \times N_Y(\bar{y})$; the second coordinate of this inclusion is (4.1.11). This ends the proof of Theorem 4.1.1. ∎

Example 4.1.2. Let V be a metric space, $f, g : V \to \mathbb{R}_+ \cup \{+\infty\}$ two nonnegative lower semicontinuous functions. Then $h(u) = (f(u)^2 + g(u)^2)^{1/2}$ (positive square root) is lower semicontinuous. In fact, let $u \in V, \{u_n\}$ a sequence with $u_n \to u, h(u_n) = (f^2(u_n) + g^2(u_n))^{1/2} \to c$. Passing if necessary to a subsequence, we may assume $f(u_n)^2 \to a^2, g(u_n)^2 \to b^2$ with $a, b \geq 0, (a^2 + b^2)^{1/2} = c$. Since f, g are nonnegative, $f(u_n) \to a, g(u_n) \to b$, and it follows that $f(u) \leq a, g(u) \leq b$, so that $h(u) = (f(u)^2 + g(u)^2)^{1/2} \leq c$. The result is false if the functions are not nonnegative; for instance, take $V = [-1, 1], g(u) \equiv 0, f(u) = 0$ for $u \neq 0, f(0) = -1$.

Example 4.1.3. (a) Let $\phi(t)$ be a function having a right-sided derivative at $t = 0$. Then $\psi(t) = \max(\phi(t), c)$ has a right-sided derivative at $t = 0$ for any c. This result is obvious if $c > \phi(0)$ or if $c < \phi(0)$. In case $c = \phi(0)$, we have: (i) if $\phi'(0) > 0$ then $\psi'(0) = \phi'(0)$, (ii) if $\phi'(0) < 0$ then $\psi'(0) = 0$, (iii) if $\phi'(0) = 0$ then $\psi'(0) = 0$. (b) If $h_k \to 0$ and $\lim_{k\to\infty} h_k^{-1}(\phi(h_k) - \phi(0))$ exists, then $\lim_{k\to\infty} h_k^{-1}(\psi(h_k) - \psi(0))$ exists as well. The proof is the same as in (a).

4.2. The Minimum Principle for Problems with Fixed Terminal Time

We do optimal control problems for

$$y'(t) = f(t, y(t), u(t)), \qquad y(0) = \zeta \qquad (4.2.1)$$

with target condition $y(\bar{t}) \in Y$ and cost functional

$$y_0(t, u) = \int_0^t f_0(\tau, y(\tau, u), u(\tau))d\tau + g_0(t, y(t, u)) \qquad (4.2.2)$$

in a fixed time interval $0 \leq t \leq \bar{t}$. The control constraint $u(t) \in U$ is included in the definition of the space $C_{ad}(0, \bar{t}; U)$.

Theorem 4.2.1. *Assume $f(t, y, u)$ (resp. $f_0(t, y, u)$) satisfy Hypothesis III with $K(\cdot, c), L(\cdot, c)$ independent of $u(\cdot) \in C_{ad}(0, T; U)$ (resp. Hypothesis III^0 with $K_0(\cdot, c), L_0(\cdot, c)$ independent of $u(\cdot)$) in $0 \leq t \leq \bar{t}$, and let the derivatives $\partial_k g_0(t, y)$ exist. Let the target set Y be closed. Further, assume that $C_{ad}(0, T; U)$ is spike complete and saturated and that $f(t, y(t, u), v), f_0(t, y(t, u), v)$ are regular in U for every $u(\cdot) \in C_{ad}(0, \bar{t}; U)$. Then, if $\bar{u}(\cdot)$ is an optimal control in $0 \leq t \leq \bar{t}$ there exists $(z_0, z) \in \mathbb{R} \times E, z_0^2 + \|z\|^2 = 1, z_0 \geq 0, z \in N_Y(\bar{y}) =$ normal cone to Y at $\bar{y} = y(\bar{t}, \bar{u})$ such that, if $\bar{z}(s)$ is the solution of*

$$\bar{z}'(s) = -\partial_y f(s, y(s, \bar{u}), \bar{u}(s))^* \bar{z}(s) - z_0 \partial_y f_0(s, y(s, \bar{u}), \bar{u}(s)),$$
$$\bar{z}(\bar{t}) = z + z_0 \partial_y g_0(\bar{t}, y(\bar{t}, \bar{u})), \tag{4.2.3}$$

then

$$z_0 f_0(s, y(s, \bar{u}), \bar{u}(s)) + \langle \bar{z}(s), f(s, y(s, \bar{u}), \bar{u}(s)) \rangle$$
$$= \min_{v \in U}\{z_0 f_0(s, y(s, \bar{u}), v) + \langle \bar{z}(s), f(s, y(s, \bar{u}), v) \rangle\} \tag{4.2.4}$$

a.e. in $0 \leq s \leq \bar{t}$.

To apply Theorem 4.1.1, we rig $C_{ad}(0, \bar{t}; U)$ with the distance

$$d(u, v) = |\{t \in [0, \bar{t}]; u(t) \neq v(t)\}| \tag{4.2.5}$$

(see **3.4**). Saturation implies completeness (Lemma 3.4.1). The functions f, f_0 in the abstract minimization problem are

$$f(u) = y(\bar{t}, u), \qquad f_0(u) = y_0(\bar{t}, u) \tag{4.2.6}$$

defined in $V = B(\bar{u}, \delta), \delta$ the constant in Lemma 3.5.1 in the interval $0 \leq t \leq \bar{t}$. To check the hypotheses on f, f_0, it is enough to show that both are continuous in $B(\bar{u}, \delta)$. That f is continuous has been proved in Lemma 3.5.1. Continuity of f_0 is proved below.

Lemma 4.2.2. *Let u, δ be as in Lemma 3.5.1. Then, under the assumptions on $f_0(t, y, u)$ in Theorem 4.2.1, the map*

$$u \to y_0(\cdot, u)$$

is continuous from $B(u, \delta)$ into $C(0, \bar{t})$.

Proof. We have

$$
y_0(t, v) - y_0(t, u) = \int_0^t \{f_0(\tau, y(\tau, v), v(\tau)) - f_0(\tau, y(\tau, u), v(\tau))\}d\tau
$$

$$
+ \int_0^t \{f_0(\tau, y(\tau, u), v(\tau)) - f_0(\tau, y(\tau, u), u(\tau))\}d\tau
$$

$$
+ g_0(t, y(t, v)) - g_0(t, y(t, u))
$$

so that

$$
|y_0(t, v) - y_0(t, u)| \leq \int_{[\{\tau \in [0, \bar{t}]; u(\tau) \neq v(\tau)\}]} K_0(\tau, c + 1)d\tau
$$

$$
+ \int_0^t L_0(\tau, c + 1)\|y(\tau, v) - y(\tau, u)\|d\tau
$$

$$
+ |g_0(t, y(t, v)) - g_0(t, y(t, u))| \tag{4.2.7}
$$

where $[\cdot]$ denotes measurable envelope (see **3.5**) and c is a bound[1] for $\|y(\tau, u)\|$ in $0 \leq \tau \leq \bar{t}$. We estimate and use continuity of the solution map. ∎

It remains to identify elements of $\liminf_{n \to \infty} \text{Var}\,(f_0, f)(\tilde{u}^n)$ or, more generally, of $\liminf_{n \to \infty} \text{Var}\,(f_0, f)(u^n)$, where $u^n \to \bar{u}$. We use the directional derivatives $\xi(t, s, u, v) \in \text{Der}f(u)$ in (2.3.4) and the directional derivatives $\xi_0(t, s, u, v) \in \text{Der}f_0(u)$ in (2.3.3). Note that in **2.3** these derivatives are computed at the optimal control $\bar{u}$; here, to apply Theorem 4.1.1, they will be computed at $\tilde{u}^n$.

Lemma 4.2.3. *Let $\bar{u} \in C_{ad}(0, \bar{t}; U)$ be such that $y(t, \bar{u})$ exists in $0 \leq t \leq \bar{t}$, $\{u^n\} \subseteq C_{ad}(0, \bar{t}; U)$. Assume that*

$$
\sum_{n=1}^{\infty} d(u^n, \bar{u}) < \infty. \tag{4.2.8}
$$

Then there exists a set e of full measure in $0 \leq t \leq \bar{t}$ such that if $s \in e$ then the directional derivatives $\xi(t, s, u^n, v)$ and $\xi_0(t, s, u^n, v)$ exist and

$$
\xi(\bar{t}, s, u^n, v) \to \xi(\bar{t}, s, \bar{u}, v), \qquad \xi_0(\bar{t}, s, u^n, v) \to \xi_0(\bar{t}, s, \bar{u}, v)
$$

with $\xi(\bar{t}, s, \bar{u}, v)$ given by (2.3.14),

$$
\xi(t, s, \bar{u}, v) = \begin{cases} 0 & (0 \leq t < s) \\ S(t, s; \bar{u})\{f(s, y(s, \bar{u}), v) - f(s, y(s, \bar{u}), \bar{u}(s))\} \\ & (s \leq t \leq \bar{t}) \end{cases} \tag{4.2.9}
$$

---

[1] Recall that, in Lemma 3.5.1, $\|y(t, u)\|, \|y(t, v)\| \leq c + 1$.

and $\xi_0(\bar{t}, s, \bar{u}, v)$ given by (2.4.6),

$$
\begin{aligned}
\xi_0(t, s, \bar{u}, v) = \{f_0(s, y(s, \bar{u}), v) - f_0(s, y(s, \bar{u}), u(s))\} \\
+ \int_s^t \langle \partial_y f_0(\tau, y(\tau, \bar{u}), \bar{u}(\tau)), \xi(\tau, s, \bar{u}, v) \rangle d\tau \\
+ \langle \partial_y g_0(t, y(t, \bar{u})), \xi(t, s, \bar{u}, v) \rangle.
\end{aligned}
\tag{4.2.10}
$$

Proof. The statement for $\xi(\bar{t}, s, u^n, v)$ is just Lemma 3.7.2 with $t_n = \bar{t}$. The statement for $\xi_0(\bar{t}, s, u^n, v)$ is proved in a similar way. ∎

Proof of Theorem 4.2.1. We use Theorem 4.1.1 in the particular case $\bar{u}^n = \bar{u} =$ optimal control and an associated sequence $\{\bar{y}^n\}$ such that (4.2.8) is guaranteed, for instance $\varepsilon_n = 1/n^3$ (see the proof of Theorem 3.7.1). We then use (4.1.10) for the elements $(\xi_0(t, s, u, v), \xi(t, s, u, v))$ of $\liminf_{n \to \infty} \operatorname{Der}(f_0, f)(\bar{u}^n)$ constructed above and obtain a multiplier $(z_0, z) \in \mathbb{R} \times E$ satisfying

$$
\begin{aligned}
z_0^2 + \|z\|^2 = 1, \qquad z_0 \geq 0, \\
z_0 \xi_0(\bar{t}, s, \bar{u}, v) + \langle z, \xi(\bar{t}, s, \bar{u}, v) \rangle \geq 0,
\end{aligned}
\tag{4.2.11}
$$

that is,

$$
\begin{aligned}
z_0 \{f_0(s, y(s, \bar{u}), v) - f_0(s, y(s, \bar{u}), \bar{u}(s))\} \\
+ z_0 \int_s^{\bar{t}} \langle \partial_y f_0(\tau, y(\tau, \bar{u}), \bar{u}(\tau)), \\
S(\tau, s; \bar{u})\{f(s, y(s, \bar{u}), v) - f(s, y(s, \bar{u}), \bar{u}(s))\}\rangle d\tau \\
+ z_0 \langle \partial_y g_0(\bar{t}, y(\bar{t}, \bar{u})), S(\bar{t}, s; \bar{u})\{f(s, y(s, \bar{u}), v) - f(s, y(s, \bar{u}), \bar{u}(s))\}\rangle \\
+ \langle z, S(\bar{t}, s; \bar{u})\{f(s, y(s, \bar{u}), v) - f(s, y(s, \bar{u}), \bar{u}(s))\}\rangle \geq 0
\end{aligned}
\tag{4.2.12}
$$

a.e. in $0 \leq s \leq \bar{t}$. The manipulations to obtain (4.2.4) and (4.2.5) are essentially the same as those in Theorem 2.5.1 and are omitted. ∎

Remark 4.2.4. For an essential use of the transversality condition $z \in N_Y(\bar{y}(\bar{t}, \bar{u}))$, see the solution of the rocket landing problem in **4.7** and **4.8**.

We reformulate Theorem 4.2.1 in terms of the Hamiltonian $H(t, \mathbf{y}, \mathbf{z}, u) = H(t, y, \mathbf{z}, u) = \langle \mathbf{z}, \mathbf{f}(t, y, u) \rangle = z_0 f_0(t, y, u) + \langle z, f(t, y, u) \rangle$. The canonical equations for the state and the costate are

$$
\mathbf{y}'(t, \bar{u}) = \nabla_{\mathbf{z}} H(t, \mathbf{y}(t, \bar{u}), \bar{\mathbf{z}}(t), \bar{u}(t)), \qquad \mathbf{y}(0) = (0, \zeta)
\tag{4.2.13}
$$

$$
\bar{\mathbf{z}}'(s) = -\nabla_{\mathbf{y}} H(s, \mathbf{y}(s, \bar{u}), \bar{\mathbf{z}}(s), \bar{u}(s)), \qquad \bar{\mathbf{z}}(\bar{t}) = (z_0, z + z_0 \nabla_y g(\bar{t}, y(\bar{t}, \bar{u}))).
\tag{4.2.14}
$$

The first is (4.2.1) plus (4.2.2);[2] the second is (4.2.3). The minimum principle becomes

$$H(s, \mathbf{y}(s, \bar{u}), \bar{\mathbf{z}}(s), \bar{u}(s)) = \min_{v \in U} H(s, \mathbf{y}(s, \bar{u}), \bar{\mathbf{z}}(s), v) \qquad (4.2.15)$$

almost everywhere in $0 \le s \le \bar{t}$.

Compare Theorem 4.2.1 with Theorem 2.5.1 (for simplicity, take $g_0 = 0$). It is astonishing that, in spite of the enormously simpler nature of the targetless problem, the only difference in the minimum principle is that the final condition for the adjoint vector is known to be $(1, 0)$, while in Theorem 4.2.1 it is the unknown vector (z_0, z). This loss of information is not as bad as it seems, since even in Theorem 2.5.1 the adjoint variational equation involves the unknown optimal control $\bar{u}(s)$; explicit knowledge of the final condition does not make the costate explicit.

Example 4.2.5. Let $f(t, y, u)$, $f_0(t, y, u)$ be continuous in all three variables and continuously differentiable with respect to y in $[0, T] \times \mathbb{R}^m \times U$, where $U \subseteq \mathbb{R}^k$ is closed and bounded, and let $C_{ad}(0, T; U)$ be the set of all measurable $\mathbb{R}^k$-valued functions taking values in U almost everywhere. Then $C_{ad}(0, T; U)$, $f(t, y, u)$, $f_0(t, y, u)$ satisfy all the assumptions of Theorem 4.2.1.

4.3. Optimal Capture of a Wandering Particle in Fixed Time, I

We look at a variant of the problem in **3.8**. This time the control interval $0 \le t \le \bar{t}$ is fixed, and we minimize

$$y_0(\bar{t}, u) = \int_0^{\bar{t}} |u(\tau)| d\tau. \qquad (4.3.1)$$

The control constraint is the same:

$$|u(t)| \le 1, \qquad (4.3.2)$$

and $C_{ad}(0, \bar{t}, U)$ consists of all measurable functions satisfying (4.3.2). The first order system is

$$x'(t) = y(t), \qquad y'(t) = u(t) \qquad (4.3.3)$$

with initial condition $(x(0), y(0)) = (\xi, \eta)$: the target condition is

$$x(\bar{t}) = y(\bar{t}) = 0. \qquad (4.3.4)$$

---

[2] At least when $g_0 = 0$. In the general case, since g_0 acts only at the final point of the trajectory we may replace $f_0(t, y, u)$ by $f_0(t, y, u) + \delta(\bar{t} - t)g_0(t, y)$.

As in **3.8**, existence of optimal controls is guaranteed by Theorem 2.2.6 if we can drive to equilibrium in time $\bar{t}$ with a control satisfying (4.3.2). Obviously, this will not be possible if (ξ, η) is too far from the origin; precise conditions will be obtained in Lemma 4.3.5 below. We get a provisional necessary condition integrating the second equation (4.3.3),

$$y(t) = \eta + \int_0^t u(\tau) d\tau. \tag{4.3.5}$$

Since $y(\bar{t}) = 0$,

$$|\eta| \le \int_0^{\bar{t}} |u(\tau)| d\tau = y_0(\bar{t}, u) \le \bar{t}. \tag{4.3.6}$$

Applying the maximum principle, we call $(\mu, p(t), q(t))$ the costate; since we use the maximum (rather than minimum) principle, $\mu \le 0$.[1] The Hamiltonian is $H(x, y, \mu, p, q, u) = \mu|u| + py + qu$ and the canonical equations for the adjoint vector are

$$p' = 0, \qquad q' = -p, \tag{4.3.7}$$

so that

$$q(s) = as + b. \tag{4.3.8}$$

Eliminating terms not depending on u, the maximum principle is

$$\mu|\bar{u}(s)| + q(s)\bar{u}(s) = \max_{|v| \le 1}(\mu|v| + q(s)v) \tag{4.3.9}$$

with $(\mu, p(\bar{t}), q(\bar{t})) \ne 0$; this last condition is equivalent to

$$(\mu, a, b) \ne 0. \tag{4.3.10}$$

For $0 \le v \le 1$, $\mu|v| + q(s)v = (\mu + q(s))v$, which reaches its maximum at $v = 1$ if $q(s) + \mu > 0$, at $v = 0$ if $q(s) + \mu < 0$. On the other hand, if $-1 \le v \le 0$, $\mu|v| + q(s)v = (-\mu + q(s))v$, which reaches its maximum at $v = -1$ if $q(s) - \mu < 0$, at $v = 0$ if $q(s) - \mu > 0$. Accordingly, we have

$$\bar{u}(s) = \begin{cases} \text{sign}(q(s) - \mu) & \text{if } |q(s)| > |\mu| \\ 0 & \text{if } |q(s)| < |\mu|. \end{cases} \tag{4.3.11}$$

If $a \ne 0$ then $q(s)$ is not constant and the optimal control $\bar{u}(\cdot)$ must have one of the following nine forms:

$$\bar{u}(t) = -1 \qquad \bar{u}(t) = 0 \qquad \bar{u}(t) = 1 \quad (0 \le t \le \bar{t}) \tag{4.3.12}$$

---

[1] Replacing minimum by maximum not only changes the sign of μ but also that of the final condition $z = (p(\bar{t}), q(\bar{t}))$, so that z belongs to $-N_Y(\bar{y}) = -N_Y((x(\bar{t}), y(\bar{t})))$ rather than to $N_Y(\bar{y})$. This makes no difference for a point target problem since $N_Y(\bar{y})$ is the whole space. The only significant information is that $(\mu, z) \ne 0$.

$$\bar{u}(t) = \begin{cases} 1 \\ 0 \end{cases} \quad \bar{u}(t) = \begin{cases} -1 \\ 0 \end{cases} \quad \bar{u}(t) = \begin{cases} 0 \\ 1 \end{cases} \quad \bar{u}(t) = \begin{cases} 0 & (0 \le t \le t_1) \\ -1 & (t_1 < t \le \bar{t}) \end{cases} \quad (4.3.13)$$

$$\bar{u}(t) = \begin{cases} -1 \\ 0 \\ 1 \end{cases} \quad \bar{u}(t) = \begin{cases} 1 & (0 \le t \le t_1) \\ 0 & (t_1 < t \le t_2) \\ -1 & (t_2 < t \le \bar{t}) \end{cases} \quad (4.3.14)$$

where the switching times t_1, t_2 are unknown. In case $\mu = 0$, the switching times coalesce; thus, we may add for clarity the following controls to the list:

$$\bar{u}(t) = \begin{cases} -1 \\ 1 \end{cases} \quad \bar{u}(t) = \begin{cases} 1 & (0 \le t \le t_1) \\ -1 & (t_1 < t \le \bar{t}). \end{cases} \quad (4.3.15)$$

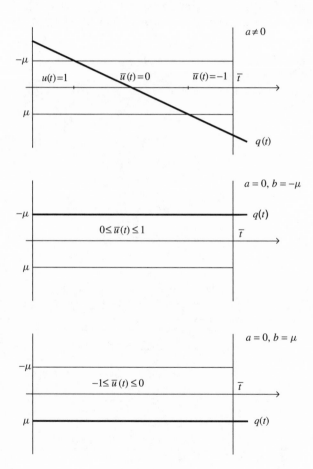

Figure 4.1.

If $a = 0$, we have $q(t) \equiv b$. This leads to controls of the form (4.3.12), except if $|b| = |\mu|$. If $b = \mu$ we must maximize $\mu(|v| + v)$ in $|v| \le 1$. Since $\mu < 0$, any v

with $-1 \leq v \leq 0$ will do the job. Similarly, if $b = -\mu$ we maximize $\mu(|v| - v)$ and the corresponding condition is $0 \leq v \leq 1$. Thus we have to add the following two (numerous families of) controls:

$$0 \leq \bar{u}(t) \leq 1 \qquad -1 \leq \bar{u}(t) \leq 0 \quad (0 \leq t \leq \bar{t}) \qquad (4.3.16)$$

to the list of candidates. Note that the controls in (4.3.12) and (4.3.13) are particular cases of (4.3.16), thus we can eliminate them from consideration and restrict ourselves to the families (4.3.14) and (4.3.16) (controls (4.3.15) are degenerate cases of the first). Even after weeding out redundancies, the list of aspirants is uncomfortably large. We will identify the optimal controls in the end, but not on the only basis of the maximum principle.

We figure out the trajectories corresponding to the possible values $-1, 0, 1$ of optimal controls; some of the work has been done in **3.8**. For $\bar{u}(t) = -1$,

$$y = -t + K_-, \qquad x = -\frac{t^2}{2} + K_- t + L_- = -\frac{y^2}{2} + C_-, \qquad (4.3.17)$$

K_-, L_-, C_- constants. For $\bar{u}(t) = 0$,

$$y = K_0, \qquad x = K_0 t + L_0 = yt + L_0. \qquad (4.3.18)$$

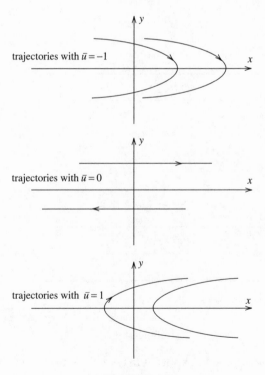

trajectories with $\bar{u} = -1$

trajectories with $\bar{u} = 0$

trajectories with $\bar{u} = 1$

Figure 4.2.

If $y \neq 0$, these are horizontal lines over which we move from left to right on the upper-half plane, from right to left on the lower-half plane. For $y = 0$, every point on the x-axis is a (constant) trajectory. Finally, for $\bar{u}(t) = 1$,

$$y = t + K_+, \qquad x = \frac{t^2}{2} + K_+ t + L_+ = \frac{y^2}{2} + C_+, \qquad (4.3.19)$$

K_+, L_+, C_+ constants. For controls of the form (4.3.16) we have the following result.

Lemma 4.3.1. *Let* $(x(t), y(t))$ *be the trajectory of* (4.3.3) *starting at* (ξ, η) *corresponding to a control* $u(t)$ *with* $0 \leq u(t) \leq 1$. *Then the graph of* $(x(t), y(t))$ *lies in the region bounded by the trajectories with the same starting point corresponding to controls* $u(t) = 0$ *and* $u(t) = 1$. *A similar result holds for trajectories with* $-1 \leq u(t) \leq 0$.

The result is obvious if $\bar{u}(\cdot)$ alternates between the values 0 and 1 in a finite number of subintervals. We may use the fact that such controls are weakly dense in $L^2(0, \bar{t})$ and the formulas (4.3.5) and

$$x(t) = \xi + \eta t + \int_0^t (t - \tau)u(\tau)d\tau \qquad (4.3.20)$$

to pass to the limit. ∎

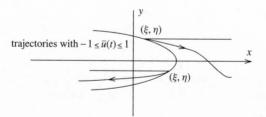

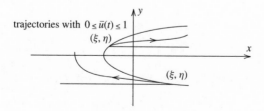

Figure 4.3.

Call a control $u(t)$ **flip-flop** in the interval $0 \leq t \leq t_0$ if $u(t) = -1$ in $0 \leq t \leq t_0/2$, $u(t) = 1$ in $t_0/2 \leq t \leq t_0$, or $u(t) = 1$ in $0 \leq t \leq t_0/2$, $u(t) = -1$ in $t_0/2 \leq t \leq t_0$

(flip-flop is a particular case of bang-bang; the control jumps *once* between two values *and* stays at each value the same time). In view of (4.3.5), driving with a flip-flop control in an interval $0 \leq t \leq t_0$ returns us to the same y coordinate; thus, we can also arrive at the same final point by coasting ($u(t) = 0$ in $0 \leq t \leq t_0$).

Lemma 4.3.2. *Coasting takes more time than flip-flop driving.*

Proof. It follows from (4.3.17) and (4.3.19) that the time elapsed over a trajectory where $u(t) = 1$ or $u(t) = -1$ equals the length of the projection of the trajectory on the y axis. Accordingly, the time elapsed on the flip-flop trajectory is $2(y_2 - y_1)$. On the other hand, in view of (4.3.14), the time elapsed during coasting is $2(x_2 - x_1)/y_1 = (y_2^2 - y_1^2)/y_1$. Our contention is then that $(y_2^2 - y_1^2)/y_1 > 2(y_2 - y_1)$ which reduces to the obvious inequality $y_2 + y_1 > 2y_1$. ∎

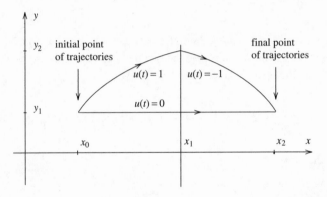

Figure 4.4.

Coasting trajectories $y = $ constant get slower as $y > 0$ decreases. The time elapsed on the coasting trajectory in Figure 4.4 is an increasing function of y_1 and tends to infinity as $y_1 \to 0+$. A symmetric observation applies to coasting trajectories with $y < 0$.

It follows immediately from (4.3.5) that

Lemma 4.3.3. *For any admissible control driving (ξ, η) to zero,*

$$y_0(\bar{t}, u) \geq |\eta|. \tag{4.3.21}$$

If $u(\cdot)$ is of the first (resp. the second) form (4.3.16), then

$$y_0(\bar{t}, u) = -\eta \quad (\text{resp. } y_0(\bar{t}, u) = \eta). \tag{4.3.22}$$

We divide below the aspirants $u(\cdot)$ to optimality in Class 1, given by (4.3.14) and their degenerate cousins (4.3.15), and Class 2, given by (4.3.16). Lemma 4.3.3 implies

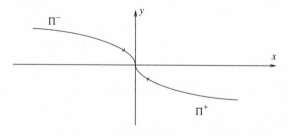

Figure 4.5.

Corollary 4.3.4. *Any control of Class 2 driving (ξ, η) to the origin is optimal.*

Final approach to the origin (from any point in the plane) under a control of Class 1 must be along one of the two half-parabolas.

$$\Pi^+ = \{(x, y); x = y^2/2 \; (y \leq 0)\}, \qquad \Pi^- = \{(x, y); x = -y^2/2 \; (y \geq 0)\}.$$
$$(4.3.22)$$

Note, however, that the option also exists of getting to $(0, 0)$ early and resting there until the deadline $\bar{t}$. Of course, if this is done with a control of Class 1, the resulting control would have a zero interval at the end and thus would not belong to Class 1 or Class 2; on the other hand, if prior to arrival the control is of Class 2, the final zero interval does keep the control in Class 2.

Lemma 4.3.5. *Let (ξ, η) be an initial state such that the time optimal problem (with constraint (4.3.2)) has a solution with $\tilde{t} =$ optimal time $\leq \bar{t}$. Then there exists an optimal control for (4.3.1)-(4.3.2) in time $\bar{t}$. If $\tilde{t} > \bar{t}$ then the problem has no solution; there are no admissible controls driving to the target in time $\bar{t}$.*

Proof. If we can reach the origin in time $\tilde{t} \leq \bar{t}$, we can rest there until time $\bar{t}$. As pointed out above, the resulting control may not belong to any of the two classes but satisfies the control constraint and drives (ξ, η) to the origin. This is enough for application of Theorem 2.2.6.[2]   ∎

We figure out the optimal controls for the lower half-plane $\eta \leq 0$. If we are below Π^-, it is clear from Lemma 4.3.1 that no control of Class 2 will drive us to the origin, thus the optimal control must be of Class 1. We obtain it as follows. First, we drive time-optimally to the origin according to the instructions in **3.8**. It we arrive to the target too early, we insert a coasting segment by "flattening out" the top of the trajectory (Figure 4.6). Move the coasting segment down; by

---

[2] It must be shown that the cost functional (4.3.1) is weakly lower semicontinuous. This can be done applying Example 2.2.16 or directly.

Lemma 4.3.2 the driving time increases without bounds, thus it will eventually equal $\bar{t}$. Optimality of this control follows from the fact that no other control, either of Class 1 or of Class 2 will drive us to the origin at all (this needs a moment's thought).

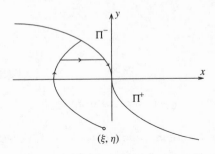

Figure 4.6.

The flattening out recipe also works in the region between Π^+ and the positive x-axis (with everything upside down; see the first Figure 4.7), but we can only raise the coasting segment until its y-coordinate equals η. It we are still too early at the target, we try another route; we initially coast to the entry parabola Π^+ ($\bar{u}(t) = 0$), then drive along Π^+ ($\bar{u}(t) = 1$) and, if we get to the target at time $t_0 < \bar{t}$ we rest there until time $\bar{t}$. This control is of Class 2 and optimality results from Corollary 4.3.4.

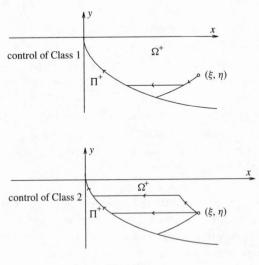

Figure 4.7.

If $t_0 < \bar{t}$ there are other ways of driving optimally, such as the one indicated in the second Figure 4.7, where the coasting segment is slightly raised (with a corresponding slight increase of the coasting time). Of course, scores of other controls of Class 2 driving optimally to the origin can be constructed simply by looking at (4.3.5)-(4.3.20) (with $t = \bar{t}$) as a system of integral equations and solving with $0 \le u(t) \le 1$.

A completely symmetric argument takes case of the upper half-plane.

Remark 4.3.6. Unlike in all previous examples, there is *no* feedback law

$$\bar{u}(t) = P(\bar{x}(t), \bar{y}(t))$$

giving the optimal control in terms of the optimal trajectory. This can be easier seen for (ξ, η) in the lower half-plane below Π^-, where optimal trajectories are unique; there are different optimal trajectories that cross at points where one is driven by $u(t) = 1$, the other by $u(t) = 0$.

4.4. Singular Intervals and Singular Arcs

The maximum principle applied to the problem in **4.3** fails to give information as precise as that in the time optimal capture problem in **3.8**. However, the information is sufficient to find the solutions, and more precision is precluded by lack of uniqueness. We look at a problem where optimal controls exist and are unique, but the maximum principle doesn't see them.

Consider minimizing

$$x_0(t, u) = \int_0^t x(t, u)^2 dt$$

among all solutions of the initial value problem

$$x'(t) = u(t), \qquad x(0) = \alpha$$

$(0 \le \alpha \le 1)$ in the interval $0 \le t \le 1$, with control constraint

$$|u(t)| \le 1$$

and no target condition. All assumptions in Theorem 2.2.6 are satisfied, thus an optimal control exists. This is one of the problems without target condition in **2.5**, so that the final condition for the second component p of the costate (μ, p) is $p(1) = 0$. If the maximum (rather than minimum) principle is used, $\mu = -1$. The Hamiltonian is then $H(x, p, q, u) = -x^2 + pu$ and the canonical equation for p is

$$p' = 2x.$$

The maximum principle says

$$p(s)\bar{u}(s) = \max_{|v| \le 1} p(s)v.$$

By inspection (not from the maximum principle) the optimal control is

$$\bar{u}(t) = \begin{cases} -1 & (0 \le t \le \alpha), \\ 0 & (\alpha < t \le 1), \end{cases}$$

corresponding to the solution

$$x(t) = \begin{cases} \alpha - t & (0 \le t \le \alpha), \\ 0 & (\alpha < t \le 1). \end{cases}$$

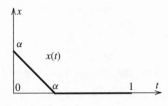

Figure 4.8.

For this $\bar{u}$ we have

$$p(t) = \begin{cases} -(\alpha - t)^2 & (0 \le t \le \alpha) \\ 0 & (\alpha < t \le 2) \end{cases}$$

so that, although the maximum principle characterizes correctly the optimal control in $0 \le t \le \alpha$, it says nothing at all in the interval $\alpha < t \le 1$. This interval is called a **singular interval** of the solution; the solution itself in the singular interval is a **singular arc**. In case $\alpha = 0$, the singular interval is the entire control interval.

4.5. Optimal Capture of a Wandering Particle in Fixed Time, II

Formulas (3.9.9) for the linear oscillator and (3.9.21) for nonlinear oscillators (the latter in combination with the argument on the zeros of $q(t)$) show that the time optimal controls in **3.9** have no singular arcs and are bang-bang with a finite number of switchings. The same is true of the systems in **3.10** if, say, $U = [-1, 1]$ and the rank condition (3.10.9) is satisfied. The conjecture that controls optimal for cost functionals other than time must have a finite number of switching points is natural but false, as the counterexample below (due to Fuller [1963]) shows. The equation and the control constraint are the same as in **4.3**:

$$x''(t) = u(t), \tag{4.5.1}$$

$$|u(t)| \le 1. \tag{4.5.2}$$

So is the admissible control space $C_{ad}(0, T; U)$. The objective is to drive the particle to equilibrium in fixed time $\bar{t}$ from $(x(0), x'(0)) = (\xi, \eta)$ minimizing the cost functional

$$y_0(\bar{t}, u) = \int_0^{\bar{t}} x(\tau, u)^2 d\tau \tag{4.5.3}$$

This is a linear-quadratic problem related to (2.7.5)-(2.7.6) but with a target condition and $c(t) \equiv 0$. The first order system is

$$x'(t) = y(t), \qquad y'(t) = u(t), \tag{4.5.4}$$

with initial condition $(x(0), y(0)) = (\xi, \eta)$ and target condition $(x(\bar{t}), y(\bar{t})) = 0$. Theorem 2.2.6 shows that an optimal control exists if we can drive (ξ, η) to zero in time $\bar{t}$.

The Hamiltonian is $H(x, y, \mu, p, q, u) = \mu x^2 + py + qu$ and the canonical equations for p, q,

$$p'(t) = -2\mu x(t), \qquad q'(t) = -p(t), \tag{4.5.5}$$

which give

$$q''(t) = 2\mu x(t). \tag{4.5.6}$$

Eliminating terms not depending on v, the maximum principle is

$$q(s)\bar{u}(s) = \max_{|v| \leq 1} q(s)v \tag{4.5.7}$$

with $(\mu, p(\bar{t}), q(\bar{t})) \neq 0$, $\mu \leq 0$. Hence,

$$\bar{u}(t) = \operatorname{sign} q(t) \tag{4.5.8}$$

where $q(t) \neq 0$.

Lemma 4.5.1. Let $(x(t, \bar{u}), y(t, \bar{u}))$ be a trajectory of (4.5.4) in $0 \leq t \leq \bar{t}$ corresponding to a control $\bar{u}(t) \in C_{ad}(0, \bar{t}; U)$ satisfying (4.5.8), with $q(t)$ given by (4.5.6). Assume that there exists an increasing sequence of zeros $\{t_n\}$ of the switching function $q(t)$ with $t_n \to \bar{t}$. Then $x(\bar{t}) = y(\bar{t}) = 0$.

Proof. If $\mu = 0$ then (4.5.6) mandates that $q(t) = at + b$ with $(\mu, p(\bar{t}), q(\bar{t})) = (0, a\bar{t} + b, -a) \neq 0$; thus, either $a \neq 0$ or $b \neq 0$. Accordingly, the sequence $\{t_n\}$ can only exist if $\mu < 0$. If it does exist, select a sequence $\{t_n^1\}$, $t_n \leq t_n^1 \leq t_n$ with $q'(t_n^1) = 0$. Eliminating, say, the terms with odd subindex, we may assume that $t_n^1 < t_{n+1}^1$. Proceeding in the same way, we construct increasing sequences $\{t_n^2\}$ and $\{t_n^3\}$, both with limit $\bar{t}$ and with $q''(t_n^2) = 0$, $q'''(t_n^3) = 0$ (note that (4.5.6) and (4.5.1) imply that $q(t)$ is three times continuously differentiable). It follows that $x(\bar{t}) = q''(\bar{t})/2\mu = 0$ and $y(\bar{t}) = x'(\bar{t}) = q'''(\bar{t})/2\mu = 0$. ∎

Lemma 4.5.2. *Let $(x(t, \bar{u}), y(t, \bar{u}))$ be a trajectory of (4.5.1) in $0 \le t \le \bar{t}$ corresponding to a control $\bar{u}(t) \in C_{ad}(0, \bar{t}; U)$ satisfying (4.5.8) and (4.5.6). Assume that $\mu \ne 0$. Then $\bar{u}(t)$ drives optimally $(x(0, \bar{u}), y(0, \bar{u}))$ to $(x(\bar{t}, \bar{u}), y(\bar{t}, \bar{u}))$.*

Proof. Let $u(\cdot) \in C_{ad}(0, \bar{t}; U)$ be such that $(x(0, u), y(0, u)) = (x(0, \bar{u}), y(0, \bar{u}))$, $(x(\bar{t}, u), y(\bar{t}, u)) = (x(\bar{t}, \bar{u}), y(\bar{t}, \bar{u}))$. Then

$$y_0(\bar{t}, u) - y_0(\bar{t}, \bar{u}) = 2 \int_0^{\bar{t}} x(\tau, \bar{u})(x(\tau, u) - x(\tau, \bar{u}))d\tau$$

$$+ \int_0^{\bar{t}} (x(\tau, u) - x(\tau, \bar{u}))^2 d\tau. \tag{4.5.9}$$

We have $x(t, \bar{u}) = q''(t)/2\mu$. Integrating by parts twice,

$$2 \int_0^{\bar{t}} x(\tau, \bar{u})(x(\tau, u) - x(\tau, \bar{u}))d\tau = \frac{1}{\mu} \int_0^{\bar{t}} q''(\tau)(x(\tau, u) - x(\tau, \bar{u}))d\tau$$

$$= -\frac{1}{\mu} \int_0^{\bar{t}} q(\tau)(\bar{u}(\tau) - u(\tau))d\tau = -\frac{1}{\mu} \int_0^{\bar{t}} q(\tau)(\text{sign}\, q(\tau) - u(\tau))d\tau$$

$$= -\frac{1}{\mu} \int_0^{\bar{t}} (|q(\tau)| - q(\tau)u(\tau))d\tau \ge 0.$$

since $\mu < 0$. This shows that $y_0(\bar{t}, u) \ge y_0(\bar{t}, \bar{u})$ as claimed. Note that we have used the fact that $q(\tau)\bar{u}(\tau) = q(\tau)\text{sign}\, q(\tau) = |q(\tau)|$ even if $q(\tau) = 0$, where (4.5.8) may not hold. ∎

Lemma 4.5.1 and Lemma 4.5.2 give a roundabout way to construct optimal controls. Let $q(t)$ be a three times continuously differentiable function with absolutely continuous third derivative satisfying the fourth order initial value problem

$$q''''(t) = -2 \,\text{sign}\, q(t),$$
$$q(0) = 0, \quad q'(0) = \zeta, \quad q''(0) = -2\xi, \quad q'''(0) = -2\eta \tag{4.5.10}$$

in $t \ge 0$. Assuming that $q(t)$ is zero at most in a null set, define

$$\bar{u}(t) = \text{sign}\, q(t), \tag{4.5.11}$$

$$y(t) = \eta + \int_0^t \bar{u}(\tau)d\tau, \qquad x(t) = \xi + \int_0^t y(\tau)d\tau. \tag{4.5.12}$$

Then $x(t) = x(t, \bar{u})$, $y(t) = y(t, \bar{u})$ and $x(0) = \xi$, $y(0) = \eta$; moreover, integrating (4.5.10) twice and minding the last two initial conditions we obtain (4.5.6) with $\mu = -1$ so that $\bar{u}(\cdot)$ satisfies the maximum principle and, in view of Lemma

4.5.2, it drives optimally (ξ, η) to $(x(\bar{t}, \bar{u}), y(\bar{t}, \bar{u}))$ for every $\bar{t} > 0$. Finally, if $q(t)$ has a sequence of zeros in $[0, \bar{t})$ tending to $\bar{t}$, then Lemma 4.5.1 will imply that $(x(t, \bar{u}), y(t, \bar{u})) = (0, 0)$ and we have the counterexample. Note that the initial condition ζ in (4.5.10) is totally arbitrary; this is decisive in obtaining the right $q(t)$.

Assume $\zeta < 0$. Then $q(t) < 0$ in the interval $0 < t < t_1^- =$ smallest positive zero of $q(t)$ so that $\bar{u}(t) = -1$ there, and

$$y(t) = -t + \eta, \qquad x(t) = -\frac{t^2}{2} + \eta t + \xi, \tag{4.5.13}$$

$$q'_-(t) = \frac{t^3}{3} - \eta t^2 - 2\xi t + \zeta, \qquad q_-(t) = \frac{t^4}{12} - \frac{\eta t^3}{3} - \xi t^2 + \zeta t, \tag{4.5.14}$$

where the first equality in (4.5.14) comes from the second in (4.5.13) and the equation (4.5.6), with $\mu = -1$; the subindex $-$ on $q(t)$ indicates association with the value $\bar{u}(t) = -1$ of the control. Note that the third and fourth initial conditions in (4.5.10) are a consequence of the initial conditions for $x(t)$ and $y(t)$.

On the other hand, if $\zeta > 0$ then $q(t) > 0$ in the interval $0 < t < t_1^+ =$ smallest positive zero of $q(t)$, so that $\bar{u}(t) = 1$ there and

$$y(t) = t + \eta, \qquad x(t) = \frac{t^2}{2} + \eta t + \xi, \tag{4.5.15}$$

$$q'_+(t) = -\frac{t^3}{3} - \eta t^2 - 2\xi t + \zeta, \qquad q_+(t) = -\frac{t^4}{12} - \frac{\eta t^3}{3} - \xi t^2 + \zeta t, \tag{4.5.16}$$

with subindex $+$ correspondingly explained. The functions $q(t)$ and $\bar{u}(t)$ in (4.5.11) will be constructed together by pieces as follows. Starting with $\zeta < 0$, we define $\bar{u}(t) = -1, q(t) = q_-(t)$ in $0 < t < t_1^- =$ smallest positive zero of $q_-(t)$. Assuming that $q'(t_1^-) > 0$, we use

$$\zeta = q'(t_1^-), \qquad \xi = x(t_1^-, \bar{u}) = -q''_-(t_1^-)/2, \qquad \eta = y(t_1^-, \bar{u}) = -q'''_-(t_1^-)/2$$

in (4.5.16) and define $\bar{u}(t) = 1, q(t) = q_1(t_1^- + t)$ in $t_1^- < t < t_1^- + t_1^+$, where $t_1^+ =$ smallest positive zero of $q_+(t)$. Assuming then that $q'(t_1^- + t_1^+) < 0$, the cycle recommences and goes on forever. The switching points are $t_1^-, t_1^- + t_1^+$, $t_1^- + t_1^+ + t_2^-, t_1^- + t_1^+ + t_2^- + t_2^+, \ldots$ so that if $\Sigma(t_n^- + t_n^+) < \infty$ we are done. It is far from clear (but shown below) that all of this can actually work.

Example 4.5.3. Let

$$h = \sqrt{\frac{\sqrt{33} - 1}{24}}, \qquad t_1 = \left(1 + \sqrt{\frac{1 - 2h}{1 + 2h}}\right)|\eta|. \tag{4.5.17}$$

(a) Let $\eta > 0$ and

$$\xi = -h\eta^2, \qquad \zeta = -h^2\eta^3. \tag{4.5.18}$$

Then $t_1 = t_1^-$ is the only zero of $p_-(t)$ in $t > 0$ and

$$y(t_1, \bar{u}) = -\sqrt{\frac{1 - 2h}{1 + 2h}} \; \eta < 0,$$

$$x(t_1, \bar{u}) = hy(t_1, \bar{u})^2, \qquad q'(t_1, \bar{u}) = -h^2 y(t_1, \bar{u})^3. \tag{4.5.19}$$

(b) Let $\eta < 0$ and

$$\xi = h\eta^2, \qquad \zeta = -h^2\eta^3. \tag{4.5.20}$$

Then $t_1 = t_1^+$ is the only zero of $p_+(t)$ in $t > 0$ and

$$y(t_1, \bar{u}) = -\sqrt{\frac{1 - 2h}{1 + 2h}} \; \eta > 0,$$

$$x(t_1, \bar{u}) = -hy(t_1, \bar{u})^2, \qquad q'(t_1, \bar{u}) = -h^2 y(t_1, \bar{u})^3. \tag{4.5.21}$$

We invite the reader to check the algebra with the help of Fuller [1963] or of a computer algebra program.

Switchings in the trajectory occur at intersections with the curve

$$x = -hy|y|. \tag{4.5.22}$$

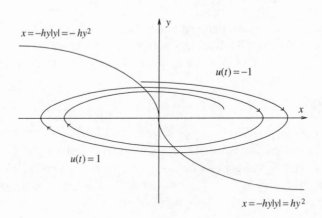

Figure 4.9.

By virtue of the second equality (4.5.17), (4.5.19), and (4.5.21) the time elapsed in the first bang-bang cycle is

$$t_1^- + t_1^+ = \left(1 + \sqrt{\frac{1 - 2h}{1 + 2h}}\right)^2 \left(\frac{1 - 2h}{1 + 2h}\right) |\eta|$$

(note that the parameter η for the second half of the cycle is $y(t_1, \bar{u})$, given by (4.5.19)). The same formula holds for all cycles, hence $t_n^+ + t_n^+ = k^n |\eta|$ with

$$k = \left(1 + \sqrt{\frac{1 - 2h}{1 + 2h}}\right)^2 \left(\frac{1 - 2h}{1 + 2h}\right) = 0.090\ldots \qquad (4.5.23)$$

If $\{t_n\}$ is the sequence of switching points, we have $\bar{t} = \lim_{n \to \infty} t_n < \infty$, so that $\bar{u}(t)$ has an infinite number of switchings in the finite interval $[0, \bar{t})$.

4.6. The Minimum Principle for Problems with Variable Terminal Time

We study optimal control problems for the autonomous system

$$y'(t) = f(y(t), u(t)), \qquad y(0) = \zeta \qquad (4.6.1)$$

with target condition $y(\bar{t}) \in Y$ and autonomous cost functional

$$y_0(t, u) = \int_0^t f_0(y(\sigma, u), u(\sigma)) d\sigma + g_0(y(t, u)) \qquad (4.6.2)$$

in a variable time interval. This control system will be named S.

The control space $C_{\text{ad}}(0, T; U)$ has the **reparametrization property** if for every $\bar{t} \in (0, T)$ there exists $\varepsilon > 0$ such that, if $t(\cdot)$ is an absolutely continuous function in $0 \leq \tau \leq \bar{t}$ with

$$|t'(\tau) - 1| \leq \varepsilon \quad (0 \leq \tau \leq \bar{t}), \qquad (4.6.3)$$

then $u(t'(\cdot)) \in C_{\text{ad}}(0, t(\bar{t}); U)$ if and only if $u(\cdot) \in C_{\text{ad}}(0, \bar{t}; U)$.

Theorem 4.6.1. (*Addendum to Theorem* 4.2.1) *Let the assumptions of Theorem 4.2.1 be satisfied and let $C_{\text{ad}}(0, T; U)$ have the reparametrization property. Then, if $\bar{u}(\cdot)$ is an optimal control for the free arrival time problem, there exists a multiplier (z_0, z) such that all the statements in Theorem 4.2.1 are satisfied and, in addition, the Hamiltonian vanishes:*

$$z_0 f_0(y(s, \bar{u}), \bar{u}(s)) + \langle \bar{z}(s), f(y(s, \bar{u}), \bar{u}(s)) \rangle = 0 \qquad (4.6.4)$$

a.e. in $0 \leq s \leq \bar{t}$.

Proof. Let ε be the parameter in (4.6.3) corresponding to $\bar{t} = $ arrival time of $\bar{u}(\cdot)$. Let $u_0(\cdot)$ be a measurable scalar function satisfying $|u_0(\tau)| \leq \varepsilon$ in $0 \leq \tau \leq \bar{t}$. Define

$$t(\tau, u_0) = \int_0^t (1 + u_0(\tau)) d\tau \quad (0 \leq \tau \leq \bar{t}). \qquad (4.6.5)$$

Then $t(\cdot, u_0)$ is an absolutely continuous function in $0 \le t \le \bar{t}$ with derivative $t'(\tau, u_0) = 1 + u_0(\tau)$ satisfying (4.6.3); thus, by the (Lebesgue style) inverse function theorem, the inverse map $\tau = \tau(t, u_0)$ is as well absolutely continuous in $0 \le t \le t(\bar{t}, u_0)$ Natanson [1955, Chapter IX]. If $y(t, u)$ satisfies (4.6.1) in $0 \le t \le t(\bar{t}, u_0)$ for some $u(\cdot) \in C_{\mathrm{ad}}(0, t(\bar{t}, u_0); U)$, then $y(t(\tau, u_0), u)$ satisfies

$$y'(t(\tau, u_0), u) = (1 + u_0(\tau)) f(y(t(\tau, u_0), u), u(t(\tau, u_0)))$$

in $0 \le \tau \le \bar{t}$ as a function of τ. Moreover, changing variables with $t(\tau, u_0)$ given by (4.6.5),

$$y_0(t(\bar{t}, u_0), u) = \int_0^{t(\bar{t}, u_0)} f_0(y(t, u), u(t)) dt + g_0(y(t(\bar{t}, u_0)))$$

$$= \int_0^{\bar{t}} (1 + u_0(\tau)) f_0(y(t(\tau, u_0)), u(t(\tau, u_0))) d\tau + g_0(y(t(\bar{t}, u_0))).$$

This motivates the introduction of a second control system **S**,

$$\mathbf{y}'(\tau) = \mathbf{f}(\mathbf{y}(\tau, \mathbf{u}), \mathbf{u}(\tau))$$

$$= (1 + u_0(\tau)) f(\mathbf{y}(\tau, \mathbf{u}), u_1(\tau)), \quad \mathbf{y}(0) = \zeta, \qquad (4.6.6)$$

whose trajectories $\mathbf{y}(\tau, \mathbf{u})$ live in the same state space $E = \mathbb{R}^m$. Time in this system is τ, and the (fixed) control interval is $0 \le \tau \le \bar{t}$. The control set is $\mathbf{U} = [-\varepsilon, \varepsilon] \times U$, and the space $C_{\mathrm{ad}}(0, \bar{t}; \mathbf{U})$ of admissible controls consists of all functions of the form

$$\mathbf{u}(\tau) = (u_0(\tau), u_1(\tau)),$$

where $u_1(\cdot) \in C_{\mathrm{ad}}(0, \bar{t}; U)$ and $u_0(\cdot)$ is a measurable scalar function with

$$|u_0(\tau)| \le \varepsilon \quad (0 \le \tau \le \bar{t}).$$

The target condition $\mathbf{y}(\bar{t}, \mathbf{u}) \in Y$ is the same and the cost functional

$$\mathbf{y}_0(\bar{t}, \mathbf{u}) = \int_0^{\bar{t}} \mathbf{f}_0(\mathbf{y}(\tau, \mathbf{u}), \mathbf{u}(\tau)) d\tau + g_0(\mathbf{y}(\bar{t}, \mathbf{u}))$$

$$= \int_0^{\bar{t}} (1 + u_0(\tau)) f_0(\mathbf{y}(\tau, \mathbf{u}), u_1(\tau)) d\tau + g_0(\mathbf{y}(\bar{t}, \mathbf{u})). \quad (4.6.7)$$

Lemma 4.6.2. *Let $\mathbf{y}(\tau, \mathbf{u})$ be a trajectory of* **S** *in $0 \le \tau \le \bar{t}$ with $\mathbf{u}(\cdot) = (u_0(\cdot), u_1(\cdot)) \in \mathbf{C}_{\mathrm{ad}}(0; \bar{t}; \mathbf{U})$, and let $t(\tau, u_0)$ be given by (4.6.5), $\tau(t, u_0)$ the function inverse to $t(\tau, u_0)$. Let*

$$u(t) = u_1(\tau(t, u_0)), \qquad y(t) = y(t, u) = \mathbf{y}(\tau(t, u_0), \mathbf{u}). \quad (4.6.8)$$

Then $y(t, u)$ is a trajectory of S in $0 \leq t \leq t_0 = t(\bar{t}, u_0)$. Moreover,

$$y_0(t_0, u) = \mathbf{y}_0(\bar{t}, \mathbf{u}).$$

Conversely, let $u(\cdot) \in C_{\text{ad}}(0, t_0; U)$, where

$$(1 - \varepsilon)\bar{t} \leq t_0 \leq (1 + \varepsilon)\bar{t}. \tag{4.6.9}$$

Then there exists a measurable function $u_0(\cdot)$ defined in $0 \leq \tau \leq \bar{t}$, satisfying $|u_0(\tau)| \leq \varepsilon$ and such that $t_0 = t(\bar{t}, u_0)$. Define

$$\mathbf{u}(\tau) = (u_0(\tau), u(t(\tau, u_0))), \qquad \mathbf{y}(\tau, \mathbf{u}) = y(t(\tau, u_0), u). \tag{4.6.10}$$

Then $\mathbf{y}(\tau, \mathbf{u})$ is a trajectory of S in $0 \leq \tau \leq \bar{t}$. Moreover,

$$\mathbf{y}_0(\bar{t}, \mathbf{u}) = y_0(t_0, u).$$

Finally,

$$y(0, u) = \mathbf{y}(0, \mathbf{u}), \qquad y(t_0, u) = \mathbf{y}(\bar{t}, \mathbf{u}). \tag{4.6.11}$$

Most of Lemma 4.5.2 has already been proved. If (4.6.9) holds, we may take $u_0(\cdot) \equiv (t_0 - \bar{t})/\bar{t}$. ∎

Lemma 4.6.3. *Let $\bar{u}(t)$ be an optimal control with arrival time $\bar{t}$ for the variable time optimal control problem for S, and*

$$\bar{\mathbf{u}}(\tau) = (0, \bar{u}(\tau)) \quad (0 \leq \tau \leq \bar{t}). \tag{4.6.12}$$

Then $\bar{\mathbf{u}}(\tau)$ is an optimal control for the fixed time optimal control problem for S in $0 \leq s \leq \bar{t}$.

Proof. Assume that our claim on optimality of $\bar{\mathbf{u}}(\cdot)$ is bogus. Then there exists another control $\mathbf{u}(\tau) = (u_0(\tau), u_1(\tau))$ in the same interval $0 \leq t \leq \bar{t}$ with

$$\mathbf{y}_0(\bar{t}, \mathbf{u}) < \mathbf{y}_0(\bar{t}, \bar{\mathbf{u}}).$$

By Lemma 4.6.2, the control (4.6.8) will drive the system S from ζ to the target set in some time t_0 and

$$y_0(t_0, u) < y_0(\bar{t}, \bar{u})$$

contradicting the alleged optimality of $\bar{u}$. ∎

End of proof of Theorem 4.6.1. We apply Theorem 4.2.1 to the auxiliary control problem S, all hypotheses readily verified. The adjoint variational equation of S

coincides with that of S for a control of the form $\bar{\mathbf{u}}(\cdot) = (0, \bar{u})$. The minimum principle (2.2.4) for $\mathbf{S}$ reads

$$z_0 f_0(y(s, \bar{u}), \bar{u}(s)) + \langle \bar{z}(s), f(y(s, \bar{u}), \bar{u}(s)) \rangle$$

$$= \min_{|v_0| \leq \varepsilon, v \in U} (1 + v_0)\{z_0 f_0(y(s, \bar{u}), v) + \langle \bar{z}(s), f(y(s, \bar{u}), v) \rangle\} \quad (4.6.13)$$

a.e. in $0 \leq s \leq \bar{t}$. Let $h(s) = z_0 f_0(y(s, \bar{u}), \bar{u}(s)) + \langle z(s), f(y(s, \bar{u}), \bar{u}(s)) \rangle$. Setting $v = \bar{u}(s)$ on the right-hand side of (4.6.13), we obtain $(1 + v_0)h(s) \geq h(s)$ a.e. for $|v_0| \leq \varepsilon$, which is absurd unless $h(s) = 0$. a.e. Since the expression in curly brackets on the right side must then be nonnegative, (4.6.13) implies the minimum principle (4.2.4).   ∎

Remark 4.6.4. Since ε can be taken as small as we wish, $\bar{u}$ need only be *locally* optimal with respect to arrival time.

Example 4.6.5. Let $f(y, u)$, $f_0(y, u)$ be continuous in y, u and continuously differentiable with respect to y in $[0, T] \times \mathbb{R}^m \times U$, where $U \subseteq \mathbb{R}^k$ is closed and bounded, and let $C_{\text{ad}}(0, T; U)$ be the set of all measurable $\mathbb{R}^k$-valued functions taking values in U almost everywhere. Then all assumptions in Theorem 4.6.1 are satisfied.

4.7. Fuel Optimal Soft Landing of a Space Vehicle: Existence and Identification of the Optimal Control

The control system is

$$h'(t) = v(t), \qquad v'(t) = -g + \frac{u(t)}{m(t)}, \qquad m'(t) = -Ku(t) \quad (4.7.1)$$

where the components of $(h(t), v(t), m(t))$ are, respectively, the instantaneous height, velocity, and mass of the vehicle (see **1.4**). Initial conditions are

$$h(0) = h_0 \geq 0, \qquad v(0) = v_0, \qquad m(0) = m_0 > 0. \quad (4.7.2)$$

The target conditions are $h(\bar{t}) = v(\bar{t}) = 0$ (soft landing); besides, the final mass of the rocket cannot be less than the "empty mass" m_e, thus

$$(h(\bar{t}), v(\bar{t}), m(\bar{t})) \in Y = \{0\} \times \{0\} \times [m_e, \infty). \quad (4.7.3)$$

The control set is $U = [0, R]$ and $C_{\text{ad}}(0, T; U)$ consists of all measurable U-valued functions. The cost functional is

$$y_0(t, u) = K \int_0^t u(\tau) d\tau, \quad (4.7.4)$$

and the problem contains two state constraints:

$$h(t) \geq 0, \qquad m(t) \geq m_e > 0 \quad (0 \leq t \leq \bar{t}). \qquad (4.7.5)$$

The second is automatic for any control satisfying the target condition (4.7.3), since $m(t)$ is a nonincreasing function of t. We ignore the first state constraint for the moment; it will be checked *a posteriori*.

Theorem 4.7.1. *Assume there exists an admissible control $u(\cdot) \in C_{\mathrm{ad}}(0, T; U)$ whose trajectory satisfies the target condition* (4.7.3). *Then there exists an optimal control $\bar{u}(\cdot)$.*

This follows from Theorem 2.2.6 with minor changes. Under the hypotheses, a minimizing sequence $\{u^n(\cdot)\}$, $u^n \in C_{\mathrm{ad}}(0, t_n; U)$ exists. Integrating the second equation (4.7.1) and using the third and the target condition for $u^n(\cdot)$,

$$gt_n = v_0 + \int_0^{t_n} \frac{u^n(\tau)}{m(\tau)} d\tau \leq v_0 + \frac{m_0 - m_e}{K m_e}$$

so that $\{t_n\}$ is bounded. The system is of the required form (2.2.7), since the right-hand side is $(v, -g, 0) + (0, 1/m, -K)u = f(h, v, m) + F(h, v, m)u$. The function f satisfies the assumptions of Theorem 2.2.6, but F does not, since it is not even defined for $m = 0$. However, this is irrelevant, since $m(t) \geq m_e > 0$.

We apply the maximum principle, calling the costate variables μ, p, q, r. The Hamiltonian is

$$H(h, v, m, \mu, p, q, r, u) = \mu K u + pv + q(-g + u/m) - rKu$$

and the canonical equations for the costate,

$$p'(t) = 0, \qquad q'(t) = -p(t), \qquad r'(t) = q(t)\frac{u(t)}{m^2(t)}. \qquad (4.7.6)$$

Since we have exchanged minimum by maximum, $\mu \leq 0$ and

$$(p(\bar{t}), q(\bar{t}), r(\bar{t})) \in -N_Y(0, 0, m(\bar{t})). \qquad (4.7.7)$$

Also,

$$(\mu, p(\bar{t}), q(\bar{t}), r(\bar{t})) \neq 0. \qquad (4.7.8)$$

We have $N_Y(0, 0, m(\bar{t})) = (-\infty, \infty) \times (-\infty, \infty) \times (-\infty, 0]$, thus (4.7.7) implies

$$r(\bar{t}) \geq 0. \qquad (4.7.9)$$

After the customary elimination of terms that do not depend on u, the maximum principle gives

$$\zeta(t)\bar{u}(t) = \max_{0 \leq u \leq R} \zeta(t)u \qquad (4.7.10)$$

where

$$\zeta(t) = K\mu + \frac{q(t)}{m(t)} - Kr(t) \qquad (4.7.11)$$

is the **switching function**.

Lemma 4.7.2. $\zeta(\cdot)$ *vanishes at most at one point in* $0 \leq t \leq \bar{t}$.

Proof. We have

$$\zeta'(t) = \frac{q'(t)}{m(t)} - \frac{m'(t)q(t)}{m(t)^2} - Kr'(t) = \frac{q'(t)}{m(t)} \qquad (4.7.12)$$

cancellation due to the third equation (4.7.1) and the third equation (4.7.6). The first and second equations (4.7.6) yield

$$p(t) = p(\bar{t}), \qquad q(t) = -p(\bar{t})t + (q(\bar{t}) + p(\bar{t})\bar{t}). \qquad (4.7.13)$$

There are two possibilities.

(a) $p(\bar{t}) \neq 0$. Since $m(t) \geq m_e > 0$, (4.7.12) shows that $\zeta(t)$ is either strictly increasing or strictly decreasing, which fulfills the claims of Lemma 4.7.2.

(b) $p(\bar{t}) = 0$. Then $q'(t) = 0$ thus $\zeta(t) = c$ and we must show that $c \neq 0$. There are two subalternatives.

(b1) $q(\bar{t}) = 0$. Since $p(\bar{t}) = 0$, it follows from (4.7.13) that $q(t) \equiv 0$. The third equation (4.7.6) then shows that $r(t)$ is constant so that $r(t) = r \geq 0$ in view of (4.7.9). Hence $\zeta(t) \equiv K\mu - Kr < 0$, since μ and r cannot be zero at the same time due to (4.7.8).

(b2) $q(\bar{t}) \neq 0$. Since $p(\bar{t}) = 0$, if $\zeta(t) \equiv 0$ the Hamiltonian at $\bar{t}$ equals $-q(\bar{t})g \neq 0$, which contradicts Theorem 4.6.1 (the terminal time is free).   ∎

Corollary 4.7.3. *Any optimal control* $\bar{u}(\cdot)$ *must be of the form*

$$\bar{u}(t) = \begin{cases} 0 & (0 \leq t \leq t_0) \\ R & (t_0 \leq t \leq \bar{t}). \end{cases} \qquad (4.7.14)$$

The switching cannot be from R to 0; the rocket would be doing free fall in the interval $[t_0, \bar{t}]$ and crash.

4.8. Fuel Optimal Soft Landing of a Space Vehicle: Identification of the Optimal Trajectory

We solve (4.7.1) for a control of the form (4.7.14). In the full thrust interval $t_0 \le t \le \bar{t}$ we have $\bar{u}(t) = R$, thus we obtain from the third equation (4.7.1) that

$$m(t) = m_0 - KR(t - t_0) \quad (t_0 \le t \le \bar{t}). \tag{4.8.1}$$

The second equation becomes

$$v'(t) = -g + \frac{R}{m_0 - KR(t - t_0)} \quad (t_0 \le t \le \bar{t}),$$

with solution

$$v(t) = v(t_0) - g(t - t_0) - \frac{1}{K} \ln \frac{m_0 - KR(t - t_0)}{m_0} \quad (t_0 \le t \le \bar{t}). \tag{4.8.2}$$

Replacing in the first equation and integrating,

$$h(t) = h(t_0) + v(t_0)(t - t_0) - \frac{1}{2} g(t - t_0)^2 + \frac{t - t_0}{K}$$
$$+ \frac{m_0 - KR(t - t_0)}{K^2 R} \ln \frac{m_0 - KR(t - t_0)}{m_0} \quad (t_0 \le t \le \bar{t}). \tag{4.8.3}$$

Assume the trajectory ends in a soft landing, that is, that $v(\bar{t}) = h(\bar{t}) = 0$. Then, setting $s = \bar{t} - t_0$ we have

$$\phi(s) = gs + \frac{1}{K} \ln \frac{m_0 - KRs}{m_0} = v(t_0), \tag{4.8.4}$$

$$\frac{1}{2} gs^2 - \frac{m_0 - KRs}{K^2 R} \ln \frac{m_0 - KRs}{m_0} - \frac{s}{K} - v(t_0)s = h(t_0). \tag{4.8.5}$$

Replacing $v(t_0)$ given by (4.8.4) in (4.8.5),

$$\psi(s) = -\frac{1}{2} gs^2 - \frac{m_0}{K^2 R} \ln \frac{m_0 - KRs}{m_0} - \frac{s}{K} = h(t_0). \tag{4.8.6}$$

The amount of fuel used up in the interval $[t_0, \bar{t}]$ is Rs, so that by (4.8.1),

$$s \le s_m = \frac{m_0 - m_e}{KR} < \frac{m_0}{KR}. \tag{4.8.7}$$

We assume

$$\frac{R}{m_0} > g \tag{4.8.8}$$

which means that, as the engine is started, the full thrust of the rocket can offset the acceleration of gravity. This condition implies

$$\phi'(s) = g - \frac{R}{m_0 - K R s} < 0$$

so that $\phi(s)$ is strictly decreasing in $0 \le s < m_0/(K R)$. On the other hand,

$$\psi'(s) = -gs + \frac{m_0}{K(m_0 - K R s)} - \frac{1}{K}, \tag{4.8.9}$$

$$\psi''(s) = -g + \frac{m_0 R}{(m_0 - K R s)^2} > -g + \frac{R}{m_0} > 0. \tag{4.8.10}$$

Inequality (4.8.9) implies that $\psi'(s)$ is strictly increasing; since $\psi'(0) = 0$, then $\psi'(s) > 0$ in $0 \le s < m_0/(K R)$. Hence, $\psi(s)$ itself is strictly increasing there.

In the initial interval $0 \le t \le t_0$ of the trajectory, the rocket performs free fall, so that we have

$$v(t) = -gt + v_0 \tag{4.8.11}$$

$$h(t) = -\tfrac{1}{2}gt^2 + v_0 t + h_0. \tag{4.8.12}$$

Accordingly, the switching point t_0 and the terminal time $\bar{t}$ can be simultaneously calculated by solving the system of two nonlinear equations with two unknowns

$$\phi(\bar{t} - t_0) = -gt_0 + v_0 \tag{4.8.13}$$

$$\psi(\bar{t} - t_0) = -\tfrac{1}{2}gt_0^2 + v_0 t_0 + h_0. \tag{4.8.14}$$

We do this graphically as follows. Plot the curve

$$v = \phi(s), \qquad h = \psi(s) \tag{4.8.15}$$

in the (v, h) plane for $0 \le s \le s_m$, s_m given by (4.8.7); call it the **full thrust curve**. Then,

(a) Find the initial position (v_0, h_0) in the (v, h) plane and draw the **free fall curve**, determined by (4.8.11) and (4.8.12) starting at (v_0, h_0) for $t = 0$.

(b) Follow the free fall curve until hitting the full thrust curve; the time t_0 where this happens is the switching time. The value s_0 of the parameter s on the full thrust curve where the two curves intersect gives the final time: $\bar{t} = s_0 + t_0$.

(c) After the intersection, the trajectory of the rocket is given by

$$v = \phi(\bar{t} - t), \qquad h = \psi(\bar{t} - t) \quad (t_0 \le t \le \bar{t}).$$

From all controls singled out by Corollary 4.7.3, the one we have constructed is the only one achieving a soft landing; thus, it must be the optimal control. The trajectories obey automatically the state constraints (4.7.5).

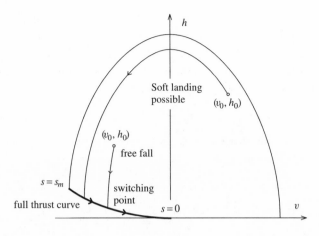

Figure 4.10.

4.9. Unbounded Control Sets: The Linear-Quadratic Problem and the Minimum Drag Nose Shape Problem

We take a last look at the linear-quadratic problem in **2.7–2.9**. The equation is (2.7.5), the cost functional (2.7.6), and all the assumptions in **2.7** are in force; as in there, the control space consists of all the functions in $L^2(0, T; \mathbb{R}^k)$ taking values in U a.e. We include a target condition

$$y(\bar{t}, u) \in Y \tag{4.9.1}$$

with Y closed in a fixed or variable time interval $0 \le t \le \bar{t}$. If U is convex and closed, Theorem 2.7.1 covers existence for this problem, and Theorem 2.7.2 proves uniqueness of optimal controls if the target set Y is convex. The Hamiltonian is $H(s, \mathbf{y}, \mathbf{z}, u) = z_0 \langle R(s)y, y \rangle + z_0 \|u\|^2 + \langle z, A(s)y + B(s)u \rangle$, and the minimum principle gives

$$z_0 \|\bar{u}(s)\|^2 + \langle B(s)^* \bar{z}(s), \bar{u}(s) \rangle = \min_{v \in U}\{z_0\|v\|^2 + \langle B(s)^*\bar{z}(s), v \rangle\}$$

a.e. for any optimal control $\bar{u}(\cdot)$, where $\bar{z}(\cdot)$ is the costate. However, this application of the minimum principle is purely formal; the equation satisfies Assumption III but *not* with $K(t, c)$, $L(t, c)$ independent of $u(\cdot)$, and the same applies to the cost functional. One may suspect (correctly) that this is not a serious problem, as the proof of the minimum principle is *local*; it only compares the optimal control with neighboring controls.

To understand things better, we consider a system

$$y'(t) = f(t, y(t), u(t)), \qquad y(0) = \zeta \tag{4.9.2}$$

with $f(t, y, u)$ continuous in all three variables and continuously differentiable with respect to y in $[0, T] \times \mathbb{R}^m \times U$, where U is a closed subset of $\mathbb{R}^k$. If U is bounded and $C_{\mathrm{ad}}(0, T; U) = F_m(0, T; U)$ (the space of all measurable functions defined in $0 \le t \le T$ and taking values in U a.e.), then the admissible control space and the system satisfy all the assumptions in Theorem 3.7.1, Theorem 4.2.1, and Theorem 4.6.1 (see Examples 3.7.5, 4.2.5, and 4.6.5). If U is unbounded, summability assumptions must be incorporated in the definition of $C_{\mathrm{ad}}(0, T; U)$ and on the growth of $f(t, y, u)$. To fix ideas, we take

$$C_{\mathrm{ad}}(0, T; U) = F_m(0, T; U) \cap L^p(0, T; \mathbb{R}^k), \qquad (4.9.3)$$

with $p \ge 1$, and assume that, for every $c > 0$ there exist $P(\cdot, c), Q(\cdot, c) \in L^q(0, T)$ $(1/q + 1/p = 1)$ such that

$$\|f(t, y, u)\| \le P(t, c)\|u\| \quad (0 \le t \le T, \|y\| \le c, u \in U), \quad (4.9.4)$$

$$\|\partial_y f(t, y, u)\| \le Q(t, c)\|u\| \quad (0 \le t \le T, \|y\| \le c, u \in U). \quad (4.9.5)$$

If we use a cost functional

$$y_0(t, u) = \int_0^t f_0(\tau, y(\tau, u), u(\tau))d\tau + g_0(t, y(t, u)), \qquad (4.9.6)$$

we likewise assume $f_0(t, y, u)$ continuous in all three variables and continuously differentiable with respect to y in $[0, T] \times \mathbb{R}^m \times U$ and such that, for every $c > 0$ there exist $P_0(\cdot, c), Q_0(\cdot, c) \in L^q(0, T)$ such that

$$\|f_0(t, y, u)\| \le P_0(t, c)\|u\| \quad (0 \le t \le T, \|y\| \le c, u \in U) \quad (4.9.7)$$

$$\|\partial_y f_0(t, y, u)\| \le Q_0(t, c)\|u\| \quad (0 \le t \le T, \|y\| \le c, u \in U) \quad (4.9.8)$$

These conditions insure that $f(t, y, u)$ satisfies Hypothesis III for every $u(\cdot) \in C_{\mathrm{ad}}(0, T; U)$ but not with the same $K(\cdot, c), L(\cdot, c)$ for all $u(\cdot) \in C_{\mathrm{ad}}(0, T; U)$; the same observation applies to $f_0(t, y, u)$. We assume that $g(t, y)$ is differentiable with respect to y. The Ekeland distance $d(u(\cdot), v(\cdot)) = |\{t \in [0, T]; u(t) \ne v(t)\}|$ makes sense in $C_{\mathrm{ad}}(0, T; U)$, but the space is not complete equipped with d; for instance, if $k = 1, U = \mathbb{R}$, the sequence $\{u_n(\cdot)\}$ with $u_n(t) = t^{-1}\chi_n(t), \chi_n(\cdot)$ the characteristic function of $[1/n, 1]$ belongs to $C_{\mathrm{ad}}(0, 1; U)$ and converges to the function $u(t) = t^{-1}$, which does not belong to $C_{\mathrm{ad}}(0, 1; U)$. However, the abstract theory will be applied below in suitable subspaces.

Let $\bar{u}(\cdot) \in C_{\mathrm{ad}}(0, T; U)$ and let $a > 0$. The space $C_{\mathrm{ad}}(0, T; U; \bar{u}; a) \subseteq C_{\mathrm{ad}}(0, T; U)$ consists of all $u(\cdot) \in C_{\mathrm{ad}}(0, T; U)$ such that

$$\|u(t)\| \le a \qquad (4.9.9)$$

in the set $\{t \in [0, T]; u(t) \ne \bar{u}(t)\}$; it is equipped with the distance d.

Lemma 4.9.1. $C_{\mathrm{ad}}(0, T; U; \bar{u}; a)$ *is saturated and spike complete for spike perturbations* $u_{s,h,v}$ *satisfying* $\|v\| \le a$.

Proof. Let $\{u_n(\cdot)\} \subset C_{\mathrm{ad}}(0, T; U; \bar{u}; a)$ be a stationary sequence, that is, let $u_n(t) = u_{n+1}(t) = \cdots$ for $n = n(t), t$ outside of a null set e. Let $u(t) =$

$\lim_{n \to \infty} u_n(t)$. If $u(t) \neq \bar{u}(t)$ then $u_n(t) \neq \bar{u}(t)$, hence, $\|u_n(t)\| \leq a$, thus the same is true of $u(t)$. The statement about spikes is obvious. $\blacksquare$

Theorem 4.9.2. *Let $\bar{u}(t)$ be an optimal control in $0 \leq t \leq \bar{t}$. Then there exists $(z_0, z) \in \mathbb{R} \times E, z_0^2 + \|z\|^2 = 1, z_0 \geq 0, z \in N_Y(\bar{y})$ such that, if $\bar{z}(s)$ is the solution of*

$$\begin{aligned}
\bar{z}'(s) &= -\partial_y f(s, y(s, \bar{u}), \bar{u}(s))^* \bar{z}(s) - z_0 \partial_y f_0(s, y(s, \bar{u}), \bar{u}(s)), \\
\bar{z}(\bar{t}) &= z + z_0 \partial_y g_0(\bar{t}, y(\bar{t}, \bar{u})),
\end{aligned} \tag{4.9.10}$$

then

$$\begin{aligned}
z_0 f_0(s, y(s, \bar{u}), \bar{u}(s)) &+ \langle \bar{z}(s), f(s, y(s, \bar{u}), \bar{u}(s)) \rangle \\
&= \min_{v \in U} \{ z_0 f_0(s, y(s, \bar{u}), v) + \langle \bar{z}(s), f(s, y(s, \bar{u}), v) \rangle \} \tag{4.9.11}
\end{aligned}$$

a.e. in $0 \leq s \leq \bar{t}$.

Proof. Let n be a fixed positive integer. In the control space $C_{\mathrm{ad}}(0, T; U; \bar{u}; n)$, $f(t, y, u)$ satisfies Hypothesis III with $K(\cdot, c), L(\cdot, c)$ independently of $u(\cdot) \in C_{\mathrm{ad}}(0, T; U; \bar{u}; n)$; likewise, $f_0(t, y, u)$ satisfies Hypothesis III^0 with $K_0(\cdot, c)$, $L_0(\cdot, c)$ independently of $u(\cdot) \in C_{\mathrm{ad}}(0, T; U; \bar{u}; n)$. Moreover the space $C_{\mathrm{ad}}(0, T; U; \bar{u}; n)$ is spike closed (as long as $\|v\| \leq n$) and saturated, and the regularity assumptions on $f(t, y, u), f_0(t, y, u)$ are satisfied. Obviously, $\bar{u}(\cdot)$ is as well optimal in the smaller control space $C_{\mathrm{ad}}(0, \bar{t}; U; \bar{u}; n)$, thus we may apply Theorem 4.2.1 and obtain a sequence $\{e_n\}$ of sets of full measure in $0 \leq t \leq \bar{t}$ and a sequence $(z_{0n}, z_n) \in \mathbb{R} \times \mathbb{R}^m, z_{0n}^2 + \|z_n\|^2 = 1, z_{0n} \geq 0, z_n \in N_Y(\bar{y})$ such that if $\bar{z}_n(s)$ is given by

$$\begin{aligned}
\bar{z}_n'(s) &= -\partial_y f(s, y(s, \bar{u}), \bar{u}(s))^* \bar{z}_n(s) - z_{0n} \partial_y f_0(s, y(s, \bar{u}), \bar{u}(s)), \\
\bar{z}_n(\bar{t}) &= z_n + z_{0n} \partial_y g_0(\bar{t}, y(\bar{t}, \bar{u})) \tag{4.9.12}
\end{aligned}$$

then

$$\begin{aligned}
z_{0n} f_0(s, y(s, \bar{u}), \bar{u}(s)) &+ \langle \bar{z}_n(s), f(s, y(s, \bar{u}), \bar{u}(s)) \rangle \\
&= \min_{v \in U, \|v\| \leq n} \{ z_{0n} f_0(s, y(s, \bar{u}), v) + \langle \bar{z}_n(s), f(s, y(s, \bar{u}), v) \rangle \} \tag{4.9.13}
\end{aligned}$$

for $s \in e_n$. Using the Bolzano–Weierstrass theorem, we may assume that the sequence $\{(z_{0n}, z_n)\}$ is convergent to $(z_0, z) \in \mathbb{R} \times \mathbb{R}^m$, so that $\bar{z}_n(s)$ is convergent to $\bar{z}(s)$, the solution of the adjoint variational equation (4.9.10) corresponding to (z_0, z). Let e be the intersection of the e_n, and let $v \in U$, so that $\|v\| \leq n$ for n large enough; then

$$\begin{aligned}
z_{0n} f_0(s, y(s, \bar{u}), \bar{u}(s)) &+ \langle \bar{z}_n(s), f(s, y(s, \bar{u}), \bar{u}(s)) \rangle \\
&\leq z_{0n} f_0(s, y(s, \bar{u}), v) + \langle \bar{z}_n(s), f(s, y(s, \bar{u}), v) \rangle
\end{aligned}$$

for $s \in e$. Taking limits, we obtain (4.9.13). $\blacksquare$

The minimum principle for the time optimal problem generalizes in the same way, as does the minimum principle for problems with variable terminal time; in the last case, we apply Theorem 4.6.1 in the space $C_{ad}(0, \bar{t}; U; \bar{u}; n)$ and obtain a multiplier (z_{0n}, z) such that

$$z_{0n} f_0(s, y(s, \bar{u}), \bar{u}(s)) + \langle \bar{z}_n(s), f(s, y(s, \bar{u}), \bar{u}(s)) \rangle = 0.$$

a.e. in $0 \leq s \leq t$. Taking limits, we perceive that the Hamiltonian of the original problem vanishes.

The minimum drag nose shape problem in **1.7** is another one that does not fit in Theorem 4.2.1. The cost functional is

$$y_0(t, u, v) = \int_0^t \frac{x(\tau)u(\tau)^3}{u(\tau)^2 + v(\tau)^2} d\tau. \tag{4.9.14}$$

Since $u(t), v(t) \geq 0$, the control set is $U = [0, \infty) \times [0, \infty)$. We have $xu^3/(u^2 + v^2) \leq xu$, thus the natural choice of admissible control space would appear to be

$$C_{ad}(0, \bar{t}; U) = F_m(0, T; U) \cap L^1(0, T; \mathbb{R}^2), \tag{4.9.15}$$

consisting of all pairs of nonnegative functions in L^1. With this space, the curves C in competition for the minimum (except for the nonnegativity condition) are simply those of parametric equations $x = x(t), y = y(t)$, with $x(\cdot), y(\cdot)$ absolutely continuous in an interval $0 \leq t \leq T$. This is the largest class of curves for which the arclength $s = s(t)$ can be reasonably defined:

$$s(t) = \int_0^t \sqrt{x'(\tau)^2 + y'(\tau)^2} d\tau, \tag{4.9.16}$$

and we shall call them **rectifiable curves**.[1] However, the space (4.9.15) is a bad place for existence (see Example 2.2.7), and to change *a priori* $L^1(0, T; \mathbb{R}^2)$ by $L^p(0, T; \mathbb{R}^2), p > 1$ is unnatural. It turns out, however, that due to the possibility of reparametrizing the curves in competition for the minimum, we may formulate the problem with a bounded control set. This is a consequence of the result below.

Reparametrization Lemma 4.9.3. *Every rectifiable curve in $0 \leq t \leq T$ can be parametrized with respect to arclength s in $0 \leq s \leq L = $ length of the curve, with $x(\cdot), y(\cdot)$ absolutely continuous and*

$$x'(s)^2 + y'(s)^2 = 1 \quad a.e. \tag{4.9.17}$$

The functional (4.9.14) does not change; $y_0(L, u, v)$ for the new parametrization and $y_0(T, u, v)$ for the old one coincide.

[1] This definition is a little more demanding than the classical one, which, due to a theorem of C. Jordan, amounts to requiring that $x(\cdot), y(\cdot)$ be continuous functions of bounded variation. See Tonelli [1921, p. 44].

Proof. Our first task is to show that (after a modification of the original parametrization) the function $s(t)$ defined by (4.9.16) may be assumed strictly increasing. If this is not the case there exists a (at most countable) family $\{[a_j, b_j]\}$ of disjoint intervals of constancy $[a_j, b_j] \subseteq [0, T]$ where $x'(\tau) = y'(\tau) = 0$. These are "dead" intervals where nothing happens and must be removed. If the family $\{[a_j, b_j]\}$ is finite it is obvious that we can do this without changing the total length of the curve or the value of the functional $y_0(T, x, y, u, v)$. We show below what to do in the case of a countable family $\{[a_j, b_j]\}$. Let s_j the value of $s(t)$ in each dead interval. The inverse function $t(s)$ is increasing and continuous except at $s = s_j$, where it has a jump $b_j - a_j$. Let $J(s) = \Sigma(b_j - a_j)v(s - s_j)$, $v(\cdot)$ the Heaviside function $v(t) = 0$ for $t < 0$, $v(t) = 0$ for $t \geq 0$. Then $t(s) - J(s)$ is continuous and strictly increasing. Its inverse $s^*(t)$ is $s(t)$ without its dead intervals, thus is strictly increasing in the new parametrization interval $[0, T']$, $T' = T - \Sigma(b_j - a_j)$.

Consider the continuous, increasing function $x(t)$. It is constant in all the intervals of $\{[a_j, b_j]\}$ (and perhaps in other intervals) so that its inverse $t(x)$ has jumps at a countably family of points including $\{s_j\}$; the same comment applies to $y(t)$ and $t(y)$. Accordingly, if $x^*(t)$ is the inverse of $t^*(x) = t(x) - J(x)$, and $y^*(t)$ is the inverse of $t^*(y) = t(y) - J(y)$, $(x^*(t), y^*(t))$ is a parametrization of the same curve with

$$s^*(t) = \int_0^t \sqrt{x^{*\prime}(\tau)^2 + y^{*\prime}(\tau)^2}\, d\tau$$

strictly increasing. Thus, we may assume from the word go that the family $\{[a_j, b_j]\}$ of dead intervals is empty.

Under this condition $s(t)$ is a (absolutely continuous) strictly increasing function and thus has a strictly increasing inverse function $t(s)$ with $t(0) = 0$, $t(L) = t(s(T))$ $= T$. As an increasing function, $t(s)$ has a derivative $t'(s)$ in a set $c \subseteq [0, L]$ of full measure in $[0, L]$ (Example 4.9.9 below). We use now the fact that an absolutely continuous function maps null sets into null sets (Example 4.9.7) so that there exists a set $e \subseteq [0, T]$ of full measure in $[0, T]$ such that, if $t \in e$ then $s'(t)$ and $t'(s(t))$ exist, thus by elementary calculus, $t'(s(t))s'(t) = 1$. The reparametrization is of course $x = x(t(s))$, $y = y(t(s))$, and it is plain that it will satisfy (4.9.17) a.e., but it is not obvious that the functions $x(t(s))$ and $y(t(s))$ are absolutely continuous. It is enough to show that $x(t(s))$ and $y(t(s))$ are the indefinite integrals of their derivatives $x'(t(s))t'(s)$, $x'(t(s))t'(s)$, which we do for $x(t(s))$. Using the fact that $s(t)$ is strictly increasing and absolutely continuous, and the general change of variable formula (Example 4.9.12) we obtain

$$\int_0^s x'(\tau(\sigma))\tau'(\sigma)d\sigma = \int_0^{t(s)} x'(\tau)\tau'(s(\tau))s'(\tau)d\tau$$

$$= \int_0^{t(s)} x'(\tau)d\tau = x(t(s)),$$

where τ (resp. σ) is the integration surrogate of t (resp. s). A similar change of variable formula shows the statement about $y_0(T, u, v)$. This ends the proof. ■

The Reparametrization Lemma makes it possible to reduce the control space to $C_{ad}(0, \bar{t}; U) = F_m(0, T; U)$, where U is the intersection of the unit disk in $\mathbb{R}^2$ with the positive quadrant. Even with this improvement, existence of a solution is not immediate. Example 2.2.16 fails to apply since $f_0(t, x, y, u, v) = xu^3/(u^2 + v^2)$ is not a convex function of u, v in the positive quadrant (in fact $\partial^2 f_0/\partial u^2 = 2uv^2(3v^2 - u^2)(u^2 + v^2)^{-3}$ is not everywhere nonnegative); existence will be postponed to **13.7**. In the meantime, we assume a solution $(\bar{u}(\cdot), \bar{v}(\cdot))$ exists and (using Lemma 4.9.3) that it is parametrized by arclength in the interval $0 \le s \le L$ so that $\bar{u}(s)^2 + \bar{v}(s)^2 = 1$ a.e. We call $(\bar{x}(t), \bar{y}(t))$ the optimal curve and apply Theorem 4.9.2 with the original control set $U = [0, \infty) \times [0, \infty)$ (we are actually using Theorem 4.9.2 "at half power" since we have a bounded optimal control). The Hamiltonian is $H(x, y, \mu, p, q, u, v) = \mu xu^3/(u^2 + v^2) + pu - qv$ and the canonical equations for the costate,

$$p'(s) = -\mu \frac{\bar{u}^3(s)}{\bar{u}(s)^2 + \bar{v}(s)^2}, \qquad q'(s) = 0 \qquad (4.9.18)$$

with $\mu \le 0$, $(\mu, p(L), q(L)) \ne 0$. The maximum principle is

$$\mu\bar{x}(s)\frac{\bar{u}(s)^3}{\bar{u}(s)^2 + \bar{v}(s)^2} + p(s)\bar{u}(s) - q\bar{v}(s)$$

$$= \max_{0 \le u, v \le \infty} \left(\mu\bar{x}(s)\frac{u^3}{u^2 + v^2} + p(s)u - qv\right) = 0 \qquad (4.9.19)$$

a.e. in $0 \le s \le L$. The vanishing of the Hamiltonian is a consequence of the remarks after Theorem 4.9.2; the problem is free terminal time.

If $\mu = 0$ then p is constant and either $p \ne 0$ or $q \ne 0$. Since the maximum in (4.9.19) must be finite, we must have $p \le 0$ and $q \ge 0$. If $p < 0$ the maximum is attained at $u = 0$ for all s. This implies $\bar{x}'(s) \equiv \bar{u}(s) \equiv 0$, which preempts the target condition for $x(\cdot)$. Likewise, if $q > 0$, the maximum is attained for $v = 0$ for all s and $\bar{y}'(s) \equiv 0$, absurd as well. It then follows that $\mu < 0$ and (multiplying by a constant) we may assume $\mu = -1$.

Suppose that $q = 0$. Then the maximum in (4.9.19) near L (where $\bar{x}(s) > 0$ since $x(L) = r > 0$) is attained at $v = \infty$, a contradiction. We may then assume that

$$\mu = -1, \qquad q > 0. \qquad (4.9.20)$$

The arclength parametrization insures that

$$\bar{x}(s) > 0 \quad (0 < s \le L). \qquad (4.9.21)$$

In fact, if $\bar{x}(s) \equiv 0$ in $0 \leq s \leq s_0 > 0$ then $\bar{u}(s) \equiv 0$ in the same interval, and (4.9.19) gives $\bar{v}(s) \equiv 0$ in $0 \leq s \leq s_0$, which is impossible.

For fixed u, the Hamiltonian in (4.9.19) has a local maximum $v > 0$ in $0 \leq v < \infty$ if

$$\frac{\partial}{\partial v}\left(-\frac{\bar{x}(s)u^3}{u^2 + v^2} - qv\right) = \frac{2\bar{x}(s)u^3 v}{(u^2 + v^2)^2} - q = 0. \tag{4.9.22}$$

The function $2u^3 v/(u^2 + v^2)^2$ is homogeneous of degree zero, taking the value $\phi(\lambda) = 2\lambda/(1 + \lambda^2)^2$ on the ray $v = \lambda u$. Since $\phi'(\lambda) = 2(1 - 3\lambda^2)/(1 + \lambda^2)^3$, $\phi(\lambda)$ reaches the maximum $3^{3/2}/8$ at $\lambda = (1/3)^{1/2}$. It then follows that if $a = 8q/3^{3/2}$ and $\bar{x}(s) < a$, (4.9.22) does not have a solution, so that the maximum of the Hamiltonian in v occurs for $v = 0$. We then deduce that $\bar{v}(s) = 0$ and thus that $\bar{u}(s) = 1$ in the interval $0 \leq s \leq a$.

If $s > a$ then $\bar{x}(s) \geq \bar{x}(a) = a$ so, that (4.9.22) has two rays $v = \lambda u$ as solutions, λ any of the two solutions of $\phi(\lambda) = q/2\bar{x}(s)$. One ray corresponds to a local minimum, the other to a maximum. This means that, for each s the Hamiltonian has a unique maximum $v = \bar{v}(s) > 0$ in $v \geq 0$. Taking into account that $\bar{u}(s)^2 + \bar{v}(s)^2 = 1$, (4.9.22) becomes

$$2\bar{x}(s)\bar{x}'(s)^3 \bar{y}'(s) = -q \tag{4.9.23}$$

(recall that $\bar{y}'(s) = -\bar{v}(s)$). This equation implies that $\bar{x}'(s)$ is bounded below in $a \leq s \leq L$, so that the inverse function $\bar{s}(x)$ is Lipschitz continuous and we can put $\bar{y}$ as a function of x in $x \geq a$ in an increasing, Lipschitz continuous fashion: $\bar{y}(x) = \bar{y}(\bar{s}(x))$ with derivative $\bar{y}'(x) = \bar{y}'(\bar{s}(x))\bar{s}'(x)$ a.e. We obtain from (4.9.23) that $2\bar{x}(s)\bar{y}'(s)/\bar{x}'(s) = -q\bar{x}'(s)^{-4} = -q(\bar{x}'(s)^2 + \bar{y}'(s)^2)^2\bar{x}'(s)^{-4}$, so that $\bar{y}'(x)$ solves

$$2x\bar{y}'(x) = -q(1 + \bar{y}'(x)^2)^2 \tag{4.9.24}$$

which is not a differential equation but just an algebraic equation for $y' = \bar{y}'(x) = \bar{y}'(s)/\bar{x}'(s)$ in $x \geq a$. The function $x(y') = -(1 + y'^2)^2/2y'$ $(y' < 0)$ reaches its minimum $8/3^{3/2} = a/q$ at $y' = -1/3^{1/2}$ and is decreasing for $y' \leq -1/3^{1/2}$, thus (4.9.24) defines univocally $\bar{y}'(x)$ as an infinitely differentiable, negative, decreasing function of x in $x \geq a$, and we obtain $y(x)$ integrating and imposing the final condition $y(r) = 0$. Note that $\bar{y}'(a) = -1/3^{1/2}$, so that the solution has a corner at $x = a$ whose angle does not depend on any of the parameters in the problem (see Figure 1.9).

Parametric equations for the solution in $x \geq a$ can be easily obtained taking $\tau = -\bar{y}'(x)$ as a parameter (this is justified since $\bar{y}'(x)$ is decreasing). Putting $\bar{x} = \bar{x}(\tau)$, $\bar{y} = \bar{y}(\tau)$, (4.9.24) becomes

$$\bar{x}(\tau) = \frac{q}{2\tau}(1 + \tau^2)^2. \tag{4.9.25}$$

The parametrization interval is $[\tau_0, \tau_1]$, where $\tau_0 = -\bar{y}'(a) = 1/3^{1/2}$ and τ_1 satisfies $x(\tau_1) = r$. We have

$$\bar{x}'(\tau) = \frac{q}{2}\left(-\frac{1}{\tau^2} + 2 + 3\tau^2\right)$$

and $\bar{y}'(\tau) = \bar{y}'(x)\bar{x}'(\tau) = -\tau\bar{x}'(\tau)$, thus

$$\bar{y}'(\tau) = \frac{q}{2}\left(\frac{1}{\tau} - 2\tau - 3\tau^3\right), \qquad \bar{y}(\tau) = c + \frac{q}{2}\left(\log\tau - \tau^2 - \frac{3}{4}\tau^4\right). \qquad (4.9.26)$$

Theorem 4.9.4. *The solution of the minimum drag nose shape problem is unique.*
We have $\bar{x}(\tau) = 8q\tau/3$, $\bar{y}(\tau) = h$ in $0 \le \tau \le 1/\sqrt{3}$; in $1/\sqrt{3} \le \tau \le \tau_1$, $\bar{x}(\tau)$ and
$\bar{y}(\tau)$ are given by (4.9.25)-(4.9.26), where q, τ_1, c solve $\bar{y}(1/3^{1/2}) = h$, $\bar{x}(\tau_1) = r$,
$y(\tau_1) = 0$, that is,

$$c - \frac{q}{2}\left(\frac{\log 3}{2} + \frac{5}{12}\right) = h, \qquad (4.9.27)$$

$$\frac{q}{2\tau_1}(1 + \tau_1^2)^2 = r, \qquad (4.9.28)$$

$$c + \frac{q}{2}\left(\log\tau_1 - \tau_1^2 - \frac{3}{4}\tau_1^4\right) = 0. \qquad (4.9.29)$$

We have reparametrized the initial segment for neatness.   ∎

Remark 4.9.5. The key to the possibility of using reparametrization is that the cost functional $y_0(T, u, v)$ is independent of the parametrization of the curve $(x(t), y(t))$ (T the upper endpoint of the parameter interval). Any cost functional (4.9.6) that has this property is called **parametric** (Cesari [1983, p. 432], Tonelli [1921, p. 218]). A necessary and sufficient condition for (4.9.6) to be parametric is that f_0 be time independent and positively homogeneous in u, that is, $f_0(y, ku) = kf_0(y, u)$ for $k \ge 0$. Of course, we cannot take advantage of a parametric cost functional unless the system (4.9.6) itself is parametric in the sense that a reparametrized trajectory is a trajectory itself.

Remark 4.9.6. The statement that the minimum in the right side of (4.9.11) (or the maximum if the maximum principle is used) is finite is obvious in case U is bounded, but it becomes highly nontrivial when U is unbounded. It played an important role in the proof of Theorem 4.9.4. We note, however that the use of unbounded controls in this example can be avoided.

Example 4.9.7. Let $f(\cdot)$ be an absolutely continuous function in $[0, T]$. Then, if e is a null set, $f(e)$ is a null set. To show this, use regularity of the Lebesgue measure (if e is a null set there exists an open set $d \supseteq e$ such that $|d| \leq \varepsilon$) and the fact that every open set in $\mathbb{R}$ is a finite or countable union of disjoint open intervals. See Natanson [1955, Theorems 2, 3, pp. 248–249].

Example 4.9.8. Let $f(\cdot)$ be nondecreasing in $[0, T]$. Then f is continuous except at a finite or countable number of points.

Example 4.9.9. Let $f(\cdot)$ be nondecreasing in $[0, T]$. Then f has a derivative $f'(t)$ almost everywhere. See Natanson [1955, Theorem 4, p. 211].

Example 4.9.10. Under the conditions of Example 4.9.9, $f'(t)$ is measurable and

$$\int_0^T f'(t)dt \leq f(T) - f(0)$$

with equality if the function is absolutely continuous. See Natanson [1955, Theorem 5, p. 212, and Theorem 2, p. 253].

Example 4.9.11. There exists a continuous strictly increasing function $\phi(t)$ in $0 \leq t \leq 1$ with $\phi'(t) = 0$ a.e; hence, $\phi(t)$ is not the indefinite integral of its derivative. See Natanson [1955, p. 213].

Example 4.9.12. (Integration by substitution in the style of Lebesgue). Let $f(\cdot) \in L^1(a, b)$ and let $\phi(t)$ be a strictly increasing absolutely continuous function mapping $[c, d]$ onto $[a, b]$. Then

$$\int_a^b f(s)ds = \int_c^d f(\phi(t))\phi'(t)dt.$$

See Natanson [1960, p. 236]. Note that we *do not* have to assume that $f(\phi(\cdot))\phi'(\cdot) \in L^1(c, d)$; this is part of the result.

4.10. Nonlinear Programming Problems:
The Kuhn–Tucker Theorem

Results on the abstract problem (4.1.1)–(4.1.2) can be applied to classical nonlinear programming problems

$$\text{minimize} \quad f_0(x) \tag{4.10.1}$$

$$\text{subject to} \quad f_j(x) = 0 \quad (j = 1, 2, \ldots, m),$$
$$f_j(x) \leq 0 \quad (j = m+1, m+2, \ldots, p) \tag{4.10.2}$$

where the functions f_j are continuously differentiable in $\mathbb{R}^k$. This problem can be put in the form (4.1.1)-(4.1.2) taking $V = \mathbb{R}^k$, $E = \mathbb{R}^p$ with their usual Euclidean norms, and

$$
\begin{aligned}
\mathbf{f}(x) &= (\mathbf{f}_1(x), \mathbf{f}_2(x)) \\
&= (f_1(x), \dots, f_m(x), f_{m+1}(x), \dots, f_p(x)), \qquad (4.10.3) \\
Y &= \underbrace{\{0\} \times \cdots \times \{0\}}_{m \text{ times}} \times \underbrace{(-\infty, 0] \times \cdots \times (-\infty, 0]}_{p-m \text{ times}}.
\end{aligned}
$$

If $x \in \mathbb{R}^k$, Der $(f_0, f)(x)$ contains all ordinary two-sided directional derivatives of (f_0, f) at x. Taking into account continuity of partial derivatives, if $x^n \to x$, $\liminf_{n\to\infty}$ Der $(f_0, f)(x^n)$ contains as well all two-sided directional derivatives of (f_0, f) at x.

All the hypotheses of Theorem 4.1.1 are satisfied. If $\bar{x} = (\bar{x}_1, \dots, \bar{x}_k)$ is a solution of (4.10.1)-(4.10.2), and

$$
\begin{aligned}
\bar{y} &= (\bar{y}_1, \dots, \bar{y}_m, \bar{y}_{m+1}, \dots, \bar{y}_p) \\
&= \mathbf{f}(\bar{x}) = (\mathbf{f}_1(\bar{x}), \mathbf{f}_2(\bar{x})) \\
&= (f_1(\bar{x}), \dots, f_m(\bar{x}), f_{m+1}(\bar{x}), \dots, f_p(\bar{x})) \\
&= (0, \dots, 0, f_{m+1}(\bar{x}), \dots, f_p(\bar{x}))
\end{aligned}
$$

there exists a nonzero multiplier $(z_0, \mathbf{z}_1, \mathbf{z}_2) = (z_0, z_1, \dots, z_m, z_{m+1}, \dots, z_p) \in \mathbb{R} \times \mathbb{R}^p$ such that $z_0 \geq 0$ and

$$
z_0 \partial f_0(\bar{x}) + \mathbf{z}_1 \partial \mathbf{f}_1(\bar{x}) + \mathbf{z}_2 \partial \mathbf{f}_2(\bar{x}) = 0, \qquad (4.10.4)
$$

where ∂ indicates gradient or Jacobian matrix and the products are matrix multiplications. The inequality in (4.1.10) becomes an equality since we can take the negatives of the directional derivatives. The tangent cone $K_Y(\bar{y})$ to Y at $\bar{y} = (0, \dots, 0, \bar{y}_{m+1}, \dots, \bar{y}_p) \in Y$ is

$$
K_Y(\bar{y}) = \underbrace{\{0\} \times \cdots \times \{0\}}_{m \text{ times}} \times \underbrace{M \times \cdots \times M}_{p-m \text{ times}}
$$

with $M = (-\infty, \infty)$ if $\bar{y}_j < 0$, $M = (-\infty, 0]$ if $\bar{y}_j = 0$. Hence

$$
N_Y(\bar{y}) = \underbrace{(-\infty, \infty) \times \cdots \times (-\infty, \infty)}_{m \text{ times}} \times \underbrace{N \times \cdots \times N}_{p-m \text{ times}},
$$

where $N = \{0\}$ if $\bar{y}_j < 0$, $N = [0, \infty)$ if $\bar{y}_j = 0$. It follows that

$$
z_{m+1}, \dots, z_p \geq 0 \qquad (4.10.5)
$$

and that $z_j = 0$ if $f_j(\bar{x}) < 0$ (that is, if the constraint is not **saturated**). These relations can be written

$$
\mathbf{z}_2 \mathbf{f}_2(\bar{x}) = 0. \qquad (4.10.6)
$$

Equations (4.10.4) and (4.10.6) are the Kuhn-Tucker theorem (Luenberger [1984, p. 314]) except for the statement that $z_0 \neq 0$. For this, an additional condition is needed (traditionally incorporated in the statement of the theorem); this condition is the linear independence of the set of gradient vectors

$$\partial f_1(\bar{x}), \dots, \partial f_m(\bar{x}), \partial f_{m+1}(\bar{x}), \dots, \partial f_p(\bar{x}). \qquad (4.10.7)$$

In fact, if $z_0 = 0$, (4.10.4) implies that $\mathbf{z}_1 = \mathbf{z}_2 = 0$, which is a contradiction.

This approach to nonlinear programming is much more general than the classical one, since it applies to nonsmooth functions; see Ekeland [1979].

Miscellaneous Comments for Part I

Control as stabilization by feedback.   The rotary steam engine, one of the key components of the early industrial revolution, was perfected by the Scottish inventor James Watt at the end of the eighteenth century. Early engines tended to change speed in response to variations in load and to race under no load, undesirable traits whose correction was undertaken by Watt himself by means of the *Watt centrifugal governor*, the first feedback device widely used in industry. However, some engines operating under the governor tended to settle into a periodic regime rather than into a steady one—a limit cycle instead of an equilibrium state. It was empirically found that raising the friction between the movable sleeve and the spindle eliminated the objectionable periodic limit regime, an insight that was mathematically justified by the British physicist J. Clerk Maxwell [1867/8] and by the Russian engineer J. Vyshnegradski [1876]; these papers are the first works in control theory. Vyshnegradski's analysis of the Watt governor is in Pontryagin [1961, §5.27]. See also Bellman-Kalaba [1964] for a collection of early papers in control theory including Maxwell's.

Towards the end of the century the British physicist E. J. Routh [1877], inadvertently overlapping with an 1856 paper by Hermite, provided an algorithm to decide whether the real parts of the roots an algebraic polynomial lie in the left half of the complex plane, a stability criterion known to Maxwell and Vyshnegradski only for polynomials of degree ≤ 3. A second solution of the problem was given independently by A. Hurwitz [1895], based on the work of Hermite. For a complete treatment of the Hermite-Routh-Hurwitz theory, together with references to several more recent works see Gantmajer [1959, Chapter 5]. Work along the same lines is A. M. Lyapunov's dissertation [1892], where a necessary and sufficient condition for the eigenvalues of a matrix to have negative real parts is given (a finite dimensional forerunner of the general Hilbert space theory of dissipative operators) along with various methods (now in common use) in the theory of stability of nonlinear equations.

The study of stabilization by feedback continued into the twentieth century, a highlight being the introduction by H. Nyquist [1932] and others of the *transfer function* of a feedback circuit and the analysis of the stability of the circuit by the *Nyquist plot* and other graphical methods based on complex function theory.

The *stabilization problem* for the linear system

$$y'(t) = Ay(t) + Bu(t) \tag{1}$$

is that of finding a matrix K of appropriate dimension such that, if we provide the system with the feedback law $u(t) = Ky(t)$, the resulting (uncontrolled) system

$$y'(t) = (A + BK)y(t)$$

is asymptotically stable, that is, all eigenvalues of $A + BK$ have negative real part. More general is the *pole assignment problem*, which demands finding $K = K(\Lambda)$

in such a way that the eigenvalues of $A + BK$ coincide with a given set Λ of real or complex numbers of appropriate number of elements (multiplicities taken into account). It was shown by Wonham [1967] that pole assignment is possible if (1) is controllable (as defined in **3.10**). This line of research is still very active and includes now many other related questions for (1) and for its *discrete time* companion

$$y_{n+1} = Ay_n + Bu_n,$$

as well as for nonlinear continuous time and discrete time systems. The theory is known by various names, one of them (multivariable) *control theory* or *system theory*; among its tools are matrix analysis, the Laplace transform, abstract algebra and algebraic geometry (see Rosenbrock [1970] and Wonham [1979] for two quite different approaches; for a more elementary account, see Barnett-Cameron [1984]).

Intimately related to control theory is *filtering theory*. A typical problem in (discrete time) filtering theory is that of recovering a signal $\{y_n\}$ propagating according to an equation $y_{n+1} = Ay_n + \varepsilon_n$ and being measured by $b_n = By_n + e_n$, where $\{e_n\}$ represents measurement error and ε_n comes from errors in the model. The *Kalman filter* (Kalman [1960]) recovers the signal recursively under suitable statistical assumptions on $\{\varepsilon_n\}$ and $\{e_n\}$; the *Kalman-Bucy filter* (Kalman-Bucy [1961]) performs the same job for continuous time systems. These works (whose influence in modern technology has been enormous) mark the beginnings of the theory of *stochastic control and filtering*, closely related to the *deterministic* theory in this book but now a vast subject with its own tools and methods. For an introduction see Fleming-Rishel [1975].

Optimal control. Until the second half of the twentieth century, control was practiced by engineers; "control engineer" was synonymous with "designer of servomechanisms and feedback devices". By mid-century, control theory began to consider the idea of *quality of control*; among the (possibly many) ways to reach a certain objective, to select the one minimizing cost, time or other performance criterion. Some mathematicians became interested in these problems; for instance, time optimality was investigated by Hestenes [1950]. Many optimal control problems were solved using ad-hoc arguments, or, later, versions of Pontryagin's maximum principle. Among these works we mention Bellman-Glicksberg-Gross [1956] on the time optimal problem for linear systems and the example on infinitely many switching points due to Fuller [1963]. The latter type of control behavior is now called *Fuller's phenomenon*; for some indications on its geometrical meaning, see Kupka [1990]. An important highlight was the Hamilton-Jacobi approach to construction of optimal controls by feedback in Bellman [1957] and other works of the same author. Except for particular cases where rigorous justification was possible (Boltyanski [1966]) much of the theory remained at a heuristic level, although it produced valuable results in many applications. A more solid mathematical justification had to wait until the late eighties and early nineties, where such ideas as viscosity solutions of nonlinear equations and nonsmooth analysis could be used to put the whole approach on a firmer footing.

Interest in control theory among the mathematical community seems to have been at best sporadic until the late fifties; for instance, Hestenes [1950] appeared only as an internal report and was probably not well known until the early sixties. Awareness increased due to an announcement of Pontryagin's maximum principle by Boltyanski-Gamkrelidze-Pontryagin [1956] combined with a talk by Pontryagin at the 1958 International Congress of Mathematicians in Edinburgh. Considerable interest (and perhaps confusion, see Markus [1994]) was caused by the involvement of a topologist like Pontryagin and a geometer like Boltyanski in "engineering matters"[1], and there was expectation for full presentation of the results. This came about with the publication of Pontryagin-Boltyanski-Gamkrelidze-Mischenko [1962] (the Russian version appeared in 1961), which clarified the connections of control theory with many aspects of classical calculus of variations. Research in control theory boomed and fueled a renewed interest in calculus of variations, whose popularity was on the wane in the early twentieth century.

Modern finite dimensional control theory has made contact with many other areas of pure and applied mathematics, such as functional analysis and differential geometry, and Lie theory. For an account of the first see Hermes-LaSalle [1969]; for the second see Brockett [1973] and Jurdjevic-Sussmann [1972]. As was the case with many other developments in applied mathematics, science and technology, existing results that originated well within pure mathematics (such as A. A. Lyapunov's 1940 theorem on the range of vector measures) found unexpected and meaningful applications. The reader willing to go in more depth into finite dimensional control theory may consult Berkowitz [1974], Cesari [1983], Fleming-Rishel [1975], Lee-Markus [1967], Warga [1972] and Young [1969]. For an elementary introduction containing many applications of the maximum principle, see Hocking [1991]. An excellent historical account and many examples can be found in McShane [1978]; for more on the history see Markus [1994].

It is not entirely clear whether control theory in full generality, in particular Pontryagin's maximum principle, can be entirely assembled from preexisting results in calculus of variations. Mayer (see Bliss [1946]), Bolza [1914] and Valentine [1937] worked in areas later incorporated in control theory, and Hestenes [1950] proved what can now be recognized as a particular case of Pontryagin's maximum principle. However, it was the fundamental achievement of control theory as developed by Pontryagin and collaborators to admit closed constraint sets and to characterize (without extraneous assumptions) extremals where the variable may lie on the boundary of the admissible set. On the other hand, it is clear that many,

---

[1] Control theory was apparently considered a subject of national interest (and actively encouraged) by the Soviet government of the time.

if not all of the tools essential in control theory were already in use in calculus of variations. For instance, spike and multispike perturbations were introduced by McShane [1939], who also pioneered, with others, the use of separation theorems for the sets of variations produced by these perturbations, a direct ancestor of Pontryagin's approach.

All the examples in Part I are well known in the literature; in particular, the solution of the soft landing problem is due to A. Miele (see Pontryagin *et al.* [1962] and Fleming-Rishel [1975] for further details and historical data). The minimum drag nose shape problem was formulated and completely solved by Newton in 1686; for Newton's original solution, see Goldstine [1980, pp. 7-29]. There exist earlier contributions in the field later called calculus of variations, notable Fermat's formulation and solution of the problem leading to Snell's law and the *brachistochrone problem* of minimizing the travel time of a particle sliding under gravity down a frictionless wire, posed by Johann Bernouilli and solved by Newton, Leibnitz, L'Hôpital, Bernouilli himself and his brother Jacob. However, these problems essentially reduce to what we would call today calculus minimization; according to Goldstine [1980, p. 7], Newton's problem is "the first genuine problem in the calculus of variations". Our treatment of the problem follows McShane [1978]. It is a challenging problem indeed; proving existence of solutions without unnatural ad-hoc devices needs the machinery of relaxed controls (treated in Part III) which postdates Newton's solution by almost three centuries.

Abstract nonlinear programming problems.   The idea (central to this work) that optimal control problems can be studied as abstract nonlinear programming problems originated in the work of Dubovitski-Milyutin [1965], Halkin [1966], Neustadt [1966] and Gamkrelidze-Jaratishvili [1967]. Other nonlinear programming theories that apply to control problems were developed by Ioffe and Tijomirov [1979] and Zowe and Kurcyuz [1979]. The abstract nonlinear programming theory in use in this book is due to the author and Frankowska and will be commented on Part II of this work, along with other approaches.

Part II

Infinite Dimensional Control Problems

5

Differential Equations in Banach Spaces and Semigroup Theory

5.0. Banach Spaces and Their Duals. Linear Operators. Integration of Vector Valued Functions

Let E, F be real normed spaces, $B : E \to F$ a linear operator. B is **bounded** if $\|By\|_F \leq C\|y\|_E$ $(y \in E)$ for some constant C. The minimum C satisfying this inequality,

$$\|B\| = \sup_{y \in E, \|y\| \leq 1} \|By\|,$$

is called the **norm** of B. The space of all linear operators from E into F is denoted (E, F) and is a Banach space equipped with $\| \cdot \|$.

Theorem 5.0.1 (Uniform Boundedness Principle). *Let E be a Banach space, F a normed space, $\mathcal{B}$ a subset of (E, F) such that $\{By; B \in \mathcal{B}\}$ is bounded in F for every $y \in E$. Then $\{\|B\|; B \in \mathcal{B}\}$ is bounded.*

For a proof see Goffman–Pedrick [1983, p. 76].

In the particular case where $F = \mathbb{R}$, the space $(E, \mathbb{R})$ (of all bounded linear functionals in E) is denoted by E^*; for a complex space, $E^* = (E, \mathbb{C})$. Application of a linear functional y^* to an element $y \in E$ is denoted by

$$\langle y^*, y \rangle \qquad \text{or} \qquad \langle y, y^* \rangle.$$

The map $(y^*, y) \to \langle y^*, y \rangle$ is called the **duality pairing** of E^* and E. It is bilinear and satisfies

$$|\langle y^*, y \rangle| \leq \|y^*\|_{E^*} \|y\|_E. \tag{5.0.1}$$

Existence of nontrivial (that is, nonzero) linear functionals in an arbitrary Banach space E is not obvious. One of the uses of the following result is to produce nontrivial functionals.

Theorem 5.0.2 (Hahn-Banach). *Let X be a real linear space, $p : X \to \mathbb{R}$ a functional such that*

$$p(x + y) \leq p(x) + p(y), \quad p(\lambda x) = \lambda p(x) \quad (x, y \in E, \lambda \geq 0).$$

Let X_0 be a subspace of X, $f : X_0 \to \mathbb{R}$ a linear functional satisfying

$$f(x) \leq p(x) \quad (x \in X_0). \tag{5.0.2}$$

Then there exists a linear functional $F : X \to \mathbb{R}$ that coincides with f in X_0 and satisfies

$$F(x) \leq p(x) \quad (x \in X). \tag{5.0.3}$$

The proof of the Hahn-Banach theorem has two parts. The first, the "one-step" result is elementary. The second (the construction of the full extension) is esoteric, but can be brought down to earth for certain functionals and spaces (see below). Note that Theorem 5.0.2 is a purely algebraic result; no topology is involved.

Lemma 5.0.3 (One-step Hahn-Banach Lemma). *Let X, f, X_0 be as in Theorem 5.0.2 and let $x_1 \notin X_0$, X_1 the subspace generated by X_0 and x_1. Then there exists a linear functional $F : X_1 \to \mathbb{R}$ that coincides with f in X_0 and satisfies (5.0.3).*

Proof. Any such extension must be of the form

$$F(x + \lambda x_1) = f(x) + \lambda F(x_1) = f(x) + \lambda a,$$

so we only have to show that a may be chosen such that

$$f(x) + \lambda a = F(x + \lambda x_1) \leq p(x + \lambda x_1) \quad (x \in X_0, \lambda \in \mathbb{R}). \tag{5.0.4}$$

For $\lambda > 0$, (5.0.4) holds if $a \leq p(x/\lambda + x_1) - f(x/\lambda)$ for $x \in X_0$, that is, if

$$a \leq M = \inf_{x \in X_0} \{p(x + x_1) - f(x)\}.$$

For $\lambda = -\mu < 0$, (5.0.4) holds if $a \geq -p(x/\mu - x_1) + f(x/\mu)$ for $x \in X_0$, that is, if

$$a \geq m = \sup_{x \in X_0} \{-p(x - x_1) + f(x)\}$$

so that the choice of a is possible if $m \leq M$. That this inequality is always true can be seen noting that $f(x) + f(y) = f(x + y) \leq p(x + y) = p(x + x_1 + y - x_1) \leq p(x + x_1) + p(y - x_1)$ so that $f(y) - p(y - x_1) \leq -f(x) + p(x + x_1)$ for all $x, y \in X_0$. ∎

A relation "$\leq$" in an arbitrary set A is called an **order** if $u \leq u$, and $u \leq v$, $v \leq w$ imply $u \leq w$ for $u, v, w \in A$; the set A equipped with $\leq$ is an **ordered set**.

An element $u \in A$ is **maximal** if, for each $v \in A$, $u \leq v$ implies $v \leq u$. An **upper bound** of a subset $B \subseteq A$ is any element v such that $u \leq v$ for $u \in B$. Finally, a subset $B \subseteq A$ is **linearly ordered** if and only if, given $u, v \in A$, we have either $u \leq v$ or $v \leq u$.

Lemma 5.0.4 (Zorn). *Assume that every linearly ordered subset of A has an upper bound. Then A has a maximal element.*

Zorn's lemma can be taken as a set theoretical postulate or can be deduced from other axioms (such as the Axiom of Choice). For such a proof see Dunford–Schwartz [1958, p. 6].

Proof of Theorem 5.0.2 By definition, an **extension** of f is any pair (Y, g) where $Y \supseteq X_0$ is a subspace of X and g is a linear functional that coincides with f in X_0 and satisfies $g(x) \leq p(x)$ $(x \in Y)$. Order the extensions (Y, g) of (X_0, f) as follows: $(Y_1, g_1) \leq (Y_2, g_2)$ if $Y_1 \subseteq Y_2$ and g_2 coincides with g_1 in Y_1. If (Y_α, g_α) is an arbitrary linearly ordered set of extensions, then (Y, g) is an upper bound, where $Y = \bigcup Y_\alpha$ and $g = g_\alpha$ in Y_α, so that by Zorn's lemma the set of all extensions has a maximal element (X', F). We must have $X' = X$; otherwise, we can obtain an extension (X'', G) with X' strictly contained in X'' using the one-step lemma. ∎

Theorem 5.0.2 and Lemma 5.0.3 are for real spaces. The complex version of Theorem 5.0.2 is

Theorem 5.0.5 (Bohnenblust-Sobczyk). *Let X be a complex linear space, $p : X \to \mathbb{R}$ a functional such that*

$$p(x + y) \leq p(x) + p(y), \quad p(\lambda x) = |\lambda| p(x) \quad (x, y \in E, \lambda \in \mathbb{C}).$$

Let X_0 be a subspace of X, $f : X_0 \to \mathbb{C}$ a linear functional satisfying

$$|f(x)| \leq p(x) \quad (x \in X_0). \tag{5.0.5}$$

Then there exists a linear functional $F : X \to \mathbb{C}$ that coincides with f in X_0 and satisfies

$$|F(x)| \leq p(x) \quad (x \in X). \tag{5.0.6}$$

The proof is an immediate consequence of the real theorem. We prove first the one-step Lemma, where $X = X_1$ is the subspace generated by X_0 and $x_1 \in X_0$. The assumptions on f and p imply the assumptions in Lemma 5.0.3 for $f_r = \mathrm{Re}\, f$, p now considered as functionals in the real space X_r (product limited to real scalars). Let F_r be the extension of f_r provided by Lemma 5.0.3. We have $f(x) = \mathrm{Re}\, f(x) + i\, \mathrm{Im}\, f(x) = \mathrm{Re}\, f(x) - i\, \mathrm{Re}\,(if(x)) = \mathrm{Re}\, f(x) - i\, \mathrm{Re}\, f(ix) = f_r(x) - if_r(ix),$

thus $F(x) = F_r(x) - i F_r(ix)$ coincides with $f(x)$ in X_0 and is easily checked to be a complex linear functional. Given $x \in X$ let θ be such that $e^{i\theta} F(x) = |F(x)|$. Then, since (5.0.3) holds for F_r we have $|F(x)| = F(e^{i\theta}x) = F_r(e^{i\theta}x) \le p(e^{i\theta}x) = |e^{i\theta}|p(x) = p(x)$. The passage from the one-step Lemma to Theorem 5.0.5 is exactly the same as for the real case.   ∎

Let E be a normed space, f a bounded linear functional in a subspace E_0 (not necessarily closed) with norm $\|f\|_0$ in this subspace. Then applying Theorem 5.0.2 (Theorem 5.0.5 in the complex case) with $p(y) = \|f\|_0\|y\|$ we can find an element $y^* \in E^*$ such that

$$\langle y^*, y \rangle = f(y) \quad (y \in E_0), \qquad \|y^*\| = \|f\|_0. \tag{5.0.7}$$

In particular, let E_0 be the space generated by a closed subspace F and an element $y_0 \notin F$. The functional $f(y + \lambda y_0) = \lambda$ is linear; moreover, $\|f(y + \lambda y_0)\| = |\lambda| = (1/\|y/\lambda + y_0\|)\|y + \lambda y_0\| \le (1/\text{dist}(F, y_0))\|y + \lambda y_0\|$. Accordingly,

Theorem 5.0.6. *Let E be a normed space, F a closed subspace of E, $y_0 \notin F$. Then there exists an element $y^* \in E^*$ such that*

$$\langle y^*, y \rangle = 0 \quad (y \in F), \qquad \langle y^*, y_0 \rangle = 1, \quad \|y^*\| = 1/\text{dist}(y_0, F). \tag{5.0.8}$$

Taking $F = \{0\}$ we obtain

Corollary 5.0.7. *Let $y \in E$. Then there exists $y^* \in E^*$ such that*

$$\|y^*\|_{E^*} = 1, \qquad \langle y^*, y \rangle = \|y\|_E. \tag{5.0.9}$$

Theorem 5.0.2 has incredible power, obviously coming from Zorn's lemma. One striking application (Corollary 5.0.8 below) will provide very weird counterexamples in Part III.

Let ℓ^∞ be the Banach space of all real bounded sequences $s = \{s_j; j = 1, 2, \ldots\}$ equipped with the supremum norm $\|s\| = \sup|s_j|$. A **Banach limit** in ℓ^∞ is a linear functional (denoted $\text{LIM}_{n\to\infty}$) having the following properties:

$$\text{LIM}_{n\to\infty} s_n = \text{LIM}_{n\to\infty} s_{n+1} \tag{5.0.10}$$

$$\text{LIM}_{n\to\infty} s_n \ge 0 \quad \text{if } s_n \ge 0 \tag{5.0.11}$$

It is easy to see that linearity, (5.0.10), and (5.0.11) imply that

$$\liminf_{n\to\infty} s_n \le \text{LIM}_{n\to\infty} s_n \le \limsup_{n\to\infty} s_n \tag{5.0.12}$$

for any $\{s_j\} \in \ell^\infty$; thus, in particular

$$\text{LIM}_{n\to\infty} s_n = \lim_{n\to\infty} s_n \tag{5.0.13}$$

whenever the latter exists; a Banach limit extends the ordinary limit.

Corollary 5.0.8. *Banach limits exist.*

Proof. Let ℓ_0 be the closed subspace of ℓ^∞ generated by all sequences of the form $\{s_0, s_1 - s_0, s_2 - s_1, \ldots\}$, where $s = \{s_j\} \in \ell^\infty$. If $\mathbf{1} = \{1, 1, \ldots\}$ we check that $\text{dist}(\mathbf{1}, \ell_0) = 1$ so that by Theorem 5.0.6 there exists a linear functional s^* in ℓ^∞ such that $\|s^*\| = 1$, $s^* = 0$ in ℓ_0 and $\langle s^*, \mathbf{1} \rangle = 1$. This functional obviously satisfies (5.0.10); as for (5.0.11), if $s_n \geq 0$ then $\| \|\{s_n\}\| \mathbf{1} - \{s_n\}\| \leq \|\{s_n\}\|$, so that, applying the functional, $\|\{s_n\}\| - \langle s^*, \{s_n\} \rangle \leq \|\{s_n\}\|$. ∎

Corollary 5.0.8 (proof from Dunford–Schwartz [1958, p. 73]) gives the full flavor of the applications of the Hahn–Banach theorem: something (a Banach limit) exists, but cannot be explicitly constructed. Some mathematicians find this unbearable. For other equally striking applications, among them an integral that integrates everything (the Banach integral), there is no better source than Banach [1932] where another proof of Corollary 5.0.8 is also given.

Many of the usual spaces in analysis are separable. For these, Theorem 5.0.5 and Corollary 5.0.6 may be proved without recourse to Zorn's lemma.

Example 5.0.9. Let E be a separable real normed space, $\{y_n\}$ a countable dense sequence in E. To show Theorem 5.0.6 and Corollary 5.0.7 we only need Theorem 5.0.2 in the case $p(y) = c\|y\|$. Using the one-step lemma repeatedly we can extend f to an F satisfying (5.0.3) in the subspace E_1 consisting of all finite linear combinations of elements of E_0 and elements of $\{y_k\}$. The final extension F is then constructed as $F(y) = \lim F(z_n)$, where $\{z_n\}$ is a sequence in E_1 converging to y.

This separable version of the Hahn–Banach theorem has important applications even in classical analysis, such as to show existence of conformal maps and Green functions. See for instance Garabedian [1964, p. 312] for the Green function application, where the manipulations in Lemma 5.0.3 are seen to have a bearing on the Dirichlet problem.

Corollary 5.0.7 implies a very useful relation between a normed space E and the dual $E^{**} = (E^*)^*$ of E^* (the "double dual" of E). An element $y \in E$ defines a continuous linear functional in E^* (that is, an element y^{**} of E^{**}) by

$$y^{**}(y^*) \to \langle y^*, y \rangle.$$

In view of (5.0.1), $\|y^{**}\|_{E^{**}} \leq \|y\|_E$. Applying Corollary 5.0.7 we construct $y^* \in E^*$ with $\|y^*\|_{E^*} = 1$ and $y^{**}(y^*) = \|y\|_E$, thus we have $\|y^{**}\|_{E^{**}} = \|y\|_E$. Accordingly, E can be isometrically imbedded in E^{**}, and it is easy to see that the imbedding is also algebraic; we can write informally $E \subseteq E^{**}$. The space E is called **reflexive** if

$$E = E^{**} \tag{5.0.14}$$

under this embedding.

A sequence $\{y_n\} \subseteq E$ is E^*-**weakly convergent** to $y \in E$ if

$$\langle y^*, y_n \rangle \to \langle y^*, y \rangle \tag{5.0.15}$$

for all $y^* \in E^*$. Correspondingly, a sequence $\{y_n^*\} \subseteq E^*$ is E^{**}-**weakly convergent** to $y^* \in E^*$ if

$$\langle y_n^*, y^{**} \rangle \to \langle y^*, y^{**} \rangle \tag{5.0.16}$$

for all $y^{**} \in E^*$; if (5.0.16) occurs for $y \in E \subseteq E^{**}$, $\{y_n^*\}$ is E-**weakly convergent** to y^*. Obviously, E-weak convergence is a weaker notion than E^{**}-weak convergence; the two coincide in a reflexive space. All three notions of weak convergence are weaker than the corresponding norm convergence.

For a Hilbert space H, duality is very special (**2.0**, Theorem 2.0.10); we are justified in writing $H^* = H$. Hilbert spaces are obviously reflexive. Some properties of weak convergence in Hilbert spaces were explicited in **2.0**.

The spaces of control theory (indeed of much of applied mathematics) are *real*; their elements represent positions in physical space, temperatures, deviations from equilibrium positions. However, to work in real spaces is mathematically inconvenient (for instance, eigenvalue-eigenvector theory of real matrices is crippled by this restriction). Accordingly, when dealing with matrices as operators or with partial differential operators, it is many times convenient to define them in spaces of complex-valued functions, and we shall do so when necessary.

We do below integration theory of functions f with values in a (real or complex) Banach space; the resulting integral is called the **Lebesgue-Bochner** integral. The good news is: only very few modifications to the finite dimensional theory in **2.0** are needed. For simplicity, we only look at the case where f is defined in a measurable subset Ω of $\mathbb{R}^m$, and integration is with respect to Lebesgue measure. A function $f : \Omega \to E$ is **countably valued** if it can be written as a (finite or infinite) sum

$$f(x) = \sum \chi_n(x) f_n \tag{5.0.17}$$

where each f_n belongs to E, each $\chi_n(\cdot)$ is the characteristic function of a measurable set e_n and the sets e_n are pairwise disjoint with union Ω. The function $f : \Omega \to E$ is **strongly measurable** if there exists a sequence $\{f_n(\cdot)\}$ of countably valued functions such that $f_n(x) \to f(x)$ uniformly outside a null set.

The countably valued function (5.0.17) is **integrable** if $\Sigma \mu(e_n) \|f_n\| < \infty$; its **integral** is

$$\int_\Omega f(x)dx = \sum \mu(e_n) f_n \tag{5.0.18}$$

In both sums, if $f_n = 0$ we take $f_n \mu(e_n) = \|f_n\| \mu(e_n) = 0$ even if $\mu(e_n) = +\infty$. Although a given countably valued function can be written in the form (5.0.17) in many ways, $\int_\Omega f dx$ does not depend on the particular representation.

A strongly measurable function $f : \Omega \rightarrow E$ is **integrable** if there exists a sequence $\{f_n\}$ of integrable countably valued functions such that $f_n(x) \rightarrow f(x)$ uniformly outside of a null set e and

$$\int_\Omega \|f_n(x) - f_m(x)\|dx \rightarrow 0 \quad \text{as } m, n \rightarrow \infty. \tag{5.0.19}$$

By definition,

$$\int_\Omega f(x)dx = \lim_{n \rightarrow \infty} \int_\Omega f_n(x)dx. \tag{5.0.20}$$

The definition of integral (or of integrability) does not depend on the particular sequence $\{f_n\}$ chosen; two functions such that $f(x) = g(x)$ a.e. are integrable or nonintegrable together and, if integrable, have the same integral. The class of Lebesgue–Bochner integrable functions is a subspace over which $\int_\Omega$ is linear. Additivity of domains holds as well. Relations between the integral of $f(\cdot)$ and the integral of (the scalar-valued function) $\|f(\cdot)\|$ follow from the triangle inequality and its consequence $\big| \|y\| - \|z\| \big| \leq \|y - z\|$. For instance, if $f(\cdot)$ is strongly measurable then $\|f(\cdot)\|$ is measurable, and $f(\cdot)$ is Lebesgue–Bochner integrable if and only if $\|f(\cdot)\|$ is Lebesgue integrable,[1] both integrals standing in the relation

$$\left\| \int_\Omega f(x)dx \right\| \leq \int_\Omega \|f(x)\|dx. \tag{5.0.21}$$

As for most definitions in a linear space, there is a version of measurability connected with linear functionals: a function $f(\cdot)$ is E^*-**weakly measurable** (or simply **weakly measurable**) if $\langle y^*, f(\cdot)\rangle$ is measurable for every $y^* \in E^*$. An E^* valued function $g(\cdot)$ may be E^{**}-weakly measurable (if $\langle y^{**}, f(\cdot)\rangle$ is measurable for $y^{**} \in E^{**}$) or E-weakly measurable (if $\langle y, f(\cdot)\rangle$ is measurable for $y \in E$). The first notion is in general stronger than the second. Strong measurability implies every type of weak measurability. The converse is not true:

Example 5.0.10. Let $E = L^1(0, 1)$. Then $E^* = L^\infty(0, 1)$ and the E^*-valued function $t \rightarrow \chi(\cdot - t)$ (χ the characteristic function of $[0, 1]$) is E-weakly measurable but not strongly measurable.

However, we have

Theorem 5.0.11. *Let E be separable. Then $f(\cdot)$ is weakly measurable if and only if it is strongly measurable. If E^* is separable, E-weak measurability is equivalent to strong measurability.*

---

[1] The symbol $\|f(\cdot)\|$ indicates the function $t \rightarrow \|f(\cdot)\|$, *not* the norm of $f(\cdot)$ in some function space.

See Example 5.0.38 or, for the first statement Hille–Phillips [1957, p. 72].

Below, $m = 1$ and $\Omega = [0, T]$; $f(\cdot)$ is a function integrable in $[0, T]$. A point $t \in (0, T)$ is a **Lebesgue point** of $f(\cdot)$ if

$$\lim_{h \to 0} \frac{1}{h} \int_t^{t+h} \| f(\tau) - f(t) \| d\tau = 0 \tag{5.0.22}$$

(the limit is two-sided). As in the scalar case, we have

Theorem 5.0.12. *Almost every point of $[0, T]$ is a Lebesgue point of f.*

For a proof, see Dunford–Schwartz [1958, p. 217] or Hille–Phillips [1958, p. 86]. At a Lebesgue point,

$$\frac{1}{h} \int_t^{t+h} f(\tau) d\tau \to f(t). \tag{5.0.23}$$

Theorem 5.0.12 implies that, if $g(\cdot)$ is integrable in $0 \leq t \leq T$ and

$$f(t) = C + \int_0^t g(\tau) d\tau, \tag{5.0.24}$$

then $f'(t)$ exists almost everywhere and

$$f'(t) = g(t).$$

As in the scalar case, it follows from equicontinuity of the integral (of $\|g(\cdot)\|$) that $f(\cdot)$ is absolutely continuous; for every $\varepsilon > 0$ there exists $\delta > 0$ such that $\Sigma \| f(b_j) - f(a_j) \| \leq \varepsilon$ for every finite collection $\{(a_j, b_j)\}$ of pairwise disjoint subintervals of the interval $[0, T]$ such that $\Sigma(b_j - a_j) \leq \delta$. However, not every Banach space valued absolutely continuous functions is the indefinite integral (5.0.24) of an integrable function:

Example 5.0.13. Let E be a Banach space, $g(\cdot)$, a E-weakly measurable E^*-valued function which is not strongly measurable (such as the one in Example 5.0.11) and let $h(t)$ be defined by

$$\langle h(t), y \rangle = \int_0^t \langle g(\tau), y \rangle d\tau \quad (y \in E).$$

Then $h(t)$ is absolutely continuous but cannot be expressed in the form (5.0.24) with $g(\cdot)$ strongly measurable.

A Banach space is called a **Gelfand space** (Diestel–Uhl [1977, p. 107]) if every E-valued absolutely continuous function is differentiable almost everywhere. The result below (Diestel–Uhl [1977, Theorem 1, p. 79, and Corollary 4, p. 82]) identifies some Gelfand spaces through the equivalent Radon–Nikodým property (Diestel–Uhl [1977, p. 61 and Theorem 2, p. 107]).

Theorem 5.0.14. *Let E be either* (a) *reflexive or* (b) *separable and the dual of another Banach space. Then E is a Gelfand space.*

A function $f(\lambda)$ with values in a complex Banach space E is **analytic** in a domain $D \subseteq \mathbb{C}$ if $f'(\lambda)$ exists (as the limit of the quotient of increments) in the norm of E. Equivalently, f is analytic if the function $\langle y^*, f(\lambda) \rangle$ is analytic for every $y^* \in E^*$. Vector-valued analytic functions enjoy the usual properties of ordinary analytic functions; for instance, they can be developed in convergent power series whose radius of convergence is the distance to the boundary of the domain of analyticity. See Hille–Phillips [1957, 3.10] for additional details.

Let E be a Banach space. We consider below linear operators A which are neither bounded nor defined in all of E but in a (not necessarily closed) subspace $D(A)$ depending on A. We write $A \subseteq B$ for two such operators if $D(A) \subseteq D(B)$ and the restriction of B to $D(A)$ coincides with A. A scalar λ belongs to $\rho(A)$, the **resolvent set** of A if and only if $\lambda I - A$ has a bounded inverse $R(\lambda; A)$, that is, if there exists a bounded, everywhere defined operator $R(\lambda; A)$ such that $R(\lambda; A)E \subseteq D(A)$ and $(\lambda I - A)R(\lambda; A) = I$, $R(\lambda; A)(\lambda I - A)y = y$ for $y \in D(A)$. The **spectrum** of A is the complement $\sigma(A)$ of $\rho(A)$; either set may be empty. The resolvent set is always open, and the **resolvent** $R(\lambda; A)$ is an analytic $(E; E)$-valued function on $\rho(A)$. When A is bounded, both sets are nonempty. For proofs of these and other results on the spectrum and the resolvent see the author [1983, p. 8]. We emphasize that these and other results below (unless otherwise stated) do not require the domain $D(A)$ to be dense in E; when it is, we call A **densely defined.** Although in this work A usually stands for a differential operator restricted by boundary conditions and the space E is chosen in such a way that $D(A)$ is dense, the adjoints A^* (see below) may not be densely defined.

The operator A is **closed** if and only if, for every sequence $\{y_n\} \in D(A)$ such that $y_n \to y \in E$ and $Ay_n \to z \in E$ we have $y \in D(A)$ and $Ay = z$. If $\rho(A) \neq \emptyset$, then A is closed. The converse is not true:

Example 5.0.15. Let A be the derivative operator $(Ay)(x) = y'(x)$ in $L^2(0, 1)$ with domain $D(A)$ consisting of all absolutely continuous functions in $[0, 1]$ with $y'(x) \in L^2(0, 1)$. Let A_1 be the restriction of A to $D(A_1) = \{y(\cdot) \in D(A); y(0) = 0\}$, A_2 the restriction of A to $D(A_2) = \{y(\cdot) \in D(A); y(0) = y(1) = 0\}$. Then A, A_1 and A_2 are densely defined and closed with $\rho(A_1) \neq \emptyset$, $\rho(A) = \rho(A_2) = \emptyset$.

Theorem 5.0.16 (Closed graph theorem). *Let A be closed and everywhere defined. Then A is bounded.*

For a proof, see Banach [1932, p. 38] or Dunford–Schwartz [1958, p. 55]. Theorem 5.0.16 has number of striking consequences. For instance, if A is a closed operator with an everywhere defined inverse A^{-1}, the latter must be bounded

$(A^{-1}$ is closed). As a particular case, let E be a linear space equipped with two norms $\| \cdot \|_1$, $\| \cdot \|_2$ and such that E is a Banach space with respect to both. Then, if $\|y\|_1 \leq C\|y\|_2$ the opposite estimate also holds (consider the identity map from $(E, \| \cdot \|_2)$ into $(E, \| \cdot \|_1)$).

Lemma 5.0.17. *Let A be closed, $f(\cdot)$ such that $f(t) \in D(A)$ almost everywhere, and let $f(\cdot)$, $Af(\cdot)$ be (strongly measurable and) integrable. Then $\int_\Omega f dx \in D(A)$ and $A \int_\Omega f(x)dx = \int_\Omega Af(x)dx$.*

For a proof see Hille–Phillips [1957, p. 83].

Let A be a densely defined linear operator. The **adjoint** A^* of A is defined as follows. An element $y^* \in E^*$ belongs to $D(A^*)$ if and only if $y \to \langle y^*, Ay \rangle$ is a continuous linear functional of y (that is, if there exists $z^* \in E^*$ such that $\langle z^*, y \rangle = \langle y^*, Ay \rangle$ ($y \in D(A)$), and we define $A^*y^* = z^*$. Since $D(A)$ is dense, z^* is uniquely defined and A^* is always closed. In general, $D(A^*)$ may not be dense in E^*, although this is the case if E^* is reflexive and A is closed. If B is bounded and everywhere defined, then B^* is bounded as well and everywhere defined; moreover, $\|B^*\| = \|B\|$. For the proofs of these statements and general rules of calculation with adjoints, see for instance, Hille–Phillips [1957, Ch. 2] and Riesz–Nagy [1955, Chapter VIII]. For unbounded densely defined operators A, B, we check easily that

$$(A + B)^* \supseteq A^* + B^*, \qquad (BA)^* \supseteq A^*B^* \qquad (5.0.25)$$

whenever $A + B$ or BA have dense domains (so that adjoints can be defined). When B is bounded and everywhere defined, the opposite inclusions hold as well (see Example 5.0.43). Hence

$$(A + B)^* = A^* + B^*, \qquad (BA)^* = A^*B^*. \qquad (5.0.26)$$

In particular, it follows that if A is densely defined and A^{-1} is bounded and everywhere defined, then A^* is invertible with

$$(A^*)^{-1} = (A^{-1})^* \qquad (5.0.27)$$

so that if $\lambda \in \rho(A)$ then $\lambda \in \rho(A^*)$ and $R(\lambda; A^*) = R(\lambda; A)^*$.

When E is a Hilbert space, adjoints are defined using the scalar product $(\cdot, \cdot)$ instead of the dual canonical pairing $\langle \cdot, \cdot \rangle$; these are the same in a real space but not in a complex space. An operator A is **symmetric** if and only if $(Ay, z) = (y, Az)$ ($y, z \in D(A)$); this relation is equivalent to $A \subseteq A^*$. The operator A is **self adjoint** if it is densely defined and

$$A^* = A. \qquad (5.0.28)$$

A bounded, everywhere defined symmetric operator is automatically self adjoint.

The operator A is **normal** if it is densely defined and

$$AA^* = A^*A. \tag{5.0.29}$$

Finally, a bounded operator B is **unitary** if and only if

$$B^* = B^{-1}. \tag{5.0.30}$$

A **projection** is a bounded symmetric operator P with

$$P^2 = P. \tag{5.0.31}$$

A **spectral measure** (denoted $P(d\lambda)$) is a set function defined in the σ-field $\Phi_b(\mathbb{C})$ of all Borel sets in the complex plane whose values are projections and which satisfies (a) $P(\emptyset) = 0$, (b) $P(\mathbb{C}) = I$, (c)

$$P\left(\bigcup e_j\right)y = \sum P(e_j)y$$

for every sequence $e_1, e_2, \ldots$ of disjoint Borel sets and every $y \in E$, (d)

$$P(e_1 \cap e_2) = P(e_1)P(e_2)$$

for any two Borel sets e_1, e_2.

Theorem 5.0.18. *Let A be normal. Then there exists a spectral measure $P(d\lambda)$ such that (a) $P(e) = 0$ if $e \cap \sigma(A) = \emptyset$, (b) $P(e)H \subseteq D(A)$, and $AP(e) \supseteq P(e)A$ for any Borel set e, (c) $P(e)$ commutes with all bounded operators that commute with A, (d) $D(A)$ consists of all $y \in H$ such that $\int_{\sigma(A)} |\lambda|^2 \|P(d\lambda)y\|^2 < \infty$ and*

$$(Ay, z) = \int_{\sigma(A)} \lambda(P(d\lambda)y, z) \quad (y \in D(A), z \in H). \tag{5.0.32}$$

Note that, if $\{e_j\}$ is a countable family of disjoint Borel sets, then $\Sigma\|P(e_j)y\|^2 = \Sigma(P(e_j)y, P(e_j)y) = \Sigma(P(e_j)y, y) = (P(e)y, y) = (P(e)y, P(e)y) = \|P(e)y\|^2$, with $e = \cup e_j$, so that $\|P(d\lambda)y\|^2$ is a countably additive measure, and the same is true of $(P(d\lambda)y, z)$. That gives sense to the integral (5.0.32) (and makes unnecessary at this point a theory of integration with respect to vector-valued measures). For a proof of Theorem 5.0.18, see Dunford–Schwartz [1963]. The spectral measure $\{P(d\lambda)\}$ (called the **resolution of the identity** for A) is also **regular** in the sense that, given $y \in E$, $\varepsilon > 0$ and a Borel set e, there exists a closed set $d \subseteq e$ with $\|E(c)y\| \leq \varepsilon$ for every Borel set $c \subseteq e \backslash d$.

   Theorem 5.0.18 applies in particular to self adjoint operators. In this case the spectral measure vanishes for any e that does not intersect $\mathbb{R}$, thus, the integral (5.0.32) is taken over the real axis. It can be proved from Theorem 5.0.18 or directly that $\sigma(A) \in [\omega, \infty)$ if and only if $(Ay, y) \geq \omega\|y\|^2$ for $y \in D(A)$. Another popular particular case is that where A is unitary; here, $\sigma(A)$ is contained in the boundary of the unit circle in $\mathbb{C}$ and, after a change of variables, (5.0.32) can be expressed as an integral in $[0, 2\pi]$.

Theorem 5.0.18 is the basis of a **functional calculus** for an arbitrary normal operator A. Let $f(\cdot)$ be a Borel measurable function defined in $\sigma(A)$ (see Example 5.0.44) and such that $|f(\lambda)|$ is bounded except in a set e with $P(e) = 0$. We define a linear, bounded operator $f(A)$ by

$$(f(A)y, z) = \int_{\sigma(A)} f(\lambda)(P(d\lambda)y, z). \tag{5.0.33}$$

The functional calculus enjoys the following properties:

$$(f + g)(A) = f(A) + g(A), \qquad (fg)(A) = f(A)g(A),$$
$$\|f(A)\| = \|f\|, \tag{5.0.34}$$

where $\|f\|$ indicates the essential supremum of f with respect to $P(d\lambda)$ ($\|f\|$ is the least constant C with $|f(\lambda)| \leq C$ except in a set e with $P(e) = 0$). We also have $f(A)^* = \bar{f}(A)$. Finally,

$$\|f(A)y\|^2 = \int_{\sigma(A)} |f(\lambda)|^2 \|P(d\lambda)y\|^2. \tag{5.0.35}$$

The functional calculus can be extended to unbounded functions: if $f(\lambda)$ is a Borel measurable a. e. finite function in $\sigma(A)$, the operator $f(A)$ is again defined by (5.0.33), the domain of A consisting of all $y \in E$ such that the right side of (5.0.35) is finite. This extended functional calculus can be used, for instance, to compute square roots: if A is a self adjoint operator with $\sigma(A) \subseteq [\omega, \infty)$, $\omega > 0$, then the function $f(\lambda) = \sqrt{\lambda}$ (positive square root) produces a self adjoint square root $A^{1/2}$ satisfying $A = A^{1/2}A^{1/2}$ and $\sigma(A) \subseteq [\sqrt{\omega}, \infty)$.

Let E be a Banach space, $\Omega \subseteq \mathbb{R}^m$ a measurable set. The space $L^p(\Omega; E)$ ($1 \leq p \leq \infty$) consists of all E-valued strongly measurable functions $f(\cdot)$ defined in Ω and such that

$$\|f\|_p = \left(\int_\Omega \|f(x)\|^p dx \right)^{1/p} < \infty. \tag{5.0.36}$$

Since there are nonzero functions with $\|f\|_p = 0$ (namely, any function zero almost everywhere) the correct definition of $L^p(\Omega; E)$ is by equivalence classes with respect to the equivalence relation: $y(\cdot) \approx z(\cdot)$ if and only if $y(x) = z(x)$ a.e. The space with $E = \mathbb{R}$ or $E = \mathbb{C}$ is simply called $L^p(\Omega)$. These spaces have been defined in **2.0** for $E = \mathbb{R}^m$. Hölder's and Minkowski's inequalities look just like (2.0.44) and (2.0.43) in **2.0**. The norm in $L^\infty(\Omega; E)$ is

$$\|f\|_\infty = \operatorname*{ess.\,sup}_{x \in \Omega} \|f(x)\| \tag{5.0.37}$$

where, as in **2.0**, ess. sup is the least C such that $\| f(x) \| \leq C$ a.e. in Ω. The equivalence relation in $L^\infty(\Omega\ E)$ is the same as the one in $L^p(\Omega; E)$, and the fact that $\| \cdot \|_\infty$ is a norm is elementary. For a proof of the fact that $L^p(\Omega; E)(1 \leq p \leq \infty)$ is complete (thus a Banach space), see Dunford–Schwartz [1958, p. 119]. If E is a Hilbert space, the space $L^2(\Omega; E)$ is a Hilbert space under the inner product

$$(g(\cdot), f(\cdot)) = \int_\Omega (g(x), f(x))dx. \tag{5.0.38}$$

Let $g(\cdot) \in L^q(\Omega; E^*)$, $1/p + 1/q = 1$. Then we show using Hölder's inequality that

$$\langle g(\cdot), f(\cdot) \rangle = \int_\Omega \langle g(x), f(x) \rangle dx \tag{5.0.39}$$

defines a bounded linear functional in $L^p(\Omega; E)$ (the integrand is Lebesgue measurable). However, depending on the space E, there may be linear functionals not representable by (5.0.39) with $g(\cdot) \in L^q(\Omega; E^*)$. The characterization of the dual $L^p(\Omega; E)^*$ is subtle and turns out to be essential for relaxed control theory in Part III. However, (5.0.39) does represent all linear functionals when $p < \infty$ and $E = \mathbb{R}, \mathbb{C}$ or, more generally, $E = \mathbb{R}^k, \mathbb{C}^k$.

Theorem 5.0.19 (F. Riesz). *Let* $1 \leq p < \infty$. *Then we have* $L^p(\Omega; \mathbb{R}^k)^* = L^q(\Omega; \mathbb{R}^k)$ *in the sense that every* $\phi \in L^p(\Omega; \mathbb{R}^k)^*$ *admits the representation* (5.0.39) *with (a unique)* $g(\cdot) \in L^q(\Omega; \mathbb{R}^k)$; *the norm* $\|\phi\|$ *and the* $L^q(\Omega; \mathbb{R}^k)$ *norm of* $g(\cdot)$ *are the same. Moreover, the map* $\phi \to g(\cdot)$ *is linear. Same statement for* $\mathbb{C}^k$.

For a proof, see Dunford–Schwartz [1958, p. 286] or Goffman–Pedrick [1983, p. 142]. A much more general result (including a complete identification of $L^p(\Omega; E)$ for E arbitrary) will be found in Chapter 12, although it uses Theorem 5.0.19.

We also use the space $C(K)$ of continuous functions in K (K a compact subset of $\mathbb{R}^m$) outfitted with the supremum norm (see **2.0**). The dual $C(K)^*$ of $C(K)$ (and of more general spaces) will be characterized in Chapter 12, but we outline its construction here. A **bounded Borel measure** μ in K is a bounded countably additive measure defined in the σ-field $\Phi_c(K)$ of all Borel sets of K. Its **total variation** is the positive measure $|\mu|$ defined by $|\mu|(e) = \sup \sum |\mu(e_j)|$, the supremum taken over all finite families $e_1, \ldots, e_n$ of pairwise disjoint Borel subsets of e. We say that μ is **regular** if, given a Borel set $e \subseteq K$ and $\varepsilon > 0$, there exist an open set $d \supseteq e$ and a closed set $c \subseteq e$ with $|\mu|(d/c) \leq \varepsilon$.

Theorem 5.0.20 (F. Riesz). *The dual* $C(K)^*$ *can be identified algebraically and metrically with the space* $\Sigma(K)$ *of all bounded regular Borel measures in* K

endowed with the norm $\|\mu\| = |\mu|(K)$. *The duality pairing is*

$$\langle \mu, f \rangle = \int_K f(x)\mu(dx). \tag{5.0.40}$$

There are approaches to Theorem 5.0.20 that are faster than that in Chapter 12; for instance, see Rudin [1966, Theorem 6.19, p. 131].

A countably valued function (5.0.17) is **simple** if the sum is *finite*, and the χ_n are characteristic functions of *bounded* pairwise disjoint sets e_n.

Theorem 5.0.21. *Simple functions are dense in* $L^p(\Omega; E)$, $1 \le p < \infty$.

Proof. Given $f(\cdot) \in L^p(\Omega; E)$, $\varepsilon > 0$, we choose N such that $\int_{\|x\| \ge N} \|f(x)\|^p dx \le \varepsilon$. The function $\chi(\cdot)f(\cdot)$ (χ the characteristic function of the ball $B(0, N)$) is integrable; thus, by definition of the Lebesgue–Bochner integral there exists an integrable countably valued function $f_\varepsilon(\cdot)$ such that $\|f_\varepsilon(x) - f(x)\| \le \varepsilon$ in $B(0, N)$ except in a null set, the functions χ_n making up f_ε the characteristic functions of a countable family of pairwise disjoint measurable sets $e_j \subseteq B(0, N)$. Using equicontinuity of the integral of $\|f(\cdot)\|$, select $\delta > 0$ such that $\int_e \|f(x)\|^p dx \le \varepsilon$ whenever $|e| \le \delta$, and then select n so large that $|e_{n+1} \cup e_{n+2} \cup \cdots| \le \delta$. If $g_\varepsilon(\cdot)$ is the simple function obtained dropping terms of index $> n$ from the sum (5.0.17) for f_ε, we have $\|g_\varepsilon(x) - f(x)\| \le \varepsilon$ in $e = B(0, N) \setminus (e_{n+1} \cup e_{n+2} \cup \cdots) \subseteq B(0, N)$; in the complement of e, $\int \|f(x)\|^p dx \le 2\varepsilon$. $\blacksquare$

Dense sets of smooth functions will be obtained using convolutions,

$$(f * g)(x) = \int_{\mathbb{R}^m} f(x - \xi)g(\xi)d\xi.$$

Theorem 5.0.22 (Young). *Let* $1 \le p \le \infty$, $f(\cdot) \in L^p(\mathbb{R}^m)$, $g(\cdot) \in L^{p'}(\mathbb{R}^m)$ *with* $1/p + 1/p' \ge 1$. *Then the convolution* $f * g$ *exists for almost all* $x \in \mathbb{R}^m$, $f * g \in L^r(\mathbb{R}^m)$ *for* $1/r = 1/p + 1/p' - 1$ *and*

$$\|f * g\|_r \le \|f\|_p \|g\|_{p'}. \tag{5.0.41}$$

See Stein–Weiss [1971, p. 178]. We only use the case $p' = 1$ ($r = p$), whose proof is easier. Write $|f(x - \xi)g(\xi)| = |f(x - \xi)||g(\xi)|^{1/p}|g(\xi)|^{1/q}$, $1/p + 1/q = 1$. Apply (formally) Hölder's inequality:

$$|(f * g)(x)| \le \left(\int_{\mathbb{R}^m} |f(x - \xi)|^p |g(\xi)| d\xi \right)^{1/p} \left(\int_{\mathbb{R}^m} |g(\xi)| d\xi \right)^{1/q}. \tag{5.0.42}$$

Then take the pth power, integrate with respect to x, and switch the order of integration:

$$\int_{\mathbb{R}^m} |(f * g)(x)|^p dx \le \left(\int_{\mathbb{R}^m} |g(\xi)| d\xi \right)^{1+p/q} \int_{\mathbb{R}^m} |f(x)|^p dx = (\|g\|_1)^p (\|f\|_p)^p. \tag{5.0.43}$$

We now argue backwards applying Tonelli's Theorem 2.0.17. Since the integral on the right side of (5.0.43) exists, so does the integral with the order of integration reversed. This one bounds the integral on the left side so that $(f * g)(\cdot) \in L^p$, and $(f * g)(x)$ exists for almost all x. The case $p = \infty$ is elementary. ∎

Let $\eta(\cdot)$ be a nonnegative infinitely differentiable function defined in $\mathbb{R}^m$ with support in $\|x\| \leq 1$ and integral 1. The sequence $\{\eta_n(x)\} = \{n^m \eta(nx)\}$ is called a **mollifier** or a δ-**sequence**; each $\eta_n(\cdot)$ is infinitely differentiable and nonnegative with support in $\|x\| \leq 1/n$, and has integral 1. It follows from Young's theorem that the operator

$$J_n f = \eta_n * f \tag{5.0.44}$$

maps $L^p(\mathbb{R}^m; E)$ into $L^p(\mathbb{R}^m; E)$ for $1 \leq p \leq \infty$ with norm $\|J_n\|_p \leq 1$. Moreover, clearly justifiable repeated differentiations under the integral sign show that each $J_n f$ is an infinitely differentiable E-valued function.

Theorem 5.0.23. *Let* $y(\cdot) \in L^p(\mathbb{R}^m; E)$, $1 \leq p < \infty$. *Then* (a) *The integrals of the sequence* $\{\|(J_n f)(\cdot)\|^p\}$ *are equicontinuous; moreover, given* $\varepsilon > 0$ *there exists* $N \geq 0$ *such that*

$$\int_{\|x\| \geq N} \|(J_n f)(x)\|^p dx \leq \varepsilon \tag{5.0.45}$$

for all n, (b) $(J_n f)(x) \to y(x)$ *in* E *for almost all* x, (c)

$$\|J_n f - f\|_p \to 0 \quad as \ n \to \infty. \tag{5.0.46}$$

For the proof of (a), let $e \subseteq \mathbb{R}^m$ be a measurable set and χ its characteristic function. We estimate as in (5.0.43) writing $\chi(x)\|\eta_n(\xi) f(x - \xi)\| = \chi(x)\|f(x - \xi)\| \eta_n^{1/p}(\xi) \eta_n^{1/q}(\xi)$ and using Hölder's inequality. The result is

$$\int_e \|(J_n f)(x)\|^p dx$$

$$= \int_{\mathbb{R}^m} \chi(x)\|(\eta_n * f)(x)\|^p dx$$

$$\leq \left(\int_{\mathbb{R}^m} \eta_n(\xi) d\xi \right)^{p/q} \int_{\mathbb{R}^m} \left(\eta_n(\xi) \int_{\mathbb{R}^m} \chi(x)\|f(x - \xi)\|^p dx \right) d\xi$$

$$= \int_{\mathbb{R}^m} \eta_n(\xi) \left(\int_{e - \{\xi\}} \|f(x)\|^p dx \right) d\xi$$

which shows equicontinuity (taking e with small measure) and condition (5.0.45) (taking $e = \{x \in \mathbb{R}^m; \|x\| \geq N\}$). By virtue of Vitali's theorem, L^p convergence is a consequence of a.e. convergence, which itself follows from Theorem 5.0.24 below.

Let $f(\cdot)$ be **locally integrable** in $\mathbb{R}^m$, that is, integrable over any bounded measurable subset of $\mathbb{R}^m$. Call a point $x \in \mathbb{R}^m$ a **symmetric Lebesgue point**[2] of $f(\cdot)$ if

$$\lim_{h \to 0} \frac{1}{h^m} \int_{\|\xi - x\| \le h} \|f(\xi) - f(x)\| d\xi = 0. \tag{5.0.47}$$

Theorem 5.0.24. *Let $f(\cdot)$ be locally integrable. Then almost every $x \in \mathbb{R}^m$ is a symmetric Lebesgue point of $f(\cdot)$.*

For a proof in the scalar case, see Stein–Weiss [1971, p. 12, and Corollary 3.14, p. 71]; the proof for vector-valued functions is the same. For $m = 1$, the definition of left Lebesgue point is also the same as in the scalar case (see (2.3.9)). Theorem 5.0.24 holds as well for left Lebesgue points.

End of Proof of Theorem 5.0.23. We have

$$\|(J_n f)(x) - f(x)\| \le \int_{\mathbb{R}^m} \eta_n(\xi) \|f(x - \xi) - f(x)\| d\xi$$

$$\le C n^m \int_{\|\xi\| \le 1/n} \|f(x - \xi) - f(x)\| d\xi, \tag{5.0.48}$$

and the right side tends to zero at a symmetric Lebesgue point of $y(\cdot)$. ∎

The approximation in Theorem 5.0.23 can be practiced in $L^p(\Omega; E)$ spaces with $\Omega \subseteq \mathbb{R}^m$ an arbitrary measurable set. It suffices to extend functions $f(\cdot) \in L^p(\Omega; E)$ to $L^p(\mathbb{R}^m; E)$ setting $f(x) = 0$ outside of Ω. The same observation applies to other results below.

If $p > 1$, pointwise estimates for mollified functions are available with the help of the **maximal operator** M, defined in $L^p(\mathbb{R}^m)$ by

$$(Mf)(x) = \sup_{r > 0} \frac{1}{b_m r^m} \int_{\|\xi - x\| \le r} |f(\xi)| d\xi \tag{5.0.49}$$

with b_m the volume of the unit ball on $\mathbb{R}^m$ (so that $b_m r^m$ is the volume of the ball of radius r). We have

Theorem 5.0.25. *Let $1 < p \le \infty$. Then M maps $L^p(\mathbb{R}^m)$ into itself and there exists a constant $C_{m,p}$ such that*

$$\|Mf\|_p \le C_{m,p} \|f\|_p \quad (f \in L^p(\Omega)). \tag{5.0.50}$$

---

[2] Particularized to $m = 1$, this definition does not duplicate definition (2.0.25) but the version with integral from $t - h$ to $t + h$. Hence the label "symmetric." There exist true generalizations of (2.0.25) to m-dimensional space; see Stein–Weiss [1971] or Dunford–Schwartz [1958].

For a proof, see Stein–Weiss [1971, p. 5]. Theorem 5.0.27 does not hold for $p = 1$.

Corollary 5.0.26. *Let $\{\eta_n(\cdot)\}$ be a mollifier. Then*

$$\|(J_n f)(x)\| \le C(M\|f\|)(x) \quad (x \in \mathbb{R}^m). \tag{5.0.51}$$

Proof.

$$\|J_n f(x)\| \le n^m \int_{\|\xi-x\|\le 1/n} \eta(n(x-\xi))\|f(\xi)\|d\xi$$

$$\le C'n^m \int_{\|\xi-x\|\le 1/n} \|f(\xi)\|d\xi,$$

where C' is the maximum of $\eta(x)$. ∎

Let $F(\Omega)$ be a space of functions in Ω. We denote by $F(\Omega) \otimes E$ the space of all finite linear combinations $\Sigma f_j(x)y_j$ $(f_j(\cdot) \in F(\Omega), y_j \in E)$. The space $\mathcal{D}(\mathbb{R}^m)$ consists of all infinitely differentiable functions with compact support in $\mathbb{R}^m$; elements of $\mathcal{D}(\mathbb{R}^m)$ are called **test functions** in $\mathbb{R}^m$.

Theorem 5.0.27. *$\mathcal{D}(\mathbb{R}^m) \otimes E$ is dense in $L^p(\Omega; E)$ for $1 \le p < \infty$; more generally, if F is a dense subspace of E then $\mathcal{D}(\mathbb{R}^m) \otimes F$ is dense in $L^p(\Omega; E)$.*

Proof. It is enough to do $\Omega = \mathbb{R}^m$. Let $\Sigma\chi_j(\cdot)y_j$ be one of the integrable simple functions in Theorem 5.0.21. Each $\chi_j(\cdot)$ is the characteristic function of a bounded set, thus belongs to $L^p(\mathbb{R}^m)$; we obtain a dense set in $L^p(\mathbb{R}^m; E)$ taking $y_j \in F$. Using Theorem 5.0.23, we approximate each $\chi_j(\cdot)$ in the L^p norm by an infinitely differentiable function $\eta_j(\cdot) \in L^p(\mathbb{R}^m)$; the approximations η_j constructed there have compact support. ∎

Weierstrass' approximation theorem states that any continuous function in a compact set can be uniformly approximated there by a multivariate polynomial. For a proof (of a more general result), see Dunford–Schwartz [1958, p. 272].

Corollary 5.0.28. *Let E be separable, $1 \le p < \infty$. Then $L^p(\Omega; E)$ is separable.*

Proof. Let Z be a countable dense set in E. Approximate each y_j in the sum $\Sigma\eta_j(x)y_j$ by an element $z_j \in Z$ and each $\eta_j(\cdot)$ uniformly by a multivariate polynomial $p_j(x)$ in a closed ball $B(0, n+1)$ such that $B(0, n)$ contains the support of η_j. Then approximate the polynomial uniformly by one with rational coefficients and multiply it by $\rho_n(x)$, where $\rho_n(x) \equiv 1 (\|x\| \le n)$, $\rho_n(x) \equiv 0 (\|x\| \ge n+1)$, $0 \le \rho_n(x) \le 1$. The set of all functions used for approximation is countable. ∎

Example 5.0.29. Simple functions are dense in $L^\infty(\Omega; E)$ if and only if E is finite dimensional. The finite dimensional case reduces to the scalar case $L^\infty(\Omega)$; to approximate $f(\cdot) \in L^\infty(\Omega)$ within $\varepsilon > 0$ by a simple function f_ε, partition the range of f in a finite number of sufficiently small disjoint intervals I_j and then define $f_\varepsilon(x) = f \in I_j$ in each $f^{-1}(I_j)$. If E is infinite dimensional, construct an infinite sequence $\{y_n\}$ in E with $\|y_n\| = 1$, $\|y_n - y_m\| \geq 1/2$ for $m \neq n$ (see Riesz–Nagy [1955, p. 218]) and define $f(x) = y_n$ in $(1/(n+1), 1/n], n = 1, 2, \ldots$; this function cannot be closely approximated in the L^∞ norm by a simple function.

Example 5.0.30. Continuous functions are not dense in L^∞; the function $f(x) = 0$ $(0 \leq x < 1/2)$ $f(x) = 1$ $(1/2 \leq x \leq 1)$ cannot be approximated within $\varepsilon < 1/2$ by any continuous function.

Example 5.0.31. The space $L^\infty(0, 1)$ is not separable. In fact, let M be the subset of $L^\infty(0, 1)$ consisting of functions taking the values $+1$ or -1 in each of the intervals $(1/(n+1), 1/n], n = 1, 2, \ldots$. The set of all such functions is uncountable, and any two different elements of M lie at a distance $= 2$. If Z is a countable dense set there must be an element of Z in each ball $B(y, 1/2), y \in M$, a contradiction since the set of such balls is uncountable and pairwise disjoint.

Example 5.0.32. Let Ω_1, Ω_2 be measurable subsets of $\mathbb{R}^m$. Then (a) if $1 \leq p < \infty$ then

$$L^p(\Omega_1 \times \Omega_2) = L^p(\Omega_1; L^p(\Omega_2)) \tag{5.0.52}$$

with equality of norms, (b) except in trivial cases, (5.0.52) does not extend to $p = \infty$; rather,

$$L^\infty(\Omega_1 \times \Omega_2) = L_w^\infty(\Omega_1; L^\infty(\Omega_2)), \tag{5.0.53}$$

where the space on the right consists of all $L^1(\Omega_2)$-weakly measurable $L^\infty(\Omega_2)$-valued functions. There is also equality of norms; for the definition of the norm in L_w^∞ spaces see **12.2**. Both equalities (5.0.52) and (5.0.53) can be proved by constructing a space F contained in the spaces on the left and right side of each equality and dense in the two spaces, both norms coinciding in F. One such space is that generated by functions of the form $f(x_1)g(x_2)$ with both f and g integrable simple functions.

Approximation and separability in the spaces $C(K)$ are handled in a similar but more elementary way. Weierstrass' theorem and (some of) the arguments used in Corollary 5.0.28 provide

Theorem 5.0.33. *The space $C(K)$ is separable. Multivariate polynomials are dense in $C(K)$.*

As in L^p spaces, approximation results may be proved by extending functions to all of $\mathbb{R}^m$. In fact, using **Tietze's extension theorem** (Dunford–Schwartz [1958, p. 15]), a function $f(\cdot)$ continuous in K can be extended to a continuous function in $\mathbb{R}^m$ that has the same supremum norm and vanishes for $\|x\| \geq N + 1$, where $B(0, N)$ is a ball containing K.

Theorem 5.0.34. *Let $f(\cdot)$ be a bounded continuous E-valued function defined in $\mathbb{R}^m$. Then*

$$\|(J_n f)(x) - f(x)\| \to 0 \tag{5.0.54}$$

uniformly on compact subsets of $\mathbb{R}^m$; if $f(\cdot)$ is uniformly continuous in $\mathbb{R}^m$ then convergence in (5.0.54) is uniform. If $\|f(x)\| \leq C$ $(x \in \mathbb{R}^m)$, then we have $\|(J_n f)(x)\| \leq C$ $(x \in \mathbb{R}^m)$.

The estimate for bounded f is plain; the convergence statements follow from (5.0.48). ∎

A function $f(\cdot)$ is **separably valued** if the range $f(\Omega) \subseteq E$ is separable (that is, it contains a countable dense set). The function is **almost separably valued** if there exists a null set $e \subseteq \Omega$ such that f, restricted to $\Omega\backslash e$, is separably valued.

Example 5.0.35. Assume that $f(\cdot)$ is the pointwise limit a.e. of a sequence of countably valued functions $f_n(\cdot)$. Then $f(\cdot)$ is almost separably valued. Moreover, there exists a sequence $\{g_n(\cdot)\}$ of countably valued functions such that $g_n(x) \to f(x)$ uniformly outside of a null set, so that $f(\cdot)$ is strongly measurable.

The first statement is plain; we take G as the closure of the (countable) set of all values of all the f_n, and it is obvious that $f(x) \in G$ at every point where $f(x) = \lim f_n(x)$. To show the second claim, we may replace E by $F =$ closed subspace generated by all values of all the f_n, and thus assume that E is separable. Let $y_1, y_2, \ldots$ be a countable dense set in E. Since $\|f_n(x) - y\| \to \|f(x) - y\|$ a.e., the norm $\|f(\cdot) - y\|$ is measurable for any y so that, given $\varepsilon > 0$ the sets $d_n = \{x \in \Omega; \|f(x) - y_j\| \leq \varepsilon\}$ are measurable; denseness of $\{y_j\}$ implies that $\cup d_n = \Omega\backslash(\text{null set})$. Define $e_1 = d_1, e_2 = d_2 \backslash e_1, e_3 = d_3 \backslash e_2, \ldots, g(x) = \Sigma \chi_n(x) y_j$, where χ_n is the characteristic function of e_j. Then $\|g(x) - f(x)\| \leq \varepsilon$ a.e. in Ω.

As a very particular case, the result implies that *every strongly measurable function is almost separably valued*. The same argument also proves

Example 5.0.36. Let $f(\cdot)$ be almost separably valued with $\|f(\cdot) - y\|$ measurable for all y. Then $f(\cdot)$ is strongly measurable.

Example 5.0.37. Let E be a separable Banach space. Then there exists a countable subset $\{y_n^*\}$ of the unit ball of E^* such that

$$\|y\| = \sup|\langle y_n^*, y\rangle|. \tag{5.0.55}$$

In fact, let $\{y_n\}$ be a countable dense set in the unit ball of E. Using Corollary 5.0.7 select a sequence $\{y_n^*\}$ in E^* with $\|y_n^*\|_{E^*} = 1$, $\langle y_n^*, y_j\rangle = 1$. If $y \in E$ with $\|y\| = 1$ there exists y_n with $\|y - y_n\| \leq \varepsilon$; thus, $\langle y_n^*, y\rangle \geq \langle y_n^*, y_n\rangle - \varepsilon = 1 - \varepsilon$.

This result does *not* imply that E^* is separable; consider $E = L^1(0, 1)$, $E^* = L^\infty(0, 1)$.

Example 5.0.38. Let the dual E^* of a Banach space E be separable. Then E is separable. To see this, select a countable dense set $\{y_n^*\}$ in E^*. Pick a sequence $\{y_n\}$ in E with $\|y_n\| = 1$ and $\langle y_n^*, y_j\rangle \geq \|y_n^*\|_{E^*}/2$. If the (countable) set of all finite linear combinations with rational coefficients of elements of $\{y_n\}$ is not dense in E, then the closed subspace of E generated by $\{y_n\}$ is strictly contained in E and, by Theorem 5.0.6, there exists a nonzero $y^* \in E^*$ with $\langle y^*, y_n\rangle = 0$ for all x. Taking y_n^* sufficiently close to y^* we obtain a contradiction.

Example 5.0.39. (*a*) If E is separable, any E^*-weakly measurable function is strongly measurable. (*b*) If E^* is separable, any E-weakly measurable function is strongly measurable.

To show (*a*) we note that (5.0.55) implies that $\|f(\cdot) - y\|_E$ is measurable for all y (the supremum of a countable number of measurable functions is measurable), so that Example 5.0.36 applies. The proof of (*b*) is the same: $\|f(\cdot) - y\|_{E^*}$ is the supremum of a countable number of measurable functions, hence measurable.

Example 5.0.40. In absence of separability, the equivalence of weak and strong measurability breaks down. Consider the space E of all families $y = \{y_\alpha; 0 \leq \alpha \leq 1\}$ of real numbers equipped with

$$\|y\|^2 = \sum_{0 \leq \alpha \leq 1} |y_\alpha|^2 < \infty$$

(finiteness of the sum implies $y_\alpha = 0$ except for a finite or countable number of indices depending on y). Equipped with pointwise operations and the norm $\|\cdot\|$, E is a Hilbert space. Define a function $y(t) = \{y_\alpha(t)\}$ by $y_\alpha(t) = \eta(t)$ if $\alpha = t$, $y_\alpha(t) = 0$ if $\alpha \neq t$ ($\eta(t)$ an arbitrary function). Then $\|y(t)\| = \eta(t)$. If $y \in E$ then $t \to \langle y, y(t)\rangle$ is measurable (in fact, zero almost everywhere) but if $\eta(t)$ is not measurable, $y(\cdot)$ cannot be strongly measurable (its norm would have to be measurable). We note that this function is equivalent (in a suitable sense) to the zero function; see **12.2**.

Remark 5.0.41. Strongly measurable functions can be defined as pointwise a.e. limits of countably valued functions. Likewise, in the definition (5.0.19)-(5.0.20) of the Lebesgue–Bochner integral, we may only require convergence a.e. The definition here is seemingly more restrictive, but it yields the same integrable functions and the same integral. This follows from the fact that, since $||\,||f(\cdot)|| - ||f_n(\cdot)||\,|| \leq ||f(\cdot) - f_n(\cdot)||$ we may use Example 2.0.31 to show that $||f(\cdot)||$ is integrable, and from the following result.

Example 5.0.42. Let $f(t)$ be a strongly measurable function with $||f(\cdot)||$ integrable. Then $f(\cdot)$ is integrable and (5.0.21) holds. To see this, write Ω as a disjoint union of subsets Ω_j of finite measure and do the approximation in Example 5.0.35 in each Ω_j but with $\varepsilon_j = \varepsilon/2^j|\Omega_j|$. We construct in this way a sequence of countably valued functions converging uniformly to $f(\cdot)$ outside of a null set and satisfying (5.0.19).

Example 5.0.43. If A is densely defined and B is everywhere defined and bounded, we have (a) $(A^* + B^*) \subseteq A^* + B^*$, (b) $(BA)^* \subseteq A^*B^*$. To show (a) apply the inclusion (5.0.25) to $A + B$ and $-B$. To show (b), pick $y^* \in D((BA)^*)$. We have $\langle B^*y^*, Ay \rangle = \langle y^*, BAy \rangle$, so that $B^*y^* \in D(A^*)$ and $(BA)^*y^* = A^*B^*y^*$.

Example 5.0.44. Let $\Omega \subseteq \mathbb{R}^m$ be a Borel set. A function $f : \Omega \to \mathbb{R}^m$ is **Borel measurable** if $f^{-1}((-\infty, a))$ is a Borel set for all a. Equivalently, f is Borel measurable if it is the uniform limit of countably valued functions (2.0.15) where the e_n are Borel sets (compare with the definition of measurable function in **2.0**). If μ is a **Borel measure** (that is, a countably additive measure defined in the field of all Borel subsets of Ω) then integrals $\int_\Omega f(x)\mu(dx)$ can be defined in the same way as Lebesgue integrals for any Borel measurable function f. See Chapters 10 and 12 for more on integration against general measures.

5.1. Partial Differential Equations as Ordinary Differential Equations in Banach Spaces

Example 5.1.1. Consider the heat equation

$$\frac{\partial y(t, x)}{\partial t} = \frac{\partial^2 y(t, x)}{\partial x^2} \quad (t \geq 0) \tag{5.1.1}$$

in the interval $0 \leq x \leq \pi$, with initial condition

$$y(0, x) = \zeta(x) \quad (0 \leq x \leq \pi), \tag{5.1.2}$$

and boundary conditions

$$y(t, 0) = y(t, \pi) = 0 \quad (t \geq 0). \tag{5.1.3}$$

We assume that $y(t, x)$ is a **classical solution** of the initial-boundary value problem; here, this means that $y(t, x)$ is continuous in $[0, \pi] \times [0, \infty)$ and satisfies the initial and boundary conditions, $y_t(t, x)$ exists and is continuous in the same region, $y_x(t, x)$ and $y_{xx}(t, x)$ exist in $(0, \pi) \times [0, \infty)$ and $y(t, x)$ satisfies the equation (1.1) there. Note that, since $y_{xx} = y_t$, y_{xx} is continuous in $[0, \pi] \times [0, \infty)$ and the same is true of y_x (by integration). Obviously, $\zeta(x)$ itself must be continuous in $[0, \pi]$ and satisfy the boundary conditions (5.1.3).

We can write this initial-boundary value problem as a pure initial value problem for an "ordinary differential equation" in a Banach space as follows. Let $E = C_0[0, \pi]$ be the space of all continuous functions $y(\cdot)$ in $0 \leq x \leq \pi$ with $y(0) = y(\pi) = 0$, equipped with its supremum norm. Define an operator A in E by

$$Ay(x) = y''(x) \tag{5.1.4}$$

with domain $D(A)$ consisting of all functions $y(\cdot) \in C_0[0, \pi]$ twice continuously differentiable in $0 < x < \pi$ with $y'(x), y''(x)$ continuous in $0 \leq x \leq \pi$ and $u(0) = u(\pi) = 0$. Let $y(t, x)$ be a classical solution of (5.1.1)-(5.1.2)-(5.1.3). Define a function $t \to y(t) \in E$ by

$$y(t)(x) = y(t, x) \quad (0 \leq x \leq \pi, \, t \geq 0).$$

Clearly, $y(t) \in D(A)$ for all t. Moreover, if $t \to z(t) \in E$ is defined by

$$z(t)(x) = y_t(t, x)$$

(note that $z(t, 0) = z(t, \pi) = 0$) then, using the uniform continuity of $y_t(t, x)$ on compact subsets of $[0, \pi] \times [0, \infty)$ we show that, for $t \geq 0$,

$$\left\| \frac{y(t + h) - y(t)}{h} - z(t) \right\| \to 0 \tag{5.1.5}$$

as $h \to 0$ (as $h \to 0+$ when $t = 0$). It turns out that $y(t)$ is continuously differentiable in the sense of norm of E and that

$$y'(t) = Ay(t) \quad (t \geq 0), \quad y(0) = \zeta, \tag{5.1.6}$$

where ζ is the initial function in (5.1.2). Conversely, let $y(\cdot)$ be a E-valued function defined and continuously differentiable in $t \geq 0$ and such that $y(t) \in D(A)$ and (5.1.7) are satisfied. Then it is plain that

$$y(t, x) = y(t)(x) \tag{5.1.7}$$

is a classical solution of the initial-boundary value problem (5.1.1)-(5.1.2)-(5.1.3). Accordingly, we have shown equivalence of that problem with the abstract initial value problem (5.1.6).

More information about (5.1.6) can be obtained from this equivalence. Assume the Fourier sine series $\Sigma a_n \sin nx$ of ζ satisfies $\Sigma n^2 |a_n| < \infty$. Then

$$y(t, x) = \sum_{n=1}^{\infty} a_n e^{-n^2 t} \sin nx \tag{5.1.8}$$

is a classical solution of the initial-boundary value problem (hence the function defined by (5.1.7) is a solution of the abstract differential equation (5.1.6)). On the other hand, let $y(\cdot)$ be an arbitrary classical solution of (5.1.1)-(5.1.2)-(5.1.3) and let $y(t, x)$ be defined by (5.1.8). Then it follows from the maximum principle for the heat equation that

$$|y(t, x)| \leq \max_{0 \leq x \leq \pi} |\zeta(x)|$$

or, in Banach space language,

$$\|y(t)\|_E \leq \|\zeta\|_E \quad (t \geq 0). \tag{5.1.9}$$

We have proved that solutions of (5.1.6) exist at least for ζ in the subspace D of E whose elements have Fourier sine series satisfying $\Sigma n^2 |a_n| < \infty$. This subspace is dense in E (Example (5.1.3). Moreover, (5.1.9) and linearity of the equation show that arbitrary solutions of (5.1.7) that coincide for $t = 0$ will coincide for all t. These properties will be used in the next section as motivation to set up a theory of *abstract Cauchy problems*.

The idea of converting partial differential equations into ordinary differential equations in Banach spaces works also for time-dependent equations.

Example 5.1.2. Consider the equation

$$\frac{\partial y(t, x)}{\partial t} = a(t, x) \frac{\partial^2 y(t, x)}{\partial^2 x} + b(t, x) \frac{\partial y(t, x)}{\partial x} \quad (t \geq 0) \tag{5.1.10}$$

where $a(t, x), b(t, x)$ are continuous in $[0, \pi] \times [0, \infty)$, combined with the initial and boundary conditions (5.1.2)-(5.1.3). This equation can be written in the form

$$y'(t) = A(t)y(t) \quad (t \geq 0), \quad y(0) = \zeta \tag{5.1.11}$$

where $A(t)$ is the operator

$$A(t)y(x) = a(t, x)y''(x) + b(t, x)y'(x) \tag{5.1.12}$$

with domain $D(A(t))$ defined as the domain of A in (5.1.4). All the remarks about equivalence of classical solutions of (5.1.1)-(5.1.2)-(5.1.3) and solutions of (5.1.6) apply here.

A partial differential equation like (5.1.1) can be cast as a differential equation (5.1.6) in different Banach spaces. For instance, we may use $E = L^p(0, \pi)$ instead of $C_0(0, \pi)$. The operator A is still defined by (5.1.4), but $D(A)$ consists now of all continuously differentiable functions $y(\cdot)$ having an absolutely continuous derivative $y'(\cdot)$ with $y''(\cdot) \in L^p(0, \pi)$ and such that $y(0) = y(\pi) = 0$ (note that, in this case, the space $E = L^p(0, \pi)$ cannot "carry" the boundary conditions, thus they must be incorporated in the domain of A). Of course, the solutions of the initial-boundary value problem obtained in this way will no longer be classical solutions, but there is nothing sacred about one or other definition of solution. The choice of the space E may be dictated by convenience (for instance, $L^2(0, \pi)$ is a Hilbert space, thus it is a simpler space than $C_0[0, \pi]$) or by physical considerations; if (5.1.1) models a diffusion process, the natural norm would be the $L^1(0, \pi)$ norm since, for positive solutions, it represents the amount of diffusing matter. See the author [1983, Chapter 1] for more on these matters.

Example 5.1.3. Functions having Fourier sine series with $\Sigma n^2 |a_n| < \infty$ are dense in $C_0[0, \pi]$. One way to show this is: given $y(\cdot) \in C_0[0, \pi]$, extend it as an odd 2π-periodic function to $\mathbb{R}$ and then apply Theorem 5.0.34 with the kernel of J_n *even*. In this way, we approximate $y(\cdot)$ in the norm of $C_0[0, \pi]$ by infinitely differentiable functions $z(\cdot)$, odd about $x = 0$ and $x = \pi$. Integrating repeatedly by parts in the integrals giving the Fourier coefficients a_n of z we obtain $\Sigma n^k |a_n| < \infty$ for all k. Dropping a tail of the series, we can even show that the elements of $C_0[0, \pi]$ having finite Fourier series are dense in $C_0[0, \pi]$.

5.2. Abstract Cauchy Problems in $t \geq 0$

Let A be a densely defined linear operator in a Banach space E. A **strong solution** of the abstract differential equation

$$y'(t) = Ay(t) \tag{5.2.1}$$

in $t \geq 0$ is, by definition, an E-valued function $y(t)$ defined and continuously differentiable in the norm of E and such that $y(t) \in D(A)$ and (5.2.1) is satisfied for $t \geq 0$. We say that the **Cauchy** or **initial value problem** for (5.2.1) is **well posed in** $t \geq 0$ if (a) there exists a dense subspace D of E such that for every $\zeta \in D$ there is a solution $y(\cdot)$ of (5.2.1) in $t \geq 0$ with

$$y(0) = \zeta, \tag{5.2.2}$$

(b) there exists a positive function $C(t)$ bounded on bounded subsets of $t \geq 0$ and such that

$$\|y(t)\| \leq C(t)\|y(0)\| \tag{5.2.3}$$

for *any* solution of (5.2.1).

Obviously, (b) and linearity of the equation imply uniqueness of solutions of the initial value problem. Assuming the Cauchy problem for (5.2.1) is well posed in $t \geq 0$, the **propagator** or **solution operator** of (5.2.1) is defined by

$$S(t)y = y(t) \quad (t \geq 0),$$

for $y \in D$, where $y(\cdot)$ is the (unique) solution of (5.2.1) with $y(0) = y$. The operator $S(t)$ is extended to the whole space E by

$$S(t)y = \lim_{n \to \infty} S(t)y_n \tag{5.2.4}$$

where $\{y_n\} \subseteq D, y_n \to y$. By virtue of (5.2.3) we have $\|S(t)(y_n - y_m)\| \leq C(t)\|y_n - y_m\|$ so that the limit exists and it does not depend on the particular sequence $\{y_n\}$. The extended operator is linear and satisfies

$$\|S(t)\| \leq C(t). \tag{5.2.5}$$

Finally, $S(t)$ is **strongly continuous** in $t \geq 0$, that is, $t \to S(t)y$ is continuous in the norm of E in $t \geq 0$ for every $y \in E$. For $y \in D$ this is obvious, since solutions $y(\cdot)$ are continuously differentiable; for arbitrary y, we approximate by elements of D and note that (5.2.3) implies uniform convergence on compact subsets of $t \geq 0$.

If $y(\cdot)$ is an arbitrary solution of (5.2.1) with $y(0) = y$, then

$$y(t) = S(t)y. \tag{5.2.6}$$

This is merely the definition of $S(t)$ if $y \in D$; for general y, we approximate y by elements of D and take limits.

For arbitrary $y \in E$, the function $y(\cdot)$ defined by (5.2.6) may not be a strong solution of (5.2.1). We define $y(\cdot)$ to be a **weak** or **generalized solution**. In general, it may not be differentiable anywhere or belong to $D(A)$. Since many control problems admit only generalized solutions, we call these simply **solutions**.

We have

$$S(0) = I = \text{ identity operator in } E. \tag{5.2.7}$$

Equation (5.2.1) does not depend explicitly on time; hence, if $y(\cdot)$ is a solution in $t \geq 0$ and $s \geq 0$, then

$$z(t) = y(s + t)$$

is also a solution in $t \geq 0$ (with $z(0) = y(s)$). Due to the representation (5.2.6) for arbitrary solutions, we have $S(t + s)y(0) = y(s + t) = z(t) = S(t)z(0) = S(t)y(s) = S(t)S(s)y(0)$. Taking $y(0) \in D$ and using denseness,

$$S(s + t) = S(s)S(t) \quad (s, t \geq 0). \tag{5.2.8}$$

Let $t \geq 0$, and let n be the largest integer not surpassing t. Then we have $S(t) = S(t - n)S(1)^n$, which implies $\|S(t)\| \leq \|S(t - n)\|\|S(1)\|^n \leq C \cdot C^n = C\exp(n \log C) \leq C \exp(t \log C)$, where C is a bound for $C(t)$ in $0 \leq t \leq 1$. It follows that there exists C, ω with

$$\|S(t)\| \leq Ce^{\omega t} \quad (t \geq 0). \tag{5.2.9}$$

An operator valued function satisfying (5.2.7) and (5.2.8) is called a **semigroup**; in this jargon, we have just proved that the solution operator $S(t)$ of a well posed Cauchy problem is a **strongly continuous semigroup**. The converse is as well true.

Theorem 5.2.1. *Let $S(\cdot)$ be a strongly continuous semigroup. Then there exists a (unique) closed, densely defined operator A such that the Cauchy problem for (5.2.1) is well posed in $t \geq 0$ and $S(t)$ is the solution operator of (5.2.1).*

Proof. We define A, the **infinitesimal generator** of S by

$$Ay = \lim_{h \to 0+} \frac{1}{h}(S(h) - I)y, \qquad (5.2.10)$$

the domain of A consisting of all y for which the limit exists. To check that A is densely defined, let

$$y^\alpha = \frac{1}{\alpha} \int_0^\alpha S(s)y\,ds.$$

Using (5.2.8) and rearranging integrals, we show that

$$\frac{1}{h}(S(h) - I)y^\alpha = \frac{1}{\alpha}\left(\frac{1}{h}\int_\alpha^{\alpha+h} S(s)y\,ds - \frac{1}{h}\int_0^h S(s)y\,ds\right)$$

for $h \leq \alpha$. Since $S(\cdot)y$ is continuous, the limit of the right-hand side as $h \to 0+$ exists and equals $\alpha^{-1}(S(\alpha)y - y)$. Accordingly, $y^\alpha \in D(A)$ and

$$Ay^\alpha = \frac{1}{\alpha}(S(\alpha)y - y). \qquad (5.2.11)$$

A similar averaging argument shows that $y^\alpha \to y$ as $\alpha \to 0+$; this establishes the denseness of $D(A)$.

We prove that A is closed. Note that

$$\frac{1}{h}(S(h) - I)S(t)y = S(t)\frac{1}{h}(S(h) - I)y;$$

thus, if we take $y \in D(A)$ and let $h \to 0+$, $S(t)y \in D(A)$ and

$$AS(t)y = S(t)Ay. \qquad (5.2.12)$$

We prove in the same way that if $y \in D(A)$, then $S(t)y$ is differentiable in $t \geq 0$ and

$$(S(t)y)' = S(t)Ay = AS(t)y \qquad (5.2.13)$$

(commutativity of A and $S(t)$ coming from (5.2.12)). This equality is obvious for the right-sided derivative, since $h^{-1}(S(t+h) - S(t))y = S(t)h^{-1}(S(h) - I)y$; for the left-sided derivative, we put to use the equality $h^{-1}(S(t) - S(t-h))y = S(t-h)\{h^{-1}(S(h) - I)y - Ay\} + S(t-h)Ay$. Now, (5.2.11) and (5.2.13) imply

$$(Ay)^h = Ay^h = \frac{1}{h}(S(h)y - y)$$

for $y \in D(A)$. Let $\{y_n\}$ be a sequence in $D(A)$ with $y_n \to y$, $Ay_n \to z$ for some $y, z \in E$. Then

$$\frac{1}{h}(S(h) - I)y = \lim_{n \to \infty} \frac{1}{h}(S(h) - I)y_n = \lim_{n \to \infty} (Ay_n)^h = z^h.$$

Taking limits as $h \to 0+$, we see that $y \in D(A)$ and that $Ay = z$, which completes the proof that A is closed.

Part (a) of the definition of well-posed Cauchy problem has already been verified in (5.2.13), with $D = D(A)$. In the matter of (b) observe that, since $S(\cdot)y$ is strongly continuous, $\|S(t)y\|$ must be bounded on bounded subsets of $t \geq 0$. By the uniform boundedness principle (Theorem 5.0.1) for every $t > 0$ there exists a constant $C(t)$ such that

$$\|S(s)\| \leq C(t) \quad (0 \leq s \leq t). \tag{5.2.14}$$

If $y(\cdot)$ is an arbitrary strong solution of (5.2.1) then $z(s) = S(t - s)y(s)$ is continuously differentiable in $0 \leq s \leq t$ and $z'(s) = S(t - s)Ay(s) - AS(t - s)y(s) = 0$. Hence we obtain $z(t) = z(0)$, or $y(t) = S(t)y(0)$. The continuous dependence assumption is satisfied as a consequence of (5.2.14). This ends the proof of Theorem 5.2.1. ∎

Theorem 5.2.2 (Hille-Yosida). *The operator A is the infinitesimal generator of a strongly continuous semigroup $S(\cdot)$ satisfying*

$$\|S(t)\| \leq Ce^{\omega t} \quad (t \geq 0) \tag{5.2.15}$$

if and only if A is closed and densely defined, $R(\lambda; A) = (\lambda I - A)^{-1}$ exists in $\lambda > \omega$, and

$$\|R(\lambda; A)^n\| \leq C(\lambda - \omega)^{-n} \quad (\lambda > \omega, n = 1, 2, \ldots). \tag{5.2.16}$$

Proofs of Theorem 5.2.2 can be found in most functional analysis textbooks, for instance Dunford–Schwartz [1958, p. 624], or in Examples 5.2.16–5.2.20 below. We note that inequalities (5.2.16) can be extended to complex λ in the half plane $\text{Re } \lambda > \omega$:

$$\|R(\lambda; A)^n\| \leq C(\text{Re } \lambda - \omega)^{-n} \quad (\text{Re } \lambda > \omega, n = 1, 2, \ldots) \tag{5.2.17}$$

Most proofs are based on Laplace transform techniques and imply that $R(\lambda; A)$ can be obtained as the Laplace transform of $S(t)$; precisely,

$$R(\lambda; A)y = \int_0^\infty e^{-\lambda t} S(t)y \, dt \quad (\text{Re } \lambda > \omega). \tag{5.2.18}$$

for every $y \in E$.

Using Theorem 5.2.2 we characterize completely the operators A such that the Cauchy problem for (5.2.1) is well posed in $t \geq 0$.

Theorem 5.2.3. *Let A be densely defined and closed. Then the Cauchy problem for (5.2.1) is well posed in $t \geq 0$ if and only if there exists ω such that $R(\lambda; A)$ exists for $\lambda > \omega$ and inequalities (5.2.16) (equivalently, (5.2.17)) hold.*

Proof. If $R(\lambda; A)$ exists in $\lambda > \omega$ and (5.2.16) holds, Theorem 5.2.2 shows that A is the infinitesimal generator of a strongly continuous semigroup $S(t)$ and, by Theorem 5.2.1, the Cauchy problem for (5.2.1) is well posed with $D = D(A)$. Conversely, let the Cauchy problem for (5.2.1) be well posed with $D \subseteq D(A)$ (recall that D must be dense in E). We begin by showing that we may take $D = D(A)$. Let $S(t)$ be the solution operator of (5.2.1), ω the constant in (5.2.9). For $\lambda > \omega$ define

$$Q(\lambda)y = \int_0^\infty e^{-\lambda t} S(t)y \, dt.$$

The norm of the integrand is bounded by $C\|y\| \exp(\omega - \lambda)t$, thus $Q(\lambda)$ is a bounded operator.

Let $T > 0$, $y \in D$. Using Lemma 5.0.17 we obtain

$$A \int_0^T e^{-\lambda t} S(t)y \, dt = \int_0^T e^{-\lambda t} AS(t)y \, dt = \int_0^T e^{-\lambda t} (S(t)y)' dt$$

$$= e^{-\lambda T} S(T)y - y + \lambda \int_0^T e^{-\lambda t} S(t)y \, dt.$$

Letting $T \to \infty$ and using closedness, we see that $Q(\lambda)y \in D(A)$ and $AQ(\lambda)y = \lambda Q(\lambda)y - y$. Let now $y \in E$, $\{y_n\}$ a sequence in D with $y_n \to y$. Then we have $Q(\lambda)y_n \to Q(\lambda)y$, $AQ(\lambda)y_n = \lambda Q(\lambda)y_n - y_n \to \lambda Q(\lambda)y - y$. Accordingly, $Q(\lambda)y \subseteq D(A)$ and $AQ(\lambda)y = \lambda Q(\lambda)y - y$; in other words, $Q(\lambda)E \subseteq D(A)$ and $(\lambda I - A)Q(\lambda) = I$, which, in particular, implies that $(\lambda I - A) : D(A) \to E$ is onto. It is also one-to-one; for, if y is an element of $D(A)$ such that $Ay = \lambda y$, then $y(t) = e^{\lambda t}y$ is a solution of (5.2.1) with $\|y(t)\| = e^{\mathrm{Re}\lambda t}\|y\|$, which contradicts the estimate (5.2.9) (recall that $\lambda > \omega$). It then follows that $\lambda \in \rho(A)$ with $R(\lambda; A) = (\lambda I - A)^{-1} = Q(\lambda)$.

Let $y \in D$. Arguing as in the proof of Theorem 5.2.1 we deduce that $(S(t)y)' = AS(t)y = S(t)Ay$, so that $(\lambda I - A)S(t)y = S(t)(\lambda I - A)y$; writing $y = R(\lambda; A)z$ and applying $R(\lambda; A)$ to both sides we obtain $S(t)R(\lambda; A)z = R(\lambda; A)S(t)z$ for $z \in R(\lambda; A)D$ and, since the operators on both sides are bounded, $S(t)R(\lambda; A) = R(\lambda; A)S(t)$. This implies that $S(t)D(A) \subseteq D(A)$ and

$$AS(t)y = S(t)Ay. \tag{5.2.19}$$

In view of this equality, we have

$$S(t)y - y = \int_0^t S(s)Ay \, ds \tag{5.2.20}$$

for $y \in D$. Applying $R(\lambda; A)$ to both sides,

$$R(\lambda; A)(S(t)y - y) = \int_0^t S(s)R(\lambda; A)Ay\,ds.$$

Again, the operators on both sides are bounded so that the equality extends to $y \in D(A)$; since $R(\lambda; A)$ is one-to-one, (5.2.20) itself holds for $y \in D(A)$. This, combined with (5.2.19), shows that we may take $D = D(A)$ as claimed. It is then clear that if B is the infinitesimal generator of $S(t)$ we have $A \subseteq B$.

We already know from Theorem 5.2.2 that λ also belongs to $\rho(B)$. Thus, $(\lambda I - B) : D(B) \to E$ is also onto. If the inclusion $A \subseteq B$ is strict, let $y \in D(B)$, $y \notin D(A)$. Then there must exist $z \in D(B)$ with $(\lambda I - B)z = (\lambda I - B)y$, so that $(\lambda I - B)(y - z) = 0$, which means $\lambda I - B$ is not one-to-one, a contradiction. This ends the proof. ∎

The following result is useful when a formula for $S(t)$ can be guessed.

Corollary 5.2.4. *Let $S(\cdot)$ be a (E, E)-valued function defined in $t \geq 0$ satisfying (5.2.15) and such that $S(\cdot)y$ is E^*-weakly measurable in E for each $y \in E$. Let A be a densely defined operator such that $R(\lambda; A)$ exists for $\lambda > \omega$ and*

$$\langle y^*, R(\lambda; A)y \rangle = \int_0^\infty e^{-\lambda t} \langle y^*, S(t)y \rangle dt \quad (\lambda > \omega, y \in E, y^* \in E^*). \quad (5.2.21)$$

Then $S(\cdot)$ is a strongly continuous semigroup and A its infinitesimal generator. Equivalently, the Cauchy problem for (5.2.1) is well posed in $t \geq 0$ and $S(\cdot)$ is the solution operator.

Proof. We estimate the integral (5.2.21) using (5.2.15) in the integrand; since y, y^* are arbitrary we obtain the sequence of inequalities (5.2.16). Applying Theorem 5.2.2, A is identified as the infinitesimal generator of a strongly continuous semigroup $T(t)$, which satisfies (5.2.21) as well. Accordingly, for any $y \in E$, $y^* \in E^*$, both $\langle y^*, S(t)y \rangle$ and $\langle y^*, T(t)y \rangle$ have $\langle y^*, R(\lambda; A)y \rangle$ as Laplace transform; thus by uniqueness of these, $\langle y^*, S(t)y \rangle = \langle y^*, T(t)y \rangle$, which implies $S(t) = T(t)$. This ends the proof. ∎

Theorem 5.2.2 is not particularly easy to apply due to the need of verifying the infinite sequence of inequalities (5.2.16). When $C = 1$, all inequalities are consequence of the first one,

$$\|R(\lambda; A)\| \leq 1/(\lambda - \omega) \quad (\lambda > \omega), \quad (5.2.22)$$

corresponding to

$$\|S(t)\| \leq e^{\omega t} \quad (t \geq 0). \quad (5.2.23)$$

This particular case can be put in a very transparent form (Theorem 5.2.5 below)

Let E^* be the dual space of E, $y \in E$. The **duality set** $\Theta(y) \subseteq E^*$ of y consists of all $y^* \in E^*$ such that $\|y^*\|^2 = \|y\|^2 = \langle y^*, y \rangle$, where $\langle \cdot, \cdot \rangle$ is the duality pairing of E and E^*. That $\Theta(y)$ is nonempty for every y is a consequence of Corollary 5.0.7. An operator A in E with domain $D(A)$ is **dissipative** if

$$\operatorname{Re} \langle y^*, Ay \rangle \leq 0 \quad (y \in D(A), y^* \in \Theta(y)). \tag{5.2.24}$$

Theorem 5.2.5 (Lumer-Phillips). *The operator A is the infinitesimal generator of a strongly continuous semigroup satisfying (5.2.23) if and only if $A - \omega I$ is dissipative (that is,* $\operatorname{Re} \langle y^*, Ay \rangle \leq \omega \|y\|^2$ *for $y \in D(A)$, $y^* \in \Theta(y)$) and*

$$(\lambda I - A)D(A) = E \tag{5.2.25}$$

for $\lambda > \omega$.

Proof. Replacing, if necessary, A by $A - \omega I$ we may take $\omega = 0$. Assume A generates a strongly continuous semigroup satisfying (5.2.23) with $\omega = 0$ (that is, with norm uniformly bounded by 1). Let $y \in D(A)$, $y^* \in \Theta(y)$, $h > 0$. Since $\|S(h)y\| \leq \|y\|$ we have $\operatorname{Re} \langle y^*, S(h)y - y \rangle \leq \|y\|^2 - \|y\|^2 = 0$. Dividing by h and taking limits, the dissipativity condition follows. That (5.2.25) is true follows from the fact that $R(\lambda; A)$ exists for $\lambda > 0$.

Conversely, assume that A is dissipative and that (5.2.25) holds. Pick $\lambda > 0$, $y \in D(A)$, $y^* \in \Theta(y)$. Then we have $\lambda \|y^*\| \|y\| = \langle y^*, \lambda y \rangle = \operatorname{Re} \langle y^*, \lambda y \rangle \leq \operatorname{Re} \langle y^*, \lambda y - Ay \rangle \leq \|y^*\| \|\lambda y - Ay\|$, so that $\|\lambda I - Ay\| \geq \lambda \|y\|$. Combining this with (5.2.25), $R(\lambda; A) = (\lambda I - A)^{-1}$ exists for $\lambda > 0$ and satisfies $\|R(\lambda; A)\| \leq 1/\lambda$; this implies $\|R(\lambda; A)^n\| \leq \lambda^{-n}$ and by, Theorem 5.2.2, the operator A is the infinitesimal generator of a strongly continuous semigroup satisfying (5.2.23). ∎

The **adjoint equation** is

$$z'(t) = A^* z(t) \tag{5.2.26}$$

and its solutions live in the dual space E^*. The full theory of (5.2.26) in relation to (5.2.1) will be postponed to **5.5**; for now, we only need

Theorem 5.2.6. *Let A be the infinitesimal generator of a strongly continuous semigroup $S(t)$ in a reflexive space E. Then A^* is the infinitesimal generator of the semigroup $S^*(t) = S(t)^*$ in E^*.*

Proof. If E is reflexive then A^* is densely defined, $\rho(A) = \rho(A^*)$, and $R(\lambda; A^*) = R(\lambda; A)^*$ (Example 5.2.14 below) so that inequalities (5.2.16) imply the corresponding inequalities for A^*, and it follows from Theorem 5.2.2 that A^* is the

infinitesimal generator of a strongly continuous semigroup $S^*(t)$. It only remains
to identify $S^*(t)$. From (5.2.18) we obtain

$$\langle R(\lambda; A^*)y^*, y\rangle = \langle R(\lambda; A)^*y^*, y\rangle = \langle y^*, R(\lambda; A)y\rangle$$
$$= \int_0^\infty e^{-\lambda t}\langle y^*, S(t)y\rangle dt = \int_0^\infty e^{-\lambda t}\langle S(t)^*y^*, y\rangle dt;$$

thus, the result follows from Corollary 5.2.4 and the fact that $E^{**} = E$.   ∎

We take a look at the inhomogeneous version of (5.2.1),

$$y'(t) = Ay(t) + f(t), \quad y(0) = \zeta \tag{5.2.27}$$

where $f(\cdot)$ takes values in E. If f is continuous, strong solutions are defined in
the same way as strong solutions of (5.2.1). Since the difference of two solutions
is a solution of the homogeneous equation, well posedness of (5.2.1) implies that
there exists only one strong solution of the initial value problem (5.2.27).

Let $y(\cdot)$ be a strong solution of (5.2.27) and let $S(\cdot)$ be the solution operator of
the homogeneous equation. We show easily that the function $\tau \to S(t - \tau)y(\tau)$
is continuously differentiable in $0 \leq \tau \leq t$ with derivative $-S'(t - \tau)y(\tau) +$
$S(t - \tau)y'(\tau) = S(t - \tau)(y'(\tau) - Ay(\tau)) = S(t - \tau)f(\tau)$. Integrating, we obtain the familiar "variation-of-constants" formula

$$y(t) = S(t)\zeta + \int_0^t S(t - \tau)f(\tau)d\tau. \tag{5.2.28}$$

In general, mere continuity of $f(\cdot)$ does not guarantee that (5.2.28) is a strong
solution. Stronger conditions do:

Lemma 5.2.7. *Assume that $\zeta \in D(A)$ and that either (a) $f(t) \in D(A)$ a.e., $f(\cdot)$
is continuous and $Af(\cdot) \in L^1(0, T; E)$, or (b) $f(\cdot)$ is the indefinite integral of a
function $f'(\cdot) \in L^1(0, T; E)$. Then (5.2.28) is a strong solution of the initial value
problem (5.2.27).*

For a proof, see Example 5.2.15. In control problems, $f(\cdot)$ is often merely
integrable; if $f \in L^1(0, T; E)$, (5.2.28) is **defined** to be the solution of (5.2.27).
Formula (5.2.28) defines a linear operator

$$(\zeta, f(\cdot)) \to y(\cdot). \tag{5.2.29}$$

Lemma 5.2.8. *The operator (5.2.29) is bounded from $E \times L^1(0, T; E)$ into
$C(0, T; E)$.*

Proof. Continuity is obvious for the first term, thus we may assume $\zeta = 0$. If $t < t'$ we have

$$\|y(t') - y(t)\| \leq \int_t^{t'} \|S(t' - \tau)f(\tau)\|d\tau$$

$$+ \int_0^t \|S(t' - \tau)f(\tau) - S(t - \tau)f(\tau)\|d\tau.$$

Let M be a bound for $\|S(t)\|$ in $0 \leq t \leq T$. The first integrand is bounded by $M\|f(\tau)\|$; thus, it tends to zero as $t' - t \to 0$ on account of equicontinuity of the integral of $\|f(\cdot)\|$. The second integrand tends to zero due to strong continuity of $S(\cdot)$ and is bounded by $2M\|f(\cdot)\|$; thus, we can apply the dominated convergence Theorem 2.0.5.

That (5.2.29) is a bounded operator follows from the obvious estimate

$$\|y\|_{C(0,T;E)} \leq M(\|\zeta\|_E + \|f\|_{L^1(0,T;E)}). \tag{5.2.30}$$

∎

Example 5.2.9. Let A be a bounded operator. Then A generates the strongly continuous semigroup $S(t) = e^{tA}$, the exponential computed, for instance, with the power series

$$e^{tA} = \sum_{k=0}^{\infty} \frac{t^k}{k!}A^k. \tag{5.2.31}$$

This semigroup is defined for all t and infinitely differentiable in the norm of (E, E).

Example 5.2.10. Let A be a normal operator in a Hilbert space H. Then the Cauchy problem for (5.2.1) is well posed in $t \geq 0$ (equivalently, A generates a strongly continuous semigroup) if and only if $\omega = \sup \mathrm{Re}\, \sigma(A) < \infty$. The semigroup $S(t)$ satisfies (5.2.23) and ω is the least possible constant for this inequality. We have $S(t) = e^{tA}$, the exponential computed according to the bounded functional calculus in **5.0.**

Example 5.2.11. Let A be a self adjoint operator in a Hilbert space H. Then A generates a strongly continuous semigroup if and only if $\omega = \sup \sigma(A) < \infty$. The semigroup $S(t)$ satisfies (5.2.23) and ω is the least possible constant for this inequality. We have $S(t) = e^{tA}$, the exponential computed according to the bounded functional calculus in **5.0.**

Example 5.2.12. In a complex Hilbert space, duality sets and dissipativity are defined using the scalar product rather than $\langle \cdot, \cdot \rangle$; in this case $\Theta(y) = \{y\}$ and the dissipativity condition (5.2.24) is

$$\mathrm{Re}\,(y, Ay) \leq 0 \quad (y \in D(A)). \tag{5.2.32}$$

Example 5.2.13. The first inequality (5.2.16) yields a useful approximation of the identity; noting that $\lambda R(\lambda; A) = I + R(\lambda; A)A$ for $\lambda > \omega$,

$$\lambda R(\lambda; A)y \to y \quad \text{as } \lambda \to \infty \tag{5.2.33}$$

for $u \in D(A)$; since $\|\lambda R(\lambda; A)\| \leq C$, (5.2.33) can be extended to arbitrary $y \in E$. The first inequality (5.2.16) does not imply the others (unless $C = 1$), so that (5.2.33) applies in a larger class than that of semigroup generators.

Example 5.2.14. Let A be a densely defined operator with $\rho(A) \neq \emptyset$ in a reflexive space E. Then $D(A^*)$ is dense in E^*. In fact, if this is not true and $\lambda \in \rho(A)$, there exists $y \in E^{**} = E$, $y \neq 0$ such that $\langle(\lambda I - A)y, y^*\rangle = \langle y, (\lambda I - A^*)y^*\rangle = 0$ for all $y^* \in D(A^*)$, so that $(\lambda I - A^*)D(A^*) \neq E^*$. This contradicts the fact that $\lambda \in \rho(A^*)$ (see (5.0.27) and following comments).

Example 5.2.15. Lemma 5.2.7 is proved as follows. In case (a) define $z(t)$ as the formal derivative of (5.2.28):

$$z(t) = S(t)A\zeta + f(t) + \int_0^t S(t - \tau)Af(\tau)d\tau.$$

This function is continuous by Lemma 5.2.8. Then show (integrating $z(\cdot)$) that $y'(t) = z(t)$. In case (b), define

$$z(t) = S(t)A\zeta + S(t)f(0) + \int_0^t S(t - \tau)f'(\tau)d\tau$$

and do the same. See the author [1983, p. 87].

The following Examples give the proof of the Hille-Yosida theorem in five easy steps.

Example 5.2.16. Let A be the infinitesimal generator of a strongly continuous semigroup $S(t)$ satisfying (5.2.15) (A is closed and densely defined by Theorem 5.2.1). Then $R(\lambda; A)$ exists for $\lambda > \omega$, is given by (5.2.18) and satisfies the inequalities (5.2.16).

That (5.2.18) holds is proved in very much the same way as the formula $Q(\lambda) = R(\lambda; A)$ in Theorem 5.2.3. To establish inequalities (5.2.16) we take derivatives with respect to λ on both sides of (5.2.18). On the right, we differentiate under the integral sign (this justified on the basis of the dominated convergence theorem; see Dunford–Schwartz [1958, p. 623]); on the left, we use the resolvent equation $R(\mu; A) - R(\lambda; A) = -(\mu - \lambda) \, R(\lambda; A)R(\mu; A)$, which implies

$\partial_\lambda R(\lambda; A) = -R(\lambda; A)^2$, and corresponding formulas for the successive derivatives. After n differentiations, we obtain

$$R(\lambda; A)^n y = \frac{1}{(n-1)!} \int_0^\infty t^{n-1} e^{-\lambda t} S(t) y \, dt \quad (\text{Re } \lambda > \omega). \tag{5.2.34}$$

Inequalities (5.2.16) follow using (5.2.15) to estimate the integrand.

Example 5.2.17. Let A be an operator such that $R(\lambda; A)$ exists in $\lambda > \omega$ and satisfies (5.2.16). Then, if $A_n = nAR(n; A)$ $(n > \omega)$ and $S_n(t) = e^{tA_n}$,

$$\|S_n(t)\| \le C e^{n\omega t / (n-\omega)}. \tag{5.2.35}$$

To prove (5.2.35) we write $A_n = n^2 R(n; A) - nI$, $S_n(t) = e^{-tn} e^{tn^2 R(n;A)}$ and estimate the operator exponential with the power series (5.2.31). We use the inequality $\|(n^2 R(n; A))^k\| \le C(n^2/(n-\omega))^k$ (consequence of (5.2.16)) in each term.

Example 5.2.18. Show

$$S_m(t) - S_n(t) = \int_0^t S_n(t-s) S_m(s) (A_m - A_n) ds. \tag{5.2.36}$$

Hint: $\partial_s S_n(t-s) S_m(s) = S_n(t-s) S_m(s) (A_m - A_n)$ (derivative taken in the norm of (E, E)). Integrate in $0 \le s \le t$.

Example 5.2.19. For each y, $\{S_n(t)y\}$ converges on bounded subsets of $t \ge 0$ to a strongly continuous semigroup $S(t)$ satisfying (5.2.15). Hint: Use Example 5.2.13 to show that $A_n y \to Ay$ for $y \in D(A)$. Then use (5.2.35), (5.2.36) and denseness of $D(A)$. Convergence implies both strong continuity and the semigroup equations (5.2.7)-(5.2.8).

Example 5.2.20. The infinitesimal generator of $S(t)$ coincides with A. In fact, as a strongly continuous semigroup, $S(\cdot)$ has an infinitesimal generator B. On the other hand, we have

$$S(t)y - y = \int_0^t S(s) Ay \, ds \quad (y \in A) \tag{5.2.37}$$

as we see writing the corresponding equality for $S_n(s)$, A_n and letting $n \to \infty$. It follows from (5.2.37) that $y \in D(B)$ and $Ay = By$, so that $A \subseteq B$. Since $R(\lambda; A)$ and $R(\lambda; B)$ both exist for $\lambda > \omega$ we must have $A = B$ (see the end of the proof of Theorem 5.2.3).

Remark 5.2.21. Approximation of the semigroup $S(t)$ by the uniformly continuous semigroups $S_n(t)$ is called the **Yosida approximation** of $S(t)$.

Remark 5.2.22. The above proof of the Hille-Yosida Theorem is "real" (it only uses the inequalities (5.2.16) for λ real) and thus produces versions of all its consequences (such as Theorem 5.2.5) for real spaces. Dissipativity in real spaces is defined by

$$\langle y^*, Ay \rangle \le 0 \quad (y \in D(A), \ y^* \in \Theta(y)). \tag{5.2.38}$$

5.3. Abstract Cauchy Problems in $-\infty < t < \infty$

Example 5.3.1. Consider the wave equation

$$\frac{\partial^2 y(t, x)}{\partial t^2} = \frac{\partial^2 y(t, x)}{\partial x^2} \tag{5.3.1}$$

in the entire axis $-\infty < x < \infty$, with initial conditions

$$y(0, x) = \zeta_0(x), \qquad y_t(0, x) = \zeta_1(x) \quad (-\infty < x < \infty). \tag{5.3.2}$$

Our aim is to rewrite this initial value problem as an ordinary differential equation in a suitable Banach space. We reduce the second order equation to a first order system for two functions, $y(t, x)$ and $y_1(t, x) = y_t(t, x)$:

$$y_t(t, x) = y_1(t, x), \tag{5.3.3}$$

$$y_{1t}(t, x) = y_{xx}(t, x). \tag{5.3.4}$$

The space **E** consists of all function pairs $\mathbf{y} = (y, y_1)$, where y is absolutely continuous with $y'(\cdot) \in L^2(-\infty, \infty)$, and $y_1(\cdot) \in L^2(-\infty, \infty)$. The norm in **E** is $\|\mathbf{y}\| = \|(y, y_1)\|_{\mathbf{E}} = (\|y'\|^2 + \|y_1\|^2)^{1/2}$ where both norms on the right are in $L^2(-\infty, \infty)$. This is not a true norm (the element $(c, 0)$ has norm zero for any constant c) but it becomes one under the equivalence relation: $(y, y_1) \sim (z, z_1)$ if and only if $z_1 = y_1, z - y = $ constant. In fact, **E** is a Hilbert space with the scalar product

$$((y, y_1), (z, z_1))_{\mathbf{E}} = (y', z') + (y_1, z_1), \tag{5.3.5}$$

$(\cdot, \cdot)$ the scalar product in $L^2(-\infty, \infty)$. The choice of space is dictated by physics: the **E**-norm of a solution $(y(t, \cdot), y_t(t, \cdot))$ is the energy at time t.

The system (5.3.3)-(5.3.4) can be written in the form

$$\mathbf{y}'(t) = \mathbf{A}\mathbf{y}(t) \tag{5.3.6}$$

where the operator **A** is

$$\mathbf{A}\mathbf{y} = \mathbf{A}(y, y_1) = (y_1, y'') \tag{5.3.7}$$

with domain $D(\mathbf{A})$ consisting of all pairs $\mathbf{y} = (y, y_1) \in E$ with y_1, y' absolutely continuous and $y_1'(\cdot), y''(\cdot) \in L^2(-\infty, \infty)$. The system (5.3.3)-(5.3.4) (thus the equation (5.3.6)) can be solved explicitly in terms of the initial conditions using D'Alembert's formula for the wave equation,

$$y(t, x) = \frac{\zeta_0(x + t) + \zeta_0(x - t)}{2} + \frac{1}{2} \int_{x-t}^{x+t} \zeta_1(\xi) d\xi, \qquad (5.3.8)$$

which gives

$$y_t(t, x) = \frac{\zeta_0'(x + t) - \zeta_0'(x - t)}{2} + \frac{\zeta_1(x + t) + \zeta_1(x - t)}{2},$$

$$y_x(t, x) = \frac{\zeta_0'(x + t) + \zeta_0'(x - t)}{2} + \frac{\zeta_1(x + t) - \zeta_1(x - t)}{2}. \qquad (5.3.9)$$

It follows that if the initial conditions ζ_0, ζ_1 are sufficiently smooth, (5.3.9) provides a strong solution of (5.3.6), which is (a) in the definition of well posed problem. To show (b), note that if $\mathbf{y}(\cdot) = (y(\cdot), y_1(\cdot))$ is an arbitrary strong solution of (5.3.6), we have

$$(\|\mathbf{y}(t)\|^2)' = (\mathbf{y}(t), \mathbf{y}(t))' = 2(\mathbf{A}\mathbf{y}(t), \mathbf{y}(t)) = 2((y, y_1), (y_1, y'')) = 0,$$

the last equality coming from integration by parts. This "conservation-of-energy" equation obviously implies (b) with $C(t) \equiv 1$, and then shows that the Cauchy problem for (5.3.6) is well posed. What is new is that our arguments are just as valid for positive or for negative t, so that assumption (a) on existence of solutions and inequality (b) are valid in $-\infty < t < \infty$. This is far from true for the parabolic equation in Example 5.1.1. We may obtain solutions of this equation in $-\infty < t < \infty$ taking as D, for instance, the space of all elements of $C_0[0, \pi]$ with finite sine series. The solution is again given by (5.1.8). However, if we take $\zeta(x) = \sin nx$, the solution is $u(t)(x) = e^{-n^2 t} \sin nx$, and we have $\|u(t)\| = e^{n^2|t|}$ for $t < 0$, whereas $\|u(0)\| = 1$. This shows that (5.2.3) must be false for t negative.

We set up an abstract model for "reversible" equations such as (5.3.6). Let A be a densely defined linear operator in a Banach space E. The Cauchy problem or initial value problem for

$$y'(t) = Ay(t) \qquad (5.3.10)$$

is **well posed in** $-\infty < t < \infty$ if (a) and (b) of **5.2** hold with the following modifications: in (a) we assume existence of a strong solution in $-\infty < t < \infty$ for every $u \in D$, and in (b) we assume the existence of a function $C(t)$ bounded on compacts of $(-\infty, \infty)$ and such that

$$\|y(t)\| \leq C(t)\|y(0)\| \quad (-\infty < t < \infty). \qquad (5.3.11)$$

In this language, what has been proved in Example 5.3.1 is that the Cauchy problem for (5.3.6) is well posed **in** $-\infty < t < \infty$.

Let the Cauchy problem for (5.3.10) be well posed in $-\infty < t < \infty$. The solution operator $S(t)$ is now defined and strongly continuous for all t and satisfies (5.2.7) and (5.2.8), the latter for $-\infty < s, t < \infty$. The argument after (5.2.8) applied to both $S(t)$ and $S(-t)$ produces the bound

$$\|S(t)\| \le Ce^{\omega|t|} \quad (-\infty < t < \infty). \tag{5.3.12}$$

An operator valued function satisfying equalities (5.2.7) and (5.2.8) the latter for $-\infty < s, t < \infty$ is called a **group**; the solution operator of a Cauchy problem well posed in $-\infty < t < \infty$ is a strongly continuous group.

The theory of the Cauchy problem in $-\infty < t < \infty$ can be wholly reduced to that of the Cauchy problem in $t \ge 0$ through this simple result, whose proof we leave to the reader.

Theorem 5.3.2. *The Cauchy problem for (5.3.10) is well posed in $-\infty < t < \infty$ if and only if the Cauchy problems for (5.3.10) and for*

$$y'(t) = -Ay(t) \tag{5.3.13}$$

are both well posed in $t \ge 0$.

Via Theorem 5.3.2, all results in **5.2** reappear in "reversible" versions.

Theorem 5.3.3. *Let $S(\cdot)$ be a strongly continuous group. Then there exists a (unique) closed, densely defined operator A such that the Cauchy problem for (5.3.10) is well posed in $-\infty < t < \infty$ and $S(t)$ is the solution operator of (5.3.10).*

Theorem 5.3.4. *The operator A is the infinitesimal generator of a strongly continuous group $S(\cdot)$ satisfying*

$$\|S(t)\| \le Ce^{\omega|t|} \quad (-\infty < t < \infty) \tag{5.3.14}$$

if and only if A is closed and densely defined, $R(\lambda) = (\lambda I - A)^{-1}$ exists in $|\lambda| > \omega$ and

$$\|R(\lambda; A)^n\| \le C(|\lambda| - \omega)^{-n} \quad (|\lambda| > \omega, n = 1, 2, \ldots). \tag{5.3.15}$$

Inequalities (5.3.15) can be extended to $|\operatorname{Re} \lambda| > \omega$:

$$\|R(\lambda; A)^n\| \le C \, (|\operatorname{Re} \lambda| - \omega)^{-n} \quad (|\operatorname{Re} \lambda| > \omega, n = 1, 2, \ldots). \tag{5.3.16}$$

Theorem 5.3.5. *Let A be densely defined and closed. Then the Cauchy problem for (5.3.10) is well posed in $-\infty < t < \infty$ if and only if there exists ω such that $R(\lambda; A)$ exists for $|\lambda| > \omega$ and (5.3.16) holds.*

Corollary 5.3.6. *Let $S(\cdot)$ be an (E, E)-valued function defined in $-\infty < t < \infty$ satisfying (5.3.14) and such that $S(\cdot)y$ is E^*-weakly measurable in E for each $y \in E$. Let A be a densely defined operator such than (5.2.21) holds for A, $S(t)$ and $-A$, $S(-t)$ for $\lambda > \omega$. Then $S(t)$ is a strongly continuous group and A its infinitesimal generator. Equivalently, the Cauchy problem for (5.3.10) is well posed in $-\infty < t < \infty$ and $S(t)$ is the solution operator.*

As in the semigroup case, Theorem 5.3.4 has an "easy" case ($C = 1$) where only the first inequality (5.3.15) has to be checked:

$$\|R(\lambda; A)\| \leq 1/(|\lambda| - \omega) \quad (|\lambda| > \omega) \tag{5.3.17}$$

corresponding to

$$\|S(t)\| \leq e^{\omega|t|} \quad (-\infty < t < \infty). \tag{5.3.18}$$

Theorem 5.3.7. *The operator A is the infinitesimal generator of a strongly continuous group satisfying (5.3.18) if and only if*

$$|\mathrm{Re}\,\langle y^*, Ay \rangle| \leq \omega \|y\|^2 \quad (y \in D(A), y^* \in \Theta(y)) \tag{5.3.19}$$

$$(\lambda I - A)D(A) = E \quad (|\lambda| > \omega). \tag{5.3.20}$$

Theorem 5.3.7 results from Theorem 5.2.5 applied to A and to $-A$; inequality (5.3.19) is equivalent to dissipativity of $A - \omega I$ ($\mathrm{Re}\,\langle y^*, (A - \omega I)y \rangle \leq 0$) and dissipativity of $-A - \omega I$ ($\mathrm{Re}\,\langle y^*, (-A - \omega I)y \rangle \leq 0$).

The condition for $\omega = 0$ is

$$\mathrm{Re}\,\langle y^*, Ay \rangle = 0 \quad (y \in D(A), y^* \in \Theta(y)). \tag{5.3.21}$$

Theorem 5.2.6 implies the following.

Theorem 5.3.8. *Let A be the infinitesimal generator of a strongly continuous group $S(t)$ in a reflexive space E. Then A^* is the infinitesimal generator of the strongly continuous group $S^*(t) = S(t)^*$ in E^*.*

The variation-of-constants formula (5.2.28) can now be used for t of any sign, and Lemma 5.2.7 and Lemma 5.2.8 have obvious counterparts.

Example 5.3.9. The semigroup $S(t)$ in Example 5.2.9 is actually a group, continuous in the norm of (E, E).

Example 5.3.10. Let A be a normal operator in a Hilbert space H. Then the Cauchy problem for (5.3.10) is well posed in $-\infty < t < \infty$ (equivalently, A generates a strongly continuous group) if and only if $-\omega_- = \inf \mathrm{Re}\,\sigma(A)$,

$\omega_+ = \sup \operatorname{Re} \sigma(A) < \omega$. The group $S(t)$ satisfies (5.3.14) with $\omega = \max(\omega_-, \omega_+)$, and ω is the least possible constant for this inequality. We have $S(t) = e^{tA}$, the exponential computed according to the bounded functional calculus in **5.0.**

Example 5.3.11. Let A be a self adjoint operator in a Hilbert space H. Then A generates a strongly continuous group if and only if $-\omega_- = \inf \sigma(A)$, $\omega_+ = \sup \sigma(A) < \infty$. The group $S(t)$ satisfies (5.3.14) with $\omega = \max(\omega_-, \omega_+)$, and ω is the least possible constant for this inequality. We have $S(t) = e^{tA}$, the exponential computed according to the bounded functional calculus in **5.0.**

Example 5.3.12. Let A be **skew adjoint** in a Hilbert space H, that is, let $D(A)$ be dense in H and $A^* = -A$. Then A generates a unitary strongly continuous group. We have $S(t) = e^{tA}$, via the bounded functional calculus in **5.0.** Note that if H is complex then A is skew adjoint if and only if iA is self adjoint.

Example 5.3.13. In a complex Hilbert space, $\Theta(y) = \{y\}$ and condition (5.3.19) reduces to

$$|\operatorname{Re}(y, Ay)| \leq \omega \|y\|^2 \quad (y \in D(A)). \tag{5.3.22}$$

For $\omega = 0$ the condition is $\operatorname{Re}(y, Ay) = 0$ $(y \in D(A))$. See Example 5.2.12.

Example 5.3.14. In a real space, (5.3.19) and (5.3.21) are

$$|\langle y^*, Ay \rangle| \leq \omega \|y\|^2 \quad (y \in D(A), y^* \in \Theta(y)), \tag{5.3.23}$$

$$\langle y^*, Ay \rangle = 0 \quad (y \in D(A), y^* \in \Theta(y)). \tag{5.3.24}$$

5.4. Evolution Equations

Motivated by Example 5.1.2, we consider the Cauchy problem for time-dependent equations

$$y'(t) = A(t)y(t), \tag{5.4.1}$$

where $\{A(t); 0 \leq t < \infty\}$ is a family of densely defined linear operators in E. We can handle this equation in a way similar to (5.2.1), assuming strong solutions for a dense subspace D of initial conditions y and continuous dependence of arbitrary solutions on their initial conditions. Instead, we shall postulate existence of solutions and, directly, existence of the solution operator. Further, unlike in the time-invariant case, it is not reasonable to set initial conditions only at $t = 0$.

We postulate a set S of continuous E-valued functions defined in intervals $[s, \infty)$, with $s \geq 0$. Elements of this set will be called **solutions** of (5.4.2).

We say that the Cauchy problem for (5.4.1) is **well posed in** $t \geq 0$ if (a) for each $s \geq 0$ and each $\zeta \in E$ there exists a unique solution $y(\cdot, s; \zeta) \in S$ of (5.4.1) in $t \geq s$ with

$$y(s, s; \zeta) = \zeta, \tag{5.4.2}$$

(b) if $s' \geq s$ then

$$y(t, s'; y(s', s; \zeta)) = y(t, s; \zeta) \quad (t \geq s'), \tag{5.4.3}$$

(c) there exists a strongly continuous operator-valued function $S(t, s)$, defined in $0 \leq s \leq t \leq \infty$ and such that

$$y(t, s; \zeta) = S(t, s)\zeta. \tag{5.4.4}$$

The operator function $S(t, s)$ is called the **propagator** or **solution operator** of (4.1).

So far, we have not specified what we mean by "solution" of (5.4.1); continuity is the only requirement. Insistence on strong solutions (continuously differentiable E-valued functions $y(\cdot)$ such that $y(t) \in D(A(t))$ and (5.4.1) is satisfied in $t \geq s$) leads to numerous complications later; besides, strong solutions have no special meaning in control theory. Thus we shall not commit ourselves to any particular type of solution. Of course, a precise definition must be given each time the theory of (5.4.1) is applied to a particular equation.

The operator $S(t, s)$ is, in general, not defined for $t < s$. It follows immediately from the postulates that

$$S(s, s) = I \qquad (0 \leq s < \infty), \tag{5.4.5}$$

$$S(t, s)S(s, r) = S(t, r) \quad (0 \leq r \leq s \leq t < \infty). \tag{5.4.6}$$

Strong continuity of $S(t, s)$ and the uniform boundedness Theorem 5.0.1 imply that $\|S(t, s)\|$ is bounded on bounded subsets of $0 \leq s \leq t < \infty$, although there is no specific growth rate at infinity like (5.2.9).

It might be objected that refusing to *define* solutions and only *postulating* them has exploded the relation between the solution operator $S(t, s)$ and the equation (5.4.1), and we might as well work with a strongly continuous operator-valued function $S(t, s)$ satisfying (5.4.5) and (5.4.6) without even mentioning any equation. This observation is on target. First, showing that the Cauchy problem is well posed usually begins rather than ends with the construction of $S(t, s)$; moreover, in many of our results not even (5.4.5) or (5.4.6) play any role.

An operator-valued function $S(t, s)$ satisfying (5.4.5) and (5.4.6) will be called an **evolution operator**.

The definition of well-posed problem for (5.4.1) can be applied to the time invariant equation (5.2.1), thus it is important to show that it is consistent with the definition in **5.2**.

Example 5.4.1. Assume that the Cauchy problem for

$$y'(t) = Ay(t) \tag{5.4.7}$$

is well posed in $t \geq 0$ as defined in **5.2**. Then it is well posed in the sense of this section; "solution" means "weak solution" and $S(t, s) = S(t - s)$.

For the nonhomogeneous equation

$$y'(t) = A(t)y(t) + f(t), \tag{5.4.8}$$

we simply follow the time-invariant case and *define*

$$y(t) = S(t, s)\zeta + \int_s^t S(t, \tau)f(\tau)d\tau \tag{5.4.9}$$

as the solution of (5.4.8) with initial condition $y(s) = \zeta$.

Lemma 5.4.2. *The operator* $(\zeta, f(\cdot)) \to y(\cdot)$ *defined by* (5.4.9) *is bounded from* $E \times L^1(s, T; E)$ *into* $C(s, T; E)$ *and*

$$\|y\|_{C(s,T;E)} \leq M(\|\zeta\|_E + \|f\|_{L^1(s,T;E)}), \tag{5.4.10}$$

where M is a bound for $\|S(t, \tau)\|$ *in* $s \leq \tau \leq t \leq T$.

The proof is the same as that of Lemma 5.2.8.

The Cauchy problem for (5.4.1) in arbitrary intervals (say, $[0, T]$) is similarly set up; $s \in [0, T]$, solutions $y(t, s; \zeta)$ satisfying (5.4.2) exist in $s \leq t \leq T$, (5.4.3) is required for $s \leq s' \leq t \leq T$ and (5.4.4) in $0 \leq s \leq t \leq T$; the solution operator is assumed to be strongly continuous and satisfies (5.4.5) and (5.4.6), the latter in $0 \leq r \leq s \leq t \leq T$. The Cauchy problem in other intervals is treated in the same way.

We say that the Cauchy problem for (5.4.1) is well posed **forwards and backwards** or **in both senses of time** in an interval, say $[0, T]$, if given $s \in [0, T]$ solutions satisfying (5.4.2) exist in $0 \leq t \leq T$ and (5.4.3) holds for arbitrary $s, s', t \in [0, T]$; (5.4.4) holds for arbitrary $s, t \in [0, T]$, and $S(t, s)$ is required to be strongly continuous for $s, t \in [0, T]$. The solution operator satisfies (5.4.5) and (5.4.6), the latter for $r, s, t \in [0, T]$. The definitions are similar for other intervals. When the interval is $(-\infty, \infty)$, forwards and backwards well posedness corresponds to well posedness in $(-\infty, \infty)$ as defined in **5.3** for time-independent equations.

In Chapter 7 we shall need an extension of the notion of well-posed problem. Let $F \subseteq E$ be a Banach space. We say that the Cauchy problem for (5.4.1) is F-**well posed** if (a) holds for every $\zeta \in F$, the solution $y(t)$ required to take values in F. Assumption (b) does not change. In Assumption (c) the solution operator is assumed to belong to (F, F) and to be F-strongly continuous in $0 \leq s \leq t < \infty$. We do not require existence of solutions for initial data in E or E-strong continuity of the solution operator.

5.5. Semilinear Equations in Banach Spaces.
Perturbation Theory

We consider the equation

$$y'(t) = A(t)y(t) + f(t, y(t)), \qquad y(s) = \zeta \tag{5.5.1}$$

where $\{A(t); 0 \leq t \leq T\}$ is a family of densely defined operators in a Banach space E. The assumption on $A(\cdot)$ is that the Cauchy problem for the linear equation

$$y'(t) = A(t)y(t) \tag{5.5.2}$$

is well posed in $0 \leq t \leq T$, as defined in **5.4**. Below, $S(t, s)$ denotes the solution operator of (5.5.2), defined and strongly continuous in the triangle $0 \leq s \leq t \leq T$. Generally speaking, the results in this section are infinite dimensional counterparts of those in **2.1**, and many of the proofs are almost the same.

On the basis of (5.4.9), we *define* a solution of (5.5.1) as a solution of the integral equation

$$y(t) = S(t, s)\zeta + \int_s^t S(t, \tau)f(\tau, y(\tau))d\tau \tag{5.5.3}$$

(the integral understood in the sense of Lebesgue–Bochner). Solutions of (5.5.3) may not have a derivative anywhere or belong to $D(A(t))$ for any t. Many of the results do not even use the functional equations (5.4.5)-(5.4.6) for the solution operator $S(t, s)$, thus, they are valid for general integral equations with a strongly continuous kernel $S(t, s)$ not necessarily associated with a differential equation.

We summarize in this section the necessary existence-uniqueness theory of (5.5.1). Results will be proved under two hypotheses on $f(t, y)$, the second stronger than the first, both counterparts of corresponding assumptions in **2.2**.

Hypothesis I. $f(t, y)$ is strongly measurable in t for y fixed and continuous in y for t fixed. For every $c > 0$ there exists $K(\cdot, c) \in L^1(0, T)$ such that

$$\|f(t, y)\| \leq K(t, c) \quad (0 \leq t \leq T, \|y\| \leq c). \tag{5.5.4}$$

Hypothesis II. $f(t, y)$ is strongly measurable in t for y fixed. For every $c > 0$ there exist $K(\cdot, c), L(\cdot, c) \in L^1(0, T)$ such that (5.5.4) holds and

$$\|f(t, y') - f(t, y)\| \leq L(t, c)\|y' - y\| \quad (0 \leq t \leq T, \|y\|, \|y'\| \leq c). \tag{5.5.5}$$

Under Hypothesis I, the function $t \to f(t, y(t))$ is strongly measurable for any $y(\cdot) \in C(0, T; E)$ (approximate $y(\cdot)$ by a piecewise constant function and use y-continuity of $f(t, y)$), hence it belongs to $L^1(0, T; E)$ by (5.5.4). This and local boundedness of $\|S(t, s)\|$ give sense to (5.5.3); the integrand is easily seen to be strongly measurable.

Theorem 5.5.1. *Assume Hypothesis II holds in* $0 \le t \le T$. *Then the integral equation* (5.5.3) (*or, more generally, the integral equation*

$$y(t) = \zeta(t) + \int_s^t S(t, \tau) f(\tau, y(\tau)) d\tau \qquad (5.5.6)$$

where $\zeta(\cdot)$ *is continuous in* $0 \le t \le T$ *and* $S(\cdot, \cdot)$ *is strongly continuous in* $0 \le s \le t \le T$) *has a unique solution in some interval* $s \le t \le T'$, *where* $s < T' \le T$.

Proof. Consider the operator

$$(\mathcal{G}y)(t) = \zeta(t) + \int_s^t S(t, \tau) f(\tau, y(\tau)) d\tau \qquad (5.5.7)$$

in the closed ball $B(\zeta(\cdot), 1) \subseteq C(s, T'; E)$ of center $\zeta(\cdot)$ and radius 1. Let M be a bound for $\|S(t, s)\|$ in $0 \le s \le t \le T$ and c a bound for $\|\zeta(t)\|$ in $0 \le t \le T$. Then

$$\|(\mathcal{G}y)(t) - \zeta(t)\| \le M \int_s^{T'} K(\tau, c + 1) d\tau.$$

Thus, by absolute continuity of the integral, $\mathcal{G}$ maps $B(\zeta(\cdot), 1)$ into itself for $T' - s \le \delta_1$ sufficiently small. On the other hand,

$$\|(\mathcal{G}z)(t) - (\mathcal{G}y)(t)\| \le M \sup_{s \le t \le T'} \|z(t) - y(t)\| \int_s^{T'} L(\tau, c + 1) d\tau$$

so that, again by absolute continuity of the integral, $\mathcal{G}$ is a contraction for $T' - s \le$ a sufficiently small $\delta_2 \le \delta_1$. Theorem 5.5.1 then follows from the Cacciopoli–Banach Theorem 2.0.13. ∎

Theorem 5.5.2. *Let* $y_1(\cdot)$ (*resp.* $y_2(\cdot)$) *be solutions of* (5.5.6) *in* $s \le t \le T'$ *with* $\zeta(t) = \zeta_1(t)$. (*resp. with* $\zeta(t) = \zeta_2(t)$). *Let* c *be a bound for* $\|y_1(t)\|, \|y_2(t)\|$ *in* $s \le t \le T'$. *Then*

$$\|y_1(t) - y_2(t)\| \le \sup_{s \le t \le T'} \|\zeta_1(t) - \zeta_2(t)\| \exp\left(M \int_s^t L(\tau, c) d\tau\right) \quad (s \le t \le T).$$
$$(5.5.8)$$

In particular, if $y_1(\cdot)$ (*resp.* $y_2(\cdot)$) *are solutions of* (5.5.3) *with* $\zeta = \zeta_1$ (*resp. with* $\zeta = \zeta_2$) *then*

$$\|y_1(t) - y_2(t)\| \le M \|\zeta_1 - \zeta_2\| \exp\left(M \int_s^t L(\tau, c) d\tau\right) \quad (s \le t \le T). \quad (5.5.9)$$

The proof is the same as that of Corollary 2.1.4.

The following two results are counterparts of Lemma 2.1.5 and Corollary 2.1.6.

Lemma 5.5.3. *Let* $y(t)$ *be a solution of* (5.5.6) *in an interval* $[s, T')$. *Assume that*

$$\|y(t)\| \le c \quad (s \le t < T'). \tag{5.5.10}$$

Then $y(\cdot)$ *can be extended to an interval* $[s, T'']$ *with* $T'' > T'$ *(that is, a solution of* (5.5.6) *coinciding with* $y(\cdot)$ *in* $s \le t < T'$ *exists in* $[0, T'']$).

Proof. (5.5.10) implies that $f(t, y(\cdot)) \in L^1(s, T'; E)$. Then it follows from the integral equation (5.5.6) and from Lemma 5.4.2 that $y(t)$ is continuous in (strictly speaking, possesses a continuous extension to) $s \le t \le T'$. We can then solve

$$y(t) = \zeta(t) + \int_s^{T'} S(t, \tau) f(\tau, y(\tau)) d\tau + \int_{T'}^t S(t, \tau) f(\tau, y(\tau)) d\tau$$

in some interval $[T', T''], T'' > T'$. The function obtained piecing together both solutions is a solution in $[0, T'']$. ∎

Corollary 5.5.4. *The solution* $y(\cdot)$ *of* (5.5.6) *exists in* $s \le t \le T$ *or in an interval* $[s, T_m), T_m \le T$ *and*

$$\limsup_{t \to T_m-} \|y(t)\| = \infty. \tag{5.5.11}$$

Lemma 2.1.7 has a counterpart (Theorem 5.8.3) whose proof is rather technical due to the fact that, unlike in the finite dimensional case, $y(t)$ may not be differentiable anywhere. The following (much weaker) result is much easier to prove.

Corollary 5.5.5. *Assume there exists* $K(\cdot) \in L^1(0, T)$ *such that*

$$\|f(t, y)\| \le K(t)(1 + \|y\|) \quad (0 \le t \le T, y \in E). \tag{5.5.12}$$

Then (5.5.10) *holds in every interval where the solution* $y(t)$ *of* (5.5.6) *exists; accordingly,* $y(t)$ *exists in* $s \le t \le T$.

In fact, we obtain from (5.5.12) that

$$\|y(t)\| \le c + M \int_s^t K(\tau)(1 + \|y(\tau)\|) d\tau,$$

where c, M are as in Theorem 5.5.1. Thus, a *priori* boundedness of $y(\cdot)$ follows from Corollary 2.1.3.

Theorem 5.5.6. *Let the Cauchy problem for* (5.5.2) *be well posed in* $0 \le t \le T$, *and let* $\{B(t); 0 \le t \le T\}$ *be a family of linear bounded operators in* E *such that* (*a*) *for each* $y \in E, t \to B(t)y$ *is strongly measurable,* (b) *there exists* $\alpha(\cdot) \in L^1(0, T)$ *such that*

$$\|B(t)\| \leq \alpha(t) \quad (0 \leq t \leq T). \tag{5.5.13}$$

Then the Cauchy problem for

$$y'(t) = (A(t) + B(t))y(t) \tag{5.5.14}$$

is well posed in $0 \leq t \leq T$, *solutions of* (5.5.14) *with* $y(s) = \zeta$ *understood as solutions of the integral equation*

$$y(t) = S(t, s)\zeta + \int_s^t S(t, \tau)B(\tau)y(\tau)d\tau. \tag{5.5.15}$$

If $U(t, s)$ *is the solution operator of* (5.5.14), *solutions of the inhomogeneous equation*

$$y'(t) = (A(t) + B(t))y(t) + f(t), \qquad y(s) = \zeta \tag{5.5.16}$$

with $f(\cdot) \in L^1(s, T; E)$, *understood as solutions of the integral equation*

$$y(t) = S(t, s)\zeta + \int_s^t S(t, \tau)(B(\tau)y(\tau) + f(\tau))d\tau, \tag{5.5.17}$$

can be expressed by the variation-of-constants [1] *formula*

$$y(t) = U(t, s)\zeta + \int_s^t U(t, \tau)f(\tau)d\tau. \tag{5.5.18}$$

Proof. Condition (a) in **5.4** on global existence of solutions results from Corollary 5.5.5 applied to $f(t, y) = B(t)y$. In view of the definition of the solution $y(t)$ of (5.5.14) with $y(s) = \zeta$ by the integral equation (5.5.15), the solution operator $U(t, s)$ of (5.5.14) is given by the integral equation

$$U(t, s)\zeta = S(t, s)\zeta + \int_s^t S(t, \tau)B(\tau)U(\tau, s)\zeta d\tau \tag{5.5.19}$$

$(0 \leq s \leq t \leq T, \zeta \in E)$. Condition (b) in the definition of well posed problem follows from manipulations with the integral equation (5.5.15) similar to those in the proof of Lemma 5.5.3. Finally, condition (c) means we must show boundedness of each $U(t, s)$ and strong continuity of the operator-valued function $U(t, s)$ in $0 \leq s \leq t \leq T$. Estimating (5.5.19) we obtain

$$\|U(t, s)\zeta\| \leq M\|\zeta\| + M \int_s^t \alpha(\tau)\|U(\tau, s)\zeta\|d\tau,$$

---

[1] Unlike in **5.4**, the variation-of-constants formula (5.5.18) cannot be taken as *definition* of solutions of the equation (5.5.16), since these solutions have already been defined otherwise (as solutions of the integral equation (5.5.17)).

thus, we obtain from Corollary 2.1.3 that

$$\|U(t,s)\| \le M \exp\left(M \int_s^t \alpha(\tau) d\tau \right) \le N. \tag{5.5.20}$$

As for continuity, if $t, t' \ge s$ and $t \le t'$,

$$U(t',s)\zeta - U(t,s)\zeta = S(t',s)\zeta - S(t,s)\zeta$$
$$+ \int_t^{t'} S(t',\tau)B(\tau)U(\tau,s)\zeta d\tau$$
$$+ \int_s^t (S(t',\tau) - S(t,\tau))B(\tau)U(\tau,s)\zeta d\tau,$$

so that

$$\|U(t',s)\zeta - U(t,s)\zeta\| \le \|S(t',s')\zeta - S(t,s)\zeta\|$$
$$+ MN\|\zeta\| \int_t^{t'} \alpha(\tau) d\tau + \int_s^t \|(S(t',\tau) - S(t,\tau))B(\tau)U(\tau,s)\zeta\| d\tau, \tag{5.5.21}$$

(N comes from (5.5.20)) with a similar estimate (switching t and t') when $t \ge t'$. On the other hand, if $s, s' \le t$ and $s \le s'$,

$$U(t,s')\zeta - U(t,s)\zeta = S(t,s')\zeta - S(t,s)\zeta$$
$$- \int_s^{s'} S(t,\tau)B(\tau)U(\tau,s)\zeta d\tau$$
$$+ \int_{s'}^t S(t,\tau)B(\tau)(U(\tau,s')\zeta - U(\tau,s)\zeta) d\tau$$

so that

$$\|U(t,s')\zeta - U(t,s)\zeta\| \le \|S(t,s')\zeta - S(t,s)\zeta\|$$
$$+ MN\|\zeta\| \int_s^{s'} \alpha(\tau) d\tau + M \int_{s'}^t \alpha(\tau)\|U(\tau,s')\zeta - U(\tau,s)\zeta\| d\tau \tag{5.5.22}$$

with a similar estimate (switching s and s') when $s \ge s'$. These estimates are exploited as follows.

(5.5.21): We replace the lower limit of integration in the second integral on the right by $s = 0$ so as to make it independent of s. The integrand in the second integral tends to zero (by strong continuity of $S(t,s)$) and is bounded by $2MN\alpha(\tau)$; thus, by the dominated convergence theorem, the integral tends to zero as $t' \to t$ as long as $t, t' \ge s$; convergence is independent of s. The first integral tends to zero by absolute continuity.

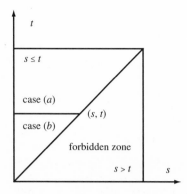

Figure 5.1.

(5.5.22): We use Corollary 2.1.3 and strong continuity of $S(t, s)$ and deduce that $\|U(t, s')\zeta - U(t, s)\zeta\| \to 0$ as $s' \to s$ as long as $s, s' \le t$; we check easily that convergence is independent of t.

If $t > s$ and $(t', s') \to (t, s)$, then eventually $t' > s'$, $t' > s$, and we can write

$$\|U(t', s')\zeta - U(t, s)\zeta\| \le \|U(t', s')\zeta - U(t', s)\zeta\| + \|U(t', s)\zeta - U(t, s)\zeta\|$$

$$(5.5.23)$$

showing that $U(t', s')\zeta \to U(t, s)\zeta$ as $(t', s') \to (t, s)$. On the other hand, let $s = t$ and let (t', s') be another point in $0 \le s \le t \le T$. There are two possibilities: (a) $t' \ge t = s$, (b) $t' < t = s$. In case (a) we estimate with (5.5.23); in case (b), we have $s' \le t' < t$ and the bound is

$$\|U(t', s')\zeta - U(s, s)\zeta\| \le \|U(t', s')\zeta - U(t, s')\zeta\| + \|U(t, s')\zeta - U(s, s)\zeta\|.$$

This completes the proof of strong continuity of $U(t, s)$. The reader will perceive the similitude with Lemma 2.1.10, modulo some acrobatics to prevent (s', t') from falling in the "forbidden zone" $s > t$ (see Figure 5.1). These contortions can actually be avoided (see Example 5.5.26) but this would compromise Theorem 5.5.6 as a model for other results in Chapter 7.

It remains to show that the solution $y(\cdot)$ of the integral equation (5.5.17) is given by formula (5.5.18). To do this we replace $y(t)$, as given by (5.5.18), in (5.5.17) using the integral equation (5.5.19) for $U(t, s)$:

$$S(t, s)\zeta + \int_s^t S(t, \tau)B(\tau)y(\tau)d\tau + \int_s^t S(t, \tau)f(\tau)d\tau$$

$$= S(t, s)\zeta + \int_s^t S(t, \sigma)B(\sigma)U(\sigma, s)\zeta \, d\sigma$$

$$+ \int_s^t S(t, \sigma)B(\sigma)\int_s^\sigma U(\sigma, \tau)f(\tau)d\tau d\sigma + \int_s^t S(t, \tau)f(\tau)d\tau$$

$$= S(t, s)\zeta + \int_s^t S(t, \sigma)B(\sigma)U(\sigma, s)\zeta \, d\sigma$$

$$+ \int_s^t \left(S(t, \tau) + \int_\tau^t S(t, \sigma)B(\sigma)U(\sigma, \tau) \, d\sigma \right) f(\tau) \, d\tau$$

$$= U(t, s)\zeta + \int_s^t U(t, \tau)f(\tau) \, d\tau = y(t) \tag{5.5.24}$$

after an interchange in the order of integration. This ends the proof of Theorem 5.5.6. ∎

Remark 5.5.7. If the Cauchy problem for (5.5.2) is well posed in both senses of time, then the same is true of (5.5.14), and we are free to use $U(t, s)$ (in particular, the variation-of-constants formula (5.5.18)) for s, t in arbitrary position.

The autonomous case deserves a nod. Applied to the equation

$$y'(t) = Ay(t), \tag{5.5.25}$$

and to the perturbed equation

$$y'(t) = (A + B)y(t), \tag{5.5.26}$$

Theorem 5.5.6 and Remark 5.5.7 can be interpreted in the following way.

Corollary 5.5.8. *Let A be the infinitesimal generator of a strongly continuous semigroup (resp. strongly continuous group), and let B be a bounded operator. Then $A + B$ generates a strongly continuous semigroup (resp. a strongly continuous group).*

It is possible to express the semigroup (group) generated by $A + B$ by means of a perturbation series involving B and the semigroup generated by A (see Examples 5.5.24 and 5.5.25 for this and other information). Perturbation theory of semigroups addresses the problem of which "perturbations" B (in general, unbounded operators) preserve the semigroup generating properties of A. It also keeps track of the relations between spectral properties of A (such as behavior of eigenvalues and eigenvectors) and the corresponding properties of the perturbed operator $A + B$. This is a vast subject with many applications in physics. For a treatment of the linear case, see Kato [1995].

Control theory needs the **reversed** or **backwards adjoint equation**

$$z'(s) = -A(s)^* z(s), \qquad z(t) = z \tag{5.5.27}$$

where the role of the variables t and s in (5.5.2) has been switched for later convenience; t is the time at which the initial (or "final") condition is given. The initial value problem for (5.5.27) is the "reversed" or "backwards" initial value problem or "final value problem" where solutions are required in the past $s \leq t$ rather than in the future (the usual Cauchy problem of requiring solutions in $s \geq t$ is called "direct" or "forward.") Note that, if the forward Cauchy problem for (5.5.2) is well posed, the backwards problem may not be well posed; an example is the reversed heat equation $u_s = -u_{xx}$ (see the comments in **5.3**).

The theory of the backwards Cauchy problem is a mirror image of the theory of the forward problem, and we leave it (mostly) to the reader. Well posedness (as defined in **5.4**) will be, in general, postulated in a finite interval $0 \leq t \leq T$. For the backwards problem, it means existence and uniqueness of solutions $z(s)$ of (5.5.27) in $0 \leq s \leq t$ for arbitrary initial condition z and existence of a strongly continuous solution operator $S^*(s, t)$, defined in $0 \leq s \leq t \leq T$ and propagating backwards the solutions: $z(s) = S^*(s, t)z(t)$. This leads to the equations

$$S^*(t, t) = I, \qquad S^*(r, s)S^*(s, t) = S(r, t), \qquad (5.5.28)$$

valid for $0 \leq t \leq T$ and $0 \leq r \leq s \leq t \leq T$ respectively. In this context, the variation-of-constants formula (5.4.9) for an E^*-valued $f(\cdot)$ becomes

$$z(s) = S^*(s, t)z + \int_s^t S^*(s, \sigma)f(\sigma)d\sigma, \qquad (5.5.29)$$

and $z(s)$ is (by definition) the solution of

$$z'(s) = -A(s)^*z(s) - f(s), \qquad y(t) = z \qquad (5.5.30)$$

in $s \leq t$; $z(\cdot)$ is continuous in $0 \leq s \leq t$ if $f(\cdot) \in L^1(0, t; E^*)$. Theorem 5.5.6 has the following counterpart.

Theorem 5.5.9. *Let the reversed Cauchy problem for* (5.5.27) *be well posed in* $0 \leq t \leq T$ *and let* $B^*(s)$ *be a family of linear bounded operators in* E^* *such that* (a) $s \to B^*(s)z$ *is strongly measurable for all* $z \in E^*$, (b) *there exists* $\alpha(\cdot) \in L^1(0, T)$ *such that*

$$\|B^*(s)\| \leq \alpha(s) \quad (0 \leq s \leq T). \qquad (5.5.31)$$

Then the backwards Cauchy problem for the perturbed equation

$$z'(s) = -(A(s)^* + B^*(s))z(s) \qquad (5.5.32)$$

is well posed.

The integral equation giving the propagator $U^*(s, t)$ of (5.5.32) in terms of the propagator $S^*(s, t)$ of (5.5.30) is

$$U^*(s, t)z = S^*(s, t)z + \int_s^t S^*(s, \sigma)B^*(\sigma)U^*(\sigma, t)zd\sigma. \qquad (5.5.33)$$

In the time-independent case (5.5.25), the backwards adjoint equation is

$$z'(t) = -A^*z(t). \tag{5.5.34}$$

The solution operator of (5.5.25) is $S(t, s) = S(t - s)$, $S(\cdot)$ the semigroup generated by A. On the other hand, if E is reflexive, Theorem 5.2.6 implies that A^* generates the semigroup $S^*(t) = S(t)^*$, thus the solution operator of (5.5.30) is $S^*(s, t) = S^*(t - s) = S(t - s)^*$. This leads to the formula

$$S^*(s, t) = S(t, s)^*, \tag{5.5.35}$$

which extends to time-dependent equations. To simplify, we assume the "base equation" time independent.

Theorem 5.5.10. *Let E be reflexive and separable, and let A be the infinitesimal generator of a strongly continuous semigroup, that is, let the Cauchy problem for (5.5.25) be well posed in $t \geq 0$. Let $B(\cdot)$ satisfy the assumptions of Theorem 5.5.6 in $0 \leq t \leq T$. Then the forward Cauchy problem for*

$$y'(t) = (A + B(t))y(t) \tag{5.5.36}$$

and the backwards Cauchy problem for

$$z'(s) = -(A^* + B(s)^*)z(s) \tag{5.5.37}$$

are well posed in $0 \leq t \leq T$. If $S(t, s)$ is the solution operator of the forward equation and $S^*(t, s)$ the solution operator of the backwards equation, then (5.5.35) holds.

Proof. Forward well posedness for (5.5.36) follows from Theorem 5.5.6. To show backwards well posedness for (5.5.37), we prove that $B(\cdot)^*$ satisfies the assumptions in Theorem 5.5.9. The bound (5.5.13) implies (5.5.31). On the other hand, if $y \in E, z \in E^*$ then $\langle B(t)^*z, y \rangle = \langle z, B(t)y \rangle$ so that $B(t)^*$ is weakly measurable, then strongly measurable by separability of E^* (Example 5.0.39).

To prove (5.5.35) we write (5.5.25) as the "twice perturbed" equation

$$y'(t) = (A + B(t))y(t) - B(t)y(t) = Ay(t)$$

and apply Theorem 5.5.6 to construct $S(t, s) = S(t - s)$, the solution operator of (5.5.25) from the solution operator $S(t, s)$ of (5.5.36) by means of the integral equation (5.5.19). We obtain

$$S(t - s)y = S(t, s)y - \int_s^t S(t, \sigma)B(\sigma)S(\sigma - s)y \, d\sigma.$$

After applying an arbitrary $z \in E^*$ to both sides and manipulating adjoints, we obtain

$$S(s, t)^* z = S(t - s)^* z + \int_s^t S(\sigma - s)^* B(\sigma)^* S(t, \sigma)^* z \, d\sigma.$$

But this is precisely the integral equation (5.5.33) giving the propagator $S^*(t, s)$ of (5.5.37) in terms of the propagator of (5.5.34). Hence, (5.5.35) follows by uniqueness. $\blacksquare$

Remark 5.5.11. Theorem 5.5.10 admits several obvious extensions. First, the assumption that E is separable can be dropped if we postulate $B(\cdot)^* z$ strongly measurable for every $z \in E^*$. We may also deal with a time-dependent $A(t)$ if we assume that the Cauchy problems for (5.5.2) and (5.5.27) are well posed and that (5.35) holds for the respective solution operators; the conclusion is that the equality extends to the perturbed problems (5.5.14) and (5.5.32).

Well posedness of the Cauchy problem for the reverse adjoint equation is not essential in control theory, and we take below a more general point of view than in Theorem 5.5.10, dropping all ad-hoc assumptions on E (such as reflexivity or separability). Assuming only that the Cauchy problem for (5.5.2) is well posed in $0 \leq t \leq T$ and taking (5.5.35) as inspiration, we *define* the expression

$$z(s) = S(t, s)^* z + \int_s^t S(\sigma, s)^* f(\sigma) d\sigma = z_h(s) + z_i(s) \qquad (5.5.38)$$

to be the solution of (5.5.30) (h = homogeneous, i = inhomogeneous). The integral $z_i(s)$ is not a Lebesgue–Bochner integral and thus requires explanation. It is defined as the only element of E^* satisfying

$$\langle z_i(s), y \rangle = \int_s^t \langle f(\sigma), S(\sigma, s) y \rangle d\sigma. \qquad (5.5.39)$$

Theorem 5.5.12. *Let $z \in E$ be arbitrary and $f(\cdot)$ an E-weakly measurable E^*-valued function with $\| f(s) \|_{E^*} < \alpha(s) \; (0 \leq s \leq t), \alpha(\cdot) \in L^1(0, t)$. Then $z(s)$ is a E-weakly continuous E^*-valued function in $0 \leq s \leq t$ and*

$$\| z(s) \|_{E^*} < M(\| z \|_{E^*} + \| \alpha \|_{L^1(0,t)}) \quad (0 \leq s \leq t), \qquad (5.5.40)$$

M a bound for $\| S(\sigma, s) \|$ in $0 \leq s \leq \sigma \leq t$.

Proof. The bound (5.5.40) is obvious. The continuity claim is plain for $z_h(s)$, since $\langle z_h(s), y \rangle = \langle z, S(t, s) y \rangle$.

We deal with $z_i(s)$ by means of an approximation argument. Extend $S(\sigma, s)$ to $0 \leq \sigma, s \leq t$ setting $S(\sigma, s) = I$ in the "forbidden zone" $\sigma < s$; the extended S is strongly continuous, thus, by compactness of $[0, t] \times [0, t]$, $S(\sigma, s)y$ is uniformly continuous for each $y \in E$. A finite collection of functions $\{\phi_j(\cdot); j = 1, 2, \ldots, n\}$ is a **partition of unity** for the interval $[0, t]$ if $\Sigma\phi_j(s) = 1$ in $0 \leq s \leq t$. For each n select a partition of unity $\{\phi_{nj}(\cdot)\}$ such that each $\phi_{nj}(\cdot)$ is smooth, nonnegative and has support in an interval of length $\leq 1/n$. Choose $s_{nj} \in [0, t]$ in the support of $\phi_{nj}(\cdot)$ and define $S_n(\sigma, s) = \Sigma\Sigma\phi_{nj}(\sigma)\phi_{nk}(s)S(s_{nj}, s_{nk})$. Then

$$\|S(\sigma, s)y - S_n(\sigma, s)y\|$$

$$= \left\| \sum\sum \phi_{nj}(\sigma)\phi_{nk}(s)S(\sigma, s)y - \sum\sum \phi_{nj}(\sigma)\phi_{nk}(s)S(s_{nj}, s_{nk})y \right\|$$

$$\leq \sum\sum \phi_{nj}(\sigma)\phi_{nk}(s)\|S(\sigma, s)y - S(s_{nj}, s_{nk})y\| \tag{5.5.41}$$

so that $S_n(\sigma, s)y \to S(\sigma, s)y$ uniformly in $0 \leq \sigma, s \leq \bar{t}$. If $z_{in}(s)$ is defined from $S_n(\sigma, s)$ in the same way $z_i(s)$ is defined from $S(\sigma, s)$, we have

$$|\langle z_{in}(s), y\rangle - \langle z_i(s), y\rangle| \leq \int_s^t \alpha(\sigma)\|S_n(\sigma, s)y - S(\sigma, s)y\|d\sigma$$

so that $\langle z_{in}(s), y\rangle \to \langle z_i(s), y\rangle$ uniformly in $0 \leq s \leq t$ and it is enough to show that $z_{in}(s)$ is E-weakly continuous; this is obvious since

$$\langle z_{in}(s), y\rangle = \sum\sum \phi_{nk}(s)\int_s^t \phi_{nj}(\sigma)\langle f(\sigma), S(s_{nj}, s_{nk})y\rangle d\sigma. \qquad \blacksquare$$

Theorem 5.5.12 gives sense to the "final condition" $z(t) = z$ in (5.5.30); it means

$$\lim_{s \to t-} \langle z(s), y\rangle = \langle z, y\rangle. \tag{5.5.42}$$

Corollary 5.5.13. *Let E^* be separable. Then $z(s)$ is strongly measurable.* $\qquad \blacksquare$

Proof. E-weakly continuous implies E-weakly measurable. For the strong measurability statement, use Example 5.0.39. $\qquad \blacksquare$

The following result is a relative of Fubini's theorem pertinent to the solutions in Theorem 5.5.12.

Lemma 5.5.14. *Let $f(\cdot)$ be an E-weakly measurable E^*-valued function with $\|f(s)\| \leq \alpha(s)$ $(0 \leq s \leq \bar{t})$, $\alpha(\cdot) \in L^1(0, \bar{t})$, and let $g(\cdot) \in L^1(0, \bar{t}; E)$. Then*

$$\int_0^{\bar{t}} \left\langle f(t), \int_0^t S(t, s)g(s)ds \right\rangle dt = \int_0^{\bar{t}} \left\langle \int_s^{\bar{t}} S(t, s)^*f(t)dt, g(s) \right\rangle ds, \tag{5.5.43}$$

the inner integral on the right side understood as in Theorem 5.5.12; the inner integral on the left side is a Lebesgue–Bochner integral.

Proof. Both outer integrals make sense; in fact, if $f(\cdot)$ (resp. $h(\cdot)$) is a E-weakly measurable E^*-valued function (resp. a E-valued strongly measurable function) then $\langle f(\cdot), h(\cdot) \rangle$ is measurable, as we see approximating $g(\cdot)$ by countably valued functions. Due to (5.4.10) and (5.5.40), the integrand on the left is bounded by $C\alpha(t)$, the one on the right by $C\|g(s)\|$.

We approximate $g(\cdot)$ in the $L^1(0, t; E)$-norm by a sequence $g_n(s) = \Sigma \chi_{nk}(s)g_{nk}$ of simple functions, that is, finite sums where the χ_{nk} are characteristic functions of pairwise disjoint sets (Theorem 5.0.21). Assuming we can establish (5.5.43) for $g_n(\cdot)$, we pass to the limit as follows. By Lemma 5.4.2, $\int_0^t S(t, \sigma)g_n(\sigma)d\sigma \to \int_0^t S(t, \sigma)g(\sigma)d\sigma$ in $C(0, T; E)$, which justifies taking limits on the left side using the dominated convergence theorem. On the right side we use (5.5.40) and L^1-convergence of $g_n(\cdot)$.

We note next that we may reduce the case of a finite sum $\Sigma \chi_{nk}(s)g_{nk}$ to one term: $g(s) = \chi_e(s)g$, χ_e the characteristic function of a measurable set e. The equality then becomes

$$\int_0^{\bar{t}} \left\langle f(t), \int_0^t S(t, s)\chi_e(s)g\, ds \right\rangle dt = \int_0^{\bar{t}} \left\langle \int_s^{\bar{t}} S(t, s)^* f(t)dt, g \right\rangle \chi_e(s)ds$$
(5.5.44)

and to prove it, it is enough to do first the corresponding equality for $S_n(t, s)$,

$$\int_0^{\bar{t}} \left\langle f(t), \int_0^t S_n(t, s)\chi_e(s)g\, ds \right\rangle dt = \int_0^{\bar{t}} \left\langle \int_s^{\bar{t}} S_n(t, s)^* f(t)\, dt, g \right\rangle \chi_e(s)ds$$
(5.5.45)

and take limits. Finally, we only have to prove (5.5.45) for each of the terms making up S_n, that is

$$\int_0^{\bar{t}} \phi_{nj}(t) \left(\int_0^t \phi_{nk}(s)\chi_e(s)ds \right) \langle f(t), S(s_{nj}, s_{nk})g \rangle dt$$

$$= \int_0^{\bar{t}} \phi_{nk}(s)\chi_e(s) \int_s^{\bar{t}} \phi_{nj}(t)\langle f(t), S(s_{nj}, s_{nk})g \rangle dt\, ds,$$

and this is just a straight application of Fubini's theorem.

It only remains to show that we obtain (5.5.44) taking limits in (5.5.45). For the inner integral on the right we note that $S_n(\cdot, s)g \to S(\cdot, s)g$ in $C(0, T; E)$, and use the dominated convergence theorem in the outer integral. On the left side we use the (easily shown) fact that $\int_0^t S_n(t, s)g\chi_e(s)ds \to \int_0^t S(t, s)g\chi_e(s)ds$ in $C(0, T; E)$. This ends the proof. ∎

The following material (up to Example 5.5.21) is on existence on the only basis of Hypothesis I. Existence results such as those in **2.1** are not to be expected; in fact, generalizing earlier efforts of Dieudonné, Godunov [1974] proved.

Example 5.5.15. There exists a continuous function $f : (-\infty, \infty) \times H \to H$ (H a separable Hilbert space) bounded on bounded subsets and such that the initial value problem

$$y'(t) = f(t, y(t)), \qquad y(s) = \zeta$$

(or, equivalently, the integral equation (5.5.3) with $S(t, \tau) \equiv I$) has no solution in any interval containing s.

Compactness cannot be supplied by the space (as when E is finite dimensional); it has to come from the equation.

A bounded operator $B \in (E, F)$ (E, F two arbitrary Banach spaces) is **compact** if, for every bounded sequence $\{y_n\} \subseteq E$ the sequence $\{By_n\}$ contains a convergent subsequence. A strongly continuous semigroup $S(\cdot)$ is called **compact** if $S(t)$ is compact for every $t > 0$.

Example 5.5.16. Let E, F, G Banach spaces, $B \in (E, F)$, $A \in (F, G)$. If either A or B is compact, AB is compact.

Example 5.5.17. Let $\{B_n\} \subseteq (E, F)$ be a sequence of compact operators such that $B_n \to B$ in the norm of (E, F). Then B is compact.

Lemma 5.5.18. *Let $S(t)$ be a compact semigroup. Then $S(t)$ is continuous in the norm of (E, F) for $t > 0$.*

Proof. If $S(t)$ is not (E, F)-continuous at $t = t_0 > 0$ there exists a sequence $\{y_n\} \subseteq E$, $\|y_n\| = 1$ and a sequence $\{h_n\} \subseteq \mathbb{R}_+$ with $h_n \to 0$ such that

$$\|S(t_0 + h_n)y_n - S(t_0)y_n\| \geq \varepsilon > 0. \qquad (5.5.46)$$

Select a subsequence such that $S(t_0/2)y_n \to z$ and note that

$$(S(t_0 + h_n) - S(t_0))y_n = (S(t_0/2 + h_n) - S(t_0/2))S(t_0/2)y_n$$

$$= (S(t_0/2 + h_n) - S(t_0/2))z + (S(t_0/2 + h_n) - S(t_0/2))(S(t_0/2)y_n - z) \to 0,$$

a contradiction with (5.5.46). ∎

Theorem 5.5.19. *Let $S(\cdot)$ be a strongly continuous semigroup in E. Then (a) the operator*

$$(\Lambda g)(t) = \int_0^t S(t - \tau)g(\tau)d\tau \qquad (5.5.47)$$

is bounded from $L^1(0, T; E)$ *into* $C(0, T; E)$. (b) *if* $S(t)$ *is compact and the sequence* $\{g_n(\cdot)\} \subseteq L^1(0, T; E)$ *is such that the integrals of the* $\|g_n(\cdot)\|$ *are equicontinuous in* $0 \le t \le T$, *then* $\{(\Lambda g_n)(\cdot)\}$ *has a subsequence convergent in* $C(0, T; U)$.

Proof. That Λ maps $L^1(0, T; E)$ into $C(0, T; E)$ follows from Lemma 5.2.8; we have

$$\|\Lambda g\|_{C(0,T;E)} \le M \|g\|_{L^1(0,T;E)}, \tag{5.5.48}$$

where M is a bound for $\|S(t)\|$ in $0 \le t \le T$. To prove (b), let $\delta > 0$. Write

$$(\Lambda_\delta g)(t) = \int_0^{t-\delta} S(t - \tau)g(\tau)d\tau = S(\delta) \int_0^{t-\delta} S(t - \delta - \tau)g(\tau)d\tau \tag{5.5.49}$$

for $t \ge \delta$, $(\Lambda_\delta y)(t) = 0$ for $t < \delta$. We have

$$\|(\Lambda - \Lambda_\delta)g_n\|_{C(0,T;E)} \le M \sup_{\delta \le t \le T} \int_{t-\delta}^t \|g_n(\tau)\|d\tau \tag{5.5.50}$$

where the right-hand side tends to zero uniformly in $0 \le t \le T$ independently of n, due to equicontinuity of the integrals.

We show that, for every $\delta > 0$, $\{(\Lambda_\delta g_n)(\cdot)\}$ has a subsequence convergent in $C(0, T; E)$. To do this, note that it follows from the second expression (5.5.49) for Λ_δ and from compactness of $S(\delta)$ that for every t fixed $\{(\Lambda_\delta g_n)(t)\}$ has a convergent subsequence. To apply the Arzelà-Ascoli Theorem 2.0.15, it is enough to show that the sequence $\{(\Lambda_\delta g_n)(t)\}$ is equicontinuous, which we do next. Let $0 \le s \le t \le T$. If $s \ge \delta$ we have

$$\|(\Lambda_\delta g_n)(t) - (\Lambda_\delta g_n)(s)\| \le \int_{s-\delta}^{t-\delta} \|S(t - \tau)\|\|g_n(\tau)\|d\tau$$

$$+ \int_0^{s-\delta} \|S(t - \tau) - S(s - \tau)\|\|g_n(\tau)\|d\tau, \tag{5.5.51}$$

which tends to zero as $t - s \to 0$ independently of n on account of the equicontinuity assumption and the (E, E)-continuity of $S(t)$ in $t \ge \delta$. The same is true if $s < \delta$, since the second integral disappears, and the domain of integration in the first becomes $0 \le \tau \le t - \delta$. The case $t < \delta$ is obvious.

Let now $\{\delta_m\}$ be a sequence of positive numbers tending to zero. We can use the diagonal sequence trick to construct a subsequence (call it also $\{g_n\}$) such that $\{\Lambda_{\delta_m} g_n\}$ is convergent in $C(0, T; E)$ for all m. If $\varepsilon > 0$, using (5.5.50) we may insure $\|\Lambda g_n - \Lambda_{\delta_m} g_n\| < \varepsilon/3$ for m large enough. Accordingly, $\|\Lambda g_n - \Lambda g_k\| < 2\varepsilon/3 + \|\Lambda_{\delta_m} g_n - \Lambda_{\delta_m} g_k\| < \varepsilon$ for $n, k \ge n_0$, and it follows that $\{\Lambda g_n\}$ is Cauchy, hence convergent. This ends the proof. $\blacksquare$

Remark 5.5.20. Λ is not a compact operator from $L^1(0, T; E)$ into $C(0, T; E)$ even if dim $E = 1$. In fact, take a δ-sequence $\{\phi_n(\cdot)\}, \phi_n(t) = n\phi(nt)$ ($\phi(\cdot)$

a continuous nonnegative function with support in $0 \le t \le 1$). For each n let $g_n(t) = \phi_n(t - t_0)y$, $0 < t_0 < T$. Then Λg_n tends to the discontinuous function $G(t) = 0$ $(t < t_0)$, $G(t) = S(t - t_0)y$ $(t > t_0)$.

However, we have

Theorem 5.5.21. *If $S(t)$ is compact for $t > 0$ then (a) Λ is compact from $L^p(0, T; E)$ into $C(0, T; E)$ for $p > 1$, (b) Λ is compact from $L^1(0, T; E)$ into $L^q(0, T; E)$ for $q > 1$.*

To show (a) it is enough to use Theorem 5.5.18 and note that, if $\{g_n(\cdot)\}$ is a bounded set in $L^p(0, T; E)$ then, by Hölder's inequality,

$$\int_e \|g_n(\tau)\|d\tau \le |e|^{1-1/p}\|g_n\|_{L^p(0,T;E)} \tag{5.5.52}$$

for any measurable e. To prove (b), extend all functions g in $L^1(0, T; E)$ setting $g(t) = 0$ in $t < 0$ and $t > T$, and extend $S(t)$ by $S(t) = 0$ in $t < 0$. Let $\phi_n(\cdot)$ be a δ-sequence as in Remark 5.5.20, and consider the operators $\Phi_n g = \phi_n * g$, $\Lambda_n g = $ restriction of $S * \Phi_n g = S * \phi_n * g = \phi_n * S * g$ to $0 \le t \le T$. By Young's Theorem 5.0.22, the operator Φ_n is bounded from $L^1(0, T; E)$ into $L^p(0, \infty; E)$ for any $p \ge 1$; thus, taking $p > 1$ and using part (a), Λ_n is a compact operator from $L^1(0, T; E)$ into $C(0, T; E)$, *a fortiori* into $L^q(0, T; E)$. Finally,

$$\|(\Lambda_n g)(t) - (\Lambda g)(t)\| \le \int_0^t \|(\phi_n * S)(t - \tau) - S(t - \tau)\|\|g(\tau)\|d\tau. \tag{5.5.53}$$

Using continuity of $S(t)$ in $t > 0$ in the norm of $L(E, E)$ and the dominated convergence theorem, we show that

$$\int_0^T \|(\phi_n * S)(\tau) - S(\tau)\|^q d\tau \to 0 \tag{5.5.54}$$

as $n \to \infty$; then, again by Young's theorem, $\Lambda_n \to \Lambda$ in the norm of (L^1, L^q), and it results that Λ is compact. This ends the proof. ∎

Theorem 5.5.22. *Assume that f satisfies Hypothesis I in $0 \le t \le T$ and that (a) the semigroup $S(t)$ is compact or (b) for each bounded sequence $\{y_n(\cdot)\} \in C(0, T; E)$ the sequence $\{f(\cdot, y_n(\cdot))\}$ has a subsequence convergent in $L^1(0, T; E)$. Then, if $g(\cdot)$ is continuous in $0 \le t \le T$, the integral equation*

$$y(t) = \zeta(t) + \int_s^t S(t - \tau)f(\tau, y(\tau))d\tau \tag{5.5.55}$$

has a continuous solution in some interval $s \le t \le T'$, where $s < T' \le T$.

Proof. As in the proof of Theorem 5.5.1 we pick T' such that the operator $\mathcal{G}$ in (5.5.7) maps $B(\zeta(\cdot), 1)$ into itself. The fact that $K(\cdot, c + 1) \in L^1$ permits application of Theorem 5.5.19 in case (a) so that $\mathcal{G}$ maps $B(\zeta(\cdot), 1)$ into a precompact subset thereof. We then apply Schauder's Theorem 2.0.14. In case (b), we use the boundedness of Λ to show that $\mathcal{G}$ maps $B(\zeta(\cdot), 1)$ into a precompact subset. ∎

When $S(\cdot)$ is a group part (a) of Corollary 5.5.17 is inapplicable; in fact,

Example 5.5.23. Let $S(\cdot)$ be a group in a Banach space E, and let $S(t)$ be compact for a single t. Then E is finite dimensional (just note that $I = S(t)S(-t)$).

Example 5.5.24. Assume that A generates a strongly continuous semigroup $S(t)$ with

$$\|S(t)\| \leq Ce^{\omega t} \quad (t \geq 0). \tag{5.5.56}$$

Then, if B is a bounded operator, $A + B$ (domain $D(A)$) generates a strongly continuous semigroup $U(t)$ with

$$\|U(t)\| \leq Ce^{(\omega + C\|B\|)t} \quad (t \geq 0). \tag{5.5.57}$$

The same holds if A generates a strongly continuous group, the estimates replaced by $\|S(t)\| \leq Ce^{\omega|t|}$ and $\|U(t)\| \leq Ce^{(\omega + C\|B\|)|t|}$ $(-\infty < t < \infty)$ respectively. To obtain these estimates, note that the integral equation (5.5.19) for $s = 0, U(t, s) = U(t), S(t, s) = S(t)$ gives, after taking norms and multiplying by $e^{-\omega t}$,

$$e^{-\omega t}\|U(t)\| \leq C + C\|B\| \int_0^t e^{-\omega \tau}\|U(\tau)\|d\tau$$

so that the result follows from Corollary 2.1.3 with $\eta(t) = e^{-\omega t}\|U(t)\|$. The group estimation is the same. The case $C = 1$ can be handled in an easier way using the Lumer-Phillips Theorem 5.2.5.

Example 5.5.25. The semigroup $U(t)$ in Example 5.5.24 can be be computed as the sum of the perturbation series

$$U(t)y = \sum_{n=1}^{\infty} U_n(t)y, \tag{5.5.58}$$

where $U_0(t) = S(t)$ and

$$U_n(t)y = \int_0^t S(t - \tau)BU_{n-1}(\tau)y \, d\tau \quad (y \in E). \tag{5.5.59}$$

For each $y \in E$, the series converges absolutely in norm, uniformly in bounded subsets of $t \geq 0$. This formula holds as well for certain unbounded operators (Dunford–Schwartz [1958, p. 631]).

Example 5.5.26. The need to keep $s < t$ in the estimations in Theorem 5.5.6 can be obviated by (a) extending $S(t, s)$ by $S(t, s) = I$ in $s > t$, (b) extending $U(t, s)$ to $s > t$ using the integral equation (5.5.19).

5.6. Wave Equations

The equation is

$$y''(t) = A(\beta)y(t). \tag{5.6.1}$$

The operator $A(\beta)$ is the restriction of the partial differential operator

$$Ay = \sum_{j=1}^{m} \sum_{k=1}^{m} \partial^j (a_{jk}(x)\partial^k y) + \sum_{j=1}^{m} b_j(x)\partial^j y + c(x)y \tag{5.6.2}$$

$(x = (x_1, x_2, \ldots, x_m), \partial_j = \partial/\partial x_j)$ obtained by means of a boundary condition, either of Dirichlet type

$$y(x) = 0 \quad (x \in \Gamma) \tag{5.6.3}$$

or of variational type

$$\partial^\nu y(x) = \gamma(x)y(x) \quad (x \in \Gamma) \tag{5.6.4}$$

on the boundary Γ of a domain $\Omega \subseteq \mathbb{R}^m$, where ∂^ν denotes the conormal derivative (defined later). The coefficients of A are real valued and defined in Ω, and the matrix $\{a_{jk}\}$ is symmetric: $a_{jk} = a_{kj}$. If the a_{jk} have first order partials, we can write A in the more natural form $Ay = \Sigma\Sigma a_{jk}\partial^j\partial^k y + \Sigma \tilde{b}_j\partial^j y + cy$. Otherwise, passing from one form to the other is not possible; since we only assume the a_{jk}, b_j, c measurable and bounded we stick with (5.6.2), the **divergence** or **variational** form of the operator. We assume A uniformly elliptic:

$$\sum_{j=1}^{m} \sum_{k=1}^{m} a_{jk}(x)\xi_j\xi_k \geq \kappa \|\xi\|^2 \quad (x \in \Omega, \xi \in \mathbb{R}^m) \tag{5.6.5}$$

for some $\kappa > 0$.

We prove that the Cauchy problem for the first-order system (or vector equation) $y'(t) = y_1(t), y_1'(t) = A(\beta)y(t)$ obtained from the equation (5.6.1) is well posed in $-\infty < t < \infty$ in a suitable space of 2-vector functions $y = (y, y_1)$, disposing first of the Dirichlet boundary condition (5.6.3). No special assumptions are placed on the domain, the boundary, or the coefficients of A, and on this level of generality, application of $A(\beta)$ will have to be understood in a suitably generalized sense.

We introduce some function spaces based on an arbitrary domain (open connected set) $\Omega \subseteq \mathbb{R}^m$. Higher order derivatives are written as usual: if

$\alpha = (\alpha_1, \alpha_2, \ldots, \alpha_m)$ is a m-dimensional vector of integers $\alpha_j \geq 0$, then $\partial^\alpha = (\partial^1)^{\alpha_1}(\partial^2)^{\alpha_2}\cdots(\partial^m)^{\alpha_m}$; the order of the derivative is $|\alpha| = \alpha_1 + \alpha_2 + \cdots + \alpha_m$.

Let $1 \leq p < \infty$, k an integer ≥ 1. The **Sobolev space** $W^{k,p}(\Omega)$ consists of all functions $y(\cdot)$ having derivatives $\partial^\alpha y$ of order $|\alpha| \leq k$ in $L^p(\Omega)$ (derivatives understood in the sense of distributions). The space $W^{k,p}(\Omega)$ is equipped with the norm

$$\|y\|_{W^{k,p}(\Omega)} = \left(\sum_{|\alpha| \leq k} \|\partial^\alpha y\|_{L^p(\Omega)}^p \right)^{1/p},$$

and $W^{k,p}(\Omega)$ is a Banach space (Adams [1975, p. 45]). We denote by $\mathcal{D}(\Omega)$ the space of all test functions on Ω (infinitely differentiable functions with compact support contained in Ω) and by $W_0^{k,p}(\Omega)$, the closure of $\mathcal{D}(\Omega)$ in $W^{k,p}(\Omega)$. Since the space $\mathcal{D}(\Omega)$ is dense in $L^p(\Omega)$ (Adams [1975, p. 31]), so is $W_0^{k,p}(\Omega)$, a fortiori $W^{k,p}(\Omega)$. For $p = 2$, the alternate notation $W^{2,p}(\Omega) = H^k(\Omega)$, $W_0^{k,p}(\Omega) = H_0^k(\Omega)$ is usual. The spaces $H^k(\Omega)$, $H_0^k(\Omega)$ are Hilbert spaces equipped with the inner product

$$(y, z)_{H^k(\Omega)} = \sum_{|\alpha| \leq k} (\partial^\alpha y, \partial^\alpha z)$$

where $(\cdot, \cdot)$ is the inner product in $L^2(\Omega)$.

We construct first the operator $A_0(\beta)$, where

$$A_0(\beta)y = \sum_{j=1}^{m} \sum_{k=1}^{m} \partial^j(a_{jk}(x)\partial^k y) - \lambda y \qquad (5.6.6)$$

and then incorporate the lower order terms by a perturbation argument:

$$A(\beta) = A_0(\beta) + P, \qquad (5.6.7)$$

where

$$Py = \sum_{j=1}^{m} b_j(x)\partial^j y + c(x)y + \lambda y. \qquad (5.6.8)$$

We do the construction in real spaces. For the Dirichlet boundary condition we take $\lambda = \kappa$ (κ the ellipticity constant in (5.6.5)) and introduce a new inner product in $H_0^1(\Omega)$ by

$$(y, z)_\lambda = \lambda \int_\Omega y(x)z(x)dx + \int_\Omega \left(\sum_{j=1}^{m} \sum_{k=1}^{m} a_{jk}(x)\partial^j y(x)\partial^k z(x) \right)dx \qquad (5.6.9)$$

We have $\lambda\|y\|_{H^1(\Omega)} \leq (y, y)_\lambda \leq C\|y\|_{H^1(\Omega)}$ with $C < \infty$, and it is obvious that (5.6.9) has all properties of an inner product. The norm $\|y\|_\lambda = \sqrt{(y, y)_\lambda}$ corresponding to (5.6.9) is equivalent to the original norm of $H_0^1(\Omega)$. The operator

$A_0(\beta)$ is defined as follows: $y \in H_0^1(\Omega)$ belongs to $D(A_0(\beta))$ if and only if the linear functional $z \to (y, z)_\lambda$, defined in $H_0^1(\Omega)$, is continuous *in the norm of* $L^2(\Omega)$; if this is the case, we extend the functional to all of $L^2(\Omega)$ (since $H_0^1(\Omega)$ is dense in $L^2(\Omega)$ in the norm of $L^2(\Omega)$ this extension is unique) and using Theorem 2.0.10 we construct an element $w \in L^2(\Omega)$ such that $(y, z)_\lambda = (w, z)$ $(z \in H_0^1(\Omega))$. Then we define $A_0(\beta)y = -w$. All these steps can be fused into the single formula

$$(A_0(\beta)y, z) = -(y, z)_\lambda \quad (y \in D(A_0(\beta))). \tag{5.6.10}$$

Motivation for this definition is: if the coefficients a_{jk} and the boundary Γ are smooth and y, z are smooth functions with $y = z = 0$ on Γ, then (5.6.10) follows using the divergence theorem.

If $y, z \in D(A_0(\beta))$ then $(A_0(\beta)y, z) = -(y, z)_\lambda = -(z, y)_\lambda = (y, A_0(\beta)z)$, so that $A_0(\beta)$ is symmetric. On the other hand, setting $y = z \in D(A_0(\beta))$ in (5.6.10) we obtain $(A_0(\beta)y, y) = -\sqrt{(y, y)_\lambda}$, so that $A_0(\beta)$ is one-to-one. Finally, $A_0(\beta)D(A_0(\beta)) = L^2(\Omega)$. In fact, let w be an arbitrary element of $L^2(\Omega)$. Define a linear functional by $z \to -(w, z)$, where $(\cdot, \cdot)$ is the inner product of $L^2(\Omega)$. Since this functional is bounded with norm $\|w\|$ in $L^2(\Omega)$, it is bounded in $H_0^1(\Omega)$ (normed with $\|\cdot\|_\lambda$) with norm $\leq \lambda^{-1}\|w\|$; thus, there exists $y \in H_0^1(\Omega)$ such that $(y, z)_\lambda = -(w, z)$. Accordingly, $y \in D(A_0(\beta))$ and $A_0(\beta)y = w$. In particular, $\|y\| \leq \lambda^{-1}\|w\|$, so that $A_0(\beta)^{-1}$ is bounded as an operator from $L^2(\Omega)$ into itself (with norm $\leq \lambda^{-1}$).

To show that $A_0(\beta)$ is densely defined in $L^2(\Omega)$ it is enough to show that $D(A_0(\beta))$ is dense in $H_0^1(\Omega)$ in the norm of $H_0^1(\Omega)$. If the latter is not true we can find an element $z \in H_0^1(\Omega)$ with $(y, z)_\lambda = 0$ for all $y \in D(A_0(\beta))$. By (5.6.10), z would be orthogonal in $L^2(\Omega)$ to $A_0(\beta)D(A_0(\beta)) = L^2(\Omega)$, hence $z = 0$.

To prove that $A_0(\beta)$ is self adjoint we use

Lemma 5.6.1. *Let A be a densely defined, symmetric operator with a bounded inverse in the Hilbert space H. Then A is self adjoint.*

Proof. We have $(A^{-1}y, z) = (A^{-1}y, A(A^{-1}z)) = (AA^{-1}y, A^{-1}z) = (y, A^{-1}z)$ so that A^{-1} is symmetric; since it is bounded, it is self adjoint. Using the calculus of inverses and adjoints (see **5.0**), we obtain $A^* = ((-A)^{-1})^{-1})^* = ((-A)^{-1})^*)^{-1} = (A^{-1})^{-1} = A$. ∎

It follows from (5.6.10) that $(A_0(\beta)y, y) \leq -\kappa\|y\|^2$, so that $\sigma(A) \subseteq (-\infty, -\kappa]$. We denote by $R = (-A_0(\beta))^{1/2}$ the positive definite root of $-A_0(\beta)$ obtained through the extended functional calculus in **5.0**; this root R has a bounded inverse $R^{-1} = (-A_0(\beta))^{-1/2}$. Although R is not easy to identify from A_0, its domain is.

Lemma 5.6.2.

$$D(R) = D((-A_0(\beta))^{1/2}) = H_0^1(\Omega), \quad \|Ry\| = \|y\|_\lambda \quad (y \in H_0^1(\Omega)). \tag{5.6.11}$$

Proof. We begin by showing that $D(A_0(\beta))$ is dense in $D(R)$ in the norm $\|y\|_R = \|Ry\|$. To see this, let $y \in D(R)$, so that $y = R^{-1}z, z \in L^2(\Omega)$. Select a sequence $\{z_n\} \in D(R)$ with $z_n \to z$. Then, $y_n = R^{-1}z_n \in D(R^2) = D(A_0(\beta))$ and $Ry_n = z_n \to z$, so that $\|Ry_n - z\| = \|R(y_n - y)\| = \|y_n - y\|_R \to 0$.

Let y be an element of $D(A_0(\beta))$. Then we have $\|Ry\|^2 = \|(-A_0(\beta))^{1/2}y\|^2 = -(A_0(\beta)^{1/2}y, A_0(\beta)^{1/2}y) = -(A_0(\beta)y, y) = (y, y)_\lambda$, so that the second equality (5.6.11) is obtained. Since $D(A_0(\beta))$ is dense both in $D(R)$ and in $H_0^1(\Omega)$, it follows that both spaces coincide and that the equality of norms can be extended to $D(A_0(\beta)^{1/2}) = H_0^1(\beta)$. ∎

In order to reduce the wave equation (5.6.1) to a first order system we introduce the operator

$$\mathbf{A}_0(\beta) = \begin{bmatrix} 0 & I \\ A_0(\beta) & 0 \end{bmatrix} \tag{5.6.12}$$

in the space $\mathbf{E} = H_0^1(\Omega) \times L^2(\Omega)$, equipped with the inner product

$$(\mathbf{y}, \mathbf{z})_\mathbf{E} = ((y, y_1), (z, z_1))_\mathbf{E} = (y, z)_\lambda + (y_1, z_1)_{L^2(\Omega)}. \tag{5.6.13}$$

The domain is $D(\mathbf{A}_0(\beta)) = D(A_0(\beta)) \times H_0^1(\Omega)$.

Theorem 5.6.3. *The Cauchy problem for*

$$\mathbf{y}'(t) = \mathbf{A}_0(\beta)\mathbf{y}(t)$$

is well posed in $-\infty < t < \infty$; *precisely,* $\mathbf{A}_0(\beta)$ *is a skew adjoint operator* $(\mathbf{A}_0(\beta)^* = -\mathbf{A}_0(\beta))$, *hence it generates a strongly continuous unitary group* $\mathbf{S}_0(t; \beta)$ *in* $\mathbf{E}$.

The proof of Theorem 5.6.3 is a consequence of an abstract operator result (Theorem 5.6.4 below). Let H be an arbitrary Hilbert space and A_0 a self adjoint operator with spectrum in $(-\infty, -\kappa]$ $(\kappa > 0)$. We set $R = (-A_0)^{1/2}$, so that R is self adjoint and $R^{-1} = (-A_0)^{-1/2}$ is bounded. Define $H^1 = D(R)$ equipped with the inner product $(y, z)_1 = (Ry, Rz)$, corresponding to the norm $\|y\|_1 = \|Ru\|$, and let $\mathbf{H} = H^1 \times H$. Elements of $\mathbf{H}$ are denoted $\mathbf{y} = (y, y_1), \mathbf{z} = (z, z_1), \ldots$, and $\mathbf{H}$ is equipped with the inner product[1]

$$(\mathbf{y}, \mathbf{z})_\mathbf{H} = ((y, y_1), (z, z_1))_\mathbf{H} = (y, z)_1 + (y_1, z_1)_H, \tag{5.6.14}$$

which product corresponds to the norm $\|\mathbf{y}\| = \|(y, y_1)\| = (\|y\|_1^2 + \|y_1\|^2)^{1/2} = (\|Ry\|^2 + \|y_1\|^2)^{1/2}$. Finally,

$$\mathbf{A}_0 = \begin{bmatrix} 0 & I \\ A_0 & 0 \end{bmatrix} \tag{5.6.15}$$

with domain $D(\mathbf{A}_0) = D(A_0) \times H^1$.

[1] Parentheses are used here both for pairs in a Cartesian product and for inner products. To avoid confusion, we subindex inner products.

Theorem 5.6.4. A_0 *is skew adjoint.*

Proof. Let $\mathbf{z} = (z, z_1) \in D(\mathbf{A}_0^*)$. Then the linear functional $\mathbf{y} \to (\mathbf{z}, \mathbf{A}_0\mathbf{y})_\mathbf{H}$ is bounded in the norm of $\mathbf{H}$ in $D(\mathbf{A}_0)$. We have

$$(\mathbf{z}, \mathbf{A}_0\mathbf{y})_\mathbf{H} = ((z, z_1), \mathbf{A}_0(y, y_1))_\mathbf{H} = ((z, z_1), (y_1, A_0y))_\mathbf{H}$$
$$= (z, y_1)_1 + (z_1, A_0y)_H = (Rz, Ry_1)_H + (z_1, A_0y)_H. \quad (5.6.16)$$

It follows that $y_1 \to (Rz, Ry_1)_H$ is bounded in the norm of H, so that $Rz \in D(R^*)$ $= D(R)$, *a fortiori* $z \in D(A_0)$; on the other hand, $y \to (z, A_0y)_H =$ $-(z_1, R(Ry))_H$ is bounded in the norm of H^1, so that $w \to -(z_1, Rw)_H$ is bounded in the norm of H, and we deduce that $z_1 \in D(R^*) = D(R)$. We have then shown that $(z, z_1) \in D(\mathbf{A}_0)$. Using (5.6.16),

$$((z, z_1), \mathbf{A}_0(y, y_1))_\mathbf{H} = (Rz, Ry_1)_H + (z_1, A_0y)_H$$
$$= -(A_0z, y_1)_H - (Rz_1, Ry)_H = -(\mathbf{A}_0(z, z_1), (y, y_1))_\mathbf{H}.$$

This ends the proof of Theorem 5.6.4 and thus also that of Theorem 5.6.3. ∎

We incorporate the lower order terms in (5.6.7). The perturbation operator is

$$\mathbf{P} = \begin{bmatrix} 0 & 0 \\ P & 0 \end{bmatrix}. \quad (5.6.17)$$

Obviously, $\mathbf{P}$ is bounded in $\mathbf{E}$, thus Corollary 5.5.7 applies, producing

Theorem 5.6.5. *Let*

$$\mathbf{A}(\beta) = \begin{bmatrix} 0 & I \\ A(\beta) & 0 \end{bmatrix} \quad (5.6.18)$$

with domain $D(\mathbf{A}(\beta)) = D(A_0(\beta)) \times H_0^1(\Omega)$. *Then the Cauchy problem for*

$$\mathbf{y}'(t) = \mathbf{A}(\beta)\mathbf{y}(t) \quad (5.6.19)$$

is well posed in $-\infty < t < \infty$; *equivalently,* $\mathbf{A}(\beta)$ *is the infinitesimal generator of a strongly continuous group* $\mathbf{S}(t; \beta)$.

To motivate the treatment of variational boundary conditions, assume that the domain Ω is bounded and that the boundary and the coefficients of the differential operator are smooth. Denote by $(\eta_1(x), \ldots, \eta_m(x))$ the outer normal vector on Γ. The **conormal vector** is $(\nu_1(x), \ldots, \nu_m(x))$, with components

$$\nu_j(x) = \sum_{j=1}^{m} a_{jk}(x)\eta_k(x).$$

The **conormal derivative** ∂^ν is the derivative in the direction of the conormal vector. Assume the coefficients of the operator A_0 in (5.6.6) are smooth and that $y(\cdot)$, $z(\cdot)$ are smooth; if $y(\cdot)$ satisfies (5.6.4) on the boundary we obtain, applying the divergence theorem,

$$
\begin{aligned}
\int_\Omega (A_0 y) z \, dx &= \int_\Omega \left(\sum \sum \partial^j (a_{jk}(x) \partial^k y(x)) - \lambda y(x) \right) z(x) \, dx \\
&= \int_\Omega \left(-\sum \sum a_{jk}(x) \partial^j y(x) \partial^k z(x) - \lambda y(x) \right) z(x) \, dx \\
&\quad + \int_\Gamma \left(\sum \sum a_{jk}(x) \eta_j(x) \partial^k y(x) z(x) \right) d\sigma \\
&= -\int_\Omega \left(\sum \sum a_{jk}(x) \partial^j y(x) \partial^k z(x) + \lambda y(x) \right) z(x) \, dx \\
&\quad + \int_\Gamma \gamma(x) y(x) z(x) \, d\sigma,
\end{aligned}
\tag{5.6.20}
$$

where $d\sigma$ is the hyperarea differential on Γ. Accordingly, if we are to define $A_0(\beta)$ as in (5.6.10) by $(A_0(\beta)) y, z) = -(y, z)_\lambda$ from an inner product $(\cdot, \cdot)_\lambda$ in $H^1(\Omega)$, this inner product must be

$$
(y, z)_\lambda = \lambda \int_\Omega y(x) z(x) \, dx + \int_\Omega \left(\sum_{j=1}^m \sum_{k=1}^m a_{jk}(x) \partial^j y(x) \partial^k z(x) \right) dx
$$
$$
- \int_\Gamma \gamma(x) y(x) z(x) \, d\sigma.
\tag{5.6.21}
$$

Due to the presence of the integral over Γ, Ω cannot be unrestricted any longer; some boundary regularity is needed.

Denote by B the open unit ball of $\mathbb{R}^m$. A domain $\Omega \subseteq \mathbb{R}^m$ with boundary Γ is said to be **of class** $C^{(k)}$ (k an integer ≥ 0) if, given any $\bar{x} \in \Gamma$, there exists an open neighborhood V of $\bar{x}$ in $\mathbb{R}^m$ and a map $\phi : \overline{V} \to \overline{B}$ such that

(i) ϕ is one-to-one and onto $\overline{B}$ with $\phi(\bar{x}) = 0$,

(ii) ϕ (resp. ϕ^{-1}) possesses partial derivatives of order $\leq k$ in V (resp. in B) continuous in the respective closures,

(iii) $\phi(V \cap \Omega) = \{x \in B; x_m \geq 0\}$, $\phi(V \cap \Gamma) = \{x \in B; x_m = 0\}$.

We outline the construction of $A_0(\beta)$; full details can be found, for instance in the author [1975, Chapter IV]. Let Ω be a domain of class $C^{(1)}$ with bounded boundary Γ. Then (Adams [1975, Theorem 5.22, p. 114]), if $y \in \mathcal{D}(\mathbb{R}^m)$ there exists a constant C such that $\|y\|_{L^1(\Gamma)} \leq C \|y\|_{W^{1,1}(\Omega)}$. Accordingly, if $\gamma(\cdot) \in L^\infty(\Gamma)$ we have

$$
\int_\Gamma |\gamma(x) y(x) z(x)| \, d\sigma \leq C \|yz\|_{W^{1,1}(\Omega)}
$$

for $y, z \in \mathcal{D}(\mathbb{R}^m)$ (the constants in this and following inequalities may change with the inequality). Writing down the norm on the right side in terms of the L^1 norm of $\|yz\|$ and of its first derivatives and using Schwarz's inequality, we end up with

$$\int_\Gamma |\gamma(x)y(x)z(x)|d\sigma \leq C\|y\|_{L^2(\Omega)}\|z\|_{L^2(\Omega)}$$

$$+C\left(\sum \|\partial^j y\|^2_{L^2(\Omega)}\right)^{1/2}\|z\|_{L^2(\Omega)} + C\|y\|_{L^2(\Omega)}\left(\sum \|\partial^j z\|^2_{L^2(\Omega)}\right)^{1/2}. \quad (5.6.22)$$

Taking $y = z$ and using the inequality $ab = (\varepsilon^{-1}a)(\varepsilon b) \leq (\varepsilon^{-2}a^2 + \varepsilon^2 b^2)/2$ in the last two terms with $a = \|y\|, b = (\Sigma\|\partial^j y\|)^{1/2}$ we obtain

$$\int_\Gamma |\gamma(x)||y(x)|^2 d\sigma \leq C\varepsilon^{-2}\|y\|^2_{L^2(\Omega)} + C\varepsilon^2 \sum \|\partial^j y\|^2_{L^2(\Omega)}$$

$$\leq C\varepsilon^{-2}\int_\Omega |y(x)|^2 dx + C\varepsilon^2\kappa^{-1}\int_\Omega \left(\sum_{j=1}^m \sum_{k=1}^m a_{jk}(x)\partial^j y(x)\partial^k y(x)\right)dx. \quad (5.6.23)$$

We deduce that, for λ large enough the inner product (5.6.21) will satisfy

$$c\|y\|^2_{H^1(\Omega)} \leq (y, y)_\lambda \leq C\|y\|^2_{H^1(\Omega)} \quad (5.6.24)$$

$(y \in \mathcal{D}(\mathbb{R}^m))$ with $c > 0$. We use now the fact (Adams [1975]) that in a domain Ω of class $C^{(0)}$ the space $\mathcal{D}(\mathbb{R}^m)$ is dense in $W^{k,p}(\Omega)$ for any integer k and $1 \leq p < \infty$; this, combined with inequality (5.6.24) allows us to extend the inner product (5.6.21) to $H^1(\Omega)$, and (5.6.24) holds for the extended product. Once the space $H^1(\Omega)$ is equipped with $(\cdot, \cdot)_\lambda$, the construction of the operator $A_0(\beta)$ is carried out exactly in the same way as for the Dirichlet boundary condition; $A_0(\beta)$ is self adjoint and $\sigma(A) \subseteq (-\infty, -\kappa]$. We define again $R = (-A_0(\beta))^{1/2}$ and we have the following counterpart of Lemma 5.6.2:

Lemma 5.6.6.

$$D(R) = D((-A_0(\beta))^{1/2}) = H^1(\Omega), \quad \|Ry\| = \|y\|_\lambda \quad (y \in H^1(\Omega)) \quad (5.6.25)$$

The proof is based on the fact that $D(A_0(\beta))$ is dense in $H^1(\Omega)$ and in $D(R)$ (which follows from the definition of $A_0(\beta)$). ∎

The operator $\mathbf{A}_0(\beta)$ is defined by (5.6.12) in the space $\mathbf{E} = H^1(\Omega) \times L^2(\Omega)$ endowed with the inner product (5.6.13). Theorem 5.6.4 applies to show that $\mathbf{A}_0(\beta)$ is skew adjoint, and the perturbation argument to incorporate the lower order terms is the same. We limit ourselves to stating the final result, companion of Theorem 5.6.5.

Theorem 5.6.7. *Let Ω be a domain of class $C^{(1)}$ with bounded boundary Γ, $\gamma(\cdot) \in L^\infty(\Gamma)$, $A_0(\beta)$ the operator defined by*

$$(A_0(\beta)y, z) = -(y, z)_\lambda$$

with respect to the scalar product (5.6.21). Let $A(\beta)$ be defined by (5.6.7) and

$$\mathbf{A}(\beta) = \begin{bmatrix} 0 & I \\ A(\beta) & 0 \end{bmatrix} \tag{5.6.26}$$

with domain $D(\mathbf{A}(\beta)) = D(A_0(\beta)) \times H^1(\Omega)$. Then the Cauchy problem for

$$\mathbf{y}'(t) = \mathbf{A}(\beta)\mathbf{y}(t) \tag{5.6.27}$$

is well posed in $-\infty < t < \infty$; equivalently, $\mathbf{A}(\beta)$ is the infinitesimal generator of a strongly continuous group $S(t; \beta)$.

Remark 5.6.8. The construction of the operator $\mathbf{A}(\beta)$ in complex spaces is similar; it suffices to put the complex conjugate $\bar{y}(x)$ in the scalar products (5.6.9) and (5.6.21).

Example 5.6.9. If $\mathbf{A}_0$ is the operator in Theorem 5.6.4, we have

$$\mathbf{S}_0(t) = \exp(t\mathbf{A}_0) = \begin{bmatrix} \cos(tR) & R^{-1}\sin(tR) \\ -R\sin(tR) & \cos(tR) \end{bmatrix} \tag{5.6.28}$$

Example 5.6.10. The group $\mathbf{S}(t, \beta)$ in Theorems 5.6.5 and 5.6.7 satisfies

$$\|\mathbf{S}(t, \beta)\| \leq e^{\omega|t|} \quad (-\infty < t < \infty) \tag{5.6.29}$$

for some ω. This follows from the fact that

$$\|\mathbf{S}_0(t, \beta)\| = 1 \quad (-\infty < t < \infty) \tag{5.6.30}$$

and from Example 5.5.24.

Example 5.6.11. The **energy** of a solution of (5.6.1) is

$$E(t) = \int_\Omega \left(\partial_t y(t, x)^2 + \sum\sum a_{jk}(x)\partial_j y(x, t)\partial_k y(t, x) \right) dx. \tag{5.6.31}$$

It is obvious that the solutions constructed in this section have finite energy for all t. More precisely, the energy increases (at most) exponentially as $|t| \to \infty$.

Example 5.6.12. Consider the wave equation

$$\partial_{tt} y = \sum_{j=1}^{m} \sum_{k=1}^{m} \partial^j (a_{jk}(x) \partial^k y) \tag{5.6.32}$$

without lower order terms. Let β the Dirichlet boundary condition (5.6.3) or the Neumann boundary condition ((5.6.4) with $\gamma(x) = 0$). Then $E(t)$ is constant in $-\infty < t < \infty$ (Hint: In both cases the norm $(\cdot, \cdot)_\lambda$ may be defined with $\lambda > 0$ arbitrarily small.)

5.7. Semilinear Wave Equations: Local Existence

We apply the theory in **5.5** to

$$y_{tt}(t, x) = \sum_{j=1}^{m} \sum_{k=1}^{m} \partial^j (a_{jk}(x) \partial^k y(t, x)) + \sum_{j=1}^{m} b_j(x) \partial^j y(t, x)$$
$$+ c(x) y(t, x) - \phi(t, x, y(t, x)) + u(t, x) \tag{5.7.1}$$

in a domain $\Omega \subseteq \mathbb{R}^m$. Assumptions on Ω, the linear part of the operator and the boundary condition are the same as in **5.6**. If β is the Dirichlet boundary condition Ω is arbitrary; for the variational boundary condition (5.6.4) we assume Ω of class C and Γ bounded.

We handle (5.7.1) as the abstract semilinear equation

$$\mathbf{y}'(t) = \mathbf{A}(\beta) \mathbf{y}(t) + \mathbf{f}(t, \mathbf{y}(t)), \quad \mathbf{y}(0) = \zeta \tag{5.7.2}$$

for the vector function $\mathbf{y}(t) = (y(t), y_1(t)) \in \mathbf{E}$ and the operator

$$\mathbf{A}(\beta) = \begin{bmatrix} 0 & I \\ A(\beta) & 0 \end{bmatrix} \tag{5.7.3}$$

in Theorem 5.6.5 for the Dirichlet boundary condition (5.6.3) (with $\mathbf{E} = H_0^1(\Omega) \times L^2(\Omega)$) and in Theorem 5.6.7 for a variational boundary condition (with $\mathbf{E} = H^1(\Omega) \times L^2(\Omega)$). In both cases, $\mathbf{f}(t, \mathbf{y})$ is

$$\mathbf{f}(t, \mathbf{y})(x) = f(t, (y, y_1))(x) = \begin{bmatrix} 0 \\ -\phi(t, x, y(x)) + u(t, x) \end{bmatrix}. \tag{5.7.4}$$

The function $\phi(t, x, y)$ is real valued. For any dimension, we require $(t, x) \to \phi(t, x, y)$ to be measurable in $[0, T] \times \Omega$ for every $y \in \mathbb{R}$ fixed. Otherwise, the assumptions on $\phi(t, x, y)$ depend on the dimension m. All functions K, L, ξ, η below are nonnegative.

Dimension $m \geq 3$: Let $\rho = m/(m-2)$. There exist $K(\cdot), L(\cdot) \in L^1(0, T)$ and $\xi(\cdot) \in L^2(\Omega), \eta(\cdot) \in L^m(\Omega)$ such that

$$|\phi(t, x, y)| \leq K(t)(\xi(x) + |y|^\rho) \quad (0 \leq t \leq T, x \in \Omega, y \in \mathbb{R}), \qquad (5.7.5)$$

$$|\phi(t, x, y') - \phi(t, x, y)| \leq L(t)(\eta(x) + |y|^{\rho-1} + |y'|^{\rho-1})|y' - y|$$
$$(0 \leq t \leq T, x \in \Omega, y, y' \in \mathbb{R}). \quad (5.7.6)$$

Dimension $m = 2$: There exist $\rho > 1$, $K(\cdot), L(\cdot) \in L^1(0, T)$ and $\xi(\cdot) \in L^2(\Omega)$, $\eta(\cdot) \in L^{2\rho/(\rho-1)}(\Omega)$ such that (5.7.5) and (5.7.6) hold.

Dimension $m = 1$: There exist $\xi(\cdot), \eta(\cdot) \in L^2(\Omega)$ such that for every $c > 0$ there exist $K(\cdot, c), L(\cdot, c) \in L^1(0, T)$ with

$$|\phi(t, x, y)| \leq K(t, c)\xi(x) \quad (0 \leq t \leq T, x \in \Omega, |y| \leq c), \qquad (5.7.7)$$

$$|\phi(t, x, y') - \phi(t, x, y)| \leq L(t, c)\eta(x)|y' - y|$$
$$(0 \leq t \leq T, x \in \Omega, |y|, |y'| \leq c). \quad (5.7.8)$$

The assumptions on $u(t, x)$ are the same in all dimensions: $u(t, x)$ is measurable in $(0, T) \times \Omega$, for almost all $t \in [0, T]$, $u(t, \cdot) \in L^2(\Omega)$, and the function $t \to \|u(t, \cdot)\|_{L^2(\Omega)}$ belongs to $L^1(0, T)$. This assumption is satisfied to excess if

$$u(\cdot, \cdot) \in L^2((0, T) \times \Omega). \qquad (5.7.9)$$

The treatment below depends on (a particular case of) Sobolev's imbedding theorem. A domain Ω has the **cone property** (Adams [1975, p. 66]) if there exists a finite cone $\mathcal{C}$ such that every point $x \in \Omega$ is the vertex of a cone $\mathcal{C}_x \subseteq \Omega$ congruent to $\mathcal{C}$ by rigid motion.

Theorem 5.7.1. *Let Ω have the cone property. Then, if $kp < m, 1 \leq p \leq q \leq mp/(m - kp)$, we have $W^{k,p}(\Omega) \subseteq L^q(\Omega)$ and the inclusion is a bounded operator. In case $m = kp$, the inclusion is bounded for $1 \leq p \leq q < \infty$.*

The proof (of a more general theorem) is in Adams [1975, p. 97].

Theorem 5.7.2. $\mathbf{f}(t, \mathbf{y})$ *satisfies Hypothesis II in* **5.5.**

We note first that all properties of $\mathbf{f}(t, \mathbf{y})$ as an operator from $\mathbf{E}$ into $\mathbf{E}$ to be proved are equivalent to corresponding properties of the operator $y(\cdot) \to \phi(t, \cdot, y(\cdot)) + u(t, \cdot)$ from $H^1(\Omega)$ into $L^2(\Omega)$. Obviously, the term $u(t, \cdot)$ satisfies Assumption II, adds a L^1 function to the right-hand side of (5.5.4) (independently of the size of y) and adds nothing to the right side of (5.5.5), thus we may and will assume that $u = 0$.

We begin by showing that for every $y(\cdot)$ measurable in Ω the function $x \to \phi(t, x, y(x))$ is measurable. This is an immediate consequence of measurability of $(t, x) \to \phi(t, x, y)$ if $y(x)$ is countably valued. The general result follows approximating $y(x)$ by a sequence of countably valued functions and using y-continuity of $\phi(t, x, y)$ (guaranteed by (5.7.6) for $m \geq 2$ and by (5.7.8) for $m = 1$).

Dimension $m \geq 3$. By Theorem 5.7.1 $H^1(\Omega) \subseteq L^{2m/(m-2)}(\Omega)$ with bounded inclusion, so that if $y(\cdot) \in H^1(\Omega)$ we have[1]

$$\|\phi(t, \cdot, y(\cdot))\|_{L^2(\Omega)} \leq K(t) \left(\int_\Omega (\xi(x) + |y(x)|^{m/(m-2)})^2 dx \right)^{1/2}$$

$$\leq K(t) \left\{ \|\xi\|_{L^2(\Omega)} + \||y|^{m/(m-2)}\|_{L^2(\Omega)} \right\}$$

$$\leq K(t) \left\{ \|\xi\|_{L^2(\Omega)} + \left(\|y\|_{L^{2m/(m-2)}(\Omega)} \right)^{m/(m-2)} \right\}$$

$$\leq K(t) \left\{ \|\xi\|_{L^2(\Omega)} + C \left(\|y\|_{H^1(\Omega)} \right)^{m/(m-2)} \right\} \quad (5.7.10)$$

and it follows that $\mathbf{f}(t, \mathbf{y})$ satisfies (5.5.4) of Hypothesis II. On the other hand, by (5.7.6), if $y(\cdot), y'(\cdot) \in H^1(\Omega)$, then

$$\|\phi(t, \cdot, y'(\cdot)) - \phi(t, \cdot, y(\cdot))\|_{L^2(\Omega)}$$

$$\leq L(t) \left(\int_\Omega (\eta(x) + |y(x)|^{2/(m-2)} + |y'(x)|^{2/(m-2)})^2 |y'(x) - y(x)|^2 dx \right)^{1/2}$$

$$\leq L(t) \left(\int_\Omega (\eta(x) + |y(x)|^{2/(m-2)} + |y'(x)|^{2/(m-2)})^m dx \right)^{1/m}$$

$$\times \left(\int_\Omega |y'(x) - y(x)|^{2m/(m-2)} dx \right)^{(m-2)/2m} \quad (5.7.11)$$

after applying Hölder's inequality with $p = m/2, q = p/(p-1) = m/(m-2)$; note that the three functions in the first integrand belong to $L^m(\Omega)$, the first by assumption, the second and the third because $y(\cdot), y'(\cdot) \in L^{2m/(m-2)}(\Omega)$. Using again the imbedding, we end up with the estimate

$$\|\phi(t, \cdot, y'(\cdot)) - \phi(t, \cdot, y(\cdot))\|_{L^2(\Omega)}$$

$$\leq L(t) \left\{ \|\eta\|_{L^m(\Omega)} + \||y|^{2/(m-2)}\|_{L^m(\Omega)} \right.$$

$$\left. + \||y'|^{2/(m-2)}\|_{L^m(\Omega)} \right\} \|y' - y\|_{L^{2m(m-2)}(\Omega)}$$

$$\leq L(t) \left\{ \|\eta\|_{L^m(\Omega)} + \left(\|y\|_{L^{2m/(m-2)}(\Omega)} \right)^{2/(m-2)} \right.$$

$$\left. + \left(\|y'\|_{L^{2m/(m-2)}(\Omega)} \right)^{2/(m-2)} \right\} \|y' - y\|_{L^{2m(m-2)}(\Omega)}$$

[1] In this section, $\|f(\cdot)\|$ indicates the norm of $f(\cdot)$ in some function space rather than the norm $t \to \|f(t)\|$ as a function.

$$\leq CL(t) \Big\{ \|\eta\|_{L^m(\Omega)} + C\big(\|y\|_{H^1(\Omega)}\big)^{2/(m-2)}$$

$$+ C\big(\|y'\|_{H^1(\Omega)}\big)^{2/(m-2)} \Big\} \|y' - y\|_{H^1(\Omega)} \qquad (5.7.12)$$

good for (5.5.5) of Hypothesis II. It only remains to show that $t \to \mathbf{f}(t, \mathbf{y})$ is strongly measurable as an **E**-valued function or, equivalently, that $t \to \phi(t, \cdot, y(\cdot))$ is a strongly measurable $L^2(\Omega)$-valued function for each $y(\cdot) \in H^1(\Omega)$. Since $L^2(\Omega)$ is separable, Example 5.0.39 authorizes us to check only weak measurability, that is, measurability of

$$t \to (z(\cdot), \phi(t, \cdot, y(\cdot))) = \int_\Omega z(x)\phi(t, x, y(x))dx$$

for $y(\cdot), z(\cdot) \in L^2(\Omega)$ arbitrary. This is obvious if $y(\cdot) \in \mathcal{D}(\mathbb{R}^m)$ (we approximate $y(\cdot)$ uniformly by a function taking a finite number of values); thus, it follows for $y(\cdot) \in H^1(\Omega)$ by denseness of $\mathcal{D}(\mathbb{R}^m)$ and (5.7.12).

Dimension $m = 2$. Everything is the same, except that $\rho \geq 1$ is not related to the dimension m. We use Theorem 5.7.1 in the case $kp = m$ with $k = 1, p = m = 2$, where the imbedding $H^1(\Omega) \to L^q(\Omega)$ is valid for every $q \geq 2$. We define a fictitious dimension $m' = 2\rho/(\rho - 1)$ so that $\rho = m'/(m' - 2)$. Although m' is no longer an integer, the estimations (5.7.10), (5.7.11) and (5.7.12) work in exactly the same way.

Dimension $m = 1$. If $y(\cdot) \in H^1(\Omega)$ then $y(\cdot)^2$ is absolutely continuous with derivative $2y(\cdot)y'(\cdot)$, which is integrable in Ω by Schwarz's inequality. It then follows that $H^1(\Omega) \subseteq BC(\overline{\Omega})$ (the space of all bounded continuous functions in $\overline{\Omega}$ equipped with the supremum norm) with bounded inclusion. Accordingly, if $\|y(\cdot)\|_{H^1(\Omega)} \leq c$ then

$$|\phi(t, x, y(x))| \leq K(t, c)\xi(x),$$
$$|\phi(t, x, y'(x)) - \phi(t, x, y(x))| \leq L(t, c)\eta(x)|y'(x) - y(x)|.$$

Integrating,

$$\|\phi(t, \cdot, y(\cdot))\|_{L^2(\Omega)} \leq K(t, c)\|\xi\|_{L^2(\Omega)}, \qquad (5.7.13)$$

$$\|\phi(t, \cdot, y'(\cdot)) - \phi(t, \cdot, y(\cdot))\|_{L^2(\Omega)} \leq L(t, c)\|\eta\|_{L^2(\Omega)}\|y'(\cdot) - y(\cdot)\|_{L^2(\Omega)}$$

$$\leq CL(t, c)\|\eta\|_{L^2(\Omega)}\|y'(\cdot) - y(\cdot)\|_{H^1(\Omega)}. \qquad (5.7.14)$$

Strong measurability of $t \to \phi(t, \cdot, y(\cdot))$ is proved as in the case $m \geq 2$. This ends the proof of Theorem 5.7.2. ∎

Applying Theorem 5.5.1 we obtain

Theorem 5.7.3. *The equation (5.7.2) is locally solvable for an arbitrary initial condition $\zeta = (\zeta, \zeta_1)$ in **E**.*

The results in this section apply to the interesting nonlinearity $\phi(y) = cy^3$ (c of any sign) for $n = 3$; for $n = 2$, arbitrary polynomial growth is fine, and no growth conditions are needed for $m = 1$.

For Hypothesis I, we assume that $(t, x) \to \phi(t, x, y)$ is measurable in $[0, T] \times \Omega$ for every $y \in \mathbb{R}$ fixed and that $y \to \phi(t, x, y)$ is continuous for every (t, x) fixed. In dimension ≥ 3 we only require (5.7.5) for $\rho = m/(m-2)$ and $\xi(\cdot) \in L^2(\Omega)$, and in dimension 2 for some $\rho > 1$. In dimension 1 (7.7) is enough. In these conditions we have

Theorem 5.7.4. $\mathbf{f}(t, \mathbf{y})$ *satisfies Assumption I in* **5.5.**

We start with dimension ≥ 3. That $\mathbf{f}(t, \mathbf{y})$ satisfies (5.5.4) of Assumption I is proved in the same way as in Theorem 5.7.2; likewise, the proof of strong measurability of $t \to \mathbf{f}(t, \mathbf{y})$ is the same. Thus, it only remains to prove that $\mathbf{y} \to \mathbf{f}(t, \mathbf{y})$ is continuous for t fixed or, equivalently, that $y(\cdot) \to \phi(t, \cdot, y(\cdot))$ is continuous from $H^1(\Omega)$ into $L^2(\Omega)$. If this is not true, there exists a sequence $\{h_n(\cdot)\} \subseteq H^1(\Omega)$ such that $h_n(\cdot) \to 0$ in $H^1(\Omega)$ and

$$\|\phi(t, \cdot, y(\cdot) + h_n(\cdot)) - \phi(t, \cdot, y(\cdot))\|_{L^2(\Omega)}^2$$
$$= \int_\Omega |\phi(t, x, y(x) + h_n(x)) - \phi(t, x, y(x))|^2 dx \qquad (5.7.15)$$

does not converge to zero. We assume (taking a subsequence) that $h_n(x) \to 0$ a.e. so that the integrand $g_n(x)$ of (5.7.15) tends to zero a.e. We will obtain a contradiction to Vitali's Theorem 2.0.4 if we show that the sequence $\{g_n(\cdot)\}$ has equicontinuous integrals in $x \in \Omega$ and (in case Ω is unbounded) that for every $\varepsilon > 0$ there exists $e \subseteq \Omega$ with $|e| < \infty$ and $\int_{\Omega \setminus e} g_n(x) dx \leq \varepsilon$ for all n. This second condition is obvious from the bound

$$g_n(x) \leq C K(t)\left(\xi(x)^2 + |y(x)|^{2m/(m-2)} + |h_n(x)|^{2m/(m-2)}\right) \qquad (5.7.16)$$

coming from (5.7.5) (see Remark 5.7.5). It suffices to estimate the integral of $|h_n(x)|^{2m/(m-2)}$ in $\Omega \setminus e$ for $n = n_0$ large enough by its integral in all of Ω and then select the set e for the functions $|y(x)|^{2m/(m-2)}, |h_n(x)|^{2m/(m-2)}$ $(n < n_0)$. The same trick can be used to estimate the integral over a small set e, which shows equicontinuity of the integrals and then completes the proof of $\mathbf{y}$-continuity of $\mathbf{f}(t, \mathbf{y})$.

The proof is similar for $m = 2, 1$, and we omit the details. ∎

Remark 5.7.5. To deduce (5.7.16) from (5.7.5) we have used the inequality $(a_1 + \cdots + a_k)^p \leq C_{p,k}(a_1^p + \cdots a_k^p)$ $(a_j \geq 0, p \geq 1)$ three times; the first two with $k = 2$ and $p = 2$, the third with $k = 2$ and $p = 2m/(m-2)$.

5.8. Semilinear Equations in Banach Spaces: Global Existence

The objective is to solve

$$y'(t) = Ay(t) + f(t, y(t)), \quad y(0) = \zeta \tag{5.8.1}$$

in the entire interval $0 \le t \le T$ using Lemma 5.5.3 (that is, obtaining a priori estimates). The strategy is to generalize Lemma 2.1.7 to infinite dimensional spaces; the basic difficulty is that solutions of (8.1) may not have derivatives (or may not belong to $D(A)$) anywhere.

The first result lays the basis for approximation of $f(t, y)$ by other functions $f_n(t, y)$. The approximating system is

$$y_n'(t) = Ay_n(t) + f_n(t, y_n(t)), \quad y_n(0) = \zeta_n \tag{5.8.2}$$

and we shall assume that each $f_n(t, y)$ satisfies Hypothesis II; the corresponding functions in inequalities (5.5.4) and (5.5.5) are named $K_n(t, c)$ and $L_n(t, c)$, and those for $f(t, y)$ in (5.8.1) are $K(t, c)$ and $L(t, c)$. The setting is that of a general Banach space.

Lemma 5.8.1. *Let f satisfy Hypothesis I and let f_n satisfy Hypothesis II in $0 \le t \le T$ with $\{L_n(\cdot, c)\}$ bounded in $L^1(0, T)$ and $\{K_n(\cdot, c)\}$ with equicontinuous integrals (thus also bounded in $L^1(0, T)$). Assume that $\zeta_n \to \zeta$, and that for each $y \in E$*

$$f_n(t, y) \to f(t, y) \quad a.e. \text{ in } \quad 0 \le t \le T \tag{5.8.3}$$

Finally, assume that the solution $y(t)$ of (5.8.1) exists in $0 \le t \le T$. Then there exists n_0 such that, if $n \ge n_0$ the solution $y_n(t)$ of (5.8.2) exists in $0 \le t \le T$ and

$$\|y_n(t) - y(t)\| \to 0 \tag{5.8.4}$$

uniformly in $0 \le t \le T$.

Proof. For each $n = 1, 2 \ldots$ let $[0, t_n] \subseteq [0, T]$ be the largest interval such that $y_n(t)$ exists in $0 \le t \le t_n$ and $\|y_n(t) - y(t)\| \le 1$ (since $y_n(0) = \zeta_n \to \zeta = y(0)$ this interval must be nonempty for n large enough). Subtract the integral equations for $y_n(t)$ and for $y(t)$, use the equality

$$f_n(\tau, y_n(\tau)) - f(\tau, y(\tau))$$
$$= \{f_n(\tau, y_n(\tau)) - f_n(\tau, y(\tau))\} + \{f_n(\tau, y(\tau)) - f(\tau, y(\tau))\}$$

in the integrand and estimate:

$$\|y_n(t) - y(t)\| \le M\|\zeta_n - \zeta\| + M \int_0^t L_n(\tau, c+1)\|y_n(\tau) - y(\tau)\|d\tau$$

$$+ M \int_0^t \|f_n(\tau, y(\tau)) - f(\tau, y(\tau))\|d\tau$$

$$= M \int_0^t L_n(\tau, c+1)\|y_n(\tau) - y(\tau)\|d\tau$$

$$+ M\|\zeta_n - \zeta\| + \rho_n(t) \qquad (0 \le t \le t_n), \qquad (5.8.5)$$

where M (resp. c) is a bound for $\|S(t)\|$ (resp. $\|y(t)\|$) in $0 \le t \le T$. Approximating $y(\cdot)$ uniformly by piecewise constant functions and using (5.8.3) we show that

$$f_n(\tau, y(\tau)) \to f(\tau, y(\tau)) \quad \text{a.e. in} \quad 0 \le t \le T. \qquad (5.8.6)$$

On the other hand, we have

$$\|f_n(\tau, y(\tau)) - f(\tau, y(\tau))\| \le K_n(\tau, c) + K(\tau, c), \qquad (5.8.7)$$

which implies that $\rho_n(t) \to 0$ as $n \to \infty$ uniformly in $0 \le t \le T$. In fact, if this claim is false we can find $\varepsilon > 0$ and a sequence $\{s_n\} \subseteq [0, \overline{t}]$ such that $\rho(s_n) \ge \varepsilon$; passing if necessary to a subsequence, we may assume that $t_n \to s \in [0, T]$. We then get a contradiction to Vitali's Theorem 2.0.4 using (5.8.6) and (5.8.7) (the latter estimate implies that the integrals of the $\|f_n(\tau, y(\tau)) - f(\tau, y(\tau))\|$ are equicontinuous), and noting that

$$\rho_n(t) = M \int_0^t \chi_n(\tau)\|f_n(\tau, y(\tau)) - f(\tau, y(\tau))\|d\tau,$$

χ_n the characteristic function of $0 \le t \le t_n$. Using (5.8.5) and Gronwall's Lemma 2.1.2 (recall that the L^1 norms of the L_n are bounded), we deduce that for sufficiently large n, $\|y(t) - y_n(t)\| < 1$ in $[0, t_n]$. We can then apply Lemma 5.5.3 and extend $y_n(t)$ to a larger interval, where we have $\|y_n(t) - y(t)\| < 1$. This contradicts the assumed maximality of the interval $[0, t_n]$ and thus ends the proof. ∎

The functions $f_n(t, y)$ in (5.8.2) will be obtained mollifying $f(t, y)$. Let $\eta(\cdot)$ be a nonnegative test function with support in $|t| \le 1$ and integral 1, $\{\eta_n(t)\}$ the mollifier $\eta_n(t) = n\eta(nt)$ (see **5.0**).

Lemma 5.8.2. Let $f : [0, T] \times E \to E$ satisfy Hypothesis II with $K(\cdot, c), L(\cdot, c) \in L^1(0, T)$. Define $f(t, y) = f(0, y)$ for $t < 0$, $f(t, y) = f(T, y)$ for $t > T$ and

$$f_n(t, y) = \int_{-\infty}^{\infty} \eta_n(\tau)f(\tau, y)d\tau \quad (0 \le t \le T, y \in E). \qquad (5.8.8)$$

Then the $f_n(t, y)$ *satisfy Hypothesis II with* $\{K_n(\cdot, c)\}, \{L_n(\cdot, c)\}$ *continuous and having bounded* L^1 *norm and equicontinuous integrals in* $0 \leq t \leq T$. *Moreover,* $f_n(t, y)$ *is infinitely differentiable in* t *for* y *fixed, jointly continuous in* t, y, *and* (5.8.3) *holds.*

Proof. It is clear that $f_n(t, y)$ satisfies Hypothesis II with

$$K_n(t, c) = \int_{t-1/n}^{t+1/n} \eta_n(t - \tau) K(\tau, c) d\tau,$$

$$L_n(t, c) = \int_{t-1/n}^{t+1/n} \eta_n(t - \tau) L(\tau, c) d\tau.$$

Infinite differentiability of each $f_n(t, y)$, the fact that (5.8.3) holds, uniform L^1 boundedness of the K_n, L_n and equicontinuity of the integrals all follow from Theorem 5.0.23. If $\|y\|, \|y'\| \leq c$ we have

$$\|f_n(t', y') - f_n(t, y)\| \leq \|f_n(t', y') - f_n(t', y)\| + \|f(t', y) - f(t, y)\|$$
$$\leq L_n(t', c)\|y' - y\| + \|f(t', y) - f(t, y)\|.$$

Thus, the continuity claim is checked. ∎

In case $K(\cdot, c), L(\cdot, c) \in L^p(0, T)$ for $p > 1$ we can use the theory of the maximal operator

$$(Mf)(t) = \sup_{r>0} \frac{1}{2r} \int_{t-r}^{t+r} f(\tau) d\tau$$

in **5.0** to deduce that the $f_n(t, y)$ satisfy Hypothesis II with $K_n(t, c) = CM(K(\cdot, c))(t), L_n = CM(L(\cdot, c))(t)$ independent of n and in $L^p(0, T)$. See Theorem 5.0.25.

Theorem 5.8.3. *Let* A *be a densely defined linear operator in a real Banach space* E. *Assume that* $A - \omega I$ *is dissipative for some* ω,

$$\langle y^*, Ay \rangle \leq \omega \|y\|^2 \quad (y \in D(A), y^* \in \Theta(y)), \tag{5.8.9}$$

($\Theta(y)$ *the duality set of* y) *and that*

$$(\lambda I - A)D(A) = E \quad (\lambda > \omega). \tag{5.8.10}$$

Further, assume that Hypothesis II holds for $f : [0, T] \times E \to E$ *with* $K(\cdot, c),$ $L(\cdot, c) \in L^1(0, T)$ *and that there exists* c *with*

$$\langle y^*, f(t, y) \rangle \leq c(1 + \|y\|^2) \quad (0 \leq t \leq T, y \in E, y^* \in \Theta(y)). \tag{5.8.11}$$

Let $y(\cdot)$ be a solution of (8.1) in an interval $0 \le t < \bar{t}$ with $\zeta \ne 0$. Then, if $\theta = \omega + c(1 + \|\zeta\|^{-2})$, $y(\cdot)$ satisfies

$$\|y(t)\| \le e^{\theta t}\|\zeta\| \qquad (0 \le t \le \bar{t}). \tag{5.8.12}$$

Proof. The assumptions on A imply (Theorem 5.2.5) that A is the infinitesimal generator of a strongly continuous semigroup; thus, the theory hitherto developed on (5.8.1) applies. In particular, Lemma 5.8.1 gives us the opportunity of proving (5.8.12) not directly for the equation (5.8.1), but for a sequence of approximating equations (5.8.2). We shall in fact approximate twice. The first time we use Lemma 5.8.2 to construct the sequence $\{f_n(t, y)\}$ by mollification of $f(t, y)$; this sequence automatically satisfies the conditions in Lemma 5.8.1. Moreover,

$$\langle y^*, f_n(t, y) \rangle = \int_{-\infty}^{\infty} \eta_n(\tau)\langle y^*, f(\tau, y)\rangle d\tau$$

$$\le c(1 + \|y\|^2) \int_{-\infty}^{\infty} \eta_n(\tau)d\tau = c(1 + \|y\|^2)$$

$$(0 \le t \le \bar{t}, y \in E, y^* \in \Theta(y)) \quad (5.8.13)$$

so that each $f_n(t, y)$ satisfies the assumptions of Theorem 5.8.3. We may then assume from the beginning that $f(t, y)$ in (5.8.1) is jointly continuous. For the second approximation, put $P_n = nR(n; A) = n(nI - A)^{-1}$ (see Example 5.2.13) so that $P_n E \subseteq D(A)$, AP_n is bounded, and $P_n y \to y$ as $n \to \infty$ for every $y \in E$. The sequence of approximating equations

$$y_n'(t) = Ay_n(t) + P_n f(t, y_n(t)), \qquad y_n(0) = P_n\zeta \tag{5.8.14}$$

satisfies the assumptions of Lemma 5.8.1, so that the solution of (5.8.14) exists in $0 \le t \le \bar{t}$ for n large enough and $y_n(t) \to y(t)$ uniformly in $0 \le t \le \bar{t}$. Hence, it is enough to show that for every $\alpha > \theta$ and every $t \in [0, \bar{t}]$ we have

$$\|y_n(t)\| \le e^{\alpha t}\|P_n\zeta\| \tag{5.8.15}$$

for n large enough (in principle, n may depend on α and t). Since $P_n\zeta \in D(A)$ and $P_n f(t, y(t))$, $AP_n f(t, y_n(t))$ are continuous, Lemma 5.2.7 (a) determines that solutions of (5.8.14) are strong solutions (they are continuously differentiable, belong to $D(A)$ for all t and $Ay_n(t)$ is continuous).

Let $\zeta \ne 0$, $\alpha > \theta = \omega + c(1 + \|\zeta\|^{-2})$. Then

$$\alpha > \omega + c(1 + \|P_n\zeta\|^{-2}) \tag{5.8.16}$$

for n large enough. Let t_n be a point where $e^{-\alpha t}\|y_n(t)\|$ reaches its maximum in $0 \le t \le \bar{t}$. Consider the function

$$\phi_n(t) = e^{-\alpha t}\langle y_n^*, y_n(t)\rangle \tag{5.8.17}$$

in $0 \leq t \leq \bar{t}$, where $y_n^* \in \Theta(y_n(t_n))$. Since

$$\phi(t) \leq e^{-\alpha t} \|y_n(t)\| \|y_n^*\| \leq e^{-\alpha t_n} \|y_n(t_n)\| \|y_n^*\| \leq$$
$$= e^{-\alpha t_n} \|y_n(t_n)\|^2 = e^{-\alpha t_n} \langle y_n^*, y_n(t_n) \rangle = \phi_n(t_n),$$

$\phi_n(t)$ also reaches its maximum at $t = t_n$. Its derivative there equals

$$\phi_n'(t_n) = -\alpha e^{-\alpha t_n} \|y_n(t_n)\|^2 + e^{-\alpha t_n} \langle y_n^*, Ay_n(t_n) \rangle$$
$$+ e^{-\alpha t_n} \langle y_n^*, f(t_n, y_n(t_n)) \rangle + e^{-\alpha t_n} \langle y_n^*, (P_n - I)f(t_n, y_n(t_n)) \rangle$$
$$\leq (\omega + c - \alpha) e^{-\alpha t_n} \|y_n(t_n)\|^2 + c e^{-\alpha t_n} + e^{-\alpha t_n} \langle y_n^*, (P_n - I)f(t_n, y_n(t_n)) \rangle$$
$$\leq e^{\alpha t_n}((\omega + c - \alpha)\|P_n \zeta\|^2 + c) + e^{-\alpha t_n} \langle y_n^*, (P_n - I)f(t_n, y_n(t_n)) \rangle \quad (5.8.18)$$

where we have used the facts that $e^{-\alpha t_n} \leq e^{\alpha t_n}$, that $\omega + c - \alpha < 0$ (consequence of (5.8.16)) and that, since $e^{-\alpha t} \|y_n(t)\|$ reaches its maximum at $t = t_n$ we have $e^{-\alpha t_n} \|y(t_n)\| \geq \|y(0)\| = \|P_n \zeta\|$; hence, $e^{-\alpha t_n} \|y_n(t_n)\|^2 = e^{\alpha t_n} (e^{-\alpha t_n} \|y_n(t_n)\|)^2 \geq e^{\alpha t_n} \|P_n \zeta\|^2$. Passing if necessary to a subsequence, we may assume that $t_n \to t$. Since $y_n(t) \to y(t)$ uniformly we have $y_n(t_n) \to y(t)$. Accordingly,

$$(P_n - I)f(t_n, y_n(t_n)) = (P_n - I)f(t, y(t))$$
$$+ (P_n - I)\{f(t_n, y_n(t_n)) - f(t, y(t))\} \to 0 \quad (5.8.19)$$

as $n \to \infty$. On the other hand, $\|y_n^*\| = \|y_n(t_n)\|$ is bounded, thus

$$e^{-\alpha t_n} \langle y_n^*, (P_n - I)f(t_n, y_n(t_n)) \rangle \to 0, \quad (5.8.20)$$

and it follows from (5.8.16) (which implies the coefficient of $e^{\alpha t_n}$ in (5.8.18) is < 0) that $\phi_n(t_n) < 0$ for n large enough, a contradiction unless $t_n = 0$, which is exactly (5.8.15). We have then completed the proof of Theorem 5.8.3. It can actually be shown (using a compactness argument) that convergence in (5.8.19) and (5.8.20) is uniform in $0 \leq t \leq \bar{t}$, so that n does not depend on t, although this is not demanded by the argument.

Note that the result extends to an interval $[s, \bar{t}]$, initial condition given at $t = s$; t is replaced by $t - s$ on the right side of (5.8.12). ∎

Using Lemma 5.5.4, we deduce

Corollary 5.8.4. *Under the hypotheses of Theorem 5.8.3, the solution of (5.8.1) exists in $0 \leq t \leq T$.*

In fact, Theorem 5.8.3 provides an *a priori* bound for every solution $y(t)$ with $y(0) \neq 0$. If $y(0) = 0$ then, either $y(t)$ is identically zero or $\zeta = y(s) \neq 0$ for some s; we apply Theorem 5.8.3 to the right of s.

Remark 5.8.5. This is why the strange exponent $\theta = \omega + c(1 + \|\zeta\|^{-2})$ in (5.8.12) (the exponent tends to infinity as $\|\zeta\| \to 0$). For E finite dimensional (with Euclidean norm) and the operator $A = 0$ we have $(\|y(t)\|^2)' = 2(y(t), y'(t)) = 2(y(t), f(t, y(t))) \le 2c(1 + \|y(t)\|^2)$, so that

$$\|y(t)\|^2 \le \|\zeta\|^2 + 2ct + 2c \int_0^t \|y(\tau)\|^2 d\tau.$$

We apply Lemma 2.1.2 in the interval $0 \le t \le \bar{t}$. In the notation there, we have $\alpha(t) = \|\zeta\|^2 + 2ct$, $\beta(t) = 2c$, $B(t) = e^{-2ct}$, so that inequality (2.1.8) becomes

$$\|y(t)\|^2 \le \|\zeta\|^2 + 2ct + 2ce^{2ct} \int_0^t (\|\zeta\|^2 + 2c\tau)e^{-2c\tau} d\tau = (\|\zeta\|^2 + 1)e^{2ct} - 1.$$

If this is to be bounded by $\|\zeta\|^2 e^{\gamma t}$, then $\gamma \to +\infty$ as $\|\zeta\| \to 0$.

In the following result, E is a real Hilbert space.

Theorem 5.8.6. *Let A be a densely defined linear operator in E. Assume that $A - \omega I$ is dissipative for some ω, that is*

$$(y, Ay) \le \omega \|y\|^2 \quad (y \in D(A)), \tag{5.8.21}$$

and that (5.8.10) holds. Further assume that f satisfies Hypothesis II, and let $y(t)$ be a solution of (5.8.1) in an interval $0 \le t \le \bar{t}$ such that

$$\int_0^{\bar{t}} (y(t), f(t, y(t)))dt \le C. \tag{5.8.22}$$

Then

$$\|y(t)\| \le \sqrt{\|\zeta\|^2 + 2C}\, e^{\omega t} \quad (0 \le t \le \bar{t}), \tag{5.8.23}$$

so that $y(t)$ can be extended beyond $\bar{t}$. If the bound C can be established with C independent of the interval, then $y(t)$ exists globally.

Proof. Again, we use the approximating equation (5.8.14) and assume (via the first approximation with Lemma 5.8.2) that $f(t, y)$ is jointly continuous, so that solutions of (5.8.14) are strong solutions. By Lemma 5.8.1, for sufficiently large n the solution $y_n(\cdot)$ of (5.8.14) exists in $0 \le t \le \bar{t}$ and $y_n(t) \to y(t)$ uniformly. If $y_n(t)$ is such a solution,

$$\begin{aligned}(\|y_n(t)\|^2)' &= 2(y_n(t), y_n'(t)) \\ &= 2(y_n(t), Ay_n(t)) + 2(y_n(t), P_n f(t, y_n(t))). \end{aligned} \tag{5.8.24}$$

Integrating and estimating

$$\|y_n(t)\|^2 \le \|P_n\zeta\|^2 + 2\omega \int_0^t \|y_n(\tau)\|^2 d\tau + 2\int_0^t (y(\tau), f(\tau, y(\tau)))d\tau$$

$$+2\int_0^t \|(y_n(\tau), P_n f(\tau, y_n(\tau))) - (y(\tau), f(\tau, y(\tau)))\|d\tau$$

so that, letting $n \to \infty$,

$$\|y(t)\|^2 \le \|\zeta\|^2 + 2C + 2\omega \int_0^t \|y(\tau)\|^2 d\tau,$$

and it follows from Corollary 2.1.3 that $\|y(t)\|^2 \le (\|\zeta\|^2 + 2C)e^{2\omega t}$.   ∎

Example 5.8.7. In the complex version of Theorem 5.8.3 we require that Re $\langle y^*, Ay\rangle \le 0$ ($y \in D(A)$, $y^* \in \Theta(y)$) and that Re $\langle y^*, f(t, y)\rangle \le c(1 + \|y\|^2)$ ($0 \le t \le T$, $y \in E$, $y^* \in \Theta(y)$). In the proof we use the function $\phi_n(t) = e^{-\alpha t}$ Re $\langle y_n^*, y_n(t)\rangle$. Similar modifications work in Theorem 5.8.6; we require Re $(y, Ay) \le 0$ ($y \in D(A)$) and we take real part on the left side of (5.8.22).

Example 5.8.8. Let E be a Banach space. The duality set $\Theta(y) \subseteq E^*$ of any $y \in E$ is convex.

Example 5.8.9. Let E^* be strictly convex ($\|y^*\| + \|z^*\| < 2$ if $\|y^*\|, \|z^*\| = 1$, $y^* \ne z^*$). Then the map $y \to \Theta(y)$ is single valued.

Example 5.8.10. Let E^* be uniformly convex (if $\|y_n^*\| = \|z_n^*\| = 1$, $\|y_n^* + z_n^*\| \to 2$ then $\|y_n^* - z_n^*\| \to 0$). Then Θ is continuous.

For the following two examples, see **5.0** for the necessary information on duals.

Example 5.8.11. Let $\Omega \subseteq \mathbb{R}^m$ be measurable, $E = L^p(\Omega)$. Then (a) if $1 < p < \infty$ Θ is single valued (and continuous): $\Theta(y) = \{y^*\}$ with

$$y^*(x) = \begin{cases} (\|y\|_{L^p(\Omega)})^{2-p}|y(x)|^{p-2}\bar{y}(x) & \text{if } y(x) \ne 0 \\ 0 & \text{if } y(x) = 0, \end{cases}$$

(b) if $p = 1$ then $\Theta(y)$ consists of all $y^*(\cdot) \in L^\infty(\Omega)$ with

$$y^*(x) = \begin{cases} \|y\|_{L^1(\Omega)}|y(x)|^{-1}\bar{y}(x) & \text{if } y(x) \ne 0 \\ \text{arbitrary with } |y^*(x)| \le \|y\|_{L^1(\Omega)} & \text{if } y(x) = 0. \end{cases}$$

Example 5.8.12. Let $K \subseteq \mathbb{R}^m$ be compact. If $y(\cdot) \in C(K)$, let

$$m(y) = \{x \in K; |y(x)| = \|y\|\}.$$

Then $m(y)$ is closed, $m(y) \neq \emptyset$ and $\Theta(y)$ consists of all measures $\mu \in C(K)^* = \Sigma(K)$ with support in $m(y)$ (that is, $\mu(e) = 0$ if e is a Borel set with $m(y) \cap e = \emptyset$) and satisfying

$$\int_{m(y)} y(x)\mu(dx) \geq 0, \qquad \int_{m(y)} |\mu|(dx) = \|y\|.$$

5.9. Semilinear Wave Equations: Global Existence

We complete the study of the semilinear wave equation

$$y_{tt}(t, x) = \sum_{j=1}^{m} \sum_{k=1}^{m} \partial^j (a_{jk}(x)\partial^k y(t, x)) + \sum_{j=1}^{m} b_j(x)\partial^j y(t, x)$$
$$+ c(x)y(t, x) - \phi(x, y(t, x)) + u(t, x) \tag{5.9.1}$$

with a global existence result for the autonomous case (ϕ does not depend on t). The assumptions on the equation are the the same in **5.7** but some requirements on ϕ will be added below. We apply Theorem 5.8.6 to the equation

$$\mathbf{y}'(t) = \mathbf{A}(\beta)\mathbf{y}(t) + \mathbf{f}(\mathbf{y}(t)) \tag{5.9.2}$$

($\mathbf{A}(\beta)$ and $\mathbf{f}(\mathbf{y})$ given by (5.7.3) and (5.7.4) respectively). Working in real spaces, everything reduces to show that

$$(\mathbf{A}(\beta)\mathbf{y}, \mathbf{y}) \leq \omega\|\mathbf{y}\|^2 \quad (\mathbf{y} \in D(\mathbf{A}(\beta))) \tag{5.9.3}$$

and that, for a solution $\mathbf{y}(\cdot)$ of (5.9.2) in an arbitrary interval $0 \leq t \leq \bar{t}$ we have

$$\int_0^t (\mathbf{y}(\tau), \mathbf{f}(\mathbf{y}(\tau)))d\tau \leq C \tag{5.9.4}$$

where C does not depend on the interval.

To prove (5.9.3) recall that $\mathbf{A}(\beta) = \mathbf{A}_0(\beta) + \mathbf{P}$, where the operator $\mathbf{A}_0(\beta)$ given by (5.6.12) is skew adjoint, so that $(\mathbf{A}_0(\beta)\mathbf{y}, \mathbf{y}) = 0$. Accordingly, we only have to show (5.9.3) for $\mathbf{P}$, given by (5.6.17). This is a consequence of the definition of the norm of $\mathbf{E}$ and of the fact that the coefficients $b_j(\cdot)$, $c(\cdot)$ are in $L^\infty(\Omega)$.

Theorem 5.9.1. *Assume $\phi(x, y)$ satisfies the assumptions in **5.7** (with $K(t) \equiv K$, $L(t) \equiv L$) and that $\phi(x, 0) = 0$ and*

$$y\phi(x, y) \geq 0 \quad (x \in \Omega, -\infty < y < \infty). \tag{5.9.5}$$

Then (5.9.4) holds for any solution $\mathbf{y}(\cdot)$ (with a constant C depending on $\mathbf{y}(\cdot)$) in $0 \leq t \leq \bar{t}$ independently of $\bar{t}$. Hence, the solution of (5.9.2) exists globally.

Proof. We do only the case $m \geq 3$; $m = 2$ is exactly the same and $m = 1$ is simpler. ∎

Lemma 5.9.2. *Let* $(y(t), y_1(t))$ *be a solution of* (5.9.2). *Then*

$$y'(t) = y_1(t) \quad (0 \leq t \leq \bar{t}) \tag{5.9.6}$$

(quotients of increments convergent in the $L^2(\Omega)$*-norm); precisely,*

$$y(t) = y(0) + \int_0^t y_1(\tau)d\tau, \tag{5.9.7}$$

the integral understood as a Lebesgue–Bochner integral in $L^2(\Omega)$.

If $(y(t), y_1(t))$ is a strong solution, then Lemma 5.9.2 is satisfied to excess; in fact, (5.9.6) is the first coordinate of the equation (5.9.2) and the derivative can even be computed taking the limit of the quotient of increments in the H^1 norm; the integral in (5.9.7) is an integral in $H^1(\Omega)$ and $y_1(\cdot)$ is a continuous $H^1(\Omega)$-valued function.

In the general case we approximate twice, the first using Lemma 5.8.2 and the second the equation (5.8.14). Although (5.9.6) is no longer the first coordinate of the approximating equation, the (strong) solution $(y_n(\cdot), y_{1n}(\cdot))$ satisfies $y_n'(t) = y_{n1}(t) + \rho_n(t)$ with $\rho_n(t) \to 0$; integrating,

$$y_n(t) = y_n(0) + \int_0^t y_{1n}(\tau)d\tau + \int_0^t \rho_n(\tau)d\tau.$$

Interpreting the integrals in $L^2(\Omega)$, we show that the second tends to zero as in Lemma 5.8.2 and Theorem 5.8.3; to take limit in the first, we use that $y_{1n}(\tau) \to y_1(\tau)$ uniformly in $L^2(\Omega)$. ∎

Define

$$\Phi(x, y) = \int_0^y \phi(x, \upsilon)d\upsilon.$$

Then it follows from (5.9.5) that $\Phi(x, y) \geq 0$ $(x \in \Omega, -\infty < y < \infty)$.

To check the estimate (5.9.4), we note that $(\mathbf{y}, \mathbf{f}(\mathbf{y}))_{\mathbf{H}} = ((y, y_1), \mathbf{f}(y, y_1))_{\mathbf{H}} = -(y_1, \phi(\cdot, y(\cdot)))_{L^2(\Omega)}$. The *formal* proof goes as follows:

$$\begin{aligned}
(\mathbf{y}(t)), \mathbf{f}(\mathbf{y}(t)) &= -\int_\Omega y_1(t, x)\phi(x, y(t, x))dx \\
&= -\int_\Omega y_t(t, x)\Phi_y(x, y(t, x))dx \\
&= -\int_\Omega \Phi_t(x, y(t, x))dx, \tag{5.9.8}
\end{aligned}$$

so that

$$\int_0^{\bar{t}} (\mathbf{y}(t), \mathbf{f}(\mathbf{y}(t))) dt = - \int_\Omega \int_0^{\bar{t}} \Phi_t(x, y(t, x)) dt dx$$

$$= - \int_\Omega \{\Phi(x, y(\bar{t}, x)) - \Phi(x, y(0, x))\} dx$$

$$\leq \int_\Omega \Phi(x, y(0, x)) dx. \tag{5.9.9}$$

To justify rigorously this computation, we use approximation by smooth functions. The first step is to extend y to $(0, \bar{t}) \times \mathbb{R}^m$ setting $y = 0$ outside of Ω, then to $\mathbb{R} \times \mathbb{R}^m$ defining $y(t, x) = y(0, x)$ for $t \leq 0$, $y_n(t, x) = y(\bar{t}, x)$ for $t \geq t$. We then extend y_1 in the same way. The conclusions of Lemma 5.9.2 hold as well for the extended functions, which are given the same names.

Let $\{\eta_n(x)\}$ be a mollifier in $\mathbb{R}^m$ and let $2 \leq p \leq 2m/(m-2)$. Since $t \to y(t, \cdot)$ is an $H^1(\Omega)$-valued continuous function, by Theorem 5.7.1 it is also a $L^p(\Omega)$-valued continuous function. The extended function is a $L^p(\mathbb{R}^m)$-valued continuous function, so that applying Theorem 5.0.23 with $E = L^p(\mathbb{R}^m)$ we deduce that $t \to \eta_n *_x y(t, \cdot)$ is also a $L^p(\mathbb{R}^m)$-valued continuous function[1] with

$$\|\eta_n *_x y(t, \cdot)\|_{L^p(\mathbb{R}^m)} \leq \|y(t, \cdot)\|_{L^p(\mathbb{R}^m)},^{[2]} \tag{5.9.10}$$

$$\eta_n *_x y(t, \cdot) \to y(t, \cdot) \tag{5.9.11}$$

in $L^p(\mathbb{R}^m)$ for each $t, 0 \leq t \leq \bar{t}$. An argument based on the fact that $\{y(t, \cdot); -\infty < t < \infty\}$ is compact in $L^p(\mathbb{R}^m)$ (consequence of L^p-continuity of $y(\cdot)$) proves that convergence in (5.9.11) is uniform in t. Accordingly, if $\{\psi_n(t)\}$ is a mollifier in $\mathbb{R}$ and we define

$$y_n = (\psi_n \eta_n) *_{t,x} y = \psi_n *_t \eta_n *_x y,$$

we have

$$y_n(t, \cdot) \to y(t, \cdot) \tag{5.9.12}$$

in $L^p(\mathbb{R}^m)$, uniformly in t, and

$$\|y_n(t, \cdot)\|_{L^p(\mathbb{R}^m)} \leq \max_{0 \leq t \leq \bar{t}} \|y(t, \cdot)\|_{L^p(\mathbb{R}^m)}. \tag{5.9.13}$$

---

[1] Subindices indicate variables in convolutions.

[2] In this section, as in **5.7**, $\|f(\cdot)\|$ indicates the norm of $f(\cdot)$ in some function space rather than the norm as a function.

for $2 \le p \le 2m/(m-2)$. We have

$$y_n(t+h, x) - y_n(t, x)$$

$$= \int_{-\infty}^{\infty} \int_{\mathbb{R}^m} (\psi_n(t+h-\tau) - \psi_n(t-\tau))\eta_n(x-\xi)y(\tau, \xi)d\xi d\tau$$

$$= \int_{-\infty}^{\infty} \psi_n(t-\tau) \int_{\mathbb{R}^m} \eta_n(x-\xi)(y(\tau+h, \xi) - y(\tau, \xi))d\xi d\tau$$

so that it follows from Lemma 5.9.2 that if we define

$$y_{1n} = (\psi_n \eta_n) *_{t,x} y_1 = \psi_n *_t \eta_n *_x y_1$$

then

$$\partial_t y_n(t, x) = y_{1n}(t, x), \qquad (5.9.14)$$

this time the derivative understood classically (both functions are infinitely differentiable). Accordingly, (5.9.8) and (5.9.9) are legitimate for y_n and $y_{1n} = \partial_t y_n$ and we only have to pass to the limit in the inequality

$$-\int_0^{\bar{t}} \int_{\Omega} y_{1n}(t, x)\phi(x, y_n(t, x))dx\, dt \le \int_{\Omega} \Phi(x, y_n(0, x))dx. \qquad (5.9.15)$$

The function $t \to y_1(t, \cdot)$ belongs to $C(0, T; L^2(\mathbb{R}^m))$, thus by Theorem 5.0.23, $\eta_n *_x y_1(t, \cdot) \to y_1(t, \cdot)$ in $L^2(\Omega)$ uniformly in t and

$$\|\eta_n *_x y_1(t, \cdot)\|_{L^2(\mathbb{R}^m)} \le \|y_1(t, \cdot)\|_{L^2(\mathbb{R}^m)}.$$

This and the dominated convergence theorem imply

$$y_{1n}(t, \cdot) \to y_1(t, \cdot) \qquad (5.9.16)$$

in $L^2(\mathbb{R}^m)$, uniformly in t, and

$$\|y_{1n}(t, \cdot)\|_{L^2(\mathbb{R}^m)} \le \max_{0 \le t \le \bar{t}} \|y_1(t, \cdot)\|_{L^2(\mathbb{R}^m)}, \qquad (5.9.17)$$

so that to pass to the limit in the integral on the left side of (5.9.15) it is enough to show that $\phi(\cdot, y_n(\cdot, \cdot)) \to \phi(\cdot, y(\cdot, \cdot))$ in $L^2((0, T) \times \mathbb{R}^m)$. For this we use (5.7.6):

$$|\phi(x, y_n(t, x)) - \phi(x, y(t, x))|$$

$$\le L(\eta(x) + |y_n(t, x)|^{2/(m-2)} + |y(t, x)|^{2/(m-2)})|y_n(t, x) - y(t, x)|,$$

and estimate exactly as in (5.7.11), using (5.9.10) and (5.9.11) for $p = 2m/(m-2)$; after integrating in x we use the dominated convergence theorem for the t-integral. For the integral on the right side of (5.9.14), note that (5.7.5) and the definition of Φ imply

$$|\Phi(x, y)| \le K(\xi(x)|y| + |y|^{(2m-2)/(m-2)}). \qquad (5.9.18)$$

Since $2 \leq (2m-2)/(m-2) \leq 2m/(m-2)$, Theorem 5.7.1 guarantees that $\|y\|_{L^{(2m-2)/(m-2)}(\mathbb{R}^m)} \leq C\|y\|_{H^1(\Omega)}$ so the integral on the right side of (5.9.15) makes sense for $y(0, x)$. By the mean value theorem there exists a point $v(x, y, y')$ in the segment joining the points y and y' such that $\Phi(x, y') - \Phi(x, y) = \phi(x, v(x, y, y'))(y' - y)$, so that

$$|\Phi(x, y_n(0, x)) - \Phi(x, y(0, x))|$$
$$\leq K(\xi(x) + |y_n(0, x)|^{m/(m-2)} + |y(0, x)|^{m/(m-2)})|y_n(0, x) - y(0, x)|.$$

We integrate and use (5.9.10) and (5.9.11) for $p = 2$ and Schwarz's inequality. This ends the proof of Theorem 5.9.1.

6

Abstract Minimization Problems in Hilbert Spaces: Applications to Hyperbolic Control Systems

6.1. Control Systems: Continuity of the Solution Map

Most of this chapter is on extension of the results in Chapters 3 and 4 to infinite dimensional systems, and we shall take advantage (when possible) of similarity of proofs.

The control systems are of the form

$$y'(t) = A(t)y(t) + f(t, y(t), u(t)), \quad y(0) = \zeta \tag{6.1.1}$$

where $\{A(t); 0 \leq t < T\}$ is a family of densely defined operators in a Banach space E. As in **5.5**, we assume that the Cauchy problem for the linear equation

$$y'(t) = A(t)y(t) \tag{6.1.2}$$

is well posed in an interval $0 \leq t \leq T$ and denote by $S(t, s)$ the solution operator of (6.1.2), which is defined and strongly continuous in the triangle $0 \leq s \leq t \leq T$. Solutions of (6.1.1) are solutions of the integral equation

$$y(t) = S(t, s)\zeta + \int_0^t S(t, \tau) f(\tau, y(\tau), u(\tau)) d\tau, \tag{6.1.3}$$

and most of the results are independent of the evolution operator properties of $S(t, s)$ (that is, equalities (5.4.5) and (5.4.6)). In other results $A(t) \equiv A$ is the infinitesimal generator of a strongly continuous semigroup $S(t)$, so that the evolution operator is given by $S(t, s) = S(t - s)$.

The **control set** U is an arbitrary set. We denote by $F(0, T; U)$ the set of all functions $u(\cdot)$ defined in $0 \leq t \leq T$ with values in U. The function $f(t, y, u)$ is defined in $[0, T] \times E \times U$ and takes values in E. We postulate a subspace $C_{\mathrm{ad}}(0, T; U) \subseteq F(0, T; U)$ (the **admissible control space**) such that Hypothesis II holds for $f(t, y) = f(t, y, u(t))$ for every $u(\cdot) \in C_{\mathrm{ad}}(0, T; U)$.[1]

---

[1] In some results, with $K(\cdot, c)$ and $L(\cdot, c)$ independent of $u(\cdot)$.

This allows us to apply the existence-uniqueness results in **5.5**. We name $y(t, u)$ the solution of (6.1.1) corresponding to the control $u(\cdot)$. Under Hypothesis II, each $y(t, u)$ is uniquely defined in a maximal interval of existence (depending on $u(\cdot)$). The **solution map** of (6.1.1) is $u \to y(t, u)$. The space $C_{\mathrm{ad}}(0, T; U)$ is equipped with the distance

$$d(u(\cdot), v(\cdot)) = |\{t \in [0, T]; u(t) \neq v(t)\}| \tag{6.1.4}$$

introduced in **3.4**, where $| \cdot |$ denotes outer Lebesgue measure or Lebesgue measure when the set is measurable. Recall that if e is an arbitrary set, a *measurable envelope* of e is a measurable set $[e]$ satisfying $e \subseteq [e]$, $|e| = |[e]|$ (see **3.5**).

Lemma 6.1.1. *Assume* $f(t, y, u(t))$ *satisfies Hypothesis II for each* $u(\cdot) \in C_{\mathrm{ad}}(0, \bar{t}; U)$ *with* $K(\cdot, c)$ *and* $L(\cdot, c)$ *independent of* $u(\cdot)$. *Let* $y(t, \bar{u})$ *exist in* $0 \leq t \leq \bar{t}$. *Then there exists* $\delta > 0$ *such that, if* $u(\cdot), v(\cdot) \in B(\bar{u}, \delta) = \{u \in C_{\mathrm{ad}}(0, \bar{t}; U); d(u, \bar{u}) \leq \delta\}$ *then* $y(t, u)$ *and* $y(t, v)$ *exist as well in* $0 \leq t \leq \bar{t}$ *and*

$$\|y(t, v) - y(t, u)\| \leq C \int_{[\{\tau \in [0, \bar{t}]; u(\tau) \neq v(\tau)\}]} K(\tau, c + 1)d\tau \tag{6.1.5}$$

in $0 \leq t \leq \bar{t}$, *where* c *is a bound for* $\|y(t, \bar{u})\|$ *in* $0 \leq t \leq \bar{t}$ *and* C *does not depend on* u, v.

Lemma 6.1.1 is proved by minor modifications of the proof of Lemma 3.5.1. Denote by $[0, t_u]$ the largest interval where $y(t, u)$ exists and satisfies $\|y(t, u) - y(t, \bar{u})\| \leq 1$. Then we have

$$y(t, u) - y(t, \bar{u}) = \int_0^t S(t, s)\{f(\tau, y(\tau, u), u(\tau)) - f(\tau, y(\tau, u), \bar{u}(\tau))\}d\tau$$

$$+ \int_0^t S(t, s)\{f(\tau, y(\tau, u), \bar{u}(\tau)) - f(\tau, y(\tau, \bar{u}), \bar{u}(\tau))\}d\tau \tag{6.1.6}$$

thus

$$\|y(t, u) - y(t, \bar{u})\| \leq 2M \int_{[\{\tau \in [0, \bar{t}]; u(\tau) \neq v(\tau)\}]} K(\tau, c + 1)d\tau$$

$$+ M \int_0^t L(\tau, c + 1)\|y(\tau, u) - y(\tau, \bar{u})\|d\tau, \tag{6.1.7}$$

where M is a bound for $\|S(t, s)\|$ in $0 \leq s \leq t \leq \bar{t}$. We use Corollary 2.1.3 and obtain (6.1.5) for $u, \bar{u}$ in $0 \leq t \leq t_u$; taking $d(u, \bar{u}) \leq \delta$ sufficiently small, the maximality of $[0, t_u]$ implies $t_u = \bar{t}$ (the extension result now is Lemma 5.5.3). We argue in the same way for $v(\cdot)$ and deduce that $y(t, v)$ exists in $0 \leq t \leq \bar{t}$. Finally, we write (6.1.6) for u and v and estimate in the same way. A look at Lemma 3.5.1 reveals that the only difference between its proof and this one is the presence of the operator $S(t, s)$ as a kernel in the two integrals in (6.1.6), which merely adds the factor M in the estimate (6.1.7). This is typical of most proofs in this chapter.

6.2. Patch Perturbations and Directional Derivatives

Spike perturbations are harsh; their directional derivatives have a jump at the spike, and convergence of quotients of increments cannot be uniform. In some of the systems considered in Chapter 7, even singularities at the spike appear. Kinder, gentler perturbations were introduced by Li and Yao in [1985]. These perturbations give less information than spike perturbations on the set where controls are modified, but have decisive advantages. Quotients of increments converge uniformly to the directional derivatives, and this is essential in problems with state constraints (Chapters 10 and 11).

On the other hand, it is quite feasible to get to the maximum principle (without state constraints) by simply mimicking the arguments in Chapters 3 and 4; **6.1** already gave a hint on how these arguments generalize. The reader wishing to follow this route is given pointers in Examples at the end of each section.

Let $u(\cdot) \in C_{\mathrm{ad}}(0, \bar{t}; U)$, $\mathbf{v}(\cdot) = (v_1(\cdot), v_2(\cdot), \ldots, v_m(\cdot))$ an arbitrary collection of m elements of $C_{\mathrm{ad}}(0, \bar{t}; U)$, $\mathbf{e} = (e_1, e_2, \ldots, e_m)$ a collection of m pairwise disjoint measurable sets in $[0, \bar{t}]$. The **patch perturbation** of $u(\cdot)$ corresponding to $\mathbf{e}, \mathbf{v}(\cdot)$ is

$$u_{\mathbf{e},\mathbf{v}}(t) = \begin{cases} v_j(t) & (t \in e_j, j = 1, 2, \ldots, m) \\ u(t) & \text{elsewhere.} \end{cases} \tag{6.2.1}$$

If d is the distance (6.1.4) we have

$$d(u_{\mathbf{e},\mathbf{v}}, u) \leq |e| \tag{6.2.2}$$

where $e = \cup e_j$.

In the following result we assume that $f(t, y, u(t))$ in the system

$$y'(t) = A(t)y(t) + f(t, y(t), u(t)), \quad y(0) = \zeta \tag{6.2.3}$$

satisfies Hypothesis II in $0 \leq t \leq \bar{t}$ for every $u(\cdot) \in C_{\mathrm{ad}}(0, \bar{t}; U)$ fixed ($K(t, c)$ and $L(t, c)$ may depend on $u(\cdot)$).

Lemma 6.2.1. *Let $\bar{u}(\cdot) \in C_{\mathrm{ad}}(0, \bar{t}; U)$ be such that $y(t, \bar{u})$ exists in $0 \leq t \leq \bar{t}$. Then, given $\mathbf{v}$, there exists $\delta > 0$ such that if $|e| \leq \delta$ then $y(t, \bar{u}_{\mathbf{e},\mathbf{v}})$ exists in $0 \leq t \leq \bar{t}$. Moreover, there exist $C > 0$ and $K(\cdot) \in L^1(0, \bar{t})$ such that*

$$\|y(t, \bar{u}_{\mathbf{e},\mathbf{v}}) - y(t, \bar{u})\| \leq C \int_e K(\tau) d\tau \quad (0 \leq t \leq \bar{t}). \tag{6.2.4}$$

The constants δ, C and the function $K(\cdot)$ depend on $\bar{u}(\cdot)$ and $\mathbf{v}(\cdot)$ but not on $\mathbf{e}$.

This result follows directly from Lemma 6.1.1. To apply the latter, we replace the space $C_{\mathrm{ad}}(0, \bar{t}; U)$ by the set of controls consisting of $\bar{u}$ and all of its patch perturbations with $\mathbf{v}(\cdot) = (v_1(\cdot), \ldots, v_m(\cdot))$ fixed; in this new control space,

Hypothesis II is satisfied with $K(t, c)$, $L(t, c)$, independent of the control (it suffices to sum all the $K(t, c)$ corresponding to $\bar{u}, v_1, \ldots, v_m$ and do the same with the $L(t, c)$).

A function g defined in an open subset D of a normed space E with values in a normed space F is **Fréchet differentiable** at $y \in D$ if and only if there exists a bounded linear operator $\partial g(y) \in (E, F)$ such that

$$g(y + h) = g(y) + \partial g(y)h + \rho(y, h) \quad (y, y + h \in D)$$

with

$$\lim_{\|h\| \to 0} \frac{\|\rho(y, h)\|}{\|h\|} = 0.$$

The operator $\partial g(y)$ is called the **Fréchet differential** or **Fréchet derivative** of g at u. Obviously, Fréchet differentiability implies continuity. The **mean value theorem** holds for Fréchet differentiable maps: we need it in the form

$$\|g(y') - g(y)\|_F \leq \|y' - y\|_E \sup_{z \in I} \|\partial g(z)\|_{(E,F)} \tag{6.2.5}$$

(I the segment joining y and y') valid for D convex (see Example 6.2.9). The Fréchet differential is of course the calculus differential if $E = \mathbb{R}^m$.

Business in the rest of the section will be transacted under Hypothesis III below which, as customary, we state for the uncontrolled system

$$y'(t) = A(t)y(t) + f(t, y(t)).$$

Hypothesis III. The Fréchet derivative $\partial_y f(t, y)$ exists in $[0, T] \times E$. The function $f(t, y)$ is strongly measurable in t for y fixed and continuous in y for t fixed. For each $z \in E$, $\partial_y f(t, y)z$ is strongly measurable in t for y fixed and continuous in y for t fixed.[1] For every $c > 0$ there exist $K(\cdot, c), L(\cdot, c) \in L^1(0, T)$ such that

$$\|f(t, y)\|_E \leq K(t, c) \quad (0 \leq t \leq T, \|y\| \leq c) \tag{6.2.6}$$

$$\|\partial_y f(t, y)\|_{(E,E)} \leq L(t, c) \quad (0 \leq t \leq T, \|y\| \leq c). \tag{6.2.7}$$

Hypothesis III, via the mean value theorem (6.2.5), implies Hypothesis II; thus, we may use any of the existence-uniqueness results in in **5.5** and Lemma 6.2.1 above.

Let $u(\cdot) \in C_{\mathrm{ad}}(0, \bar{t}; U)$, and assume the trajectory $y(t, u)$ of (6.2.3) exists in $0 \leq t \leq \bar{t}$. The **variational equation** at u is

$$\xi'(t) = \{A(t) + \partial_y f(t, y(t, u), u(t))\}\xi(t), \quad \xi(s) = \zeta. \tag{6.2.8}$$

---

[1] As in the finite dimensional case, there are redundancies in Hypothesis III; y-continuity of $f(t, y)$ and t-strong measurability of $\partial_y f(t, y)z$ need not be assumed (see Footnote 1, **2.3**). The same observation applies to Hypothesis III^0 in **6.6**.

By Hypothesis III and a simple approximation argument, the operator function $B(t) = \partial_y f(t, y(t, u), u(t))$ satisfies the assumptions of Theorem 5.5.6. Since the Cauchy problem for $\xi'(t) = A(t)\xi(t)$ is well posed, the same is true of (6.2.8) and, according to Theorem 5.5.6, solutions of (6.2.8) are given by the integral equation

$$y(t) = S(t, s)\zeta + \int_s^t S(t, \tau)B(\tau)y(\tau)d\tau \qquad (6.2.9)$$

or by

$$y(t) = S(t, s; u)\zeta, \qquad (6.2.10)$$

$S(t, s; u)$ the solution operator of (6.2.8). As shown in Theorem 5.5.6, $S(t, s; u)$ is strongly continuous in $0 \leq s \leq t \leq \bar{t}$. More properties of $S(t, s; u)$ will be found in next section.

Handling of patch perturbations depends on the following measure theoretic result, where $\mathbf{p} = (p_1, p_2, \ldots, p_m)$ is a probability vector: $p_j \geq 0$, $\Sigma p_j = 1$.

Theorem 6.2.2. *Let E be an arbitrary Banach space, $\phi(t, \tau)$ an E-valued function defined in $0 \leq \tau, t \leq T$ such that $\phi(t, \cdot)$ is strongly measurable in $0 \leq \tau \leq T$ for all t and*[(2)]

$$\int_0^T \|\phi(t, \tau)\|d\tau < \infty \quad (0 \leq t \leq T), \qquad (6.2.11)$$

$$\int_0^T \|\phi(t', \tau) - \phi(t, \tau)\|d\tau \to 0 \qquad (6.2.12)$$

as $t' \to t$. Then, given $0 < h \leq 1, \varepsilon > 0$ and a probability vector $\mathbf{p}$ there exist disjoint measurable sets $e_1(h), e_2(h), \ldots, e_m(h) \subseteq [0, T]$ with $|e_j(h)| = p_j hT$ such that

$$\left\|\frac{1}{h}\int_{(0,t)\cap e_j(h)} \phi(t, \tau)d\tau - p_j\int_0^t \phi(t, \tau)d\tau\right\| \leq \varepsilon$$

$$(0 \leq t \leq T, j = 1, 2, \ldots, m). \qquad (6.2.13)$$

We prove a few auxiliary results first. We use the fact that if e is a measurable set in $[0, T]$ with $|e| > 0$ and $0 < h \leq 1$ we may always find a measurable $d \subseteq e$ such that $|d| = h|e|$; it suffices to note that $\phi(t) = \int_{(0,t)\cap e} d\tau$ is absolutely continuous, hence continuous, thus it must take all intermediate values in $0 \leq t \leq T$. Refining this argument a little bit, we show that given a probability vector $\mathbf{p} = (p_1, p_2, \ldots, p_m)$ we can find pairwise disjoint measurable sets $d_j \subseteq e$ such that $|d_j| = p_j h|e|$.

[(2)] We only assume the integral (6.2.11) to be finite for each t, not bounded in $0 \leq t \leq T$.

Lemma 6.2.3. *Let* $e \subseteq [0, T]$ *be a measurable set,* $0 < h \leq 1, \varepsilon > 0,$ **p** $=$ $(p_1, p_2, \ldots, p_m)$ *a probability vector. Then there exist disjoint measurable sets* $e_1(h), e_2(h), \ldots, e_m(h) \subseteq e$ *with* $|e_j(h)| = p_j h |e|$ *and such that*

$$\left| |(0, t) \cap e_j(h)| - p_j h |(0, t) \cap e| \right| \leq \varepsilon$$

$$(0 \leq t \leq T, j = 1, 2, \ldots, m). \quad (6.2.14)$$

Proof. Since $|(0, t) \cap e_j(h)| - p_j h |(0, t) \cap e|$ (as a function of t) may vary at most by ε in an interval of length ε, it is enough to show that $|(0, t) \cap e_j(h)| = p_j h |(0, t) \cap e|$ for, say, $t = t_k = (k/n)T$ ($k = 0, 1, \ldots, n$), where n is such that $T/n \leq \varepsilon$. To do the latter, it suffices to take $e_{1k}(h), e_{2k}(h), \ldots, e_{mk}(h) \subseteq$ $[t_k, t_{k+1}) \cap e$ pairwise disjoint and such that $|e_{jk}(h)| = p_j h |[t_k, t_{k+1}) \cap e|$ and define $e_j(h) = \cup_k e_{jk}(h)$ for $j = 1, \ldots, m$; we have $|e_j(h)| = p_j h \Sigma |[t_k, t_{k+1}) \cap e|$ $= p_j h |e|$. ∎

Corollary 6.2.4. *Let* $f(\cdot)$ *be integrable in* $0 \leq t \leq T, 0 < h \leq 1, \varepsilon > 0,$ **p** *a probability vector. Then there exist disjoint measurable sets* $e_1(h), e_2(h), \ldots,$ $e_m(h) \subseteq [0, T]$ *with* $|e_j(h)| = p_j h T$ *and*

$$\left\| \frac{1}{h} \int_{(0,t) \cap e_j(h)} f(\tau) d\tau - p_j \int_0^t f(\tau) d\tau \right\| \leq \varepsilon \quad (0 \leq t \leq T, j = 1, 2, \ldots, m).$$

$$(6.2.15)$$

Proof. A strongly measurable function can be uniformly approximated by a countably valued function outside of a null set (this is in fact the definition of strongly measurable function adopted in **5.0**); thus we may assume $f(\cdot)$ itself is a countably valued function. Let $\{e_n; n = 1, 2, \ldots\}$ be a (finite or countable) partition of $[0, T]$ in pairwise disjoint measurable sets where $f(\tau) = f_n$ is constant; we have $M = \Sigma |e_n| \| f_n \| = \int \| f(\tau) \| d\tau < \infty$. For each e_n use Lemma 6.2.3 to construct disjoint measurable sets $e_{n1}(h), e_{n2}(h), \ldots, e_{nm}(h) \subseteq e_n$ with $|e_{nj}(h)| = p_j h |e_n|$ and

$$\left| |(0, t) \cap e_{nj}(h)| - p_j h |(0, t) \cap e_n| \right| \leq \varepsilon h |e_n| / M$$

$$(0 \leq t \leq T, j = 1, 2, \ldots, m). \quad (6.2.16)$$

(the appearance of h on the right side does not affect this construction since ε and h in (6.2.14) are unrelated). Finally, define $e_j(h) = \cup_n e_{nj}(h)$. Then we have $|e_j(h)| = p_j h T$ and

$$\left\| \frac{1}{h} \int_{(0,t) \cap e_j(h)} f(\tau) d\tau - p_j \int_0^t f(\tau) d\tau \right\|$$

$$\leq \sum_n \left| h^{-1} |(0, t) \cap e_{nj}(h)| - p_j |(0, t) \cap e_n| \right| \| f_n \| \leq \varepsilon$$

as claimed. ∎

Corollary 6.2.5. *Let* $f_1(\cdot), f_2(\cdot), \ldots, f_n(\cdot)$ *be integrable in* $0 \le t \le T$ *and let* $0 < h \le 1, \varepsilon > 0$, $\mathbf{p}$ *a probability vector. Then there exist disjoint measurable sets* $e_1(h), e_2(h), \ldots, e_m(h) \subseteq [0, T]$ *with* $|e_j(h)| = p_j hT$ *such that* (6.2.15) *holds for* $f = f_k$ $(k = 1, 2, \ldots, n)$, $j = 1, 2, \ldots, m$.

Proof. Apply Corollary 6.2.4 to the integrable $(E \times E \times \cdots \times E)$-valued function $\mathbf{f} = (f_1, \ldots, f_n)$. ∎

Proof of Theorem 6.2.2. Assumption (6.2.12) and compactness of the interval $[0, T]$ implies the existence of $\delta > 0$ such that

$$\int_0^T \|\phi(t', \tau) - \phi(t, \tau)\| d\tau \le \frac{\varepsilon}{2T} \qquad (6.2.17)$$

whenever $0 \le t' - t \le \delta$. Let $T/n \le \delta, t_k = (k/n)T$. Using Corollary 6.2.5, select $e_1(h), e_2(h), \ldots, e_m(h)$ such that (6.2.15) holds with $\varepsilon/2$ for $f_k(\tau) = \phi(t_k, \tau)$, $k = 1, 2, \ldots, n$. Write then $\phi(t, \tau) = \phi(t_k, \tau) + (\phi(t, \tau) - \phi(t_k, \tau))$, where t_k is such that $|t - t_k| \le \delta$, and combine (6.2.17) and (6.2.15). ∎

Corollary 6.2.6. *Let* $\phi_1(t, \tau), \phi_2(t, \tau), \ldots, \phi_n(t, \tau)$ *be* E-valued functions satisfying the assumptions of Theorem 6.2.2. Then, given $0 < h \le 1$ and a probability vector $\mathbf{p}$ there exist disjoint measurable sets $e_1(h), e_2(h), \ldots, e_m(h) \subseteq [0, T]$ with $|e_j(h)| = p_j hT$ such that (2.13) holds for each $\phi_k(t, \tau)$.

Proof. Apply Theorem 6.2.2 to $\phi(t, \tau) = (\phi_1(t, \tau), \phi_2(t, \tau), \ldots, \phi_n(t, \tau))$ in the product space $E \times E \times \cdots \times E$. ∎

We construct below directional derivatives of the solution map

$$\mathbf{f}(u) = y(\bar{t}, u) \qquad (6.2.18)$$

of the control system (6.2.3).

Theorem 6.2.7. *Assume that* $f(t, y) = f(t, y, u(t))$ *satisfies Hypothesis III for every* $u(\cdot) \in C_{ad}(0, \bar{t}; U)$. *Let* $\bar{u}(\cdot) \in C_{ad}(0, \bar{t}; U)$ *be such that* $y(t, \bar{u})$ *exists in* $0 \le t \le \bar{t}$. *Let* $\mathbf{p} = (p_1, p_2, \ldots, p_m)$ *be a probability vector and* $\mathbf{v}(\cdot) = (v_1(\cdot), v_2(\cdot), \ldots, v_m(h))$ *an arbitrary collection of elements of* $C_{ad}(0, \bar{t}; U)$. *Then, for each* $h \in (0, 1]$ *there exists* $\mathbf{e}(h) = (e_1(h), e_2(h), \ldots, e_m(h))$ *such that* $|e_j(h)| = p_j h\bar{t}$, $y(t, \bar{u}_{\mathbf{e}(h), \mathbf{v}})$ *exists in* $0 \le t \le \bar{t}$ *for sufficiently small* h *and*

$$\xi(t, \bar{u}, \mathbf{p}, \mathbf{v}) = \lim_{h \to 0+} \frac{1}{h}\left(y(t, \bar{u}_{\mathbf{e}(h), \mathbf{v}}) - y(t, \bar{u})\right) \qquad (6.2.19)$$

uniformly in $0 \leq t \leq \bar{t}$, where $\xi(t) = \xi(t, \bar{u}, \mathbf{p}, \mathbf{v})$ is the solution of the variational equation

$$\xi'(t) = \{A(t) + \partial_y f(t, y(t, \bar{u}), \bar{u}(t))\}\xi(t)$$

$$+ \sum_{j=1}^{m} p_j \{f(t, y(t, \bar{u}), v_j(t)) - f(t, y(t, \bar{u}), \bar{u}(t))\}, \quad \xi(0) = 0. \quad (6.2.20)$$

Proof. We use Corollary 6.2.6 for $T = \bar{t}, m = n, \varepsilon = h$ and functions $\phi_j(t, \tau) = S(t, \tau)\{f(\tau, y(\tau, \bar{u}), v_j(\tau)) - f(\tau, y(\tau, \bar{u}), \bar{u}(\tau))\}$, continued to $t < \tau$ setting $\phi_j(t, \tau) = 0$ there. These functions are easily seen to satisfy (6.2.11) and (6.2.12), the first by (6.2.6), the second as a consequence of strong continuity of $S(t, s)$ and the dominated convergence theorem. We obtain via Corollary 6.2.6 disjoint measurable sets $e_1(h), e_2(h), \ldots, e_m(h) \subseteq [0, \bar{t}]$ with $|e_j(h)| = p_j h \bar{t}$ and such that

$$\left\| \frac{1}{h} \int_{(0,t) \cap e_j(h)} S(t, \tau)\{f(\tau, y(\tau, \bar{u}), v_j(\tau)) - f(\tau, y(\tau, \bar{u}), \bar{u}(\tau))\}d\tau \right.$$

$$\left. - p_j \int_0^t S(t, \tau)\{f(\tau, y(\tau, \bar{u}), v_j(\tau)) - f(\tau, y(\tau, \bar{u}), \bar{u}(\tau))\}d\tau \right\| \leq h.$$

$$(0 \leq t \leq \bar{t}). \quad (6.2.21)$$

Existence of $y(t, \bar{u}_{e(h),\mathbf{v}})$ in $0 \leq t \leq \bar{t}$ for sufficiently small h is a consequence of Lemma 6.2.1; also, it follows from its proof that, having fixed $\mathbf{v}(\cdot)$, we may use the same $K(t, c), L(t, c)$ in Hypothesis III for all the functions $f(t, y, \bar{u}_{e(h),\mathbf{v}}(t))$, independently of h.

Subtract the integral equations for $y(t, \bar{u}_{e(h),\mathbf{v}})$ and for $y(t, \bar{u})$. Rearrange the integrand in the form

$$S(t, \tau)\{f(\tau, y(\tau, \bar{u}_{e(h),\mathbf{v}}), \bar{u}_{e(h),\mathbf{v}}(\tau)) - f(\tau, y(\tau, \bar{u}), \bar{u}(\tau))\}$$

$$= S(t, \tau)\{f(\tau, y(\tau, \bar{u}_{e(h),\mathbf{v}}), \bar{u}_{e(h),\mathbf{v}}(\tau)) - f(\tau, y(\tau, \bar{u}), \bar{u}_{e(h),\mathbf{v}}(\tau))\}$$

$$+ S(t, \tau)\{f(\tau, y(\tau, \bar{u}), \bar{u}_{e(h),\mathbf{v}}(\tau)) - f(\tau, y(\tau, \bar{u}), \bar{u}(\tau))\}$$

and estimate,[3] using the local Lipschitz condition (coming from (6.2.7) and inequality (6.2.5)) in the first term and the fact that $\|S(t, \tau)\|$ is bounded:

$$\|y(t, \bar{u}_{e(h),\mathbf{v}}) - y(t, \bar{u})\| \leq C \int_0^t L(\tau, c)\|y(\tau, \bar{u}_{e(h),\mathbf{v}}) - y(\tau, \bar{u})\|d\tau$$

$$+ C' \sum_{j=1}^{m} \left\| \int_{(0,t) \cap e_j(h)} S(t, \tau)\{f(\tau, y(\tau, \bar{u}), v_j(\tau)) - f(\tau, y(\tau, \bar{u}), \bar{u}(\tau))\}d\tau \right\|.$$

$$(6.2.22)$$

---

[3] The constants C, C' may be different in different inequalities.

Noting that the second term between norm bars in (6.2.21) is independent of h,

$$\left\| \int_{(0,t)\cap e_j(h)} S(t,\tau)\{f(\tau, y(\tau,\bar{u}), v_j(\tau)) - f(\tau, y(\tau,\bar{u}), \bar{u}(\tau))\}d\tau \right\| \leq Ch,$$

thus it follows from (6.2.22) and Corollary 2.1.3 that

$$\|y(t, \bar{u}_{e(h),\mathbf{v}}) - y(t,\bar{u})\| \leq Ch \quad (0 \leq t \leq \bar{t}). \tag{6.2.23}$$

Define

$$\rho(t, y, y', u) = f(t, y', u) - f(t, y, u) - \partial_y f(t, y, u)(y' - y)$$
$$(y, y' \in E, u \in U).$$

By Assumption III and the mean value theorem we have

$$\|\rho(t, y, y', u(t))\| \leq CL(t, c)\|y' - y\| \quad (\|y\|, \|y'\| \leq c) \tag{6.2.24}$$

where we have the right to use $L(\cdot, c)$ independent of h if $u = \bar{u}_{e(h),\mathbf{v}}$ (*vide supra*). On the other hand, for each y we have

$$\|\rho(t, y', y, u)\| \to 0 \quad \text{as } y' \to y \quad (u \in U). \tag{6.2.25}$$

Define

$$\eta(t, h) = \frac{1}{h}\big(y(t, \bar{u}_{e(h),\mathbf{v}}) - y(t,\bar{u})\big) - \xi(t, \bar{u}, \mathbf{p}, \mathbf{v}).$$

Taking into account that $\xi(t, \bar{u}, \mathbf{p}, \mathbf{v})$ satisfies the integral equation

$$\xi(t, \bar{u}, \mathbf{p}, \mathbf{v}) = \int_0^t S(t, \tau)\partial_y f(\tau, y(\tau, \bar{u}), \bar{u}(\tau))\xi(\tau, \bar{u}, \mathbf{p}, \mathbf{v})d\tau$$
$$+ \sum_{j=1}^m p_j \int_0^t S(t, \tau)\{f(\tau, y(\tau,\bar{u}), v_j(\tau)) - f(\tau, y(\tau,\bar{u}), \bar{u}(\tau))\}d\tau$$

we obtain

$$\eta(t, h) = \int_0^t S(t, \tau)\frac{1}{h}\{f(\tau, y(\tau, \bar{u}_{e(h),\mathbf{v}}), \bar{u}_{e(h),\mathbf{v}}(\tau)) - f(\tau, y(\tau,\bar{u}), \bar{u}(\tau))\}d\tau$$
$$- \int_0^t S(t, \tau)\partial_y f(\tau, y(\tau,\bar{u}), \bar{u}(\tau))\xi(\tau, \bar{u}, \mathbf{p}, \mathbf{v})d\tau$$
$$- \sum_{j=1}^m p_j \int_0^t S(t, \tau)\{f(\tau, y(\tau,\bar{u}), v_j(\tau)) - f(\tau, y(\tau,\bar{u}), \bar{u}(\tau))\}d\tau$$

$$
= \int_0^t S(t, \tau) \partial_y f(\tau, y(\tau, \bar{u}), \bar{u}(\tau)) \eta(\tau, h) d\tau
$$

$$
+ \int_0^t S(t, \tau) \frac{1}{h} \rho(\tau, y(\tau, \bar{u}), y(\tau, \bar{u}_{\mathbf{e}(h),\mathbf{v}}), \bar{u}(\tau)) d\tau
$$

$$
+ \frac{1}{h} \int_0^t S(t, \tau) \{ f(\tau, y(\tau, \bar{u}_{\mathbf{e}(h),\mathbf{v}}), \bar{u}_{\mathbf{e}(h),\mathbf{v}}(\tau)) - f(\tau, y(\tau, \bar{u}_{\mathbf{e}(h),\mathbf{v}}), \bar{u}(\tau)) \} d\tau
$$

$$
- \sum_{j=1}^m p_j \int_0^t S(t, \tau) \{ f(\tau, y(\tau, \bar{u}), v_j(\tau)) - f(\tau, y(\tau, \bar{u}), \bar{u}(\tau)) \} d\tau
$$

$$
= \int_0^t S(t, \tau) \partial_y f(\tau, y(\tau, \bar{u}), \bar{u}(\tau)) \eta(\tau, h) d\tau
$$

$$
+ \int_0^t S(t, \tau) \frac{1}{h} \rho(\tau, y(\tau, \bar{u}), y(\tau, \bar{u}_{\mathbf{e}(h),\mathbf{v}}), \bar{u}(\tau)) d\tau
$$

$$
+ \frac{1}{h} \int_0^t S(t, \tau) \{ f(\tau, y(\tau, \bar{u}_{\mathbf{e}(h),\mathbf{v}}), \bar{u}_{\mathbf{e}(h),\mathbf{v}}(\tau)) - f(\tau, y(\tau, \bar{u}), \bar{u}_{\mathbf{e}(h),\mathbf{v}}(\tau)) \} d\tau
$$

$$
- \frac{1}{h} \int_0^t S(t, \tau) \{ f(\tau, y(\tau, \bar{u}_{\mathbf{e}(h),\mathbf{v}}), \bar{u}(\tau)) - f(\tau, y(\tau, \bar{u}), \bar{u}(\tau)) \} d\tau
$$

$$
+ \frac{1}{h} \int_0^t S(t, \tau) \{ f(\tau, y(\tau, \bar{u}), \bar{u}_{\mathbf{e}(h),\mathbf{v}}(\tau)) - f(\tau, y(\tau, \bar{u}), \bar{u}(\tau)) \} d\tau
$$

$$
- \sum_{j=1}^m p_j \int_0^t S(t, \tau) \{ f(\tau, y(\tau, \bar{u}), v_j(\tau)) - f(\tau, y(\tau, \bar{u}), \bar{u}(\tau)) \} d\tau
$$

$$
= \int_0^t S(t, \tau) \partial_y f(\tau, y(\tau, \bar{u}), \bar{u}(\tau)) \eta(\tau, h) d\tau
$$

$$
+ \int_0^t S(t, \tau) \frac{1}{h} \rho(\tau, y(\tau, \bar{u}), y(\tau, \bar{u}_{\mathbf{e}(h),\mathbf{v}}), \bar{u}(\tau)) d\tau
$$

$$
+ \frac{1}{h} \int_0^t S(t, \tau) \{ f(\tau, y(\tau, \bar{u}_{\mathbf{e}(h),\mathbf{v}}), \bar{u}_{\mathbf{e}(h),\mathbf{v}}(\tau)) - f(\tau, y(\tau, \bar{u}), \bar{u}_{\mathbf{e}(h),\mathbf{v}}(\tau)) \} d\tau
$$

$$
- \frac{1}{h} \int_0^t S(t, \tau) \{ f(\tau, y(\tau, \bar{u}_{\mathbf{e}(h),\mathbf{v}}), \bar{u}(\tau)) - f(\tau, y(\tau, \bar{u}), \bar{u}(\tau)) \} d\tau
$$

$$
+ \sum_{j=1}^m \frac{1}{h} \int_{(0,t) \cap e_j(h)} S(t, \tau) \{ f(\tau, y(\tau, \bar{u}), v_j(\tau)) - f(\tau, y(\tau, \bar{u}), \bar{u}(\tau)) \} d\tau
$$

$$
- \sum_{j=1}^m p_j \int_0^t S(t, \tau) \{ f(\tau, y(\tau, \bar{u}), v_j(\tau)) - f(\tau, y(\tau, \bar{u}), \bar{u}(\tau)) \} d\tau.
$$

$$(6.2.26)$$

Write this equation in the form

$$
\eta(t, h) = \int_0^t S(t, \tau) \partial_y f(\tau, y(\tau, \bar{u}), \bar{u}(\tau)) \eta(\tau, h) d\tau
$$

$$
+ \kappa_1(t, h) + \kappa_2(t, h) + \kappa_3(t, h), \tag{6.2.27}
$$

where $\kappa_1(t, h)$ is the second integral on the right side of the last equality and $\kappa_2(t, h)$ (resp. $\kappa_3(t, h)$) is the combination of the third and fourth (resp. of the fifth and sixth sums of) integrals. Estimating,

$$\|\eta(t, h)\| \leq \kappa(t, h) + C \int_0^t L(\tau, c + 1)\|\eta(\tau, h)\| d\tau, \tag{6.2.28}$$

(c a bound for $\|y(t, \bar{u})\|$ in $0 \leq t \leq \bar{t}$) with $\kappa(t, h) = \|\kappa_1(t, h)\| + \|\kappa_2(t, h)\| + \|\kappa_3(t, h)\|$, so that Theorem 6.2.7 will follow if we show that each $\|\kappa_j(t, h)\|$ tends to zero as $h \to 0+$ uniformly in $0 \leq t \leq \bar{t}$. This is done as follows:

$\kappa_1(t, h)$: We use (6.2.25) in combination with (6.2.23) to show that the integrand tends to zero and then (6.2.24), again in combination with (6.2.23), to estimate it by a fixed summable function independently of h. Hence, that $\kappa_1(t, h) \to 0$ follows from the dominated convergence theorem. If the convergence is not uniform in $0 \leq t \leq \bar{t}$ there exists a sequence $\{t_n\} \subseteq [0, t]$ (which we may assume convergent) with $\|\kappa(t_n, h_n)\| \geq \varepsilon > 0$ for some sequence $\{h_n\} \subseteq \mathbb{R}_+, h_n \to 0$. We then obtain a contradiction to the dominated convergence theorem writing the integral in $(0, t_n)$ as an integral in $(0, \bar{t})$ with the characteristic function $\chi_n(t)$ of $[0, t_n]$ multiplying the integrand (see the argument after (2.3.20)).

$\kappa_2(t, h)$: We use the local Lipschitz condition for $f(t, y, u)$ coming from (6.2.7) in combination with (6.2.24) to show that the integrands in both integrals are bounded by a summable function independently of h. Combining both integrals into one, the integrands cancel out outside of $e(h) = \cup e_j(h)$. We then estimate the integral for any t by the integral over e and use equicontinuity of the integral of the bounding function.

$\kappa_3(t, h)$: It tends to zero uniformly in $0 \leq t \leq \bar{t}$ because of (6.2.21).

The proof of Theorem 6.2.7 then ends applying Corollary 2.1.3. ∎

Remark 6.2.8. If $e = \cup_j e_j$, then $d(\bar{u}, \bar{u}_{e(h), v}) \leq |e| \leq \bar{t}h$, hence $\bar{t}^{-1}\xi(t, \bar{u}, \mathbf{p}, \mathbf{v})$ (not $\xi(t, \bar{u}, \mathbf{p}, \mathbf{v})$) is a directional derivative.

Example 6.2.9. The mean value theorem (6.2.5) follows taking $y^* \in F^*$ arbitrary and applying the calculus mean value theorem to the function $t \to \langle y^*, g((1 - t)y + ty')\rangle$, which has the derivative $\langle y^*, \partial g((1 - t)y + ty')(y' - y)\rangle$ in $0 \leq t \leq 1$. It follows that

$$|\langle y^*, g(y')\rangle - \langle y^*, g(y)\rangle| \leq \|y^*\|_{F^*}\|y' - y\|_E \sup_{0 \leq t \leq 1} \|\partial g((1 - t)y + ty')\|_{(E,F)},$$

which implies (5.2.5).

Remark 6.2.10. Setting $m = 1$ (so that $\mathbf{p} = \{1\}$, $\mathbf{e}(h) = \{e(h)\}$) and $t = T$, one cannot help noticing that (6.2.15) becomes a sort of approximation of the Lebesgue-Bochner integral of $f(\cdot)$ in $0 \leq t \leq T$ by a "smeared Riemann sum", the set $e(h)$ playing the role of the finite set of partition points defining the Riemann sum.

Example 6.2.11. *Spike perturbations and directional derivatives.* Let $u_{s,h,v}(t)$ be the spike perturbation (2.3.2), and assume $C_{ad}(0, \bar{t}; U)$ contains all constants $v(t) \equiv v \in U$ and that the assumptions of Theorem 6.2.7 are satisfied. Let $\bar{u}(\cdot) \in U_{ad}(0, \bar{t}; U)$ be such that $y(t, \bar{u})$ exists in $0 \leq t \leq \bar{t}$, and let e be the set of all left Lebesgue points of both functions

$$f(\cdot, y(\cdot, \bar{u}), \bar{u}(\cdot)), \qquad f(\cdot, y(\cdot, \bar{u}), v)$$

in $(0, \bar{t}]$. Then, if $s \in e$ the limit

$$\xi(t, s, \bar{u}, v) = \lim_{h \to 0+} \frac{1}{h}(y(t, \bar{u}_{s,h,v}) - y(t, \bar{u}))$$

exists uniformly in $s \leq t \leq \bar{t}$ and is given by

$$\xi(t, s, \bar{u}, v) = \begin{cases} 0 & (0 \leq t < s) \\ S(t, s; \bar{u})\{f(s, y(s, \bar{u}), v) - f(s, y(s, \bar{u}), \bar{u}(s))\} & (s \leq t \leq \bar{t}) \end{cases}$$

(see (6.2.10)). Equivalently, $\xi(t) = \xi(t, s, \bar{u}, v)$ solves

$$\xi'(t) = \{A(t) + \partial_y f(t, y(t, \bar{u}), \bar{u}(t))\}\xi(t)$$

$$+ \delta(t - s)\{f(s, y(s, \bar{u}), v) - f(s, y(s, \bar{u}), \bar{u}(s))\}, \quad \xi(0) = 0.$$

The proof is essentially the same as that of Theorem 2.3.3.

6.3. Continuity of the Solution Operator of the Variational Equation

Lemma 6.3.1. *Let the control system* (6.1.1) *satisfy Hypothesis III in an interval* $0 \leq t \leq \bar{t}$ *for each* $u(\cdot) \in C_{ad}(0, \bar{t}; U)$ *with* $K(\cdot, c)$ *and* $L(\cdot, c)$ *independent of* $u(\cdot)$, *and assume that* $y(t, \bar{u})$ *exists in* $0 \leq t \leq \bar{t}$. *Let* $S(t, s; u)$ *be the solution operator of the variational equation*

$$\xi'(t) = \{A(t) + \partial_y f(t, y(t, u), u(t))\}\xi(t). \tag{6.3.1}$$

Then $u \to S(t, s; u)\zeta$ *is continuous in* $B(\bar{u}, \delta) \subseteq C_{ad}(0, \bar{t}; U)$ *for each* $\zeta \in E$ (δ *the constant in Lemma* 6.1.1), *uniformly in* $0 \leq s \leq t \leq \bar{t}$.

Corollary 6.3.2. *Under the assumptions of Lemma* 6.3.1, $(t, s, u) \to S(t, s; u)\zeta$ *is continuous in* $\{(s, t); 0 \leq s \leq t \leq \bar{t}\} \times B(\bar{u}, \delta)$ *for each* $\zeta \in E$ (δ *the constant in Lemma* 6.1.1).

To see that Lemma 6.3.1 implies this result, take $u, v \in B(\bar{u}, \delta)$ and $(s, t), (s', t')$ in the triangle $0 \leq s \leq t \leq \bar{t}$. Write

$$\|S(t', s'; v)\zeta - S(t, s; u)\zeta\|$$

$$\leq \|S(t', s'; v)\zeta - S(t', s'; u)\zeta\| + \|S(t', s'; u)\zeta - S(t, s; u)\zeta\|,$$

and apply Theorem 5.5.6 on (t, s)-continuity of $S(t, s; u)\zeta$ for u fixed. ∎

Proof of Lemma 6.3.1. The assumptions and Lemma 6.1.1 imply

$$\|\partial_y f(t, y(t, u), u(t))\| \le L(t, c + 1) \tag{6.3.2}$$

(c a bound for $\|y(t, \bar{u})\|$) for all $u(\cdot) \in B(\bar{u}, \delta)$. In view of Theorem 5.5.6,

$$\|S(t, s; u)\| \le C. \tag{6.3.3}$$

Using the integral equation (5.5.19) for $S(t, s; u)$ and $S(t, s; v)$,

$$S(t, s; v)\zeta - S(t, s; u)\zeta$$

$$= \int_s^t S(t, \tau)\{\partial_y f(\tau, y(\tau, v), v(\tau)) - \partial_y f(\tau, y(\tau, v), u(\tau))\}S(\tau, s; v)\zeta d\tau$$

$$+ \int_s^t S(t, \tau)\{\partial_y f(\tau, y(\tau, v), u(\tau)) - \partial_y f(\tau, y(\tau, u), u(\tau))\}S(\tau, s; u)\zeta d\tau$$

$$+ \int_s^t S(t, \tau)\partial_y f(\tau, y(\tau, v), u(\tau))\{S(\tau, s; v) - S(\tau, s; u)\}\zeta d\tau$$

thus

$$\|(S(t, s; v)\zeta - S(t, s; u)\zeta\|$$

$$\le 2CM\|\zeta\| \int_{\{[\tau \in [0,\bar{t}]; u(\tau) \ne v(\tau)\}]} L(\tau, c + 1)d\tau$$

$$+ CM \int_s^t \|\{\partial_y f(\tau, y(\tau, v), u(\tau)) - \partial_y f(\tau, y(\tau, u), u(\tau))\}S(\tau, s; u)\zeta\|d\tau$$

$$+ M \int_s^t L(\tau, c + 1)\|S(\tau, s; v)\zeta - S(\tau, s; u)\zeta\|d\tau \tag{6.3.4}$$

The first two integrals on the right tend to zero as $v \to u$, the first on account of equicontinuity of the integral of $K(\cdot, c + 1)$, the second by Lemma 6.1.1, y-continuity of $\partial_y f(\tau, y, y)z$ and the dominated convergence theorem. Note that if $y \to \partial_y f(\tau, y, u)$ is continuous in the norm of (E, E) (this occurs in most applications) an obvious modification of (6.3.4) implies that $u \to S(t, s; u)$ is continuous in the norm of (E, E), uniformly in $0 \le s \le t \le \bar{t}$. ∎

6.4. Abstract Minimization Problems Again

We bring into play again the two problems in Part I:

Abstract minimization problem (see **4.1**): Let V be a metric space, E a Banach space, $Y \subseteq E$. Given functions $f : D(f) \to E, f_0 : D(f_0) \to \mathbb{R}$ ($D(f), D(f_0) \subseteq V$) characterize the $\bar{u} \in D(f) \cap D(f_0)$ satisfying

$$\text{minimize} \quad f_0(u) \tag{6.4.1}$$

$$\text{subject to} \quad f(u) \in Y \tag{6.4.2}$$

or, more generally, the **approximate** or **suboptimal** solutions of (6.4.1)-(6.4.2), that is, the sequences $\{\bar{u}^n\} \subseteq D(f) \cap D(f_0)$ satisfying

$$\limsup_{n \to \infty} f_0(\bar{u}^n) \leq m, \qquad \lim_{n \to \infty} \text{dist}(f(\bar{u}^n), Y) = 0, \qquad (6.4.3)$$

where m is the minimum of (6.4.1) under the constraint (6.4.2); we assume that $-\infty < m < \infty$.

Abstract time optimal problem (see **3.1**): Let $\{V_n\}$ be a sequence of metric spaces, E a Banach space, $Y \subseteq E$. Given a sequence of functions $f_n : D(f_n) \to E$ $(D(f_n) \subseteq V_n)$ such that $f_n(D(f_n)) \cap Y = \emptyset \, (n = 1, 2, \ldots)$ characterize the **solutions** of the problem, that is, the sequences $\{\bar{u}^n\}, \bar{u}^n \in D(f_n)$ satisfying

$$\lim_{n \to \infty} \text{dist}(f_n(\bar{u}^n), Y) = 0 \qquad (6.4.4)$$

We assume that E is a Hilbert space, that V is complete and that Y is closed. For the time optimal problem, the functions $\Phi_n(u, y) = \|f_n(u) - y\| \, (u \in D(f_n))$, $\Phi_n(u, y) = +\infty \, (u \notin D(f_n))$ are assumed lower semicontinuous in V for each $y \in E$. For the abstract minimization problem, the same is asked of f; for f_0 we assume that $\Phi_0(u, y) = \max(f_0(u), m - \varepsilon) \, (u \in D(f_0))$, $\Phi_0(u, y) = +\infty$ $(u \notin D(f_0))$ is lower semicontinuous for $\varepsilon > 0$ sufficiently small and that $D(f) \cap D(f_0) \neq \emptyset$.

Theorem 6.4.1. *Let $\{\bar{u}_n\}$ be a solution of the abstract time optimal problem, $\{\bar{y}^n\} \subseteq Y$ a sequence associated with $\{\bar{u}^n\}$, that is, such that*

$$\varepsilon_n = \|f_n(\bar{u}^n) - \bar{y}^n\| \to 0 \quad as \; n \to \infty. \qquad (6.4.5)$$

Then there exist sequences $\{\tilde{u}^n\}, \tilde{u}^n \in D(f_n), \{\tilde{y}^n\} \subseteq Y$ such that

$$\|f_n(\tilde{u}^n) - \tilde{y}^n\| \leq \|f_n(\bar{u}^n) - \bar{y}^n\|, \qquad (6.4.6)$$

$$d_n(\tilde{u}^n, \bar{u}^n) + \|\tilde{y}^n - \bar{y}^n\| \leq \sqrt{\varepsilon_n}, \qquad (6.4.7)$$

and a sequence $\{z_n\} \subseteq E$ such that

$$\|z_n\| = 1, \qquad \langle z_n, \xi^n - w^n \rangle \geq -\sqrt{\varepsilon_n}(1 + \|w^n\|) \qquad (6.4.8)$$

for $\xi^n \in \text{Var} \, f_n(\tilde{u}^n)$ and $w^n \in I_Y(\tilde{y}^n)$ or for $\xi^n \in \text{Der} \, f_n(\tilde{u}^n)$ and $w^n \in K_Y(\tilde{y}^n)$. If z is a weak limit point of $\{z_n\}$ then we have

$$\langle z, \xi \rangle \geq 0 \qquad (6.4.9)$$

for $\xi \in \liminf_{n \to \infty} \text{Var} \, f_n(\tilde{u}^n)$. If $\bar{y}^n \to \bar{y}$, then

$$z \in N_Y(\bar{y}). \qquad (6.4.10)$$

Theorem 6.4.2. *Let* $\{\bar{u}^n\}$ *be a suboptimal solution of* (6.4.1)-(6.4.2) *and* $\{\bar{y}^n\} \subseteq Y$ *a sequence associated with* $\{\bar{u}^n\}$, *that is, such that*

$$f_0(\bar{u}_n) \le m + \varepsilon_n, \qquad \|f(\bar{u}^n) - \bar{y}^n\| \le \varepsilon_n \qquad (6.4.11)$$

with $\varepsilon_n \to 0$. *Then there exist sequences* $\{\tilde{u}^n\} \subseteq D(f) \cap D(f_0), \{\tilde{y}^n\} \subseteq Y$ *with*

$$d(\tilde{u}^n, \bar{u}^n) + \|\tilde{y}^n - \bar{y}^n\| \le \sqrt[4]{5}\sqrt{\varepsilon_n} \qquad (6.4.12)$$

and a sequence $\{(z_{0n}, z_n)\} \subseteq \mathbb{R} \times E$ *such that*

$$z_{0n}^2 + \|z_n\|^2 = 1, \qquad z_{0n} \ge 0, \qquad (6.4.13)$$

$$z_{0n}\xi_0^n + \langle z_n, \xi^n - w^n \rangle \ge -\sqrt[4]{5}\sqrt{\varepsilon_n}(1 + \|w^n\|), \qquad (6.4.14)$$

for $(\xi_0^n, \xi^n) \in \mathrm{Var}\,(f_0, f)(\tilde{u}^n)$ *and* $w^n \in I_Y(\tilde{y}^n)$ *or* $(\xi_0^n, \xi^n) \in \mathrm{Der}\,(f_0, f)(\tilde{u}^n)$ *and* $w^n \in K_Y(\tilde{y}^n)$. *If* (z_0, z) *is a weak limit point of* $\{(z_{0n}, z_n)\}$ *then we have*

$$z_0 \ge 0, \qquad z_0\xi_0 + \langle z, \xi \rangle \ge 0 \qquad (6.4.15)$$

for $(\xi_0, \xi) \in \liminf_{n\to\infty} \mathrm{Var}\,(f_0, f)(\tilde{u}^n)$. *If* $\bar{y}^n \to \bar{y}$, *then*

$$z \in N_Y(\bar{y}). \qquad (6.4.16)$$

There is no need to reprove these results. Theorem 6.4.1 (resp. Theorem 6.4.2) is Theorem 3.3.2 (resp. Theorem 4.1.1), but there are two important distinctions. In Theorem 6.4.1,

(*i*) z is a weak (not strong) limit point of $\{z_n\}$,

(*ii*) there is no statement about the norm of z in (6.4.9); in Theorem 3.3.2 we had $\|z\| = 1$.

The reason for (*i*) and (*ii*) is the lack of the Bolzano-Weierstrass theorem in infinite dimensional spaces; we can be sure that $\{z_n\}$ has weak limit points (Theorem 2.0.11) but not strong limit points. Unlike strong limit points, weak limit points of a sequence $\{z_n\}$ with $\|z_n\| = 1$ may vanish (the classical example is $z_n(x) = \pi^{-1/2} \sin nx$ in $L^2(0, \pi)$). This means that we cannot exclude the possibility that $z = 0$, so that (6.4.9) may be trivial. This is basically the difference between finite and infinite dimensional control problems; in the former, the multipliers z (for the time optimal problem) and (z_0, z) (for the general control problem) are automatically nonzero. In the latter, we need additional conditions (on the system or the target set) to insure that $z \ne 0$ in Theorem 6.4.1 and that $(z_0, z) \ne 0$ in Theorem 6.4.2.

A result in this direction is Lemma 6.4.3 below. In view of future use, we present this result in a Banach space version.

A sequence $\{Q_n\}$ of subsets of E is **precompact** if any sequence $\{q_n\}, q_n \in Q_n$ has a subsequence convergent to some $q \in E$.

Lemma 6.4.3. *Let E be a Banach space, $\{\Delta_n\}$ a sequence of sets in E, $\{z_n\}$ a sequence in the dual space E^* such that*

$$0 < c \leq \|z_n\| \leq C, \quad \langle z_n, y \rangle \geq -\varepsilon_n \quad (y \in \Delta_n) \tag{6.4.17}$$

with $\varepsilon_n \to 0$. Assume that there exists a precompact sequence $\{Q_n\}$ such that

$$\Delta = \bigcap_{n=n_0}^{\infty} (\Delta_n + Q_n) \tag{6.4.18}$$

contains an interior point for n_0 large enough. Then every weakly convergent subsequence of $\{z_n\}$ has a nonzero limit.

Proof. Assume there exists a subsequence of $\{z_n\}$ (denoted by the same symbol) with $z_n \to 0$ weakly. Let $B(\bar{x}, \delta)$ ($\delta > 0$) be a ball contained in Δ. Choose $x_n \in E$ such that $\|x_n\| = 1$, $\langle z_n, x_n \rangle \geq c/2$. For each $n \geq n_0$ there exists $y_n \in \Delta_n$ and $q_n \in Q_n$ such that $\bar{x} - \delta x_n = y_n + q_n$; by hypothesis, we may assume that $\{q_n\}$ is convergent to some $q \in E$. Using the second inequality (6.4.17) we obtain

$$\langle z_n, \bar{x} \rangle = \langle z_n, y_n \rangle + \delta \langle z_n, x_n \rangle + \langle z_n, q_n \rangle \geq -\varepsilon_n + \delta c/2 + \langle z_n, q_n \rangle.$$

However, $\langle z_n, q_n \rangle \to \langle z, q \rangle = 0$, which contradicts the fact that $\{z_n\}$ is weakly convergent to zero. ∎

Corollary 6.4.4. *Let $\{\tilde{y}^n\}$, $\{z_n\}$ be the sequences in Theorem 6.4.1. Assume that there exist $\rho > 0$ and a precompact sequence $\{Q_n\}$ such that the set Δ in (6.4.18) contains an interior point for n_0 large enough, where*

$$\Delta_n = \mathrm{Var}\, f_n(\tilde{u}^n) - I_Y(\tilde{y}^n) \cap B(0, \rho) \quad or \quad \Delta_n = \mathrm{Der}\, f_n(\tilde{u}^n) - K_Y(\tilde{y}^n) \cap B(0, \rho).$$

Then $\{z_n\}$ cannot have zero weak limit points, so that the multiplier z in (6.4.9) is not zero.

Proof. Inequality (6.4.8) for $\xi^n \in \mathrm{Var}\, f_n(\tilde{u}^n)$ and $w^n \in I_Y(\tilde{y}^n)$ with $\|w^n\| \leq \rho$ implies the second inequality (6.4.17) for $\Delta_n = \mathrm{Var}\, f_n(\tilde{u}^n) - I_Y(\tilde{y}^n) \cap B(0, \rho)$, so that Lemma 6.4.3 applies. The other half of the statement is proved in the same way. ∎

    The comments on Theorem 6.4.1 extend to Theorem 6.4.2; z is just a weak limit point of $\{z_n\}$, and we cannot exclude the possibility that $(z_0, z) = 0$ in (6.4.15) although each (z_{0n}, z_n) in (6.4.14) has norm 1. The counterpart of Corollary 6.4.4 is

Corollary 6.4.5. *Let $\{\tilde{y}^n\}$, $\{(z_{0n}, z_n)\}$ be the sequences in Theorem 6.4.2. Assume that there exist $\rho > 0$ and a precompact sequence $\{Q_n\}$ such that the set Δ in (6.4.18) contains an interior point for n_0 large enough, where*

$$\Delta_n = \Pi(\mathrm{Var}\,(f_0, f)(\tilde{u}^n)) - I_Y(\tilde{y}^n) \cap B(0, \rho)$$

or

$$\Delta_n = \Pi(\mathrm{Der}\,(f_0, f)(\tilde{u}^n)) - K_Y(\tilde{y}^n) \cap B(0, \rho),$$

and Π is the projection from $\mathbb{R} \times E$ into E. Then $\{(z_{0n}, z_n)\}$ cannot have zero weak limit points, so that the multiplier (z_0, z) in (6.4.15) is not zero.

Proof. Inequality (6.4.14) implies the second inequality (6.4.17) for the sequence $\{(z_{0n}, z_n)\} \subseteq \mathbf{E} = \mathbb{R} \times E$ and the sequence $\{\boldsymbol{\Delta}_n\}$ of subsets of $\mathbf{E}$ defined by $\boldsymbol{\Delta}_n = \mathrm{Var}\,(f_0, f)(\tilde{u}^n) - (\{0\} \times (I_Y(\tilde{y}^n) \cap B(0, \rho)))$. Assume that the intersection Δ in (6.4.18) with $\Delta_n = \Pi(\mathrm{Var}\,(f_0, f)(\tilde{u}^n)) - I_Y(\tilde{y}^n) \cap B(0, \rho)$ contains an interior point. Then the intersection

$$\bigcap_{n=n_0}^{\infty} \left(\boldsymbol{\Delta}_n + [-1, 1] \times Q_n\right)$$

contains an interior point in $\mathbf{E}$ and the sequence $\{\mathbf{Q}_n\} = \{[-1, 1] \times Q_n\}$ is precompact in $\mathbf{E}$, so that Lemma 6.4.3 applies. The other half of the statement is proved in the same way. ∎

Remark 6.4.6. If $D \subseteq E$ then $\overline{\mathrm{conv}}\,D$ denotes the **closed convex envelope** of D, the intersection of all convex closed subsets of E that contain D. The second inequality (6.4.17) obviously extends to $y \in \overline{\mathrm{conv}}\,\Delta_n$, thus we may replace Δ_n by $\overline{\mathrm{conv}}\,\Delta_n$ in (6.4.18).

In many control systems the sets $\mathrm{Var}\,(f_0, f)(\tilde{u}^n)$ and $\mathrm{Der}\,(f_0, f)(\tilde{u}^n)$ are so small as to contribute next to nothing to the right-hand side of (6.4.18). In such cases, we write these sets off and require that for every sequence $\{y^n\} \subseteq Y$ such that $y^n \to \bar{y} \in Y$ there exist $\rho > 0$ and a precompact sequence $\{Q_n\}$ such that

$$\bigcap_{n=n_0}^{\infty} \{\overline{\mathrm{conv}}(I_Y(y^n) \cap B(0, \rho)) + Q_n\} \tag{6.4.19}$$

contains an interior point for n_0 large enough.[1] We call Y I-**full** at $\bar{y}$ if this condition is satisfied; Y is I-**full** if it is I-full at every $\bar{y} \in Y$. Fullness can also

---

[1] Multiplying each term by a scalar $\lambda > 0$ we see that, if the fullness condition is satisfied for some $\rho > 0$ then it is satisfied for every $\rho > 0$. The same observation applies to the other cones.

be defined with respect to the contingent cone K or the tangent cone T. Since $T_Y(y) \subseteq I_Y(y) \subseteq K_Y(y)$, we have the implications

$$
\begin{aligned}
T\text{-full at } \bar{y} &\Rightarrow I\text{-full at } \bar{y} \Rightarrow K\text{-full at } \bar{y} \\
T\text{-full} &\Rightarrow I\text{-full} \Rightarrow K\text{-full.}
\end{aligned}
\tag{6.4.20}
$$

Due to convexity of the tangent cone, (6.4.19) becomes

$$
\bigcap_{n=n_0}^{\infty} \{T_Y(y^n) \cap B(0, \rho) + Q_n\}
\tag{6.4.21}
$$

in the definition of T-full sets.[2]

Lemma 6.4.7. *Let Y be closed and convex with* $\mathrm{Int}(Y) \neq \emptyset$. *Then Y is T-full.*

In fact, let $B(y_0, \varepsilon)$ $(\varepsilon > 0)$ be a ball contained in $\mathrm{Int}(Y)$, and let $y \in Y$. Then (Example 3.3.4),

$$
T_Y(y) \supseteq \bigcup_{\lambda \geq 0} \lambda(B(y_0, \varepsilon) - y) \supseteq B(y_0, \varepsilon) - y = B(y_0 - y, \varepsilon)
$$

so that $T_Y(y^n) \cap B(0, \rho) \supseteq B(y_0 - y^n, \varepsilon)$ for ρ large enough. The precompact sequence $\{Q_n\}$ is not needed; we may take $Q_n = \{0\}$.

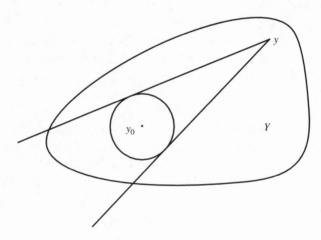

Figure 6.1.

[2] To take the closure of the convex set $T_Y(y^n) \cap B(0, \rho)$ in (6.4.21) is unnecessary since $T_Y(y^n)$ (hence $T_Y(y^n) \cap B(0, \rho)$) is closed (see Example 3.3.5). Note also that, by requiring the sequence $\{Q_n\}$ to be bounded we may eliminate $B(0, \rho)$ in (6.4.21) (but there are unbounded precompact sequences, for instance $E, \{0\}, E, \{0\}, \dots$). Finally, "for n_0 large enough" can be omitted replacing Q_n by E for $n < n_0$.

A map $\Phi : D \to F$ (D an open set in a Banach space E, F another Banach space) is called **continuously Fréchet differentiable** if the Fréchet derivative $\partial \Phi(y) \in (E, F)$ exists for all $y \in D$ and is continuous with respect to y in the norm of (E, F). Φ is called a **diffeomorphism** if $\Phi(D)$ is open, the inverse Φ^{-1} exists in $\Phi(D)$ and both Φ and Φ^{-1} are continuously Fréchet differentiable. If all of this holds we have

$$\partial \Phi^{-1}(\Phi(y)) = (\partial \Phi(y))^{-1} \qquad (6.4.22)$$

(see Example 6.4.12). If $\Phi : D \to F$ is continuously Fréchet differentiable and $\partial \Phi(\bar{y})$ has a bounded inverse for some $\bar{y} \in D$ then the **inverse function theorem** guarantees that Φ is a diffeomorphism in some open ball $B_0(\bar{y}, \varepsilon)$ with center $\bar{y}$ (Schwartz [1969, p. 14]).

Lemma 6.4.8. *Let $\bar{y} \in Y \subseteq E$ (D an open set containing $\bar{y}$), $\Phi : D \to F$ a continuously Fréchet differentiable map such that $\partial \Phi(\bar{y})$ has a bounded inverse. Then Y is T-full at $\bar{y}$ if and only if $Z = \Phi(D \cap Y)$ is T-full at $\Phi(\bar{y})$.*

Proof. Using the implicit function theorem we may assume (exchanging D by an open ball $B_0(\bar{y}, \varepsilon)$) that Φ is a diffeomorphism and that $\|\partial \Phi(y)\|$, $\|\partial \Phi(\Phi(y))^{-1}\|$ are uniformly bounded in D. It follows from the definitions involved (Example 6.4.14) that

$$T_{\Phi(D \cap Y)}(\Phi(\bar{y})) = \partial \Phi(\bar{y})T_{D \cap Y}(\bar{y}) = \partial \Phi(\bar{y})T_Y(\bar{y}) \qquad (6.4.23)$$

Let $\{z^n\} \subseteq Z$ such that $z^n \to \Phi(\bar{y})$, $y^n = \Phi^{-1}(z_n)$. Then $y^n \to \bar{y}$ and, since Y is T-full at $\bar{y}$, there exist $\rho > 0$ and a precompact sequence $\{Q_n\}$ such that

$$\bigcap_{n=n_0}^{\infty} \{T_Y(y^n) \cap B(0, \rho) + Q_n\} \qquad (6.4.24)$$

contains an interior point for n_0 sufficiently large. Since $\|\partial \Phi(\Phi(y))^{-1}\|$ is uniformly bounded in D, if $B(y_1, \delta)$ ($\delta > 0$) is an arbitrary ball, the intersection of all the sets $\partial \Phi(y)B(y_1, \delta)$ over $y \in D$ contains a ball of positive radius. Accordingly,

$$\bigcap_{n=n_0}^{\infty} \partial \Phi(y^n)\{T_Y(y^n) \cap B(0, \rho) + Q_n\}$$

contains as well an interior point for n_0 sufficiently large. Note now that if A is an arbitrary linear operator in E then $A(C + D) = A(C) + A(D)$ and $A(C \cap D) \subseteq A(C) \cap A(D)$ for arbitrary sets $C, D \subseteq E$. Hence, using (6.4.23),

$$\partial \Phi(y^n)\{T_Y(y^n) \cap B(0, \rho) + Q_n\}$$
$$\subseteq \partial \Phi(y^n)T_Y(y^n) \cap \partial \Phi(y^n)B(0, \rho) + \partial \Phi(y^n)Q_n$$
$$\subseteq T_{\Phi(D \cap Y)}(\Phi(y^n)) \cap B(0, \rho') + \partial \Phi(y^n)Q_n,$$

where $\rho' > 0$ is such that $\partial\Phi(y^n)B(0, \rho) \subseteq B(0, \rho')$ (recall that $\|\partial\Phi(y)\|$ is uniformly bounded in D). It follows that

$$\bigcap_{n=n_0}^{\infty} \{T_{\Phi(D\cap Y)}(z^n) \cap B(0, \rho') + \partial\Phi(y^n)Q_n\}$$

contains an interior point. Since $\{Q_n\}$ is precompact and $\partial\Phi(y^n)$ is convergent in the norm of (E, E), the sequence $\{\partial\Phi(y^n)Q_n\}$ is precompact, and we have shown that $\Phi(D \cap Y)$ is T-full at $\Phi(\bar{y})$. The reciprocal statement is obtained interchanging D, Φ, by $\Phi(D)$, Φ^{-1}, and we omit the details.   ∎

Lemma 6.4.9. *Let the set* $Z \subseteq E$ *be convex and closed and let the functionals* $y_1^*, y_2^*, \ldots, y_n^* \in E^*$ *be linearly independent. Define*

$$N = \{y \in E; \langle y_j^*, y \rangle = 0 \ (j = 1, 2, \ldots, n)\}.$$

Then, if $\text{Int}(Z) \cap N \neq \emptyset, Y = Z \cap N$ *is* T*-full.*

Proof. Since the $\{y_j^*\}$ are linearly independent we may select $\{y_j\} \subseteq E$ such that $\langle y_j^*, y_k \rangle = \delta_{jk}$. Let F be the subspace generated by the $\{y_j\}$. Then we have

$$E = N \oplus F \tag{6.4.25}$$

algebraically and topologically; the latter means the projection P into F is bounded (so that $I - P$, the projection into N is also bounded). Boundedness of P is obvious since $Py = \Sigma\langle y_j^*, y \rangle y_j \in F$ for $y \in E$.

Let $B(y_0, \varepsilon)$ $(\varepsilon > 0)$ be a ball contained in Z with center $y_0 \in N$. Then $Y = Z \cap N$ contains $B_N(y_0, \varepsilon)$, the ball of center y_0 and radius ε in N. It follows as in Lemma 6.4.7 that if $y \in Y$ then the tangent cone to Y at y in N, *a fortiori* the tangent cone $T_Y(y)$ to Y at y in E, contains $B_N(y_0, \varepsilon) - y = B_N(y_0 - y, \varepsilon)$. Now, if C is a bound for $\|P\|$ and $\|I - P\|$ we have $B_E(y_0 - y, C^{-1}\varepsilon) \subseteq B_N(y_0 - y, \varepsilon) + B_F(0, \varepsilon)$ (subindices indicating in what space balls are taken). Accordingly, if $\{y^n\}$ is a sequence in Y with $y^n \to \bar{y} \in Y$ and ρ is large enough,

$$T_Y(y^n) \cap B(0, \rho) + B_F(0, \varepsilon)$$
$$\supseteq B_N(y_0 - y^n, \varepsilon) + B_F(0, \varepsilon) \supseteq B_E(y_0 - y^n, C^{-1}\varepsilon),$$

where the (constant) sequence $\{B_F(0, \varepsilon)\}$ is precompact (F is finite dimensional). This ends the proof.   ∎

Other T-full sets can be constructed by application of Lemma 6.4.9 and the "deformation Lemma" 6.4.8. One example is

Lemma 6.4.10. *Let $\phi_1, \phi_2, \ldots, \phi_n$ be continuously Fréchet differentiable functionals in E. Define*

$$\Sigma = \{y; \phi_j(y) = c_j \ (j = 1, 2, \ldots, n)\}. \tag{6.4.26}$$

Then, if $\bar{y} \in \Sigma$ and $\partial \phi_1(\bar{y}), \partial \phi_2(\bar{y}), \ldots, \partial \phi_n(\bar{y})$ are linearly independent in E^*, Σ is T-full at $\bar{y}$.

Proof. Select $\{y_j\} \subseteq E$ such that $\langle \partial \phi_j(\bar{y}), y_k \rangle = \delta_{jk}$, and denote by P, as above, the projection operator $Py = \Sigma \langle \partial \phi_j(\bar{y}), y \rangle y_j$ into the subspace F generated by $y_1, y_2, \ldots, y_n$. Define

$$N = \{z \in E; \langle \partial \phi_j(\bar{y}), z \rangle = 0, j = 1, 2, \ldots, n\}$$

and do the decomposition (6.4.25); the projections P and $I - P$ are the same. Then define

$$\Phi(y) = (I - P)y + \Sigma(\phi_j(y) - c_j)y_j.$$

Obviously $\Phi : \Sigma \to N$ and Φ is continuously Fréchet differentiable with derivative $\partial \Phi(y)z = (I - P)z + \Sigma \langle \partial \phi_j(y), z \rangle y_j = ((I - P) + P)z = z$. By the inverse function theorem Φ is a diffeomorphism in an open ball $B_0(\bar{y}, \varepsilon)$, and we only have to show that N is full at $\Phi(\bar{y})$, which is a consequence of Lemma 6.4.8 with $Z = E$. ∎

Corollary 6.4.11. *Let Σ be defined as in Lemma 6.4.9 but with equalities and inequalities,*

$$\Sigma = \{y; \phi_j(y) = c_j \ (j = 1, 2, \ldots, k), \phi_j(y) \le c_j \ (j = k + 1, \ldots, n)\}, \tag{6.4.27}$$

$\phi_1, \phi_2, \ldots, \phi_n$ *continuously Fréchet differentiable functionals in E. Then, if $\partial \phi_1(\bar{y}), \partial \phi_2(\bar{y}), \ldots, \partial \phi_n(\bar{y})$ are linearly independent in E^*, Σ is T-full at $\bar{y}$.*

Proof. Let Σ_e be the set defined as in (6.4.26) (all equalities); then $\Sigma_e \subseteq \Sigma$. If $\phi_j(\bar{y}) = c_j$ ($1 \le j \le n$) then Σ_e is T-full at $\bar{y}$ by Lemma 6.4.9, thus Σ itself is T-full at $\bar{y}$. If $\phi_j(\bar{y}) < c_j$ for some set of indices, we replace Σ_e by $\Sigma' \cap B(\bar{y}, \delta)$ for δ sufficiently small, where Σ' is the set defined by the constraints corresponding to the other indices. Since T-fullness at $\bar{y}$ is a local property, $\Sigma' \cap B(\bar{y}, \delta)$ is T-full at $\bar{y}$ by Lemma 6.4.10, thus Σ is T-full at $\bar{y}$. ∎

Corollary 6.4.12. *Let $\phi_1, \phi_2, \ldots, \phi_n$ be continuously Fréchet differentiable functionals in E such that $\partial \phi_1(y), \partial \phi_2(y), \ldots, \partial \phi_n(y)$ are linearly independent in E^* for each y, and let Σ be given by (6.4.26) or (6.4.27). Then $\bar{y}$ is T-full.*

Example 6.4.13. Putting suitable conditions on domains and ranges prove: (a) if $\partial\Phi(y), \partial\Psi(y)$ exist then $\partial(\Phi + \Psi)(y)$ exists and $\partial(\Phi + \Psi)(y) = \partial\Phi(y) + \partial\Psi(y)$, (b) if $\partial\Phi(y), \partial\Psi(\Phi(y))$ exist then $\partial(\Psi \circ \Phi)(y)$ exists and $\partial(\Psi \circ \Phi)(y) = \partial\psi(\Phi(y))\partial\Phi(y)$, (c) if Φ is invertible and $\partial\Phi(y), \partial\Phi^{-1}(\Phi(y))$ exist then $(\partial\Phi^{-1})(\Phi(y)) = \partial\Phi(y)^{-1}$.

Example 6.4.14. If $\Phi(y)$ is differentiable, we may apply the mean value theorem to $\Psi(y) = \Phi(y) - \partial\Phi(\tilde{y})y$ and obtain the following estimate for the term $\rho(\tilde{y}, h)$ in the definition of derivative:

$$\|\rho(\tilde{y}, h)\|_F = \|\Phi(\tilde{y} + h) - \Phi(\tilde{y}) - \partial\Phi(\tilde{y})h\|_F$$

$$\leq \|h\|_E \max_{\xi \in I} \|\partial\Phi(\xi) - \partial\Phi(\tilde{y})\|_{(E,F)}, \qquad (6.4.28)$$

where I is the segment joining $\tilde{y}$ and $\tilde{y} + h$.

We use this estimate to prove (6.4.23). Let $w \in T_Y(\tilde{y})$, $\{\bar{z}_k\}$ a sequence in $\Phi(D \cap Y)$ with $\bar{z}_k \to \Phi(\tilde{y})$ and $\{h_k\} \subseteq R_+$ a sequence with $h_k \to 0$. Then there exists a sequence $\{y_k\} \subseteq D \cap Y$ with

$$y_k = \Phi^{-1}(\bar{z}_k) + h_k w + o(h_k).$$

Applying Φ on both sides and using the Fréchet derivative at $\bar{y}_k = \Phi^{-1}(\bar{z}_k)$,

$$\Phi(y_k) = \bar{z}_k + \partial\Phi(\bar{y}_k)(h_k w + o(h_k)) + \rho(\bar{y}_k, h_k w + o(h_k)). \qquad (6.4.29)$$

Using continuity of the derivative we see that $\partial\Phi(\bar{y}_k)h_k w = h_k \partial\Phi(\tilde{y})w + o(h_k)$. For the third term on the right of (6.4.29) we use (6.4.28) and the equality $\partial\Phi(\xi_k) - \partial\Phi(\bar{y}_k) = (\partial\Phi(\xi_k) - \partial\Phi(\tilde{y})) + (\partial\Phi(\tilde{y}) - \partial\Phi(\bar{y}_k))$ (ξ_k a point in the segment joining $\bar{y}_k$ and $\bar{y}_k + h_k w + o(h_k)$) to deduce that $\|\rho(\bar{y}_k, h_k w + o(h_k))\| = o(h_k)$. We have then shown that $\partial\Phi(\tilde{y}) \in T_{\Phi(D \cap Y)}(\Phi(\tilde{y}))$. The converse statement follows interchanging D, Φ by $\Phi(D)$, Φ^{-1}.

Analogues of (6.4.23) hold for the contingent and intermediate cones, and are much easier to prove since, in these, $\bar{z}_k = z$. There is also an analogue of Lemma 6.4.7 for "I-full" and "K-full"; the proof below is slightly different than that for "K-full" since we use $\overline{\text{conv}}$ in the definition (see (6.4.19)). Note that if A is bounded then $A(\overline{\text{conv}}D) \subseteq \overline{\text{conv}}A(D)$, with equality if A is invertible. Then

$$\partial\Phi(y^n)\{\overline{\text{conv}}(I_Y(y^n) \cap B(0, \rho)) + Q_n\}$$

$$\subseteq \overline{\text{conv}}(\partial\Phi(y^n)I_Y(y^n) \cap \partial\Phi(y^n)B(0, \rho)) + \partial\Phi(y^n)Q_n$$

$$\subseteq \overline{\text{conv}}(I_{\Phi(D \cap Y)}(\Phi(y^n)) \cap B(0, \rho')) + \partial\Phi(y^n)Q_n,$$

where $\rho' > 0$ is such that $\partial\Phi(y^n)B(0, \rho) \subseteq B(0, \rho')$. This is used to show that

$$\bigcap_{n=n_0}^{\infty} \{\overline{\text{conv}}(I_{\Phi(D \cap Y)}(\Phi(y^n)) \cap B(0, \rho')) + \partial\Phi(y^n)Q_n\}$$

contains an interior point for n_0 large enough. Same argument for K.

6.5. The Minimum Principle for the Time Optimal Problem

As in **6.1**, we assume that the Cauchy problem for the linear part $y'(t) = A(t)y(t)$ of

$$y'(t) = A(t)y(t) + f(t, y(t), u(t)), \quad y(0) = \zeta \tag{6.5.1}$$

is well posed in $0 \le t \le T$ with solution operator $S(t, s)$. E is a Hilbert space and $f(t, y, u(t))$ satisfies Hypothesis III for each $u(\cdot) \in C_{\text{ad}}(0, T; U)$ with $K(\cdot, c)$, $L(\cdot, c)$ independent of $u(\cdot)$. The target condition is $y(\bar{t}) \in Y$.

Recall (see **3.4**) that a sequence $\{u_n(\cdot)\} \subseteq C_{\text{ad}}(0, \bar{t}; U)$ is *stationary* if there exists a null set e such that for every $t \in [0, T] \backslash e$ there exists n (depending on t) such that $u_n(t) = u_{n+1}(t) = u_{n+2}(t) = \cdots$. The admissible control space $C_{\text{ad}}(0, \bar{t}; U)$ is *saturated* if the limit of every stationary sequence belongs to $C_{\text{ad}}(0, \bar{t}; U)$. As shown in Theorem 3.4.1, saturation implies completeness with respect to the distance

$$d(u(\cdot), v(\cdot)) = |\{t \in [0, T]; u(t) \ne v(t)\}|. \tag{6.5.2}$$

Finally, $C_{\text{ad}}(0, T; U)$ is **patch complete** if every patch perturbation of a control $u(\cdot) \in C_{\text{ad}}(0, T; U)$ with arbitrary m and $\mathbf{v}(\cdot) = (v_1(\cdot), v_2(\cdot), \ldots, v_m(\cdot))$ $(v_j(\cdot) \in C_{\text{ad}}(0, T; U))$ belongs to $C_{\text{ad}}(0, T; U)$.

Let $\phi(t, v)$ be a Banach space valued function defined in $[0, T] \times U$ (U an arbitrary set) such that $\phi(t, v)$ is integrable in t for v fixed. The *left Lebesgue set* of $\phi(t, U)$ is the set d of all $t \in [0, T]$ such that t is a left Lebesgue point of $\phi(t, v)$ for every $v \in U$. We say that $\phi(t, v)$ is *regular* in U if its left Lebesgue set has full measure in $[0, T]$.

Theorem 6.5.1. (*The minimum principle for the time optimal problem.*) *Assume that the target set Y is closed and that $C_{\text{ad}}(0, T; U)$ is patch complete and saturated. Then, if $\bar{u}(\cdot)$ is a time optimal control in $0 \le t \le \bar{t}$ there exists $z \in N_Y(\bar{y}) = $ normal cone to Y at $\bar{y} = y(\bar{t}, \bar{u})$ such that, if $\bar{z}(s)$ is the solution of*

$$\bar{z}'(s) = -\{A(s)^* + \partial_y f(s, y(s, \bar{u}), \bar{u}(s))^*\}\bar{z}(s), \quad z(\bar{t}) = z \tag{6.5.3}$$

in $0 \le s \le \bar{t}$ then

$$\int_0^{\bar{t}} \langle \bar{z}(\sigma), f(\sigma, y(\sigma, \bar{u}), v(\sigma)) - f(\sigma, y(\sigma, \bar{u}), \bar{u}(\sigma)) \rangle d\sigma \ge 0 \tag{6.5.4}$$

*for all $v(\cdot) \in C_{\text{ad}}(0, \bar{t}; U)$. If E is separable,[1] $f(t, y(t, \bar{u}), v)$ is regular in U, and $C_{\text{ad}}(0, \bar{t}; U)$ is spike complete (see **2.3**), then*

$$\langle \bar{z}(s), f(s, y(s, \bar{u}), \bar{u}(s)) \rangle = \min_{v \in U} \langle \bar{z}(s), f(s, y(s, \bar{u}), v) \rangle \tag{6.5.5}$$

a.e. in $0 \le s \le \bar{t}$.

---

[1] See Remark 6.5.11.

Proof. The result follows from Theorem 6.4.1 for the abstract time optimal problem. By Lemma 6.1.1 there is a ball $B(\bar{u}, \delta)$ $(\delta > 0)$ such that $y(t, v)$ exists in $0 \le t \le \bar{t}$ for $u(\cdot) \in B(\bar{u}, \delta)$ and $u \to y(\cdot, u)$ is continuous in $B(\bar{u}, \delta)$ as a $C(0, \bar{t}; E)$-valued function. Choose a sequence $\{t_n\}$, $t_n < \bar{t}$ with $t_n \to \bar{t}$, and take $V_n = B_n(\bar{u}, \delta) \subseteq C_{\text{ad}}(0, t_n; U)$ ($B_n(\bar{u}, \delta)$ the ball of center $\bar{u}$ and radius δ in $C_{\text{ad}}(0, t_n; U)$). The functions $f_n : V_n \to E$ in the abstract time optimal problem are

$$\mathbf{f}_n(u) = y(t_n, u) \tag{6.5.6}$$

and are thus continuous. The solution of the abstract time optimal problem (in the sense of Theorem 6.4.1) is $\{\bar{u}^n\}$, $\bar{u}^n(\cdot)$ the restriction of $\bar{u}(\cdot)$ to $0 \le t \le t_n$. Theorem 6.4.1 provides sequences $\{\tilde{u}^n(\cdot)\}$, $\tilde{u}^n(\cdot) \in B_n(\bar{u}, \delta) \subseteq C_{\text{ad}}(0, t_n; U)$ with $d_n(\tilde{u}^n(\cdot), \bar{u}(\cdot)) \to 0$ (d_n the distance in $C_{\text{ad}}(0, t_n; U)$), $\{\tilde{y}^n\} \subseteq Y$ with $\tilde{y}^n \to \bar{y} = y(\bar{t}, \bar{u})$ and $\{z_n\} \subseteq E$ such that $\|z_n\| = 1$ and that (6.4.8) holds for any directional derivative $\xi^n \in \text{Der } f_n(\tilde{u}^n)$ and any $w^n \in K_Y(\tilde{y}^n)$. We use this for the directional derivatives $\bar{t}^{-1}\xi(t_n, \tilde{u}^n, \mathbf{p}, \mathbf{v})$ produced in Theorem 6.2.7 by means of patch perturbations. To graduate to the Kuhn-Tucker inequality (6.4.9), we must compute limits of these directional derivatives as $n \to \infty$. ∎

Lemma 6.5.2. *Let $\bar{u} \in C_{\text{ad}}(0, \bar{t}; U)$ be such that $y(t, \bar{u})$ exists in $0 \le t \le \bar{t}$, $\{u^n\} \subseteq C_{\text{ad}}(0, \bar{t}; U)$ with $d(u^n, \bar{u}) \to 0$. Then*

$$\xi(t, u^n, \mathbf{p}, \mathbf{v}) \to \xi(t, \bar{u}, \mathbf{p}, \mathbf{v}) \text{ as } n \to \infty \tag{6.5.7}$$

uniformly in $0 \le t \le \bar{t}$.

Proof. We use the variation-of-constants formula to express the solution of (6.2.20). After some rearrangements,

$$\xi(t, u^n, \mathbf{p}, \mathbf{v})$$

$$= \sum_{j=1}^{m} p_j \int_0^t S(t, \tau; u^n)\{f(\tau, y(\tau, u^n), v_j(\tau)) - f(\tau, y(\tau, u^n), u^n(\tau))\}d\tau$$

$$= \sum_{j=1}^{m} p_j \int_0^t \left(S(t, \tau; u^n) - S(t, \tau, \bar{u})\right)$$

$$\times \{f(\tau, y(\tau, u^n), v_j(\tau)) - f(\tau, y(\tau, u^n), \bar{u}(\tau))\}d\tau$$

$$+ \sum_{j=1}^{m} p_j \int_0^t S(t, \tau, \bar{u})\{f(\tau, y(\tau, u^n), v_j(\tau)) - f(\tau, y(\tau, u^n), \bar{u}(\tau))\}d\tau$$

$$+ \int_0^t S(t, \tau; u^n)\{f(\tau, y(\tau, u^n), \bar{u}(\tau)) - f(\tau, y(\tau, u^n), u^n(\tau))\}d\tau. \tag{6.5.8}$$

We now use Lemma 6.1.1 for $y(\tau, u^n)$, so that all terms between curly brackets can be estimated by $2K(\tau, c + 1)$ independently of n. In the first integral on the right we use boundedness and strong continuity of the solution operator as function of u (Lemma 6.3.1) and the dominated convergence theorem; in the second the dominated convergence theorem and in the third, again, boundedness of $S(t, \tau; u^n)$ and the fact that the term between curly brackets vanishes outside of a set of measure $d(u^n, \bar{u}) \to 0$. Uniform convergence is extorted from the dominated convergence theorem as customary (see the argument after (2.3.20)). ∎

To apply Lemma 6.5.2 to the time optimal problem we extend $u^n(\cdot)$ to $[t_n, \bar{t}]$ setting $u^n(t) = \bar{u}(t)$ there. We have $d(u^n, \bar{u}) = d_n(u^n, \bar{u}) \to 0$ so that Lemma 6.1.1 guarantees that $y(t, u^n)$ exists in $0 \le t \le \bar{t}$. Uniform convergence of $\xi(t, u^n, \mathbf{p}, \mathbf{v})$ in $0 \le t \le \bar{t}$ implies $\xi(t_n, u^n, \mathbf{p}, \mathbf{v}) \to \xi(\bar{t}, u^n, \mathbf{p}, \mathbf{v})$, which is all we need here.

Proof of Theorem 6.5.1. We use the Kuhn-Tucker inequality (6.4.9) only for "one-patch" perturbations $(m = 1)$,[2] so that $\mathbf{e} = \{e\}$ and $\mathbf{v}(\cdot) = \{v(\cdot)\}$, $\mathbf{p} = \{1\}$. For these $\mathbf{p}$ and $\mathbf{v}(\cdot)$ we use the shorthand $\xi(t, u, \mathbf{p}, \mathbf{v}) = \xi(t, u, v)$. The resulting particular case of (6.4.9) is

$$\langle z, \xi(\bar{t}, \bar{u}, v) \rangle \ge 0 \qquad (6.5.9)$$

where z is the weak limit of some subsequence of the $\{z_n\}$ and $v = v(\cdot)$ is an arbitrary element of $C_{ad}(0, \bar{t}; U)$. This inequality can be written

$$\int_0^{\bar{t}} \langle z, S(\bar{t}, \tau; \bar{u})\{f(\tau, y(\tau, \bar{u}), v(\tau)) - f(\tau, y(\tau, \bar{u}), \bar{u}(\tau))\}\rangle d\tau \ge 0.$$

Putting $S(\bar{t}, \tau; \bar{u})$ on the other side of the duality product we obtain (6.5.4) for all $v(\cdot) \in C_{ad}(0, \bar{t}; U)$. We take here the point of view in Theorem 5.5.12 that

$$\bar{z}(s) = S(\bar{t}, s; \bar{u})^* z \qquad (6.5.10)$$

is *by definition* the solution of (6.5.3); *nothing* is assumed of $A(s)^*$ or of $\partial_y f(s, y(s, \bar{u}), \bar{u}(s))^*$, and $\bar{z}(\cdot)$ is just an E-weakly continuous function.

Inequality (6.5.4) may be called the **integral form** of the minimum principle. If E is separable then $\bar{z}(\cdot)$ is strongly measurable (Example 5.0.39; recall that E is Hilbert, so that $E^* = E$ is separable). Since $\bar{z}(\cdot)$ is bounded, it is integrable. Assuming $f(\cdot, y(\cdot, \bar{u}), v)$ regular in U, select s in the set $d \cap e \cap e^*$, where d is the left Lebesgue set of $f(\sigma, y(\sigma, \bar{u}), U)$ and e (resp. e^*) is the set of left Lebesgue points of $f(\cdot, y(\cdot, \bar{u}), \bar{u}(\cdot))$ (resp. of $\bar{z}(\cdot)$). Use as $v(\sigma)$ in (6.5.4) the spike perturbation $u_{s,h,v}(\sigma)$, divide by h and take limits. The result is

$$\langle \bar{z}(s), f(s, y(s, \bar{u}), v)) - f(s, y(s, \bar{u}), \bar{u}(s))\rangle \ge 0 \qquad (6.5.11)$$

in $d \cap e \cap e^*$ which is (6.5.5) (see Example 6.5.6 below). This ends the proof of Theorem 6.5.1. ∎

---

[2] Patch perturbations here (and in the rest of the chapter) are used with only one patch. The multipatch construction will be essential in general Banach spaces (see Chapter 7).

Compared with Theorem 3.7.1, Theorem 6.5.1 lacks any assurance that $z \neq 0$ (see the comments after Theorem 6.4.2). We need additional conditions (those in Corollary 6.4.4) to achieve a nonzero multiplier.

Consider the inhomogeneous variational equation

$$\xi'(t) = \{A(t) + \partial_y f(t, y(t, \tilde{u}), \tilde{u}(t))\}\xi(t) + w(t), \quad \xi(0) = 0 \qquad (6.5.12)$$

for $\tilde{u} \in C_{ad}(0, \tilde{t}; U)$ such that $y(t, \tilde{u})$ exists in $0 \leq t \leq \tilde{t}$. Equation (6.5.12) is viewed as a control system with control $w(\cdot)$; the *admissible control space* $C_{ad}(0, \tilde{t}; U, \tilde{u})$ consists of all functions of the form

$$w(t) = f(t, y(t, \tilde{u}), u(t)) - f(t, y(t, \tilde{u}), \tilde{u}(t)), \qquad (6.5.13)$$

where $u(\cdot)$ belongs to the admissible control space $C_{ad}(0, \tilde{t}; U)$ of the system (6.5.1). We denote by $R(0, \tilde{t}; U, \tilde{u})$ the **reachable set** or **reachable space** (at time $\tilde{t}$) of (6.5.12), that is, the set of all $\xi(\tilde{t}, v)$ $(v(\cdot) \in C_{ad}(0, \tilde{t}; U, \tilde{u}))$. Equivalently, $R(0, \tilde{t}; U, \tilde{u})$ consists of all elements of the form

$$\int_0^{\tilde{t}} S(\tilde{t}, s; \tilde{u})\{f(s, y(s, \tilde{u}), u(s)) - f(s, y(s, \tilde{u}), \tilde{u}(s))\}ds, \qquad (6.5.14)$$

where $S(t, s; \tilde{u})$ is the solution operator of (6.5.12) and $u(\cdot) \in C_{ad}(0, \tilde{t}; U)$.

Lemma 6.5.3. *Let $\{t_n\}$, $\{\tilde{u}^n\}$, $\{\tilde{y}^n\}$ be the sequences in Theorem 6.5.1. Assume that there exist $\rho > 0$ and a precompact sequence $\{Q_n\}$, $Q_n \subseteq E$ such that[3]*

$$\bigcap_{n=n_0}^{\infty} \{t_n^{-1} R(0, t_n; U, \tilde{u}^n) - K_Y(\tilde{y}^n) \cap B(0, \rho) + Q_n\} \qquad (6.5.15)$$

contains an interior point for n_0 large enough. Then the multiplier z in Theorem 6.5.1 is not zero.

Proof. To apply Corollary 6.4.4 it is enough to check that

$$t_n^{-1} R(0, t_n; U, \tilde{u}^n) \subseteq \mathrm{Der} f_n(\tilde{u}^n). \qquad (6.5.16)$$

This is obvious writing an arbitrary one-patch directional derivative $t_n^{-1}\xi(t_n, \tilde{u}^n, v)$ by means of the variation-of-constants formula and comparing with (6.5.14). Note that (at the cost of changing ρ) since $t_n \to \tilde{t} > 0$ and $K_Y(\tilde{y}^n)$ is a cone, we may get rid of the factor t_n^{-1} in (6.5.15). Of course a precompact sequence remains precompact multiplied by a factor or a convergent sequence. ∎

---

[3] Note the exclusive appearance of the contingent rather than the intermediate cone. This is due to the use of directional derivatives rather than general variations; see Corollary 6.4.5.

Remark 6.5.4. Checking the hypotheses of Lemma 6.5.3 for the actual sequences $t_n, \{\tilde{u}^n\}, \{\tilde{y}^n\}$ is impossible since we don't actually know them; they make their entrance in a nonconstructive manner. Accordingly, we must do the checking for arbitrary sequences $\{t_n\}, \{\tilde{u}^n\}, \{\tilde{y}^n\}$ such that $t_n \to \bar{t}, d_n(\tilde{u}^n, \bar{u}) \to 0, \|y_n - \bar{y}\| \to 0$ where $\bar{u}(\cdot)$ (resp. $\bar{y}$) is an arbitrary element of $C_{\mathrm{ad}}(0, \bar{t}; U)$ (resp. of Y). We may, however, assume that $t - t_n \leq \delta_n, d_n(\tilde{u}^n, \bar{u}) \leq \delta_n, \|y^n - \bar{y}\| \leq \delta_n$, where $\{\delta_n\}$ is an arbitrary sequence with $\delta_n \to 0$.

Example 6.5.5. Assume $A(t) \equiv A =$ infinitesimal generator of a strongly continous semigroup. Then the reversed Cauchy problem for $z'(s) = -A^*z(s)$ is well posed (Theorem 5.2.6). We have

$$\langle \partial_y f(s, y(s, \bar{u}), \bar{u}(s))^*z, y \rangle = \langle z, \partial_y f(s, y(s, \bar{u}), \bar{u}(s))y \rangle$$

so that if E (thus E^*) is separable, $s \to \partial_y f(s, y(s, \bar{u}), \bar{u}(s))^*z$ is strongly measurable. Theorem 5.5.10 can be applied and the costate $\bar{z}(s)$ results continuous.

Example 6.5.6. If the E-valued function $f(\cdot)$ and the E^*-valued function $z(\cdot)$ are strongly measurable with $z(\cdot)$ bounded and $f(\cdot)$ integrable, then, if s is a (left) Lebesgue point of both, s is a (left) Lebesgue point of the scalar function $g(s) = \langle z(s), f(s) \rangle$. To see this, note the equality $g(s) - g(\tau) = \langle z(s) - z(\tau), f(s) \rangle + \langle z(\tau), f(s) - f(\tau) \rangle$ and estimate.

Remark 6.5.7. Taking advantage of Remark 6.4.6 we can replace the set $t_n^{-1}R(0, t_n; U, \tilde{u}^n) - K_Y(\tilde{y}^n) \cap B(0, \rho)$ in the intersection (6.5.15) by

$$\overline{\mathrm{conv}}\big(t_n^{-1}R(0, t_n; U, \tilde{u}^n) - K_Y(\tilde{y}^n) \cap B(0, \rho)\big) \tag{6.5.17}$$

or by the (presumably easier to compute) set

$$\overline{\big(t_n^{-1}\mathrm{conv}R(0, t_n; U, \tilde{u}^n) - \mathrm{conv}K_Y(\tilde{y}^n) \cap B(0, \rho)\big)}, \tag{6.5.18}$$

where the **convex envelope** $\mathrm{conv}\,D$ is the intersection of all convex sets containing D. To convince ourselves that the second replacement makes sense it is enough to show that

$$\mathrm{conv}(C) - \mathrm{conv}(D) \subseteq 2\,\mathrm{conv}(C - D) \tag{6.5.19}$$

for arbitrary subsets of E, both containing 0. To prove this last inclusion, consider a finite collection $\{y_j\}$ (resp. $\{z_k\}$) of elements of C (resp. of D). Then we have $\Sigma_j\alpha_j y_j - \Sigma_k\beta_k z_k = 2\{\Sigma_j(\alpha_j/2)(y_j - 0) + \Sigma_k(\beta_k/2)(0 - z_k)\}$, where both $y_j - 0$ and $0 - z_k$ belong to $C - D$.

Applying this to (6.5.17) and (6.5.18) the factor 2 on the right-hand side of (6.5.19) does not matter; we have the right to modify ρ.

Although it is obvious that using either (6.5.17) or (6.5.18) in the intersection fortifies Lemma 6.5.3, the advantage seems marginal in the examples studied in this work.

Example 6.5.8. Let $f(t, y, u)$ be defined and continuous in all three variables in $[0, T] \times E \times U$, where U is a closed and bounded subset of another Banach space F. Assume the Fréchet derivative $\partial_y f(t, y, u)$ with respect to y exists, that $\partial_y f(t, y, u)z$ is continuous in all three variables for z fixed, and that $f(t, y, u), \partial_y f(t, y, u)$ are bounded for y on bounded subsets of E and $u \in U$. Let $C_{ad}(0, T; U)$ be the set of all strongly measurable U-valued functions taking values in U almost everywhere. Then $C_{ad}(0, T; U)$ and $f(t, y, u)$ satisfy all the assumptions in Theorem 6.5.1.

Example 6.5.9. Prove the minimum principle using Example 6.2.11 on spike perturbations and generalizing the arguments in Theorem 3.7.1.

Example 6.5.10. The present proof of Theorem 6.5.1 makes almost irresistible to formulate (and prove) a version of the minimum principle for a time-dependent control constraint $u(t) \in U(t)$. Here, $\{U(t); 0 \le t \le T\}$ is a family of arbitrary sets (a set-valued function in the terminology of Part III), and $C_{ad}(0, T; U(\cdot))$ is a space of functions defined in $0 \le t \le T$ with $u(t) \in U(t)$. Assumptions I, II and III on the control system (6.5.1) stand formulation with such a control space, as does the construction of patch perturbations in **6.2** for arbitrary $\bar{u}(\cdot) \in C_{ad}(0, T; U(\cdot))$, $\mathbf{v}(\cdot) = (v_1(\cdot), v_2(\cdot), \dots, v_m(\cdot)), v_j(\cdot) \in C_{ad}(0, T; U(\cdot))$. Patch completeness and stationarity are extended verbatim, the latter implying completeness of the space $C_{ad}(0, T; U(\cdot))$ equipped with the distance (6.5.2). The argument in Theorem 6.5.1 can be applied, and (6.5.4) is then obtained. If we wish to proceed to the pointwise minimum principle

$$\langle \bar{z}(s), f(s, y(s, \bar{u}), \bar{u}(s)) \rangle = \min_{v \in U(s)} \langle \bar{z}(s), f(s, y(s, \bar{u}), v) \rangle, \qquad (6.5.20)$$

we need E to be separable and the following very *ad hoc*

Assumption. There exists a set $e \subseteq [0, T]$ of full measure such that, for every $v \in U(s)$ $(s \in e)$ there exists a control $v(\cdot) \in C_{ad}(0, T; U(\cdot))$ with $v(s) = v$ and such that s is a left Lebesgue point of $f(\tau, y(\tau, \bar{u}), v(\tau))$.

Remark 6.5.11. Separability is unnecessary; in fact, since E is a Hilbert space, E^* is a Gelfand space (see **5.0**, especially Theorem 5.0.14) and every weakly measurable E^*-valued function $z(\cdot)$ is equivalent to a strongly measurable function $g(\cdot)$ in the sense that $\langle z(t), y \rangle = \langle g(t), y \rangle$ a.e. for every $y \in E$ (see Example 12.2.13); this implies that $\langle z(t), f(t) \rangle$ is measurable for any strongly measurable $f(\cdot)$.

6.6. The Minimum Principle for General Control Problems

The assumptions on the control system

$$y'(t) = A(t)y(t) + f(t, y(t), u(t)), \quad y(0) = \zeta \tag{6.6.1}$$

are the same in **6.5**. The target condition is $y(\bar{t}) \in Y$ and the cost functional

$$y_0(t, u) = \int_0^t f_0(\tau, y(\tau, u), u(\tau))d\tau + g_0(t, y(t, u)), \tag{6.6.2}$$

where $f_0 : [0, T] \times E \times U \to \mathbb{R}$; for each $u(\cdot) \in C_{\mathrm{ad}}(0, T; U)$, the function $f_0(t, y) = f_0(t, y, u(t))$ satisfies

Hypothesis III^0. The Fréchet derivative $\partial_y f_0(t, y)$ exists in $[0, T] \times E$. The function $f_0(t, y)$ is measurable in t for y fixed and continuous in y for t fixed. For each $z \in E$, $\partial_y f(t, y)z$ is measurable in t for y fixed and continuous in y for t fixed. For every $c > 0$ there exists $K_0(\cdot, c), L_0(\cdot, c) \in L^1(0, T)$ such that

$$|f_0(t, y)| \leq K_0(t, c) \quad (0 \leq t \leq T, \|y\| \leq c) \tag{6.6.3}$$

$$\|\partial_y f_0(t, y)\|_{E^*} \leq L_0(t, c) \quad (0 \leq t \leq T, \|y\| \leq c). \tag{6.6.4}$$

Finally, we assume that the Fréchet derivative $\partial_y g_0(t, y)$ exists in $[0, T] \times E$.

Lemma 6.6.1. *Let $\bar{u} \in C_{\mathrm{ad}}(0, \bar{t}; U)$ be such that $y(t, \bar{u})$ exists in $0 \leq t \leq \bar{t}$. Then the sets $e_j(h)$ in Theorem 6.2.7 can be chosen in such a way that*

$$\left| \frac{1}{h} \int_{(0,t) \cap e_j(h)} \left\{ f_0(\tau, y(\tau, \bar{u}), v_j(\tau)) - f_0(\tau, y(\tau, \bar{u}), \bar{u}(\tau)) \right\} d\tau \right.$$

$$\left. - p_j \int_0^t \left\{ f_0(\tau, y(\tau, \bar{u}), v_j(\tau)) - f_0(\tau, y(\tau, \bar{u}), \bar{u}(\tau)) \right\} d\tau \right| \leq h \tag{6.6.5}$$

and we have

$$\xi_0(t, \bar{u}, \mathbf{p}, \mathbf{v}) = \lim_{h \to 0+} \frac{1}{h} \left(y_0(t, \bar{u}_{\mathbf{e}(h), \mathbf{v}}) - y_0(t, \bar{u}) \right) \tag{6.6.6}$$

uniformly in $0 \leq t \leq \bar{t}$, where

$$\xi_0(t, \bar{u}, \mathbf{p}, \mathbf{v}) = \int_0^t \left\langle \partial_y f_0(\tau, y(\tau, \bar{u}), \bar{u}(\tau)), \xi(\tau, \bar{u}, \mathbf{p}, \mathbf{v}) \right\rangle d\tau$$

$$+ \sum_{j=1}^m p_j \int_0^t \left\{ f_0(\tau, y(\tau, \bar{u}), v_j(\tau)) - f_0(\tau, y(\tau, \bar{u}), \bar{u}(\tau)) \right\} d\tau$$

$$+ \left\langle \partial_y g_0(t, y(t, \bar{u})), \xi(t, \bar{u}, \mathbf{p}, \mathbf{v}) \right\rangle. \tag{6.6.7}$$

Proof. A direct proof in the style of Theorem 2.4.1 is not difficult to do. A more cunning approach is to expand the space E to $\mathbb{R} \times E$ and the family $\{A(t)\}$ to $\{0, A(t)\}$, so that the expanded solution operator is $\mathbf{S}(t, s) = (I, S(t, s))$; then the statement of Lemma 6.6.1 is the "first coordinate" of the corresponding statement in Theorem 6.2.7.   ∎

We assume below that $f(t, y, u(t))$ (resp. $f_0(t, y, u(t))$) satisfies Hypothesis III (resp. Hypothesis III^0) in $0 \leq t \leq \bar{t}$ with $K(\cdot, c)$, $L(\cdot, c)$ (resp. $K_0(\cdot, c)$, $L_0(\cdot, c)$) independent of $u(\cdot) \in C_{\mathrm{ad}}(0, \bar{t}; U)$. Regarding the separability assumption, see Remark 6.5.11.

Theorem 6.6.2. (*The minimum principle for general control problems.*) *Assume that the target set Y is closed and that $C_{\mathrm{ad}}(0, T; U)$ is patch complete and saturated. Then, if $\bar{u}(\cdot)$ is an optimal control in $0 \leq \bar{t} \leq \bar{t}$ there exists $(z_0, z) \in \mathbb{R} \times E$, $z_0 \geq 0$, $z \in N_Y(\bar{y}) = $ normal cone to Y at $\bar{y} = y(\bar{t}, \bar{u})$ such that, if $z(s)$ is the solution of*

$$\bar{z}'(s) = -\{A(s)^* + \partial_y f(s, y(s, \bar{u}), \bar{u}(s))^*\}\bar{z}(s) - z_0 \partial_y f_0(s, y(s, \bar{u}), \bar{u}(s)),$$

$$\bar{z}(\bar{t}) = z + z_0 \partial_y g_0(\bar{t}, y(\bar{t}, \bar{u})) \tag{6.6.8}$$

in $0 \leq s \leq \bar{t}$ then

$$z_0 \int_0^{\bar{t}} \{f_0(\sigma, y(\sigma, \bar{u}), v(\sigma)) - f_0(\sigma, y(\sigma, \bar{u}), \bar{u}(\sigma))\}d\sigma$$

$$+ \int_0^{\bar{t}} \langle \bar{z}(\sigma), f(\sigma, y(\sigma, \bar{u}), v(\sigma)) - f(\sigma, y(\sigma, \bar{u}), \bar{u}(\sigma)) \rangle d\sigma \geq 0 \tag{6.6.9}$$

for all $v(\cdot) \in C_{\mathrm{ad}}(0, \bar{t}; U)$. If E is separable, $f_0(t, y(t, \bar{u}), v)$ and $f(t, y(t, \bar{u}), v)$ are regular in U and $C_{\mathrm{ad}}(0, \bar{t}; U)$ is spike complete, then

$$z_0 f_0(s, y(s, \bar{u}), \bar{u}(s)) + \langle z(s), f(s, y(s, \bar{u}), \bar{u}(s)) \rangle$$

$$= \min_{v \in U}\{z_0 f_0(s, y(s, \bar{u}), v) + \langle z(s), f(s, y(s, \bar{u}), v) \rangle\} \tag{6.6.10}$$

a.e. in $0 \leq s \leq \bar{t}$.

Proof. As in **6.5**, "solution of (6.6.8)" is understood in the spirit of Remark 5.5.11; $\bar{z}(s)$ is defined by formula (5.5.38), in this case

$$\bar{z}(s) = S(\bar{t}, s; \bar{u})^*\{z + z_0 \partial g_0(\bar{t}, y(\bar{t}, \bar{u}))\}$$

$$+ z_0 \int_s^{\bar{t}} S(\sigma, s; \bar{u})^* \partial_y f_0(\sigma, y(\sigma, \bar{u}), \bar{u}(\sigma))d\sigma, \tag{6.6.11}$$

the integral understood as in the comments following (5.5.38) and enjoying the properties granted by Theorem 5.5.12, among them weak continuity and the bound (5.5.40). To apply Theorem 5.5.12 we must show that the inhomogeneous term $\partial_y f_0(s, y(s, \bar{u}), \bar{u}(s))$ (which takes values in $E^* = E$) is weakly measurable and bounded in norm by an integrable function. This is satisfied to excess since $\partial_y f_0(s, y(s, \bar{u}), \bar{u}(s))$ is assumed strongly measurable in Hypothesis III^0 and

$$\|\partial_y f_0(s, y(s, \bar{u}), \bar{u}(s))\| \le L_0(t, c), \tag{6.6.12}$$

c a bound for $\|y(s, \bar{u})\|$ in $0 \le s \le \bar{t}$.

To prove Theorem 6.6.2 we apply Theorem 6.4.2 in $V = B(\bar{u}, \delta)$, $B(\bar{u}, \delta)$ the ball in Lemma 6.1.1. The functions f, f_0 in the abstract minimization problem are

$$\mathbf{f}(u) = y(\bar{t}, u), \qquad \mathbf{f}_0(u) = y_0(\bar{t}, u) \tag{6.6.13}$$

and the "suboptimal solution" is $\{\bar{u}^n\}$, $\bar{u}^n = \bar{u}$. That $\mathbf{f}$ is continuous has been proved in Lemma 6.1.1. The corresponding result for $\mathbf{f}_0$ is

Lemma 6.6.3. *Let $\bar{u}, \delta$ be as in Lemma* 6.1.1. *Then the map*

$$u \to y_0(\cdot, u) \tag{6.6.14}$$

is continuous from $B(\bar{u}, \delta)$ into $C(0, \bar{t})$.

Its proof is exactly the same as that of Lemma 4.2.2 and thus omitted. All assumptions verified, Theorem 6.4.2 produces sequences $\{\bar{u}^n(\cdot)\} \subseteq C_{\mathrm{ad}}(0, \bar{t}; U)$ with $d(\bar{u}^n(\cdot), \bar{u}(\cdot)) \to 0$, $\{\tilde{y}^n\} \subseteq Y$ with $\tilde{y}^n \to \bar{y} = y(\bar{t}, \bar{u})$ and $\{(z_{0n}, z_n)\} \in \mathbb{R} \times E$ such that $z_{0n}^2 + \|z_n\|^2 = 1$, $z_{0n} \ge 0$ and (6.4.14) holds for any directional derivative $(\xi_0^n, \xi^n) \in \mathrm{Der}\,(f_0, f)(\bar{u}^n)$ and $w^n \in K_y(\tilde{y}^n)$. To pass to (6.4.15) we use the following addendum to Lemma 6.5.2.

Lemma 6.6.4. *Let $\bar{u} \in C_{\mathrm{ad}}(0, \bar{t}; U)$ be such that $y(t, \bar{u})$ exists in $0 \le t \le \bar{t}$, $\{u^n\} \subseteq C_{\mathrm{ad}}(0, \bar{t}; U)$ with $d(u^n, \bar{u}) \to 0$. Then*

$$\xi_0(t, u^n, \mathbf{p}, \mathbf{v}) \to \xi_0(t, \bar{u}, \mathbf{p}, \mathbf{v}), \qquad \xi(t, u^n, \mathbf{p}, \mathbf{v}) \to \xi(t, \bar{u}, \mathbf{p}, \mathbf{v}) \tag{6.6.15}$$

uniformly in $0 \le t \le \bar{t}$.

The second limit is that in Lemma 6.5.2. We can prove the first directly from (6.6.7) or both at once using the "expansion-of-the-space" trick in Lemma 6.6.1. ∎

End of Proof of Theorem 6.6.2. Using Lemma 6.6.4 we obtain

$$z_0 \xi_0(\bar{t}, \bar{u}, v) + \langle z, \xi(\bar{t}, \bar{u}, v) \rangle \ge 0, \tag{6.6.16}$$

where (z_0, z) is the weak limit of (a subsequence of) $\{(z_{0n}, z_n)\}$ and we have used the shorthand in **6.5** for one-patch perturbations. Using the variation-of-constants formula and (6.6.7), this inequality becomes

$$z_0 \int_0^{\bar{t}} \left\{ f_0(\sigma, y(\sigma, \bar{u}), v(\sigma)) - f_0(\sigma, y(\sigma, \bar{u}), \bar{u}(\sigma)) \right\} d\sigma$$

$$+ z_0 \int_0^{\bar{t}} \left\langle \partial_y f_0(\tau, y(\tau, \bar{u}), \bar{u}(\tau)), \right.$$

$$\left. \int_0^{\tau} S(\tau, \sigma, \bar{u}) \{ f(\sigma, y(\sigma, \bar{u}), v(\sigma)) - f(\sigma, y(\sigma, \bar{u}), \bar{u}(\sigma)) \} d\sigma \right\rangle d\tau$$

$$+ z_0 \left\langle \partial_y g_0(\bar{t}, y(\bar{t}, \bar{u})), \right.$$

$$\left. \int_0^{\bar{t}} S(\bar{t}, \sigma; \bar{u}) \{ f(\sigma, y(\sigma, \bar{u}), v(\sigma)) - f(\sigma, y(\sigma, \bar{u}), \bar{u}(\sigma)) \} d\sigma \right\rangle$$

$$+ \left\langle z, \int_0^{\bar{t}} S(\bar{t}, \sigma; \bar{u}) \{ f(\sigma, y(\sigma, \bar{u}), v(\sigma)) - f(\sigma, y(\sigma, \bar{u}), \bar{u}(\sigma)) \} d\sigma \right\rangle \geq 0.$$
$$(6.6.17)$$

In the third and fourth terms, we simply pull the integral to the left of the duality product $\langle \cdot, \cdot \rangle$ and then switch $S(\bar{t}, \sigma; \bar{u})$ to the other side taking adjoints (as we did in the proof of Theorem 6.5.1). In the second term we interchange order of integration on the basis of Lemma 5.5.14 (which is just cut to measure for this task). Using then (6.6.11), we obtain the integral form (6.6.9) of the maximum principle. We go from there to (6.6.10) as in the proof of Theorem 6.5.1.   ∎

Lemma 6.6.5. *Let $\{\tilde{u}_n\}, \{\tilde{y}^n\}$ be the sequences in Theorem 6.6.2. Assume that there exist $\rho > 0$ and a precompact sequence $\{Q_n\}, Q_n \subseteq E$ such that*

$$\bigcap_{n=n_0}^{\infty} \left\{ \bar{t}^{-1} R(0, \bar{t}; U, \tilde{u}^n) - K_Y(\tilde{y}^n) \cap B(0, \rho) + Q_n \right\} \tag{6.6.18}$$

contains an interior point for n_0 large enough. Then the multiplier (z_0, z) in Theorem 6.6.2 is not zero.

Proof. Same as that of Lemma 6.5.3; this time we use Corollary 6.4.5. See also Remark 6.5.7.   ∎

Remark 6.6.6. Theorem 6.6.2 does not distinguish between the fixed endpoint and variable endpoint problems. In finite dimensional problems, the difference is detected by the vanishing of the Hamiltonian (Theorem 4.6.1). In infinite dimensional problems, we run into trouble even defining the Hamiltonian $H(t, \mathbf{y}, \mathbf{z}, u) = H(t, (y_0, y), (z_0, z), u)$, which should be

$$H(t, \mathbf{y}, \mathbf{z}, u) = z_0 f_0(t, y, u) + \langle z, Ay \rangle + \langle z, f(t, y, u) \rangle. \tag{6.6.19}$$

In fact, the term $\langle z, Ay \rangle$ may not be defined for a solution $y(t, u)$ of (6.6.1) ($y(t)$ may not belong to the domain of A anywhere). "Putting A on the other side" may not work either: the solution $\bar{z}(t)$ of (6.6.8) may not belong to $D(A^*)$. Thus, vanishing of the Hamiltonian can only be proved in some particular cases. See Miscellaneous Comments to Part II.

Example 6.6.7. Example 6.5.5 has a counterpart; if $A(t) \equiv A =$ infinitesimal generator of a strongly continuous semigroup and E is separable, then the costate $\bar{z}(s)$ is continuous. The argument is the same; the inhomogeneous term $s \to \partial_y f_0(s, y(s, \bar{u}), \bar{u}(s))$ satisfies the assumptions of Lemma 5.2.7.

Example 6.6.8. Let $f(t, y, u)$ and $C_{ad}(0, T; U)$ be as in Example 6.5.8. Let $f_0(t, y, u)$ be defined and continuous in all three variables in $[0, T] \times E \times U$, and assume the Fréchet derivative $\partial_y f_0(t, y, u)$ with respect to y exists, that $\partial_y f_0(t, y, u)z$ is continuous in all three variables for z fixed and that $f_0(t, y, u)$, $\partial_y f_0(t, y, u)$ are bounded for y on bounded subsets of E and $u \in U$. Then all the assumptions in Theorem 6.6.2 are satisfied.

Example 6.6.9. Prove the minimum principle using Example 6.2.11 on spike perturbations and generalizing Theorem 4.2.1.

Example 6.6.10. The present proof of the minimum principle generalizes to time-dependent control constraints $u(t) \in U(t)$. For the statement of Assumption III in this context and of the notion of stationarity, see Example 6.5.10. Without further assumptions, the integral form (6.6.9) of the minimum principle is obtained. The pointwise statement

$$z_0 f_0(s, y(s, \bar{u}), \bar{u}(s)) + \langle z(s), f(s, y(s, \bar{u}), \bar{u}(s)) \rangle$$
$$= \min_{v \in U(s)} \{ z_0 f_0(s, y(s, \bar{u}), v) + \langle z(s), f(s, y(s, \bar{u}), v) \rangle \} \quad (6.6.20)$$

needs the following companion of the assumption in Example 6.5.10.

Assumption. There exists a set $e \subseteq [0, \bar{t}]$ of full measure such that, for every $v \in U(s)$ ($s \in e$) there exists a control $v(\cdot) \in C_{ad}(0, T; U(\cdot))$ with $v(s) = v$ and such that s is a left Lebesgue point of $f(\tau, y(\tau, \bar{u}), v(\tau))$ and $f_0(\tau, y(\tau, \bar{u}), v(\tau))$.

6.7. Optimal Problems for Some Linear and Semilinear Equations

To see what can be done with the minimum principle, we look at examples, the first one

$$y'(t) = Ay(t) + f(t, y(t)) + Bu(t), \quad y(0) = \zeta, \quad (6.7.1)$$

with control constraint $u(t) \in U \subseteq E$ and target condition $y(\bar{t}) \in Y$ in a real separable Hilbert space E. The operator A generates a strongly continuous semigroup

$\{S(t); t \geq 0\}$, $B : E \to E$ is linear and bounded, U is bounded and closed and Y is closed. The function $f(t, y)$ satisfies Hypothesis III in $0 \leq t \leq T$. The admissible control space $C_{ad}(0, T; U)$ consists of all strongly measurable E-valued functions such that $u(t) \in U$ almost everywhere; $C_{ad}(0, T; U)$ is patch and spike complete and saturated. Existence questions are postponed until Part III of this work.

We obtain straight from Theorem 6.5.1 the following

Theorem 6.7.1. *Assume that $\bar{u}(\cdot)$ is a time optimal control. Then there exists* $z \in N_Y(\bar{y}) = $ *normal cone to Y at $\bar{y} = y(\bar{t}, \bar{u})$ such that*[1]

$$(B^* S(\bar{t}, s; \bar{u})^* z, \bar{u}(s)) = \min_{v \in U}(B^* S(\bar{t}, s; \bar{u})^* z, v) \qquad (6.7.2)$$

a.e. in $0 \leq s \leq \bar{t}$, where $S(t, s; \bar{u})$ is the solution operator of the variational equation

$$\xi'(t) = \{A + \partial_y f(t, y(t, \bar{u}))\}\xi(t). \qquad (6.7.3)$$

The multiplier z is not zero if the target set is K-full (see the comments after Corollary 6.4.5). If Y is not K-full (in particular, in the point target case $Y = \{\bar{y}\}$) we use Lemma 6.5.3. Given $\tilde{u} \in C_{ad}(0, \tilde{t}; U)$ such that $y(t, \tilde{u})$ exists in $0 \leq t \leq \tilde{t}$, the reachable space $R(0, \tilde{t}; U, \tilde{u})$ of the inhomogeneous variational equation

$$\xi'(t) = \{A + \partial_y f(t, y(t, \tilde{u}))\}\xi(t) + B(u(t) - \tilde{u}(t)), \quad \xi(0) = 0 \qquad (6.7.4)$$

consists of all elements of the form

$$\xi(\tilde{t}, v) = \int_0^{\tilde{t}} S(\tilde{t}, \tau; \tilde{u}) Bu(\tau) d\tau - \int_0^{\tilde{t}} S(\tilde{t}, \tau; \tilde{u}) B\tilde{u}(\tau) d\tau$$

$$= \int_0^{\tilde{t}} S(\tilde{t}, \tau; \tilde{u}) Bu(\tau) d\tau - \tilde{\xi}, \qquad (6.7.5)$$

with $u(\cdot) \in C_{ad}(0, T; U)$. If the assumptions of Lemma 6.5.3 are to be satisfied (with $Q_n \equiv \emptyset$), single reachable spaces $R(0, \tilde{t}; U, \tilde{u})$ should contain interior points. As seen in Chapter 11, this is never the case for parabolic equations, and the compact sets Q_n don't help. On the other hand,

Lemma 6.7.2. *Assume that A is the infinitesimal generator of a strongly continuous group $\{S(t); -\infty < t < \infty\}$, that B is invertible, that 0 is an interior point of the control set U, and let $\bar{u}(\cdot) \in C_{ad}(0, \tilde{t}; U)$ be such that $y(t, \bar{u})$ exists in $0 \leq t \leq \tilde{t}$. Then, if $\{t_n\} \subseteq [0, \tilde{t}]$ with $t_n \to \tilde{t}$ and $\tilde{u}^n \in C_{ad}(0, t_n; U)$ are such that $d_n(\tilde{u}^n, \bar{u}) \to 0$, the spaces $R(0, t_n, U, \tilde{u}^n)$ contain a common ball for n large enough.*[2]

---

[1] When in a Hilbert space, there are two forms of the minimum principle: one uses the duality pairing $\langle \cdot, \cdot \rangle$ of E and E^* and the other the scalar product $(\cdot, \cdot)$. Of course, the adjoint operators in the adjoint variational equation are different. If the space is real, both forms coincide.

[2] In the next few lines, the reader is cautioned against confusing the operator B (applied to something) with balls $B(0, r)$.

Proof. We use (6.7.5) for $\tilde{t} = t_n, \tilde{u} = \tilde{u}^n$,

$$\xi(t_n, v) = \int_0^{t_n} S(t_n, \tau; \tilde{u}^n) Bu(\tau) d\tau - \tilde{\xi}_n \qquad (6.7.6)$$

with the obvious definition of $\tilde{\xi}_n$. Extend $\tilde{u}^n$ to $t_n \leq t \leq \tilde{t}$ setting $\tilde{u}^n(t) = u =$ fixed element of U there. Lemma 6.1.1 implies that $y(t, \tilde{u}^n)$ exists in $0 \leq t \leq \tilde{t}$ for n large enough. Since $S(t)$ is a group, the solution operator $S(t, s; \tilde{u}^n)$ is defined for all $s, t \in [0, \tilde{t}]$ (Remark 5.5.7), and $\|S(t, s; \tilde{u}^n)\| \leq M$ independently of n. Let $B(0, \varepsilon)$ be a ball of positive radius contained in U. Then the control $u_n(\tau) = B^{-1} S(\tau, t_n; \tilde{u}^n) y$ belongs to $C_{ad}(0, t_n; U)$ if $\|y\| \leq \varepsilon / M \|B^{-1}\|$. We have

$$\xi(t_n, v_n) = \int_0^{t_n} S(t_n, \tau; \tilde{u}^n) Bu_n(\tau) d\tau - \tilde{\xi}_n$$

$$= \int_0^{t_n} S(t_n, \tau; \tilde{u}^n) S(\tau, t_n; \tilde{u}^n) y d\tau - \tilde{\xi}_n = t_n y - \tilde{\xi}_n, \qquad (6.7.7)$$

so that $t_n^{-1} R(0, t_n; U, \tilde{u}^n) \supseteq B(0, \varepsilon / M \|B^{-1}\|) - t_n^{-1} \tilde{\xi}_n$. We use Lemma 5.3.1 on strongly continuous dependence of $S(t, s; \tilde{u})$ on all parameters to show that $\tilde{\xi}_n \to \bar{\xi}, \bar{\xi}$ defined from (6.7.5) with $\tilde{u} = \bar{u}$. This ends the proof. ∎

Assume we have shown (either by Lemma 6.7.2 or by fullness of the target set) that $z \neq 0$ in Theorem 6.7.1. Let $B = I$ and $U = B(0, 1)$. Then we obtain from the maximum principle (6.7.2) that

$$\bar{u}(t) = -\frac{S(\bar{t}, t; \bar{u})^* z}{\|S(\bar{t}, t; \bar{u})^* z\|} \qquad (6.7.8)$$

(recall that each $S(\bar{t}, s; \bar{u})$ is invertible if A generates a group, so that $S(\bar{t}, s; \bar{u})^* z \neq 0$ for all t). This is not an explicit formula for the optimal control; $\bar{u}$ is on both sides. However, it gives nontrivial information. For instance, existence theorems only produce measurable optimal controls; on the other hand, (6.7.8) implies that the optimal control $\bar{u}(\cdot)$ is "as smooth as" $S(\bar{t}, s; \bar{u})$.

Example 6.7.3. Consider the control system

$$y'(t) = Ay(t) + u(t) \quad y(0) = \zeta \qquad (6.7.9)$$

where A is *skew adjoint*, so that the semigroup $S(\cdot)$ generated by A is unitary $(S(t)^* = S(t)^{-1} = S(-t)$; see **5.0**). We consider the time optimal problem with control constraint $\|u(t)\| \leq 1$, point target condition $y(\bar{t}) = \bar{y} \in E$ and $C_{ad}(0, T; U)$ as above. Theorem 6.7.1 and Lemma 6.7.2 apply and there are extra dividends. According to (6.7.8) (normalizing z if necessary),

$$\bar{u}(t) = S(\bar{t} - t)^* z = S(t - \bar{t}) z \qquad (6.7.10)$$

with an unknown z with $\|z\| = 1$. It follows that $S(\bar{t} - t)\bar{u}(t) \equiv z$, hence the trajectory is $y(t, \bar{u}) = S(t)\zeta + tz$, and the target condition

$$S(\bar{t})\zeta + \bar{t}z = \bar{y}. \tag{6.7.11}$$

Taking norms we get an equation for the optimal time,

$$\bar{t} = \|\bar{y} - S(\bar{t})\zeta\|, \tag{6.7.12}$$

which must have at least one solution since $\|\bar{y} - S(t)\zeta\|$ is nonnegative and bounded. Once $\bar{t}$ has been found from (6.7.12), we obtain z from (6.7.11) and then the (unique) optimal control from (6.7.10). Of course, if (6.7.12) has more than one solution, the optimal time must be the least of these. For $\zeta = 0$ we simply have $\bar{t} = \|\bar{y}\|, z = \bar{y}/\|\bar{y}\|$.

Linear control problems do not need the full power of the minimum principle; we can use separation of convex sets. Given two sets A, B in a real Banach space E we say that a functional $z \in E^*$ **separates** A and B if $z \neq 0$ and

$$\langle z, x \rangle \leq \langle z, y \rangle \quad (x \in A, y \in B). \tag{6.7.13}$$

We say that z **separates strictly** A and B if $z \neq 0$ and

$$\langle z, y \rangle + \varepsilon \leq \langle z, z' \rangle \quad (x \in A, y \in B) \tag{6.7.14}$$

for some $\varepsilon > 0$. Geometrically, to separate A and B means to put them on opposite sides of a hyperplane $\langle z, y \rangle = a$; to separate strictly A and B means to put them on opposite sides of the "hyperstrip" $a \leq \langle z, y \rangle \leq a + \varepsilon$. It follows from the

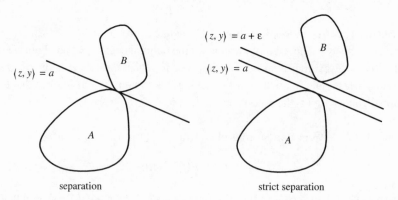

Figure 6.2.

forthcoming Theorem 12.0.7 that disjoint convex sets A, B can be separated by a nonzero functional if one of them contains interior points. This implies that if A is closed and $B = \{y_0\}$ with $y_0 \notin A$ we can separate A from a ball containing y_0, which gives strict separation of A and $\{y_0\}$. For details, see **12.0**.

Example 6.7.4. Consider the control system

$$y'(t) = Ay(t) + u(t), \quad y(0) = \zeta \tag{6.7.15}$$

in a Banach space E with A the generator of a strongly continuous group $S(t)$ and U a closed convex set with the origin as an interior point. Then, if $\bar{u}(\cdot)$ is a time optimal control for a point target condition, there exists $z \in E$, $\|z\| = 1$ such that

$$\langle S(\bar{t} - t)^*z, \bar{u}(t) \rangle = \max_{v \in U} \langle S(\bar{t} - t)^*z, v \rangle \tag{6.7.16}$$

a.e. in $0 \leq t \leq \bar{t}$. In fact, let $R(t)$ be the set of all elements of the form

$$y(t, u) = S(t)\zeta + \int_0^t S(t - \tau)u(\tau)d\tau \tag{6.7.17}$$

for $u(\cdot) \in C_{\mathrm{ad}}(0, T; U)$. Each $R(t)$ is convex and closed. Select a sequence $\{t_n\}$ with $t_n < \bar{t}, t_n \to \bar{t}$. Obviously, $\bar{y} = y(\bar{t}, \bar{u})$ cannot belong to $R(t_n)$, thus there exists z_n, $\|z_n\| = 1$ such that

$$\langle z_n, y(t_n, u) \rangle \leq \langle z_n, y(\bar{t}, \bar{u}) \rangle \quad (u(\cdot) \in C_{\mathrm{ad}}(0, \bar{t}; U)). \tag{6.7.18}$$

Select a subsequence (same name) such that $z_n \to z$ weakly. If $z = 0$ then it follows from (6.7.18) that $\langle z_n, y(t_n, u) \rangle \leq \delta_n \to 0$ $(u \in C_{\mathrm{ad}}(0, \bar{t}; U))$, thus a contradiction ensues using Lemma 6.4.3 and the fact (particular case of Lemma 6.7.2) that the $R(t_n)$ contain a common ball for n large enough. It follows then that $z \neq 0$ and, taking limits in (6.7.18) we obtain $\langle z, y(\bar{t}, u) \rangle \leq \langle z, y(\bar{t}, \bar{u}) \rangle$ $(u(\cdot) \in C_{\mathrm{ad}}(0, \bar{t}; U))$, or

$$\int_0^{\bar{t}} \langle S(t - \tau)^*z, u(\tau) \rangle d\tau \leq \int_0^{\bar{t}} \langle S(t - \tau)^*z, \bar{u}(\tau) \rangle d\tau \tag{6.7.19}$$

which translates into (6.7.16). Note that E is a Banach rather than a Hilbert space so that, at this stage, Example 6.7.4 is not a particular case of the general minimum principle (6.7.2).

This separation argument was used by Bellman–Glicksberg–Gross [1956] in finite dimensional spaces, where the argument on z being nonzero is moot since one can use the Bolzano–Weierstrass theorem on the sequence $\{z_n\}$. For more sophisticated applications of separation arguments in infinite dimensional spaces, see Chapters 9 and 11.

Remark 6.7.5. The existence arguments in **2.2** extend to systems like (6.7.15), in fact to the more general model $y'(t) = Ay(t) + f(t, y(t)) + F(t, y(t))u(t)$, with special results in the linear case $y'(t) = Ay(t) + Bu(t)$. For these and other existence questions see Chapter 14.

6.8. Semilinear Wave Equations Again

We return to the semilinear wave equation

$$y_{tt}(t, x) = \sum_{j=1}^{m} \sum_{k=1}^{m} \partial^j (a_{jk}(x) \partial^k y(t, x))$$

$$+ \sum_{j=1}^{m} b_j(x) \partial^j y(t, x) + c(x) y(t, x) - \phi(t, x, y(t, x)) + u(t, x) \qquad (6.8.1)$$

in a domain $\Omega \subseteq \mathbb{R}^m$. Assumptions on the coefficients a_{jk}, b_j, c are the same as in **5.6** and **5.7**. If we use the Dirichlet boundary condition, Ω is totally arbitrary. For a variational boundary condition, Ω is of class $C^{(1)}$ with bounded boundary. As in **5.7**, we reduce (6.8.1) to the abstract semilinear equation

$$\mathbf{y}'(t) = \mathbf{A}(\beta) \mathbf{y}(t) + \mathbf{f}(t, \mathbf{y}(t)) \qquad (6.8.2)$$

for $\mathbf{y}(t) = (y(t, \cdot), y_1(t, \cdot)) = (y(t, \cdot), y_t(t, \cdot)) \in \mathbf{E}$, where $\mathbf{E} = H_0^1(\Omega) \times L^2(\Omega)$ for the Dirichlet boundary condition, $\mathbf{E} = H^1(\Omega) \times L^2(\Omega)$ for a variational boundary condition. The operators are

$$\mathbf{A}(\beta) = \begin{bmatrix} 0 & I \\ A(\beta) & 0 \end{bmatrix}, \qquad (6.8.3)$$

$$\mathbf{f}(t, \mathbf{y})(x) = \mathbf{f}(t, (y, y_1))(x) = \begin{bmatrix} 0 \\ -\phi(t, x, y(x)) + u(t, x) \end{bmatrix}. \qquad (6.8.4)$$

We assume that $\phi(t, x, y)$ is continuously differentiable in $\mathbb{R}$ with respect to y and that $(t, x) \to \phi(t, x, y)$ and $(t, x) \to \partial_y \phi(t, x, y)$ are measurable for y fixed. The rest, as in **5.7**, depends on the space dimension m.

Dimension $m \geq 3$: Let $\rho = m/(m - 2)$. There exist $K(\cdot), L(\cdot) \in L^1(0, T)$ and $\xi(\cdot) \in L^2(\Omega), \eta(\cdot) \in L^m(\Omega)$ such that

$$|\phi(t, x, y)| \leq K(t)(\xi(x) + |y|^\rho) \qquad (0 \leq t \leq T, x \in \Omega, y \in \mathbb{R}), \qquad (6.8.5)$$

$$|\partial_y \phi(t, x, y)| \leq L(t)(\eta(x) + |y|^{\rho-1}) \qquad (0 \leq t \leq T, x \in \Omega, y \in \mathbb{R}). \qquad (6.8.6)$$

Dimension $m = 2$: There exist $\rho > 1$, $K(\cdot), L(\cdot) \in L^1(0, T)$ and $\xi(\cdot) \in L^2(\Omega)$, $\eta(\cdot) \in L^{2\rho/(\rho-1)}(\Omega)$ such that (6.8.5) and (6.8.6) hold.

Dimension $m = 1$: There exist $\xi(\cdot), \eta(\cdot) \in L^2(\Omega)$ such that for every $c > 0$ there exist $K(\cdot, c), L(\cdot, c) \in L^1(0, T)$ with

$$|\phi(t, x, y)| \leq K(t, c)\xi(x) \qquad (0 \leq t \leq T, x \in \Omega, y \in \mathbb{R}), \qquad (6.8.7)$$

$$|\partial_y \phi(t, x, y)| \leq L(t, c)\eta(x) \qquad (0 \leq t \leq T, x \in \Omega, y \in \mathbb{R}). \qquad (6.8.8)$$

The control $u(t, x)$ is measurable in $(0, T) \times \Omega$ and, for almost all $t \in [0, T]$ $u(t, \cdot) \in L^2(\Omega)$ and $t \to \|u(t, \cdot)\|_{L^2(\Omega)}$ is in $L^1(0, T)$.

Theorem 6.8.1. $\mathbf{f}(t, \mathbf{y})$ *satisfies Hypothesis III in* **6.2**.

Proof. Inequality (6.8.6) and the mean value theorem imply the Lipschitz condition (5.7.6) so that everything done in **5.7** can be used here. Again, the term $u(t, \cdot)$ satisfies Hypothesis III, adds an integrable function to (6.2.6) and nothing to (6.2.7), thus we may and will assume that $u = 0$.

We show existence of $\partial_y \mathbf{f}(t, \mathbf{y})$. Formally, we have

$$\partial_y \mathbf{f}(t, \mathbf{y})(x) = \partial_y \mathbf{f}(t, (y, y_1))(x) = \begin{bmatrix} 0 & 0 \\ -\partial_y \phi(t, x, y(x)) & 0 \end{bmatrix}. \tag{6.8.9}$$

Call $\partial(t, \mathbf{y})$ the linear operator defined by the matrix (6.8.9). We check first (beginning with the case $m = 3$) that $\partial(t, \mathbf{y})$ defines a bounded linear operator in $\mathbf{E}$ or, equivalently, that the operator of multiplication by $\partial_y \phi(t, x, y(x))$ is a bounded operator from $H^1(\Omega)$ into $L^2(\Omega)$. For $y(\cdot), z(\cdot) \in H^1(\Omega)$ we have

$$\|\partial_y \phi(t, \cdot, y(\cdot))z(\cdot)\|_{L^2(\Omega)} = \left(\int_\Omega |\partial_y \phi(t, x, y(x))|^2 |z(x)|^2 dx \right)^{1/2}$$

$$\leq L(t) \left(\int_\Omega \left(\eta(x) + |y(x)|^{2/(m-2)} \right)^m dx \right)^{1/m} \left(\int_\Omega |z(x)|^{2m/(m-2)} dx \right)^{(m-2)/2m}$$

$$\leq L(t) \left\{ \|\eta\|_{L^m(\Omega)} + \| |y|^{2/(m-2)} \|_{L^m(\Omega)} \right\} \|z\|_{L^{2m(m-2)}(\Omega)}$$

$$\leq L(t) \left\{ \|\eta\|_{L^m(\Omega)} + \left(\|y\|_{L^{2m(m-2)}(\Omega)} \right)^{2/(m-2)} \right\} \|z\|_{L^{2m(m-2)}(\Omega)}$$

$$\leq CL(t) \left\{ \|\eta\|_{L^m(\Omega)} + C \left(\|y\|_{H^1(\Omega)} \right)^{2/(m-2)} \right\} \|z\|_{H^1(\Omega)} \tag{6.8.10}$$

where we have applied Hölder's inequality for $p = m/2, q = m/(m - 2)$ and exploited the continuous imbedding $L^{2m/(m-2)}(\Omega) \subseteq H^1(\Omega)$ (see **5.7**). Boundedness of $\partial(t, \mathbf{y})$ out of the way we show that for t fixed, $\partial(t, \mathbf{y})$ depends continuously on $\mathbf{y}$ in the operator norm. Using Hölder's inequality as above

$$\|\{\partial_y \phi(t, \cdot, y(\cdot) + h(\cdot)) - \partial_y \phi(t, \cdot, y(\cdot))\}z(\cdot)\|_{L^2(\Omega)}$$

$$= \left(\int_\Omega |\partial_y \phi(t, x, y(x) + h(x)) - \partial_y \phi(t, x, y(x))|^2 |z(x)|^2 dx \right)^{1/2}$$

$$\leq \left(\int_\Omega |\partial_y \phi(t, x, y(x) + h(x)) - \partial_y \phi(t, x, y(x))|^m dx \right)^{1/m}$$

$$\times \left(\int_\Omega |z(x)|^{2m/(m-2)} dx \right)^{(m-2)/2m}$$

$$\leq C \left(\int_\Omega |\partial_y \phi(t, x, y(x) + h(x)) - \partial_y \phi(t, x, y(x))|^m dx \right)^{1/m} \|z(\cdot)\|_{H^1(\Omega)}$$

$$= C \|\partial_y \phi(t, \cdot, y(\cdot) + h(\cdot)) - \partial_y \phi(t, \cdot, y(\cdot))\|_{L^m(\Omega)} \|z(\cdot)\|_{H^1(\Omega)}, \tag{6.8.11}$$

and we must show that

$$\|\partial_y\phi(t,\cdot,y(\cdot)+h(\cdot)) - \partial_y\phi(t,\cdot,y(\cdot))\|_{L^m(\Omega)} \to 0 \qquad (6.8.12)$$

as $h(\cdot) \to 0$ in $H^1(\Omega)$, which we do as in the proof of Theorem 5.7.4. Assume that (6.8.12) is bogus. Then there exists a sequence $\{h_n(\cdot)\} \subseteq H^1(\Omega)$ such that $h_n(\cdot) \to 0$ in $H^1(\Omega)$ and (6.8.12) does not tend to zero. Selecting if necessary a subsequence we may assume that $h_n(x) \to 0$ almost everywhere, so that

$$g_n(x) = |\partial_y\phi(t,x,y(x)+h_n(x)) - \partial_y\phi(t,x,y(x))|^m \to 0 \qquad (6.8.13)$$

almost everywhere. We shall obtain a contradiction (to Vitali's Theorem 2.0.4) if we show that the sequence $\{g_n(\cdot)\}$ has equicontinuous integrals in $x \in \Omega$, and (in case Ω is unbounded) that for every $\varepsilon > 0$ there exists $e \subseteq \Omega$ with $|e| < \infty$ and $\int_{\Omega\setminus e} g_n(x)dx \le \varepsilon$ for all n. This second condition is obvious from the bound

$$g_n(x) \le CL(t)\big(\eta(x)^m + |y(x)|^{2m/(m-2)} + |h_n(x)|^{2m/(m-2)}\big) \qquad (6.8.14)$$

coming from (6.8.6) (see Remark 5.7.5). It suffices to estimate the integral of $|h_n(x)|^{2m/(m-2)}$ for $n \ge$ a sufficiently large n_0 by its integral in all of Ω and then select the set e for the functions $\eta(x)^m$, $|y(x)|^{2m/(m-2)}$, $|h_n(x)|^{2m/(m-2)} (n \le n_0)$. The same trick shows equicontinuity and completes the proof of $\mathbf{y}$-continuity of $\partial(t,\mathbf{y})$. It also shows that $\|\partial(t,\mathbf{y})\|$ is locally bounded.

We verify finally that $\partial(t,\mathbf{y}) = \partial_y\mathbf{f}(t,\mathbf{y})$. Take $y(\cdot), h(\cdot) \in H^1(\Omega)$. By the mean value theorem,

$$\phi(t,x,y(x)+h(x)) - \phi(t,x,y(x)) - \partial_y\phi(t,x,y(x))h(x)$$

$$= \int_0^1 \big\{\partial_y\phi(t,x,y(x)+\tau h(x)) - \partial_y\phi(t,x,y(x))\big\}h(x)d\tau. \quad (6.8.15)$$

We estimate the integrand as in (6.8.11):

$$\|\{\partial_y\phi(t,\cdot,y(\cdot)+\tau h(\cdot)) - \partial_y\phi(t,\cdot,y(\cdot))\}h(\cdot)\|_{L^2(\Omega)}$$

$$\le C\|\partial_y\phi(t,\cdot,y(\cdot)+\tau h(\cdot)) - \partial_y\phi(t,\cdot,y(\cdot))\|_{L^m(\Omega)}\|h(\cdot)\|_{H^1(\Omega)}.$$

Taking norms in (6.8.15) we then obtain

$$\|\phi(t,\cdot,y(\cdot)+h(\cdot)) - \phi(t,\cdot,y(\cdot)) - \partial_y\phi(t,\cdot,y(\cdot))h(\cdot)\|_{L^2(\Omega)}$$

$$\le \int_0^1 \big\|\{\partial_y\phi(t,\cdot,y(\cdot)+\tau h(\cdot)) - \partial_y\phi(t,\cdot,y(\cdot))\}h(\cdot)\big\|_{L^2(\Omega)}d\tau$$

$$\le \|h(\cdot)\|_{H^1(\Omega)} \int_0^1 \big\|\partial_y\phi(t,\cdot,y(\cdot)+\tau h(\cdot)) - \partial_y\phi(t,\cdot,y(\cdot))\big\|_{L^m(\Omega)}d\tau, \quad (6.8.16)$$

and it only remains to show that the integral on the right-hand side of (6.8.16) tends to zero as $\|h\|_{H^1(\Omega)} \to 0$. That the integrand converges to zero for all τ follows from (6.8.12). We then use local boundedness of $\partial(t, \mathbf{y})$ to bound the integrand, and the dominated convergence theorem. To complete the proof of Hypothesis III it only remains to check that $t \to \partial_y \mathbf{f}(t, \mathbf{y})\mathbf{z}$ is strongly measurable in $\mathbf{E}$ for $\mathbf{y}, \mathbf{z}$ fixed or, equivalently, that $t \to \partial_y \phi(t, \cdot, y(\cdot))z(\cdot)$ is strongly (or just weakly) measurable in $L^2(\Omega)$. This is done as in the proof of weak measurability of $t \to \phi(t, y(\cdot))$ in **5.7** and we omit the details. This ends the proof of Theorem 6.8.1 for $m = 3$. The proof for $m = 2$ is the same using the ersatz dimension $m' = 2\rho/(\rho - 1)$, and we leave the proof for $m = 1$ to the reader. ∎

6.9. The Time Optimal Problem for a Semilinear Wave Equation, I: Dampening Nonlinear Vibrations in Minimum Time

The control system is

$$y_{tt}(t, x) = \sum_{j=1}^{m} \sum_{k=1}^{m} \partial^j (a_{jk}(x)\partial^k y(t, x))$$

$$+ \sum_{j=1}^{m} b_j(x)\partial^j y(t, x) + c(x)y(t, x) - \phi(t, x, y(t, x)) + u(t, x) \quad (6.9.1)$$

and all assumptions the same in the last section. We consider two different time optimal problems: one with control constraint

$$\int_\Omega |u(t, x)|^2 dx \leq 1 \quad \text{a.e. in } 0 \leq t \leq \bar{t} \quad (6.9.2)$$

and the other (when Ω is bounded) with control constraint

$$|u(t, x)| \leq 1 \quad \text{a.e. in } 0 \leq t \leq \bar{t}, x \in \Omega. \quad (6.9.3)$$

We write (6.9.1) in the customary vector form

$$\mathbf{y}'(t) = \mathbf{A}(\beta)\mathbf{y}(t) + \mathbf{f}(t, \mathbf{y}(t)) + \mathbf{B}u(t), \quad (6.9.4)$$

where

$$\mathbf{B} = \begin{bmatrix} 0 \\ I \end{bmatrix}.$$

For the constraint (6.9.2) the control space is the unit ball

$$U = U_2 = \{u \in L^2(\Omega), \|u\| \leq 1\} \quad (6.9.5)$$

and, since $L^2((0, T) \times \Omega) = L^2(0, T; L^2(\Omega))$ (Example 5.0.32), $C_{ad}(0, T; U)$ consists of all strongly measurable $L^2(\Omega)$-valued functions defined in $0 \leq t \leq T$ such that $u(t) \in U$ a.e. For the constraint (6.9.3) we take

$$U = U_\infty = \{u \in L^2(\Omega); |u(x)| \leq 1 \text{ a.e. in } \Omega\}. \tag{6.9.6}$$

Again, $C_{ad}(0, T; U)$ consists of all strongly measurable $L^2(\Omega)$-valued functions defined in $0 \leq t \leq T$ such that $u(t) \in U$ a.e. It is clear that in both cases $U = U_2$, U_∞, the space $C_{ad}(0, T; U)$ is patch and spike complete and saturated. The assumptions on $\phi(t, x, y)$ are the same in **6.8**, and we have shown that $\mathbf{f}(t, \mathbf{y}, u)(x) = \mathbf{f}(t, \mathbf{y}) + \mathbf{B}u$ satisfies Hypothesis III with $K(\cdot, c), L(\cdot, c)$ independent of $u \in C_{ad}(0, T; U)$. For the pointwise maximum principle, it remains to show that $\mathbf{f}(t, \mathbf{y}(t, \bar{u}), u)$ is regular in U, which means that, given $\mathbf{y} \in \mathbf{E}$ there should exist a set d of full measure in $0 \leq t \leq T$ such that each point in d is a Lebesgue point of the $\mathbf{E}$-valued function $t \to \mathbf{f}(t, \mathbf{y}, u)$. Since we don't know $\mathbf{y}(t, \bar{u})$, we do the checking with an arbitrary continuous $\mathbf{E}$-valued function $\mathbf{y}(t)(\cdot) = (y(t, \cdot), y_1(t, \cdot))$. Regularity of $t \to \mathbf{f}(t, \mathbf{y}(t), u)$ is equivalent to regularity of the $L^2(\Omega)$-valued function $t \to \phi(t, y(t, \cdot)) + u(\cdot)$, thus we may take d as the set of all left Lebesgue points of $t \to \phi(t, y(t, \cdot))$.

Finally, the target set $\mathbf{Y} \in \mathbf{E}$ is assumed closed. With all hypotheses checked, we have

Theorem 6.9.1. *Let* $\bar{u} = \bar{u}(\cdot, \cdot)$ *be a time optimal control. Then there exists* $\mathbf{z} \in N_Y(\bar{\mathbf{y}}) =$ *normal cone to* $\mathbf{Y}$ *at* $\bar{\mathbf{y}} = \bar{\mathbf{y}}(\bar{t}, \bar{u})$ *such that, if* $\bar{\mathbf{z}}(s)$ *is the solution of*

$$\bar{\mathbf{z}}'(s) = -\{\mathbf{A}(\beta)^* + \partial_y \mathbf{f}(s, \mathbf{y}(s, \bar{u}))^*\}\bar{\mathbf{z}}(s), \quad \mathbf{z}(\bar{t}) = \mathbf{z} \tag{6.9.7}$$

in $0 \leq s \leq \bar{t}$, *then*

$$(\mathbf{B}^*\bar{\mathbf{z}}(s), \bar{u}(s)) = \min_{v \in U}(\mathbf{B}^*\bar{\mathbf{z}}(s), v). \tag{6.9.8}$$

The question of whether $\mathbf{z} \neq 0$ is open, and it will be decided in the following sections via Lemma 6.5.3 and results on the reachable spaces $R(0, \tilde{t}; U, \tilde{u})$ of the nonhomogeneous variational equation

$$\xi'(t) = \{\mathbf{A}(\beta) + \partial_y \mathbf{f}(t, \mathbf{y}(t, \tilde{u}))\}\xi(t) + \mathbf{B}(u(t) - \tilde{u}(t)), \quad \xi(0) = 0. \tag{6.9.9}$$

For the time being, we simply recall that if the set $\mathbf{Y}$ is K-full in $\mathbf{E}$, then $\mathbf{z} \neq 0$ (Corollary 6.4.4). This allows, for instance, the target set

$$\mathbf{Y} = \mathbf{B}(0, \varepsilon), \tag{6.9.10}$$

the ball of center 0 and (arbitrarily small) positive radius ε. We may also impose a finite number of additional conditions

$$\int_\Omega \{\phi_j(x)y(\bar{t}, x) + \psi_j(x)y_t(\bar{t}, x)\}dx = 0 \quad (j = 1, 2, \ldots, n) \tag{6.9.11}$$

on the final value $y(\bar{t}, \cdot)$ of the solution, where the $\phi_j(\cdot)$ and the $\psi_j(\cdot)$ belong to $L^2(\Omega)$. We can apply Lemma 6.4.9 to show that the set chiseled out from $\mathbf{B}(0, \varepsilon)$ by the exact target conditions (6.9.11) is T-full if the (ϕ_j, ψ_j) are linearly independent as linear functionals in $\mathbf{E}$; a sufficient condition for this is that (ϕ_j, ψ_j) be linearly independent in $L^2(\Omega) \times L^2(\Omega)$. We note that the ϕ_j can be chosen in the larger space $H^{-1}(\Omega)$ (see Example 6.10.6).

When $A = \Delta$, $b_j = c = 0$, (6.9.1) is an actual wave equation, and the problem is that of controlling the vibrations of an oscillating nonlinear string $(m = 1)$, membrane $(m = 2)$ or body $(m = 3)$ (however, control applied on the whole body may be unrealistic in the latter case). The target set (6.9.10) corresponds to damping the oscillations to a prescribed small energy level, while (6.9.11) imposes exact damping according to the observations of a finite number of integral sensors. Using as kernels of the sensors the functions $(\phi_j, 0)$ and $(0, \phi_j)$, ϕ_j the eigenfunctions of the Laplacian, (6.9.11) requires exact vanishing of a finite number of "vibration modes." We shall see in **6.12** that for the control constraint (6.9.2) we can actually use a point target.

To understand the minimum principle (6.9.8), we must obtain an independent equation for the second coordinate $\mathbf{B}^*\bar{\mathbf{z}}(s)$ of the solution of (6.9.7). This is done in next section.

6.10. Some Remarks on Adjoint Equations

We digress on the linear second order equation

$$y''(t) = (A_0 + P(t))y(t) + u(t) \tag{6.10.1}$$

in a Hilbert space H. The operator A_0 is self adjoint with spectrum in $(-\infty, -\kappa]$ $(\kappa > 0)$ so that negative fractional powers of $-A_0$ are bounded self adjoint operators. We set $R = (-A_0)^{1/2}$, $H^1 = D(R)$ equipped with the norm $\|y\|_1 = \|Ry\|$ and assume that each $P(t) : H^1 \to H$ is bounded, that $t \to P(t)y$ is strongly measurable for each $y \in H^1$ and that there exists $\alpha(\cdot) \in L^1(0, T)$ such that

$$\|P(t)\|_{(H^1, H)} \le \alpha(t) \quad (0 \le t \le T). \tag{6.10.2}$$

Of course, we have in mind the variational equation (6.9.9). There, A_0 is the operator $A_0(\beta)$ generated by

$$A_0 y(x) = \sum_{j=1}^{m} \sum_{k=1}^{m} \partial^j (a_{jk}(x)\partial^k y(x)) - \lambda y(x) \tag{6.10.3}$$

and the boundary condition β (see **5.6**), and H^1 is either $H_0^1(\Omega)$ or $H^1(\Omega)$ depending on the boundary condition. We have $P(t) = P + P_1(t)$, where

$$Py(x) = \sum_{j=1}^{m} b_j(x)\partial^j y(x) + c(x)y(x) + \lambda y(x) \tag{6.10.4}$$

and
$$P_1(t)y(x) = \partial_y \phi(t, x, \eta(t, x))y(x) \qquad (6.10.5)$$

with $t \to \eta(t, \cdot)$ a continuous $H^1(\Omega)$-valued function. We have shown in **6.8** that $P_1(t)y$ is strongly measurable in $0 \le t \le T$ for all $y \in H^1(\Omega)$ and that (6.10.2) holds; P is bounded and independent of t.

Arguing as in **5.6** we reduce the abstract equation (6.10.1) to the system

$$\mathbf{y}'(t) = (\mathbf{A}_0 + \mathbf{P}(t))\mathbf{y}(t) + \mathbf{B}u(t) \qquad (6.10.6)$$

for $\mathbf{y}(t) = (y(t), y'(t)) \in \mathbf{E} = H^1 \times H$. The inner product in the space $\mathbf{E}$ is $(\mathbf{y}, \mathbf{z}) = ((y, y_1), (z, z_1)) = (Ry, Rz) + (y_1, z_1)$, corresponding to the norm $\|\mathbf{y}\| = \|(y, y_1)\| = (\|Ry\|^2 + \|y_1\|)^{1/2}$; $\mathbf{E}$ is a Hilbert space.[1] In matrix form the operators are

$$\mathbf{A}_0 = \begin{bmatrix} 0 & I \\ A_0 & 0 \end{bmatrix} \qquad (6.10.7)$$

with domain $D(\mathbf{A}_0) = D(A_0) \times E_1$, and

$$\mathbf{P}(t) = \begin{bmatrix} 0 & 0 \\ P(t) & 0 \end{bmatrix} \qquad (6.10.8)$$

$$\mathbf{B} = \begin{bmatrix} 0 \\ I \end{bmatrix}. \qquad (6.10.9)$$

The two latter operators are bounded and everywhere defined, and $\|\mathbf{P}(t)\| \le C\alpha(t)$. Using Theorem 5.6.4 we prove that $\mathbf{A}_0$ is skew adjoint, thus it generates a unitary group $\mathbf{S}_0(t)$ (explicitly given by formula (5.6.28)). We then obtain from Theorem 5.5.6 and Remark 5.5.7 the following.

Theorem 6.10.1. *The Cauchy problem for*

$$\mathbf{y}'(t) = (\mathbf{A}_0 + \mathbf{P}(t))\mathbf{y}(t) \qquad (6.10.10)$$

is well posed (forwards and backwards) in $0 \le t \le T$.

Of course, the result holds in intervals other than $[0, T]$, in particular in $(-\infty, \infty)$ if (6.10.2) holds with a locally integrable function. The same observation applies to other results below.

Since no claims are made on t-smoothness of $P(t)$, solutions of (6.10.10) will in general be no more than continuous in $\mathbf{E}$; this includes continuity of $Ry(t)$ and of $y'(t)$ as H-valued functions.

[1] The reader is again cautioned that parentheses are used to indicate elements in a Cartesian product and also scalar product. Also, note that $y \in H^1$ and $y_1 \in H$ (our apologies).

We figure out the adjoint equation. $\mathbf{A}_0$ is skew adjoint, so that $\mathbf{A}_0^* = -\mathbf{A}_0$. Consider $P(t)$ as an operator from H into H (say, with domain H^1). Since $P(t)$ is densely defined, its adjoint $P(t)^*$ is well defined. Now, it follows from the second relation (5.0.25) that $R^{-1}P(t)^* \subseteq (P(t)R^{-1})^*$; since $P(t)R^{-1}$ is bounded, $R^{-1}P(t)^*$ is bounded as well, and possesses a bounded extension to all of H, namely $(P(t)R^{-1})^*$. On the other hand, if $\mathbf{y}, \mathbf{z} \in \mathbf{E}$ we have

$$
\begin{aligned}
(\mathbf{z}, \mathbf{P}(t)\mathbf{y}) &= ((z, z_1), \mathbf{P}(t)(y, y_1)) = ((z, z_1), (0, P(t)y)) = (z_1, P(t)y) \\
&= (z_1, P(t)R^{-1}Ry) = ((P(t)R^{-1})^* z_1, Ry) \\
&= ((R^{-1}(P(t)R^{-1})^* z_1, 0), (y, y_1)),
\end{aligned}
$$

thus we have shown that

$$
\mathbf{P}(t)^* = \begin{bmatrix} 0 & R^{-1}(P(t)R^{-1})^* \\ 0 & 0 \end{bmatrix}. \tag{6.10.11}
$$

Accordingly,

$$
(\mathbf{A}_0 + \mathbf{P}(t))^* = \begin{bmatrix} 0 & -I + R^{-1}(P(t)R^{-1})^* \\ -A_0 & 0 \end{bmatrix}. \tag{6.10.12}
$$

Note that we have avoided on the one hand, giving $P(t)$ its maximal domain, and on the other hand, computing the domain of $P(t)^*$ or even showing that it is dense in E. These two tasks are harder than they seem due to lack of smoothness of the coefficients b_j, c.[2]

The reversed (or backwards) adjoint equation is

$$
\mathbf{z}'(s) = -(\mathbf{A}_0 + \mathbf{P}(s))^* \mathbf{z}(s) = (\mathbf{A}_0 - \mathbf{P}(s)^*)\mathbf{z}(s). \tag{6.10.13}
$$

Theorem 6.10.2. *Let H be separable. The Cauchy problem for (6.10.13) is well posed (forward and backwards) in $0 \le s \le T$.*

Proof. The operator function $\mathbf{P}(s)^*$ qualifies for Theorem 5.5.6; $\|\mathbf{P}(s)^*\| = \|\mathbf{P}(s)\| \le C\alpha(s)$ and strong measurability of $s \to \mathbf{P}(s)^* \mathbf{z}$ follows from measurability of $s \to (\mathbf{y}, \mathbf{P}(s)^* \mathbf{z}) = (\mathbf{P}(s)\mathbf{y}, \mathbf{z})$; in fact, we are in a separable Hilbert space and weak and strong measurability coincide (note that $\mathbf{E}$ is separable if H is). ∎

Coordinate by coordinate, (6.10.13) is the system

$$
z'(s) = (I - R^{-1}(P(s)R^{-1})^*)z_1(s), \qquad z_1'(s) = A_0 z(s), \tag{6.10.14}
$$

[2] On the other hand, there is never any doubt about the domain of $\mathbf{P}(t)^*$: $\mathbf{P}(t)$ is bounded.

which does not produce a simple second order equation for $z(s)$. However, in the minimum principle (6.9.8) we only use $\mathbf{B}^*(z(s), z_1(s))$. Since $(\mathbf{B}u, (z, z_1)) = ((0, u), (z, z_1)) = (u, z_1)$,

$$\mathbf{B}^*(z, z_1) = z_1 \tag{6.10.15}$$

so that only $z_1(s)$ is relevant. Formally, the equation for $z_1(s)$ is

$$z_1''(s) = (A_0 + R(P(s)R^{-1})^*)z_1(s), \tag{6.10.16}$$

or, even more formally,

$$z_1''(s) = (A_0 + P(s)^*)z_1(s). \tag{6.10.17}$$

Neither equation should be taken seriously; $z_1(t)$ will always be the second coordinate of the corresponding solution $\mathbf{z}(t)$ of (6.10.13). Since $\mathbf{P}(s)^*$ is not assumed smooth as a function of s, solutions of (6.10.13) are no more than continuous in $\mathbf{E}$ and, in particular, their second coordinates $z_1(s)$ will be no more than continuous in H. However, we have the following serious version of the second equation (6.10.14).

Lemma 6.10.3. *Let $(z(s), z_1(s))$ be a solution of (6.10.13). Then $A_0^{-1}z_1(s)$ is continuously differentiable in the norm of H^1 with*

$$\left(A_0^{-1}z_1(s)\right)' = z(s). \tag{6.10.18}$$

Proof. Use the variation-of-constants formula for $\mathbf{z}'(s) = \mathbf{A}_0\mathbf{z}(s) - \mathbf{P}(s)^*\mathbf{z}(s)$ considering $-\mathbf{P}(s)^*\mathbf{z}(s)$ as the inhomogeneous term and taking the initial condition at s_0 arbitrary:[3]

$$\mathbf{z}(s) = \mathbf{S}_0(s - s_0)\mathbf{z}(s_0) - \int_{s_0}^{s} \mathbf{S}_0(s - \sigma)\mathbf{P}(\sigma)^*\mathbf{z}(\sigma)d\sigma. \tag{6.10.19}$$

Then apply $\mathbf{A}_0^{-1}$ to both sides:

$$\mathbf{A}_0^{-1}\mathbf{z}(s) = \mathbf{S}_0(s - s_0)\mathbf{A}_0^{-1}\mathbf{z}(s_0) - \int_{s_0}^{s} \mathbf{S}_0(s - \sigma)\mathbf{A}_0^{-1}\mathbf{P}(\sigma)^*\mathbf{z}(\sigma)d\sigma. \tag{6.10.20}$$

The first term on the right-hand side of (6.10.20) is continuously differentiable with derivative $\mathbf{S}_0(s - s_0)\mathbf{z}(s_0)$. As for the integral, the function $\mathbf{S}_0(s)\mathbf{A}_0^{-1}$ is continuously differentiable in the norm of $(\mathbf{E}, \mathbf{E})$, thus we easily check that the integral is continuously differentiable (derivative in the sense of the norm of $\mathbf{E}$) and that

$$\left(\mathbf{A}_0^{-1}\mathbf{z}(s)\right)' = \mathbf{z}(s) - \mathbf{A}_0^{-1}\mathbf{P}(s)^*\mathbf{z}(s) = \left(\mathbf{I} - \mathbf{A}_0^{-1}\mathbf{P}(s)^*\right)\mathbf{z}(s). \tag{6.10.21}$$

---

[3] Since this is the adjoint variational equation, we are tempted to use the backwards variation-of-constant formula. Obviously, this makes no difference since the Cauchy problem is well posed forwards and backwards.

The inverse $\mathbf{A}_0^{-1}$ is

$$\mathbf{A}_0^{-1} = \begin{bmatrix} 0 & A_0^{-1} \\ I & 0 \end{bmatrix},$$

The first coordinate of $\mathbf{A}_0^{-1}\mathbf{z}(s)$ is $A_0^{-1}z_1(s)$, thus it follows from (6.10.21) that $A_0^{-1}z_1(s)$ is continuously differentiable in the norm of H^1. The first coordinate of the right-hand side is $z(s)$, so that (6.10.18) is proved. ∎

Corollary 6.10.4. *If* $(z(\cdot), z_1(\cdot))$ *is a solution of* (6.10.13) *not identically zero, the set of zeros of* $z_1(s)$ *in a finite interval* $0 \leq s \leq \bar{t}$ *is finite.*

Proof. Assume that $z_1(s)$ has an infinite sequence $\{s_n\}$ of different zeros; take a subsequence and assume that $s_n \to \tilde{s} \in [0, \bar{t}]$. Each s_n is a zero of $A_0^{-1}z_1(s)$ as well. By Lemma 6.10.3, $A_0^{-1}z_1(s)$ is continuously differentiable with derivative $z(s)$. By continuity, $A_0^{-1}z_1(\tilde{s}) = 0$; on the other hand,

$$(A_0^{-1}z_1(\tilde{s}))' = \lim_{n \to \infty} \frac{A_0^{-1}z_1(s_n) - A_0^{-1}z_1(\tilde{s})}{s_n - \tilde{s}} = 0.$$

Using (6.10.18) we deduce that $z(\tilde{s}) = z_1(\tilde{s}) = 0$; by uniqueness of solutions of (6.10.13), $z(s) \equiv z_1(s) \equiv 0$. ∎

Remark 6.10.5. Our handling of the equation (6.10.1) by means of the vector equation (6.10.6) and of the adjoint equation (6.10.16) (or (6.10.17)) by means of the vector equation (6.10.13) is obviously asymmetric. The equation (6.10.1) is read off the first coordinate of (6.10.6). In view of the fact that the Cauchy problem for the homogeneous equation (6.10.10) is well posed and of the variation-of-constants formula, (6.10.1) can be uniquely solved with initial conditions $y(s) = \zeta \in H^1$, $y'(s) = \zeta_1 \in H$ and $u(\cdot) \in L^1(0, T; H)$: the solution $y(t)$ is the first coordinate of the function

$$\mathbf{y}(t) = \mathbf{S}(t, s)\zeta + \int_s^t \mathbf{S}(t, \tau)\mathbf{B}u(\tau)d\tau, \tag{6.10.22}$$

with $\zeta = (\zeta, \zeta_1)$, where $\mathbf{S}(t, s)$ is the solution operator of (6.10.10). The solution $y(t)$ takes values in H^1 with derivative $y'(t)$ in H.

On the other hand, equation (6.10.16) comes off the second coordinate of (6.10.13), which means that $z_1(t)$ is merely continuous and takes values in H. Pressing questions are: What is the derivative $z_1'(s)$? Where is the second initial condition $z_1'(s) = z$? Answers to these can be worked out looking at Lemma 6.10.3. In fact, although $z_1'(s)$ may not exist, $(A_0^{-1}z_1(t))'$ does and equals $z(s)$. Thus, the missing initial condition is $(A_0^{-1}z_1(s))' = z(s)$ or, formally,

$$z_1'(s) = A_0 z(s), \tag{6.10.23}$$

which is purely formal since $z(s)$ belongs to H^1 but not necessarily to $D(A_0)$. However, (6.10.23) can be given literal sense as follows.

Example 6.10.6. The space H^{-1} is the completion of H under the norm $\|z\|_{-1} = \|R^{-1}z\|$ ($z \in H$); see Example 7.4.17 for the definition of completion. The operator R^{-1} can be extended to H^{-1} and $R^{-1} : H^{-1} \to H$ is an isometric isomorphism. A_0 is extended to H^{-1} by $A_0 = RA_0R^{-1}$. Equality (6.10.23) holds; $z(\cdot) \in D(A_0)$ and $z_1(\cdot)$ is continuously differentiable in the norm of H^{-1}.

6.11. Some Remarks on Controllability

We continue to work on the equation

$$y''(t) = (A_0 + P(t))y(t) + u(t) \tag{6.11.1}$$

under the assumptions in **6.10**.

Theorem 6.11.1. *The equation* (6.11.1) *is* **controllable** *in the following sense: for every* $\bar{t} > 0$ *and every* $\boldsymbol{\zeta} = (\zeta, \zeta_1)$, $\mathbf{y} = (y, y_1) \in \mathbf{E}$ *there exists a strongly measurable E-valued function* $u(t)$ *defined in* $0 \leq t \leq \bar{t}$ *with*

$$\|u(t)\| \leq M(\bar{t})(1 + \alpha(t))(\|\zeta\|_1 + \|\zeta_1\| + \|y\|_1 + \|y_1\|) \quad (0 \leq t \leq \bar{t}) \tag{6.11.2}$$

and such that the solution of (6.11.1) *with* $y(0) = \zeta$, $y'(t) = \zeta$ *satisfies*

$$y(\bar{t}) = y, \quad y'(\bar{t}) = y_1. \tag{6.11.3}$$

Here, $\alpha(\cdot)$ *is the function in* (6.10.2)*. The constant* $M(\bar{t})$ *does not depend on* ζ, ζ_1, y, y_1.

Proof. We begin with the particular case $P(t) \equiv 0$,

$$y''(t) = A_0y(t) + u(t), \tag{6.11.4}$$

where the bound is

$$\|u(t)\| \leq N(\bar{t})(\|\zeta\|_1 + \|\zeta_1\| + \|y\|_1 + \|y_1\|). \tag{6.11.5}$$

The proof is clumsy, but we don't know a better one. Let

$$C_0(t) = \cos(tR), \qquad S_0(t) = R^{-1}\sin(tR),$$

where $R = (-A_0)^{1/2}$; $C_0(t)$ and $S_0(t)$ are the two entries in the first row of the matrix of the semigroup $\mathbf{S}_0(t)$ generated by the operator $\mathbf{A}_0$ given by (6.10.7) in the space $\mathbf{E} = H^1 \times H$:

$$\mathbf{S}_0(t) = \exp(t\mathbf{A}_0) = \begin{bmatrix} \cos(tR) & R^{-1}\sin(tR) \\ -R\sin(tR) & \cos(tR) \end{bmatrix} \tag{6.11.6}$$

(see Example 5.6.9, in particular (5.6.28)). It follows from the functional calculus for the self adjoint operator R (or from parallel properties of the semigroup $\mathbf{S}_0(t)$) that $C_0(\cdot)$, $S_0(\cdot)$ are strongly continuous bounded operator functions with $C_0(0) = I$, $S_0(0) = 0$. Also, $S_0(t)H \subseteq D(R)$, $S_0(t)H^1 \subseteq D(A_0)$ and the estimates

$$\|C_0(t)\| \le 1, \quad \|S_0(t)\| \le 1/\kappa, \quad \|RS_0(t)\| \le 1 \quad (-\infty < t < \infty) \quad (6.11.7)$$

hold ($-\kappa$ an upper bound for $\sigma(A_0)$). Moreover, $C_0(\cdot)y$ and $S_0(\cdot)y$ are twice continuously differentiable for $y \in D(A_0)$ with $C_0''(t)y = A_0C_0(t)y$, $S_0''(t)y = A_0S_0(t)y$; $S_0(t)y$ is continuously differentiable for $y \in H$ and $S_0'(t)y = C_0(t)y$. It follows from this last inequality and a simple approximation argument that we have $\int_0^t S_0(\tau)y d\tau \in D(A_0)$ for every $y \in E$ and

$$A_0 \int_0^t S_0(\tau)y d\tau = C_0(t)y - y. \quad (6.11.8)$$

Finally, we have the **cosine** and **sine functional equations**, respectively

$$C_0(s+t) + C_0(s-t) = 2C_0(s)C_0(t) \quad (6.11.9)$$

$$S_0(s+t) + S_0(s-t) = 2S_0(s)C_0(t), \quad (6.11.10)$$

Before tackling the controllability problem, two observations are opportune. (a) To find a control $u(t)$ that drives from (ζ, ζ_1) to (y, y_1) in a time interval $0 \le t \le \bar{t}$ we may simply superpose two controls, one driving from (ζ, ζ_1) to $(0, 0)$ and the other from $(0, 0)$ to (y, y_1). (b) The equation is reversible (if $y(t)$ is a solution with control $u(t)$ then $y(-t)$ is a solution with control $u(-t)$). Thus, all we have to do is to drive from $(0, 0)$ to an arbitrary $(y, y_1) \in H^1 \times H$ in an arbitrary time interval $0 \le t \le \bar{t}$, with due attention paid to the bound (6.11.5). In view of formula (6.11.6) for the semigroup $\mathbf{S}_0(t)$, the expression (6.10.9) for **B** and the variation-of-constants formula, this means finding a control $u(\cdot)$ such that

$$\int_0^{\bar{t}} S_0(\bar{t} - \tau)u(\tau)d\tau = y, \qquad \int_0^{\bar{t}} C_0(\bar{t} - \tau)u(\tau)d\tau = y_1 \quad (6.11.11)$$

We do this in three steps. In the first, $y, y_1 \in D(A_0)$; in the second, $y, y_1 \in H^1$. Full generality ($y \in H^1$, $y_1 \in H$) is achieved in the third step.

Step 1. Let $y, y_1 \in D(A_0)$, and let $\phi(t)$, $\psi(t)$ be scalar functions, twice continuously differentiable in $0 \le t \le \bar{t}$, with

$$\phi(0) = 0 \quad \phi(\bar{t}) = -1 \quad \phi'(0) = 0 \quad \phi'(\bar{t}) = 0$$

$$\psi(0) = 0 \quad \psi(\bar{t}) = 0 \quad \psi'(0) = 0 \quad \psi'(\bar{t}) = -1$$

(for instance, the cubic interpolation polynomials $\phi(t) = t^2\{2\bar{t}^{-3}(t - \bar{t}) - \bar{t}^{-2}\}$, $\psi(t) = -\bar{t}^{-2}t^2(t - \bar{t})$). Define

$$u_1(t) = \phi(t)A_0y - \phi''(t)y + \psi(t)A_0y_1 - \psi''(t)y_1. \quad (6.11.12)$$

Then both equalities (6.11.11) hold, as we see integrating by parts twice. This control satisfies

$$\|u_1(t)\| \le K(\bar{t})(\|A_0 y\| + \|A_0 y_1\|). \tag{6.11.13}$$

Step 2. Let $z, z_1 \in H^1$, and let

$$u_2(t) = C_0(t)z + A_0 S_0(t)z_1. \tag{6.11.14}$$

Then we obtain, using the cosine and sine functional equations,

$$\int_0^{\bar{t}} S_0(\bar{t} - \tau)u_2(\tau)d\tau = C_0(\bar{t})(\bar{t}z_1/2) + S_0(\bar{t})(\bar{t}z/2) - S_0(\bar{t})(z_1/2), \tag{6.11.15}$$

$$\int_0^{\bar{t}} C_0(\bar{t} - \tau)u_2(\tau)d\tau = A_0 S_0(\bar{t})(\bar{t}z_1/2) + C_0(\bar{t})(\bar{t}z/2) + S_0(\bar{t})(z/2). \tag{6.11.16}$$

Let $y, y_1 \in H^1$. We claim that we can find $z, z_1 \in H^1$ such that

$$C_0(\bar{t})(\bar{t}z_1/2) + S_0(\bar{t})(\bar{t}z/2) = y,$$

$$A_0 S_0(\bar{t})(\bar{t}z_1/2) + C_0(\bar{t})(\bar{t}z/2) = y_1.$$

To see this, notice that, in view of (6.11.6), these two equations are equivalent to the single vector equation

$$\mathbf{S}_0(\bar{t})(\bar{t}z_1/2, \bar{t}z/2) = (y, y_1)$$

with solution $(z_1, z) = (2/\bar{t})\mathbf{S}_0(-\bar{t})(y, y_1)$. We can then find a control $u_2(\cdot)$ of the form (6.11.14) driving $(0, 0)$ to

$$(y - S_0(\bar{t})(z_1/2), y_1 + S_0(\bar{t})(z/2)).$$

In view of (6.11.6), if $y, y_1 \in H^1 = D(R)$ then $z, z_1 \in D(R)$ with $\|Rz\| \le C(\|Ry\| + \|Ry_1\|)$ and $\|Rz_1\| \le C(\|Ry\| + \|Ry_1\|)$; using the estimates (6.11.7) we obtain

$$\|u_2(t)\| \le L(\bar{t})(\|Ry\| + \|Ry_1\|). \tag{6.11.17}$$

We get rid of the "residual" $(-S_0(\bar{t})(z_1/2), S_0(\bar{t})(\bar{t}z/2))$ superposing a control of the form (6.11.12); this is possible since $S_0(t)D(R) \subseteq D(A_0)$. This auxiliary control also satisfies (6.11.17).

Step 3. For every $y \in E$ define

$$e(y) = \left\{ t \in [0, \bar{t}]; \|C_0(t)y\| < \frac{\|y\|}{3} \right\}.$$

Since $C_0(0)y = y$, this set may be empty if $\bar{t}$ is small; this we don't mind since we precisely want to show that $e(y)$ cannot be unduly large. In view of the cosine functional equation, $2C_0(t/2)^2 = C_0(t) + I$. On the other hand, $\|C_0(t/2)^2 y\| \leq \|C_0(t/2)\| \|C_0(t/2)y\| \leq \|C_0(t/2)y\|$, hence

$$\|C_0(t/2)y\| \geq \frac{\|y\| - \|C_0(t)y\|}{2} > \frac{\|y\|}{3}$$

which shows that if $t \in e(y)$ then $t/2 \notin e(y)$, that is, $e(y) \cap (e(y)/2) = \emptyset$; hence $|e(y)| + |e(y)/2| = 3|e(y)|/2 \leq \bar{t}$, or $|e(y)| \leq 2\bar{t}/3$. Thus, if we define

$$N(\bar{t})y = \int_0^{\bar{t}} C_0(t)^2 y \, dt,$$

the operator $N(\bar{t})$ is bounded and self adjoint. If $d(y) \subseteq [0, \bar{t}]$ is the complement of $e(y)$ in $[0, \bar{t}]$ we have

$$(N(\bar{t})y, y) = \int_0^{\bar{t}} \|C_0(t)y\|^2 dt \geq \int_{d(y)} \|C_0(t)y\|^2 dt \geq \frac{\bar{t}}{3} \left(\frac{1}{3}\right)^2 \|y\|^2 = \frac{\bar{t}}{27} \|y\|^2,$$

so that $N(\bar{t})$ is invertible with

$$\|N(\bar{t})^{-1}\| \leq \frac{27}{\bar{t}}. \tag{6.11.18}$$

We undertake the drive from $(0, 0)$ to $(y, y_1) \in \mathbf{E}$, that is, the construction of a control satisfying both equalities (6.11.11). The second is fulfilled if we take

$$u_3(\tau) = C_0(\bar{t} - \tau) N(\bar{t})^{-1} y_1.$$

We calculate the element z produced by this control in the first equality (6.11.11) using the sine functional equation (6.11.10):

$$\int_0^{\bar{t}} S_0(\bar{t} - \tau) u_3(\tau) d\tau = \int_0^{\bar{t}} S_0(\bar{t} - \tau) C_0(\bar{t} - \tau) N(\bar{t})^{-1} y_1 d\tau$$

$$= \frac{1}{2} \int_0^{\bar{t}} S_0(2\bar{t} - 2\tau) N(\bar{t})^{-1} y_1 d\tau = \frac{1}{4} A_0^{-1} (C_0(2\bar{t}) - I) N(\bar{t})^{-1} y_1 d\tau.$$

This shows that $z \in D(A_0)$ with $\|A_0 z\| \leq C \|y_1\|$. We add then to $u_3(\cdot)$ one of the controls constructed in Step II driving $(0, 0)$ to $(y - z, 0)$. Obviously, this control satisfies (6.11.5)

End of Proof of Theorem 6.11.1. We have

$$y''(t) = (A_0 + P(t))y(t) + u(t) = A_0 y(t) + (P(t)y(t) + u(t)) = A_0 y(t) + v(t).$$

Construct a control $v(\cdot)$ for the equation (6.11.4) steering (ζ, ζ_1) to (y, y_1) in time $\bar{t}$ and satisfying (6.11.2); then define $u(t) = v(t) - P(t)y(t)$. The first coordinate of the variation-of-constants formula reads

$$y(t) = C_0(t)\zeta + S_0(t)\zeta_1 + \int_0^t S_0(t - \tau)v(\tau)d\tau. \tag{6.11.19}$$

Applying $R = (-A_0)^{1/2}$ to both sides, we deduce that $Ry(t)$ is bounded by a multiple of the right side of (6.11.5); writing then $P(t)y(t) = (P(t)R^{-1})(Ry(t))$ we obtain

$$\|P(t)y(t)\| \le C\alpha(t)N(\bar{t})(\|\zeta\|_1 + \|\zeta_1\| + \|y\|_1 + \|y_1\|).$$

This completes the proof.   ∎

Theorem 6.11.1 refers to a single equation and a single time interval $0 \le t \le \bar{t}$. We need a "uniform" version, where both the equation and the time interval vary.

Corollary 6.11.2. *Let $\{P_\nu(t)\}$ be a family of operator-valued functions each satisfying the assumptions in* **6.10** *with $\alpha(\cdot)$ independent of ν, that is*

$$\|P_\nu(t)\|_{(H^1, H)} \le \alpha(\cdot) \quad (0 \le t \le T) \tag{6.11.20}$$

for all ν, and let $0 < a < b \le T$. Then Theorem 6.11.1 holds with (6.11.2) independent of ν and of $\bar{t} \in [a, b]$. In other words, given $\nu, \bar{t} \in [a, b]$, $\zeta = (\zeta, \zeta_1), \mathbf{y} = (y, y_1)$ there exists a strongly measurable E-valued function $u_{\nu,\bar{t}}(\cdot)$ defined in $0 \le t \le \bar{t}$ with

$$\|u_{\nu,\bar{t}}(t)\| \le M(\bar{t})(1 + \alpha(t))(\|\zeta\|_1 + \|\zeta_1\| + \|y\|_1 + \|y_1\|) \qquad (0 \le t \le \bar{t}) \tag{6.11.21}$$

such that the solution of

$$y''(t) = (A_0 + P_\nu(t))y(t) + u_{\nu,\bar{t}}(t) \tag{6.11.22}$$

with $y(0) = \zeta, y'(0) = \zeta_1$ satisfies

$$y(\bar{t}) = y, \qquad y'(\bar{t}) = y_1. \tag{6.11.23}$$

The constant $M(\bar{t})$ does not depend on ν or on $\bar{t} \in [a, b]$.

We prove Corollary 6.2.2 reviewing the proof of Theorem 6.11.1. We argue away first dependence on ν. The control $v(\cdot)$ constructed at the end of the proof of Theorem 6.11.1 is independent of ν; that $u_{\nu,\bar{t}}(t) = v(t) - P_\nu(t)y(t)$ satisfies (6.11.21) independently of ν is obvious. To check independence on $\bar{t}$ we go over the three steps in the proof of Theorem 6.11.1. In Step 1 it is plain that the specific ϕ, ψ suggested are bounded independently of $\bar{t} \in [a, b]$. In Step 2 note that the control (6.11.14) is bounded by $C/\bar{t}$ and apply the previous remark to the control used to dispose of the residual term $(-S_0(\bar{t})(z_1/2), S_0(\bar{t})(z/2))$. The argument for Step 3 is obvious.

6.12. The Time Optimal Problem for a Semilinear Wave Equation, II: Stopping Nonlinear Vibrations in Minimum Time

We go back to **6.9** to prove that the time optimal problem with control constraint (6.9.2) and a point target set $\mathbf{Y} = \{\bar{\mathbf{y}}\} = \{(y, y_1)\}$ admits a nontrivial minimum principle; this means that $\mathbf{B}^*\bar{\mathbf{z}}(s)$ in (6.9.8) is nonzero (except for finitely many values). The assumptions in **6.9** are slightly reinforced; we require $K(t) \equiv K$ and $L(t) \equiv L$ for $m \geq 2$ and $K(t, c) = K(c)$, $L(t, c) = L(c)$ for $m = 1$. The theory in **6.5** and **6.6** is applied to the vector alias of (6.9.1),

$$\mathbf{y}'(t) = \mathbf{A}(\beta)\mathbf{y}(t) + \mathbf{f}(t, \mathbf{y}(t)) + \mathbf{B}u(t). \tag{6.12.1}$$

The first result is a companion of Lemma 6.7.2 dealing with the inhomogeneous variational equation

$$\boldsymbol{\xi}'(t) = \{\mathbf{A}(\beta) + \partial_y \mathbf{f}(t, \mathbf{y}(t, \tilde{u}))\}\boldsymbol{\xi}(t) + \mathbf{B}v(t), \quad \boldsymbol{\xi}(0) = 0 \tag{6.12.2}$$

in $\mathbf{E} = H_0^1(\Omega) \times L^2(\Omega)$ or $H^1(\Omega) \times L^2(\Omega)$ depending on the boundary condition. The control $\tilde{u}$ belongs to the admissible control space $C_{\text{ad}}(0, \tilde{t}; U)$, consisting of all $L^2(\Omega)$-valued strongly measurable functions taking values in the unit ball U of $L^2(\Omega)$ (see **6.9**). We assume that $\tilde{u} \in C_{\text{ad}}(0, \tilde{t}; U)$ is such that the solution $\mathbf{y}(t, \tilde{u})$ of (6.12.1) exists in $0 \leq t \leq \tilde{t}$. The admissible control space for (6.12.2) consists of all controls

$$v(t) = u(t) - \tilde{u}(t) \quad (u(\cdot) \in C_{\text{ad}}(0, \tilde{t}; U)), \tag{6.12.3}$$

and the reachable space $R(0, \tilde{t}; U, \tilde{u}) \subseteq \mathbf{E}$ (at time $\tilde{t}$) of (6.12.2) consists of all elements $\boldsymbol{\xi}(\tilde{t}) = (\xi(\tilde{t}), \xi_1(\tilde{t}))$, where $\boldsymbol{\xi}(\cdot) = (\xi(\cdot), \xi_1(\cdot))$ is a solution of (6.12.2) with a control of the form (6.12.3).

Lemma 6.12.1. *Let* $\{t_n\} \subseteq \mathbb{R}_+$ *with* $t_n \to \tilde{t}$, $\tilde{u} \in C_{\text{ad}}(0, \tilde{t}; U)$ *such that* $\mathbf{y}(t, u)$ *exists in* $0 \leq t \leq \tilde{t}$, $\tilde{u}^n \in C_{\text{ad}}(0, t_n; U)$ *such that* $d_n(\tilde{u}^n, \tilde{u}) \to 0$. *Then the spaces* $R(0, t_n; U, \tilde{u}^n)$ *contain a common ball for n large enough.*

We prove this using the controllability results in **6.11**. It follows from Lemma 6.1.1 that $\mathbf{y}(t, \tilde{u}^n)$ exists in $0 \leq t \leq \tilde{t}$ for n large enough (we extend $\tilde{u}^n$ to $t_n \leq t \leq \tilde{t}$ setting $\tilde{u}^n(t) = u = $ fixed element of U there) and that $\{\mathbf{y}(\cdot, \tilde{u}^n)\}$ is bounded in $C(0, \tilde{t}; \mathbf{E})$. Then, after the work in **6.8** and the assumed time independence of $K(\cdot)$ and $L(\cdot)$,

$$\|\partial_y \mathbf{f}(t, \mathbf{y}(t, \tilde{u}^n))\| \leq C \quad (0 \leq t \leq \tilde{t}). \tag{6.12.4}$$

Let $\boldsymbol{\xi}(t) = (\xi(\cdot), \xi_1(\cdot))$ be a solution of (6.12.2) with a control of the form (6.12.3). Let $(\eta(\cdot), \eta_1(\cdot))$ (resp. $(\eta_n(\cdot), \eta_{n1}(\cdot))$, $(\bar{\eta}(\cdot), \bar{\eta}_1(\cdot))$) be the solution corresponding to $v(\cdot) = u(\cdot)$ (resp. to $v(\cdot) = \tilde{u}^n(\cdot)$, $v(\cdot) = \bar{u}(\cdot)$). Then we have

$$(\xi(\cdot), \xi_1(\cdot)) = (\eta(\cdot), \eta_1(\cdot)) - (\eta_n(\cdot), \eta_{n1}(\cdot))$$

and we show that

$$(\eta_n(\cdot), \eta_{n1}(\cdot)) \to (\bar{\eta}(\cdot), \bar{\eta}_1(\cdot)),$$

as in Lemma 6.7.2, using u-continuity of the solution operator of (6.12.1). On the other hand, using Corollary 6.11.2 and noting that $\alpha(t) \equiv \alpha$ is constant in the bound (6.11.21), we deduce that there exists $\varepsilon > 0$ such that for every $(\eta, \eta_1) \in \mathbf{E}$ with $\|(\eta, \eta_1)\| \le \varepsilon$ there exists $u(\cdot) \in C_{\mathrm{ad}}(0, \bar{t}; U)$ such that $(\eta(t_n), \eta_1(t_n)) = (\eta, \eta_1)$. It then follows that, for n_0 large enough, the intersection of the $R(0, t_n; U, \tilde{u}^n)$ for $n \ge n_0$ contains an open ball of positive radius and center $(\bar{\eta}(\bar{t}), \bar{\eta}_1(\bar{t}))$. This ends the proof. ∎

Remark 6.12.2. The inhomogeneous variational equation (6.12.2) is shorthand for the equation

$$\xi_{tt}(t, x) = \sum_{j=1}^{m} \sum_{k=1}^{m} \partial^j (a_{jk}(x) \partial^k \xi(t, x)) + \sum_{j=1}^{m} b_j(x) \partial^j \xi(t, x)$$
$$+ \{c(x) - \partial_y \phi(t, x, y(t, x))\} \xi(t, x) + v(t, x). \qquad (6.12.5)$$

Theorem 6.12.3. *Let the target set* **Y** *be closed but otherwise arbitrary. Let* $\bar{u}(\cdot) = \bar{u}(t, x)$ *be a time optimal control for the control constraint*

$$\int_{\Omega} |u(t, x)|^2 dx \le 1 \quad a.e. \text{ in } 0 \le t \le \bar{t}. \qquad (6.12.6)$$

Then there exists $\mathbf{z} = (z, z_1) \in N_Y(\bar{\mathbf{y}}) = $ *normal cone to* **Y** *at* $\bar{\mathbf{y}} = \bar{\mathbf{y}}(\bar{t}, \bar{u})$, $(z, z_1) \ne 0$ *such that if* $\bar{\eta}(t, x)$ *is the solution of*

$$\bar{\eta}_{ss}(s, x) = \sum_{j=1}^{m} \sum_{k=1}^{m} \partial^j \left(a_{jk}(x) \partial^k \bar{\eta}(s, x) \right)$$
$$- \sum_{j=1}^{m} \partial^j (b_j(x) \bar{\eta}(s, x)) + \{c(x) - \partial_y \phi(s, x, y(s, x, \bar{u}))\} \bar{\eta}(s, x)$$

$$(x \in \Omega, 0 \le s \le \bar{t}), \qquad (6.12.7)$$

$$\bar{\eta}(\bar{t}, x) = z_1(x), \qquad \bar{\eta}_t(\bar{t}, x) = A_0 z(x) \quad (x \in \Omega), \qquad (6.12.8)$$

then we have

$$\int_{\Omega} \bar{\eta}(s, x) \bar{u}(s, x) dx = \min_{v \in L^2(\Omega), \|v\| \le 1} \int_{\Omega} \bar{\eta}(s, x) v(x) dx. \qquad (6.12.9)$$

The function $s \to \bar{\eta}(s, \cdot) \in L^2(\Omega)$ *is nonzero for all s except possibly for a finite set* $\{s_j\}, s_1 < \cdots < s_n$, *so that, completing the set if necessary with* $s_0 = 0, s_{n+1} = \bar{t}$,

$$\bar{u}(s) = -\frac{\bar{\eta}(s)}{\|\bar{\eta}(s)\|} \quad (s_j < s < s_{j+1}). \tag{6.12.10}$$

Proof. The *actual* definition of $\bar{\eta}(s)$ is: $\bar{\eta}(s) = \bar{z}_1(s) =$ second coordinate of the solution of the adjoint variational equation (6.9.7). Equation (6.12.7) is (6.10.17) in longhand, and (the same as the latter) is interpreted in a purely formal way (see the comments after (6.10.17)). In the same vein, the taking of "final conditions" in (6.12.8) is formal, although it can be given literal sense (see Remark 6.10.5 and Example 6.10.6).

The fact that the multiplier (z, z_1) is nonzero follows from Lemma 6.12.1, which makes possible the application of Lemma 6.5.3 (the contingent cones $K_Y(\bar{y}^n)$ and the precompact sequence $\{Q_n\}$ can be thrown away). That $(z, z_1) \neq 0$ makes the solution $(\bar{z}(s), \bar{\eta}(s)) = (\bar{z}(s), \bar{z}_1(s))$ not identically zero and gives then opportunity for application of Corollary 6.10.4; we prove in this way our claim on nonvanishing of $\bar{\eta}(s)$. ∎

Corollary 6.12.4. *Let* $\bar{u}(\cdot)$ *be an optimal control. Then the constraint* (6.12.6) *is saturated, that is*

$$\int_\Omega |\bar{u}(s, x)|^2 dx = 1 \quad \text{a.e. in } 0 \le s \le \bar{t}. \tag{6.12.11}$$

Proof. Obvious from (6.12.9). ∎

In contrast, the results in **6.9** for the time-optimal problem with constraint

$$|u(t, x)| \le 1 \quad \text{a.e. in } 0 \le t \le \bar{t}, x \in \Omega \tag{6.12.12}$$

do not admit much improvement. In absence of an analogue of Lemma 6.12.1, we cannot refine the result to point target sets. The available result reads

Theorem 6.12.5. *Let the target set* **Y** *be closed and K-full (see* **6.4**) *and let the assumptions in* **6.9** *be satisfied. Let* $\bar{u}(\cdot) = \bar{u}(t, x)$ *be a time optimal control for* (6.12.12). *Then there exists* $(z, z_1) \in N_Y(\bar{y}) =$ *normal cone to* **Y** *at* $\bar{y} = \bar{y}(\bar{t}, \bar{u})$, $(z, z_1) \neq 0$ *such that if* $\bar{\eta}(s)$ *is the solution of* (6.12.7)-(6.12.8) *we have*

$$\int_\Omega \bar{\eta}(s, x)\bar{u}(s, x)dx = \min_{v \in L^\infty(\Omega), \|v\| \le 1} \int_\Omega \bar{\eta}(s, x)v(x)dx. \tag{6.12.13}$$

At first sight, the conclusion is more precise than that of Theorem 6.12.3 and Corollary 6.12.4; in fact, while (6.12.11) is an integral condition (it bears on the L^2 norm of $\bar{u}(s, \cdot)$), (6.12.13) implies that

$$\bar{u}(s, x) = -\text{sign}\,\bar{\eta}(s, x) \tag{6.12.14}$$

wherever $\bar{\eta}(s, x) \neq 0$, so that $\bar{u}(s, x) = \pm 1$. However, information about the set where $\bar{\eta}(s, x) = 0$ is difficult to come by, so that (6.12.14) may provide no information on the optimal control in large subsets of $(0, \bar{t}) \times \Omega$. Target sets such as $\mathbf{B}(0, \varepsilon)$ in (6.9.10) are admissible, even if combined with a finite number of exact sensings (see the comments after (6.9.10)).

Remark 6.12.6. A prerequisite to dampening vibrations in minimum time (as in **6.9**) or according to any other optimality criterion is the existence of an admissible control $u(t, x)$ that dampens the vibrations at all, that is, that drives the initial condition $\zeta = (\zeta, \zeta_1)$ into the ball $\mathbf{B}(0, \varepsilon)$ in (6.9.10) in some time $\bar{t}$. This is an **approximate controllability** problem. One way (not the only one) to solve it is by means of energy decay estimates showing that, in the uncontrolled case $u = 0$, every solution $\mathbf{y}(t) = (y(t), y_1(t))$ of (6.9.4) satisfies

$$\|\mathbf{y}(t)\|_{\mathbf{E}} = \sqrt{\|y(t)\|^2_{H^1(\Omega)} + \|y_1(t)\|^2_{L^2(\Omega)}} \to 0 \quad \text{as } t \to \infty. \tag{6.12.15}$$

This of course depends on dissipation of energy by the nonlinear term ϕ or by the boundary conditions. On the other hand, if vibrations are to be actually stopped, we need to construct a control u such that

$$\mathbf{y}(\bar{t}) = (y(\bar{t}), y_1(\bar{t})) = 0, \tag{6.12.16}$$

for some time $\bar{t}$, which is an **exact controllability** problem. Via the energy decay estimate (6.12.15), we will be able to attain (6.12.16) at some time $\bar{t}$ using Corollary 6.12.10 below, which requires some auxiliary results for the system

$$y'(t) = A(t)y(t) + f(t, y(t)) + Bu(t), \quad y(0) = \zeta. \tag{6.12.17}$$

We impose on $A(\cdot)$ the assumptions in **6.1**; the Cauchy problem for the linear equation $y'(t) = A(t)y(t)$ is well posed with solution operator $S(t, s)$. We assume that $B \in (F, E)$ (F another Banach space) and that $f(t, y)$ satisfies Hypothesis III in $0 \leq t \leq \bar{t}$. The control space is $L^p(0, \bar{t}; F)$, $1 \leq p \leq \infty$.

We denote by $\Phi : L^p(0, \bar{t}; F) \to C(0, \bar{t}; E)$ the map $u(\cdot) \to \Phi u = y(\cdot, u)$, $y(t, u)$ the solution of (6.12.17) corresponding to $u(\cdot)$ (ζ is fixed). Since solutions may not exist in the whole interval, the map is not in general everywhere defined in $L^p(0, \bar{t}; F)$; we call $D(\Phi)$ its domain of definition. Given $\bar{u}$ such that $y(t, \bar{u})$ exists in $0 \leq t \leq \bar{t}$, the (linear) map $\Psi(\bar{u}): L^p(0, \bar{t}; F) \to C(0, \bar{t}; U)$ is defined by $\Psi(\bar{u})h = \xi(\cdot, \bar{u}, h), \xi(t, \bar{u}, h)$ the solution of the variational equation

$$\xi'(t, \bar{u}, h) = \{A(t) + \partial_y f(t, y(t, \bar{u}))\}\xi(t, \bar{u}, h) + Bh(t), \quad \xi(0, \bar{u}, h) = 0. \tag{6.12.18}$$

Since this is a linear equation, $\Psi(\bar{u})h$ is defined for all $h(\cdot) \in L^p(0, \bar{t}; F)$.

Theorem 6.12.7. (a) $D(\Phi)$ is open in $L^p(0, \bar{t}; F)$. (b) The map Φ is continuously Fréchet differentiable at any $u \in D(\Phi)$ with derivative

$$\partial\Phi(u)h = \Psi(u)h = \xi(\cdot, u, h). \tag{6.12.19}$$

The proof is roughed out in Examples 6.12.11 and 6.12.12 below. We write $\Phi(\bar{t}; u) = y(\bar{t}, u)$; this map has the same domain as Φ and takes values in E. Since it is the composition of Φ and the evaluation at $\bar{t}$ (which is a bounded linear operator from $C(0, \bar{t}; E)$ into E), Theorem 6.12.7 implies

Corollary 6.12.8. The map $u \to \Phi(\bar{t}, u)$ is continuously Fréchet differentiable at any $\bar{u} \in D(\Phi)$ with derivative

$$\partial_u \Phi(\bar{t}, \bar{u})h = (\Psi(\bar{u})h)(\bar{t}) = \xi(\bar{t}, \bar{u}, h). \tag{6.12.20}$$

Theorem 6.12.9. Let $\bar{u} \in D(\Phi)$ be such that $\partial_u \Phi(\bar{t}, \bar{u})L^p(0, \bar{t}; F) = E$ (that is, such that the reachable space of (6.12.18) is the whole space E). Then there exists a ball $B(\bar{u}, \delta) \subseteq D(\Phi)$ such that the image $\Phi(\bar{t}, B(\bar{u}, \delta))$ contains a ball $B(y(\bar{t}, \bar{u}), \varepsilon)$ $(\varepsilon > 0)$ in E.

Theorem 6.12.9 is an immediate consequence of one of the forms of the implicit function theorem (Dieudonné [1960, Chapter X, Exercise 8, p. 263]). This result states that a continuously Fréchet differentiable map Φ between Banach spaces whose Fréchet derivative at a point $\bar{u}$ is *onto* the range space will be *locally onto* in the sense that the image of some ball around $B(\bar{u}, \delta)$ contains a ball $B(\Phi(\bar{u}), \varepsilon)$ of positive radius around $\Phi(\bar{u})$. It also shows (and this is used in Theorem 6.12.9) that, taking ε sufficiently small we may assume that δ is as small as we wish. ∎

In order to apply the result to (6.9.4) we assume below that the function ϕ in (6.9.1) is independent of t,

$$\phi(t, x, y) = \phi(x, y) \tag{6.12.21}$$

and that

$$\phi(x, 0) = 0. \tag{6.12.22}$$

The first condition makes the equation (6.9.4) autonomous. The second insures that (6.9.4) has the equilibrium solution $\mathbf{y}(t) \equiv (0, 0)$; this in turn implies that $\Phi(0) = (0, 0)$ (functions identically zero on both sides) and $\Phi(\bar{t}, 0) = (0, 0) \in \mathbf{E}$.

Corollary 6.12.10. Assume the wave equation (6.9.1) possesses the energy decay property (6.12.15) for uncontrolled motion or, more generally, that the solution can be driven to arbitrarily small energy with some admissible control. Then there exists an admissible control u such that (6.12.16) holds for $\mathbf{y}(t) = \mathbf{y}(t, u)$.

Proof. Written in the form (6.9.4), the equation fits into the model (6.12.17). We apply Theorem 6.12.9 with $E = \mathbf{E}$, $F = L^2(\Omega)$, $B = \mathbf{B}$, $p = \infty$ and $\zeta = 0$, taking as $\bar{u}$ the identically zero control. The variational equation (6.12.18) is (6.12.2); condition (6.12.22) implies that, for this particular control, the variational equation is

$$\xi'(t) = \mathbf{A}(\beta)\xi(t) + \mathbf{B}u(t), \qquad \xi(0) = 0, \qquad (6.12.23)$$

and the fact that the reachable space for this equation is all of $\mathbf{E}$ is a particular case of Lemma 6.12.1. If follows that $\Phi(\bar{t}, B(0, \delta)) \supseteq \mathbf{B}(0, \varepsilon) = $ ball of center $(0, 0)$ and sufficiently small radius ε in $\mathbf{E}$. All of this means that for every $\mathbf{y} \in \mathbf{E}$ with $\|\mathbf{y}\|_{\mathbf{E}} \leq \varepsilon$ there exists a control $u(\cdot)$ in $C_{\mathrm{ad}}(0, \bar{t}; U) = $ unit ball of $L^\infty(0, \bar{t}; F)$ such that, for the solution of (6.9.4) with $\mathbf{y}(0, u) = \zeta = 0$ we have $\mathbf{y}(\bar{t}, u) = \mathbf{y}$. However, the equation is reversible (if $\mathbf{y}(t, u)$ is a solution, $\mathbf{y}(-t, u) = \mathbf{y}(t, u^-)$, where $u^-(t) = u(-t)$). Thus, the above implies: there exists $\varepsilon > 0$ such that for every $\zeta \in \mathbf{E}$ with $\|\zeta\|_{\mathbf{E}} \leq \varepsilon$ there is a control $u(\cdot) \in C_{\mathrm{ad}}(0, \bar{t}; U)$ with $\mathbf{y}(\bar{t}, u) = 0$. This ends the proof of Corollary 6.12.10. ∎

Example 6.12.11. The proof of (*a*) in Theorem 6.12.7 is essentially the same as that of Lemma 6.1.1. If $\bar{u} \in D(\Phi)$ and $u \in L^p(0, \bar{t}; F)$, we have

$$y(t, u) - y(t, \bar{u}) = \int_0^t S(t, \tau)B(u(\tau) - \bar{u}(\tau))d\tau$$

$$+ \int_0^t S(t, \tau)\{f(\tau, y(\tau, u)) - f(\tau, y(\tau, \bar{u}))\}d\tau$$

in the interval $[0, t_u]$ where $y(t, u)$ exists and satisfies $\|y(t, u) - y(t, \bar{u})\| \leq 1$; estimating,

$$\|y(t, u) - y(t, \bar{u})\| \leq C\|u - \bar{u}\|_{L^p(0, \bar{t}; F)}$$

$$+ M \int_0^t K(\tau, c + 1)\|y(\tau, u) - y(\tau, \bar{u})\|d\tau.$$

The proof ends like that of Lemma 6.1.1, and implies that Φ is locally Lipschitz continuous.

Example 6.12.12. The arguments to prove (*b*) in Theorem 6.12.6 are similar to (but much simpler than) those in the proof of Theorem 6.2.7. We begin by observing that the solution of (6.12.18) obeys the integral equation

$$\xi(t, u, h) = \int_0^t S(t, \tau)\{\partial_y f(\tau, y(\tau, u))\xi(\tau, u, h) + Bh(\tau)\}d\tau.$$

Hence,

$$y(t, u + h) - y(t, u) - \xi(t, u, h)$$
$$= \int_0^t S(t, \tau)\{f(\tau, y(\tau, u + h)) - f(\tau, y(\tau, u)) - \partial_y f(\tau, y(\tau, u))\xi(\tau, u, h)\}d\tau$$
$$= \int_0^t S(t, \tau)\{f(\tau, y(\tau, u + h)) - f(\tau, y(\tau, u))$$
$$- \partial_y f(\tau, y(\tau, u))(y(\tau, u + h) - y(\tau, u)\}d\tau$$
$$+ \int_0^t S(t, \tau)\partial_y f(\tau, y(\tau, u))\{y(\tau, u + h) - y(\tau, u) - \xi(\tau, u, h)\}d\tau$$

To show that $\|y(t, u + h) - y(t, u) - \xi(t, u, h)\| = o(\|h\|)$ we divide by $\|h\|$ and estimate using Gronwall's inequality. In the first integral on the right, we use the dominated convergence theorem; the integrand is bounded using the local Lipschitz continuity of Φ. Continuity of $\partial_u \Phi$ in the norm of $(L^p(0, \bar{t}; F), E)$ is proved writing

$$\xi(t, u, h) = \int_0^t S(t, \tau; u)h(\tau)d\tau$$

($S(t, \tau; u)$ the solution operator of the equation (6.12.18)) and then showing u-continuity of $S(t, \tau; u)$ (u measured in the L^p norm) in the uniform operator topology. This continuity property follows from estimations very similar to those in the proof of Lemma 6.3.1.

7

Abstract Minimization Problems in Banach Spaces. Abstract Parabolic Linear and Semilinear Equations

7.1. Some Geometry of Banach Spaces

Every two norms in a finite dimensional space are equivalent (Example 2.0.18), thus one can always use a nice one, usually the Euclidean norm. In infinite dimensional spaces, replacement of nasty or indifferent norms by better ones may not feasible.

Let E, F be Banach spaces. A function g, defined in an open subset D of a normed space E with values in a normed space F is **Gâteaux differentiable** at y if and only if there exists an operator $\partial g(y) \in (E, F)$ such that, for each $\xi \in E$,

$$\lim_{h \to 0} \frac{g(y + h\xi) - g(y)}{h} = \frac{d}{dh}\bigg|_{h=0} g(y + h\xi) = \partial g(y)\xi \qquad (7.1.1)$$

(the limit is two-sided). Equivalently,

$$g(y + h\xi) = g(y) + h\partial g(y)\xi + r(y, \xi, h) \qquad (7.1.2)$$

with

$$\frac{\|r(y, \xi, h)\|}{h} \to 0 \quad \text{as } h \to 0 \qquad (7.1.3)$$

for each ξ fixed. The operator $\partial g(y) \in L(E, F)$ is the **Gâteaux derivative** or the **Gâteaux differential** of g at y.[1] A look at the definition of Fréchet differentiability in **6.2** reveals Fréchet differentiability implies Gâteaux differentiability, but the converse is not true even in finite dimension.

Example 7.1.1. Gâteaux differentiability at y means (*a*) existence of directional derivatives in any direction, (*b*) linear continuous dependence of the derivative on

---

[1] There seems to be more than one definition of Gâteaux differentiability in the literature. Compare Hille–Phillips [1957, p. 109] and Aubin–Ekeland [1984, p. 23].

the direction. The function $g(x, y) = y/x$ $(x \neq 0)$, $g(x, y) = 0$ $(x = 0)$ has directional derivatives in any direction but is not Gâteaux differentiable. The function $g(x, y) = y^2/x$ $(x \neq 0)$, $g(x, y) = 0$ $(x = 0)$ is Gâteaux differentiable at $(0, 0)$ with $\partial g(0) = 0$. However, it is not even continuous at $(0, 0)$, thus it cannot be Fréchet differentiable there.

Let E be an arbitrary Banach space. Given $y, \xi \in E$ define

$$[y, \xi] = \lim_{h \to 0+} = \frac{\|y + h\xi\| - \|y\|}{h} = \frac{d}{dh}\bigg|_{h=0+} \|y + h\xi\|. \qquad (7.1.4)$$

That $[y, \xi]$ exists for all y, ξ follows from the (geometrically obvious) result:

Example 7.1.2. Let $\phi(t)$ be a real-valued convex function $(\phi(\alpha s + (1 - \alpha)t) \leq \alpha\phi(s) + (1 - \alpha)\phi(t)$ for $0 \leq \alpha \leq 1)$ defined in $t \geq 0$. Then the quotient $h^{-1}(\phi(h) - \phi(0))$ is a nondecreasing function of $h > 0$.

Convexity of $t \to \|y + t\xi\|$ follows from the triangle inequality:

$$\|y + (\alpha s + (1 - \alpha)t)\xi\| = \|\alpha(y + s\xi) + (1 - \alpha)(y + t\xi)\|$$

$$\leq \alpha\|y + s\xi\| + (1 - \alpha)\|y + t\xi\| \quad (0 \leq \alpha \leq 1).$$

The triangle inequality also implies

$$|[y, \xi]| \leq \|\xi\|, \qquad (7.1.5)$$

and we check easily that

$$[\alpha y, \xi] = [y, \xi], \qquad [y, \alpha\xi] = \alpha[y, \xi] \quad (\alpha \geq 0). \qquad (7.1.6)$$

Since $\|2y + h(\xi + \eta)\| - \|2y\| \leq (\|y + h\xi\| - \|y\|) + (\|y + h\eta\| - \|y\|)$,

$$[y, \xi + \eta] = [2y, \xi + \eta] \leq [y, \xi] + [y, \eta]. \qquad (7.1.7)$$

If $y = 0$ then $[y, \xi] = \|\xi\|$. If E is a Hilbert space we have

$$[y, \xi] = \frac{(y, \xi)}{\|y\|} \quad (y \neq 0). \qquad (7.1.8)$$

More generally, if E is a Banach space with a Gâteaux differentiable norm off the origin, then

$$[y, \xi] = \langle \partial\|y\|, \xi \rangle \quad (y \neq 0), \qquad (7.1.9)$$

where $\partial\|y\| \in E^*$ is the Gâteaux derivative; the angled brackets indicate the duality of E and E^*. Conversely,

Example 7.1.3. Assume $\xi \to [y, \xi]$ is linear in ξ for every $y \neq 0$. Then the norm of E is Gâteaux differentiable off the origin.

In fact, if $\xi \to [y, \xi]$ is linear it is bounded by (7.1.5). That (7.1.1) occurs for $h > 0$ is the definition of $[y, \xi]$; on the other hand $[y, \xi] = -[y, -\xi]$, so that (7.1.1) follows as well for $h < 0$. We note the qualification "off the origin": the norm of *any* nontrivial normed space fails to be Gâteaux differentiable at the origin. In fact, the limit on the left side of (7.1.4) is $\|\xi\|$ for $h > 0$, $-\|\xi\|$ for $h < 0$.

Example 7.1.4. Assume the limit (7.1.4) is two sided. Then $[y, \xi]$ is linear in ξ for every $y \neq 0$.

Equality of the left-hand and the right-hand limit implies $[y, \xi] = -[y, -\xi]$. By (7.1.7), $[y, \xi + \eta] \leq [y, \xi] + [y, \eta]$ so that $[y, \xi + \eta] = -[y, -\xi - \eta] \geq -[y, -\xi]$ $- [y, -\eta] = [y, \xi] + [y, \eta]$. On the other hand, by (7.1.6), $[y, \alpha\xi] = \alpha[y, \xi]$ and $[y, -\alpha\xi] = -[y, \alpha\xi] = -\alpha[y, \xi]$ for $\alpha \geq 0$.

The L^1 and the L^∞ norm are not well behaved, even in finite dimension.

Example 7.1.5. (a) Let $E = \mathbb{R}^2$ equipped with the supremum norm $\|(y_1, y_2)\|_\infty = \max(|y_1|, |y_2|)$. Then the norm is (Gâteaux or Fréchet) differentiable at $y = (y_1, y_2)$ if and only if $y_1 \neq y_2$. (b) Let $E = \mathbb{R}^2$, equipped with the norm $\|(y_1, y_2)\|_1 = |y_1| + |y_2|$. Then the norm is (Gâteaux or Fréchet) differentiable at $y = (y_1, y_2)$ if and only if $y_1 \neq 0$, $y_2 \neq 0$.

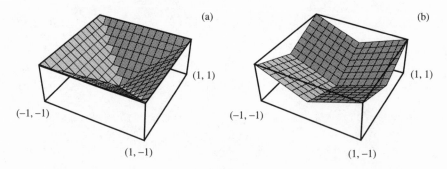

Figure 7.1.

This result can be easily generalized to higher dimensions. If $\mathbb{R}^m$ is equipped with the L^∞ norm, the norm is differentiable at $y = (y_1, \ldots, y_m)$ if and only if all the y_j are different. For the L^1 norm the requirement is that $y_j \neq 0$ for all j.

The L^p norm, $1 < p < \infty$ is much better; in fact,

Example 7.1.6. Let Ω be a measurable subset of $\mathbb{R}^m$, $E = L^p(\Omega)$ with $1 < p < \infty$. Then the norm $\| \cdot \|_p$ of $L^p(\Omega)$ is Fréchet differentiable off the origin with derivative

$$\partial \|y\|_p \xi = \|y\|_p^{1-p} \langle |y|^{p-2} y, \xi \rangle \tag{7.1.10}$$

(note that $|y|^{p-2} y \in L^q(\Omega)$, where $1/p + 1/q = 1$.)

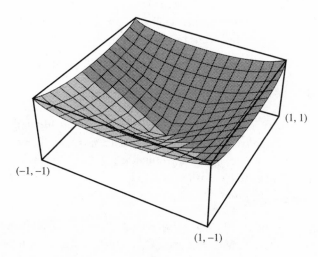

$(1, 1)$

$(-1, -1)$

$(1, -1)$

Figure 7.2.

We go back to a general Banach space. By (7.1.6) and (7.1.7), $[y, \xi]$ is a convex function of ξ, thus for every real a and every $y \in E$ the set $P_a(y) = \{\xi; [y, \xi] \leq a\}$ is convex. It is never empty if $y \neq 0$, since

$$[y, \alpha y] = \alpha \|y\| \quad (-\infty < \alpha < \infty). \tag{7.1.11}$$

(note that (7.1.4) gives $[y, \alpha y] = \|y\| (d/dh)_{h=0+} |1 + \alpha h|$). Finally, $P_a(y)$ has nonempty interior; in fact, $[y, \xi] \leq [y, \xi - \xi_0] + [y, \xi_0] \leq \|\xi - \xi_0\| + [y, \xi_0]$, thus any $\xi_0 \in P_a(y)$ with $[y, \xi_0] < a$ is an interior point. That there are such ξ_0 follows from (7.1.11).

Lemma 7.1.7. *Let $y \neq 0, \delta > 0$, and let D be a convex set with*

$$D \subseteq Q_{-\delta}(y) = \{\xi; [y, \xi] \geq -\delta\}. \tag{7.1.12}$$

Then there exists $z \in E^*$ such that

$$\|z\| = 1, \qquad \langle z, \xi \rangle \geq -\delta \quad (\xi \in D). \tag{7.1.13}$$

Proof. Let $\xi_0 \in Q_{-\delta}(y) \cap P_{-\delta}(y)$ so that

$$[y, \xi_0] = -\delta. \tag{7.1.14}$$

If ξ_0 is an interior point of $P_{-\delta}(y)$ then there exists a ball $B(\xi_0, \varepsilon)$ of center ξ_0 and radius ε contained in $P_{-\delta}$. Hence, $\alpha\xi_0 \in P_{-\delta}(y)$ for α sufficiently close to 1. This and (7.1.6) imply that $[y, \xi_0] < -\delta$, which contradicts (7.1.14). A completely symmetric argument can be applied to $Q_{-\delta}(y)$; accordingly the intersection $Q_{-\delta}(y) \cap P_{-\delta}(y)$ may contain only boundary points of either set, and the same applies to $P_{-\delta}(y) \cap D$. On the other hand, $P_{-\delta}(y)$ has nonempty interior, thus it can be separated from D by means of a functional $z \in E^*, z \neq 0$:

$$\langle z, \xi \rangle \geq \alpha \geq \langle z, \eta \rangle \quad (\xi \in D, \eta \in P_{-\delta}(y)) \tag{7.1.15}$$

(see Remark 7.1.10). Let α be the largest number such that (7.1.15) holds. If $\alpha \geq 0$, we have $\langle z, \xi \rangle \geq 0$ ($\xi \in D$), which implies the second condition (7.1.13); for the first we simply rescale z. If $\alpha < 0$ we may assume (multiplying z by $\delta/|\alpha|$) that $\alpha = -\delta$. Now, the set $P_{-\delta}(y)$ contains the point $\eta = -\delta y/\|y\|$ (see (7.1.11)) with norm $\|\eta\| = \delta$ and $\langle z, \eta \rangle \leq -\delta$, so that $\|z\| \geq 1$. Since $-\delta/\|z\| \geq -\delta$, the inequality in (7.1.13) will be satisfied as well by $z/\|z\|$, thus we may assume that $\|z\| = 1$. This ends the proof.   ∎

When the norm of E is Gâteaux differentiable off the origin, then by (7.1.9) $Q_\delta(y)$ is convex (in fact, both $P_{-\delta}$ and $Q_{-\delta}$ are half-spaces). We may then take $D = Q_{-\delta}$ in Lemma 7.1.7. Even better, we may throw the Lemma away, since (7.1.13) will be achieved by the vector

$$z = \partial \|y\|. \tag{7.1.16}$$

That the inequality in (7.1.13) holds is obvious. It follows from (7.1.5) that $\|z\| \leq 1$; on the other hand, $\langle z, y \rangle = \langle \partial \|y\|, y \rangle = [y, y] = \|y\|$, so that $\|z\| = 1$.

Example 7.1.8. As in Example 7.1.5 let $E = \mathbb{R}^2$ equipped with the supremum norm $\|(y_1, y_2)\|_\infty = \max(|y_1|, |y_2|)$. Let $y = (y_1, y_2)$, and assume $y_1, y_2 \geq 0$. If

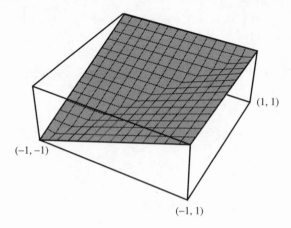

Figure 7.3.

$y_1 > y_2$ (resp. $y_1 < y_2$) then $[(y_1, y_2), (\xi_1, \xi_2)] = \xi_1$ (resp. $[(y_1, y_2), (\xi_1, \xi_2)] = \xi_2$)
so that $Q_{-\delta}(y)$ is a half space. On the other hand, if $y = (y_1, y_2)$ with $y_1 = y_2 > 0$,
then $[(y_1, y_2), (\xi_1, \xi_2)] = \max(\xi_1, \xi_2)$; the latter case shows that $Q_{-\delta}(y)$ may not
be convex. Figure 7.3 shows the function $(\xi_1, \xi_2) \to [(y_1, y_2), (\xi_1, \xi_2)]$ for, say,
$y = (1, 1)$; the sets $P_{-\delta}(y)$ and $Q_{-\delta}(y)$ are in Figure 7.4.

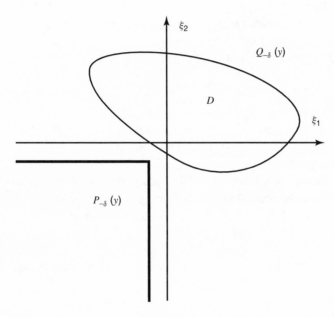

Figure 7.4.

We apply Lemma 7.1.7 in the space $\mathbf{E} = \mathbb{R} \times E$ equipped with the "Euclidean
norm" $\|\mathbf{y}\| = \|(t, y)\| = (t^2 + \|y\|^2)^{1/2}$. We check easily that if $\mathbf{y} = (t, y) \neq 0$
and $\boldsymbol{\xi} = (\tau, \xi)$ then

$$[\mathbf{y}, \boldsymbol{\xi}] = \frac{t\tau + \|y\|[y, \xi]}{(t^2 + \|y\|^2)^{1/2}} = \frac{|t|[t, \tau] + \|y\|[y, \xi]}{(t^2 + \|y\|^2)^{1/2}} \qquad (7.1.17)$$

(note that in the space $\mathbb{R}$ we have $[t, \tau] = (t/|t|)\tau$ if $t \neq 0$). On the other hand,
the dual $E^* = (\mathbb{R} \times E)^*$ can be identified with $\mathbb{R} \times E^*$ equipped with the same
norm $\|\mathbf{z}\| = \|(\mu, z)\| = (|\mu|^2 + \|z\|^2)^{1/2}$; the element $\mathbf{z} = (\mu, z) \in \mathbb{R} \times E^*$ acts
on elements $\boldsymbol{\xi} = (\tau, \xi) \in \mathbb{R} \times E$ in the form

$$\langle \mathbf{z}, \boldsymbol{\xi} \rangle = \langle (\mu, z), (\tau, \xi) \rangle = \mu\tau + \langle z, \xi \rangle. \qquad (7.1.18)$$

Lemma 7.1.7, applied to a convex subset $\mathbf{D}$ of $Q_{-\delta}(\mathbf{y})$, produces a $\mathbf{z} = (\mu, z)$ with

$$|\mu|^2 + \|z\|^2 = 1, \qquad \mu\tau + \langle z, \xi \rangle \geq -\delta \quad ((\tau, \xi) \in \mathbf{D}). \qquad (7.1.19)$$

Lemma 7.1.9. *Let* $\mathbf{z} = (\mu, z)$ *be the vector in Lemma* 7.1.7 *corresponding to* $\mathbf{y} = (t, y) \neq 0$ *and to a convex subset* $\mathbf{D}$ *of* $Q_{-\delta}(\mathbf{y}), \delta > 0$. *Then* (a) *if* $t > 0$ *we must have* $\mu \geq 0$, (b) *if* $t = 0$ *we may replace* (μ, z) *by a vector satisfying the conclusion of Lemma* 7.1.7 *with* $\mu = 0$.

Proof. (a) Let $t > 0$. In view of (7.1.17) the set $P_{-\delta}(\mathbf{y}) = P_{-\delta}((t, y))$ contains all vectors of the form $(\tau, 0)$ with $\tau \leq -\delta$. The second inequality (7.1.15) implies that $\mu\tau + \langle z, \xi \rangle = \mu\tau$ is bounded above for all such vectors so that $\mu \geq 0$ as claimed. (b) If $t = 0$ it results from (7.1.17) that $[\mathbf{y}, \boldsymbol{\xi}] = [(t, y), (\tau, \xi)] = [y, \xi]$ so that (7.1.12) determines that all $(\tau, \xi) \in \mathbf{D}$ must satisfy

$$[y, \xi] \geq -\delta. \tag{7.1.20}$$

We apply Lemma 7.1.7 in the space E to the convex set $\Pi(\mathbf{D})$ (Π the projection from $\mathbb{R} \times E$ into E). We obtain a $z \in E^*$ such that (7.1.13) holds for $\Pi(\mathbf{D})$; plainly, $\mathbf{z} = (0, z)$ will do the job for $\mathbf{D}$. This ends the proof.   ∎

Remark 7.1.10. It follows from the forthcoming Theorem 12.0.7 that the separation (7.1.15) can be effected for two disjoint convex sets one of which has interior points. The sets in Lemma 7.1.7 are not disjoint, but it suffices to separate D from the interior of $P_{-\delta}(y)$ and then use

Example 7.1.11. Let A be a convex set with $\mathrm{Int}(A) \neq \emptyset$. Then $\mathrm{Int}\,(A)$ is dense in A.

7.2. Abstract Minimization Problems for the Last Time

Abstract minimization problem (**4.1** and **6.4**): Let V be a metric space, E a Banach space, $Y \subseteq E$. Given functions $f : D(f) \to E$, $f_0 : D(f_0) \to \mathbb{R}$ $(D(f), D(f_0) \subseteq V)$ characterize the $\bar{u} \in D(f) \cap D(f_0)$ satisfying

$$\text{minimize} \quad f_0(u) \tag{7.2.1}$$

$$\text{subject to} \quad f(u) \in Y \tag{7.2.2}$$

or, more generally, the **approximate** or **suboptimal solutions** of (7.2.1)-(7.2.2), that is, the sequences $\{\bar{u}^n\} \subseteq D(f) \cap D(f_0)$ satisfying

$$\limsup_{n \to \infty} f_0(\bar{u}^n) \leq m, \qquad \lim_{n \to \infty} \text{dist}(f(\bar{u}^n), Y) = 0 \tag{7.2.3}$$

where m is the minimum in (7.2.1) under the constraint (7.2.2). We assume that $-\infty < m < \infty$.

Abstract time optimal problem (**3.3** and **6.4**): Let $\{V_n\}$ be a sequence of metric spaces, E a Banach space, $Y \subseteq E$. Given a sequence of functions $f_n : D(f_n) \to E$

$(D(f_n) \subseteq V_n)$ such that $f_n(D(f_n)) \cap Y = \emptyset$ $(n = 1, 2, \ldots)$, characterize the **so-lutions** of the problem, that is, the sequences $\{\bar{u}^n\}$, $\bar{u}^n \in D(f_n)$ satisfying

$$\text{dist}(f_n(\bar{u}^n), Y) \to 0 \quad \text{as } n \to \infty. \tag{7.2.4}$$

All results in this section have a version valid for general Banach spaces and a **smooth version** valid in spaces whose norm is Gâteaux differentiable off the origin. The results (in their smooth versions) contain the Hilbert space results.

The first result is a generalization of Lemma 3.3.1.

Lemma 7.2.1. *Let V be a complete metric space, E a real Banach space, $f : D(f) \to E$, where $D(f) \subseteq V$, $D(f) \neq \emptyset$. Assume that $Y \subseteq E$ is closed, that $f(D(f)) \cap Y = \emptyset$ and that for each $y \in Y$ the real-valued function*

$$\Phi(u, y) = \begin{cases} \|f(u) - y\| & (u \in D(f)) \\ +\infty & (u \notin D(f)) \end{cases} \tag{7.2.5}$$

is lower semicontinuous with respect to u. Finally, let $\bar{u} \in D(f)$, $\bar{y} \in Y$,

$$\varepsilon = \|f(\bar{u}) - \bar{y}\|. \tag{7.2.6}$$

Then there exist $\tilde{u} \in D(f)$, $\tilde{y} \in Y$ such that

$$\|f(\tilde{u}) - \tilde{y}\| \le \|f(\bar{u}) - \bar{y}\|, \tag{7.2.7}$$

$$d(\tilde{u}, \bar{u}) + \|\tilde{y} - \bar{y}\| \le \sqrt{\varepsilon}, \tag{7.2.8}$$

and such that for every convex set $D \subseteq \text{Var} f(\tilde{u})$ and every $\rho > 0$ there exists $z \in E^*$ with[1]

$$\|z\|_{E^*} = 1, \qquad \langle z, \xi - w \rangle \ge -\sqrt{\varepsilon}(1 + \rho) \tag{7.2.9}$$

for $\xi \in D$ and $w \in T_Y(\tilde{y}) \cap B(0, \rho)$.

Smooth version: *Assume the norm of E is Gâteaux differentiable off the origin. Then there exist $\tilde{u} \in D(f)$, $\tilde{y} \in Y$ satisfying (7.2.7), (7.2.8) and $z \in E^*$ such that*

$$\|z\|_{E^*} = 1, \qquad \langle z, \xi - w \rangle \ge -\sqrt{\varepsilon}(1 + \|w\|) \tag{7.2.10}$$

for $\xi \in \text{Var} f(\tilde{u})$ and $w \in I_Y(\tilde{y})$, or for $\xi \in \text{Der} f(\tilde{u})$ and $w \in K_Y(\tilde{y})$. The multiplier z is given by $z = \partial \|f(\tilde{u}) - \tilde{y}\|$.

---

[1] The reader is cautioned against confusing the convex set D with the domain of definition $D(f)$ of f. The same observation applies to the sets D_n and the domains $D(f_n)$ in Theorem 7.2.2 below.

Proof. The function Φ defined by (7.2.5) in the space $V \times Y$ equipped with the distance $d((u, y), (u', y')) = d(u, u') + \|y - y'\|$ (which distance makes $V \times Y$ complete) is jointly lower semicontinuous in $V \times Y$ (Example 3.3.13). We have

$$\Phi(\bar{u}, \bar{y}) = \|f(\bar{u}) - \bar{y}\| = \varepsilon \leq 0 + \varepsilon \leq \inf_{u \in V, y \in Y} \Phi(u, y) + \varepsilon,$$

thus, applying Ekeland's variational principle (Theorem 3.2.2) to $\Phi(u, y)$ in $V \times Y$ we obtain $(\tilde{u}, \tilde{y}) \in V \times Y$ satisfying (7.2.7), (7.2.8) and

$$\Phi(u, y) \geq \Phi(\tilde{u}, \tilde{y}) - \sqrt{\varepsilon}\{d(u, \tilde{u}) + \|y - \tilde{y}\|\} \quad ((u, y) \in V \times Y). \quad (7.2.11)$$

Inequality (7.2.7) implies that $\tilde{u} \in D(f)$. To show (7.2.9), take $w \in I_Y(\tilde{y}) \cap B(0, \rho)$, $\xi \in \text{Var } f(\tilde{u})$. Let $\{h_k\}, \{u_k\}$ be the sequences employed in (3.3.6) for the definition of ξ, so that $d(u_k, \tilde{u}) \leq h_k$,

$$f(u_k) = f(\tilde{u}) + h_k\xi + o(h_k), \quad (7.2.12)$$

and let $\{y_k\}$ be the sequence in (3.3.2) in the definition of w (with the same sequence $\{h_k\}$), so that

$$y_k = \tilde{y} + h_k w + o(h_k). \quad (7.2.13)$$

Using (7.2.11) for each u_k, y_k we obtain

$$\|f(\tilde{u}) + h_k\xi - \tilde{y} - h_k w + o(h_k)\| - \|f(\tilde{u}) - \tilde{y}\|$$
$$\geq -\sqrt{\varepsilon}h_k(1 + \|w + o(1)\|).$$

Dividing by h_k and letting $k \to \infty$,

$$[f(\tilde{u}) - \tilde{y}, \xi - w] \geq -\sqrt{\varepsilon}(1 + \|w\|), \quad (7.2.14)$$

so that

$$[f(\tilde{u}) - \tilde{y}, \xi - w] \geq -\sqrt{\varepsilon}(1 + \rho) \quad (7.2.15)$$

for $\xi - w \in \text{Var } f(\tilde{u}) - I_Y(\tilde{y}) \cap B(0, \rho)$, in particular for $\xi - w$ in the convex set $D - T_Y(\tilde{y}) \cap B(0, \rho)$ (the tangent cone $T_Y(\tilde{y})$ is convex; Example 3.3.10). Since $f(D(f)) \cap Y = \emptyset$ we have $f(\tilde{u}) - \tilde{y} \neq 0$ and Lemma 7.1.7 provides a vector $z \in E^*$ such that (7.2.9) holds. End of proof.   ∎

Proof of the smooth version. We use (7.1.9), so that (7.2.14) is exactly (7.2.10) with $z = \partial\|f(\tilde{u}) - \tilde{y}\|$ and $\xi \in \text{Var } f(\tilde{u})$, $w \in I_Y(\tilde{y})$. That $\|z\|_{E^*} = 1$ follows from the considerations after (7.1.16). The statement for $\xi \in \text{Der } f(\tilde{u})$ and $w \in K_Y(\tilde{y})$ results from the fact that the sequence $\{h_k\}$ in (7.2.12) is now arbitrary; we use in (7.2.12) the same sequence employed in (7.2.13) to define w (for more details on this see the proof of Lemma 3.3.1).   ∎

Given a solution $\{\bar{u}^n\}$ of the abstract time optimal problem, a sequence $\{\bar{y}^n\} \subseteq Y$ is **associated** to $\{\bar{u}^n\}$ if

$$\|f_n(\bar{u}^n) - \bar{y}^n\| = \varepsilon_n \to 0 \quad \text{as } n \to \infty. \tag{7.2.16}$$

The assumptions for the abstract time optimal problem are:

(a) V_n is complete for each n and E is a real Banach space,

(b) for each n and each $y \in Y$, the real valued function

$$\Phi_n(u, y) = \begin{cases} \|f_n(u) - y\| & (u \in D(f_n)) \\ +\infty & (u \notin D(f_n)) \end{cases} \tag{7.2.17}$$

is lower semicontinuous with respect to u,

(c) the target set Y is closed.

Theorem 7.2.2. *Let $\{\bar{u}^n\}$ be a solution of the abstract time optimal problem, $\{\bar{y}^n\} \subseteq Y$ an associated sequence,*

$$\varepsilon_n = \|f(\bar{u}^n) - \bar{y}^n\|. \tag{7.2.18}$$

Then there exist sequences $\{\tilde{u}^n\}$, $\tilde{u}^n \in D(f_n)$, $\{\tilde{y}^n\} \subseteq Y$ such that

$$\|f_n(\tilde{u}^n) - \tilde{y}^n\| \le \|f_n(\bar{u}^n) - \bar{y}^n\|, \tag{7.2.19}$$

$$d_n(\tilde{u}^n, \bar{u}^n) + \|\tilde{y}^n - \bar{y}^n\| \le \sqrt{\varepsilon_n} \tag{7.2.20}$$

(d_n the distance in V_n) and such that, for every sequence of convex sets $\{D_n\}$, $D_n \subseteq \text{Var} f_n(\tilde{u}^n)$ and every $\rho > 0$ there exists a sequence $\{z_n\} \subset E^*$ such that

$$\|z_n\|_{E^*} = 1, \qquad \langle z_n, \xi^n - w^n \rangle \ge -\sqrt{\varepsilon_n}(1 + \rho) \tag{7.2.21}$$

for $\xi^n \in D_n$, $w^n \in T_Y(\tilde{y}^n) \cap B(0, \rho)$. If z is a E-weak limit point of $\{z_n\}$ in E^* then we have

$$\langle z, \xi \rangle \ge 0 \tag{7.2.22}$$

for $\xi \in \liminf_{n \to \infty} D_n$. Finally,

$$z \in \left(\liminf_{n \to \infty} T_Y(\tilde{y}^n) \right)^-. \tag{7.2.23}$$

Smooth version: *Assume the norm of E is Gâteaux differentiable off the origin. Then there exist sequences $\{\tilde{u}^n\}$, $\tilde{u}^n \in D(f_n)$, $\{\tilde{y}^n\} \subseteq Y$ such that (7.2.19) and (7.2.20) hold and a sequence $\{z_n\} \in E^*$ such that*

$$\|z_n\|_{E^*} = 1, \qquad \langle z_n, \xi^n - w^n \rangle \ge -\sqrt{\varepsilon_n}(1 + \|w^n\|) \tag{7.2.24}$$

for $\xi^n \in \text{Var} f_n(\tilde{u}^n)$ and $w \in I_Y(\tilde{y}^n)$, or for $\xi^n \in \text{Der} f_n(\tilde{u}^n)$ and $w \in K_Y(\tilde{y}^n)$; z_n is given by $z_n = \partial \|f_n(\tilde{u}^n) - \tilde{y}^n\|$. If $\tilde{y}^n \to \bar{y}$ and $z \in E^*$ is a E-weak limit point of z then

$$z \in N_Y(\bar{y}). \tag{7.2.25}$$

Proof. We apply Lemma 7.2.1 to each term $\bar{u}^n$ of a solution of the abstract time optimal problem and each term $\bar{y}^n$ of the associated sequence. We obtain in this way an element $(\bar{u}^n, \bar{y}^n)$ of $V_n \times Y$ satisfying (7.2.19) and (7.2.20) and a sequence $\{z_n\} \subseteq E^*$ (depending on ρ and on the sequence $\{D_n\}$ of convex sets) such that (7.2.21) holds for $\xi^n \in D_n$ and $w^n \in T_Y(\bar{y}^n) \cap B(0, \rho)$. Clearly, (7.2.22) holds for $\xi \in \liminf_{n\to\infty} D_n$. To prove (7.2.23) let $w \in \liminf_{n\to\infty} T_Y(\bar{y}^n)$, so that there exists a sequence $\{w^n\}$, $w^n \in T_Y(\bar{y}^n)$ with $w^n \to w$. Using (7.2.21) for $\xi^n = 0$ and $w^n = (\rho/\|w^n\|)w^n$ and letting $n \to \infty$, (7.2.23) results. ∎

Proof of the smooth version. The statements for Gâteaux differentiable norm have already been proved (via the smooth part of Lemma 7.2.1) except for the claim that $z \in N_Y(\bar{y})$ when $\bar{y}^n \to \bar{y}$. To do this, note that (7.2.20) implies that $\bar{y}^n \to \bar{y}$ as well. If $y \neq 0$ we have

$$\|y + h\xi\| = \|y\| + h\langle \partial\|y\|, \xi\rangle + r(y, \xi, h) \tag{7.2.26}$$

where, for each y, ξ fixed $\|r(y, \eta, h)\| = o(h)$ as $h \to 0$. Pick $w \in T_Y(\bar{y})$ and, for each n select $\{h_n\}$ such that

$$\frac{1}{h_n}\|r(f_n(\bar{u}^n) - \bar{y}^n, w, -h_n)\| \to 0 \tag{7.2.27}$$

as $n \to \infty$. By definition of the tangent cone $T_Y(\bar{y})$ there exists $\{y^n\} \subseteq Y$ such that

$$y^n = \bar{y}^n + h_n w + o(h_n) \tag{7.2.28}$$

as $n \to \infty$ (see (3.3.4)). We write (7.2.11) for $\Phi_n(u, y) = \|f_n(u) - y\|$ and $u = \bar{u}^n$, $y = y^n$ (y^n the sequence in (7.2.28)); since the distance from $(\bar{u}^n, y^n)$ to $(\bar{u}^n, \bar{y}^n)$ in $V_n \times Y$ is $\|y^n - \bar{y}^n\|$,

$$\|f(\bar{u}^n) - y^n\| \geq \|f(\bar{u}^n) - \bar{y}^n\| - \sqrt{\varepsilon_n}\|y^n - \bar{y}^n\|. \tag{7.2.29}$$

We insert the expression (7.2.28) for y^n in the left-hand side of (7.2.29). Since $\|x^n + o(h_n)\| = \|x^n\| + o(h_n)$ for any sequence $\{x^n\}$, we have

$$\|f(\bar{u}^n) - \bar{y}^n - h_n w + o(h_n)\| = \|f(\bar{u}^n) - \bar{y}^n - h_n w\| + o(h_n).$$

Accordingly, (7.2.29) yields

$$\|f(\bar{u}^n) - \bar{y}^n - h_n w\| \geq \|f(\bar{u}^n) - \bar{y}^n\| - \sqrt{\varepsilon_n}\|h_n w + o(h_n)\| + o(h_n)$$
$$= \|f(\bar{u}^n) - \bar{y}^n\| - \sqrt{\varepsilon_n}(\|h_n w\| + o(h_n)) + o(h_n)$$
$$= \|f(\bar{u}^n) - \bar{y}^n\| - \sqrt{\varepsilon_n}h_n(\|w\| + o(1)) + o(h_n).$$

We then use (7.2.26) on the left side, obtaining

$$\|f_n(\bar{u}^n) - \bar{y}^n\| - h_n\langle \partial\|f_n(\bar{u}^n) - \bar{y}^n\|, w\rangle + r(f_n(\bar{u}^n) - \bar{y}^n, w, -h_n) + o(h_n)$$
$$\geq \|f_n(\bar{u}^n) - \bar{y}^n\| - \sqrt{\varepsilon_n}h_n(\|w\| + o(1)).$$

Knocking off $\| f_n(\tilde{u}^n) - \tilde{y}^n \|$ from both sides and dividing by h_n,

$$\langle z_n, w \rangle = \langle \partial \| f(\tilde{u}^n) - \tilde{y}^n \|, w \rangle$$
$$\leq \frac{1}{h_n} r(f(\tilde{u}^n) - \tilde{y}^n, w, -h_n) + \sqrt{\varepsilon}_n (\|w\| + o(1)) + o(1).$$

Letting $n \to \infty$, we obtain $\langle z, w \rangle \leq 0$ as claimed.   ∎

Note that in the smooth case the sequence $\{z_n\}$ depends only on the solution $\{\tilde{u}^n\}$ of the abstract time optimal problem; in the general case, $\{z_n\}$ depends on $\{D_n\}$ and on ρ.

As for Hilbert spaces in **6.4**, we need additional conditions to insure that $z \neq 0$. These are based on Lemma 6.4.3.

Corollary 7.2.3. (*of Lemma* 6.4.3). *Let* $\{D_n\}$ *and* ρ *be as in Theorem* 7.2.2. *Assume that there exists a precompact sequence* $\{Q_n\}$ *such that the set*

$$\Delta = \bigcap_{n=n_0}^{\infty} (\Delta_n + Q_n) \tag{7.2.30}$$

contains an interior point for n_0 *large enough, where*

$$\Delta_n = D_n - T_Y(y^n) \cap B(0, \rho).$$

Then the sequence $\{z_n\}$ *in Theorem* 7.2.2 *corresponding to* $\{D_n\}$ *and* ρ *cannot have zero E-weak limit points, so that the vector* z *in* (7.2.22) *is not zero.*

Smooth version: *Assume the norm of* E *is Gâteaux differentiable off the origin. Then the same result holds with* $\Delta_n = \text{Var} f_n(\tilde{u}^n) - I_Y(\tilde{y}^n) \cap B(0, \rho)$ *or* $\Delta_n = \text{Der} f_n(\tilde{u}^n) - K_Y(\tilde{y}^n) \cap B(0, \rho)$ *for arbitrary* ρ.

Remark 7.2.4. A sufficient (but not necessary) condition for lower semicontinuity of (7.2.5) for any y is weak E^*-continuity of $f_n(u)$. In fact, if this is the case, $u^n \to u$ in V implies $f(u^n) - y \to f(u) - y$ E^*-weakly in E, so that $\| f(u) - y \| \leq \liminf_{n \to \infty} \| f(u^n) - y \|$. Similarly, if $E = F^*$ is the dual of another Banach space E, F-weak continuity of $f_n(u)$ will do.

We do now the abstract nonlinear programming problem (7.2.1)-(7.2.2). If $\{\tilde{u}^n\}$ is a suboptimal solution, a sequence $\{\tilde{y}^n\} \subseteq Y$ is **associated** with $\{\tilde{u}^n\}$ if there exists $\{\varepsilon_n\} \subset \mathbb{R}_+, \varepsilon_n \to 0$ with

$$f_0(\tilde{u}_n) \leq m + \varepsilon_n, \qquad \| f(\tilde{u}^n) - \tilde{y}^n \| \leq \varepsilon_n. \tag{7.2.31}$$

We write $\mathbf{f}(u) = (f_0, f)(u) = (f_0(u), f(u)) \in \mathbf{E} = \mathbb{R} \times E$. The assumptions are

(a) V is complete and E is a real Banach space,

(b) $D(f) \cap D(f_0) \neq \emptyset$ and the real valued functions

$$\Phi(u, y) = \begin{cases} \|f(u) - \bar{y}\| & (u \in D(f)) \\ +\infty & (u \notin D(f)) \end{cases} \tag{7.2.32}$$

$$\Phi_0(u, y) = \begin{cases} \max(f_0(u), m - \varepsilon) & (u \in D(f_0)) \\ +\infty & (u \notin D(f_0)) \end{cases} \tag{7.2.33}$$

are lower semicontinuous with respect to to u, the first for any $y \in Y$ and the second for ε sufficiently small,

(c) the target set Y is closed.

Theorem 7.2.5. *Let* $\{\bar{u}^n\}$ *be a suboptimal solution of* (7.2.1)-(7.2.2) *and* $\{\bar{y}^n\} \subseteq Y$ *an associated sequence. Then there exists sequences* $\{\tilde{u}^n\} \subseteq D(f) \cap D(f_0)$, $\{\tilde{y}^n\} \subseteq Y$ *with*

$$d(\tilde{u}^n, \bar{u}^n) + \|\tilde{y}^n - \bar{y}^n\| \leq \sqrt[4]{5}\sqrt{\varepsilon_n} \tag{7.2.34}$$

and such that, for every sequence of convex sets $\{\mathbf{D}_n\}, \mathbf{D}_n \subseteq \mathrm{Var}\,\mathbf{f}(\tilde{u}^n) = \mathrm{Var}\,(f_0, f)(\tilde{u}^n)$ *and every* $\rho > 0$ *there exists a sequence* $\{(z_{0n}, z_n)\} \subseteq \mathbb{R} \times E^*$ *with*

$$z_{0n}^2 + \|z_n\|^2 = 1, \qquad z_{0n} \geq 0, \tag{7.2.35}$$

$$z_{0n}\xi_0^n + \langle z_n, \xi^n - w^n \rangle \geq -\sqrt[4]{5}\sqrt{\varepsilon_n}(1 + \rho) \tag{7.2.36}$$

for $(\xi_0^n, \xi^n) \in \mathbf{D}_n$, $w^n \in T_Y(\tilde{y}^n) \cap B(0, \rho)$. *If* (z_0, z) *is a* $\mathbb{R} \times E$-*weak limit point of* $\{(z_{0n}, z_n)\}$ *then we have*

$$z_0\xi_0 + \langle z, \xi \rangle \geq 0, \tag{7.2.37}$$

$$z_0 \geq 0, \quad z \in \left(\liminf_{n \to \infty} T_Y(\tilde{y}^n)\right)^- \tag{7.2.38}$$

for $(\xi_0, \xi) \in \liminf_{n \to \infty} \mathbf{D}_n$.

Smooth version: *Assume the norm of* E *is Gâteaux differentiable off the origin. Then there exist sequences* $\{\tilde{u}^n\} \subseteq D(f) \cap D(f_0)$, $\{\tilde{y}^n\} \subseteq Y$ *satisfying* (7.2.34) *and a sequence* $\{(z_{0n}, z_n)\} \subseteq \mathbb{R} \times E^*$ *such that*

$$z_{0n}^2 + \|z_n\|^2 = 1, \qquad z_{0n} \geq 0, \tag{7.2.39}$$

$$z_{0n}\xi_0^n + \langle z_n, \xi^n - w^n \rangle \geq -\sqrt[4]{5}\sqrt{\varepsilon_n}(1 + \|w^n\|) \tag{7.2.40}$$

for $(\xi_0^n, \xi^n) \in \mathrm{Var}\,(f_0, f)(\tilde{u}^n)$ *and* $w^n \in I_Y(\tilde{y}^n)$, *or for* $(\xi_0^n, \xi^n) \in \mathrm{Der}\,(f_0, f)(\tilde{u}^n)$ *and* $w^n \in K_Y(\tilde{y}^n)$ *(an explicit expression for* z_n *is given below). If* $\tilde{y}^n \to \bar{y}$ *and* z *is a* E-*weak limit point of* $\{z_n\}$, *then*

$$z \in N_Y(\bar{y}). \tag{7.2.41}$$

Proof. We reduce this problem to the abstract time optimal problem setting $V_n = V$ and $\mathbf{E} = \mathbb{R} \times E$ equipped with the "Euclidean" norm $\|\mathbf{y}\| = \|(t, y)\| = (t^2 + \|y\|^2)^{1/2}$. The functions $\mathbf{f}_n : V \to \mathbf{E}$ are

$$\mathbf{f}_n(u) = (\max(f_0(u) + \varepsilon_n, m), f(u))$$

with $D(\mathbf{f}_n) = D(f) \cap D(f_0)$, and the target set is $\mathbf{Y} = \{m\} \times Y$, which is closed if Y is closed. Obviously, $\mathbf{f}_n(D(\mathbf{f}_n)) \cap \mathbf{Y} = \emptyset$ (otherwise we would have $u \in U$ with $f(u) \in Y$ and $f_0(u) \leq m - \varepsilon_n < m$, contradicting the definition of m). The sequence $\{\bar{u}^n\}$ is a solution of the abstract time optimal problem; a sequence associated to $\{\bar{u}^n\}$ is $\{\bar{\mathbf{y}}^n\} = \{(m, \bar{y}^n)\}$, since

$$\|\mathbf{f}_n(\bar{u}^n) - \bar{\mathbf{y}}^n\|$$
$$= \{(\max(f_0(\bar{u}^n) + \varepsilon_n, m) - m)^2 + \|f(\bar{u}^n) - \bar{y}^n\|^2\}^{1/2} \leq \sqrt{5}\varepsilon_n. \quad (7.2.42)$$

Finally, if $\Phi_n(u, \mathbf{y})$ is the function (7.2.17) corresponding to $\mathbf{f}_n(u)$ we have

$$\Phi_n(u, \mathbf{y}) = \Phi_n(u, (m, y)) = \|\mathbf{f}_n(u) - (m, y)\|$$
$$= \{(\max(f_0(u) + \varepsilon_n, m) - m)^2 + \|f(u) - y\|^2\}^{1/2}$$

for $(m, y) \in \{m\} \times Y$, thus lower semicontinuity of $\Phi_n(u, \mathbf{y})$ follows from the hypotheses on f and f_0 (see Example 4.1.2). All assumptions verified, we apply Theorem 7.2.2 for $\{\sqrt{5}\varepsilon_n\}$ obtaining sequences $\{\bar{u}^n\} \subseteq D(f) \cap D(f_0)$ and $\{\bar{\mathbf{y}}_n\} = \{(m, \bar{y}^n)\} \in \{m\} \times Y = \mathbf{Y}$ and a sequence $\{\mathbf{z}_n\} = \{(z_{0n}, z_n)\} \subseteq \mathbb{R} \times E^*$ (depending on $\{\mathbf{D}_n\}$ and on ρ) satisfying the conclusions of Theorem 7.2.2, which must now be translated into those of Theorem 7.2.5. Inequality (7.2.20) is (7.2.34); (7.2.19) implies that $\bar{u}^n \in D(\mathbf{f}_n) = D(f) \cap D(f_0)$ (and produces an inequality that we have omitted from the statement of Theorem 7.2.5). The first equality in (7.2.35) is obvious. The first coordinate of $\mathbf{f}(\bar{u}^n) - (m, y^n)$ equals $\max(f_0(\bar{u}^n) + \varepsilon_n, m) - m \geq 0$, so that Lemma 7.1.9 guarantees that $z_{0n} \geq 0$. Note that $T_{\mathbf{Y}}((m, \bar{y}^n)) = \{0\} \times T_Y(\bar{y}^n)$, so that the last inequality in (7.2.21) becomes

$$\langle (z_{0n}, z_n), (\xi_0^n, \xi^n) - (0, w^n) \rangle$$
$$= z_{0n}\xi_0^n + \langle z_n, \xi^n - w^n \rangle \geq -\sqrt[4]{5}\sqrt{\varepsilon}_n(1 + \rho) \quad (7.2.43)$$

for all $(\xi_0^n, \xi^n) \in \mathbf{D}_n$ and all $w^n \in T_Y(\bar{y}^n)$ with $\|w^n\| \leq \rho$, which is (7.2.36). However, the variations (ξ_0^n, ξ^n) are in $\mathrm{Var}\,\mathbf{f}_n(\bar{u}^n)$ rather than in $\mathrm{Var}\,\mathbf{f}(\bar{u}^n)$. To clarify this point, note that there are two possibilities:

(i) $f_0(\bar{u}^n) + \varepsilon_n > m$,   (ii) $f_0(\bar{u}^n) + \varepsilon_n \leq m$.

In case (i), every variation (ξ_0^n, ξ^n) of $\mathbf{f}$ at $\bar{u}^n$ is also a variation of $\mathbf{f}_n$ at $\bar{u}^n$. In case (ii), the first coordinate of $\mathbf{f}_n(\bar{u}^n) - \bar{\mathbf{y}}^n$ is zero, so again by virtue of Lemma 7.1.9

we may take $z_{0n} = 0$; here, it is irrelevant whether we use variations of $\mathbf{f}_n$ or of $\mathbf{f}$. However, we must show that for every $(\xi_0^n, \xi^n) \in \text{Var}\,\mathbf{f}(\tilde{u}^n)$ there exists an element of $\text{Var}\,\mathbf{f}_n(\tilde{u}^n)$ with the same second coordinate. This is done as in the end of the proof of Theorem 4.1.1, and we omit the details.

Finally, that $z_0 \geq 0$ follows from the fact that $z_{0n} \geq 0$; the second statement in (7.2.38) results from (7.2.23) for the abstract time optimal problem and from the characterization of $T_Y(\tilde{\mathbf{y}}^n) = T_Y((m, \tilde{y}^n))$. ∎

Proof of the smooth version. If the norm of E is Gâteaux differentiable off the origin, then the Euclidean norm of $\mathbf{E} = \mathbb{R} \times E$ is as well Gâteaux differentiable off the origin with

$$\langle \partial \|(t, y)\|, (\tau, \xi) \rangle = \frac{t\tau + \|y\| \langle \partial \|y\|, \xi \rangle}{(t^2 + \|y\|^2)^{1/2}}. \tag{7.2.44}$$

Inequality (7.2.24) translates to

$$\langle (z_{0n}, z_n), (\xi_0^n, \xi^n) - (0, w^n) \rangle$$
$$= z_{0n}\xi_0^n + \langle z_n, \xi^n - w^n \rangle \geq -\sqrt[4]{5}\sqrt{\varepsilon_n}(1 + \|w^n\|)$$

(which is (7.2.40)) with

$$(z_{0n}, z_n) = \partial \|\mathbf{f}_n(\tilde{u}^n) - \tilde{\mathbf{y}}^n\|$$
$$= \partial \| \max(f_0(\tilde{u}^n) + \varepsilon_n, m) - m, f(\tilde{u}^n) - \tilde{y}^n \|. \tag{7.2.45}$$

The arguments on comparison of $\text{Var}\,\mathbf{f}_n(\tilde{u}^n)$ with $\text{Var}\,\mathbf{f}(\tilde{u}^n)$ and of $\text{Der}\,\mathbf{f}_n(\tilde{u}^n)$ with $\text{Der}\,\mathbf{f}(\tilde{u}^n)$ are the same as in Theorem 4.1.1 (and in the nonsmooth version of this theorem). ∎

Corollary 7.2.6. (*of Lemma* 6.4.3). *Let* $\{\mathbf{D}_n\}$ *and* ρ *be as in Theorem* 7.2.5. *Assume there exists a precompact sequence* $\{Q_n\}$ *such that the set* Δ *in* (7.2.30) *contains an interior point in* E *for* n_0 *large enough, where*

$$\Delta_n = \Pi(\mathbf{D}_n) - T_Y(\tilde{y}^n) \cap B(0, \rho), \tag{7.2.46}$$

Π *the projection from* $\mathbb{R} \times E$ *into* E. *Then the sequence* (z_{0n}, z_n) *corresponding to* $\{\mathbf{D}_n\}$ *and* ρ *cannot have zero weak limit points, so that the vector* (z_0, z) *in* (7.2.36) *is not zero.*

Smooth version: *Assume the norm of* E *is Gâteaux differentiable off the origin. Then the same result holds with* $\Delta_n = \Pi(\text{Var}\,(f_0, f)(\tilde{u}^n)) - I_Y(\tilde{y}^n) \cap B(0, \rho)$ *or* $\Delta_n = \Pi(\text{Der}\,(f_0, f)(\tilde{u}^n)) - K_Y(\tilde{y}^n) \cap B(0, \rho)$.

Proof. If Δ contains an interior point in E for $\Delta_n = \Pi(\mathbf{D}_n) - T_Y(\tilde{y}^n) \cap B(0, \rho)$, then

$$\bigcap_{n=n_0}^{\infty} \left\{ \mathbf{D}_n - T_Y(\tilde{y}^n) \cap B(0, \rho) + [-1, 1] \times Q_n \right\} \tag{7.2.47}$$

contains an interior point in $\mathbf{E} = \mathbb{R} \times E$ for n_0 large enough and the argument in Corollary 6.4.5 can be applied. The other statements are proved in the same way. ∎

Remark 7.2.7. Using the same arguments in Remark 6.4.6, one may replace Δ_n in the intersection (7.2.30) by $\overline{\text{conv}}(\Delta_n)$; of course, when the tangent cone is used, only closure is necessary.

Theorem 7.2.2 has some curious geometric consequences that we explore below.

Example 7.2.8. Let $X \subseteq E$ be a closed set, $f(x) = x$ ($x \in X$). Then, if $x \in X$ we have

$$\text{Var } f(x) = K_X(x) \cap B(0, 1), \qquad \text{Der } f(x) = I_X(x) \cap B(0, 1).$$

The inclusions $\subseteq$ are obvious. For the converse, note that if $(x_k - x)/h_k \to w$ with $\|w\| = 1$ then $(x_k - x)/\rho_k \to w$, where $\rho_k = h_k \|x_k - x\|/h_k = \|x_k - x\|$. For Der $f(x)$ use the characterization given in Example 3.3.3.

Example 7.2.9. Let $X, Y \subseteq E$ be closed and disjoint. Let $\{\bar{x}^n\} \subseteq X$, $\{\bar{y}^n\} \subseteq Y$, $\varepsilon_n = \|\bar{x}^n - \bar{y}^n\|$. We apply Theorem 7.2.2 with convex sets $D_n = T_X(x) \cap B(0, 1)$ $\subseteq K_X(x) \cap B(0, 1) = \text{Var} f(x)$ and obtain sequences $\{\tilde{x}^n\} \subseteq X$, $\{\tilde{y}^n\} \subseteq Y$ such that

$$\|\tilde{x}^n - \tilde{y}^n\| \le \|\bar{x}^n - \bar{y}^n\|, \qquad \|\tilde{x}^n - \bar{x}^n\| + \|\tilde{y} - \bar{y}^n\| \le \sqrt{\varepsilon_n}$$

and a sequence $\{z_n\} \subseteq E^*$ such that

$$\|z_n\| = 1, \qquad \langle z_n, \xi^n - w^n \rangle \ge -2\sqrt{\varepsilon_n}$$

for $\xi^n \in T_X(\tilde{x}^n) \cap B(0, 1)$ and $w^n \in T_Y(\tilde{y}^n) \cap B(0, 1)$, a sort of "local approximate separation theorem" for arbitrary sets.

7.3. The Minimum Principle in Banach Spaces

We apply the results in **7.2** to a control system

$$y'(t) = A(t)y(t) + f(t, y(t), u(t)), \qquad y(0) = \zeta \qquad (7.3.1)$$

in an arbitrary Banach space E. As in Chapter 6, we assume that the Cauchy problem for the linear part $y'(t) = A(t)y(t)$ of (7.3.1) is well posed in $0 \le t \le T$ with solution operator $S(t, \tau)$. The target condition is $y(\bar{t}) \in Y$ with Y closed, and the nonlinear term $f(t, y, u(t))$ satisfies Hypothesis III for each $u(\cdot) \in C_{ad}(0, T; U)$ with $K(\cdot, c), L(\cdot, c)$ independent of $u(\cdot)$. Finally, the admissible control space $C_{ad}(0, \bar{t}; U)$ is saturated and patch complete. Statements of theorems are exactly as in **6.5** and **6.6**; thus, we just list these theorems and point out differences in statements and proofs (if any).

Theorem 6.5.1. The adjoint variational equation (6.5.3)

$$\bar{z}'(s) = -\{A(s)^* + \partial_y f(s, y(s, \bar{u}), \bar{u}(s))^*\}z(s), \qquad z(\bar{t}) = z \qquad (7.3.2)$$

lives in the dual space E^* and the final condition z satisfies

$$z \in \left(\liminf_{n \to \infty} T_Y(\bar{y}^n)\right)^-, \qquad (7.3.3)$$

$\{\bar{y}^n\} \subseteq Y$ the sequence produced in Theorem 7.2.2; in the smooth case (norm of E Gâteaux differentiable off the origin), $z \in N_Y(\bar{y})$. Solutions of (7.3.2) are understood, as in Remark 5.5.11, as

$$\bar{z}(s) = S(\bar{t}, s; \bar{u})^* z, \qquad (7.3.4)$$

$S(t, s; \bar{u})$ the solution operator of

$$\xi'(t) = \{A(t) + \partial_y f(t, y(t, \bar{u}), \bar{u}(t))\}\xi(t). \qquad (7.3.5)$$

No assumptions are placed on $A(s)^*$ or on $\partial_y f(s, y(s, \bar{u}), \bar{u}(s))^*$. To prove the integral form

$$\int_0^{\bar{t}} \langle \bar{z}(\sigma), f(\sigma, y(\sigma, \bar{u}), v(\sigma)) - f(\sigma, y(\sigma, \bar{u}), \bar{u}(\sigma)) \rangle d\sigma \geq 0 \qquad (7.3.6)$$

of the minimum principle, we select a sequence $\{t_n\}$, $t_n < \bar{t}$ with $t_n \to \bar{t}$ and use Theorem 7.2.2 for the abstract time optimal problem in the spaces $V_n = B_n(\bar{u}, \delta)$ where $B(\bar{u}, \delta)$ is the ball provided by Lemma 6.1.1 and $B_n(\bar{u}, \delta) \subseteq C_{ad}(0, t_n; U)$ is the ball of center $\bar{u}$ and radius δ in $C_{ad}(0, t_n; U)$). The functions $\mathbf{f}_n : V_n \to E$ are

$$\mathbf{f}_n(u) = y(t_n, u), \qquad (7.3.7)$$

continuous on account of Lemma 6.1.1. The sets $D_n \subseteq \text{Der } f_n(\bar{u}^n)$ required in Theorem 7.2.2 are $D_n = \{t_n^{-1}\xi(t_n, \bar{u}^n, \mathbf{p}, \mathbf{v})\}$, where $t_n^{-1}\xi(t_n, \bar{u}^n, \mathbf{p}, \mathbf{v})$ are the directional derivatives produced in Theorem 6.2.7 by means of patch perturbations, with $\mathbf{p}$ a probability vector (of arbitrary length m) and $\mathbf{v}(\cdot)$ an m-vector of elements of $C_{ad}(0, t_n; U)$. That the D_n are convex follows from formula (6.2.20). We fix $\rho > 0$ and obtain from Theorem 7.2.2 a sequence $\{z_n\}$ satisfying (7.2.21) for $\xi^n \in D_n$ and $w^n \in T_Y(\bar{y}^n) \cap B(0, \rho)$, thus if z is a E-weak limit point of $\{z_n\}$ we have inequality (7.2.22) for $\xi \in \liminf_{n \to \infty} D_n$. These limits computed in Lemma 6.5.2, we end up with

$$\langle z, \xi(\bar{t}, \bar{u}, v) \rangle \geq 0 \qquad (7.3.8)$$

as a particular case of (7.2.22), where $\xi(\bar{t}, \bar{u}, v)$ denotes a one-patch directional derivative ($m = 1$, $\mathbf{p} = \{1\}$, $\mathbf{v}(\cdot) = \{v(\cdot)\}$). From then on, the proof of the minimum principle proceeds just as that of Theorem 6.5.1. To pass to the pointwise minimum principle

$$\langle \bar{z}(s), f(s, y(s, \bar{u}), \bar{u}(s)) \rangle = \min_{v \in U} \langle \bar{z}(s), f(s, y(s, \bar{u}), v) \rangle \tag{7.3.9}$$

a.e. in $0 \le s \le \bar{t}$, we assume E^* separable, so that the costate is strongly measurable (Example 5.0.39) and again do as in Theorem 6.5.1. If E is reflexive and separable and $A(t) \equiv A =$ infinitesimal generator of a strongly continuous semigroup, we argue along the lines of Example 6.5.5 and show that the costate $\bar{z}(s)$ is continuous.

The crucial difference between the Hilbert space (or smooth Banach space) case and the general Banach space result is that in the latter we need to work with a convex subset D_n of Der $f(\tilde{u}^n)$ rather than with Der $f(\tilde{u}^n)$ itself; hence the need for multipatch perturbations. Another difference is that, in the general case, the parameter ρ has an influence on the choice of the sequence $\{z_n\}$. None of these dissimilarities is "seen" in the final result.

Lemma 6.5.3. We use this time Corollary 7.2.3, which amounts to replacing (6.5.15) by

$$\bigcap_{n=n_0}^{\infty} \left\{ t_n^{-1} R(0, t_n; U, \tilde{u}^n) - T_Y(\tilde{y}^n) \cap B(0, \rho) + Q_n \right\} \tag{7.3.10}$$

In the smooth case, we may use the contingent cones as in Lemma 6.5.3.

Remark 6.5.4 and Remark 6.5.7. Valid without changes.

Example 6.5.10. Valid without changes; to prove the pointwise minimum principle with variable control set

$$\langle \bar{z}(s), f(s, y(s, \bar{u}), \bar{u}(s)) \rangle = \min_{v \in U(s)} \langle \bar{z}(s), f(s, y(s, \bar{u}), v) \rangle \tag{7.3.11}$$

we need E^* to be separable and the ad-hoc assumption there.

For the general optimal control problem with cost functional

$$y_0(t, u) = \int_0^t f_0(\tau, y(\tau, u), u(\tau)) d\tau + g_0(t, y(t, u)), \tag{7.3.12}$$

we require Hypothesis III^0 for f_0 and existence of the Fréchet derivative $\partial_y g_0(t, y)$ in $[0, T] \times E$. Lemma 6.6.1 was proved for Banach spaces and needs no retouching. The formula for $\xi_0(t, u, \mathbf{p}, \mathbf{v})$ is again (6.6.7),

$$\xi_0(t, u, \mathbf{p}, \mathbf{v}) = \int_0^t \left\langle \partial_y f_0(\tau, y(\tau, u), u(\tau)), \xi(\tau, u, \mathbf{p}, \mathbf{v}) \right\rangle d\tau$$

$$+ \sum_{j=1}^m p_j \int_0^t \left\{ f_0(\tau, y(\tau, u), v_j(\tau)) - f_0(\tau, y(\tau, u), u(\tau)) \right\} d\tau$$

$$+ \left\langle \partial_y g_0(t, y(t, u)), \xi(t, u, \mathbf{p}, \mathbf{v}) \right\rangle, \tag{7.3.13}$$

but now $\langle \cdot, \cdot \rangle$ indicates E, E^*-duality; E^* is the home space for $\partial_y f_0$ and $\partial_y g_0$.

Theorem 6.6.2. The adjoint variational equation

$$\bar{z}'(s) = -\{A(s)^* + \partial_y f(s, y(s, \bar{u}), \bar{u}(s))^*\}z(s) - z_0 \partial_y f_0(s, y(s, \bar{u}), \bar{u}(s)),$$

$$\bar{z}(\bar{t}) = z + z_0 \partial_y g_0(\bar{t}, y(\bar{t}, \bar{u})) \tag{7.3.14}$$

lives in the dual space E^*. Solutions are again understood as in Remark 5.5.11,

$$\bar{z}(s) = S(\bar{t}, s; \bar{u})^*\{z + z_0 \partial_y g_0(\bar{t}, y(\bar{t}, \bar{u}))\}$$

$$+ z_0 \int_s^{\bar{t}} S(\sigma, s; \bar{u})^* \partial_y f_0(\sigma, y(\sigma, \bar{u}), \bar{u}(\sigma))d\sigma, \tag{7.3.15}$$

the integral interpreted as in Theorem 5.5.12; by Hypothesis III^0, the inhomogeneous term $\partial_y f_0(\cdot, y(\cdot, \bar{u}), \bar{u}(\cdot))$ is an E^*-valued strongly measurable function. To prove the integral form

$$z_0 \int_0^{\bar{t}} \{f_0(\sigma, y(\sigma, \bar{u}), v(\sigma)) - f_0(\sigma, y(\sigma, \bar{u}), \bar{u}(\sigma))\}d\sigma$$

$$+ \int_0^{\bar{t}} \langle \bar{z}(\sigma), f(\sigma, y(\sigma, \bar{u}), v(\sigma)) - f(\sigma, y(\sigma, \bar{u}), \bar{u}(\sigma)) \rangle d\sigma \geq 0 \tag{7.3.16}$$

of the maximum principle, we apply this time Theorem 7.2.2 in the space $V = B(\bar{u}, \delta)$, $B(\bar{u}, \delta)$ the ball in Lemma 6.1.1. The functions $\mathbf{f}, \mathbf{f}_0$ in the abstract minimization problem are

$$\mathbf{f}(u) = y(\bar{t}, u), \qquad \mathbf{f}_0(u) = y_0(\bar{t}, u). \tag{7.3.17}$$

That $\mathbf{f}$ (resp. $\mathbf{f}_0$) is continuous has been proved in full Banach space generality in Lemma 6.1.1 (resp. Lemma 6.6.3). The sets $\mathbf{D}_n \subseteq \mathrm{Der}\,(f_0, f)(\bar{u}^n) \subseteq \mathbb{R} \times E$ required in Theorem 7.2.5 are $\mathbf{D}_n = \{t^{-1}\xi_0(\bar{t}, \bar{u}^n, \mathbf{p}, \mathbf{v}), \bar{t}^{-1}\xi(\bar{t}, \bar{u}^n, \mathbf{p}, \mathbf{v})\}$, where $\bar{t}^{-1}\xi(t, \bar{u}^n, \mathbf{p}, \mathbf{v})$ (resp. $\bar{t}^{-1}\xi_0(t, \bar{u}^n, \mathbf{p}, \mathbf{v})$) are the directional derivatives produced in Theorem 6.2.7 (resp. Lemma 6.6.1) with $\mathbf{p}$ a probability vector (of any length m) and $\mathbf{v}(\cdot)$ an arbitrary m-vector of elements of $C_{ad}(0, t; U)$. That each $\mathbf{D}_n$ is convex follows from formulas (6.2.20) and (6.6.7). We fix $\rho > 0$ and obtain from Theorem 7.2.5 a sequence $\{(z_{0n}, z_n)\}$ satisfying (7.2.35) and (7.2.36) for $(\xi_0^n, \xi^n) \in \mathbf{D}_n$ and $w^n \in T_Y(\tilde{y}^n) \cap B(0, \rho)$; thus if (z_0, z) is an E-weak limit point of $\{(z_{0n}, z_n\}$ we have inequality (7.2.37) for $(\xi_0, \xi) \in \liminf_{n \to \infty} \mathbf{D}_n$. These limits computed in Lemma 6.5.2 and Lemma 6.6.4, we end up with

$$z_0\xi_0(\bar{t}, \bar{u}, v) + \langle z, \xi(\bar{t}, \bar{u}, v) \rangle \geq 0, \tag{7.3.18}$$

where $\xi(\bar{t}, \bar{u}, v), \xi_0(\bar{t}, \bar{u}, v)$ denote one-patch directional derivatives. From then on, the proof of the minimum principle is just like that of Theorem 6.6.2.

For the pointwise minimum principle

$$z_0 f_0(s, y(s, \bar{u}), \bar{u}(s)) + \langle \bar{z}(s), f(s, y(s, \bar{u}), \bar{u}(s)) \rangle$$

$$= \min_{v \in U} \{z_0 f_0(s, y(s, \bar{u}), v) + \langle \bar{z}(s), f(s, y(s, \bar{u}), v) \rangle\} \tag{7.3.19}$$

a.e. in $0 \leq s \leq \bar{t}$, we assume E^* separable, so that the costate is strongly measurable, and do as in Theorem 6.6.2.

Lemma 6.6.5. Proof on the basis of Corollary 7.2.6; we replace (6.6.18) by

$$\bigcap_{n=n_0}^{\infty} \{\bar{t}^{-1} R(0, \bar{t}; U, \tilde{u}^n) - T_Y(\tilde{y}^n) \cap B(0, \rho) + Q_n\}. \tag{7.3.20}$$

Example 6.6.7. Valid without changes.

Example 6.6.8. Valid without changes.

Example 6.6.10. Valid without changes; to prove the pointwise minimum principle with variable control set

$$z_0 f_0(s, y(s, \bar{u}), \bar{u}(s)) + \langle \bar{z}(s), f(s, y(s, \bar{u}), \bar{u}(s)) \rangle$$
$$= \min_{v \in U(s)} \{z_0 f_0(s, y(s, \bar{u}), v) + \langle \bar{z}(s), f(s, y(s, \bar{u}), v) \rangle\}, \tag{7.3.21}$$

we require E^* to be separable and the ad-hoc Assumption.

7.4. Fractional Powers of Infinitesimal Generators. Analytic Semigroups. Duality

Let $S(t)$ be a strongly continuous semigroup with infinitesimal generator A. Assume that

$$\|S(t)\| \leq Ce^{-ct} \quad (t \geq 0) \tag{7.4.1}$$

for some $c > 0$. We can define fractional powers of the infinitesimal generator (or, rather, of $-A$) by the formula

$$(-A)^{-\alpha} y = \frac{1}{\Gamma(\alpha)} \int_0^{\infty} t^{\alpha-1} S(t) y \, dt \tag{7.4.2}$$

for $\alpha > 0$. Obviously, $(-A)^{-\alpha}$ is a bounded operator. Fractional powers with $\alpha \leq 0$ are defined by composition with A. We need mostly the range $0 \leq \alpha \leq 1$, where the definition is

$$(-A)^{\alpha} = (-A)(-A)^{\alpha-1}, \tag{7.4.3}$$

the domain $D((-A)^{\alpha})$ consisting of all $y \in E$ such that $(-A)^{\alpha-1} y \in D(A)$. The operator $(-A)^{\alpha}$, being the composition of a bounded operator and a closed operator, is closed. We refer the reader to Yosida [1978, p. 260] for properties of fractional powers. In particular, if A is self adjoint with $\sigma(A) \subseteq (-\infty, -\kappa]$ (or, more generally, normal with spectrum contained in $\text{Re } \lambda \leq -\kappa$) where $\kappa > 0$, fractional powers coincide with those defined by the extended functional calculus in **5.0** (at least if the function employed therein is the principal value of $\lambda^{-\alpha}$). Finally, we note that for a general semigroup generator A we can define fractional powers $(\lambda I - A)^\alpha$, with $\lambda > \omega$, ω the constant in (5.2.9); in fact, $A - \lambda I$ generates the semigroup $e^{-\lambda t} S(t)$.

Let $S(\cdot)$ be a strongly continuous semigroup. The semigroup is **analytic** if $S(\cdot)$ can be extended to an (E, E)-valued function $S(z)$ defined and analytic in the sector

$$\Sigma_0(\phi) = \{z; |\arg z| < \phi, z \neq 0\} \tag{7.4.4}$$

and strongly continuous in

$$\Sigma(\phi) = \{z; |\arg z| \leq \phi\}, \tag{7.4.5}$$

where $0 < \phi \leq \pi/2$ (if ϕ must be precised in the definition, we say that the semigroup $S(\cdot)$ **belongs to the class** $\mathcal{A}_\phi$). Using an analytic continuation argument (and continuity up to the boundary of the sector) we prove that the semigroup equation (5.2.8) extends to s, t, complex: $S(z + \zeta) = S(z)S(\zeta)$ $(z, \zeta \in \Sigma(\phi))$. It follows from the uniform boundedness principle (Theorem 5.0.1) that $\|S(z)\|$ is bounded in $|\arg z| \leq \phi$, $|z| \leq 1$. We use then the extended semigroup equation and an argument similar to that following (5.2.8) applied to $S(te^{i\psi})$, $t \geq 0$, $|\psi| \leq \phi$ to show that

$$\|S(z)\| \leq Ce^{a|z|} \quad (z \in \Sigma(\phi)). \tag{7.4.6}$$

Lemma 7.4.1. *Let $S(\cdot)$ be analytic. Then $S(z)E \subseteq D(A)$ for $z \in \Sigma_0(\phi)$ and $AS(z)$ is bounded. Moreover, there exists b such that*

$$\|AS(t)\| \leq Ct^{-1}e^{bt} \quad (t > 0). \tag{7.4.7}$$

Proof. Since the function $S(\cdot)$ is analytic in $\Sigma_0(\phi)$, the derivative $S'(z)$ exists. On the real axis we have $S'(t)y = AS(t)y$ for $y \in D(A)$, thus by analytic continuation $S'(z)y = AS(z)y$ for $z \in \Sigma_0(\phi)$. If $y \in E$, select a sequence $\{y_n\} \in D(A)$ with $y_n \to y$; since $S'(z)$ is a bounded operator and $S'(z)y_n = AS(z)y_n$, it follows from closedness of A that $S(z)y \in D(A)$ and that $AS(z)$ is bounded. The estimate (7.4.7) follows bounding Cauchy's formula for the derivative,

$$S'(t) = \frac{1}{2\pi i} \int_{|\zeta - t| = t \sin \phi} \frac{S(\zeta)}{(\zeta - t)^2} d\zeta, \tag{7.4.8}$$

$t \sin \phi$ the radius of the largest circle of center t contained in $\Sigma_0(\phi)$. ∎

A similar argument replacing the positive real axis by a ray $\text{Arg}(\zeta) = \psi$ shows that (7.4.7) can be extended to $z \in \Sigma(\psi)$ with $0 \leq \psi < \phi$,

$$\|S(z)\| \leq C_\psi |z|^{-1} e^{b|z|} \quad (z \in \Sigma(\psi), 0 \leq \psi \leq \phi), \tag{7.4.9}$$

with $b < 0$ if $a < 0$ in (7.4.6).

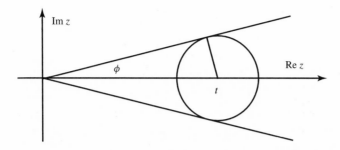

Figure 7.5.

Consider formula (5.2.18),

$$R(\lambda; A)y = \int_0^\infty e^{-\lambda t} S(t)y \, dt, \qquad (7.4.10)$$

applied to a semigroup in $\mathcal{A}_\phi$. Using analyticity and the bound (7.4.6) we may deform the path of integration for λ large enough into either of the rays $t e^{\pm i\phi}$, $0 \le t \le \infty$. This extends the resolvent $R(\lambda; A)$ to the half planes $\mathrm{Re}(\lambda e^{i\phi}) > a$ and $\mathrm{Re}(\lambda e^{-i\phi}) > a$ and produces the two estimates

$$\|R(\lambda; A)\| \le C(\mathrm{Re}(\lambda e^{i\phi}) - a)^{-1} \quad (\mathrm{Re}(\lambda e^{i\phi}) > a),$$

$$\|R(\lambda; A)\| \le C(\mathrm{Re}(\lambda e^{-i\phi}) - a)^{-1} \quad (\mathrm{Re}(\lambda e^{-i\phi}) > a).$$

These imply that if $\alpha = a/\cos\phi$ and $0 \le \psi < \phi$ then $R(\lambda; A)$ exists in the sector

$$\Sigma(\alpha, \pi/2 + \psi) = \{\lambda; \arg(\lambda - \alpha) \le \pi/2 + \psi, \lambda \ne \alpha\}, \qquad (7.4.11)$$

and

$$\|R(\lambda; A)\| \le C_\psi |\lambda - \alpha|^{-1} \quad (\lambda \in \Sigma(\alpha, \pi/2 + \psi)). \qquad (7.4.12)$$

The converse is also true.

Theorem 7.4.2. *Assume $R(\lambda; A)$ exists in a sector $\Sigma(\alpha, \pi/2 + \phi)$ and that (7.4.11) holds there. Then A is the infinitesimal generator of a semigroup of class $\mathcal{A}_\psi$ for $0 \le \psi < \phi$.*

The proof can be found in many functional analysis books, for instance the author [1983, p. 184] and includes the statement that

$$S(z) = \frac{1}{2\pi i} \int_{\Gamma(\alpha, \pi/2 + \psi)} e^{\lambda z} R(\lambda; A) \, d\lambda, \qquad (7.4.13)$$

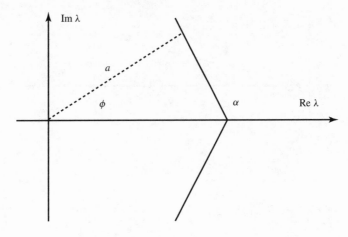

Figure 7.6.

where the contour $\Gamma(\alpha, \pi/2 + \psi)$ is the boundary of $\Sigma(\alpha, \pi/2 + \psi)$ oriented counterclockwise with respect to $\sigma(A)$; (7.4.13) is understood as an integral in (E, E) and exists for z in any sector $\Sigma_0(\psi')$ with $\psi' < \psi$. One consequence of (7.4.13) is that $\|S(t)\| \le C e^{\omega t}$ for $t \ge 0$ if $\omega > \omega_0$, where

$$\omega_0 = \max_{\lambda \in \sigma(A)} \mathrm{Re}\,\lambda, \tag{7.4.14}$$

a fact that is notoriously false for general semigroups (see Hille–Phillips [1957, p. 665]). In the realm of analytic semigroups, (7.4.14) has numerous applications: one is that an analytic semigroup satisfying (7.4.1) with $c > 0$ will satisfy (7.4.6) with $a < 0$ in some sector $\Sigma(\phi)$ (Example 7.4.19). It follows from this and from Lemma 7.4.1 that (7.4.1) with $c > 0$ implies (7.4.7) with $b < 0$.

Lemma 7.4.3. *Let A be an analytic semigroup satisfying* (7.4.1) *with $c > 0$, and let* $(-A)^\alpha$ *be the fractional powers defined in* (7.4.3) *with $0 \le \alpha \le 1$. Then $S(t)E \subseteq D((-A)^\alpha)$, $(-A)^\alpha S(t)$ is continuous in (E, E) in $t > 0$ and there exists $k > 0$ such that*

$$\|(-A)^\alpha S(t)\| \le C_\alpha t^{-\alpha} e^{-kt} \quad (t > 0). \tag{7.4.15}$$

Proof. It follows from the observations above that (7.4.7) holds with $b = -k < 0$. Applying Lemma 5.0.17 to (7.4.2) we deduce that if $y \in D(A)$ then $(-A)^{\alpha-1} y \in D(A)$, so that $D((-A)^\alpha) \supseteq D(A)$. We have

$$(-A)^\alpha S(t)y = \frac{1}{\Gamma(\alpha)} \int_0^\infty s^{-\alpha} A S(t+s)y\,ds. \tag{7.4.16}$$

Since the operator $(-A)^\alpha$ is closed and the operator on the right-hand side is bounded, $(-A)^\alpha S(t)$ is bounded for $t > 0$. Continuity in the norm of (E, E) is obvious. Finally,

$$\|(-A)^\alpha S(t)\| \le Ce^{-kt} \int_0^\infty s^{-\alpha}(t+s)^{-1}e^{-ks}ds,$$

which produces the bound (7.4.15).   ∎

If $S(\cdot)$ is an arbitrary analytic semigroup, we can apply the preceding theory to construct fractional powers of $\lambda I - A$ for λ real and large enough.

The adjoint $S^*(t) = S(t)^*$ of a strongly continuous semigroup in E is a semigroup in E^* (that is, it satisfies the semigroup equations (5.2.7) and (5.2.8)). If E is reflexive, then $S(\cdot)^*$ is strongly continuous with infinitesimal generator A^* (Theorem 5.2.6), but in general $S(\cdot)^*$ may not be strongly continuous; also, the adjoint A^* of the infinitesimal generator A may not be densely defined.

Example 7.4.4. Let $S(\cdot)$ be the translation group

$$S(t)y(x) = y(x+t) \quad (-\infty < t < \infty)$$

in $L^1(-\infty, \infty)$. It is easy to show that $S(\cdot)$ is strongly continuous (using for instance Vitali's theorem) and that its infinitesimal generator is

$$Ay(x) = y'(x),$$

where $D(A)$ consists of all absolutely continuous $y(\cdot) \in L^1(-\infty, \infty)$ with $y'(\cdot) \in L^1(-\infty, \infty)$. We have $L^1(-\infty, \infty)^* = L^\infty(-\infty, \infty)$ and

$$S^*(t)y(x) = y(x-t) \quad (-\infty < t < \infty).$$

The adjoint semigroup is not strongly continuous; for instance, if χ is the characteristic function of $[0, \infty)$ then $\|S^*(t')\chi - S^*(t)\chi\|_{L^\infty(-\infty,\infty)} = 1$ for $t \ne t'$. The adjoint A^* of the infinitesimal generator is

$$Ay(x) = -y'(x), \tag{7.4.17}$$

where $D(A^*)$ consists of all absolutely continuous $y(\cdot) \in L^\infty(\cdot)$ such that $y'(\cdot) \in L^\infty(-\infty, \infty)$; $D(A^*)$ is not dense in $L^\infty(-\infty, \infty)$.

For nonreflexive spaces, the dual E^* and the adjoint semigroup can be "replaced" by nicer things. Let E be a Banach space, $S(t)$ a strongly continuous semigroup in E. Define

$$E^\odot = \overline{D(A^*)} \subseteq E^*.$$

Theorem 7.4.5. (a) *The restriction $S^\odot(t)$ of $S(t)^*$ to $E^\odot$ is a strongly continuous semigroup in $E^\odot$ with infinitesimal generator $A^\odot$ defined by*

$$A^\odot y^* = A^* y^*, \qquad D(A^\odot) = \{y \in D(A^*); A^*y \in E^\odot\}. \tag{7.4.18}$$

(*b*) *We have*

$$E^{\circ} = \{y^* \in E^*; S(t)^* y^* \text{ is continuous in } t \geq 0\}. \qquad (7.4.19)$$

Proof. We take below $\lambda > \omega$, ω the constant in (5.2.9). We have $AR(\lambda; A) = \lambda R(\lambda; A) - I$; taking adjoints and keeping in mind that A and $R(\lambda; A)$ commute we obtain

$$A^* R(\lambda; A^*) y^* = \lambda R(\lambda; A^*) y^* - y^*. \qquad (7.4.20)$$

We use this equality for $y^* \in E^{\circ}$; since $R(\lambda; A^*) y^* \in D(A^*) \subseteq E^{\circ}$ we deduce that $A^* R(\lambda; A^*) y^* \in E^{\circ}$. In other words,

$$R(\lambda; A^*) E^{\circ} \subseteq D(A^{\circ}). \qquad (7.4.21)$$

Using (7.4.20) for $y^* \in D(A^*)$ and commuting operators on the left side we deduce that

$$\lambda R(\lambda; A^*) y^* \to y^*. \qquad (7.4.22)$$

On the other hand, $\|\lambda R(\lambda; A^*)\| = \|\lambda R(\lambda; A)\| = 0(1)$ as $\lambda \to \infty$, so that (7.4.22) extends to $y^* \in \overline{D(A^*)} = E^{\circ}$. This relation and the inclusion (7.4.21) show that $D(A^{\circ})$ is dense in E°. The inclusion also implies that if $y^* \in E^{\circ}$ then

$$(\lambda I - A^{\circ}) R(\lambda; A^*) y^* = (\lambda I - A^*) R(\lambda; A^*) y^* = y^*.$$

On the other hand, if $y^* \in D(A^{\circ})$,

$$R(\lambda; A^*)(\lambda I - A^{\circ}) y^* = R(\lambda; A^*)(\lambda I - A^*) y^* = y^*$$

so that $(\lambda I - A^{\circ})^{-1} = R(\lambda; A^{\circ})$ exists and coincides with the restriction of $R(\lambda; A^*)$ to E°, that is,

$$R(\lambda; A^{\circ}) = R(\lambda; A^*)|_{E^{\circ}}. \qquad (7.4.23)$$

Accordingly,

$$\|R(\lambda; A^{\circ})^n\| \leq \|R(\lambda; A^*)^n\| = \|R(\lambda; A)^n\| \quad (\lambda \geq \omega, n = 1, 2, \ldots)$$

and the operator A° satisfies the premises of the Hille-Yosida Theorem 5.2.2 in E°, thus it generates a strongly continuous semigroup $S^{\circ}(\cdot)$ in E°. If $y^* \in E^{\circ}$ and $y \in E$ we have

$$\int_0^{\infty} e^{-\lambda t} \langle S^{\circ}(t) y^*, y \rangle dt$$

$$= \langle R(\lambda; A^{\circ}) y^*, y \rangle = \langle R(\lambda; A^*) y^*, y \rangle = \langle y^*, R(\lambda; A) y \rangle$$

$$= \int_0^{\infty} e^{-\lambda t} \langle y^*, S(t) y \rangle dt = \int_0^{\infty} e^{-\lambda t} \langle S(t)^* y^*, y \rangle dt.$$

By uniqueness of Laplace transforms, $S(t)^* y^* = S^{\odot}(t) y^*$. In particular, it follows that $S(t)^* E^{\odot} \subseteq E^{\odot}$ and, *a fortiori*, that $S^{\odot}(t)$ is the restriction of $S(t)^*$ to $E^{\odot}$. Conversely, let $y^* \in E^*$ such that $S(t)^* y^*$ is continuous in $t \geq 0$. Take $\varepsilon > 0$ and δ so small that $\|S(t)^* y^* - y^*\| \leq \varepsilon$ in $0 \leq t \leq \delta$. Then

$$\|\lambda R(\lambda; A^*) y^* - y^*\| \leq \lambda \int_0^{\infty} e^{-\lambda t} \|S(t)^* y^* - y^*\| \, dt \leq \varepsilon + o(1)$$

as $\lambda \to \infty$ dividing the interval of integration in $(0, \delta)$ and (δ, ∞). Accordingly, $\lambda R(\lambda; A^*) y^* \to y^*$ as $\lambda \to \infty$, so that $y^* \in E^{\odot}$. This ends the proof of Theorem 7.4.5. ∎

We call $E^{\odot}$ the **Phillips dual** of E with respect to A (or with respect to $S(\cdot)$) and $S^{\odot}(\cdot)$ the **Phillips adjoint** of the semigroup $S(\cdot)$. In many control theory situations where E is not reflexive, $S^{\odot}(\cdot)$ plays the role of the adjoint semigroup in the adjoint variational equation.

Define a norm $\| \cdot \|_0$ in E by

$$\|y\|_0 = \sup_{y^* \in E^{\odot}, \|y^*\| \leq 1} |\langle y^*, y \rangle|. \tag{7.4.24}$$

Lemma 7.4.6. *Let C be the constant in (5.2.15) or (5.2.16). Then*

$$\|y\|_0 \leq \|y\| \leq C \|y\|_0. \tag{7.4.25}$$

Proof. Since $|\langle y^*, y \rangle| \leq \|y^*\|_{E^*} \|y\|_E$, the first inequality is obvious. As for the second, let $\varepsilon > 0$, $y^* \in E^*$, $\|y^*\| = 1$ such that $\langle y^*, y \rangle \geq \|y\| - \varepsilon$. Then $y_\lambda^* = \lambda R(\lambda; A^*) y^* \in D(A^*) \subseteq E^{\odot}$ with $\|y_\lambda^*\| \leq C$ and $\langle y_\lambda^*, y \rangle = \langle y_\lambda^*, \lambda R(\lambda; A) y \rangle \to \langle y^*, y \rangle$ as $\lambda \to \infty$; accordingly, $\|y\|_0 \geq (\|y\| - \varepsilon)/C$. ∎

Example 7.4.7. In Example 7.4.4, $E^{\odot}$ consists of all $y(\cdot) \in L^{\infty}(-\infty, \infty)$ uniformly continuous in $-\infty < x < \infty$ and $A^{\odot}$ is given by (7.4.17), the domain $D(A^{\odot})$ consisting of all $y(\cdot) \in E^{\odot}$ continuously differentiable with bounded uniformly continuous derivative $y'(\cdot)$.

Analytic semigroups are particular examples of semigroups **uniformly continuous in** $t > 0$ (that is, continuous in the operator norm in $t > 0$). Phillips adjoints enjoy special privileges for these semigroups.

Lemma 7.4.8. *Let $S(t)$ be continuous in (E, E) for $t > 0$. Then*

$$S(t)^* E^* \subseteq E^{\odot} \qquad (t \geq 0), \tag{7.4.26}$$

$$S(s + t)^* = S^{\odot}(s) S(t)^* \qquad (s \geq 0, t > 0). \tag{7.4.27}$$

Proof. The adjoint semigroup $S^*(t)$ is as well uniformly continuous in $t > 0$. We deduce that $s \to S(s)^* S(t)^* y^* = S(s + t)^* y^*$ is continuous in $s \geq 0$ for every $t > 0$, which implies (7.4.26); since $S^{\odot}(t)$ is the restriction of $S(t)^*$ to $E^{\odot}$, (7.4.27) follows. ∎

We close this section with a few results on fractional powers of the infinitesimal generator of a strongly continuous (not necessarily (E, E)-continuous) semigroup, under assumption (7.4.1), which of course implies that $0 \in \rho(A)$, that is, that A^{-1} is a bounded everywhere defined operator. We begin with the equality

$$(-A)^{-\alpha-\beta} = (-A)^{-\alpha}(-A)^{-\beta} \tag{7.4.28}$$

valid for $\alpha, \beta \geq 0$, whose proof is left to the reader (Example 7.4.13). Next, we extend the definition (7.4.3) to the full range $-\infty < \alpha < \infty$ setting

$$(-A)^{\alpha} = (-A)^n(-A)^{\alpha-n}, \tag{7.4.29}$$

n the smallest integer with $n \geq \alpha$, with $D((-A)^{\alpha})$ consisting of all $y \in E$ such that $(-A)^{\alpha-n} y \in D(A^n)$. We use the fact that A^n is closed (Example 7.4.14) combined with Lemma 5.0.17 and the fact that A^n commutes with $S(t)$ to show that, if $y \in D(A^n)$ then $(-A)^{\alpha-n} y \in D(A^n)$ with $A^n(-A)^{\alpha-n} y = (-A)^{\alpha-n} A^n y$, so that $D((-A)^{\alpha}) \supseteq D(A^n)$, and, again by Example 7.4.14, $(-A)^{\alpha}$ is densely defined. The operator $(-A)^{\alpha}$ is the composition of the bounded operator $(-A)^{\alpha-n}$ with the closed operator $(-A)^n$, thus it is closed. If $\alpha > 0$, we have, using (7.4.28) and Example 7.4.15,

$$(-A)^{\alpha}(-A)^{-\alpha} = (-A)^n(-A)^{\alpha-n}(-A)^{-\alpha} = (-A)^n(-A)^{-n} = I.$$

On the other hand, if $y \in D((-A)^{\alpha})$, (7.4.28) implies

$$\begin{aligned}
(-A)^{-\alpha}(-A)^{\alpha} y &= (-A)^{-\alpha}(-A)^n(-A)^{\alpha-n} y \\
&= (-A)^n(-A)^{-\alpha}(-A)^{\alpha-n} y = y.
\end{aligned}$$

We have then shown that

$$(-A)^{\alpha} = ((-A)^{-\alpha})^{-1} \quad (\alpha \geq 0), \tag{7.4.30}$$

which is a better characterization of $(-A)^{\alpha}$ than (7.4.29). For instance,

Lemma 7.4.9. *Let* $-\infty < \alpha, \beta < \infty$. *Then*

$$(-A)^{\alpha+\beta} = (-A)^{\alpha}(-A)^{\beta}. \tag{7.4.31}$$

Proof. For $\alpha, \beta \geq 0$ it is enough to write $(-A)^{-\alpha-\beta} = (-A)^{-\alpha}(-A)^{-\beta}$ (equality (7.4.28)) and take inverses on both sides. If, say, $\alpha \geq 0$ and $\beta < 0$ with $\alpha + \beta < 0$

we use (7.4.28) to show that $(-A)^{-\alpha}(-A)^{\alpha+\beta} = (-A)^\beta$ and then apply $(-A)^\alpha$ to both sides. The other cases follow in the same way.

For $\alpha \geq 0$ we define

$$E_\alpha = D((-A)^\alpha) \tag{7.4.32}$$

equipped with the norm

$$\|y\|_\alpha = \|y\|_{E_\alpha} = \|(-A)^\alpha y\|. \tag{7.4.33}$$

The fact that $((-A)^\alpha)^{-1}$ is bounded implies that E_α is a Banach space (Example 7.4.16). For $\alpha < 0$, E_α is the completion of E with respect to the norm (7.4.33) (Example 7.4.17), thus it is a space larger than E. We have $E_\alpha \subseteq E = E_0 \subseteq E_\beta$ if $\alpha \geq 0, \beta \leq 0$; more generally,

$$E_\beta \subseteq E_\alpha, \qquad \|y\|_\alpha \leq C\|y\|_\beta \quad (-\infty < \alpha \leq \beta < \infty) \tag{7.4.34}$$

with a constant C depending on α, β. This follows from (7.4.31) written in the form $(-A)^\alpha = (-A)^{-(\beta-\alpha)}(-A)^\beta$.   ∎

Lemma 7.4.10. *Let* $-\infty < \alpha < \infty, \rho \geq 0$. *Then*

$$(-A)^{-\rho} E_\alpha = E_{\alpha+\rho} \tag{7.4.35}$$

algebraically and metrically.

Proof. For $\alpha > 0$ the proof is a consequence of the equality

$$(-A)^{\rho+\alpha}(-A)^{-\rho}y = (-A)^\alpha y \tag{7.4.36}$$

which follows from (7.4.31), and shows that $(-A)^{-\rho} E_\alpha = (-A)^{-\rho} D((-A)^\alpha) \subseteq D((-A)^{\rho+\alpha})$ and that

$$\|(-A)^{-\rho}y\|_{\rho+\alpha} = \|y\|_\alpha.$$

To show the opposite inclusion note that $(-A)^{-\rho-\alpha} = (-A)^{-\rho}(-A)^{-\alpha}$, so that every $z \in D((-A)^{\rho+\alpha})$ can be written $z = (-A)^\rho y$ with $y \in D((-A)^\alpha)$.

For $\alpha = -\beta < 0$, the operator $(-A)^{-\rho}$ must be extended to E_α to give sense to (7.4.35). This is done as follows in two separate cases.

(i) $\rho \geq \beta$. If $y \in E_{-\beta}$ then there exists a sequence $\{y_n\} \subseteq E$ such that $\|y_n - y\|_{-\beta} \to 0$. It follows that $\{(-A)^{-\beta} y_n\}$ is Cauchy in E, hence $(-A)^{-\beta} y_n \to z \in E$. Then $(-A)^{-\rho} y_n = (-A)^{-(\rho-\beta)}(-A)^{-\beta} y_n \to (-A)^{-(\rho-\beta)}z$, so that we define $(-A)^{-\rho}y = (-A)^{-(\rho-\beta)}z$. It is plain that this definition does not depend on the particular sequence $\{y_n\}$. We have then demonstrated that $(-A)^{-\rho} E_{-\beta} \subseteq E_{-\beta+\rho} = D((-A)^{\rho-\beta})$ and

$$\|(-A)^{-\rho}y\|_{-\beta+\rho} = \|(-A)^{-\beta+\rho}(-A)^{-\rho}y\|$$
$$= \|(-A)^{-\beta+\rho}(-A)^{-(\rho-\beta)}z\| = \|z\|$$
$$= \lim\|(-A)^{-\beta}y_n\| = \|y\|_{-\beta}.$$

To show the opposite inclusion, take $w \in E_{-\beta+\rho} = D((-A)^{\rho-\beta})$ arbitrary. Write $w = (-A)^{-(\rho-\beta)}z$, construct a sequence $\{y_n\} \subseteq E$ with $(-A)^{-\beta}y_n \to z$ and take $y = \lim y_n$ in $E_{-\beta}$. That this sequence can be constructed is obvious since $D((-A)^{\beta})$ is dense in E.

(ii) $\rho < \beta$. The sequence $\{y_n\} \subseteq E$ is chosen in the same way as in (i), that is, it satisfies $\|y_n - y\|_{-\beta} \to 0$. Now, $\{(-A)^{-\beta+\rho}(-A)^{-\rho}y_n\} = \{(-A)^{-\beta}y_n\}$ is convergent, so that there exists $w \in E_{-\beta+\rho}$ such that $\|(-A)^{-\rho}y_n - w\|_{-\beta+\rho} \to 0$; we define $w = (-A)^{-\rho}y$, and show again that the definition does not depend on $\{y_n\}$, so that the inclusion $(-A)^{-\rho}E_{-\beta} \subseteq E_{-\beta+\rho}$ holds with

$$\|(-A)^{-\rho}y\|_{-\beta+\rho} = \lim\|(-A)^{-\beta+\rho}(-A)^{-\rho}y_n\|$$
$$= \lim\|(-A)^{-\beta}y_n\| = \|y\|_{-\beta}.$$

To show (7.4.35), take $w \in E_{-\beta+\rho}$ and a sequence $\{z_n\} \subseteq E$ such that $\|z_n - w\|_{-\beta+\rho} \to 0$. Then use denseness of $D((-A)^{\rho})$ in E to construct a sequence $\{y_n\} \subseteq E$ with $\|(-A)^{-\rho}y_n - z_n\| \to 0$. This implies

$$\|(-A)^{-\rho}y_n - w\|_{-\beta+\rho} \le \|(-A)^{-\rho}y_n - z_n\|_{-\beta+\rho} + \|z_n - w\|_{-\beta+\rho}$$
$$\le C\|(-A)^{-\rho}y_n - z_n\| + \|z_n - w\|_{-\beta+\rho} \to 0,$$

(see (7.4.34)) and ends the proof.   ∎

If E is reflexive, the adjoint A^* is the infinitesimal generator of the semigroup $S^*(\cdot) = S(\cdot)^*$, and the spaces $(E^*)_\alpha$ are defined in the same way with respect to the adjoint A^*; in particular, we have the companion of (7.4.35),

$$((-A)^{-\rho})^*(E^*)_\alpha = (E^*)_{\alpha+\rho}. \tag{7.4.37}$$

If E^* is not reflexive, A^* may not even be a semigroup generator, and the fractional power theory does not apply; we use $((-A)^{-\alpha})^*$ instead of $(-A^*)^{-\alpha}$ in the definition.

Combining (7.4.30) with (5.0.27) we obtain

$$((-A)^{\alpha})^* = (((-A)^{-\alpha})^{-1})^* = (((-A)^{-\alpha})^*)^{-1}. \tag{7.4.38}$$

Lemma 7.4.11. *Let* $-\infty < \alpha, \beta < \infty$. *Then*

$$((-A)^{\alpha+\beta})^* = ((-A)^{\alpha})^*((-A)^{\beta})^*. \tag{7.4.39}$$

Proof. We take adjoints in (7.4.31). If $\alpha \le 0$ $(-A)^\alpha$ is bounded, so that (7.4.39) follows from (5.0.26). If $\beta \le 0$ we obtain in the same way that $((-A)^{\alpha+\beta})^* = ((-A)^{\beta})^*((-A)^{\alpha})^*$; since $(-A)^{\beta}$ and $(-A)^{\alpha}$ commute, so do $((-A)^{\alpha})^*$ and $((-A)^{\alpha})^*$. For $\alpha, \beta > 0$ we write (7.4.39) for $-\alpha, -\beta$ and then take inverses using (7.4.38).   ∎

Lemma 7.4.12. *Let* $-\infty < \alpha < \infty, \rho \geq 0$. *Then*

$$((-A)^{-\rho})^*(E^*)_\alpha \subseteq (E^*)_{\alpha+\rho} \tag{7.4.40}$$

with algebraic and isometric inclusion.

Proof. The proof of (7.4.40) is based on (7.4.39) and is essentially the same as that of Lemma 7.4.9. Note that we have to stop short of equality in (7.4.40) since $D(((-A)^\rho)^*)$ may not be dense in E^*. However, equality holds in a certain range of parameters (Example 7.4.18). ∎

Example 7.4.13. We can prove (7.4.28) as follows. First, we write the definition (7.4.2) for α and β using different integration variables (say, s and t). Then we multiply out, use the semigroup equation $S(s)S(t) = S(s + t)$ and do the change of variables $s + t = \tau$.

Example 7.4.14. A^n is densely defined, closed and invertible with $(A^n)^{-1} = (A^{-1})^n$. In fact, annihilating one A with one A^{-1} at a time, we obtain $(A^n)(A^{-1})^n y = y$ for arbitrary $y \in E$ and $(A^{-1})^n A^n y = y$ for $y \in D(A^n)$, so that $(A^n)^{-1} = (A^{-1})^n$ as claimed. The inverse of a bounded operator is closed, so that A^n is closed. As for denseness of the domain, $A^{-1}E$ is dense in E, thus $A^{-1}D$ is dense if D is dense in E. Taking $D = A^{-1}E$ and repeating the argument, we are through.

Example 7.4.15. If $\alpha = -n$, definition (7.4.2) gives the "right result." This follows from formula (5.2.34) applied for $\lambda = 0$.

Example 7.4.16. Let B be a linear operator with domain $D(B) \subseteq E$. The **graph norm** of $D(B)$ is

$$\|y\|_{D(B)} = \|y\| + \|By\|. \tag{7.4.41}$$

If B is closed, this norm makes $D(B)$ a Banach space; if B is invertible, the first term on the right can be omitted.

Example 7.4.17. Let E be a normed space. Show that there exists a space $\bar{E}$ (the **completion** of E) such that (a) $\bar{E}$ is a Banach space, (b) E "is" a dense subspace of $\bar{E}$ and the norms of E and $\bar{E}$ coincide in E. Hint: $\bar{E}$ is the set of all Cauchy sequences $\{y_n\}$ in E with the equivalence relation $\{y_n\} \approx \{z_n\}$ if $y_n - z_n \to 0$. The space $\bar{E}$ is endowed with obvious operations and norm $\|\{y_n\}\| = \lim \|y_n\|$. The space $E = \{y, \ldots\}$ is identified with the subspace of all equivalence classes in $\bar{E}$ containing the constant sequence $\{y_n\}, y_n = y$.

Example 7.4.18. In general, there is no equality in (7.4.40). For instance, let A be such that $D(A^*)$ is not dense in E^*. Then

$$(A^{-1})^*(E^*)_{-1} \subseteq \overline{D(A^*)} \neq E^*. \tag{7.4.42}$$

To see this, take $z \in (E^*)_{-1}$ and select a sequence $\{z_n\} \subseteq E^*$ such that $\|z_n - z\|_{(E^*)_{-1}} \to 0$. Then $(A^{-1})^* z_n \to w = (A^{-1})^* z$ in E^* and $(A^{-1})^* z_n \in D(A^*)$, so that $(A^{-1})^* z$ belongs to the closure of $D(A^*)$.

However, equality in (7.4.40) obtains when $\alpha \geq 0$, as we can see imitating the proof of Lemma 7.4.10 (denseness of $D(A)$ was not used there). In case (*ii*) ($\alpha < 0, \rho < -\alpha$), we don't need denseness of $D(((-A)^\rho)^*)$ in E^*, just in $(E^*)_{\alpha+\rho}$.

Example 7.4.19. If (7.4.1) holds with $c > 0$, then (7.4.6) holds with $a < 0$ in some sector $\Sigma(\phi)$. In fact, let $0 < c' < c'' < c$. Inequality (7.4.1) implies that $\sigma(A)$ is contained in the half-plane Re $\lambda \leq -c$, so that the contour of integration in (7.4.13) can be deformed into the contour Γ coming from $e^{-i(\pi/2+\psi)}\infty$ along the ray $\arg \lambda = -\pi/2 - \psi$, then switching to Re $\lambda = -c''$ and finally going off to $e^{i(\pi/2+\psi)}\infty$ along the ray $\arg \lambda = \pi/2 + \psi$. Rotations $e^{i\phi}\Gamma$ of Γ stay in the half-plane Re $\lambda \leq c'$ for ϕ small enough, thus we only have to switch to $e^{i\phi}\Gamma$ in (7.4.13) and estimate.

7.5. Elliptic Operators in L^2 Spaces

We consider the parabolic equation

$$y'(t) = A(\beta)y(t), \tag{7.5.1}$$

where A is the operator in **5.6**,

$$Ay = \sum_{j=1}^{m} \sum_{k=1}^{m} \partial^j (a_{jk}(x)\partial^k y) + \sum_{j=1}^{m} b_j(x)\partial^j y + c(x)y, \tag{7.5.2}$$

and $A(\beta)$ indicates the restriction obtained by means of a boundary condition β on the boundary Γ of a domain Ω, either the Dirichlet boundary condition $y(x) = 0$ ($x \in \Gamma$), or the variational boundary condition $\partial^\nu y(x) = \gamma(x)y(x)$ ($x \in \Gamma$), where ∂^ν is the conormal derivative, *i.e.* the derivative in the direction of the conormal vector $\{\nu_j(x)\} = \{\Sigma a_{jk}(x)\eta_k(x)\}$ ($\{\eta_k\}$ the outer normal vector). We do in this section the $L^2(\Omega)$ theory; the $L^p(\Omega)$ and $C(\overline{\Omega})$ treatment are in next section. Assumptions are the same in **5.6**; $a_{jk}(x), b_j(x), c(x)$ are real valued, measurable and bounded in Ω, the matrix $\{a_{jk}(\cdot)\}$ is symmetric and A is uniformly elliptic: $\Sigma\Sigma a_{jk}(x)\xi_j\xi_k \geq \kappa \|\xi\|^2$ ($x \in \Omega, \xi \in \mathbb{R}^m$). For the Dirichlet boundary condition there are no other assumptions; for a variational boundary condition we assume the domain Ω of class $C^{(1)}$ with bounded boundary and $\gamma(\cdot)$ measurable and bounded on Γ. The operators $A(\beta), A_0(\beta), P$, are the same in **5.6**, and the approach is indirect; we begin with another reduction of the wave equation (5.6.1) to a first-order system, this time in the product space $\mathbf{E} = L^2(\Omega) \times L^2(\Omega)$ (complex spaces: see Remark 5.6.8). The first result is for an arbitrary complex

Hilbert space and a self adjoint operator A_0 with spectrum in $(-\infty, -\kappa]$, $\kappa > 0$. We set $\mathbf{E} = H \times H$ with inner product $(\mathbf{y}, \mathbf{z}) = ((y, y_1), (z, z_1)) = (y, z) + (y_1, z_1)$, take $R = (-A_0)^{1/2}$ and define

$$\mathbf{A}_0 = \begin{bmatrix} 0 & iR \\ iR & 0 \end{bmatrix} \tag{7.5.3}$$

with domain $D(\mathbf{A}_0) = D(R) \times D(R) = D((-A_0)^{1/2}) \times D((-A_0)^{1/2})$.

Lemma 7.5.1. $\mathbf{A}_0$ *is skew adjoint.*

The proof is the same as that of Theorem 5.6.4 and is omitted.

We apply Lemma 7.5.1 to the operator $A_0(\beta)$ in $H = L^2(\Omega)$ (the operator matrix is called $\mathbf{A}_0(\beta)$) and incorporate the lower order terms in (5.2) by means of

$$\mathbf{P} = \begin{bmatrix} 0 & -iPR^{-1} \\ 0 & 0 \end{bmatrix}. \tag{7.5.4}$$

It follows from the results in **5.6** that PR^{-1} is bounded in H. By Corollary 5.5.8, we have

Theorem 7.5.2. *The operator*

$$\mathbf{A}(\beta) = \mathbf{A}_0(\beta) + \mathbf{P} = \begin{bmatrix} 0 & iR - iPR^{-1} \\ iR & 0 \end{bmatrix} \tag{7.5.5}$$

is the infinitesimal generator of a strongly continuous group $\mathbf{S}(t; A, \beta)$ *satisfying*

$$\|\mathbf{S}(t; A, \beta)\| \leq e^{\omega|t|} \quad (-\infty < t < \infty). \tag{7.5.6}$$

We write $\mathbf{S}(t; A, \beta)$ in matrix form

$$\mathbf{S}(t; A, \beta) = \begin{bmatrix} S_{11}(t; A, \beta) & S_{12}(t; A, \beta) \\ S_{21}(t; A, \beta) & S_{22}(t; A, \beta) \end{bmatrix}, \tag{7.5.7}$$

each $S_{jk}(t; A, \beta)$ a strongly continuous operator function in $-\infty < t < \infty$. The resolvent $R(\lambda; \mathbf{A}_0(\beta) + P)$ must exist for $\lambda > \omega$. Write

$$R(\lambda; \mathbf{A}_0(\beta) + P) = \begin{bmatrix} R_{11}(\lambda) & R_{12}(\lambda) \\ R_{21}(\lambda) & R_{22}(\lambda) \end{bmatrix}. \tag{7.5.8}$$

Looking at the first column of the matrix equation

$$(\lambda\mathbf{I} - \mathbf{A}_0(\beta) - \mathbf{P})R(\lambda; \mathbf{A}_0(\beta) + P)$$

$$= \begin{bmatrix} \lambda I & -iR + iPR^{-1} \\ -iR & \lambda I \end{bmatrix} \begin{bmatrix} R_{11}(\lambda) & R_{12}(\lambda) \\ R_{21}(\lambda) & R_{22}(\lambda) \end{bmatrix} = \begin{bmatrix} I & 0 \\ 0 & I \end{bmatrix} = \mathbf{I} \tag{7.5.9}$$

we obtain

$$\lambda R_{11}(\lambda) - (iR - iPR^{-1})R_{21}(\lambda) = I, \qquad -iRR_{11}(\lambda) + \lambda R_{21}(\lambda) = 0.$$

Eliminating $R_{21}(\lambda)$ from these equalities,

$$(\lambda^2 I + R^2 - P)R_{11}(\lambda) = (\lambda^2 I - A_0(\beta) - P)R_{11}(\lambda) = \lambda I. \qquad (7.5.10)$$

We work in the same way with the equation

$$R(\lambda; \mathbf{A}_0(\beta) + \mathbf{P})(\lambda \mathbf{I} - \mathbf{A}_0(\beta) - \mathbf{P})\mathbf{y} = \mathbf{y}$$

valid for $\mathbf{y} = (y, z) \in D(\mathbf{A}_0(\beta) + \mathbf{P}) = D(\mathbf{A}_0(\beta)) = D(R) \times D(R)$. Taking $z = y$ we obtain

$$R_{11}(\lambda)(\lambda^2 I + R^2 - P)y = R_{11}(\lambda)(\lambda^2 I - A_0(\beta) - P)y = \lambda y \qquad (7.5.11)$$

for $y \in D(A_0(\beta)) = D(A_0(\beta) + P)$.

It follows from (7.5.10) and (7.5.11) that

$$R_{11}(\lambda) = \lambda R(\lambda^2; A_0(\beta) + P). \qquad (7.5.12)$$

Finally, observing the element in the upper left corner of the matrix equation

$$R(\lambda; \mathbf{A}_0(\beta) + \mathbf{P})\mathbf{y} = \int_0^\infty e^{-\lambda t} \mathbf{S}(t; A, \beta)\mathbf{y}\, dt$$

$(\mathbf{y} \in \mathbf{E})$ we obtain from (7.5.12) that

$$\lambda R(\lambda^2; A_0(\beta) + P)y = \int_0^\infty e^{-\lambda t} S_{11}(t; A, \beta)y\, dt. \qquad (7.5.13)$$

Theorem 7.5.3. *The operator $A(\beta) = A_0(\beta) + P$ generates an analytic semigroup $S(\zeta; A, \beta)$ given by*

$$S(\zeta; A, \beta)y = \frac{1}{\sqrt{\pi \zeta}} \int_0^\infty e^{-s^2/4\zeta} S_{11}(s; A, \beta)y\, ds \quad (\mathrm{Re}\, \zeta > 0, y \in E) \tag{7.5.14}$$

(precisely, $A(\beta) \in \mathcal{A}_\phi$ for $0 < \phi < \pi/2$).

Proof. In view of inequality (7.5.6), (7.5.16) is defined for $\mathrm{Re}\, \zeta > 0$. We rewrite (7.5.6) in the form

$$\|S_{11}(t; A, \beta)\| \le 2\cosh(\omega t) \quad (-\infty < t < \infty) \qquad (7.5.15)$$

so that the integral equality

$$\frac{1}{\sqrt{\pi t}} \int_0^\infty e^{-s^2/4t} \cosh(\omega s)\, ds = e^{\omega^2 t} \quad (t > 0) \qquad (7.5.16)$$

and the fact that $|e^{-s^2/4\zeta}| = e^{-s^2 \operatorname{Re}\zeta/4|\zeta|^2}$ provide the bound

$$\|S(\zeta; A, \beta)\| \leq 2\sqrt{\frac{|\zeta|}{\operatorname{Re}\zeta}} e^{\omega^2(|\zeta|^2/\operatorname{Re}\zeta)} \quad (\operatorname{Re}\zeta > 0). \tag{7.5.17}$$

Differentiation of (7.5.14) under the integral sign is clearly permissible in $\operatorname{Re}\zeta > 0$, so that for every $\phi, 0 \leq \phi < \pi/2$, $S(\cdot; A, \beta)$ is analytic in any sector $\Sigma_0(\phi)$ and strongly continuous in $\Sigma(\phi)$; strong continuity at the origin is proved via the equality

$$S(\zeta; A, \beta)y - y = \frac{1}{\sqrt{\pi\zeta}} \int_0^\infty e^{-s^2/4\zeta} (S_{11}(s; A, \beta)y - y)ds$$

and the subsequent estimate

$$\|S(\zeta; A, \beta)y - y\| \leq \frac{1}{\sqrt{\pi|\zeta|}} \int_0^\delta e^{-s^2 \operatorname{Re}\zeta/4|\zeta|^2} \|S_{11}(s; A, \beta)y - y\|ds$$
$$+ \frac{1}{\sqrt{\pi|\zeta|}} \int_\delta^\infty e^{-s^2 \operatorname{Re}\zeta/4|\zeta|^2} e^{\omega s}ds$$

in any sector $\Sigma(\phi)$. In the first integral we use strong continuity of $S(t; A, \beta)$ at $t = 0$ to deduce that $\|S_{11}(s; A, \beta)y - y\| \leq \varepsilon$ in $0 \leq s \leq \delta$ for δ small enough; in the second we make the change of variables $\sigma = s\sqrt{\operatorname{Re}\zeta/4|\zeta|^2}$. The reader will recognize the similarity of the manipulations with those in the construction of Weierstrass' solution of the heat equation.

We show finally that $S(\cdot)$ is a semigroup with infinitesimal generator $A(\beta) = A_0(\beta) + P$ using Lemma 5.2.4. This amounts to proving

$$\int_0^\infty e^{-\lambda t} S(\zeta; A, \beta)y\, dt = \int_0^\infty \frac{e^{-\lambda t}}{\sqrt{\pi t}} dt \int_0^\infty e^{-s^2/4t} S_{11}(s; A, \beta)y\, ds$$
$$= R(\lambda; A(\beta)) = R(\lambda; A_0(\beta) + P) \tag{7.5.18}$$

fot $\lambda > \omega^2$. We interchange the order of integration on the basis of Tonelli's theorem, estimating the integrand with (7.5.15), integrating first in t and using (7.5.16). Formula (7.5.18) then follows from (7.5.14) and the scalar formula

$$\int_0^\infty \frac{e^{-s^2/4t} e^{-\lambda t}}{\sqrt{\pi t}} dt = \frac{e^{-\sqrt{\lambda}s}}{\sqrt{\lambda}} \quad (t > 0). \qquad \blacksquare$$

7.6. Elliptic Operators in L^p and C Spaces. Duality

If A is the operator

$$Ay = \sum_{j=1}^m \sum_{k=1}^m \partial^j (a_{jk}(x)\partial^k y) + \sum_{j=1}^m b_j(x)\partial^j y + c(x)y \tag{7.6.1}$$

in last section, the **formal adjoint** A' is

$$A'y = \sum_{j=1}^{m}\sum_{k=1}^{m} \partial^j(a_{jk}(x)\partial^k y) - \sum_{j=1}^{m} \partial_j(b_j(x)y) + c(x)y. \tag{7.6.2}$$

The adjoint boundary condition β' is $\beta' = \beta$ for the Dirichlet boundary condition. For a variational boundary condition (7.2.3), β' is

$$\partial^\nu y(x) = (\gamma(x) + b(x))y(x) \tag{7.6.3}$$

with $b(x) = \Sigma b_j(x)\eta_j(x)$, $\{\eta_j\}$ the outer normal vector. The motivation is: if everything is smooth, then the divergence theorem gives $(A(\beta)y, z) = (z, A'(\beta')z)$; conversely, this property determines A' and β' uniquely. If we insist on the level of generality in last section (measurable bounded coefficients), the first order term of A' is not of the same form as that of A. To bypass this, we assume from now on that the $a_{jk}(x)$ and the b_j are continuously differentiable in Ω and that c is continuous in $\overline{\Omega}$. The domain Ω is bounded and of class $C^{(2)}$.

The $L^2(\Omega)$-duality is: defining operators as in **7.5**, $A(\beta)^* = A'(\beta')$. This is a particular case of the $L^p(\Omega)$-duality theory that we sketch below. The proofs of all results (except analyticity of the semigroups) can be found in the author [1983, Chapter 6] with historical notes. Analyticity (and most of the other statements) can be found in Stewart [1974].

Theorem 7.6.1. *Let $1 < p < \infty$. Then there exists an operator $A_p(\beta)$ that can be characterized in any of the two equivalent forms.*

Strong form: $D(A_p(\beta)) = W_\beta^{2,p}(\Omega)$, *the space of all y in the Sobolev space $W^{2,p}(\Omega)$ that satisfy the boundary condition β, and $A_p(\beta)y = Ay$.*

Weak form: $D(A_p(\beta))$ *consists of all elements $y \in L^p(\Omega)$ such that there exists z $(= A_p(\beta)y)$ in $L^p(\Omega)$ with*

$$\int_\Omega y(x)(A'(\beta')v)(x)dx = \int_\Omega z(x)v(x)dx \tag{7.6.4}$$

for every $v \in C_{\beta'}^{(2)}(\overline{\Omega})$ or, equivalently, for every $v \in W_{\beta'}^{2,q}(\Omega)$ with $1/p + 1/q = 1$ ($C_{\beta'}^{(2)}(\overline{\Omega})$ is the subspace of $C^{(2)}(\overline{\Omega})$ consisting of all functions y that satisfy the boundary condition β'). Moreover, we have

$$A_p(\beta)^* = A_q'(\beta') \tag{7.6.5}$$

The operator $A_p(\beta)$ generates a compact analytic semigroup $S_p(t; A, \beta)$ and

$$S_p(t; A, \beta)^* = S_q(t; A', \beta'). \tag{7.6.6}$$

This statement needs some explanation. For the definition of the Sobolev spaces $W^{2,p}(\Omega)$ see **5.6**. The fact that the boundary conditions make sense for functions in $W^{2,p}(\Omega)$ is explained in Adams [1975, Theorem 5.22, p. 114]. For more on assumption of boundary values by members of Sobolev spaces, see **8.3**.

For $p = 1$, we have

Theorem 7.6.2. *The operator $A_1(\beta)$ in $L^1(\Omega)$ characterized in the weak form above generates a compact analytic semigroup $S_1(t; A, \beta)$.*

The operator $A_1(\beta)$ cannot be characterized in the strong form.

In the realm of continuous functions, we need two different spaces. For variational boundary conditions the space is $C(\overline{\Omega})$. We use the subspace $C_0(\overline{\Omega}) \subseteq C(\overline{\Omega})$ consisting of all y which are zero at the boundary Γ for the Dirichlet boundary condition.

Theorem 7.6.3. *Let β be a variational boundary condition. Then there exists an operator $A_c(\beta)$ that can be characterized in any of the two equivalent forms:*

Strong form:

$$D(A_c(\beta)) = \left\{ y \in \bigcap_{p \geq 1} W_\beta^{2,p}(\Omega); \, Ay \in C(\overline{\Omega}) \right\} \tag{7.6.7}$$

and $A_c(\beta)y = Ay$.

Weak form: $D(A_c(\beta))$ *consists of all elements* $y \in C(\overline{\Omega})$ *such that there exists* $z \,(= A_c(\beta)y)$ *in* $C(\overline{\Omega})$ *with*

$$\int_\Omega y(x)(A'(\beta')v)(x)dx = \int_\Omega z(x)v(x)dx \tag{7.6.8}$$

for every $v \in C_{\beta'}^{(2)}(\overline{\Omega})$. *The operator $A_c(\beta)$ generates a compact analytic semigroup $S_c(t; A, \beta)$.*

Theorem 7.6.4. *Let β be the Dirichlet boundary condition. Then there exists an operator $A_c(\beta)$ that can be characterized in any of the two equivalent forms:*

$$D(A_c) = \left\{ y \in \bigcap_{p \geq 1} W_\beta^{2,p}(\Omega); \, Ay \in C_0(\overline{\Omega}) \right\} \tag{7.6.9}$$

and $A_c(\beta)y = Ay$.

Weak form: $D(A_c(\beta))$ *consists of all elements* $y \in C(\overline{\Omega})$ *such that there exists* $z \,(= A_c(\beta)y)$ *in* $C_0(\overline{\Omega})$ *with*

$$\int_\Omega y(x)(A'(\beta')v)(x)dx = \int_\Omega z(x)v(x)dx \tag{7.6.10}$$

for every $v \in C_0^{(2)}(\overline{\Omega})$. *The operator* $A_c(\beta)$ *generates a compact analytic semigroup* $S_c(t; A, \beta)$.

Duality theory in the spaces $L^1(\Omega), C(\overline{\Omega}), C_0(\overline{\Omega})$ is handled via Phillips adjoints. We have $L^1(\Omega)^* = L^\infty(\Omega)$ and $C(\overline{\Omega})^* = \Sigma(\overline{\Omega})$, the latter space consisting of all finite regular Borel measures in $\overline{\Omega}$ equipped with the total variation norm (see **5.0**). The duality pairing of $C(\overline{\Omega})$ and $\Sigma(\overline{\Omega})$ is

$$\langle \mu, y \rangle = \int_\Omega y(x)\mu(dx) \tag{7.6.11}$$

(for a proof of this statement and of much more general results see **12.4**). The space $C_0(\overline{\Omega})^*$ can be identified with $\Sigma_0(\overline{\Omega})$, the subspace of $\Sigma(\overline{\Omega})$ consisting of all $\mu \in \Sigma(\overline{\Omega})$ with $|\mu|(\Gamma) = 0$; the duality pairing is again given by (7.6.11).

Theorem 7.6.5. *Let β the Dirichlet boundary condition or a variational boundary condition, and let $A_1(\beta)$ and $S_1(t; A, \beta)$ be the operator and semigroup in the space $L^1(\Omega)$ defined in Theorem 7.6.2. Then*

$$L^1(\Omega)^\odot = C(\overline{\Omega}) \quad \text{(variational boundary condition)}$$
$$L^1(\Omega)^\odot = C_0(\overline{\Omega}) \quad \text{(Dirichlet boundary condition)}$$
$$A_1(\beta)^\odot = A_c'(\beta'), \quad S_1^\odot(t; A, \beta) = S_c(t; A', \beta').$$

Theorem 7.6.6. *Let β the Dirichlet boundary condition or a variational boundary condition, and let $A_c(\beta)$ and $S_c(t; A, \beta)$ be the operator and semigroup in the spaces $C_0(\overline{\Omega})$ or $C(\overline{\Omega})$ defined in Theorems 7.6.4 and 7.6.3. Then*

$$C(\overline{\Omega})^\odot = L^1(\Omega) \quad \text{(variational boundary condition)}$$
$$C_0(\overline{\Omega})^\odot = L^1(\Omega) \quad \text{(Dirichlet boundary condition)}$$
$$A_c(\beta)^\odot = A_1'(\beta'), \quad S_c^\odot(t; A, \beta) = S_1(t; A', \beta')$$

It will be essential later to know if the semigroups $S(t)$ generated by $A_p(\beta)$ or $A_c(\beta)$ satisfy

$$\|S(t)\| \le e^{\omega t} \quad (t \ge 0) \tag{7.6.12}$$

for some ω, rather than the weaker inequality $\|S(t)\| \le Ce^{\omega t}$; as shown in **5.2**, (7.6.12) is equivalent to

$$\langle y^*, Ay \rangle \le \omega \|y\|^2 \quad (y \in D(A), y^* \in \Theta(y)), \tag{7.6.13}$$

where $\Theta(y)$ denotes the duality set of y (see **5.2**) and A denotes one of the operators $A_p(\beta)$ or $A_c(\beta)$.

Theorem 7.6.7. *Let $1 \le p < \infty$. Then (7.6.12) holds for some ω.*

Theorem 7.6.8. *If β is the Dirichlet boundary condition, $A_c(\beta)$ satisfies (7.6.12) for some ω. If β is a variational boundary condition, there exists a twice continuously differentiable function $\rho(x)$ in $\overline{\Omega}$ such that $\rho(x) > 0$ and (7.6.12) holds for some ω after renorming the space $C(\overline{\Omega})$ with the norm*

$$\|y\|_\rho = \max_{x \in \overline{\Omega}} |y(x)| \rho(x). \tag{7.6.14}$$

For a proof of this particular result, see the author [1983, Chapter 6].

Remark 7.6.9. In order to keep our footing through the following sections, we try to clarify what adjoints and Phillips adjoints mean in the applications under study. Outside of reflexive spaces (where Phillips adjoints do not appear), the only application in mind is to the differential operators $A = A_c(\beta)$, $A_1(\beta)$ in this section, so that $E = C(\overline{\Omega})$, $C_0(\overline{\Omega})$ or $L^1(\Omega)$. We have

$$S_1(t; A, \beta) = S_c^\odot(t; A', \beta') \subseteq S_c(t; A', \beta')^* = S_1^\odot(t; A, \beta)^*, \tag{7.6.15}$$

the first semigroup acting in $L^1(\Omega)$, the second in $\Sigma(\overline{\Omega}) \supseteq L^1(\Omega)$. Accordingly, $S_1^\odot(t; A, \beta)^* = S_c(t; A', \beta')^*$ extends $S_1(t; A, \beta)$ from $L^1(\Omega)$ to $C(\overline{\Omega})^* = \Sigma(\overline{\Omega})$ and could be informally named "the semigroup generated by A and β in $\Sigma(\overline{\Omega})$." It is not strongly continuous at the origin; in fact, (7.4.26) implies that $S_c(t; A', \beta')^* \Sigma(\overline{\Omega}) \subseteq L^1(\Omega)$ for $t > 0$, thus $S_c(t; A', \beta')^* \mu$ cannot converge to μ in $\Sigma(\overline{\Omega})$ if, say, μ is a Dirac delta (or, in general, any measure μ not of the form $\mu(dx) = y(x) dx$, $y(\cdot) \in L^1(\Omega)$).

Completely symmetric arguments apply to $C(\overline{\Omega})$. Here we have

$$S_c(t; A, \beta) = S_1^\odot(t; A', \beta') \subseteq S_1(t; A', \beta')^* = S_c^\odot(t; A, \beta)^*, \tag{7.6.16}$$

so that $S_c^\odot(t; A, \beta)^* = S_1(t; A', \beta')^*$ extends $S_c(t; A, \beta)$ from $C(\overline{\Omega})$ to $L^1(\Omega) = L^\infty(\Omega)$ and deserves to be called "the semigroup generated by A, β in $L^\infty(\Omega)$." By (7.4.26), $S_1(t; A', \beta')^* L^\infty(\Omega) \subseteq C(\overline{\Omega})$, so that $S_1(t; A', \beta')^*$ is not strongly continuous at the origin; $S_1(t; A', \beta')^* y$ cannot converge to $y \in L^\infty(\Omega)$ if y is discontinuous.

The usual modifications apply for the Dirichlet boundary condition.

7.7. Semilinear Abstract Parabolic Equations

Consider the semilinear distributed parameter control system

$$y_t(t, x) = \Delta y(t, x) + f(t, y(t, x), \nabla y(t, x)) + u(t, x) \tag{7.7.1}$$

in a domain Ω with, say, the boundary condition $y(t, x) = 0$. If (7.7.1) describes a heat process, the supremum norm (the norm of $C(\overline{\Omega})$) is more natural that the $L^p(\Omega)$ norm—the latter allows for large temperature spikes possibly exceeding the melting point of the material and invalidating (7.7.1) as a model. Estimates in the supremum norm can be obtained from estimates in Sobolev norms and imbedding theorems, but in some instances it is more natural[1] to look at (7.7.1) directly in $C(\overline{\Omega})$. This introduces a complication: if we insist on existence of optimal controls, it is inconvenient to take controls $u(t, x)$ such that $u(t, \cdot) \in C(\overline{\Omega})$; the "right" class is $L^\infty((0, T) \times \Omega)$. Thus, instantaneous values of the controls belong to $L^\infty(\Omega)$, a space larger that $C(\overline{\Omega})$.

We do in this section a theory for the equation

$$y'(t) = Ay(t) + f(t, y(t)), \qquad y(s) = \zeta \tag{7.7.2}$$

that can be applied to (7.7.1) in the spaces $C(\overline{\Omega})$ or in $L^p(\Omega)$ (it is much simpler in the latter case). The basic ingredient is a Banach space E and a strongly continuous semigroup $S(t)$ in E with infinitesimal generator A. As in **7.4**, $E^\odot \subseteq E^*$ denotes the Phillips dual of E relative to A, and $S^\odot(t)$ is the Phillips adjoint semigroup. Since $S^\odot(t)$ is a strongly continuous semigroup in $E^\odot$, we can apply to it the Phillips adjoint theory; we obtain in this way the double Phillips adjoint semigroup $S^{\odot\odot}(t) = (S^\odot)^\odot(t)$, restriction of the adjoint semigroup $S^\odot(t)^*$ to the closure $E^{\odot\odot} = (E^\odot)^\odot$ of $D((A^\odot)^*)$ in $(E^\odot)^*$.

Lemma 7.7.1. (a) *There exists a bicontinuous linear imbedding from E into $E^{\odot\odot}$, that is, modulo a change of equivalent norms,*

$$E \subseteq E^{\odot\odot} = (E^\odot)^\odot. \tag{7.7.3}$$

(b) *We have*

$$A \subseteq A^{\odot\odot} = (A^\odot)^\odot, \qquad S(t) \subseteq S^{\odot\odot}(t) = (S^\odot)^\odot(t) \quad (t > 0). \tag{7.7.4}$$

Proof. We have already seen that $E \subseteq (E^\odot)^*$ with equivalence of the norms (Lemma 7.4.6). If $y \in D(A)$ and $y^* \in D(A^\odot)$, then $\langle A^\odot y^*, y \rangle = \langle y^*, Ay \rangle$ so that $y \in D((A^\odot)^*)$ and $(A^\odot)^* y = Ay$; this means $A \subseteq (A^\odot)^*$. In particular, $D(A) \subseteq E^{\odot\odot}$; thus, since $D(A)$ is dense in E, (7.7.3) follows. Finally, if $y \in D(A)$ we have $(A^\odot)^* y = Ay \in E \subseteq E^{\odot\odot}$, so that the first inequality (7.7.4) results. For the second, note that if $y \in E \subseteq E^{\odot\odot}$ and $y^* \in E^\odot$ we have $\langle y^*, S(t)y \rangle = \langle S^\odot(t)y^*, y \rangle = \langle y^*, S^{\odot\odot}(t)y \rangle$. ∎

The space E is $^\odot$-**reflexive** (relative to the operator A or to the semigroup $S(t)$) if

$$E = E^{\odot\odot}$$

---

[1] One such instance is found in next section. Global existence results for parabolic equations are essentially related to the duality of $C(\overline{\Omega})$ rather than that of $L^p(\Omega)$.

Lemma 7.7.2. *Let E be $^\odot$-reflexive. Then*

$$A = A^{\odot\odot}, \qquad S(t) = S^{\odot\odot}(t) \quad (t > 0). \tag{7.7.5}$$

Proof. The second inequality (7.7.5) is obvious from (7.7.3) and (7.7.4). The first follows from the fact that A (resp. $A^{\odot\odot}$) is the infinitesimal generator of $S(t)$ (resp. $S^{\odot\odot}(t)$). ∎

Lemma 7.7.3. *Let $S(t)$ be continuous in the norm of (E, E) in $t > 0$. Then*

$$S(t)^* E^* \subseteq E^\odot, \qquad S^\odot(t)^*(E^\odot)^* \subseteq E^{\odot\odot} \qquad (t > 0) \tag{7.7.6}$$

$$S(s + t)^* = S^\odot(s)S(t)^* \qquad\qquad (s \geq 0, t > 0) \tag{7.7.7}$$

$$S^\odot(s + t)^* = S^{\odot\odot}(s)S^\odot(t)^* \qquad (s \geq 0, t > 0). \tag{7.7.8}$$

If E is $^\odot$-reflexive with respect to $S(t)$ we have

$$S^\odot(t)^*(E^\odot)^* \subseteq E, \qquad S^\odot(s + t)^* = S(s)S^\odot(t)^* \quad (s \geq 0, t > 0). \tag{7.7.9}$$

Proof. The first inclusion (7.7.6) and the equality (7.7.7) are proved in Lemma 7.4.8; the second inclusion (7.7.6) and (7.7.8) follow from Lemma 7.4.8 applied to $S^\odot(t)$. Everything gets upgraded to (7.7.9) in the $^\odot$-reflexive case where $E^{\odot\odot} = E$, $S^{\odot\odot}(t) = S(t)$. ∎

To guarantee existence of the fractional powers $(-A)^\alpha$ defined in **7.4** we assume that

$$\|S(t)\| \leq Ce^{-ct} \quad (t \geq 0). \tag{7.7.10}$$

This can always be achieved replacing A by $A - kI$ (and f by $f - kI$) in (7.7.2). Since $S^\odot(t)$ satisfies the same estimate, the fractional powers $(-A^\odot)^\alpha$ exist as well and all results in **7.4** apply to them. The spaces E_α, $(E^*)_\alpha$ are constructed as in **7.4**, the first using $(-A)^\alpha$, the second $((-A)^\alpha)^*$; so are the spaces $(E^\odot)_\alpha$ (using $(-A^\odot)^\alpha$) and $((E^\odot)^*)_\alpha$ (using $((-A^\odot)^\alpha)^*$). Application of Lemma 7.4.10 and Lemma 7.4.12 produces

$$(-A)^{-\rho} E_\alpha = E_{\alpha+\rho}, \qquad (-A^\odot)^{-\rho}(E^\odot)_\alpha = (E^\odot)_{\alpha+\rho} \tag{7.7.11}$$

$$((-A)^{-\rho})^*(E^*)_\alpha \subseteq (E^*)_{\alpha+\rho}, \qquad ((-A^\odot)^{-\rho})^*((E^\odot)^*)_\alpha \subseteq ((E^\odot)^*)_{\alpha+\rho} \tag{7.7.12}$$

for $-\infty < \alpha < \infty$ and $\rho \geq 0$ all inclusions and equalities being algebraic and metric.

Lemma 7.7.4. *Let $S(t)$ be continuous in the norm of (E, E) and let E be $^\odot$-reflexive with respect to $S(t)$. Then, for $\alpha > 0$ we have*

$$((-A)^{-\alpha})^* E^* \subseteq E^\odot, \qquad ((-A^\odot)^{-\alpha})^*(E^\odot)^* \subseteq E. \tag{7.7.13}$$

$$((-A)^{-\alpha})^*|_{E^\odot} = (-A^\odot)^{-\alpha}, \qquad ((-A^\odot)^{-\alpha})^*|_E = (-A)^{-\alpha} \tag{7.7.14}$$

Proof. To compute $((-A)^{-\alpha})^*$, we simply take adjoints inside the integral sign in (7.4.2); this can be easily justified on the basis of the boundedness of the adjoint operation and the (E, E)-continuity of the integrand. Then we divide the interval of integration at $t = h > 0$ and use (7.7.9). The result is

$$((-A)^{-\alpha})^* y^* = \frac{1}{\Gamma(\alpha)} \int_0^\infty (t + h)^{\alpha-1} S(t)^\odot S(h)^* y^* \, dt$$

$$+ \frac{1}{\Gamma(\alpha)} \int_0^h t^{\alpha-1} S(t)^* y^* dt = I_1(h) + I_2(h)$$

for any $h > 0$. Now, $I_2(h) \to 0$ as $h \to \infty$, hence $((-A)^{-\alpha})^* y^* = \lim_{h\to 0+} I_1(h)$ in E^*. Since $I_1(h) \in E^\odot$ and $E^\odot$ is closed, $((-A)^{-\alpha})^* \in E^\odot$ as claimed. For $y^* \in E^\odot$, we apply the dominated convergence theorem and obtain

$$((-A)^{-\alpha})^* y^* = \frac{1}{\Gamma(\alpha)} \int_0^\infty t^{\alpha-1} S(t)^\odot y^* \, dt, \tag{7.7.15}$$

valid whenever $y \in E^\odot$, which proves the first restriction relation (7.7.14). The same argument applied in $E^\odot$ to $A^\odot$ takes care of the rest of Lemma 7.7.4. ∎

Good examples of $^\odot$-reflexive spaces were seen in **7.6**. In fact, let A, β be, respectively, an operator and a boundary condition. Then it follows from Theorem 7.6.5 and Theorem 7.6.6 that if β is variational, the space $E = L^1(\Omega)$ is $^\odot$-reflexive relative to the operator $A(\beta)$ with $E^\odot = C(\overline{\Omega})$; likewise, $C(\overline{\Omega})$ is $^\odot$-reflexive with respect to $A_c(\beta)$ with $E^\odot = L^1(\Omega)$. For the Dirichlet boundary condition, $C(\overline{\Omega})$ is replaced by $C_0(\overline{\Omega})$. Of course, in the reflexive spaces $L^p(\Omega)$ $(1 < p < \infty)$, adjoints and Phillips adjoints coincide. All semigroups are analytic and compact.

The opening remarks on (7.7.1) make clear that any useful theory on (7.7.2) must contemplate the possibility that $f(t, y)$ is defined in a space smaller than E, and that it takes values in a space larger than E. We account for this requiring only that

$$f : [0, T] \times E_\alpha \to ((E^\odot)^*)_{-\rho} \tag{7.7.16}$$

for some $\alpha, \rho \geq 0$. We say that f satisfies **Hypothesis I$_{\alpha,\rho}$** (resp. **Hypothesis II$_{\alpha,\rho}$**) in $0 \leq t \leq T$ if the function[2]

$$g(t, \eta) = ((-A^\odot)^{-\rho})^* f(t, (-A)^{-\alpha}\eta) \tag{7.7.17}$$

maps E into E and satisfies Hypothesis I (resp. Hypothesis II) in $0 \leq t \leq T$ (see **5.5**) with $K(t, c) = K(c)$, $L(t, c) = L(c)$ independent of t.[3] We also assume

---

[2] In many applications, this "front-and-back" smoothing is unnecessary; one can take either $\rho = 0$ or $\alpha = 0$. However, it is useful in the Navier–Stokes equations (see **8.6**). Also, it unifies the theory of the variational equation and the adjoint variational equation (see **7.9**).

[3] The constants $K(c), L(c)$ can be replaced by functions of t; see Example 7.7.15.

from now on that the semigroup $S(\cdot)$ generated by A is analytic, that (7.7.10) holds and that the space E is $\odot$-reflexive with respect to A.[4] Informally, a solution of (7.7.2) is a function $y(\cdot)$ satisfying the integral equation

$$y(t) = S(t-s)\zeta + \int_s^t S(t-\tau)f(\tau, y(\tau))d\tau.$$

This equation actually stands for

$$\eta(t) = (-A)^\alpha S(t-s)\zeta$$
$$+ \int_s^t (-A)^{\alpha+\rho} S(t-\tau)((-A^\odot)^{-\rho})^* f(\tau, (-A)^{-\alpha}\eta(\tau))d\tau. \quad (7.7.18)$$

Once (7.7.18) is solved, $y(t) = (-A)^{-\alpha}\eta(t)$ is, by definition, the solution of (7.7.2).

Theorem 7.7.5. *Let $f(t, y)$ satisfy Hypothesis $II_{\alpha,\rho}$, with $\alpha + \rho < 1$. Then, if $\zeta \in E_\alpha$ the integral equation (7.7.18) or, more generally, the integral equation*

$$\eta(t) = \zeta(t) + \int_s^t (-A)^{\alpha+\rho} S(t-\tau)((-A^\odot)^{-\rho})^* f(\tau, (-A)^{-\alpha}\eta(\tau))d\tau$$
$$= \zeta(t) + \int_s^t (-A)^{\alpha+\rho} S(t-\tau)g(\tau, \eta(\tau))d\tau, \quad (7.7.19)$$

where $\zeta(t) \in C(s, T; E)$, has a unique solution in some interval $s \le t \le T'$, where $s < T' \le T$ (g the function in (7.7.17)).

Proof. Much the same as that of Theorem 5.5.1. The kernel of the integral equation is now singular, but the singularity is integrable since, by Lemma 7.4.3,

$$\|(-A)^{\alpha+\rho} S(t)\| \le C_{\alpha+\rho} t^{-\alpha-\rho} \quad (0 \le t \le T). \quad (7.7.20)$$

The details are as follows. Consider the operator

$$(\mathcal{G}\eta)(t) = \zeta(t) + \int_s^t (-A)^{\alpha+\rho} S(t-\tau)g(\tau, \eta(\tau))d\tau \quad (7.7.21)$$

in the closed ball $B(\zeta(\cdot); 1) \subseteq C(s, T'; E)$ of center $\zeta(\cdot)$ and radius 1. We have

$$(\mathcal{G}\eta)(t') - (\mathcal{G}\eta)(t)$$
$$= \int_s^{T'} \{\chi_{t'}(\tau)(-A)^{\alpha+\rho} S(t'-\tau) - \chi_t(\tau)(-A)^{\alpha+\rho} S(t-\tau)\} g(\tau, \eta(\tau))d\tau$$

---

[4] If $E^\odot$ is $\odot$-reflexive and separable then $E^\odot$ is separable as well (Example 7.7.19) so that the assumptions place E and $E^\odot$ on a totally symmetric footing.

where $\chi_t(\cdot)$ is the characteristic function of $[s, t]$. On account of continuity of $(-A)^{\alpha+\rho}S(t - \tau)$ in $\tau < t$ in the uniform operator topology (Lemma 7.4.3) the integrand tends to zero almost everywhere as $t' \to t$. On the other hand, by (7.7.20) the integrand is bounded by a constant times $\chi_t(\tau)(t - \tau)^{-\alpha-\rho} + \chi_{t'}(\tau)(t' - \tau)^{-\alpha-\rho}$, a family of functions with equicontinuous integrals; thus Vitali's theorem applies to show that $\mathcal{G}\eta(\cdot)$ is continuous. If c is a bound for $\|\eta(\tau)\|$ in $s \leq \tau \leq t$ we have

$$\|(\mathcal{G}\eta)(t) - \zeta(t)\| \leq C_{\alpha+\rho}K(c + 1) \int_s^t (t - \tau)^{-\alpha-\rho}d\tau$$

so that $\mathcal{G}$ maps $B(\zeta(\cdot); 1)$ into itself for T' sufficiently small. On the other hand,

$$\|(\mathcal{G}\eta')(t) - (\mathcal{G}\eta)(t)\| \leq C_{\alpha+\rho}L(c + 1) \sup_{s \leq \tau \leq t} \|\eta'(\tau) - \eta(\tau)\| \int_s^t (t - \tau)^{-\alpha-\rho}d\tau.$$

Thus, $\mathcal{G}$ is a contraction for T' sufficiently small. The result then follows from the Cacciopoli-Banach Theorem 2.0.13.   ∎

To obtain uniqueness in the large for solutions of (7.7.19) we need a new version of Gronwall's inequality (Henry [1981, p. 188, Lemma 7.1.1])

Lemma 7.7.6. *Let $b \geq 0, \gamma > -1$ and $a(t), u(t)$ nonnegative and integrable in $0 \leq s \leq T$ with*

$$u(t) \leq a(t) + b \int_s^t (t - \tau)^\gamma u(\tau)d\tau \quad (s \leq t \leq T).^{(5)}$$

Then there exists C depending only on γ, T such that

$$u(t) \leq a(t) + Cb \int_s^t (t - \tau)^\gamma a(\tau)d\tau \quad (s \leq t \leq T).$$

The most direct proof of Lemma 7.7.6 is: iterate the starting inequality again and again. One obtains a series that does not contain $u(\cdot)$ and a residual term seen to converge to zero with the help of the beta integral

$$\int_s^t (t - \tau)^{-\alpha}\tau^{-\beta}d\sigma = \frac{\Gamma(1 + \alpha)\Gamma(1 + \beta)}{\Gamma(2 + \alpha + \beta)}(t - s)^{1-\alpha-\beta}, \quad (7.7.22)$$

valid for $\alpha, \beta < 1$. The series is shown to be convergent using (7.7.22).   ∎

Theorem 7.7.7. *Let $\eta_1(\cdot)$ (resp. $\eta_2(\cdot)$) be a solution of (7.7.19) in $s \leq t \leq T'$ with $\zeta(t) = \zeta_1(t)$ (resp. with $\zeta(t) = \zeta_2(t)$). Then*

$$\|\eta_1(t) - \eta_2(t)\| \leq C \sup_{s \leq \tau \leq t} \|\zeta_1(\tau) - \zeta_2(\tau)\| \quad (s \leq t \leq \bar{t}). \quad (7.7.23)$$

---

[5] That this integral and the next are finite is a consequence of Young's Theorem.

For the proof, estimate using the local Lipschitz condition and apply Lemma 7.7.6. ∎

Corollary 7.7.8. *Let* $y_1(\cdot), y_2(\cdot)$ *be solutions of* (7.7.2) *in* $s \leq t \leq T'$ *with* $y_1(0) = \zeta_1 \in E_\alpha$ *(resp.* $y_2(0) = \zeta_2 \in E_\alpha$*). Then*

$$\|y_1(t) - y_2(t)\|_\alpha \leq C\|\zeta_1 - \zeta_2\|_\alpha \quad (s \leq t \leq \bar{t}). \tag{7.7.24}$$

Extension of solutions and maximal intervals of existence are treated as in Chapter 5. If we have a solution $y(t)$ such that

$$\|y(t)\|_\alpha \leq C \quad (s \leq t \leq T'), \tag{7.7.25}$$

then $y(\cdot)$ can be extended to an interval $[s, T'']$ with $T'' > T$. We limit ourselves to an analogue of Corollary 5.5.4.

Corollary 7.7.9. *The solution* $y(\cdot)$ *of* (7.7.2) *exists in* $s \leq t \leq T$ *or in an interval* $[s, T_m), T_m \leq T$ *and*

$$\limsup_{t \to T_m-}\|y(t)\|_\alpha = \infty.$$

The proofs are the same as those of Lemma 5.5.3 and Corollary 5.5.4.

The reader interested only in the $L^p(\Omega)$ treatment of equations like (7.7.1) with $1 < p < \infty$ may stop here. In the abstract treatment of (7.7.2) we may assume from the beginning that E is reflexive, which produces the following simplifications: $E^\odot = E^*, A^\odot = A^*, ((-A^\odot)^{-\rho})^* = (-A)^{-\rho}$, so that Phillips adjoints (and some adjoints) can be entirely eliminated. For a complete list of simplifications pertaining to E reflexive, see Remark 7.7.18.

The second inclusion in the result below shows that the requirement that the function g in (7.7.17) should take values in E is redundant when $\rho > 0$.

Lemma 7.7.10. *Let* $\rho > 0$. *Then*

$$((-A)^{-\rho})^*(E^*)_{-\rho} \subseteq E^\odot, \quad ((-A^\odot)^{-\rho})^*((E^\odot)^*)_{-\rho} \subseteq E. \tag{7.7.26}$$

Proof. Obviously, we only have to show the first inclusion; for the second, we replace E, A by $E^\odot, A^\odot$. Let $y^* \in (E^*)_{-\rho}$, $\{y_n^*\}$ a sequence in E^* such that $\|y_n^* - y^*\|_{(E^*)_{-\rho}} \to 0$. Then $((-A)^{-\rho})^* y_n^* \to z^* = ((-A)^{-\rho})^* y^* \in E^*$. Now, $((-A)^{-\rho})^* y^* \in E^\odot$ due to the first inclusion (7.7.14) and $E^\odot$ is closed, so that $z^* \in E^\odot$ as claimed. ∎

When $\rho = 0$, there are important examples where g satisfies Hypothesis $I_{\alpha,0}$ or Hypothesis $II_{\alpha,0}$ with $K(t, c) \equiv K(c)$ and $L(t, c) \equiv L(c)$ as required, except that

$f(t, y)$ takes values in $(E^{\odot})^* \supseteq E$ and that $t \to f(t, y)$ is merely $E^{\odot}$-weakly measurable, that is, $t \to \langle y^*, f(t, y) \rangle$ is measurable for every $y^* \in E^{\odot}$. If this happens, we say that $f(t, y)$ satisfies **Hypothesis** $\text{I}_{\alpha,0}(wm)$ or **Hypothesis** $\text{II}_{\alpha,0}(wm)$. An example is the equation (7.7.1) in the space $C(\overline{\Omega})$ with $u(t, x) \in L^{\infty}((0, T) \times \Omega)$; for instance, if $f(t, y) = f(y) + u$ does not depend on t or on the gradient, $f(y)$ is Lipschitz continuous and $u(\cdot, \cdot) \in L^{\infty}((0, T) \times \Omega)$, then $f(t, y)$ satisfies Hypothesis $\text{II}_{\alpha,0}(wm)$ in the space $C_0(\overline{\Omega})$ but, except for certain choices of u, it does not satisfy the corresponding "strong" hypothesis. We show below (Corollary 7.7.12 and Lemma 7.7.13) that this does not make a difference in the treatment of (7.7.2).

Lemma 7.7.11. *Let the* $(E^{\odot})^*$-*valued function* $g(\cdot)$ *be* $E^{\odot}$-*weakly measurable, and let* $\delta > 0$. *Then the function*

$$g_{\delta}(\tau) = ((-A^{\odot})^{-\delta})^* g(\tau)$$

is a strongly measurable E-*valued function.*

Proof. That $g_{\delta}(\tau)$ takes values in E follows from the second inclusion (7.7.13). Since E is separable, it is enough to show that $g_{\delta}(\cdot)$ is E^*-weakly measurable (Example 5.0.39). Let $y^* \in E^*$. Using (7.4.38) we can write

$$y^* = ((-A)^{\delta/2})^*((-A)^{-\delta/2})^* y^* = ((-A)^{\delta/2})^* z^*$$

where, in view of the first inclusion (7.7.13), $z^* \in E^{\odot}$. On the other hand, again by (7.4.39) and then by the second inclusion (7.7.13) and the second equality (7.7.14),

$$g_{\delta}(\tau) = ((-A^{\odot})^{-\delta/2})^*((-A^{\odot})^{-\delta/2})^* g(\tau) = (-A)^{-\delta/2}((-A^{\odot})^{-\delta/2})^* g(\tau),$$

so that

$$\langle y^*, g_{\delta}(\tau) \rangle = \langle ((-A)^{\delta/2})^* z^*, (-A)^{-\delta/2}((-A^{\odot})^{-\delta/2})^* g(\tau) \rangle$$

$$= \langle z^*, ((-A^{\odot})^{-\delta/2})^* g(\tau) \rangle = \langle (-A^{\odot})^{-\delta/2} z^*, g(\tau) \rangle,$$

which is measurable by hypothesis since z^*, thus $(-A^{\odot})^{-\delta/2} z^*$, belongs to $E^{\odot}$. ∎

Corollary 7.7.12. *Let* $E, S(t)$ *be as in Lemma* 7.1.11. *If* $f(t, y)$ *satisfies Hypothesis* $\text{II}_{\alpha,0}(wm)$ *then it satisfies Hypothesis* $\text{II}_{\alpha,\rho}$ *for any* $\rho > 0$. *Same statement for Hypothesis* $\text{I}_{\alpha,0}$.

Corollary 7.7.12 shows that, in the case $\rho = 0$, Theorem 7.7.5 can be proved weakening Hypothesis $\text{II}_{\alpha,0}$ to Hypothesis $\text{II}_{\alpha,0}(wm)$. In some cases this "inflation of ρ" is inconvenient, and we prefer to do the smoothing with the semigroup itself rather than with a negative fractional power of A. This is explained below: before stating the result, we note that if X is a separable Banach space and $g(\cdot)$ is a X^*-valued X-weakly measurable function, then $t \to \|g(t)\|_{X^*}$ is measurable; it suffices to note that $\|g(t)\| = \sup\{\langle g(t), y_n \rangle\}$, where $\{y_n\}$ is a countable dense set in the unit ball of X. Measurability of the norm gives sense to the definition of the spaces $L_w^p(0, T; X^*)$ of X-weakly measurable X^*-valued functions with $\|g(\cdot)\|_{X^*} \in L^p(0, T)$; the norm of $g(\cdot)$ is the L^p norm of $\|g(\cdot)\|_{X^*}$.

Lemma 7.7.13. (a) Let the $(E^{\odot})^*$-valued function $g(\cdot)$ be $E^{\odot}$-weakly measurable. Then

$$\tau \to (-A)^{\alpha} S^{\odot}(t - \tau)^* g(\tau) \qquad (7.7.27)$$

is a strongly measurable E-valued function in $0 \leq \tau \leq t$. (b) If $0 \leq \alpha < 1$ then the operator

$$(\Lambda_{\alpha} g)(t) = \int_0^t (-A)^{\alpha} S^{\odot}(t - \tau)^* g(\tau) d\tau \qquad (7.7.28)$$

is bounded from $L_w^p(0, T; (E^{\odot})^*)$ into $C(0, T; E)$ for $p > 1/(1 - \alpha)$.

Proof. It is enough to show that (7.7.27) is strongly measurable for $\tau \leq t - \varepsilon$ for every $\varepsilon > 0$. By (7.7.9) we have

$$(-A)^{\alpha} S^{\odot}(t - \tau)^* g(\tau) = (-A)^{\alpha} S(t - \tau - \varepsilon) S^{\odot}(\varepsilon)^* g(\tau);$$

since $(-A)^{\alpha} S(\cdot)$ is continuous in (E, E) we only have to show that $S^{\odot}(\varepsilon)^* g(\tau)$ is strongly measurable for $\varepsilon > 0$ fixed. If $y^* \in E^*$ we have, again using (7.7.9),

$$\langle y^*, S^{\odot}(\varepsilon)^* g(\tau) \rangle = \langle y^*, S(\varepsilon/2) S^{\odot}(\varepsilon/2)^* g(\tau) \rangle = \langle S(\varepsilon/2)^* y^*, S^{\odot}(\varepsilon/2)^* g(\tau) \rangle$$

$$= \langle S(\varepsilon/2)^{\odot} S(\varepsilon/2)^* y^*, g(\tau) \rangle = \langle S(\varepsilon)^* y^*, g(\tau) \rangle.$$

Since $S(\varepsilon)^* E^* \subseteq E^{\odot}$, $\langle y^*, S^{\odot}(\varepsilon)^* g(\tau) \rangle$ is measurable for all $y \in E^*$ so that $S^{\odot}(\varepsilon)^* g(\cdot)$ is E^*-weakly measurable, thus strongly measurable since E is separable. We prove (b). By Lemma 7.4.3 we have $\|(-A)^{\alpha} S(t)\| \leq C_{\alpha} t^{-\alpha}$, hence

$$\|(-A)^{\alpha} S^{\odot}(t)^*\| \leq C t^{-\alpha} \quad (0 \leq t \leq T), \qquad (7.7.29)$$

as we see writing $(-A)^{\alpha} S^{\odot}(t)^* = (-A)^{\alpha} S(t - \varepsilon) S^{\odot}(\varepsilon)^*$, estimating and letting $\varepsilon \to 0$. Accordingly, the integrand of (7.7.28) is bounded by $C(t - \tau)^{-\alpha} \|g(\tau)\|$ and continuity of $\Lambda_{\alpha} g(\cdot)$ is proved using an argument similar to that in the proof of continuity of $\mathcal{G}\eta(\cdot)$ in (7.7.21). In the present case, equicontinuity of the integrals of the bounding functions follows from Hölder's inequality and the fact that $t^{-\alpha}$ belongs to L^q for $q < 1/\alpha$. Hölder's inequality is also used to prove boundedness of Λ_{α}. ∎

Remark 7.7.14. In case $f(t, y)$ satisfies Hypothesis $II_{\alpha,0}(wm)$ with $\alpha < 1$, Theorem 7.7.5 can be proved without modifying ρ. It suffices to replace the integral equation (7.7.18) by

$$\eta(t) = (-A)^\alpha S(t - s)\zeta$$
$$+ \int_s^t (-A)^\alpha S^\odot(t - \tau)^*((-A^\odot)^{-\rho})^* f(\tau, (-A)^\alpha \eta(\tau))d\tau \quad (7.7.30)$$

whose treatment is the same as that of (7.7.18), but this time on the basis of Lemma 7.7.13; the required mapping and contraction properties of $\mathcal{G}$ are proved in the same way. We can also solve (7.7.30) with $\zeta \in ((E^\odot)^*)_\alpha$. In this case the first term on the right is $(-A)^\alpha S^\odot(t - s)^*\zeta$. To obtain a local existence-uniqueness theorem under these conditions it suffices to replace $C(s, T; E)$ by the space $C(s+, T; E)$ of all bounded continuous functions in $(s, T]$ endowed with the supremum norm, and argue in the same way. All results on uniqueness and extension are similar.

We note for the record the particular case where Assumption $II_{0,0}(wm)$ is satisfied (that is, $\alpha = \rho = 0$), corresponding roughly to Hypothesis II (bounded Lipschitz continuous nonlinearity) but not completely, as strong measurability is replaced by weak measurability. Fractional powers disappear, $y(t) = \eta(t)$, and the equation becomes

$$y(t) = S^\odot(t - s)^*\zeta + \int_s^t S^\odot(t - \tau)^* f(\tau, y(\tau))d\tau. \quad (7.7.31)$$

Example 7.7.15. Hypotheses $II_{\alpha,\rho}$ and $II_{\alpha,\rho}(wm)$ can be weakened replacing $K(c), L(c)$ by $K(\cdot, c)$ and $L(\cdot, c) \in L^q(0, T)$ with $q > 1/(1 - \alpha - \rho)$; taking into account that $t^{-\alpha-\rho}$ belongs to L^p for $p < 1/(\alpha + \rho)$, any q as above is the conjugate index of such a p. Under the new hypotheses, Theorems 7.7.5 and its variant in Remark 7.7.14 undergo no changes, except for two uses of Hölder's inequality. Corollary 7.7.8, however, cannot be squeezed out of Lemma 7.7.6; the version of Gronwall's inequality required must use the kernel $(t - \tau)^{-\alpha-\rho} L(\tau, c)$. A version of Gronwall's lemma that applies to this case can be found in the author–Sritharan [1995 : 2, Lemma 3.1].

As in **5.6**, existence results can be obtained on the basis of Hypothesis $I_{\alpha,\rho}$ or $I_{\alpha,\rho}(wm)$ if compactness is supplied by the semigroup. This is particularly interesting here as all semigroups generated by uniformly elliptic operators in **7.6** are compact.

Theorem 7.7.16. *Assume all the hypotheses in Lemma 7.7.13 are satisfied, and that $S(t)$ is compact for every $t > 0$. Then, if $p > 1/(1 - \alpha)$ the operator Λ_α is compact from $L_w^p(0, T; (E^\odot)^*)$ into $C(0, T; E)$.*

Proof. Similar to that of Theorem 5.5.19. For $\delta > 0$ we write

$$(\Lambda_{\alpha,\delta} g)(t) = \int_0^{t-\delta} (-A)^\alpha S^\odot(t-\tau)^* g(\tau) d\tau$$

$$= S(\delta) \int_0^{t-\delta} (-A)^\alpha S^\odot(t-\delta-\tau)^* g(\tau) d\tau \qquad (7.7.32)$$

for $t \geq \delta$, $(\Lambda_{\alpha,\delta} g)(t) = 0$ for $t < \delta$. It easy to show on the basis of Hölder's inequality that

$$\Lambda_{\alpha,\delta} \to \Lambda_\alpha \quad \text{in } (L_w^p(0, T; (E^\odot)^*), C(0, T; E))$$

as $\delta \to 0$, thus we only have to show that each $\Lambda_{\alpha,\delta}$ is a compact operator from $L_w^p(0, T; (E^\odot)^*)$ into $C(0, T; E)$. To do this, pick a bounded sequence $\{g_n(\cdot)\} \subseteq L_w^p(0, T; (E^\odot)^*)$. The second expression (7.7.32) implies that $\{(\Lambda_{\alpha,\delta})(t)\}$ has a convergent subsequence for each t fixed, so that to apply the Arzelà–Ascoli Theorem 2.0.15 it is enough to show that the sequence $\{(\Lambda_{\alpha,\delta})(\cdot)\}$ is equicontinuous. If $0 \leq s \leq t \leq T$ and $s \geq \delta$ we have

$$\|(\Lambda_{\alpha,\delta} g_n)(t) - (\Lambda_{\alpha,\delta} g_n)(s)\|$$

$$\leq \int_{s-\delta}^{t-\delta} \|(-A)^\alpha S^\odot(t-\tau)^*\| \, \|g_n(\tau)\| d\tau$$

$$+ \int_0^{s-\delta} \|(-A)^\alpha (S^\odot(t-\tau)^* - S^\odot(s-\tau)^*)\| \, \|g_n(\tau)\| d\tau$$

$$\leq C \left(\int_{s-\delta}^{t-\delta} (t-\tau)^{-\alpha q'} d\tau \right)^{1/q'}$$

$$+ C \left(\int_0^{s-\delta} \|(-A)^\alpha S^\odot(t-\tau)^* - S^\odot(s-\tau)^*)\|^{q'} d\tau \right)^{1/q'}, \qquad (7.7.33)$$

where we have applied Hölder's inequality to both integrals with $p > p' > 1/(1-\alpha)$ and $1/q' + 1/p' = 1$, so that $q' < 1/\alpha$. In the first term, we estimate with the worst case $\delta = 0$; in the second, we note that $(-A)^\alpha S^\odot(t)^*$ is continuous in the norm of (E, E) in $t > 0$ (consequence of the equality $(-A)^\alpha S^\odot(t)^* = (-A)^\alpha S(t-\varepsilon) S^\odot(\varepsilon)^*$) and use Vitali's convergence theorem, whose application is justified since the integrand is bounded by the function $\chi_t(\tau)(t-\tau)^{-\alpha q'} + \chi_s(\tau)(s-\tau)^{-\alpha q'}$, χ_t the characteristic function of $(0, t)$. The estimate (7.7.33) is also good in the range $s < \delta$ (where the second integral disappears) and the case $t < \delta$ is obvious.   ■

Corollary 7.7.17. *Assume that $S(t)$ is compact. Let $f(t, y)$ satisfy Hypothesis $I_{\alpha,\rho}$, or $I_{\alpha,\rho}(wm)$ with $\alpha + \rho < 1$. Then, if $\zeta \in E_\alpha$ the integral equation (7.7.18), or more generally, the integral equation (7.7.19) has a solution in some interval $s \leq T' \leq T$.*

The proof is based on Schauder's theorem and is essentially the same as that of Theorem 5.5.22.

Remark 7.7.18. Here are the simplifications pertaining to the case E reflexive (see also the comments after Corollary 7.7.9). We have

$$E^\odot = E^*, \qquad (E^\odot)^* = E, \qquad A^\odot = A^*, \qquad (A^\odot)^* = A,$$
$$((-A^\odot)^{-\rho})^* = (-A)^{-\rho}, \qquad S^\odot(t) = S(t)^*, \qquad S^\odot(t)^* = S(t), \tag{7.7.34}$$

the function $g(t, \eta)$ in (7.7.17) is

$$g(t, \eta) = (-A)^{-\rho} f(t, (-A)^{-\alpha}\eta), \tag{7.7.35}$$

and the integral equation (7.7.18) becomes

$$\eta(t) = (-A)^\alpha S(t - s)\zeta$$
$$+ \int_s^t (-A)^{\alpha+\rho} S(t - \tau)(-A)^{-\rho} f(\tau, (-A)^{-\alpha}\eta(\tau))d\tau. \tag{7.7.36}$$

Example 7.7.19. Let E be a Banach space, $S(\cdot)$ a strongly continuous semigroup in E. Then, if the Phillips adjoint $E^\odot$ is separable, so is E. The proof is similar to that of Example 5.0.38. Given a countable dense set $\{y_n^*\}$ in $E^\odot$, pick a sequence $\{y_n\} \subseteq E$ with

$$\|y_n\| = 1, \qquad \langle y_n^*, y_n \rangle \geq \|y_n^*\|/2. \tag{7.7.37}$$

Fix λ large enough and let K be the closed subspace of E generated by all elements of the form $R(\lambda; A)^m y_n$ $(n = 1, 2, \ldots, m = 0, 1, \ldots)$. Clearly K is separable and $R(\lambda; A)K \subseteq K$; also, $y_n \in K$ for all n. If K is not dense in E there exists $y^* \in E^*$, $y^* \neq 0$ such that

$$\langle y^*, y \rangle = 0 \quad (y \in K).$$

Let $z^* = R(\lambda; A)^* y^*$. Then $z^* \in D(A^*) \subseteq E^\odot, z^* \neq 0$ and

$$\langle z^*, y \rangle = \langle R(\lambda; A)^* y^*, y \rangle = \langle y^*, R(\lambda; A)y \rangle = 0 \quad (y \in K).$$

Pick n so large that $\|z^* - y_n^*\| < \|y_n^*\|/2$. Then

$$\langle y_n^*, y_n \rangle = \langle z^* - y_n^*, y_n \rangle + \langle z^*, y_n \rangle = \langle z^* - y_n^*, y_n \rangle < \|y_n^*\|/2,$$

a contradiction with (7.7.37).

If $E^\odot$ is $\odot$-reflexive with respect to $S(\cdot)$, the preceding argument shows that $E^\odot$ is separable if $E = E^{\odot\odot}$ is separable.

7.8. Semilinear Abstract Parabolic Equations: Global Existence

We study global existence for

$$y'(t) = Ay(t) + f(t, y(t)), \quad y(0) = \zeta \tag{7.8.1}$$

using (and generalizing) results in **5.8**. The first (corresponding to Lemma 5.8.1) approximates solutions of (7.8.1) by solutions of

$$y_n'(t) = Ay_n(t) + f_n(t, y_n(t)), \quad y_n(0) = \zeta_n. \tag{7.8.2}$$

As in **7.7**, the standing assumptions are: A generates an analytic semigroup $S(\cdot)$ and E is separable and $^\odot$-reflexive with respect to A.

Lemma 7.8.1. *Let f satisfy Hypothesis $I_{\alpha,\rho}(wm)$ and let f_n satisfy Hypothesis $II_{\alpha,\rho}(wm)$ in $0 \le t \le T$ with α, ρ fixed, $\alpha + \rho < 1$ and $L_n(c)$ uniformly bounded. Assume that $\zeta_n \to \zeta$ in E_α and that, for every $y(\cdot) \in C(0, T; E_\alpha)$ we have*

$$\|\kappa_n(t)\| \to 0 \tag{7.8.3}$$

uniformly in $0 \le t \le T$, where[1]

$$\kappa_n(t) = \int_0^t (-A)^{\alpha+\rho} S^\odot(t - \tau)^* ((-A^\odot)^{-\rho})^* \{f_n(\tau, y(\tau)) - f(\tau, y(\tau))\} d\tau.$$

Finally, assume that the solution $y(t)$ of (7.8.1) exists in $0 \le t \le \bar{t}$. Then there exists n_0 such that, if $n \ge n_0$ the solution $y_n(t)$ of (7.8.2) exists in $0 \le t \le \bar{t}$ and

$$\|y_n(t) - y(t)\|_\alpha \to 0 \tag{7.8.4}$$

uniformly in $0 \le t \le \bar{t}$.

The reader will note that the expected analogue of (5.8.3) in Lemma 5.8.1,

$$((-A^\odot)^{-\rho})^* f_n(t, y) \to ((-A^\odot)^{-\rho})^* f(t, y) \quad \text{a.e. in } 0 \le t \le T, \tag{7.8.5}$$

has been replaced by the more general (7.8.3). This will be decisive later, since (7.8.3) takes advantage of smoothing by the semigroup $S^\odot(t)^*$ while (7.8.5) does not; see Remark 7.8.3.

The integral equations corresponding to (7.8.1) and (7.8.2) are

$$\eta(t) = (-A)^\alpha S(t - s)\zeta$$
$$+ \int_0^t (-A)^{\alpha+\rho} S^\odot(t - \tau)^* ((-A^\odot)^{-\rho})^* f(\tau, (-A)^\alpha \eta(\tau)) d\tau, \tag{7.8.6}$$

$$\eta_n(t) = (-A)^\alpha S(t - s)\zeta_n$$
$$+ \int_0^t (-A)^{\alpha+\rho} S^\odot(t - \tau)^* ((-A^\odot)^{-\rho})^* f_n(\tau, (-A)^{-\alpha} \eta_n(\tau)) d\tau, \tag{7.8.7}$$

---

[1] Hypothesis $II_{\alpha,\rho}(wm)$ makes its appearance only when $\rho = 0$ (see Lemma 7.7.10 and following comments); this is the single case where we need $S^\odot(t - \tau)^*$ rather than $S(t - \tau)$ in the integrand.

where $\eta(t) = (-A)^{\alpha} y(t)$ and $\eta_n(t) = (-A)^{\alpha} y_n(t)$. For each $n = 1, 2, \ldots$ let $[0, t_n] \subseteq [0, \bar{t}]$ be the largest interval such that $\eta_n(t)$ exists in $0 \leq t \leq t_n$ and $\|\eta_n(t) - \eta(t)\| \leq 1$. Subtract (7.8.6) from (7.8.7), write

$$
f_n(\tau, (-A)^{-\alpha}\eta_n(\tau)) - f(\tau, (-A)^{-\alpha}\eta(\tau))
$$
$$
= f_n(\tau, (-A)^{-\alpha}\eta_n(\tau)) - f_n(\tau, (-A)^{-\alpha}\eta(\tau))
$$
$$
+ f_n(\tau, (-A)^{-\alpha}\eta(\tau)) - f(\tau, (-A)^{-\alpha}\eta(\tau)),
$$

and estimate:

$$
\|\eta_n(t) - \eta(t)\| \leq \|S(t)(\zeta_n - \zeta)\|_{\alpha}
$$
$$
+ CL_n(c+1) \int_0^t (t - \tau)^{-\alpha-\rho} \|\eta_n(\tau) - \eta(\tau)\| d\tau
$$
$$
+ \left\| \int_0^t (-A)^{\alpha+\rho} S^{\odot}(t - \tau)^* ((-A^{\odot})^{-\rho})^*
$$
$$
\times \left\{ f_n(t, (-A)^{\alpha}\eta(\tau)) - f(t, (-A)^{\alpha}\eta(\tau) \right\} d\tau \right\|
$$
$$
= M\|\zeta_n - \zeta\|_{\alpha} + \|\kappa_n(t)\| + CL_n(c+1) \int_0^t (t - \tau)^{-\alpha-\rho}\|\eta_n(\tau) - \eta(\tau)\| d\tau,
$$

M a bound for $\|S(t)\|$. We end as in Lemma 5.8.1, only using this time the generalized Gronwall Lemma 7.7.6. ∎

Global existence is based on Theorem 5.8.3. This result requires

$$
\langle y^*, Ay \rangle \leq \omega \|y\|^2 \quad (y \in D(A), y^* \in \Theta(y)), \tag{7.8.8}
$$

($\Theta(y)$ the duality set of y),

$$
(\lambda I - A)D(A) = E \quad (\lambda > \omega), \tag{7.8.9}
$$

and Hypothesis II for $f : [0, T] \times E \to E$, plus

$$
\langle y^*, f(t, y) \rangle \leq c(1 + \|y\|^2) \quad (0 \leq t \leq T, y \in E, y^* \in \Theta(y)). \tag{7.8.10}
$$

Theorem 5.8.3 produces the *a priori* bound

$$
\|y(t)\| \leq e^{\theta t}\|\zeta\| \tag{7.8.11}
$$

for $\zeta \neq 0$, with $\theta = \omega + c(1 + \|\zeta\|^{-2})$, thus it implies global existence; for $\zeta = 0$ we apply the estimate in a suitable subinterval (see Corollary 5.8.4). As a sample application, we show global existence for (7.8.1) and an application to a partial differential equation.

Corollary 7.8.2. *Assume that $f(t, y)$ satisfies Hypothesis $II_{0,0}$, that (7.8.8), (7.8.9) and (7.8.10) hold and that $u(\cdot) \in L_w^\infty(0, T; (E^\odot)^*)$. Then the solution of*

$$y'(t) = Ay(t) + f(t, y(t)) + u(t), \quad y(0) = \zeta \tag{7.8.12}$$

exists in $0 \le t \le T$; if $\|\zeta\| \ne 0$ it satisfies

$$\|y(t)\| \le e^{\theta t}\|\zeta\| \quad (0 \le t \le T') \tag{7.8.13}$$

with $\theta = \omega + (c + M\|u\|)(1 + \|\zeta\|^{-2})$, $\|u\|$ the $L_w^\infty(0, T; (E^\odot)^*)$-norm of $u(\cdot)$ and M a bound for $\|S(t)\|$ in $0 \le t \le T$.

Proof. It is obvious that (7.8.12) satisfies Hypothesis $II_{0,0}(wm)$. We use the approximating equation

$$y_h'(t) = Ay_h(t) + f(t, y_h(t)) + S^\odot(h)^*u(t), \quad y(s) = \zeta \tag{7.8.14}$$

and check that it fits into Lemma 7.8.1. It has been proved in Lemma 7.7.13 that $S^\odot(h)^*u(\cdot)$ is an E-valued strongly measurable function. Thus, the nonlinear term $f(t, y) + S^\odot(h)^*u(t)$ in (7.8.14) satisfies Hypothesis $II = II_{0,0}$ with $K_n(c)$ and $L_n(c)$ independent of n. Moreover, if c is the constant in (7.8.10) we have

$$\langle y^*, f(t, y) + S^\odot(h)^*u(t)\rangle \le c(1 + \|y\|^2) + M\|y\| \|u(t)\|$$
$$\le (c + M\|u\|)(1 + \|y\|^2) \quad (0 \le t \le T, y \in E, y^* \in \Theta(y))$$

modifying $u(\cdot)$ in a set of null measure (so that $\|u(t)\| \le \|u\|$) and noting that $\|y\| \le 1 + \|y\|^2$. Theorem 5.8.3 then shows that $y_h(t)$ exists in $0 \le t \le T$ and satisfies (7.8.11). If we prove that

$$\kappa_h(t) = \int_0^t S^\odot(t - \tau)^*\{S^\odot(h)^*u(\tau) - u(\tau)\}d\tau \to 0 \tag{7.8.15}$$

uniformly in $0 \le t \le T$, then Lemma 7.8.1 applies to show that if the solution of (7.8.1) exists in $0 \le t \le \bar{t}$ it must satisfy the *a priori* bound (7.8.11), thus it must actually exist in $0 \le t \le T$. Rewrite (7.8.15) in the form

$$\kappa_h(t) = \int_0^T \chi_t(\tau)\{S^\odot(t - \tau + h)^* - S^\odot(t - \tau)^*\}u(\tau)d\tau, \tag{7.8.16}$$

χ_t the characteristic function of $[0, t]$, and assume $\kappa_h(t)$ does not converge to zero uniformly in $0 \le t \le T$. Then there exists a sequence $\{h_n\} \subseteq \mathbb{R}_+, h_n \to 0$ and a sequence $\{t_n\} \subseteq [0, T]$ with $t_n \to \bar{t} \in [0, T]$ and $\|\kappa_{h_n}(t_n)\| \ge \varepsilon \ge 0$. A contradiction with the dominated convergence theorem ensues if we note that the integrand of (7.8.16) (for $h = h_n, t = t_n$) is uniformly bounded and converges to zero a.e. on account of the (E, E)-continuity of $S^\odot(t)^*$ in $t > 0$. ∎

Remark 7.8.3. That (7.8.5) is much more demanding than (7.8.3) is plain in the proof above: since $S^\odot(h)^*u(\tau)$ belongs to E it cannot in general tend to $u(\tau) \in (E^\odot)^*$ as $h \to 0$.

We apply the results to the parabolic equation

$$y_t(t, x) = \sum_{j=1}^{m}\sum_{k=1}^{m} \partial^j(a_{jk}(x)\partial^k y(t, x)) + \sum_{j=1}^{m} b_j(x)\partial^j y(t, x)$$

$$+ c(x)y(t, x) - \phi(t, x, y(t, x)) + u(t, x) \qquad (7.8.17)$$

in a bounded domain Ω of class $C^{(2)}$. The assumptions on the linear part of the operator are those in Theorems 7.6.3 and 7.6.4, and the space is $E = C(\overline{\Omega})$ or $E = C_0(\overline{\Omega})$ according to whether the boundary condition is variational or Dirichlet. Controls $u(t, x)$ are taken in the space $L^\infty((0, T) \times \Omega) = L_w^\infty(0, T; L^\infty(\Omega)) = L_w^\infty(0, T; (E^\odot)^*)$ (see Example 5.0.32). The nonlinear term $\phi(t, x, y)$ in (7.8.17) is measurable in t for x, y fixed and continuous in x, y for t fixed; for every $c > 0$ there exist $K(c), L(c)$ such that

$$|\phi(t, x, y)| \le K(c) \quad (0 \le t \le T, x \in \Omega, |y| \le c), \qquad (7.8.18)$$

$$|\phi(t, x, y') - \phi(t, x, y)| \le L(c)|y' - y|$$
$$(0 \le t \le T, x \in \Omega, |y|, |y'| \le c). \qquad (7.8.19)$$

Equation (7.8.17) is viewed as the abstract equation

$$y'(t) = A_c(\beta)y(t) + f(t, y(t)) + u(t), \qquad (7.8.20)$$

where $A_c(\beta)$ is the operator introduced in **7.6** and $f(t, y)$ is the nonlinear operator

$$f(t, y(\cdot))(x) = -\phi(t, x, y(x)). \qquad (7.8.21)$$

We check that $f(t, y)$ satisfies the assumptions of Corollary 7.8.2 first for a variational boundary condition, keeping in mind that

$$E^\odot = L^1(\Omega), \qquad (E^\odot)^* = L^\infty(\Omega).$$

Obviously, $f(t, y)$ takes values in $C(\overline{\Omega})$, and (7.8.18) and (7.8.19) translate into the two inequalities needed in Hypothesis $\mathrm{II}_{0,0}$, namely

$$\|f(t, y)\| \le K(c), \qquad \|f(t, y') - f(t, y)\| \le L(c)\|y' - y\| \quad (0 \le t \le T)$$

for $y, y' \in C(\overline{\Omega})$, $\|y\|, \|y'\| \le c$. The space $C(\overline{\Omega})$ is separable, hence, to show that $t \to \phi(t, x, y(x))$ is strongly measurable in $C(\overline{\Omega})$ it is enough to show that $t \to \int_{\overline{\Omega}} \phi(t, x, y(x))\mu(dx)$ is measurable for $\mu \in C(\overline{\Omega})^* = \Sigma(\overline{\Omega})$. Since linear combinations of Dirac delta measures are $C(\overline{\Omega})$-weakly dense in $\Sigma(\overline{\Omega})$

(Theorem 10.6.4) this reduces to measurability of $t \to \phi(t, x, y(x))$ for x fixed, which is postulated.

The fact that (7.8.8) and (7.8.9) hold for the operator $A_c(\beta)$ has been established in Theorem 7.6.8, modulo a renorming of the space with

$$\|y(\cdot)\|_\rho = \max_{x \in \overline{\Omega}} |y(x)| \rho(x) \qquad (7.8.22)$$

where $\rho(x)$ is positive and twice continuously differentiable in $\overline{\Omega}$. The dual space is $\Sigma(\overline{\Omega})$, and the the the duality pairing of $C(\overline{\Omega})$ and $\Sigma(\overline{\Omega})$ is

$$\langle \mu, y \rangle = \int_{\overline{\Omega}} y(x) \rho(x) \mu(dx).$$

If $y(\cdot) \in C(\overline{\Omega})$ (see Example 5.8.12) then the duality set $\Theta(y) \subseteq \Sigma(\overline{\Omega})$ consists of all measures μ supported by the set

$$m(y) = \{x \in \overline{\Omega}; |y(x)| \rho(x) = \|y\|\} = m_+(y) \cup m_-(y)$$
$$= \{x \in \overline{\Omega}; y(x) \rho(x) = \|y\|\} \cup \{x \in \overline{\Omega}; y(x) \rho(x) = -\|y\|\}$$

with $\|\mu\|_{\Sigma(\overline{\Omega})} = \|y\|_{C(\overline{\Omega})}$, $\mu \geq 0$ in $m_+(y)$, $\mu \leq 0$ in $m_-(y)$. Assume that

$$(\text{sgn } y)\phi(t, x, y) \geq -C(1 + |y|) \quad (x \in \overline{\Omega}, -\infty < y < \infty), \qquad (7.8.23)$$

and let $y(\cdot) \in C(\overline{\Omega})$ and $\mu \in \Theta(y)$. Then we have

$$\langle \mu, f(t, \cdot, y(\cdot)) \rangle = -\int_{m(y)} \phi(t, x, y(x)) \mu(dx)$$
$$= -\int_{m(y)} (\text{sgn } y(x)) \phi(t, x, y(x)) |\mu|(dx)$$
$$\leq C \int_{m(y)} (1 + |y(x)|) |\mu|(dx)$$
$$= C(\|y\| + \|y\|^2) \leq 2C(1 + \|y\|^2).$$

This inequality is (7.8.10), thus we may apply Corollary 7.8.2 and assure global existence. Condition (7.8.23) is satified for instance by $\phi(y) = cy^3$ or, more generally, by any polynomial

$$\phi(t, x, y) = \sum_{k \text{ odd}} \phi_k(x) y^k \qquad (7.8.24)$$

with $\phi_k(x)$ continuous and nonnegative; in fact, $(\text{sign } y)y^k \geq 0$ for k odd. The reader should note that inequality (7.8.23) only limits the growth of $\phi(t, x, y)$ in one direction; for instance, it is satisfied by $\phi(t, x, y) = \phi(x)e^y$ with $\phi(x)$ continuous and nonnegative or, more generally, by any function of linear growth for $y < 0$.

We leave to the reader the modifications pertaining to the Dirichlet boundary condition.

7.9. Linear Abstract Parabolic Equations. Duality

To understand both variational equations, we digress on the linear equation

$$y'(t) = (A + B(t))y(t) + f(t), \qquad y(s) = \zeta. \tag{7.9.1}$$

Control theory requires us to solve the variational equation with the same type of initial condition and right-hand side as (7.7.2). However, the adjoint variational equation will have to be solved with initial condition and nonhomogeneous term $f(\cdot)$ in larger spaces, thus we extend the theory in **7.7**. As in the last two sections, we assume that A is the infinitesimal generator of an analytic semigroup $S(\cdot)$ satisfying

$$\|S(t)\| \le Ce^{-ct} \quad (t \ge 0) \tag{7.9.2}$$

for some $c > 0$, and that E is separable and $\odot$-reflexive with respect to A. The requirements on $B(t)$ and $f(t)$ are:

$B(t)$ is densely defined in E and there exist α, ρ with $\alpha + \rho < 1$ such that

$$B(t) \in (E_\alpha, ((E^\odot)^*)_{-\rho}). \tag{7.9.3}$$

The function

$$t \to F(t)\eta = ((-A^\odot)^{-\rho})^* B(t)(-A)^{-\alpha}\eta \tag{7.9.4}$$

is $E^\odot$-weakly measurable for every $\eta \in E$ and there exists K such that

$$\|((-A^\odot)^{-\rho})^* B(t)(-A)^{-\alpha}\|_{(E,(E^\odot)^*)} \le K \quad (0 \le t \le T).^{(1)} \tag{7.9.5}$$

There exists γ, δ such that $\alpha + \gamma < 1, \alpha + \delta < 1$ and

$$\zeta \in ((E^\odot)^*)_{-\gamma}, \tag{7.9.6}$$

$$f(t) \in ((E^\odot)^*)_{-\delta}. \tag{7.9.7}$$

The function

$$t \to ((-A^\odot)^{-\delta})^* f(t) \tag{7.9.8}$$

is $E^\odot$-weakly measurable and there exists L such that

$$\|((-A^\odot)^{-\delta})^* f(t)\|_{(E^\odot)^*} \le L \quad (0 \le t \le T). \tag{7.9.9}$$

These boundedness conditions in (7.9.5) and (7.9.9) are assumed for simplicity; L^q conditions for suitable q suffice (see Example 7.7.15).

Note that the condition that $B(t)$ be densely defined as an operator in E (which is necessary to define $B(t)^*$ later) does *not* follow from (7.9.3), since if $\eta \in E_\alpha$, $B(t)\eta$ belongs to the larger space $((E^\odot)^*)_{-\rho}$, not necessarily to E. As an operator

---

(1) If $\rho > 0$ then $F(t)\eta \in E$, thus the $(E^\odot)^*$-norm is used only for $\rho = 0$. See Lemma 7.7.10.

in E, $B(t)$ is neither everywhere defined nor bounded unless $\alpha = \rho = 0$; in this case, $B(t) \in (E, (E^\odot)^*)$.

To avoid having to contend with weakly measurable functions in the case $\rho = 0$ we use Lemma 7.7.11 and raise ρ and δ (without disturbing the inequalities $\alpha + \rho < 1$, $\alpha + \delta < 1$ and possible changing K, L) in such a way that $F(t)\eta$ and $((-A^\odot)^{-\delta})^* f(t)$ are actually strongly measurable E-valued functions). Minding the new type of initial condition, the integral equation for $\eta(t) = (-A)^\alpha y(t)$ defining the solution $y(t)$ of (7.9.1) is

$$\eta(t) = (-A)^{\alpha+\gamma} S^\odot(t-s)^*((-A^\odot)^{-\gamma})^* \zeta$$
$$+ \int_s^t (-A)^{\alpha+\rho} S(t-\tau)((-A^\odot)^{-\rho})^* B(\tau)(-A)^{-\alpha} \eta(\tau) d\tau$$
$$+ \int_s^t (-A)^{\alpha+\delta} S(t-\tau)((-A^\odot)^{-\delta})^* f(\tau) d\tau. \tag{7.9.10}$$

The second term on the right requires showing that

$$t \to F(t)\eta(t) = ((-A^\odot)^{-\rho})^* B(t)(-A)^{-\alpha}\eta(t) \tag{7.9.11}$$

is strongly measurable for $\eta(\cdot)$ continuous. That this is true for $\eta(\cdot)$ constant is assumed; in general, we approximate $\eta(\cdot)$ by functions constant in intervals and use the bound (7.9.5). We note the use of $S^\odot(t-s)^*$ in the first term; it is only necessary when $\gamma = 0$.

Equation (7.9.10) can be locally solved by methods close to those in **7.7**. We consider the operator $\eta(\cdot) \to (\mathcal{G}\eta)(\cdot)$ given by the right side of (7.9.10) in the space $C_{\alpha+\gamma}(s+, T'; E)$ of all E-valued functions continuous in $s < t \le T'$ and satisfying $\|\eta(t)\| = 0((t-s)^{-\alpha-\gamma})$ as $t \to s+$, endowed with the norm

$$\|\eta\| = \sup_{s < t \le T'} (t-s)^{\alpha+\gamma} \|\eta(t)\|. \tag{7.9.12}$$

Arguing as in Theorem 7.7.8 but using this time the beta formula (7.7.29) we show that $\mathcal{G}$ maps the ball $B(\zeta(\cdot), 1) \subseteq C_{\alpha+\gamma}(s+, T'; E)$ ($\zeta(\cdot)$) the first term on the right side of (7.9.10)) into itself and is a contraction for $T' - s$ small enough. Existence of solutions is in fact global; estimating (7.9.10) with (7.7.20),

$$\|\eta(t)\| \le C_{\alpha+\gamma}(t-s)^{-\alpha-\gamma} \|\zeta\|_{-\gamma}$$
$$+ C_{\alpha+\rho} K \int_s^t (t-\tau)^{-\alpha-\rho} \|\eta(\tau)\| d\tau$$
$$+ C_{\alpha+\delta} L \int_s^t (t-\tau)^{-\alpha-\delta} \|((-A^\odot)^{-\delta})^* f(\tau)\| d\tau, \tag{7.9.13}$$

and it follows from the generalized Gronwall Lemma 7.7.9 that

$$\|\eta(t)\| \le C'(t-s)^{-\alpha-\gamma} \tag{7.9.14}$$

where C' depends only on K, L and $\|\zeta\|_{-\gamma}$; this guarantees global existence (see the comments after (7.7.30)). Global existence can also be proved solving (7.9.10) directly by successive approximations and estimating terms with the beta formula (7.7.22).[2] If we insist on using weakly measurable functions, $S(t - \tau)$ must be replaced by $S^{\odot}(t - \tau)^*$ in both integrands and we use Lemma 7.7.13; otherwise, everything is the same.

We show below that the Cauchy problem for $y'(t) = (A + B(t))y(t)$ is E_α-well posed as defined in **5.4**. Conditions (a) and (b) are obvious. The solution operator is $S(t, s)\zeta = y(t)$, where $y(t)$ is the solution of the homogeneous equation with $y(s) = \zeta$ or, equivalently, $S(t, s) = (-A)^{-\alpha} R_\alpha(t, s)$, where $R_\alpha(t, s)$ is the solution of the operator integral equation

$$R_\alpha(t, s)\zeta = (-A)^\alpha S(t - s)\zeta$$
$$+ \int_s^t (-A)^{\alpha + \rho} S(t - \tau)((-A^{\odot})^{-\rho})^* B(\tau)(-A)^{-\alpha} R_\alpha(\tau, s)\zeta \, d\tau \quad (7.9.15)$$

The continuity properties of $S(t, s)$ are embodied in the result below, a descendant of Theorem 5.5.6.

Theorem 7.9.1. *The solution operator $S(t, s)$ of*

$$y'(t) = (A + B(t))y(t) \tag{7.9.16}$$

is bounded from E_α to E_α, strongly continuous for $0 \le s \le t \le T$ and continuous in the (E_α, E_α)-norm for $s < t$; we have

$$\|S(t, s)\|_{(E_\alpha, E_\alpha)} \le C \quad (0 \le s \le t \le T). \tag{7.9.17}$$

The estimate (7.9.17) for $S(t, s)$ and the claimed continuity properties in the norm of (E_α, E_α) are equivalent to the corresponding estimate and continuity properties of $R_\alpha(t, s)$ in the norm of (E_α, E), thus we limit ourselves to studying the latter operator.

Theorem 7.9.2. *The operator $R_\alpha(t, s)$ is bounded from E_α to E, strongly continuous for $0 \le s \le t \le T$ and continuous in the (E_α, E)-norm for $s < t$; we have*

$$\|R_\alpha(t, s)\|_{(E_\alpha, E)} \le C \quad (0 \le s \le t \le T). \tag{7.9.18}$$

Proof. Theorem 7.9.2 is the first of several results on operator integral equations, all similar to (7.9.15). The "guiding principle" in all proofs is: the solution is just as good as the first term on the right-hand side (the one not under an integral), since the subsequent integrals smooth out singularities. Boundedness and continuity estimates are carried out using the generalized Gronwall lemma and the beta integral.

---

[2] In these solutions, the initial condition $y(s) = \zeta$ must be satisfied in a larger space since $\|y(\cdot)\|_E$ blows up as $t \to s+$ if $\gamma > 0$. See Example 7.9.12.

See the table later in the section comparing all operators to those corresponding to the equation with $B(t) = 0$.

We estimate the integral equation (7.9.15); here and below, $C, C', \ldots$ denote constants that may differ in different inequalities. We have

$$\|R_\alpha(t, s)\zeta\| \le M\|\zeta\|_\alpha + C_{\alpha+\rho}K \int_s^t (t - \tau)^{-\alpha-\rho}\|R_\alpha(\tau, s)\zeta\|d\tau, \quad (7.9.19)$$

(M a bound for $\|S(t)\|$), thus the estimate (7.9.18) follows from Lemma 7.7.6; global existence follows from Corollary 7.7.9 or directly. The estimates used to prove the strong continuity statement are direct descendants of (5.5.21) and (5.5.22). If $t, t' \ge s$ and $t \le t'$ we have

$$
\begin{aligned}
R_\alpha(t', s)\zeta - R_\alpha(t, s)\zeta &= (-A)^\alpha S(t' - s)\zeta - (-A)^\alpha S(t - s)\zeta \\
&+ \int_t^{t'} (-A)^{\alpha+\rho} S(t' - \tau)((-A^\odot)^{-\rho})^* B(\tau)(-A)^{-\alpha} R_\alpha(\tau, s)\zeta d\tau \\
&+ \int_s^t \{(-A)^{\alpha+\rho} S(t' - \tau) - (-A)^{\alpha+\rho} S(t - \tau)\} \\
&\quad \times ((-A^\odot)^{-\rho})^* B(\tau)(-A)^{-\alpha} R_\alpha(\tau, s)\zeta d\tau, \quad (7.9.20)
\end{aligned}
$$

which produces the estimate

$$
\begin{aligned}
\|R_\alpha(t', s)\zeta - R_\alpha(t, s)\zeta\| &\le \|(-A)^\alpha S(t' - s)\zeta - (-A)^\alpha S(t - s)\zeta\| \\
&+ C\|\zeta\|_\alpha \int_t^{t'} (t' - \tau)^{-\alpha-\rho} d\tau \\
&+ C'\|\zeta\|_\alpha \int_s^t \|(-A)^{\alpha+\rho} S(t' - \tau) - (-A)^{\alpha+\rho} S(t - \tau)\| d\tau. \quad (7.9.21)
\end{aligned}
$$

We use the same estimate for $t \ge t'$ switching t and t'. For $s, s' \le t, s \le s'$ we have

$$
\begin{aligned}
R_\alpha(t, s')\zeta - R_\alpha(t, s)\zeta &= (-A)^\alpha S(t - s')\zeta - (-A)^\alpha S(t - s)\zeta \\
&- \int_s^{s'} (-A)^{\alpha+\rho} S(t - \tau)((-A^\odot)^{-\rho})^* B(\tau)(-A)^{-\alpha} R_\alpha(\tau, s)\zeta d\tau \\
&+ \int_{s'}^t (-A)^{\alpha+\rho} S(t - \tau)((-A^\odot)^{-\rho})^* B(\tau)(-A)^\alpha (R_\alpha(\tau, s')\zeta - R_\alpha(\tau, s)\zeta)d\tau, \\
&\quad (7.9.22)
\end{aligned}
$$

so that

$$
\begin{aligned}
\|R_\alpha(t, s')\zeta - R_\alpha(t, s)\zeta\| &\le \|(-A)^\alpha S(t - s')\zeta - (-A)^\alpha S(t - s)\zeta\| \\
&+ C\|\zeta\|_\alpha \int_s^{s'} (t - \tau)^{-\alpha-\rho} d\tau \\
&+ C' \int_{s'}^t (t - \tau)^{-\alpha-\rho} \|R_\alpha(\tau, s')\zeta - R_\alpha(\tau, s)\zeta\| d\tau, \quad (7.9.23)
\end{aligned}
$$

and we switch s and s' when $s' \le s$. We handle these estimates as follows.

(7.9.21): In the second integral we change the lower limit of integration to $s = 0$, use (E, E)-continuity of $(-A)^{\alpha+\rho} S(t - \tau)$ for $\tau \leq t$ and, as customary, pass to the limit under the integral with help from Vitali's theorem; we obtain

$$\|R_\alpha(t', s)\zeta - R_\alpha(t, s)\zeta\| \leq \varepsilon(t, t')\|\zeta\|_\alpha \qquad (7.9.24)$$

with $\varepsilon(t, t') \to 0$ as $|t' - t| \to 0$ with $t, t' \geq s$; $\varepsilon(t, t')$ is independent of s.

(7.9.23): In the first integral we replace $(t - \tau)^{-\alpha-\rho}$ by $(s' - \tau)^{-\alpha-\rho}$ or by $(s - \tau)^{-\alpha-\rho}$ according to whether $s \leq s'$ or $s \geq s'$; in the second integral we use Lemma 7.7.6. The result is

$$\|R_\alpha(t', s')\zeta - R_\alpha(t, s)\zeta\| \leq \delta(s, s')\|\zeta\|_\alpha \qquad (7.9.25)$$

with $\delta(s, s') \to 0$ as $|s' - s| \to 0$ with $s, s' \leq t$; $\delta(s, s')$ is independent of t.

(E_α, E)-continuity for $t > s$ is handled as follows. If $t > s$ and $(t', s') \to (t, s)$, then eventually $t' > s', t' > s$, and

$$\|R_\alpha(t', s')\zeta - R_\alpha(t, s)\zeta\| \leq \|R_\alpha(t', s')\zeta - R_\alpha(t', s)\zeta\| + \|R_\alpha(t', s)\zeta - R_\alpha(t, s)\zeta\|,$$

thus we can combine (7.9.24) and (7.9.25).

We must also prove strong continuity when $s = t$. This is handled exactly as in Theorem 5.5.6 and we omit the details. ∎

Theorem 7.9.3. *Let $\gamma \geq 0, \alpha + \gamma < 1$. Then the operator $R_\alpha(t, s)(-A)^\gamma$ (with domain $E_\gamma = D((-A)^\gamma)$) has an extension $R_{\alpha,\gamma}(t, s) : (E^\odot)^* \to E$ bounded and continuous in the norm of $((E^\odot)^*, E)$ for $s < t$ and*

$$\|R_{\alpha,\gamma}(t, s)\|_{((E^\odot)^*, E)} \leq C(t - s)^{-\alpha-\gamma} \quad (0 \leq s < t \leq T). \qquad (7.9.26)$$

Proof. We write equation (7.9.15) for $(-A)^\gamma \zeta$ ($\zeta \in E_\gamma$),

$$R_\alpha(t, s)(-A)^\gamma \zeta = (-A)^{\alpha+\gamma} S(t - s)\zeta$$
$$+ \int_s^t (-A)^{\alpha+\rho} S(t - \tau)((-A^\odot)^{-\rho})^* B(\tau)(-A)^{-\alpha} R_\alpha(\tau, s)(-A)^\gamma \zeta d\tau,$$

and take it as *definition* of $R_{\alpha,\gamma}(t, s)$:

$$R_{\alpha,\gamma}(t, s)\zeta = (-A)^{\alpha+\gamma} S^\odot(t - s)^* \zeta$$
$$+ \int_s^t (-A)^{\alpha+\rho} S(t - \tau)((-A^\odot)^{-\rho})^* B(\tau)(-A)^{-\alpha} R_{\alpha,\gamma}(\tau, s)\zeta d\tau. \qquad^{(3)} \qquad (7.9.27)$$

---

[3] The use of $S^\odot(t - s)^*$ in the first term is unavoidable since we want to define $R_{\alpha,\gamma}(t, s)$ in $(E^\odot)^*$, not just in E.

This equation is a particular case of (7.9.10) and can be solved globally in the same way. The bound (7.9.26) results estimating with (7.7.29),

$$\|R_{\alpha,\gamma}(t,s)\zeta\|_E \le C(t-s)^{-\alpha-\gamma}\|\zeta\|_{(E^{\odot})^*} + C'\int_s^t (t-\tau)^{-\alpha-\rho}\|R_{\alpha,\gamma}(\tau,s)\zeta\|_E d\tau,$$
(7.9.28)

and applying Lemma 7.7.6; the result is

$$\|R_{\alpha,\gamma}(t,s)\|_E \le C(t-s)^{-\alpha-\gamma}\|\zeta\|_{(E^{\odot})^*}$$
$$+ C'\|\zeta\|_{(E^{\odot})^*}\int_s^t (t-\tau)^{-\alpha-\rho}(\tau-s)^{-\alpha-\gamma}d\tau$$
$$= C''\{(t-s)^{-\alpha-\gamma} + (t-s)^{1-2\alpha-\rho-\gamma}\}\|\zeta\|_{(E^{\odot})^*}$$
$$\le C'''(t-s)^{-\alpha-\gamma}\|\zeta\|_{(E^{\odot})^*}$$
(7.9.29)

by the gamma formula (7.7.22); in the last inequality, we have used the fact that $1 - 2\alpha - \rho - \gamma > -\alpha - \gamma$, which follows from $\alpha + \rho < 1$. To show continuity, the estimates corresponding to (7.9.21) and (7.9.23) are, in view of (7.9.26),

$$\|R_{\alpha,\gamma}(t',s)\zeta - R_{\alpha,\gamma}(t,s)\zeta\|$$
$$\le \|(-A)^{\alpha+\gamma}S^{\odot}(t'-s)^*\zeta - (-A)^{\alpha+\gamma}S^{\odot}(t-s)^*\zeta\|$$
$$+ C\|\zeta\|_{(E^{\odot})^*}\int_t^{t'} (t'-\tau)^{-\alpha-\rho}(\tau-s)^{-\alpha-\gamma}d\tau$$
$$+ C'\|\zeta\|_{(E^{\odot})^*}\int_s^t \|(-A)^{\alpha+\rho}S(t'-\tau) - (-A)^{\alpha+\rho}S(t-\tau)\|(\tau-s)^{-\alpha-\gamma}d\tau$$
(7.9.30)

for $t, t' > s, t \le t'$; when $t \ge t'$ we switch t and t'. If $s, s' < t, s \le s'$ the estimate is

$$\|R_{\alpha,\gamma}(t,s')\zeta - R_{\alpha,\gamma}(t,s)\zeta\|$$
$$\le \|(-A)^{\alpha+\gamma}S^{\odot}(t-s')^*\zeta - (-A)^{\alpha+\gamma}S^{\odot}(t-s)^*\zeta\|$$
$$+ C\|\zeta\|_{(E^{\odot})^*}\int_s^{s'} (t-\tau)^{-\alpha-\rho}(\tau-s)^{-\alpha-\gamma}d\tau$$
$$+ C'\int_{s'}^t (t-\tau)^{-\alpha-\rho}\|R_{\alpha,\gamma}(\tau,s')\zeta - R_{\alpha,\gamma}(\tau,s)\zeta\|d\tau,$$
(7.9.31)

with the correspondingly modified estimate when $s' \ge s$. The only significant difference between these estimates and (7.9.21)-(7.9.23) is that in (7.9.30) t, t' must be kept away from s to prevent blowup of the integrals; for the same reason, in (7.9.31), s, s' must be kept away from t. Thus, the statements corresponding to (7.9.24) and (7.9.25) are

$$\|R_{\alpha,\gamma}(t',s) - R_{\alpha,\gamma}(t,s)\| \le \varepsilon(t,t')\|\zeta\|_{(E^{\odot})^*},$$
$$\|R_{\alpha,\gamma}(t,s') - R_{\alpha,\gamma}(t,s)\| \le \delta(s,s')\|\zeta\|_{(E^{\odot})^*},$$

where, for $r > 0$ fixed, $\varepsilon(t, t') \to 0$ as $|t' - t| \to 0$, $t, t' \geq s + r$ and $\delta(s, s') \to 0$ as $|s' - s| \to 0$, $s, s' \leq t - r$. Since only continuity in $s < t$ is in question, this is enough. A little more work is needed to show that $\varepsilon(t, t')$ is independent of s and that $\delta(t, t')$ is independent of s; we omit the details. This ends the proof. ∎

We note that the conclusion of Theorem 7.9.3 is news even when $\gamma = 0$. In that case $R_{\alpha,0}(t, s)$ is an extension of the operator $R_\alpha(t, s)$ (analyzed in Theorem 7.9.2) from E_α to $(E^\odot)^*$. As a particular case, Theorem 7.9.3 exposes the behavior of $R_\alpha(t, s)$ in the norm of (E, E) rather than in that of (E_α, E).

We show next that the solution of (7.9.1) is given by the (suitably interpreted) variation-of-constants formula

$$y(t) = S(t, s)\zeta + \int_s^t S(t, \tau) f(\tau) d\tau. \tag{7.9.32}$$

Theorem 7.9.4. *The solution $\eta(t)$ of (7.9.10) is given by*

$$\eta(t) = R_{\alpha,\gamma}(t, s)((-A^\odot)^{-\gamma})^*\zeta + \int_s^t R_{\alpha,\delta}(t, \tau)((-A^\odot)^{-\delta})^* f(\tau) d\tau. \tag{7.9.33}$$

Proof. We replace $\eta(\cdot)$, given by (7.9.33), into the integral equation (7.9.10), and use the integral equation (7.9.27) for $R_{\alpha,\gamma}(t, s)$ and $R_{\alpha,\delta}(t, s)$; to save space, we use the shorthand (7.9.4) for $((-A^\odot)^{-\rho})^* B(t)(-A)^\alpha$. The computation is a counterpart of (5.5.24):

$$(-A)^{\alpha+\gamma} S^\odot(t - s)^*((-A^\odot)^{-\gamma})^*\zeta + \int_s^t (-A)^{\alpha+\rho} S^\odot(t - \tau)^* F(\tau)\eta(\tau)d\tau$$

$$+ \int_s^t (-A)^{\alpha+\delta} S^\odot(t - \tau)^*((-A^\odot)^{-\delta})^* f(\tau)d\tau$$

$$= (-A)^{\alpha+\gamma} S^\odot(t - s)^*((-A^\odot)^{-\gamma})^*\zeta$$

$$+ \int_s^t (-A)^{\alpha+\rho} S^\odot(t - \sigma)^* F(\sigma)$$

$$\times \left(R_{\alpha,\gamma}(\sigma, s)((-A^\odot)^{-\gamma})^*\zeta + \int_s^\sigma R_{\alpha,\delta}(\sigma, \tau)((-A^\odot)^{-\delta})^* f(\tau)d\tau \right) d\sigma$$

$$+ \int_s^t (-A)^{\alpha+\delta} S^\odot(t - \tau)^*((-A^\odot)^{-\delta})^* f(\tau)d\tau$$

$$= \left((-A)^{\alpha+\gamma} S^\odot(t - s)^* \right.$$

$$\left. + \int_s^t (-A)^{\alpha+\rho} S^\odot(t - \sigma)^* F(\sigma) R_{\alpha,\gamma}(\sigma, s)d\sigma \right)((-A^\odot)^{-\gamma})^*\zeta$$

$$+ \int_s^t \left((-A)^{\alpha+\delta} S^{\odot}(t-\tau)^* \right.$$

$$\left. + \int_\tau^t (-A)^{\alpha+\rho} S^{\odot}(t-\sigma)^* F(\sigma) R_{\alpha,\delta}(\sigma,\tau) d\sigma \right) ((-A^{\odot})^{-\delta})^* f(\tau) d\tau$$

$$= R_{\alpha,\gamma}(t,s)((-A^{\odot})^{-\gamma})^* \zeta + \int_s^t R_{\alpha,\delta}(t,\tau)((-A)^{\odot})^{-\delta})^* f(\tau) d\tau = \eta(t) \tag{7.9.34}$$

after an excusable interchange in the order of integration. To simplify, we have written $S^{\odot}(\cdot)^*$ even when $S(\cdot)$ will do, taking advantage that the first semigroup is an extension of the latter. ∎

The reversed adjoint equation is

$$z'(s) = -(A^{\odot} + B(s)^*) z(s) - g(s), \qquad z(t) = z. \tag{7.9.35}$$

Since $B(s)$ is densely defined in E the adjoint $B(s)^*$ is well defined. Using the fact that $K \subseteq K^{**}$ for any bounded operator and the rules of computation with adjoints ((5.0.25) and (5.0.26)) we obtain

$$((-A)^{-\alpha})^* B(s)^* (-A^{\odot})^{-\rho} \subseteq ((-A)^{-\alpha})^* B(s)^* ((-A^{\odot})^{-\rho})^{**}$$

$$= ((-A)^{-\alpha})^* \{ ((-A^{\odot})^{-\rho})^* B(s) \}^* \subseteq \{ ((-A^{\odot})^{-\rho})^* B(s)(-A)^{-\alpha} \}^*. \tag{7.9.36}$$

Assuming that[4]

$$B(s)^* : (E^{\odot})_\rho \to (E^*)_{-\alpha}, \tag{7.9.37}$$

the function $G(s) = ((-A)^{-\alpha})^* B(s)^* (-A^{\odot})^{-\rho}$ takes values in the space $(E^{\odot}, E^{\odot})$ (in $(E^{\odot}, E^*)$ if $\alpha = 0$) and shares the bound (7.9.5). To do a theory of (7.9.35) that is completely symmetric with that of (7.9.1) we only need to check that (i) $E^{\cdot}$ is separable, and (ii) $G(\cdot)z$ is a strongly measurable $E^{\odot}$-valued function for $z \in E^{\odot}$ (a E-weakly measurable E^*-valued function when $\alpha = 0$). That (i) holds has been proved in Example 7.7.19; to show (ii) note that $\langle G(s)z, y \rangle = \langle z, F(s)y \rangle$ for $z \in E^{\odot}, y \in E$.

All assumptions verified, everything proved on (7.9.1) can be automatically translated to (7.9.35) interchanging A, α, ρ with $A^{\odot}, \rho, \alpha$. For the record, we state the companion of each assumption and result. The final value z in (7.9.35) belongs to $(E^*)_{-\gamma}$, and $g(s) \in (E^*)_{-\delta}$ with

$$s \to ((-A)^{-\delta})^* g(\cdot) \tag{7.9.38}$$

E-weakly measurable and

$$\| ((-A)^{-\delta})^* g(s) \|_{E^*} \leq L \quad (0 \leq s \leq T). \tag{7.9.39}$$

---

[4] All extra assumptions on $B(s)^*$ can be eliminated; see Example 7.9.13.

The counterpart of (7.9.10) is the integral equation

$$
\upsilon(s) = (-A^\odot)^{\rho+\gamma} S(t-s)^* ((-A)^{-\gamma})^* z
$$
$$
+ \int_s^t (-A^\odot)^{\alpha+\rho} S^\odot(\tau-s)((-A)^{-\alpha})^* B(\tau)^* (-A^\odot)^{-\rho} \upsilon(\tau) d\tau
$$
$$
+ \int_s^t (-A^\odot)^{\alpha+\delta} S^\odot(\tau-s)((-A)^{-\delta})^* g(\tau) d\tau, \tag{7.9.40}
$$

and the solution of (7.9.35) is given by $z(t) = (-A^\odot)^{-\rho}\upsilon(t)$. The solution operator $S^\odot(s,t)$ of the homogeneous equation

$$
z'(s) = -(A^\odot + B(s)^*)z(s) \tag{7.9.41}
$$

is defined[5] by $S^\odot(s,t) = (-A^\odot)^{-\rho} R^\odot_\rho(s,t)$, where $R^\odot_\rho(s,t)$ is the solution of the operator integral equation

$$
R^\odot_\rho(s,t)z = (-A^\odot)^\rho S^\odot(t-s)z
$$
$$
+ \int_s^t (-A^\odot)^{\alpha+\rho} S^\odot(\sigma-s)((-A)^{-\alpha})^* B(\sigma)^* (-A^\odot)^{-\rho} R^\odot_\rho(\sigma,t)z\, d\sigma.
$$
$$
\tag{7.9.42}
$$

Theorem 7.9.5. *The solution operator $S^\odot(s,t)$ of (7.9.41) is bounded from $(E^\odot)_\rho$ to $(E^\odot)_\rho$, strongly continuous for $0 \le s \le t \le T$ and continuous in the $((E^\odot)_\rho, (E^\odot)_\rho))$-norm for $s < t$; we have*

$$
\|S^\odot(s,t)\|_{((E^\odot)_\rho,(E^\odot)_\rho)} \le C \quad (0 \le s \le t \le T). \tag{7.9.43}
$$

Theorem 7.9.5 is a consequence of

Theorem 7.9.6. *$R^\odot_\rho(s,t)$ is bounded from $(E^\odot)_\rho$ to $E^\odot$, strongly continuous for $0 \le s \le t \le T$ and continuous in the $((E^\odot)_\rho, E^\odot)$-norm for $s < t$; we have*

$$
\|R^\odot_\rho(s,t)\|_{((E^\odot)_\rho,E^\odot)} \le C \quad (0 \le s \le t \le T). \tag{7.9.44}
$$

Theorem 7.9.7. *Let $\gamma \ge 0$, $\rho + \gamma < 1$. Then the operator $R^\odot_\rho(s,t)(-A^\odot)^\gamma$ (domain $(E^\odot)_\gamma = D((-A^\odot)^\gamma))$ has an extension $R^*_{\rho,\gamma}(s,t) : E^* \to E^\odot$ given by the integral equation*

$$
R^*_{\rho,\gamma}(s,t)z = (-A^\odot)^{\rho+\gamma} S(t-s)^* z
$$
$$
+ \int_s^t (-A^\odot)^{\alpha+\rho} S^\odot(\sigma-s)((-A)^{-\alpha})^* B(\sigma)^* (-A^\odot)^{-\rho} R^*_{\rho,\gamma}(\sigma,t)z\, d\sigma.
$$
$$
\tag{7.9.45}
$$

---

[5] The superindex $\odot$ on $S^\odot(t,s)$ and $R^\odot_\rho(s,t)$ is simply a label and does not indicate the taking of any Phillips adjoint. Likewise, the superindex $*$ in the operator $R^*_{\rho,\gamma}(s,t)$ below does not indicate adjoint.

The operator $R^*_{\rho,\gamma}(s,t)$ is bounded and continuous in the $(E^*, E^\odot)$-norm for $s < t$ and

$$\|R^*_{\rho,\gamma}(s,t)\|_{(E^*,E^\odot)} \le C(t-s)^{-\rho-\gamma}. \tag{7.9.46}$$

The variation-of-constants formula

$$z(s) = S^\odot(s,t)z + \int_s^t S^\odot(s,\sigma)g(\sigma)d\sigma \tag{7.9.47}$$

is interpreted as follows.

Theorem 7.9.8. *The solution $\upsilon(\cdot)$ of (7.9.40) is given by*

$$\upsilon(s) = R^*_{\rho,\gamma}(s,t)((-A)^{-\gamma})^* z + \int_s^t R^*_{\rho,\delta}(s,\sigma)((-A)^{-\delta})^* g(\sigma)d\sigma. \tag{7.9.48}$$

Theorem 7.9.9.

$$S^\odot(s,t) = S(t,s)^*|_{E^\odot} \quad (0 \le s \le t \le T) \tag{7.9.49}$$

(* *indicates adjoint in* (E,E)).

Proof. Since we have $S(t,s) = (-A)^{-\alpha}R_\alpha(t,s)$ and $S^\odot(s,t) = (-A^\odot)^{-\rho}R^\odot_\rho(s,t)$, (7.9.49) amounts to showing

$$(-A^\odot)^{-\rho}R^\odot_\rho(s,t)z = R_\alpha(t,s)^*(-A^\odot)^{-\alpha}z \tag{7.9.50}$$

for $s < t, z \in E^\odot$. Adjoints are taken in (E,E); all operators are bounded, since it follows from their respective integral equations that $R_\alpha(t,s), S(t,s) \in (E,E)$ for $s < t$. The function $y(t) = S(t-s)\zeta$ ($\zeta \in E_\alpha = D((-A)^\alpha)$) satisfies the equation

$$\begin{aligned} y'(t) = Ay(t) &= (A+B(t))y(t) - B(t)y(t) \\ &= (A+B(t))y(t) - f(t), \qquad y(s) = \zeta, \end{aligned}$$

where $((-A^\odot)^{-\rho})^* f(t) = ((-A^\odot)^{-\rho})^* B(t)(-A)^{-\alpha}S(t-s)(-A)^\alpha\zeta$ is strongly measurable and bounded. We apply the variation-of-constants formula (7.9.33) (with $\gamma = 0$), obtaining the following equation for $\eta(t) = (-A)^\alpha S(t-s)\zeta$:

$$\begin{aligned} (-A)^\alpha S(t-s)\zeta = R_\alpha(t,s)\zeta \\ - \int_s^t R_{\alpha,\rho}(t,\sigma)((-A^\odot)^{-\rho})^* B(\sigma)(-A)^{-\alpha}S(\sigma-s)(-A)^\alpha\zeta d\sigma. \end{aligned} \tag{7.9.51}$$

We note next that

$$R_{\alpha,\rho}(t,\sigma)^* = ((-A)^\rho)^* R_\alpha(t,\sigma)^*. \tag{7.9.52}$$

To see this, recall that $R_{\alpha,\rho}(t, \sigma)$ has been defined in (7.9.27) as an extension of $R_\alpha(t, \sigma)(-A)^\rho$, so that $R_{\alpha,\rho}(t, \sigma)^* = (R_\alpha(t, \sigma)(-A)^\rho)^*$; (7.9.52) then follows from the second rule (5.0.26) on adjoints, and implies that $R_\alpha(t, \sigma)^* E^* \subseteq D(((-A)^\rho)^*)$.

We apply a functional $(-A^\odot)^{-\alpha}z, z \in E^\odot$ to both sides of (7.9.51) and then put adjoints one by one on the other side of the duality product, using (7.9.52) in one of the steps. The result is

$$R_\alpha(t, s)^*(-A^\odot)^{-\alpha}z = S^\odot(t - s)z$$
$$+ \int_s^t (-A^\odot)^\alpha S^\odot(\sigma - s)$$
$$\times ((-A)^{-\alpha})^* B(\sigma)^*(-A^\odot)^{-\rho}((-A)^\rho)^* R_\alpha(t, \sigma)^*(-A^\odot)^{-\alpha}z d\sigma \quad (7.9.53)$$

Then we apply $((-A)^\rho)^*$ to both sides of this equation and notice that the function $((-A)^\rho)^* R_\alpha(t, s)^*((-A^\odot)^\alpha)z$ satisfies the same integral equation (7.9.42) defining $R_\rho^\odot(s, t)z$; by uniquess, both functions must coincide. Applying $(-A^\odot)^{-\rho}$ on both sides, (7.9.50) results.   ∎

The following table compares all the operators with the corresponding operators for the unperturbed equation $(B(t) \equiv 0)$.

$y'(t) = Ay(t)$	$y'(t) = (A + B(t))y(t)$
$S(t - s)$	$S(t, s)$
$(-A)^\alpha S(t - s)$	$R_\alpha(t, s)$
$(-A)^{\alpha+\gamma} S^\odot(t - s)^*$	$R_{\alpha,\gamma}(t, s)$

$z'(t) = -A^\odot y(t)$	$z'(t) = -(A^\odot + B(t)^*)z(t)$
$S^\odot(t - s)$	$S^\odot(s, t)$
$(-A^\odot)^\rho S^\odot(t - s)$	$R_\rho^\odot(s, t)$
$(-A^\odot)^{\rho+\gamma} S(t - s)^*$	$R_{\rho,\gamma}^*(s, t)$

Example 7.9.10. Let A be the infinitesimal generator of a strongly continuous semigroup in an arbitrary Banach space E. Consider the equation

$$y'(t) = Ay(t) + f(t), \qquad y(0) = \zeta. \qquad (7.9.54)$$

Assume that $\zeta \in D(A)$ and that there exists $\alpha, 0 < \alpha \le 1$ such that, either (a) $f(t) \in D((-A)^\alpha)$ and $(-A)^\alpha f(t)$ is continuous in $0 \le t \le T$, or (b) $f(\cdot)$ is Hölder continuous with exponent α, that is,

$$\|f(t) - f(s)\| \le C|t - s|^\alpha \quad (0 \le s \le t \le T). \qquad (7.9.55)$$

Then the variation-of-constants formula

$$y(t) = S(t - s)\zeta + \int_0^t S(t - \tau)f(\tau)d\tau$$

gives a genuine solution of (7.9.54). We may obviously assume that $\zeta = 0$. For the proof of case (a), we write

$$Ay(\sigma) = \int_0^\sigma (-A)^{1-\alpha} S(\sigma - \tau)(-A)^\alpha f(\tau) d\tau,$$

so that $Ay(\sigma)$ is continuous. Integrating in $0 \le \sigma \le t$ and using Fubini's theorem,

$$\int_0^t Ay(\sigma) d\sigma = \int_0^t (-A)^{1-\alpha} \int_\tau^t S(\sigma - \tau)(-A)^\alpha f(\tau) d\sigma d\tau$$

$$= \int_0^t (-A)^{1-\alpha} A^{-1} (S(t - \tau) - I)(-A)^\alpha f(\tau) d\tau$$

$$= y(t) - \int_0^t f(\tau) d\tau.$$

In case (b), let $\phi_n(t) = n\phi_n(nt)$ be a mollifier (ϕ a nonnegative test function with support in $|t| \le 1$ and integral 1). Extend $f(t)$ to $t \le 0$ by $f(t) = f(0)$ and to $t \ge T$ by $f(t) = T$. The functions $f_n = \phi_n * f$ are infinitely differentiable so that the solution $y_n(t)$ corresponding to $f_n(t)$ is a genuine solution (Example 5.2.15), and we check that the f_n satisfy (7.9.55) with C independent of n. We have

$$\int_0^t Ay_n(\sigma) d\sigma = y_n(t) - \int_0^t f_n(\sigma) d\sigma. \tag{7.9.56}$$

When $n \to \infty$, $f_n(t) \to f(t)$ and $y_n(t) \to y(t)$ uniformly in $0 \le t \le T$, thus we can take limits on the right side of (7.9.56). In the left side, we note that

$$Ay_n(\sigma) = \int_0^\sigma AS(\sigma - \tau)(f_n(\tau) - f_n(\sigma)) d\tau + (S(\sigma) - I) f_n(\sigma) \tag{7.9.57}$$

so we can take limits first in (7.9.57) and then in the left side of (7.9.56), in both cases using the dominated convergence theorem.

Remark 7.9.11. The simplifications corresponding to the reflexive case (see Remark 7.7.18) are as follows. All relations (7.7.34) hold. Conditions (7.9.3) and (7.9.37) become

$$B(t) \in (E_\alpha, E_{-\rho}), \qquad B(t)^* \in (E_\rho, E_{-\alpha}), \tag{7.9.58}$$

and $f(t) \in E_{-\delta}, g(s) \in E_{-\delta}$. The integral equation (7.9.10) defining $\eta(t) = (-A)^\alpha y(t)$ ($y(\cdot)$ the solution of (7.9.1)) becomes

$$\eta(t) = (-A)^{\alpha + \gamma} S(t - s)(-A)^{-\gamma} \zeta$$

$$+ \int_s^t (-A)^{\alpha + \rho} S(t - \tau)(-A)^{-\rho} B(\tau)(-A)^{-\alpha} \eta(\tau) d\tau$$

$$+ \int_s^t (-A)^{\alpha + \delta} S(t - \tau)(-A)^{-\delta} f(\tau) d\tau, \tag{7.9.59}$$

and the integral equation (7.9.40) giving $\upsilon(t) = (-A^*)^\rho z(t)$, $z(\cdot)$ the solution of the reversed adjoint equation

$$z'(s) = -(A^* + B(s)^*)z(s) - g(s), \qquad z(t) = z \tag{7.9.60}$$

is

$$\begin{aligned}
\upsilon(s) = &\ (-A^*)^{\rho+\gamma} S(t-s)^* (-A^*)^{-\gamma} z \\
&+ \int_s^t (-A^*)^{\alpha+\rho} S^*(\tau - s)(-A^*)^{-\alpha} B(\tau)^* (-A^*)^{-\rho} \upsilon(\tau) d\tau \\
&+ \int_s^t (-A^*)^{\alpha+\delta} S^*(\tau - s)(-A^*)^{-\delta} g(\tau) d\tau.
\end{aligned} \tag{7.9.61}$$

The variation-of-constants formula (7.9.33) simplifies to

$$\eta(t) = R_{\alpha,\gamma}(t,s)(-A)^{-\gamma}\zeta + \int_s^t R_{\alpha,\delta}(t,\tau)(-A)^{-\delta} f(\tau) d\tau \tag{7.9.62}$$

with the operator $R_{\alpha,\gamma}(t,s)$ answering to the simplified relative of (7.9.27),

$$\begin{aligned}
R_{\alpha,\gamma}(t,s)\zeta = &\ (-A)^{\alpha+\gamma} S(t-s)\zeta \\
&+ \int_s^t (-A)^{\alpha+\rho} S(t-\tau)(-A)^{-\rho} B(\tau)(-A)^{-\alpha} R_{\alpha,\gamma}(\tau,s)\zeta \, d\tau.
\end{aligned} \tag{7.9.63}$$

The same equation with $\gamma = 0$ takes the place of (7.9.15). The estimate (7.9.26) is now in the (E, E)-norm. For the adjoint equation, (7.9.48) needs no translation (although we may replace $((-A)^{-\gamma})^*$ by $(-A^*)^{-\gamma}$ and $((-A)^{-\delta})^*$ by $(-A^*)^{-\delta}$); in (7.9.45) we do the same and replace $A^\odot$ by A^*. The estimation (7.9.46) is in the (E^*, E^*)-norm. Finally, (7.9.49) becomes the more familiar equality

$$S^*(t,s) = S(t,s)^* \tag{7.9.64}$$

for $S^*(t,s) = S^\odot(t,s)$.

Remark 7.9.12. Solutions $y(\cdot)$ of (7.9.10) "assume their initial condition" in various senses. One is $\langle y^*, (-A)^{-\gamma} y(t)\rangle \to \langle y^*, ((-A^\odot)^{-\gamma})^* \zeta\rangle$ $(y^* \in E^\odot)$ as $t \to s+$. Another is $(-A)^{-\delta} y(t) \to ((-A^\odot)^{-\delta})^* \zeta$ $(\delta > \gamma)$. Similar observations apply to the adjoint equation.

Remark 7.9.13. A condition on $B(s)^*$ such as (7.9.37) may be inopportune to verify in practice. At the risk of introducing a slight asymmetry in the treatment of (7.9.1) and (7.9.35) we may dispense with *any* additional condition on $B(s)^*$ by simply *defining*

$$F(s)^\odot = \{((-A^\odot)^{-\rho})^* B(s)(-A)^{-\alpha}\}^*|_{E^\odot}$$

(restriction of $F(s)^*$ to $E^\odot$) and carrying out the theory of the adjoint equation (7.9.35) using $F(s)^\odot$ instead of $G(s) = ((-A)^{-\alpha})^* B(s)^* (-A^\odot)^{-\rho}$ in the same way $F(\cdot)$ was used in the theory of (7.9.1). This calls for replacement of $G(s)$ by $F(s)^\odot$ in the integral equations (7.9.40), (7.9.42) and (7.9.45). However, a slight complication ensues in that $F(s)^\odot$ belongs to $(E^\odot, E^*)$ rather than to $(E^\odot, E^\odot)$. This is insubstantial, since we can use smoothing by the semigroup (Lemma 7.7.13). Only modifications needed: in the first integrand in (7.9.40) replace $S^\odot(\tau - s)$ by $S(\tau - s)^*$. Do the same in (7.9.42) and in (7.9.45). The proof of equality (7.9.49) is the same; in the process of taking adjoints one uses $F(s)^\odot$ without disassembling into components and then replaces $S^\odot(\sigma - s)$ by $S(\sigma - s)^*$.

7.10. Patch Perturbations and Directional Derivatives

We generalize here the results in **6.1** and **6.2** to the abstract parabolic world. Theorem 7.10.4 below constructs directional derivatives of the solution map

$$\mathbf{f}(u) = y(\bar{t}, u) \tag{7.10.1}$$

of a control system

$$y'(t) = Ay(t) + f(t, y(t), u(t)), \qquad y(0) = \zeta. \tag{7.10.2}$$

As in last section, we assume that A generates an analytic semigroup satisfying (7.9.2) and that E is separable and $\odot$-reflexive with respect to A. The control space $C_{\text{ad}}(0, T; U)$ is equipped with the usual distance

$$d(u(\cdot), v(\cdot)) = |\{t \in [0, T]; u(t) \neq v(t)\}| \tag{7.10.3}$$

where $|\cdot|$ denotes outer Lebesgue measure.

Lemma 7.10.1. *Let $\zeta \in E_\alpha$ and assume that $f(t, y, u(t))$ satisfies Hypothesis $II_{\alpha,\rho}$ in **7.7** with $\alpha + \rho < 1$ for $u(\cdot) \in C_{\text{ad}}(0, \bar{t}; U)$ with $K(c)$, $L(c)$ independent of $u(\cdot)$. Let $\bar{u}(\cdot) \in C_{\text{ad}}(0, \bar{t}; U)$ be such that $y(t, \bar{u})$ exists in $0 \leq t \leq \bar{t}$. Then there exists $\delta > 0$ such that if $u(\cdot), v(\cdot) \in B(\bar{u}, \delta) = \{u \in C_{\text{ad}}(0, \bar{t}; U); d(u, \bar{u}) \leq \delta\}$ then $y(t, u)$ and $y(t, v)$ exist as well in $0 \leq t \leq \bar{t}$ and*

$$\|y(t, v) - y(t, u)\|_\alpha \leq CL(c + 1) \int_{[\{\tau \in [0, \bar{t}]; u(\tau) \neq v(\tau)\}]} (t - \tau)^{-\alpha - \rho} d\tau \tag{7.10.4}$$

*in $0 \leq t \leq \bar{t}$ where $[\cdot]$ denotes measurable envelope (see **6.1**), c is a bound for $\|y(t, \bar{u})\|$ in $0 \leq t \leq \bar{t}$ and C does not depend on u, v.*

Lemma 7.10.1 is a generalization of Lemma 6.1.1 and the proof is similar; this time we estimate $\eta(t, u) = (-A)^\alpha y(t, u)$ in the norm of E. Let

$$g(t, \eta, u) = ((-A^\odot)^{-\rho})^* f(t, (-A)^{-\alpha}\eta, u). \tag{7.10.5}$$

If $[0, t_u]$ is the largest interval where $\eta(t, u)$ exists and satisfies the estimate $\|y(t, u) - y(t, \bar{u})\|_\alpha = \|\eta(t, u) - \eta(t, \bar{u})\| \leq 1$ we obtain from the integral equation (7.7.18) that

$$
\begin{aligned}
\eta(t, u) &- \eta(t, \bar{u}) \\
&= \int_0^t (-A)^{\alpha+\rho} S(t - \tau)\{g(\tau, \eta(\tau, u), u(\tau)) - g(\tau, \eta(\tau, u), \bar{u}(\tau))\}d\tau \\
&\quad + \int_0^t (-A)^{\alpha+\rho} S(t - \tau)\{g(\tau, \eta(\tau, u), \bar{u}(\tau)) - g(\tau, \eta(\tau, \bar{u}), \bar{u}(\tau))\}d\tau;
\end{aligned}
\tag{7.10.6}
$$

thus,

$$
\begin{aligned}
\|\eta(t, u) - \eta(t, \bar{u})\| &\leq C'K(c+1) \int_{[\{\tau \in [0, \bar{t}]; u(\tau) \neq \bar{u}(\tau)\}]} (t - \tau)^{-\alpha-\rho}d\tau \\
&\quad + C''L(c+1) \int_0^t (t - \tau)^{-\alpha-\rho}\|\eta(\tau, u) - \eta(\tau, \bar{u})\|d\tau.
\end{aligned}
\tag{7.10.7}
$$

We use then Lemma 7.7.6 and obtain (7.10.6) for $u, \bar{u}$ in $0 \leq t \leq t_u$; taking $d(u, \bar{u}) \leq \delta$ sufficiently small, maximality of $[0, t_u]$ implies $t_u = \bar{t}$. The rest is the same as in Lemma 6.1.1 and we omit the details. ∎

When $\rho = 0$ Lemma 7.10.1 can be equally proved under Hypothesis $\text{II}_{\alpha,\rho}(wm)$. We can either raise ρ so as to make $g(t, \eta, u(t))$ E-valued and strongly measurable, or we can use Lemma 7.7.13; the latter is better since ρ shows up in the estimates.

We recall various definitions in **6.2**. Let $u(\cdot)$ be a control in $C_{\text{ad}}(0, \bar{t}; U)$, $\mathbf{v}(\cdot) = (v_1(\cdot), v_2(\cdot), \ldots, v_m(\cdot))$ an arbitrary collection of m elements of $C_{\text{ad}}(0, \bar{t}; U)$, $\mathbf{e} = (e_1, e_2, \ldots, e_m)$ a collection of m pairwise disjoint measurable sets in $[0, \bar{t}]$. The *patch perturbation* of $u(\cdot)$ corresponding to $\mathbf{e}, \mathbf{v}(\cdot)$ is

$$u_{\mathbf{e},\mathbf{v}}(t) = \begin{cases} v_j(t) & (t \in e_j, j = 1, 2, \ldots, m) \\ u(t) & \text{elsewhere.} \end{cases} \tag{7.10.8}$$

If d is the distance (7.10.3) we have

$$d(u_{\mathbf{e},\mathbf{v}}, u) \leq |e| \tag{7.10.9}$$

where $e = \cup e_j$. The result below is the counterpart of Lemma 6.2.1. We only need Hypothesis $\text{II}_{\alpha,\rho}$ or $\text{II}_{\alpha,\rho}(wm)$ with $\alpha + \rho < 1$ for each $u(\cdot) \in C_{\text{ad}}(0, \bar{t}; U)$; $K(c)$ and $L(c)$ may depend on $u(\cdot)$.

Lemma 7.10.2. *Let* $\bar{u}(\cdot) \in C_{\mathrm{ad}}(0, \bar{t}; U)$ *be such that* $y(t, \bar{u})$ *exists in* $0 \leq t \leq \bar{t}$. *Then, given* $\mathbf{v}(\cdot)$, *there exists* $\delta > 0$ *such that, if* $0 \leq |e| \leq \delta$ *then* $y(t, \bar{u}_{\mathbf{e},\mathbf{v}})$ *exists in* $0 \leq t \leq \bar{t}$. *Moreover, there exists* $C > 0$ *such that*

$$\|y(t, \bar{u}_{\mathbf{e},\mathbf{v}}) - y(t, \bar{u})\|_\alpha \leq C \int_{[\{\tau \in [0, \bar{t}]; \, u(\tau) \neq v(\tau)\}]} (t - \tau)^{-\alpha-\rho} d\tau \quad (7.10.10)$$

The constants δ, C *depend on* $\bar{u}(\cdot)$ *and* $\mathbf{v}(\cdot)$ *but not on* $\mathbf{e}$.

This result follows directly from Lemma 7.10.1. We replace the space $C_{\mathrm{ad}}(0, \bar{t}; U)$ by the set of controls consisting of $\bar{u}$ and all of its patch perturbations with $\mathbf{v}$ fixed; in this new control space, Assumption $\mathrm{II}_{\alpha,\rho}$ is satisfied with $K(c)$, $L(c)$ independent of the control. ∎

In the rest of the section, we function under Hypothesis $\mathrm{III}_{\alpha,\rho}$ or $\mathrm{III}_{\alpha,\rho}(wm)$ below (the latter for $\rho = 0$). These are the counterparts of Hypothesis III in **6.2**. As usual, we state the hypotheses for the uncontrolled system $y'(t) = Ay(t) + f(t, y(t))$. We assume that

$$f : [0, T] \times E_\alpha \to ((E^\odot)^*)_{-\rho} \quad (7.10.11)$$

and that f has a Fréchet derivative $\partial_y f(t, y) \in (E_\alpha, ((E^\odot)^*)_{-\rho})$ in $[0, T] \times E_\alpha$. This means that the function

$$g(t, \eta) = ((-A^\odot)^{-\rho})^* f(t, (-A)^{-\alpha} \eta) \quad (7.10.12)$$

maps $[0, T] \times E$ into E (into $(E^\odot)^*$ if $\rho = 0$)[1] and has a Fréchet derivative

$$\partial_\eta g(t, \eta) = ((-A^\odot)^{-\rho})^* \partial_y f(t, (-A)^{-\alpha} \eta)(-A)^{-\alpha}. \quad (7.10.13)$$

We say that $f(t, y)$ satisfies Hypothesis $\mathrm{III}_{\alpha,\rho}$, if $g(t, \eta)$ takes values in E and satisfies Hypothesis III with $K(t, c) \equiv K(c)$ and $L(t, c) \equiv L(c)$. This means: $g(t, \eta)$ is strongly measurable in t for η fixed and continuous in η (in the norm of E) for t fixed. For each $z \in E$, $\partial_\eta g(t, \eta)z$ is strongly measurable in t for η fixed and continuous in η (in the norm of E) for t fixed. Finally, for every $c > 0$ there exist $K(c)$, $L(c)$ such that

$$\|g(t, \eta)\|_E = \|((-A^\odot)^{-\rho})^* f(t, (-A)^{-\alpha} \eta)\|_E \leq K(c)$$

$$(0 \leq t \leq T, \|\eta\|_E \leq c) \quad (7.10.14)$$

$$\|\partial_\eta g(t, \eta)\|_{(E,E)} = \|((-A^\odot)^{-\rho})^* \partial_y f(t, (-A)^{-\alpha} \eta)(-A)^{-\alpha}\|_{(E,E)} \leq L(c)$$

$$(0 \leq t \leq T, \|\eta\|_E \leq c). \quad (7.10.15)$$

---

[1] See Lemma 7.7.10.

On the other hand, for $\rho = 0$, Hypothesis $III_{\alpha,\rho}(wm)$ only postulates that the function g in (7.10.12) maps E into $(E^{\odot})^*$; $g(t, \eta)$ is $E^{\odot}$-weakly measurable in t for η fixed and continuous in η for t fixed. For each $z \in (E^{\odot})^*$, $\partial_{\eta}g(t, \eta)z$ is $E^{\odot}$-weakly measurable in t for η fixed and continuous in η for t fixed. Inequalities (7.10.14) and (7.10.15) become

$$\|g(t, \eta)\|_{(E^{\odot})^*} = \|f(t, (-A)^{-\alpha}\eta)\|_{(E^{\odot})^*} \leq K(c)$$

$$(0 \leq t \leq T, \|\eta\|_E \leq c) \quad (7.10.16)$$

$$\|\partial_{\eta}g(t, \eta)\|_{(E,(E^{\odot})^*)} = \|\partial_y f(t, (-A)^{-\alpha}\eta)(-A)^{-\alpha}\|_{(E,(E^{\odot})^*)} \leq L(c)$$

$$(0 \leq t \leq T, \|\eta\|_E \leq c). \quad (7.10.17)$$

Lemma 7.10.3. *Assume $f(t, y)$ satisfies Hypothesis $III_{\alpha,0}(wm)$. Then $f(t, y)$ satisfies Hypothesis $III_{\alpha,\rho}$ for any $\rho > 0$.*

The proof, as that of Corollary 7.7.12, is a consequence of Lemma 7.7.11. We omit the details.

The following result is the companion of Theorem 6.2.7.

Theorem 7.10.4. *Assume that $f(t, y) = f(t, y, u(t))$ satisfies Hypothesis $III_{\alpha,\rho}$ with $\alpha + \rho < 1$ for every $u(\cdot) \in C_{ad}(0, \bar{t}; U)$. Let $\bar{u}(\cdot) \in C_{ad}(0, \bar{t}; U)$ be such that $y(t, \bar{u})$ exists in $0 \leq t \leq \bar{t}$. Let $\mathbf{p} = (p_1, p_2, \ldots, p_m)$ be a probability vector and $\mathbf{v}(\cdot) = (v_1(\cdot), v_2(\cdot), \ldots, v_m(\cdot))$ a collection of elements of $C_{ad}(0, \bar{t}; U)$. Then, for each $h \in (0, 1]$ there exists $\mathbf{e}(h) = (e_1(h), e_2(h), \ldots, e_m(h))$ (the $e_j(h)$ pairwise disjoint) such that $|e_j(h)| = p_j h\bar{t}$, $y(t, \bar{u}_{\mathbf{e}(h),\mathbf{v}})$ exists in $0 \leq t \leq \bar{t}$ for sufficiently small h and*

$$\xi(t, \bar{u}, \mathbf{p}, \mathbf{v}) = \lim_{h \to 0+} \frac{1}{h}\left(y(t, \bar{u}_{\mathbf{e}(h),\mathbf{v}}) - y(t, \bar{u})\right) \quad (7.10.18)$$

exists uniformly in $0 \leq t \leq \bar{t}$ in the norm of E_{α}, where $\xi(t) = \xi(t, \bar{u}, \mathbf{p}, \mathbf{v})$ is the solution of the variational equation

$$\xi'(t) = \{A + \partial_y f(t, y(t, \bar{u}), \bar{u}(t))\}\xi(t)$$

$$+ \sum_{j=1}^{m} p_j\{f(t, y(t, \bar{u}), v_j(t)) - f(t, y(t, \bar{u}), \bar{u}(t))\}, \quad \xi(0) = 0. \quad (7.10.19)$$

The proof is much the same as that of Theorem 6.2.7, only that (as in Lemma 7.10.1 and Lemma 7.10.2) we estimate $\eta(t, u)$, not $y(t, u)$, in the norm of E. Using Lemma 7.10.3 and modifying ρ if necessary we may assume that $g(t, \eta, u(t))$ ($g(t, \eta, u)$ the function in (7.10.5)) satisfies Hypothesis III for each control $u(\cdot) \in C_{ad}(0, \bar{t}; U)$.

Existence of $y(t, \bar{u}_{e(h), \mathbf{v}})$ for h small enough follows from Lemma 7.10.2. The linear equation (7.10.19) is understood as (7.9.1) with $\zeta = 0$,

$$B(t) = \partial_y f(t, y(t, \bar{u}), \bar{u}(t)) \tag{7.10.20}$$

$$f(t) = \sum_{j=1}^{m} p_j \{ f(t, y(t, \bar{u}), v_j(t)) - f(t, y(t, \bar{u}), \bar{u}(t)) \}. \tag{7.10.21}$$

This means the solution is $\xi(t, \bar{u}, \mathbf{p}, \mathbf{v}) = (-A)^{-\alpha} v(t, \bar{u}, \mathbf{p}, \mathbf{v})$, $v(t) = v(t, \bar{u}, \mathbf{p}, \mathbf{v})$ given by the integral equation corresponding to (7.9.10),

$$v(t) = \int_0^t (-A)^{\alpha + \rho} S(t - \tau) F(\tau) v(\tau) d\tau$$

$$+ \int_0^t (-A)^{\alpha + \rho} S(t - \tau)((-A^\odot)^{-\rho})^* f(\tau) d\tau.$$

with

$$F(t) = ((-A^\odot)^{-\rho})^* \partial_y f(t, (-A)^{-\alpha} \eta(t; \bar{u}), \bar{u}(t))(-A)^{-\alpha}$$

$$= \partial_\eta g(t, \eta(t, \bar{u}), \bar{u}(t)) \tag{7.10.22}$$

and $f(\cdot)$ given by (7.10.21).

We use Corollary 6.2.6 for $T = \bar{t}$, $m = n$, $\varepsilon = h$ and

$$\phi_j(t, \tau) = (-A)^{\alpha + \rho} S(t - \tau) \{ g(\tau, \eta(\tau, \bar{u}), v_j(\tau)) - g(\tau, \eta(\tau, \bar{u}), \bar{u}(\tau)) \},$$

which satisfy both assumptions (6.2.11) and (6.2.12), the first due to the bounds on f and (7.7.20), the second by (E, E)-continuity of $(-A)^{\alpha + \rho} S(t - \tau)$ and the dominated convergence theorem. Corollary 6.2.6 provides disjoint measurable sets $e_1(h), e_2(h), \ldots, e_m(h) \subseteq [0, \bar{t}]$ with $|e_j(h)| = p_j h \bar{t}$ and such that

$$\left\| \frac{1}{h} \int_{(0, t) \cap e_j(h)} (-A)^{\alpha + \rho} S(t - \tau) \{ g(\tau, \eta(\tau, \bar{u}), v_j(\tau)) - g(\tau, \eta(\tau, \bar{u}), \bar{u}(\tau)) \} d\tau \right.$$

$$\left. - p_j \int_0^t (-A)^{\alpha + \rho} S(t - \tau) \{ g(\tau, \eta(\tau, \bar{u}), v_j(\tau)) - g(\tau, \eta(\tau, \bar{u}), \bar{u}(\tau)) \} d\tau \right\| \leq h.$$

$$(0 \leq t \leq \bar{t}) \tag{7.10.23}$$

We subtract the integral equations (7.7.18) for $\eta(t, \bar{u}_{e(h), \mathbf{v}})$ and $\eta(t, \bar{u})$, writing

$$g(\tau, \eta(\tau, \bar{u}_{e(h), \mathbf{v}}), \bar{u}_{e(h), \mathbf{v}}(\tau)) - g(\tau, \eta(\tau, \bar{u}), \bar{u}(\tau))$$

$$= g(\tau, \eta(\tau, \bar{u}_{e(h), \mathbf{v}}), \bar{u}_{e(h), \mathbf{v}}(\tau)) - g(\tau, \eta(\tau, \bar{u}), \bar{u}_{e(h), \mathbf{v}}(\tau))$$

$$+ g(\tau, y(\tau, \bar{u}), \bar{u}_{e(h), \mathbf{v}}(\tau)) - g(\tau, \eta(\tau, \bar{u}), \bar{u}(\tau)),$$

and estimate using the local Lipschitz condition coming from (7.10.15) and the mean value theorem in the first term; the constants may be different in different inequalities.

$$\|\eta(t, \bar{u}_{\mathbf{e}(h),\mathbf{v}}) - \eta(\tau, \bar{u})\| \leq C \int_0^t (t - \tau)^{-\alpha-\rho} \|\eta(\tau, \bar{u}_{\mathbf{e}(h),\mathbf{v}}) - \eta(\tau, \bar{u})\| d\tau$$

$$+ C' \sum_{j=1}^m \left\| \int_{(0,t) \cap e_j(h)} (-A)^{\alpha+\rho} S(t - \tau) \right.$$

$$\left. \times \{g(\tau, \eta(\tau, \bar{u}), v_j(\tau)) - g(\tau, \eta(\tau, \bar{u}), \bar{u}(\tau))\} d\tau \right\|. \quad (7.10.24)$$

The second term between bars in (7.10.23) is independent of h, thus each term in the sum on the right-hand side of (7.10.24) is $O(h)$. It then follows from Lemma 7.7.6 that

$$\|\eta(t, \bar{u}_{\mathbf{e}(h),\mathbf{v}}) - \eta(t, \bar{u})\| \leq Ch \quad (0 \leq t \leq \bar{t}). \quad (7.10.25)$$

Define

$$\rho(t, \eta, \eta', u) = g(t, \eta', u) - g(t, \eta, u) - \partial_\eta g(t, \eta, u)(\eta' - \eta)$$

for $\eta, \eta' \in E, u \in U$. By the mean value theorem,

$$\|\rho(t, \eta', \eta, \bar{u}(t))\| \leq CL(c)\|\eta' - \eta\| \quad (0 \leq t \leq \bar{t}, \|\eta\|, \|\eta'\| \leq c). \quad (7.10.26)$$

where we have the right to use $L(c)$ independent of h if $u = \bar{u}_{\mathbf{e}(h),\mathbf{v}}$ (see the comments following (6.2.4)) Also, for each η,

$$\|\rho(t, \eta', \eta, \bar{u}(t))\| \to 0 \quad \text{as } \eta' \to \eta \quad (0 \leq t \leq \bar{t}). \quad (7.10.27)$$

Let $\psi(t, h) = h^{-1}(\eta(t, \bar{u}_{\mathbf{e}(h),\mathbf{v}}) - \eta(t, \bar{u})) - \upsilon(t, \bar{u}, \mathbf{p}, \mathbf{v})$. Taking into account the integral equations satisfied by each function, we obtain the following counterpart of (6.2.26), where we omit the intermediate steps:

$$\psi(t, h)$$

$$= \int_0^t (-A)^{\alpha+\rho} S(t - \tau) \frac{1}{h} \{g(\tau, \eta(\tau, \bar{u}_{\mathbf{e}(h),\mathbf{v}}), \bar{u}_{\mathbf{e}(h),\mathbf{v}}(\tau)) - g(\tau, \eta(\tau, \bar{u}), \bar{u}(\tau))\} d\tau$$

$$- \int_0^t (-A)^{\alpha+\rho} S(t - \tau) \partial_\eta g(t, \eta(t, \bar{u}), \bar{u}(t)) \upsilon(\tau, \bar{u}, \mathbf{p}, \mathbf{v}) d\tau$$

$$- \sum_{j=1}^m p_j \int_0^t (-A)^{\alpha+\rho} S(t - \tau) \{g(\tau, \eta(\tau, \bar{u}), v_j(\tau)) - g(\tau, \eta(\tau, \bar{u}), \bar{u}(\tau))\} d\tau$$

$$= \int_0^t (-A)^{\alpha+\rho} S(t - \tau) \partial_\eta g(\tau, \eta(\tau, \bar{u}), \bar{u}(\tau)) \psi(\tau, h) d\tau$$

$$+ \int_0^t (-A)^{\alpha+\rho} S(t - \tau) \frac{1}{h} \rho(\tau, \eta(\tau, \bar{u}), \eta(\tau, \bar{u}_{\mathbf{e}(h),\mathbf{v}}), \bar{u}(\tau)) d\tau$$

$$+ \frac{1}{h} \int_0^t (-A)^{\alpha+\rho} S(t - \tau)$$

$$\times \big\{ g(\tau, \eta(\tau, \bar{u}_{e(h),v}), \bar{u}_{e(h),v}(\tau)) - g(\tau, \eta(\tau, \bar{u}), \bar{u}_{e(h),v}(\tau)) \big\} d\tau$$

$$- \frac{1}{h} \int_0^t (-A)^{\alpha+\rho} S(t - \tau) \big\{ g(\tau, \eta(\tau, \bar{u}_{e(h),v}), \bar{u}(\tau)) - g(\tau, \eta(\tau, \bar{u}), \bar{u}(\tau)) \big\} d\tau$$

$$+ \sum_{j=1}^m \frac{1}{h} \int_{(0,t)\cap e_j(h)} (-A)^{\alpha+\rho} S(t - \tau)$$

$$\times \big\{ g(\tau, \eta(\tau, \bar{u}), v_j(\tau)) - g(\tau, \eta(\tau, \bar{u}), \bar{u}(\tau)) \big\} d\tau$$

$$- \sum_{j=1}^m p_j \int_0^t (-A)^{\alpha+\rho} S(t - \tau)$$

$$\times \big\{ g(\tau, \eta(\tau, \bar{u}), v_j(\tau)) - g(\tau, \eta(\tau, \bar{u}), \bar{u}(\tau)) \big\} d\tau. \qquad (7.10.28)$$

Write this equation in the form

$$\psi(t, h) = \int_0^t (-A)^{\alpha+\rho} S(t - \tau) \partial_\eta g(\tau, \eta(\tau, \bar{u}), \bar{u}(\tau)) \psi(\tau, h) d\tau$$
$$+ \kappa_1(t, h) + \kappa_2(t, h) + \kappa_3(t, h) \qquad (7.10.29)$$

where $\kappa_1(t, h)$ is the second integral on the right side of the last equality and $\kappa_2(t, h)$ (resp. $\kappa_3(t, h)$) is the combination of the third and fourth (resp. of the fifth and sixth sums of) integrals. Estimating,

$$\|\psi(t, h)\| \leq \delta(t, h) + C \int_0^t (t - \tau)^{-\alpha-\rho} \|\psi(\tau, h)\| d\tau \qquad (7.10.30)$$

with $\delta(t, h) = \|\kappa_1(t, h)\| + \|\kappa_2(t, h)\| + \|\kappa_3(t, h)\|$, so that Theorem 7.10.4 will follow if we show that each $\kappa_j(t, h)$ tends to zero uniformly in $0 \leq t \leq T$. The proof ends like that of Theorem 6.2.7 but with some modifications.

$\kappa_1(t, h)$: If it does not converge uniformly to zero as $h \to 0$ pick a sequence $\{t_n\}$ (which we may assume convergent) with $\|\kappa(t_n, h_n)\| \geq \varepsilon > 0$ for some $\{h_n\} \in R_+, h_n \to 0$. Write the integral as an integral in $(0, \bar{t})$ with the characteristic function $\chi_n(\tau)$ of $0 \leq t \leq t_n$ multiplying the integrand. Use (7.7.20), (7.10.26) and (7.10.25) to bound the integrand; the bound implies the equicontinuity-of-integrals condition needed to apply Vitali's theorem. Then use (7.10.25), (7.10.27) and (E, E)-continuity of $(-A)^{\alpha+\rho} S(t - \tau)$ for $\tau < t$ to show that the integrand tends to zero a.e., which contradicts Vitali's theorem.

$\kappa_2(t, h)$: Use the local Lipschitz condition coming from (7.10.15) to bound both integrands by $C(t - \tau)^{-\alpha-\rho}$. Then notice that the integrands cancel out outside of $e(h)$ and bound with the worst possible case $e(h) = (t - h\bar{t}, t)$.

$\kappa_3(t, h)$: Use (7.10.23).   ∎

Applied to $\upsilon(t, u, \mathbf{p}, \mathbf{v}) = (-A)^\alpha \xi(t, u, \mathbf{p}, \mathbf{v})$, the variation-of-constants formula (7.9.33) reads

$$\upsilon(t, u, \mathbf{p}, \mathbf{v}) = \sum_{j=1}^{m} p_j \int_0^t R_{\alpha,\rho}(t, \tau; \bar{u})((-A^\odot)^{-\rho})^*$$

$$\times \{ f(\tau, (-A)^{-\alpha} \eta(\tau, \bar{u}), v_j(\tau)) - f(\tau, (-A)^{-\alpha} \eta(\tau, \bar{u}), \bar{u}(\tau)) \} d\tau \quad (7.10.31)$$

where $R_{\alpha,\rho}(t, s; \bar{u})$ is the extension of $(-A)^\alpha S(t, s; \bar{u})(-A)^\rho$ constructed in Theorem 7.9.3, $S(t, s; u) = (-A)^{-\alpha} R_\alpha(t, s; u)$ the solution operator of the variational equation

$$\xi'(t) = \{A(t) + \partial_y f(t, y(t, u), u(t))\} \xi(t). \quad (7.10.32)$$

The following two results roughly correspond to Lemma 6.3.1.

Lemma 7.10.5. *Let the control system* (7.10.1) *satisfy Hypothesis* $III_{\alpha,\rho}$ *with* $\alpha + \rho < 1$ *in an interval* $0 \le t \le \bar{t}$ *for each* $u(\cdot) \in C_{\mathrm{ad}}(0, \bar{t}; U)$ *with* $K(c)$ *and* $L(c)$ *independent of* $u(\cdot)$, *and let* $y(t, \bar{u})$ *exist in* $0 \le t \le \bar{t}$. *Then* $R_\alpha(t, s; u)\zeta$ *is continuous with respect to* $u \in B(\bar{u}, \delta)$ *for each* $\zeta \in E_\alpha$ *(δ the constant in Lemma 7.10.1) in the norm of* E, *uniformly in* $0 \le s \le t \le \bar{t}$.

Lemma 7.10.6. *Under the same hypotheses of Lemma* 7.10.5, *let* $\alpha + \gamma < 1$. *Then* $R_{\alpha,\gamma}(t, s; u)\zeta$ *is continuous with respect to* $u \in B(\bar{u}, \delta)$ *for each* $\zeta \in (E^\odot)^*$ *in the norm of* E *in* $0 \le s < t \le \bar{t}$.

The proof of both results apes that of Lemma 6.3.1, the only difference being that estimations are based on Lemma 7.7.6 rather than on Gronwall's inequality. The key estimate is the counterpart of (6.3.4) and implies, just as in Lemma 6.3.1, that if $\eta \to \partial_\eta g(\tau, \eta, u)$ is continuous in the norm of $(E, (E^\odot)^*)$ then $u \to R_{\alpha,\gamma}(t, s; u)$ is continuous in the norm of (E, E) in $0 \le s < t \le \bar{t}$.

Lemma 7.10.5 and Lemma 7.10.6 imply corresponding results on continuity of $R_\alpha(t, s; u)$ and $R_{\alpha,\gamma}(t, s; u)$ in all variables of the type of Corollary 6.3.2.

Remark 7.10.7. We have $d(\bar{u}, \bar{u}_{e(h),\mathbf{v}}) \le |e(h)| \le \bar{t}h$, hence $\bar{t}^{-1}\xi(t, \bar{u}, \mathbf{p}, \mathbf{v})$ (not $\xi(t, \bar{u}, \mathbf{p}, \mathbf{v})$) is a directional derivative. See Remark 6.2.8.

Remark 7.10.8. Note that we are free to replace h on the right side of (7.10.23) by any function $\varepsilon(h) > 0$ converging to zero as $h \to 0+$ as fast as we wish. However, this does not improve (7.10.25), thus the rate of convergence of the quotient of increments is the same.

8

Interpolation and Domains of Fractional Powers

8.1. Trace Spaces and Semigroups

Let E, F be two normed spaces. We use the notation $F \hookrightarrow E$ to indicate that F is contained in E and that there exists a constant C such that

$$\|y\|_E \leq C\|y\|_F \quad (y \in F).$$

Colloquially, "F is imbedded in E" or "there is an imbedding of F into E." We write $E \approx F$ or $F \approx E$ and say that E and F are "equivalent" when $F \hookrightarrow E$ and $E \hookrightarrow F$; this means that E and F coincide as sets and that there exist two constants $C \geq c > 0$ such that

$$c\|y\|_E < \|y\|_F \leq C\|y\|_E. \quad (y \in E = F).$$

In some cases, equivalence follows from set equality without having to compare norms. For instance, we have

Lemma 8.1.1. *Let $E = F$ and let $(E, \| \cdot \|_E)$ and $(F, \| \cdot \|_F)$ be Banach spaces. Assume further that for every sequence $\{y_n\} \in E = F$ with $\|y_n - y\|_E \to 0$ and $\|y_n - z\|_F \to 0$ we have $y = z$. Then $E \approx F$.*

In view of the closed graph Theorem 5.0.16, it is enough to show that the identity maps from $(E, \| \cdot \|_E)$ into $(F, \| \cdot \|_F)$ and back are closed. This is elementary and we omit the details. ∎

An E-valued function $f(\cdot)$ defined in an interval $I \subseteq (-\infty, \infty)$ is (Lebesgue-Bochner) **locally integrable** there if it is integrable over any compact interval contained in I. A second E-valued function $g(\cdot)$ is the **derivative** of $f(\cdot)$ if it is itself locally integrable and, for every $s \in I$

$$f(t) = f(s) + \int_s^t g(\tau)d\tau \quad \text{a.e. in } I. \tag{8.1.1}$$

It follows that, after eventual modification in a null set, $f(\cdot)$ is continuous in $t \in I$ and the derivative (as limit of the quotient of increments in the norm of E) exists at any Lebesgue point of $g(\cdot)$, that is, almost everywhere (Theorem 5.0.12). We write $f'(t) = g(t)$. This definition is equivalent to differentiability in the sense of distributions (see Example 8.1.6).

The only version of interpolation theory we need begins with two Banach spaces E, F with $F \hookrightarrow E$. Given v, p with $-\infty < v < \infty$ and $1 \le p \le \infty$ we denote by $W(p, v; F, E)$ the space of all E-valued functions such that $f(\cdot)$ has a derivative $f'(t)$ in E in $t > 0$ and

$$t^v f \in L^p(0, \infty; F) \qquad t^v f' \in L^p(0, \infty; E). \tag{8.1.2}$$

Note that the first relation demands that $f(t)$ take values in F a.e. The space $W(p, v; F, E)$ becomes a Banach space equipped with the norm

$$\|f\|_{W(p,v;F,E)} = \left(\|t^v f\|_{L^p(0,\infty,F)}^p + \|t^v f'\|_{L^p(0,\infty,E)}^p\right)^{1/p} \tag{8.1.3}$$

(for $p = \infty$ we take the maximum of the two L^∞ norms).

Lemma 8.1.2. *Let* $1 \le p \le \infty$, $v + 1/p < 1$. *Then*

$$f(0) = \lim_{t \to 0+} f(t) \tag{8.1.4}$$

exists in E for every $f(\cdot) \in W(p, v; F, E)$.

Proof. Let $1 < p < \infty$. If $0 < s < t$ we have, using (8.1.1) and Hölder's inequality,

$$
\begin{aligned}
\|f(t) - f(s)\|_E &= \left\| \int_s^t f'(\tau) d\tau \right\|_E \\
&\le \int_s^t \|\tau^v f'(\tau)\|_E \tau^{-v} d\tau \\
&\le \|\tau^v f'\|_{L^p(0,\infty,F)} \left(\int_s^t \tau^{-vp/(p-1)} d\tau \right)^{(p-1)/p}
\end{aligned} \tag{8.1.5}
$$

with $vp/(p-1) < 1$, thus the right side tends to zero as $t, s \to 0$. The estimation works as well for $p = \infty$. For $p = 1$ the inequality $v + 1/p < 1$ is $v < 0$, thus we use (8.1.5) up to the second integral and then estimate τ^{-v} by t^{-v}. $\blacksquare$

We call the limit (8.1.4) the **trace** of $f(t)$ at 0. Always under the assumption that $v + 1/p < 1$, the **trace space** $T(p, v; F, E) \subseteq E$ consists of all traces $y = f(0)$ with $f(\cdot) \in W(p, v; F, E)$ and is equipped with the norm

$$\|y\|_{T(p,v;F,E)} = \inf \|f\|_{W(p,v;F,E)} \tag{8.1.6}$$

the infimum taken over all $f(\cdot) \in W(p, v; F, E)$ with $f(0) = y$. It follows from the definition that the map $f(\cdot) \to f(0)$ of $W(p, v; F, E)$ into $T(p, v; F, E)$ is linear and bounded.

Let A be the infinitesimal generator of a strongly continuous semigroup $S(t)$ satisfying

$$\|S(t)\| \le C_0 e^{-ct} \quad (t > 0) \tag{8.1.7}$$

with $c > 0$. Given v, p with $1 \le p < \infty$, $v + 1/p < 1$, the space $T(p, v; E, S)$ is defined as the set of all $y \in E$ with

$$\|y\|_{T(p,v;E,S)} = \left(\int_0^\infty (t^{v-1} \|S(t)y - y\|_E)^p dt \right)^{1/p} < \infty. \tag{8.1.8}$$

Note that, since $(v - 1)p < -1$, the integral is always convergent at infinity; moreover, in view of (8.1.7), $\|S(t)y - y\|_E \ge \|y\|_E/2$ for t large enough (independently of y) so that there exists a constant C with

$$\|y\|_E \le C\|y\|_{T(p,v;E,S)} \quad (y \in T(p, v; E, S)). \tag{8.1.9}$$

For $p = \infty$ we take $v \le 1$; $T(\infty, v; E, S)$ consists of all $y \in E$ with

$$\|y\|_{T(\infty,v;E,S)} = \sup_{0<t<\infty} t^{v-1} \|S(t)y - y\|_E < \infty. \tag{8.1.10}$$

We show in a similar way that (8.1.9) holds. The spaces $T(p, v; E, S)$, equipped with the norm (8.1.9) ((8.1.10) for $p = \infty$) are Banach spaces, and (8.1.9) means that $T(p, v; E, S) \hookrightarrow E$.

Theorem 8.1.3. *Let* $1 \le p < \infty$, $v + 1/p < 1$, *or* $p = \infty$, $0 \le v \le 1$, *and let* $F = D(A)$ *be equipped with the norm* $\|y\|_{D(A)} = \|Ay\|_E$. *Then*

$$T(p, v; E, S) \approx T(p, v; D(A), E). \tag{8.1.11}$$

We need an auxiliary result:

Lemma 8.1.4. (Hardy) *Let* f *be real valued and locally integrable in* $t > 0$ *for all* t. *Define*

$$g(t) = \frac{1}{t} \int_0^t f(s)ds. \tag{8.1.12}$$

Let $1 \le p < \infty$ *and* $\theta = v + 1/p < 1$. *Then*

$$\int_0^\infty (t^v |g(t)|)^p \, dt \le \frac{1}{(1-\theta)^p} \int_0^\infty (t^v |f(t)|)^p \, dt. \tag{8.1.13}$$

Proof. We assume the right side is finite. Under the transformation $t = e^\tau$, the inequality becomes

$$\int_{-\infty}^\infty (e^{\theta\tau} |\tilde{g}(\tau)|)^p d\tau \le \frac{1}{(1-\theta)^p} \int_{-\infty}^\infty (e^{\theta\tau} |\tilde{f}(\tau)|)^p d\tau \tag{8.1.14}$$

for the functions $\tilde{f}(\tau) = f(e^\tau)$, $\tilde{g}(\tau) = g(e^\tau)$. The relation (8.1.12) among f and g translates to

$$\tilde{g}(\tau) = e^{-\tau} \int_{-\infty}^{\tau} \tilde{f}(\sigma)e^\sigma d\sigma$$

for the transformed functions. Equivalently,

$$e^{\theta\tau}\tilde{g}(\tau) = \int_{-\infty}^{\tau} (\tilde{f}(\sigma)e^{\theta\sigma})e^{(\theta-1)(\tau-\sigma)} d\sigma,$$

that is, $e^{\theta\tau}\tilde{g}(\tau)$ is the convolution of $\tilde{f}(\tau)e^{\theta\tau}$ and $\chi(\tau)e^{(\theta-1)\tau}$, χ the characteristic function of $[0, \infty)$. But then (8.1.14) (or, rather, its $1/p^{\text{th}}$ power) is nothing but Young's Theorem 5.0.22, since $1/(1-\theta)$ is the L^1 norm of $\chi(\tau)e^{(\theta-1)\tau}$.   ∎

Proof of Theorem 8.1.3. Let $y \in T(p, v; D(A), E)$. Select $f \in W(p, v; D(A), E)$ such that $f(0) = y$. Consider first the range $1 \leq p < \infty$ in the two cases

$$(a) \quad v \leq 0, \qquad (b) \quad v > 0.$$

In case (a) $1 \leq t^v$ near zero, so that $f'(t)$ and $Af(t)$ are locally integrable in $t \geq 0$. Using the equality

$$f(t) = S(t)f(0) + \int_0^t S(t-\tau)(f'(\tau) - Af(\tau)) d\tau \tag{8.1.15}$$

(see Example 8.1.11) we obtain

$$S(t)y - y = \int_0^t f'(\tau)d\tau - \int_0^t S(t-\tau)(f'(\tau) - Af(\tau)) d\tau \tag{8.1.16}$$

hence

$$\left\| \frac{S(t)y - y}{t} \right\|_E \leq \frac{1 + C_0}{t} \int_0^t \|f'(\tau)\|_E \, d\tau + \frac{C_0}{t} \int_0^t \|Af(\tau)\|_E \, d\tau. \tag{8.1.17}$$

By Hardy's inequality,

$$\int_0^\infty (t^{v-1}\|S(t)y - y\|_E)^p \, dt \leq C \int_0^\infty (t^v(\|f'(t)\|_E + \|Af(t)\|_E))^p \, dt$$

$$\leq C'\|f\|^p_{W(p,v;D(A),E)},$$

where C, C' (in this inequality and in others) are constants that depend only on v, p. In case (b), $\|f'(\cdot)\|_E$ and $\|Af(\cdot)\|_E$ may not be integrable near zero, and (8.1.15) is unjustified. We define instead

$$f_\varepsilon(t) = f(t + \varepsilon) \quad (t \geq 0).$$

Since $t^v \leq (t + \varepsilon)^v$, $f_\varepsilon \in W(p, v; D(A), E)$ with

$$\|f_\varepsilon\|_{W(p,v;D(A),E)} \leq \|f\|_{W(p,v;D(A),E)},$$

and $\|f'_\varepsilon(\cdot)\|_E$, $\|Af_\varepsilon(\cdot)\|_E$ are integrable near zero. We argue with f_ε in the same way as with f in (a), obtaining

$$\int_0^\infty (t^{\nu-1}\|S(t)f(\varepsilon) - f(\varepsilon)\|_E)^p \, dt \le C\|f_\varepsilon\|_{W(p,\nu;D(A),E)}^p \le C\|f\|_{W(p,\nu;D(A),E)}^p.$$

We then use then Lemma 8.1.2 and Fatou's theorem (Example 8.1.7), obtaining

$$\int_0^\infty (t^{\nu-1}\|S(t)y - y\|_E)^p \, dt \le C\|f\|_{W(p,\nu;D(A),E)}^p. \tag{8.1.18}$$

When $p = \infty$, we must show that

$$\sup_{0\le t<\infty} t^{\nu-1}\|S(t)y - y\|_E \le C\|f\|_{W(\infty,\nu;D(A),E)}. \tag{8.1.19}$$

This estimate is obvious from (8.1.17) for $\nu = 0$. For $0 \le \nu < 1$ we avail ourselves of the right of taking $f(\cdot)$ with support in $[0, a]$ (see Example 8.1.12), write (8.1.18) for p so large that $\nu + 1/p < 1$, and then let $p \to \infty$ (see Example 8.1.13); note that the constant $\theta = \nu + 1/p$ in Hardy's inequality tends to ν as $p \to \infty$.

At this point, the imbedding

$$T(p, \nu; D(A), E) \hookrightarrow T(p, \nu; E, S) \tag{8.1.20}$$

has been proved in all cases. We show now the opposite imbedding

$$T(p, \nu; E, S) \hookrightarrow T(p, \nu; D(A), E). \tag{8.1.21}$$

Let $y \in T(p, \nu; E, S)$. Define

$$g(t) = \frac{1}{t} \int_0^t S(\tau)y \, d\tau. \tag{8.1.22}$$

Then $g(t) \in D(A)$ for $t > 0$, and

$$g(0) = \lim_{t\to 0+} g(t) = y, \qquad Ag(t) = t^{-1}(S(t)y - y) \tag{8.1.23}$$

(this is in fact true for every $y \in E$; see **5.2**). Let $\phi(\cdot)$ be a test function such that $0 \le \phi(t) \le 1, \phi(0) = 1, \phi(t) = 0$ for $t \ge 1$. To show that $y \in T(p, \nu; D(A), E)$ it is enough to show that $f(\cdot) = \phi(\cdot)g(\cdot) \in W(p, \nu; D(A), E)$ with norm bounded by (a constant times) $\|y\|_{T(p,\nu;E,S)}$; this reduces to proving that $t^\nu\|g'\|_E$ and $t^\nu\|Ag\|_E$ belong to $L^p(0, 1)$ with a similar bound on both norms.

Let $p < \infty, \nu + 1/p < 1$. Using the second relation (8.1.23),

$$\int_0^1 (t^\nu\|Ag(t)\|_E)^p \, dt = \int_0^1 (t^{\nu-1}\|S(t)y - y\|_E)^p \, dt.$$

On the other hand,

$$g'(t) = \frac{1}{t} S(t)y - \frac{1}{t^2} \int_0^t S(\tau)y \, d\tau$$
$$= \frac{1}{t}(S(t)y - y) - \frac{1}{t} \int_0^t \frac{S(\tau)y - y}{t} \, d\tau = h_1(t) + h_2(t).$$

We have

$$\int_0^1 (t^\nu \|h_1(t)\|_E)^p \, dt = \int_0^1 (t^{\nu-1} \|S(t)y - y\|_E)^p \, dt.$$

For $h_2(t)$ we apply Lemma 8.1.4:

$$\int_0^1 (t^\nu \|h_2(t)\|_E)^p \, dt \leq C \int_0^1 (t^{\nu-1} \|S(t)y - y\|_E)^p \, dt.$$

For $p = \infty$ we obviously have

$$\sup_{0\leq t\leq 1} t^\nu \|Ag(t)\|_E \leq \sup_{0\leq t<\infty} t^{\nu-1} \|S(t)y - y\|_E,$$

and we obtain from (8.1.23) that

$$\sup_{0\leq t\leq 1} t^\nu \|h_j(t)\|_E \leq \sup_{0\leq t<\infty} t^{\nu-1} \|S(t)y - y\|_E.$$

This is obvious for $j = 1$; for $j = 2$ introduce t^ν into the integral and note that, since $\nu - 1 \leq 0$ we have $t^{\nu-1} \leq \tau^{\nu-1}$. This shows (8.1.21) in all cases and thus ends the proof of Theorem 8.1.3. ∎

A surprising consequence of Theorem 8.1.3 is that the spaces $T(p, \nu; E, S)$ depend only on the domain $D(A)$ of the infinitesimal generator A, not on the semigroup $S(\cdot)$. This follows from

Lemma 8.1.5. *Let A, B be two closed operators such that $D(A) = D(B) = F$. Then the graph norms $\|y\|_{D(A)} = \|y\| + \|Ay\|$ and $\|y\|_{D(B)} = \|y\| + \|By\|$ of F are equivalent; in other words, $D(A) \approx D(B)$.*

The proof is a consequence of Lemma 8.1.1. Recall that (8.1.7) implies that A^{-1} exists, thus the graph norm of $D(A)$ is equivalent to the norm $\|y\|_{D(A)} = \|Ay\|$ (Example 7.4.16).

It follows from Lemma 8.1.2 that, if $p < \infty$, then $T(p, \nu; F, E) = \{0\}$ if $\nu p \leq -1$ (the first norm in (8.1.3) is infinite unless $f(0) = 0$). Since the condition that $\nu + 1/p < 1$ is required in all the results above, the permissible range for the pair (ν, p) is

$$0 < \nu + 1/p < 1 \quad (1 < p < \infty) \tag{8.1.24}$$
$$0 \leq \nu \leq 1 \quad\quad (p = \infty). \tag{8.1.25}$$

It is plain that $T(\infty, 1; D(A), E) \approx E$ and almost equally plain that

$$D(A) \hookrightarrow T(\infty, 0; E, S) \approx T(\infty, 0; D(A), E) \qquad (8.1.26)$$

The imbedding opposite to (8.1.26) is true when E is reflexive but not in general (Examples 8.1.9 and 8.1.10).

The following imbeddings are obvious from either definition, more so from the semigroup definition (8.1.8):

$$T(p', \nu; D(A), E) \hookrightarrow T(p, \nu; D(A), E)$$
$$(1 \le p \le p' < \infty, 0 < \nu + 1/p', \nu + 1/p < 1$$
$$\text{or } 1 \le p < p' = \infty, 0 < \nu, \nu + 1/p < 1). \qquad (8.1.27)$$
$$T(p, \nu; D(A), E) \hookrightarrow T(p, \nu'; D(A), E)$$
$$(1 \le p < \infty, \nu \le \nu', 0 < \nu + 1/p, \nu' + 1/p < 1$$
$$\text{or } p = \infty, 0 \le \nu \le \nu' < 1). \qquad (8.1.28)$$

Example 8.1.6. An E-valued locally integrable function $f(\cdot)$ defined in an open interval I **has a derivative** $g(\cdot)$ **in the sense of distributions** if $g(\cdot)$ is itself locally integrable and for every test function $\phi(\cdot)$ with support in I we have

$$\int_I \phi'(t) f(t)\, dt = -\int_I \phi(t) g(t)\, dt.$$

If $f(\cdot)$ has a derivative $g(\cdot)$ in this sense, then (8.1.1) holds. For a scalar function, the result is well-known (Dunford–Schwartz [1963, p. 1291]). We may then apply it to the scalar functions $\langle y^*, f(t) \rangle$, $\langle y^*, g(t) \rangle$ $(y^* \in E^*)$ obtaining (8.1.1) with y^* applied to both sides. Arbitrariness of y^* proves (8.1.1) itself.

Example 8.1.7 (Fatou's Theorem). Let $\{\phi_n(\cdot)\}$ be a sequence of nonnegative Lebesgue integrable functions in a measurable set $\Omega \subseteq \mathbb{R}^m$ of finite measure such that $\int_\Omega \phi_n(x)\, dx$ is uniformly bounded and $\phi_n(x) \to \phi(x)$ a.e. in Ω. Then $\phi(\cdot)$ is integrable in Ω; precisely,

$$\int_\Omega \phi(x)\, dx \le \liminf_{n \to \infty} \int_\Omega \phi_n(x)\, dx. \qquad (8.1.29)$$

To prove this statement, note first that, given $\varepsilon > 0$ there exists a subsequence such that $M = \sup \int_\Omega \phi_n(x)\, dx$ does not exceed the lim inf on the right side of (8.1.29) by more than ε, thus we may replace lim inf by sup. Given $m = 1, 2, \ldots$ and applying the Lebesgue dominated convergence theorem to the sequence $\{\min(\phi_n(x), m); n = 1, 2, \ldots\}$, we deduce that $\int_\Omega \min(\phi(x), m) dx \le M$ for all m. This implies that if $e_m = \{x \in \Omega; m < \phi(x) \le m + 1\}$, then $\Sigma m |e_m| < \infty$, so that $\Sigma(m + 1)|e_m| < \infty$ as well and $\phi(\cdot)$ is integrable. We have $\phi(x) = \lim_{m \to \infty} \min(\phi(x), m)$ with $\min(\phi(x), m) \le \phi(x)$, thus (8.1.29) follows from the dominated convergence theorem.

The result can be considerably generalized at no extra cost. For instance, if we do not assume that $\lim \phi_n(x)$ exists, (8.1.29) is still true with $\phi(x) = \liminf_{n \to \infty} \phi(x)$; to see this, we just replace the sequence $\{\phi_n(x)\}$ by $\{\inf_{k \geq n} \phi_k(x)\}$. See Natanson [1955/1960, p. 140] or Dunford–Schwartz [1958, p. 152].

Example 8.1.8. An operator A with domain $D(A) \subseteq E$ is E^***-weakly closed** if for every (generalized) sequence[1] $\{y_\kappa\} \subseteq E$ such that $y_k \to y$ and $Ay_\kappa \to z$ we have $y \in D(A)$ and $Ay = z$. Equivalently, A is weakly closed if the graph $\Gamma(A) = \{(y, Ay); y \in D(A)\} \subseteq E \times E$ is $(E \times E)^*$-weakly closed $((E \times E)^*$ can be identified with $E^* \times E^*$). Obviously, a E^*-weakly closed operator is closed, but the reciprocal is not true (see Example 8.1.10 below). However, in a reflexive space, A is closed if and only if it is E^*-weakly closed. To see this, note that $\Gamma(A)$ is a subspace of $E \times E$, thus it is convex; then (Theorem 12.0.11) it is closed if and only if is E-weakly closed. As a consequence, we have

Example 8.1.9. If E is reflexive we have $T(\infty, 0; D(A), E) \hookrightarrow D(A)$, so that, combining with (8.1.26),

$$T(\infty, 0; D(A), E) \approx D(A).$$

In fact, assume that $y \in T(\infty, 0; D(A), E)$, and let $y_t = g(t)$ be defined by (8.1.22). Then $y_t \to y$ as $t \to 0+$. On the other hand, $Ay_t = t^{-1}(S(t)y - y)$; hence, $\|Ay_t\|$ is bounded as $t \to 0$ and specializing t to a (generalized) subsequence tending to zero we may assume that $Ay_t \to z$ E^*-weakly (this follows from Alaoglu's Theorem 12.0.17). Since A is weakly closed, $y \in D(A)$.

Example 8.1.10. Let $S(t)$ be the translation semigroup $S(t)y(x) = y(x + t)$ in the space E of all continuous 1-periodic functions $y(x)$ in $-\infty < x < \infty$ equipped with the supremum norm. The infinitesimal generator of $S(t)$ is $(Ay)(x) = y'(x)$ with domain $D(A)$ consisting of all continuously differentiable functions in E. We have $T(\infty, 0; D(A), E) \supset D(A)$ with strict inclusion; in fact, $T(\infty, 0; D(A), E)$ consists of all Lipschitz continuous functions in E.

Example 8.1.11. Equality (8.1.15) is intuitively obvious since $f'(t) = Af(t) + (f'(t) - Af(t))$. To establish it rigorously, it is sufficient to note that

$$\int_0^t A^{-1}S(t - \tau)f'(\tau)d\tau = \int_0^t S(t - \tau)f(\tau)d\tau = A^{-1}f(t) - A^{-1}S(t)f(0).$$

This equality can be shown integrating both sides with respect to t in an interval $[0, s]$ and switching the order of integration in the left-hand side.

[1] For generalized sequences and weak topologies, see **12.0**.

Remark 8.1.12. The trace spaces $T(p, v; F, E)$ can be defined using functions $f(\cdot) \in W(p, v; F, E)$ with support in some fixed interval $[0, a]$, $a > 0$. To see this pick a test function $\eta(t)$ with $0 \leq \eta(t) \leq 1$, $\eta(t) = 1$ $(0 \leq t \leq a)$, $\eta(t) = 0$ $(t \geq a + 1)$. Then, if $g(t) = \eta(t)f(t)$ we have $g'(t) = \eta'(t)f(t) + \eta(t)f'(t)$, so that

$$\|t^v g\|_{L^p(0,\infty; F)} \leq \|t^v f\|_{L^p(0,\infty; F)},$$
$$\|t^v g'\|_{L^p(0,\infty; E)} \leq C\big(\|t^v f'\|_{L^p(0,\infty; E)} + \|t^v f\|_{L^p(0,\infty; F)}\big),$$

the replacement of E by F in the last norm justified on the basis that $F \hookrightarrow E$. Accordingly, the norms defined by (8.1.6) with or without the support restriction are equivalent.

Example 8.1.13. Let $\eta(\cdot) \in L^\infty(0, a)$. Then $\|\eta\|_{L^p(0,a)} \to \|\eta\|_{L^\infty(0,a)}$ as $p \to \infty$. To see this, note that, on the one hand $\|\eta\|_{L^p(0,a)} \leq a^{1/p}\|\eta\|_{L^\infty(0,a)}$. On the other hand, if $\varepsilon < \|\eta\|_{L^\infty(0,a)}$ the set $e \subseteq [0, a]$ where $|\eta(t)| \geq \|\eta\|_{L^\infty(0,a)} - \varepsilon$ has positive measure and $\|\eta\|_{L^p(0,a)} \geq |e|^{1/p}(\|\eta\|_{L^\infty(0,a)} - \varepsilon)$.

8.2. Interpolation and Fractional Powers

As in **8.1**, F and E are two Banach spaces with $F \hookrightarrow E$. Let $0 < \alpha < 1, 1 \leq p \leq \infty$. The **interpolation space** $(E, F)_{\alpha, p}$ is defined by

$$(E, F)_{\alpha, p} = T(p, 1 - \alpha - 1/p; F, E). \tag{8.2.1}$$

Note that $v = 1 - \alpha - 1/p$ is within the range (8.1.24); when $p = \infty$, $\alpha = 0, 1$ are admissible (see (8.1.25)). In case $F = D(A)$, A the infinitesimal generator of a strongly continuous semigroup $S(\cdot)$, we may use the alternative definition

$$(E, F)_{\alpha, p} = T(p, 1 - \alpha - 1/p; E, S) \tag{8.2.2}$$

Unless otherwise stated, we assume in this section that $S(\cdot)$ is analytic and satisfies (8.1.7) with $c > 0$, so that the fractional powers obey the bound

$$\|(-A)^\alpha S(t)\| \leq C_\alpha t^{-\alpha} e^{-ct} \quad (t > 0) \tag{8.2.3}$$

for $0 < \alpha \leq 1$, possibly with a different $c > 0$ (see Example 7.4.19); for $\alpha = 1$ this bound is (7.4.7).

Given v, p with $1 \leq p < \infty, 0 < v + 1/p < 1$, the space $T'(p, v; E, S) \subseteq E$ consists of all $y \in E$ with

$$\|y\|_{T'(p,v;E,S)} = \left(\int_0^\infty (t^v \|AS(t)y\|)^p \, dt\right)^{1/p} < \infty. \tag{8.2.4}$$

For $p = \infty$ we take $0 \leq v \leq 1$ and define $T'(\infty, v; E, S)$ as the set of all $y \in E$ with

$$\|y\|_{T'(\infty,v;E,S)} = \sup_{0 < t < \infty} t^v \|AS(t)y\| < \infty. \tag{8.2.5}$$

Theorem 8.2.1. *Let* $1 \leq p < \infty, 0 < v + 1/p < 1$, *or* $p = \infty, 0 \leq v \leq 1$. *Then*

$$T(p, v; E, S) \approx T'(p, v; E, S) \tag{8.2.6}$$

The proof needs a chain of auxiliary results.

Lemma 8.2.2. *Assume that*

$$\|AS(t)y\| \leq K \quad (0 \leq t < \infty). \tag{8.2.7}$$

Then

$$\|S(t)y - y\| \leq Ct \quad (0 \leq t < \infty) \tag{8.2.8}$$

with $C = K$. *Conversely, assume that* (8.2.8) *holds. Then* (8.2.7) *holds with* $K = C_0 C$ (C_0 *the constant in* (8.1.7)).

Proof. To obtain (8.2.8) from (8.2.7), note that

$$S(t)y - y = \lim_{\varepsilon \to 0+} \int_{\varepsilon}^{t} AS(\tau)y \, d\tau. \tag{8.2.9}$$

Conversely let (8.2.8) be true. Then $AS(t)y = \lim_{h \to 0+} h^{-1}(S(h) - I)S(t)y = \lim_{h \to 0+} S(t)h^{-1}(S(h) - I)y$, which implies (8.2.7). ∎

The next result (except for precision in the constants) generalizes Lemma 8.2.2 to the range $0 \leq \alpha < 1$.

Lemma 8.2.3. *Let* $0 < \alpha < 1$. *Assume that*

$$\|AS(t)y\| \leq Kt^{\alpha - 1} \quad (0 \leq t < \infty). \tag{8.2.10}$$

Then

$$\|S(t)y - y\| \leq Ct^{\alpha} \quad (0 \leq t < \infty) \tag{8.2.11}$$

with $C = K/\alpha$. *Conversely, there exists a constant* $L = L(\alpha)$ *such that, if* (8.2.11) *holds then* (8.2.10) *holds with* $K = L(\alpha)C$.

Proof. That (8.2.10) implies (8.2.11) is proved using (8.2.9) exactly as in Lemma 8.2.2.

To show the converse, we note several simplifications. Multiplying y by $1/C$ we may assume that $C = 1$ in (8.2.11). Choosing then t with $C_0 e^{-ct} < 1$, (8.2.11) implies $(1 - C_0 e^{-ct})\|y\| \leq \|y - S(t)y\| \leq t^{\alpha}$, so we may restrict ourselves to bounded y. Thus, we assume

$$C = 1, \qquad \|y\| \leq B. \tag{8.2.12}$$

Finally, due to exponential decay at infinity, we only need to show (8.2.10) in the interval $0 < t \leq 1$.

Assume then that (8.2.11) holds. Let $t_n = 2^{-n}$, $n = 0, 1, 2, \ldots$. Write

$$A(S(t_{n-1}) - S(t_n)) = AS(t_n)(S(t_{n-1}) - I) - AS(t_{n-1})(S(t_n) - I),$$

and estimate using (8.2.3) (for $\alpha = 1$, dropping the exponential) and (8.2.11):

$$\begin{aligned}
\|A(S(t_{n-1}) - S(t_n))y\| &\leq \|AS(t_n)\|\|S(t_{n-1})y - y\| + \|AS(t_{n-1})\|\|S(t_n)y - y\| \\
&\leq C_1(2^n 2^{-(n-1)\alpha} + 2^{n-1}2^{-n\alpha}) \\
&= C_1 2^{n(1-\alpha)}(2^\alpha + 2^{-1}) \leq 3C_1 2^{n(1-\alpha)},
\end{aligned}$$

so that

$$\begin{aligned}
\|AS(t_n)y\| &\leq \|AS(1)y\| + \sum_{j=1}^{n} \|A(S(t_{j-1}) - S(t_j))y\| \\
&\leq \|AS(1)y\| + 3C_1 \sum_{j=1}^{n} 2^{j(1-\alpha)} \\
&\leq \|AS(1)\|B + 3C_1 2^{1-\alpha} \frac{2^{n(1-\alpha)} - 1}{2^{1-\alpha} - 1}.
\end{aligned}$$

Now, we have $\|AS(1)\|B \leq C_1 B \leq C_1 t_n^{\alpha-1} = C_1 2^{n(1-\alpha)}$ if $n > n_0$ large enough. Since $2^{n(1-\alpha)} - 1 \leq 2^{n(1-\alpha)} = t_n^{\alpha-1}$, we have

$$\|AS(t_n)y\| \leq C_1 \left(1 + 3\frac{2^{1-\alpha}}{2^{1-\alpha} - 1}\right) t_n^{\alpha-1} \tag{8.2.13}$$

for $n \geq n_0$, thus, with a different constant, for $n > 1$ (recall that n_0 is independent of y). Inequality (8.2.13) is (8.2.10), but only for $t = t_n$. To extend the bound to arbitrary t we write

$$A(S(t) - S(t_n)) = AS(t)(I - S(t_n)) - AS(t_n)(I - S(t))$$

for $t_n \leq t \leq t_{n-1}$, and bound in a similar way:

$$\begin{aligned}
\|A(S(t) - S(t_n))y\| &\leq \|AS(t)\|\|S(t_n)y - y\| + \|AS(t_n)\|\|S(t)y - y\| \\
&\leq C_1\{t^{-1}t_n^\alpha + t_n^{-1}t^\alpha\} \\
&\leq C_1\{t^{-1}t^\alpha + 2t^{-1}t^\alpha\} \leq 3C_1 t^{\alpha-1}. \tag{8.2.14}
\end{aligned}$$

We combine (8.2.13) and (8.2.14) writing

$$\|AS(t)y\| \leq \|AS(t_n)y\| + \|A(S(t) - S(t_n))y\|.$$

This ends the proof. ∎

Lemma 8.2.4. *Let* $1 \le p < \infty, 0 < v + 1/p < 1$. *Assume that* $y \in T'(p, v; E, S)$.
Then $y \in T(p, v; E, S)$ *and we have*

$$\int_0^\infty (t^{v-1} \|S(t)y - y\|)^p \, dt \le \frac{C_0^p}{(1-\theta)^p} \int_0^\infty (t^v \|AS(t)y\|)^p \, dt \qquad (8.2.15)$$

with $\theta = v + 1/p$. *Conversely, if* $y \in T(p, v; E, S)$ *then* $y \in T'(p, v; E, S)$ *and*
there exists a constant $L(p, v)$ *depending only on* p *and* v *such that*

$$\int_0^\infty (t^v \|AS(t)y\|)^p \, dt \le L(p, v) \int_0^\infty (t^{v-1} \|S(t)y - y\|)^p \, dt. \qquad (8.2.16)$$

Proof. Again, we have to consider separately the cases $v \le 0$ and $v > 0$. In the
first, $\|AS(\tau)y\|$ is integrable near the origin. Using (8.2.9) we have

$$\int_0^\infty (t^{v-1} \|S(t)y - y\|)^p \, dt \le \int_0^\infty \left(t^v \frac{1}{t} \int_0^t \|AS(\tau)y\| d\tau \right)^p dt \qquad (8.2.17)$$

so that (8.2.15) (with $C_0 = 1$) is a consequence of Lemma 8.1.4. If $v > 0$ we obtain
in the same way inequality (8.2.15), but for $S(\varepsilon)y$ instead of y. In the right side
we use the fact that $\|AS(t)S(\varepsilon)y\| \le C_0\|AS(t)y\|$; in the left, we take limits as
$\varepsilon \to 0$ using Fatou's theorem.

To show the opposite inequality, write

$$S(t) = \sum_{j=0}^n (S(2^j t) - S(2^{j+1} t)) + S(2^{n+1} t)$$

$$= \sum_{j=0}^n S(2^j t)(I - S(2^j t)) + S(2^{n+1} t).$$

Apply A to both sides and then apply them to $y \in E$. Estimating,

$$\|AS(t)y\| \le \sum_{j=0}^n \|AS(2^j t)\| \|S(2^j t)y - y\| + \|AS(2^{n+1} t)\| \|y\|$$

so that using (8.2.3) for $\alpha = 1$ (and again dropping the exponential),

$$t^v \|AS(t)y\| \le C_1 \sum_{j=0}^n 2^{-j} t^{v-1} \|S(2^j t)y - y\| + C_1 2^{-n-1} t^{v-1} \|y\|$$

$$= C_1 \sum_{j=0}^n 2^{-vj} (2^j t)^{v-1} \|S(2^j t)y - y\| + C_1 2^{-n-1} t^{v-1} \|y\|.$$

Assume now that $y \in T(p, v; E, S)$ and take the $L^p(\varepsilon, \infty)$ norm on both sides,
using the triangle inequality on the right; then perform the substitution $2^j t = \tau$ in

the first integral on the right. The result is

$$
\left(\int_\varepsilon^\infty (t^\nu \|AS(t)y\|)^p \, dt \right)^{1/p}
$$

$$
\le C_1 \sum_{j=0}^n 2^{-(\nu+1)j} \left(\int_{2^j\varepsilon}^\infty (\tau^{\nu-1}\|S(\tau)y - y\|)^p \, d\tau \right)^{1/p}
$$

$$
+ C_1 2^{-n-1}\|y\| \left(\int_\varepsilon^\infty t^{(\nu-1)p} \, dt \right)^{1/p}
$$

$$
\le \frac{C_1}{1 - 2^{-(\nu+1)}} \left(\int_0^\infty (\tau^{\nu-1}\|S(\tau)y - y\|)^p \, d\tau \right)^{1/p} + \|y\| \quad (8.2.18)
$$

if $n = n(\varepsilon)$ is so large that the coefficient of $\|y\|$ in the last term does not exceed 1. Using finally (8.1.9) and the fact that (8.2.18) holds for ε arbitrary, the proof is over. $\blacksquare$

Proof of Theorem 8.2.1. For $1 \le p < \infty, 0 < \nu + 1/p < 1$ we use Lemma 8.2.4. For $p = \infty, 0 \le \nu < 1$ (resp. $\nu = 0$), we use Lemma 8.2.3 (resp. Lemma 8.2.2). Finally, if $p = \infty, \nu = 1$ then $T(\infty, 1; E, S) \approx E$ (see the comments before (8.1.26)) and we show just as easily that $T'(\infty, 1; E, S) \approx E$. $\blacksquare$

We use below the equivalent definition of real interpolation spaces,

$$
(E, D(A))_{\alpha,p} = T'(p, 1 - \alpha - 1/p; E, S), \quad (8.2.19)
$$

consequence of Theorem 8.2.1.

Lemma 8.2.5. *Let* $0 < \alpha < 1$, $E_\alpha = D((-A)^\alpha)$ *endowed with the norm* $\|y\|_\alpha = \|(-A)^\alpha y\|$. *Then*

$$
(E, D(A))_{\alpha,1} \hookrightarrow D((-A)^\alpha) \hookrightarrow (E, D(A))_{\alpha,\infty}. \quad (8.2.20)
$$

Proof. Assume $y \in D((-A)^\alpha)$. Then $(-A)S(t)y = (-A)^{1-\alpha}S(t)(-A)^\alpha y$; in view of (8.2.3) we have $\|AS(t)y\| \le C_\alpha \|y\|_\alpha t^{\alpha-1}e^{-ct}$ $(t > 0)$. It then follows that $y \in T'(\infty, 1 - \alpha; E, S) = (E, D(A))_{\alpha,\infty}$ with norm $\le C_\alpha \|y\|_\alpha$.

Conversely, assume $y \in (E, D(A))_{\alpha,1} = T'(1, -\alpha; E, S)$. Then

$$
t^{-\alpha}\|AS(t)y\| \in L^1(0, \infty) \quad (8.2.21)
$$

with L^1 norm $\|y\|_{T'(1,-\alpha;E,S)}$. We have

$$
(-A)^{-(1-\alpha)}y = \frac{1}{\Gamma(1-\alpha)} \int_0^\infty t^{-\alpha}S(t)y \, dt \quad (8.2.22)
$$

(see **(7.4)**), thus (8.2.21) and Lemma 5.0.17 imply that $(-A)$ can be applied to both sides of (8.2.22) and inserted under the integral on the right,

$$(-A)(-A)^{-(1-\alpha)}y = \frac{1}{\Gamma(1-\alpha)} \int_0^\infty t^{-\alpha}(-A)S(t)y\,dt,$$

so that $y \in D((-A)^\alpha)$ with $\|y\|_\alpha = \|(-A)^\alpha y\| \le C\|y\|_{T'(1,-\alpha;E,S)}$. This ends the proof.   ∎

Lemma 8.2.6. *Let E be a Hilbert space and A a self adjoint operator with $\sigma(A) \subseteq (-\infty, -\kappa]$, $\kappa > 0$. Then*

$$E_\alpha \approx (E, D(A))_{\alpha,2} \quad (0 < \alpha < 1). \tag{8.2.23}$$

Proof. Let $P(d\lambda)$ be the resolution of the identity for A (Theorem 5.0.18). Then, using the (bounded) functional calculus for A in **5.0** we have

$$\|AS(t)y\|^2 = \int_{\sigma(A)} (\lambda e^{\lambda t})^2 \|P(d\lambda)y\|^2 \quad (y \in E)$$

so that, formally,

$$\int_0^\infty (t^\nu \|AS(t)y\|)^2\, dt = \int_{\sigma(A)} \left(\int_0^\infty (\lambda t^\nu e^{\lambda t})^2\, dt \right) \|P(d\lambda)y\|^2$$

$$= 2^{-(2\nu+1)}\Gamma(2\nu+1) \int_{\sigma(A)} (-\lambda)^{1-2\nu} \|P(d\lambda)y\|^2$$

$$= 2^{-(2\nu+1)}\Gamma(2\nu+1) \int_{\sigma(A)} ((-\lambda)^\alpha)^2 \|P(d\lambda)y\|^2$$

$$= 2^{-(2\nu+1)}\Gamma(2\nu+1) \|(-A)^\alpha y\|^2$$

with $\alpha = 1/2 - \nu$. This equality is justified on the basis of Tonelli's theorem if either the integral on the left side or the integral on the right side are finite.   ∎

Example 8.2.7. Another semigroup with strict inclusion $T(\infty, 0; D(A), E) \supset D(A)$ (see Example 8.1.10), but this one is analytic. Let $E = C_0(-\infty, \infty)$ be the space of all functions continuous in $-\infty < x < \infty$ with $\lim_{|x|\to\infty} y(x) = 0$ equipped with the supremum norm, and let $(Ay)(x) = y''(x)$, $D(A)$ the subspace of all twice continuously differentiable $y(\cdot) \in E$ with $y''(\cdot) \in E$. Then A generates the (analytic) Weierstrass semigroup

$$S(t)y(x) = \frac{1}{2\sqrt{\pi t}} \int_{-\infty}^\infty e^{-|x-\xi|^2/4t} y(\xi)\, d\xi$$

and $T(\infty, 0; D(A), E)$ contains all continuously differentiable $y(\cdot) \in E$ with $y'(x)$ absolutely continuous and $y''(\cdot) \in L^\infty(-\infty, \infty)$. To see this it is enough to show that any such function can be approximated pointwise by a sequence $y_n(\cdot) \in D(A)$ in such a way that $\|y_n\|_{L^\infty(-\infty,\infty)} \le (1 + o(1))\|y\|_{L^\infty(-\infty,\infty)}$. This can be achieved by chopping off and mollifying.

8.3. Interpolation and Sobolev Spaces

Recall (5.6) that, given a domain $\Omega \subseteq \mathbb{R}^m$ the *Sobolev space* $W^{k,p}(\Omega)$ ($k = 1, 2, \ldots$, $1 \leq p < \infty$) consists of all functions $y(\cdot)$ having distributional derivatives $\partial^\alpha y$ of order $|\alpha| \leq k$ in $L^p(\Omega)$, equipped with the norm

$$\|y\|_{W^{k,p}(\Omega)} = \left(\sum_{|\alpha| \leq k} \|\partial^\alpha y\|^p_{L^p(\Omega)} \right)^{1/p}. \tag{8.3.1}$$

The space $W^{k,p}(\Omega)$ is a Banach space. This definition is extended to fractional differentiation orders with the **Sobolev-Slobodetzki** spaces $W^{s,p}(\Omega)$, where $s \geq 0$ and $1 \leq p < \infty$. By definition, $W^{o,p}(\Omega) = L^p(\Omega)$ and $W^{s,p}(\Omega) = W^{k,p}(\Omega)$ if $s = k =$ integer ≥ 1. For $0 < s < 1$, $W^{s,p}(\Omega)$ consists of all $y(\cdot) \in L^p(\Omega)$ such that

$$\|u\|^p_{W^{s,p}(\Omega)} = \|u\|^p_{L^p(\Omega)} + I_{s,p}(y) = \|u\|^p_{L^p(\Omega)} + \int_{\Omega \times \Omega} \frac{|y(x) - y(\xi)|^p}{|x - \xi|^{m+sp}} dx d\xi < \infty \tag{8.3.2}$$

outfitted with the norm $\| \cdot \|_{W^{s,p}(\Omega)}$. For $s > 1$ not an integer, $W^{s,p}(\Omega)$ consists of all $y(\cdot) \in W^{[s],p}(\Omega)$ ($[s] =$ integer part of s) such that

$$\|u\|^p_{W^{s,p}(\Omega)} = \|u\|^p_{W^{[s],p}(\Omega)} + \sum_{|\alpha| \leq [s]} I_{s-[s],p}(\partial^\alpha y) < \infty \tag{8.3.3}$$

equipped with $\|u\|_{W^{s,p}(\Omega)}$. The spaces $W^{s,p}(\Omega)$ are Banach spaces for $1 \leq p < \infty$ and $s > 0$. See Adams [1975, p. 204 ff.] for details.

We recall (5.6) that a domain $\Omega \subseteq \mathbb{R}^m$ is *of class* $C^{(k)}$ (k an integer ≥ 0) if, given any point $\bar{x}$ on the boundary Γ there exists an open neighborhood V of $\bar{x}$ in $\mathbb{R}^m$ and a map $\phi : \overline{V} \to \overline{B} =$ closed unit ball of radius 1 in $\mathbb{R}^m$ such that ϕ is one-to-one and onto $\overline{B}$ with $\phi(\bar{x}) = 0$, ϕ (resp. ϕ^{-1}) possesses partial derivatives of order $\leq k$ in V (resp. in S) continuous in the respective closures, and

$$\phi(V \cap \Omega) = \{x \in S; x_m \geq 0\}, \quad \phi(V \cap \Gamma) = \{x \in S; x_m = 0\}. \tag{8.3.4}$$

The set V is called a **boundary patch** and ϕ the corresponding **boundary map**. The domain Ω is **of class** $C^{(\infty)}$ if it is of class $C^{(k)}$ for all $k \geq 0$. Working in a bounded domain of class $C^{(\infty)}$ makes it possible to apply the theory of elliptic operators in **7.6** as well as all the interpolation and imbedding results needed below. The "template" for the latter is Theorem 8.3.1. It follows from the definitions that $W^{s_1,p}(\Omega) \hookrightarrow W^{s_0,p}(\Omega)$ if $s_1 \geq s_0$, so that the definition of the interpolation spaces $(E, F)_{\alpha,p}$ applies to $E = W^{s_0,p}(\Omega)$, $F = W^{s_1,p}(\Omega)$.

Theorem 8.3.1. *Let* $0 \leq s_0 < s_1 < \infty$, $s = (1 - \alpha)s_0 + \alpha s_1$. *Then*

$$(W^{s_0,p}(\Omega), W^{s_1,p}(\Omega))_{\alpha,p} = W^{s,p}(\Omega) \tag{8.3.5}$$

if s is not an integer.

For the proof of a more general theorem, see Triebel [1978, pp. 242, 318]; when $s_0 = 0$, $s_1 = 1$, (8.3.5) is often taken as definition of the spaces $W^{s,p}(\Omega)$ (see Adams [1975, p. 204]). We need versions of this result that apply to spaces including boundary conditions (such as domains of differential operators), thus we must pay attention to Sobolev-Slobodetzki spaces on the boundary Γ, which spaces were implicit in the taking of boundary conditions in Theorems 7.6.1 and 7.6.3.

Using the existence of a boundary patch containing every $\bar{x} \in \Gamma$ and the compactness of Γ we deduce the existence of a finite collection of boundary patches $V_1, V_2, \ldots, V_n$ covering Γ. Using Adams [1975, Theorem 3.14, p. 51] we construct a **partition of unity** $\{\omega_1(\cdot), \omega_2(\cdot), \ldots, \omega_n(\cdot)\}$ **subordinated** to the covering $\{V_1, V_2, \ldots, V_m\}$, that is, a collection of nonnegative infinitely differentiable functions $\omega_j(\cdot)$ in $\mathbb{R}^m$ such that $\text{supp}(\omega_j) \subset V_j$ and $\Sigma \omega_j(x) = 1$ on Γ. The operator θ_j, acting on functions $y(x)$ defined on the boundary Γ, is

$$(\theta_j y)(\hat{x}) = \begin{cases} \omega_j(\phi_j^{-1}(\hat{x}, 0)) y(\phi_j^{-1}(\hat{x}, 0)) & (\|\hat{x}\| \leq 1) \\ 0 & (\|\hat{x}\| > 1) \end{cases} \tag{8.3.6}$$

where $\hat{x} = (x_1, x_2, \ldots, x_{m-1})$ is the point x bereft of its last coordinate. For $s \geq 0$, $1 \leq p < \infty$ the space $W^{s,p}(\Gamma)$ consists of all $y(\cdot)$ defined a.e. on Γ and such that each $\theta_j y$ belongs to $W^{s,p}(\mathbb{R}^{m-1})$; the norm of $W^{s,p}(\Gamma)$ is

$$\|u\|_{W^{s,p}(\Gamma)} = \left(\sum_{j=1}^{n} \|\theta_j u\|_{W^{s,p}(\mathbb{R}^{m-1})}^p \right)^{1/p}, \tag{8.3.7}$$

and we check that the definition (modulo equivalent norms) does not depend on the particular collection of boundary patches, associated maps and subordinated partitions of unity. It can also be shown that restrictions of functions in the space $\mathcal{D}(\mathbb{R}^m)$ (of test functions in $\mathbb{R}^m$) to Ω (resp. to Γ) are dense in $W^{s,p}(\Omega)$ (resp. in $W^{s,p}(\Gamma)$) (Adams [1975, p. 215]).

We consider two **trace operators**, namely

$$\gamma_0 y = y|_\Gamma, \quad \gamma_1 y = \partial^\nu y. \tag{8.3.8}$$

In the first, $|_\Gamma$ indicates restriction to Γ. The second trace operator is the conormal derivative $\partial^\nu y = \Sigma \Sigma a_{jk}(x) \eta_k(x) \partial_j y$ associated to one of the uniformly elliptic operators in **7.6**. The trace operators are defined in (the restriction to Ω of) $\mathcal{D}(\mathbb{R}^m)$.

Theorem 8.3.2. Let $1 < p < \infty$, $0 < s \leq 2$. (a) If $s > 1/p$ then

$$\|\gamma_0 y\|_{W^{s-1/p,p}(\Gamma)} \leq C \|y\|_{W^{s,p}(\Omega)} \quad (y \in \mathcal{D}(\mathbb{R}^m)). \tag{8.3.9}$$

(b) If $s > 1 + 1/p$ then

$$\|\gamma_1 y\|_{W^{s-1-1/p,p}(\Gamma)} \leq C \|y\|_{W^{s,p}(\Omega)} \quad (y \in \mathcal{D}(\mathbb{R}^m)). \tag{8.3.10}$$

The proof is in Adams [1975, p. 216].

Theorem 8.3.2 and the denseness of $\mathcal{D}(\mathbb{R}^m)$ in $W^{s,p}(\Omega)$ make it possible to extend the trace operators γ_0 and γ_1 to $W^{s,p}(\Omega)$ preserving their boundedness properties. Accordingly, the Dirichlet boundary condition $y = 0$ can be meaningfully imposed if $s > 1/p$. For a variational boundary condition we require $s > 1 + 1/p$.

A version of Theorem 8.3.1 that heeds boundary conditions is the result below, where $W_\beta^{s,p}(\Omega)$ indicates the subspace of $W^{s,p}(\Omega)$ whose elements satisfy a boundary condition β. If β is a variational boundary condition $\partial^\nu y(x) = \gamma(x)y(x)$, we assume the coefficient $\gamma(x)$ infinitely differentiable on Γ.

Theorem 8.3.3. *Let* $0 < \alpha < 1$, $k = 1, 2, \ldots$, $k\alpha$ *not an integer,* $1 < p < \infty$.
(a) If β *is the Dirichlet boundary condition,*

$$
\left(L^p(\Omega), W_\beta^{k,p}(\Omega)\right)_{\alpha,p} \approx \begin{cases} W_\beta^{k\alpha,p}(\Omega) & \text{if } k\alpha > 1/p \\ W^{k\alpha,p}(\Omega) & \text{if } k\alpha < 1/p, \end{cases} \tag{8.3.11}
$$

(b) if β *is a variational boundary condition,*

$$
\left(L^p(\Omega), W_\beta^{k,p}(\Omega)\right)_{\alpha,p} \approx \begin{cases} W_\beta^{k\alpha,p}(\Omega) & \text{if } k\alpha > 1 + 1/p \\ W^{k\alpha,p}(\Omega) & \text{if } k\alpha < 1 + 1/p. \end{cases} \tag{8.3.12}
$$

For (references to) a proof of a result much more general than Theorem 8.3.3 see Triebel [1978, p. 321].[1]

For k integer, define $C^{(k)}(\overline{\Omega})$ as the space of all k times differentiable functions with partial derivatives continuous in $\overline{\Omega}$, equipped with the norm

$$
\|y\|_{C^{(k)}(\overline{\Omega})} = \sum_{|\alpha| \leq k} \max_{x \in \overline{\Omega}} |\partial^\alpha y(x)|. \tag{8.3.13}
$$

For $s > 0$ not an integer, $C^{(s)}(\overline{\Omega})$ is the space of all functions $y(\cdot) \in C^{([s])}(\overline{\Omega})$ such that

$$
\|y\|_{C^{(s)}(\overline{\Omega})} = \|y\|_{C^{([s])}(\overline{\Omega})} + \sum_{|\alpha|=[s]} \sup_{x,y \in \overline{\Omega}, x \neq y} \frac{|\partial^\alpha y(x) - \partial^\alpha y(\xi)|}{|x - \xi|^{s-[s]}} < \infty \tag{8.3.14}
$$

(see Triebel [1978, pp. 200, 324]). The spaces $C^{(s)}(\overline{\Omega})$ are Banach spaces for any $s > 0$.

---

[1] This result requires the boundary condition to be *normal* as defined in Triebel [1978, p. 321]. The Dirichlet boundary condition is normal. For a variational boundary condition β, normality means that the conormal vector (in whose direction the derivative is taken) should be nowhere tangential, that is, should have nonzero scalar product with the normal vector. This is an immediate consequence of ellipticity.

Theorem 8.3.4. *Let* $0 \leq r \leq s, 1 < p \leq q$. *Then* (a) *if* $s - m/p \geq r - m/q$
we have

$$W^{s,p}(\Omega) \hookrightarrow W^{r,q}(\Omega), \tag{8.3.15}$$

(b) *if* $s - m/p > r$ *we have*

$$W^{s,p}(\Omega) \hookrightarrow C^{(r)}(\overline{\Omega}). \tag{8.3.16}$$

The last imbedding holds with $s - m/p = r$ *if* r *is not an integer.*

The proof of part (*a*) and of part (*b*) (the latter with r integer) is in Adams [1975, p. 217]. For a complete proof see Triebel [1978, p. 327–328].

8.4. Parabolic Equations

We consider

$$y'(t) = A_c(\beta)y(t) + f(t, y(t)), \quad y(0) = \zeta \tag{8.4.1}$$

where

$$Ay = \sum_{j=1}^{m} \sum_{k=1}^{m} \partial^j (a_{jk}(x)\partial^k y) + \sum_{j=1}^{m} b_j(x)\partial^j y + c(x)y \tag{8.4.2}$$

and $A_c(\beta)$ is the restriction in $C_0(\overline{\Omega})$ or in $C(\overline{\Omega})$ associated with the boundary condition β. We apply the theory of semilinear abstract parabolic equations in Chapter 7 to the space $E = C(\overline{\Omega})$ and the operator $A = A_c(\beta)$; recall that $E^* = \Sigma(\overline{\Omega})$, $E^{\odot} = L^1(\Omega), A_c(\beta)^{\odot} = A'_1(\beta'), S(t) = S_c(t; A, \beta), S^{\odot}(t) = S_1(t, A', \beta')$, where A' and β' are the adjoints of the operator and the boundary condition respectively (see **7.6** and **7.7**). We may assume (8.1.7) by translating $A_c(\beta)$ (and antitranslating $f(t, y(t))$). This makes possible the construction of fractional powers. Although the spaces $E_\alpha = D((-A_c(\beta))^\alpha)$ are not easy to characterize, only imbeddings are needed.

Our starting point are the imbeddings (8.2.20) combined with the result below, where A is the infinitesimal generator of a strongly continuous semigroup satisfying (8.1.7).

Lemma 8.4.1. *We have*

$$(E, D(A))_{\alpha+\varepsilon,p} \hookrightarrow (E, D(A))_{\alpha,1}$$
$$(1 \leq p \leq \infty, 0 < \alpha < \alpha + \varepsilon < 1), \tag{8.4.3}$$
$$(E, D(A))_{\alpha,\infty} \hookrightarrow (E, D(A))_{\alpha-\varepsilon,p}$$
$$(1 \leq p \leq \infty, 0 < \alpha - \varepsilon < \alpha < 1). \tag{8.4.4}$$

Proof. In view of (8.2.2) the two imbeddings are, respectively,

$$T(p, 1 - \alpha - \varepsilon - 1/p; E, S) \hookrightarrow T(1, -\alpha; E, S) \tag{8.4.5}$$

$$T(\infty, 1 - \alpha; E, S) \hookrightarrow T(p, 1 - \alpha + \varepsilon - 1/p; E, S). \tag{8.4.6}$$

To prove the first for $1 < p < \infty$ we use Hölder's inequality:

$$\int_0^1 t^{-\alpha - 1} \|S(t)y - y\| dt$$

$$= \int_0^1 t^{\varepsilon - (1 - 1/p)} t^{-\alpha - \varepsilon - 1/p} \|S(t)y - y\| dt$$

$$\leq \left(\int_0^1 t^{\varepsilon q - 1} dt \right)^{1/q} \left(\int_0^1 (t^{-\alpha - \varepsilon - 1/p} \|S(t)y - y\|)^p dt \right)^{1/p}$$

for $y \in T(p, 1 - \alpha - \varepsilon - 1/p; E, S)$ with $1/p + 1/q = 1$ (see the proof of Lemma 8.2.3 on irrelevance of the integral in $t \geq 1$). The proof for $p = \infty$ is similar and the statement is obvious for $p = 1$.

On the other hand, let $\varepsilon > 0$ and $1 \leq p < \infty$. If $y \in T(\infty, 1 - \alpha; E, S)$ then $t^{-\alpha} \|S(t)y - y\|$ is bounded in $0 \leq t < \infty$. We have

$$(t^{-\alpha + \varepsilon - 1/p} \|S(t)y - y\|)^p = (t^{-\alpha} \|S(t)y - y\|)^p t^{\varepsilon p - 1}$$

Thus we obtain the imbedding (8.4.6) integrating in $0 \leq t \leq 1$. For $p = \infty$ the statement is obvious. ∎

Combining Lemma 8.4.1 with Lemma 8.2.5 we obtain

Corollary 8.4.2. *Let* $0 < \alpha - \varepsilon < \alpha < \alpha + \varepsilon < 1, 1 \leq p \leq \infty$. *Then*

$$(E, D(A))_{\alpha + \varepsilon, p} \hookrightarrow D((-A)^\alpha) \hookrightarrow (E, D(A))_{\alpha - \varepsilon, p}. \tag{8.4.7}$$

The result below is an immediate consequence of the first definition of the trace spaces $T(p, v; F, E)$ and $T(p, v; F, E)$ in **8.1**. The first two imbeddings in (8.4.8) are to permit the definition of these spaces.

Lemma 8.4.3. *Let* E, F, E_1, F_1 *be four Banach spaces such that*

$$F \hookrightarrow E, F_1 \hookrightarrow E_1, E_1 \hookrightarrow E, F_1 \hookrightarrow F. \tag{8.4.8}$$

Then

$$T(p, v; F_1, E_1) \hookrightarrow T(p, v; F, E). \tag{8.4.9}$$

Corollary 8.4.4. *Let* $0 < \alpha - \varepsilon < \alpha < 1, 2\alpha - 2\varepsilon$ *not an integer,* $1 < p < \infty$. *Then (a) if* β *is the Dirichlet boundary condition,*

$$D((-A_c(\beta))^\alpha) \hookrightarrow \begin{cases} W_\beta^{2\alpha - 2\varepsilon, p}(\Omega) & \text{if } 2\alpha - 2\varepsilon > 1/p \\ W^{2\alpha - 2\varepsilon, p}(\Omega) & \text{if } 2\alpha - 2\varepsilon < 1/p, \end{cases} \tag{8.4.10}$$

(b) *if β is a variational boundary condition,*

$$D((-A_c(\beta))^\alpha) \hookrightarrow \begin{cases} W_\beta^{2\alpha-2\varepsilon,p}(\Omega) & \text{if } 2\alpha - 2\varepsilon > 1 + 1/p \\ W^{2\alpha-2\varepsilon,p}(\Omega) & \text{if } 2\alpha - 2\varepsilon < 1 + 1/p. \end{cases} \qquad (8.4.11)$$

Proof. We have $E = C(\overline{\Omega}) \hookrightarrow L^p(\overline{\Omega}) = E_1$; on the other hand, it follows from the definitions of operators and domains that $F = D(A_c(\beta)) \hookrightarrow D(A_p(\beta)) = W_\beta^{2,p}(\Omega) = F_1$. By Lemma 8.4.2,

$$D((-A_c(\beta))^\alpha) \hookrightarrow (C(\overline{\Omega}), D(A_c(\beta)))_{\alpha-\varepsilon,p} \hookrightarrow (L^p(\Omega), W_\beta^{2,p}(\Omega))_{\alpha-\varepsilon,p}$$

so that the result is a direct consequence of Theorem 8.3.3 with $k = 2$. ∎

Corollary 8.4.7 below is an imbedding in the opposite sense, and it will not be used until Chapter 11. It needs a couple of auxiliary results. In the first, $1 < p < \infty$ and, as before, E, F are two Banach spaces with $F \hookrightarrow E$. If $0 < \alpha < \delta < 1$ we have

$$(t^{-\alpha-1/p}\|S(t)y - y\|)^p \, dt = t^{(\delta-\alpha)p}(t^{-\delta-1/p}\|S(t)y - y\|)^p,$$

so that, arguing as in the proof of Lemma 8.4.1 we deduce that

$$(E, F)_{\delta,p} \hookrightarrow (E, F)_{\alpha,p} \qquad (8.4.12)$$

and intermediate spaces between $(E, F)_{\alpha,p}$ and $(E, F)_{\delta,p}$ can be defined. The result below, called the Reiteration Theorem, identifies these new spaces as intermediate between E and F.

Theorem 8.4.5. *Let $0 < \alpha < \delta < 1$, $0 < \theta < 1$. Then*

$$((E, F)_{\alpha,p}, (E, F)_{\delta,p})_{\theta,p} \approx (E, F)_{\theta\delta+(1-\theta)\alpha,p}.$$

The proof of more general theorems can be found in Triebel [1978, p. 63] or Butzer–Berens [1967, p. 177].

Corollary 8.4.6. *Let $0 < \alpha < 1$, $0 < \varepsilon < 1$, $2\alpha + \varepsilon$ not an integer. (a) If β is the Dirichlet boundary condition and $0 < \varepsilon < 1/p$, then*

$$\left(W^{\varepsilon,p}(\Omega), W_\beta^{2+\varepsilon,p}(\Omega)\right)_{\alpha,p} \approx \begin{cases} W_\beta^{2\alpha+\varepsilon,p}(\Omega) & \text{if } 2\alpha + \varepsilon > 1/p \\ W^{2\alpha+\varepsilon,p}(\Omega) & \text{if } 2\alpha + \varepsilon < 1/p. \end{cases} \qquad (8.4.13)$$

(b) *If β is the Dirichlet boundary condition and $1/p < \varepsilon < 1$, then*

$$\left(W_\beta^{\varepsilon,p}(\Omega), W_\beta^{2\alpha+\varepsilon,p}(\Omega)\right)_{\alpha,p} \approx W_\beta^{2\alpha+\varepsilon}(\Omega). \qquad (8.4.14)$$

(c) *If β is a variational boundary condition,*

$$\left(W^{\varepsilon,p}(\Omega), W_\beta^{2+\varepsilon,p}(\Omega)\right)_{\alpha,p} \approx \begin{cases} W_\beta^{2\alpha+\varepsilon,p}(\Omega) & \text{if } 2\alpha + \varepsilon > 1 + 1/p \\ W^{2\alpha+\varepsilon,p}(\Omega) & \text{if } 2\alpha + \varepsilon < 1 + 1/p. \end{cases} \qquad (8.4.15)$$

Proof. We use Theorem 8.3.3 to express the spaces $W^{\varepsilon,p}(\Omega)$, $W_\beta^{2+\varepsilon,p}(\Omega)$ as intermediate spaces between $L^p(\Omega)$ and $W_\beta^{3,p}(\Omega)$. If β is the Dirichlet boundary condition,

$$W^{\varepsilon,p}(\Omega) \approx \left(L^p(\Omega), W_\beta^{3,p}(\Omega)\right)_{\varepsilon/3,p} \quad (0 < \varepsilon < 1/p)$$

$$W_\beta^{\varepsilon,p}(\Omega) \approx \left(L^p(\Omega), W_\beta^{3,p}(\Omega)\right)_{\varepsilon/3,p} \quad (1/p < \varepsilon < 1)$$

(see (8.4.10); the relevant parameter is $k\alpha = 3(\varepsilon/3) = \varepsilon$) and

$$W_\beta^{2+\varepsilon,p}(\Omega) \approx \left(L^p(\Omega), W_\beta^{3,p}(\Omega)\right)_{(2+\varepsilon)/3,p}$$

in all cases, since $k\alpha = 3((2+\varepsilon)/3) = 2+\varepsilon > 1/p$.

On the other hand, if β is a variational boundary condition,

$$W^{\varepsilon,p}(\Omega) \approx \left(L^p(\Omega), W_\beta^{3,p}(\Omega)\right)_{\varepsilon/3,p}$$

in all cases, since $k\alpha = 3(\varepsilon/3) = \varepsilon < 1 + 1/p$; also

$$W_\beta^{2+\varepsilon,p}(\Omega) \approx \left(L^p(\Omega), W_\beta^{3,p}(\Omega)\right)_{(2+\varepsilon)/3,p}$$

in all cases, since $3((2+\varepsilon)/3) = 2+\varepsilon > 1 + 1/p$. Applying then the Reiteration Theorem 8.4.5 we obtain

$$\left(W_\beta^{\varepsilon,p}(\Omega), W_\beta^{2+\varepsilon,p}(\Omega)\right)_{\alpha,p}$$
$$\approx \left(\left(L^p(\Omega), W_\beta^{3,p}(\Omega)\right)_{\varepsilon/3,p}, \left(L^p(\Omega), W_\beta^{3,p}(\Omega)\right)_{(2+\varepsilon)/3,p}\right)_{\alpha,p}$$
$$\approx \left(\left(L^p(\Omega), W_\beta^{3,p}(\Omega)\right)_{(2\alpha+\varepsilon)/3,p} \approx W_\beta^{2\alpha+\varepsilon,p}(\Omega) \quad (8.4.16)$$

for a variational boundary condition; in the last step, Theorem 8.3.3 has been applied again and preservation of the boundary condition results from the inequality $3((2\alpha + \varepsilon)/3) = 2\alpha + \varepsilon > 1 + 1/p$ postulated in the first line of (8.4.15). When $2\alpha + \varepsilon < 1 + 1/p$ the boundary condition is lost, and the space on the right side of (8.4.16) is $W^{2\alpha+\varepsilon,p}(\Omega)$.

If β is the Dirichlet boundary condition and $1/p < \varepsilon < 1$, (8.4.16) applies verbatim. When $2\alpha + \varepsilon > 1/p$ the boundary condition is preserved, and is lost when $2\alpha + \varepsilon < 1/p$, where the space on the right hand side is replaced by $W^{2\alpha+\varepsilon,p}(\Omega)$.

Finally, if β is the Dirichlet boundary condition and $\varepsilon < 1/p$, the space $W_\beta^{\varepsilon,p}(\Omega)$ on the left side of (8.4.16) is replaced by $W^{\varepsilon,p}(\Omega)$ and the space on the right hand side by $W^{2\alpha+\varepsilon,p}(\Omega)$ when $2\alpha + \varepsilon < 1/p$. ∎

Corollary 8.4.7. *Let* $0 < \alpha < \alpha + \varepsilon < 1$, $2\alpha + 2\varepsilon$ *not an integer,* $\varepsilon > m/p$. *Then*
(a) If β *is the Dirichlet boundary condition,*

$$W_\beta^{2\alpha+2\varepsilon,p}(\Omega) \hookrightarrow D((-A_c(\beta))^\alpha). \quad (8.4.17)$$

(b) If β is a variational boundary condition,

$$W_\beta^{2\alpha+2\varepsilon,p}(\Omega) \hookrightarrow D((-A_c(\beta))^\alpha) \quad \text{if } 2\alpha + 2\varepsilon > 1 + 1/p$$
$$W^{2\alpha+2\varepsilon,p}(\Omega) \hookrightarrow D((-A_c(\beta))^\alpha) \quad \text{if } 2\alpha + 2\varepsilon < 1 + 1/p \tag{8.4.18}$$

Proof. We do first the Dirichlet boundary condition. Since $\varepsilon - m/p > 0$, the imbedding (8.3.16) applies, that is $W^{\varepsilon,p}(\Omega) \hookrightarrow C(\overline{\Omega})$ and $W^{2+\varepsilon,p}(\Omega) \hookrightarrow C^{(2)}(\overline{\Omega})$. We can refine both imbeddings to

$$W_\beta^{\varepsilon,p}(\Omega) \hookrightarrow C_0(\overline{\Omega}), \qquad W_\beta^{2+\varepsilon,p}(\Omega) \hookrightarrow C_\beta^{(2)}(\overline{\Omega}) \tag{8.4.19}$$

(if the boundary condition β is satisfied in the sense of the traces (8.3.8) it is satisfied classically.) Since $C_\beta^{(2)}(\overline{\Omega}) \hookrightarrow D(A_c(\beta))$, we also have $W_\beta^{2+\varepsilon,p}(\Omega) \hookrightarrow D(A_c(\beta))$ and it follows from Corollary 8.4.6 (*b*) and Corollary 8.4.2 that

$$W_\beta^{2\alpha+2\varepsilon,p}(\Omega) \approx \left(W^{\varepsilon,p}(\Omega), W_\beta^{2+\varepsilon,p}(\Omega)\right)_{\alpha+\varepsilon/2,p}$$
$$\hookrightarrow (C_0(\overline{\Omega}), D(A_c(\beta)))_{\alpha+\varepsilon/2,p} \hookrightarrow D((-A_c(\beta))^\alpha)$$

(note that, since $\varepsilon > m/p$ we are automatically in the range $2(\alpha + \varepsilon/2) + \varepsilon = 2\alpha + 2\varepsilon > 1/p$). For a variational boundary condition the first imbedding (8.4.19) becomes

$$W^{\varepsilon,p}(\Omega) \hookrightarrow C(\overline{\Omega});$$

the second is the same. The only difference is that now we use the first imbedding (8.4.15) if $2\alpha + 2\varepsilon > 1 + 1/p$, the second if $2\alpha + 2\varepsilon < 1 + 1/p$. ∎

We consider a nonlinear term in (8.4.1) of the form

$$f(t, y(\cdot))(x) = \phi(t, x, y(x), \partial_1 y(x), \ldots, \partial_m y(x))$$
$$= \phi(t, x, y(x), \nabla y(x)), \tag{8.4.20}$$

the function $\phi(t, x, y, y_1, \ldots, y_m) = \phi(t, x, y, \mathbf{y})$ defined in $[0, T] \times \overline{\Omega} \times \mathbb{R} \times \mathbb{R}^m$. Unless ϕ is independent of the last m variables, f is not defined in all of $C(\overline{\Omega})$. All we do in this section is to figure out conditions on ϕ which guarantee that the nonlinear operator (8.4.20) satisfies Hypotheses $\text{II}_{\alpha,\rho}$, $\text{II}_{\alpha,\rho}(wm)$ in **7.7** or their counterparts Hypotheses $\text{III}_{\alpha,\rho}$, $\text{III}_{\alpha,\rho}(wm)$ in **7.10** for suitable α, ρ. We only need consider the (wm)-hypotheses, since the others are the same modulo a change in ρ (see Corollary 7.7.12 and Lemma 7.10.3).

Sometimes there is some freedom in the choice of α, ρ, and there are two competing interests at play. One is to take α as large as possible, at least large enough that constraints on the gradient of the state are tractable. However, assuming the state constraints do not affect the gradient, the other is to take $\alpha = 0$, since this is the case where global existence results are more natural (this may be due to the "physical" nature of the supremum norm; see the opening lines of **7.7**).

The first two results are for the case $\rho = 0$.

Theorem 8.4.8. *Assume that* $\phi(t, x, y, y_1, \ldots, y_m) = \phi(t, x, y, \mathbf{y})$ *is measurable in* t, x *for* y, $\mathbf{y}$ *fixed and that for every* $c > 0$ *there exist* $K(c)$, $L(c)$ *such that*

$$|\phi(t, x, y, \mathbf{y})| \leq K(c) \quad (0 \leq t \leq T, x \in \Omega, |y|, \|\mathbf{y}\| \leq c)^{(1)} \tag{8.4.21}$$

$$|\phi(t, x, y', \mathbf{y}') - \phi(t, x, y, \mathbf{y})| \leq L(c)(|y' - y| + \|\mathbf{y}' - \mathbf{y}\|)$$
$$(0 \leq t \leq T, x \in \Omega, |y|, |y'|, \|\mathbf{y}\|, \|\mathbf{y}'\| \leq c) \tag{8.4.22}$$

and let $\alpha > 1/2$. *Then the operator* $f : [0, T] \times E_\alpha \to L^\infty(\Omega)$ *satisfies Hypothesis* $II_{\alpha,0}(wm)$. *Equivalently, the operator*

$$g(t, \eta)(\cdot) = f(t, (-A_c(\beta))^{-\alpha}\eta) \tag{8.4.23}$$

from $[0, T] \times C(\overline{\Omega})$ *or* $[0, T] \times C_0(\overline{\Omega})$ *into* $L^\infty(\Omega)$ *is an* $L^1(\Omega)$-*weakly measurable* $L^\infty(\Omega)$-*valued function for each* η *and, for every* $c > 0$ *there exist* $K(c)$, $L(c)$ *such that*

$$\|g(t, \eta)\|_{L^\infty(\Omega)} \leq K(c) \quad (0 \leq t \leq T, \|\eta\|_{C(\overline{\Omega})} \leq c) \tag{8.4.24}$$

$$\|g(t, \eta') - g(t, \eta)\|_{L^\infty(\Omega)} \leq L(c)\|\eta' - \eta\|_{C(\overline{\Omega})}$$
$$(0 \leq t \leq T, \|\eta\|_{C(\overline{\Omega})}, \|\eta'\|_{C(\overline{\Omega})} \leq c). \tag{8.4.25}$$

Proof. If β is a variational boundary condition, $(-A_c(\beta))^{-\alpha}$ is a bounded operator from $C(\overline{\Omega})$ onto $D((-A_c(\beta))^\alpha)$. By Corollary 8.4.4 (a), $D((-A_c(\beta))^\alpha) \hookrightarrow W^{2\alpha-2\varepsilon,p}(\Omega)$ for $\varepsilon > 0$, $2\alpha - 2\varepsilon$ not an integer, $1 < p < \infty$. Choose ε so small and p so large that $2\alpha - 2\varepsilon - m/p > 1$. Then by Theorem 8.3.5, $W^{2\alpha-2\varepsilon,p}(\Omega) \hookrightarrow C^{(1)}(\overline{\Omega})$, so that $(-A_c(\beta))^{-\alpha}$ is a bounded operator from $C(\overline{\Omega})$ into $C^{(1)}(\overline{\Omega})$. This makes obvious both estimates (8.4.24) and (8.4.25). The weak measurability statement is that $t \to \int_\Omega z(x)g(t, \eta(\cdot))(x)\,dx$ is measurable for every $\eta(\cdot) \in C(\overline{\Omega})$ and every $z(\Omega) \in L^1(\Omega)$, so it is sufficient to show that

$$t \to \int_\Omega z(x)\phi(t, x, y(x), y_1(x), \ldots, y_m(x))\,dx \tag{8.4.26}$$

is measurable for $z(\cdot) \in L^1(\Omega)$, $y(\cdot)$, $y_j(\cdot) \in C(\overline{\Omega})$. An approximation argument reveals that it is enough to prove this for $z(\cdot)$ countably valued and $y(\cdot)$, $y_j(\cdot)$ finitely valued, thus we may assume that $z(\cdot)$, $y(\cdot)$, $y_1(\cdot)$ are constant, in which case (8.4.26) is obviously measurable. For the Dirichlet boundary condition we replace $C(\overline{\Omega})$ by $C_0(\overline{\Omega})$. ∎

Theorem 8.4.9. *Assume that* $\phi(t, x, y, y_1, \ldots, y_m) = \phi(t, x, y, \mathbf{y})$ *is differentiable with respect to* $y, y_1, \ldots, y_m$ *for* t, x *fixed, with* $\nabla_{(y,\mathbf{y})}\phi(t, x, y, \mathbf{y})$ *continuous*

---

(1) $\|\mathbf{y}\|$ is the Euclidean norm of $\mathbf{y}$. Same for other finite dimensional vectors.

in $y, \mathbf{y}$ *uniformly with respect to* $x \in \Omega$ *for* t *fixed. Moreover, let* $\phi(t, x, y, \mathbf{y})$, $\nabla_{(y,\mathbf{y})}\phi(t, x, y, \mathbf{y})$ *be measurable in* t, x *for* $y, \mathbf{y}$ *fixed. Finally, assume that for every* $c > 0$ *there exists* $K(c), L(c)$ *such that*

$$|\phi(t, x, y, \mathbf{y})| \le K(c) \qquad (0 \le t \le T, x \in \Omega, |y|, \|\mathbf{y}\| \le c) \qquad (8.4.27)$$

$$\|\nabla_{(y,\mathbf{y})}\phi(t, x, y', \mathbf{y}')\| \le L(c) \qquad (0 \le t \le T, x \in \Omega, |y|, \|\mathbf{y}\| \le c). \qquad (8.4.28)$$

Then the operator (8.4.20) *satisfies Hypothesis* $III_{\alpha,\rho}(wm)$. *Equivalently, the operator* (8.4.23) *from* $C(\overline{\Omega})$ *into* $L^\infty(\Omega)$ *is* $L^1(\Omega)$-*weakly measurable in* t *for* y *fixed and has a Fréchet derivative* $\partial_\eta g(t, \eta)$ *given by*

$$(\partial_\eta g(t, \eta(\cdot))h(\cdot))(x)$$
$$= \partial_0 \phi(t, x, ((-A_c(\beta))^{-\alpha}\eta)(x), \nabla((-A_c(\beta))^{-\alpha}\eta)(x))((-A_c(\beta))^{-\alpha}h)(x)$$
$$+ \Sigma_j \partial_j \phi(t, x, ((-A_c(\beta))^{-\alpha}\eta)(x), \nabla((-A_c(\beta))^{-\alpha}\eta)(x))\partial_j((-A_c(\beta))^{-\alpha}h)(x),$$
$$(8.4.29)$$

where $\partial_0, \partial_1, \ldots, \partial_m$ *are the derivatives with respect to the last* $m + 1$ *variables in* ϕ. *For each* $\upsilon \in C(\overline{\Omega})$ $\partial_\eta g(t, \eta)\upsilon$ *is* $L^1(\Omega)$-*weakly measurable in* t *for* η *fixed,* $L^\infty(\Omega)$-*continuous in* η *for* t *fixed, and*

$$\|g(t, \eta)\|_{L^\infty(\Omega)} \le K(c) \qquad (0 \le t \le T, \|\eta\|_{C(\overline{\Omega})} \le c), \qquad (8.4.30)$$

$$\|\partial_\eta g(t, \eta)\|_{(C(\overline{\Omega}), L^\infty(\Omega))} \le L(c) \qquad (0 \le t \le T, \|\eta\|_{C(\overline{\Omega})} \le c). \qquad (8.4.31)$$

Proof. Consider the operator $\Phi : C(\overline{\Omega}) \times C(\overline{\Omega})^{m+1} \to L^\infty(\overline{\Omega})$

$$\Phi(t, y(\cdot), y_1(\cdot), \ldots, y_m(\cdot))(x) = \Phi(t, y(\cdot), \mathbf{y}(\cdot))(x)$$
$$= \phi(t, x, y(x), y_1(x), \ldots, y_m(x)).$$

The fact that $(-A_c(\beta))^{-\alpha}$ is a bounded operator from $C(\overline{\Omega})$ into $C^{(1)}(\overline{\Omega})$ reduces the differentiability statement in Lemma 8.4.5 to the proof that Φ is Fréchet differentiable with derivative

$$\partial_{(y,\mathbf{y})}\Phi(t, y(\cdot), \mathbf{y}(\cdot))(h(\cdot), \mathbf{h}(\cdot)) = \partial_0 \phi(t, x, y(x), \mathbf{y}(x))h(x)$$
$$+ \Sigma_j \partial_j \phi(t, x, y(x), \mathbf{y}(x))h_j(x). \qquad (8.4.32)$$

To show this, notice that the mean value theorem implies

$$\Phi(t, x, y(x) + h(x), \mathbf{y}(x) + \mathbf{h}(x)) - \Phi(t, x, y(x), \mathbf{y}(x))$$
$$= \partial_0 \phi(t, x, y(x), \mathbf{y}(x))h(x) + \Sigma_j \partial_j \phi(t, x, y(x), \mathbf{y}(x))h_j(x)$$
$$+ \{\partial_0 \phi(t, x, \theta(x), \boldsymbol{\theta}(x)) - \partial_0 \phi(t, x, y(x), \mathbf{y}(x))\}h(x)$$
$$+ \Sigma_j \{\partial_j \phi(t, x, \theta(x), \boldsymbol{\theta}(x)) - \partial_j \phi(t, x, y(x), \mathbf{y}(x))\}h_j(x),$$

where $(\theta(x), \boldsymbol{\theta}(x))$ belongs to the segment joining the points $(y(x), \mathbf{y}(x))$ and $(y(x) + h(x), \mathbf{y}(x) + \mathbf{h}(x))$ in $\mathbb{R} \times \mathbb{R}^m$. Given the continuity hypothesis on the gradient, the sum of the last two terms is $o(\|h\|_{C(\overline{\Omega})} + \|\mathbf{h}\|_{C(\overline{\Omega})^m})$, so that (8.4.32) is justified and so is (8.4.29). The fact that $\partial_\eta g(t, \eta)v$ is continuous for v fixed is obvious; in fact, $\eta \to \partial_\eta g(t, \eta)$ is continuous in the norm of $(C(\overline{\Omega}), L^\infty(\Omega))$. The modifications for the Dirichlet boundary condition are standard. ∎

Remark 8.4.10. Condition (8.4.21) is of course a consequence of (8.4.22). If we ask only (8.4.21) *and* continuity of $\phi(t, x, y, \mathbf{y})$ in $y, \mathbf{y}$ uniformly with respect to x for t fixed (plus the measurability condition) then f satisfies Hypothesis $I_{\alpha,0}(wm)$. Note finally that, here and in Theorem 8.4.9, "continuity in $y, \mathbf{y}$ uniformly with respect to x" is *not* meant to require uniform continuity in $y, \mathbf{y}$.

Below, $\partial_{j,p}$ is the partial derivative operator $\partial/\partial x_j$ in $L^p(\Omega)$, $1 < p < \infty$, with domain $D(\partial_{j,p}) = W^{1,p}(\Omega)$. The operator $\partial_{j,p,0}$ is the restriction of $\partial_{j,p}$ whose domain consists of all $y(\cdot) \in W^{1,p}(\Omega)$ with

$$y(x)\eta_j(x) = 0 \quad (x \in \Gamma), \tag{8.4.33}$$

$\eta_j(x)$ the j-th component of the outer unit normal vector at Γ (note that Theorem 8.3.2 (a) legalizes the taking of traces on the boundary).

Lemma 8.4.11. *Let Ω be a bounded domain of class $C^{(1)}$, $1/p + 1/q = 1$. Then*

$$\partial_{j,p,0} = -\partial_{j,q}^*. \tag{8.4.34}$$

Proof. Assume that $y(\cdot) \in D(\partial_{j,q}^*)$. Then there exists $z(\cdot) \in L^p(\Omega)$ such that

$$\int_\Omega z(x)v(x)\,dx = \int_\Omega y(x)\partial_j v(x)\,dx \tag{8.4.35}$$

for every $v(\cdot) \in W_0^{1,q}(\Omega)$. Accordingly, (8.4.35) holds for a test function with support in Ω and, by definition, $\partial_j y(\cdot)$ exists in the sense of distributions and equals $-z(\cdot)$. Using (formally) $1/m^{\text{th}}$ of the divergence theorem,

$$\int_\Omega \{\partial_j y(x)v(x) + y(x)\partial_j v(x)\}dx = \int_\Gamma y(x)v(x)\eta_j(x)\,d\sigma \tag{8.4.36}$$

so that the boundary condition must be satisfied since we can take $v(\cdot) \in \mathcal{D}(\mathbb{R}^m)$ arbitrary. Conversely, if we write (8.4.36) for $y(\cdot) \in D(\partial_{j,p,0})$ and $v(\cdot) \in D(\partial_{j,q})$ the right side vanishes. Instead of trying to justify this calculation directly, we recall that $\mathcal{D}(\mathbb{R}^m)$ is dense in $W^{1,q}(\Omega)$ and that $\mathcal{D}(\Omega)$ is dense in $W_0^{1,p}(\Omega)$ (Adams [1975, p. 66]) so that we can use (8.4.36) for sequences of functions in these spaces converging to $y(\cdot)$ and $v(\cdot)$. Theorem 8.3.2 makes it possible to take limits in the boundary integral. ∎

Corollary 8.4.12. *Let $1 < p < \infty$, $\rho > 1/2$.*[2] *Then the operator*

$$(-A_p(\beta))^{-\rho} \partial_{j,p} \tag{8.4.37}$$

(with domain $W_0^{1,p}(\Omega)$) is bounded in $L^p(\Omega)$, that is,

$$\|(-A_p(\beta))^{-\rho} \partial_{j,p} y\|_{L^p(\Omega)} < C\|y\|_{L^p(\Omega)} \quad (y \in W_0^{1,p}(\Omega)). \tag{8.4.38}$$

Proof. Let $1/p + 1/q = 1$, $A_q'(\beta')$ the operator defined from the formal adjoint A' and the adjoint boundary condition β' in $L^q(\Omega)$; we have

$$A_p(\beta) = A_q'(\beta')^* \tag{8.4.39}$$

(see **7.6**). In view of (8.4.7),

$$(-A_q'(\beta'))^{-\rho} L^q(\Omega) = D((-A_q'(\beta'))^\rho) \hookrightarrow (L^q(\Omega), D(A_q'(\beta')))_{\rho-\varepsilon,q}$$
$$\hookrightarrow (L^q(\Omega), W^{2,q}(\Omega))_{\rho-\varepsilon,q} \approx W^{2(\rho-\varepsilon),q} \hookrightarrow W^{1,q}(\Omega)$$

for ε sufficiently small, so that $\partial_{j,q}(-A_q'(\beta'))^{-\rho}$ is a bounded operator in $L^q(\Omega)$. Using the fact that $((-A_q'(\beta'))^{-\rho})^* = (-A_p(\beta))^{-\rho}$ (consequence of (8.4.39)) and the second equality (5.0.25), we obtain

$$(-A_p(\beta))^{-\rho} \partial_{j,p,0} = -((-A_q'(\beta'))^{-\rho})^* \partial_{j,q}^* \subseteq -(\partial_{j,q}(-A_q'(\beta'))^{-\rho})^*. \tag{8.4.40}$$

Since the domain of $\partial_{j,p,0}$ contains $W_0^{1,p}(\Omega)$, (4.38) follows. ∎

Taking into account that $W_0^{1,p}(\Omega)$ is dense in $L^p(\Omega)$, Corollary 8.4.11 shows that the operator (8.4.37) admits a unique extension as a bounded operator (denoted with the same symbol) from $L^p(\Omega)$ into $L^p(\Omega)$. To construct the corresponding extension in $L^\infty(\Omega)$, we proceed in an indirect way. It follows from the definition of both operators that

$$A_c(\beta) \subseteq A_p(\beta) \tag{8.4.41}$$

for $1 \leq p < \infty$, that is, that $A_c(\beta)y = A_p(\beta)y$ if $y \in D(A_c(\beta))$. Hence, in the same sense,

$$S_c(t; A, \beta) \subseteq S_p(t; A, \beta) \tag{8.4.42}$$

and (from formula (7.4.2)),

$$(-A_c(\beta))^{-\rho} \subseteq (-A_p(\beta))^{-\rho}. \tag{8.4.43}$$

Given $\rho > 1/2$ choose ν with $1/2 < \nu < \rho$, and then p so large that

$$2(\rho - \nu) - m/p > 0. \tag{8.4.44}$$

---

[2] The theorem is in fact true with $\alpha = 1/2$. See next section.

Due to Corollary 8.4.11,

$$\|(-A_p(\beta))^{-\nu}\partial_{j,p}y\|_{L^p(\Omega)} \leq C\|y\|_{L^p(\Omega)} \quad (y \in L^p(\Omega)). \tag{8.4.45}$$

Now, by Corollary 8.4.2, Theorem 8.3.3 and Theorem 8.3.4 (b),

$$(-A_p(\beta))^{-(\rho-\nu)}L^p(\Omega) = D((-A_p(\beta))^{\rho-\nu})$$
$$\hookrightarrow W^{2(\rho-\nu)-2\varepsilon,p}(\Omega) \hookrightarrow C(\overline{\Omega}). \tag{8.4.46}$$

For the first imbedding we need $2(\rho - \nu) - 2\varepsilon \neq$ integer, and for the second $2(\rho - \nu) - 2\varepsilon - m/p > 0$. Both these conditions hold for ε sufficiently small, the first since $0 < \rho - \nu < 1/2$, the second as a consequence of (8.4.44).

It follows from (8.4.45) and (8.4.46) that

$$\|(-A_p(\beta))^{-\rho}\partial_{j,p}y\|_{C(\overline{\Omega})} \leq C\|y\|_{L^p(\Omega)} \quad (y \in L^p(\Omega)). \tag{8.4.47}$$

Denote by $\partial_{j,c}$ the partial derivative operator with domain $C^{(1)}(\overline{\Omega})$. Obviously $\partial_{j,c} \subseteq \partial_{j,p}$; hence, we deduce from Corollary 8.4.12 that

Corollary 8.4.13. *Assume that* $\rho > 1/2$, $p > m/(2\rho - 1)$. *Then the operator* $(-A_p(\beta))^{-\rho}\partial_{j,c}$ *(domain* $C^{(1)}(\overline{\Omega})$*) has a unique extension as a bounded operator from* $L^p(\Omega)$ *into* $C(\overline{\Omega})$*; in particular, this extension is a bounded operator from* $L^\infty(\Omega)$ *into* $C(\overline{\Omega})$*, that is,*

$$\|(-A_p(\beta))^{-\rho}\partial_{j,c}y\|_{C(\overline{\Omega})} \leq C\|y\|_{L^\infty(\Omega)} \quad (y \in L^\infty(\Omega)). \tag{8.4.48}$$

In the following result we stretch a bit Hypothesis $III_{0,\rho}$. Instead of assuming that $f(t, y)$ maps E into $D(((-A^\odot)^{-\rho})^*)$ we ask $f(t, y)$ to be defined in a dense subspace $D \subseteq E$, and to take values in E. Moreover, we require the function

$$g(t, y) = (-A)^{-\rho}f(t, y) \tag{8.4.49}$$

(domain D) to be continuous in the norm of E, so that $g(t, y)$ can be uniquely extended (with the same name) to E. We then assume that $\partial_y g(t, y)$ exists and that it satisfies the assumptions pertaining to Hypothesis $III_{0,\rho}$ (see (7.10.13) and following lines). All of this amounts to the actual Hypothesis $III_{0,\rho}$ for an extension of f. In fact, let $\{y_n\} \subseteq E$ be such that $y_n \to y$. Then $(-A)^{-\rho}f(t, y_n) \to g(t, y)$; it follows that $\{f(t, y_n)\}$ is convergent in $E_{-\rho}$. We can then extend $f(t, y)$ (with the same name) to a function $f : E \to E_{-\rho}$ satisfying Hypothesis $III_{0,\rho}$; (8.4.49) is good for the extensions.

This is the way Hypothesis $III_{0,\rho}$ is understood in the result below, where we also take advantage of (8.4.43).

Corollary 8.4.14. *Let* $\phi_j : [0, T] \times \mathbb{R} \to \mathbb{R}$ $(1 \leq j \leq m)$ *be differentiable with respect to* y. *Moreover, assume that* $\phi_j(t, y)$, $\partial_y \phi_j(t, y)$ *are continuous with respect* *to y for t fixed, measurable in t for y fixed, and*

$$|\phi_j(t, y)| \leq K(c) \quad (0 \leq t \leq T, |y| \leq c, j = 1, \ldots, m) \qquad (8.4.50)$$

$$|\partial_y \phi_j(t, y)| \leq L(c) \quad (0 \leq t \leq T, |y|, |y'| \leq c, j = 1, 2, \ldots, m). \qquad (8.4.51)$$

Consider the operator

$$f(t, y(\cdot))(x) = \Sigma_j \partial_{x_j} \phi_j(t, y(x)) = \Sigma_j \partial_y \phi_j(t, y(x)) \partial_{x_j} y(x) \qquad (8.4.52)$$

with domain $C^{(1)}(\overline{\Omega})$. Then $f(t, y)$ satisfies Assumption $III_{0,\rho}$ for $\rho > 1/2$. Precisely, the operator

$$g(t, y) = (-A_c(\beta))^{-\rho} f(t, y) \qquad (8.4.53)$$

(domain $C^{(1)}(\overline{\Omega})$) has an extension to $C(\overline{\Omega})$ (denoted with the same symbol) taking values in $C(\overline{\Omega})$. The operator $g(t, y)$ is strongly measurable in t for y fixed, continuous in y for t fixed and has a Fréchet derivative $\partial_y g(t, y)$ such that for each z, $\partial_y g(t, y)z$ is strongly measurable in t for y fixed and continuous in y for t fixed, and

$$\|g(t, y)\|_{C(\overline{\Omega})} \leq K(c) \quad (0 \leq t \leq T, \|y\|_{C(\overline{\Omega})} \leq c) \qquad (8.4.54)$$

$$\|\partial_y g(t, y)\|_{(C(\overline{\Omega}), C(\overline{\Omega}))} \leq L(c) \quad (0 \leq t \leq T, \|y\|_{C(\overline{\Omega})} \leq c). \qquad (8.4.55)$$

Proof. The operator $g(t, y)$ is a sum of operators of the form $(-A_p(\beta))^{-\rho} \partial_{j,c} \Phi_j$, where Φ_j is the operator $\Phi_j(t, y(\cdot))(x) = \phi_j(t, y(x))$ mapping $C(\overline{\Omega})$ into $C(\overline{\Omega})$. In view of Corollary 8.4.12, it is enough to show corresponding properties for each $\Phi_j : C(\overline{\Omega}) \to C(\overline{\Omega})$. This is elementary and left to the reader. Note that (as customary) $y \to \partial_y g(t, y)$ is continuous in the norm of $(C(\overline{\Omega}), C(\overline{\Omega}))$. ∎

8.5. Fractional Powers and the Complex Interpolation Method

Fractional powers of certain infinitesimal generators can be exactly characterized using a different interpolation method. We don't need this in the $C(\overline{\Omega})$-treatment of parabolic equations, but it has important applications to the Navier-Stokes equations in next section.

As in **8.1** and **8.2**, E and F are Banach spaces with $F \hookrightarrow E$; we also assume F dense in E. Denote by $\mathcal{F}(E, F)$ the space of all E-valued functions $f(z)$ analytic in $0 < \mathrm{Re}\, z < 1$, E-continuous and bounded in $0 \leq \mathrm{Re}\, z \leq 1$ and such that $f(1 + i\tau) \in F$ and is continuous and bounded in F. The space $\mathcal{F}(E, F)$ is equipped with

$$\|f\| = \max \left\{ \sup_{-\infty \leq \tau \leq \infty} \|f(i\tau)\|_E, \sup_{-\infty \leq \tau \leq \infty} \|f(1 + i\tau)\|_F \right\}. \qquad (8.5.1)$$

All properties of a norm are obvious except that $\|f\| = 0$ implies $f = 0$. This follows from

Theorem 8.5.1 (Hadamard's three lines theorem). *Let $f(\cdot)$ be a Banach space valued function analytic in $0 < \operatorname{Re} z < 1$ and continuous and bounded in $0 \leq \operatorname{Re} z \leq 1$. Let $M(\theta) = \sup_{-\infty \leq \tau \leq \infty} \|f(\theta + i\tau)\|$. Then*

$$M(\theta) \leq M(0)^{1-\theta} M(1)^{\theta} \quad (0 \leq \theta \leq 1).$$

For a proof in the scalar case see Dunford–Schwartz [1958, p. 520]; it applies equally well to vector valued functions. The space $\mathcal{F}_0(E, F) \subseteq \mathcal{F}(E, F)$ consists of all $f(\cdot) \in \mathcal{F}(E, F)$ with $\lim_{|\tau| \to \infty} \|f(\tau)\|_E = \lim_{|\tau| \to \infty} \|f(1 + i\tau)\|_F = 0$ and inherits the norm of $\mathcal{F}(E, F)$. Both spaces contain more than the zero element; $e^{bz}y$ belongs to $\mathcal{F}(E, F)$ for b real and $y \in F$ and $e^{az^2} e^{bz} y$ belongs to $\mathcal{F}_0(E, F)$ for $a > 0$, b real and $y \in F$.

The three lines theorem implies that $\mathcal{F}(E, F), \mathcal{F}_0(E, F)$ are Banach spaces. In fact if $\{f_n(\cdot)\}$ is a Cauchy sequence in $\mathcal{F}(E, F)$ then $f_n(\theta + i\tau) - f_m(\theta + i\tau) \to 0$ uniformly in the norm of E for $\theta = 0$, uniformly in the norm of F for $\theta = 1$; in view of the embedding $F \hookrightarrow E$ and the three lines theorem, the Cauchy condition holds uniformly in the norm of E in $0 \leq \operatorname{Re} z \leq 1$, and we conclude that $f_n(\cdot)$ converges to an element $f(\cdot) \in \mathcal{F}(E, F)$ in the norm of $\mathcal{F}(E, F)$. Since $\mathcal{F}_0(E, F)$ is a closed subspace of $\mathcal{F}(E, F)$ it is also a Banach space.

Let $0 \leq \theta \leq 1$. We define a space $[E, F]_\theta$ and its norm as follows:

$$[E, F]_\theta = \{y \in E; y = f(\theta) \text{ for some } f \in \mathcal{F}_0(E, F)\}, \tag{8.5.2}$$

$$\|y\|_{[E,F]_\theta} = \inf\{\|f\|; f(\cdot) \in \mathcal{F}_0(E, F); f(\theta) = y\}. \tag{8.5.3}$$

The space $[E, F]_\theta$ is a Banach space. In fact, all properties required of a norm are evident, and completeness follows from completeness of $\mathcal{F}_0(E, F)$.

Lemma 8.5.2. *We have*

$$F \hookrightarrow [E, F]_\theta \hookrightarrow E \quad (0 \leq \theta \leq 1). \tag{8.5.4}$$

Proof. It follows from the three lines theorem and the imbedding $F \hookrightarrow E$ that if $f(\cdot) \in \mathcal{F}_0(E, F)$ satisfies $f(\theta) = y$ then $\|y\|_E = \|f(\theta)\|_E \leq M(\theta) \leq M(0)^{1-\theta} M(1)^{\theta} \leq C\|f\|$, so that $\|y\|_E \leq C\|y\|_{[E,F]_\theta}$. To verify the first imbedding note that if $y \in F$ then the function $f(z) = e^{(z-\theta)^2} y$ belongs to $\mathcal{F}_0(E, F)$ with $\|f\| \leq C(\|y\|_E + \|y\|_F) \leq C'\|y\|_F$ and satisfies $f(\theta) = y$ so that $\|y\|_{[E,F]_\theta} \leq C'\|y\|_F$. ∎

Lemma 8.5.3. *We have*

$$[E, F]_0 \approx E, \quad [E, F]_1 \approx F. \tag{8.5.5}$$

Proof. We have the imbedding $F \hookrightarrow [E, F]_1$ from Lemma 8.5.2; the opposite imbedding $[E, F]_1 \hookrightarrow F$ is obvious. Lemma 8.5.2 also provides $[E, F]_0 \hookrightarrow E$.

For the opposite imbedding, since F is dense in E it is enough to show inequality of norms for $y \in F$. To do this we take $f(z) = e^{az^2} y$ with $a > 0$, $y \in F$. Then $f(0) = y$ and $\|f\| \le \|y\|_E + c(a)\|y\|_F$ with $c(a) \to 0$ as $a \to \infty$, so that $\|y\|_{[E,F]_0} \le \|y\|_E$.   ∎

If $\mathcal{H}$ is a space of (complex valued) functions and X a complex linear space, $\mathcal{H} \otimes X$ indicates the space of all finite sums

$$\sum h_j(z)y_j \quad (h_j(\cdot) \in \mathcal{H}, \, y_j \in X).$$

Below, $\mathcal{H}$ is the space of all finite complex linear combinations $\sum c_j e^{a_j z^2} e^{b_j z}$ with $a_j > 0$, b_j real.

Lemma 8.5.4. *$\mathcal{H} \otimes F$ is dense in $\mathcal{F}_0(E, F)$.*

The proof is technical and can be found in Krein–Petunin–Semenov [1982, p. 220].

Corollary 8.5.5. *Let $0 < \theta < 1$. Then F is dense in $[E, F]_\theta$.*

Proof. Let $y \in [E, F]_\theta$, $f \in \mathcal{F}_0(E, F)$ with $f(\theta) = y$, $g(\cdot) \in \mathcal{H} \otimes F$ with $\|f - g\| \le \varepsilon$. Then $\|y - g(\theta)\|_{[E,F]_\theta} \le \varepsilon$.   ∎

We write $\mathcal{F}(\mathbb{C}, \mathbb{C}) = \mathcal{F}$, $\mathcal{F}_0(\mathbb{C}, \mathbb{C}) = \mathcal{F}_0$ for the spaces of scalar-valued functions. Note that Lemma 8.5.4 and the three lines theorem implies that, given $\varepsilon > 0$ and $f(\cdot) \in \mathcal{F}_0$ there exists $g(\cdot) \in \mathcal{H}$ such that

$$\sup_{0 \le \theta \le 1, -\infty < \tau < \infty} |f(\theta + i\tau) - g(\theta + i\tau)| \le \varepsilon$$

Lemma 8.5.6. *Let $y \in F$, $0 \le \theta \le 1$. Then*

$$\|y\|_{[E,F]_\theta} = \inf\{\|f\|; \, f(\cdot) \in \mathcal{F}_0 \otimes F, \, f(\theta) = y\}. \tag{8.5.6}$$

Proof. Denote provisionally by $\|y\|'_\theta$ the norm defined by the right-hand side of (8.5.6). Since $\mathcal{F}_0 \otimes F \subseteq \mathcal{F}_0(E, F)$, $\|y\|_{[E,F]_\theta} \le \|y\|'_\theta$.

Given $\varepsilon > 0$, let $f(\cdot) \in \mathcal{F}_0(E, F)$ be such that $f(\theta) = y$, $\|f\| \le \|y\|_{[E,F]_\theta} + \varepsilon$. Then choose a function $g(z)$ mapping conformally the strip $0 \le \operatorname{Re} z \le 1$ into the unit disk $|z| \le 1$ and satisfying $g(\theta) = 0$, $g'(\theta) = 1$. The modulus $|g(z)|$ is bounded away from zero outside any disk with center θ and positive radius, so that the function

$$g(z)^{-1}(f(z) - e^{a(z-\theta)^2} y)$$

belongs to $\mathcal{F}_0(E, F)$. Use Lemma 8.5.4 to construct a function $h(\cdot) \in \mathcal{F}_0 \otimes F$ such that

$$\|g(\cdot)^{-1}(f(\cdot) - e^{a(\cdot -\theta)^2}y) - h(\cdot)\| \le \varepsilon.$$

Multiplying by $g(z)$ we obtain

$$\|(f(\cdot) - e^{a(\cdot -\theta)^2}y) - g(\cdot)h(\cdot)\| \le \varepsilon.$$

Set $f_0(z) = e^{a(z-\theta)^2}y + g(z)h(z)$, so that $f_0(\cdot) \in \mathcal{F}_0 \otimes F$. Moreover, $f_0(\theta) = y$ and

$$\|f_0\| \le \|f\| + \|f_0 - f\| \le \|y\|_{[E,F]_\theta} + 2\varepsilon.$$

This ends the proof. ∎

Lemma 8.5.7. *Let X be a subspace of F dense in F (in the norm of F), $0 \le \theta \le 1$. Then, for every $y \in X$,*

$$\|y\|_{[E,F]_\theta} = \inf\{\|f\|;\, f(\cdot) \in \mathcal{H} \otimes X,\, f(\theta) = y\}. \tag{8.5.7}$$

Proof. By Lemma 8.5.6 there exists $f(\cdot) \in \mathcal{F}_0 \otimes F$ such that $f(\theta) = y$, $\|f\| \le \|y\|_{[E,F]_\theta} + \varepsilon$. Write $f(z) = \Sigma f_j(z)y_j$ $(f_j(\cdot) \in \mathcal{F}_0, y_j \in F)$, choose $x_j \in X$ and (using Lemma 8.5.4) $g_j \in \mathcal{H}$ such that, if $g(z) = \Sigma g_j(z)x_j$ then we have $\|f(\cdot) - g(\cdot)\| \le \varepsilon$. Set $f_0(z) = g(z) + e^{a(z-\theta)^2}(f(\theta) - g(\theta))$ with $a > 0$. Then $f_0 \in \mathcal{H} \otimes F$, $f_0(\theta) = y$ and $\|f_0(\cdot)\| \le \|g(\cdot)\| + C\varepsilon$ for a large enough. ∎

To work with the fractional powers $(-A)^\alpha$ in the present context, we need them defined for α complex. This presents no problem; if $S(\cdot)$ is an arbitrary strongly continuous semigroup the integral (7.4.2) makes sense for $\text{Re}\,\zeta > 0$ and

$$\|(-A)^{-\zeta}\|_{(E,E)} \le C_\varepsilon \quad (\text{Re}\,\zeta \ge \varepsilon > 0). \tag{8.5.8}$$

Complex fractional powers are extended to the range $0 \le \text{Re}\,\zeta < 1$ by (7.4.3). In particular, the imaginary powers $(-A)^{i\tau}$ are defined, although they are not necessarily bounded operators. We shall need the equality $(-A)^{-z-\zeta} = (-A)^{-z}(-A)^{-\zeta}$ for $\text{Re}\,z, \text{Re}\,\zeta > 0$; in this range, it is an immediate consequence of the definition (7.4.2). It extends to $\text{Re}\,z = \text{Re}\,\zeta = 0$ if the imaginary powers $(-A)^{i\tau}$ are bounded operators. Finally, it also follows from formula (7.4.2) that $\zeta \to (-A)^{-\zeta}$ is a (E, E)-valued analytic function in $\text{Re}\,\zeta > 0$.

In the following result, $D(A)$ is endowed with its customary norm $\|y\|_1 = \|Ay\|_E$; we have the imbedding $D(A) \hookrightarrow E$ and $D(A)$ is dense in E, so that the complex interpolation spaces $[E, D(A)]_\theta$ can be defined.

Theorem 8.5.8. *Let A be the infinitesimal generator of a strongly continuous semigroup satisfying (8.1.7). Then, if $0 \le \alpha - \varepsilon < \alpha + \varepsilon \le 1$,*

$$[E, D(A)]_{\alpha+\varepsilon} \hookrightarrow D((-A)^\alpha) \hookrightarrow [E, D(A)]_{\alpha-\varepsilon}. \tag{8.5.9}$$

If, in addition, $(-A)^{i\tau}$ *is a bounded operator and*

$$\|(-A)^{i\tau}\|_{(E,E)} \leq C \quad (|\tau| \leq \delta > 0),$$ (8.5.10)

then

$$D((-A)^{\alpha}) \approx [E, D(A)]_{\alpha} \quad (0 \leq \alpha \leq 1).$$ (8.5.11)

Proof. Let $y \in D(A)$. Then $y = (-A)^{-\alpha}z$ for some $z \in E$. Approximate z by a sequence $\{z_n\} \subseteq D(A)$ in the norm of E. Then, if $y_n = (-A)^{-\alpha}z_n$, $y_n \in D(A)$ and $\|y_n - y\|_{\alpha} = \|z - z_n\|_E$, which shows that $D(A)$ is dense in $E_{\alpha} = D((-A)^{\alpha})$. On the other hand, $D(A)$ is dense in $[E, D(A)]_{\theta}$ (Corollary 8.5.5). Accordingly, to prove both imbeddings (8.5.9) it is enough to show that

$$\|y\|_{\alpha} = \|(-A)^{\alpha}y\| \leq C\|y\|_{[E,D(A)]_{\alpha+\varepsilon}} \quad (y \in D(A))$$ (8.5.12)

$$\|y\|_{[E,D(A)]_{\alpha-\varepsilon}} \leq \|y\|_{\alpha} = C\|(-A)^{\alpha}y\| \quad (y \in D(A)).$$ (8.5.13)

We prove first the second bound. Define

$$\begin{aligned}
f(z) &= e^{a(z-\alpha+\varepsilon)^2}(-A)^{-(z-\alpha+\varepsilon+1)}(-A)y \\
&= e^{a(z-\alpha+\varepsilon)^2}(-A)^{-(z-\alpha+\varepsilon)}y \\
&= e^{a(z-\alpha+\varepsilon)^2}(-A)^{-(z-\alpha+\varepsilon-1)}(-A)^{-1}y
\end{aligned}$$ (8.5.14)

for $a > 0$ and $y \in D(A)$. We have $\mathrm{Re}(z - \alpha + \varepsilon + 1) \geq 1 - (\alpha - \varepsilon) > 0$, so that it follows from the first expression for $f(\cdot)$ that $f(\cdot) \in \mathcal{F}_0(E, D(A))$. Moreover using the second expression to calculate the maximum of $\|f(i\tau)\|_E$ and the third for the maximum of $\|f(1 + i\tau)\|_{D(A)}$ we obtain

$$\|f\| \leq C \sup_{-\infty < \tau < \infty} e^{-a\tau^2}\|(-A)^{-\varepsilon-i\tau}(-A)^{\alpha}y\|_E \leq C'\|(-A)^{\alpha}y\|_E$$

after (8.5.8). Finally, $f(\alpha - \varepsilon) = y$, so that (8.5.13) stands proved.

To show (8.5.12) we use Lemma 8.5.6 to define the norm $\|y\|_{[E,D(A)]_{\alpha+\varepsilon}}$ in $D(A)$ by means of (8.5.7) with $X = D(A)$. Select $f(\cdot) = \Sigma f_j(\cdot)y_j \in \mathcal{H} \otimes D(A)$ such that $f(\alpha + \varepsilon) = y$ and $\|f\| \leq \|y\|_{[E,D(A)]_{\alpha+\varepsilon}} + \delta$. Then set

$$\begin{aligned}
g(z) &= \Sigma f_j(z)(-A)^{z-\varepsilon}y_j \\
&= \Sigma f_j(z)(-A)^{z-\varepsilon-1}(-A)y_j = (-A)^{z-\varepsilon}f(z).
\end{aligned}$$ (8.5.15)

The estimate (8.5.8) shows that $g(z)$ is bounded in $0 \leq \mathrm{Re}\, z \leq 1$. Also, we have

$$\|g(i\tau)\|_E \leq \|(-A)^{i\tau-\varepsilon}\|_{(E,E)}\|f(i\tau)\|_E$$ (8.5.16)

$$\|g(1 + i\tau)\|_{D(A)} = \|(-A)^{i\tau-\varepsilon}\|_{(E,E)}\|f(1 + i\tau)\|_{D(A)}$$ (8.5.17)

and $(-A)^{\alpha}y = g(\alpha + \varepsilon)$, thus $\|(-A)^{\alpha}y\|_E \leq \|g\| \leq C\|f\| \leq C(\|y\|_{[E,D(A)]_{\alpha-\varepsilon}} + \delta)$. This completes the proof of (8.5.12).

Assume that (8.5.10) holds. Let $\tau \in \mathbb{R}$ and write $\tau = \delta n + \sigma$, n an integer, $|\sigma| \leq \delta$. Then $(-A)^{i\tau} = (-A)^{i\delta n}(-A)^{i\sigma}$, so that

$$\|(-A)^{i\tau}\| \leq \|(-A)^{i\delta}\|^n \|(-A)^{i\sigma}\| \leq C^n \cdot C \leq C' e^{c|\tau|}$$

for suitable c. It follows that: (a) we may take $\varepsilon = 0$ in the definition (8.5.14) of the function $f(\cdot)$, (b) we may take $\varepsilon = 0$ in the definition (8.5.15) of the function $g(z)$. Replacing $\|(-A)^{i\tau-\varepsilon}\|$ by $\|(-A)^{i\tau})\|$ introduces an exponential $e^{c|\tau|}$ in the estimations (8.5.16) and (8.5.17), which is of no consequence in view of the definition of the clase $\mathcal{H}$. The end result reduces to (8.5.9) for $\varepsilon = 0$, which is (8.5.11). ■

Conditions on the domain and on the coefficients of the operator $A_p(\beta)$ $(1 < p < \infty)$ that guarantee that (8.5.10) holds can be found in Prüss–Sohr [1993] and Amman–Hieber–Simonett [1994]. If these hold, we have

$$D((-A_p(\beta))^\alpha) \approx [L^p(\Omega), W_\beta^{2,p}(\Omega)]_\alpha \quad (0 \leq \alpha \leq 1). \tag{8.5.18}$$

It should be noted, however, that complex interpolation between $L^p(\Omega)$ and $W^{2,p}(\Omega)$ does not yield the intermediate spaces $W^{s,p}(\Omega)$ but new spaces $H^{s,p}(\Omega)$. These coincide with $W^{s,p}(\Omega)$ if $p = 2$ (Triebel [1978, p. 179]). Also, if $1 < p < \infty$, $H^{s,p}(\Omega) = W^{s,p}(\Omega)$ for $s = 1, 2, \ldots$ (see Triebel [1978, p. 310]) thus we have

$$D((-A_p(\beta))^{1/2}) \approx [L^p(\Omega), W_\beta^{2,p}(\Omega)]_{1/2} \approx W^{1,p}(\Omega), \tag{8.5.19}$$

($W_0^{1,p}(\Omega)$ for the Dirichlet boundary conditions). This justifies

Corollary 8.5.9. *Assume that the operator $A_p(\beta)$ satisfies (8.5.10). Then Corollary 8.4.12 is true not only for $\rho > 1/2$ but also for $\rho = 1/2$.*

Complex interpolation offers an alternative way to prove the imbeddings between domains of fractional powers and Sobolev-Slobodetski spaces of fractional order used in **8.5**. In fact, we have

$$H^{s+\varepsilon,p}(\Omega) \hookrightarrow W^{s,p}(\Omega) \hookrightarrow H^{s-\varepsilon,p}(\Omega). \tag{8.5.20}$$

See Triebel [1978, p. 180], for these imbeddings. Full details on complex interpolation and the spaces $H^{s,p}(\Omega)$, in particular their intrinsic (interpolation-independent) definitions can also be found in Triebel [1978], especially Chapters 2 and 4. We note that the spaces $H^{s,p}(\Omega)$ have interpolation properties similar to those of the spaces $W^{s,p}$. In particular, there is a twin of Theorem 8.3.4, whose proof can be found in Triebel [1978, pp. 327–328]:

Theorem 8.5.10. *Let* $0 \le r \le s$, $1 < p \le q$. *Then* (a) *if* $s - m/p \ge r - m/q$ *we have*

$$H^{s,p}(\Omega) \hookrightarrow H^{r,q}(\Omega), \qquad (8.5.21)$$

(b) *if* $s - m/p > r$ *we have*

$$H^{s,p}(\Omega) \hookrightarrow C^{(r)}(\overline{\Omega}). \qquad (8.5.22)$$

The imbedding holds with $s - m/p = r$ *if* r *is not an integer.*

There is also an identical twin of Theorem 8.3.3 on interpolation between spaces incorporating boundary conditions; real interpolation is replaced by complex interpolation and the spaces $H^{s,p}(\Omega)$ and $H_\beta^{s,p}(\Omega)$ take the places of $W^{s,p}(\Omega)$ and $W_\beta^{s,p}(\Omega)$. (see Triebel [1978, pp. 320–321]). The interplay among the parameters k, p, α is the same. It catches one's eye that in results that depend on real interpolation (like Theorem 8.3.3) the parameter p in the real interpolation functor must match the parameter in the spaces; on the other hand, the complex interpolation functor does the matching automatically.

8.6. The Navier-Stokes Equations

In suitable units, the velocity $\mathbf{y}(t, x) = (y_1(t, x), \ldots, y_m(t, x))$ and the pressure $p(t, x)$ of a viscous fluid moving in a domain in m-dimensional space $\mathbb{R}^m$ are governed by the **Navier-Stokes equations**

$$\mathbf{y}_t(t, x) = \Delta\mathbf{y}(t, x) - (\mathbf{y}(t, x) \cdot \nabla)\mathbf{y}(t, x)$$
$$+ \mathbf{u}(t, x) - \nabla p(t, x) \quad (x \in \Omega) \qquad (8.6.1)$$
$$\nabla \cdot \mathbf{y}(t, x) = 0 \qquad (x \in \Omega) \qquad (8.6.2)$$
$$\mathbf{y}(t, x) = 0 \qquad (x \in \Gamma) \qquad (8.6.3)$$
$$\mathbf{y}(0, x) = \zeta(x) \quad (x \in \Omega), \qquad (8.6.4)$$

where $\Delta\mathbf{y} = (\Delta y_1, \ldots, \Delta y_m)$, and $\mathbf{u}(t, x) = (u_1(t, x), \ldots, u_m(t, x))$ denotes an external force, considered as a control. The physically significant dimensions are $m = 2$ (plane flow) and $m = 3$ (space flow); dimensions $m > 3$ have apparently only mathematical interest. Since $\int_\Omega \|\mathbf{y}(t, x)\|^2 \, dx$ represents energy, the L^2 norm is natural for the Navier-Stokes equations. However, there are dividends from the L^p treatment, so we keep our options open.

We assume Ω a bounded domain of class $C^{(\infty)}$ in $\mathbb{R}^m$, its boundary denoted by Γ. The space $\mathcal{D}(\Omega; \mathbb{R}^m)$ consists of all m-vectors of test functions in Ω (infinitely differentiable functions with support contained in Ω). We denote by $X^p(\Omega)$ the closure in $L^p(\Omega; \mathbb{R}^m)$ of all $\mathbf{y} \in \mathcal{D}(\Omega; \mathbb{R}^m)$ with $\nabla \cdot \mathbf{y} = 0$. Finally, $G^p(\Omega)$ consists of all vectors ∇p with $p \in W^{1,p}(\Omega)$.

Elaborating on the classical Helmholtz decomposition of vector analysis, Fujiwara–Morimoto [1977] have proved that

$$L^p(\Omega; \mathbb{R}^m) = X^p(\Omega) \oplus G^p(\Omega), \tag{8.6.5}$$

where the direct sum is orthogonal if $p = 2$; if $p \neq 2$ the sum is norm-direct, which means that the projection $\mathbf{P}_p$ of $L^p(\Omega; \mathbb{R}^m)$ into $X^p(\Omega)$ is a bounded operator (then, so is the projection $\mathbf{I} - \mathbf{P}_p$ into $G^p(\Omega)$).

We use below Sobolev spaces $W^{k,p}(\Omega; \mathbb{R}^m)$ of vector-valued functions $\mathbf{y} = (y_1, \ldots, y_m)$, each of whose components y_j belong to the scalar space $W^{k,p}(\Omega)$; these spaces are equipped with the natural product norms such as

$$\|\mathbf{y}\|_{W^{k,p}(\Omega;\mathbb{R}^m)} = \left(\|y\|^p_{W^{k,p}(\Omega)} + \cdots + \|y_m\|^p_{W^{k,p}(\Omega)}\right)^{1/p}.$$

Other spaces are "vectorized" in the same way. The imbedding and interpolation results in the previous sections extend in an obvious manner.

The operator $\mathbf{\Delta}_p$ is the (m-vector) Laplacian in $L^p(\Omega; \mathbb{R}^m)$ with the Dirichlet boundary condition $\mathbf{y} = 0$ on Γ; in other words,

$$\mathbf{\Delta}_p \mathbf{y} = \mathbf{\Delta}_p(y_1, \ldots, y_m) = (\Delta_p y_1, \ldots, \Delta_p y_m)$$

with $D(\mathbf{\Delta}_p) = W_0^{2,p}(\Omega; \mathbb{R}^m) = \{\mathbf{y} \in W^{2,p}(\Omega; \mathbb{R}^m); \ y_j = 0 \text{ on } \Gamma\}$. Finally, the **Stokes operator** $\mathbf{A}_p$ is defined by

$$\mathbf{A}_p = \mathbf{P}_p \mathbf{\Delta}_p \tag{8.6.6}$$

with domain $D(\mathbf{A}_p) = X^p(\Omega) \cap D(\mathbf{\Delta}_p)$. Applying the projection $\mathbf{P}_p$ to the equation (8.6.1) we obtain the semilinear equation

$$\mathbf{y}'(t) = \mathbf{A}_p \mathbf{y}(t) - \mathbf{F}(\mathbf{y}(t)) + \mathbf{B}\mathbf{u}(t), \quad \mathbf{y}(0) = \zeta \tag{8.6.7}$$

for the velocity $\mathbf{y}(t)(x) = \mathbf{y}(t, x)$ in the space $X^p(\Omega)$, where $\mathbf{B} = \mathbf{P}_p$ and (8.6.2) and (8.6.3) are automatically taken care of. We may (and will) make things a little more general by taking $\mathbf{B} : H \to X^p(\Omega)$ (H another Banach space) and assuming the control $\mathbf{u}(t)$ with values in H; $\mathbf{B} = \mathbf{I}$ if $H = X^p(\Omega)$. The nonlinear term is

$$\mathbf{F}(\mathbf{y}) = \mathbf{P}_p(\mathbf{y} \cdot \nabla)\mathbf{y}. \tag{8.6.8}$$

For control problems based on actual physical problems (where control is applied on boundaries by suction and blowing) see the author–Sritharan [1992] and the author–Sritharan [1994]. After some preliminary work, these control systems can be written in the form (8.6.7).

Results on the Stokes operator $\mathbf{A}_p$ turned up first for $p = 2$. It was shown by Sobolevski [1959] (see also Fujita–Kato [1962], Fujita–Kato [1964]) that the Stokes operator $\mathbf{A}_2$ is self adjoint (with negative spectrum) in the Hilbert space

$X^2(\Omega)$. For $1 < p < \infty$, the operator $\mathbf{A}_p$ generates a bounded analytic semi-group $\mathbf{S}_p(\cdot)$ in $X^p(\Omega)$ (Sobolevski [1964], Solonnikov [1977], Von Wahl [1980], Miyakawa [1981], Giga [1981]). We also have (Fujiwara–Morimoto [1977]) the following duality relations:

$$X^p(\Omega)^* = X^q(\Omega), \quad \mathbf{A}_p^* = \mathbf{A}_q, \quad \mathbf{P}_p^* = \mathbf{P}_q, \tag{8.6.9}$$

with $1/p + 1/q = 1$. By means of a translation of $\mathbf{A}_p$ (and the opposite translation in the nonlinear term) we may assume that

$$\|\mathbf{S}_p(t)\| \le M e^{-ct} \quad (t \ge 0)$$

(norm in $(X^p(\Omega), X^p(\Omega))$) so that the theory in **7.7** is automatically applicable if we can show that the nonlinear term $\mathbf{F}(\mathbf{y})$ satisfies Hypothesis II$_{\alpha,\rho}$ for suitable α, ρ. Since we are now in a reflexive space, the theory "cleans up" considerably. Weak measurability bows out, and the Phillips adjoint $\odot$ is replaced by the adjoint $*$. For Hypothesis II$_{\alpha,\rho}$ the nonlinear operator $(-\mathbf{A}_p)^{-\rho}\mathbf{F}(y)$ must be defined in $D((-\mathbf{A}_p)^\alpha)$ and satisfy

$$\|(-\mathbf{A}_p)^{-\rho}\mathbf{F}(\mathbf{y})\| \le K(c) \quad (\|(-\mathbf{A}_p)^\alpha \mathbf{y}\| \le c) \tag{8.6.10}$$

$$\|(-\mathbf{A}_p)^{-\rho}(\mathbf{F}(\mathbf{y}') - \mathbf{F}(\mathbf{y}))\| \le L(c)\|(-\mathbf{A}_p)^\alpha(\mathbf{y}' - \mathbf{y})\|$$
$$(\|(-\mathbf{A}_p)^\alpha \mathbf{y}'\|, \|(-\mathbf{A}_p)^\alpha \mathbf{y}\| \le c) \tag{8.6.11}$$

with $\alpha, \rho \ge 0$, $\alpha + \rho < 1$ (the norms here and below are in $X^p(\Omega)$). Assume we show that

$$\|(-\mathbf{A}_p)^{-\rho}\mathbf{P}_p(\mathbf{y} \cdot \nabla)\mathbf{z}\| \le C\|(-\mathbf{A}_p)^\beta \mathbf{y}\|\|(-\mathbf{A}_p)^\gamma \mathbf{z}\| \tag{8.6.12}$$

for $\mathbf{y} \in D((-\mathbf{A}_p)^\beta)$, $\mathbf{z} \in D((-\mathbf{A}_p)^\gamma)$, with $\beta, \gamma, \rho \ge 0$, $\alpha + \rho < 1$ ($\alpha = \max(\beta, \gamma)$). Then the two estimates above (and more) follow. In fact, (8.6.12) implies (8.6.10) putting $\mathbf{y} = \mathbf{z}$. On the other hand

$$\|(-\mathbf{A}_p)^{-\rho}\mathbf{P}_p(\mathbf{y}' \cdot \nabla)\mathbf{y}' - (-\mathbf{A}_p)^{-\rho}\mathbf{P}_p(\mathbf{y} \cdot \nabla)\mathbf{y}\|$$
$$\le \|(-\mathbf{A}_p)^{-\rho}\mathbf{P}_p((\mathbf{y}' - \mathbf{y}) \cdot \nabla)\mathbf{y}'\| + \|(-\mathbf{A}_p)^{-\rho}\mathbf{P}_p(\mathbf{y} \cdot \nabla)(\mathbf{y}' - \mathbf{y})\|$$
$$\le C\|(-\mathbf{A}_p)^\beta(\mathbf{y}' - \mathbf{y})\|\|(-\mathbf{A}_p)^\gamma \mathbf{y}'\| + C\|(-\mathbf{A}_p)^\beta \mathbf{y}\|\|(-\mathbf{A}_p)^\gamma(\mathbf{y}' - \mathbf{y})\|$$
$$\le C'(\|(-\mathbf{A}_p)^\alpha \mathbf{y}\| + \|(-\mathbf{A}_p)^\alpha \mathbf{y}'\|)\|(-\mathbf{A}_p)^\alpha(\mathbf{y}' - \mathbf{y})\|.$$

The proof of (8.6.12) depends on several auxiliary results. The first (Giga [1985]) is

Theorem 8.6.1. $(-\mathbf{A}_p)^{i\tau}$ *is a bounded operator for all real* τ *and for every* $\varepsilon > 0$ *there exists* C_ε *such that*

$$\|(-\mathbf{A}_p)^{i\tau}\| \le C_\varepsilon e^{\varepsilon|\tau|} \quad (-\infty < \tau < \infty). \tag{8.6.13}$$

This result, combined with Theorem 8.5.8 shows that the domains of the fractional powers $(-\mathbf{A}_p)^\alpha$ can be exactly characterized by complex interpolation,

$$D((-\mathbf{A}_p)^\alpha) \approx [X^p(\Omega), D(\mathbf{A}_p)]_\alpha. \tag{8.6.14}$$

A much more useful expression stemming from general interpolation results (Triebel [1978, p. 118]) and the fact that $D(\mathbf{A}_p) = X^p(\Omega) \cap D(\mathbf{\Delta}_p)$ is given in Giga [1985]:

$$D((-\mathbf{A}_p)^\alpha) \approx X^p(\Omega) \cap [L^p(\Omega; \mathbb{R}^m), D(\mathbf{\Delta}_p)]_\alpha. \tag{8.6.15}$$

The operator $\mathbf{\Delta}_p$ is a particular case of those in Prüss–Sohr [1993] or Amann–Hieber–Simonett [1994], thus it satisfies (8.6.13) and the domains of its fractional powers can also be characterized by complex interpolation. Using (8.6.5) and the $H^{s,p}$-companion of Theorem 8.3.3 (see the comments after Theorem 8.5.10) we obtain

$$D((-\mathbf{A}_p)^\alpha) \approx \begin{cases} X_p(\Omega) \cap H^{2\alpha, p}(\Omega; \mathbb{R}^m) & (0 < \alpha < 1/p) \\ X_p(\Omega) \cap H_0^{2\alpha, p}(\Omega; \mathbb{R}^m) & (1/p < \alpha < 1). \end{cases} \tag{8.6.16}$$

In particular,

$$D((-\mathbf{A}_p)^{1/2}) \approx X_p(\Omega) \cap W_0^{1,p}(\Omega; \mathbb{R}^m). \tag{8.6.17}$$

As in **8.4**, we denote by $\partial_{j,p}$ the partial derivative operator in $L^p(\Omega; \mathbb{R}^m)$, with $D(\partial_{j,p}) = W^{1,p}(\Omega; \mathbb{R}^m)$.

Theorem 8.6.2. *The operator* $(-\mathbf{A}_p)^{-1/2}\mathbf{P}_p\partial_{j,p}$ *(domain* $W_0^{1,p}(\Omega; \mathbb{R}^m)$*) is bounded from* $L_p(\Omega; \mathbb{R}^m)$ *into* $X^p(\Omega)$, *that is,*

$$\|(-\mathbf{A}_p)^{-1/2}\mathbf{P}_p\partial_{j,p}\mathbf{y}\|_{X^p(\Omega)} \le C\|\mathbf{y}\|_{L^p(\Omega; \mathbb{R}^m)}. \tag{8.6.18}$$

The proof is essentially the same as that of Corollary 8.4.11. We generalize first Lemma 8.4.10; as in there, $\partial_{j,p,0}$ denotes the restriction of $\partial_{j,p}$ obtained imposing the boundary condition (8.4.33) in each coordinate, and we have $\partial_{j,p,0} = -\partial_{j,q}^*$. The imbedding (8.6.17) determines that the operator $\partial_{j,q}\mathbf{P}_q(-\mathbf{A}_q)^{-1/2} = \partial_{j,q}(-\mathbf{A}_q)^{-1/2}$ is a bounded operator from $X^q(\Omega)$ into $L^q(\Omega; \mathbb{R}^m)$, so the first duality relation (8.6.9) pins down $(\partial_{j,q}\mathbf{P}_q(-\mathbf{A}_q)^{-1/2})^*$ as a bounded operator from $L^p(\Omega; \mathbb{R}^m)$ into $X^p(\Omega)$. The other duality relations (8.6.9) and the second relation (5.0.25) imply

$$\begin{aligned} (-\mathbf{A}_p)^{-1/2}\mathbf{P}_p\partial_{j,p,0} &= -(\mathbf{P}_q(-\mathbf{A}_q)^{-1/2})^*\partial_{j,q}^* \\ &= -((-\mathbf{A}_q)^{-1/2})^*\partial_{j,q}^* \subseteq -(\partial_{j,q}(-\mathbf{A}_q)^{-1/2})^*. \end{aligned}$$

This ends the proof. ∎

Theorem 8.6.2 allows us to extend $(-\mathbf{A}_p)^{-1/2}\mathbf{P}_p\partial_{j,p}$ to an operator in $(L^p(\Omega; \mathbb{R}^m), X^p(\Omega))$ (which will be given the same name).

The estimations below (following Giga–Miyakawa [1985]) aim to establish (8.6.12). Note first that, by definition of $X^p(\Omega)$ the space $\mathcal{J}(\Omega) = \{\mathbf{y} \in \mathcal{D}(\Omega; \mathbb{R}^m); \nabla \cdot \mathbf{y} = 0\}$ is dense in $X^p(\Omega)$, thus $(-\mathbf{A}_p)^{-1}\mathcal{J}(\Omega)$ is dense in $D(\mathbf{A}_p)$; this means that we may take $\mathbf{y}, \mathbf{z}$ smooth functions in $D(\mathbf{A}_p)$. If $(\mathbf{y} \cdot \nabla)z_k$ is one of the components of $(\mathbf{y} \cdot \nabla)\mathbf{z}$,

$$(\mathbf{y} \cdot \nabla)z_k = \sum_{j=1}^{m} y_j \partial_j z_k = \sum_{j=1}^{m} \partial_j(y_j z_k)$$

since $\nabla \cdot \mathbf{y} = \partial_1 y_1 + \cdots + \partial_m y_m = 0$. Accordingly,

$$\|(-\mathbf{A}_p)^{-1/2}\mathbf{P}_p(\mathbf{y} \cdot \nabla)\mathbf{z}\|^p \leq C \int_\Omega \left(\sum \sum |y_j(x)z_k(x)|^p \right) dx$$
$$\leq C' \left(\int_\Omega \|\mathbf{y}(x)\|^{pr'} dx \right)^{1/r'} \left(\int_\Omega \|\mathbf{z}(x)\|^{pr} dx \right)^{1/r}$$
$$(8.6.19)$$

where we have used Theorem 8.6.2 in the first inequality and Hölder's inequality with $1 < r < \infty$, $1/r + 1/r' = 1$ in the second. We then bring into play (8.6.16), the imbedding

$$H^{m/pr,p}(\Omega) \hookrightarrow L^{pr'}(\Omega) \qquad (8.6.20)$$

coming from Theorem 8.5.10 (a), the fact that $H^{0,p}(\Omega) = L^p(\Omega)$ and the equality

$$s - \frac{m}{p} = \frac{m}{pr} - \frac{m}{p} = \frac{m}{p}\left(\frac{1}{r} - 1 \right) = -\frac{m}{pr'} = 0 - \frac{m}{pr'}.$$

Combining (8.6.20) with (8.6.16) for $\alpha = m/2pr$,

$$D((-\mathbf{A}_p)^{m/2pr}) \hookrightarrow X_p(\Omega) \cap H^{m/pr,p}(\Omega; \mathbb{R}^m) \hookrightarrow L^{pr'}(\Omega; \mathbb{R}^m).$$

Thus we can bound the first integral on the right side of (8.6.19) by a constant times $\|(-\mathbf{A}_p)^{m/2pr}\mathbf{y}\|$. The second integral is estimated in the same way interchanging r by r'. The result is

$$\|(-\mathbf{A}_p)^{-1/2}\mathbf{P}_p(\mathbf{y} \cdot \nabla)\mathbf{z}\| \leq C\|(-\mathbf{A}_p)^{m/2pr}\mathbf{y}\|\|(-\mathbf{A}_p)^{m/2pr'}\mathbf{z}\|. \qquad (8.6.21)$$

This bound is usable, but we can do better. Using Hölder's inequality directly we obtain

$$\|\mathbf{P}_p(\mathbf{y} \cdot \nabla)\mathbf{z}\|^p \leq C \left(\int_\Omega \|\mathbf{y}(x)\|^{pr'} dx \right)^{1/r'} \left(\int_\Omega \|\nabla \mathbf{z}(x)\|^{pr} dx \right)^{1/r}. \qquad (8.6.22)$$

We now use the imbedding $\hookrightarrow$ in (8.6.17), which yields the inequality $\|\nabla \mathbf{z}\| \leq C\|(-\mathbf{A}_p)^{1/2}\mathbf{z}\|$ and estimate (8.6.22) in the same way as (8.6.19):

$$\|\mathbf{P}_p(\mathbf{y} \cdot \nabla)\mathbf{z}\| \leq C\|(-\mathbf{A}_p)^{m/2pr}\mathbf{y}\|\|(-\mathbf{A}_p)^{m/2pr'+1/2}\mathbf{z}\|. \qquad (8.6.23)$$

Consider

$$\Phi(z) = e^{z^2}(-\mathbf{A}_p)^{(z-1)/2}\mathbf{P}_p(\mathbf{y}\cdot\nabla)(-\mathbf{A}_p)^{-(m/2pr'+z/2)}\mathbf{z}.$$

for $z = \sigma + i\tau$. The function e^{z^2} tames the exponential produced by $\|(-\mathbf{A}_p)^{i\tau/2}\|$ and $\|(-\mathbf{A}_p)^{-i\tau/2}\|$ (see (8.6.13)) thus $\Phi(z)$ is holomorphic and bounded in the strip $0 \le \operatorname{Re} z \le 1$. Moreover. taking into account (8.6.21) and (8.6.23) (and using the same name for different constants), $\Phi(z)$ is bounded by $C\|(-A)^{m/2pr}\mathbf{y}\|\|\mathbf{z}\|$ for $\operatorname{Re} z = 0$ and $\operatorname{Re} z = 1$. It then follows from the three lines theorem that it is bounded by $C\|(-A)^{m/2pr}\mathbf{y}\|\|\mathbf{z}\|$ in $0 \le \operatorname{Re} z \le 1$; this implies

$$\|(-\mathbf{A}_p)^{(\theta-1)/2}\mathbf{P}_p(\mathbf{y}\cdot\nabla)\mathbf{z}\| \le C\|(-\mathbf{A}_p)^{m/2pr}\mathbf{y}\|\|(-\mathbf{A}_p)^{m/2pr'+\theta/2}\mathbf{z}\|, \quad (8.6.24)$$

where the parameters θ, p, r can be arbitrarily (and independently) chosen in the range $0 \le \theta \le 1, 1 < p < \infty, 1 < r < \infty$.

Particular cases of this inequality (Giga–Miyakawa [1985]) were previously known for $p = 2$; some are in the following table, whose first two lines are due to Sobolevski [1959] (see also Fujita–Kato [1962], [1964]); the third is in Fujita–Kato [1962].

m	ρ	β	γ
2	1/4	1/4	1/2
3	1/4	1/2	1/2
3	0	1/2	3/4

The parameters ρ, β, γ are those in (8.6.12). The three lines in the table are obtained from (8.6.24) taking $\theta = 1/2, r = 2, \theta = 1/2, r = 3/2$ and $\theta = 1, r = 3/2$. Note that the last line follows from the elementary inequality (8.6.23). For other particular cases, see Giga–Miyakawa [1985].

With (8.6.12) established, the theory in **7.7** can be applied straight to the equation (8.6.7). The integral equation (7.7.18) for $\eta(t) = (-\mathbf{A}_p)^\alpha \mathbf{y}(t)$ is

$$\eta(t) = (-\mathbf{A}_p)^\alpha \mathbf{S}_p(t)\zeta - \int_0^t (-\mathbf{A}_p)^{\alpha+\rho}\mathbf{S}_p(t-\tau)(-\mathbf{A}_p)^{-\rho}\mathbf{F}((-\mathbf{A}_p)^{-\alpha}\eta(\tau))\,d\tau$$

$$+ \int_0^t (-\mathbf{A}_p)^\alpha \mathbf{S}_p(t-\tau)\mathbf{Bu}(\tau)\,d\tau. \quad (8.6.25)$$

Since $\|(-\mathbf{A}_p)^\alpha \mathbf{S}_p(t)\| \le Ct^{-\alpha}$ the last term makes sense if the control $\mathbf{u}(\cdot)$ belongs to $L^q(0, T; H)$ with $q > 1/(1-\alpha)$. However, in the Hilbert space case $p = 2$ certain flow control problems call for $\alpha = 2$ and controls in $L^2(0, T; H)$ (see the author–Sritharan [1992]). In this case, the last integral in (8.6.25) can be given sense according to the following result due to Masuda [1975], whose part (b) adds a footnote to the theory of the operator Λ_α in (7.7.28).

Lemma 8.6.3. *Let A be a self adjoint operator in a Hilbert space H with $\sigma(A) \subseteq (-\infty, 0]$, $S(t) = e^{At}$ the semigroup generated by A. Then (a) for every $y \in H$*

$(-A)^{1/2}S(\cdot)y \in L^2(0, \infty; H)$ and

$$\|(-A)^{1/2}S(\cdot)y\|_{L^2(0,\infty;H)} = \frac{1}{\sqrt{2}}\|y\|, \qquad (8.6.26)$$

(b) the operator

$$(\Lambda f)(t) = \int_0^t S(t - \tau)f(\tau)\,d\tau \quad (0 \le t \le T) \qquad (8.6.27)$$

is bounded from $L^2(0, T; H)$ into $C(0, T; H_{1/2})$ ($H_{1/2} = D((-A)^{1/2})$ equipped with the graph norm $\|y\|_{1/2} = (\|y\|^2 + \|(-A)^{1/2}y\|^2)^{1/2}$).[(1)] In particular,

$$\|(-A)^{1/2}\Lambda f\|_{C(0,T;H_{1/2})} \le \frac{1}{\sqrt{2}}\|f\|_{L^2(0,T;H)}. \qquad (8.6.28)$$

Proof. Let $\{P(d\lambda)\}$ be the resolution of the identity for A (Theorem 5.0.18). Using the functional calculus for self adjoint operators in **5.0**, we have

$$(-A)^{1/2}S(t)y = \int_{-\infty}^0 (-\lambda)^{1/2}e^{\lambda t}P(d\lambda)y,$$

so that

$$\|(-A)^{1/2}S(t)y\|^2 = \int_{-\infty}^0 (-\lambda)e^{2\lambda t}(P(d\lambda)y, y),$$

Integrating with respect to t in $0 \le t \le \infty$ and using Tonelli's theorem,

$$\|(-A)^{1/2}S(\cdot)y\|_{L^2(0,\infty;H)}^2 = \frac{1}{2}\int_{-\infty}^0 (P(d\lambda)y, y) = \frac{1}{2}\|y\|^2$$

which shows (a). To prove (b), let $z \in H_{1/2}$, $y(\cdot) = \Lambda f(\cdot)$. We have

$$(y(t), (-A)^{1/2}z) = \int_0^t \left(f(\tau), (-A)^{1/2}S(t - \tau)z\right)d\tau. \qquad (8.6.29)$$

By Schwarz's inequality,

$$|(y(t), (-A)^{1/2}z)| \le \|f\|_{L^2(0,T;H)}\|(-A)^{1/2}S(\cdot)z\|_{L^2(0,\infty;H)}$$
$$\le \frac{1}{\sqrt{2}}\|f\|_{L^2(0,T;H)}\|z\|_H$$

so that $y(t) \in D((-A)^{1/2})$ for $0 \le t \le T$ and the bound (8.6.28) holds. We must show that $(-A)^{1/2}y(t)$ is continuous. This is done as follows. Let $t < t'$. We have

$$\left(y(t') - y(t), (-A)^{1/2}z\right) = \int_t^{t'} \left(f(\tau), (-A)^{1/2}S(t' - \tau)z\right)d\tau$$
$$+ \int_0^t \left((S(t' - t) - I)f(\tau), (-A)^{1/2}S(t - \tau)z\right)d\tau$$

[(1)] We use the graph norm instead of $\|(-A)^{1/2}y\|$ since the assumptions do not imply that $0 \in \rho(A)$.

so that, arguing as after (8.6.29), we deduce that

$$\|y(t') - y(t)\|_{H^{1/2}} \leq \frac{1}{\sqrt{2}} \|f\|_{L^2(t,t';H)} + \frac{1}{\sqrt{2}} \|(S(t'-t) - I)f\|_{L^2(0,T;H)}.$$

Both integrals on the right converge to zero, the first by equicontinuity of integrals, the second by the dominated convergence theorem. This ends the proof. ∎

Example 8.6.4. According to Theorem 7.7.16, if $S(t)$ is compact for $t > 0$ and $p > 2$ then $\Lambda_{1/2} = (-A)^{1/2}\Lambda$ is a compact operator from $L^p(0, T; H)$ into $C(0, T; H)$. This does not extend to $p = 2$. To see this take H separable, $\{\phi_n\}$ a complete orthonormal sequence in H, $\{\lambda_n\}$ an increasing sequence of positive numbers tending to infinity and $A(\Sigma c_n \phi_n) = -\Sigma \lambda_n c_n \phi_n$, the domain of A consisting of all $y = \Sigma c_n \phi_n$ such that $\Sigma(\lambda_n |c_n|)^2 < \infty$. Then A is self adjoint with spectrum $\{\lambda_n\}$ and $S(t)(\Sigma c_n \phi_n) = \Sigma e^{-\lambda_n t} c_n \phi_n$ is compact for every $t > 0$. If

$$f_n(t) = \sqrt{\lambda_n} \, e^{-\lambda_n t} \phi_n \quad (n = 1, 2, \dots.)$$

then $\|f_n\|_{L^2(0,T;H)} \leq 1/\sqrt{2}$, and we check easily that $f_n(\cdot) \to 0$ weakly in $L^2(0, T; H)$. On the other hand,

$$((-A)^{1/2}\Lambda f_n)(t) = \left(\int_0^t \lambda_n e^{-\lambda_n t} \, d\tau \right) \phi_n = \lambda_n t e^{-\lambda_n t} \phi_n.$$

The maximum of $\lambda_n t e^{-\lambda_n t}$ occurs at $t = 1/\lambda_n$ and equals e^{-1}, thus no subsequence of $\{((-A)^{1/2}\Lambda f_n)(\cdot)\}$ may converge to zero in $C(0, T; H)$.

9

Linear Control Systems

9.1. Linear Systems: The Minimum Principle

Numerous results for linear control systems

$$y'(t) = Ay(t) + Bu(t), \qquad y(0) = \zeta \qquad (9.1.1)$$

were obtained prior to general theories of nonlinear systems. Some of these results have never been completely superseded by their nonlinear counterparts.

We assume in this section that A is the infinitesimal generator of an analytic semigroup $S(\cdot)$ in a Banach space E. In order to deal with parabolic equations in the most interesting cases—L^1 spaces and spaces of continuous functions—we use the $\odot$-adjoint theory in **7.4**. For existence reasons we assume

$$F = X^*, \qquad (9.1.2)$$

where X is a separable Banach space. B is a bounded operator with

(a) $B : X^* \to (E^\odot)^*$, $B^* : E^\odot \to X$,

(b) E is $\odot$-reflexive and separable.

The control constraint for (9.1.1) is

$$u(t) \in U \subseteq F = X^*. \qquad (9.1.3)$$

Admissible controls for (9.2.1) are X^*-valued, X-weakly measurable functions satisfying (9.1.3) a.e.; in other words, $C_{ad}(0, T; U)$ is the subset of $L_w^\infty(0, T; X^*)$ characterized by the control constraint (9.1.3). In other results, the spaces $L_w^p(0, T; X^*)$, $1 \le p < \infty$ will be used. These spaces were introduced in Chapter 7 (see the comments before Lemma 7.7.13) and are studied at length in Chapter 12. We limit ourselves to quote a few results:

Theorem 9.1.1. *Let* $1 \le p < \infty$, $1/p + 1/q = 1$. *Then*

$$L^p(0, T; X)^* = L_w^q(0, T; X^*).$$

For the proof, see Theorems 12.2.11 and 12.9.2. In all cases, the duality pairing of $L^p(0, T; X)^*$ and $L^q_w(0, T; X^*)$ is given by $\langle u(\cdot), f(\cdot) \rangle = \int \langle u(t), f(t) \rangle dt$ for $f(\cdot) \in L^p(0, T; X)$ and $u(\cdot) \in L^q_w(0, T; X^*)$; the integral is given sense approximating $f(\cdot)$ by countably valued functions.

If X is not separable, the spaces $L^q_w(0, T; X^*)$ are rather esoteric (see Chapter 12). Fortunately, the applications in this chapter do not call for this degree of generality. Further simplifications ensue if X^* is separable.

Lemma 9.1.2. (*i*) *Let* X^* *be separable. Then*

$$L^p_w(0, T; X^*) = L^p(0, T; X^*). \tag{9.1.4}$$

(*ii*) *Let* X *be separable. Then functions* $u(\cdot) \in L^p_w(0, T; X^*)$, $1 \le p \le \infty$ *have measurable norm.*

For proofs, see Chapter 12 and also **5.0** for definitions of weak measurability and results connecting it with strong measurability. Part (*ii*) of Lemma 9.1.2 justifies the usual norm in $L^p_w(0, T; X^*)$.

Given $\zeta \in E$ and $u(\cdot) \in L^p_w(0, T; X^*)$, $1 \le p \le \infty$, the solution of (9.1.1) is

$$y(t, \zeta, u) = S(t)\zeta + \int_0^t S^\odot(t - \tau)^* Bu(\tau) d\tau. \tag{9.1.5}$$

If $z \in E^\odot$ then $\langle z, Bu(\tau) \rangle = \langle B^*z, u(\tau) \rangle$, so that $Bu(\cdot)$ is a $(E^\odot)^*$-valued, $E^\odot$-weakly measurable function; then the integrand is a strongly measurable E-valued function (Lemma 7.7.13), so that the integral in (9.1.5) is a Lebesgue-Bochner integral. The solution $y(t)$ takes values in E for $0 \le t \le T$. The integral in (9.1.5) is continuous by Lemma 7.7.13 so that trajectories $y(t, \zeta, u)$ are continuous in $0 \le t \le T$.

The **admissible control space** $C_{ad}(0, \bar{t}; U)$ is the subset of $L^\infty_w(0, T; X^*)$ whose elements satisfy the control constraint (9.1.3) a.e.

We take a look at the time optimal problem. The first result (Theorem 9.1.4) is on existence, and we add the assumption

(*c*) The admissible control space $C_{ad}(0, T; U)$ is $L^1(0, T; X)$-weakly compact in $L^\infty_w(0, T; X^*)$.

Hypothesis (*c*) is satisfied, for instance, if U is the unit ball of X^*; in this case $C_{ad}(0, T; U)$ is the unit ball of $L^\infty_w(0, T; X^*)$, which is $L^1(0, T; X)$-weakly compact by Alaoglu's Theorem 12.0.17. It is easy to see that Hypothesis (*c*) implies convexity of U. In fact,

Example 9.1.3. Assume Hypothesis (*c*) is satisfied. Then U must be convex and X^*-weakly closed. In fact, assume U is not convex; then there exist $u, v \in U$ such that $w = \alpha u + (1 - \alpha)v \notin U$ for some α, $0 < \alpha < 1$. Given $n = 1, 2, \ldots$

partition $[0, T]$ into the subintervals $I_{nk} = [t_{n,k}, t_{n,k+1}) = [kT/n, (k+1)T/n)$
$(k = 0, 1, \ldots, k - 1)$ and define

$$u_n(t) = \begin{cases} u & t_{n,k} \leq t \leq t_{n,k} + \alpha(t_{n,k+1} - t_{n,k}) \\ v & t_{n,k} + \alpha(t_{n,k+1} - t_{n,k}) < t < t_{n,k+1}. \end{cases}$$

Then $u_n(\cdot) \to w(\cdot)$ $L^1(0, T; X)$-weakly in $L_w^\infty(0, T; X^*)$, where $w(\cdot) \equiv w$.

If U is not X-weakly closed there exists a (generalized) sequence $\{u_n\} \subseteq U$ such
that $u_n \to u$ X-weakly, $u \notin U$. Then the constant controls $u_n(\cdot) \equiv u_n$ converge
to the constant control $u(\cdot) \equiv u$ $L^1(0, T; X)$-weakly in $L_w^\infty(0, T; X^*)$.

Theorem 9.1.4. *Assume that $S(t)$ is compact for $t > 0$ and that the target set Y
is closed. Then, if there exists a control driving ζ to Y for some $t > 0$ the time
optimal problem has a solution.*

Proof. The hypotheses make clear that a minimizing sequence $\{u_n(\cdot)\}, u_n(\cdot) \in$
$C_{\text{ad}}(0, t_n; U)$ exists, where $t_n \geq \bar{t} =$ optimal time. Extend $u_n(\cdot)$ (with the same
name) to $[0, T]$ $(T = t_1)$ setting $u_n(t) = u =$ fixed element of U in $t \geq t_n$. Passing
to a subsequence,[1] we may assume that $\{u_n(\cdot)\}$ is $L^1(0, T; X)$-weakly convergent
in $L^\infty(0, T; X^*)$ to $\bar{u}(\cdot) \in C_{\text{ad}}(0, T; U)$. We use Theorem 7.7.16 to show that the
sequence $\{y(t, \zeta, u_n)\}$ is uniformly convergent in $0 \leq t \leq T$ to $y(t, \zeta, \bar{u})$, and is
clear that this trajectory satisfies the target condition. ∎

The **reachable set** $R(\bar{t}, \zeta; U)$ consists of the instantaneous values $y(\bar{t}, \zeta; u)$
of trajectories for all $u(\cdot) \in C_{\text{ad}}(0, \bar{t}; U)$. We also use the **reachable subspace**
$R^p(\bar{t})$ $(1 \leq p \leq \infty)$ of all elements of the form (9.1.5) with $\zeta = 0$ and
$u(\cdot) \in L^p(0, \bar{t}; X^*)$ (no control constraint). $R^p(\bar{t})$ is a Banach space equipped
with the norm

$$\|y\|_{R^p(\bar{t})} = \inf \|u\|_{L_w^p(0,\bar{t};X^*)}, \tag{9.1.6}$$

the infimum taken over all $u \in L^p(0, \bar{t}; X^*)$ with

$$y = \int_0^{\bar{t}} S^\circ(\bar{t} - \tau)^* Bu(\tau)d\tau. \tag{9.1.7}$$

Hölder's inequality implies

$$\|y\|_E \leq C(\bar{t})\|y\|_{R^p(\bar{t})} \tag{9.1.8}$$

for all p; in the jargon of Chapter 8, $R^p(\bar{t}) \hookrightarrow E$ $(\bar{t} > 0, 1 \leq p \leq \infty)$.

---

[1] Recall that, since X is separable, $L^1(0, T; X)$ is separable as well; thus, the $L^1(0, T; E)$-weak
topology of bounded sets in $L_w^\infty(0, T; X^*)$ is given by a metric. We may then avoid generalized
sequences in weak compactness arguments (see **12.0** for details).

Lemma 9.1.5. $R(\bar{t}, \zeta; U)$ *is convex and* E^*-*weakly compact, and the imbedding* $R^p(\bar{t}) \hookrightarrow E$ $(p > 1)$ *is* E^*-*weakly compact. If* $S(t)$ *is compact for* $t > 0$, *then* $R(\bar{t}, \zeta; U)$ *is compact and the imbedding* $R^p(\bar{t}) \hookrightarrow E$ $(p > 1)$ *is compact.*

Proof. Let $\{u_n(\cdot)\} \subseteq L_w^\infty(0, T; X^*)$; select a subsequence $L^1(0, \bar{t}; X)$-weakly convergent to $\bar{u}(\cdot) \in L_w^\infty(0, T; X^*)$. Given $y^* \in E^*$, we have

$$\langle y^*, y(\bar{t}, \zeta, u_n) \rangle = \langle y^*, S(\bar{t})\zeta \rangle + \int_0^{\bar{t}} \langle g^*(\tau), u_n(\tau) \rangle d\tau$$

where, since $S^\odot(\bar{t} - \tau)^* = S(\delta)S^\odot(\bar{t} - \tau - \delta)^*$ for $\delta > 0$ and $t < \bar{t} - \delta$ (see (7.7.9)), the equality $g^*(\tau) = B^* S^\odot(\bar{t} - \tau - \delta)S(\delta)^* y^*$ holds in $t < \bar{t} - \delta$. On account of (a) and of the fact that $\delta > 0$ is arbitrary, the function on the left side of the angled brackets in the integrand is X-valued, bounded and continuous in $\tau < \bar{t}$, thus an element of $L^q(0, \bar{t}; X)$ for any $q \geq 1$; the taking of limits is obvious. The argument for $R^p(\bar{t})$ exploits in a similar way the duality of $L^q(0, \bar{t}; X)$ and $L_w^p(0, \bar{t}; X^*)$. The proof of the statements pertaining to $S(t)$ compact is the same as that of Theorem 9.1.4. ∎

In Lemma 9.1.6 below, H is an arbitrary Banach space and R a linear, possibly unbounded operator with domain $D(R)$ (not necessarily dense in H) and range in H. We assume that R^{-1} exists and is everywhere defined and bounded. This means that R^n (defined in $D(R^n) = (R^{-1})^n H$) is invertible with $R^{-n} = (R^n)^{-1} = (R^{-1})^n$. We define $H_{-n}(R)$ as the completion of H under the norm $\|y\|_{H_{-n}(R)} = \|R^{-n}y\|$; boundedness of R^{-1} implies that

$$H \hookrightarrow H_{-1}(R) \hookrightarrow H_{-2}(R) \hookrightarrow \cdots. \tag{9.1.9}$$

The reader will note that all of this is related to the spaces E_α constructed in Chapter 7; see in particular Example 7.4.17 for the precise meaning of completion.

The operator R^{-1} can be extended to $H_{-n}(R)$ as follows. Let $z \in H_{-n}(R)$. Then there exists a sequence $\{z_k\} \subseteq H$ such that $\|z_k - y\|_{H_{-n}(R)} \to 0$. This implies that $\{R^{-(n-1)}R^{-1}z_k\} = \{R^{-n}z_k\}$ is Cauchy in H so that $\{R^{-1}z_k\}$ converges in $H_{-(n-1)}(R)$ to $w \in H_{-(n-1)}(R)$. We then define $w = R^{-1}z$, and we have

$$\|w\|_{H_{-(n-1)}(R)} = \lim \|R^{-(n-1)}R^{-1}z_k\| = \lim \|R^{-n}z_k\| = \|z\|_{H_{-n}(R)}$$

so that the definition of the extended operator R^{-1} (that shall keep the same name) does not depend on the sequence $\{z_k\}$ and R^{-1} is an isometry from $H_{-n}(R)$ into $H_{-(n-1)}(R)$.

$$\cdots \xrightarrow{R^{-1}} H_{-2}(R) \xrightarrow{R^{-1}} H_{-1}(R) \xrightarrow{R^{-1}} H. \tag{9.1.10}$$

If R is not densely defined, R^{-1} is not onto, as (i) of the result below shows.

Lemma 9.1.6. (i) $R^{-1}(H_{-n}(R)) = Cl_{-(n-1)}(D(R))$ $(Cl_{-(n-1)}$ indicates closure in $H_{-(n-1)}(R))$. (ii) Assume $D(R^2)$ is dense in $D(R)$ in the norm of H. Then $R^{-1}(H_{-2}(R)) \supseteq H$.

Proof. (i) Let $z \in H_{-n}(R)$. Then there exists a sequence $\{z_k\} \in H$ with $\|z_k - z\|_{H_{-n}(R)} \to 0$, so that the sequence $\{R^{-n}z_k\}$ is Cauchy in E. Now, $\{R^{-n}z_k\} = \{R^{-(n-1)}R^{-1}z_k\}$, so that $\{R^{-1}z_k\}$ converges in $H_{-(n-1)}(R)$ to $w \in H_{-(n-1)}(R)$; since $R^{-1}z_k \in D(R)$, $w \in Cl_{-(n-1)}(D(R))$.

Conversely, let $w \in Cl_{-(n-1)}(D(R))$, so that there exists $\{w_k\} \in D(R)$ with $\|w_k - w\|_{H_{-(n-1)}(R)} \to 0$. Then $\{R^{-(n-1)}w_k\} = \{R^{-n}Rw_k\} = \{R^{-n}z_k\}$ is Cauchy in H, so that $z_k \to z \in H_{-n}(R)$ in $H_{-n}(R)$. Finally, we have $\{R^{-1}z_k\} = \{w_k\}$, hence $R^{-1}z = w$.

(ii) Using (i), it is enough to show that $Cl_{-1}(D(R)) \supseteq E$. To check this, let $w \in E$. Using denseness of $D(R^2)$ in $D(R)$ select a sequence $\{w_k\} \subseteq E$ such that $R^{-1}w - R^{-1}R^{-1}w_k \to 0$. Then $\|w - R^{-1}w_k\|_{H_{-1}(R)} \to 0$, so that $w \in Cl_{-1}(D(R))$. This ends the proof.   ∎

Below, K is a linear operator with dense domain $D(K)$ in E and range in E. We assume that

(d) $S(t)E \subseteq D(K)$ and $KS(t)$ is bounded for $t > 0$,

(e) K has a bounded everywhere defined inverse K^{-1} that commutes with $S(t)$ $(t > 0)$, and $D(K^*) \subseteq E^{\odot}$.

Note that we have $KS(t) = S(t-h)KS(h)$ $(t > h > 0)$ hence $KS(\cdot)$ is continuous in the norm of (E, E) in $t > 0$. Since K^{-1} is bounded and everywhere defined, $K^*(K^{-1})^* = I$; on the other hand, $(K^{-1})^*K^* \subseteq I$, so that, although K^* may not be everywhere defined, $(K^{-1})^* = (K^*)^{-1}$ (see (5.0.27)). Taking adjoints,

$$S(t)^*(K^*)^{-1} = (K^*)^{-1}S(t)^*. \qquad (9.1.11)$$

Computing the adjoint of $S(t)K$ using (5.0.26) we obtain that $S(t)^*E^* \subseteq D(K^*)$ and $K^*S(t)^* = (S(t)K)^*$ is bounded for $t > 0$, so that it follows from (9.1.11) that $S(t)^*K^* \subseteq K^*S(t)^*$. Writing $K^*S(t)^* = S(t-h)^*K^*S(h)^*$ we deduce that $K^*S(t)^*$ is continuous in the norm of (E^*, E^*) for $t > 0$; noting then that $(K^*)^n S(t)^* = (K^*S(t/n)^*)^n$ we obtain

$$S(t)^*E^* \subseteq D((K^*)^n). \qquad (9.1.12)$$

The considerations above (in particular, Lemma 9.1.6) on H, R will be applied to $H = E^*$, $R = K^*$, but we must check first that $D((K^*)^2)$ is dense in $D(K^*)$ in the norm of E^*. Since, by hypothesis $D(K^*) \subseteq E^{\odot}$ and we have (9.1.12), it is enough to prove that $S(t)^*E^*$ is dense in $E^{\odot}$ for $t > 0$. This is a particular case of the result below.

Lemma 9.1.7. $S^\circ(t)E^\circ$ is dense in E° for $t > 0$.

Proof. If this is not true for a certain t, there exists $y \in (E^\circ)^*$ with

$$\langle z, S^\circ(t)^*y \rangle = \langle S^\circ(t)z, y \rangle = 0 \qquad (z \in E^\circ)$$

so that $S^\circ(t)^*y = 0$. Accordingly, $S^\circ(s)^*y = 0$ for $s \geq t$ and, by analyticity, $S^\circ(s)^*y = 0$ for $s > 0$. Now, if $z \in E^\circ$,

$$\langle z, y \rangle = \lim_{s \to 0^+} \langle S^\circ(s)z, y \rangle = \lim_{s \to 0^+} \langle z, S^\circ(s)^*y \rangle = 0,$$

and we conclude that $y = 0$. This ends the proof. ∎

We define $E_1(K) = D(K)$ equipped with its usual norm $\|y\|_{E_1(K)} = \|y\|_{D(K)} = \|Ky\|$. The versions of the minimum principle in this chapter will be obtained applying separation theorems in $E_1(K)$, thus it is imperative to pin down the dual space $E_1(K)^*$. A reasonable guess might be

$$E_1(K)^* = (E^*)_{-1}(K^*) \tag{9.1.13}$$

with canonical pairing $\langle (K^*)^{-1}z, Ky \rangle$. This turns out to be right when $D(A^*)$ is dense in E^* but not in general. To see why, note that Lemma 9.1.6 (i) implies that $(K^*)^{-1}(E^*)_{-1}(K^*) = \overline{D(K^*)}$, thus if $y^* \in E^* \setminus \overline{D(K^*)}$, $\langle y^*, Ky \rangle$ gives a functional in $E_1(K)$ not representable in the form above. However, a little fine tuning of the argument provides a suitable identification. Note first that Lemma 9.1.6 (ii) certifies that an arbitrary element $y^* \in E^*$ can be written in the form

$$y^* = (K^*)^{-1}z \qquad (z \in (E^*)_{-2}(K^*)). \tag{9.1.14}$$

We denote by $K^*E^* \subseteq (E^*)_{-2}(K^*)$ the set of all $z \in (E^*)_{-2}(K^*)$ such that (9.1.14) holds for some $y^* \in E^*$. Since $(K^*)^{-1} : (E^*)_{-2}(K^*) \to (E^*)_{-1}(K^*)$ is an isometry, $(K^*)^{-1}$ is $1-1$ as an operator from $(E^*)_{-2}(K^*)$ into $(E^*)_{-1}(K^*)$. On the other hand, $(K^*)^{-1}$ is an isometry from $(E^*)_{-1}(K^*)$ into E^*, thus an element of E^* that is zero as an element of $(E^*)_{-1}(K^*)$ must be zero itself. It finally results that

$$(K^*)^{-1} : K^*E^* \to E^* \tag{9.1.15}$$

is $1-1$, and it is onto by definition. We renorm K^*E^* with

$$\|z\|_{K^*E^*} = \|(K^*)^{-1}z\|_{E^*}, \tag{9.1.16}$$

under which norm it graduates to a Banach space. We have

$$(E^*)_{-1}(K^*) \hookrightarrow K^*E^* \hookrightarrow (E^*)_{-2}(K^*), \tag{9.1.17}$$

The inclusions are obvious. The first imbedding is isometric; if $z \in (E^*)_{-1}(K^*)$ then, by (9.1.9) $\|z\|_{(E^*)_{-1}(K^*)} = \|(K^*)^{-1}z\|_{E^*} = \|z\|_{K^*E^*}$. In the matter of the second imbedding, let $z \in K^*E^*$. Then $\|z\|_{(E^*)_{-2}(K^*)} = \|(K^*)^{-1}z\|_{(E^*)_{-1}(K^*)} \leq C\|(K^*)^{-1}z\|_{E^*}$ by (9.1.9).

Lemma 9.1.8. K^*E^* *is algebraically and metrically isomorphic to the dual space* $E_1(K)^*$; *the canonical pairing is*

$$\langle z, y \rangle_{K^*E^* \times E_1(K)} = \langle (K^*)^{-1}z, Ky \rangle_{E^* \times E}. \tag{9.1.18}$$

If $D(K^*)$ *is dense in* E^* *then* (9.1.13) *holds algebraically and metrically.*

Proof. There is little doubt that (9.1.18) provides a bounded linear functional in $E_1(K) = D(K)$ for every $z \in K^*E^*$. On the other hand, if ϕ is a bounded linear functional in $E_1(K)$ then $y \to \phi(K^{-1}y)$ is a bounded linear functional in E, so that there exists $y^* \in E^*$ with $\phi(K^{-1}y) = \langle y^*, y \rangle$; this means that $\phi(y)$ is given by (9.1.18). The metric isomorphism follows from definition of the norms of $E_1(K)$ and of K^*E^*. If $D(K^*)$ is dense in E^* then $(K^*)^{-1} : (E^*)_{-1}(K^*) \to E^*$ is onto, so that $K^*E^* = (E^*)_{-1}(K^*)$. Clearly, the norms of both spaces are the same. ∎

Lemma 9.1.9. $S(t)^*$ *can be extended to* K^*E^* *and* $S(t)^*(K^*E^*) \subseteq E^\odot$ *for* $t > 0$.

Proof. The extension is

$$S(t)^*z = K^*S(t)^*(K^{-1})^*z. \tag{9.1.19}$$

If $t > h > 0$ then we have $K^*S(t)^*(K^{-1})^*z = S(t-h)^*K^*S(h)^*(K^{-1})^*z$. That $S(t)^*(K^*E^*) \subseteq E^\odot$ follows from the fact that $S(t)^*E^* \subseteq E^\odot$. ∎

The space $(K^*E^*)^1(B^*)$ consists of all $z \in K^*E^*$ such that

$$\int_0^1 \|B^*S(t)^*z\|_X \, dt < \infty \tag{9.1.20}$$

($S(t)^*$ extended as in Lemma 9.1.9; recall that $B^* : E^\odot \to X$).

Theorem 9.1.10. *Let* $S(t)$ *be compact for* $t > 0$, *and let* $U \subseteq X^*$ *and* $Y \subseteq D(K)$ *be convex and closed (the first in* X^*, *the second in* E). *Moreover, assume that there exists* $\rho > 0$ *such that*

$$B(0, \rho) \subseteq U, \tag{9.1.21}$$

$B(0, \rho)$ *the ball of center* 0 *and radius* ρ *in* X^*, *and that*

$$R(\tilde{t}, \zeta; U) - Y \tag{9.1.22}$$

contains a ball of positive radius in $D(K)$ *for some* $\tilde{t} < \bar{t}$. *Let* $\bar{u}(\cdot)$ *be a time optimal control with optimal time* $\bar{t}$. *Then there exists* $z \in (K^*E^*)^1(B^*)$, $z \neq 0$ *with* $z \in N_Y(y(\bar{t}, \zeta, \bar{u}))$ *(the normal cone defined according to the duality* (9.1.11)) *and such that*

$$\langle B^*S(\bar{t}-s)^*z, \bar{u}(s) \rangle = \min_{v \in U} \langle B^*S(\bar{t}-s)^*z, v \rangle \tag{9.1.23}$$

a.e. in $0 \le s \le \bar{t}$.

We need

Lemma 9.1.11. *Assume $R(\tilde{t}, \zeta; U) - Y$ contains a ball in $D(K)$ for $\tilde{t} < \bar{t}$. Then, if $\{t_n\}$ is a sequence with $t_n \to \bar{t}$, the sets $R(t_n, \zeta; U) - Y$ contain a common ball in $D(K)$ for n large enough.*

Proof. Let $t < t'$. Then

$$
\begin{aligned}
y(t, \zeta, u) &= S(t)\zeta + \int_0^t S^\odot(t - \tau)^* Bu(\tau)d\tau \\
&= (S(t) - S(t'))\zeta + S(t')\zeta \\
&\quad + \int_{t'-t}^{t'} S^\odot(t' - \tau)^* Bu(\tau - (t' - t))d\tau.
\end{aligned}
\tag{9.1.24}
$$

Since the function defined by $\tilde{u}(\tau) = 0$ $(0 \le \tau < t' - t)$, $\tilde{u}(\tau) = u(\tau + (t' - t))$ $(t' - t \le \tau \le t')$ belongs to $C_{\text{ad}}(0, t'; U)$, we obtain

$$
R(t, \zeta; U) \subseteq (S(t) - S(t'))\zeta + R(t', \zeta; U) \quad (t < t'),
\tag{9.1.25}
$$

hence

$$
R(t, \zeta; U) - Y \subseteq (S(t) - S(t'))\zeta + (R(t', \zeta; U) - Y) \quad (t < t'),
$$

and the result follows from continuity of $KS(t)$ in $t > 0$, which implies that $S(t)$ is continuous in $D(K)$. ∎

Proof of Theorem 9.1.10. We denote by $C_{\text{ad},K}(0, \tilde{t}; U)$ the set of all $u(\cdot) \in C_{\text{ad}}(0, \tilde{t}; U)$ with $y(\tilde{t}, \zeta, u) \in D(K)$ and by $R_K(\tilde{t}, \zeta; U) \subseteq E$ the set of all elements of the form

$$
Ky(\tilde{t}, \zeta, u) \quad (u(\cdot) \in C_{\text{ad},K}(0, \tilde{t}, u)).
$$

Select a sequence $\{t_n\} \subseteq [0, \bar{t}), \bar{t} = \text{optimal time}, t_n \to \bar{t}$. Due to optimality of $\bar{t}$, the sets KY and $R_K(t_n, \zeta; U)$ are disjoint. More is true; if $\tilde{t} < \bar{t}$ the distance from KY to $R_K(\tilde{t}, \zeta; U)$ is positive. In fact, if this were not the case we could construct a sequence $\{u_m(\cdot)\} \subseteq C_{\text{ad},K}(0, \tilde{t}; U)$ and a sequence $y_m \in Y$ such that

$$
Ky(\tilde{t}, \zeta, u_m) - Ky_m \to 0
\tag{9.1.26}
$$

as $m \to \infty$. Applying K^{-1} to both sides of (9.1.26) and using Lemma 9.1.5, we deduce that $y(\tilde{t}, \zeta, \tilde{u}) \in Y$ for some $\tilde{u}(\cdot) \in C_{\text{ad}}(0, \tilde{t}; U)$, which contradicts the optimality of $\bar{t}$.

Since $\text{dist}(KY, R_K(\tilde{t}, \zeta; U)) > 0$ the sets $(KY)_\varepsilon = \{y \in E; \text{dist}(y, KY) \le \varepsilon\}$ and $R_K(\tilde{t}, \zeta; U)$ are disjoint for ε small enough and the first contains interior points, thus we can apply the separation theorem (Corollary 12.0.8) and construct $w_n \in E^*$, $\|w_n\| = 1$ such that

$$\langle w_n, Ky \rangle \le \langle w_n, Ky(t_n, \zeta, u) \rangle \tag{9.1.27}$$

or

$$\langle w_n, -Ky(t_n, \zeta, u) + Ky \rangle \le 0 \tag{9.1.28}$$

for $u \in C_{\mathrm{ad},K}(0, t_n, u)$ and $y \in Y$. The assumptions on the set (9.1.22) and Lemma 9.1.11 imply that the sets

$$\{Ky(t_n, \zeta, u) - Ky; u \in C_{\mathrm{ad},K}(0, t_n; U), y \in K\}$$

contain a common ball in E for n large enough, so that it follows from Lemma 6.4.3 that if w is the E-weak limit of (a subsequence of) $\{w_n\}$, then $w \neq 0$.

If $\delta > 0$ and $u(t) = 0$ in $\tilde{t} - \delta \le t \le \tilde{t}$ we have

$$y(t, \zeta, u) = S(t)\zeta + S(\delta) \int_0^{t-\delta} S^{\odot}(t - \delta - \tau)^* Bu(\tau)d\tau \tag{9.1.29}$$

so that every control $u(\cdot) \in C_{\mathrm{ad}}(0, \tilde{t} - \delta; U)$, extended by $u(t) = 0 \, (\tilde{t} - \delta \le t \le \tilde{t})$ belongs to $C_{\mathrm{ad},K}(0, \tilde{t}; U)$. In particular, if $B(0, \rho) \subseteq U$, inequality (9.1.28) will hold for $u(\cdot) \in C_{\mathrm{ad}}(0, t_n - \delta; B(0, \rho))$ and δ arbitrary. Since the space $C_{\mathrm{ad}}(0, t_n - \delta; B(0, \rho))$ is invariant through product by scalars of modulus ≤ 1, we have

$$|\langle w_n, Ky(t_n, \zeta, u) \rangle| \le C \tag{9.1.30}$$

for $u(\cdot) \in C_{\mathrm{ad}}(0, t_n - \delta; B(0, \rho))$, with C independent of n and of $\delta > 0$. In other words,

$$\begin{aligned}
\Big\langle w_n, & \int_0^{t_n-\delta} KS^{\odot}(t_n - \tau)^* Bu(\tau)d\tau \Big\rangle \\
&= \Big\langle w_n, S(\delta/2)K \int_0^{t_n-\delta} S^{\odot}(t_n - \delta/2 - \tau)^* Bu(\tau)d\tau \Big\rangle \\
&= \int_0^{t_n-\delta} \langle K^* S(\delta/2)^* w_n, S(\delta/2) S^{\odot}(t_n - \delta - \tau)^* Bu(\tau) \rangle d\tau \\
&= \int_0^{t_n-\delta} \langle S(\delta/2)^* K^* S(\delta/2)^* w_n, S^{\odot}(t_n - \delta - \tau)^* Bu(\tau) \rangle d\tau \\
&= \int_0^{t_n-\delta} \langle S^{\odot}(t_n - \delta - \tau) S(\delta/2)^* K^* S(\delta/2)^* w_n, Bu(\tau) \rangle d\tau \\
&= \int_0^{t_n-\delta} \langle B^* K^* S(t_n - \tau)^* w_n, u(\tau) \rangle d\tau \tag{9.1.31}
\end{aligned}$$

is bounded for $u(\cdot) \in C_{\mathrm{ad}}(0, t_n - \delta; B(0, \rho))$ independently of n and of $\delta > 0$. In the various steps we have used (7.7.6), (7.7.7) and the facts that $S(\delta)^* E^* \subseteq D(K^*)$

and that $S(\delta)^*$ and K^* commute. One of the intermediate steps on (9.1.31) was the equality $B^*K^*S(t_n - \tau)^*w_n = B^*S^\odot(t_n - \varepsilon - \tau)S(\varepsilon/2)^*K^*S(\varepsilon/2)^*w_n$ (valid for ε arbitrary) which, by hypothesis (a) shows that $B^*K^*S(t_n - \tau)^*w_n$ is a continuous X-valued function for $\tau < t_n$. We may then approximate it uniformly in $[0, t_n - \delta]$ by a piecewise constant X-valued function $h(\cdot)$ and then select $u(\cdot) \in C_{ad}(0, t_n; B(0, \rho))$ with $\langle h(t), u(t) \rangle \geq (\rho/2)\|h(t)\|$. In view of the arbitrariness of δ,

$$\int_0^1 \|B^*K^*S(t)^*w_n\|_X dt \leq C,$$

where C does not depend on $n = 1, 2, \ldots$. Now, $w_n \to w$ E-weakly and $S(t)^*$ is compact, thus $S(t)^*w_n \to S(t)^*w$ in the norm of E^* for $t > 0$ (Dunford–Schwartz [1958, Theorem 6, p. 486]). Since $B^*K^*S(t)^* = B^*S(t - h)^*K^*S(h)^*$ the same is true of $B^*K^*S(t)^*w_n$ and it follows from Fatou's Theorem (Example 8.1.7) that

$$\int_0^1 \|B^*K^*S(t)^*w\|_X \, dt \leq C. \tag{9.1.32}$$

The next task is to take limits in inequality (9.1.27) with the objective of obtaining

$$\langle w, Ky \rangle \leq \langle w, Ky(\bar{t}, \zeta, u) \rangle \tag{9.1.33}$$

for all $y \in Y$ and $u(\cdot) \in C_{ad, K}(0, \bar{t}; U)$. The limit on the left side is obvious, so that only the right side is in question. For an arbitrary control $u(\cdot)$ we have

$$y(\bar{t}, \zeta, u) = S(\bar{t})\zeta + \int_0^{\bar{t}} S^\odot(\bar{t} - \tau)^* Bu(\tau)d\tau$$

$$= S(t_n)\zeta + \int_0^{t_n} S^\odot(t_n - \tau)^* Bu(\tau + (\bar{t} - t_n))d\tau$$

$$+ (S(\bar{t}) - S(t_n)\zeta) + S(t_n)\int_0^{\bar{t} - t_n} S^\odot(\bar{t} - t_n - \tau)^* Bu(\tau)d\tau. \tag{9.1.34}$$

Accordingly, if $u(\cdot) \in C_{ad, K}(0, \bar{t}; U)$ and $u_n(\tau) = u(\tau + (\bar{t} - t_n))$ $(0 \leq t \leq t_n)$ then $u_n(\cdot) \in C_{ad, K}(0, t_n; U)$ and

$$Ky(t_n, \zeta, u_n) \to Ky(\bar{t}, \zeta, u).$$

Plugging $Ky(t_n, \zeta, u_n)$ in (9.1.27) and taking limits on the right side, (9.1.33) results.

Let $u(\cdot) \in C_{ad, K}(0, \bar{t}; U)$. Then any control $v(\cdot)$ that coincides with $u(\cdot)$ in $\bar{t} - \delta \leq t \leq \bar{t}$ also belongs to $C_{ad, K}(0, \bar{t}; U)$; this follows from the equality

$$y(\bar{t}, \zeta, v) = S(\bar{t})\zeta + \int_0^{\bar{t}} S^\odot(\bar{t} - \tau)^* Bv(\tau)d\tau$$

$$= S(\bar{t})\zeta + \int_0^{\bar{t}} S^\odot(\bar{t} - \tau)^* Bu(\tau)d\tau$$

$$+ S(\delta)\int_0^{\bar{t} - \delta} S^\odot(\bar{t} - \delta - \tau)^* B(v(\tau) - u(\tau))d\tau. \tag{9.1.35}$$

Keeping it in mind, we use (9.1.33) for $y = y(\bar{t}, \zeta, \bar{u})$ and $u(\cdot) \in C_{ad}(0, \bar{t}; U)$ such that $u(t) = \bar{u}(t)$ $(\bar{t} - \delta \le t \le \bar{t})$. After some manipulations similar to those in (9.1.31),

$$\int_0^{\bar{t}-\delta} \langle B^* K^* S(\bar{t} - \tau)^* w, \bar{u}(\tau)\rangle d\tau \le \int_0^{\bar{t}-\delta} \langle B^* K^* S(\bar{t} - \tau)^* w, u(\tau)\rangle d\tau, \quad (9.1.36)$$

or

$$\int_0^{\bar{t}-\delta} \langle B^* S(\bar{t} - \tau)^* z, \bar{u}(\tau)\rangle d\tau \le \int_0^{\bar{t}-\delta} \langle B^* S(\bar{t} - \tau)^* z, u(\tau)\rangle d\tau, \quad (9.1.37)$$

for every $u(\cdot) \in C_{ad}(0, \bar{t} - \delta; U)$ with arbitrary $\delta > 0$, where the element $z \in K^* E^*$ is defined by $w = (K^{-1})^* z$; see Lemma 9.1.9, especially (9.1.19). The estimate (9.1.32) implies that $z \in (K^* E^*)^1(B^*)$ and, since $w \not\ni 0$, we have $z \ne 0$.

We exploit (9.1.37) as follows. Since $\bar{u}(\cdot)$ is X-weakly measurable, the function $\tau \to \langle B^* S(\bar{t} - \tau)^* z, \bar{u}(\tau)\rangle$ is bounded and measurable. Call e the set of all its Lebesgue points in $0 < t < \bar{t}$. Given $s \in e$ take $u = \bar{u}_{s,h,v}$ a spike perturbation of $u(\cdot)$ for $v \in U$. Since $\bar{u}(\cdot)$ and $\bar{u}_{s,h,v}(\cdot)$ coincide in $s \le t \le \bar{t}$, (9.1.37) can be applied and yields

$$\frac{1}{h} \int_{s-h}^s \langle B^* S(\bar{t} - \tau)^* z, \bar{u}(\tau)\rangle d\tau \le \frac{1}{h} \int_{s-h}^s \langle B^* S(\bar{t} - \tau)^* z, v\rangle d\tau.$$

Taking limits, (9.1.23) results.

Finally, we take $u = \bar{u}$ in (9.1.33), obtaining

$$\langle w, Ky - Ky(\bar{t}, \zeta, \bar{u})\rangle \le 0. \quad (9.1.38)$$

hence, by (9.1.18),

$$\langle z, y - y(\bar{t}, \zeta, \bar{u})\rangle_{K^* E^* \times E_1(K)} = \langle (K^{-1})^* z, Ky - Ky(\bar{t}, \zeta, \bar{u})\rangle_{E^* \times E} \le 0 \quad (9.1.39)$$

for all $y \in Y$, by (9.1.38). In view of the characterization of the tangent cone of convex sets (Example 3.3.4), this shows that $z \in N_Y(y(\bar{t}, \zeta, \bar{u}))$ and thus completes the proof of Theorem 9.1.10. ∎

Remark 9.1.12. When E is a *reflexive* space some assumptions may be discarded and numerous simplifications ensue. In the first place, $D(K^*)$ is dense in E^* (see the comments before (5.0.25)) so that, by Lemma 9.1.8 we have $E_1(K)^* = (E^*)_{-1}(K^*)$ with equality of norms. The semigroup $S^*(t) = S(t)^*$ is strongly continuous, so that $\odot$-adjoints are unnecessary. Analyticity of the semigroup can be given up if one assumes X^* separable so that controls are strongly measurable (Lemma 9.1.2); formula (9.1.5) becomes the familiar one,

$$y(t, \zeta, u) = S(t)\zeta + \int_0^t S(t - \tau) Bu(\tau) d\tau. \quad (9.1.40)$$

A version of Theorem 9.1.4 can be proved without compactness (see Theorem 14.5.4).

Example 9.1.13. Denote by $E = \ell^1$ the space of all numerical sequences $y = \{y_n; \ n = 1, 2, \cdots\}$ with $\|y\|_1 = \Sigma|y_n| < \infty$ equipped with $\|\cdot\|_1$. Then E^* is the space ℓ^∞ of all $z = \{z_n\}$ with $\|z\|_\infty = \sup|z_n|$ equipped with $\|\cdot\|_\infty$, and $\langle z, y \rangle = \Sigma z_n y_n$. Let K be the operator in ℓ^1 defined by

$$Ky = K(\{y_n\}) = \{ny_n\} \tag{9.1.41}$$

with maximal domain. The adjoint K^* is given by the same formula, also with maximal domain, so that $D(K^*)$ is contained in the subspace $c_0 \subseteq \ell^\infty$ characterized by $|z_n| \to 0$, which subspace is not dense in E^*. The inclusion $(E^*)_{-1}(K^*) \subset A^*E^*$ is strict; in fact, A^*E^* can be identified with the space of all sequences $\{z_n\}$ such that $\|z\|_{A^*E^*} = \sup n^{-1}|z_n| < \infty$, whereas elements of $(E^*)_{-1}(K^*)$ satisfy $n^{-1}|z_n| \to 0$.

9.2. The Minimum Principle with Full Control

We use the name "full control" in connection with the system

$$y'(t) = Ay(t) + u(t), \qquad y(0) = \zeta \tag{9.2.1}$$

with $X = E^\odot$, so that $U \subseteq (E^\odot)^*$ and $u(\cdot) \in L_w^\infty(0, T; (E^\odot)^*)$. Hypothesis (a) in last section (with $B = I$) is obvious; we assume (b) and (c). We show below that in this particular case, Theorem 9.1.10 always applies with $K = A$ (we may assume via a translation that A^{-1} exists) if 0 is an interior point of U. We can deal with the point target condition

$$y(t, \zeta, u) = \bar{y} \in D(A) \tag{9.2.2}$$

if we show that $R(\tilde{t}, \zeta; U)$ contains a ball in $D(A)$ for any $\tilde{t} > 0$, and we may limit ourselves to $\zeta = 0$. For this, it suffices to note that

$$y = \int_0^{\tilde{t}} S(\tilde{t} - \tau)\frac{1}{\tilde{t}}(y - \tau Ay)d\tau \tag{9.2.3}$$

for $y \in D(A)$; the control $u(\tau) = (y - \tau Ay)/\tilde{t}$ has norm $\leq C\|y\|_{D(A)}$. We restrict ourselves below to the point target condition (9.2.2).

Theorem 9.2.1. *Let $\bar{u}(\cdot)$ be a time optimal control for (9.2.1) with optimal time $\tilde{t}$ and $\bar{y} \in D(A)$. Assume that 0 is an interior point of U. Then there exists $z \in (A^*E^*)^1(I), z \neq 0$ such that*

$$\langle S(\tilde{t} - s)^*z, \bar{u}(s) \rangle = \min_{v \in U}\langle S(\tilde{t} - s)^*z, v \rangle \tag{9.2.4}$$

a.e. in $0 \leq s \leq \tilde{t}$.

There exists a more general version of this result where $S(\cdot)$ is an arbitrary semigroup; see Example 9.2.8.

When E is reflexive, $X = E^*$. A subclass of the class of reflexive spaces is that of ζ-*convex* spaces (Example 9.2.3). In these, something can be said about *any* optimal control $\bar{u}(\cdot)$ even if $y(\bar{t}, \zeta, u) \notin D(A)$.

Example 9.2.2. (De Simon [1964]) Let A be the infinitesimal generator of an analytic semigroup in a Hilbert space H, and let $f(\cdot) \in L^p(0, T; H), 1 < p < \infty$. Then the function

$$y(t) = \int_0^t S(t - \tau)u(\tau)d\tau \tag{9.2.5}$$

has a derivative $y'(t)$ in $L^p(0, T; H)$ (see **8.1**) Moreover, $y(t) \in D(A)$ a.e., and

$$\|y\|_{L^p(0,T;D(A))} \leq C\|u\|_{L^p(0,T;H)}, \tag{9.2.6}$$

$$\|y'\|_{L^p(0,T;H)} \leq C\|u\|_{L^p(0,T;H)}, \tag{9.2.7}$$

$$y'(t) = Ay(t) + f(t) \quad \text{a.e. in } 0 \leq t \leq T. \tag{9.2.8}$$

Example 9.2.3. ζ-**convex** Banach spaces E were introduced by Burkholder [1983]; these are exactly the spaces where the E-valued Hilbert transform is a bounded operator from $L^p(-\infty, \infty; E)$ into itself for $1 < p < \infty$ (Bourgain [1983]). Hilbert spaces are ζ-convex, as are $L^p(\Omega)$ spaces, $1 < p < \infty$. The spaces $L^1(\Omega)$ and $C(\overline{\Omega})$ are not ζ-convex.

It has been proved by Dore and Venni [1987] that the result in Example 9.2.2 extends to ζ-convex spaces if we assume in addition that

$$\|(-A)^{i\tau}\| \leq Ce^{c|\tau|} \quad (-\infty < \tau < \infty) \tag{9.2.9}$$

with $c < \pi/2$. See **8.3** for other situations where condition (9.2.9) is relevant.

Example 9.2.4. Let the control $\bar{u}(\cdot) \in C_{\text{ad}}(0, \bar{t}; U)$ be time optimal for (9.1.1) in $0 \leq t \leq \bar{t}$, and let $\tilde{t} < \bar{t}$. Then $\bar{u}(\cdot)$ drives time optimally ζ to $y(\tilde{t}, \zeta, \bar{u})$; in other words, $\bar{u}(\cdot)$ is time optimal in any subinterval $[0, \tilde{t}]$.

In fact, if some other admissible control $u(\cdot)$ drives ζ to $y(\tilde{t}, \zeta, \bar{u})$ in time $t_0 < \tilde{t}$, the control that coincides with $u(\cdot)$ in $[0, t_0]$ and equals $\bar{u}(t + (\tilde{t} - t_0))$ in $[t_0, \bar{t} - (\tilde{t} - t_0)]$ drives ζ to $\bar{y} = y(\bar{t}, \zeta, u)$ but gets there in time $\bar{t} - (\tilde{t} - t_0) < \bar{t}$, a contradiction. We can show in the same way that $\bar{u}(\cdot)$ is time optimal in any interval $[a, b], 0 < a < b < \bar{t}$.

This result is a particular case of the so-called **optimality principle.** It depends essentially on time invariance of the control system (9.1.1) and of the control set U. It extends to other cost functionals as long as the functional is time invariant.

Theorem 9.2.5. *Assume E is a Hilbert space or a ζ-convex Banach space and, in the latter case, that (9.2.9) holds, and let U be convex with nonempty interior.*

Let $\bar{u}(\cdot)$ be a time optimal control with optimal time $\bar{t}$ for (9.2.1). Then there exists a sequence $\{t_n\} \subseteq [0, \bar{t}]$, $t_n \to \bar{t}$ and a sequence $\{z_n\} \subseteq (A^*E^*)^1(I)$, $z_n \neq 0$ such that

$$\langle S(t_n - s)^*z_n, \bar{u}(s)\rangle = \min_{v \in U}\langle S(t_n - s)^*z_n, v\rangle \quad a.e. \ in \quad 0 \leq s \leq t_n. \quad (9.2.10)$$

In fact, using Examples 9.2.2 or 9.2.3 we can pick an increasing sequence $\{t_n\} \subseteq [0, \bar{t})$ with $t_n \to \bar{t}$ and such that $y_n = y(t_n, \zeta, \bar{u}) \in D(A)$ and, using the optimality principle, apply Theorem 9.2.1 in each subinterval. It does not seem obvious that one can "take limits as $n \to \infty$" in (9.2.10) in any useful way. ∎

In a space which is not ζ-convex, we cannot say anything about optimal trajectories that do not end up in $D(A)$. It is not immediately obvious that such trajectories exist. If they do, we must have $R^\infty(\bar{t}) \neq D(A)$ for some $\bar{t}$. To see what this means, assume

$$R^\infty(\bar{t}) = D(A) \quad (9.2.11)$$

for some $\bar{t} > 0$. Then $R^\infty(\bar{t}) = D(A)$ is a Banach space under $\| \cdot \|_{R^\infty(\bar{t})}$ in (9.1.6) and under $\| \cdot \|_{D(A)}$; since $D(A) \hookrightarrow R^\infty(\bar{t})$ (as shown by (9.2.3)), Lemma 8.1.1 proves that both norms in $R^\infty(\bar{t})$ are equivalent. In particular, the operator $u(\cdot) \to A \int_0^{\bar{t}} S^\odot(\bar{t} - \tau)^*u(\tau)d\tau$ from $L_w^\infty(0, \bar{t}; (E^\odot)^*)$ into E is bounded, so that the functional

$$u(\cdot) \to \left\langle z, A \int_0^{\bar{t}} S^\odot(\bar{t} - \tau)^*u(\tau)d\tau \right\rangle \quad (9.2.12)$$

is bounded in $L_w^\infty(0, T; (E^\odot)^*)$ for all $z \in E^*$. Arguing then with controls vanishing in $[\bar{t} - \delta, \bar{t}]$ as after (9.1.31) we deduce that

$$\int_0^t \|A^*S(t)^*z\|_{E^*}dt < \infty \quad (9.2.13)$$

for all $z \in E^*$.

Example 9.2.6. (A. Pazy) Here is an unbounded infinitesimal generator satisfying (9.2.13). The spaces c_0 and ℓ^1 are the same in Example 9.1.13, and we have $c_0^* = \ell^1$. The operator

$$Ay = A\{y_n\} = \{-ny_n\} \quad (9.2.14)$$

(maximal domain) is the infinitesimal generator of the (analytic, compact) semigroup

$$S(t)\{y_n\} = \{e^{-nt}y_n\}. \quad (9.2.15)$$

in c_0. The adjoint semigroup $S(t)^*$ in ℓ^1 is given by the same formula, and it clearly satisfies (9.2.13) for every $z \in \ell^1$.

It has been proved in the author [1974] that relation between (9.2.11) and (9.2.13) holds in much more generality; the semigroup need not be analytic or compact. See Example 9.2.9, and also Example 9.2.10 for a related result.

Example 9.2.7. There are semigroups not satisfying (9.2.13). For instance, consider the (self adjoint) operator $Ay(x) = -xy(x)$ in $L^2(1, \infty)$. A is the infinitesimal generator of the analytic semigroup $S(t)y(x) = e^{-tx}y(x)$. If $y(x) = 1/x$, $\|AS(t)y\| \geq ct^{-1}$ as $t \to \infty$.

The following problem seems to be open; can there be unbounded infinitesimal generators in "smooth" spaces (i.e., reflexive, ζ-convex . . .) such that (9.2.13) holds for all $z \in E^*$?

In the following three Examples, the system is

$$y'(t) = Ay(t) + u(t) \tag{9.2.16}$$

with $F = E$ and A is the infinitesimal generator of a (merely) strongly continuous semigroup $S(t)$; as usual, we assume that A^{-1} exists and is bounded. Solutions of (9.2.1) are given by (9.1.40), with controls in $L^\infty(0, T; F)$.

Example 9.2.8. Theorem 9.2.1 admits the following (partial) generalization. Let $\bar{u}(\cdot)$ be a time optimal control for (9.2.16) with optimal time $\bar{t}$ and with $\zeta, \bar{y} \in D(A)$. Assume that U is convex and that 0 is an interior point of U. Then there exists $z \in E^*$ such that $S(t)^*z \in D(A^*)$ for $t > 0$, $\|A^*S(t)^*z\|_{E^*}$ is integrable near zero and

$$\langle A^*S(\bar{t} - s)^*z, \bar{u}(s) \rangle = \min_{v \in U} \langle A^*S(\bar{t} - s)^*z, v \rangle \tag{9.2.17}$$

a.e. in $0 \leq s \leq \bar{t}$. For a proof, see the author [1974, Theorem 4.1].

Example 9.2.9. We define $R^\infty(\bar{t})$ here using controls in $L^\infty(0, \bar{t}; F)$. If (9.2.11) holds, then $S(t)^*E^* \subseteq D(A^*)$ for $t > 0$ and (9.2.13) holds for all $z \in E^*$. See the author [1974, Lemma 4.3]. The converse is as well true.

Example 9.2.10. Assume that, for every $u(\cdot)$ continuous in $0 \leq t \leq T$, the function $y(t, 0, u)$ is Lipschitz continuous in $0 \leq t \leq T$. Then $S(t)^*E^* \subseteq D(A^*)$ for $t > 0$ and (9.2.13) holds for all $z \in E^*$. The converse is as well true. See the author–Frankowska [1990 : 2, Theorem 3.2].

We find out below what the minimum principle means in various examples. In the first result (Lemma 9.2.11 below) we place ourselves under the assumptions of Theorem 9.1.10, so that K is an operator complying with (d) and (e), and optimal controls satisfy (9.1.23). Assuming that U is the unit ball of X^* (so that hypothesis

(c) is automatic) and exchanging minimum by maximum we have

$$\langle B^* S(\bar{t} - s)^* z, \bar{u}(t) \rangle = \max_{\|v\| \leq 1} \langle B^* S(\bar{t} - s)^* z, v \rangle \tag{9.2.18}$$

a.e. in $0 \leq s \leq \bar{t}$. It then follows from the definition of the duality set $\Theta(v)$ for elements $y \in X$ (see **5.2**) that if t belongs to the set

$$n(\bar{t}, z) = \{t \in [0, \bar{t}]; B^* S(\bar{t} - t)^* z \neq 0\} \tag{9.2.19}$$

then

$$\|B^* S(\bar{t} - t)^* z\| \bar{u}(t) \in \Theta(B^* S(\bar{t} - t)^* z). \tag{9.2.20}$$

This is also true (but trivial) if $B^* S(\bar{t} - t)^* z = 0$. In spaces where the duality set $\Theta(v)$ contains exactly one element, (9.2.20) characterizes uniquely the optimal control (in terms of z) in $n(\bar{t}, z)$.

Lemma 9.2.11. *Let $S(t)$ be an analytic semigroup.* (i) *Let $z \in K^* E^*$. Then, either $n(\bar{t}, z)$ is empty or it consists of the complement of a sequence $\{t_n\} \subseteq [0, \bar{t}]$; if infinite, this sequence can only accumulate at $\bar{t}$.* (ii) *In particular, if $B = I$ then $n(\bar{t}, z) = [0, \bar{t}]$ if $z \in K^* E^*$, $z \neq 0$.*

The proof of (i) is a simple consequence of analyticity of the function $t \rightarrow B^* S(\bar{t} - t)^* z = B^* K^* S(\bar{t} - t)^* (K^{-1})^* z$. To show ($ii$), let $z \in K^* E^*$ and assume that $S(t_0)^* z = 0$ for a single $t_0 > 0$. Then the semigroup equations show that $S(t)^* z = 0$ for $t \geq t_0$, and analyticity of $S(t)^*$ that $S(t)^* z = 0$ for all $t > 0$. We have $S(t)^* z = K^* S(t)^* (K^{-1})^* z$, thus the statement reduces to: if $S(t)^* w = 0$ in $t > 0$ for $w \in E^*$, then $w = 0$. To see this, take $y \in E$ and note that $\langle w, y \rangle = \lim_{t \to 0+} \langle w, S(t)y \rangle = \lim_{t \to 0+} \langle S(t)^* w, y \rangle = 0$. ∎

For a general strongly continuous semigroup, none of this is valid; $n(\bar{t}, z)$ may be a complicated set for arbitrary B and, even if $B = I$ and E is a reflexive space we only know that, if $z \neq 0$, $n(\bar{t}, z)$ contains an interval $(\bar{t} - \delta, \bar{t})$, $\delta > 0$.

In the rest of the section, $K = A$ and we combine Theorem 9.2.1 (and Theorem 9.2.5 when appropriate) with Lemma 9.2.11 (ii). The equations are

$$y'(t) = A_p(\beta)y(t) + u(t, x) \qquad y(0) = \zeta \tag{9.2.21}$$

in the space $L^p(\Omega)$, $1 \leq p < \infty$, and

$$y'(t) = A_c(\beta)y(t) + u(t, x) \qquad y(0) = \zeta. \tag{9.2.22}$$

in the space $C(\overline{\Omega})$ or $C_0(\overline{\Omega})$, depending on whether β is variational or Dirichlet (for details on the definition of the operators and on the hypotheses see **7.6**). Both equations are of the form (9.2.1). For (9.2.21), $1 < p < \infty$, we take $E = L^p(\Omega)$, $X = L^q(\Omega)$ with $1/p + 1/q = 1$, so that $X^* = L^p(\Omega)$. The control set is $U =$

unit ball of $L^p(\Omega)$, thus Assumption (c) holds. The operator $A_p(\beta)$ generates a compact analytic semigroup $S_p(t; A, \beta)$ in $L^p(\Omega)$.

For $p = 1$, $E = L^1(\Omega)$ and we use the $\odot$-theory. If β is a variational boundary condition we take $X = C(\overline{\Omega})$, so that $X^* = \Sigma(\overline{\Omega})$. The admissible control space is then the unit ball of $L_w^\infty(0, T; \Sigma(\overline{\Omega}))$, thus the equation actually is

$$y'(t) = A_1(\beta)y(t) + \mu(t), \qquad y(0) = \zeta \tag{9.2.23}$$

with $\mu(t)$ a measure in $\Sigma(\overline{\Omega})$ for each t. We have $L^1(\Omega)^\odot = C(\overline{\Omega})$. The same is true for the Dirichlet boundary condition as long as we take $X = C_0(\overline{\Omega}) = L^1(\Omega)^\odot$; then $X^* = \Sigma_0(\overline{\Omega})$, the control set is the unit ball of $\Sigma_0(\overline{\Omega})$ and the admissible control space is the unit ball of $L_w^\infty(0, T; \Sigma_0(\overline{\Omega}))$. The operator $A_1(\beta)$ generates a compact analytic semigroup $S_1(t; A, \beta)$ in $C(\overline{\Omega})$ or $C_0(\overline{\Omega})$.

Finally, we precise (9.2.22). For a variational boundary condition β we take $E = C(\overline{\Omega})$, $X = C(\overline{\Omega})^\odot = L^1(\Omega)$, so that $X^* = L^\infty(\Omega)$; for the Dirichlet boundary condition, $E = C_0(\overline{\Omega})$. The control set is the unit ball of $L^\infty(\Omega)$ and the admissible control space the unit ball of $L_w^\infty(0, T; L^\infty(\Omega)) = L^\infty((0, T) \times \Omega)$. The operator $A_c(\beta)$ generates a compact analytic semigroup $S_c(t; A, \beta)$ in $C(\overline{\Omega})$ or $C_0(\overline{\Omega})$. It is clear that (a), (b) and (c) are satisfied in all these examples. Since A and β are fixed, we use below the following abbreviations: $A_p = A_p(\beta)$, $A_c = A_c(\beta)$, $S_p(t) = S_p(t; A, \beta)$ and $S_c(t) = S_c(t; A, \beta)$.

We apply to each system Theorem 9.2.1 (with $A = A_p, A_c$) combined with Lemma 9.2.11 (b); $n(\bar{t}, z) = [0, \bar{t}]$ in all cases.

Example 9.2.12. The system is (9.2.21) with $1 < p < \infty$. We have (9.2.20) everywhere in $0 \le t \le \bar{t}$ and, in view of Example 5.8.11 $\Theta(y)$ is single valued with $\Theta(y) = \|y\|^{2-p}|y(x)|^{p-2}y(x)$. It follows that if $\bar{u}(\cdot)$ is time optimal in an interval $0 \le t \le \bar{t}$ for the target condition $y(t, \zeta, \bar{u}) = \bar{y} \in D(A_p)$ then there exists $z \in (A^*E^*)^1(B^*) = (A_p^*L^p(\Omega)^*)^1(I) = (A_p^*L^q(\Omega))^1(I)$, $z \ne 0$ such that

$$\bar{u}(t, x) = \frac{|(S_p(\bar{t} - t)^*z)(x)|^{p-2}(S_p(\bar{t} - t)^*z)(x)}{\|S_p(\bar{t} - t)^*z\|^{p-1}} \tag{9.2.24}$$

in $0 \le t \le \bar{t}$. In the Hilbert case $p = 2$ this reduces to the familiar formula

$$\bar{u}(t, x) = \frac{(S_2(\bar{t} - t)^*z)(x)}{\|S_2(\bar{t} - t)^*z\|}. \tag{9.2.25}$$

It follows from (9.2.24) that

$$\|\bar{u}(t)\|_{L^p(\Omega)} = 1 \tag{9.2.26}$$

a.e. in $0 \le t \le \bar{t}$.

If the target $\bar{y}$ does not belong to $D(A_p)$, Theorem 9.2.1 does not apply. However, $L^p(\Omega)$, $1 < p < \infty$ is ζ-convex (Example 9.2.3) so that Theorem 9.2.5 can be used, and we can then construct a sequence $\{t_n\} \subseteq [0, \bar{t})$ with $t_n \to \bar{t}$ and a sequence $z_n \subseteq (A_p^* L^q(\Omega))^1(I)$ such that $z_n \neq 0$ and

$$\bar{u}(t, x) = \frac{|(S_p(\bar{t} - t)^* z_n)(x)|^{p-2}(S_p(\bar{t} - t)^* z_n)(x)}{\|S_p(\bar{t} - t)^* z_n\|^{p-1}} \tag{9.2.27}$$

in $0 \leq t \leq t_n$. This, incidentally, is enough to prove (9.2.26).

Example 9.2.13. The system is (9.2.21) with $p = 1$ (or, rather, (9.2.23)). Duality sets in $\Sigma(\overline{\Omega}) = C(\overline{\Omega})^*$ are characterized in Example 5.8.12. If $\bar{\mu}(\cdot) \in C_{\mathrm{ad}}(0, \bar{t}; U)$ is a time optimal control and $\bar{y} = y(\bar{t}, \zeta, \bar{\mu}) \in D(A_1)$ then there exists an element $z \in (A_1^* L^1(\Omega)^*)^1(I) = (A_1^* L^\infty(\Omega))^1(I), z \neq 0$ such that if

$$\bar{z}(t, x) = (S_1(\bar{t} - t)^* z)(t, x) \tag{9.2.28}$$

then the measure $\bar{\mu}(\cdot)$ is supported in the set $m(\bar{z}(t, \cdot))$, where

$$m(z(\cdot)) = \{x \in \overline{\Omega}; |z(x)| = \|z\|_{C(\overline{\Omega})}\}$$

for any $z(\cdot) \in C(\overline{\Omega})$. Moreover, $\bar{z}(t, x)\mu(t, dx)$ is a positive measure and $\|\bar{\mu}(t)\| = 1$.

This information can be precised under additional hypotheses. For instance, assume that the boundary Γ is real analytic and the coefficients of the operator A are real analytic in $\overline{\Omega}$. Then $\bar{z}(t, \cdot)$ is real analytic in $\overline{\Omega}$. Since $n(\bar{t}, z) = [0, \bar{t}]$, the function $\bar{z}(t, \cdot)$ is not identically zero for any t. For each t, the functions

$$\|\bar{z}(t)\|_{C(\overline{\Omega})} - \bar{z}(t, x), \qquad \|\bar{z}(t)\|_{C(\overline{\Omega})} + \bar{z}(t, x) \tag{9.2.29}$$

are real analytic in $\overline{\Omega}$, and the union of the sets where either one vanishes is $m(\bar{z}(t, \cdot))$. However, a nonzero real analytic function can vanish only in a set of measure zero (Example 9.2.14 below), thus, $m(\bar{z}(t, \cdot))$ has Lebesgue measure zero for all t. We deduce that, for each t the optimal measure $\bar{\mu}(t)$ is supported by a set of measure zero (of course, the set of zeros of an analytic function enjoys many other privileges besides being of Lebesgue measure zero).

Example 9.2.14. The system is (9.2.22). Theorem 9.2.1 shows that there exists a nonzero $z \in (A_c^* C(\overline{\Omega})^*)^1(I) = (A_c^* \Sigma(\overline{\Omega}))^1(I) ((A_c^* \Sigma_0(\overline{\Omega}))^1(I)$ for the Dirichlet boundary condition) such that if

$$\bar{z}(t, x) = (S_1(\bar{t} - t)^* z)(t, x) \tag{9.2.30}$$

then

$$\bar{u}(t, x) = \mathrm{sign}\,\bar{z}(t, x). \tag{9.2.31}$$

Arguing exactly as in Example 9.2.12 we deduce that, if the boundary Γ and coefficients of A are real analytic the set $e(t) \subseteq \overline{\Omega}$ where $\bar{z}(t, x) = 0$ is of measure zero for all t; thus, the set

$$e(\bar{t}, z) = \{(t, x) \in [0, \bar{t}] \times \overline{\Omega}; \bar{z}(t, x) = 0\} \tag{9.2.32}$$

has measure zero (to show, this, we integrate the characteristic function of $e(\bar{t}, z)$ first in x, then in t). It follows that (9.2.31) characterizes $\bar{u}(t, x)$ (in function of $\bar{z}$) a.e. in $[0, \bar{t}] \times \overline{\Omega}$.

Example 9.2.15. A function $f(x) = f(x_1, \ldots, x_m)$ defined in a domain Ω is **real analytic** if, for every $\bar{x} = (\bar{x}_1, \ldots, \bar{x}_m) \in \Omega$ it can be expressed as a multiple power series

$$f(x_1, \ldots, x_m) = \sum_{j_1=0}^{\infty} \cdots \sum_{j_m=0}^{\infty} a_{j_1 \cdots j_m} (x_1 - \bar{x}_1)^{j_1} \cdots (x_m - \bar{x}_m)^{j_m} \tag{9.2.33}$$

convergent in some square $S = |x_j - \bar{x}_j| \leq \delta$, $j = 1, \ldots, m$. If the function is not identically zero, the set $e = \{x; f(x) = 0\}$ has measure zero in $\overline{\Omega}$. We check this in the particular case $m = 2$. Since the boundary has measure zero, we only look at the set of zeros in Ω. Let S be a square $|x_j - \bar{x}_j| \leq \delta$ where the power series development (9.2.33) holds. For each x_1, $|x_1 - \bar{x}_1| \leq \delta$ the function $f(x_1, x_2)$ is an analytic function of x_2. Thus, either (a) $f(x_1, x_2) \equiv 0$ in $|x_2 - \bar{x}_2| \leq \delta$, or (b) the zeros of $f(x_1, x_2)$ are a finite subset of $|x_2 - \bar{x}_2| \leq \delta$. Option (a) cannot occur more than a finite number of times; in fact, if this were the case the function $f(x_1, x_2)$, as analytic function of x_1 for x_2 fixed would have an infinite number of zeros in $|x_1 - \bar{x}_1| \leq \delta$ and would then be identically zero. It then suffices to cover Ω with a countable number of squares where (9.2.33) holds (a connectedness argument is implicit here).

9.3. Bang-Bang Theorems and Approximate Controllability

The terminology comes from finite dimensional control theory (see **3.8–3.10**). For instance, if the control set is $U = [-1, 1]$, controls are **bang-bang** if they only take the values -1 and 1; a little more generally, if U is a polyhedron in $\mathbb{R}^m$, a control is bang-bang if it takes values on the vertices. The name "bang-bang control" has come to mean "a control that takes values on the set of extremal points of the control set," for instance a control with $\|u(t)\| = 1$ when the control set U is an Euclidean ball in $\mathbb{R}^m$. We use this meaning of the word also in infinite dimensional situations. Sometimes, "bang-bang" is understood as "taking values on the boundary of the control set," whether or not these values are extremal points. To avoid confusion, we call these controls "weakly bang-bang."

Ordinarily, bang-bang and weakly bang-bang theorems for a system

$$y'(t) = Ay(t) + Bu(t), \qquad y(0) = \zeta \tag{9.3.1}$$

(A the generator of a strongly continuous semigroup) are a consequence of the maximum principle

$$\langle B^* S(\bar{t} - t)^* z, \bar{u}(t) \rangle = \max_{v \in U} \langle B^* S(\bar{t} - t)^* z, v \rangle. \tag{9.3.2}$$

Let $n(\bar{t}, z)$ be the set (9.2.19) where $B^* S(\bar{t} - t)^* z \neq 0$. The minimum principle (9.3.2) always implies that $\bar{u}(t)$ is weakly bang-bang in $n(\bar{t}, z)$; in fact, $v \to \langle z, v \rangle$ always reaches its minimum (or maximum) at the boundary of U. In some cases, (9.3.2) implies that $\bar{u}(t)$ is extremal; this happens, for instance, in Example 9.2.12 (where the extremal points of U coincide with its boundary points) and in Example 9.2.14, under additional analyticity assumptions (note that functions taking only the values ± 1 a.e. are extremals in the unit ball of $L^\infty(\Omega)$). Note, however, Example 9.2.13. Extremals in the unit ball of $\Sigma(\overline{\Omega})$ are $\pm$ Dirac deltas with mass anywhere in $\overline{\Omega}$ (Lemma 12.6.6). Thus, to prove that the optimal control $\bar{u}(\cdot)$ is bang-bang rather than just weakly bang-bang we would need to show that $|\bar{z}(t, \cdot)|$ assumes its maximum at a single point.

The "worst possible case" for the minimum principle (9.3.2) as proved in Theorem 9.1.10 is that where $n(\bar{t}, z)$ is empty; in this case (9.3.2) says nothing. This eventuality is prevented by the condition

$$B^* S(\bar{t} - t)^* z = 0 \quad (0 \leq t < \bar{t}) \quad \text{implies } z = 0 \tag{9.3.3}$$

for $z \in K^* E^*$ (K the operator in Theorem 9.1.10). We look into (9.3.3) for $z \in E^*$; results for $z^* \in K^* E^*$ can be squeezed out of this case (see Lemma 9.3.3).

In the results below, unless otherwise stated, we just assume that E and F are Banach spaces, that $B : F \to E$ is a bounded operator and that $S(t)$ is a strongly continuous semigroup in E.

Call the system (9.3.1) **approximately controllable in time** $\bar{t}$ if, given $\zeta \in E$, $\bar{y} \in E$ and $\varepsilon > 0$ there exists $u(\cdot) \in C(0, \bar{t}; F)$ such that

$$\|y(\bar{t}, \zeta, u) - \bar{y}\| \leq \varepsilon. \tag{9.3.4}$$

Equivalently, (9.3.1) is approximately controllable in time $\bar{t}$ if

$$\overline{R_c(\bar{t})} = E, \tag{9.3.5}$$

$R_c(\bar{t})$ the set of all $y(\bar{t}, 0, u)$ with $u(\cdot) \in C(0, \bar{t}; F)$. In fact, it is clear that (9.3.5) holds if (9.3.1) is approximately controllable in time $\bar{t}$; conversely, if $R_c(\bar{t})$ is dense in E we can find $u(\cdot) \in C(0, \bar{t}; F)$ with $\|y(\bar{t}, 0; u) - (\bar{y} - S(\bar{t})\zeta)\| \leq \varepsilon$, which is (9.3.4).

Lemma 9.3.1. *Let* $S^*(t) = S(t)^*$ *be a strongly continuous semigroup. Then the implication (9.3.3) holds for* $z \in E^*$ *if and only if (9.3.1) is approximately controllable in time* $\bar{t}$.

Proof. If $R_c(\bar{t})$ is not dense in E, by the Hahn-Banach theorem there exists $z \in E^*$ such that $z \neq 0$ and

$$\langle z, y \rangle = 0 \qquad (y \in R_c(\bar{t})). \tag{9.3.6}$$

This is

$$\int_0^t \langle B^* S(t - \tau)^* z, u(\tau) \rangle d\tau = 0 \tag{9.3.7}$$

for all $u(\cdot) \in C(0, \bar{t}, F)$, which implies $B^* S(\bar{t} - t)^* z = 0$ in $0 \leq t \leq \bar{t}$ and thus provides a rebuttal to (9.3.3). Conversely, if $z \neq 0$ is such that $B^* S(\bar{t} - t)^* z = 0$ in $0 \leq t < \bar{t}$ (9.3.7) (hence (9.3.6)) holds, so that $R_c(\bar{t})$ is not dense in E. ∎

The choice of control space in the definition of approximate controllability is not critical; simple approximation arguments show we may replace $C(0, \bar{t}; F)$ by $L^p(0, \bar{t}; F)$ or (in the other direction) by the space of all polynomials with coefficients in F. Note also that (9.3.3) (for $z \in E^*$ or for $z \in K^* E^*$) can be restated as

$$B^* S(t)^* z = 0 \quad (0 < t \leq \bar{t}) \quad \text{implies } z = 0. \tag{9.3.8}$$

Lemma 9.3.1 has an obvious counterpart for an analytic semigroup under assumptions (a) and (b) in **9.1**; although $S^*(t)$ may not be strongly continuous at $t = 0$ it is continuous for $t > 0$ so that the proof applies verbatim. Under the same assumptions, we have the following three results for analytic semigroups:

Lemma 9.3.2. (9.3.8) (*or* (9.3.3)) *are independent of $\bar{t}$.*

In fact, by analyticity, vanishing of $B^* S(t)^* z$ in any interval $0 < t \leq \bar{t}$ implies that $B^* S(t)^* z$ is identically zero in $t > 0$.

Lemma 9.3.3. *Assume the implication* (9.3.3) (*or* (9.3.8)) *hold for $z \in E^*$. Then they also hold for $z \in K^* E^*$.*

Proof. Assume (9.3.8) holds for $w \in E^*$, and let $z \in K^* E^*$. Then if $h > 0$, $B^* S(t)^* z = B^* K^* S(t)^* (K^{-1})^* z = B^* S(t - h)^* K^* S(h)^* (K^{-1})^* z$ in $t \geq h$; by analyticity, $B^* S(t)^* K^* S(h)^* (K^{-1})^* z = 0$ for all $t > 0$. Using (9.3.8) for $K^* S(h)^* (K^{-1})^* z \in E^*$ we obtain $K^* S(h)^* (K^{-1})^* z = 0$ and, since K^* is one-to-one, $S(h)^* (K^{-1})^* z = 0$ for all $h > 0$. If $y \in E$ we have $\langle (K^{-1})^* z, y \rangle = \lim_{h \to 0} \langle (K^{-1})^* z, S(h) y \rangle = \lim_{h \to 0} \langle S(h)^* (K^{-1})^* z, y \rangle = 0$, so that $(K^{-1})^* z = 0$ in E^*; this implies $z = 0$ in $K^* E^*$ (see the definition of $K^* E^*$ in **9.1**). ∎

Corollary 9.3.4. *If* (9.3.1) *is approximately controllable in any time $\bar{t} > 0$ and $z \in K^* E^*, z \neq 0$, then $n(\bar{t}, z)$ is the complement of a sequence $\{t_n\} \subseteq [0, \bar{t}]$ which, if infinite, can only accumulate at $\bar{t}$.*

In certain cases, the assumptions of Theorem 9.1.10 imply approximate controllability (and thus guarantee, via Corollary 9.3.4, that the adjoint vector is nontrivial). One case is that where Y is a point, where $R(\tilde{t}, \zeta; U)$ must contain a ball of positive radius in $D(K)$; this is easily seen to imply that

$$R^\infty(\tilde{t}) \supseteq D(K)$$

and thus, since $D(K)$ is dense in E, that (9.3.5) holds for $\tilde{t}$, thus for any $t > 0$. On the other hand, if Y is a large set, (9.1.15) does not demand much of $R(\tilde{t}, \zeta; U)$, thus approximate controllability may not follow. For reference, we state

Corollary 9.3.5. *Let the assumptions of Theorem 9.1.10 be satisfied and (9.1.22) reinforced to: $R(\tilde{t}, \zeta; U)$ contains a ball of positive radius in $D(K)$. Then the conclusions hold, plus: $\bar{z}(t) = B^*S(\tilde{t} - t)^*z$ is nonzero in the complement of a sequence $\{t_n\} \subseteq [0, \tilde{t}]$ which, if infinite, can only accumulate at $\tilde{t}$.*

Results of this type give information on the *switching points* of an optimal control $\bar{u}(\cdot)$. To fix ideas, assume that F is a Hilbert space and that the control set U is the unit ball of F. Then (9.3.2) gives

$$\bar{u}(t) = \frac{B^*S(\tilde{t} - t)^*z}{\|B^*S(\tilde{t} - t)^*z\|} \tag{9.3.9}$$

in the intervals (t_n, t_{n+1}) between points of the sequence $\{t_n\}$ in Corollary 9.3.5. The optimal control is then analytic between **switching points** t_n. We have encountered these points in **3.8, 3.9** and **4.3, 4.5, 4.7, 4.8** in the finite dimensional setting (in fact, the computation of optimal controls reduced, for the most part, to figuring out their switching points). For some general statements on switchings in the finite dimensional case, see **3.10**.

If $S(\cdot)$ is not an analytic semigroup, not much seems to be known about $n(\tilde{t}, z)$, except in particular examples. To extract a bang-bang or weak bang-bang theorem from the maximum principle we would need to know that $n(\tilde{t}, z)$ has full measure in $[0, \tilde{t}]$ for $z \neq 0$. This has apparently never been proved in full generality, even for the equation

$$y'(t) = Ay(t) + u(t), \qquad y(0) = \zeta. \tag{9.3.10}$$

Moreover, it seems to be unknown at present whether (some version of) the minimum principle holds for an arbitrary time optimal trajectory of (9.3.10), without extraneous assumptions on E, A or the initial and final points of the trajectory. Hence the interest of the following result, where a bang-bang result is obtained without intercession of the minimum principle. The system (9.3.10) lives in an arbitrary Banach space E, with control set $U = $ unit ball of E, and A is the generator of a strongly continuous semigroup $S(t)$. We assume the controls strongly measurable.

Theorem 9.3.6. *Let $\bar{u}(t)$ be a time optimal control. Then*

$$\|\bar{u}(t)\| = 1 \qquad (9.3.11)$$

a.e. in $0 \le t \le \bar{t} =$ optimal time.

The proof follows from a series of auxiliary results. A **density point** t of a measurable set $e \subseteq \mathbb{R}$ is a point where

$$\lim_{h \to 0} \frac{|(t - h, t + h) \cap e|}{2h} = 1.$$

In the definition of **left** density point one replaces $(t - h, t + h)$ by $(t - h, t)$ and $2h$ by h, with corresponding adjustments for a right density point.

Lemma 9.3.7. *Almost every point of e is a density point.*

Proof. The characteristic function $\chi_e(\cdot)$ of e is measurable, thus almost every $t \in \mathbb{R}$ is a Lebesgue point of χ_e (Theorem 2.0.6). A Lebesgue point t of χ_e where $\chi_e(t) = 1$ (that is, with $t \in e$) is a density point of e. ∎

Lemma 9.3.8. *Let e be a set of positive measure in $\mathbb{R}$. Then, for almost every $t \in e$ there exist $c, \rho > 0$ and a sequence $\{t_n\}$, $t_1 < t_2 < \cdots < t$, $t_n \to t$ such that*

$$|(t_n, t_{n+1}) \cap e| \ge \rho(t_{n+1} - t_n), \qquad (9.3.12)$$

$$\frac{t_n - t_{n-1}}{t_{n+1} - t_n} \le c, \qquad (9.3.13)$$

for $n = 1, 2, \ldots$.

Proof. Let

$$e_m = \{t \in e; \ |e \cap (t - 1/k, t)| \ge 1/2k, k \ge m\}.$$

If t is a left density point of e then $t \in e_m$ for m large enough, so that

$$e = \left(\bigcup e_m \right) \cup b,$$

with $|b| = 0$. If d_m is the set of left density points of e_m, each d_m has full measure in e_m, thus

$$e = \left(\bigcup d_m \right) \cup b', \qquad (9.3.14)$$

with $|b'| = 0$, and it is enough to prove existence of the sequence $\{t_n\}$ for $t \in d_m$ with m arbitrary. As long as we take $\{t_n\} \subseteq e_m$, and t_1 sufficiently near t, inequality (9.3.12) is automatic (with $\rho = 1/2m$), thus we only have to take care of (9.3.13). We take $2/3 < r < 1$ and pick t_1 in such a way that

$$|(s, t) \cap e_m| \ge r(t - s) \quad (t_1 \le s \le t).$$

It follows that, if we divide the interval (t_1, t) in three equal subintervals, the middle one will contain some point of e_m; we pick then t_2 in that subinterval, so that $t_1 + (t - t_1)/3 \le t_2 \le t_1 + 2(t - t_1)/3$. Once t_2 has been chosen, we select $t_3, t_4, \ldots$ in the same way; we have

$$\frac{1}{3}(t - t_n) \le t_{n+1} - t_n \le \frac{2}{3}(t - t_n),$$

so that

$$\frac{t_n - t_{n-1}}{t_{n+1} - t_n} \le 2 \frac{t - t_{n-1}}{t - t_n} \le 6. \qquad \blacksquare$$

Figure 9.1.

Lemma 9.3.9. *Let* $t, t' > 0$. *Then*

$$R^\infty(t) = R^\infty(t').$$

Proof. Assume that $t < t'$. That $R^\infty(t) \subseteq R^\infty(t')$ follows from the following particular case of (9.1.24),

$$\int_0^t S(t - \tau)u(\tau)d\tau = \int_{t'-t}^{t'} S(t' - \tau)u(\tau - (t' - t))d\tau \qquad (9.3.15)$$

and the opposite inclusion from the equality

$$\int_0^{t'} S(t' - \tau)u(\tau)d\tau$$

$$= S(t) \int_0^{t'-t} S(t' - t - \tau)u(\tau)d\tau + \int_{t'-t}^{t'} S(t' - \tau)u(\tau)d\tau$$

$$= \int_0^t S(t - \tau)\left(\frac{1}{t}S(\tau)\int_0^{t'-t} S(t' - t - \tau)u(\tau)d\tau + u(\tau + (t' - t))\right)d\tau$$

$$= \int_0^t S(t - \tau)v(\tau)d\tau. \qquad (9.3.16)$$

The second part of the proof can be reinterpreted thus: if $y \in R^\infty(t)$, then for every $h > 0$, y may be written in the form

$$y = \int_{t-h}^t S(t - \tau)u(\tau)d\tau \qquad (9.3.17)$$

with

$$\|u\|_{L^\infty(t-h,t)} \le C(h)\|y\|_{R^\infty(t)}. \qquad (9.3.18)$$

$\blacksquare$

Lemma 9.3.10. *Let $u(\cdot) \in C_{ad}(0, \bar{t}; U)$ satisfy $\|u(t)\| \leq 1 - \varepsilon \ (0 \leq t \leq \bar{t})$. Then $u(\cdot)$ cannot be time optimal in $0 \leq t \leq \bar{t}$.*

For the proof, we just make $t' = \bar{t}$ in (9.3.16); clearly, $\|v(\tau)\| \leq 1$ if $t' - t = \bar{t} - t$ is small enough.   ∎

Given a measurable set $e \subseteq \mathbb{R}$, we denote by $R^\infty(t; e)$ the subspace of $R^\infty(t)$ corresponding to controls $u(\cdot) \in L^\infty((0, t) \cap e)$ (or, rather, controls in $L^\infty(0, t; E)$ supported by e). We equip $R^\infty(t; e)$ with the norm

$$\|y\|_{R^\infty(t;e)} = \inf \|u\|_{L^\infty((0,t)\cap e; E)}, \qquad (9.3.19)$$

the infimum taken over all $u \in L^\infty((0, t) \cap e)$ with

$$y = \int_0^t S(t - \tau)Bu(\tau)d\tau. \qquad (9.3.20)$$

Obviously, $R^\infty(t; e) \subseteq R^\infty(t)$, and

$$\|y\|_{R^\infty(t)} \leq \|y\|_{R^\infty(t;e)}$$

(in other words, $R^\infty(t; e) \hookrightarrow R^\infty(t)$). In general, $R^\infty(t; e)$ may be smaller than $R^\infty(t)$; for instance, if $e \subseteq (0, t'), t' < t, R^\infty(t; e) \subseteq S(t - t')R^\infty(t)$, which is *much* smaller that $R^\infty(t)$ if $S(t)$ is analytic. However, we have

Lemma 9.3.11. *For almost all $t \in e, t \geq 0$ we have*

$$R^\infty(t; e) \approx R^\infty(t). \qquad (9.3.21)$$

Proof. We only have to show that $R^\infty(t) \hookrightarrow R^\infty(t; e)$. Let $t \in e$ be such that the sequence $\{t_n\}$ in Lemma 9.3.8 exists, and let $y \in R^\infty(t)$. Write y in the form (9.3.17) with $t - h = t_1$. We have

$$y = \int_{t_1}^t S(t - \sigma)u(\sigma)d\sigma$$

$$= \sum_{n=1}^\infty S(t - t_{n+1}) \int_{t_n}^{t_{n+1}} S(t_{n+1} - \sigma)u(\sigma)d\sigma$$

$$= \sum_{n=1}^\infty \int_{(t_{n+1}, t_{n+2}) \cap e} S(t - \tau)\frac{1}{|(t_{n+1}, t_{n+2}) \cap e|}$$

$$\times S(\tau - t_{n+1})\left(\int_{t_n}^{t_{n+1}} S(t_{n+1} - \sigma)u(\sigma)d\sigma\right)d\tau$$

$$= \int_0^t S(t - \tau)v(\tau)d\tau,$$

with

$$v(\tau) = \frac{1}{|(t_{n+1}, t_{n+2}) \cap e|} S(\tau - t_{n+1}) \left(\int_{t_n}^{t_{n+1}} S(t_{n+1} - \sigma)u(\sigma)d\sigma \right)$$

for $\tau \in (t_{n+1}, t_{n+2}) \cap e$ $(n = 1, 2, \ldots)$, $v(t) = 0$ elsewhere. Obviously, $v(\cdot)$ is zero outside of e, and we have

$$\|v(\tau)\|_E \leq \frac{C(t_{n+1} - t_n)}{|(t_{n+1}, t_{n+2}) \cap e|} \|u\|_{L^\infty(0, t; E)} \leq C' \|u\|_{L^\infty(0, t; E)}.$$

in $(t_{n+1}, t_{n+2}) \cap e$, where C' does not depend on n. This ends the proof. ∎

Proof of Theorem 9.3.6. Let $\bar{u}(\cdot)$ be a time optimal control, $\bar{t}$ the optimal time. Assume that the set where $\|\bar{u}(t)\| < 1$ has positive measure. Then there exists $\varepsilon > 0$ and a set e of positive measure with

$$\|u(t)\| \leq 1 - \varepsilon \quad (t \in e). \tag{9.3.22}$$

Let $t' \in e$ be such that $R^\infty(t'; e) = R^\infty(t')$. Using Lemma 9.3.10 construct a control $w(\cdot)$ zero outside of e and such that $y(t', w) = y(t', u)$, and then set $v(t) = (1 - \delta)u(t) + \delta w(t)$. It is clear that $\|v(t)\| \leq 1 - \varepsilon/2$ in $0 \leq t \leq t'$ if δ is small enough. By Lemma 9.3.10 $\bar{u}(\cdot)$ is not time optimal in the interval $0 \leq t \leq t'$, thus by the optimality principle (Example 9.2.4) it is not optimal in $0 \leq t \leq \bar{t}$, a contradiction. ∎

Remark 9.3.12. If one works under the setup in **9.1** the controls live in $L^\infty_w(0, T; X^*)$ and are not strongly measurable. Extension of Theorem 9.3.6 and auxiliary results to controls $\bar{u}(\cdot)$ in $L^\infty_w(0, T; X^*)$ presents no problems; strong measurability of $u(\cdot)$ is nowhere used. Measurability of the norm $\|u(\cdot)\|$ comes into play selecting the set e in (9.3.22) and is insured by separability of X.

Remark 9.3.13. Remark 9.3.12, combined with Lemma 9.3.11 has a surprising consequence. Applying the result in an arbitrary interval $[0, \bar{t}]$ with $e = [0, \bar{t}]$, the control $v(\cdot)$ manufactured in Lemma 9.3.11 is *continuous in the norm of E* in each interval $[t_n, t_{n+1}]$. As a very particular case we obtain that if $u(\cdot) \in L^\infty_w(0, \bar{t}; X^*)$ then there exists $u(\cdot) \in L^\infty(0, \bar{t}; X^*)$ with

$$y(\bar{t}, \zeta, u) = y(\bar{t}, \zeta, v) \tag{9.3.23}$$

In other words, defining $R^\infty(\bar{t})$ using $L^\infty_w(0, \bar{t}; X^*)$ or $L^\infty(0, \bar{t}; X^*)$ gives the same space, although the norms are not the same: just equivalent. At first sight, this may seem to imply that, in setting up control problems, we may always limit ourselves to controls in $L^\infty(0, \bar{t}; X^*)$. This is illusory; for instance, if $\bar{u}(\cdot) \in L^\infty_w(0, \bar{t}; X^*)$ is a time optimal control there exists $v(\cdot) \in L^\infty(0, \bar{t}; X^*)$ such that (9.3.23) holds, but $v(\cdot)$ may not satisfy the control constraint. Example 9.2.14 confirms that strongly measurable controls are not enough; unless $\bar{z}(t, x)$ does not change sign, the optimal control does not belong to $L^\infty(0, \bar{t}; X^*)$.

9.4. Exact and Approximate Controllability

Assume the linear system

$$y'(t) = Ay(t) + Bu(t), \qquad y(0) = \zeta \qquad (9.4.1)$$

is equipped with the control space $C_{ad}(0, \bar{t}; F) = L_w^\infty(0, \bar{t}; X^*)$ (no control constraint). Granted that trajectories $y(t, \zeta, u)$ are suitably defined (this is the case under the hypotheses in **9.1** or **9.2**), the system (9.4.1) is **(exactly) controllable in time** $\bar{t} > 0$ if

$$R(\bar{t}, \zeta; F) = E \qquad (9.4.2)$$

for every $\zeta \in E$. (the qualifier "exactly" avoids confusion with *approximate* controllability). Equivalently, the system is exactly controllable if, given $\zeta, \bar{y} \in E$ there exists $u(\cdot) \in C_{ad}(0, \bar{t}; F)$ such that

$$y(\bar{t}, \zeta, u) = \bar{y}. \qquad (9.4.3)$$

It is enough to check (9.4.2) with $\zeta = 0$; in fact, if $R(\bar{t}, 0; F) = E$ there exists a control $u = u(\cdot) \in C_{ad}(0, \bar{t}; F)$ such that $y(\bar{t}, 0, u) = \bar{y} - S(\bar{t})\zeta$, which is $y(\bar{t}, \zeta, u) = S(\bar{t})\zeta + y(\bar{t}, 0, \bar{u}) = \bar{y}$.

We say that (9.4.1) is **exactly reachable** if, for $\zeta = 0$ and $\bar{y}$ arbitrary there exists $\bar{t}$ (depending of $\bar{y}$) such that (9.4.3) holds; this amounts to

$$\bigcup_{\bar{t}>0} R(\bar{t}, 0; F) = E. \qquad (9.4.4)$$

We may also define exact controllability and reachability with different control spaces, for instance, $L_w^p(0, T; X^*)$, $1 \le p < \infty$ instead of $L_w^\infty(0, T; X^*)$. To precise, we may speak then of "p-exact controllability," "p-exact reachability," etc., so that (9.4.1) is p-exactly controllable in time $\bar{t}$ if

$$R^p(\bar{t}) = E, \qquad (9.4.5)$$

and p-exactly reachable if

$$\bigcup_{\bar{t}>0} R^p(\bar{t}) = E. \qquad (9.4.6)$$

Finally, we complement our definition of approximate controllability in time $\bar{t}$ in last section. We say that the system (9.4.1) is **approximately reachable** if

$$\overline{\bigcup_{\bar{t}>0} R_c(\bar{t})} = E. \qquad (9.4.7)$$

As noted in **9.3**, the choice of the control space in the approximate definitions is not critical. Obviously, all are weaker than their exact companions.

When $S(t)$ is merely a strongly continuous semigroup, trajectories may not be defined for $u(\cdot) \in L_w^p(0, T; X^*)$; in this case, we take $C_{ad}(0, T; F) = L^p(0, T; F)$, and there is no need to assume that F is a dual space. In this situation, or under the hypotheses in **9.1** and **9.2** we have

Lemma 9.4.1. (9.4.1) *is approximately reachable if and only if*

$$B^*S(t)^*z = 0 \quad (t > 0) \text{ implies } z = 0. \tag{9.4.8}$$

The proof is much the same as that of Lemma 9.3.1.

Example 9.4.2. Approximate reachability does not necessarily imply approximate controllability in any time $\bar{t} > 0$. Consider the equation

$$
\begin{aligned}
& y_t(t, x) = -y_x(t, x) + \chi(x)u(t) \quad (x \geq 0) \\
& y(t, 0) = 0 \\
& y(0, x) = \zeta(x) \quad (x \geq 0)
\end{aligned}
$$

in the space $E = L^2(0, \infty)$, where χ is the characteristic function of the interval $[0, 1]$. We put this equation in the form (9.4.1) defining $Ay(\cdot) = y'(\cdot)$ with domain $D(A)$ consisting of all absolutely continuous functions $y(\cdot) \in L^2(0, \infty)$ with $y'(\cdot) \in L^2(0, \infty)$ and $y(0) = 0$. The operator A generates the translation semigroup

$$S(t)y(x) = \begin{cases} y(x - t) & (x \geq t) \\ 0 & (0 \leq x < t) \end{cases} \tag{9.4.9}$$

We take controls $u(\cdot)$ in L^∞. This system cannot be approximately reachable for any finite time $\bar{t}$; in fact, the influence of the control term at $x = x_0$ is not felt until $\bar{t} = x_0 - 1$. However, it is approximately reachable. To see this, note that

$$S(t)^*z(x) = z(x + t) \quad (x \geq 0) \tag{9.4.10}$$

and $B^*z = \int_0^1 z(x)dx$. If the system were not approximately reachable there would exist $z(\cdot) \in L^2(0, \infty)$ such that $\int_0^1 z(x + t)dx = 0$ for all t, or

$$\phi(t) = \int_t^{t+1} z(x)dx = 0 \quad (t \geq 0)$$

so that, differentiating, $z(t + 1) - z(t) = 0$ a.e. in $t \geq 0$: since $z(\cdot)$ is square summable, $z(x) = 0$ a.e.

The situation is different when "approximate" is replaced by "p-exact." We show this in the setting of **9.1**. The control space is $L_w^p(0, T; X^*)$, $1 < p \leq \infty$.

Theorem 9.4.3. *Let $1 < p \leq \infty$. Assume the system (9.4.1) is p-exactly reachable. Then (9.4.1) is p-exactly controllable for some $\bar{t} > 0$.*

Theorem 9.4.3 follows from

Theorem 9.4.4 (The Baire category theorem). *If a complete metric space V is the union of a sequence of closed subsets, at least one of these subsets has nonempty interior.*

For a proof, see Dunford–Schwartz [1958, p. 20]. Arguing in a closed ball $B(x, r) \subseteq V$ we can squeeze the seemingly more general result: if the union of a sequence of closed sets in V has an interior point, so does some element of the sequence.

Theorem 9.4.4 motivates a definition. A subset $W \subseteq V$ is **of the first category** (or of **Baire's first category**) if it is the countable union of a sequence of closed sets, each with empty interior. Colloquially, Lemma 9.4.4 says that "sets of the first category are thin." In modern analysis, a statement $S(x)$ depending on an element x of a metric space is sometimes called **generic** if it holds in the complement of a set of the first category. The word "generic" is also used in a more general way as "holding outside a small set" where "small" may have other meanings (for instance, a set of measure zero if a measure is at play).

We prove Theorem 9.4.3 under the assumptions in **9.1**. We note first that we can "make countable" the union (9.4.6). In fact, using the analogue of (9.3.15) for controls in L^p we show that $R^p(t) \subseteq R^p(t')$ if $t \leq t'$, so that (9.4.6) holds if and only if

$$\bigcup_{n=1}^{\infty} R^p(n) = E.$$

Next, we define $R_m^p(\bar{t})$ as the subset of $R^p(\bar{t})$ corresponding to controls $u(\cdot)$ in the closed ball $B(0, m)$ of $L_w^p(0, \bar{t}; X^*)$. We have

Lemma 9.4.5. *Let $1 < p \leq \infty$. Then $R_m^p(\bar{t})$ is closed in E for every $\bar{t} > 0$.*

In fact, the arguments in Lemma 9.1.5 show that $R_m^p(\bar{t})$ is E^*-weakly closed, *a fortiori* closed.

End of Proof of Theorem 9.4.3. Since each $R^p(\bar{t})$ is the union of the $R_m^p(\bar{t})$ for $m = 1, 2, \ldots$, we have

$$\bigcup_{n=1}^{\infty} \bigcup_{m=1}^{\infty} R_m^p(n) = E,$$

and, by Baire's theorem, one of the $R_m^p(n)$ must have an interior point. This is easily seen to imply that $R^p(n) = E$, and the proof is complete. ∎

There are situations where approximate reachability and approximate controllability in (any) finite time $\bar{t} > 0$ are equivalent. One example is

Lemma 9.4.6. *Let A be the infinitesimal generator of an analytic semigroup $S(\cdot)$. Then (9.4.1) is approximately reachable if and only if it is approximately controllable in any time $\bar{t} > 0$.*

Obviously, approximate controllability implies approximate reachability. On the other hand, assume (9.4.1) is approximately reachable. If $S(\bar{t} - t)^*z = 0$ in $0 \le t < \bar{t}$ for some $z = 0$ then, by analyticity, $S(t)^*z = 0$ in $t > 0$ and Lemma 9.4.1 implies $z = 0$. ∎

The most elementary result on exact controllability is probably this one:

Example 9.4.7. Let A generate a strongly continuous group $S(\cdot)$. Then

$$y'(t) = Ay(t) + u(t) \tag{9.4.11}$$

is exactly controllable in any time $\bar{t} > 0$, that is, $R^\infty(\bar{t}) = E$. The proof is essentially contained in the results in **6.7**.

Example 9.4.8. Consider the second order equation

$$y''(t) = A_0 y(t) + u(t) \tag{9.4.12}$$

in a Hilbert space H, where A_0 is self adjoint with spectrum in $(-\infty, -\kappa]$, $\kappa > 0$. The *energy space* $\mathbf{E}$ for (9.4.12) is the set of all pairs $(y, y_1) \in D(R) \times H$, where $R = (-A_0)^{1/2}$; $\mathbf{E}$ is fitted with the inner product $(y, z) = (Ry, Rz) + (y_1, z_1)$ corresponding to the norm $\|y\| = \|(y, y_1)\| = (\|Ry\| + \|y_1\|^2)^{1/2}$. The equation (9.4.12) is exactly controllable in the energy space $\mathbf{E}$ for any $\bar{t} > 0$ in the following sense: the first order equation

$$\mathbf{y}'(t) = \mathbf{A}_0 \mathbf{y}(t) + \mathbf{B}u(t)$$

for $\mathbf{y}(t) = (y(t), y'(t)) \in \mathbf{E}$, with

$$\mathbf{A}_0 = \begin{bmatrix} 0 & I \\ A_0 & 0 \end{bmatrix}, \qquad \mathbf{B} = \begin{bmatrix} 0 \\ I \end{bmatrix}, \tag{9.4.13}$$

$(D(\mathbf{A}_0) = D(A_0) \times D(R))$ is exactly controllable in any time $\bar{t} > 0$. The proof (of a more general result) was the job done in **6.11**.

The next result goes in the opposite direction.

Example 9.4.9. (the author [1974, p. 170]). Let E be a Hilbert space. Assume that, for some $t > 0$ we have

$$R^2(t) = E. \tag{9.4.14}$$

Then

$$S(t)E = E \tag{9.4.15}$$

for all $t \geq 0$. Equality (9.4.15) does *not* imply that $S(\cdot)$ is a group; the left translation semigroup (9.4.10) in $L^2(0, \infty)$ is a counterexample.

Example 9.4.10. Let F, E be Banach spaces, E infinite dimensional, and let $L : F \to E$ be a compact operator. Then $L(F)$ has empty interior in E. To see this, note that

$$L(F) \subseteq \bigcup_{n=1}^{\infty} \overline{L(B(0, n))},$$

where each set in the union is compact, thus it cannot have nonempty interior. Apply Theorem 9.4.4.

Example 9.4.11. (Triggiani [1977]) Let $S(t)$ be compact for $t > 0$. Then (9.4.6) cannot hold for any $1 < p \leq \infty$, that is, the system cannot be exactly reachable. In fact, by Theorem 9.4.4 we would have $R^p(\bar{t}) = E$ for some $t > 0$, which contradicts Example 9.4.10 since the operator $u(\cdot) \to y(\bar{t}, 0, u)$ is compact (the latter is a particular case of Theorem 5.5.19).

For proving the maximum principle exact controllability is not necessary; Theorem 9.1.10 shows that the relevant condition is

$$R^{\infty}(\bar{t}) \supseteq H \tag{9.4.16}$$

where H is a subspace of E (in Theorem 9.1.10 the domain of an operator K satisfying certain assumptions). We may call (9.4.16) **exact controllability to H in time $\bar{t}$**, with a corresponding definition of p-**exact controllability to H in time $\bar{t}$**. As we see in the following section this sort of controllability can be attained in interesting examples.

Approximate controllability can be studied (for special operators) using methods similar to those used in finite dimensional spaces. Call a closed, densely defined operator A a **Jordan operator** if (a) $R(\lambda; A)$ exists for all λ except for a finite or countable sequence $\{\lambda_n\}$ in the complex plane, (b) $R(\lambda; A)$ has a pole (as opposed to an essential singularity) at each λ_n; this amounts to requiring

$$\|R(\lambda; A)\| = O(|\lambda - \lambda_n|^{-m_n}) \quad \text{as } \lambda \to \lambda_n \tag{9.4.17}$$

for some integer m_n.

The (local) structure of Jordan operators is not difficult to elucidate. Condition (9.4.17) implies that λ_n is an eigenvalue of A, that is, that there exists a nonzero solution $y \in D(A)$ of the **eigenvalue equation**

$$(\lambda_n I - A)y = 0. \tag{9.4.18}$$

The space of all solutions of (9.4.18) is the **eigenvector space** E_n corresponding to λ_n; the **multiplicity** of λ_n is the dimension of E_n. A **generalized eigenvector** of A corresponding to the eigenvalue λ_n is any nonzero solution of the equation

$$(\lambda_n I - A)^k y = 0 \tag{9.4.19}$$

for some $k = 1, 2, \ldots$ The **index** of an eigenvalue λ_n is the smallest integer m_n such that

$$(\lambda I - A)^{m_n+1} y = 0 \quad \text{implies} \quad (\lambda I - A)^{m_n} y = 0. \tag{9.4.20}$$

The index of an eigenvalue coincides with m_n in (9.4.17) if the latter is the least integer making the estimate work. The implication (9.4.20) shows that we only need consider $k \le m_n$ in the **generalized eigenvalue equation** (9.4.19). The space E_n^g of all solutions of (9.4.19) is the **generalized eigenvector space** corresponding to λ_n and contains E_n. When E_n^g is finite dimensional one can choose an adequate basis and express the action of A in E_n^g in matrix form by a Jordan block, hence the name "Jordan operator." Note also that when $m_n = 1$ we have $E_n^g = E_n$; all generalized eigenvectors are actually eigenvectors.

"Local" approximate controllability of (9.4.1) in each E_n^g is studied essentially in the same way as for finite dimensional systems. If the sequence $\{\lambda_n\}$ is finite, we have

$$E = E_1^g \oplus E_2^g \oplus \cdots \oplus E_n^g, \tag{9.4.21}$$

and the local results can be assembled to give approximate controllability results in the whole space as in the finite dimensional case in **3.10** (Jordan operators with a finite set of eigenvalues are not very different from finite dimensional operators). The same can be done when the sequence is infinite, but only if we show that (the generalization of) (9.4.21) holds in the sense that every $y \in E$ can be approximated by linear combinations of generalized eigenvectors. For details on this approach see the author [1966 : 2]; for all facts on Jordan operators quoted, see Dunford–Schwartz [1958, Chapter VII], especially Section 3.

If the operator A is self adjoint, approximate controllability can be studied using its resolution of the identity (see the author [1967 : 2]) for details) whether the spectrum is discrete or continuous. If the control space F is assumed to be finite dimensional, the minimum dimension needed for approximate controllability coincides with the multiplicity of the operator; for operators with pure point spectrum the multiplicity is the maximum of the dimensions of all eigenvector spaces and may of course be infinite, thus precluding the use of finite dimensional controls. For the definition of multiplicity when continuous spectrum is present, see Dunford–Schwartz [1963, Chapter XII] or Halmos [1957]. For simple examples, see the first three in next section.

9.5. Controllability with Finite Dimensional Controls

It might be argued that any control that can be implemented in practice must be finite dimensional (or, at best, a finite dimensional approximation to a hypothetical infinite dimensional control). Accordingly, a look at controllability properties of the system

$$y'(t) = Ay(t) + \sum_{j=1}^{m} b_j u_j(t), \qquad y(0) = \zeta \tag{9.5.1}$$

$(b_1, \ldots, b_m \in E)$ is justified. Here $F = \mathbb{R}^m$ and $B : \mathbb{R}^m \to E$ is $B((u_1, \ldots, u_m))$ $= \Sigma u_j b_j$. In the first three examples we look into approximate controllability. Most examples in this section use operators of the form

$$A(\Sigma c_n y_n) = \Sigma \lambda_n c_n y_n \tag{9.5.2}$$

in a separable Hilbert space E, $\{y_n\}$ a complete orthonormal sequence in H, $\{\lambda_n\}$ a sequence of complex numbers. The domain $D(A)$ of A consists of all $\Sigma c_n y_n$ such that $\Sigma |\lambda_n|^2 |c_n|^2 < \infty$ (this is a necessary and sufficient condition for convergence in E of the series on the right side of (9.5.2); we have $\|\Sigma \lambda_n c_n y_n\|^2 = \Sigma |\lambda_n|^2 |c_n|^2$). The operator A is always normal and it is bounded (with $\|A\| = \sup|\lambda_n|$) if $|\lambda_n|$ is bounded. If Re λ_n is bounded, A generates a strongly continuous semigroup $S(t)$ given by

$$S(t)(\Sigma c_n y_n) = e^{\lambda_n t} c_n y_n. \tag{9.5.3}$$

Example 9.5.1. The operator is (9.5.2) with negative eigenvalues ordered by non-decreasing modulus: $\lambda_j = -\mu_j$ with $0 < \mu_1 \le \mu_2 \le \ldots, \mu_n \to \infty$. The multiplicity of an eigenvalue λ_n is the number of elements in the eigenvalue sequence that coincide with λ_n, and the multiplicity of A is the supremum (possibly infinite) of all the multiplicities of its eigenvalues. The system (9.5.1) is approximately controllable in arbitrary time $\bar{t} > 0$ with some $b_1, \ldots, b_m$ if and only if $m \ge$ multiplicity of A. In particular, (9.5.1) cannot be approximately controllable for any m if A has infinite multiplicity.

To see this, we use Lemma 9.4.1 and Lemma 9.4.6, since $S(\cdot)$ is an analytic semigroup. Taking into account that $B^* y = ((b_1, y), \ldots, (b_m, y))$, the condition $B^* S(t)^* z \equiv B^* S(t)(\Sigma c_n y_n) \equiv 0$ is

$$\sum_{n=1}^{\infty} c_n e^{-\mu_n t} (b_j, y_n) \equiv 0 \quad (j = 1, 2, \ldots, m). \tag{9.5.4}$$

We group together terms with the same eigenvalue and use a well known result, (Bernstein [1933, p. 12]), to the effect that if a Dirichlet series vanishes identically all of its coefficients must vanish. Accordingly, if $\mu_{k-1} < \mu_k = \mu_{k+1} = \cdots = \mu_{k+r} < \mu_{k+r+1}$, (9.5.4) implies

$$\sum_{n=k}^{k+r} c_n (b_j, y_n) \equiv 0 \quad (j = 1, 2, \ldots, m). \tag{9.5.5}$$

The requirements in Lemma 9.4.6 are satisfied if we can choose the b_j in such a way that (9.5.5) implies that all the c_n vanish, and this is obviously possible if and only if no eigenvalue has multiplicity $> m$.

Example 9.5.2. Same operator as in Example 9.5.1, but we assume that the eigenvalues are all distinct. We know that a one dimensional control $bu(t)$ suffices for approximate controllability, which will occur if and only if

$$\sum_{n=1}^{\infty} c_n e^{-\mu_n t}(b, y_n) \equiv 0 \quad \text{implies} \quad c_n = 0 \quad (n = 1, 2, \ldots) \tag{9.5.6}$$

or

$$(b, y_n) \neq 0 \quad (n = 1, 2, \ldots). \tag{9.5.7}$$

The systems in Examples 9.5.1 and 9.5.2 are realizable as differential systems. Let $H = L^2(\Omega)$, where $\Omega \subseteq \mathbb{R}^2$ is the rectangle $0 < x_1 < \pi/a_1, 0 < x_2 < \pi/a_2$, $A = \Delta$ the Laplacian, and β the Dirichlet boundary condition on the boundary Γ. We bring into play $A_2(\beta)$ as constructed in **7.5**; $A_2(\beta)$ is self adjoint with pure point spectrum and the eigenvalues are

$$-\lambda_{n_1,n_2} = \mu_{n_1,n_2} = a_1^2 n_1^2 + a_2^2 n_2^2 \quad (n_1, n_2 = 1, 2, \ldots), \tag{9.5.8}$$

corresponding to (nonnormalized) eigenfunctions

$$y_{n_1,n_2}(x_1, x_2) = \sin a_1 n_1 x_1 \sin a_2 n_2 x_2. \tag{9.5.9}$$

The system (9.5.1) is

$$
\begin{aligned}
y_t(t, x_1, x_2) &= \Delta y(t, x_1, x_2) \\
&+ b_1(x_1, x_2)u_1(t) + \cdots + b_m(x_1, x_2)u_m(t) \quad ((x_1, x_2) \in \Omega), \\
y(t, x_1, x_2) &= 0 \quad ((x_1, x_2) \in \Gamma). \quad (9.5.10)
\end{aligned}
$$

If a_1^2/a_2^2 is an irrational number, then a_1^2 and a_2^2 are linearly independent over the integers, so that a given number can be written as the right side of (9.5.8) for only one pair (n_1, n_2). It then follows that eigenvalues are all distinct, and we have an operator as in Example 9.5.2. Approximate controllability with a one dimensional control $b(x_1, x_2)u(t)$ obtains as long as

$$\int_0^{\pi/a_1} \int_0^{\pi/a_2} b(x_1, x_2) \sin a_1 n_1 x_1 \sin a_2 n_2 x_2 \, dx_1 \, dx_2 \neq 0 \tag{9.5.11}$$

for all possible combinations n_1, n_2 with $n_k = 1, 2, \ldots$. On the other hand, assume that Ω is a square: $a_1 = a_2$. It follows from Landau [1927, Satz 162] that the number of ways an integer n can be decomposed in a sum of two squared integers is unbounded as $n \to \infty$, so there are eigenvalues of arbitrarily high multiplicity, and it is not possible to attain approximate controllability with finite dimensional controls.

One wonders what this "application" means. In the first place, the condition that a_1^2/a_2^2 is irrational is meaningless in practice. In the second, the controllability condition (9.5.11) is highly unstable: it can be destroyed by slightly varying $b(x_1, x_2)$ in the L^2 norm (or, for that matter, in any of the other usual norms).

Example 9.5.3. The operator is

$$A(\Sigma c_n y_n) = \sum_{n=-\infty}^{\infty} (-in) c_n y_n. \tag{9.5.12}$$

The semigroup generated by A is no longer analytic (in fact, is a group) so that Lemma 9.4.6 does not apply. Condition (9.3.8) for approximate controllability in time $\bar{t}$ is

$$\sum_{n=-\infty}^{\infty} c_n e^{-int}(b, y_n) = 0 \quad (0 \le t \le \bar{t}) \quad \text{implies} \quad c_n = 0$$

$$(n = \ldots, -1, 0, 1, \ldots) \tag{9.5.13}$$

so that

$$b_n = (b, y_n) \ne 0 \quad (n = \ldots, -1, 0, 1, \ldots) \tag{9.5.14}$$

is again a necessary condition. It follows from elementary Fourier series theory that it is also sufficient if $\bar{t} \ge 2\pi$. In fact, assume that the series in (9.5.13) vanishes in $0 \le t \le 2\pi$. Then the function $\phi(t) = \Sigma c_n e^{-int}(b, y_n)$ has an absolutely convergent Fourier series whose sum vanishes in $0 \le t \le 2\pi$, thus every one of its Fourier coefficients must be zero.

On the other hand, the implication (9.5.13) may hold if $\bar{t} < 2\pi$ for particular sequences $\{b_n\}$. For instance, if

$$b_n = 0(e^{-\varepsilon n}) \quad \text{as } n \to \infty \tag{9.5.15}$$

then the function on the left side of (9.5.13) can be extended analytically to the strip $|\text{Im } t| < \varepsilon$, thus it will vanish identically if it vanishes in $[0, \bar{t}], \bar{t} > 0$. However, other choices of $\{b_n\}$ preclude approximate controllability if $\bar{t} < 2\pi$ (see below).

Example 9.5.3 is also realizable as a differential system; we take $\Omega = (0, 2\pi)$, $Ay = y'$ the first derivative operator with domain consisting of all absolutely continuous functions $y(\cdot)$ with $y'(\cdot) \in L^2(0, 2\pi)$ satisfying the periodic boundary condition $y(0) = y(2\pi)$. The system is then

$$y_t(t, x) = y_x(t, x) + b(x)u(t) \quad (0 < x < 2\pi),$$
$$y(t, 0) = y(t, 2\pi) \tag{9.5.16}$$

and the (nonnormalized) eigenvectors $y_n(x) = e^{-inx}$. The fact that a time $\bar{t} \geq 2\pi$ may be needed for approximate controllability can now be seen in terms of velocity of propagation of control effects as in Example 9.4.2; for instance, if $b(x)$ has support in $[0, \varepsilon]$ there cannot be approximate controllability in time $< 2\pi - \varepsilon$. A more transparent example can be produced by Fourier analysis:

Example 9.5.4. Let $b_n = 1/n$. Then (9.5.16) is not approximately controllable in any time $\bar{t} < 2\pi$.

To see this, let $\psi(\cdot)$ be a nonzero test function with support in $(\bar{t}, 2\pi]$. Repeated integration by parts shows that the Fourier series $\psi(t) = \Sigma a_n e^{-int}$ has coefficients satisfying $|a_n| = O(|n|^{-p})$ for all p and not all zero, thus if we define $c_n = na_n$, we obtain a serious violation of (9.5.13). Of course, the counterexample works with $b_n = 1/n^k$, $c_n = n^k a_n$.

In the rest of the section we deal with "exact controllability to H" as defined in the previous section. We require that H be dense in E and begin with a reexamination of Example 9.5.3 or, equivalently, of the system (9.5.16). Since exact controllability to H in time $\bar{t}$ implies approximate controllability in the same time we assume that $\bar{t} \geq 2\pi$ and that (9.5.14) holds. Equation

$$y(\bar{t}, 0, u) = \bar{y} = \Sigma \bar{c}_n y_n \tag{9.5.17}$$

for the system (9.5.16) is, coordinate by coordinate,[1]

$$\int_0^{\bar{t}} e^{-int} v(\tau) d\tau = \frac{\bar{c}_n}{b_n} \quad (n = -1, 0, 1, \ldots) \tag{9.5.18}$$

for $v(\tau) = u(\bar{t} - \tau)$. This is an example of a **moment problem** for a sequence of functions $\phi_n(\cdot)$ in an interval $[a, b]$; we want to solve simultaneously the infinite set of equations

$$\int_a^b \phi_n(t) v(t) dt = a_n. \tag{9.5.19}$$

One of the nicest methods (when it works) is that of **biorthogonal sequences**. To fix ideas, assume that each $\phi_n(\cdot)$ belongs to $L^2(a, b)$. A sequence $\{\psi_n(\cdot)\}$ of functions in $L^2(a, b)$ is **biorthogonal** to $\{\phi_n(\cdot)\}$ if

$$\int_a^b \phi_j(t) \psi_k(t) dt = \delta_{jk}, \tag{9.5.20}$$

δ_{jk} the Kronecker delta $\delta_{jk} = 0$ if $j \neq k$, $\delta_{kk} = 1$. Then, if the series

$$v(t) = \sum a_n \psi_n(t) \tag{9.5.21}$$

---

[1] Here and in other places below the bar does not indicate complex conjugate: the $\bar{c}_n$ are the Fourier coefficients of $\bar{y}$.

is convergent in $L^2(a, b)$, direct replacement in (9.5.19) reveals $v(\cdot)$ as a solution in $L^2(a, b)$ of the moment problem. One may seek solutions in different spaces (say, $C(a, b)$ or $L^p(a, b)$), and, depending on the space, conditions must be put on the sequence a_n that guarantee convergence of (9.5.21). The scalar product (9.5.20) is modified as usual for complex spaces. The moment problem can also be considered with measures μ instead of functions v.

Example 9.5.5. The moment problem (9.5.18) is one of the simplest, at least if $\bar{t} = 2\pi$. In fact, the sequence $\phi_n(t) = e^{-int}$; $n = \ldots, -1, 0, 1, \ldots$ is orthogonal, so that, except for a factor, is its own biorthogonal sequence: $\psi_n(t) = e^{-int}/2\pi$. Accordingly, if one assumes that $\Sigma |\bar{c}_n/b_n|^2 < \infty$, the (unique) solution of (9.5.18) is given by

$$v(t) = \frac{1}{2\pi} \sum_{n=-\infty}^{\infty} \frac{\bar{c}_n}{b_n} e^{-int}. \tag{9.5.22}$$

We solve (9.5.17) with the control $u(\bar{t} - \tau) = v(\tau)$ in $L^2(0, \bar{t})$. We obtain controls in $C(0, \bar{t})$ in the form (9.5.22) if $\bar{y}$ belongs to the space H defined by

$$\|\bar{y}\|_H = \sum_{n=-\infty}^{\infty} \left| \frac{\bar{c}_n}{b_n} \right| < \infty \tag{9.5.23}$$

and we have

$$\|u\|_{C(0,\bar{t})} \leq C \|\bar{y}\|_H. \tag{9.5.24}$$

All the L^2 arguments (except uniqueness) apply to $\bar{t} > 2\pi$; we just solve the moment problem in $[0, 2\pi]$ and extend the solution by $u(t) = 0$ in $t > 2\pi$.

Example 9.5.6. We consider the operator

$$A(\Sigma c_n y_n) = \sum_{n=1}^{\infty} -n^2 c_n y_n \tag{9.5.25}$$

generating the self adjoint analytic semigroup

$$S(t)(\Sigma c_n y_n) = \sum_{n=1}^{\infty} e^{-n^2 t} c_n y_n \tag{9.5.26}$$

and realizable as the one dimensional version of (9.5.10):

$$y_t(t, x) = y_{xx}(t, x) + b(x)u(t) \quad (0 < x < \pi),$$
$$y(t, 0) = y(t, \pi) = 0. \tag{9.5.27}$$

In this impersonation, the operator A is the second derivative in $L^2(0, \pi)$ with domain consisting of all continuously differentiable $y(\cdot)$ with absolutely continuous $y'(\cdot)$ and $y''(\cdot) \in L^2(0, \pi)$, satisfying the Dirichlet boundary condition

$y(0) = y(\pi) = 0$. The (nonnormalized) eigenfunctions are $y_n(x) = \sin nx$. The *a priori* restriction to 1-dimensional control is at least heuristically justified by the fact that this operator fits Example 9.5.2, so that approximate controllability in any time $\bar{t} > 0$ can be achieved with 1-dimensional controls. We assume the approximate controllability condition (9.5.7), in this case

$$b_n = \int_0^\pi b(x) \sin nx \neq 0 \quad (n = 1, 2, \ldots), \qquad (9.5.28)$$

which is the 1-dimensional companion of (9.5.11).

Coordinate by coordinate, the equation (9.5.17) is

$$\int_0^{\bar{t}} e^{-n^2 \tau} v(\tau) d\tau = \frac{\bar{c}_n}{b_n} \quad (n = 1, 2, \ldots), \qquad (9.5.29)$$

for $v(\tau) = u(\bar{t} - \tau)$, a moment problem having little in common with (9.5.18), since the functions $\{e^{-n^2 t}\}$ are anything but orthogonal in $L^2(0, \bar{t})$. Biorthogonal sequences will be constructed using the following result.

Theorem 9.5.7. *Let $\{\phi_n\}$ be a sequence in a Hilbert space H, and denote by H_n the closed subspace of H generated by $\{\phi_j; j \neq n\}$. Then (a) if there exists a biorthogonal sequence $\{\psi_n\}$,*

$$(\psi_j, \phi_k) = \delta_{jk}, \qquad (9.5.30)$$

then $\phi_n \notin H_n$ for all n. (b) If $\phi_n \notin H_n$ for all n, a biorthogonal sequence $\{\psi_n\}$ exists and satisfies

$$\|\psi_n\| = 1/\text{dist}(\phi_n, H_n), \qquad (9.5.31)$$

(c) the biorthogonal sequence constructed in (b) is norm optimal in the sense that, if $\{\tilde{\psi}_n\}$ is an arbitrary sequence biorthogonal to $\{\phi_n\}$ then $\|\tilde{\psi}_n\| \geq 1/\text{dist}(\phi_n, H_n)$.

Proof. (a) If $\phi_j \in H_j$ then $(\psi_j, \phi_k) = 0$ for $j \neq k$ implies $(\psi_j, \phi_j) = 0$ and there cannot be any biorthogonal sequence. (b) By intercession of the projection theorem there exists $y_n \in H_n$ minimizing the distance $\|\phi_n - y\|$, $y \in H_n$. We define $\psi_n = (\phi_n - y_n)/d_n^2$. Since $\phi_n - y_n$ must be orthogonal to H_n (see (2.0.36)), it is clear that $\{\psi_n\}$ is a biorthogonal sequence satisfying (9.5.31). (c) If $\{\tilde{\psi}_n\}$ is biorthogonal to ϕ_n then $\tilde{\psi}_n - \psi_n$ is orthogonal to the closed subspace H generated by all ϕ_n; since $\psi_n \in H$ we have $\|\tilde{\psi}_n\|^2 = \|\psi_n\|^2 + \|\tilde{\psi}_n - \psi_n\|^2 \geq \|\psi_n\|^2$. This ends the proof. ∎

Theorem 9.5.8. *Let $\{\mu_n\}$ be a sequence $0 < \mu_1 < \mu_2 < \cdots$ and let*

$$\phi_n(t) = e^{-\mu_n t} \qquad (9.5.32)$$

(a) If

$$\sum_{n=1}^\infty \frac{1}{\mu_n} < \infty \qquad (9.5.33)$$

*then, for $T > 0$ arbitrary there exists a biorthogonal sequence $\{\psi_n(\cdot)\}$ in $L^2(0, T)$
with*

$$\|\psi_n\|_{L^2(0,T)} \leq C\sqrt{2\mu_n} \prod_{j=1}^{\infty}{}' \left|\frac{\mu_j + \mu_n}{\mu_j - \mu_n}\right| = C\sqrt{\frac{\mu_n}{2}} \frac{\prod_{j=1}^{\infty}\left(1 + \dfrac{\mu_n}{\mu_j}\right)}{\prod_{j=1}^{\infty}{}'\left|1 - \dfrac{\mu_n}{\mu_j}\right|} \qquad (9.5.34)$$

*where C is a constant depending on T and $\{\mu_n\}$ and the prime in the infinite
products indicates elimination of the factor corresponding to $j = n$. (b) If (9.5.33)
does not hold, there exists no sequence biorthogonal to $\{\phi_n(\cdot)\}$, and the subspace
H generated by all $\phi_n(\cdot)$ coincides with $L^2(0, T)$.*

Proof. Consider the exponentials (9.5.32) as elements of the Hilbert space
$L^2(0, \infty)$. Kaczmarz and Steinhaus develop in [1935, p. 86] an explicit expression for the distance d_∞ in $L^2(0, \infty)$ from $\phi_n(\cdot)$ to the subspace H_n generated by
the rest of the ϕ_n. This distance is zero if $\Sigma 1/\mu_n = \infty$ (so that no biorthogonal
sequences exist); if $\Sigma 1/\mu_n < \infty$, the distance is given by

$$d_\infty^2 = \frac{1}{2\mu_n} \prod_{j=1}^{\infty}{}' \left(\frac{\mu_j - \mu_n}{\mu_j + \mu_n}\right)^2 = \frac{2}{\mu_n} \frac{\prod_{j=1}^{\infty}{}'\left(1 - \dfrac{\mu_n}{\mu_j}\right)^2}{\prod_{j=1}^{\infty}\left(1 + \dfrac{\mu_n}{\mu_j}\right)^2} \qquad (9.5.35)$$

The proof is elementary and the reader is invited (Example 9.5.10) to reconstruct
it. Another proof can be found in Schwartz [1959, p. 70].

To relate the interval $[0, \infty)$ with the finite interval $[0, T]$ Schwartz [1959,
p. 54] shows that, under condition (9.5.33) there exists a constant C depending on
T and $\{\mu_n\}$ such that

$$\|p\|_{L^2(0,\infty)} \leq C\|p\|_{L^2(0,T)} \qquad (9.5.36)$$

for arbitrary finite linear combinations $p(t)$ of exponentials (9.5.32). It results from
this that $d_\infty \leq Cd_T$ (subindices on d indicate the interval), thus existence of the
biorthogonal sequence and the bound for it follow from Theorem 9.5.7. We note
that if $\Sigma 1/\mu_n = \infty$ the subspace E of $L^2(0, \infty)$ generated by all exponentials
(9.5.32) coincides with $L^2(0, \infty)$. This is equivalent through a simple change
of variable to the classical Müntz-Szász Theorem that gives $\Sigma 1/\mu_n = \infty$ as a
necessary and sufficient condition for the linear combinations of the powers t^{μ_n} to
be dense in $L^2[0, T]$ (Kaczmarz–Steinhaus [1935, p. 86]).   ∎

We go back to the moment problem (9.5.29), where the sequence is $\{\mu_n\} = \{n^2\}$.
To compute the norms $\|\psi_n\|$ of the biorthogonal sequence manufactured in

Theorem 9.5.8 we use the infinite products for $\sin x$ and $\sinh x$,

$$\sin x = x \prod_{j=1}^{\infty} \left(1 - \frac{x^2}{j^2 \pi^2} \right), \qquad \sinh x = x \prod_{j=1}^{\infty} \left(1 + \frac{x^2}{j^2 \pi^2} \right).$$

The first one gives

$$\prod_{j=1}^{\infty}{}' \left(1 - \frac{x^2}{j^2 \pi^2} \right) = -\frac{n^2 \pi^2 \sin x}{x(x - n\pi)(x + n\pi)}. \tag{9.5.37}$$

We calculate the denominator in the last expression (9.5.34) putting $x = \pi n$ in (9.5.37) and taking absolute value; the result equals $1/2$. For the numerator we plug πn in the infinite product for the sinh. We end up with the estimate

$$\|\psi_n\|_{L^2(0,T)} \le C e^{\pi n} \quad (n = 1, 2, \ldots) \tag{9.5.38}$$

and thus deduce that the moment problem (9.5.29) has a solution as long as $\Sigma e^{\pi n} |\bar{c}_n / b_n| < \infty$. However, in applications to the proof of the maximum principle we need an L^∞ bound on the controls, and this does not come as cheap as in Example 9.5.5, since we have no information whatsoever on the L^∞ norm of the ψ_n. To deal with this problem we construct a different sequence $\{\eta_n\}$ biorthogonal to the exponentials $\{e^{-n^2 t}\}$ using this time as $\{\mu_n\}$ the sequence

$$\{\varepsilon + n^2; n = 1, 2, \ldots\} \tag{9.5.39}$$

where $0 < \varepsilon < 1$. Then we define

$$\psi_{n,\varepsilon}(t) = n^2 \int_0^t \eta_n(t) e^{-\varepsilon t} dt \quad (0 \le t \le T, \ n = 1, 2, \ldots). \tag{9.5.40}$$

Obviously $\psi_{n,\varepsilon}(0) = 0$, and it follows from one of the biorthogonality relations that $\psi_{n,\varepsilon}(T) = 0$. Now we have, integrating by parts,

$$\delta_{jn} = \int_0^T \eta_j(t) e^{-(n^2 + \varepsilon)t} dt$$

$$= \int_0^T (\eta_j(t) e^{-\varepsilon t}) e^{-n^2 t} dt = \int_0^T \psi_j(t) e^{-n^2 t} dt$$

so that the sequence $\{\psi_n(\cdot)\}$ is biorthogonal to $\{e^{-n^2 t}\}$ in $[0, T]$. We obtain from (9.5.40) and Schwarz's inequality that

$$\|\psi_n\|_{C(0,T)} \le \frac{n^2}{\sqrt{2\varepsilon}} \|\eta_n\|_{L^2(0,T)}. \tag{9.5.41}$$

Thus, it only remains to calculate the L^2 norm of the η_n. Since $\mu_j - \mu_n = j^2 - n^2$, no changes are necessary in the denominator of (9.5.34) except for an additional factor $n^2 - \varepsilon$. In the numerator, we use the bound

$$1 + \frac{n^2 + \varepsilon}{j^2 + \varepsilon} \le 1 + \frac{n^2 + \varepsilon}{j^2}$$

and estimate as in the construction of the ψ_n, with an additional factor $n^2 + \varepsilon$. We finally obtain

$$\|\eta_n\|_{L^2(0,T)} \le C e^{\pi\sqrt{n^2+\varepsilon}} \le C' e^{\pi n}$$

which we combine with (9.5.41):

$$\|\Psi_n\|_{C(0,T)} \le C(\varepsilon, T) n^2 e^{\pi n}. \tag{9.5.42}$$

We then obtain a solution of (9.5.29) in the form

$$u(t) = \sum_{n=1}^{\infty} \frac{\bar{c}_n}{b_n} \psi_n(t) \tag{9.5.43}$$

if $\|\bar{y}\|$ belongs to the space H defined by

$$\|\bar{y}\|_H = \sum_{n=1}^{\infty} n^2 e^{n\pi} \left| \frac{\bar{c}_n}{b_n} \right| < \infty. \tag{9.5.44}$$

The solution satisfies

$$\|u\|_{C(0,\bar{t})} \le M(\bar{t}) \|\bar{y}\|_H. \tag{9.5.45}$$

Example 9.5.9. Similar arguments work in different spaces. For instance, consider the system (9.5.27) in the space $C_0[0, \pi]$ of all continuous functions in $[0, \pi]$ that vanish at 0 and π, equipped with the supremum norm. The operator A is the second derivative operator with domain consisting of all twice continuously differentiable functions such that y and Ay vanish at 0 and π. The operator A generates an analytic compact semigroup given by (9.5.26) for arbitrary $y \in C_0[0, \pi]$, y_n the eigenfunctions $\sin nx$, c_n the sine Fourier coefficients of y (the series is convergent in $L^2(0, \pi)$ but not necessarily in $C_0[0, \pi]$). The control $u(\cdot)$ constructed above solves the problem $y(\bar{t}, 0, u) = \bar{y}$ for every $\bar{y} \in H$ (H is obviously a subspace of $C_0[0, \pi]$).

Example 9.5.10. The **Gram determinant** of a finite set $\phi_1, \ldots, \phi_n$ of elements of a Hilbert space H is

$$G(\phi_1, \ldots, \phi_n) = \det\{(\phi_j, \phi_k)\}.$$

(a) $G(\phi_1, \ldots, \phi_n) \ne 0$ if and only if $\phi_1, \ldots, \phi_n$ are linearly independent. (b) If $\phi_1, \ldots, \phi_n$ are linearly independent, the distance d of an element $y \in H$ to the subspace generated by $\phi_1, \ldots, \phi_n$ is given by

$$d^2 = \frac{G(y, \phi_1, \ldots, \phi_n)}{G(\phi_1, \ldots, \phi_n)}$$

(see Kaczmarz–Steinhaus [1935, p. 77]) This formula can be used to calculate the distance from one exponential (9.5.32) to the subspace generated by a finite subset of the others and then, taking limits, to the subspace generated by all the others. For the actual computation of the Gram determinants the following theorem of Cauchy is used: if $\{a_{jk}\} = \{1/(p_j + q_k)\}\ (1 \le j, k \le n)$ then $\det\{a_{jk}\}$ is

$$\frac{\Pi_{j>k}(p_j - p_k)(q_j - q_k)}{\Pi_{j,k}(p_j + q_k)}$$

(Kaczmarz–Steinhaus [1935, p. 87]).

9.6. Controllability and the Minimum Principle

We apply the minimum principle (Theorem 9.1.10) to the control system (9.5.27),

$$y_t(t, x) = y_{xx}(t, x) + b(x)u(t) \quad (0 < x < \pi),$$
$$y(t, 0) = y(t, \pi) = 0. \tag{9.6.1}$$

To cover later applications, we work in the space $E = C_0[0, \pi]$ as in Example 9.5.9. The function $b(\cdot)$ satisfies

$$b_n = \int_0^\pi b(x) \sin nx\, dx \ne 0 \quad (n = 1, 2, \ldots) \tag{9.6.2}$$

as in **9.5**. Theorem 9.1.10 is applied in the point target case

$$y(\bar{t}, \zeta, u) = \bar{y} \tag{9.6.3}$$

with the following cast: $F = X^* = \mathbb{R}$, $U = [-1, 1]$, K the operator

$$Ky = K(\Sigma c_n \sin nx) = \sum_{n=1}^\infty \frac{n^{3+\rho} e^{n\pi}}{b_n} c_n \sin nx \tag{9.6.4}$$

($\rho > 0$ fixed), the domain $D(K)$ consisting of all $y \in C_0[0, \pi]$ such that the right side of (9.6.4) belongs to $C_0[0, \pi]$. There is no simple characterization of the domain in terms of summability conditions on the Fourier coefficients, but there is little doubt that K has a bounded everywhere defined inverse given by

$$K^{-1}y = K^{-1}(\Sigma c_n \sin nx) = \sum_{n=1}^\infty n^{-3-\rho} e^{-n\pi} b_n c_n \sin nx \tag{9.6.5}$$

and that $S(t)E \subseteq D(K)$ and $S(t)K^{-1} = K^{-1}S(t)$. Moreover, if $y \in D(K)$ then the $\{n^{3+\rho} e^{n\pi} c_n/b_n\}$ are the sine Fourier coefficients of the continuous function Ky, thus they are bounded by a constant times $\|Ky\|_E$. Defining the norm $\|\cdot\|_H$ as in (9.5.44) we have

$$\|y\|_H = \sum_{n=1}^\infty n^2 e^{n\pi} \left| \frac{\bar{c}_n}{b_n} \right| \le \sum_{n=1}^\infty n^{-1-\rho} n^{3+\rho} e^{n\pi} \left| \frac{\bar{c}_n}{b_n} \right|$$

$$\le \left(\sum_{n=1}^\infty n^{-1-\rho} \right) \sup_{n \ge 1} \left(n^{3+\rho} e^{n\pi} \left| \frac{\bar{c}_n}{b_n} \right| \right) \le C \|Ky\|_E. \tag{9.6.6}$$

Combining with (9.5.45), we deduce that for every $\bar{t} > 0$ and every $\bar{y} \in D(K)$ there exists a control in $C(0, \bar{t})$ such that

$$\|u\|_{C(0,\bar{t})} \le C(\bar{t})\|\bar{y}\|_{D(K)} \tag{9.6.7}$$

and $y(\bar{t}, 0, u) = \bar{y}$. It follows that $R(\bar{t}, 0; U)$ contains a ball in $D(K)$, thus the same is true of $R(\bar{t}, \zeta; U)$ for any ζ since $S(\bar{t})\zeta \in D(K)$.

Theorem 9.6.1. *Let $\bar{u}(\cdot)$ be a time optimal control for the system* (9.6.1) *with $\bar{y} \in D(K)$. Then there exists $z \in (K^*C_0[0, \pi]^*)^1(B^*) = (K^*\Sigma_0[0, \pi])^1(B^*)$, $z \ne 0$ such that*

$$\bar{u}(t) = \frac{\kappa(t)}{|\kappa(t)|} \quad \text{where } \kappa(t) \ne 0, \tag{9.6.8}$$

with

$$\kappa(t) = \int_0^\pi b(x)\bar{z}(t, x)dx, \tag{9.6.9}$$

$\bar{z}(t, x)$ *the solution of the backwards heat equation*

$$\begin{aligned} \bar{z}_t(t, x) &= -\bar{z}_{xx}(t, x) \quad (0 < x < \pi), \\ \bar{z}(t, 0) &= \bar{z}(t, \pi) = 0 \end{aligned} \tag{9.6.10}$$

with final condition

$$\bar{z}(\bar{t}, x) = z(x). \tag{9.6.11}$$

The set where $\kappa(t) \ne 0$ is the complement of a (possibly empty) increasing sequence $\{t_n\}$ which, if infinite, can only accumulate at the terminal time $\bar{t}$.

The proof is a direct application of Theorem 9.1.10. Note that the operator B^* is given by $(B^*z)(x) = \int b(x)z(x)dx$; we only need to apply it to $z(\cdot) \in L^1(0, \pi)$ since $S(t)^* : K^*\Sigma_0[0, \pi] \to C_0[0, \pi]^\odot = L^1(0, \pi)$ (Lemma 9.1.9). The last statement on $\kappa(t)$ follows from Corollary 9.3.5. It implies that the optimal control $\bar{u}(\cdot)$ is bang-bang on the entire control interval $0 \le t \le \bar{t}$ with switching points contained in the sequence $\{t_n\}$. Note also that $\bar{z}(t, x)$ is smooth for $t < \bar{t}$ but it may become singular as $t \to \bar{t}$; however, the fact that $z \in (K^*\Sigma_0[0, \pi])^1(B^*)$ means that the **switching function** $\kappa(t)$ satisfies

$$\int_0^{\bar{t}} |\kappa(t)|dt < \infty. \tag{9.6.12}$$

For some choices of $b(\cdot)$ the requirement that $\bar{y} \in D(K)$ can be checked without actually computing Fourier coefficients:

Example 9.6.2. Assume that the sine Fourier coefficients $\{b_n\}$ of $b(\cdot)$ satisfy $1/|b_n| = 0(n^k)$ for some k. Then $\bar{y} \in D(K)$ if (but not only if) it can be extended to a function $\bar{y}(x + i\xi)$, (a) 2π-periodic and odd in x, (b) analytic in the strip $|\xi| < \pi$, (c) infinitely differentiable in the strip $|\xi| \le \pi$.

To show this we calculate the Fourier coefficients a_n with respect to $\{e^{inx}\}$ and, invoking analyticity, periodicity and oddness, we displace the domain of integration to the line $\xi = \pi$ for $n \geq 0$ and to $\xi = -\pi$ for $n < 0$; then we use infinite differentiability on the lines and integrate by parts as much as necessary. We obtain $|a_n| = 0(n^{-j}e^{-\pi|n|})$ for j arbitrary; of course the same is true for the Fourier sine coefficients.

We close this section with a few remarks on the general system

$$y'(t) = Ay(t) + \sum_{j=1}^{m} b_j u(t), \qquad y(0) = \zeta, \tag{9.6.13}$$

A the infinitesimal generator of a strongly continuous semigroup $S(t)$ in a Banach space E; some of these remarks (but not all) are related to the minimum principle.

Assume that we want to drive an arbitrary initial condition ζ to the origin $\bar{y} = 0$ in time $\bar{t}$ (in many practical problems, this "driving to equilibrium" is what is needed). This motivates the following definition, which we may formulate for the general system (9.1.1). We say that (9.1.1) (in particular, (9.6.13)) is **exactly controllable to zero** in time $\bar{t}$ if, given $\zeta \in E$ arbitrary there exists $u(\cdot) \in L_w^\infty(0, \bar{t}; F)$ with

$$y(\bar{t}, \zeta, u) = 0, \tag{9.6.14}$$

with a corresponding definition of p-exact controllability.

Exact controllability to zero is a much weaker concept than exact controllability. For instance, the system (9.6.13) can never be exactly controllable when E is infinite dimensional and A the generator of a compact semigroup (see Example 9.4.11). On the other hand, $y(\bar{t}, \zeta, u) = 0$ is $y(\bar{t}, 0, u) = -S(\bar{t})\zeta$, thus p-exact controllability to zero in time $\bar{t}$ is equivalent to

$$R^p(\bar{t}) \supseteq S(\bar{t})E. \tag{9.6.15}$$

This holds, for instance, for the system (9.5.27) as a particular case of the results in **9.5**, since $S(\bar{t})E \subseteq D(K)$ for every $\bar{t} > 0$.

The question of what the inclusion (9.6.15) means for a system (9.6.13) driven by finite dimensional controls has been addressed in the author [1975 : 3] in a somewhat more general formulation. First, measure-valued controls are admitted, that is, we use the equation

$$dy(t) = Ay(t)dt + \sum_{j=1}^{m} b_j \mu_j(dt), \qquad y(0) = \zeta, \tag{9.6.16}$$

with μ_j in the space $\Sigma[0, T]$ of regular, bounded Borel measures in the interval $[0, T]$. The solution is, by definition,

$$y(t) = S(t)\zeta + \sum_{j=1}^{m} \int_0^t S(t - \tau)b_j \mu(d\tau), \tag{9.6.17}$$

where the integrals in (9.6.17) are understood as in Bartle [1956]; for more details on these integrals and more general ones, see Chapter 10, especially **10.2**. The set of all elements (9.6.17) for $t = T$ is named $R_\Sigma(T)$. This formula leads to a functional calculus for the operator A. The class of functions in this calculus is $\mathcal{E}[0, T]$, consisting of all Fourier-Laplace transforms $f(\lambda)$ of measures in $\Sigma[0, T]$,

$$f(\lambda) = \int_0^T e^{\lambda t} \mu(dt). \tag{9.6.18}$$

Each function in $\mathcal{E}[0, T]$ is entire and satisfies $|f(\lambda)| \leq \|\mu\| e^{(\text{Re}\lambda)T}$ ($\lambda \in \mathbb{C}$), where $\|\mu\|$ is the total variation norm in $\Sigma[0, T]$. Clearly, $\mathcal{E}[0, T]$ is a linear space. It follows from properties of convolution of measures (Dunford–Schwartz [1958, p. 643]) that, if $f(\cdot) \in \mathcal{E}[0, T]$ and $g(\cdot) \in \mathcal{E}[0, T']$ then the product $f(\cdot)g(\cdot)$ belongs to $\mathcal{E}[0, T + T']$.

By definition, $f(A)$ is the operator given by

$$f(A)y = \int_0^T S(t) y \mu(dt) \tag{9.6.19}$$

and is well defined due to uniqueness of Laplace transforms of measures. Moreover, if $f, g \in \mathcal{E}[0, T]$ then $(f + g)(A) = f(A) + g(A)$ and $(fg)(A) = f(A)g(A)$.

Example 9.6.3. Let $g(\cdot) \in \mathcal{E}[0, \bar{t}]$ such that $g \neq 0$,

$$g(A)E \subseteq R_\Sigma(\bar{t}). \tag{9.6.20}$$

Then there exists $f(\cdot) \in \mathcal{E}[0, T]$ for some $T > \bar{t}$ such that $f \neq 0$ and

$$f(A) = 0. \tag{9.6.21}$$

For a proof see the author [1975 : 3, Lemma 2.1 and Corollary 9.2.2].

Example 9.6.3 gives (the beginning of) an answer to our question about whether the exact controllability inclusion (9.6.15) is possible; if $S(\bar{t})E \subseteq R^1(\bar{t})$ then (9.6.20) holds (with $\mu = \delta(t - \bar{t})$, $f(\lambda) = e^{\lambda \bar{t}}$) so that we have (9.6.21) for some nonzero $f(\cdot) \in \mathcal{E}[0, T]$. It remains to clarify the consequences of annihilation by an element of $\mathcal{E}[0, T]$.

We note that (9.6.21) does not depend sharply on the class of functions f in use. In fact, if it holds for the Laplace transform of a measure μ in $\Sigma[0, T]$ then, by the convolution theorem, it holds as well for the Laplace transform of $\phi * \mu$, ϕ a test function with support in $t \geq 0$. It follows then that (9.6.21) will hold with the Laplace transform of an infinitely differentiable function with compact support contained in $t \geq 0$. It turns out, however, that the function space most convenient for the study of this problem is H^2, consisting of all functions $f(\lambda) = f(x + i\xi)$ analytic in the left half plane Re $\lambda < 0$ and such that

$$\|f\|_{H^2}^2 = \sup_{x<0} \int_{-\infty}^{\infty} |f(x + i\xi)|^2 d\xi < \infty.$$

The space H^2, endowed with $\| \cdot \|_{H^2}$ is a Hilbert space; the Paley-Wiener theorem (Hoffman [1962, p. 113]) establishes that H^2 is algebraically isomorphic to $L^2(0, \infty)$ through the Laplace transform

$$f(\lambda) = \int_0^\infty e^{\lambda t} \phi(t) dt.$$

The isomorphism is also metric in the sense that $\| f \|_{H^2} = 2\pi \|\phi\|_{L^2}$. The functional calculus is based on the integral

$$f(A)y = \int_0^\infty S(t) y \phi(t) dt \qquad (9.6.22)$$

which exists if we assume (always possible via a translation) that

$$\|S(t)\| \le Ce^{-ct}. \qquad (9.6.23)$$

For properties of this functional calculus, see the author [1966 : 1]. Both functional calculi used in this section are particular cases of that introduced in Dunford–Schwartz [1958, VIII.2, p. 641].

An answer to the question of what (9.6.21) means for an infinitesimal generator was given in the author [1966 : 1] for A the infinitesimal generator of an analytic semigroup satisfying (9.6.23). Under these conditions (see Theorem 7.4.2) the spectrum $\sigma(A)$ is contained in a sector $|\arg(\lambda - \alpha)| \le \pi/2 + \psi$ with $\psi > 0$ and we have

Theorem 9.6.4. *Assume there exists a nonzero $f \in H^2$ such that $f(A) = 0$. Then A is a Jordan operator and*

$$\sum \frac{m_n}{|Re\ \lambda_n|} < \infty \qquad (9.6.24)$$

where $\{\lambda_n\}$ are the eigenvalues of A and m_n is the multiplicity of the eigenvalue λ_n.

For a proof of Theorem 9.6.5, see the author [1966 : 1, Proposition 2.1]. This result is a slight variant of several results where (9.6.21) with f in a given function class, $f \ne 0$ implies properties "of Jordan type" for an operator A; for other examples see Dunford–Schwartz [1958, p. 571].

Another question connected with the system (9.6.13) is illustrated by

Example 9.6.5. The system is (9.5.10) with 1-dimensional control:

$$y_t(t, x_1, x_2) = \Delta y(t, x_1, x_2) + b(x_1, x_2)u(t) \quad ((x_1, x_2) \in \Omega),$$
$$y(t, x_1, x_2) = 0 \qquad\qquad ((x_1, x_2) \in \Gamma). \qquad (9.6.25)$$

We assume that a_1^2/a_2^2 is an irrational number, so that we have an increasing sequence $\{-\mu_n\}$ of distinct eigenvalues. The moment problem corresponding to the equation $y(\bar{t}, 0, u) = \bar{y}$ is

$$\int_0^{\bar{t}} e^{-\mu_n t} v(t) dt = \frac{\bar{c}_n}{b_n} \qquad (n = 0, 1, \ldots), \qquad (9.6.26)$$

for $v(\tau) = u(\bar{t} - \tau)$. Using the integral test for the double series it is easy to show that $\Sigma 1/\mu_n = \infty$ so that, according to Theorem 9.5.8, linear combinations of the exponentials $\{e^{-\mu_n t}\}$ are dense in $L^2(0, T)$. Hence, solutions of (9.6.26) in $L^2(0, \bar{t})$, *a fortiori* in $L^\infty(0, \bar{t})$ (when they exist) are unique in any interval $0 \le t \le \bar{t}$. This means

$$y(\bar{t}, 0, u_1) = y(\bar{t}, 0, u_2) \quad \text{implies} \quad u_1(t) = u_2(t) \text{ a.e. in } 0 \le t \le \bar{t}. \quad (9.6.27)$$

Systems with this property were baptized **rigid** (in $0 \le t \le \bar{t}$) in the author [1967 : 1]; the definition applies as well to the general system (9.1.1). Rigid systems are interesting, if on no other grounds, because they give counterexamples to the minimum principle with point target:

Example 9.6.6. Let the system

$$y'(t) = Ay(t) + Bu(t), \qquad y(0) = \zeta \qquad (9.6.28)$$

be rigid. Then, *any* control $\bar{u}(\cdot)$ in $0 \le t \le \bar{t}$ that does not vanish a.e. in an interval $0 \le t \le \delta > 0$ drives $\zeta = 0$ time optimally to $y(\bar{t}, 0, \bar{u})$. In fact, if $u(\cdot)$ gets to the target in time $\tilde{t} < \bar{t}$ then $v(\cdot)$, defined by $u(t) = 0$ $(0 \le \bar{t} - \tilde{t})$, $u(t) = u(t + (\bar{t} - \tilde{t})) (\bar{t} - \tilde{t} \le t \le \bar{t})$ satisfies $y(\bar{t}, 0, \bar{u}) = y(\bar{t}, 0, u)$; by rigidity $\bar{u} = u$, which is impossible.

Example 9.6.6 makes possible the construction of optimal controls that are not bang-bang and thus precludes any result like Theorem 9.6.1 for the system (9.6.25); note that the summability requirement in Theorem 9.6.1 for the sine Fourier coefficients of the target can be achieved by giving the control a zero of sufficiently high order at $\bar{t}$.

Rigidity of the system (9.6.13) was studied in the author [1966 : 1], [1967 : 1] with controls in $L^2(0, \bar{t}; \mathbb{R}^m)$, assuming A generates an analytic semigroup, using the H^2-functional calculus for A. It was shown there that if A is not a Jordan operator, or if it is and (9.6.24) is violated, then the system (9.6.13) will be always rigid except for $(b_1, b_2, \ldots, b_m)$ in a set N of first category in $E \times E \times \cdots \times E$, that is, in a countable union of closed sets each having nonempty interior. In other words, rigidity of (9.6.13) is generic with respect to $(b_1, \ldots, b_m) \in E \times \ldots \times E$. In particular, there exist vectors $(b_1, \ldots, b_m)$ that make (9.6.13) rigid, and this preempts the minimum principle with point targets, although set targets are still tractable.

Let Ω be a bounded domain, $A_2(\beta)$ one of the elliptic operators in $L^2(\Omega)$ in **6.5**. Under suitable smoothness assumptions, if $A' = A$ and $\beta' = \beta$, $A_2(\beta)$ is self adjoint with compact resolvent, thus its spectrum is real and punctual. It has been shown by Gårding [1953] that

$$\lambda_n = -cn^{2/m}(1 + o(1)) \quad \text{as } n \to \infty \qquad (9.6.29)$$

so that (9.6.13) will be generically rigid except in dimension 1. We note, however, that (9.6.29) takes multiplicities into account, while the condition $\Sigma 1/|\lambda_n| < \infty$ does not.

10

Optimal Control Problems with State Constraints

10.1. Optimal Control Problems with State Constraints

We study in this chapter optimal control problems for the system

$$y'(t) = A(t)y(t) + f(t, y(t), u(t)), \qquad y(0) = \zeta \tag{10.1.1}$$

in a Banach space E. There is a state constraint and a target condition, respectively

$$y(t) \in M(t) \subseteq E \quad (0 \le t < \bar{t}), \qquad y(\bar{t}) \in Y \tag{10.1.2}$$

in a fixed or variable time interval $0 \le t \le \bar{t}$. We may subsume the target condition into the state constraint replacing $M(\bar{t})$ by $M(\bar{t}) \cap Y$ but it is convenient not to do so (among other things, because the terminal time $\bar{t}$ may be unknown).

We have already met these problems. One was the rocket landing problem in **1.4**, where the instantaneous height $h(t)$ obeys the constraint $h(t) \ge 0$. Another problem including state constraints is the minimum surface problem in **1.1**, where the radius $r(x)$ must be nonnegative. We can set this problem up as an optimal control problem writing the curve $r = r(x)$ in parametric form $x = x(t), r = r(t)$; the equations are

$$x'(t) = u(t), \qquad r'(t) = v(t), \qquad x(0) = a, \qquad r(0) = r_a. \tag{10.1.3}$$

It seems reasonable to assume that $x(t)$ is nondecreasing (if it isn't, the surface has kinks and the area can be reduced by ironing them out). The only control constraint is then $u(t) \ge 0$, so that the control set is

$$U = [0, \infty) \times (-\infty, \infty). \tag{10.1.4}$$

The target conditions are

$$x(\bar{t}) = b, \qquad r(\bar{t}) = r_b, \tag{10.1.5}$$

and the state constraint,

$$r(t) \geq 0, \tag{10.1.6}$$

corresponding to $Y = (b, r_b)$ and[1] $M(t) = (-\infty, \infty) \times [0, \infty)$ in (10.1.2). Finally, the cost functional is

$$y_0(t, u, v) = \int_0^t r(t) \sqrt{u(t)^2 + v(t)^2} dt. \tag{10.1.7}$$

A strategy to deal with problems with state constraints was outlined in Example 3.1.4. Let $C(0, T; E)$ be, as usual, the space of all continuous functions $y : [0, T] \to E$ equipped with the supremum norm, and

$$\mathbf{M}(\bar{t}) = \{y(\cdot) \in C(0, \bar{t}; E); y(t) \in M(t) \ (0 \leq t < \bar{t})\}. \tag{10.1.8}$$

Let $C_{ad}(0, \bar{t}; U)$ be the admissible control space for (10.1.1). Define functions $\mathbf{f} : C_{ad}(0, \bar{t}; U) \to C(0, T; E) \times E$ and $\mathbf{f}_0 : C_{ad}(0, \bar{t}; U) \to \mathbb{R}$ by

$$\mathbf{f}(u) = (y(\cdot, u), y(\bar{t}, u)), \qquad \mathbf{f}_0(u) = y_0(\bar{t}, u), \tag{10.1.9}$$

where $y_0(t, u)$ is the cost functional in use. Then the optimal control problem with state constraints is

$$\text{minimize} \quad \mathbf{f}_0(u) \tag{10.1.10}$$

$$\text{subject to} \quad \mathbf{f}(u) \in \mathbf{Y} = \mathbf{M}(\bar{t}) \times Y \tag{10.1.11}$$

with the state constraints appearing as (part of) a target condition. This is a nonlinear programming problem of the type already studied in Chapter 7. To bring into play Kuhn–Tucker inequalities such as (7.2.37) we need to know the dual space $C(0, \bar{t}; E)^*$ which, as we may suspect, is a space of measures with values in the dual E^* and is constructed in next section.

10.2. Integration with Respect to Vector-Valued Measures

A set function μ defined in a field Φ of subsets of a set S with values in a Banach space E is called a (**finitely additive**) **vector measure** if $\mu(\emptyset) = 0$ and

$$\mu\left(\bigcup e_j\right) = \sum \mu(e_j) \tag{10.2.1}$$

for every finite collection $e_1, e_2, \ldots, e_n$ of pairwise disjoint sets in Φ. When necessary to be precise, we speak of an E-**valued measure**. The vector measure μ

---

[1] We are neglecting the obvious state constraint $a \leq x(t) \leq b$. This is immaterial, as this constraint is a consequence of the control constraint and the initial and target conditions.

is σ-**additive** or **countably additive** if (10.2.1) can be extended to countable collections of pairwise disjoint sets in Φ whose union belongs to Φ; the series on the right is assumed to converge in the norm of E.

A finitely additive vector measure μ is **of bounded variation** if

$$\|\mu\|(e) = \sup \sum \|\mu(e_j)\| < \infty, \tag{10.2.2}$$

for every $e \in \Phi$, the supremum taken over arbitrary finite collections $\{e_j\}$ of pairwise disjoint sets in Φ with $e_j \subseteq e$. The **variation** $\|\mu\|$ is itself a positive, finitely additive measure defined in Φ. In fact, if $d, e \in \Phi$ are disjoint and $\{d_j\}, \{e_j\}$ are pairwise disjoint collections in d, e with $\Sigma \|\mu(d_j)\| \geq \|\mu\|(d) - \varepsilon$, $\Sigma \|\mu(e_j)\| \geq \|\mu\|(e) - \varepsilon$, it follows that $\|\mu\|(d \cup e) \geq \Sigma \|\mu(d_j)\| + \Sigma \|\mu(e_j)\| \geq \|\mu\|(d) + \|\mu\|(e)\| - 2\varepsilon$. On the other hand, if c_j is a pairwise disjoint collection in $d \cup e$ giving $\Sigma \|\mu(c_j)\| \geq \|\mu\|(d \cup e) - \varepsilon$, take $d_j = d \cap c_j, e_j = e \cap c_j$; we have $\|\mu\|(d \cup e) \leq \Sigma \|\mu(c_j)\| + \varepsilon = \Sigma \|\mu(d_j \cup e_j)\| + \varepsilon = \Sigma \|\mu(d_j) + \mu(e_j)\| + \varepsilon \leq \Sigma \|\mu(d_j)\| + \Sigma \|\mu(e_j)\| + \varepsilon \leq \|\mu\|(d) + \|\mu\|(e) + \varepsilon$.

The finitely additive measure $\|\mu\|$ is a **majorant** of μ in the sense that

$$\|\mu(e)\| \leq \|\mu\|(e) \quad (e \in \Phi). \tag{10.2.3}$$

We outline below a theory of integration of E-valued functions with respect to E^*-valued finitely additive measures μ of bounded variation defined in a field Φ of subsets of the interval $0 \leq t \leq T$. It is a very particular case of a general theory of Bartle [1956]. Denote by $S(0, T; E, \Phi)$ the space of all **simple** functions, that is, finite linear combinations

$$y(t) = \sum_{j=1}^{m} \chi_j(t) y_j, \tag{10.2.4}$$

with $y_j \in E$ and $\chi_j(t)$ characteristic functions of sets e_j in Φ. We may and will assume that the e_j are pairwise disjoint. If μ is an E^*-valued finitely additive measure then we define

$$\int_0^T \langle y(t), \mu(dt) \rangle = \sum_{j=1}^{m} \langle y_j, \mu(e_j) \rangle \tag{10.2.5}$$

for $y(\cdot) \in S(0, T; E; \Phi)$ (the integrand may also be written $\langle \mu(dt), y(t) \rangle$). The integral does not depend on the representation (10.2.4) for the function $y(t)$. Since $|\langle y_j, \mu(e_j) \rangle| \leq \|y_j\| \|\mu(e_j)\| \leq \|y_j\| \|\mu\|(e_j)$ we have[2]

$$\left| \int_0^T \langle y(t), \mu(dt) \rangle \right|$$

$$\leq \int_0^T \|y(t)\| \|\mu\|(dt) \leq \|\mu\|([0, T]) \max_{0 \leq t \leq T} \|y(t)\|. \tag{10.2.6}$$

---

[2] Norms in different spaces are not always subindexed in this section.

Using this estimate the integral can be extended in an obvious way to functions $y(\cdot)$ which are the uniform limits of functions in $S(0, T; E, \Phi)$; its value does not depend on the particular sequence used in the definition. Finally, the integral satisfies (10.2.6). See Bartle [1956] for details.

We shall (mostly) use the above theory for functions $y(\cdot) \in C(0, T; E)$. Let Φ_0 be the field generated by all intervals $[a, b]$, $[a, b)$, $(a, b]$, $(a, b) \subseteq [0, T]$. Then any $y(\cdot) \in C(0, T; E)$ is the uniform limit of functions in $S(0, T; E, \Phi_0)$, so that the integral $\int \langle y(t), \mu(t) \rangle$ exists.

The measure $\|\mu\|$ on the right-hand side can be upgraded to a better one, as shown below.

Lemma 10.2.1. *Let μ be a E^*-valued finitely additive measure of bounded variation in Φ_0. Then there exists a bounded, regular, countably additive measure $\|\mu\|_0$ defined in a σ-field containing the Borel field $\Phi_b(C)$ in $[0, 1]$ and such that*

$$\left| \int_0^T \langle y(t), \mu(dt) \rangle \right| \leq \int_0^T \|y(t)\| \|\mu\|_0 (dt) \qquad (10.2.7)$$

for every $y(\cdot) \in C(0, T; E)$.

The proof needs

Lemma 10.2.2. *Let μ be an E^*-valued finitely additive measure defined in a field Φ_0 of subsets of $[0, T]$. Assume μ is of bounded variation in Φ_0. Then μ can be extended to the field Φ of all subsets of $[0, T]$ (as a finitely additive measure) with the same variation.*

Proof. Furnish the linear space $B(0, T; E)$ of all bounded E-valued functions defined in $[0, T]$ with the supremum norm, and define a linear functional μ^* in the subspace $S(0, T; E, \Phi_0) \subseteq B(0, T; E)$ by

$$\langle \mu^*, y(\cdot) \rangle = \int_0^T \langle y(t), \mu(dt) \rangle. \qquad (10.2.8)$$

Then, by (10.2.6), $\|\mu^*\|_{S(0,T;E,\Phi_0)^*} \leq \|\mu\|_{\Phi_0}([0, T])$, where $\|\mu\|_{\Phi_0}$ is the total variation of μ computed with sets in Φ_0. On the other hand, if $\{e_j\}$ is a pairwise disjoint finite collection in Φ_0 with $\Sigma \|\mu(e_j)\| \geq \|\mu\|_{\Phi_0}([0, T]) - \varepsilon$, the set $\{y_j\} \subseteq E$ is such that $\|y_j\| = 1$ and $\langle \mu(e_j), y_j \rangle \geq \Sigma \|\mu(e_j)\| - 2^{-j}\varepsilon$, and $f(t) = \Sigma \chi_j(t) y_j$ (χ_j the characteristic function of e_j), then $\langle \mu^*, f \rangle = \|\mu\|_{\Phi_0}([0, T]) - 2\varepsilon$. Hence

$$\|\mu^*\|_{S(0,T;E,\Phi_0)^*} = \|\mu\|_{\Phi_0}([0, T]). \qquad (10.2.9)$$

We then apply the Hahn–Banach theorem and extend μ^* from $S(0, T; E, \Phi_0)$ to $B(0, T; E)$ with the same norm. The measure μ is extended setting

$$\langle \mu(e), y \rangle = \langle \mu^*, \chi_e(\cdot) y \rangle, \tag{10.2.10}$$

for e an arbitrary subset of $[0, T]$. Note that the extended functional μ^* and the extended measure μ do not necessarily stand in the relation (10.2.8) in the whole space $B(0, T; E)$; in fact, if E is infinite dimensional, not every function $y(\cdot)$ in $B(0, T; E)$ is the uniform limit of functions in $S(0, T; E, \Phi)$, which is a condition for the definition of the integral (take for instance $y(0) = y_0$, $y(t) = y_n$ in $(1/(n+1), 1/n]$, $n = 1, 2, \ldots$, where $\{y_k\}$ is a sequence with $\|y_j - y_k\| \geq \varepsilon > 0$ for $i \neq j$). Nevertheless, (10.2.8) holds in the space $B(0, T; E, \Phi)$, and an argument identical to the one for $B(0, T; E, \Phi_0)$ shows that

$$\|\mu^*\|_{S(0,T;E,\Phi)^*} = \|\mu\|_{\Phi}([0, T]). \tag{10.2.11}$$

We have $\|\mu^*\|_{S(0,T;E,\Phi_0)^*} \leq \|\mu^*\|_{S(0,T;E,\Phi)^*} \leq \|\mu^*\|_{B(0,T;E)^*} = \|\mu^*\|_{S(0,T;E,\Phi_0)^*}$, thus $\|\mu\|_{\Phi}([0, T]) = \|\mu\|_{\Phi_0}([0, T])$. ∎

Proof of Lemma 10.2.1. Extend μ as in Lemma 10.2.2 to the field Φ of all subsets of $[0, T]$, so that the variation $\|\mu\|$ is a finitely additive measure defined in Φ. Applying Dunford–Schwartz [1958, p. 262, Theorem 2] or Theorem 12.4.7 in this book, we deduce that there exists a *regular* finitely additive measure $\|\mu\|_0$ defined in the field $\Phi(C)$ generated by the closed subsets of $[0, T]$ and such that

$$\int_0^T \phi(t) \|\mu\|(dt) = \int_0^T \phi(t) \|\mu\|_0(dt) \tag{10.2.12}$$

for scalar continuous functions $\phi(t)$ and then use Alexandrov's theorem (Dunford–Schwartz [1958, p. 138] or Theorems 12.4.10 and 12.4.11 in this book) to deduce that $\|\mu\|_0$ is countably additive and can be extended from $\Phi(C)$ to a σ-field containing the Borel field $\Phi_b(C)$ (the σ-field generated by all closed sets $\subseteq [0, T]$). For a definition of regular measure see **12.4**, and all of Chapter 12 for other information on measures of various kinds. ∎

Lemma 10.2.1 shows that (as far as integration of continuous functions goes), a finitely additive E^*-valued measure is "controlled" by a countably additive measure $\|\mu\|_0$.

The following is a "quick-and-dirty" identification of the dual $C(0, T; E)^*$. We improve on it later.[3]

---

[3] The reader exclusively interested in the particular cases $E = C(\overline{\Omega})$ and $E = L^p(\Omega)$ ($1 \leq p < \infty$) may skip over most of this section and proceed to Example 10.2.20.

Theorem 10.2.3. *Let μ be an E^*-valued finitely additive measure of bounded variation defined in the field Φ_0 generated by the subintervals of $[0, T]$. Then (10.2.8) defines an element of the dual space $C(0, T; E)^*$ satisfying*

$$\|\mu^*\|_{C(0,T;E)^*} \leq \|\mu\|([0, T]) \tag{10.2.13}$$

with $\|\mu\|$ computed in Φ_0. Conversely, let μ^* be a bounded linear functional in $C(0, T; E)$. Then there exists a finitely additive E^*-valued measure μ of bounded variation, defined in the field of all subsets of $[0, T]$ and satisfying (10.2.8) and (10.2.13), the latter with $\|\mu\|$ computed in Φ_0.

Proof. That such a measure defines an element $\mu^* \in C(0, T; E)^*$ is obvious and it follows from (10.2.6) that $\|\mu^*\|_{C(0,T;E)^*} \leq \|\mu\|([0, T])$. Conversely, let μ^* be an arbitrary bounded linear functional in the space $C(0, T; E)$. Invoking the Hahn–Banach theorem, extend this functional (with the same name and norm) to the space $B(0, T; E)$ of all bounded E-valued functions endowed with the supremum norm. Define an E^*-valued (finitely additive) measure μ in the field of all subsets of $[0, T]$ by (10.2.10). We have already seen (in the proof of Lemma 10.2.2) that μ is of bounded variation, with

$$\|\mu\|_{\Phi_0}([0, T]) \leq \|\mu\|_\Phi([0, T]) = \|\mu^*\|_{S(0,T;E,\Phi)^*} \leq \|\mu^*\|_{B(0,T;E)^*}.$$

The extended functional and the extended measure satisfy (10.2.8) for $y(\cdot) \in S(0, T; E, \Phi)$, in particular for $y(\cdot) \in S(0, T; E, \Phi_0)$. Given $y(\cdot) \in C(0, \bar{t}; E)$ choose a sequence $\{y_n(\cdot)\} \in S(0, T; E, \Phi_0)$ such that $y_n(\cdot) \to y(\cdot)$ in $B(0, T; E)$ and take limits on both sides of (10.2.8); on the left side we use continuity of the extended functional, on the left the definition of integral. This ends the proof. ∎

Theorem 10.2.3 is unsatisfactory on two grounds: the measures μ in the dual $C(0, \bar{t}; E)$ are merely finitely additive, and their norm as functionals may not be the same as their total variation (there is only inequality in (10.2.13)). However, by Lemma 10.2.1 each measure μ is "controlled" by the countably additive measure $\|\mu\|_0$, and this alone permits proving interesting results. One is

Lemma 10.2.4. *Let $\{y_n(\cdot)\}$ be a bounded sequence in $C(0, T; E)$ such that $y_n(t) \to y(t)$ everywhere in $0 \leq t \leq T$. Then $y_n(\cdot) \to y(\cdot)$ $C(0, T; E)^*$-weakly.*

Proof. Use the representation (10.2.8), then (10.2.7) for $y - y_n$ and the dominated convergence theorem. Note that only countable additivity of $\|\mu\|_0$ is used; countable additivity of μ is not essential.[4] ∎

---

[4] We are applying the dominated convergence theorem for the measure $\|\mu\|_0$. Since we don't know the null sets for $\|\mu\|_0$, *everywhere* convergence is essential. The same observation applies in many other places.

We show below that, as far as continuous functions go, our integration theory needs only countably additive measures.

Theorem 10.2.5. *Let μ be a finitely additive E^*-valued measure of bounded variation defined in Φ_0. Then there exists an E^*-valued countably additive measure μ_0 of bounded variation, defined in Φ_0 and such that*

$$\int_0^T \langle y(t), \mu_0(dt) \rangle = \int_0^T \langle y(t), \mu(dt) \rangle \qquad (10.2.14)$$

for every $y(\cdot) \in C(0, T; E)$. Moreover, $\|\mu\|_0$ is a majorant of $\|\mu_0\|$:

$$\|\mu_0\|(e) \le \|\mu\|_0(e) \quad (e \in \Phi_0), \qquad (10.2.15)$$

and $\|\mu_0\|$ is countably additive in Φ_0.

The proof of Theorem 10.2.5 depends on some auxiliary results.

Lemma 10.2.6. *Every element of Φ_0 can be written as a finite union of disjoint intervals.*

Lemma 10.2.7. *Let $y(t) \in S(0, T; E, \Phi_0)$. Then there exists a sequence $\{y_n(\cdot)\} \subseteq C(0, T; E)$ such that (a)*

$$\|y_n(t)\| \le \max_{0 \le \tau \le T} \|y(\tau)\| \quad (0 \le t \le T), \qquad (10.2.16)$$

(b) $y_n(t) \to y(t)$ for all $t \in [0, T]$.

The proof of Lemma 10.2.6 is an immediate consequence of the definition of "field generated by." Using this result we may reduce the proof of Lemma 10.2.7 to the case where $y(t)$ is of the form (10.2.4), with each $\chi_j(t)$ the characteristic function of an interval. The approximation of each characteristic function $\chi_j(t)$ is effected by nonnegative "trapezoidal" functions $f_{jn}(t)$, from above or from below according to whether the corresponding endpoint belongs to the interval or not.

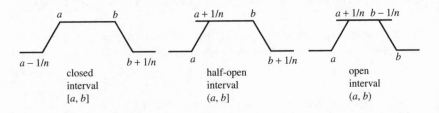

Figure 10.1.

The approximating function is then

$$y_n(t) = \sum_{j=1}^{m} f_{jn}(t) y_j. \tag{10.2.17}$$

Due to the fact that the number of intervals is finite, if we take n large enough the supports of the trapezoidal functions can only overlap for adjacent intervals e_j, e_{j+1}. If t is a point where $f_{jn}(t)$ and $f_{j+1,n}(t)$ are not zero, then $f_{jn}(t) + f_{j+1,n}(t) \le 1$, so that $\|y_n(t)\| \le f_{jn}(t)\|y_j\| + f_{j+1,n}(t)\|y_{j+1}\| \le \max(\|y_j\|, \|y_{j+1}\|)$, and (10.2.16) follows. ∎

As a particular case, we have

Lemma 10.2.8. *Let $\chi(\cdot)$ be the characteristic function of a set $e \in \Phi_0$. Then there exists a sequence $\{f_n(\cdot)\}$ of continuous functions such that $0 \le f_n(t) \le 1$ and $f_n(t) \to \chi(t)$ for all $t \in [0, T]$.*

Lemma 10.2.9 (Lebesgue). *Let $f_n(t), f(t)$ be Borel measurable functions in $0 \le t \le T$ with $f_n(t) \to f(t)$ everywhere in $0 \le t \le T$, v a finite positive countably additive measure defined in $\Phi_b(C)$. Then*

$$v(\{t; |f_n(t) - f(t)| \ge \varepsilon\}) \to 0 \quad as\ n \to \infty \tag{10.2.18}$$

for every $\varepsilon > 0$.

To prove this, let $e_n(\varepsilon) = \{t; |f_n(t) - f(t)| \ge \varepsilon\}$, $d_n(\varepsilon) = e_n(\varepsilon) \cup e_{n+1}(\varepsilon) \dots$. Then $d_1(\varepsilon) \supseteq d_2(\varepsilon) \supseteq \dots$ and

$$\bigcap_{n=1}^{\infty} d_n(\varepsilon) = \emptyset.$$

Since $v(\emptyset) = 0$ and v is countably additive, $v(d_n(\varepsilon)) \to 0$, which is stronger than (10.2.18). ∎

Proof of Theorem 10.2.5. The vector measure μ_0 is defined by

$$\langle \mu_0(e), y \rangle = \lim_{n \to \infty} \int_0^T \langle f_n(t)y, \mu(dt) \rangle \tag{10.2.19}$$

for $e \in \Phi_0$, where $f_n(t)$ is a bounded sequence in $C(0, T)$ with $f_n(t) \to \chi(t)$ for all t, χ the characteristic function of e (for instance, the sequence in Lemma 10.2.8). Applying (10.2.7) and the dominated convergence theorem to $f_n(t)y - f_m(t)y$

we deduce that the limit exists. The same argument proves that the limit does not depend on the particular sequence $\{f_n(\cdot)\}$.

Let $e_1, e_2 \in \Phi_0$ be disjoint sets, $\chi_1(t), \chi_2(t)$ their characteristic functions, $\{f_{1n}(\cdot)\}, \{f_{2n}(\cdot)\}$ the respective sequences for $\chi_1(t), \chi_2(t)$. Then, if $e = e_1 \cup e_2$ and $\chi(t)$ is the characteristic function of e, $f_{1n}(t) + f_{2n}(t) \to \chi(t)$ for all t, thus $\langle \mu_0(e), y \rangle = \langle \mu_0(e_1), y \rangle + \langle \mu_0(e_2), y \rangle$. This shows that μ_0 is a (for the time being, finitely additive) E^*-valued measure. We have

$$|\langle \mu_0(e), y \rangle| \le \lim_{n \to \infty} \|y\| \int_0^T |f_n(t)| \|\mu\|_0(dt),$$

hence it follows from the dominated convergence theorem that $|\langle \mu_0(e), y \rangle| \le \|y\| \|\mu\|_0(e)$, a fortiori $\|\mu_0(e)\| \le \|\mu\|_0(e)$, which implies (10.2.15) by definition of total variation (note that $\|\mu\|_0$ is positive, then is its own total variation).

To show (10.2.14), pick $y(\cdot) \in C(0, T; E)$ and then select

$$y_0(\cdot) = \sum_{j=1}^m \chi_j(\cdot) y_j \in S(0, T; E, \Phi_0)$$

in such a way that $\|y(t) - y_0(t)\| \le \varepsilon$ in $0 \le t \le T$. Applying (10.2.6) to the measure μ_0 and then using (10.2.8) we obtain

$$\left| \int_0^T \langle y(t), \mu_0(dt) \rangle - \int_0^T \langle y_0(t), \mu_0(dt) \rangle \right|$$
$$\le \varepsilon \|\mu_0\|([0, T]) \le \varepsilon \|\mu\|_0([0, T]). \tag{10.2.20}$$

Next, use Lemma 10.2.8 to construct sequences $\{f_{jn}(\cdot)\} \subseteq C(0, T)$ with $0 \le f_{jn}(t) \le 1$ and $f_{jn}(t) \to \chi_j(t)$ as $n \to \infty$ in $0 \le t \le T$. Using then Lemma 10.2.9 select $n(\varepsilon)$ so large that

$$|f_{j,n(\varepsilon)}(t) - \chi_j(t)| \le \varepsilon/m \quad (0 \le t \le T, \ j = 1, \ldots, m) \tag{10.2.21}$$

except in a Borel set $e_{j\varepsilon} \subseteq [0, T]$ with

$$\|\mu\|_0(e_{j\varepsilon}) \le \varepsilon/m \quad (j = 1, \ldots, m) \tag{10.2.22}$$

and

$$\left| \langle y_j, \mu_0(e_j) \rangle - \int_0^T \langle f_{j,n(\varepsilon)}(t) y_j, \mu(dt) \rangle \right| \le \frac{\varepsilon}{m} \quad (j = 1, \ldots, m) \tag{10.2.23}$$

(the latter by definition of $\mu_0(e_j)$). Adding up the m inequalities (10.2.23) we obtain

$$\left| \int_0^T \langle y_0(t), \mu_0(dt) \rangle - \int_0^T \langle y_\varepsilon(t), \mu(dt) \rangle \right| \le \varepsilon. \tag{10.2.24}$$

If $y_\varepsilon(t) = \Sigma_j f_{j,n(\varepsilon)}(t) y_j$ we have

$$\|y(t) - y_\varepsilon(t)\| \le \|y(t) - y_0(t)\| + \|y_0(t) - y_\varepsilon(t)\|$$

$$\le \|y(t) - y_0(t)\| + \sum_{j=1}^{m} |f_{j,n(\varepsilon)}(t) - \chi_j(t)| \|y_j\| \le (K+1)\varepsilon \qquad (10.2.25)$$

except in the set $e_\varepsilon = \cup e_{j\varepsilon}$ with $\|\mu\|_0(e) \le \varepsilon$, where $K = \max(\|y_1\|, \ldots, \|y_m\|) = \max_{0 \le \tau \le T} \|y_0(\tau)\| \le \max_{0 \le \tau \le T} \|y(\tau)\| + \varepsilon = C + \varepsilon$. On the other hand, (10.2.25) and the fact that $|f_{j,n(\varepsilon)}(t)|$, $|\chi_j(t)| \le 1$ imply

$$\|y(t) - y_\varepsilon(t)\| \le \varepsilon + 2K \quad (0 \le t \le T). \qquad (10.2.26)$$

Combining (10.2.25) and (10.2.26) with (10.2.7), we obtain

$$\left| \int_0^T \langle y(t), \mu(dt) \rangle - \int_0^T \langle y_\varepsilon(t), \mu(dt) \rangle \right|$$

$$\le \int_0^T \|y(t) - y_\varepsilon(t)\| \|\mu\|_0(dt) \le \varepsilon(K+1)T + \varepsilon(\varepsilon + 2K) \le K'\varepsilon. \qquad (10.2.27)$$

Finally, the fact that μ_0 is of bounded variation means that (10.2.6) can be applied:[5]

$$\left| \int_0^T \langle y(t), \mu_0(dt) \rangle - \int_0^T \langle y_0(t), \mu_0(dt) \rangle \right|$$

$$\le \int_0^T \|y(t) - y_0(t)\| \|\mu_0\|(dt) \le \varepsilon \|\mu_0\|([0, T]). \qquad (10.2.28)$$

Putting together (10.2.24), (10.2.27) and (10.2.28) and letting $\varepsilon \to 0$, the proof of (10.2.14) is over.

It only remains to show that the vector measure μ_0 and the scalar measure $\|\mu_0\|$ are countably additive in Φ_0. Let $\{e_j\}$ be a sequence of pairwise disjoint sets in Φ_0 such that $e = \cup_j e_j \in \Phi_0$, e^k the union of the first k of the e_j, $\chi(\cdot)$ (resp. $\chi^k(\cdot)$) the characteristic function of e (resp. e^k). We have

$$\sum_{j=1}^{k} \langle y, \mu_0(e_j) \rangle - \langle y, \mu_0(e) \rangle = \int_0^T \langle (\chi^k(t) - \chi(t)) y, \mu_0(dt) \rangle$$

so that, after (10.2.6),

$$\left| \sum_{j=1}^{k} \langle y, \mu_0(e_j) \rangle - \langle y, \mu_0(e) \rangle \right| \le \|y\| \int_0^T |\chi^k(t) - \chi(t)| \|\mu_0\|(dt).$$

[5] Recall that (10.2.6) is valid for uniform limits of functions in $S(0, T; E, \Phi)$. This is clearly the case for $y(t) - y_0(t)$.

Hence, using the estimate (10.2.15),

$$\left\| \sum_{j=1}^{k} \mu_0(e_j) - \mu_0(e) \right\| \le \int_0^T |\chi^k(t) - \chi(t)| \|\mu\|_0(dt),$$

and countable additivity of μ_0 follows from the dominated convergence theorem. The result for $\|\mu_0\|$ is a consequence of

Lemma 10.2.10. *Let the E-valued measure μ be countably additive and of bounded variation in a field Φ. Then $\|\mu\|$ is countably additive in Φ.*

To see this, let $\{e_j\}$ be a sequence of pairwise disjoint sets in Φ such that $e = \cup_j e_j \in \Phi$. Applying finite additivity of $\|\mu\|$ for $e_1, \dots, e_n$ and letting $n \to \infty$ we obtain $\Sigma_j \|\mu\|(e_j) \le \|\mu\|(e)$. On the other hand, let $\{d_k\}$ be a pairwise disjoint collection of subsets of e in Φ with $\Sigma_k \|\mu(d_k)\| \ge \|\mu\|(e) - \varepsilon$. Then

$$\|\mu\|(e) \le \Sigma_k \|\mu(d_k)\| + \varepsilon = \Sigma_k \|\Sigma_j \mu(d_k \cap e_j)\| + \varepsilon$$
$$\le \Sigma_k \Sigma_j \|\mu(d_k \cap e_j)\| + \varepsilon = \Sigma_j \Sigma_k \|\mu(d_k \cap e_j)\| + \varepsilon \le \Sigma_j \|\mu\|(e_j) + \varepsilon. \qquad \blacksquare$$

Remark 10.2.11. We do *not* claim that $\mu_0(e) = \mu(e)$ for all $e \in \Phi_0$. Accordingly, we do not claim that (10.2.14) holds for functions in $S(0, T; E, \Phi_0)$. To see why, consider the case $E = E^* = \mathbb{R}, T = 1$ and a measure μ defined by $\mu([a, b)) = \mu((a, b]) = \mu((a, b)) = \mu([a, b]) = \phi(b) - \phi(a)$, where ϕ is a bounded nondecreasing function in $0 \le t \le T$, continuous except at $t = 1$, where it is discontinuous to the left ($\phi(1-) < \phi(1)$). The measure μ is not countably additive, as can be seen with the sequence $\{e_j\} = \{[1 - 1/j, 1 - 1/(j+1))\}$ of disjoint sets in Φ_0, whose union is $e = [0, 1)$; we have $\Sigma \mu(e_j) = \phi(1-) - \phi(0)$, while $\mu(e) = \phi(1) - \phi(0)$. However, μ is countably additive in $[0, 1 - \varepsilon]$ for every $\varepsilon > 0$. It follows from the definition (10.2.19) and the dominated convergence theorem that $\mu_0(e) = \mu(e)$ if $e \subseteq [0, 1 - \varepsilon]$. On the other hand, as a consequence of the definition we obtain $\mu_0((a, 1)) = \phi(1-) - \phi(a), \mu_0(\{1\}) = \phi(1) - \phi(1-)$, whereas $\mu((a, 1)) = \phi(1) - \phi(a), \mu(\{1\}) = \phi(1) - \phi(1) = 0$.

However (and this is the gist of Theorem 10.2.5) continuous functions integrated against μ or μ_0 are impervious to the discrepancy between the two measures; this follows from the fact that $\mu((a, 1]) = \mu_0((a, 1])$ for any $a < 1$.

We call $\Sigma(0, T; E^*, \Phi_0)$ (or $\Sigma(0, T; E^*)$ for short) the space of all E^*-valued countably additive regular measures of bounded variation defined in Φ_0. The space $\Sigma(0, T; E^*)$ is equipped with the total variation norm over Φ_0; $\|\mu\|_{\Sigma(0,T;E^*)} = \|\mu\|([0, T])$.

Lemma 10.2.12. *Let $\mu \in \Sigma(0, T; E^*)$. Then $\|\mu\|$ can be extended as a countably additive function to a σ-field containing the Borel field $\Phi_b(C)$.*

Proof. The extension is done on the basis of Carathéodory's Theorem 2.0.2, and is not unlike the construction of the Lebesgue measure in **2.0.** First, we define an outer measure λ in the σ-field of all subsets of $[0, T]$ by

$$\lambda(e) = \inf \sum \|\mu\|(d_n),$$

the infimum taken over all finite or countable families $\{d_n\}$ of sets in Φ_0 with $\cup d_n \supseteq e$. Then we apply Theorem 2.0.2 and obtain the desired extension to a σ-field containing Φ_0; obviously, this field must contain $\Phi_b(C)$. We denote the extension with the same name $\|\mu\|$. ∎

The next result says that the construction (10.2.19), if applied to a measure already in $\Sigma(0, T; E^*)$, produces the same measure.

Corollary 10.2.13. *Let* $\mu \in \Sigma(0, T; E^*)$. *Then, if* $e \in \Phi_0$, χ *is the characteristic function of* e *and* $f_n(t)$ *is a bounded sequence in* $C(0, T)$ *with* $f_n(t) \to \chi(t)$ *for all* t, *then*

$$\langle \mu(e), y \rangle = \lim_{n \to \infty} \int_0^T \langle f_n(t)y, \mu(dt) \rangle. \tag{10.2.29}$$

Proof. Inequality (10.2.6) is valid for uniform limits of functions in $S(0, T; E, \Phi_0)$, thus is valid for $\chi(t) - f_n(t)$. Hence,

$$\left| \langle \mu(e), y \rangle - \int_0^T \langle f_n(t)y, \mu(dt) \rangle \right| = \left| \int_0^T \langle \chi(t) - f_n(t), \mu(dt) \rangle \right|$$

$$\leq \int_0^T |\chi(t) - f_n(t)| \|\mu\|(dt), \tag{10.2.30}$$

so the result follows from σ-additivity of $\|\mu\|$ and the dominated convergence theorem. ∎

Theorem 10.2.14. *Let* E *be an arbitrary Banach space. The dual space* $C(0, \bar{t}; E)^*$ *is algebraically and metrically isomorphic to the space* $\Sigma(0, T; E^*)$. *The duality relation between the two spaces is* (10.2.8).

Theorem 10.2.14 is an improvement on Theorem 10.2.3. That the duties of a finitely additive measure can be assumed by a countably additive measure is insured by Theorem 10.2.5, thus we only have to upgrade (10.2.13) to

$$\|\mu^*\|_{C(0,T;E)^*} = \|\mu\|([0, T]). \tag{10.2.31}$$

This is done as follows. We use the fact (consequence of Lemma 10.2.6) that there exists a finite collection $\{e_j\}$ of pairwise disjoint intervals e_j such that

$\Sigma \|\mu(e_j)\| \geq \|\mu\|([0, T]) - \varepsilon$. Choose $\{y_j\} \subseteq E$, $\|y_j\| = 1$ with $\langle \mu(e_j), y_j \rangle \geq \|\mu(e_j)\| - 2^{-j}\varepsilon$ and let χ_j be the characteristic function of e_j, $y_0(t) = \Sigma \chi_j(t) y_j$. Then we have $\int \langle y_0(t), \mu(dt) \rangle = \Sigma \langle \mu(e_j), y_j \rangle \geq \Sigma \|\mu(e_j)\| - \varepsilon \geq \|\mu\|([0, T]) - 2\varepsilon$. Finally, approximate $y_0(\cdot)$ pointwise with a sequence $\{y_n(\cdot)\} \subseteq C(0, T; E)$ with $\|y_n(t)\| \leq 1$ (Lemma 10.2.7); by virtue of Corollary 10.2.13,

$$\sum \langle \mu(e_j), y_j \rangle = \int_0^T \langle y_0(t), \mu(dt) \rangle = \lim_{n \to \infty} \int_0^T \langle y_n(t), \mu(dt) \rangle$$

with each of the integrals on the right side $\leq \|\mu^*\|_{C(0,T;E)}$ in absolute value. This ends the proof. ∎

The material below aims to fitting the adjoint variational equation with a "measure forcing term" as required by the minimum principle proved in next section. We do this for an arbitrary linear equation

$$y'(t) = A(t)y(t), \tag{10.2.32}$$

$\{A(t); 0 \leq t \leq \bar{t}\}$ a family of densely defined operators in a Banach space E, and we take the point of view in 5.5 of requiring nothing of the adjoint equation (Theorem 5.5.12). We assume that the Cauchy problem for (10.2.32) is well posed in $0 \leq t \leq T$ in the sense of **5.4** and denote by $S(t, s)$ the (strongly continuous) solution operator. The "measure driven" (reversed) adjoint equation is

$$dz(s) = -A(s)^* z(s) ds - \mu(ds), \qquad z(\bar{t}) = z \tag{10.2.33}$$

with $\mu \in \Sigma(0, T; E^*)$. By definition, the solution $z(s)$ of (10.2.33) is

$$z(s) = S(\bar{t}, s)^* z + \int_s^{\bar{t}} S(\sigma, s)^* \mu(d\sigma) = z_h(s) + z_i(s), \tag{10.2.34}$$

where the integral on the right-hand side is understood as follows: $z_i(s)$ is the unique element of E^* such that

$$\langle y, z_i(s) \rangle = \int_s^{\bar{t}} \langle S(\sigma, s)y, \mu(d\sigma) \rangle.^{(6)} \tag{10.2.35}$$

See formulas (5.5.35) and (5.5.39) for motivation. The present definition uses no properties of $S(\sigma, s)^*$; the assumptions are on $S(\sigma, s)$. It follows from (10.2.6) that

$$\|z(s)\|_{E^*} \leq C(\|z\|_{E^*} + \|\mu\|_{\Sigma(0,\bar{t};E^*)}) \quad (0 \leq s \leq \bar{t}), \tag{10.2.36}$$

where C is a bound for $\|S(\sigma, s)\|$ in $0 \leq s \leq \sigma \leq \bar{t}$.

---

(6) The integral includes both endpoints (see Example 10.2.18).

Theorem 10.2.15. *Let* $z \in E^*$, $\mu \in \Sigma(0, \bar{t}; E^*)$. *Then* $z(s)$ *is an E-weakly left continuous E^*-valued function in* $0 \le s \le \bar{t}$.

Proof. Theorem 10.2.15 is a descendant of Theorem 5.5.12, and we show it in a very similar way. The same goes for Corollary 10.2.16 and Lemma 10.2.17, relatives of Corollary 5.5.13 and Lemma 5.5.14.

Note first that $\langle z_h(s), y \rangle = \langle z, S(\bar{t}, s)y \rangle$ so that $z_h(s)$ is E-weakly continuous. To do $z_i(s)$ extend $S(\sigma, s)$ to $0 \le \sigma, s \le \bar{t}$ setting $S(\sigma, s) = I$ for $\sigma < s$ and approximate $S(\sigma, s)$ by $S_n(\sigma, s) = \Sigma\Sigma\phi_{nj}(\sigma)\phi_{nk}(s)S(s_{nj}, s_{nk})$, $\{\phi_{nj}(\cdot)\}$ a smooth partition of unity in $[0, T]$. Then, as shown in Theorem 5.5.12, $S_n(\sigma, s)y \to S(\sigma, s)y$ uniformly in $0 \le \sigma, s \le \bar{t}$ (inequality (5.5.41)). If $z_{in}(s)$ is defined from $S_n(\sigma, s)$ in the same way $z_i(s)$ is defined from $S(\sigma, s)$ then we have

$$|\langle y, z_{in}(s) \rangle - \langle y, z_i(s) \rangle| \le \|S_n(\cdot, s)y - S(\cdot, s)y\|_{C(0,\bar{t};E)}\|\mu\|_{\Sigma(0,\bar{t};E^*)},$$

thus it is enough to show that each $z_{in}(s)$ is E-weakly left continuous. Since

$$\langle y, z_{in}(s) \rangle = \sum\sum \phi_{nk}(s) \int_s^{\bar{t}} \phi_{nj}(\sigma)\langle S(s_{nj}, s_{nk})y, \mu(d\sigma) \rangle, \qquad (10.2.37)$$

it is finally enough to show that each of the integrals in (10.2.37) is left continuous, which is elementary (see Example 10.2.18 below).   ∎

The "final condition" $z(t) = z$ in (10.2.33) cannot be taken seriously, even in the weak sense $\lim_{s \to t-}\langle z(s), y \rangle = \langle z, y \rangle$; for instance, if $z = 0$, $\mu(dt) = \delta_{\bar{t}}(dt)w$ ($\delta_{\bar{t}}$ the Dirac delta with mass at $\bar{t}$) we have $z(s) = S(\bar{t}, s)^*w$.

Corollary 10.2.16. *Let E^* be separable. Then $z(s)$ is strongly measurable.*

Proof. E-weakly left continuous implies E-weakly measurable. Use Example 5.0.39.   ∎

The following result is a variant of Fubini's theorem related to the solutions in Theorem 10.2.15.

Lemma 10.2.17. *Let $g(\cdot) \in L^1(0, \bar{t}; E)$. Then*

$$\int_0^{\bar{t}} \left\langle \int_0^t S(t, s)g(s)ds, \mu(dt) \right\rangle = \int_0^{\bar{t}} \left\langle g(s), \int_s^{\bar{t}} S(t, s)^*\mu(dt) \right\rangle ds. \quad (10.2.38)$$

Proof. As in Lemma 5.5.14, we approximate $g(\cdot)$ in the $L^1(0, \bar{t}; E)$-norm by a sequence $g_n(s) = \Sigma\chi_{nk}(s)g_{nk}$ of simple functions. If we can establish (10.2.38) for each one of these functions, we pass to the limit as follows. By Lemma 5.4.2 $\int_0^t S(t, \sigma)g_n(\sigma)d\sigma \to \int_0^t S(t, \sigma)g(\sigma)d\sigma$ in $C(0, \bar{t}; E)$, which justifies taking limit on the left side. On the right side we use the fact that $s \to \int_s^{\bar{t}} S(t,s)^* \mu(ds) \in L_w^\infty(0, \bar{t}; E^*)$ and L^1-convergence of $g_n(\cdot)$.

It is then enough to show (10.2.38) for a one-term sum $g(s) = \chi_e(s)g$, with χ_e the characteristic function of a measurable set e. The equality becomes

$$\int_0^{\bar{t}} \left\langle \int_0^t S(t,s)\chi_e(s)g\,ds, \mu(dt) \right\rangle = \int_0^{\bar{t}} \int_s^{\bar{t}} \langle S(t,s)g, \mu(dt)\rangle \chi_e(s)ds,$$
(10.2.39)

and to prove it we show first the corresponding equality for $S_n(t,s)$

$$\int_0^{\bar{t}} \left\langle \int_0^t S_n(t,s)\chi_e(s)g\,ds, \mu(dt) \right\rangle = \int_0^{\bar{t}} \int_s^{\bar{t}} \langle S_n(t,s)g, \mu(dt)\rangle \chi_e(s)ds$$
(10.2.40)

and take limits. Finally, it is enough to prove (10.2.40) for each of the terms making up S_n, that is,

$$\int_0^t \phi_{nj}(t)\left(\int_0^t \phi_{nk}(s)\chi_e(s)ds\right)\langle S(s_{nj}, s_{nk})g, \mu(dt)\rangle$$
$$= \int_0^{\bar{t}} \phi_{nk}(s)\chi_e(s)\int_0^{\bar{t}} \phi_{nj}(t)\langle S(s_{nj}, s_{nk})g, \mu(dt)\rangle ds,$$

and this is a tautology. It only remains to show that we can obtain (10.2.39) taking limits in (10.2.40). In the inner integral on the right we exploit the fact that $S_n(t,s)g \to S(t,s)g$ uniformly in $0 \le s, t \le \bar{t}$ and (10.2.6); then we use the dominated convergence theorem in the outer integral. On the left side we use the fact that $\int_0^t S_n(t,s)g\chi_e(s)ds \to \int_0^t S(t,s)g\chi_e(s)ds$ in the norm of $C(0, \bar{t}; E)$, consequence of (10.2.6). ∎

Example 10.2.18. Let ν be a (scalar-valued) bounded countably additive measure defined in Φ_0. Then, if

$$\phi(t) = \int_0^t \nu(d\tau) \qquad \psi(s) = \int_s^{\bar{t}} \nu(d\sigma),$$

(endpoints included in the integrals) $\phi(t)$ is continuous from the right and $\psi(s)$ is continuous from the left.

To prove this, extend the integrals to $[0, \bar{t}]$, putting characteristic functions in the integrand, and use the dominated convergence theorem for $\|\nu\|$. Obvious examples with Dirac deltas show that two-sided continuity is not to be expected.

Example 10.2.19. Show that the integral (10.2.5) of a simple function $y(t)$ does not depend on the particular expression (10.2.4).

Example 10.2.20. The dual of $C(0, T; E)$ can be characterized in an elementary way if $E = L^p(\Omega)$ $(1 \leq p < \infty)$, $\Omega \subset \mathbb{R}^m$ a bounded domain, or if $E = C(K), K$ compact; these two cases suffice for most applications in this work. For the latter space we have $C(0, T; C(K)) = C([0, T] \times K)$, whose dual is $\Sigma([0, T] \times K)$ (Theorem 5.0.20), and

$$\Sigma([0, T] \times K) = \Sigma(0, T; \Sigma(K)) \tag{10.2.41}$$

with equality of norms. To see this, let $\mu(dt\, dx)$ be an element of $\Sigma([0, T] \times K)$. We define an element $\tilde{\mu}(dt) \in \Sigma(0, T; \Sigma(K)) = \Sigma(0, T; C(K)^*)$ by

$$\langle \tilde{\mu}(e), y \rangle = \int_{e \times K} y(x)\mu(dt\, dx) \quad (y(\cdot) \in C(K)) \tag{10.2.42}$$

(e a Borel set in $[0, T]$). Obviously, this assignment is linear, 1-1 and onto. It also follows that $\|\tilde{\mu}(e)\|_{\Sigma(K)} = \int_{e \times K} |\mu|(dt\, dx)$, thus equality of norms results. It remains to show that

$$\int_{e \times K} y(t, x)\mu(dt\, dx) = \int_0^T \langle y(t, \cdot), \tilde{\mu}(dt) \rangle \tag{10.2.43}$$

for $y(\cdot, \cdot) \in C(0, T; C(K))$. This is obvious for a function $y(\cdot) = \Sigma_j \chi_j(\cdot)y_j \in S(0, T; C(K), \Phi_0)$ (χ_j the characteristic functions of a finite collection $\{I_j\}$ of pairwise disjoint intervals) where both sides of (10.2.43) reduce to $\Sigma_j \langle y_j, \tilde{\mu}(I_j) \rangle$; we extend the equality to $y(\cdot, \cdot) \in C(0, T; C(K))$ approximating uniformly by functions in $S(0, T; C(K), \Phi_0)$. Note the gratuity: $\tilde{\mu}$ is defined in the Borel field $\Phi_b(C)$ rather than in Φ_0.

We have $C(\overline{\Omega}) \hookrightarrow L^p(\Omega)$, so that $C(0, T; C(\overline{\Omega})) \hookrightarrow C(0, T; L^p(\Omega))$, and every element of the dual space $C(0, T; L^p(\Omega))^*$ is a bounded linear functional in $C(0, T; C(\overline{\Omega}))$. Let $\mu \in \Sigma([0, T] \times \overline{\Omega})$ be the measure impersonating this functional; since $C(\overline{\Omega})$ is dense in $L^p(\Omega)$, $C(0, T; C(\overline{\Omega}))$ is dense in $C(0, T; L^p(\Omega))$ and μ is uniquely defined. Let $J = \cup_j I_j$ a finite union of disjoint intervals in $[0, T]$, $f_{jn}(\cdot)$ sequences of trapezoidal functions converging pointwise to the characteristic function $\chi_j(\cdot)$ of I_j, $0 \leq \Sigma_j f_{jn}(t) \leq 1$. (Lemma 10.2.7), $y_j(x)$ arbitrary elements of $C(\overline{\Omega})$. We have $\|\Sigma_j f_{jn} \otimes y_j\|_{C(0,T;L^p(\Omega))} \leq \max_j \|y_j\|_{L^p(\Omega)}$, so that $|\langle \mu, \Sigma_j f_{jn} \otimes y_j \rangle| \leq \|\mu\|_{C(0,T;L^p(\Omega))^*} (\max_j \|y_j\|_{L^p(\Omega)})$ for all n. Taking limits as $n \to \infty$ and using the dominated convergence theorem,

$$\left| \sum_j \int_\Omega y_j(x)\mu(I_j \times dx) \right| \leq \|\mu\|_{C(0,T;L^p(\Omega))^*} \left(\max_j \|y_j\|_{L^p(\Omega)} \right), \tag{10.2.44}$$

and, taking $y_j = y$,

$$\int_\Omega y(x)\mu(J, dx) = \sum_j \int_\Omega y(x)\mu(I_j, dx). \tag{10.2.45}$$

Using this equality and (10.2.44) for one interval $I_j = I$ it follows that the restriction of μ to $I \times \overline{\Omega}$ is of the form

$$\mu(I \times dx) = \tilde{\mu}(I, x)dx \qquad (10.2.46)$$

with $\tilde{\mu}(I, \cdot) \in L^q(\Omega)$, $1/q + 1/p = 1$. Choosing then $y_j(\cdot)$ in (10.2.43) in such a way that $\|y_j\|_{L^p(\Omega)} = 1$, $\langle \tilde{\mu}(I_j), y_j \rangle \geq \|\tilde{\mu}(I_j)\|_{L^q(\Omega)} - \varepsilon$ we obtain

$$\sum \|\tilde{\mu}(I_j)\|_{L^q(\Omega)} \leq \|\mu\|_{C(0,T;L^p(\Omega))^*}. \qquad (10.2.47)$$

We can rewrite (10.2.44) in the form $\langle \tilde{\mu}(J), y \rangle = \Sigma_j \langle \tilde{\mu}(I_j), y \rangle$, so that $\tilde{\mu}(J) = \Sigma_j \tilde{\mu}(I_j)$. If $\{I_j\}$ is a countable sequence of intervals with $\cup I_j = I =$ interval we obtain from (10.2.47) applied to a tail of the series that

$$\tilde{\mu}(I) = \sum \tilde{\mu}(I_j), \qquad (10.2.48)$$

so that the assignation (10.2.46) produces an element $\tilde{\mu} \in \Sigma(0, T; L^q(\Omega))$. We have

$$\|\tilde{\mu}\|_{C(0,T;L^p(\Omega))^*} = \|\tilde{\mu}\|_{\Sigma(0,T;L^q(\Omega))} = \|\mu\|_{C(0,T;L^p(\Omega))^*} \qquad (10.2.49)$$

where we have used (10.2.31) in the first equality. It remains to show that (10.2.43) holds for $y(\cdot, \cdot) \in C(0, T; L^p(\Omega))$; for this, the same approximation argument used for $C(K)$ works since $C(\overline{\Omega})$ is dense in $L^p(\Omega)$.

10.3. The Minimum Principle with State Constraints

We prove the minimum principle for the control system.

$$y'(t) = A(t)y(t) + f(t, y(t), u(t)), \qquad y(0) = \zeta \qquad (10.3.1)$$

in an arbitrary Banach space E with a state constraint and a target condition, respectively

$$y(t) \in M(t) \subseteq E \quad (0 \leq t < t), \qquad y(\bar{t}) \in Y, \qquad (10.3.2)$$

with Y and $M(t)$ closed. The cost functional is

$$y_0(t, u) = \int_0^t f_0(\tau, y(\tau, u), u(\tau))d\tau + g(t, y(t, u)), \qquad (10.3.3)$$

and we assume the Cauchy problem for $y'(t) = A(t)y(t)$ well posed in $0 \leq t \leq T$ with evolution operator $S(t, \tau)$. The admissible control space $C_{ad}(0, T; U)$ is requested to be patch complete and saturated and for each $u(\cdot) \in C_{ad}(0, T; E)$, $f(t, y, u(\cdot))$ satisfies Assumption III (resp. $f_0(t, y, u(\cdot))$ satisfies Assumption III^0) with $K(t, c), L(t, c)$ independent of $u(\cdot)$ (resp. with $K_0(t, c), L_0(t, c)$ independent of $u(\cdot)$) in $0 \leq t \leq T$. The function $g(t, y)$ has a Fréchet derivative with respect to y. As in **10.1**,

$$\mathbf{M}(\bar{t}) = \{y(\cdot) \in C(0, \bar{t}; U); y(t) \in M(t) \ (0 \leq t < \bar{t})\} \subseteq C(0, T; E). \quad (10.3.4)$$

Theorem 10.3.1. *Let all the assumptions above be satisfied and let the control* $\bar{u}(\cdot) \in C_{ad}(0, \bar{t}; U)$ *be optimal. Then there exist sequences* $\{\tilde{y}^n(\cdot)\} \subseteq \mathbf{M}(\bar{t})$ *with* $\tilde{y}^n(\cdot) \to y(\cdot, \bar{u})$ *and* $\{\tilde{y}^n\} \subseteq Y$ *with* $\tilde{y}^n \to \bar{y} = y(\bar{t}, \bar{u})$, *and a triple* $(z_0, \mu, z) \in \mathbb{R} \times \Sigma(0, \bar{t}; E^*) \times E^*$ *with*

$$z_0 \geq 0, \quad \mu \in \left(\liminf_{n \to \infty} T_{\mathbf{M}(\bar{t})}(\tilde{y}^n(\cdot)) \right)^-, \quad z \in \left(\liminf_{n \to \infty} T_Y(\tilde{y}^n) \right)^-, \quad (10.3.5)$$

and such that, if $\bar{z}(s)$ *is the solution of*

$$d\bar{z}(s) = -\{A(t)^* + \partial_y f(s, y(s, \bar{u}), \bar{u}(s))^*\}\bar{z}(s)ds$$
$$\qquad - z_0 \partial_y f_0(s, y(s, \bar{u}), \bar{u}(s))ds - \mu(ds),$$
$$\bar{z}(\bar{t}) = z + z_0 \partial_y g_0(\bar{t}, y(\bar{t}, \bar{u})) \qquad (10.3.6)$$

in $0 \leq t \leq \bar{t}$, *then*

$$z_0 \int_0^{\bar{t}} \left\{ f_0(\sigma, y(\sigma, \bar{u}), v(\sigma)) - f_0(\sigma, y(\sigma, \bar{u}), \bar{u}(\sigma)) \right\} d\sigma$$

$$+ \int_0^{\bar{t}} \langle \bar{z}(\sigma), f(\sigma, y(\sigma, \bar{u}), v(\sigma)) - f(\sigma, y(\sigma, \bar{u}), \bar{u}(\sigma)) \rangle d\sigma \geq 0 \quad (10.3.7)$$

for all $v(\cdot) \in C_{ad}(0, \bar{t}; U)$. *If* E^* *is separable and* $f_0(\sigma, y(\sigma, \bar{u}), u)$, $f(\sigma, y(\sigma, \bar{u}), u)$ *are regular in* U, *then*

$$z_0\{f_0(s, y(s, \bar{u}), \bar{u}(s)) + \langle \bar{z}(s), f(s, y(s, \bar{u}), \bar{u}(s)) \rangle\}$$
$$= \min_{v \in U}\{z_0 f_0(s, y(s, \bar{u}), v) + \langle \bar{z}(s), f(s, y(s, \bar{u}), v) \rangle\} \qquad (10.3.8)$$

a.e. in $0 \leq s \leq \bar{t}$.

Proof. We apply the theory in **7.2** to the abstract nonlinear programming problem (10.1.10)-(10.1.11) with $V = B(\bar{u}, \delta)$, $B(\bar{u}, \delta)$ the ball in Lemma 6.1.1, $\mathbf{f} : V \to C(0, \bar{t}; E)$ and $\mathbf{f}_0 : V \to \mathbb{R}$ defined in (10.1.9) and target set $\mathbf{M}(\bar{t}) \times Y$. Both functions are continuous (Lemma 6.1.1, Lemma 6.6.3), and $B(\bar{u}, \delta)$, as a closed subspace of $C_{ad}(0, \bar{t}; U)$, is a complete metric space. Theorem 7.2.5 produces sequences $\{\tilde{u}^n\} \subseteq B(\bar{u}, \delta)$ and $\{(\tilde{y}^n(\cdot), \tilde{y}^n)\} \in \mathbf{M}(\bar{t}) \times Y$ having the convergence properties announced in Theorem 10.3.1. We select $\rho > 0$ and[1] the following sequence $\{\mathbf{D}_n\}$ of convex sets: $\mathbf{D}_n \subseteq \mathrm{Der}\,(\mathbf{f}_0, \mathbf{f})(\tilde{u}^n) \subseteq \mathbb{R} \times C(0, \bar{t}; E) \times E$ consists of all triples

$$(\xi_0(\bar{t}, u, \mathbf{p}, \mathbf{v}), \xi(\cdot, u, \mathbf{p}, \mathbf{v}), \xi(\bar{t}, u, \mathbf{p}, \mathbf{v})) \in \mathbb{R} \times C(0, \bar{t}; E) \times E \qquad (10.3.9)$$

---

[1] For the role of ρ in this result see Theorem 7.2.5. Incidentally, the smooth version of this theorem (where ρ doesn't appear) is useless for problems with state constraints; the norm of the space $C(0, \bar{t}; E)$ is not Gâteaux differentiable off the origin even when $\dim E = 1$. The same applies to Theorem 7.2.2.

$\xi_0(\cdot, u, \mathbf{p}, \mathbf{v})$ (resp. $\xi(\cdot, u, \mathbf{p}, \mathbf{v})$) given by (6.5.8) (resp. (6.6.7)) with $u = \tilde{u}^n$. The limits of these sequences have been calculated in Lemma 6.6.4; every element of the form (10.3.9) with $u = \bar{u}$ belongs to $\liminf_{n \to \infty} \mathbf{D}_n$. Once ρ and $\{\mathbf{D}_n\}$ have been selected, Theorem 7.2.5 gives sequences $\{z_{0n}\} \subseteq \mathbb{R}$, $\{\mu_n\} \subseteq \Sigma(0, \bar{t}; E^*) = C(0, \bar{t}; E)^*$, $\{z_n\} \subseteq E^*$ such that

$$z_{0n}^2 + \|\mu_n\|_{\Sigma(0,\bar{t};E^*)}^2 + \|z_n\|_{E^*}^2 = 1, \quad z_{0n} \geq 0,$$

$$z_{0n}\xi_0(\bar{t}, \tilde{u}^n, \mathbf{p}, \mathbf{v}) + \langle \mu_n, \xi(\cdot, \tilde{u}^n, \mathbf{p}, \mathbf{v}) \rangle_c + \langle z_n, \xi(\bar{t}, \tilde{u}^n, \mathbf{p}, \mathbf{v}) \rangle \geq -\delta_n \quad (10.3.10)$$

where the first angled bracket (resp. the second) indicates the duality of $C(0, \bar{t}; E)$ and $\Sigma(0, \bar{t}; E^*)$ (resp. the duality of E and E^*) and $\delta_n \to 0$. If $(z_0, \mu, z) \in \mathbb{R} \times \Sigma(0, \bar{t}; E^*) \times E^*$ is a $(\mathbb{R} \times C(0, \bar{t}; E) \times E)$-weak limit point of (z_{0n}, μ_n, z_n), then

$$z_0\xi_0(\bar{t}, \bar{u}, \mathbf{p}, \mathbf{v}) + \langle \mu_n, \xi(\cdot, \bar{u}, \mathbf{p}, \mathbf{v}) \rangle_c + \langle z_n, \xi(\bar{t}, \bar{u}, \mathbf{p}, \mathbf{v}) \rangle \geq 0. \quad (10.3.11)$$

Using this inequality for single patches ($m = 1, \mathbf{p} = \{1\}, \mathbf{v} = \{v\}$) we obtain

$$z_0\xi_0(\bar{t}, \bar{u}, v) + \langle \mu, \xi(\cdot, \bar{u}, v) \rangle_c + \langle z, \xi(\bar{t}, \bar{u}, v) \rangle \geq 0 \quad (10.3.12)$$

in the customary shorthand. This inequality is

$$z_0 \int_0^{\bar{t}} \{f_0(\sigma, y(\sigma, \bar{u}), v(\sigma)) - f_0(\sigma, y(\sigma, \bar{u}), \bar{u}(\sigma))\} d\sigma$$

$$+ z_0 \int_0^{\bar{t}} \Big\langle \partial_y f_0(\tau, y(\tau, \bar{u}), \bar{u}(\tau)),$$

$$\int_0^\tau S(\tau, \sigma; \bar{u})\{f(\sigma, y(\sigma, \bar{u}), v(\sigma)) - f(\sigma, y(\sigma, \bar{u}), \bar{u}(\sigma))\} d\sigma \Big\rangle d\tau$$

$$+ \int_0^{\bar{t}} \Big\langle \mu(dt), \int_0^t S(t, \sigma; \bar{u})\{f(\sigma, y(\sigma, \bar{u}), v(\sigma)) - f(\sigma, y(\sigma, \bar{u}), \bar{u}(\sigma))\} d\sigma \Big\rangle$$

$$+ z_0\Big\langle \partial_y g_0(\bar{t}, y(\bar{t}, u)),$$

$$\int_0^{\bar{t}} S(\bar{t}, \sigma; \bar{u})\{f(\sigma, y(\sigma, \bar{u}), v(\sigma)) - f(\sigma, y(\sigma, \bar{u}), \bar{u}(\sigma))\} d\sigma \Big\rangle$$

$$+ \Big\langle z, \int_0^{\bar{t}} S(\bar{t}, \sigma; \bar{u})\{f(\sigma, y(\sigma, \bar{u}), v(\sigma)) - f(\sigma, y(\sigma, \bar{u}), \bar{u}(\sigma))\} d\sigma \Big\rangle \geq 0.$$

$$(10.3.13)$$

According to the theory in **10.2**, the solution of (10.3.6) is given by

$$\bar{z}(s) = S(\bar{t}, s; \bar{u})^*(z + z_0\partial_y g_0(\bar{t}, y(\bar{t}, \bar{u})))$$

$$+ z_0 \int_s^{\bar{t}} S(\tau, s; \bar{u})^* \partial_y f_0(\tau, y(\tau, \bar{u}), \bar{u}(\tau)) d\tau$$

$$+ \int_s^{\bar{t}} S(\tau, s; \bar{u})^* \mu(d\tau), \quad (10.3.14)$$

the last integral understood as in (10.2.34). We interpret the first in the same way, after noticing that $g(\cdot) = \partial_y f_0(\cdot, y(\cdot, \bar{u}), \bar{u}(\cdot))$ is an E^*-valued Lebesgue–Bochner integrable function and thus defines an element v of $\Sigma(0, \bar{t}; E^*)$ through the formula $v(e) = \int_e g(s)ds$. Obviously, we have $\|v\|_{\Sigma(0,\bar{t};E^*)} \leq \|g\|_{L^1(0,\bar{t};E^*)}$ (the first integral can also be interpreted in the spirit of Remark 5.5.11). To go from (10.3.13) to (10.3.7) we do as in previous appearances of the minimum principle. In the last two terms, the integral is pulled to the left of $\langle \cdot, \cdot \rangle$ and S is switched to the other side of $\langle \cdot, \cdot \rangle$ taking adjoints; in the second and third integral a switching of orders of integration is meted out via Lemma 10.2.17. This ends the proof of (10.3.7). We pass to (10.3.8) as in **6.5** and **6.6** and we omit the details.

As usual, we can obtain the formulation for the time optimal problem setting $f_0 \equiv 1$ in the general control problem. However, subsequent results stating that the multiplier (z_0, μ, z) is nonzero may be useless since, in the time optimal problem, we must show that $(\mu, z) \neq 0$. We apply instead Theorem 7.2.2 with $V_n = V = B(\bar{u}, \delta)$, $B(\bar{u}, \delta)$ the ball[2] in Lemma 6.1.1, and

$$\mathbf{f}_n(u) = (y(\cdot, u), y(t_n, u)), \tag{10.3.15}$$

where $\{t_n\}$ is a sequence with $t_n < \bar{t} = $ optimal time, $t_n \to \bar{t}$. The target set is $\mathbf{M}(\bar{t}) \times Y$, and the solution of the time optimal problem is the sequence $\{\bar{u}^n\} = \{\bar{u}\}$. It is clear that $\mathbf{f}_n(V) \cap (\mathbf{M}(\bar{t}) \times Y) = \emptyset$ for all n. Theorem 7.2.2 produces sequences $\{\bar{u}^n\} \subseteq B(\bar{u}, \delta)$ and $\{\bar{y}^n(\cdot), \bar{y}^n\} \in \mathbf{M}(\bar{t}) \times Y$ satisfying the convergence statements. We select $\rho > 0$ and use $\{D_n\} \subseteq \mathrm{Der}\,\mathbf{f}_n(\bar{u}^n)$ consisting of all pairs

$$(\xi(\cdot, u, \mathbf{p}, \mathbf{v}), \xi(t_n, u, \mathbf{p}, \mathbf{v})) \in C(0, \bar{t}; E) \times E, \tag{10.3.16}$$

$\xi(\cdot, u, \mathbf{p}, \mathbf{v})$ given by (6.5.8) with $u = \bar{u}^n$. By Lemma 6.6.4 every element of the form (10.3.16) with $t_n = \bar{t}$ and $u = \bar{u}$ belongs to $\liminf_{n \to \infty} D_n$. Theorem 7.2.2 provides sequences $\{\mu_n\} \subseteq \Sigma(0, \bar{t}; E^*) = C(0, \bar{t}; E)^*$, $\{z_n\} \subseteq E^*$ such that

$$\|\mu_n\|_{\Sigma(0,\bar{t};E^*)}^2 + \|z_n\|_{E^*}^2 = 1,$$

$$\langle \mu_n, \xi(\cdot, \bar{u}^n, \mathbf{p}, \mathbf{v}) \rangle_c + \langle z_n, \xi(\bar{t}, \bar{u}^n, \mathbf{p}, \mathbf{v}) \rangle \geq -\delta_n, \tag{10.3.17}$$

where $\delta_n \to 0$. If $(\mu, z) \in \Sigma(0, \bar{t}; E^*) \times E^*$ is a $(C(0, \bar{t}; E) \times E)$-weak limit point of (μ_n, z_n), then

$$\langle \mu, \xi(\cdot, \bar{u}, \mathbf{p}, \mathbf{v}) \rangle_c + \langle z, \xi(\bar{t}, \bar{u}, \mathbf{p}, \mathbf{v}) \rangle \geq 0 \tag{10.3.18}$$

and μ, z satisfy the last two relations (10.3.5). For single patches,

$$\langle \mu, \xi(\cdot, \bar{u}, v) \rangle_c + \langle z, \xi(\bar{t}, \bar{u}, v) \rangle \geq 0, \tag{10.3.19}$$

[2] See Remark 10.3.9.

which is (10.3.13) with $z_0 = 0$, so that

$$\int_0^{\bar{t}} \langle \bar{z}(\sigma), f(\sigma, y(\sigma, \bar{u}), v(\sigma)) - f(\sigma, y(\sigma, \bar{u}), \bar{u}(\sigma)) \rangle d\sigma \geq 0 \qquad (10.3.20)$$

for all $v(\cdot) \in C_{\mathrm{ad}}(0, \bar{t}; U)$. If E^* is separable and $f(\sigma, y(\sigma, \bar{u}), \bar{u}(\sigma))$ is regular in U, then

$$\langle \bar{z}(s), f(s, y(s, \bar{u}), \bar{u}(s)) \rangle = \min_{v \in U} \langle \bar{z}(s), f(s, y(s, \bar{u}), v) \rangle \qquad (10.3.21)$$

a.e. in $0 \leq s \leq \bar{t}$. This ends the proof of Theorem 10.3.1 in all cases. Note that we have ignored the fact that $\xi(t, \bar{u}, \mathbf{p}, \mathbf{v})$ and $\xi_0(t, \bar{u}, \mathbf{p}, \mathbf{v})$ must be multiplied by $\bar{t}^{-1}$ to be actual directional derivatives; this obviously has no bearing on the arguments.   ∎

Let $\tilde{u} \in C_{\mathrm{ad}}(0, \bar{t}; U)$ be such that $y(t, \tilde{u})$ exists in $0 \leq t \leq \bar{t}$. We denote by $\Xi(0, \bar{t}; U, \tilde{u}) \subseteq C(0, \bar{t}; E)$ the set of all trajectories of the inhomogeneous variational equation

$$\xi'(t) = \{A(t) + \partial_y f(t, y(t, \tilde{u}), \tilde{u}(t))\} \xi(t)$$
$$+ \{f(t, y(t, \tilde{u}), u(t)) - f(t, y(t, \tilde{u}), \tilde{u}(t))\}, \quad \xi(0) = 0 \quad (10.3.22)$$

for $u(\cdot) \in C_{\mathrm{ad}}(0, \bar{t}; U)$. The set of values of all these trajectories at time $\tilde{t} \leq \bar{t}$ is the *reachable space* $R(0, \tilde{t}; U, \tilde{u})$ defined in **6.5** (except for the minor detail that $y(t, \tilde{u})$ is only assumed there to exist in $0 \leq t \leq \tilde{t}$). Finally, we define $\Xi(0, \bar{t}; U, \tilde{u})(\tilde{t}) \subseteq C(0, \bar{t}; E) \times E$ as the set of all elements $(\xi(\cdot), \xi(\tilde{t})), \xi(\cdot)$ an arbitrary trajectory of (10.3.22). Clearly, $\mathrm{conv}\, \Xi(0, \bar{t}; U, \tilde{u})(\tilde{t})$ consists of all elements of the form

$$(\xi(\cdot, \tilde{u}, \mathbf{p}, \mathbf{v}), \xi(\tilde{t}, \tilde{u}, \mathbf{p}, \mathbf{v})) \subseteq C(0, \bar{t}; E) \times E$$

for all probability vectors $\mathbf{p}$ and all vectors $\mathbf{v}$ of elements of $C_{\mathrm{ad}}(0, \bar{t}; U)$.

Theorem 10.3.2. *Assume that for every sequence* $\{\tilde{u}^n\} \subseteq C_{\mathrm{ad}}(0, \bar{t}; E)$ *with* $\tilde{u}^n \to \bar{u}$ *and every sequence* $\{(\tilde{y}^n(\cdot), \tilde{y}^n)\} \subseteq \mathbf{M}(\bar{t}) \times Y$ *with* $(\tilde{y}^n(\cdot), \tilde{y}^n) \to (y(\cdot, \bar{u}), y(\bar{t}, \bar{u}))$ *there exists* $\rho > 0$ *and a precompact sequence* $\{\mathbf{Q}_n\}$ *in the space* $C(0, \bar{t}; E) \times E$ *such that*

$$\Delta = \bigcap_{n=n_0}^{\infty} (\Delta_n + \mathbf{Q}_n) \qquad (10.3.23)$$

contains an interior point in $C(0, T; E) \times E$ *for* n_0 *large enough, where*

$$\Delta_n = \bar{t}^{-1} \mathrm{conv}\, \Xi(0, \bar{t}; U, \tilde{u}^n)(\tilde{t}) - (T_{\mathbf{M}(\bar{t})}(\tilde{y}^n(\cdot)) \times T_Y(\tilde{y}^n)) \cap \mathbf{B}(0, \rho) \qquad (10.3.24)$$

with $\mathbf{B}(0, \rho)$ *the ball of center* 0 *and radius* ρ *in* $C(0, \bar{t}; E) \times E$. *Then any limit point* (z_0, μ, z) *of the sequence* (z_{0n}, μ_n, z_n) *(corresponding to* $\{\mathbf{D}_n\}$ *and* ρ*) is nonzero.*

The result for the time optimal problem is

Theorem 10.3.3. *Assume that for every sequence* $\{t_n\} \subseteq [0, \bar{t})$ *with* $t_n \to \bar{t}$, *every sequence* $\{\tilde{u}^n\} \in C_{ad}(0, \bar{t}; E)$ *with* $\tilde{u}^n \to \bar{u}$ *and every sequence* $\{(\tilde{y}^n(\cdot), \tilde{y}^n)\} \subseteq M(\bar{t}) \times Y$ *with* $(\tilde{y}^n(\cdot), \tilde{y}^n) \to (y(\cdot, \bar{u}), y(\bar{t}, \bar{u}))$ *there exists* $\rho > 0$ *and a precompact sequence* $\{\mathbf{Q}_n\}$ *in* $C(0, \bar{t}; E) \times E$ *such that* (10.3.23) *contains an interior point for* n_0 *large enough, where*

$$\Delta_n = t_n^{-1} \operatorname{conv} \Xi(0, \bar{t}; U, \tilde{u}^n)(t_n) - (T_{M(\bar{t})}(\tilde{y}^n(\cdot)) \times T_Y(\tilde{y}^n)) \cap \mathbf{B}(0, \rho) \quad (10.3.25)$$

$\mathbf{B}(0, \rho)$ *the ball of center* 0 *and radius* ρ *in* $C(0, \bar{t}; E)$. *Then any limit point* (μ, z) *of the sequence* (μ_n, z_n) *(corresponding to* $\{\mathbf{D}_n\}$ *and* ρ*) is nonzero.*

Theorem 10.3.3 is an immediate consequence of Corollary 7.2.3. If the reader wants a direct proof, use the sequence of inequalities (10.3.17) in combination with Lemma 6.4.3. Likewise, Theorem 10.3.2 follows from Corollary 7.2.6, or can be proved as follows. Note that the assumptions imply that the set of all $\{\xi_0(\bar{t}, \tilde{u}^n, \mathbf{p}, \mathbf{v})\}$ is bounded. If $z_{0n} \to z_0 \neq 0$ one has nothing to do; if $z_{0n} \to 0$ then we can pass to (10.3.17) (with a different δ_n) and the proof ends in the same way. We note also that $\Xi(0, \bar{t}; U, \tilde{u})(\bar{t})$ is bounded, thus if $\{\mathbf{Q}_n\}$ is bounded we have the right to knock off $\mathbf{B}(0, \rho)$ from (10.3.24) and (10.3.25). Finally, see Remark 6.5.7 on the right of taking closures.   ∎

We take a look below at diverse ways of fulfilling conditions (10.3.24) and (10.3.25); it is enough to do so for the latter since it is more demanding (in the first, $t_n \equiv \bar{t}$). The particular case $\mathbf{Q}_n = Q_{n,c} \times Q_n$ is especially interesting; here, (10.3.25) will be satisfied (with a different ρ)[3] if there exist $\rho > 0$, an integer n_0, and precompact sequences $\{Q_{n,c}\}$ in $C(0, \bar{t}; E)$ and $\{Q_n\}$ in E such that the set $\Delta_n \subseteq C(0, \bar{t}; E) \times E$ of all elements with coordinates

$$t_n^{-1} \xi(\cdot, \tilde{u}^n, \mathbf{p}, \mathbf{v}) - T_{M(\bar{t})}(\tilde{y}^n(\cdot)) \cap B_c(0, \rho) + Q_{n,c} \quad (10.3.26)$$

$$t_n^{-1} \xi(t_n, \tilde{u}^n, \mathbf{p}, \mathbf{v}) - T_Y(\tilde{y}^n) \cap B(0, \rho) + Q_n \quad (10.3.27)$$

contains a common polyball $B_c(y(\cdot), \varepsilon) \times B(y, \varepsilon)$ in $C(0, \bar{t}; E) \times E$ for $n \geq n_0$. Here (for each n) $\mathbf{p}$ ranges over all probability vectors, $\mathbf{v}$ over all vectors of elements of $C_{ad}(0, \bar{t}; U)$, and $\mathbf{p}$ and $\mathbf{v}$ are the same in (10.3.26) and (10.3.27).

---

[3] Replacing ρ means using a different sequence (z_{0n}, μ_n, z_n) in Theorem 10.3.1 and a different sequence (μ_n, z_n) for the time optimal problem. This does not affect the statements on limit points.

In the first result we set $\mathbf{p} = \{1\}$, $\mathbf{v} = \{\tilde{u}^n\}^{(4)}$ (so that $\xi(t, \tilde{u}^n, \mathbf{p}, \mathbf{v}) \equiv 0$) and satisfy (10.3.26) and (10.3.27) relying only on the tangent cones and the precompact sequences in the style of **6.4**.

Corollary 10.3.4. *Assume that* $\mathbf{M}(\bar{t})$ *is* T*-full in* $C(0, \bar{t}; E)$ *and that* Y *is* T*-full in* E. *Then* $(z_0, \mu, z) \neq 0$ $((\mu, z) \neq 0$ *for the time optimal problem*$)$.

Generally speaking, fullness of $\mathbf{M}(\bar{t})$ (or some such condition) cannot be given up (see Example 10.3.7). However, lack of fullness of the target set Y can sometimes be compensated with $R(0, t_n; U, \tilde{u}^n)$.

Theorem 10.3.5. *Assume* $\mathbf{M}(\bar{t})$ *is* T*-full in* $C(0, \bar{t}; E)$ *and that, for all sequences* $\{t_n\} \subseteq [0, \bar{t}]$ *with* $t_n \to \bar{t}$ $(t_n \equiv \bar{t}$ *for the general control problem*$)$, $\{\tilde{u}^n\} \subseteq C_{\mathrm{ad}}(0, \bar{t}; E)$ *with* $\tilde{u}^n \to \bar{u}$, *and* $\{\tilde{y}^n\} \subseteq Y$ *with* $\tilde{y}^n \to y(\bar{t}, \bar{u})$ *there exists* $\rho > 0$ *and a precompact sequence* $\{Q_n\}$ *in* E *such that*

$$\bigcap_{n=n_0}^{\infty} \{\operatorname{conv} R(0, t_n; U, \tilde{u}^n) - T_Y(\tilde{y}^n) \cap B(0, \rho) + Q_n\} \tag{10.3.28}$$

contains an interior point for n_0 *large enough. Then* $(z_0, \mu, z) \neq 0$ $((\mu, z) \neq 0$ *for the time optimal problem*$)$.

Proof. We begin by observing that, under the present assumptions, formula (6.5.8) determines that

$$\|\xi(t, \tilde{u}^n, \mathbf{p}, \mathbf{v})\| \leq C \quad (0 \leq t \leq \bar{t}) \tag{10.3.29}$$

for all $\mathbf{p}$, $\mathbf{v}$. The assumption of T-fullness of $\mathbf{M}(\bar{t})$ implies the existence of $\sigma > 0$ and of a precompact sequence $\{Q'_{n,c}\}$ in $C(0, \bar{t}; E)$ such that the sets $T_{\mathbf{M}(\bar{t})}(\tilde{y}^n(\cdot)) \cap B(0, \sigma) + Q'_{n,c}$ contain a common ball $B_c(y(\cdot), \varepsilon)$ for $n \geq n_0$ large enough. Replacing $Q'_{n,c}$ by $Q'_{n,c} - y(\cdot)$ we shift the ball to $B_c(0, \varepsilon)$, so that

$$\bigcap_{n=n_0}^{\infty} \{T_{\mathbf{M}(\bar{t})}(\tilde{y}^n(\cdot)) \cap B_c(0, r\sigma) + r(Q'_{n,c} - y(\cdot))\}$$

contains the ball $B_c(0, r\varepsilon) \subseteq C(0, \bar{t}; E)$ for $n \geq n_0$ large enough (recall that $T_{\mathbf{M}(\bar{t})}(\tilde{y}^n(\cdot))$ is a cone). Taking r so huge that $r\varepsilon > 2C\bar{t}^{-1}$ (C the constant in (10.3.29)) and defining $Q_{n,c} = r(Q'_{n,c} - y(\cdot))$ we deduce that the sets (10.3.26) contain $B_c(0, C)$ for $n \geq n_0$ *for any choice of* $\mathbf{p}$, $\mathbf{v}$, thus we are free to select $\mathbf{p}$, $\mathbf{v}$ at will in (10.3.27); all we have to do is to check that the sets (10.3.27) contain a common ball in E for n large enough. This is insured by the assumptions of Theorem 10.3.5, since we may always suppose that $r\sigma > \rho$. ∎

---

(4) Here, $\{\tilde{u}^n\}$ means the set whose only element is $\tilde{u}^n$, not the sequence $\{\tilde{u}^n\}$.

Remark 10.3.6. In the case $\dim E < \infty$, condition (10.3.28) is obviously satisfied, thus the only hypothesis needed for nonvanishing of the multipliers is T-fullness of $\mathbf{M}(\bar{t})$ in $C(0, \bar{t}; E)$.

Remark 10.3.7. Consider the finite dimensional system $y'(t) = f(t, y(t), u(t))$. Under standard assumptions, $\Xi(0, \bar{t}; U, \tilde{u}^n)$ is a uniformly bounded equicontinuous set of functions. Thus, by the Arzelà-Ascoli theorem, it is precompact in $C(0, \bar{t}; E)$. Accordingly, the contribution of $\Xi(0, \bar{t}; U, \tilde{u}^n)$ to the first coordinate of (10.3.23) is negligible; it can be offset modifying the first coordinate of the precompact sequence. A similar observation (with different justification of compactness) applies to abstract parabolic equations; see Theorem 7.7.16.

Remark 10.3.8. It must be emphasized that the condition $(z_0, \mu, z) \neq 0$ does *not* guarantee that the costate $\bar{z}(s)$ is nontrivial (i.e. not identically zero). To see this, note that the costate can be written in the form

$$\bar{z}(s) = \int_s^{\bar{t}} S(\sigma, s; \bar{u})\mu_1(ds) \tag{10.3.30}$$

with $\mu_1(ds) = \mu(ds) + z_0 f_0(s, y(s, \bar{u}), \bar{u}(s))ds + \delta_{\bar{t}}(ds)z$, $\delta_{\bar{t}}$ the Dirac delta with mass at $\bar{t}$; for the time optimal problem, $\mu_1(ds) = \mu(ds) + \delta_{\bar{t}}(ds)z$. Even if $(z_0, \mu, z) \neq 0$ we may have $\mu_1 = 0$ (an example is $(z_0, \mu, z) = (0, -\delta_{\bar{t}}(ds)z, z)$). It seems natural to conjecture that the costate is nontrivial if μ_1 is not concentrated at $\{0\}$, which reduces to the implication

$$\int_s^{\bar{t}} S(\sigma, s; \bar{u})^* v(d\sigma) = 0 \text{ a.e. in } 0 \leq s \leq \bar{t} \implies \|v\|((0, \bar{t}]) = 0 \tag{10.3.31}$$

for arbitrary $v \in \Sigma(0, \bar{t}; E)$. Integrating against a function $g(\cdot) \in C(0, \bar{t}; E)$ and applying Lemma 10.2.17, the equality on the left of (10.3.31) becomes

$$\int_0^{\bar{t}} \left\langle \int_0^t S(t, s)g(s)ds, v(dt) \right\rangle = 0, \tag{10.3.32}$$

thus everything reduces to showing that the set of all functions of the form

$$y(t) = \int_0^t S(t, s)g(s)ds \quad (g(\cdot) \in C(0, \bar{t}; E)) \tag{10.3.33}$$

with $S(t, s) = S(t, s; \bar{u})$ is dense in $C_0(0, \bar{t}; E)$. (the subindex means $y(0) = 0$). This is true in a number of particular cases (for instance, $S(t, s) = S(t - s)$, $S(\cdot)$ a strongly continuous semigroup) and is probably valid for an arbitrary (E, E)-valued function $S(t, s)$ defined and strongly continuous in $0 \leq s \leq t \leq \bar{t}$ with $S(s, s) = I$ $(0 \leq s \leq \bar{t})$, but we don't have a proof.

Remark 10.3.9. The reader has no doubt noted a slight discrepancy among the treatment of time optimal problems without state constraints (Theorem 6.5.1) and with state constraints (Theorem 10.3.1). In Chapter 6 the metric spaces are $V_n = B_n(\bar{u}, \delta) \subseteq C_{ad}(0, t_n; U)$, whereas in this chapter all the spaces coincide with $V = B(\bar{u}, \delta)$. The first choice is more natural, since it does not seem to require that $y(t, u)$ exist in $0 \le t \le \bar{t}$, but this is illusory; due to Lemma 6.1.1, global existence holds no matter what. To insist in using the V_n in problems with state constraints would need a generalization of the theory of the abstract time optimal problem to functions f_n taking values in different Banach spaces $E_n (= C(0, t_n; U) \times E)$.

10.4. Saturation of the State Constraint

It is natural to expect that, in case $y(t, \bar{u})$ does not **saturate** the state constraint (that is, when $y(t, \bar{u}) \in \text{Int}(M(t))$), the minimum principle should be the same as the one without state constraints; this means the measure $\mu \in \Sigma(0, \bar{t}; E^*)$ in the adjoint variational equation (10.3.7) should be zero. The following result shows (as a very particular case) that this is true.

Call the set valued function $t \to U(t) \subseteq E$ defined in a set $e \subseteq \mathbb{R}$ **continuous in** e if for every closed set $C \subseteq E$ the set

$$\{t \in e; U(t) \cap C \ne \emptyset\} \tag{10.4.1}$$

is closed. We denote by ∂M or $\partial(M)$ the boundary of M.

Theorem 10.4.1. *Assume that each $M(t)$ is closed and that $t \to \partial(M(t))$ is a continuous set valued function. Let $\bar{y}(\cdot) \in \mathbf{M}(\bar{t})$,*

$$e_0 = \{t \in [0, \bar{t}]; \bar{y}(t) \in \text{Int}(M(t))\}, \tag{10.4.2}$$

and $\{\tilde{y}^n(\cdot)\}$ a sequence in $\mathbf{M}(\bar{t})$ with $\tilde{y}^n(\cdot) \to \bar{y}(\cdot)$ in $C(0, \bar{t}; E)$. Then, if $\mu \in$ $(\liminf_{n \to \infty} T_{\mathbf{M}(\bar{t})}(\tilde{y}^n(\cdot)))^-$ we have

$$\|\mu\|(e_0) = 0, \tag{10.4.3}$$

with $\|\mu\|$ the total variation of μ (extended as in Lemma 10.2.12).

Proof. We show first that if $\bar{y}(\cdot) \in \mathbf{M}(\bar{t})$ then the set

$$e_+(\varepsilon) = \{t \in [0, \bar{t}]; \text{dist}(\bar{y}(t), \partial(M(t))) > \varepsilon\} \tag{10.4.4}$$

is open in $[0, \bar{t}]$. To see this, let $\bar{y} \in E$, and note that the set

$$\{t \in [0, \bar{t}]; \text{dist}(\bar{y}, \partial(M(t))) \le \varepsilon\}$$
$$= \bigcap_{j=1}^{\infty} \{t \in [0, \bar{t}]; \partial(M(t)) \cap B(\bar{y}, \varepsilon + 1/j) \ne \emptyset\}$$

is closed in $[0, \bar{t}]$. Define $t_{nj} = j\bar{t}/n$ $(j = 0, 1, \ldots, n)$ and let

$$\delta_{nj} = \max_{t_{n,j-1} \le t \le t_{nj}} \|\bar{y}(t_{nj}) - \bar{y}(t)\|, \quad j = 1, \ldots, n.$$

Then, if $t_{n,j-1} \le t \le t_{nj}$ and $\text{dist}(\bar{y}(t), \partial(M(t))) \le \varepsilon$ we have

$$\text{dist}(\bar{y}(t_{nj}), \partial((M(t))))$$
$$\le \|\bar{y}(t_{nj}) - \bar{y}(t)\| + \text{dist}(\bar{y}(t), \partial((M(t)))) \le \delta_{nj} + \varepsilon.$$

Symmetrically, if $\text{dist}(\bar{y}(t_{nj}), \partial((M(t)))) \le \delta_{nj} + \varepsilon$ and $t_{n,j-1} \le t \le t_{nj}$,

$$\text{dist}(\bar{y}(t), \partial((M(t))))$$
$$\le \|\bar{y}(t_{nj}) - \bar{y}(t)\| + \text{dist}(\bar{y}(t_{nj}), \partial((M(t)))) \le 2\delta_{nj} + \varepsilon.$$

Accordingly,

$$e_-(\varepsilon) = \{t \in [0, \bar{t}]; \ \text{dist}(\bar{y}(t), \partial(M(t))) \le \varepsilon\}$$
$$= \bigcap_{n=1}^{\infty} \bigcup_{j=1}^{n} \{t \in [t_{n,j-1}, t_{n,j}]; \ \text{dist}(\bar{y}(t_{nj}), \partial(M(t))) \le \varepsilon + \delta_{nj}\}$$

so that $e_-(\varepsilon)$ is closed and $e_+(\varepsilon)$, its complement in $[0, \bar{t}]$, is open in $[0, \bar{t}]$.

We write $e_+(\varepsilon)$ as the union of a finite or countable collection of disjoint subintervals $e_j(\varepsilon)$ of $[0, \bar{t}]$, each open in $[0, \bar{t}]$, thus of the form

$$[0, \bar{t}], \qquad [0, b) \quad (0 < a < \bar{t}), \qquad (a, \bar{t}] \quad (0 < b < \bar{t}) \qquad (10.4.5)$$

or

$$(a, b) \qquad (0 < a < b < \bar{t}), \qquad (10.4.6)$$

and use the following auxiliary result.

Lemma 10.4.2. *Assume $e_j(\varepsilon)$ has the form (10.4.6) and let $f(\cdot) \in C(0, \bar{t}; E)$ have support in the closure $\overline{e_j(\varepsilon)}$. Let $y(\cdot) \in M(\bar{t})$ with*

$$\|y - \bar{y}\|_{C(0,\bar{t};E)} < \varepsilon. \qquad (10.4.7)$$

Then

$$f(\cdot) \in T_{M(\bar{t})}(y(\cdot)). \qquad (10.4.8)$$

If $e_j(\varepsilon) = [0, b)$ (resp. $(a, t]$) (10.4.8) holds for $f(\cdot) \in C(0, \bar{t}; E)$ such that $f(t) = 0$ for $t \ge b$ (resp. $t \le a$). Finally, if $e_j(\varepsilon) = [0, \bar{t}]$, (10.4.8) holds for every $f(\cdot) \in C(0, \bar{t}; E)$.

Proof. Assume $y(\cdot) \in \mathbf{M}(\bar{t})$ satisfies (10.4.7). Let $\{\bar{y}_k(\cdot)\}$ be an arbitrary sequence in $\mathbf{M}(\bar{t})$ with $\bar{y}_k(\cdot) \to y(\cdot)$ in $C(0, \bar{t}; E)$ and let $\{h_k\} \subseteq \mathbb{R}_+$ be an arbitrary sequence with $h_k \to 0$. If k is so large that

$$\|\bar{y}_k - \bar{y}\|_{C(0,\bar{t};E)} \leq \|\bar{y}_k - y\|_{C(0,\bar{t};E)} + \|y - \bar{y}\|_{C(0,\bar{t};E)} \leq r\varepsilon \qquad (10.4.9)$$

with $0 < r < 1$, then

$$\tilde{y}_k(t) = h_k f(t) + \bar{y}_k(t) \in M(t) \quad (0 \leq t \leq \bar{t}) \qquad (10.4.10)$$

for k large enough. To see this, note that $\mathrm{dist}(\bar{y}(t), \partial(M(t))) > \varepsilon$ for $t \in e_j(\varepsilon)$, so that (10.4.9) implies

$$\mathrm{dist}(\bar{y}_k(t), \partial(M(t))) \geq (1 - r)\varepsilon \qquad (10.4.11)$$

for $t \in e_j(\varepsilon)$. Since $f(\cdot)$ is bounded in $e_j(\varepsilon)$ and zero outside of $e_j(\varepsilon)$, (10.4.10) follows. This inclusion relation means that $y_k(\cdot) \in \mathbf{M}(\bar{t})$.

We have

$$f(\cdot) = \lim_{k\to\infty} \frac{\tilde{y}_k(\cdot) - \bar{y}_k(\cdot)}{h_k} \qquad (10.4.12)$$

in $C(0, \bar{t}; E)$. Since $\{\bar{y}_k(\cdot)\}$ and $\{h_k\}$ are arbitrary, $f(\cdot)$ belongs to $T_{\mathbf{M}(\bar{t})}(y(\cdot))$ as claimed (see **3.3**). We deal with the other cases in the same way. ∎

End of Proof of Theorem 10.4.1. Let $\mu \in (\liminf_{n\to\infty} T_{\mathbf{M}}(\tilde{y}^n(\cdot)))^-$, and let $f(\cdot) \in C(0, \bar{t}; E)$ satisfy the assumptions of Lemma 10.4.2. Then $f(\cdot) \in T_{\mathbf{M}}(\bar{t})(\tilde{y}^n(\cdot))$ for n large enough, so that $f(\cdot) \in \liminf_{n\to\infty} T_{\mathbf{M}}(\tilde{y}^n(\cdot))$. It follows that

$$\int_{e_j(\varepsilon)} \langle f(t), \mu(dt) \rangle = 0, \qquad (10.4.13)$$

where the upgrade of $\leq$ to $=$ is due to the possibility of multiplying $f(\cdot)$ by -1. If d is an interval contained in $e_j(\varepsilon)$ then its characteristic function χ_d can be approximated pointwise by a sequence $\{f_n(\cdot)\}$ of continuous functions with support in $\overline{e_j(\varepsilon)}$ (with the necessary modifications when $e_j(\varepsilon)$ is of the form (10.4.5)). We then deduce from Corollary 10.2.13 that $\mu(d) = 0$. Computing the total variation of μ using partitions on intervals we deduce that $\|\mu\|(e_j(\varepsilon)) = 0$. Now, (Lemma 10.2.12) $\|\mu\|$ can be extended to a countably additive measure in the Borel field $\Phi_b(C)$ and we have $e_+(\varepsilon) = \cup e_j(\varepsilon)$, $e_0 = \cup e_+(1/j)$, so that $\|\mu\|(e_0) = 0$ as claimed. ∎

Remark 10.4.3. As pointed out in Remark 10.3.8, the fact that $(z_0, \mu, z) \neq 0$ $((\mu, z) \neq 0$ for the time optimal problem) does not guarantee nontriviality of the costate $\bar{z}(s)$. In certain situations, we may be sure that the measure μ_1 in (10.3.30) is not supported by $\{0\}$. If the implication (10.3.31) holds, this guarantees nontriviality of the costate. For instance, consider the time optimal problem with $\zeta \in \text{Int } M(0)$, $Y \subseteq \text{Int } M(\bar{t})$. Then e_0 contains nonempty intervals $[0, b)$ and $(a, \bar{t}]$, where $\mu = 0$; if $\mu_1 = \mu + \delta_{\bar{t}} z = 0$ is supported by $\{0\}$, both μ and z are zero.

Example 10.4.4. Theorem 10.4.1 admits generalizations that contemplate saturation of state constraints in subspaces. To fix ideas, let F, G be subspaces of E such that $E = F \oplus G$ algebraically and topologically, so that the projection $P : E \to F$ is bounded; then so is the projection $Q = I - P$ onto G. We have $P^* P^* = (PP)^* = P^*$, thus P^* and $Q^* = I^* - P^*$ are the projections for an (algebraic and topological) dual decomposition $E^* = F^* \oplus G^*$. Both decompositions interact in the expected way: $\langle y^*, y \rangle = \langle P^* y^*, Py \rangle + \langle Q^* y^*, Qy \rangle$.

The decomposition $E = F \oplus G$ induces in an obvious way a corresponding decomposition $C(0, \bar{t}; E) = C(0, \bar{t}; F) \oplus C(0, \bar{t}; G)$, where the projections are P and Q acting pointwise; likewise, the projections for the dual decomposition $\Sigma(0, \bar{t}; E^*) = \Sigma(0, \bar{t}; F^*) \oplus \Sigma(0, \bar{t}; G^*)$ are P^* and Q^* acting pointwise.

Let $\bar{y}(\cdot)$, $\tilde{y}^n(\cdot)$ satisfy the assumptions of Theorem 10.4.1 in the space E. To study saturation of constraints in F, let

$$e_0(F) = \{t \in [0, \bar{t}]; P\bar{y}(t) \in \text{Int}(PM(t))\}.$$

A direct sum decomposition carries over to tangent cones, hence we have

$$T_{M(\bar{t})}(\tilde{y}^n(\cdot)) = T_{PM(\bar{t})}(P\tilde{y}^n(\cdot)) \oplus T_{QM(\bar{t})}(Q\tilde{y}^n(\cdot))$$

Then, if $\mu \in (\liminf_{n \to \infty} T_{M(\bar{t})}(\tilde{y}^n(\cdot)))^-$ we have

$$P^* \mu \in \left(\liminf_{n \to \infty} T_{PM(\bar{t})}(P\tilde{y}^n(\cdot)) \right)^-, \qquad Q^* \mu \in \left(\liminf_{n \to \infty} T_{QM(\bar{t})}(Q\tilde{y}^n(\cdot)) \right)^-$$

It follows from Theorem 10.4.1 applied in F that

$$\|P^* \mu\|(e_0(F)) = 0. \tag{10.4.14}$$

but we must verify first that the set function $t \to PM(t)$ is closed. This is obvious; in fact if $C \subseteq F$, then $PM(t) \cap C \neq \emptyset$ if and only if $M(t) \cap (C \oplus G) \neq \emptyset$.

10.5. Surface of Revolution of Minimum Area as a Control Problem

Since $r\sqrt{u^2 + v^2} \leq r(|u| + |v|)$ and there are no control constraints, the natural choice of admissible control space for the minimum surface problem in **10.1** should be

$$C_{\text{ad}}(0, \bar{t}; U) = L^1(0, T; \mathbb{R}^2).$$

The curves in competition for the minimum are rectifiable (in the sense of **4.9**) and their length can be defined. However, as in the minimum drag nose shape problem in

4.9 this control space is unsuitable for existence results. We use the Reparametrization Lemma 4.9.3 to show that every rectifiable curve can be parametrized by arclength, and then the general change-of-variable formula (Example 4.9.12) to show that reparametrization does not alter the value of the cost functional. To prove existence we use Example 2.2.16 and check the convexity condition using (*b*) in Example 2.2.15 (the Hessian is zero). In proving existence (and in applying the maximum principle) it is convenient to ignore the control constraint $u \geq 0$; this introduces spurious solutions, but they can be easily spotted. We assume

$$r_a = r(0) > 0 \quad \text{and} \quad r_b = r(\bar{t}) > 0. \tag{10.5.1}$$

Theorem 10.5.1. *A surface of revolution of minimum area exists.*

We identify the solution below using the minimum principle in **10.3**. The optimal control will be parametrized by arclength, but we do *not* assume the same for other controls in competition for the minimum. Hence, the control set is unbounded and the minimum principle does not apply: Hypotheses III and III^0 are not satisfied independently of the control. However, a suitable extension of the minimum principle via the trick in **4.9** (see Theorem 4.9.2) can be easily proved and we omit the details.

Denote by $(\bar{u}(\cdot), \bar{v}(\cdot))$ a solution of the problem parametrized by arclength, and let L be the total length of the curve. To begin with, we transform the minimum into a maximum principle by assuming that $z_0 \leq 0$ and changing the sign of the measure $\mu = (\mu_1, \mu_2)$ in the adjoint variational equation (10.3.6). The problem is finite dimensional, thus the maximum principle can be formulated using the Hamiltonian formalism: using a notation similar to that in Chapters 3 and 4 the Hamiltonian is

$$H(x, r, z_0, p, q, u, v) = z_0 r \sqrt{u^2 + v^2} + pu + qv, \tag{10.5.2}$$

and the canonical equations for the costate $(p(s), q(s))$ are

$$dp = -\mu_1(ds), \qquad dq = -z_0 \sqrt{u^2 + v^2} \, ds - \mu_2(ds), \tag{10.5.3}$$

with final condition

$$(z_0, p(L), q(L)), \qquad z_0 \leq 0. \tag{10.5.4}$$

Keeping in mind the change of sign, we have

$$(\mu_1, \mu_2) \in -\left(\liminf_{n \to \infty} T_{\mathbf{M}(\bar{t})}(\tilde{x}^n(\cdot), \tilde{r}^n(\cdot)) \right)^-, \tag{10.5.5}$$

where[1] $M(t) = (-\infty, \infty) \times [0, \infty)$ and $(\tilde{x}^n(\cdot), \tilde{r}^n(\cdot))$ is a sequence of (pairs of) continuous functions with $\tilde{r}^n(t) \geq 0, \tilde{x}^n(t) \to \bar{x}(t), \tilde{r}^n(t) \to \bar{r}(t)$ uniformly in

---

[1] We are not assuming that $u(t) \geq 0$, so that the state constraint $a \leq x(t) \leq b$ is no longer in force; the surfaces may spill over the endpoints of the interval. Compare with **10.1**.

$0 \le t \le L$. Condition (10.5.5) and some work in the style of Example 10.4.4 show that

$$\mu_1 = 0, \qquad \mu_2 \ge 0, \tag{10.5.6}$$

while Theorem 10.4.1 implies that $\mu_2 = 0$ in the set where $\bar{r}(t) > 0$. The maximum principle is

$$z_0 \bar{r}(s) \sqrt{\bar{u}(s)^2 + \bar{v}(s)^2} + p\bar{u}(s) + q(s)\bar{v}(s)$$

$$= \max_{-\infty < u, v < \infty} \{ z_0 \bar{r}(s) \sqrt{u^2 + v^2} + pu + q(s)v \} \tag{10.5.7}$$

(recall that we do not assume that $u \ge 0$). We show next that $z_0 < 0$. In fact, if this fails we obtain from (10.5.7) that $p = 0$ and $q(s) \equiv 0$ a.e. (otherwise the maximum would be infinite). However, the second equality and equation (10.5.3) imply that $\mu_2 = 0$ in $0 < t \le \bar{t}$. Since the initial condition (a, r_a) belongs to the interior of the state constraint set, μ_2 cannot have mass at 0, and we deduce that $\mu_2 = 0$. Consequently,

$$(z_0, \mu_1, \mu_2, p, q(L)) = 0,$$

a contradiction with Remark 10.3.6, since $\mathbf{M}(\bar{t})$ is T-full in $C(0, \bar{t}; E) = C(0, \bar{t}; \mathbb{R}^2)$. (in fact, it is convex with interior points).

Normalizing to $z_0 = -1$ and keeping in mind that $(\bar{x}(s), \bar{r}(s))$ is parametrized by arclength we obtain

$$-\bar{r}(s) + p\bar{u}(s) + q(s)\bar{v}(s)$$

$$= \max_{-\infty < u, v < \infty} \{ -\bar{r}(s) \sqrt{u^2 + v^2} + pu + q(s)v \}. \tag{10.5.8}$$

There are two alternatives.

(a) $\bar{r}(s) > 0$ in $0 \le t \le L$. If $p = 0$, the maximum of (10.5.8) in u (for v fixed) would occur at $u = 0$; since the curve is parametrized by arclength, $\bar{u}(s) \equiv 0$, $|\bar{v}(s)| \equiv 1$ in $0 \le t \le L$, a demented "solution." It then follows that $p \ne 0$. On the other hand, we must have $|p| \le \bar{r}(s)$, otherwise the maximum in (10.5.8) would be infinite. Under these conditions, the Hamiltonian tends to zero as $|u| \to \infty$, and the maximum in u occurs where the u-derivative of the Hamiltonian is zero, so that

$$\frac{\partial}{\partial u} (-\bar{r}(s) \sqrt{u^2 + v^2} + pu) = -\frac{\bar{r}(s)u}{\sqrt{u^2 + v^2}} + p = 0, \tag{10.5.9}$$

and the equation

$$\bar{r}(s)\bar{u}(s) = p\sqrt{\bar{u}(s)^2 + \bar{v}(s)^2} = p \tag{10.5.10}$$

results. It follows that $\bar{u}(s) = \bar{x}'(s)$ is always negative or always positive according to the sign of p. The first choice produces a curve that "walks from right to left"; this is a bogus solution brought into play by ignoring the control constraint $u \geq 0$ and it cannot satisfy the target conditions (10.1.5). Hence, $p > 0$.

Obviously (10.5.10) implies that $\bar{x}'(s)$ is bounded below; hence by the inverse function theorem (Lebesgue style) we may put $s = s(x)$ ($s(\cdot)$ absolutely continuous in $a \leq x \leq b$) and then put $r = r(s(x)) = r(x)$ with $r(\cdot)$ absolutely continuous in $a \leq x \leq b$. Doing this, (10.5.10) becomes

$$\bar{r}(x) = p\sqrt{1 + \bar{r}'(x)^2}, \tag{10.5.11}$$

which is equation (1.1.10) with solutions

$$\bar{r}(x) = p \cosh\left(\frac{x - \alpha}{p}\right). \tag{10.5.12}$$

We met these solutions in **1.1**. However, we can now answer the question (left open there) of what happens when there is no curve (10.5.12) joining $(0, r_a)$ and (L, r_b), plus some other open issues.

(b) Assume that $\bar{r}(s) = 0$ for some $s \in [0, L]$. Let $e \subseteq [0, L]$ be the set where $\bar{r}(s) = 0$, $s_0 = \inf e$, $s_1 = \sup e$. We have $0 < s_0 < s_1 < L$ and, by obvious reasons, $\bar{r}(s) \equiv 0$ in $[s_0, s_1]$ (if not, we may "squeeze to a wire" the surface between s_0 and s_1, thereby reducing the area). Accordingly, $e = [s_0, s_1]$. We then apply the argument pertaining to the case $\bar{r}(s) > 0$ in each of the subintervals $[0, s_0)$ and $(s_1, L]$. We conclude in the same way that $z_0 < 0$, so we can use (10.5.8) rather than merely (10.5.7). If $p \neq 0$ we deduce that $\bar{r}(s)$ must be of the form (10.5.12) in the intervals $[0, s_0)$ and $(s_1, L]$, which, by (10.5.1), is absurd since (10.5.12) never vanishes. It follows then that $p = 0$, in which case (10.5.8) implies $\bar{u}(s) = \bar{x}'(s) = 0$ in $s \in [0, s_0] \cup [s_1, L]$. To figure out $\bar{v}(s)$ in $[0, s_0]$ and $[s_1, L]$, note that (10.5.8) reduces to

$$-\bar{r}(s) + q(s)\bar{v}(s) = \max_{-\infty < v < \infty} \{-\bar{r}(s)|v| + q(s)v\}. \tag{10.5.13}$$

If $|q(s)| > \bar{r}(s)$, the maximum would be infinite, thus we must have $|q(s)| \leq \bar{r}(s)$. However, if $|q(s)| < \bar{r}(s)$ the maximum is attained at $\bar{v}(s) = 0$, which implies $\bar{r}(s) = 0$, absurd. The only possibility left is

$$|q(s)| = \bar{r}(s). \tag{10.5.14}$$

Note that (10.5.8) implies that $p = q(s) = 0$ where $\bar{r}(s) = 0$, so that (10.5.14) extends to the whole interval $0 \leq t \leq L$. Under this equality, the only information that (10.5.13) gives is: $\bar{v}(s) > 0$ if $q(s) = \bar{r}(s) > 0$, $\bar{v}(s) < 0$ if $q(s) = -\bar{r}(s) < 0$. However, we know that $\bar{u}(s) = 0$ in $[0, s_0]$ and $[s_1, L]$, thus the parametrization by arclength implies

$$\bar{v}(s) = 1 \text{ if } q(s) = \bar{r}(s) > 0, \qquad \bar{v}(s) = -1 \text{ if } q(s) = -\bar{r}(s) < 0. \tag{10.5.15}$$

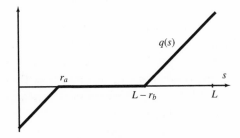

Figure 10.2.

Taking again into account the parametrization by arclenght and that $z_0 = -1$, the
second canonical equation (10.5.3) becomes $dq = ds - \mu_2(ds)$, with solution

$$q(s) = c + s + \int_s^L \mu_2(ds). \qquad (10.5.16)$$

Paying heed to (10.5.14) and to the fact that $\mu_2 = 0$ in $[0, s_0]$ and $[s_1, 0]$, we deduce that $q(s) = s - r_a \ (0 \le s \le r_a)$, $q(s) = 0 \ (s_0 \le s \le s_1)$, $q(s) = s - L + r_b$
$(L - r_b \le s \le L)$ (see Figure 10.2). Hence, (10.5.15) identifies the measure as
$\mu_2(ds) = \chi(s)ds$, χ the characteristic function of $[s_0, s_1]$. It finally follows from
(10.5.15) that $\bar{v}(s) = -1$ in $0 \le s \le s_0$, $\bar{v}(s) = 1$ in $s_1 \le s \le L$, which completes
the proof of

Theorem 10.5.2. *Let $\bar{x}(s), \bar{r}(s)$ be a surface of revolution of minimum area,
parametrized by arclength. Then (a) if $\bar{r}(s) > 0$ in $[0, L]$, $\bar{x}'(s)$ has an absolutely
continuous inverse and we can put $y = y(x)$ with $y(\cdot)$ given by (10.5.12) with
suitable constants. (b) If $\bar{r}(s) = 0$ anywhere in $[0, L]$ then*

$$\begin{aligned}
\bar{x}(s) &= 0, & \bar{r}(s) &= s - r_a & (0 \le s < r_a) \\
\bar{x}(s) &= s - r_a, & \bar{r}(s) &= 0 & (r_a \le s < L - r_b) \\
\bar{x}(s) &= 0, & \bar{r}(s) &= s - L + r_b & (L - r_b \le s \le L).
\end{aligned}$$

Following the classics, we call the latter type of solution a **broken extremal** of
the problem.

Remark 10.5.3. It is stated in (at least) one textbook that if the points (a, r_a)
and (b, r_b) can be joined by a single curve of the form (10.5.12) then this curve
must be the solution of the problem. This is an exaggeration. To see this we retake
Example 1.1.2, where $a = 0, b = 1$ and $r_a = r_b = r$. We have seen there that if
$r = m = 0.7544\ldots$ there exists a unique curve (10.5.11) joining $(0, r)$ and $(1, r)$,
corresponding to $\beta = 0.4167\ldots$. However, the area of the surface spanned by this
curve equals $4.29036\ldots$, whereas the area $2\pi r^2$ of the two disks making up the
broken extremal is $3.57626\ldots$. Using the computation for $r = 1$ in Example 1.1.2

and continuity arguments, we may argue the existence of m_0, $m < m_0 < 1$ such that: (a) if $m < r < m_0$, none of the two extremals (10.5.11) joining the endpoints is a minimum (the minimum is the broken extremal), (b) if $r = m_0$ one of the two extremals (10.5.12) is a minimum (the other minimum is the broken extremal), (c) if $r > m_0$ one of the two curves (10.5.12) is a minimum. The complete analysis where $r_a \neq r_b$ is very involved (in a classical sort of way) and can be found in Bliss [1925, p. 90] or Cesari [1983, p. 143].

10.6. Other Applications

Example 10.6.1. We take a new look at the time optimal problem for the system

$$y'(t) = Ay(t) + f(t, y(t)) + u(t), \qquad y(0) = \zeta \qquad (10.6.1)$$

in **6.7**, this time in an arbitrary Banach space, where $C_{\text{ad}}(0, \bar{t}; U)$ consists of all strongly measurable E-valued functions with $u(t) \in U \subseteq E$ a.e. The minimum principle for a state constraint and target condition

$$y(t) \in M(t) \quad (0 \le t \le \bar{t}), \qquad y(\bar{t}) \in Y, \qquad (10.6.2)$$

is

$$\langle \bar{z}(s), \bar{u}(s) \rangle = \min_{v \in U} \langle \bar{z}(s), v \rangle \qquad (10.6.3)$$

a.e. in $0 \le t \le \bar{t}$, where $\bar{z}(s)$ is the solution of

$$d\bar{z}(s) = -\{A^* + \partial_y f(s, y(s, \bar{u}))^*\} \bar{z}(s) ds - \mu(ds) \quad (0 \le s \le \bar{t}),$$
$$\bar{z}(\bar{t}) = z \qquad (10.6.4)$$

with $(\mu, z) \in \Sigma(0, \bar{t}; E^*) \times E^*$ satisfying the last two conditions in (10.3.5). By Corollary 10.3.4 $(\mu, z) \neq 0$ if $M(\bar{t})$ and Y are T-full. If A generates a strongly continuous group, then the point target condition

$$y(\bar{t}, u) = \bar{y} \qquad (10.6.5)$$

can also be accommodated; in fact, it was proved in Lemma 6.7.2 that if $\{t_n\}$ is a sequence with $t_n \le t_{n+1} \le \cdots \le \bar{t}$, $t_n \to \bar{t}$ and if $\bar{u} \in C_{\text{ad}}(0, \bar{t}; U)$, $\tilde{u}^n \in C_{\text{ad}}(0, t_n; U)$ are such that $d_n(\tilde{u}^n, \bar{u}) \to 0$, the reachable spaces $R(0, t_n; U, \tilde{u}^n)$ of the variational equation contain a common ball for n large enough, so that Lemma 10.3.5 applies and the multiplier (μ, z) is not zero. If U is the unit ball of E then (10.6.3) implies the bang-bang theorem

$$\|\bar{u}(s)\| = 1 \qquad (10.6.6)$$

in the set where $\bar{z}(s) \neq 0$. We show below that (10.6.6) is always true in the set where the state constraint is not saturated, although we must concede that the proof has nothing to do with the minimum principle with state constraints; we only invoke the "unconstrained" minimum principle in Chapters 6 and 7.

Example 10.6.2. (The optimality principle) Let $f = f(y)$ and M be independent of t and let $\bar{u}(\cdot)$ be a time optimal control in the interval $0 \leq t \leq \bar{t}$. Then $\bar{u}(\cdot)$ is time optimal in any subinterval $[t_0, t_1], 0 \leq t_0 < t_1 \leq \bar{t}$.

The proof is the same as in the linear case (see Example 9.2.4).

Theorem 10.6.3. *Assume A generates a strongly continuous group and that f and M are independent of t. Then (10.6.6) holds in the set $e_0 = \{t; y(t, \bar{u}) \in \text{Int}(M)\}$.*

Proof. The set e_0 is open, thus it is a countable union of open intervals (t_0, t_1). By the optimality principle the optimal control $\bar{u}(t)$ is time optimal (no state constraints) in $[t_0, t_1]$ and the costate $\bar{z}(\cdot)$ corresponding to this interval is guaranteed to be nonzero, so that (10.6.6) holds in $t_0 \leq t \leq t_1$. ∎

Example 10.6.4. The system is that described by the semilinear wave equation in **6.8**:

$$y_{tt}(t, x) = \sum_{j=1}^{m}\sum_{k=1}^{m} \partial^j (a_{jk}(x)\partial^k y(t, x)) + \sum_{j=1}^{m} b_j(x)\partial^j y(t, x)$$
$$+ c(x)y(t, x) - \phi(t, x, y(t, x)) + u(t, x), \qquad (10.6.7)$$

reduced to the first order equation

$$\mathbf{y}'(t) = \mathbf{A}(\beta)\mathbf{y}(t) + \mathbf{f}(t, \mathbf{y}(t)) + \mathbf{B}u(t) \qquad (10.6.8)$$

for $\mathbf{y}(t) = (y(t, \cdot), y_1(t, \cdot)) = (y(t, \cdot), y_t(t, \cdot)) \in \mathbf{E} = H_0^1(\Omega) \times L^2(\Omega)$ for the Dirichlet boundary condition, $\mathbf{E} = H^1(\Omega) \times L^2(\Omega)$ for variational boundary conditions. The control set is the unit ball of $L^2(\Omega)$.

The minimum principle (Theorem 10.3.1) for the time optimal problem with state constraint and target condition

$$\mathbf{y}(t) \in \mathbf{M}(t) \quad (0 \leq t \leq \bar{t}), \qquad y(\bar{t}) \in Y \qquad (10.6.9)$$

produces a multiplier $(\boldsymbol{\mu}, \mathbf{z}) \in \Sigma(0, T; \mathbf{E}) \times \mathbf{E}$ satisfying the last two conditions (10.3.5) and such that if $\bar{\mathbf{z}}(s) = (\bar{z}(t), \bar{z}_1(s))$ is the solution of

$$d\bar{\mathbf{z}}(s) = -\{\mathbf{A}(\beta)^* + \partial_y \mathbf{f}(s, \mathbf{y}(s, \bar{\mathbf{u}}))^*\}\mathbf{z}(s)ds - \boldsymbol{\mu}(ds),$$
$$\bar{\mathbf{z}}(\bar{t}) = \mathbf{z} \qquad (10.6.10)$$

in $0 \leq s \leq \bar{t} = $ optimal time, then

$$(\bar{z}_1(s), \bar{u}(s)) = \min_{\|v\|_{L^2(\Omega)} \leq 1} (\bar{z}_1(s), v) \qquad (10.6.11)$$

a.e. in $0 \leq t \leq \bar{t}$, since $\mathbf{B}^*(z, z_1) = z_1$ and the conditions necessary for the pointwise minimum principle (Theorem 10.3.5) are satisfied. In fact, it has been shown in

6.11 that if $\{t_n\}$ is a sequence with $t_n \leq t_{n+1} \leq \cdots \leq \bar{t}$, $t_n \to \bar{t}$, and if $\bar{u} \in$ $C_{\mathrm{ad}}(0, \bar{t}; U)$, $\tilde{u}^n \in C_{\mathrm{ad}}(0, t_n; U)$ are such that $d_n(\tilde{u}^n, \bar{u}) \to 0$, the reachable spaces $\mathbf{R}(0, t_n; U, \tilde{u}^n)$ of the variational equation contain a common ball for n large enough, so that Theorem 10.3.5 applies and the multiplier $(\mu, \mathbf{z})$ is not zero if $\mathbf{M}(\bar{t})$ is T-full. There are no conditions on Y beyond closedness, so that the point target problem is included. We have the following counterpart of Theorem 10.6.3.

Theorem 10.6.5. *Assume* ϕ, $\mathbf{M}$ *are independent of* t. *Then*

$$\|\bar{u}(t)\|_{L^2(\Omega)} = 1 \qquad \text{a.e. in } e_0 = \{t; \mathbf{y}(t, \bar{u}) \in \mathrm{Int}(\mathbf{M})\}. \tag{10.6.12}$$

The proof uses the minimum principle without state constraints in component intervals (t_0, t_1) of e_0. That (10.6.12) will hold in each of these intervals is a consequence of Corollary 6.12.4.

11

Optimal Control Problems with State Constraints: The Abstract Parabolic Case. Convergence of Suboptimal Controls

11.1. Abstract Parabolic Equations: The Measure-Driven Adjoint Variational Equation

We study in this chapter the control system

$$y'(t) = Ay(t) + f(t, y(t), u(t)), \qquad y(0) = \zeta \qquad (11.1.1)$$

within the $\odot$-theory in Chapter 7; specifically, $S(t)$ generates an analytic semigroup $S(\cdot)$ with $\|S(t)\| \leq Ce^{-ct}$, $c > 0$ (so that fractional powers of A and $A^\odot$ can be defined), E is $\odot$-reflexive, and E and $E^\odot$ are separable.[1] The nonlinear term $f(t, y, u(t))$ satisfies Hypothesis $\mathrm{III}_{\alpha,\rho}$ in **7.10** for each $u(\cdot) \in C_{\mathrm{ad}}(0, \bar{t}; U)$ with $\alpha + \rho < 1$ and constants $K(c)$, $L(c)$ independent of $u(\cdot)$. Hypothesis $\mathrm{III}_{\alpha,\rho}(wm)$ is enough modulo a change in ρ (see **7.10**).

The business of this section is to fit the adjoint variational equation

$$\bar{z}'(s) = -\{A^\odot + \partial_y f(s, y(s, \bar{u}), \bar{u}(s))^*\} \bar{z}(s)$$

with a measure forcing term $\mu(ds)$ taking values in the dual $(E_\alpha)^*$ of the space $E_\alpha = D((-A)^\alpha)$, where $0 < \alpha < 1$. Although not strictly necessary (see Remark 11.1.13) we assume that (7.9.37) holds, that is

$$\partial_y f(s, y(s, \bar{u}), \bar{u}(s))^* : (E^\odot)_\rho \to (E^*)_{-\alpha} \quad (0 \leq s \leq \bar{t}), \qquad (11.1.2)$$

so that the theory of the variational and adjoint variational equation can be fitted to the treatment in **7.9**.

We begin by identifying $(E_\alpha)^*$. Recall (**7.4**) that if $\alpha \geq 0$ the space $(E^*)_{-\alpha}$ is the completion of E^* with respect to the norm $\|z\|_{-\alpha} = \|((-A)^{-\alpha})^* z\|_{E^*}$. The first result is related to Lemma 9.1.6 and requires the extension of the operator $((-A)^{-\alpha})^*$ to spaces $(E^*)_{-\beta}$, where $\alpha, \beta \geq 0$. This is done as follows. Assume first that $\beta \geq \alpha$ and let $z \in (E^*)_{-\beta}$. Then there exists a sequence $\{z_n\} \subseteq E^*$ such that

[1] See Example 7.7.19.

$\|z_n - z\|_{-\beta} \to 0$. This implies $\{((-A)^{-\beta})^* z_n\} = \{((-A)^{-(\beta-\alpha)})^*((-A)^{-\alpha})^* z_n\}$ is Cauchy so that $\{((-A)^{-\alpha})^* z_n\}$ converges in $(E^*)_{-(\beta-\alpha)}$ to $w \in (E^*)_{-(\beta-\alpha)}$; we define $w = ((-A)^{-\alpha})^* z$. The definition does not depend on the sequence $\{z_n\}$, and we have $\|w\|_{-(\beta-\alpha)} = \|z\|_{-\beta}$, so that

$$((-A)^{-\alpha})^* : (E^*)_{-\beta} \to (E^*)_{-(\beta-\alpha)}$$

is an isometry. The proof is similar for $\beta < \alpha$; see **7.4** for this and additional details. As shown in Lemma 11.1.1 below, $((-A)^{-\alpha})^*$ is in general not onto.

Part (a) of Lemma 11.1.1 uses the fact that $D(((-A)^{\alpha})^*)$ is dense in $D(((-A)^{\delta})^*)$ for arbitrary $\alpha \geq \delta$. Since $D(((-A)^{\delta})^*) = ((-A)^{-\delta})^* E^* \subseteq E^{\odot}$ (Lemma 7.7.4, (7.7.13)), it is enough to show that $D(((-A)^n)^*) = ((-A)^{-n})^* E^*$ is dense in $E^{\odot}$ for n integer. In view of (7.7.14), it is in turn enough to prove that $D((A^{\odot})^n) = (-A^{\odot})^{-n} E^{\odot}$ is dense in $E^{\odot}$. This is of course a consequence of the fact that $A^{\odot}$ is a semigroup generator in $E^{\odot}$; under the present assumptions, it also follows from the fact that $S^{\odot}(t) E^{\odot}$ is dense in $E^{\odot}$ for any $t > 0$ (Lemma 9.1.7).

Lemma 11.1.1. *Let* $0 < \alpha \leq \beta \leq 1$. *Then we have* (a) $((-A)^{-\alpha})^*(E^*)_{-\beta} = Cl_{\alpha-\beta}(D(((-A)^{\alpha})^*))$ $(Cl_{\gamma}$ *denotes closure in the norm of* $(E^*)_{\gamma})$, (b) *if* $\beta > \alpha$ *then* $((-A)^{-\alpha})^*(E^*)_{-\beta} \supseteq E^*$.

Proof. (a) Let $z \in (E^*)_{-\beta}$, $\{z_n\} \subseteq E^*$ with $\|z_n - z\|_{-\beta} \to 0$. Then, by definition of $((-A)^{-\alpha})^* z$ we have $((-A)^{-\alpha})^* z_n \to ((-A)^{-\alpha})^* z$ in $(E^*)_{-(\beta-\alpha)}$; since $((-A)^{-\alpha})^* z_n \in D(((-A)^{\alpha})^*)$, the inclusion $\subseteq$ in (a) follows. Conversely, let $w \in Cl_{\alpha-\beta}(D(((-A)^{\alpha})^*))$, so that there exists $\{w_n\} \subseteq D(((-A)^{\alpha})^*)$ with $w_n \to w$ in $(E^*)_{\alpha-\beta}$. Then, if $\{z_n\} = \{((-A)^{\alpha})^* w_n\}$, $\{z_n\}$ is Cauchy in the norm of $(E^*)_{-\beta}$ so that it converges to z with $((-A)^{-\alpha})^* z = w$.

(b) In view of (a) it is enough to prove that $Cl_{-\delta}(D(((-A)^{\alpha})^*)) \supseteq E^*$ for $\delta > 0$, which is done as follows. Let $w \in E^*$. Using the fact that $D(((-A)^{\alpha+\delta})^*)$ is dense in $D(((-A)^{\delta})^*)$, pick $w_n \in E^*$ with $((-A)^{-\alpha-\delta})^* w_n - ((-A)^{-\delta})^*)w \to 0$. Then $\|((-A)^{-\alpha})^* w_n - w\|_{-\delta} \to 0$, so that $w \in Cl_{-\delta}(D(((-A)^{\alpha})^*))$ as claimed. ∎

The identification of $(E_{\alpha})^*$ follows the lines of **9.1**. Fix $\varepsilon > 0$. Lemma 11.1.1 (b) allows us to express an arbitrary element w of E^* in the form

$$w = ((-A)^{-\alpha})^* z \quad (z \in (E^*)_{-(\alpha+\varepsilon)}). \tag{11.1.3}$$

We denote by $((-A)^{\alpha})^* E^* \subseteq (E^*)_{-(\alpha+\varepsilon)}$ the set of all $z \in (E^*)_{-(\alpha+\varepsilon)}$ such that (11.1.3) holds for some $w \in E^*$. Since $((-A)^{-\alpha})^* : (E^*)_{-(\alpha+\varepsilon)} \to (E^*)_{-\varepsilon}$ is an isometry, $((-A)^{-\alpha})^* : ((-A)^{\alpha})^* E^* \to E^*$ is $1 - 1$; it is also onto by definition of $((-A)^{\alpha})^* E^*$. We renorm $((-A)^{\alpha})^* E^*$ with

$$\|z\|_{((-A)^{\alpha})^* E^*} = \|((-A)^{-\alpha})^* z\|_{E^*}, \tag{11.1.4}$$

under which norm $((-A)^{\alpha})^* E^*$ is a Banach space, and we have

Lemma 11.1.2. *Let* $\alpha > 0$. *Then* $((-A)^\alpha)^* E^*$ *is algebraically and metrically isomorphic to the dual space* $(E_\alpha)^*$; *the duality pairing is*

$$\langle z, y \rangle_{((-A)^\alpha)^* E^* \times E_\alpha} = \langle ((-A)^{-\alpha})^* z, (-A)^\alpha y \rangle_{E^* \times E}. \tag{11.1.5}$$

Proof. It can hardly be denied that (11.1.5) provides a bounded linear functional in E_α for every $z \in ((-A)^\alpha)^* E^*$. On the other hand, if ϕ is a bounded linear functional in E_α then $y \to \phi((-A)^\alpha y)$ is a bounded linear functional in E so that there exists $w \in E^*$ with $\phi((-A)^\alpha y) = \langle w, y \rangle$. This means that $\phi(y)$ is given by (11.1.5) with $w = ((-A)^{-\alpha})^* z$. The metric isomorphism follows from the definition of the norm of $((-A)^\alpha)^* E^*$. ∎

Corollary 11.1.3. *Let* $\varepsilon > 0$. *Then*

$$(E^*)_{-\alpha} \hookrightarrow (E_\alpha)^* = ((-A)^\alpha)^* E^* \hookrightarrow (E^*)_{-(\alpha+\varepsilon)}. \tag{11.1.6}$$

Proof. The embedding $(E^*)_{-\alpha} \hookrightarrow ((-A)^\alpha)^* E^*$ is obvious (and isometric). To show the second, note that it is plain from Lemma 11.1.1 that $((-A)^\alpha)^* E^* \subseteq (E^*)_{-(\alpha+\varepsilon)}$. If $z \in (E^*)_{-(\alpha+\varepsilon)}$, we have

$$\begin{aligned}
\|z\|_{(E^*)_{-(\alpha+\varepsilon)}} &= \|((-A)^{-(\alpha+\varepsilon)})^* z\|_{E^*} \\
&= \|((-A)^{-\varepsilon})^* ((-A)^{-\alpha})^* z\|_{E^*} \\
&\leq C \|((-A)^{-\alpha})^* z\|_{E^*} = C \|z\|_{((-A)^\alpha)^* E^*}. \quad\blacksquare
\end{aligned}$$

Lemma 11.1.4. (a) $S(t)^*$ *can be extended to* $(E^*)_{-\alpha}$. *We have*

$$S(t)^* (E^*)_{-\alpha} \subseteq E^\odot \quad (t > 0), \tag{11.1.7}$$

with $S(t)^*(E^*)_{-\alpha}$ *continuous in the norm of* $((E^*)_{-\alpha}, E^\odot)$ *in* $t > 0$ *and*

$$\|S(t)^*\|_{((E^*)_{-\alpha}, E^\odot)} \leq C_\alpha t^{-\alpha} e^{-ct} \quad (t > 0). \tag{11.1.8}$$

(b) $S(t)^*$ *can be extended to* $((-A)^\alpha)^* E^*$. *We have*

$$S(t)^* ((-A)^\alpha)^* E^* \subseteq E^\odot \quad (t > 0), \tag{11.1.9}$$

with $S(t)^*$ *continuous in the norm of* $(((-A)^\alpha)^* E^*, E^\odot)$ *in* $t > 0$ *and*

$$\|S(t)^*\|_{(((-A)^\alpha)^* E^*, E^\odot)} \leq C_\alpha t^{-\alpha} e^{-ct} \quad (t > 0). \tag{11.1.10}$$

Proof. In case (a), the extension is

$$S(t)^* = ((-A)^\alpha)^* S(t)^* ((-A)^{-\alpha})^* = (-A^\odot)^\alpha S^\odot(t)((-A)^{-\alpha})^*,$$

where, in the second equality, we have used the fact that $((-A)^{-\alpha})^*(E^*)_{-\alpha} = E^\odot$ (see Remark 11.1.11); this proves all the claims in (a), the estimate (11.1.8) a consequence of Lemma 7.4.3 applied to $A^\odot$. In case (b), the extension is

$$S(t)^* = ((-A)^\alpha)^* S(t)^* ((-A)^{-\alpha})^* = S^\odot(t-h)((-A)^\alpha)^* S(h)^* ((-A)^{-\alpha})^*,$$

where the second line attends to the continuity claims taking $h > 0$ and $t \geq h$. The estimate (11.1.10) follows taking adjoints in (7.4.15).   ∎

Instead of the adjoint variational equation, we consider the more general equation

$$z'(s) = -(A^\odot + B(s)^*)z(s) - \mu(ds), \qquad z(\bar{t}) = z \qquad (11.1.11)$$

under the assumptions in **7.9**; in particular, the operator function $s \to B(s)$ takes values in $(E_\alpha, ((E^\odot)^*)_{-\rho})$, $((-A^\odot)^{-\rho})^* B(s)(-A)^{-\alpha}\eta$ is strongly measurable in E for every $\eta \in E$, and

$$\|((-A^\odot)^{-\rho})^* B(s)(-A)^{-\alpha}\|_{(E,E)} \leq K.$$

Symetrically, the operator function $s \to B(s)^*$ takes values in $((E^\odot)_\rho, (E^*)_{-\alpha})$, $((-A)^{-\alpha})^* B(s)^* (-A^\odot)^{-\rho}\upsilon$ is strongly measurable in $E^\odot$ for every $\upsilon \in E^\odot$, and

$$\|((-A)^{-\alpha})^* B(s)^* (-A^\odot)^{-\rho}\|_{(E^\odot, E^\odot)} \leq K.$$

The measure μ belongs to the space

$$C(0, \bar{t}; E_\alpha)^* = \Sigma(0, \bar{t}; (E_\alpha)^*) = \Sigma(0, \bar{t}; ((-A)^\alpha)^* E^*).$$

Taking into account that $((-A)^{-\alpha})^* : ((-A)^\alpha)^* E^* \to E^*$ is an isometric isomorphism, $\mu \in \Sigma(0, \bar{t}; ((-A)^\alpha)^* E^*)$ if and only if $((-A)^{-\alpha})^* \mu \in \Sigma(0, \bar{t}; E^*)$. Finally, the final condition z in (11.1.11) belongs to $((-A)^\alpha)^* E^*$, so that $((-A)^\alpha)^* z \in E^*$.

By definition, the solution of (11.1.11) is $z(t) = (-A^\odot)^{-\rho}\upsilon(t)$, where

$$\upsilon(s) = R^*_{\rho,\alpha}(s, \bar{t})((-A)^{-\alpha})^* z$$
$$+ \int_s^t R^*_{\rho,\alpha}(s, \sigma)((-A)^{-\alpha})^* \mu(d\sigma) = \upsilon_h(s) + \upsilon_i(s) \qquad (11.1.12)$$

and $R^*_{\alpha,\rho}(s, t)$ is the operator defined by the integral equation (7.9.45). The precise meaning of the integral $\upsilon_i(s)$ is explicited after the next result, where the operator $R_{\alpha,\rho}(t, s)$ is defined by the integral equation (7.9.27).

Lemma 11.1.5. *Let $R_{\alpha,\rho}(t, s)$ be considered as an operator in (E, E). Then*

$$R_{\alpha,\rho}(t, s)^* = R^*_{\rho,\alpha}(s, t), \qquad R^*_{\rho,\alpha}(s, t)E^* \subseteq E^\odot \quad (s < t). \qquad (11.1.13)$$

Proof. If $y \in D((-A)^\rho)$ then $R_{\alpha,\rho}(t,s)y = R_\alpha(t,s)(-A)^\rho y$ so that, by (5.0.26) $R_{\alpha,\rho}(t,s)^* = ((-A)^\rho)^* R_\alpha(t,s)^*$. Applying both sides of this equality to $((-A^\odot)^{-\alpha})z$ and using (7.9.50), $R_{\alpha,\rho}(t,s)^*((-A^\odot)^{-\alpha})z = R_\rho^*(s,t)$. Accordingly, $R_{\alpha,\rho}(t,s)^*z = R_\rho^*(s,t)((-A^\odot)^\alpha)z = R_\rho^\odot(s,t)((-A^\odot)^\alpha)z$ for $z \in D((-A^\odot)^\alpha)$. The integral equation (7.9.42) for $R_\rho^\odot(s,t)$ then shows that $R_\rho^\odot(s,t)((-A^\odot)^\alpha)z$ satisfies the integral equation (7.9.45) for $R_{\rho,\alpha}^*(s,t)$, so that the equality in (11.1.13) results. The inclusion was proved in Theorem 7.9.7. ∎

Using (11.1.13) as justification, the integral on the right-hand side of (11.1.12) is understood as follows: $v_i(s)$ is the unique element of E^* such that

$$\langle y, v_i(s) \rangle = \int_s^{\bar{t}} \langle R_{\alpha,\rho}(\sigma,s)y, ((-A)^{-\alpha})^* \mu(d\sigma) \rangle \quad (y \in E). \tag{11.1.14}$$

Since $\nu = ((-A)^{-\alpha})^* \mu \in \Sigma(0,\bar{t}; E^*)$ we only have to elucidate the properties of functions $v(s)$ defined by

$$\langle y, v(s) \rangle = \int_s^{\bar{t}} \langle R_{\alpha,\rho}(\sigma,s)y, v(d\sigma) \rangle^{(2)} \quad (y \in E). \tag{11.1.15}$$

The only difference with (10.2.35) is that $R_{\alpha,\rho}(\sigma,s)$ is not continuous for $\sigma = s$. To study (11.1.15) we need some additions to the theory in **7.9**. The first concerns the operator $R_{\alpha,\rho}(t,s) : (E^\odot)^* \to E$ constructed in Theorem 7.9.3.

Lemma 11.1.6. *Let $\varepsilon > 0$ be such that $\alpha + \rho + \varepsilon < 1$. Then $R_{\alpha,\rho}(t,s)(E^\odot)^* \subseteq D((-A)^\varepsilon)$ for $t > s$ and $(-A)^\varepsilon R_{\alpha,\rho}(t,s)$ is continuous in the norm of $((E^\odot)^*, E)$ with*

$$\|(-A)^\varepsilon R_{\alpha,\rho}(t,s)\|_{((E^\odot)^*,E)} \le C(t-s)^{-\alpha-\rho-\varepsilon} \quad (0 \le s < t \le \bar{t}). \tag{11.1.16}$$

Proof. Apply $(-A)^\varepsilon$ to both sides of the integral equation (7.9.27). The result is

$$(-A)^\varepsilon R_{\alpha,\rho}(t,s)\zeta = (-A)^{\alpha+\rho+\varepsilon} S^\odot(t-s)^*\zeta$$

$$+ \int_s^t (-A)^{\alpha+\rho+\varepsilon} S(t-\tau)((-A^\odot)^{-\rho})^* B(\tau)(-A)^{-\alpha} R_{\alpha,\rho}(\tau,s)\zeta \, d\tau. \tag{11.1.17}$$

The estimation is the same as that following (7.9.27). ∎

Lemma 11.1.7. *Let $\varepsilon > 0$. Then*

$$\|(S(h) - I)(-A)^{-\varepsilon}\| \le Ch^\varepsilon \quad (h > 0), \tag{11.1.18}$$

where C does not depend on h.

---

(2) The integral includes both endpoints.

Proof. We have

$$(S(h) - I)(-A)^{-\varepsilon} = \int_0^h (-A)^{1-\varepsilon} S(t) dt.$$

Estimate using (7.4.15). ∎

Lemma 11.1.8. $v(s)$ *exists, belongs to* $E^\odot$ *a.e. in* $0 \le s \le \bar{t}$, *is strongly measurable and satisfies*

$$\|v(s)\|_{E^\odot} \le C\omega(s) = C \int_s^{\bar{t}} (\sigma - s)^{-\kappa} \|v\|(d\sigma) \quad a.e. \tag{11.1.19}$$

where $\kappa = \alpha + \rho$ *and* $\omega(\cdot) \in L^1(0, \bar{t})$.

Proof. The estimate (11.1.19) is clear. That $\omega(\cdot) \in L^1(0, \bar{t})$ is a consequence of Tonelli's theorem; we integrate the right side of (11.1.19) first in s and then in σ and obtain

$$\int_0^{\bar{t}} \omega(s) ds = \int_0^{\bar{t}} \|v(d\sigma)\| \int_0^\sigma (\sigma - s)^{-\kappa} ds$$

$$= C \int_0^{\bar{t}} \sigma^{1-\kappa} \|v(d\sigma)\| < \infty. \tag{11.1.20}$$

Let $\phi_n(t) = nt$ $(0 \le t \le 1/n)$, $\phi_n(t) = 1$ $(1/n \le t \le \bar{t})$, and let $v_n(s)$ be defined as above but using $R_{\alpha,\rho,n}(\sigma, s) = \phi_n(\sigma - s) R_{\alpha,\rho}(\sigma, s)$ instead of $R_{\alpha,\rho}(\sigma, s)$. In view of (11.1.16) with $\varepsilon = 0$, $R_{\alpha,\rho,n}(\sigma, s)$ is continuous in $\sigma \ge s$, hence each $v_n(s)$ is everywhere defined. Estimating as above we have

$$\|v(s) - v_n(s)\|_{E^*} \le \omega_n(s) = C \int_s^{s+1/n} (\sigma - s)^{-\kappa} \|v\|(d\sigma) \tag{11.1.21}$$

where, by Tonelli's theorem,

$$\int_0^{\bar{t}} \omega_n(s) ds = C \int_0^{\bar{t}} \|v(d\sigma)\| \int_{\max(0,\sigma-1/n)}^\sigma (\sigma - s)^{-\kappa} ds \le Cn^{\kappa-1} \int_0^{\bar{t}} \|v(d\sigma)\|,$$

so that that $\omega_n(s) \to 0$ in $L^1(0, \bar{t})$ and (passing if necessary to a subsequence), $v_n(s) \to v(s)$ a.e. It is enough to prove that $v_n(s) \in E^\odot$ for all s. To see this, note that

$$|\langle y, (S(h)^* - I) v_n(s) \rangle|$$

$$\le \|y\|_E \|v\|_{\Sigma(0,\bar{t};E)} \max_{0 \le s \le \sigma \le \bar{t}} \|(S(h) - I)(-A)^{-\varepsilon}(-A)^\varepsilon R_{\alpha,\rho,n}(\sigma, s)\|_{((E^\odot)^*,E)}, \tag{11.1.22}$$

where Lemma 11.1.6 guarantees that $\|(-A)^\varepsilon R_{\alpha,\rho,n}(\sigma, s)\|$ is uniformly bounded in $s \le \sigma$ and Lemma 11.1.7 that the right side tends to zero as $h \to 0$. Since $S(t)^* E^* \subseteq E^\odot$, $v_n(s) \in E^\odot$ for all s.

It remains to show strong measurability of $\upsilon(s)$. Continuity of $R_{\alpha,\rho,n}(\sigma, s)$ for $s \leq \sigma$ allows application of Theorem 10.2.15 so that each $\upsilon_n(\cdot)$ is E-weakly left continuous, thus E-weakly measurable; it then follows that $\upsilon(\cdot)$ is E-weakly measurable. Since $E^{\odot}$ is a separable space, it is strongly measurable. This ends the proof of Theorem 11.1.8. ∎

Remark 11.1.9. The case $\alpha = \rho = 0$ deserves recognition. The bound (11.1.19) is uniform,

$$\|\upsilon(s)\|_{E^{\odot}} \leq C\|v\|_{\Sigma(0,\bar{t};E)} \quad (0 \leq s \leq \bar{t}), \tag{11.1.23}$$

and (11.1.21) becomes

$$\|\upsilon(s) - \upsilon_n(s)\|_{E^*} \leq \omega_n(s) = C \int_s^{s+1/n} \|v\|(d\sigma) \to \|v\|(s) \tag{11.1.24}$$

as $n \to \infty$. Accordingly, if s is not an atom of $\|v\|$, that is, if $\|v\|(\{s\}) = 0$ then $\upsilon_n(s) \to \upsilon(s)$, so that $\upsilon(s) \in E^{\odot}$. If $\{s_j\}$ is a finite collection of atoms of $\|v\|$ then $\Sigma \|v(\{s_j\})\| \leq \|v\|_{\Sigma(0,\bar{t};E^*)}$, so that $\|v\|$ can have only a finite or countable collection of atoms $\{s_j\}$ and

$$\upsilon(s) \in E^{\odot} \quad (s \neq s_j). \tag{11.1.25}$$

Lemma 11.1.10. *Let* $g(\cdot) \in L^{\infty}(0, T; E)$, $v \in \Sigma(0, \bar{t}; E^*)$. *Then*

$$\int_0^{\bar{t}} \left\langle \int_0^t R_{\alpha,\rho}(t, s)g(s)ds, v(dt) \right\rangle = \int_0^{\bar{t}} \left\langle g(s), \int_s^{\bar{t}} R_{\alpha,\rho}(t, s)^* v(dt) \right\rangle ds. \tag{11.1.26}$$

Proof. Lemma 10.2.17 uses only the strong continuity of the operator function $S(t, s)$, thus it can be used to show that (11.1.26) holds for $R_{\alpha,\rho,n}(t, s)$ and leaves us the only task of taking limits as $n \to \infty$. We have

$$\int_0^t \|R_{\alpha,\rho}(t, s) - R_{\alpha,\rho,n}(t, s)\|\|g(s)\|ds \leq C \int_{t-1/n}^t (t - s)^{1-\kappa}\|g(s)\|ds,$$

so that we use uniform convergence on the left side of (11.1.26). On the right side, we use that $\upsilon_n(s) \to \upsilon(s)$ a.e. (for a subsequence) (note that the bound (11.1.19) applies equally well to the $\upsilon_n(\cdot)$), and apply the dominated convergence theorem. ∎

Remark 11.1.11. As a consequence of Lemma 11.1.1 (b) we have $((-A)^{-\rho})^*(E^*)_{-\rho} = Cl(D(((-A)^{\rho})^*))$, $Cl = Cl_0$ the closure in E^*. Since

$(D(((-A)^\rho)^*)) \subseteq E^\odot$ (see the observations before the proof of Lemma 11.1.1), we have

$$((-A)^{-\rho})^*(E^*)_{-\rho} = E^\odot, \tag{11.1.27}$$

an addendum to the first inclusion (7.7.26) (of course, there is a corresponding addendum to the second).

Example 11.1.12. If $E^\odot$ is separable, an $E^\odot$-weakly measurable E-valued function (that is, an E-valued function such that $\langle y^*, f(\cdot) \rangle$ is measurable for every $y^* \in E^\odot$) is strongly measurable.

To see this, replace the norm of E by the equivalent norm $\| \cdot \|_0$ given by (7.4.24). Since $E^\odot$ is separable, this norm can be calculated as

$$\|y\|_0 = \sup |\langle y_k^*, y \rangle|,$$

$\{y_k^*\}$ a countable dense set in the unit ball of $E^\odot$. It follows that $\| f(\cdot) - y \|_0$ is measurable for every $y \in E$, so that Example 5.0.36 applies.

Remark 11.1.13. If we insist on having (11.1.2), we only need to raise α. In fact, let $B(s) = \partial_y f(s, y(s, \bar{u}), \bar{u}(s))$. By assumption $B(s) \in (E_\alpha, ((E^\odot)^*)_{-\rho})$, so that

$$B(s)^* : (((E^\odot)^*)_{-\rho})^* \to (E_\alpha)^* \hookrightarrow (E^*)_{-(\alpha+\varepsilon)}$$

by (11.1.6). To give sense to a condition like (11.1.2) we must show that

$$(E^\odot)_\rho \hookrightarrow (((E^\odot)^*)_{-\rho})^* \tag{11.1.28}$$

Now, since for any Banach space F we have $F \hookrightarrow F^{**}$ (isometric imbedding) it is enough to show that

$$((E^\odot)_\rho)^{**} \hookrightarrow (((E^\odot)^*)_{-\rho})^* \tag{11.1.29}$$

To see this note that (11.1.6) implies $((E^\odot)^*)_{-\rho} \hookrightarrow ((E^\odot)_\rho)^*$, so that a bounded linear functional in $((E^\odot)_\rho)^*$ is a bounded linear functional in $((E^\odot)^*)_{-\rho}$ with norm less or equal to the original norm (although (11.1.29) may not be $1 - 1$). The imbedding (11.1.28) is especially transparent for $\rho = 0$, in which case it reduces to $E^\odot \hookrightarrow (E^\odot)^{**}$.

Remark 11.1.14. When E is reflexive, $D(A^*)$ is dense in E^* so that Lemma 11.1.1 gives $((-A)^{-\alpha})^*(E^*)_\alpha = E^*$. This implies

$$(E_\alpha)^* = ((-A)^{-\alpha})^* E^* = (E^*)_{-\alpha} \tag{11.1.30}$$

and clears up most of the unpleasantness coming from the discrepancy between $E^\odot$ and E^*. For instance, interchanging α by $-\alpha$ in (11.1.5) we show that (11.1.30) extends to $\alpha < 0$, so that the E_α are reflexive. We have $(((E^\odot)^*)_{-\rho})^* = (((E^*)^*)_{-\rho})^* = (E_{-\rho})^* = ((E^*)_\rho)^{**} = (E^*)_\rho$, so that (11.1.28) is actually an isometric isomorphism.

11.2. Abstract Parabolic Equations: The Minimum Principle with State Constraints

We prove the minimum principle for the control system

$$y'(t) = Ay(t) + f(t, y(t), u(t)), \qquad y(0) = \zeta \tag{11.2.1}$$

under the assumptions in **11.1**. The state constraint and the target condition are

$$y(t) \in M(t) \subseteq E_\alpha \quad (0 \le t \le \bar{t}), \qquad y(\bar{t}) \in Y \subseteq E_\alpha \tag{11.2.2}$$

with Y and $M(t)$ closed in E_α. The cost functional is

$$y_0(t, u) = \int_0^t f_0(\tau, y(\tau, u), u(\tau)) d\tau + g_0(t, y(t, u)) \tag{11.2.3}$$

with $f_0 : [0, T] \times E_\alpha \times U \to \mathbb{R}$ and $g_0 : \mathbb{R} \times E_\alpha \to \mathbb{R}$. We define $f_0(t, y) = f_0(t, y, u(t))$ and require $f_0(t, y)$ to satisfy Hypothesis III_α^0 below for all $u(\cdot) \in C_{\mathrm{ad}}(0, T; U)$ with $K_0(c), L_0(c)$ independent of $u(\cdot)$.

Hypothesis III_α^0. The Fréchet derivative $\partial_y f_0(t, y)$ exists in $[0, T] \times E_\alpha$. $f_0(t, y)$ is measurable in t for y fixed and continuous in $y \in E_\alpha$ (in the norm of E_α) for t fixed. For each $z \in E_\alpha$ the $((-A)^\alpha)^* E^* = (E_\alpha)^*$-valued function $\partial_y f_0(t, y)z$ is measurable in t for y fixed and continuous in $y \in E_\alpha$ for t fixed. For every $c > 0$ there exist $K_0(c), L_0(c)$ such that

$$|f_0(t, y)| \le K_0(c) \quad (0 \le t \le T, \|y\|_\alpha \le c) \tag{11.2.4}$$

$$\|\partial_y f_0(t, y)\|_{((-A)^\alpha)^* E^*} \le L_0(c) \quad (0 \le t \le T, \|y\|_\alpha \le c). \tag{11.2.5}$$

The function $g_0 : [0, T] \times E_\alpha \to \mathbb{R}$ is assumed Fréchet differentiable with respect to y. The first result is a complement to Theorem 7.10.4.

Lemma 11.2.1. *Let $\bar{u} \in C_{\mathrm{ad}}(0, \bar{t}; U)$ be such that $y(t, \bar{u})$ exists in $0 \le t \le \bar{t}$. Then the sets $e_j(h)$ in Theorem 7.10.4 can be chosen in such a way that*

$$\left| \frac{1}{h} \int_{(0,t) \cap e_j(h)} \{f_0(\tau, y(\tau, \bar{u}), v_j(\tau)) - f_0(\tau, y(\tau, \bar{u}), \bar{u}(\tau))\} d\tau \right.$$
$$\left. - p_j \int_0^t \{f_0(\tau, y(\tau, \bar{u}), v_j(\tau)) - f_0(\tau, y(\tau, \bar{u}), \bar{u}(\tau))\} d\tau \right| \le h, \tag{11.2.6}$$

and that the limit

$$\xi(t, \bar{u}, \mathbf{p}, \mathbf{v}) = \lim_{h \to 0+} \frac{1}{h} (y_0(t, \bar{u}_{\mathbf{e}(h), \mathbf{v}}) - y_0(t, \bar{u})) \tag{11.2.7}$$

exists uniformly in $0 \leq t \leq \bar{t}$, where

$$
\begin{aligned}
\xi_0(t, \bar{u}, \mathbf{p}, \mathbf{v}) &= \int_0^t \langle \partial_y f_0(\tau, y(\tau, \bar{u}), \bar{u}(\tau)), \xi(\tau, \bar{u}, \mathbf{p}, \mathbf{v}) \rangle_\alpha d\tau \\
&\quad + \sum_{j=1}^m p_j \int_0^t \{f_0(\tau, y(\tau, \bar{u}), v_j(\tau)) - f_0(\tau, y(\tau, \bar{u}), \bar{u}(\tau))\} d\tau \\
&\quad + \langle \partial_y g_0(t, y(t, \bar{u})), \xi(t, \bar{u}, \mathbf{p}, \mathbf{v}) \rangle_\alpha \\
&= \int_0^t \langle ((-A)^{-\alpha})^* \partial_y f_0(\tau, y(\tau, \bar{u}), \bar{u}(\tau)), \zeta(\tau, \bar{u}, \mathbf{p}, \mathbf{v}) \rangle d\tau \\
&\quad + \sum_{j=1}^m p_j \int_0^t \{f_0(\tau, y(\tau, \bar{u}), v_j(\tau)) - f_0(\tau, y(\tau, \bar{u}), \bar{u}(\tau))\} d\tau \\
&\quad + \langle ((-A)^{-\alpha})^* \partial_y g_0(t, y(t, \bar{u})), \zeta(t, \bar{u}, \mathbf{p}, \mathbf{v}) \rangle. \qquad (11.2.8)
\end{aligned}
$$

The proof is similar to that of Lemma 6.6.1 and is omitted. In this formula, $\langle \cdot, \cdot \rangle_\alpha$ indicates $E_\alpha, ((-A)^\alpha)^* E^*$-duality (recall that $((-A)^\alpha)^* E^* = (E_\alpha)^*$), $\langle \cdot, \cdot \rangle$ indicates E, E^*-duality, and $\zeta(t, \bar{u}, \mathbf{p}, \mathbf{v}) = (-A)^\alpha \xi(t, \bar{u}, \mathbf{p}, \mathbf{v})$.

The following is the integral form of Pontryagin's minimum principle. We set

$$
\mathbf{M}(\bar{t}) = \{y(\cdot) \in C(0, \bar{t}; E_\alpha); y(t) \in M(t) \ (0 \leq t < \bar{t})\}. \qquad (11.2.9)
$$

Theorem 11.2.2. *Let $\bar{u}(\cdot) \in C_{ad}(0, \bar{t}; U)$ be an optimal control. Then there exists $(z_0, \mu, z) \in \mathbb{R} \times \Sigma(0, T; ((-A)^\alpha)^* E^*) \times ((-A)^\alpha)^* E^*$ with*

$$
z_0 \geq 0, \quad \mu \in \left(\liminf_{n \to \infty} T_{\mathbf{M}(\bar{t})}(\tilde{y}^n(\cdot)) \right)^-, \quad z \in \left(\liminf_{n \to \infty} T_Y(\tilde{y}^n) \right)^-, \qquad (11.2.10)
$$

where $\tilde{y}^n(\cdot)$ (resp. $\tilde{y}^n$) is a sequence in $\mathbf{M}(\bar{t})$ (resp. in Y) such that $\tilde{y}^n(\cdot) \to y(\cdot, \bar{u})$ in $C(0, \bar{t}; E_\alpha)$ (resp. $\tilde{y}^n \to y(\bar{t}, \bar{u})$ in E_α) and such that if $\bar{z}(s)$ is the solution of

$$
\begin{aligned}
d\bar{z}(s) &= -\{A^\circ + \partial_y f(s, y(s, \bar{u}), \bar{u}(s))^*\} \bar{z}(s) ds \\
&\quad - z_0 \partial_y f_0(s, y(s, \bar{u}), \bar{u}(s)) ds - \mu(ds), \\
\bar{z}(\bar{t}) &= z + z_0 \partial_y g_0(\bar{t}, y(\bar{t}, \bar{u}))
\end{aligned} \qquad (11.2.11)
$$

in $0 \leq t \leq \bar{t}$ then

$$
\begin{aligned}
z_0 \int_0^{\bar{t}} &\{f_0(s, y(s, \bar{u}), v(s)) - f_0(s, y(s, \bar{u}), \bar{u}(s))\} ds \\
&+ \int_0^{\bar{t}} \langle \bar{z}(s), f(s, y(s, \bar{u}), v(s)) - f(s, y(s, \bar{u}), \bar{u}(s)) \rangle_\rho \, ds \geq 0 \quad (11.2.12)
\end{aligned}
$$

for all $v(\cdot) \in C_{ad}(0, \bar{t}; U)$.

Proof. The duality product inside the integral needs explanation. Note that $\bar{z}(s) = (-A^{\odot})^{-\rho}\upsilon(s) \in (E^{\odot})_{\rho}$ ($\upsilon(s)$ given by (11.1.12)), this expression for $\bar{z}(s)$ justified by the fact that $\upsilon(s) \in E^{\odot}$ a.e. (Lemma 11.1.8). On the other hand, $f(t, y, u) \in (E^*)_{-\rho} \hookrightarrow ((E^{\odot})_{\rho})^*$ (Corollary 11.1.3), thus $\langle \cdot, \cdot \rangle_{\rho}$ is simply the $(E^{\odot})_{\rho}$, $((E^{\odot})_{\rho})^*$-duality pairing, and can be computed in the form $\langle z, f \rangle = \langle (-A^{\odot})^{\rho}z, ((-A^{\odot})^{-\rho})^*f \rangle$, $\langle \cdot, \cdot \rangle$ the duality pairing of $E^{\odot}$ and $(E^{\odot})^*$. Accordingly, the integrand in (11.2.12) can be written

$$\langle \bar{z}(s), f(s, y(s, \bar{u}), v(s)) - f(s, y(s, \bar{u}), \bar{u}(s)) \rangle_{\rho}$$
$$= \langle (-A^{\odot})^{\rho}\bar{z}(s), ((-A^{\odot})^{-\rho})^*\{f(s, y(s, \bar{u}), v(s)) - f(s, y(s, \bar{u}), \bar{u}(s))\} \rangle.$$
(11.2.13)

The proof of Theorem 11.2.2 follows closely that of Theorem 10.3.1. We apply the nonlinear programming theory in Chapter 7 to the problem

$$\text{minimize} \quad \mathbf{f}_0(u) \tag{11.2.14}$$

$$\text{subject to} \quad \mathbf{f}(u) \in \mathbf{Y} = \mathbf{M}(\bar{t}) \times Y, \tag{11.2.15}$$

where $\mathbf{f}_0(u) = y_0(\bar{t}, u) \in \mathbb{R}$, $\mathbf{f}(u) = (y(\cdot, u), y(\bar{t}, u)) \in C(0, \bar{t}; E_{\alpha}) \times E_{\alpha}$. The metric space V is the ball $B(\bar{u}, \delta)$ provided by Lemma 7.10.1. Continuity of $\mathbf{f}_0(u)$ has been proved in Lemma 7.10.1 and continuity of $\mathbf{f}_0(u)$ is proved as in Lemma 4.2.2. Theorem 7.2.5 produces the sequences $\{\tilde{u}^n\} \subseteq C_{\text{ad}}(0, \bar{t}; U)$ and $\{(\tilde{y}^n(\cdot), \tilde{y}^n)\} \subseteq \mathbf{M}(\bar{t}) \times Y$ in the statement. We select $\rho > 0$ and the sequence $\{\mathbf{D}_n\}$, $\mathbf{D}_n \subseteq \text{Der}(\mathbf{f}_0, \mathbf{f})(\tilde{u}^n) \subseteq \mathbb{R} \times C(0, \bar{t}; E_{\alpha}) \times E_{\alpha}$ of convex sets consisting of all triples

$$(\bar{t}^{-1}\xi_0(\bar{t}, \tilde{u}^n, \mathbf{p}, \mathbf{v}), \bar{t}^{-1}\xi(\cdot, \tilde{u}^n, \mathbf{p}, \mathbf{v}), \bar{t}^{-1}\xi(\bar{t}, \tilde{u}^n, \mathbf{p}, \mathbf{v})) \in \mathbb{R} \times C(0, \bar{t}; E_{\alpha}) \times E_{\alpha}. \tag{11.2.16}$$

Keeping $\mathbf{v}$ fixed, the limit of (11.2.16) in $\mathbb{R} \times C(0, \bar{t}; E_{\alpha}) \times E_{\alpha}$ as $n \to \infty$ is

$$(\bar{t}^{-1}\xi_0(\bar{t}, \bar{u}, \mathbf{p}, \mathbf{v}), \bar{t}^{-1}\xi(\cdot, \bar{u}, \mathbf{p}, \mathbf{v}), \bar{t}^{-1}\xi(\bar{t}, \bar{u}, \mathbf{p}, \mathbf{v})) \in \mathbb{R} \times C(0, \bar{t}; E_{\alpha}) \times E_{\alpha}, \tag{11.2.17}$$

and this is proved essentially as in Lemma 6.6.4; the pertinent continuity properties of the solution operator are in Lemmas 7.10.5 and 7.10.6. Convergence of the second coordinate is in $C(0, \bar{t}; E_{\alpha})$ and of the third in E_{α}. Once ρ and $\{\mathbf{D}_n\}$ have been selected, Theorem 7.2.5 gives sequences $\{z_{0n}\} \subseteq \mathbb{R}$, $\{\mu_n\} \subseteq C(0, \bar{t}; E_{\alpha})^* = \Sigma(0, T; (E_{\alpha})^*) = \Sigma(0, T; ((-A)^{\alpha})^*E^*)$ and $\{z_n\} \subseteq (E_{\alpha})^* = ((-A)^{\alpha})^*E^*$ such that

$$z_{0n}^2 + \|\mu_n\|_{\Sigma(0,T;((-A)^{\alpha})^*E^*)}^2 + \|z_n\|_{((-A)^{\alpha})^*E^*}^2 = 1, \quad z_{0n} \geq 0,$$

$$z_{0n}\xi_0(\bar{t}, \tilde{u}^n, \mathbf{p}, \mathbf{v}) + \langle \mu_n, \xi(\cdot, \tilde{u}^n, \mathbf{p}, \mathbf{v}) \rangle_{\alpha,c} + \langle z_n, \xi(\bar{t}, \tilde{u}^n, \mathbf{p}, \mathbf{v}) \rangle_{\alpha} \geq -\delta_n,$$
(11.2.18)

where the angled brackets $\langle \cdot, \cdot \rangle_{\alpha,c}$ indicate $C(0, \bar{t}; E_\alpha)$, $\Sigma(0, \bar{t}; ((-A)^\alpha)^* E^*)$-duality, the second angled brackets $\langle \cdot, \cdot \rangle_\alpha$ indicate E_α, $((-A)^\alpha)^* E^*$-duality, and $\delta_n \to 0$. If (z_0, μ, z) is a $(\mathbb{R} \times C(0, \bar{t}; E_\alpha) \times E_\alpha)^*$-weak limit point of (z_{0n}, μ_n, z_n) then we have

$$z_0 \xi_0(\bar{t}, \bar{u}, \mathbf{p}, \mathbf{v}) + \langle \mu, \xi(\cdot, \bar{u}, \mathbf{p}, \mathbf{v}) \rangle_{\alpha,c} + \langle z, \xi(\bar{t}, \bar{u}, \mathbf{p}, \mathbf{v}) \rangle_\alpha \geq 0. \qquad (11.2.19)$$

For single patches,

$$z_0 \xi_0(\bar{t}, \bar{u}, v) + \langle \mu, \xi(\cdot, \bar{u}, v) \rangle_{\alpha,c} + \langle z, \xi(\bar{t}, \bar{u}, v) \rangle_\alpha \geq 0, \qquad (11.2.20)$$

which is

$$z_0 \int_0^{\bar{t}} \{f_0(\sigma, y(\sigma, \bar{u}), v(\sigma)) - f_0(\sigma, y(\sigma, \bar{u}), \bar{u}(\sigma))\} d\sigma$$

$$+ z_0 \int_0^{\bar{t}} \Big\langle ((-A)^{-\alpha})^* \partial_y f_0(\tau, y(\tau, \bar{u}), \bar{u}(\tau)),$$

$$\int_0^\tau R_{\alpha,\rho}(\tau, \sigma; \bar{u})((-A^\odot)^{-\rho})^* \{f(\sigma, y(\sigma, \bar{u}), v(\sigma)) - f(\sigma, y(\sigma, \bar{u}), \bar{u}(\sigma))\} d\sigma \Big\rangle d\tau$$

$$+ \int_0^{\bar{t}} \Big\langle ((-A)^{-\alpha})^* \mu(dt),$$

$$\int_0^t R_{\alpha,\rho}(t, \sigma; \bar{u})((-A^\odot)^{-\rho})^* \{f(\sigma, y(\sigma, \bar{u}), v(\sigma)) - f(\sigma, y(\sigma, \bar{u}), \bar{u}(\sigma))\} d\sigma \Big\rangle$$

$$+ \Big\langle ((-A)^{-\alpha})^* \{z + z_0 \partial_y g_0(\bar{t}, y(\bar{t}, \bar{u}))\},$$

$$\int_0^{\bar{t}} R_{\alpha,\rho}(\bar{t}, \sigma; \bar{u})((-A^\odot)^{-\rho})^* \{f(\sigma, y(\sigma, \bar{u}), v(\sigma)) - f(\sigma, y(\sigma, \bar{u}), \bar{u}(\sigma))\} d\sigma \Big\rangle \geq 0$$

$$(11.2.21)$$

After a switch in the order of integration (justified by Lemma 11.1.10 in the fourth integral) we obtain (11.2.12). This ends the proof. ∎

Corollary 11.2.3. *Assume that almost every $s \in [0, T]$ is a left Lebesgue point of all the functions*

$$\langle \bar{z}(s), f_0(s, y(s, \bar{u}), v) \rangle, \qquad f_0(s, y(s, \bar{u}), v)$$

(in other words, that $\langle \bar{z}(s), f(s, y(s, \bar{u}), v) \rangle$ and $f_0(s, y(s, \bar{u}), v)$ are regular in $[0, T] \times U$). Then, under the assumptions of Theorem 11.2.2, we have

$$z_0 f_0(s, y(s, \bar{u}), \bar{u}(s)) + \langle z(s), f(s, y(s, \bar{u}), \bar{u}(s)) \rangle_\rho$$

$$= \min_{v \in U} \{z_0 f_0(s, y(s, \bar{u}), v) + \langle \bar{z}(s), f(s, y(s, \bar{u}), v) \rangle_\rho\} \qquad (11.2.22)$$

a.e. in $0 \leq s \leq \bar{t}$.

The proof of the minimum principle for the time optimal problem also follows the scheme in **10.3**; this time we obtain sequences $\{\mu_n\} \subseteq \Sigma(0, T; ((-A)^\alpha)^* E^*)$ and $\{z_n\} \subseteq ((-A)^\alpha)^* E^*$ such that

$$\|\mu_n\|^2_{\Sigma(0,T;((-A)^\alpha)^* E^*)} + \|z_n\|^2_{((-A)^\alpha)^* E^*} = 1,$$

$$\langle \mu_n, \xi(\cdot, \tilde{u}^n, \mathbf{p}, \mathbf{v}) \rangle_{\alpha,c} + \langle z_n, \xi(\bar{t}, \tilde{u}^n, \mathbf{p}, \mathbf{v}) \rangle_\alpha \geq -\delta_n \qquad (11.2.23)$$

with $\delta_n \to 0$. If (μ, z) is a $(C(0, \bar{t}; E_\alpha) \times E_\alpha)^*$-weak limit point of (μ_n, z_n) then we have

$$\langle \mu, \xi(\cdot, \bar{u}, \mathbf{p}, \mathbf{v}) \rangle_{\alpha,c} + \langle z, \xi(\bar{t}, \bar{u}, \mathbf{p}, \mathbf{v}) \rangle_\alpha \geq 0 \qquad (11.2.24)$$

and operating in the same way we obtain

$$\int_0^{\bar{t}} \langle \bar{z}(\sigma), f(\sigma, y(\sigma, \bar{u}), v(\sigma)) - f(\sigma, y(\sigma, \bar{u}), \bar{u}(\sigma)) \rangle_\rho d\sigma \geq 0 \qquad (11.2.25)$$

for all $v(\cdot) \in C_{\mathrm{ad}}(0, \bar{t}; U)$, where

$$d\bar{z}(s) = -\{A^\odot + \partial_y f(s, y(s, \bar{u}), \bar{u}(s))^*\}\bar{z}(s)ds - \mu(ds), \quad \bar{z}(\bar{t}) = z. \qquad (11.2.26)$$

If $\langle \bar{z}(s), f(s, y(s, \bar{u}), v) \rangle$ is regular in $[0, T] \times U$ we obtain the pointwise minimum principle

$$\langle z(s), f(s, y(s, \bar{u}), \bar{u}(s)) \rangle_\rho = \min_{v \in U} \langle (\bar{z}(s), f(s, y(s, \bar{u}), v)) \rangle_\rho \qquad (11.2.27)$$

a.e. in $0 \leq s \leq \bar{t}$.

The counterpart of Lemma 10.3.5 is

Lemma 11.2.4. *Assume* $\mathbf{M}(\bar{t})$ *is T-full in* $C(0, \bar{t}; E_\alpha)$ *and that for all sequences* $\{t_n\} \subseteq (0, \bar{t})$ *with* $t_n \to \bar{t}$, $\{\tilde{u}^n\} \subseteq C_{\mathrm{ad}}(0, \bar{t}; E)$ *with* $\tilde{u}^n \to \bar{u}$ *and* $\{\tilde{y}^n\} \subseteq E$ *with* $\tilde{y}^n \to y(\bar{t}, \bar{u})$ *in* E_α *there exist* $\rho > 0$ *and a precompact sequence* $\{Q_n\}$, $Q_n \subseteq E_\alpha$ *such that*

$$\bigcap_{n=n_0}^{\infty} \{t_n^{-1} \mathrm{conv}\, R(0, t_n; U, \tilde{u}^n) - T_Y(\tilde{y}^n) \cap B(0, \rho) + Q_n\} \qquad (11.2.28)$$

contains an interior point in E_α *for* n_0 *large enough.*[1] *Then the multipliers* (z_0, μ, z) *in the general optimal problem and* (μ, z) *in the time optimal problem are nonzero. In particular, if* Y *is T-full in* E_α *the assumptions are automatically satisfied.*

The proof of Lemma 11.2.4 is the same as that of Lemma 10.3.5 modulo the replacement of the space E by E_α and of $C(0, \bar{t}; E)$ by $C(0, \bar{t}; E_\alpha)$ and is omitted.

---

[1] The factor t_n^{-1} could be eliminated as in (10.3.28), but we keep it here with something in mind; see the forthcoming Lemma 11.5.6.

Remark 11.2.5. The results in this section do not cover the point target case, where the tangent cone $T_Y(\bar{y}^n)$ volunteers no help in (11.2.28). Any hopes of fulfilling the assumptions of Lemma 11.2.4 on the only basis of $R(0, t_n; U, \tilde{u}^n)$ are unfounded; in fact, for the linear equation

$$y'(t) = Ay(t) + u(t)$$

with controls, say, in the unit ball of $L_w^\infty(0, T; (E^\odot)^*)$, $R(0, t_n; U, \tilde{u}^n)$ consists of all elements of the form

$$\int_0^{t_n} S^\odot(t_n - \tau)^*(v(\tau) - \tilde{u}^n(\tau))d\tau.$$

There is no possibility that $R(0, t_n; U, \tilde{u}^n)$ contain interior points in E_α for any given $\alpha < 1$ since $R(0, t_n; U, \tilde{u}^n) \subseteq E_\gamma$ for every $\gamma < 1$; to see this, it suffices to apply $(-A)^\gamma$ under the integral sign. Thinness of the set on the right side of (11.2.28) cannot be fixed with any precompact sequence $\{Q_n\}$.

Remark 11.2.6. The case where $f(t, y, u(t))$ satisfies Hypothesis $\text{III}_{\alpha,0}(wm)$ appears often in practice (as in all examples in this chapter). It fits into the theory above, since Hypothesis $\text{III}_{\alpha,0}(wm)$ implies Hypothesis $\text{III}_{\alpha,\rho}$ for any $\rho > 0$ (Lemma 7.10.3). We can get rid of the auxiliary parameter ρ in the integrands of (11.2.12) and (11.2.25) and in the pointwise maximum principles by simply observing that f takes values in $(E^\odot)^*$ and $((-A^\odot)^\rho)\bar{z}(s)$ takes values in $E^\odot$; putting $((-A^\odot)^{-\rho})^*$ on the other side of the duality product in (11.2.13) we obtain

$$\langle \bar{z}(s), f(s, y(s, \bar{u}), v(s)) - f(s, y(s, \bar{u}), \bar{u}(s)) \rangle_\rho$$
$$= \langle ((-A^\odot)^\rho)\bar{z}(s), ((-A^\odot)^{-\rho})^*\{f(s, y(s, \bar{u}), v(s)) - f(s, y(s, \bar{u}), \bar{u}(s))\}\rangle$$
$$= \langle \bar{z}(s), \{f(s, y(s, \bar{u}), v(s)) - f(s, y(s, \bar{u}), \bar{u}(s))\}\rangle. \qquad (11.2.29)$$

In fact, in this case, ρ may be cleaned up from all constructs. Note first that ρ need not be used in the definition of the solutions of (11.2.1) due to the smoothing prowess of the semigroup (Remark 7.7.14; the integral equation is (7.7.30)). Neither is it necessary to use ρ in the definition of the adjoint vector in (11.1.12). The integral equation (7.9.27) defining the operator $R_{\alpha,\rho}(t, s)$ is modified as follows when $\rho = 0$:

$$R_{\alpha,0}(t, s)\zeta = (-A)^\alpha S^\odot(t - s)^*\zeta$$
$$+ \int_s^t (-A)^\alpha S^\odot(t - \tau)^* B(\tau)(-A)^{-\alpha} R_{\alpha,0}(\tau, s)\zeta d\tau, \quad (11.2.30)$$

and the equation (7.9.45) for $R^*_{\rho,\alpha}(s,t)$ becomes

$$R^*_{0,\alpha}(s,t)z = (-A^{\odot})^{\alpha} S(t-s)^* z$$

$$+ \int_s^t (-A^{\odot})^{\alpha} S^{\odot}(\sigma - s)((-A)^{-\alpha})^* B(\sigma)^* R^*_{0,\alpha}(\sigma, t) z \, d\sigma. \quad (11.2.31)$$

The adjoint relation (11.1.13) holds. The function $\upsilon(s)$ is defined by (11.1.15) with $R_{\alpha,0}(t,s)$. There is no significant difference in any of the ulterior developments and Remark 11.1.9 applies verbatim.

11.3. Applications to Parabolic Distributed Parameter Systems

The system is

$$y_t(t,x) = A_c(\beta)y(t,x) + \phi(t,x,y(t,x),\nabla y(t,x),u(t)),$$
$$y(0,x) = \zeta(x) \quad (11.3.1)$$

in a domain $\Omega \subseteq \mathbb{R}^m$; the assumptions on Ω, the operator A and the boundary condition β are those in **8.4**. The state space is $E = C(\overline{\Omega})$ or $C_0(\overline{\Omega})$ according to the boundary condition. The control set U is arbitrary and the function $\phi(t,x,y,\mathbf{y},u)$ $= \phi(t,x,y,y_1,\ldots,y_m,u)$ satisfies the assumptions in Theorem 8.4.9 for every $u(\cdot) \in C_{\mathrm{ad}}(0,\bar{t};U)$, the constants in (8.4.27) and (8.4.28) independent of $u(\cdot)$. It has been proved in Theorem 8.4.9 that if $\alpha > 1/2$ the operator

$$f(t,y(\cdot),u(t))(x) = \phi(t,x,y(x),\nabla y(x),u(t)). \quad (11.3.2)$$

from $[0,T] \times E_{\alpha} \times U \to L^{\infty}(\Omega)$ satisfies Assumption $\mathrm{III}_{\alpha,0}(wm)$ with constants $K(c)$, $L(c)$ independent of $u(\cdot)$, thus it satisfies Assumption $\mathrm{III}_{\alpha,\rho}$ for any $\rho > 0$ with $K(c)$, $L(c)$ independent of $u(\cdot)$ (Lemma 7.10.3).

To fix ideas, we assume the state constraints are of the form

$$y(t,x) \in M_s, \qquad \nabla y(t,x) \in M_g \quad (0 \le t \le \bar{t}) \quad (11.3.3)$$

where $M_s \subseteq \mathbb{R}$, $M_g \subseteq \mathbb{R}^m$ are closed, so that $M \subseteq E_{\alpha}$ consists of all $y(\cdot) \in E_{\alpha}$ such that $y(x) \in M_s$, $\nabla y(x) \in M_g$. We fix $\alpha > 1/2$ and, to simplify, consider only the time optimal problem.

Theorem 11.3.1. *Let $\bar{u}(\cdot)$ be a time optimal control. Then there exist sequences $\{\tilde{y}^n(\cdot)\} \subseteq \mathbf{M}(\bar{t})$ and $\{\tilde{y}^n\} \subseteq Y$ such that $\tilde{y}^n(\cdot) \to y(\cdot,\bar{u})$ in $C(0,\bar{t};E_{\alpha})$ and $\tilde{y}^n \to y(\bar{t},\bar{u})$ in E_{α}, a measure $\mu \in \Sigma(0,\bar{t};((-A)^{\alpha})^* E^*)$ and an element $z \in ((-A)^{\alpha})^* E^*$ with*

$$\mu \in \left(\liminf_{n \to \infty} T_{\mathbf{M}(\bar{t})}(\tilde{y}^n(\cdot))\right)^-, \quad z \in \left(\liminf_{n \to \infty} T_Y(\tilde{y}^n)\right)^-, \quad (11.3.4)$$

and such that, if $\bar{z}(s, x)$ is the solution of

$$d\bar{z}(s, x) = -A_1'(\beta')\bar{z}(s, x)ds - \partial_y\phi(s, x, y(s, x, \bar{u}), \nabla y(s, x, \bar{u}), \bar{u}(s))\bar{z}(s, x)ds$$

$$+ \nabla \cdot (\nabla_{\mathbf{y}}\phi(s, x, y(s, x, \bar{u}), \nabla y(s, x, \bar{u}), \bar{u}(s))\bar{z}(s, x))ds - \mu(ds)^{(1)}$$

$$\bar{z}(\bar{t}, \cdot) = z \tag{11.3.5}$$

then

$$\int_\Omega \bar{z}(s, x)\phi(s, x, y(s, x, \bar{u}), \nabla y(s, x, \bar{u}), \bar{u}(s))dx$$

$$= \min_{v \in U} \int_\Omega \bar{z}(s, x)\phi(s, x, y(s, x, \bar{u}), \nabla y(s, x, \bar{u}), v)dx. \tag{11.3.6}$$

The proof is a direct consequence of Theorem 11.2.2 and Corollary 11.2.3; see also Remark 11.2.6. For conditions on M_s, M_g and on the target set Y that guarantee nontriviality of the multiplier, see Remark 11.3.4.

We note the case where $\phi = \phi(t, y, u)$ does not depend on the derivatives of y and there are only constraints on the state, not on the gradient (we only have the first constraint in (11.3.3)). Here it is natural to take $\alpha = 0$ as well; hence, $\mu \in \Sigma(0, \bar{t}; C(\overline{\Omega})^*) = \Sigma(0, \bar{t}; \Sigma(\overline{\Omega}))$ for a variational boundary condition. Since we have the obvious algebraic and metric identification

$$C(0, \bar{t}; C(\overline{\Omega})) = C([0, \bar{t}] \times \overline{\Omega}) \tag{11.3.7}$$

we have a corresponding identification of the duals

$$\Sigma(0, \bar{t}; \Sigma(\overline{\Omega})) = \Sigma([0, \bar{t}] \times \overline{\Omega}), \tag{11.3.8}$$

so that μ is just a bounded regular countably additive measure in the cylinder $\Sigma([0, \bar{t}] \times \overline{\Omega})$. If $\tilde{y}^n(t, x)$ is the sequence in the statement of Theorem 11.3.1, then the first condition (11.3.4) translates into

$$\mu \in \left(\liminf_{n \to \infty} T_{\mathbf{M}(\bar{t})}(\tilde{y}^n(\cdot, \cdot)) \right)^-, \tag{11.3.9}$$

the duality in (11.3.9) being that of $C([0, \bar{t}] \times \overline{\Omega})$ and $\Sigma([0, \bar{t}] \times \overline{\Omega})$. It then follows from a multivariate analogue of Lemma 10.4.1 (particular case of Theorem 11.3.2 below; see also Remark 11.3.7) that $\mu = 0$ in the set

$$e_0 = \{(t, x) \in [0, \bar{t}] \times \overline{\Omega}; y(t, x) \in \text{Int}(M_s)\}, \tag{11.3.10}$$

$^{(1)}$ The unindexed nablas operate on the x-variables. Application of $\nabla \cdot$ is purely formal since both $\bar{z}(s, x)$ and the expression between parentheses may not have first partials.

hence the measure μ is supported by the set where the state constraint is saturated (that is, the set where $y(t, x)$ hits the boundary of M_s).

In the case of constraints on the gradient we must take $\alpha > 1/2$ and μ cannot be identified any longer with a measure, but with a distribution. However, there is a generalization of the result above for constraints of the form (11.3.3) and a nonlinear term (11.3.2) satisfying the assumptions of Theorem 11.3.1.

Theorem 11.3.2. *Let $\mu \in \Sigma(0, \bar{t}; ((-A)^\alpha)^* E^*)$ satisfy the first condition (11.3.4). Then $\mu = 0$ in the set*

$$e_0 = \{(t, x) \in [0, \bar{t}] \times \overline{\Omega}; (y(t, x, \bar{u}), \nabla y(t, x, \bar{u})) \in \text{Int}(M_s \times M_g)\}. \quad (11.3.11)$$

Proof. We introduce some notation. If $B \subseteq \mathbb{R} \times \mathbb{R}^m$ is open, $\mathcal{D}(B)$ is the space of all test functions with support in B; $\mathcal{D}(B)|_K$ (K a closed set) means restriction to K and $\mathcal{D}(B)|_{K,\beta}$ indicates compliance of the restrictions with the boundary condition on Γ.

If $\alpha = 0$ then μ is a bounded Borel measure in $[0, \bar{t}] \times \overline{\Omega}$; hence, if e_0 is open in $[0, \bar{t}] \times \overline{\Omega}$, the statement "$\mu = 0$ in e_0" is unequivocal: it means that

$$\int_{[0,\bar{t}]\times\overline{\Omega}} \phi(t, x)\mu(dtdx) = 0 \quad (11.3.12)$$

for every $\phi \in \mathcal{D}(B)|_{[0,\bar{t}]\times\overline{\Omega},\beta}$, $B \subseteq \mathbb{R} \times \mathbb{R}^m$ with $B \cap ([0, \bar{t}] \times \overline{\Omega}) \subseteq e_0$. Nothing changes if we limit the open set B to a finite or countable family $\{B_j\}$ with

$$(\cup B_j) \cap ([0, \bar{t}] \times \overline{\Omega}) = e_0. \quad (11.3.13)$$

However, if $\alpha > 0$, μ is not a measure, thus what "$\mu = 0$ in e_0" means must be elucidated. With this in mind, we write

$$\nu = ((-A_c(\beta))^{-1})^* \mu = ((-A_c(\beta))^{\alpha-1})^* ((-A_c(\beta))^{-\alpha})^* \mu. \quad (11.3.14)$$

Now, $((-A_c(\beta))^{-\alpha})^* \mu \in \Sigma(0, \bar{t}; E^*) \subseteq \Sigma(0, \bar{t}; \Sigma(\overline{\Omega}))$, thus $((-A_c(\beta))^{-\alpha})^* \mu$ is a *bona fide* measure in $[0, \bar{t}] \times \overline{\Omega}$, and (on the only grounds that the operator $((-A_c(\beta))^{\alpha-1})^*$ is bounded) the same applies to ν. The statement that $\mu = 0$ in e_0 then means

$$A_c(\beta)^* \nu = 0 \quad \text{in } e_0, \quad (11.3.15)$$

understood in the following way:

$$\int_{[0,\bar{t}]\times\overline{\Omega}} A_c(\beta)\phi(t, x)\nu(dtdx) = 0. \quad (11.3.16)$$

for every $\phi \in \mathcal{D}(B)|_{[0,\bar{t}]\times\overline{\Omega},\beta}$ with $B \cap ([0, \bar{t}] \times \overline{\Omega}) \subseteq e_0$; we also allow verification for a family $\{B_j\}$ satisfying (11.3.13).

In order to prove (11.3.16) we recall that $(-A_c(\beta))^{-1}$ is an isometric isomorphism from $E_{\alpha-1}$ onto E_α (Lemma 7.4.10). Then, if M is the set defined by both conditions (11.3.3) in E_α and $\{\tilde{y}^n(\cdot)\}$ is the sequence in (11.3.4), we have

$$M = (-A_c(\beta))^{-1} M_1, \quad \tilde{y}^n(t) = (-A_c(\beta))^{-1} \tilde{y}_1^n(t),$$

the latter in $0 \leq t \leq \bar{t}$, where $M_1 \subseteq E_{\alpha-1}$ and $\tilde{y}_1^n(\cdot) \in C(0, \bar{t}; E_{\alpha-1})$. Tangent cones are invariant through isometric isomorphisms, thus

$$T_{M(\bar{t})}(\tilde{y}^n(\cdot)) = (-A_c(\beta))^{-1} T_{M_1(\bar{t})}(\tilde{y}_1^n(\cdot)). \tag{11.3.17}$$

Solutions $t \rightarrow y(t, \cdot, u)$ of (11.3.1) that exist for $0 \leq t \leq \bar{t}$ belong to

$$C(0, \bar{t}; E_\alpha) \hookrightarrow C(0, \bar{t}; C^{(1)}(\overline{\Omega})), \tag{11.3.18}$$

thus $y(t, x, u)$ and $\nabla y(t, x, u)$ are continuous in $[0, \bar{t}] \times \overline{\Omega}$. Hence, if $(t_0, x_0) \in e_0$ we can find $\delta > 0$ so small that

$$\text{dist}((y(t, x, \bar{u}), \nabla y(t, x, \bar{u})), \partial(M_s \times M_g)) > 2\varepsilon$$

$$((t, x) \in ([t_0 - \delta, t_0 + \delta] \cap [0, \bar{t}]) \times B(x_0, \delta)),$$

where ∂ indicates boundary; the distance is taken in $\mathbb{R} \times \mathbb{R}^m$. Using again the imbedding (11.3.18), we deduce that

$$\text{dist}((\tilde{y}^n(t, x), \nabla \tilde{y}^n(t, x), \partial(M_s \times M_g)) > \varepsilon$$

$$((t, x) \in ([t_0 - \delta, t_0 + \delta] \cap [0, \bar{t}]) \times B(x_0, \delta))$$

for n large enough, which implies that if $B = (t_0 - \delta, t_0 + \delta) \times B(x_0, \delta)$, the tangent cone $T_{M(\bar{t})}(\tilde{y}^n(\cdot))$ contains the space $\mathcal{D}(B)|_{[0, \bar{t}] \times \overline{\Omega}}$, a fortiori the subspace $\mathcal{D}(B)|_{[0, \bar{t}] \times \overline{\Omega}, \beta}$. The smoothness assumptions on Ω and the coefficients of A and the boundary condition mean that any element of $\mathcal{D}(B)|_{[0, \bar{t}] \times \overline{\Omega}, \beta}$ belongs to $D(A_c(\beta))$, thus it follows from (11.3.17) that

$$T_{M_1(\bar{t})}(\tilde{y}_1^n(\cdot)) \supseteq A_c(\beta) \mathcal{D}(B)|_{[0, \bar{t}] \times \overline{\Omega}, \beta} \tag{11.3.19}$$

for n large enough. Now, we also obtain from (11.3.17) that

$$(-A_c(\beta)) \liminf_{n \to \infty} T_{M(\bar{t})}(\tilde{y}^n(\cdot)) = \liminf_{n \to \infty} T_{M_1(\bar{t})}(\tilde{y}_1^n(\cdot)),$$

limits taken in the each cone's home space. Accordingly,

$$v = ((-A_c(\beta))^{-1})^* \mu \in \left(\liminf_{n \to \infty} T_{M_1(\bar{t})}(\tilde{y}_1^n(\cdot)) \right)^-, \tag{11.3.20}$$

which is (11.3.16) for the set B. Clearly we can do the construction above in such a way as to obtain a sequence $\{B_j\}$ satisfying (11.3.13). This ends the proof. ∎

There is an obvious analogue of Theorem 11.3.2 (Example 11.3.3 below) for the target condition, when of the form

$$y(\bar{t}, x) \in Y_s, \qquad \nabla y(\bar{t}, x) \in Y_g, \qquad (11.3.21)$$

with $Y_s \subseteq \mathbb{R}$, $Y_g \subseteq \mathbb{R}^m$ closed: the target set $Y \subseteq E_\alpha$ consists of all $y(\cdot) \in E_\alpha$ satisfying (11.3.21).

Example 11.3.3. Assume $z \in ((-A_c(\beta)^\alpha)^* E^*$ satisfies the second condition (11.3.4). Then $z = 0$ in the set $e_0(\bar{t}) \subseteq \overline{\Omega}$ defined by

$$e_0(\bar{t}) = \{x \in \overline{\Omega}; (y(\bar{t}, x, \bar{u}), \nabla y(\bar{t}, x, \bar{u})) \in \text{Int}(Y_s \times Y_g)\}. \qquad (11.3.22)$$

This result is a time-invariant companion of Theorem 11.3.2. The definition of "$z = 0$ in e_0" is (11.3.15)-(11.3.16) with $v = ((-A_c(\beta))^{-1})^* z \in E^*$, the test function $\phi(x)$ depending only on x and the integral over $\overline{\Omega}$. The proof is exactly that of Theorem 11.3.2 after "erasing t" everywhere.

Remark 11.3.4. Nontriviality of the multiplier (μ, z) is guaranteed by Lemma 11.2.4 if the sets $M_s \subseteq \mathbb{R}$ and $M_g \subseteq \mathbb{R}^m$ in the state constraint (11.3.3) and the sets $Y_s \subseteq \mathbb{R}$, $Y_g \subseteq \mathbb{R}^m$ in the target condition (11.3.21) are convex with nonempty interior. In fact, in this case M and Y are convex and, since $E_\alpha \hookrightarrow C^{(1)}(\overline{\Omega})$, M has nonempty interior in $C(0, \bar{t}; E_\alpha)$ and Y has nonempty interior in E_α, thus both sets are T-full in the spaces where they live (Lemma 6.4.7). Lemma 6.4.8 allows us to add a finite number of "evaluations" $y(\bar{t}, x_j) = c_j \in \mathbb{R}$, $\nabla y(\bar{t}, x_j) = d_j \in \mathbb{R}^m$. Corollary 6.4.11 admits target sets defined by a finite number of equalities and inequalities involving y and ∇y. The point target case is not included but will be treated in **11.4** and **11.5**.

Remark 11.3.5. Theorem 11.3.2 can be extended to variable state constraints

$$y(t, x) \in M_s(t, x), \qquad \nabla y(t, x) \in M_g(t, x) \qquad (11.3.23)$$

under the assumption that the set functions $(t, x) \to M_s(t, x) \subseteq \mathbb{R}$ and $(t, x) \to M_g(t, x) \in \mathbb{R}^m$ are **closed**; this means that

$$\{(t, x) \in [0, \bar{t}] \times \overline{\Omega}; M_s(t, x) \cap C \neq \emptyset\}$$

must be closed in $[0, \bar{t}] \times \overline{\Omega}$ for every closed set $C \subseteq \mathbb{R}$ and that

$$\{(t, x) \in [0, \bar{t}] \times \overline{\Omega}; M_g(t, x) \cap C \neq \emptyset\}$$

must be closed in $[0, \bar{t}] \times \overline{\Omega}$ for every closed set $C \subseteq \mathbb{R}^m$. The sets $M(t) \subseteq E_\alpha$ in the previous section consist of all $y(\cdot) \in E_\alpha$ such that $y(x) \in M_s(t, x)$ and

$\nabla g(x) \in M_g(t, x)$. We show first that if $\bar{y}(t, x)$ is a function in $C(0, \bar{t}; C^{(1)}(\overline{\Omega}))$ then the set

$$\{(t, x) \in [0, \bar{t}] \times \overline{\Omega}; \ \mathrm{dist}((\bar{y}(t, x), \nabla \bar{y}(t, x)), \partial(M_s(t, x) \times M_g(t, x))) > 0\}$$

is open in $[0, \bar{t}] \times \overline{\Omega}$ (this is a simple multidimensional analogue of the corresponding result in Theorem 10.4.1). From then on, the proof proceeds just as that of Theorem 11.3.2.

Remark 11.3.6. All results extend to state constraints of the form

$$\mathfrak{S}y(t, x) \in M_s \subseteq \mathbb{R}^k \quad (0 \le t \le \bar{t}, x \in \Omega) \tag{11.3.24}$$

where $\mathfrak{S}$ is a differential operator

$$\mathfrak{S}y(x) = \sum_{j=1}^{m} \chi_j(x) \partial_j y(x) + \chi(x) y(x); \tag{11.3.25}$$

the coefficients $\chi_j(x)$ and $\chi(x)$ are k-vector functions with entries in $C(\overline{\Omega})$. Likewise, the target condition may be of the form

$$\mathfrak{T}y(\bar{t}, x) \in M_Y \subseteq \mathbb{R}^l \quad (x \in \Omega) \tag{11.3.26}$$

where $\mathfrak{T}$ is a differential operator of the same form as $\mathfrak{S}$,

$$\mathfrak{T}y(x) = \sum_{j=1}^{m} \eta_j(x) \partial_j y(x) + \eta(x) y(x) \tag{11.3.27}$$

with $\eta_j(x)$ and $\eta(x)$ l-vector functions with entries in $C(\overline{\Omega})$. Time dependent state constraints are also tractable.

Remark 11.3.7. In case $\alpha = 0$ we may choose to prove Theorem 11.3.2 applying (11.3.12) straight (this does not necessitate the mapping of tangent cones provided by (11.3.17)) or using (11.3.16) for some $\alpha > 0$. There is no difference; in fact, putting $A_c(\beta)$ on the other side of the scalar product in (11.3.16), the equality becomes (11.3.12) with the added condition that β must be satisfied on the boundary. If β is a variational boundary condition this additional requirement does not affect the conclusion that $\mu = 0$ on e_0; for a Dirichlet boundary condition the measure μ is required *a priori* to vanish on Γ.

11.4. Parabolic Distributed Parameter Systems: The Minimum Principle with Point Target, I

The system is

$$\begin{aligned} y_t(t, x) &= A_c(\beta) y(t, x) + \phi(t, x, y(t, x), \nabla y(t, x)) + u(t, x), \\ y(0, x) &= \zeta(x), \end{aligned} \tag{11.4.1}$$

a particular case of (11.3.1). The control space U is a closed subset of $L^\infty(\Omega)$ and

$$C_{\text{ad}}(0, T; U) = \{u \in L^\infty_w(0, T; L^\infty(\Omega)) = L^\infty((0, T) \times \Omega); u(t, \cdot) \in U \text{ a.e.}\}. \tag{11.4.2}$$

We need an addendum to Lemma 8.4.8 on the operator

$$f(t, y)(x) = \phi(t, x, y(x), \partial_1 y(x), \dots, \partial_m y(x))$$

$$= \phi(t, x, y(x), \nabla y(x)) \tag{11.4.3}$$

under the following assumptions. (*a*) For t fixed $\phi(t, x, y, \mathbf{y})$ is continuously differentiable with respect to $y \in \mathbb{R}$, $\mathbf{y} \in \mathbb{R}^m$, the derivatives and ϕ locally Lipschitz continuous in $\overline{\Omega} \times \mathbb{R} \times \mathbb{R}^m$. (*b*) For x, y, $\mathbf{y}$ fixed, $\phi(t, x, y, \mathbf{y})$ is measurable.

Lemma 11.4.1. *Let β be a variational boundary condition and*

$$1/2 < \alpha < 1, \quad \delta < \alpha - 1/2. \tag{11.4.4}$$

Then $f(t, (-A_c(\beta))^{-\alpha}\eta) \in D((-A_c(\beta))^\delta)$ and

$$g(t, \eta) = (-A_c(\beta))^\delta f(t, (-A_c(\beta))^{-\alpha}\eta) \tag{11.4.5}$$

is continuously Fréchet differentiable from $C(\overline{\Omega})$ into $C(\overline{\Omega})$, with Fréchet derivative

$$\partial g(t, \eta)z(x)$$
$$= (-A_c(\beta))^\delta \partial_y \phi(t, x, (-A_c(\beta))^{-\alpha}\eta(x), \nabla(-A_c(\beta))^{-\alpha}\eta(x))(-A_c(\beta))^{-\alpha}z(x)$$
$$+ \nabla_{\mathbf{y}}(-A_c(\beta))^\delta \phi(t, x, (-A_c(\beta))^{-\alpha}\eta(x), \nabla(-A_c(\beta))^{-\alpha}\eta(x)) \cdot \nabla(-A_c(\beta))^{-\alpha}z(x)$$
$$\tag{11.4.6}$$

Moreover, $\eta \to g(t, \eta)$ is continuous from $C(\overline{\Omega})$ into $C(\overline{\Omega})$.[1]

The proof is a slight reworking of that of Theorem 8.4.8. It again uses the operator $\Phi(t, y, y_1, \dots, y_m)$ in Theorem 8.4.9 given by

$$\Phi(t, y(\cdot), y_1(\cdot), \dots, y_m(\cdot))(x) = \phi(t, x, y(x), y_1(x), \dots, y_m(x)), \tag{11.4.7}$$

and depends on two auxiliary parameters, $\varepsilon > 0$ and $p > 1$ that duly disappear from the final result. The first parameter is $\varepsilon = (\alpha - \delta)/2 - 1/4$; note that (11.4.4) implies $\alpha - \delta > 1/2$, so $\varepsilon > 0$ as required. Besides, we have

$$2\alpha - 2\varepsilon - 1 = 2\delta + 2\varepsilon, \quad 2\alpha - 2\varepsilon > 1, \quad 2\delta + 2\varepsilon < 1. \tag{11.4.8}$$

---

[1] The unindexed nablas operate on the x-variables. Except for premultiplication by $(-A_c(\beta))^{-\delta}$, formula (11.4.6) is (8.4.29) in vector disguise.

The parameter p is then chosen such that

$$\varepsilon > m/p, \quad 2\delta + 2\varepsilon > m/p. \tag{11.4.9}$$

(of course, the second inequality follows from the first). The inequality $2\alpha - 2\varepsilon > 1$ implies the imbedding

$$D((-A_c(\beta))^\alpha) \hookrightarrow W^{2\alpha - 2\varepsilon, p}(\Omega) \tag{11.4.10}$$

with no restrictions on $p > 1$. (Corollary 8.4.4; we give up the boundary condition). On the other hand, the inequalities $2\delta + 2\varepsilon < 1$ and $\varepsilon > m/p$ give (via Corollary 8.4.7) the imbedding

$$W^{2\delta + 2\varepsilon, p}(\Omega) \hookrightarrow D((-A_c(\beta))^\alpha) \tag{11.4.11}$$

These imbeddings plus the first equality in (11.4.9) suggest looking at the operator Φ (for t fixed) as

$$\Phi : W^{s,p}(\Omega) \times \cdots \times W^{s,p}(\Omega) \to W^{s,p}(\Omega)$$

with $s = 2\alpha - 2\varepsilon - 1 = 2\delta + 2\varepsilon$ $(0 < s < 1)$. It is clearly enough to show

Lemma 11.4.2. *Let*

$$s - m/p > 0. \tag{11.4.12}$$

Then, for t fixed, Φ is a continuous locally bounded operator and has a continuous locally bounded Fréchet derivative with respect to y given by

$$\partial_y \Phi(t, y, y_1, \ldots, y_m)(h, h_1, \ldots, h_m)$$
$$= \partial_0 \phi(t, x, y(x), y_1(x), \ldots, y_m(x))h(x)$$
$$+ \sum_{j=1}^{m} \partial_j \phi(t, x, y(x), y_1(x), \ldots, y_m(x))h_j(x). \tag{11.4.13}$$

Proof. The pattern of the proof emerges when $m = 1$, and we limit ourselves to this case; we may also assume that ϕ is independent of t. We have

$$\phi(y(x) + h(x), y_1(x) + h_1(x)) - \phi(y(x), y_1(x))$$
$$- \partial_0 \phi(y(x), y_1(x))h(x) - \partial_1 \phi(y(x), y_1(x))h_1(x)$$
$$= \int_0^1 \{\partial_0 \phi(y(x) + \tau h(x), y_1(x) + \tau h_1(x)) - \partial_0 \phi(y(x), y_1(x))\}h(x)d\tau$$
$$+ \int_0^1 \{\partial_1 \phi(y(x) + \tau h(x), y_1(x) + \tau h_1(x)) - \partial_1 \phi(y(x), y_1(x))\}h_1(x)d\tau,$$

so that Lemma 11.4.2 is a consequence of the following two results, where $0 < s < 1$ and (11.4.12) holds.

Lemma 11.4.3. *Let* $y, z \in W^{s,p}(\Omega)$. *Then* $yz \in W^{s,p}(\Omega)$ *and*

$$\|yz\|_{W^{s,p}(\Omega)} \leq K \|y\|_{W^{s,p}(\Omega)} \|z\|_{W^{s,p}(\Omega)} \quad (y, z \in W^{s,p}(\Omega)). \tag{11.4.14}$$

Lemma 11.4.4. *Let* $\psi(x, y, y_1)$ *be locally Lipschitz continuous in all variables,*

$$\Psi(y, y_1)(x) = \psi(x, y(x), y_1(x)). \tag{11.4.15}$$

Then Ψ *is a locally bounded, continuous operator from* $W^{s,p}(\Omega) \times W^{s,p}(\Omega)$ *into* $W^{s,p}(\Omega)$.

The proof of both results uses the norm (8.3.2) for $W^{s,p}(\Omega)$:

$$\|u\|_{W^{s,p}(\Omega)}^p = \|u\|_{L^p(\Omega)}^p + I_{s,p}(y)$$

$$= \|u\|_{L^p(\Omega)}^p + \int_{\Omega \times \Omega} \frac{|y(x) - y(\xi)|^p}{|x - \xi|^{m+sp}} dx d\xi \tag{11.4.16}$$

and the imbedding

$$W^{s,p}(\Omega) \hookrightarrow C^{(r)}(\overline{\Omega}) \hookrightarrow C(\overline{\Omega}) \tag{11.4.17}$$

valid if $s - m/p > r > 0$ (Theorem 8.3.4 (b)).

To show Lemma 11.4.3 we note that, due to (11.4.14),

$$|y(x)z(x)|^p \leq \|y\|_{C(\overline{\Omega})} |z(x)|^p \leq C \|y\|_{W^{s,p}(\overline{\Omega})} |z(x)|^p,$$

$$|y(x)z(x) - y(\xi)z(\xi)|^p$$

$$\leq C(|y(x)||z(x) - z(\xi)| + |z(\xi)||y(x) - y(\xi)|)^p$$

$$\leq C \{ \|y\|_{C(\overline{\Omega})}^p |z(x) - z(\xi)|^p + \|z\|_{C(\overline{\Omega})}^p |y(x) - y(\xi)|^p \}$$

$$\leq C' \{ \|y\|_{W^{s,p}(\overline{\Omega})}^p |z(x) - z(\xi)|^p + \|z\|_{W^{s,p}(\overline{\Omega})}^p |y(x) - y(\xi)|^p \}.$$

Computing the norm according to (11.4.16), the result follows.

For the proof of Lemma 11.4.4, note that (11.4.17) implies

$$|\psi(x, y(x), y_1(x))| \leq K(c),$$

$$|\psi(x, y(x), y_1(x)) - \psi(\xi, y(\xi), y_1(\xi))|$$

$$\leq L(c)(|x - \xi| + |y(x) - y(\xi)| + |y_1(x) - y_1(\xi)|)^{(2)}$$

for $\|(y, y_1)\|_{W^{s,p}(\Omega) \times W^{s,p}(\Omega)} \leq c$. Using the imbedding (11.4.17) and computing the norm according to (11.4.16), we obtain that

$$\|\Psi(y, y_1)\|_{W^{s,p}(\Omega)} \leq K(c) \quad (\|(y, y_1)\|_{W^{s,p}(\Omega) \times W^{s,p}(\Omega)} \leq c) \tag{11.4.18}$$

---

[2] Recall that $\overline{\Omega}$ is bounded, thus Lipschitz continuity in x is global.

which proves local boundedness. In the matter of continuity, assume Ψ is not continuous at some $(y, y_1) \in W^{s,p}(\Omega) \times W^{s,p}(\Omega)$. Then there exist sequences $\{h^n\}, \{h_1^n\} \subseteq W^{s,p}(\Omega)$ with $h^n \to 0$, $h_1^n \to 0$ in $W^{s,p}(\Omega)$ and

$$\|\Psi(y + h^n, y_1 + h_1^n) - \Psi(y, y_1)\|_{W^{s,p}(\Omega)} \geq \varepsilon > 0. \tag{11.4.19}$$

We have

$$|\psi(x, y(x) + h^n(x), y_1(x) + h_1^n(x)) - \psi(x, y(x), y_1(x))|$$
$$\leq L(c)(|h^n(x)| + |h_1^n(x)|), \tag{11.4.20}$$

so that

$$\|\Psi(y + h^n, y_1 + h_1^n) - \Psi(y, y_1)\|_{L^p(\Omega)} \to 0. \tag{11.4.21}$$

On the other hand,

$$\left| \{ \psi(x, y(x) + h^n(x), y_1(x) + h_1^n(x)) - \psi(x, y(x), y_1(x)) \} \right.$$
$$\left. - \{ \psi(\xi, y(\xi) + h^n(\xi), y_1(x) + h_1^n(\xi)) - \psi(\xi, y(\xi), y_1(\xi)) \} \right|$$
$$\leq L(c) \big\{ |x - \xi| + |y(x) - y(\xi)| + |y_1(x) - y_1(\xi)|$$
$$+ |h^n(x) - h(\xi)| + |h_1^n(x) - h_1(\xi)| \big\}. \tag{11.4.22}$$

The proof ends taking the pth power of (11.4.22), integrating against $|x - \xi|^{-m-sp}$ in Ω and showing that

$$I_{s,p}\big(\Psi(y + h^n, y_1 + h_1^n) - \Psi(y, y_1) \big) \to 0 \tag{11.4.23}$$

as $n \to \infty$; this relation, combined with (11.4.21), clashes with (11.4.19).

To show (11.4.23) we make use of Vitali's theorem. In view of (11.4.17), $W^{s,p}(\Omega)$-convergence implies convergence in $C(\overline{\Omega})$, so that there is pointwise convergence of the integrand and it is only necessary to check that the integrals are equicontinuous. Given a measurable set $e \subseteq \Omega$ we bound the integral over e as follows: first, we take n_0 so large that $I_{s,p}(h^n - h) + I_{s,p}(h_1^n - h_1) \leq \varepsilon$ for $n \geq n_0$ and bound the integral over e of this part of the integrand by the integral over Ω. The rest of the integral does not depend on n, thus equicontinuity is automatic. This ends the proof. $\blacksquare$

It is clear that Theorem 11.4.1 cannot hold for the Dirichlet boundary condition; in fact, in that case the space is $C_0(\overline{\Omega})$, so that the whole space carries the boundary condition and (11.4.11) is impossible. However, we have

Theorem 11.4.5. *Let all the assumptions in Theorem* 11.4.1 *be satisfied except that β is the Dirichlet boundary condition and*

$$\phi(t, x, 0, y_1, \ldots, y_m) = 0 \quad (x \in \Gamma), \tag{11.4.24}$$

all the other variables being arbitrary. Then the conclusions of Theorem 11.4.1 *hold in the space* $C_0(\overline{\Omega})$.

Proof. It is enough to show that the operator (11.4.7) maps into $W_\beta^{2\delta+2\varepsilon,p}(\Omega)$ and use the imbedding (8.4.17) in Corollary 8.4.7 (*a*). This is achieved as follows. If the functions $y(\cdot), y_1(\cdot), \ldots, y_m(\cdot)$ are Lipschitz continuous in $\overline{\Omega}$ and $y(\cdot)$ satisfies the Dirichlet boundary condition, then the function $\phi(t, x, y(x), y_1(x), \ldots, y_m(x))$ is Lipschitz continuous in $\overline{\Omega}$ and satisfies the Dirichlet boundary condition, thus it satisfies β as a function in $W^{s,p}(\Omega) = W^{2\delta+2\varepsilon,p}(\Omega)$. Since evaluating β is a continuous map in $W^{s,p}(\Omega)$, the operator Φ maps continuously into $W^{s,p}(\Omega)$ and (vectors of) Lipschitz continuous functions are dense in $W^{s,p}(\Omega) \times \cdots \times W^{s,p}(\Omega)$ we are done. ∎

At this point it is convenient to model what is happening in abstract form. We consider an equation

$$y'(t) = Ay(t) + f(t, y(t)) + u(t) \tag{11.4.25}$$

under the hypotheses in **7.7**; A generates an analytic semigroup $S(\cdot)$ satisfying (7.4.15) in a separable, $\odot$-reflexive Banach space E. Controls are functions in $L_w^\infty(0, T; (E^\odot)^*)$, and we require that $f(t, y) \in D((-A)^\delta)$ for some $\delta > 0$ and that $(-A)^\delta f(t, y)$ satisfy Hypothesis $\mathrm{III}_{\alpha,0}$. For the operator $f(t, y)$ in (11.4.3) this has already been verified, except for the statement that $t \to (-A)^\delta f(t, y)$ is strongly measurable. This follows from

Lemma 11.4.6. *Let* $f(\cdot)$ *be a strongly measurable E-valued function defined in* $0 \le t \le T$. *Assume that* $f(t) \in D((-A)^\delta)$ *a.e. Then* $(-A)^\delta f(\cdot)$ *is a strongly measurable E-valued function.*

Proof. Let $y^* \in D(((-A)^{-\delta})^*)$. Then $\langle y^*, (-A)^\delta f(t) \rangle = \langle ((-A)^{-\delta})^* y^*, f(t) \rangle$, so that $\langle y^*, (-A)^\delta f(\cdot) \rangle$ is measurable; since $D(((-A)^{-\delta})^*)$ is dense in $E^\odot$, $\langle y^*, (-A)^\delta f(\cdot) \rangle$ is measurable for $y^* \in E^\odot$. This means that $(-A)^\delta f(\cdot)$ is strongly measurable (Example 11.1.12). ∎

Lemma 11.4.6 is obviously applicable to the operator (11.4.3); the fact that $t \to f(t, y)$ is strongly measurable has been proved in Theorem 8.4.8.

Below, we need to improve a little on Theorem 7.10.4; the patch perturbations there will be chosen in a more refined way, on the basis of the following generalization of Corollary 6.2.5.

Lemma 11.4.7. *Let* $f_1(\cdot), f_2(\cdot), \ldots$ *be a finite number of E-valued locally integrable functions in* $0 \le t < T$, $\{t_n\}$ *a strictly increasing sequence in* $[0, T)$ *such that* $t_0 = 0$, $t_n \to T$, $0 < h \le 1$, $\varepsilon > 0$, $\mathbf{p} = (p_1, \ldots, p_m)$ *a probability* *vector. Then there exist disjoint measurable sets $e_1(h), \ldots, e_m(h) \subseteq [0, T]$ with $|e_j(h)| = p_j h T$ and*

$$\left\| \frac{1}{h} \int_{(t_n, t) \cap e_j(h)} f(\tau) d\tau - p_j \int_{t_n}^{t} f(\tau) d\tau \right\| \leq \varepsilon (t_{n+1} - t_n) \qquad (11.4.26)$$

in $t_n \leq t \leq t_{n+1}$ for $f = f_r, j = 1, \ldots, m$. In particular,

$$\left\| \frac{1}{h} \int_{(0, t) \cap e_j(h)} f(\tau) d\tau - p_j \int_{0}^{t} f(\tau) d\tau \right\| \leq \varepsilon T \qquad (11.4.27)$$

in $0 \leq t < T, f = f_r, j = 1, \ldots, m$.

Proof. Apply Corollary 6.2.5 in each interval $[t_n, t_{n+1}]$ with $\varepsilon(t_{n+1} - t_n)$ on the right side; the resulting sets $e_{nj}(h) \subseteq [t_n, t_{n+1}]$ satisfy $|e_{nj}(h)| = p_j h (t_{n+1} - t_n)$. Then put $e_j(h) = \cup_n e_{nj}(h)$; clearly (11.4.26) holds. Adding up the versions of (11.4.26) from 0 to n, where n is the largest integer with $t_n < t$, (11.4.27) results. ∎

Corollary 11.4.8. *Let $\phi_1(t, \tau), \phi_2(t, \tau), \ldots$ be a finite set of (E, E)-valued functions satisfying the assumptions of Corollary 6.2.6, $f_1(\tau), f_2(\tau) \ldots$ a finite set of E-valued locally integrable functions in $0 \leq t < T, \{t_n\}$ a strictly increasing sequence in $[0, T)$ with $t_0 = 0, t_n \to T, 0 < h \leq 1, \varepsilon > 0$, $\mathbf{p}$ a probability vector. Then there exist disjoint measurable sets $e_1(h), \ldots e_m(h) \subseteq [0, T]$ with $|e_j(h)| = p_j h T$ and such that (6.2.13) holds, that is*

$$\left\| \frac{1}{h} \int_{(0, t) \cap e_j(h)} \phi(t, \tau) d\tau - p_j \int_{0}^{t} \phi(t, \tau) d\tau \right\| \leq \varepsilon$$

$$(0 \leq t \leq T, j = 1, \ldots, m), \qquad (11.4.28)$$

for each $\phi(t, \tau) = \phi_l(t, \tau)$ and such that (11.4.26) and (11.4.27) hold for each $f(\tau) = f_r(\tau), j = 1, \ldots, m$.

Proof. Exactly like that of Theorem 6.2.2; one brings Lemma 11.4.7 to bear on the functions $\phi(t_k, \tau)$ and $f_r(\tau)$. Since the $\phi(t_k, \cdot)$ are integrable, (6.2.13) for $\phi(t_k, \cdot)$ is actually valid in $0 \leq t \leq T$. ∎

The result will be applied in an interval $0 \leq t \leq \bar{t}$ to the functions

$$\phi(t, \tau) = (-A)^{\alpha} S(t - \tau)(v_j(\tau) - \bar{u}(\tau)), \qquad (11.4.29)$$

$$f_j(\tau) = A S(\bar{t} - \tau)(v_j(\tau) - \bar{u}(\tau)), \qquad (11.4.30)$$

for $j = 1, \ldots, m$, where $\bar{u}(\cdot) \in L^\infty_w(0, \bar{t}; (E^\odot)^*)$ is such that the trajectory $y(t, \bar{u})$

exists in $0 \le t \le \bar{t}$ and $\mathbf{v}(\cdot) = (v_1(\cdot), \dots, v_m(\cdot))$ is the function vector associated with a patch perturbation of $\bar{u}(\cdot)$; the corresponding probability vector is $\mathbf{p}$.

The sets $e_j(h)$ chosen as above, the following upgrade of Theorem 7.10.4 ensues.

Theorem 11.4.9. *Let $y(\bar{t}, \bar{u}) \in E_1 = D(A)$. Then, if $\mathbf{v}$ is such that $\xi(t, \bar{u}, \mathbf{p}, \mathbf{v}) \in E_1$, we have $y(\bar{t}, \bar{u}_{\mathbf{e}(h), \mathbf{v}}) \in E_1$ and*

$$\xi(\bar{t}, \bar{u}, \mathbf{p}, \mathbf{v}) = \lim_{h \to 0+} \frac{1}{h} \left(y(\bar{t}, \bar{u}_{\mathbf{e}(h), \mathbf{v}}) - y(\bar{t}, \bar{u}) \right) \tag{11.4.31}$$

in the norm of E_1.

Proof. Under the assumptions for (11.4.25) we have

$$(-A) y(t, \bar{u}) = \int_0^t (-A)^{\alpha - \delta} S(t - \tau)(-A)^\delta f(\tau, y(\tau, \bar{u})) d\tau$$

$$+ \int_0^t (-A)^\alpha S(t - \tau) \bar{u}(\tau) d\tau \tag{11.4.32}$$

with a corresponding equation for $\bar{u}_{\mathbf{e}(h), \mathbf{v}}(\cdot)$ and $y(t, \bar{u}_{\mathbf{e}, (h), \mathbf{v}})$. Let $D(A, \bar{t})$ be the set of all $w(\cdot) \in L^\infty(0, \bar{t}; (E^{\odot})^*)$ such that

$$\int_0^{\bar{t}} S(\bar{t} - \tau) w(\tau) d\tau \in D(A) = E_1. \tag{11.4.33}$$

Pulling $(-A)^\alpha$ out of the second integral in (11.4.32) and then applying $(-A)^{1-\alpha}$ to both sides it is clear that

$$y(\bar{t}, \bar{u}) \in D(A) \quad \text{if and only if } \bar{u}(\cdot) \in D(A, \bar{t}) \tag{11.4.34}$$

(operators going in and out of integrals under the eye of Lemma 5.0.17). The same statement holds for $\bar{u}_{\mathbf{e}(h), \mathbf{v}}(\cdot)$; $y(\bar{t}, u_{\mathbf{e}(h), \mathbf{v}}) \in D(A)$ if and only if $\bar{u}_{\mathbf{e}(h), \mathbf{v}}(\cdot) \in D(A, \bar{t})$. On the other hand,

$$(-A)^\alpha \xi(t, \bar{u}, \mathbf{p}, \mathbf{v})$$

$$= \int_0^t (-A)^{\alpha - \delta} S(t - \tau)(-A)^\delta \partial_y f(\tau, y(\tau, \bar{u})) \xi(\tau, \bar{u}, \mathbf{p}, \mathbf{v}) d\tau$$

$$+ \int_0^t (-A)^\alpha S(t - \tau) \left(\sum_{j=1}^m p_j v_j(\tau) - \bar{u}(\tau) \right) d\tau, \tag{11.4.35}$$

thus, arguing like after (11.4.32) we deduce that

$$\xi(\bar{t}, \bar{u}, \mathbf{p}, \mathbf{v}) \in D(A) \quad \text{if and only if } \Sigma p_j v_j(\cdot) - \bar{u}(\cdot) \in D(A, \bar{t}). \tag{11.4.36}$$

If we assume that $y(\bar{t}, \bar{u}) \in D(A)$ then $\bar{u}(\cdot) \in D(A, \bar{t})$, thus $\xi(\bar{t}, \bar{u}, \mathbf{p}, \mathbf{v}) \in D(A)$ if and only if $\Sigma p_j v_j(\cdot) \in D(A, \bar{t})$.

We now take $\varepsilon = h$ in (11.4.27) and (11.4.28) and $R_k = kR(k; A) = k(kI - A)^{-1}$ (k large enough). We obtain from (11.4.32) and (11.4.35) that

$$\frac{1}{h}\big((-A)R_k y(t, \bar{u}_{\mathbf{e}(h),\mathbf{v}}) - (-A)R_k y(t, \bar{u})\big) - (-A)R_k \xi(t, \bar{u}, \mathbf{p}, \mathbf{v})$$

$$= R_k \int_0^t (-A)^{1-\delta} S(t - \tau)$$

$$\times (-A)^{\delta}\left\{\frac{1}{h}\big(f(\tau, y(\tau, \bar{u}_{\mathbf{e}(h),\mathbf{v}})) - f(\tau, y(\tau, \bar{u}))\big) - \partial_y f(\tau, y(\tau, u))\xi(\tau, \bar{u}, \mathbf{p}, \mathbf{v})\right\}d\tau$$

$$+(-A)\int_0^t S(t - \tau)R_k\left\{\frac{1}{h}(\bar{u}_{\mathbf{e}(h),\mathbf{v}}(\tau) - \bar{u}(\tau)) - (\Sigma p_j v_j(\tau) - \bar{u}(\tau))\right\}d\tau.$$

$$(11.4.37)$$

Since, for each k the integrand of the last integral belongs to $L^1(0, T; E)$, we can set $t = \bar{t}$ in (11.4.37) and write the integral as a sum of the integrals in $[t_n, t_{n+1}]$. Now, we have

$$\int_0^{\bar{t}} S(\bar{t} - \tau)\left\{\frac{1}{h}(\bar{u}_{\mathbf{e}(h),\mathbf{v}}(\tau) - \bar{u}(\tau)) - (\Sigma p_j v_j(\tau) - \bar{u}(\tau))\right\}d\tau$$

$$= \sum_{n=1}^{\infty} \int_{t_n}^{t_{n+1}} S(\bar{t} - \tau)\left\{\frac{1}{h}(\bar{u}_{\mathbf{e}(h),\mathbf{v}}(\tau) - \bar{u}(\tau)) - (\Sigma p_j v_j(\tau) - \bar{u}(\tau))\right\}d\tau$$

$$(11.4.38)$$

with

$$A\int_{t_n}^{t_{n+1}} S(\bar{t} - \tau)\left\{\frac{1}{h}(\bar{u}_{\mathbf{e}(h),\mathbf{v}}(\tau) - \bar{u}(\tau)) - (\Sigma p_j v_j(\tau) - \bar{u}(\tau))\right\}d\tau$$

$$= \sum_{j=1}^{m}\left\{\frac{1}{h}\int_{(t_n,t_{n+1})\cap e_j(h)} AS(\bar{t} - \tau)(v_j(\tau) - \bar{u}(\tau))\right.$$

$$\left. - p_j \int_{t_n}^{t_{n+1}} AS(\bar{t} - \tau)(v_j(\tau) - \bar{u}(\tau))d\tau\right\}.$$

$$(11.4.39)$$

The series on the right converges with sum $\leq h$ in norm; in fact, (11.4.26), with $\varepsilon = h$, has been prearranged for all functions (11.4.30).

The proof of Theorem 11.4.9 now ends as follows. (i) We write formula (11.4.37) for $t = \bar{t}$, the operator R_k taking care of summability of all integrands. (ii) We write the last integral as a sum of integrals in $[t_n, t_{n+1}]$. Combining with

(11.4.38) and (11.4.39) we obtain

$$
\left\| A \int_0^{\bar{t}} S(t-\tau)R_k \left\{ \frac{1}{h}(\bar{u}_{e(h),v}(\tau) - \bar{u}(\tau)) - (\Sigma p_j v_j(\tau) - \bar{u}(\tau)) \right\} d\tau \right\|
$$

$$
\leq \sum_{n=1}^{\infty} \|R_k\| \sum_{j=1}^{m} \left\{ \frac{1}{h} \int_{(t_n,t_{n+1}) \cap e_j(h)} AS(\bar{t}-\tau)(v_j(\tau) - \bar{u}(\tau)) d\tau \right.
$$

$$
\left. - p_j \int_{t_n}^{t_{n+1}} AS(\bar{t}-\tau)(v_j(\tau) - \bar{u}(\tau)) d\tau \right\} \right\| \leq Ch
$$

with C a bound for all the $\|R_k\|$. (*iii*) We bound the first integral on the right-hand side of (11.4.37) using Hypothesis $\text{III}_{\alpha,0}$ for $(-A)^\delta f(t, y)$ and Theorem 7.10.4; note that only E_α-convergence is needed, and also that R_k does not intervene in the estimate and just contributes the constant C above to the final bound. (*iv*) We let $k \to \infty$ and use the fact that $y(\bar{t}, \bar{u})$, $y(\bar{t}, \bar{u}_{e(h),v})$, $\xi(\bar{t}, \bar{u}, \mathbf{p}, \mathbf{v}) \in D(A)$. ∎

11.5. Parabolic Distributed Parameter Systems: The Minimum Principle with Point Target, II

We work with the abstract model (11.4.25) under the assumptions in **11.4**; U is a bounded subset of $(E^\odot)^*$, and the admissible control space $C_{ad}(0, \bar{t}; U)$ is

$$
C_{ad}(0, T; U) = \left\{ u(\cdot) \in L_w^\infty(0, T; (E^\odot)^*); u(t) \in U \text{ a.e. in } 0 \leq t \leq T \right\} \quad (11.5.1)
$$

(which is patch complete and saturated). We also assume that U is bounded and that zero is an interior point of U in $(E^\odot)^*$. The minimum principle will be obtained applying the theory in Chapter 7 to the functions $\mathbf{f} : V \to C(0, \bar{t}; E_\alpha) \times E_1$, $\mathbf{f}_0 : V \to \mathbb{R}$ defined by

$$
\mathbf{f}(u) = (y(\cdot, u), y(\bar{t}, u)), \qquad \mathbf{f}_0(u) = y_0(\bar{t}, u) \quad (11.5.2)
$$

where $y_0(t, u)$ is a cost functional. The metric space V is the subset $B_1(\bar{u}, \delta)$ of the ball $B(\bar{u}, \delta)$ in Lemma 7.10.1 consisting of all $u \in B(\bar{u}, \delta)$ with $y(\bar{t}, \bar{u}) \in E_1$; $B_1(\bar{u}, \delta)$ is outfitted with the distance

$$
\theta(u(\cdot), v(\cdot)) = d(u(\cdot), v(\cdot)) + \|y(\bar{t}, u) - y(\bar{t}, v)\|_{E_1} \quad (11.5.3)
$$

where $d(u, v) = |\{t \in [0, T]; u(t) \neq v(t)\}|$ is the usual distance in $C_{ad}(0, \bar{t}; U)$. A simple argument using Lemma 7.10.1 and closedness of A shows that $B_1(\bar{u}, \delta)$ is complete. Moreover, $\mathbf{f}$ is obviously continuous in $B_1(\bar{u}, \delta)$ (the second coordinate is Lipschitz continuous). Let $(B_1(\bar{u}, \delta), d)$ (resp. $B_1(\bar{u}, \delta), \theta)$) indicate $B_1(\bar{u}, \delta)$ is equipped with d (resp. θ). Since $d \leq \theta$, a directional derivative $(\xi(\cdot), \xi(\bar{t}))$ of $\mathbf{f}$ in $(B_1(\bar{u}, \delta), \theta)$ is also a directional derivative of $\mathbf{f}$ in $(B_1(\bar{u}, \delta), d)$. Going the other way, assume that $(\xi(\cdot), \xi(\bar{t})) \in C(0, \bar{t}; E_\alpha) \times E_1$ is a directional derivative of $\mathbf{f}$ at $\tilde{u}$ in $(B_1(\bar{u}, \delta), d)$, and let $\tilde{u}(h)$ be the function used in the definition (3.3.7); we have $d(\tilde{u}(h), \bar{u}) \le h$. If $\xi(\bar{t}) = \lim h^{-1}(y(\bar{t}, \tilde{u}(h)) - y(\bar{t}, \tilde{u}))$ in the norm of E_1 then $\|y(\bar{t}, \tilde{u}(h)) - y(\bar{t}, \tilde{u})\|_{E_1} \le (\|\xi(\bar{t})\|_{E_1} + o(h))h$. Accordingly, if $c > 0$,

$$\theta(\tilde{u}(h/c), \tilde{u}) = d(\tilde{u}(h/c), \tilde{u}) + \|y(\bar{t}, \tilde{u}(h/c)) - y(\bar{t}, \tilde{u})\|_{E_1}$$
$$\le \frac{h}{c}(1 + \|\xi(\bar{t})\|_{E_1} + o(h))$$

which inequality motivates the following result:

Lemma 11.5.1. *Let* $(\xi_0, \xi(\cdot), \xi(\bar{t})) \in \mathbb{R} \times C(0, \bar{t}; E_\alpha) \times E_1$ *be a directional derivative of the function* $(\mathbf{f}_0, \mathbf{f})$ *at* $\tilde{u} \in B_1(\bar{u}, \delta)$ *in the space* $(B_1(\bar{u}, \delta), d)$. *If* $\xi(\bar{t}) = \lim h^{-1}(y(\bar{t}, \tilde{u}(h)) - y(\bar{t}, \tilde{u}))$ *in the norm of* E_1 *and*

$$c > 1 + \|\xi(\bar{t})\|_{E_1} \tag{11.5.4}$$

then $c^{-1}(\xi_0, \xi(\cdot), \xi(\bar{t}))$ *is a directional derivative of* $(\mathbf{f}_0, \mathbf{f})$ *at* $\tilde{u}$ *in* $(B_1(\bar{u}, \delta), \theta)$.

The proof follows from the comments above and the equality

$$\frac{(\mathbf{f}_0, \mathbf{f})(\tilde{u}(h/c)) - (\mathbf{f}_0, \mathbf{f})(\tilde{u})}{h} = \frac{1}{c}\frac{(\mathbf{f}_0, \mathbf{f})(\tilde{u}(h/c)) - (\mathbf{f}_0, \mathbf{f})(\tilde{u})}{(h/c)}. \qquad \blacksquare$$

Combining the notations in **9.1** and **11.1**, we denote by $((-A)^* E^*)^1(I^*)$ or simply $((-A)^* E^*)^1$ the subspace of all $z \in (-A)^* E^*$ such that

$$\int_0^1 \|S(t)^* z\|_{E^\odot} dt = \int_0^1 \|(-A^\odot)S(t)^*((-A)^{-1})^* z\|_{E^\odot} dt < \infty \tag{11.5.5}$$

(recall that $((-A)^{-1})^* z \in E^*$).

In the minimum principle below, we assume, as in **11.4**, that $f(t, y) \in D((-A)^\delta)$ for $y \in D((-A)^\alpha)$ and that $(-A)^\delta f(t, y)$ satisfies Hypothesis $\mathrm{III}_{\alpha,0}$. The cost functional is

$$y_0(t, u) = \int_0^t f_0(\tau, y(\tau, u), u(\tau)) d\tau \tag{11.5.6}$$

where $f_0(t, y, u(t))$ satisfies Hypothesis III_α^0 in **11.2** for all $u(\cdot) \in C_{\mathrm{ad}}(0, T; U)$, with $K_0(\cdot), L_0(\cdot)$ independent of $u(\cdot)$. The state constraint and target condition are

$$y(t) \in M(t) \subseteq E_\alpha \quad (0 \le t \le \bar{t}), \qquad y(\bar{t}) \in Y \subseteq E_1, \tag{11.5.7}$$

with each $M(t)$ closed in E_α and Y closed in $D(A)$.

Theorem 11.5.2. *Let* $\bar{u}(\cdot) \in C_{\mathrm{ad}}(0, \bar{t}; U)$ *be an optimal control with* $y(\bar{t}, \bar{u}) \in D(A)$. *Then there exists* $(z_0, \mu, z) \in \mathbb{R} \times \Sigma(0, T; ((-A)^\alpha)^* E^*)) \times ((-A)^* E^*)^1$ *with*

$$z_0 \geq 0, \quad \mu \in \left(\liminf_{n \to \infty} T_{\mathbf{M}(\bar{t})}(\tilde{y}^n(\cdot)) \right)^-, \quad z \in \left(\liminf_{n \to \infty} T_Y(\tilde{y}^n) \right)^-, \tag{11.5.8}$$

where $\tilde{y}^n(\cdot)$ *(resp.* $\tilde{y}^n$*) is a sequence in* $\mathbf{M}(\bar{t})$ *(resp.* Y*) such that* $\tilde{y}^n(\cdot) \to y(\cdot, \bar{u})$ *in* $C(0, \bar{t}; E_\alpha)$ *(resp.* $\tilde{y}^n \to y(\bar{t}, \bar{u})$ *in* E_1*) and such that, if* $\bar{z}(s)$ *is the solution of*

$$d\bar{z}(s) = -\{A^\odot + \partial_y f(s, y(s, \bar{u}))^*\}\bar{z}(s)ds$$
$$- z_0 \partial_y f_0(s, y(s, \bar{u}), \bar{u}(s))ds - \mu(ds), \qquad \bar{z}(\bar{t}) = z, \tag{11.5.9}$$

then

$$z_0 \int_0^{\bar{t}} \{f_0(s, y(s, \bar{u}), v(s)) - f_0(s, y(s, \bar{u}), \bar{u}(s))\}ds$$
$$+ \int_0^{\bar{t}} \langle \bar{z}(s), v(s) - \bar{u}(s) \rangle ds \geq 0 \tag{11.5.10}$$

for all $v(\cdot) \in C_{\mathrm{ad}}(0, \bar{t}; U)$.

Proof. Equation (11.5.9) (specifically, the "final condition" $\bar{z}(\bar{t}) = z$) needs explanation, since so far we have only taken $z \in ((-A)^\alpha)^* E^*$ with $\alpha < 1$. The definition of $\bar{z}(s)$ is of course (11.1.12) with $\rho = 0$ (so that $\bar{z}(s) = v(s)$) but the first term needs modification. On the basis of the formal equality $((-A)^{-\alpha})^* z = ((-A)^{1-\alpha})^*((-A)^{-1})^* z$ we may try to palm off $R_{0,\alpha}^*(s, \bar{t})((-A)^{1-\alpha})^*((-A)^{-1})^* z$ as a definition, but this causes problems since the only fact known about z is that $((-A)^{-1})^* z \in E^*$. Instead, we define

$$\bar{z}(s) = ((-A)^{1-\alpha} R_{\alpha,0}(\bar{t}, s))^*((-A)^{-1})^* z + \int_s^{\bar{t}} R_{0,\alpha}^*(s, \sigma)((-A)^{-\alpha})^* \mu(d\sigma). \tag{11.5.11}$$

Still, since $1 - \alpha > 0$ we need to clarify what we mean by $(-A)^{1-\alpha} R_{\alpha,0}(\bar{t}, s)$. Moreover, we must show that the first term of (11.5.11) takes values in $E^\odot$ and that it belongs to $L_1(0, \bar{t}; E^\odot)$ lest the second term of (11.5.10) should lose all sense. All of this will be taken care of below.

The proof of Theorem 11.5.2 follows closely that of Theorem 11.2.2. We apply the nonlinear programming theory to the functions (11.5.2) in the metric space $(B_1(\bar{u}, \delta), \theta)$. The target set is

$$\mathbf{Y} = \mathbf{M}(\bar{t}) \times Y \subseteq C(0, \bar{t}; E_\alpha) \times E_1. \tag{11.5.12}$$

The assumptions of Theorem 7.2.5 are satisfied and we obtain in this way the sequences $\{\tilde{u}^n\} \subseteq B_1(\bar{u}, \delta)$ and $\{\tilde{y}^n(\cdot), \tilde{y}^n\} \subseteq \mathbf{M}(\bar{t}) \times Y$. We select $\rho > 0$ and the sequence $\{\mathbf{D}_n\}$, $\mathbf{D}_n \subseteq \mathrm{Der}(\mathbf{f}_0, \mathbf{f})(\tilde{u}^n)$ of convex sets consisting of all triples

$$\frac{1}{\bar{t}(\|\xi(\bar{t}, \tilde{u}^n, \mathbf{p}, \mathbf{v})\|_{E_1} + 2)}(\xi_0(\bar{t}, \tilde{u}^n, \mathbf{p}, \mathbf{v}), \xi(\cdot, \tilde{u}^n, \mathbf{p}, \mathbf{v}), \xi(\bar{t}, \tilde{u}^n, \mathbf{p}, \mathbf{v})) \quad (11.5.13)$$

where $\mathbf{v}$ is such that $\xi(\bar{t}, \tilde{u}^n, \mathbf{p}, \mathbf{v}) \in E_1 = D(A)$. Note that the normalization factor to insure that the vector is a directional derivative in the space $(B_1(\bar{u}, \delta), d)$ is just $\bar{t}^{-1}$; the rest of the factor comes from Lemma 11.5.1 (inequality (11.5.4)) as we and taking directional derivatives in $(B_1(\bar{u}, \delta), \theta)$.

Since $\tilde{u}^n(\cdot)$ belongs to $B_1(\bar{u}, \delta)$, $y(\bar{t}, \tilde{u}^n) \in D(A)$. Hence, (11.4.34) implies that $\tilde{u}^n(\cdot) \in D(A, \bar{t})$. Combining this observation with a look at (11.4.36) (applied to $\tilde{u}^n(\cdot)$) we conclude that

$$\xi(\bar{t}, \tilde{u}^n, \mathbf{p}, \mathbf{v}) \in D(A) \quad \text{if and only if } \Sigma p_j v_j(\cdot) \in D(A, \bar{t}). \qquad (11.5.14)$$

Application of Theorem 7.5.2 requires identification of elements of $\liminf \mathbf{D}_n$, that is, computation of limits in $\mathbb{R} \times C(0, \bar{t}; E_\alpha) \times E_1$ of elements of the form (11.5.11). To do this we note that convergence in $R \times C(0, \bar{t}; E_\alpha) \times E_1$ implies convergence of the factor in (11.5.13), hence the factor can be ignored for the moment. ∎

To take limits, we use

Lemma 11.5.3. *Let* $\tilde{u}^n(\cdot) \to \bar{u}(\cdot)$ *in* $(B_1(\bar{u}, \delta), \theta)$, $\Sigma p_j v_j(\cdot) \in D(A, \bar{t})$. *Then*

$$(\xi_0(\bar{t}, \tilde{u}^n, \mathbf{p}, \mathbf{v}), \xi(\cdot, \tilde{u}^n, \mathbf{p}, \mathbf{v}), \xi(\bar{t}, \tilde{u}^n, \mathbf{p}, \mathbf{v}))$$

$$\to (\xi_0(\bar{t}, \bar{u}, \mathbf{p}, \mathbf{v}), \xi(\cdot, \bar{u}, \mathbf{p}, \mathbf{v}), \xi(\bar{t}, \bar{u}, \mathbf{p}, \mathbf{v})) \qquad (11.5.15)$$

in $\mathbb{R} \times C(0, \bar{t}; E_\alpha) \in E_1$.

Proof. The statement on convergence of the first two coordinates was covered in **11.2**, thus we only have to show E_1-convergence of the last coordinate. To do this, note that, by (11.4.32),

$$(-A)y(\bar{t}, \tilde{u}^n) = \int_0^{\bar{t}} (-A)^{1-\delta} S(\bar{t} - \tau)(-A)^\delta f(\tau, y(\tau, \tilde{u}^n))d\tau$$

$$+ (-A) \int_0^{\bar{t}} S(\bar{t} - \tau)\tilde{u}^n(\tau)d\tau$$

with a corresponding equation for $y(\bar{t}, \bar{u})$. Since $y(\bar{t}, \tilde{u}^n) \to y(\bar{t}, \bar{u})$ in E_1, we obtain using Lemma 7.10.1 that

$$A \int_0^{\bar{t}} S(\bar{t} - \tau)\tilde{u}^n(\tau)d\tau \to A \int_0^{\bar{t}} S(\bar{t} - \tau)\bar{u}(\tau)d\tau. \qquad (11.5.16)$$

With this information in hand, we go back to the integral equation (11.4.35) for $\xi(\bar{t}, \tilde{u}^n, \mathbf{p}, \mathbf{v})$, apply $(-A)^{1-\alpha}$ to both sides and take limits.   ∎

End of Proof of Theorem 11.5.2. Having selected $\{\mathbf{D}_n\}$, Theorem 7.2.5 gives sequences $\{z_{0n}\} \subseteq \mathbb{R}, \{\mu_n\} \subseteq \Sigma(0, T; ((-A)^\alpha)^* E^*)$ and $\{z_n\} \subseteq ((-A)^* E^*)^1$ such that

$$z_{0n} \geq 0, \qquad z_{0n}^2 + \|\mu_n\|_{\Sigma(0,\bar{t};((-A)^\alpha)^* E^*)}^2 + \|z_n\|_{(-A)^* E^*}^2 = 1,$$

$$z_{0n}\xi_0(\bar{t}, \tilde{u}^n, \mathbf{p}, \mathbf{v}) + \langle \mu, \xi(\cdot, \tilde{u}^n, \mathbf{p}, \mathbf{v}) \rangle_{\alpha,c} + \langle z, \xi(\bar{t}, \tilde{u}^n, \mathbf{p}, \mathbf{v}) \rangle_1 \qquad (11.5.17)$$
$$\geq \tilde{\delta}_n = \delta_n \bar{t} (\|\xi(\bar{t}, \tilde{u}^n, \mathbf{p}, \mathbf{v})\|_{E_1} + 2).$$

In this inequality, the first angled bracket $\langle \cdot, \cdot \rangle_{\alpha,c}$ indicates the duality of $C(0, \bar{t}; E_\alpha)$ and $\Sigma(0, \bar{t}; ((-A)^\alpha)^* E^*)$, the second angled bracket $\langle \cdot, \cdot \rangle_1$ the duality of E_1 and $(-A)^* E^*$ and $\{\delta_n\}$ is the sequence produced by Theorem 7.2.5. The conclusion of Lemma 11.5.3 includes E_1-convergence of $\xi(\bar{t}, \tilde{u}^n, \mathbf{p}, \mathbf{v})$, thus $\tilde{\delta}_n \to 0$ also.

Assume that (z_0, μ, z) is a $(\mathbb{R} \times C(0, \bar{t}; E_\alpha) \times E_1)$-weak limit point of the sequence (z_{0n}, μ_n, z_n). Taking limits in (11.5.17) via Lemma 11.5.3 we obtain

$$z_0\xi_0(\bar{t}, \tilde{u}^n, \mathbf{p}, \mathbf{v}) + \langle \mu, \xi(\cdot, \tilde{u}^n, \mathbf{p}, \mathbf{v}) \rangle_{\alpha,c} + \langle z, \xi(\bar{t}, \tilde{u}^n, \mathbf{p}, \mathbf{v}) \rangle_1 \geq 0 \qquad (11.5.18)$$

valid whenever $\Sigma p_j v_j(\cdot) \in D(A, \bar{t})$ (see 11.5.14). For single patches with $v(\cdot) \in D(A, \bar{t})$ the inequality yields

$$z_0\xi_0(\bar{t}, \bar{u}, v) + \langle \mu, \xi(\cdot, \bar{u}, v) \rangle_{\alpha,c} + \langle z, \xi(\bar{t}, \bar{u}, v) \rangle_1 \geq 0 \qquad (11.5.19)$$

which is

$$z_0 \int_0^{\bar{t}} \{f_0(\sigma, y(\sigma, \bar{u}), v(\sigma)) - f_0(\sigma, y(\sigma, \bar{u}), \bar{u}(\sigma))\} d\sigma$$

$$+ z_0 \int_0^{\bar{t}} \Big\langle ((-A)^{-\alpha})^* \partial_y f_0(\tau, y(\tau, \bar{u}), \bar{u}(\tau)),$$

$$\int_0^\tau R_{\alpha,0}(\tau, \sigma, \bar{u})\{f(\sigma, y(\sigma, \bar{u}), v(\sigma)) - f(\sigma, y(\sigma, \bar{u}), \bar{u}(\sigma))\} d\sigma \Big\rangle d\tau$$

$$+ \int_0^{\bar{t}} \Big\langle ((-A)^{-\alpha})^* \mu(dt),$$

$$\int_0^t R_{\alpha,0}(t, \sigma; \bar{u})\{f(\sigma, y(\sigma, \bar{u}), v(\sigma)) - f(\sigma, y(\sigma, \bar{u}), \bar{u}(\sigma))\} d\sigma \Big\rangle$$

$$+ \Big\langle ((-A)^{-1})^* z,$$

$$(-A)^{1-\alpha} \int_0^{\bar{t}} R_{\alpha,0}(\bar{t}, \sigma; \bar{u})\{f(\sigma, y(\sigma, \bar{u}), v(\sigma)) - f(\sigma, y(\sigma, \bar{u}), \bar{u}(\sigma))\} d\sigma \Big\rangle \geq 0,$$

$$(11.5.20)$$

the companion of (12.2.11) for $\rho = 0$; the last angled bracket indicates the duality of E_1 and $(E_1)^*$. We now use the fact that the nonlinearity has the form

$$f(t, y, u) = f(t, y) + u, \tag{11.5.21}$$

so that

$$f(\sigma, y(\sigma, \bar{u}), v(\sigma)) - f(\sigma, y(\sigma, \bar{u}), \bar{u}(\sigma)) = v(\sigma) - \bar{u}(\sigma), \tag{11.5.22}$$

and we obtain

$$z_0 \int_0^{\bar{t}} \{f_0(\sigma, y(\sigma, \bar{u}), v(\sigma)) - f_0(\sigma, y(\sigma, \bar{u}), \bar{u}(\sigma))\}d\sigma$$

$$+ z_0 \int_0^{\bar{t}} \left\langle ((-A)^{-\alpha})^* \partial_y f_0(\tau, y(\tau, \bar{u}), \bar{u}(\tau)), \int_0^{\tau} R_{\alpha,0}(\tau, \sigma, \bar{u})(v(\sigma) - \bar{u}(\sigma))d\sigma \right\rangle d$$

$$+ \int_0^{\bar{t}} \left\langle ((-A)^{-\alpha})^* \mu(dt), \int_0^t R_{\alpha,0}(t, \sigma; \bar{u})(v(\sigma) - \bar{u}(\sigma))d\sigma \right\rangle$$

$$+ \left\langle ((-A)^{-1})^* z, (-A)^{1-\alpha} \int_0^{\bar{t}} R_{\alpha,0}(\bar{t}, \sigma; \bar{u})(v(\sigma) - \bar{u}(\sigma))d\sigma \right\rangle \geq 0. \tag{11.5.23}$$

To obtain the minimum principle we need to pull operators to the left side of all duality products, which we do as in Theorem 11.2.2 in all terms except the last. The element on the right side of the duality product is

$$(-A)^{1-\alpha} \int_0^{\bar{t}} R_{\alpha,0}(\bar{t}, \sigma; \bar{u})(v(\sigma) - \bar{u}(\sigma))d\sigma. \tag{11.5.24}$$

Bringing into play the integral equation (11.2.30) for $R_{\alpha,0}(t, s; \bar{u})$ we obtain

$$(-A)^{1-\alpha} R_{\alpha,0}(t, s; \bar{u})\zeta = (-A)S(t - s)\zeta$$

$$+ \int_s^t (-A)^{1-\delta} S(t - \tau)(-A)^{\delta} \partial_y f(\tau, y(\tau, \bar{u}))(-A)^{-\alpha} R_{\alpha,0}(\tau, s; \bar{u})\zeta d\tau$$

$$= (-A)S(t - s)\zeta + N(t, s; \bar{u})\zeta, \tag{11.5.25}$$

with $N(t, s; \bar{u})$ continuous in the norm of (E, E) for $s < t$ and satisfying

$$\|N(t, s; \bar{u})\| \leq C(t - s)^{-(\alpha-\delta)} \quad (t > s). \tag{11.5.26}$$

We turn back our attention to inequality (11.5.23). The first three terms on the left are obviously bounded for $v(\cdot) \in C_{ad}(0, \bar{t}; U)$ without any need to assume that $v(\cdot) \in C_{ad}(0, \bar{t}; U)$ belongs to $D(A, \bar{t})$. Consequently, shifting all terms except the last to the right side, the inequality implies that if $v(\cdot) \in D(A, \bar{t})$ then the expression

$$\left\langle ((-A)^{-1})^* z, (-A)^{1-\alpha} \int_0^{\bar{t}} R_{\alpha,0}(\bar{t}, \sigma; \bar{u})v(\sigma)d\sigma \right\rangle \tag{11.5.27}$$

is bounded below independently of $v(\cdot)$. By hypothesis, the control set U contains a ball $B(0, r)$ in $(E^{\odot})^*$, so that we may take $v(\cdot) \in C_{\mathrm{ad}}(0, \bar{t}; B(0, r)) \cap D(A, \bar{t})$ in (11.5.27) and the fact that we now can multiply $v(\cdot)$ by -1 upgrades "bounded below" to "bounded."

The proof ends this way. For $\varepsilon > 0$ let $C_{\mathrm{ad}, \varepsilon}(0, \bar{t}; B(0, r))$ be the subspace of $C_{\mathrm{ad}}(0, \bar{t}; B(0, r))$ consisting of all $v(\cdot)$ satisfying $v(s) = 0$ in $[\bar{t} - \varepsilon, \bar{t}]$. It is plain that $C_{\mathrm{ad}, \varepsilon}(0, \bar{t}; B(0, r)) \subseteq C_{\mathrm{ad}}(0, \bar{t}; B(0, r)) \cap D(A, \bar{t})$. Accordingly, if $v(\cdot) \in C_{\mathrm{ad}, \varepsilon}(0, \bar{t}; B(0, r))$, (11.5.25) and Lemma 5.0.17 sponsor pushing the operator $(-A)^{1-\alpha}$ inside of the integral in (11.5.27). We obtain that

$$\int_0^{\bar{t}-\varepsilon} \left\langle (-A)^* S(t-\tau)^*((-A)^{-1})^* z, v(\sigma) \right\rangle d\sigma$$

$$= \left\langle ((-A)^{-1})^* z, \int_0^{\bar{t}-\varepsilon} (-A) S(\bar{t}-\tau) v(\sigma) d\sigma \right\rangle$$

is bounded independently of $v(\cdot)$ and $\varepsilon > 0$. As in Theorem 9.1.10, our freedom in the choice of $v(\cdot)$ and ε procures

$$\int_0^{\bar{t}} \|(-A^{\odot}) S(\bar{t}-\sigma)^*((-A)^{-1})^* z\|_{E^{\odot}} d\sigma < \infty, \tag{11.5.28}$$

and this means that $z \in ((-A)^* E^*)^1$ as claimed.

Taking adjoints in (11.5.25) we get

$$((-A)^{1-\alpha} R_{\alpha,0}(t, s))^* = (-A^{\odot}) S(t-s)^* + N(t, s; \bar{u})^*, \tag{11.5.29}$$

so that, using (11.5.28),

$$\int_0^{\bar{t}} \|((-A)^{1-\alpha} R_{\alpha,0}(t, s))^*((-A)^{-1})^* z\|_{E^{\odot}} d\sigma < \infty \tag{11.5.30}$$

if we show previously that

$$((-A)^{1-\alpha} R_{\alpha,0}(t, s))^* E^* \subseteq E^{\odot} \tag{11.5.31}$$

(otherwise, we only have the right to use the norm of E^* in (11.5.30)). To check (11.5.31) we note that, after standard manipulations with the integral equation (11.5.25) we obtain

$$((-A)^{1-\alpha} R_{\alpha,0}(t, s; \bar{u}))^* z = (-A^{\odot}) S(t-s)^* z$$

$$+ \int_s^t R_{\alpha,0}(\tau, s; \bar{u})^*((-A)^{\delta} \partial_y f(\tau, y(\tau, \bar{u}))(-A)^{-\alpha})^*(-A^{\odot})^{-\delta} S(t-\tau)^* z \, d\tau$$

and use Lemma 11.1.5.

We return to (11.5.23) for the last time. If $v(\cdot) \in C_{\mathrm{ad},\varepsilon}(0, \bar{t}; U)$ we can write the last term in the form

$$\int_0^{\bar{t}} \big\langle ((-A)^{1-\alpha} R_{\alpha,0}(t, s))^* ((-A)^{-1})^* z, v(\sigma) - \bar{u}(\sigma) \big\rangle d\sigma$$

with (11.5.20) valid since $v(\cdot) \in D(A, \bar{t})$. However, being in posession of (11.5.30) gives us leave to let $\varepsilon \to 0$ with preservation of the inequality. This completes the proof of Theorem 11.5.2. ∎

For once, the modifications for the time optimal problem are slightly nontrivial. As in the proof of Theorem 10.3.3, the sequence $\{\mathbf{f}_n\}$ in the application of the abstract time optimal problem is

$$\mathbf{f}_n(u) = (y(\cdot, u), y(t_n, u)). \tag{11.5.32}$$

Each $\mathbf{f}_n$ is defined in the space $(B_n(\bar{u}, \delta), \theta_n)$, $B_n(\bar{u}, \delta)$ the set of all $u \in B(\bar{u}, \delta)$ with $y(t_n, u) \in D(A) = E_1$, $\theta_n(u, v) = d(u, v) + \|y(t_n, u) - y(t_n, v)\|_{E_1}$. We prove that $(B_n(\bar{u}, \delta), \theta_n)$ is complete and that $\mathbf{f}_n$ is continuous in the same way we did it for $(B_1(\bar{u}, \delta), \theta)$ and $\mathbf{f}$. The target set is (11.5.12) and it is plain that

$$\mathbf{f}_n(B_n(\bar{u}, \delta)) \cap \mathbf{Y} = \emptyset \tag{11.5.33}$$

for all n, as required. However, we cannot take $\{\bar{u}^n\} = \{\bar{u}\}$ as a solution of the abstract time optimal problem, since (a) $\bar{u}^n$ may not belong to $B_n(\bar{u}, \delta)$ (that is, $y(t_n, \bar{u}^n)$ may not belong to E_1), and (b) even if it does for all n, it is not guaranteed that $y(t_n, \bar{u}^n) \to y(\bar{t}, \bar{u})$ in E_1 (trajectories need not be continuous in the E_1 norm). We take instead $\{\bar{u}^n\}$ defined by

$$\bar{u}^n(t) = u(t + (\bar{t} - t_n)) \quad (0 \le t \le \bar{t}) \tag{11.5.34}$$

with $\bar{u}$ extended to $t \ge \bar{t}$ by $\bar{u}(t) = \bar{u}(\bar{t})$.

Lemma 11.5.4. (a) $y(\cdot, \bar{u}^n) \to y(\cdot, \bar{u})$ in $C(0, \bar{t}; E_\alpha)$. (b) $y(t_n, \bar{u}^n) \in E_1$ and $y(t_n, \bar{u}^n) \to y(\bar{t}, \bar{u})$ in E_1.

Proof. Combining the integral equations,

$$(-A)^\alpha y(t, \bar{u}^n) - (-A)^\alpha y(t, \bar{u})$$

$$= \int_0^t (-A)^{\alpha-\delta} S(t - \tau)(-A)^\delta \{f(\tau, y(\tau, \bar{u}^n)) - f(\tau, y(\tau, \bar{u}))\} d\tau$$

$$+ \int_{\bar{t}-t_n}^t (-A)^\alpha \{S(t - \tau + (\bar{t} - t_n)) - S(t - \tau)\} \bar{u}(\tau) d\tau$$

$$+ \int_t^{t+(\bar{t}-t_n)} (-A)^\alpha S(t - \tau + (\bar{t} - t_n)) \bar{u}(\tau) d\tau$$

$$- \int_0^{\bar{t}-t_n} (-A)^\alpha S(t - \tau) \bar{u}(\tau) d\tau, \tag{11.5.35}$$

so that (a) follows from uniform convergence to zero in $0 \le t \le \bar{t}$ of the three last integrals on the right and the generalized Gronwall lemma. (For the second integral, uniform convergence follows from the bound for $\|(-A)^\alpha S(t)\|$ and Vitali's theorem, for the third and fourth directly from the bound.) Now,

$$
\begin{aligned}
(-A)y(\bar{t}, \bar{u}) &- (-A)y(t_n, \bar{u}^n) \\
&= \int_{t_n}^{\bar{t}} (-A)^{1-\delta} S(\bar{t} - \tau)(-A)^\delta f(\tau, y(\tau, \bar{u}))d\tau \\
&+ \int_0^{t_n} (-A)^{1-\delta}\{S(\bar{t} - \tau) - S(t_n - \tau)\}(-A)^\delta f(\tau, y(\tau, \bar{u}))d\tau \\
&+ \int_0^{t_n} (-A)^{1-\delta} S(t_n - \tau)(-A)^\delta \{f(\tau, y(\tau, \bar{u})) - f(\tau, y(\tau, \bar{u}^n))\}d\tau \\
&+ \int_0^{\bar{t}-t_n} (-A)S(\bar{t} - \tau)\bar{u}(\tau)d\tau
\end{aligned}
\tag{11.5.36}
$$

Using (a) in the third integral on the right, the proof is complete. ∎

Lemma 11.5.4 implies that

$$
\text{dist}(\mathbf{f}_n(\bar{u}^n), \mathbf{Y}) \to 0 \quad \text{as } n \to \infty
\tag{11.5.37}
$$

so that $\{\bar{u}^n\}$ is a solution of the abstract time optimal problem in the sense of 7.2. Theorem 7.2.2 provides the sequences $\{\tilde{u}^n\} \subseteq B_1(\bar{u}, \delta)$ and $\{(\tilde{y}^n(\cdot), \tilde{y}^n)\} \subseteq \mathbf{M}(\bar{t}) \times Y$. We select $\rho > 0$ and use the convex sets $D_n \subseteq \text{Der } \mathbf{f}_n(\tilde{u}^n)$ given by

$$
\frac{1}{t_n(\|\xi(t_n, \tilde{u}^n, \mathbf{p}, \mathbf{v})\|_{E_1} + 2)}(\xi(\cdot, \tilde{u}^n, \mathbf{p}, \mathbf{v}), \xi(t_n, \tilde{u}^n, \mathbf{p}, \mathbf{v}))
\tag{11.5.38}
$$

where $\mathbf{v}$ is such that $\xi(t_n, \tilde{u}^n, \mathbf{p}, \mathbf{v}) \in E_1 = D(A)$. Taking now into account that $y(t_n, \tilde{u}^n) \in D(A)$ and that this condition is regulated by (11.4.34) for any $\bar{t}$ and any control $\bar{u}$, we deduce that $\tilde{u}^n(\cdot) \in D(A, t_n)$; using then (11.4.36) it is plain that

$$
\xi(t_n, \tilde{u}^n, \mathbf{p}, \mathbf{v}) \in D(A) \quad \text{if and only if } \Sigma p_j v_j(\cdot) \in D(A, t_n).
\tag{11.5.39}
$$

Theorem 7.2.2 provides the "abstract time optimal" companion of (11.5.17): there exist sequences $\{\mu_n\} \in \Sigma(0, \bar{t}; ((-A)^\alpha)^* E^*)$ and $\{z_n\} \subseteq ((-A)^* E^*)^1$ such that

$$
\|\mu_n\|^2_{\Sigma(0,\bar{t};((-A)^\alpha)^* E^*)} + \|z_n\|^2_{(-A)^* E^*} = 1,
$$

$$
\langle \mu, \xi(\cdot, \tilde{u}^n, \mathbf{p}, \mathbf{v})\rangle_{\alpha,c} + \langle z, \xi(t_n, \tilde{u}^n, \mathbf{p}, \mathbf{v})\rangle_1
\tag{11.5.40}
$$

$$
\ge \tilde{\delta}_n = \delta_n t_n(\|\xi(t_n, \tilde{u}^n, \mathbf{p}, \mathbf{v})\|_{E_1} + 2).
$$

To graduate to the Kuhn-Tucker inequality

$$
\langle \mu, \xi(\cdot, \bar{u}, \mathbf{p}, \mathbf{v})\rangle_{\alpha,c} + \langle z, \xi(\bar{t}, \bar{u}, \mathbf{p}, \mathbf{v})\rangle \ge 0
\tag{11.5.41}
$$

for weak limits, we must prove

Lemma 11.5.5. $(a)\, \xi(\cdot, \tilde{u}^n, \mathbf{p}, \mathbf{v}) \to \xi(\cdot, \bar{u}, \mathbf{p}, \mathbf{v})$ in $C(0, \bar{t}; E_\alpha)$, $(b)\, \xi(t_n, \tilde{u}^n, \mathbf{p}, \mathbf{v})$ $\to \xi(\bar{t}, \bar{u}, \mathbf{p}, \mathbf{v})$ in E_1.

This result does not reduce to Lemma 11.5.3, since it is no longer true that $\tilde{u}^n(\cdot) \to \bar{u}(\cdot)$ in $B_1(\bar{u}, \delta)$. All that we know is that $\theta_n(\tilde{u}^n, \bar{u}^n) \to 0$, which means

$$d(\tilde{u}_n, \bar{u}^n) \to 0, \qquad \|y(t_n, \tilde{u}^n) - y(t_n, \bar{u}^n)\|_{E_1} \to 0. \qquad (11.5.42)$$

Note that the two convergence relations (11.5.42) imply that

$$\begin{aligned}
\|\xi(\cdot, \tilde{u}^n, \mathbf{p}, \mathbf{v}) - \xi(\cdot, \bar{u}^n, \mathbf{p}, \mathbf{v})\|_{C(0,\bar{t};E_\alpha)} &\to 0, \\
\|\xi(t_n, \tilde{u}^n, \mathbf{p}, \mathbf{v}) - \xi(t_n, \bar{u}, \mathbf{p}, \mathbf{v})\|_{E_1} &\to 0.
\end{aligned} \qquad (11.5.43)$$

Thus, it is enough to show Lemma 11.5.5 with $\bar{u}^n$ instead of $\tilde{u}^n$, that is,

$$(\xi(\cdot, \bar{u}^n, \mathbf{p}, \mathbf{v}), \xi(t_n, \bar{u}^n, \mathbf{p}, \mathbf{v})) \to (\xi(\cdot, \bar{u}, \mathbf{p}, \mathbf{v}), \xi(\bar{t}, \bar{u}, \mathbf{p}, \mathbf{v}))$$

in $C(0, \bar{t}; E_\alpha) \times E_1$. Using the integral equations for $(-A)^\alpha \xi(t, \bar{u}^n, \mathbf{p}, \mathbf{v})$ and $(-A)^\alpha \xi(t, \bar{u}, \mathbf{p}, \mathbf{v})$, convergence of the first coordinate reduces to

$$\int_0^t (-A)^\alpha S(t - \tau)((\bar{u}^n(\tau) - \bar{u}(\tau))d\tau \to 0 \qquad (11.5.44)$$

uniformly in $0 \le t \le \bar{t}$. This integral equals the combination of the last three integrals on (11.5.35), thus convergence is proved in the same way. For the second coordinate, we write

$$(-A)\xi(\bar{t}, \bar{u}, \mathbf{p}, \mathbf{v}) - (-A)\xi(t_n, \bar{u}^n, \mathbf{p}, \mathbf{v})$$

$$= \int_{t_n}^{\bar{t}} (-A)^{1-\delta} S(\bar{t} - \tau)(-A)^\delta \partial_y f(\tau, y(\tau, \bar{u}))\xi(\tau, \bar{u}, \mathbf{p}, \mathbf{v})d\tau$$

$$+ \int_0^{t_n} (-A)^{1-\delta}\{S(\bar{t} - \tau) - S(t_n - \tau)\}(-A)^\delta \partial_y f(\tau, y(\tau, \bar{u}))\xi(\tau, \bar{u}, \mathbf{p}, \mathbf{v})d\tau$$

$$+ \int_0^{t_n} (-A)^{1-\delta} S(t_n - \tau)(-A)^\delta \{\partial_y f(\tau, y(\tau, \bar{u})) - \partial_y f(\tau, y(\tau, \bar{u}))\}\xi(\tau, \bar{u}^n, \mathbf{p}, \mathbf{v})d\tau$$

$$+ \int_0^{t_n} (-A)^{1-\delta} S(t_n - \tau)(-A)^\delta \partial_y f(\tau, y(\tau, \bar{u}))\{\xi(\tau, \bar{u}, \mathbf{p}, \mathbf{v}) - \xi(\tau, \bar{u}^n, \mathbf{p}, \mathbf{v})\}d\tau$$

$$+ \int_0^{\bar{t}-t_n} (-A)S(\bar{t} - \tau)\bar{u}(\tau)d\tau. \qquad (11.5.45)$$

We then use (a) in the fourth integral and Lemma 11.5.4 (a) in the third. ∎

The following result is an obvious extension of Lemma 11.2.4 to $\alpha = 1$; if there exists $\rho > 0$ and a precompact sequence $\{Q_n\}$, $Q_n \subseteq E_1$ such that the set (11.2.28) contains an interior point in E_1 for n_0 large enough, then multipliers are nontrivial. We show below that this is true without any help from $T_Y(\tilde{y}^n)$; in particular, the point target condition

$$y(\bar{t}, u) = \bar{y} \in D(A) \qquad (11.5.46)$$

is included.

Lemma 11.5.6. *Assume* $\mathbf{M}(\bar{t})$ *is* T-*full in* $C(0, \bar{t}; E_\alpha)$ *and that the origin is an interior point of the control set* U. *Then the multiplier* (z_0, μ, z) *is nontrivial; for the time optimal principle, the multiplier* (μ, z) *is nontrivial.*

Proof. We do it first for the time optimal problem. Using the set D_n of directional derivatives in (11.5.38), all we have to show is that the subsets of $C(0, \bar{t}; E_\alpha) \times E_1$ of coordinates

$$\frac{1}{t_n(\|\xi(t_n, \tilde{u}^n, \mathbf{p}, \mathbf{v})\|_{E_1} + 2)}\xi(\cdot, \tilde{u}^n, \mathbf{p}, \mathbf{v}) - T_{\mathbf{M}(\bar{t})}(\tilde{y}^n(\cdot)) \qquad (11.5.47)$$

$$\frac{1}{t_n(\|\xi(t_n, \tilde{u}^n, \mathbf{p}, \mathbf{v})\|_{E_1} + 2)}\xi(t_n, \tilde{u}^n, \mathbf{p}, \mathbf{v}) \qquad (11.5.48)$$

contain a common ball for n large enough, where $\mathbf{p}$ and $\mathbf{v}$ take the same values in (11.5.47) and (11.5.48) and $\Sigma p_j v_j(\cdot) \in D(A, t_n)$. For this, we imitate the arguments in Theorem 10.3.3. Since $\mathbf{M}(\bar{t})$ is T-full, the tangent cones $T_{\mathbf{M}(\bar{t})}(\tilde{y}^n(\cdot))$ will take care of the first coordinate; we only need boundedness of the first vector in (11.5.47) in $C(0, \bar{t}; E_\alpha)$ and this follows from boundedness of $\xi(\cdot, \tilde{u}^n, \mathbf{p}, \mathbf{v})$, since the factor is $\leq (2t_n)^{-1}$. We only have to show that the second coordinates (11.5.48) cover a ball in $E_1 = D(A)$. On this matter, note that if $\xi \in D(A)$ and $\xi(t) = t\xi$, then $\xi(t)$ satisfies the initial value problem

$$\xi'(t) = \xi = \{A + \partial_y f(t, y(t, \tilde{u}^n))\}\xi(t) + \{\xi - tA\xi - t\partial_y f(t, y(t, \tilde{u}^n))\xi\}$$

$$= \{A + \partial_y f(t, y(t, \tilde{u}^n))\}\xi(t) + v(t), \quad y(0) = 0, \qquad (11.5.49)$$

where the function $v(\cdot)$ is given by

$$v(t) = \xi - tA\xi - t\partial_y f(t, y(t, \tilde{u}^n))\xi,$$

Keeping in mind that the origin is an interior point of U we deduce that $v(\cdot)$ satisfies the control constraint in the interval $0 \leq t \leq t_n$ if $\|\xi\|_{E_1} = \|A\xi\| \leq \delta$, with $\delta > 0$ independent of n, and it is clear that, since $\xi(t_n) \in E_1$ then $v(\cdot) \in D(A, t_n)$. It follows that, for each n the set (11.5.48) contains a ball $B(y_n, \delta)$ in E_1 of fixed radius. That the intersection of all these balls for $n \geq n_0$ will in turn contain a ball is a consequence of the second convergence relation (11.5.43) applied to $\mathbf{p} = \{1\}, \mathbf{v}(\cdot) = \{v(\cdot)\}$.

The proof is similar (but easier) for the general control problem; we simply take $t_n = \bar{t}$. ∎

We apply the results to the time optimal problem for the system (11.4.1) with control set $U = $ unit ball of $L^\infty(\Omega)$ and state constraints as in **11.3**,

$$y(t, x) \in M_s \subseteq \mathbb{R} \qquad \nabla y(t, x) \in M_g \subseteq \mathbb{R}^m \quad (0 \leq t \leq \bar{t}) \tag{11.5.50}$$

where M_s and M_g are closed and such that the set $\mathbf{M}(\bar{t})$ is T-full (for this is sufficient, but not necessary, that M_s and M_g be convex with nonempty interior). The target condition is $y(\bar{t}, \cdot) \in Y = $ closed set in E_1; in particular, the point target condition (11.5.46) is permissible.

Theorem 11.5.7. *Let $\bar{u}(t, x)$ be a time optimal control with $y(\bar{t}, \bar{u}) \in D(A_c(\beta))$. Then there exist sequences $\{\tilde{y}^n(\cdot)\} \subseteq \mathbf{M}(\bar{t})$ and $\{y^n\} \subseteq Y$ such that $\tilde{y}^n(\cdot) \to y(\cdot, \bar{u})$ in $C(0, \bar{t}; E_\alpha)$ and $y^n \to \tilde{y}$ in E_1, a measure $\mu \in \Sigma(0, \bar{t}; ((-A)^\alpha)^* E^*)$ and an element $z \in ((-A)^* E^*)^1$ with $(\mu, z) \neq 0$,*

$$\mu \in \left(\liminf_{n \to \infty} T_{\mathbf{M}(\bar{t})}(\tilde{y}^n(\cdot)) \right)^-, \qquad z \in \left(\liminf_{n \to \infty} T_Y(\tilde{y}^n) \right)^- \tag{11.5.51}$$

and such that if $\bar{z}(s, x)$ is the solution of

$$\begin{aligned} d\bar{z}(s, x) = &-A_1'(\beta')\bar{z}(s, x)ds - \partial_y \phi(s, x, y(s, x, \bar{u}), \nabla y(s, x, \bar{u}))\bar{z}(s, x)ds \\ &+ \nabla \cdot (\nabla_{\mathbf{y}} \phi(s, x, y(s, x, \bar{u}), \nabla y(s, x, \bar{u}))\bar{z}(s, x))ds - \mu(ds) \end{aligned}$$
$$\bar{z}(\bar{t}, \cdot) = z \tag{11.5.52}$$

then

$$\int_\Omega \bar{z}(s, x)\bar{u}(s, x)dx = \min_{\|v\|_{L^\infty(\Omega)} \leq 1} \int_\Omega \bar{z}(s, x)v(x)dx \tag{11.5.53}$$

a.e. in $0 \leq s \leq \bar{t}$.

11.6. Linear Systems: The Minimum Principle with State Constraints

The results in **9.1** and **9.2** on the linear system

$$y'(t) = Ay(t) + Bu(t) \tag{11.6.1}$$

have counterparts for problems with state constraints. The hypotheses are those in **9.1**; A generates an analytic semigroup $S(t)$, E is $\odot$-reflexive, E, $E^\odot$ are separable, and $B : X^* \to (E^\odot)^*$, $B^* : E^\odot \to X$, where X is separable. The state constraint is

$u(t) \in U \subseteq X^*$ a.e., and the admissible control space $C_{ad}(0, T; U)$ is $L^1(0, T; X)$-weakly compact in $L^\infty_w(0, T; X^*)$. The solution $y(t, \zeta; u)$ of (11.6.1) is given by (9.1.5) and belongs to $D((-A)^\alpha)$ if $\alpha < 1$, with

$$(-A)^\alpha y(t, \zeta; u) = (-A)^\alpha S(t)\zeta + \int_0^t (-A)^\alpha S^\odot(t - \tau)^* Bu(\tau)d\tau \quad (11.6.2)$$

The state constraints and the target condition are

$$y(t) \in M \subseteq E_\alpha, \qquad y(t) \in Y. \quad (11.6.3)$$

We also assume $S(t)$ compact. If M is closed in E_α and Y is closed, existence of time optimal controls follows as in Theorem 9.1.4. The operator K below satisfies (d) and (e) in **9.1** and $Y \subseteq D(K)$.

The following is an upgrade of Theorem 9.1.10 that takes care of state constraints.

Theorem 11.6.1. *Assume that U, M and Y are convex, M is closed in E_α with* Int(M) $\neq \emptyset$,

$$B(0, \rho) \subseteq U \quad (11.6.4)$$

for some $\rho > 0$, Y is closed, and

$$R(\tilde{t}, \zeta; U) - Y \quad (11.6.5)$$

contains a ball of positive radius in $D(K)$ (norm $\|y\|_K = \|Ky\|_E$) for some $\tilde{t} < \bar{t}$. Let $\bar{u}(\cdot)$ be a time optimal control with optimal time $\bar{t}$. Then there exists a multiplier $(\mu, z) \in \Sigma(0, \bar{t}; ((-A)^\alpha)^* E^*) \times (K^* E^*)^1 (B^*)$, $(\mu, z) \neq 0$ with

$$\mu \in N_{\mathbf{M}(\bar{t})}(y(\cdot, \zeta, \bar{u})), \qquad z \in N_Y(y(\bar{t}, \zeta, \bar{u})) \quad (11.6.6)$$

and such that

$$\langle B^* \bar{z}(s), \bar{u}(s) \rangle = \min_{v \in U} \langle B^* \bar{z}(s), v \rangle \quad (11.6.7)$$

a.e. in $0 \leq s \leq \bar{t}$, $\bar{z}(\cdot)$ the solution of

$$d\bar{z}(s) = -A^\odot \bar{z}(s)ds - \mu(ds), \qquad \bar{z}(\bar{t}) = z \quad (11.6.8)$$

in $0 \leq t \leq \bar{t}$.

Note that $\bar{z}(s)$ is now simply defined by

$$\bar{z}(s) = K^* S(\bar{t} - s)^* (K^{-1})^* z + \int_s^{\bar{t}} (-A^\odot)^\alpha S(\sigma - s)^* ((-A)^{-\alpha})^* \mu(ds). \quad (11.6.9)$$

The proof of Theorem 11.6.1 results from minor modifications of that of Theorem 9.1.10. Given $\tilde{t} > 0$, $\mathbf{M}_K(\tilde{t}, Y) \subseteq C(0, \tilde{t}; E_\alpha) \times E$ consists of all elements of the form

$$(y(\cdot), Ky) \qquad (y(\cdot) \in \mathbf{M}(\tilde{t}), y \in Y),$$

and $C_{\mathrm{ad}, K}(0, \tilde{t}; U)$ is the set of all $u(\cdot) \in C_{\mathrm{ad}}(0, \tilde{t}; U)$ with $y(\tilde{t}, \zeta, u) \in D(K)$. Finally, $\mathbf{R}_K(\tilde{t}, \zeta; U) \subseteq C(0, \tilde{t}; E_\alpha) \times E$ is the set of all elements of the form

$$(y(\cdot, \zeta, u), Ky(\tilde{t}, \zeta, u)) \qquad (u(\cdot) \in C_{\mathrm{ad}, K}(0, \tilde{t}, u)).$$

Select a sequence $\{t_n\} \subseteq [0, \bar{t})$, $\bar{t} =$ optimal time, $t_n \to \bar{t}$. If $\tilde{t} < \bar{t}$ the distance from $\mathbf{M}_K(\tilde{t}, Y)$ to $\mathbf{R}_K(\tilde{t}, \zeta; U)$ is positive. In fact, if this were not the case we could construct a sequence $\{u_j(\cdot)\} \subseteq C_{\mathrm{ad}, K}(0, \tilde{t}; U)$ and a sequence $y_j(\cdot) \in \mathbf{M}(\tilde{t})$ such that

$$\|y(t, \zeta, u_j) - y_j(t)\|_\alpha \to 0 \tag{11.6.10}$$

uniformly in $0 \le t \le \tilde{t}$ and

$$Ky(\tilde{t}, \zeta, u_j) - Ky_j \to 0. \tag{11.6.11}$$

as $j \to \infty$, where $\{y_j\} \in Y$. Using Theorem 7.7.16 we can then select a subsequence of $\{u_j(\cdot)\}$ such that $u_j(\cdot) \to \tilde{u}(\cdot) \to C_{\mathrm{ad}}(0, \tilde{t}; U)$ $L^1(0, \tilde{t}; E)$-weakly in $L_w^\infty(0, \tilde{t}; X^*)$ and $y(\cdot, \zeta, u_j)$ is convergent to $y(\cdot, \zeta, \tilde{u})$ in $C(0, \tilde{t}; E_\alpha)$; in particular, $y(\tilde{t}, \zeta, u_j) \to y(\tilde{t}, \zeta, \tilde{u})$. Applying K^{-1} to both sides of (11.6.10), we deduce that $y(\tilde{t}, \zeta, u_j) - y_j \to 0$, so that $y(\tilde{t}, \zeta, \tilde{u}) = \tilde{y} \in Y$. On the other hand, (11.6.10) implies $\mathrm{dist}(y(t, \zeta, u_j), M) \to 0$ as $j \to \infty$ in $0 \le t \le \tilde{t}$ (distance in E_α) thus $y(t, \zeta, \tilde{u})$ satisfies the state constraint and goes from ζ to $\tilde{y} \subseteq Y$ in time $\tilde{t} < \bar{t}$, which contradicts the definition of $\bar{t}$ as optimal time.

Since $\mathrm{dist}(\mathbf{M}_K(t_n, Y), \mathbf{R}_K(t_n, \zeta; U)) > 0$, $\mathbf{M}_K(t_n, Y)_\varepsilon$ and $\mathbf{R}_K(t_n, \zeta; U)$ are disjoint for $\varepsilon > 0$ sufficiently small ($A_\varepsilon = \{y; \mathrm{dist}(y, A) < \varepsilon\}$) we may apply the separation theorem and thus construct $(\mu_n, w_n) \in \Sigma(0, t_n; ((-A)^\alpha)^* E^*) \times E^*$, $\|(\mu_n, w_n)\| = 1$ such that

$$\langle (\mu_n, w_n), (y(\cdot), Ky) \rangle_{\alpha, c} \le \langle (\mu_n, w_n), (y(\cdot, \zeta, u), Ky(t_n, \zeta, u)) \rangle_{\alpha, c}$$

($\langle \cdot, \cdot \rangle_{\alpha, c}$ the duality of $C(0, t_n; E_\alpha) \times E$ and $\Sigma(0, t_n; ((-A)^\alpha)^* E^*) \times E^*))$.[1] Equivalently,

$$\int_0^{t_n} \langle \mu_n(dt), y(t) \rangle_\alpha + \langle w_n, Ky \rangle$$

$$\le \int_0^{t_n} \langle \mu_n(dt), y(t, \zeta, u) \rangle_\alpha + \langle w_n, Ky(t_n, \zeta, u) \rangle \tag{11.6.12}$$

---

[1] The symbol $\langle \cdot, \cdot \rangle_{\alpha, c}$ has a different meaning than in past sections.

for every $y(\cdot) \in \mathbf{M}(t_n)$, $y \in Y$ and $u(\cdot) \in C_{\text{ad},K}(0, t_n; U)$ ($\langle \cdot, \cdot \rangle_\alpha$ the duality of E_α and $((-A)^\alpha)^* E^*$ and $\langle \cdot, \cdot \rangle$ the duality of E and E^*). We extend μ_n to $[t_n, \bar{t}]$ setting $\mu_n = 0$ there and select a subsequence such that $(\mu_n, w_n) \to (\mu, w) \in \Sigma(0, \bar{t}; ((-A)^\alpha)^* E^*) \times E^*$ $C(0, \bar{t}; E_\alpha) \times E$-weakly.

Assume $(\mu, w) = 0$. By Theorem 7.7.16 the set

$$\{y(\cdot, \zeta, u); u \in C_{\text{ad}}(0, \bar{t}; U)\}$$

is compact in $C(0, \bar{t}; E_\alpha)$, thus the integral on the right side of (11.6.12) tends to zero uniformly in u. Sending left the second term on the right, we obtain

$$\int_0^{t_n} \langle \mu_n(dt), y(t) \rangle_\alpha - \langle w_n, y \rangle \le \varepsilon_n \to 0 \qquad (11.6.13)$$

for $(y(\cdot), y) \in \mathbf{M}(\bar{t}) \times K B_K(\tilde{y}, \delta)$, $B_K(\tilde{y}, \delta)$ a ball in $D(K)$ contained in all $R(t_n, \zeta; U) - Y$ (Lemma 9.1.11). Hence, we obtain a contradiction using Lemma 6.4.3, and we conclude that $(\mu, w) \ne 0$.

If we take $y(t) = y(t, \zeta, u)$ in (11.6.12), we obtain inequality (9.1.30), holding in the same conditions as in Theorem 9.1.10. We may then mete out the same treatment to this inequality and deduce that $w = (K^{-1})^* z$ with $z \in N_Y(y(\bar{t}, \zeta, \bar{u}))$, $z \in (K^* E^*)^1(B^*)$. Once this is done we proceed to "pull operators to the other side" in (11.6.12) and obtain the minimum principle in integral form

$$\int_0^{\bar{t}} \langle B^* \bar{z}(t), \bar{u}(t) \rangle dt \le \int_0^{\bar{t}} \langle B^* \bar{z}(t), u(t) \rangle dt$$

from which the pointwise form (11.6.7) is obtained as in Theorem 9.1.10; note that, since $(\mu, w) \ne 0$ and $(K^{-1})^*$ is an isometry from $K^* E^*$ onto E^*, we have $(\mu, z) \ne 0$.

Finally, we use again (11.6.12) with $u = \bar{u}$ on both sides, obtaining

$$\int_0^{\bar{t}} \langle \mu(dt), y(t) - y(t, \zeta; \bar{u}) \rangle \le 0 \qquad (11.6.14)$$

for all $y(\cdot) \in \mathbf{M}(\bar{t})$ so that $\mu \in N_{\mathbf{M}(\bar{t})}(y(\cdot, \zeta, u))$ as claimed. ∎

There are also results for the optimal problem for (11.6.1) with cost functional

$$y_0(\bar{t}, u) = \int_0^{\bar{t}} f_0(\tau, y(\tau), u(\tau)) d\tau \qquad (11.6.15)$$

in a fixed or variable time interval $0 \le t \le \bar{t}$, where (11.6.15) satisfies Hypothesis III_α^0. For the last step of deducing the pointwise version of the maximum principle from the integral version, we need the following result for spike perturbations $u_{s,h,v}$ of a control u.

Example 11.6.2. There exists a set e of full measure in $[0, \bar{t}]$ such that

$$\lim_{h \to 0+} \frac{1}{h}(-A)^\alpha (y(t, u_{s,h,v}) - y(t, u)) = (-A)^\alpha S^\odot (t - s)^* B(v - u(s)) \quad (t > s).$$

The proof is a simple consequence of the fact (Lemma 7.7.13) that $s \to (-A)^\alpha S^\odot (t - s)^* B(v - u(s))$ is strongly measurable. Convergence is uniform in $t \geq s + \varepsilon$, and approximations are bounded by $C(t - s)^{-\alpha}$, which bound plays a role in the proof of the result below.

Example 11.6.3. There exists a set e of full measure in $[0, \bar{t}]$ such that for every $s \in e$ we have

$$\lim_{h \to 0+} \frac{1}{h}(y_0(\bar{t}, u_{s,h,v}) - y_0(\bar{t}, u)) = f_0(s, y(s, u), v) - f_0(s, y(s, u), u(s))$$

$$+ \int_s^{\bar{t}} \langle ((-A)^{-\alpha})^* \partial_y f_0(\tau, y(\tau, u), u(\tau)), (-A)^\alpha S^\odot (\tau - s)^* B(v - u(s)) \rangle d\tau.$$

Existence results require weak lower semicontinuity of the functional, and the proof of the minimum principle demands convexity: if $u, v \in U$ and $0 \leq \alpha \leq 1$, then

$$y_0(\bar{t}, \alpha u + (1 - \alpha)v) \leq \alpha y_0(\bar{t}, u) + (1 - \alpha)y_0(\bar{t}, v).$$

The optimal control problem has a solution if M is closed in E_α, Y is closed, and there exists a minimizing sequence $\{u_n(\cdot)\} \in C_{ad}(0, t; U)$ (with t_n bounded), driving ζ to Y in time t_n and satisfying the state constraint.

Theorem 11.6.4. *Same assumptions on U, M and Y in Theorem 11.6.1. Let $\bar{u}(\cdot)$ be an optimal control in the interval $0 \leq t \leq \bar{t}$. Then there exists a multiplier $(z_0, \mu, z) \in \mathbb{R} \times \Sigma(0, \bar{t}; ((-A)^\alpha)^* E^*) \times (K^* E^*)^1 (B^*)$, $(z_0, \mu, z) \neq 0$ with*

$$z_0 \geq 0, \qquad \mu \in N_{M(\bar{t})}(y(\cdot, \zeta, \bar{u})), \qquad z \in N_Y(y(\bar{t}, \zeta, \bar{u})) \qquad (11.6.16)$$

and such that

$$z_0 f_0(s, y(s, \zeta, u), \bar{u}(s)) + \langle B^* \bar{z}(s), \bar{u}(s) \rangle$$

$$= \min_{v \in U}\{z_0 f_0(s, y(s, \zeta, u), v) + \langle B^* \bar{z}(s), v \rangle\} \qquad (11.6.17)$$

a.e. in $0 \leq s \leq \bar{t}$, $\bar{z}(\cdot)$ the solution of

$$d\bar{z}(s) = -A^\odot \bar{z}(s)ds - z_0 \partial_y f_0(s, y(s, \bar{u}), \bar{u}(s))ds - \mu(ds),$$
$$\bar{z}(\bar{t}) = z. \qquad (11.6.18)$$

Proof. Given $a \in \mathbb{R}$, the set $\mathbf{M}_K(a, \bar{t}, Y) \subseteq \mathbb{R} \times C(0, \bar{t}; E_\alpha) \times E$ consists of all elements of the form

$$(r, y(\cdot), Ky) \quad (r \leq a, y(\cdot) \in \mathbf{M}(\bar{t}), y \in Y), \tag{11.6.19}$$

and $\mathbf{R}_K(\bar{t}, \zeta; U) \subseteq \mathbb{R} \times C(0, \bar{t}; E_\alpha) \times E$ consists of all elements of the form

$$(r, y(\cdot, \zeta, u), Ky(\bar{t}, \zeta, u)) \quad (r \geq y_0(\bar{t}, u), u(\cdot) \in C_{\text{ad}, K}(0, \bar{t}, u)).^{(2)} \tag{11.6.20}$$

Both sets are convex. If m is the minimum of $y_0(\bar{t}, u)$ (under state and control constraints and target condition) then for every $\varepsilon > 0$ the sets $\mathbf{M}_K(m - \varepsilon, \bar{t}, \bar{y})$ and $\mathbf{R}_K(\bar{t}, \zeta; U)$ lie at a positive distance; if this were not true we would have four sequences $\{r_j\} \subseteq \mathbb{R}$, $\{u_j(\cdot)\} \subseteq C_{\text{ad}, K}(0, \bar{t}; U)$, $\{y_j\} \subseteq Y$ and $y_j(\cdot) \in \mathbf{M}(\bar{t})$ such that

$$y_0(\bar{t}, u_j) - r_j \to 0, \qquad r_j \leq m - \varepsilon, \tag{11.6.21}$$

$$\|y(t, \zeta, u_j) - y_j(t)\|_\alpha \to 0 \tag{11.6.22}$$

uniformly in $0 \leq t \leq \bar{t}$, and

$$Ky(\bar{t}, \zeta, u_j) - Ky_j \to 0 \tag{11.6.23}$$

as $j \to \infty$. We select a subsequence of $\{u_j\}$ such that $u_j(\cdot) \to \tilde{u}(\cdot) \in C_{\text{ad}}(0, \bar{t}; U)$ $L^1(0, \bar{t}; E)$-weakly in $L_w^\infty(0, \bar{t}; X^*)$ and $y(\cdot, \zeta, u_j) \to y(\cdot, \zeta, \tilde{u})$ in $C(0, \bar{t}; E_\alpha)$. Applying K^{-1} to both sides we deduce that $y(\bar{t}, \zeta; u_j) - y_j \to 0$, so that $y(\bar{t}, \zeta, \tilde{u}) \in Y$ with $y_0(\bar{t}, \tilde{u}) \leq m - \varepsilon$ by (11.6.21) and weak lower semicontinuity of $y_0(\bar{t}, u)$, a contradiction to optimality of $\tilde{u}$.

Let $\{\varepsilon_n\}$ be a sequence of positive numbers with $\varepsilon_n \to 0$. Having established that $\text{dist}(\mathbf{M}_K(m - \varepsilon_n, \bar{t}, Y), \mathbf{R}_K(\bar{t}, \zeta; U)) > 0$ we apply the separation theorem and obtain $(z_{0n}, \mu_n, w_n) \in \mathbb{R} \times \Sigma(0, \bar{t}; ((-A)^\alpha)^* E^*) \times E^*$, $\|(z_{0n}, \mu_n, w_n)\| = 1$ such that

$$z_{0n} r + \int_0^{\bar{t}} \langle \mu_n(dt), y(t) \rangle_\alpha + \langle w_n, Ky \rangle$$

$$\leq z_{0n} y_0(\bar{t}, u) + \int_0^{\bar{t}} \langle \mu_n(dt), y(t, \zeta, u) \rangle_\alpha + \langle w_n, Ky(\bar{t}, \zeta, u) \rangle \tag{11.6.24}$$

for $r \leq m - \varepsilon_n$, $y(\cdot) \in \mathbf{M}(\bar{t})$, $y \in Y$ and $u(\cdot) \in C_{\text{ad}, K}(0, t_n; U)$. We select a $(\mathbb{R} \times C(0, \bar{t}; E_\alpha) \times E)$-weakly convergent subsequence of (z_{0n}, μ_n, w_n) and show that $(z_0, \mu, w) \neq 0$. If $z_{0n} \to z_0 \neq 0$ there is nothing to prove. If $z_{0n} \to 0$, then the proof that $(\mu, w) \neq 0$ is exactly the same as in that in Theorem 11.6.1. Making

---

$^{(2)}$ The symbol $\mathbf{R}_K(\bar{t}, \zeta; U)$ has a different meaning than in Theorem 11.6.1.

$y(t) = y(t, \zeta, u)$ in (11.6.24) and using the fact that $y_0(t, u)$ is bounded for $u(\cdot)$ bounded in $L_w^\infty(0, T; X^*)$ (consequence of the hypotheses), we obtain

$$\langle w_n, Ky \rangle \le \langle w_n, Ky(\bar{t}, \zeta, u) \rangle + \delta_n \qquad (11.6.25)$$

with $\delta_n \to 0$. This inequality is just as good as (9.1.30) to prove that $w = (K^{-1})^* z$, $z \in (K^* E^*)^1 (B^*)$, $z \in N_Y(y(\bar{t}, \zeta, \bar{u}))$. Taking limits, we end up with

$$z_0 r + \int_0^{\bar{t}} \langle \mu(dt), y(t) \rangle_\alpha + \langle w, Ky \rangle$$

$$\le z_0 y_0(\bar{t}, u) + \int_0^{\bar{t}} \langle \mu(dt), y(t, \zeta, u) \rangle_\alpha + \langle w, Ky(\bar{t}, \zeta, u) \rangle \quad (11.6.26)$$

for $r \le m$, $y(\cdot) \in \mathbf{M}(\bar{t})$, $y \in Y$ and $u(\cdot) \in C_{\mathrm{ad}, K}(0, \bar{t}; U)$. For $y(t) = y(t, \zeta, \bar{u})$, $r = y_0(\bar{t}, \bar{u})$ and $y = y(\bar{t}, \zeta, \bar{u})$ we obtain, after some rearrangements,

$$z_0 y_0(\bar{t}, \bar{u}) + \int_0^{\bar{t}} \langle B^* \bar{z}_0(t), \bar{u}(t) \rangle dt \le z_0 y_0(\bar{t}, u) + \int_0^{\bar{t}} \langle B^* \bar{z}_0(t), u(t) \rangle dt$$

for all $u(\cdot) \in C_{\mathrm{ad}, K}(0, \bar{t}; U)$, where $\bar{z}_0$ solves

$$d\bar{z}_0(s) = -A^\circ \bar{z}_0(s) ds - \mu(ds), \qquad \bar{z}(\bar{t}) = z.$$

We use this inequality with spike perturbations, whose effect on the cost functional has been determined in Example 11.6.3, and we obtain (11.6.17). It follows from (11.6.24) and the definition of $\mathbf{M}_K(m - \varepsilon_n, \bar{t}, Y)$ that $z_{0n} \ge 0$, thus $z_0 \ge 0$. That $\mu \in N_{\mathbf{M}(\bar{t})}(y(\cdot, \zeta, \bar{u}))$ and $z \in N_Y(y(\bar{t}, \zeta, \bar{u}))$ is proved as in Theorem 11.6.1. ■

Example 11.6.5. We apply Theorem 11.6.1 to the system

$$\begin{aligned} y_t(t, x) &= y_{xx}(t, x) + b(x)u(t) \quad (0 < x < \pi), \\ y(t, 0) &= y(t, \pi) = 0 \end{aligned} \qquad (11.6.27)$$

in **9.5**. The state space is $E = C_0[0, \pi]$ as in Example 9.5.6, with $F = X^* = \mathbb{R}$, $U = [-1, 1]$ and $b(\cdot)$ satisfying

$$b_n = \int_0^\pi b(x) \sin nx \, dx \ne 0 \quad (n = 1, 2, \ldots). \qquad (11.6.28)$$

It was determined in **9.6** that the operator

$$Ky = K(\Sigma c_n \sin nx) = \sum_{n=1}^\infty \frac{n^{3+\rho} e^{n\pi}}{b_n} c_n \sin nx \qquad (11.6.29)$$

$(\rho > 0)$ satisfies all that is needed in Theorem 11.6.1. We limit ourselves here to "add state constraints" and see where the results in **9.6** have to be modified. The constraints are

$$y(t, x) \in M_s, \qquad y_x(t, x) \in M_g,$$

where M_s, M_g are bounded closed intervals with nonempty interior. If $\alpha > 1/2$, the set $M \subseteq E_\alpha$ defined by $y(x) \in M_s$, $y_x(x) \in M_g$ is convex with nonempty interior so we may apply Theorem 11.6.1, obtaining the following upgrade of Theorem 9.6.1. The only assumption on Y is closedness; thus the point target problem $y(\bar{t}, u) = \bar{y} \in D(K)$ is included.

Theorem 11.6.6. *Let $\bar{u}(\cdot)$ be a time optimal control for the system* (11.6.27) *with $\bar{y} \in D(K)$. Then there exist $\mu \in \Sigma(0, \bar{t}; ((-A)^\alpha)^* E^*)$, $z \in (K^* C_0[0, \pi]^*)^1 (B^*)$ with $(\mu, z) \neq 0$ and such that*

$$\mu \in N_{M(\bar{t})}(y(\cdot, \zeta, \bar{u})) \tag{11.6.30}$$

and

$$\bar{u}(t) = \frac{\kappa(t)}{|\kappa(t)|} \quad (\kappa(t) \neq 0), \tag{11.6.31}$$

where

$$\kappa(t) = \int_0^\pi b(x)\bar{z}(t, x)dx, \tag{11.6.32}$$

$\bar{z}(t, x)$ the solution of the backwards heat equation

$$\begin{aligned} d\bar{z}(t, x) &= -\bar{z}_{xx}(t, x)dt - \mu(dt) \quad (0 < x < \pi) \\ \bar{z}(t, 0) &= \bar{z}(t, \pi) = 0 \end{aligned} \tag{11.6.33}$$

with final condition

$$\bar{z}(\bar{t}, \cdot) = z. \tag{11.6.34}$$

As a particular case of Theorem 11.3.2 we deduce that $\mu = 0$ in the set

$$e_0 = \{(t, x); (y(t, x, \bar{u}), y_x(t, x, \bar{u})) \in \text{Int}(M_s \times M_g)\}.$$

When the only state constraint is $y(t, x) \in M_s$ (no constraints on the gradient) we may take $\alpha = 0$, so that μ is a bounded Borel measure in the rectangle $[0, \bar{t}] \times [0, \pi]$ vanishing in e_0.

Results like Theorems 11.3.1, 11.5.2 and 11.6.6 lead to the problem below, where Ω is a domain in $\mathbb{R}^m$ and β is a boundary condition on the boundary Γ.

Problem 11.6.7. Let $\mu \in \Sigma(0, \bar{t}; ((-A)^{\alpha})^* C(\overline{\Omega})^*) = \Sigma(0, \bar{t}; ((-A)^{\alpha})^* \Sigma(\overline{\Omega}))$, and let $z(t, x)$ be a nontrivial solution of

$$dz(t, x) = -A_1(\beta)z(t, x)dt - D(t)z(t, x)dt - \mu(dt) \quad (0 \le t < \bar{t}, x \in \overline{\Omega}).$$
$$(11.6.35)$$

What can be said about the *nodal set*

$$e(\bar{t}, z, \mu) = \{(t, x) \in [0, \bar{t}] \times \overline{\Omega}; z(t, x) = 0\}? \quad (11.6.36)$$

This needs clarification. We want to encompass all examples, thus the symbol $D(t)$ in (11.6.35) is a first order differential operator like those in equations (11.3.5) and (11.5.52); of course, the precise definition of these operators (and of solution of (11.6.35)) is given by the abstract theory in 11.1, and for the equation (11.6.33), $D(t) \equiv 0$. "Nontrivial" means "not identically zero" (when $\mu = 0$).

Prima facie, what we mean by solution (and thus the answer to the problem) should depend on the final condition z satisfied by $z(t, x)$. In Theorem 11.3.1, $z \in ((-A_c(\beta))^{\alpha})^* C(\overline{\Omega})^* = ((-A_c(\beta))^{\alpha})^* \Sigma(\overline{\Omega})$; on the other hand, in Theorem 11.5.2, $z \in (((-A_c(\beta))^* C(\overline{\Omega})^*)^1 = (((-A_c(\beta))^* \Sigma(\overline{\Omega}))^1$, and in Theorem 11.6.6 $z \in (K^* C_0[0, \pi]^*)^1 (B^*) = (K^* \Sigma_0[0, \pi])^1 (B^*)$. In all three cases, however, the solution $z(t, \cdot)$ is swallowed up by $L^1(\Omega)$ for $t < \bar{t}$ and if $\{t_n\} \subseteq [0, \bar{t}], t_n \to \bar{t}$,

$$e(\bar{t}, z, \mu) = \bigcup_{n=1}^{\infty} e(t_n, z(t_n), \mu)$$

so that it is enough to look at Problem 11.6.7 for $z \in L^1(\Omega)$.

The answer to Problem 11.6.7 we would most like to hear would certainly be "$e(\bar{t}, z, \mu)$ has measure zero," for this would coax maximum information from the diverse minimum principles. Even in the case $\alpha = 0$ where μ is a pure measure, this is likely to be false in the support of μ, where state constraints are saturated (although one imagines that statements like (11.6.30) could be put to use in special cases). That $|e(\bar{t}, z, \mu) \backslash \text{supp}(\mu)| = 0$ is false (for instance, consider the case where $z = 0, t_m = \max \text{supp}(\mu) < \bar{t}$, where $z(t, x) = 0$ for $t > t_m$). However, it does not seem out of line to conjecture that $|e(\bar{t}, z, \mu)| = 0$ when $\mu = 0$, shown below in a particular case; the equation is the homogeneous version of (11.6.35),

$$z_1(t, x) = -A_1(\beta)z(t, x), \qquad z(\bar{t}, \cdot) = z \in L^1(\Omega). \quad (11.6.37)$$

Theorem 11.6.8. *Assume the domain Ω is of class $C^{(\infty)}$, that the operator $A(\beta)$ is self adjoint ($A' = A, \beta' = \beta$) with coefficients infinitely differentiable in $\overline{\Omega}$, and that $\gamma(\cdot)$ is infinitely differentiable on Γ if β is a variational boundary condition. Then the set $e(\bar{t}, z, 0)$ has measure zero.*

Proof. By hypoellipticity of the operator $\partial/\partial t - A_1(\beta)$, the solution $z(t, x)$ of (11.6.37) is infinitely differentiable in $[0, \bar{t}) \times \overline{\Omega}$; in particular, $z(t, \cdot) \in C^{(\infty)}(\overline{\Omega})$

for $t < \bar{t}$, so that we may exchange $A_1(\beta)$ by $A_2(\beta)$ in (11.6.37) for $t < \bar{t}$ and (via another replacement of $\bar{t}$ by $\tilde{t} < \bar{t}$) we may assume from the beginning that $z \in L^2(\Omega)$. Let $\{\phi_n(x)\}$ be the eigenfunctions of $A_2(\beta)$ in $L^2(\Omega)$ (corresponding to eigenvalues $\{-\lambda_n\}$); they are infinitely differentiable in $\overline{\Omega}$ and

$$\|\phi_n\|_{C(\overline{\Omega})} = O(n^k) \quad \text{as } n \to \infty \tag{11.6.38}$$

(Sogge [1993]). Write $z(t, x) = \Sigma c_n(t)\phi_n(x)$ (convergence in $L^2(\Omega)$). We have

$$c_n'(t) = \int_\Omega z_t(t, x)\phi_n(x)dx = -\int_\Omega A_2(\beta)z(t, x)\phi_n(x)dx$$

$$= -\int_\Omega z(t, x)A_2(\beta)\phi_n(x)dx = \lambda_n c_n(t), \tag{11.6.39}$$

so that

$$z(\bar{t} - t, x) = \sum c_n e^{-\lambda_n(\bar{t}-t)}\phi_n(x),$$

$\{c_n\}$ the Fourier coefficients of z. In view of (11.6.38), the series is uniformly convergent in $\overline{\Omega}$ for $t > 0$. Assume now that the set $e(\bar{t}, z, 0)$ has positive measure. Then, if χ is its characteristic function,

$$\int_\Omega \int_0^{\bar{t}} \chi(t, x)dtdx = |e(\bar{t}, z, 0)| > 0,$$

hence there exists a set $d \subseteq \Omega$ of positive measure such that, for every $x \in d$, we have

$$\sum c_n e^{-\lambda_n t}\phi_n(x) = 0$$

in a set of positive measure, thus, by analiticity, in $0 < t \leq \bar{t}$. However, a Dirichlet series can vanish identically only if all its coefficients are zero (Bernstein [1933]) thus $\phi_n(x) = 0$ for each $x \in d$ for all n. If x is a density point of d then all partial derivatives of ϕ_n vanish at x; iterating the argument, all partial derivatives of all orders of ϕ_n vanish in a set of positive measure, in particular ϕ_n has a zero of infinite order at some point $x_0 \in \Omega$. Now, $\phi_n(x)$ is a solution of the elliptic equation

$$A_2(\beta)\phi_n(x) + \lambda_n\phi_n(x) = 0$$

so one can bring to bear results on unique continuation of solutions of elliptic equations having zeros of infinite order (Garofalo–Lin [1987], Jerison–Kenig [1985], Robbiano [1988], Sogge [1990], Stein [1985]. In particular, using Sogge [1990, Theorem 11.2.1] we deduce that $\phi_n(x) \equiv 0$ for all n, an obvious contradiction.   ∎

That the result applies to a system like (11.6.27) (without state constraints) goes without saying, but it may come as a mild surprise that Theorem 11.6.8 is useless outside the realm of linear systems. To see this for the system (11.4.1) note that the adjoint variational equation (11.5.52) contains two terms whose infinite smoothness cannot be trusted, namely those depending on the nonlinearity ϕ. Even if ϕ is infinitely differentiable, these two terms contain the optimal trajectory $y(s, x, \bar{u})$ which, due to the possible roughness of the control $\bar{u}(t, x)$ may not even have two continuous derivatives in x or one continuous derivative in s. For more information on Problem 11.6.7 see the Miscellaneous Comments to Part II.

11.7. Control Problems for the Navier–Stokes Equations

The arguments in **11.4**, **11.5** and **11.6** yield results on optimal control problems for the Navier–Stokes equations (8.6.1)

$$\mathbf{y}_t(t, x) = \mathbf{A}_p\mathbf{y}(t, x) - \mathbf{P}_p((\mathbf{y}(t, x) \cdot \nabla)\mathbf{y}(t, x) + \mathbf{u}(t, x)), \quad \mathbf{y}(t, 0) = \zeta(x)$$
(11.7.1)

in the space $E = X^p(\Omega)$, $1 < m < \infty$; presence of the projection operator $\mathbf{P}_p$ precludes a treatment based on spaces of continuous functions (for details on the spaces and operators involved see **8.6**). To fix ideas, we assume that the control set $\mathbf{U}$ is a closed bounded subset of $L^\infty(\Omega; \mathbb{R}^m)$, the control space $C_{\mathrm{ad}}(0, T, U)$ consisting of all $\mathbf{u}(\cdot) \in L_w^\infty(0, T; L^\infty(\Omega; \mathbb{R}^m)) = L^\infty((0, T) \times \Omega; \mathbb{R}^m)$ such that $\mathbf{u}(t) \in \mathbf{U}$ a.e. The results actually hold for the more general equation

$$\mathbf{y}_t(t, x) = \mathbf{A}_p\mathbf{y}(t, x) - \mathbf{P}_p(\phi(t, x, \mathbf{y}(t, x), \nabla\mathbf{y}(t, x)) + \mathbf{u}(t, x)),$$

$$\mathbf{y}(t, 0) = \zeta(x)$$
(11.7.2)

where $\phi : [0, T] \times \mathbb{R}^m \times \mathbb{R}^m \times \mathbb{R}^{m \times m} \to \mathbb{R}^m$, $\nabla\mathbf{y}(x)$ the Jacobian matrix of the m-vector function $\mathbf{y} = (y_1, \ldots, y_m)$:

$$\nabla\mathbf{y} = \begin{bmatrix} \partial_1 y_1 & \cdots & \partial_m y_1 \\ \cdots\cdots\cdots\cdots \\ \partial_1 y_m & \cdots & \partial_m y_m \end{bmatrix}.$$

We handle the equation (11.7.2) via the abstract model

$$y'(t) = Ay(t) + f(t, y(t)) + Bu(t),$$
(11.7.3)

(see also Remark 11.7.3). We assume that A in (11.7.3) is the infinitesimal generator of a bounded analytic semigroup in a Banach space E and that $B \in (X^*, E)$ (X a separable Banach space) with $B^* : E^* \to X$. The control set U for (11.7.3) is a subset of X^* and $C_{\mathrm{ad}}(0, T; U)$ is the set of all $u(\cdot) \in L_w^\infty(0, T; X^*)$ with $u(t) \in U$ a.e. We assume that $f(t, y)$ satisfies Hypothesis III$_{\alpha,0}$ with $0 < \alpha < 1$ and that the space E is reflexive and separable. Under this last condition, the operator B

upgrades the controls in (11.7.3) to strongly measurable; in fact, if $y^* \in E^*$ then $\langle y^*, Bu(t) \rangle = \langle B^* y^*, u(t) \rangle$, so that $Bu(\cdot)$ is E^*-weakly measurable; since E is separable, $Bu(\cdot)$ is strongly measurable.

The state constraints and cost functional associated with (11.7.3) are (11.2.2) and (11.2.3), with Y and $M(t)$ closed in E_α and $f_0(t, y, u(t))$ satisfying Hypothesis III_α^0 for all $u(\cdot)$, with $K_0(c), L_0(c)$ independent of $u(\cdot)$. We skip the existence question; under additional assumptions on $C_{ad}(0, T; U)$ and on $y_0(t, u)$, existence is covered by the forthcoming Theorem 14.5.1.

Dovetailing (11.7.2) in the abstract model (11.7.3) we have $f(t, \mathbf{y}) = \mathbf{P}_p g(t, \mathbf{y})$, with

$$g(t, \mathbf{y})(x) = \phi(t, x, \mathbf{y}(x), \nabla \mathbf{y}(x)). \tag{11.7.4}$$

We assume that $\phi(t, x, \mathbf{y}, \mathbf{Y})$ is differentiable with respect to y_j, y_{jk} for t, x fixed ($\mathbf{Y}$ denotes the matrix $\{y_{jk}\}$) with $\phi(t, x, \mathbf{y}, \mathbf{Y})$, $\nabla_{(\mathbf{y}, \mathbf{Y})}\phi(t, x, \mathbf{y}, \mathbf{Y})$ continuous in $\mathbf{y}, \mathbf{Y}$ for t, x fixed and measurable in t, x for $\mathbf{y}, \mathbf{Y}$ fixed. Finally, we assume that for every $c > 0$ there exists $K(x, c), L(x, c)$ with $K(\cdot, c), L(\cdot, c) \in L^p(\Omega)$ and

$$\|\phi(t, x, \mathbf{y}, \mathbf{Y})\| \le K(x, c) \qquad (0 \le t \le T, x \in \Omega, \|\mathbf{y}\|, \|\mathbf{Y}\| \le c) \tag{11.7.5}$$

$$\|\nabla_{(\mathbf{y}, \mathbf{y})}\phi(t, x, \mathbf{y}, \mathbf{Y})\| \le L(x, c) \qquad (0 \le t \le T, x \in \Omega, \|\mathbf{y}\|, \|\mathbf{Y}\| \le c). \tag{11.7.6}$$

Lemma 11.7.1. *Let $p > m$,*

$$1/2 + m/2p < \alpha < 1. \tag{11.7.7}$$

Then the operator function f maps $[0, T] \times X^p(\Omega)_\alpha = [0, T] \times D((-\mathbf{A}_p)^\alpha)$ into $X^p(\Omega)$ and satisfies Hypothesis $\text{III}_{\alpha,0}$.

Proof. Since $p > m \ge 2$, requesting α to satisfy (11.7.7) implies $\alpha > 1/2 \ge 1/p$, thus the domain of $(-\mathbf{A}_p)^\alpha$ is given by the second relation (8.6.16). Combining this relation with the imbedding (8.5.20) we obtain

$$D((-\mathbf{A}_p)^\alpha) \approx X_p(\Omega) \cap H_0^{2\alpha, p}(\Omega; \mathbb{R}^m)$$
$$\hookrightarrow H^{2\alpha, p}(\Omega; \mathbb{R}^m) \hookrightarrow W^{2\alpha - 2\varepsilon, p}(\Omega; \mathbb{R}^m). \tag{11.7.8}$$

On the other hand, (11.7.7) guarantees that $2\alpha - 2\varepsilon - 1 > m/p$ for $\varepsilon > 0$ small enough, hence $W^{2\alpha - 2\varepsilon - 1, p}(\Omega; \mathbb{R}^m) \hookrightarrow C(\overline{\Omega})$. Finally, the projection operator $\mathbf{P}_p$ is bounded from $L^p(\Omega; \mathbb{R}^m)$ into $X^p(\Omega)$, thus, all we have to show is that the operator

$$\Phi(t, x, \mathbf{y}, \mathbf{Y})(x) = \phi(t, x, \mathbf{y}(x), \mathbf{Y}(x)) \tag{11.7.9}$$

from $C(\overline{\Omega}; \mathbb{R}^m) \times C(\overline{\Omega}; \mathbb{R}^{m \times m})$ into $L^p(\Omega; \mathbb{R}^m)$ satisfies Assumption $\text{III}_{\alpha,0}$. This is done essentially as in Theorem 8.4.9 and we omit the details. ∎

The minimum principle for the system (11.7.3) can be read directly off Theorem 11.2.2 noting that, since E is reflexive we have $A^\circ = A^*$ and $((-A)^\alpha)^* E^* = (E^*)_{-\alpha}$ (Remark 11.1.14); saturation and patch completeness of $C_{ad}(0, T; U)$ are automatic.

In its integral version, the minimum principle for (11.7.3) takes the form

$$z_0 \int_0^{\bar{t}} \{f_0(s, y(s, \bar{u}), v(s)) - f_0(s, y(s, \bar{u}), \bar{u}(s))\} ds$$

$$+ \int_0^{\bar{t}} \langle B^* \bar{z}(s), v(s) - u(s) \rangle ds \geq 0, \tag{11.7.10}$$

for all $v(\cdot) \in C_{ad}(0, \bar{t}; U)$, the costate $\bar{z}(s)$ given by

$$d\bar{z}(s) = -\{A^* + \partial_y f(s, y(s, \bar{u}))^*\} \bar{z}(s) ds$$

$$-z_0 \partial_y f_0(s, y(s, \bar{u}), \bar{u}(s)) ds - \mu(ds),$$

$$\bar{z}(\bar{t}) = z + z_0 \partial_y g_0(\bar{t}, y(\bar{t}, \bar{u})) \tag{11.7.11}$$

in $0 \leq t \leq \bar{t}$; $(z_0, \mu, z) \in \mathbb{R} \times \Sigma(0, T; (E^*)_{-\alpha}) \times (E^*)_{-\alpha}$ satisfies

$$z_0 \geq 0, \quad \mu \in \left(\liminf_{n \to \infty} T_{M(\bar{t})}(\tilde{y}^n(\cdot)) \right)^-, \quad z \in \left(\liminf_{n \to \infty} T_Y(\tilde{y}^n) \right)^-, \tag{11.7.12}$$

where $\tilde{y}^n(\cdot)$ (resp. $\tilde{y}^n$) is a sequence in $M(\bar{t})$ (resp. in Y) such that $\tilde{y}^n(\cdot) \to y(\cdot, \bar{u})$ in $C(0, \bar{t}; E_\alpha)$ (resp. $\tilde{y}^n \to y(\bar{t}, \bar{u})$ in E_α). The version for the time optimal problem follows as in **11.5**. The conditions to pass to the pointwise minimum principle

$$\langle B^* \bar{z}(s), \bar{u}(s) \rangle = \min_{v \in U} \langle B^* \bar{z}(s), v \rangle \text{ a.e.} \tag{11.7.13}$$

are automatically satisfied; if in addition $M(\bar{t})$ is T-full in $C(0, \bar{t}; E_\alpha)$ and Y is T-full in E_α then (Lemma 11.2.4) $(z_0, \mu, z) \neq 0$ $((\mu, z) \neq 0$ for the time optimal problem).

To apply all of this to the system (11.7.1) we assume the state constraints are defined as in Remark 11.3.6,

$$\mathfrak{S} \mathbf{y}(t, x) \in M_s \subseteq \mathbb{R}^k \qquad (0 \leq t \leq \bar{t}, x \in \Omega) \tag{11.7.14}$$

where $\mathfrak{S}$ is a linear differential operator $\mathfrak{S} \mathbf{y}(x) = \Sigma \Sigma \chi_{ij}(x) \partial_i y_j(x) + \Sigma \chi_j(x) y_j(x)$ whose coefficients $\chi_{ij}(x), \chi_j(x)$ are k-vector functions with entries in $C(\bar{\Omega})$; similarly, the target condition is

$$\mathfrak{T} \mathbf{y}(\bar{t}, x) \in M_Y \subseteq \mathbb{R}^l \qquad (x \in \Omega) \tag{11.7.15}$$

where $\mathfrak{T} \mathbf{y}(x) = \Sigma \Sigma \eta_{ij}(x) \partial_i y_j(x) + \Sigma \eta_j(x) y_j(x)$, the $\eta_{ij}(x), \eta_j(x)$ l-vector functions with entries in $C(\bar{\Omega})$. If p and α satisfy (11.7.7) then the imbedding (8.5.22) implies that if M_s and M_Y are convex and have interior points then M, Y are convex and have interior points, which guarantees nontriviality of the multipliers. We have $B = \mathbf{P}_p \mathbf{I}_p$, where $\mathbf{I}_p$ is the canonical imbedding of $L^\infty(\Omega; \mathbb{R}^m)$ into $L^p(\Omega; \mathbb{R}^m)$ so that $\mathbf{B}^* = \mathbf{I}_p^* \mathbf{P}_p^* = \mathbf{I}_p^* \mathbf{P}_q$ (the last equality because of the last relation (8.6.9)). Since $\mathbf{I}_p^*$ is the canonical imbedding of $L^q(\Omega; \mathbb{R}^m)$ into $L^1(\Omega; \mathbb{R}^m) \subseteq (L^\infty(\Omega; \mathbb{R}))^*$, we can write (11.7.10) in the form

$$\int_\Omega \bar{\mathbf{z}}(s, x) \cdot \bar{\mathbf{u}}(s, x) dx = \min_{\mathbf{v} \in \mathbf{U}} \int_\Omega \bar{\mathbf{z}}(s, x) \cdot \mathbf{v}(x) dx \qquad (11.7.16)$$

(note that $\bar{\mathbf{z}}(s)$ resides in $X^p(\Omega)$, so that $\mathbf{P}_q \bar{\mathbf{z}}(s) = \bar{\mathbf{z}}(s)$).

It follows from the theory for (11.7.3) that $\mu = 0$ if there are no state constraints. Theorem 11.3.2 has also a counterpart: using the same arguments we can prove (11.3.16) for $v = ((-\mathbf{A}_p)^{-1})^* \mu$ and $\phi \in \mathcal{D}(B_j; \mathbb{R}^m)|_{[0, \bar{t}] \times \bar{\Omega}, \beta}$ with $\nabla \cdot \phi = 0$ for a family $\{B_j\}$ of open sets satisfying (11.3.13). For $m = 3$ we may take $\phi = \nabla \times \psi, \psi \in \mathcal{D}(B_j \cap ((0, \bar{t}) \times \Omega); \mathbb{R}^m)$ and obtain in this way that $\nabla \times \mu = 0$ in the set $e_0 \cap ((0, \bar{t}) \times \Omega)$, where e_0 is the set where the state constraints are not saturated, that is, where $\circledS \mathbf{y}(t, x) \in \text{int}(M_s)$.

Remark 11.7.2. For the Navier–Stokes nonlinearity. $\partial \mathbf{F}(\mathbf{y}) = \mathbf{P}_p \partial \mathbf{G}(\mathbf{y})$, where $\partial \mathbf{G}(\mathbf{y})\mathbf{h} = -(\mathbf{y} \cdot \nabla)\mathbf{h} - (\mathbf{h} \cdot \nabla)\mathbf{y}$. Coordinatewise,

$$(\partial \mathbf{G}(\mathbf{y})\mathbf{h})_k = -\Sigma_j (y_j \partial_j h_k + \partial_j y_k h_j), \qquad (11.7.17)$$

so that

$$\begin{aligned} ((\partial \mathbf{G}(\mathbf{y})^*)\mathbf{z})_k &= \Sigma_j\{(\partial_j(y_j z_k) - (\partial_j y_k)z_k)\} \\ &= \Sigma_j\{y_j \partial_j z_k - (\partial_j y_k)z_k\}, \end{aligned} \qquad (11.7.18)$$

the second inequality stemming from the fact that $\mathbf{y}$ is divergence-free. In vector form,

$$\partial \mathbf{G}(\mathbf{y})^* \mathbf{z} = (\mathbf{y} \cdot \nabla)\mathbf{z} - (\nabla \mathbf{y})^T \mathbf{z}. \qquad (11.7.19)$$

Since $\partial \mathbf{F}(\mathbf{y})^* = \mathbf{P}_q \partial \mathbf{G}(\mathbf{y})^* \mathbf{P}_q$, the adjoint variational equation is

$$\begin{aligned} d\mathbf{z}(s) = &-\mathbf{A}_q \mathbf{z}(s) ds - \mathbf{P}_q(\mathbf{y}(s) \cdot \nabla)\mathbf{z}(s) ds + \mathbf{P}_q (\nabla \mathbf{y}(s))^T \mathbf{z}(s) ds \\ &-z_0 \mathbf{P}_q \partial_\mathbf{y} f_0(s, \mathbf{y}(s), \mathbf{u}(s)) ds - \mu(ds) \end{aligned} \qquad (11.7.20)$$

in the space $X^q(\Omega)$, where $\mu \in \Sigma(0; \bar{t}; X^q(\Omega)_{-\alpha})$.

Remark 11.7.3. A maximum principle for a system of the form

$$\mathbf{y}_t(t, x) = \mathbf{A}_p \mathbf{y}(t, x) - \mathbf{P}_p \phi(t, x, \mathbf{y}(t, x), \nabla \mathbf{y}(t, x), u(t)) \qquad (11.7.21)$$

along the lines of Theorem 11.3.1 can be obtained using Theorem 11.2.2; the assumptions are on $\phi(t, x, \mathbf{y}, \mathbf{Y}, u(t))$ with $K(t, x), L(t, x)$ independent of $u(\cdot) \in C_{\text{ad}}(0, T; U)$.

11.8. Control Problems for the Navier–Stokes Equations: The Point Target Case

As in **11.7** we consider the system

$$\mathbf{y}_t(t, x) = \mathbf{A}_p \mathbf{y}(t, x) - \mathbf{P}_p(\phi(t, x, \mathbf{y}(t, x), \nabla \mathbf{y}(t, x)) + \mathbf{u}(t, x)),$$

$$\mathbf{y}(t, 0) = \zeta(x) \tag{11.8.1}$$

in $X^p(\Omega)$, modelled as the abstract control system (11.7.3). The control space $\mathbf{U}$ is a closed bounded subset of $L^p(\Omega; \mathbb{R}^m)$ and $C_{\text{ad}}(0, T; \mathbf{U})$ consists of all $\mathbf{u}(\cdot) \in L^\infty(0, T; L^p(\Omega; \mathbb{R}^m))$ with $\mathbf{u}(t, \cdot) \in \mathbf{U}$ a.e. The result below on the nonlinear operator $f(t, \mathbf{y})$ corresponds to Lemma 11.4.1 and the assumptions on ϕ are as follows. For t fixed $\phi(t, x, \mathbf{y}, \mathbf{Y})$ is differentiable with respect to y_j, y_{jk} for t, x fixed, with ϕ and its first partials locally Lipschitz continuous in $\overline{\Omega} \times \mathbb{R} \times \mathbb{R}^m$ and with $\phi(t, x, \mathbf{y}, \mathbf{Y})$ measurable in t for $x, \mathbf{y}, \mathbf{Y}$ fixed. Inequalities (11.7.5) and (11.7.6) are requested with K, L independent of x.

Lemma 11.8.1. *Let* $1/2 < \alpha < 1$, $p > m$ *so large that*

$$\frac{1}{2}\left(\alpha - \frac{1}{2p}\right) - \frac{1}{4} > \frac{m}{2p}. \tag{11.8.2}$$

Then there exists $\delta > 0$ *such that the operator* $f(t, \mathbf{y}) = \mathbf{P}_p g(t, \mathbf{y})$, $g(t, \mathbf{y})$ *given by* (11.7.4), *maps* $X^p(\Omega)_\alpha$ *into* $X^p(\Omega)_\delta$ *and* $(-\mathbf{A}_p)^\delta f(t, \mathbf{y})$ *satisfies Hypothesis* $III_{\alpha,0}$.

Proof. The first tool is the imbedding (11.7.8) (the assumptions of Lemma 11.7.1 are satisfied). To go the other way, note that (11.8.2) implies that the left side is positive, so that $\alpha - 1/2 > 1/2p$. Select then δ such that

$$0 < \delta < \frac{1}{2p}, \qquad \varepsilon = \frac{\alpha - \delta}{2} - \frac{1}{4} > \frac{m}{2p}. \tag{11.8.3}$$

We have

$$2\alpha - 2\varepsilon - 1 = 2\delta + 2\varepsilon = s > 2\varepsilon > m/p. \tag{11.8.4}$$

Applying the first relation (8.6.16) we obtain

$$X^p(\Omega) \cap W^{2\delta + 2\varepsilon, p}(\Omega; \mathbb{R}^m) \hookrightarrow X^p(\Omega) \cap H^{2\delta, p}(\Omega; \mathbb{R}^m) \approx D((-\mathbf{A}_p)^\delta) \tag{11.8.5}$$

(the boundary condition lost since $2\delta < 1/p$). This imbedding, the equality in (11.8.4) and the work in **11.4** suggest looking at the operator (11.7.9) as

$$\Phi : W^s(\Omega; \mathbb{R}^m) \times W^s(\Omega, \mathbb{R}^{m \times m}) \to W^s(\Omega; \mathbb{R}^m).$$

Lemma 11.8.2. *Let $s > m$.*

$$s - m/p > 0. \tag{11.8.6}$$

Then, for t fixed, Φ is a continuous, locally bounded operator and has a continuous, locally bounded Fréchet derivative given by

$$\partial \Phi_{(\mathbf{y},\mathbf{Y})}(t, \mathbf{y}, \mathbf{Y})(\mathbf{h}, \mathbf{H}) = \sum_{j=1}^{m} \partial_j \phi(t, x, \mathbf{y}(x), \mathbf{Y}(x)) h_j(x)$$

$$+ \sum_{j=1}^{m} \sum_{k=1}^{m} \partial_{jk} \phi(t, x, \mathbf{y}(x), \mathbf{Y}(x)) h_{jk}(x), \quad (11.8.7)$$

where $\mathbf{h}(x) = \{h_j(x)\}$, $\mathbf{H}(x) = \{h_{jk}(x)\}$, ∂_j = derivative with respect to y_j, ∂_{jk} = derivative with respect to y_{jk}.

Lemma 11.8.2 is a "vector version" of Lemma 11.4.2, and its proof is exactly the same. The result implies that $g(t, \mathbf{y})$ satisfies Assumption $\mathrm{III}_{\alpha,0}$ as an operator from $[0, T] \times X^p(\Omega) \to W^{2\delta+2\varepsilon, p}(\Omega)$, hence the claims in Lemma 11.8.1 about the operator $f(t, \mathbf{y})$ follow from the imbedding (11.8.5) and from

Lemma 11.8.3. *Let $0 < s < 1$. Then $\mathbf{P}_p$ is a bounded operator from $W^{s,p}(\Omega; \mathbb{R}^m)$ into itself.*

Proof. See von Wahl [1985, p. XXIII]. ∎

The result below is for the abstract equation (11.7.3), under the assumptions on A and B in **11.7**. We assume that $f(t, y)$ maps E_α into $D((-A)^\delta)$ and that $(-A)^\delta f(t, y)$ satisfies Hypothesis $\mathrm{III}_{\alpha,0}$. The control space is defined as in **11.7**, with the additional assumption that 0 is an interior point of U. We consider the optimal control problem with cost functional

$$y_0(t, u) = \int_0^t f_0(\tau, y(\tau, u), u(\tau)) d\tau, \tag{11.8.8}$$

where $f_0(t, y, u(t))$ satisfies Hypothesis III_α^0 in **11.2** for all $u(\cdot) \in C_{\mathrm{ad}}(0, T; U)$ with K_0, L_0 independent of u. The state constraints and target condition are (11.2.2) with each $M(t)$ closed in E_α and Y closed in $E_1 = D(A)$. Since E is reflexive, $(E_\alpha)^* = ((-A)^\alpha)^* E^* = (E^*)_{-\alpha}$. The role of the space $((-A)^* E^*)^1$ in Theorem 11.5.2 is here taken over by $(E^*)^1_{-1}(B^*)$, the subspace of all $z \in (E^*)_{-1}$ with

$$\int_0^1 \|B^* S(t)^* z\|_X dt$$

$$= \int_0^1 \|B^*(-A)^* S(t)^*((-A)^{-1})^* z\|_X dt < \infty. \tag{11.8.9}$$

(the notation follows that in **9.1**).

The minimum principle is a particular case of Theorem 11.5.2. If $\bar{u}(\cdot)$ is an optimal control then

$$z_0 \int_0^{\bar{t}} \{ f_0(s, y(s, \bar{u}), v(s)) - f_0(s, y(s, \bar{u}), \bar{u}(s)) \} ds$$

$$+ \int_0^{\bar{t}} \langle B^* \bar{z}(s), v(s) - \bar{u}(s) \rangle ds \geq 0 \qquad (11.8.10)$$

for all $v(\cdot) \in C_{\text{ad}}(0, \bar{t}; U)$, $\bar{z}(s)$ the solution of

$$d\bar{z}(s) = -\{ A^* + \partial_y f(s, y(s, \bar{u}))^* \} \bar{z}(s) ds$$
$$- z_0 \partial_y f_0(s, y(s, \bar{u}), \bar{u}(s)) ds - \mu(ds), \quad \bar{z}(\bar{t}) = z \qquad (11.8.11)$$

in $0 \leq t \leq \bar{t}$, with $(z_0, \mu, z) \in \mathbb{R} \times \Sigma(0, \bar{t}; (E^*)_{-\alpha}) \times (E^*)^1_{-1}(B^*)$ satisfying

$$z_0 \geq 0, \quad \mu \in \left(\liminf_{n \to \infty} T_{\mathbf{M}(\bar{t})}(\tilde{y}^n(\cdot)) \right)^-, \quad z \in \left(\liminf_{n \to \infty} T_Y(\tilde{y}^n) \right)^-, \qquad (11.8.12)$$

where $\{ y^n(\cdot) \}$ (resp. $\{ \tilde{y}^n \}$) is a sequence in $\mathbf{M}(\bar{t})$ (resp. Y) such that $\tilde{y}^n(\cdot) \to y(\cdot, \bar{u})$ in $C(0, \bar{t}; E_\alpha)$ (resp. $\tilde{y}^n \to y(\bar{t}, \bar{u})$ in E_1).

That $(z_0, z, \mu) \neq 0$ $((z, \mu) \neq 0$ for the time optimal problem) should not be expected in general, since we must rely on Lemma 11.5.6, where $B = I$. For the system (11.8.1) the argument generalizes handily; it suffices to note that the control defined by (11.5.49) belongs to $X^p(\Omega)$ and to write $v(t) = \mathbf{P}_p v(t)$.

11.9. Convergence of Suboptimal Controls, I

The results in these last four sections are motivated by numerical computation of optimal controls for distributed parameter systems. It is often possible (for instance, using penalty methods combined with discretization) to compute a sequence of controls, none of them optimal, but satisfying the state constraints and target condition with increasing accuracy, and giving values of the cost functional increasingly close to the minimum. A pressing question is: will this sequence of "suboptimal" controls converge to an optimal control in some norm meaningful for numerical computation?

We precise this poser below. In this section and the next, the control system is

$$y'(t) = Ay(t) + f(t, y(t), u(t)), \quad y(0) = \zeta \qquad (11.9.1)$$

The hypotheses are those in Chapter 10; we assume that A is the infinitesimal generator of a strongly continuous semigroup $S(\cdot)$ and $f(t, y, u)$ satisfies Hypothesis III with $K(\cdot, c), L(\cdot, c)$ independent of $u(\cdot) \in C_{\text{ad}}(0, T; U)$. It makes no difference if we allow $A = A(t)$ and postulate a strongly continuous solution operator $S(t, s)$. The state constraints and target condition are

$$y(t) \in M(t) \subseteq E \quad (0 \leq t \leq \bar{t}), \qquad y(\bar{t}) \in Y \qquad (11.9.2)$$

respectively, and the cost functional

$$\mathbf{f}(u) = y_0(t, u) = \int_0^t f_0(\tau, y(\tau, u), u(\tau)) d\tau \qquad (11.9.3)$$

where $f_0(t, y, u)$ satisfies Hypothesis III^0 with $K_0(\cdot, c), L_0(\cdot, c)$ independent of $u(\cdot) \in C_{ad}(0, T; U)$.

Let m be the minimum of the functional (11.9.3) subject to the constraints (11.9.2). A control $u(\cdot)$ is ε-**suboptimal** for this problem if $y(t, \zeta, u)$ exists in $0 \le t \le t_n$, $dist(y(t_n, \zeta, u), Y) \le \varepsilon$, $dist(y(t, \zeta, u), M) \le \varepsilon$ $(0 \le t \le t_n)$ and $y_0(t_n, \zeta, u) \le m + \varepsilon$. We call a sequence $\{u_n(\cdot)\}$ of ε_n-suboptimal controls with $\varepsilon_n \to 0$ a **minimizing sequence.**

Problem 11.9.1. If $\{u_n(\cdot)\}$ is a minimizing sequence with $\{t_n\}$ bounded, do we have

$$u_n(\cdot) \to \bar{u}(\cdot) \qquad (11.9.4)$$

where $\bar{u}(\cdot)$ is an optimal control?

A number of caveats have to be added. Some are:

(a) There cannot be an affirmative answer if there is more than one optimal control (take the sequence that alternates two different optimal controls). Thus, in the absence of uniqueness, (11.9.4) can at most be expected for a subsequence.

(b) It is unlikely that much can be said in the subset where the state constraints are *saturated*, that is, where $y(t, \zeta, u)$ belongs to the boundary of the state constraint set.

(c) L^∞ convergence is out of the question (a large perturbation of an optimal control in a small set will produce an ε-suboptimal control for arbitrarily small ε).

Restriction (c) is reflected in the results in this and following sections, where convergence is in L^p with $p < \infty$. We note also that an affirmative answer to Problem 11.9.1 is often cheap to obtain if we are content with weak convergence of a subsequence in (11.9.4); however, weak convergence does not make the grade as a practical tool in numerical computation. We shall consider (11.9.1) with different initial conditions, thus we call $y(t, \zeta, u)$ the solution corresponding to ζ and $u(\cdot)$.

Our first result is for the time optimal problem, where the minimum principle reads

$$\langle \bar{z}(s), f(s, y(s, \bar{\zeta}, \bar{u}), v) - f(s, y(s, \bar{\zeta}, \bar{u}), \bar{u}(s)) \rangle \ge 0 \quad (v \in U) \qquad (11.9.5)$$

in a set $e \subseteq [0, \bar{t}]$ of full measure, the costate $\bar{z}(s)$ the solution of the adjoint variational equation $d\bar{z}(s) = -\{A^* + \partial_y f(s, y(s, \bar{\zeta}, \bar{u}), \bar{u}(s))^*\} \bar{z}(s) ds - \mu(ds)$, $\bar{z}(\bar{t}) = z$. We can write $\bar{z}(s)$ in the form

$$\bar{z}(s) = \int_s^{\bar{t}} S(\sigma, s; \bar{u})^* \mu_1(d\sigma), \qquad (11.9.6)$$

where $S(t, s; \bar{u})$ is the solution operator of the homogeneous variational equation $\xi'(t) = \{A + \partial_y f(t, y(t, \bar{\zeta}, \bar{u}), \bar{u}(t))\}\xi(t)$ and $\mu_1(ds) = \mu(ds) + \delta_{\bar{t}}(ds)z$ (for proofs see **10.3**).

We assume below that the two controls $\bar{u}(\cdot), u(\cdot) \in C_{ad}(0, \bar{t}; U)$ and the two initial conditions $\zeta, \bar{\zeta}$ at play are such that the solutions exist in $0 \le t \le \bar{t}$ and

$$\|y(t, \bar{\zeta}, \bar{u})\|, \|y(t, \zeta, u)\| \le c. \tag{11.9.7}$$

Rewriting (11.9.1) in the form

$$y'(t, \zeta, u) = \{A + \partial_y f(t, y(t, \bar{\zeta}, \bar{u}), \bar{u}(t))\}y(t, \zeta, u)$$
$$+ f(t, y(t, \zeta, u), u(t)) - \partial_y f(t, y(t, \bar{\zeta}, \bar{u}), \bar{u}(t))y(t, \zeta, u) \tag{11.9.8}$$

we discover that $y(t, \zeta, u)$ solves

$$y(t, \zeta, u) = S(t, 0; \bar{u})\zeta + \int_0^t S(t, \sigma; \bar{u})f(\sigma, y(\sigma, \zeta, u), u(\sigma))d\sigma$$

$$- \int_0^t S(t, \sigma; \bar{u})\partial_y f(\sigma, y(\sigma, \bar{\zeta}, \bar{u}), \bar{u}(\sigma))y(\sigma, \zeta, u)d\sigma$$

$$= S(t, 0; \bar{u})\zeta + \int_0^t S(t, \sigma; \bar{u})f(\sigma, y(\sigma, \bar{\zeta}, \bar{u}), u(\sigma))d\sigma$$

$$+ \int_0^t S(t, \sigma; \bar{u})\{f(\sigma, y(\sigma, \zeta, u), u(\sigma)) - f(\sigma, y(\sigma, \bar{\zeta}, \bar{u}), u(\sigma))\}d\sigma$$

$$- \int_0^t S(t, \sigma; \bar{u})\partial_y f(\sigma, y(\sigma, \bar{\zeta}, \bar{u}), \bar{u}(\sigma))y(\sigma, \zeta, u)d\sigma.^{(1)} \tag{11.9.9}$$

We write the same equation for $y(t, \bar{\zeta}, \bar{u})$,

$$y(t, \bar{\zeta}, \bar{u}) = S(t, 0; \bar{u})\bar{\zeta} + \int_0^t S(t, \sigma; \bar{u})f(\sigma, y(\sigma, \bar{\zeta}, \bar{u}), \bar{u}(\sigma))d\sigma$$

$$- \int_0^t S(t, \sigma; \bar{u})\partial_y f(\sigma, y(\sigma, \bar{\zeta}, \bar{u}), \bar{u}(\sigma))y(\sigma, \bar{\zeta}, \bar{u})d\sigma,$$

and subtract from (11.9.9):

$$y(t, \zeta, u) - y(t, \bar{\zeta}, \bar{u}) = S(t, 0, \bar{u})(\zeta - \bar{\zeta})$$

$$+ \int_0^t S(t, \sigma; \bar{u})\{f(\sigma, y(\sigma, \bar{\zeta}, \bar{u}), u(\sigma)) - f(\sigma, y(\sigma, \bar{\zeta}, \bar{u}), \bar{u}(\sigma))\}d\sigma$$

$$+ \int_0^t S(t, \sigma; \bar{u})\{f(\sigma, y(\sigma, \zeta, u), u(\sigma)) - f(\sigma, y(\sigma, \bar{\zeta}, \bar{u}), u(\sigma))\}d\sigma$$

$$- \int_0^t S(t, \sigma; \bar{u})\partial_y f(\sigma, y(\sigma, \bar{\zeta}, \bar{u}), \bar{u}(\sigma))(y(\sigma, \zeta, u) - y(\sigma, \bar{\zeta}, \bar{u}))d\sigma. \tag{11.9.10}$$

---

[1] Solutions are defined by integral equations as in Chapter 5, thus equalities like this must be verified via the integral equations.

Integrate both sides of (11.9.10) against $\mu_1(dt)$ in $0 \le t \le \bar{t}$,

$$\int_0^{\bar{t}} \langle \mu_1(dt), y(t, \zeta, u) - y(t, \bar{\zeta}, \bar{u}) \rangle$$

$$= \int_0^{\bar{t}} \langle \mu_1(dt), S(t, 0, \bar{u})(\zeta - \bar{\zeta}) \rangle$$

$$+ \int_0^{\bar{t}} \left\langle \mu_1(dt), \int_0^t S(t, \sigma; \bar{u})\{f(\sigma, y(\sigma, \bar{\zeta}, \bar{u}), u(\sigma)) - f(\sigma, y(\sigma, \bar{\zeta}, \bar{u}), \bar{u}(\sigma))\}d\sigma \right\rangle$$

$$+ \int_0^{\bar{t}} \left\langle \mu_1(dt), \int_0^t S(t, \sigma; \bar{u})\{f(\sigma, y(\sigma, \zeta, u), u(\sigma)) - f(\sigma, y(\sigma, \bar{\zeta}, \bar{u}), u(\sigma))\}d\sigma \right\rangle$$

$$- \int_0^{\bar{t}} \left\langle \mu_1(dt), \int_0^t S(t, \sigma; \bar{u})\partial_y f(\sigma, y(\sigma, \bar{\zeta}, \bar{u}), \bar{u}(\sigma))(y(\sigma, \zeta, u) - y(\sigma, \bar{\zeta}, \bar{u}))d\sigma \right\rangle$$

$$(11.9.11)$$

and switch the order of integration in the last three integrals using Lemma 10.2.17. After rearranging we obtain

$$\int_0^{\bar{t}} \langle \bar{z}(\sigma), f(\sigma, y(\sigma, \bar{\zeta}, \bar{u}), u(\sigma)) - f(\sigma, y(\sigma, \bar{\zeta}, \bar{u}), \bar{u}(\sigma)) \rangle d\sigma$$

$$= -\langle \bar{z}(0), \zeta - \bar{\zeta} \rangle + \int_0^{\bar{t}} \langle \mu_1(dt), y(t, \zeta, u) - y(t, \bar{\zeta}, \bar{u}) \rangle dt$$

$$- \int_0^{\bar{t}} \langle \bar{z}(\sigma), f(\sigma, y(\sigma, \zeta, u), u(\sigma)) - f(\sigma, y(\sigma, \bar{\zeta}, \bar{u}), u(\sigma)) \rangle d\sigma$$

$$+ \int_0^{\bar{t}} \langle \bar{z}(\sigma), \partial_y f(\sigma, y(\sigma, \bar{\zeta}, \bar{u}), \bar{u}(\sigma))(y(\sigma, \zeta, u) - y(\sigma, \bar{\zeta}, \bar{u})) \rangle d\sigma. \quad (11.9.12)$$

We exploit this equality as follows. Let V be an arbitrary set with boundary ∂V in a Banach space E. We say that $z \in E^*$ is an **inner normal** to V at $\bar{v} \in \partial V$ if $\langle z, v - \bar{v} \rangle \ge 0$ $(v \in V)$. In this language, (11.9.5) says that the costate $\bar{z}(s)$ is an inner normal to $f(s, y(s, \bar{\zeta}, \bar{u}), U)$ at $f(s, y(s, \bar{\zeta}, \bar{u}), \bar{u}(s))$ for $s \in e$.

Given a cone $Z \subseteq E^*$ (this means $\lambda Z \subseteq Z$ for $\lambda > 0$) the set V is (R, p)-**round with respect to** Z if for each inner normal z to V at $\bar{v} \in \partial V$ with $z \in Z$, $\|z\|_{E^*} \ne 0$ we have

$$\|z\|_{E^*}\|v - \bar{v}\|_E^p \le 2R\langle z, v - \bar{v} \rangle \quad (v \in V). \quad (11.9.13)$$

If $Z = E^*$ the set is (R, p)-**round**.

In a Hilbert space, $(R, 2)$-round means that for every $\bar{v}$ in the boundary of V and each $z \in H$ normal to V at $\bar{v}$ with $\|z\| \ne 0$ there is a ball B of radius R such that $V \subseteq B \subseteq H(\bar{v}) = \{y \in H; \langle z, v - \bar{v} \rangle \ge 0\}$ (Figure 11.1). For an example of a set V (in $\mathbb{R}^2$) $(R, 2)$-round with respect to a cone Z but not $(R, 2)$-round take $Z = \mathbb{R} \times [0, \infty)$ and build V joining the south half of a disk with a triangle.

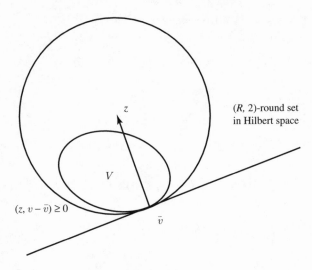

Figure 11.1.

Theorem 11.9.2. *Assume* (11.9.5) *holds a.e. in* $0 \le t \le \bar{t}$, *that* $\|z(\cdot)\|_{E^*}$ *is measurable and that*

$$f(s, y(s, \bar{\zeta}, \bar{u}), U) \qquad (11.9.14)$$

is (R, p)-*round for almost all* s. *Then*

$$\int_0^{\bar{t}} \|\bar{z}(\sigma)\|_{E^*} \| f(\sigma, y(\sigma, \bar{\zeta}, \bar{u}), u(\sigma)) - f(\sigma, y(\sigma, \bar{\zeta}, \bar{u}), \bar{u}(\sigma)) \|_E^p d\sigma$$

$$\le 2RC(\|z\|_{E^*} + \|\mu\|_{\Sigma(0,\bar{t};E^*)})\|\zeta - \bar{\zeta}\|_E$$

$$+ 2R\|z\|_{E^*} \|y(\bar{t}, \zeta, u) - y(\bar{t}, \bar{\zeta}, \bar{u})\|_E$$

$$+ 2R\|\mu\|_{\Sigma(0,\bar{t};E^*)} \max_{0 \le t \le \bar{t}} \|y(t, \zeta, u) - y(t, \bar{\zeta}, \bar{u})\|_E$$

$$+ 4RC(\|z\|_{E^*} + \|\mu\|_{\Sigma(0,\bar{t};E^*)}) \int_0^{\bar{t}} L(\sigma, c)\|y(\sigma, \zeta, u) - y(\sigma, \bar{\zeta}, \bar{u})\|_E d\sigma$$

$$(11.9.15)$$

where c *is the constant in* (11.9.7) *and* C *is a bound for* $\|S(t, s, \bar{u})\|_{(E,E)}$ *in* $0 \le s \le t \le \bar{t}$.

Proof. Inequality (11.9.13) obviously extends to $z = 0$, and we use it in (11.9.12). The different terms in (11.9.15) arise as follows. When estimating $\|\bar{z}(\sigma)\|$ and $\|\bar{z}(0)\|$ we decouple the contributions of z and μ using (10.2.34) and (10.2.36); the constant in the latter is called C (separating z and μ comes in handy when reading the estimate for problems without state constraints, where $\mu = 0$). The first term in (11.9.15) springs up from the first term in (11.9.12). In the second term we write $\mu_1 = \mu + \delta_{\bar{t}}z$, which gives rise to the second and third terms of (11.9.15). In the final term in (11.9.15) we use Hypothesis III straight in the fourth term of (11.9.12) and the mean value theorem in the third; $L(\sigma, c)$ is the function in Hypothesis III corresponding to the constant c.   ∎

The minimum principle for a general cost functional (11.9.3) is

$$z_0\{f_0(s, y(s, \bar{\zeta}, \bar{u}), v) + f_0(s, y(s, \bar{\zeta}, \bar{u}), \bar{u}(s))\}$$
$$+ \langle \bar{z}(s), f(s, y(s, \bar{\zeta}, \bar{u}), v) - f(s, y(s, \bar{\zeta}, \bar{u}), \bar{u}(s)) \rangle \geq 0 \quad (v \in U) \quad (11.9.16)$$

with $z_0 \geq 0$. The costate $\bar{z}(s)$ can be written in the form (11.9.6) with $\mu_1(ds) = \mu(ds) + z_0 \partial_y f_0(s, y(s, \bar{\zeta}, \bar{u}), \bar{u}(s))ds + \delta_{\bar{t}}(ds)z = v(ds) + \delta_{\bar{t}}(ds)z$. Define a $(\mathbb{R} \times E)$-valued function by $(f_0, f)(s, y, u) = (f_0(s, y, u), f(s, y, u))$. Then, (11.9.16) says that the vector $(z_0, \bar{z}(s)) \in \mathbb{R} \times E$ is an inner normal to the set $(f_0, f)(s, y(s, \bar{\zeta}, \bar{u}), U) \subseteq \mathbb{R} \times E$ at $(f_0, f)(s, y(s, \bar{\zeta}, \bar{u}), \bar{u}(s))$.

In the result below $\|(y_0, y)\|_{\mathbb{R} \times E}$ is any of the natural norms in $\mathbb{R} \times E$ and $\|(z_0, z)\|_{\mathbb{R} \times E^*}$ its conjugate norm in $\mathbb{R} \times E^*$ (say, $\|(y_0, y)\|_{\mathbb{R} \times E} = |y_0| + \|y\|_E$, $\|(z_0, z)\|_{\mathbb{R} \times E^*} = \max(|z_0|, \|z\|_{E^*})$).

Theorem 11.9.3. *Assume that* (11.9.16) *holds in* $0 \leq t \leq \bar{t}$, *that* $\|z(\cdot)\|_{E^*}$ *is measurable and that*

$$(f_0, f)(s, y(s, \bar{\zeta}, \bar{u}), U) \tag{11.9.17}$$

is (R, p)-*round in* $\mathbb{R} \times E$ *with respect to the cone*

$$\{(z_0, z) \in \mathbb{R} \times E^*; z_0 \geq 0\}. \tag{11.9.18}$$

Then

$$\int_0^{\bar{t}} \|(z_0, \bar{z}(\sigma))\|_{\mathbb{R} \times E^*}$$
$$\times \|(f_0, f)(\sigma, y(\sigma, \bar{\zeta}, \bar{u}), u(\sigma)) - (f_0, f)(\sigma, y(\sigma, \bar{\zeta}, \bar{u}), \bar{u}(\sigma))\|_{\mathbb{R} \times E}^p d\sigma$$
$$\leq 2z_0 R |y_0(\bar{t}, \bar{\zeta}, \bar{u}) - y_0(\bar{t}, \zeta, u)|$$
$$+ 2z_0 R \int_0^{\bar{t}} L_0(\sigma, c) \|y(\sigma, \bar{\zeta}, \bar{u}) - y(\sigma, \zeta, u)\|_E d\sigma$$
$$+ 2RC(\|z\|_{E^*} + \|v\|_{\Sigma(0,\bar{t};E^*)})\|\zeta - \bar{\zeta}\|_E)$$
$$+ 2R\|z\|_{E^*}\|y(\bar{t}, \zeta, u) - y(\bar{t}, \bar{\zeta}, \bar{u})\|_E$$
$$+ 2R\|v\|_{\Sigma(0,\bar{t};E^*)} \max_{0 \leq t \leq \bar{t}} \|y(t, \zeta, u) - y(t, \bar{\zeta}, \bar{u})\|_E$$
$$+ 4RC(\|z\|_{E^*} + \|v\|_{\Sigma(0,\bar{t};E^*)}) \int_0^{\bar{t}} L(\sigma; c)\|y(\sigma, \zeta, u) - y(\sigma, \bar{\zeta}, \bar{u})\|_E d\sigma$$
$$\tag{11.9.19}$$

where c is the constant in (11.9.7) *and C is a bound for* $\|S(t, s, \bar{u})\|_{(E,E)}$ *in*
$0 \le s \le t \le \bar{t}$.

Proof. We apply (11.9.13) to the vector $(z_0, \bar{z}(s))$ and to the set (11.9.17). On
the left we obtain the left side of (11.9.19); on the right we find ($2R$ times) the
following expression:

$$
z_0 \int_0^{\bar{t}} \{f_0(\sigma, y(\sigma, \bar{\zeta}, \bar{u}), u(\sigma)) - f_0(\sigma, y(\sigma, \bar{\zeta}, \bar{u}), \bar{u}(\sigma))\} d\sigma
$$

$$
+ \int_0^{\bar{t}} \langle \bar{z}(\sigma), f(\sigma, y(\sigma, \bar{\zeta}, \bar{u}), u(\sigma)) - f(\sigma, y(\sigma, \bar{\zeta}, \bar{u}), \bar{u}(\sigma)) \rangle d\sigma
$$

$$
= z_0 \int_0^{\bar{t}} \{f_0(\sigma, y(\sigma, \zeta, u), u(\sigma)) - f_0(\sigma, y(\sigma, \bar{\zeta}, \bar{u}), \bar{u}(\sigma))\} d\sigma
$$

$$
+ z_0 \int_0^{\bar{t}} \{f_0(\sigma, y(\sigma, \bar{\zeta}, \bar{u}), u(\sigma)) - f_0(\sigma, y(\sigma, \zeta, u), u(\sigma))\} d\sigma
$$

$$
+ \int_0^{\bar{t}} \langle \bar{z}(\sigma), f(\sigma, y(\sigma, \bar{\zeta}, \bar{u}), u(\sigma)) - f(\sigma, y(\sigma, \bar{\zeta}, \bar{u}), \bar{u}(\sigma)) \rangle d\sigma. \quad (11.9.20)
$$

Estimated, the first two terms produce the first two terms of (11.9.19), where
$L_0(\sigma, c)$ is the function in Hypothesis III^0 corresponding to c. The third term is
appraised exactly as in (11.9.15), keeping in mind that $\mu(ds)$ is to be replaced by
$\nu(ds) = \mu(ds) + z_0 \partial_y f_0(s, y(s, \bar{u}), \bar{u}(s)) ds$. ∎

11.10. Convergence of Suboptimal Controls, II

Estimates (11.9.15) and (11.9.19) add two obvious qualifications to (a), (b) and
(c) in last section:

(d) The convergence estimates are for

$$
f(\sigma, y(\sigma, \bar{\zeta}, \bar{u}), u(\sigma)) - f(\sigma, y(\sigma, \bar{\zeta}, \bar{u}), \bar{u}(\sigma)), \quad (11.10.1)
$$

not for $u(\sigma) - \bar{u}(\sigma)$.

(e) The estimates are in terms of the whole trajectories and not of the initial
and final points only.

Obviously, (d) is simply an intelligent reformulation of Problem 11.9.1, since
controls only enter the equation through $f(t, y, u)$. On the other hand, (e) precises
the reservations voiced in (a); different optimal controls may drive to the target
along different trajectories, thus the trajectories themselves must take part in the
estimate.

An unqualified affirmative answer to Problem 11.9.1 is obtainable only in the
linear case ($L(\tau, c) \equiv 0$), time optimal problem ($L_0(\tau, c) \equiv 0$), no state constraints ($\mu = 0$); this is illustrated in Examples 11.10.1 and 11.10.3 below. However, there are also interesting nonlinear applications.

Note finally that (11.9.15) produces nontrivial information on (11.10.1) only on the set where the costate $\bar{z}(s)$ does not vanish, a condition difficult to verify except in the absence of state constraints. In contrast, nontriviality of (11.9.19) only requires that $z_0 \neq 0$.

Example 11.10.1. We consider the time optimal problem for the system

$$y'(t) = Ay(t) + f(t, y(t)) + u(t), \quad y(0) = \zeta \tag{11.10.2}$$

in **10.6**, with state constraint $y(t) \in M$ $(0 \leq t \leq \bar{t})$ and target condition $y(\bar{t}) \in Y$. The space E is separable, the admissible control space $C_{ad}(0, T; U)$ consists of all strongly measurable E-valued functions with $\|u(t)\| \leq 1$ a.e. in $0 \leq t \leq T$ and the assumptions are those in Example 10.6.1. In case A is merely a semigroup generator it was proved there that the minimum principle

$$\langle \bar{z}(s), v \rangle \geq \langle \bar{z}(s), \bar{u}(s) \rangle \quad (v \in U) \tag{11.10.3}$$

holds a.e. in $0 \leq t \leq \bar{t} =$ optimal time for an arbitrary optimal control $\bar{u}(\cdot)$; if $\mathbf{M}(\bar{t})$ and Y are T-full in their home spaces then $(\mu, z) \neq 0$. All assumptions on the target set Y other than closedness can be lifted if A generates a group. (see Example 10.6.1 for details). This is essential in Theorem 11.10.2 below, where point targets must be handled.

Inequality (11.9.15) may fail to give full information, since $\bar{z}(\sigma)$ may be zero in parts of $0 \leq s \leq \bar{t}$. Convergence information in the set where the state constraint is saturated is not likely to be available, hence the following result seems a reasonable compromise.

Theorem 11.10.2. *Assume that A generates a strongly continuous group, that $f(t, y) = f(y)$ is independent of t, that M is convex and closed with $\text{Int}(M) \neq \emptyset$ and that Y is closed. Let $y(t, \bar{\zeta}, \bar{u})$ be a time optimal trajectory, $e_0 \subseteq [0, \bar{t}]$ the set where the state constraint is not saturated (that is, $y(t, u) \in \text{Int}(M)$), and let $\zeta \in E, u(\cdot) \in C_{ad}(0, \bar{t}; U)$ be such that $y(t, \zeta, u)$ exists in $0 \leq t \leq \bar{t}$ and satisfies (11.9.7). Then there exists a bounded function $\rho(\sigma, c) > 0$ in e_0 such that*

$$\int_0^{\bar{t}} \rho(\sigma, c)\|u(\sigma) - \bar{u}(\sigma)\|^2 d\sigma \leq \max_{0 \leq \sigma \leq \bar{t}} \|y(\sigma, \zeta, u) - y(\sigma, \bar{\zeta}, \bar{u})\|. \tag{11.10.4}$$

Proof. Let C be an upper bound for $\|S(t, s; \bar{u})\|_{(E,E)}$ in $0 \leq s, t \leq \bar{t}$. Since $f(t, y)$ is independent of t, $L(t, c) = L(c)$ in Hypothesis III. Write e_0 as a disjoint, at most countable union of open intervals and let (t_1, t_2) be one of these intervals. Taking into account that $f(t, y)$ (thus the equation (11.10.2)) does not contain time explicitly, the optimality principle (Example 9.2.4) implies that $y(t, \bar{u})$ is time optimal in $t_1 \leq t \leq t_2$. (initial condition $y(t_1, \bar{u})$, target set $\{y(t_2, \bar{u})\}$).

Applying the minimum principle we obtain a costate $\bar{z}(s)$ good for $[t_1, t_2]$ with $(z, \mu) \neq 0$; since the state constraint is not saturated, $\mu = 0$ (Theorem 10.4.1) so that the costate is $\bar{z}(s) = S(t_2, s; \bar{u})^* z, z \neq 0$. We then apply inequality (11.9.15) noting that the set (11.9.14) reduces to

$$f(y(s, \bar{\zeta}, \bar{u})) + U$$

(thus it is (1, 2)-round) and that we have $\|z\|_{E^*} \leq \|S(t_2, s; \bar{u})^*\|_{(E^*, E^*)} \|\bar{z}(s)\|_{E^*} \leq C \min_{t_1 \leq s \leq t_2} \|\bar{z}(s)\|$. The final result is

$$\int_{t_1}^{t_2} \|u(\sigma) - \bar{u}(\sigma)\|^2 d\sigma \leq 2C^2 \|y(t_1, \zeta, u) - y(t_1, \bar{\zeta}, \bar{u})\|$$
$$+ 2C \|y(t_2, \zeta, u) - y(t_2, \bar{\zeta}, \bar{u})\|$$
$$+ 4C^2 L(c) \int_{t_1}^{t_2} \|y(\sigma, \zeta, u) - y(\sigma, \bar{\zeta}, \bar{u})\| d\sigma$$
$$\leq C' \max_{t_1 \leq \sigma \leq t_2} \|y(\sigma, \zeta, u) - y(\sigma, \bar{\zeta}, \bar{u})\|. \qquad (11.10.5)$$

We obtain (11.10.4) multiplying each inequality (11.10.5) by 2^{-n} and adding. ∎

If the equation is linear ($f = 0$) and there is no state constraint, we can apply (11.10.5) in the whole interval $[0, \bar{t}]$, obtaining the explicit estimate

$$\int_0^{\bar{t}} \|u(\sigma) - \bar{u}(\sigma)\|^2 d\sigma$$
$$\leq 2C^2 \|\zeta - \bar{\zeta}\| + 2C \|y(\bar{t}, \zeta, u) - y(\bar{t}, \bar{\zeta}, \bar{u})\|. \qquad (11.10.6)$$

Example 11.10.3. The semilinear wave equation (6.8.1),

$$y_{tt}(t, x) = \sum_{j=1}^m \sum_{k=1}^m \partial^j (a_{jk}(x) \partial^k y(t, x))$$
$$+ \sum_{j=1}^m b_j(x) \partial^j y(t, x) + c(x) y(t, x) - \phi(t, x, y(t, x)) + u(t, x)$$
$$\qquad (11.10.7)$$

with control constraint

$$\int_\Omega |u(t, x)|^2 dx \leq 1 \qquad (11.10.8)$$

qualifies for a similar treatment, since the minimum principle holds for the point target problem (Example 10.6.4); for assumptions and proofs, see Chapter 6, Sections 8-12. Consider for instance the time optimal problem. Written as a system in $H^1(\Omega) \times L^2(\Omega)$ ($H_0^1(\Omega) \times L^2(\Omega)$ for the Dirichlet boundary condition) the equation is

$$\mathbf{y}'(t) = \mathbf{A}(\beta)\mathbf{y}(t) + \mathbf{f}(t, \mathbf{y}(t)) + \mathbf{B}u(t), \quad \mathbf{y}(0) = \zeta \qquad (11.10.9)$$

for $\mathbf{y}(t) = (y(t, \cdot), y_1(t, \cdot)) = (y(t, \cdot), y_t(t, \cdot))$. Since $\mathbf{B}$ is the projection on the second coordinate, the minimum principle is

$$\langle \bar{z}_1(s), v \rangle \geq \langle \bar{z}_1(s), \bar{u}(s) \rangle \quad (v \in U) \tag{11.10.10}$$

a.e. in $0 \leq t \leq \bar{t}$. Straight application of Theorem 11.9.2 fails, since the set (11.9.14) is $\mathbf{f}(y(s, \bar{\zeta}, \bar{u})) + \mathbf{B}U \subseteq \mathbf{f}(y(s, \bar{\zeta}, \bar{u})) + L^2(\Omega)$, as flat as can be in $\mathbf{E}$. Instead, we note that $\mathbf{f}(y(s, \bar{\zeta}, \bar{u})) + \mathbf{B}U$ is (a translate of) the unit ball $B(0, 1)$ in $L^2(\Omega)$ and is thus $(1, 2)$-round there. The minimum principle (11.10.10) implies that the $L^2(\Omega)$-valued function $\bar{z}_1(s)$ is an inner normal to $U = B(0, 1)$ at $\bar{u}(s)$. Hence, we only have to modify (11.9.15) as follows:

$$\int_0^{\bar{t}} \|\bar{z}_1(\sigma)\|_{L^2(\Omega)} \|u(\sigma) - \bar{u}(\sigma)\|^2_{L^2(\Omega)} d\sigma$$

$$\leq 2C(\|\mathbf{z}\|_{\mathbf{E}} + \|\boldsymbol{\mu}\|_{\Sigma(0,\bar{t};\mathbf{E})}) \|\zeta - \bar{\zeta}\|_{\mathbf{E}})$$

$$+ 2\|\mathbf{z}\|_{\mathbf{E}} \|\mathbf{y}(\bar{t}, \zeta, u) - \mathbf{y}(\bar{t}, \bar{\zeta}, \bar{u})\|_{\mathbf{E}}$$

$$+ 2\|\boldsymbol{\mu}\|_{\Sigma(0,\bar{t};\mathbf{E})} \max_{0 \leq t \leq \bar{t}} \|\mathbf{y}(t, \zeta, u) - \mathbf{y}(t, \bar{\zeta}, \bar{u})\|_{\mathbf{E}}$$

$$+ 4C(\|\mathbf{z}\|_{\mathbf{E}} + \|\boldsymbol{\mu}\|_{\Sigma(0,\bar{t};\mathbf{E})}) \int_0^{\bar{t}} L(\sigma, c) \|\mathbf{y}(\sigma, \zeta, u) - \mathbf{y}(\sigma, \bar{\zeta}, \bar{u})\|_{\mathbf{E}} d\sigma \tag{11.10.11}$$

with the same meaning for c, C. In the absence of state constraints the estimate is nontrivial, since, as a consequence of Corollary 6.10.4 the weight function $\|\bar{z}_1(\sigma)\|$ is a continuous function having (at most) a finite number of isolated zeros. This estimate can be exploited to obtain a result of the type of Theorem 11.10.2; since the arguments are exactly the same we omit the details. If the equation is linear and there are no state constraints, we have the following analogue of (11.10.6):

$$\int_0^{\bar{t}} \|\bar{z}_1(\sigma)\|_{L^2(\Omega)} \|u(\sigma) - \bar{u}(\sigma)\|^2_{L^2(\Omega)} d\sigma$$

$$\leq 2C\|\mathbf{z}\|_{\mathbf{E}} \|\zeta - \bar{\zeta}\|_{\mathbf{E}} + 2\|\mathbf{z}\|_{\mathbf{E}} \|\mathbf{y}(\bar{t}, \zeta, u) - \mathbf{y}(\bar{t}, \bar{\zeta}, \bar{u})\|_{\mathbf{E}}. \tag{11.10.12}$$

In some cases, strong convergence results for suboptimal controls can be obtained by more elementary means. As a sample, we look at the linear system

$$y'(t) = Ay(t) + By(t), \quad y(0) = \zeta \tag{11.10.13}$$

where A generates a strongly continuous semigroup in a Banach space E; H is a separable Hilbert space, $B : H \to E$ is bounded, U is the unit ball of H and the admissible control space $C_{\mathrm{ad}}(0, \bar{t}; U)$ consists of all strongly measurable H-valued $u(\cdot)$ with $u(t) \in U$ a.e. We consider the time optimal problem with a target condition $y(t) \in Y = E^*$-weakly closed set in E.

A convex set V with boundary ∂V in a Banach space X is **round** if for every $\bar{u} \in \partial V$ and every $\rho > 0$ we have

$$\bar{u} \notin \overline{\text{conv}}(B(\bar{u}, \rho)^c \cap V). \tag{11.10.14}$$

Example 11.10.4. Let X be a Banach space, and let $V \subseteq X$ be round. Let $\{u_n\} \subseteq V$ with $u_n \to \bar{u} \in \partial V$ X^*-weakly. Then $u_n \to \bar{u}$ strongly.

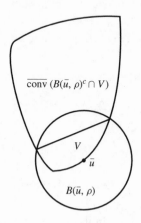

Figure 11.2.

In fact, if the conclusion is false we can assume, passing to a subsequence, that $\|u_n - \bar{u}\| \geq \rho > 0$. This means that $u_n \in \overline{\text{conv}}(B(\bar{u}, \rho)^c \cap V)$. Accordingly, $\bar{u} \in \overline{\text{conv}}(B(\bar{u}, \rho)^c \cap V)$, and we have a contradiction.

Example 11.10.5. Balls in Hilbert space are round.

Example 11.10.6. A control for (11.10.13) is *bang-bang* in a set e if $\|u(t)\| = 1$ in e. Assume that every time optimal control $\bar{u}(\cdot)$ is bang-bang in a set $e \subseteq [0, \bar{t}]$ of positive measure. Then if $\{u_n(\cdot)\}$ is a sequence of suboptimal controls in the interval $0 \leq t \leq t_n$ with $y(t_n, u_n) \in Y$, $t_n \to \bar{t}$ we have

$$u_n(\cdot) \to \bar{u}(\cdot) \quad \text{strongly in } L^2(e; H) \tag{11.10.15}$$

for a subsequence, where $\bar{u}(\cdot)$ is a time optimal control.

In fact a standard argument (very similar to Theorem 14.5.4) implies that there exists a subsequence of $\{u_n(\cdot)\}$ such that $u_n(\cdot)$ converges weakly in $L^2(0, \bar{t}; H)$ to a time optimal control $\bar{u}(\cdot) \in C_{\text{ad}}(0, \bar{t}; U)$. Obviously, $u_n(\cdot) \to \bar{u}(\cdot)$ in $L^2(e; H)$ as well, $u_n(\cdot)$ belongs to the ball of radius $|e|$ in $L^2(e; H)$ and $\bar{u}(\cdot)$ belongs to the boundary of the ball. Balls in Hilbert space are round (Example 11.10.5) so that Example 11.10.4 implies the convergence relation (11.10.15).

Interestingly enough, Example 11.10.6 can be occasionally applied bypassing the maximum principle. A notable example is (11.10.13) with $E = H$, $B = I$,

$$y'(t) = Ay(t) + u(t)$$

with an arbitrary point target $\bar{y}$ (not necessarily in $D(A)$). It is apparently unknown whether some form of the maximum principle holds, but Theorem 9.3.6 establishes the bang-bang property in $e = [0, \bar{t}]$. It follows that (11.10.15) holds in $[0, \bar{t}]$.

11.11. Parabolic Equations

To fix ideas, we study two particular cases. One is the time optimal problem for the system

$$y_t(t, x) = A_c(\beta)y(t, x) + \phi(t, x, y(t, x), \nabla y(t, x)) + u(t, x),$$
$$y(0, x) = \zeta(x),$$
$$(11.11.1)$$

with control constraint
$$|u(t, x)| \le 1 \quad \text{a.e.} \tag{11.11.2}$$

under the assumptions in **11.5**. Equation (11.1.1) is modelled as

$$y'(t) = Ay(t) + f(t, y(t)) + u(t), \quad y(0) = \zeta \tag{11.11.3}$$

in a Banach space E, with admissible control space $C_{ad}(0, \bar{t}; U) =$ unit ball of $L_w^\infty(0, T; (E^\odot)^*)$. The minimum principle for this problem takes the form

$$\langle \bar{z}(s), v - \bar{u}(s) \rangle \ge 0 \quad (v \in U) \tag{11.11.4}$$

a.e. in $0 \le s \le \bar{t}$, U the unit ball of $(E^\odot)^*$. The other model is that in **11.2**,

$$y'(t) = Ay(t) + f(t, y(t), u(t)), \quad y(0) = \zeta \tag{11.11.5}$$

coupled with a general cost functional $y_0(t, u)$; the assumptions are those in **11.2** and we take $g_0 = 0$ in (11.2.3) for simplicity. The minimum principle for this system with a target set Y in E_α is

$$z_0\{f_0(s, y(s, \bar{\zeta}, \bar{u}), v) - f_0(s, y(s, \bar{\zeta}, \bar{u}), \bar{u}(s))\}$$
$$+ \langle ((-A)^\rho)^* \bar{z}(s), ((-A^\odot)^{-\rho})^* \{f(s, y(s, \bar{\zeta}, \bar{u}), v) - f(s, y(s, \bar{\zeta}, \bar{u}), \bar{u}(s))\}\rangle \ge 0$$
$$(v \in U) \quad (11.11.6)$$

a.e. in $[0, \bar{t}]$, with costate $\bar{z}(s) = ((-A)^{-\rho})^* \upsilon(s)$, $\upsilon(s)$ given by

$$\upsilon(s) = \int_s^{\bar{t}} R_{\alpha, \rho}(\sigma, s; \bar{u})^* ((-A)^{-\alpha})^* \mu_1(d\sigma) \tag{11.11.7}$$

and $\mu_1(ds) = \mu(ds) + z_0 \partial_y f_0(s, y(s, \bar{\zeta}, \bar{u}), \bar{u}(s))ds + \delta_{\bar{t}}(ds)z = \nu(ds) + \delta_{\bar{t}}(ds) \in \Sigma(0, \bar{t}; ((-A)^{-\alpha})^* E^*)$.

Our objective in this section is the translation of Theorems 11.9.2 and 11.9.3 to the present situation, and we do only the first in detail. The manipulations below apply to two controls $\bar{u}(\cdot), u(\cdot) \in C_{ad}(0, \bar{t}; U)$ and to two initial conditions $\bar{\zeta}, \zeta$ such that the solutions exist in $0 \leq t \leq \bar{t}$ and are uniformly bounded in the E_α norm, that is,

$$\|(-A)^\alpha y(t, \bar{\zeta}, \bar{u})\|_E, \|(-A)^\alpha y(t, \zeta, u)\|_E \leq c \quad (0 \leq t \leq \bar{t}). \tag{11.11.8}$$

The solution of the variational equation

$$\xi'(t) = \{A(t) + \partial_y f(t, y(t, \bar{\zeta}, \bar{u}), \bar{u}(t))\}z(t) + g(t), \qquad \xi(0) = \eta \in E_{-\gamma}$$

where $g(t)$ is a $(E^*)_{-\rho}$-valued function is given by the variation-of-constants formula

$$(-A)^\alpha \xi(t) = R_{\alpha,\gamma}(t, 0; \bar{u})((-A^\odot)^{-\gamma})^* \eta$$

$$+ \int_0^t R_{\alpha,\rho}(t, \sigma; \bar{u})((-A^\odot)^{-\rho})^* g(\sigma)d\sigma \tag{11.11.9}$$

(see **7.9**). We use this formula for $\gamma = 0$ and a solution $y(t, \zeta, u)$ of (11.11.5) with $\zeta \in E_\alpha$:

$$(-A)^\alpha y(t, \zeta, u) = R_{\alpha,0}(t, 0; \bar{u})(-A)^{-\alpha}(-A)^\alpha \zeta$$

$$+ \int_0^t R_{\alpha,\rho}(t, \sigma; \bar{u})((-A^\odot)^{-\rho})^* f(\sigma, y(\sigma, \bar{\zeta}, \bar{u}), u(\sigma))d\sigma$$

$$+ \int_0^t R_{\alpha,\rho}(t, \sigma; \bar{u})((-A^\odot)^{-\rho})^* \{f(\sigma, y(\sigma, \zeta, u), u(\sigma)) - f(\sigma, y(\sigma, \bar{\zeta}, \bar{u}), u(\sigma))\}d\sigma$$

$$- \int_0^t R_{\alpha,\rho}(t, \sigma; \bar{u})((-A^\odot)^{-\rho})^* \partial_y f(\sigma, y(\sigma, \bar{\zeta}, \bar{u}), \bar{u}(\sigma))y(\sigma, \zeta, u)d\sigma. \tag{11.11.10}$$

Writing this equation for $y(t, \bar{\zeta}, \bar{u})$ and subtracting we obtain the following companion of (11.9.10):

$$(-A)^\alpha y(t, \zeta, u) - (-A)^\alpha y(t, \bar{\zeta}, \bar{u})$$

$$= R_{\alpha,0}(t, 0; \bar{u})(-A)^{-\alpha}(-A)^\alpha(\zeta - \bar{\zeta})$$

$$+ \int_0^t R_{\alpha,\rho}(t, \sigma; \bar{u})((-A^\odot)^{-\rho})^* \{f(\sigma, y(\sigma, \bar{\zeta}, \bar{u}), u(\sigma)) - f(\sigma, y(\sigma, \bar{\zeta}, \bar{u}), \bar{u}(\sigma))\}d\sigma$$

$$+ \int_0^t R_{\alpha,\rho}(t, \sigma; \bar{u})((-A^\odot)^{-\rho})^* \{f(\sigma, y(\sigma, \zeta, u), u(\sigma)) - f(\sigma, y(\sigma, \bar{\zeta}, \bar{u}), u(\sigma))\}d\sigma$$

$$- \int_0^t R_{\alpha,\rho}(t, \sigma; \bar{u})((-A^\odot)^{-\rho})^* \partial_y f(\sigma, y(\sigma, \bar{\zeta}, \bar{u}), \bar{u}(\sigma))(y(\sigma, \zeta, u) - y(\sigma, \bar{\zeta}, \bar{u}))d\sigma.$$

$$\tag{11.11.11}$$

Integrate both sides against $((-A)^{-\alpha})^*\mu_1(dt) \in \Sigma(0, \bar{t}; E^*)$ in $0 \le t \le \bar{t}$. The result is

$$
\int_0^{\bar{t}} \langle ((-A)^{-\alpha})^*\mu_1(dt), (-A)^\alpha y(t, \zeta, u) - (-A)^\alpha y(t, \bar{\zeta}, \bar{u}) \rangle
$$

$$
= \int_0^{\bar{t}} \langle ((-A)^{-\alpha})^*\mu_1(dt), R_{\alpha,0}(t, 0, \bar{u})(-A)^{-\alpha}(-A)^\alpha(\zeta - \bar{\zeta}) \rangle
$$

$$
+ \int_0^{\bar{t}} \Big\langle ((-A)^{-\alpha})^*\mu_1(dt),
$$

$$
\int_0^t R_{\alpha,\rho}(t, \sigma; \bar{u})((-A^\odot)^{-\rho})^*\{f(\sigma, y(\sigma, \bar{\zeta}, \bar{u}), u(\sigma)) - f(\sigma, y(\sigma, \bar{\zeta}, \bar{u}), \bar{u}(\sigma))\}d\sigma \Big\rangle
$$

$$
+ \int_0^{\bar{t}} \Big\langle ((-A)^{-\alpha})^*\mu_1(dt),
$$

$$
\int_0^t R_{\alpha,\rho}(t, \sigma; \bar{u})((-A^\odot)^{-\rho})^*\{f(\sigma, y(\sigma, \zeta, u), u(\sigma)) - f(\sigma, y(\sigma, \bar{\zeta}, \bar{u}), u(\sigma))\}d\sigma \Big\rangle
$$

$$
- \int_0^{\bar{t}} \Big\langle ((-A)^{-\alpha})^*\mu_1(dt),
$$

$$
\int_0^t R_{\alpha,\rho}(t, \sigma; \bar{u})((-A^\odot)^{-\rho})^*\partial_y f(\sigma, y(\sigma, \bar{\zeta}, \bar{u}), \bar{u}(\sigma))(y(\sigma, \zeta, u) - y(\sigma, \bar{\zeta}, \bar{u}))d\sigma \Big\rangle.
$$

$$(11.11.12)$$

We switch the order of integration using the bound (7.9.26) for $R_{\alpha,\rho}(t, s)$, Lemma 11.1.10 and the bound (11.1.19) for $\upsilon(s)$,

$$
\|\upsilon(s)\|_{E^*} \le C\{(\bar{t} - s)^{-\alpha-\rho}\|z\|_{((-A)^{-\alpha})^*E^*} + \omega(s)\}, \tag{11.11.13}
$$

where $\omega(\cdot) \in L^1(0, \bar{t})$ is given by

$$
\omega(s) = \int_s^{\bar{t}} (\sigma - s)^{-\alpha-\rho}\|v\|(d\sigma) \tag{11.11.14}
$$

(see (11.1.19)). In this way, we obtain

$$
\int_0^{\bar{t}} \langle ((-A)^\rho)^*\bar{z}(\sigma), ((-A^\odot)^{-\rho})^*\{f(\sigma, y(\sigma, \bar{\zeta}, \bar{u}), u(\sigma)) - f(\sigma, y(\sigma, \bar{\zeta}, \bar{u}), \bar{u}(\sigma))\}\rangle d\sigma
$$

$$
= - \int_0^{\bar{t}} \langle ((-A)^{-\alpha})^*\mu_1(dt), R_{\alpha,0}(t, 0; \bar{u})(-A)^{-\alpha}(-A)^\alpha(\zeta - \bar{\zeta}) \rangle
$$

$$
+ \int_0^{\bar{t}} \langle ((-A)^{-\alpha})^*\mu_1(dt), (-A)^\alpha y(t, \zeta, u) - (-A)^\alpha y(t, \bar{\zeta}, \bar{u}) \rangle
$$

$$- \int_0^{\bar{t}} \langle ((-A)^\rho)^* \bar{z}(\sigma), ((-A^\odot)^{-\rho})^* \{ f(\sigma, y(\sigma, \zeta, u), u(\sigma)) - f(\sigma, y(\sigma, \bar{\zeta}, \bar{u}), u(\sigma)) \} \rangle$$

$$+ \int_0^{\bar{t}} \langle ((-A)^\rho)^* \bar{z}(\sigma), ((-A^\odot)^{-\rho})^* \partial_y f(\sigma, y(\sigma, \bar{\zeta}, \bar{u}), \bar{u}(\sigma))(y(\sigma, \zeta, u) - y(\sigma, \bar{\zeta}, \bar{u})) \rangle$$

(11.11.1)

corresponding to (11.9.12). To estimate, we use the fact that $\| R_{\alpha,0}(t, s, \bar{u})(-A)^{-\alpha} \|$ is bounded in $0 \le s \le t \le \bar{t}$, an immediate consequence of the integral equation (7.9.27) defining $R_{\alpha,\rho}(t, s)$. We obtain in this way

$$\int_0^{\bar{t}} \langle ((-A)^*)^\rho \bar{z}(\sigma), ((-A^\odot)^{-\rho})^* \{ f(\sigma, y(\sigma, \bar{\zeta}, \bar{u}), u(\sigma)) - f(\sigma, y(\sigma, \bar{\zeta}, \bar{u}), \bar{u}(\sigma))) \} \rangle$$

$$\le C(\| z \|_{(-A)^{-\alpha} * E^*} + \| v \|_{\Sigma(0, \bar{t}; ((-A)^{-\alpha} * E^*))}) \| \zeta - \bar{\zeta} \|_{E_\alpha}$$

$$+ \| z \|_{(-A)^{-\alpha} * E^*} \| y(\bar{t}, \zeta, u) - y(\bar{t}, \bar{\zeta}, \bar{u}) \|_{E_\alpha}$$

$$+ \| v \|_{\Sigma(0, \bar{t}; ((-A)^{-\alpha} * E^*))} \max_{0 \le t \le \bar{t}} \| y(t, \zeta, u) - y(t, \bar{\zeta}, \bar{u}) \|_{E_\alpha}$$

$$+ C' L(c) \int_0^{\bar{t}} ((\bar{t} - \sigma)^{-\alpha - \rho} \| z \|_{(-A)^{-\alpha} * E^*} + \omega(\sigma)) \| y(\sigma, \zeta, u) - y(\sigma, \bar{\zeta}, \bar{u}) \|_{E_\alpha} d\sigma$$

(11.11.1)

where C, C' are constants coming from previous estimates and $L(c)$ is the constant in Hypothesis $\text{III}_{\alpha,\rho}$ corresponding to the bound c in (11.11.8); C and C' do not depend on c. Inequality (11.1.16) (with $f_0 = 1$, which implies $\nu = \mu$) is made to order for the time optimal problem; for a general cost functional the first two terms in (11.9.19) must be added.

Imitation of the arguments in **11.9** and **11.10** could be carried further, but in the most interesting applications to parabolic equations, the sets (11.9.14) and (11.9.17) are not (R, p)-round (for instance, (11.9.14) may be a translate of the unit ball of $L^\infty(\Omega)$). To fix ideas about possible uses of (11.11.16), we look at the equation (11.11.1) with admissible control space $C_{ad}(0, T; U) =$ unit ball of $L^\infty((0, T) \times \Omega)$; the assumptions are still those in **11.2** but we take $\rho = 0$. Let $\bar{u}(t, x)$ be a time optimal control in the interval $0 \le t \le \bar{t}$. The minimum principle yields

$$\int_{(0, \bar{t}) \times \Omega} \bar{z}(s, x)(u(s, x) - \bar{u}(s, x)) ds dx \ge 0 \tag{11.11.17}$$

for all $u(\cdot, \cdot)$ in the unit ball of $L^\infty((0, \bar{t}) \times \Omega)$. Hence,

$$\bar{u}(s, x) = -\text{sign } \bar{z}(s, x) \tag{11.11.18}$$

wherever $\bar{z}(s, x) \ne 0$. We deduce that the integrand is nonnegative a.e. and that $\bar{z}(s, x)(u(s, x) - \bar{u}(s, x)) = |\bar{z}(s, x)| \, |u(s, x) - \bar{u}(s, x)|$. Combining this with

(11.11.16), we obtain the following estimate for the equation (11.11.1):

$$\int_{(0,\bar{t})\times\Omega} |\bar{z}(s,x)|\, |u(s,x) - \bar{u}(s,x)|\, ds\, dx$$

$$\leq C \max_{0\leq t\leq \bar{t}} \|(-A_c(\beta))^\alpha(y(t,\zeta,u) - y(t,\bar{\zeta},\bar{u}))\|_{C(\overline{\Omega})}$$

$$+ C' \|(-A_c(\beta))^\alpha(\zeta - \bar{\zeta})\|_{C(\overline{\Omega})} \qquad (11.11.19)$$

for an arbitrary control $u(t,x)$ such that

$$\|(-A_c(\beta))^\alpha y(t,\zeta,u)\|_{C(\overline{\Omega})}, \ \|(-A_c(\beta))^\alpha y(t,\bar{\zeta},\bar{u})\|_{C(\overline{\Omega})} \leq c \qquad (11.11.20)$$

in $0 \leq t \leq \bar{t}$; the first constant depends on the bound c, the second does not. The first term on the right is $C\|(-A_c(\beta))^\alpha(y(\bar{t},\zeta,u) - y(\bar{t},\bar{\zeta},\bar{u}))\|_{C(\overline{\Omega})}$ when the equation is linear and there are no state constraints.

Theorem 11.11.1. *Let $\{u_n(\cdot)\}$ $(u_n(\cdot) \in C_{ad}(0,t_n;U))$ be a minimizing sequence for the time optimal problem corresponding to the initial condition $y(0) = \bar{\zeta}$, with $(-A_c(\beta))^\alpha y(t,\zeta,u_n)$ uniformly bounded in the norm of $C(\overline{\Omega})$. Then there exists a subsequence such that (a) $u_n \to \bar{u}$ L^1-weakly in $L^\infty((0,\bar{t}) \times \Omega)$, (b)*

$$\int_{(0,\bar{t})\times\Omega} |\bar{z}(s,x)|\, |u_n(s,x) - \bar{u}(s,x)|\, ds\, dx \to 0 \qquad (11.11.21)$$

as $n \to \infty$, where $\bar{u}(\cdot)$ is an optimal control.

Proof. We take $\tilde{t} > t$, and n so large that $t_n < \tilde{t}$. Extend $u_n(\cdot)$ (resp. $y(\cdot,\bar{\zeta},u_n)$) to $0 \leq t \leq \tilde{t}$ by $u(t) = u \in U$ (resp. by $y(t,\bar{\zeta},u_n) = y(t_n,\bar{\zeta},u_n)$), and let $\tilde{y}(t,\bar{\zeta},u_n)$ be the function defined by the integral equation (7.7.18) with the extended $y(t,\bar{\zeta},u_n)$ on the right hand side. Then $\tilde{y}(t,\bar{\zeta},u_n) = y(t,\bar{\zeta},u_n)$ in $0 \leq t \leq t_n$ and, by Theorem 7.7.16 we may select a subsequence with $u_n(\cdot) \to \bar{u}(\cdot)$ $L^1(0,\tilde{t};L^1(\Omega))$-weakly in $L^\infty_w(0,\tilde{t};L^\infty(\Omega))$ and $\tilde{y}(\cdot,\bar{\zeta},u_n) \to y(\cdot,\bar{\zeta},\bar{u})$ in $C(0,\tilde{t};E_\alpha)$; clearly $y(t,\bar{\zeta},\bar{u})$ satisfies the state constraints and target condition, so that $\bar{u}(\cdot)$ is an optimal control. Application of (11.11.19) yields the convergence relation (11.11.21). ∎

If the state constraints and target condition are defined as in Remark 11.3.6 with M_s and M_Y convex with nonempty interior, then the state constraint set M and the target set Y are convex with nonempty interior and the multiplier (z_0, μ, z) $((\mu, z)$ for the time optimal problem) is nontrivial.

The requirement that Y have a nonempty interior can be waived (still with nontrivial multipliers) using the techniques in **11.4** and **11.5**. The minimum principle for the time optimal problem is still (11.11.4), but the "final value" z of the costate $\bar{z}(s)$ merely belongs to $((-A)^* E^*)^1$. We have

$$\bar{z}(s) = ((-A)^{1-\alpha} R_{\alpha,0}(\bar{t}, s; \bar{u}))^* ((-A)^{-1})^* z$$

$$+ \int_s^{\bar{t}} R_{\alpha,0}(\sigma, s; \bar{u})^* ((-A)^{-\alpha})^* v(d\sigma) = \bar{z}_h(s) + \bar{z}_i(s) \qquad (11.11.22)$$

with $\quad v(ds) = \mu(ds) \; (v(ds) = \mu(ds) + z_0 \partial_y f_0(s, y(s, \bar{\zeta}, \bar{u}), \bar{u}(s))ds \quad$ for the general control problem). In view of (11.5.30) $\bar{z}_h(s) \in L^1(0, \bar{t}; E^{\odot})$. However, since in general $z \notin ((-A)^{\alpha})^* E^*$, the measure $\mu_1(ds) = v(ds) + \delta_{\bar{t}}(ds)z$ no longer belongs to $\Sigma(0, \bar{t}; ((-A)^{\alpha})^* E^*)$, thus the companion of the fundamental estimate (11.11.16) must be obtained by other means. We begin by writing the particular case of (11.11.11) for the system (11.1.1), assuming that $\zeta, \bar{\zeta} \in D(A)$:

$$(-A)^{\alpha} y(t, \zeta, u) - (-A)^{\alpha} y(t, \bar{\zeta}, \bar{u})$$

$$= R_{\alpha,0}(t, 0; \bar{u})(-A)^{-\alpha}(-A)^{\alpha}(\zeta - \bar{\zeta})$$

$$+ \int_0^t R_{\alpha,0}(t, \sigma; \bar{u})(u(\sigma)) - \bar{u}(\sigma))d\sigma$$

$$+ \int_0^t R_{\alpha,0}(t, \sigma; \bar{u})\{f(\sigma, y(\sigma, \zeta, u)) - f(\sigma, y(\sigma, \bar{\zeta}, \bar{u}))\}d\sigma$$

$$- \int_0^t R_{\alpha,0}(t, \sigma; \bar{u})\partial_y f(\sigma, y(\sigma, \bar{\zeta}, \bar{u}))(y(\sigma, \zeta, u) - y(\sigma, \bar{\zeta}, \bar{u}))d\sigma.$$

$$(11.11.23)$$

Integrating against $v(ds)$ and switching the order of integration as we did to obtain (11.11.15), the equality

$$\int_0^{\bar{t}} \langle \bar{z}_i(\sigma), u(\sigma) - \bar{u}(\sigma) \rangle d\sigma$$

$$= - \int_0^{\bar{t}} \langle ((-A)^{-\alpha})^* v(dt), R_{\alpha,0}(t, 0; \bar{u})(-A)^{-\alpha}(-A)^{\alpha}(\zeta - \bar{\zeta}) \rangle$$

$$+ \int_0^{\bar{t}} \langle ((-A)^{-\alpha})^* v(dt), (-A)^{\alpha} y(t, \zeta, u) - (-A)^{\alpha} y(t, \bar{\zeta}, \bar{u}) \rangle$$

$$- \int_0^{\bar{t}} \langle \bar{z}_i(\sigma), f(\sigma, y(\sigma, \zeta, u)) - f(\sigma, y(\sigma, \bar{\zeta}, \bar{u})) \rangle d\sigma$$

$$+ \int_0^{\bar{t}} \langle \bar{z}_i(\sigma), \partial_y f(\sigma, y(\sigma, \bar{\zeta}, \bar{u}))(y(\sigma, \zeta, u) - y(\sigma, \bar{\zeta}, \bar{u})) \rangle d\sigma \quad (11.11.24)$$

results. The next step is to obtain the same equality with $v = 0$ and $\bar{z}_h(s)$ instead of $\bar{z}_i(s)$. To this end we use (11.5.24),

$$(-A)^{1-\alpha} R_{\alpha,0}(t, s; \bar{u}) = (-A)S(t - s) + N_1(t, s), \qquad (11.11.25)$$

$$(-A)^{1-\alpha} R_{\alpha,0}(t, s; \bar{u})(-A)^{-\delta} = (-A)^{1-\delta} S(t - s) + N_2(t, s), \qquad (11.11.26)$$

where the operator functions $N_j(t, s)$ are continuous in the norm of (E, E) for $t > s$ and $\|N_j(t, s)\|_{(E,E)} \leq C(t - s)^{-\kappa}$ with $\kappa < 1$. By hypothesis, $y(\bar{t}, \bar{\zeta}, \bar{u}) \in Y \subseteq D(A)$, and we assume that $y(\bar{t}, \zeta, \bar{u}) \in D(A)$ as well. We take $t = \bar{t}$ in (11.11.23) and apply $(-A)^{1-\alpha}$ to both sides. In view of (11.11.26) we can introduce $(-A)^{1-\alpha}$ in the third and fourth integrals, and using (11.11.25) we can do the same in the first integral if $u(\sigma) = \bar{u}(\sigma)$ near $\bar{t}$. Next, we apply the functional $((-A)^{-1})^* z$ to both sides and put operators on the other side of the duality products. The result is

$$\int_0^{\bar{t}} \langle \bar{z}_h(\sigma), u(\sigma) - \bar{u}(\sigma) \rangle d\sigma$$

$$= -\langle ((-A)^{-1})^* z, (-A)^{1-\alpha} R_{\alpha,0}(t, 0, \bar{u})(-A)^{-\alpha}(-A)^\alpha(\zeta - \bar{\zeta}) \rangle$$

$$+ \langle ((-A)^{-1})^* z, (-A)y(\bar{t}, \zeta, u) - (-A)y(t, \bar{\zeta}, \bar{u}) \rangle$$

$$- \int_0^{\bar{t}} \langle \bar{z}_h(\sigma), f(\sigma, y(\sigma, \zeta, u) - f(\sigma, y(\sigma, \bar{\zeta}, \bar{u})) \rangle d\sigma$$

$$+ \int_0^{\bar{t}} \langle \bar{z}_h(\sigma), \partial_y f(\sigma, y(\sigma, \bar{\zeta}, \bar{u}))(y(\sigma, \zeta, u) - y(\sigma, \bar{\zeta}, \bar{u})) \rangle d\sigma. \quad (11.11.27)$$

where, since $\|\bar{z}_h(\cdot)\|_{E^*}$ is integrable, the condition that $u(\sigma) = \bar{u}(\sigma)$ near $\bar{t}$ can be jettisoned. Putting together (11.11.26) and (11.11.27) we finally obtain the companion of (11.11.16) we were pursuing.

Applied to the system (11.11.1), the estimate becomes

$$\int_{(0,\bar{t}) \times \Omega} |\bar{z}(t, x)| |u(s, x) - \bar{u}(s, x)| ds dx$$

$$\leq C \max_{0 \leq t \leq \bar{t}} \|(-A_c(\beta))^\alpha (y(t, \zeta, u) - y(t, \bar{\zeta}, \bar{u}))\|_{C(\overline{\Omega})}$$

$$+ C' \|(-A_c(\beta))^\alpha (\zeta - \bar{\zeta})\|_{C(\overline{\Omega})} + C'' \|A_c(\beta)(y(\bar{t}, \zeta, u) - y(\bar{t}, \bar{\zeta}, \bar{u}))\|_{C(\overline{\Omega})}$$

$$(11.11.28)$$

for a control $u(t, x)$ whose trajectory satisfies (11.11.20) in $0 \leq t \leq \bar{t}$ together with the optimal trajectory. The first constant depends on the bound c, the others do not (but C' depends on $\bar{t}$). The first term on the right is absent when the equation is linear and there are no state constraints.

Remark 11.11.2. Nontriviality of estimates (11.11.19) and (11.11.27) depends on showing that the nodal set

$$e(\bar{t}, z) = \{(t, x) \in [0, \bar{t}] \times \overline{\Omega}; \bar{z}(t, x) = 0\}. \tag{11.11.29}$$

is of measure zero, at least where the state constraint is not saturated. On this, see Problem 11.6.7.

Example 11.11.3. Estimates like (11.11.19) and (11.11.28) can be obtained in simpler ways for linear systems. For instance, consider the system

$$y_t(t, x) = y_{xx}(t, x) + b(x)u(t) \quad (0 < x < \pi),$$
$$y(t, 0) = y(t, \pi) = 0, \qquad y(0, x) = \zeta(x) \tag{11.11.30}$$

in Example 11.6.5, with $E = C_0[0, \pi]$, $Ay = y''$, $F = X^* = \mathbb{R}$, $U = [-1, 1]$, $b(\cdot)$ satisfying (11.6.28) and $\zeta(\cdot) \in C_0[0, \pi]$. Theorem 11.6.6 guarantees that if $\bar{u}(\cdot)$ is a time optimal control for a target condition $y(\bar{t}, \zeta, u) \in \bar{y} \in D(K)$ (K the operator (11.6.29)), then $\kappa(s)(v - \bar{u}(s)) \geq 0$ for $|v| \leq 1$ a.e. in $0 \leq s \leq \bar{t}$, with $\kappa(t) = \int b(x)\bar{z}(t, x)dx$. In absence of state constraints, the costate $\bar{z}(t, x)$ is a solution of the backwards heat equation

$$\bar{z}_t(t, x) = -\bar{z}_{xx}(t, x) \quad (0 < x < \pi),$$
$$\bar{z}(t, 0) = \bar{z}(t, \pi) = 0, \tag{11.11.31}$$

with $\bar{z}(\bar{t}, \cdot) = z(\cdot) \in (K^* C_0[0, \pi]^*)^1(B^*)$. Recall (Theorem 9.6.1) that $\kappa(\cdot) \in L^1(0, \pi)$ and that $\kappa(t) \neq 0$ except for t in a (possibly empty) sequence which, if infinite, can only accumulate at the terminal time $\bar{t}$.

Theorem 11.11.4. *Let $\bar{u}(\cdot)$ be an optimal control, $u(\cdot)$ an arbitrary admissible control. Then we have*

$$\int_0^{\bar{t}} |\kappa(t)| \, |u(t) - \bar{u}(t)| dt$$
$$\leq C \|y(\bar{t}, \zeta, u) - y(\bar{t}, \bar{\zeta}, \bar{u})\|_{D(K)} + C' \|\zeta - \bar{\zeta}\|_{C_0(0, \pi)}. \tag{11.11.32}$$

Proof. As a solution of the heat equation (11.11.31), $\bar{z}(t, x)$ is infinitely differentiable in $[0, \bar{t}) \times (0, \pi)$ and continuous in $[0, \bar{t}) \times [0, \pi]$, but it has a singularity at $t = \bar{t}$. Thus, in the computation below we use $z(t - \delta, x)$, which is smooth in $(-\infty, \bar{t}] \times [0, \pi]$ for every $\delta > 0$.

$$\int_0^{\bar{t}} \kappa(t - \delta)(u(t) - \bar{u}(t))dt$$
$$= \int_0^{\bar{t}} \int_0^{\pi} \bar{z}(t - \delta, x)b(x)(u(t) - \bar{u}(t))dxdt$$

$$= \int_0^{\bar{t}} \int_0^{\pi} \bar{z}(t - \delta, x)(y_t(t, x, \zeta, u) - y_{xx}(t, x, \zeta, u))dxdt$$

$$- \int_0^{\bar{t}} \int_0^{\pi} \bar{z}(t - \delta, x)(y_t(t, x, \bar{\zeta}, \bar{u}) - y_{xx}(t, x, \bar{\zeta}, \bar{u}))dxdt$$

$$= \int_0^{\pi} \bar{z}(\bar{t} - \delta, x)(y(\bar{t}, x, \zeta, u) - y(\bar{t}, x, \bar{\zeta}, \bar{u}))dx$$

$$- \int_0^{\pi} \bar{z}(-\delta, x)(\zeta(x) - \bar{\zeta}(x))dx \qquad (11.11.33)$$

using the divergence theorem. Taking limits in the second integral is obvious. In the first, we note that $\bar{z}(s) = K^* S_1(\bar{t} - s)^* (K^{-1})^* z$, so that $(K^{-1})^* \bar{z}(\bar{t} - \delta) = S_1(\delta)^* (K^{-1})^* z$, with $(K^{-1})^* z \in C_0[0, \pi]^*$. Accordingly, $(K^{-1})^* \bar{z}(\bar{t} - \delta) \to (K^{-1})^* z$ $C_0[0, \pi]$-weakly as $\delta \to 0$. We then have

$$\langle \bar{z}(\bar{t} - \delta), y(\bar{t}, \zeta, u) - y(\bar{t}, \bar{\zeta}, \bar{u}) \rangle$$
$$= \langle (K^{-1})^* z(\bar{t} - \delta), K(y(\bar{t}, \zeta, u) - y(\bar{t}, \bar{\zeta}, \bar{u})) \rangle$$
$$\to \langle (K^{-1})^* z(\bar{t}), K(y(\bar{t}, \zeta, u) - y(\bar{t}, \bar{\zeta}, \bar{u})) \rangle.$$

We then use the fact that

$$\int_0^{\bar{t}} |\kappa(t)| \, |u(t) - \bar{u}(t)| dt = \int_0^{\bar{t}} \kappa(t)(u(t) - \bar{u}(t))dt.$$

(proved as in (11.11.19) and take limits in (11.11.33), obtaining (11.11.32). This completes the proof. ∎

11.12. The Navier–Stokes Equations

The arguments in **11.11** apply straight to (11.7.1) with the simplification that $X^p(\Omega)$, unlike $C(\bar{\Omega})$, is reflexive. We use the assumptions in **11.7** and an L^∞ bound on the control $\mathbf{u}(t, x) = (u_1(t, x), \ldots, u_m(t, x))$:

$$|u_k(t, x)| \le 1 \quad \text{a.e.} \quad (k = 1, \ldots, m) \qquad (11.12.1)$$

so that the control space is the unit ball of $L^\infty((0, T) \times \Omega; \mathbb{R}^m)$. State constraints and target condition are as in **11.7**, and the corresponding abstract model is (11.7.3) with $X = L^1(\Omega; \mathbb{R}^m)$, $X^* = L^\infty(\Omega; \mathbb{R}^m)$ and $B = \mathbf{P}_p \mathbf{I}_p$. All the arguments in **11.11** regarding the model (11.11.3) generalize to (11.7.3) with no other change than the presence of the operator B or its adjoint B^* in the right places. Since the results on point targets do not apply to the present control space, we assume that M_s and M_Y are convex with interior points, or, in more generality that $\mathbf{M}(\bar{t})$ is T-full in $C(0, \bar{t}; E_\alpha)$ and that Y is T-full in E_α. The basic equality (11.11.15) becomes

$$\int_0^{\bar{t}} \langle B^* \bar{z}(\sigma), u(\sigma) - \bar{u}(\sigma) \rangle d\sigma$$

$$= - \int_0^{\bar{t}} \langle ((-A)^{-\alpha})^* \mu_1(dt), R_{\alpha,0}(t, 0; \bar{u})(-A)^{-\alpha}(-A)^\alpha(\zeta - \bar{\zeta}) \rangle$$

$$+ \int_0^{\bar{t}} \langle ((-A)^{-\alpha})^* \mu_1(dt), (-A)^\alpha y(t, \zeta, u) - (-A)^\alpha y(t, \bar{\zeta}, \bar{u}) \rangle$$

$$- \int_0^{\bar{t}} \langle \bar{z}(\sigma), f(\sigma, y(\sigma, \zeta, u)) - f(\sigma, y(\sigma, \bar{\zeta}, \bar{u})) \rangle d\sigma$$

$$+ \int_0^{\bar{t}} \langle \bar{z}(\sigma), \partial_y f(\sigma, y(\sigma, \bar{\zeta}, \bar{u}))(y(\sigma, \zeta, u) - y(\sigma, \bar{\zeta}, \bar{u})) \rangle d\sigma. \quad (11.12.2)$$

Applied to (11.7.1), this inequality yields

$$\sum_{k=1}^m \int_{(0,\bar{t}) \times \Omega} |\bar{z}_k(s, x)| |u_k(s, x) - \bar{u}_k(s, x)| ds dx$$

$$\leq C \max_{0 \leq t \leq \bar{t}} \|(-\mathbf{A}_p)^\alpha (\mathbf{y}(t, \zeta, \mathbf{u}) - \mathbf{y}(t, \bar{\zeta}, \bar{\mathbf{u}}))\|_{X^p(\Omega)} \quad (11.12.3)$$

for the optimal control $\bar{\mathbf{u}} = (\bar{u}_1, \ldots, \bar{u}_m)$ and an arbitrary admissible control $\mathbf{u} = (u_1, \ldots, u_m)$ such that $\mathbf{y}(t, \zeta, \mathbf{u})$ exists in $0 \leq t \leq \bar{t}$, the constants depending on a bound c for both solutions:

$$\|(-\mathbf{A}_p)^\alpha \mathbf{y}(t, \zeta, \mathbf{u})\|_{X^p(\Omega)}, \quad \|(-\mathbf{A}_p)^\alpha \mathbf{y}(t, \bar{\zeta}, \bar{\mathbf{u}})\|_{X^p(\Omega)} \leq c \quad (0 \leq t \leq \bar{t}).$$
$$(11.12.4)$$

Theorem 11.11.1 has an exact counterpart; here, convergence means $\mathbf{u}_n \to \bar{\mathbf{u}}$ $L^1((0, T) \times \Omega; \mathbb{R}^m)$-weakly in $L^\infty((0, T) \times \Omega; \mathbb{R}^m)$ and

$$\sum_{k=1}^m \int_{(0,\bar{t}) \times \Omega} |z_k(s, x)| |u_{kn}(s, x) - u_k(s, x)| ds dx \to 0 \quad (11.12.5)$$

as $n \to \infty$.

Miscellaneous Comments for Part II

Nonlinear programming in infinite dimensional spaces. In the the seventies, new proofs of Pontryagin's maximum principle based on Ekeland's variational principle were given by Ekeland [1974] and Clarke [1976:1] [1976:2] (see also Ekeland [1979]). Apart from other advantages, this new approach proved very fruitful when extended to infinite dimensions, the difference between finite and infinite dimensionality showing up in the possibility that, in the latter case, the weak limit of a sequence of elements of norm 1 may be zero. Applications of the Clarke-Ekeland approach to infinite dimensional problems are in the author [1986] [1987:1], the set target problem treated under ungainly conditions. A much improved version appears in the author and Frankowska [1990:1] [1991:1]. Among the decisive improvements due to Frankowska are the use of the function $\Phi_n(u, y)$ in **4.1** in Ekeland's variational principle, the introduction of enough nonsmooth analysis to achieve a clean treatment of the set target problem, and the jettisoning of the cumbersome "input-output" approach in the author [1986] [1987:1] in favor of the nonlinear programming setup, which is crucial for the treatment of problems with state constraints. In its Hilbert space version in Chapter 6, the theory used here can be found in the author-Frankowska [1991:1]. The general Banach space treatment of the abstract nonlinear programming problem was given by Frankowska [1990:2], who contributed the fundamental insight that convex sets of variations, rather than the full set, must be be brought into play. The present version of the theory, with more direct proofs and somewhat weaker hypotheses is taken from the author [1993:1]. Ekeland's variational principle was employed earlier for different types of optimal problems for partial differential equations; see Plotnikov-Sumin [1982].

Since its beginning in the sixties (see the Miscellaneous Notes to Part I), the theory of nonlinear programming in infinite dimensional spaces has undergone a rapid expansion and its bibliography is enormous. See Sachs [1978:2], Zowe-Kurcyusz [1979], Ioffe [1984], Barbu-Precupanu [1986], Ledzewicz-Nowakowski [1994], Mordukhovich [1994], Páles-Zeidan [1994], Bonnans-Cominetti [1996] and other papers. Most of the results on cones and their behavior under mappings in Chapter 6 are particular cases of results in the literature: see for instance Treiman [1983].

The minimum principle in Hilbert spaces. The minimum principles in **6.5** and **6.6** are in the author-Frankowska [1991:1] and supersede earlier results of the author [1986] [1987:1]. The versions presented here are somewhat more general in that the adjoint variational equation is handled without *ad hoc* assumptions, and there are no restrictions on the control set. The results in Chapter 6 pertaining to control theory of the wave equation, in particular, the proof of the minimum principle in **6.12** can be found in the author-Frankowska [1990:2]; some are of earlier data such as those on continuity of solution operators and Theorem 6.7.1

(the author [1986] [1987:1]). The proofs in Chapter 6 are a little different from those in the references since they are based on patch rather than spike perturbations. Theorems 6.2.2 and 6.2.7 are slight generalizations of Li-Yao [1985:1].

Exact controllability. In the minimum principle, controllability comes into play only when the target set is "thin" (e.g. a point) and the results needed are linear, since what must be controlled is the variational equation. The results on controllability of the linear wave equation (6.11.1) are of earlier data under somewhat different conditions (the author [1977:1]); the operator $N(\bar{t})$ is lifted from Giusti [1967]. Another, more streamlined, but less elementary proof of these results was given by Chen [1979]. The exact controllability argument underlying the proof of Lemma 11.5.6 is in the author [1974].

The only nonlinear controllability result in this work is Theorem 6.12.9 (null controllability of the linearized system implies local null controllability of the original system). It is due (in somewhat different versions) to Chewning [1976:1], Russell [1978] and Cârjă [1979]. The result is a descendant of a finite dimensional result of Lee and Markus [1961] (see also Lee-Markus [1967]). The Lee-Markus paradigm was implemented in infinite dimension by the author [1975:2], the necessary controllability properties of the linearized system available thanks to Russell [1967:1] (*vide infra*). The version of the Banach space implicit function theorem used here goes back to Lyusternik [1934]. Generalizations of the controllability result to semilinear systems $y'(t) = Ay(t) + f(t, y(t), u(t))$ that rely on Lyusternik's theorem need differentiability assumptions with respect to u. The need for these was obviated by Frankowska [1989] [1990:1] by extending Lyusternik's theory to metric spaces. For a few results on controllability of abstract differential equations see Ball-Marsden-Slemrod [1982], Kobayashi [1978] and Zhou [1984]. The problem of driving a system to a subset and keeping it there was considered by Schmidt-Stern [1980].

Moment problems and exact controllability. The controllability results in **9.5** are obtained by the *method of moments*, used by Yu. Egorov [1962] [1963:1], Butkovski [1965], Russell [1967:1] and Galchuk [1968], with Egorov, Russell and Galchuk in pursuit of exact controllability. The method of moments was used by the author [1966] with a different purpose, namely that of showing time invariance of the set of reachable states of certain linear systems and of studying rigid systems (see **9.6**). Example 9.5.5 is a very particular case of the results in Russell [1967:1]; in his paper, eigenvalues are merely asymptotic to the sequence $\{ian\}$, thus one needs to to use the theory of nonharmonic Fourier series introduced by Paley and Wiener [1934], and theorems on biorthogonal systems. Russell's paradigm "controllability via moment problems" was pursued by the author and Russell [1971] using again the method of biorthogonal sequences. The results are on boundary controllability of parabolic equations in 1 space dimension. (essentially equivalent to 1- or 2-

dimensional distributed controllability). Theorems 9.5.7 and 9.5.8 are taken from this work. The results in the author-Russell [1971] apply to parabolic equations with spatially variable coefficients, so that the sequence of eigenvalues $\{\lambda_n\}$ is only asymptotic to a sequence $\{-an^2\}$; this makes it impossible to evaluate infinite products like (9.5.35) by means of trigonometric and hyperbolic functions. The estimations improve those of Egorov [1962] [1963:2] and also those of Mizel-Seidman [1969] where the emphasis is observability rather than controllability.

It was observed by Russell that controllability problems for parabolic equations in several space variables in domains admitting separation of variables could be approached by solving sequences of moment problems in a suitable "uniform" way; this program was carried out by the author and Russell [1974/5] for balls and in the author [1975:1] for parallelepipeda, both in arbitrary dimension. The biorthogonal sequence in Theorem 9.5.8 and its uniformly bounded companion are particular cases ($m = 1$) of a construct in the author [1975:1]. Among other early applications of the method of moments: a face alone is enough for exact controllability of the heat equation in a parallelepipedon (the author [1975:1]), while one face is not enough for the wave equation (the author [1977:2]). Here, it is actually shown that, controlling only one face one can drive the system from zero to certain very smooth elements having a rapidly convergent series in the eigenfunctions, a type of behavior that one associates more with parabolic than with hyperbolic equations. Another application is in the author [1978], where the result on time invariance of the set of reachable states for parabolic linear systems in the author [1966] is extended to arbitrary space dimension. (for further extension and simplification, see Seidman [1979]). An early result in exact controllability of hyperbolic equations in one space variable was obtained by Cirinà [1969] switching the roles of the time and space variables.

At this juncture, exact controllability results in dimension > 1 (of boundary and distributed type) were limited to special domains (balls, squares, cylinders . . .). As pointed out by Seidman [1976] one can obtain boundary controllability for an arbitrary domain Ω by (a) boxing the domain in one of the special domains, (b) doing the controllability job in the special domain (initial and final conditions suitably extended), (c) using as control the restriction of the solution to the boundary of Ω. However, this leaves open the question of whether controllability can be effected applying control only in part of the boundary. The fundamental step was taken by Russell [1973:1], [1973:2] who discusses boundary controllability of the wave equation in arbitrary domains in $\mathbb{R}^m$ and then twists the results into corresponding theorems for the heat equation (this approach was generalized by Seidman [1978] to equations with spatially dependent coefficients). Russell's method underlies most of the subsequent developments in exact controllability; it was used in combination with Carleman estimates and integral transforms by Lebeau and Robbiano [1995] to solve the long standing open problem of whether boundary controllability to zero for the heat equation is possible using control in an arbitrarily small subset of the boundary. Carleman estimates have become a weapon of choice in the study of exact controllability (see Tataru [1994] and Triggiani [1997]); for hyperbolic equations sharp results have been obtained using microlocal analysis (Bardos-Lebeau-Rauch [1992] and other papers). Controllability of the Navier-Stokes equations is studied in Fursikov [1995], Fursikov-Imanuvilov [1994]. For surveys of developments in controllability not mentioned here see Russell [1978], Lions [1990], Avdonin-Ivanov [1995] and Fursikov-Imanuvilov [1996].

It should be noted that results saying that all states in a given space can be driven to zero do not imply a characterization of the spaces $\mathcal{K}$ of states to which the solution can be driven from zero. For the heat equation under boundary control a complete characterization of $\mathcal{K}$ remains elusive, even in dimension 1. On the other hand, numerous particular subspaces have been identified; for instance, it follows from Russell's results in [1973] that $\mathcal{K}$ contains all states whose eigenfunction expansion is rapidly convergent, in particular those with a finite expansion. Entirely different subspaces are identified in Sachs-Schmidt [1989], Weck [1984] and Schmidt [1986].

One of the present trends in controllability theory is to use direct partial differential equation methods in combination with functional analysis. For a survey of the subject, especially of controllability via functional moment problems see Avdonin-Ivanov [1995]. For other applications of the moment method, see Schmidt [1980], Krabs [1985], Sachs-Schmidt [1989]. The monograph by Krabs [1992] contains additional information and references. We mention an approach to the time optimal control by Krabs-Schmidt [1980], Krabs [1985] [1989]; under certain conditions, time optimal controls are solutions of the problem of minimizing the control norm in the optimal interval. This paves the way for reducing a numerical treatment of a difficult problem (the time optimal problem) to something much more tractable. Computation of optimal controls (by discretization, penalty functions, or otherwise) is an area with vast bibliography. See Balakrishnan [1968] and, for more references, Li-Yong [1995]; for functional differential equations, see Kunisch [1982], Gibson [1983].

Most exact controllability results for evolution partial differential equations, obtained by whatever method, assume time independence of the coefficients. See Seidman [1984] for an exception for the one-dimensional heat equation.

In many geometries, distributed or boundary controllability of hyperbolic equations (e.g. the wave equation) cannot be effected in arbitrarily small time due to finite speed of propagation of control effects. Things are of course different for parabolic equations, where speed of propagation is infinite. This raises the question: how expensive is (that is, how large a control we need) to control the heat equation in a small time T? The answer is: *very* expensive. In fact, for boundary control, Güichal [1985] shows that for certain targets the L^2-norm of the control is bounded below by $(C/\sqrt{T})e^{b/T}$, where $b > 0$ is a constant; in the other direction, Seidman [1984] establishes an upper bound of the same type. For more on this, see Seidman [1988].

Approximate controllability.   Under the name of *complete controllability* was introduced by the author in [1966:2] [1967:2] [1967:3] [1968:2] for differential equations in Banach spaces. These papers aim to model partial differential equations of arbitrary type, the last work incorporating control on the boundary of a domain (see **Boundary Control** below). The purpose is to generalize some of the results in Kalman-Ho-Narendra [1963] on controllability; several extensions of the rank condition (3.10.7) are given. Many other constructs of finite-dimensional multivariable control theory due to Kalman, Wonham and others (see the Miscellaneous Notes to Part I) have been generalized to infinite dimensions, and a vast literature exists in this subject, which has incorporated sophisticated operator theoretical tools; see the monographs of Curtain-Pritchard [1978], Fuhrmann [1981] and Curtain-Zwart [1995].

In the realm of partial differential equations, McCamy-Mizel-Seidman [1968] and Lions [1968] gave the first controllability results on parabolic equations without intercession of operator modelling. It should be mentioned that Russell was considering controllability of hyperbolic partial differential equations as early as in [1967:1] [1967:2] although his emphasis was on exact rather than on approximate controllability. These papers, together with Russell [1971] (the latter on approximate controllability) brought to light the role of characteristics and velocity of propagation in control of hyperbolic equations.

For other results in Banach space language see Triggiani [1975:1] [1975:2] [1976] [1977], Reghis-Hiriş [1973] Megan-Hiriş [1978], Topuzu-Topuzu [1979, Nambu [1979] and other papers. For a survey of linear controllability in infinite dimensional spaces see Conti [1982]. That approximate controllability is preserved under certain nonlinear perturbations was proved by Fujii-Sakawa [1974]; for other nonlinear results see Naito [1987] [1989], Zhou [1982] [1983] [1984] and references there. Many results on approximate controllability of linear and nonlinear equations are specific to partial differential equations; see Ahmed [1982], Lions [1990]. For parabolic equations, a type of approximate controllability that allows for a finite number of "exact sensings" has been studied by Chewning [1976:2].

An interesting subcase of approximate (and also exact) controllability is that where the controls are bounded or otherwise restricted; in fact, these questions are a preliminary to most optimal control problems. See the author [1969] for an infinite dimensional generalization of an approximate controllability result of Antosiewicz [1963]. For more recent results and references see Narukawa [1981], Ahmed [1985:1], Peichl-Schappacher [1986]; for delay equations Klamka [1993] and for boundary control, Narukawa [1982] [1984], Cârjă [1988], [1989] (where some results in the author [1969] are generalized and analogues are provided for exact controllability). For corresponding results for integrodifferential equations see Leugering [1987:1], [1987:2].

Stabilization by feedback.   This branch of control theory is closely related both
to exact and approximate controllability and also to the origins of control theory
(see Miscellaneous Notes to Part I). The stabilization problem is formulated in
the same way as for finite dimensional systems, but "stable" has more than one
meaning in infinite dimensions. If A is a finite dimensional operator then $\sigma(A)$
is contained in the left half-plane if and only if $\|e^{tA}\| = O(e^{-ct})$ as $t \to \infty$ for
some $c > 0$. This equivalence collapses in infinite dimension; one must distinguish
between *exponential stabilization* (as above) and other types of stabilization where
all one has is $\|e^{tA}\| \to 0$ as $t \to \infty$ or the even weaker condition $\|e^{tA}y\| \to 0$ as
$t \to \infty$ for all y. For some of the earlier works in evolution equation language see
Pritchard [1969], Russell [1969], Lukes-Russell [1969], Slemrod [1972], [1974:1],
[1974:3], [1976], Triggiani [1975:3], [1979], Gibson [1979], [1980], Pritchard-
Zabczyk [1981]. For more recent works see Nambu [1984] Yamamoto [1991], You
[1991], [1992]. Engineering problems motivate the theory of *modal stabilization,*
where a finite number of modes-corresponding to the first few eigenvalues of a
system-are brought to (or near) equilibrium. A key problem in this area is that of
spillover; the energy suppressed in the modes under control may be transferred
to high frequency modes. For more information and references see Wang [1986],
Balas [1988] [1991].

In partial differential equations one can apply feedback in several ways: boundary $\to$ boundary, boundary $\to$ domain, and so on. For a thorough exposition and
references see Chen-Zhou [1993]. A fascinating sidebar: some of these feedback
laws are highly unstable with respect to time delays, which delays would be unavoidable in practical implementation. See Datko [1988], Desch-Wheeler [1989],
Datko-You [1991].

Semilinear abstract differential equations.   Much of the theory in Chapters 5
and 6 is classical, and credits can be found in Pazy [1983]; we note in particular that
Theorem 5.5.1 is a slight generalization of Segal [1963]. Theorem 5.5.21 is due
to Triggiani [1977] for $p > 1$, who observes (in the addendum) that the operator
Λ is not compact with L^1 as domain; the L^1 result is in the author [1994:1]. We
note, however, that the result for $p = \infty$ is of earlier date: see Pazy [1975], who
also proves an existence-via-compactness of the type of Theorem 5.5.22. See also
Pavel [1977] for other types of existence theorems.

We note in passing that compactness of the operator Λ has negative repercus-
sions for controllability of the linear system $y'(t) = Ay(t) + Bu(t)$; for instance
see Kuperman-Repin [1971] and Triggiani [1977]. The same applies to stabilization: see Gibson [1980].

There exists an enormous literature on semilinear equations in Banach spaces;
for some of it see Pazy [1983]. The global existence Theorem 5.8.3 is generalized
from the author [1987:3]. The results in Chapter 6 pertaining to control theory of
the wave equation, in particular, the proof of the minimum principle in **6.12** can
be found in the author-Frankowska [1990:2] in a slightly less general form.

The "front-and-back smoothing" by fractional powers in Chapter 7 goes back to Sobolevski in the reflexive case, with the Navier-Stokes equations as the main objective; the L^p treatment is in [1959] [1960] for $p = 2$ and in [1964] for $1 < p < \infty$. The nonreflexive version based on Phillips adjoint theory follows essentially the author [1993:1], although the treatment there only uses front smoothing (roughly speaking, $\rho = 0$ in Hypotheses $\text{II}_{\alpha,\rho}$ and $\text{III}_{\alpha,\rho}$) so that the treatment of the variational and adjoint variational equations is much less symmetric. Similar ideas were used much earlier for linear control systems (see the author [1974]). Although formulated in a somewhat different setting, the material in the author [1993:1] and in this book has a substantial intersection with Clément-Diekmann-Gyllenberg-Heijmans-Thieme [1987] [1988] [1989] and other papers of these authors. Instances are the results on linear perturbations in **7.9**, the treatment of semilinear equations in **7.7**, the setting of $^\odot$-reflexive spaces and the use of $S^\odot(t)^*$ as a smoothing operator.

Linear control systems.   The theory of strongly continuous semigroups was first applied to linear control systems by the author [1964] and Balakrishnan [1965]. The setup in **9.1** (and in part of the following sections) is related to the author [1974/5]. Theorem 9.1.10 is a particular case of the author [1995:1] (modulo a faulty identification of the multiplier z that only works in reflexive spaces). The approach leading to Theorem 9.1.10 is related to Galchuk [1968] and also to earlier results of the author [1974/5], [1976], [1977:1]; in these, the central idea is to consider attainable sets not in the whole space E, where they may be too sparse to apply separation theorems, but in subspaces where they have nonempty interior. This places the multiplier z (the final condition for the costate) in spaces larger than E^*. For a version of Theorem 9.2.5 where the semigroup need not be analytic, see Cârjă [1984].

Existence theory of optimal controls enjoys of course many prerogatives in the linear case. The earlier result in the present setup seems to be in the author [1964]; it is a somewhat less general version of Theorem 14.5.4. See the Miscellaneous Comments to Part III for more on this and for nonlinear existence theory.

Although the result in Bellman-Glicksberg-Gross [1956] was almost immediately submerged by Pontryagin's maximum principle, the simplicity of the linear separation argument beckoned generalization to infinite dimensional spaces. This was done by Balakrishnan [1965] and Friedman [1967] [1968]; in particular Friedman proves the group case of the minimum principle (Example 6.7.4). For the time dependent case see also Conti [1968], who deals with targets with nonempty interior and the author [1968] for reversible equations. For a general semigroup the standard minimum principle (that is, the minimum principle with multiplier in E^* rather than in larger spaces) is "generic" in the sense that it holds for a set of "good targets" dense in the boundary of the reachable set. There seems to be no criteria to identify good targets except in the case where A generates a group and $B = I$, in which case all targets are good. However, as shown in Balakrishnan [1965], denseness of good targets makes possible to approximate (in weak topologies) optimal controls, which may not satisfy the maximum principle with $z \in E^*$, with controls that do satisfy the principle, a result that can be considered a remote ancestor of Example 14.10.2. For an actual example of an optimal control that does not satisfy the maximum principle with $z \in E^*$ see the author [1974/5].

It is disappointing that, 40 years into infinite dimensional control theory, its "simplest" question (the time optimal problem for $y'(t) = Ay(t) + Bu(t)$ with a point target) has never been completely settled, even in the case $B = I$. In fact, all we know about optimal controls in the latter case (apart from approximation theorems) seems to be the "bang-bang theorem" 9.3.6 in the author [1964]. (in the time-dependent case $A = A(t)$ not even this has been proved). Needless to say, the case of general B is even less understood. There are of course numerous results that require special conditions on A or on the target set. Examples are the "full control" results in **9.2**; most are in the author [1974/5], some in more general versions (strongly continuous rather than analytic semigroups). Theorem 9.2.5 exists in [1974/5] only in its Hilbert space version, although the extension to ζ-convex spaces is immediate.

Parallel to the "semigroup approach", versions of the maximum principle were derived for linear partial differential equations without intercession of semigroup theory. We mention Russell [1966] [1967:2] for a symmetric hyperbolic system under finite dimensional control. The vagaries of the maximum principle for hyperbolic systems, in particular the prevalence of "singular regions" where the vanishing adjoint state gives no information, are illustrated here. For another early paper on hyperbolic equations dealing with these questions, see Malanowski [1969]; for more recent literature see Vinter-Johnson [1977], White [1983], Robbiano [1991] and references in these papers.

The finite dimensional theory of the linear-quadratic problem in **2.8** (due to Kalman) has been generalized to infinite dimensional systems, the first step taken by Lukes-Russell [1969] for a system $y'(t) = Ay(t) + Bu(t)$, A an infinitesimal generator, B bounded, culminating in the Riccati operator equation for the feedback law. As a model for partial differential equations, it corresponds to a distributed input. For other early papers along the same lines see Datko [1974], Curtain-Pritchard [1974], [1976].

When attempting to model systems with control on the boundary as operator equations (see below) one is led to the same model, but with B unbounded. This introduces numerous complications setting up and interpreting the solutions of the Riccati equation; see Curtain-Pritchard [1977] [1978], Balakrishnan [1978], Pritchard-Salamon [1987]. As in the finite dimensional case, the linear-quadratic problem is intimately related with controllability, observability and stabilization questions (see **Stabilization by feedback** above). For some of the literature, see Gibson [1979], Banks-Kunisch [1984], Da Prato-Ichikawa [1985], Da Prato-

Lasiecka-Triggiani [1986], Yao [1985], Yao-You [1985] and, in particular, a series of papers by Lasiecka and Triggiani, two of which are [1987:1] [1987:2]. For references to the rest of this series and other bibliography, see Bensoussan-Da Prato-Delfour-Mitter [1992/3], Lasiecka-Triggiani [1991] and a forthcoming monograph by Lasiecka and Triggiani.

A parallel development for systems described by partial differential equations, especially of parabolic type was initiated by Lions [1969] and is based on direct methods. For a comparison between these and operator methods, see the introduction to Gibson [1979].

An intriguing slant on controllability (and its dual notion, observability) of partial differential equations with finite dimensional controls is to allow the spatial distribution of the control to depend on time (say, generalize the control term $u(t)\delta(x - x_0)$ to $u(t)\delta(x - x_0(t))$). This approach has certain advantages, such as elimination of some of the difficulties related with multiplicity of the eigenvalues. See Butkovski-Pustylnikov [1980], Khapalov [1995:1], [1995:2].

Another active area of research is that of *systems under uncertainty*. These contain a control parameter u and a "disturbance" parameter v over which one has little or no information. The problem is that of minimizing a cost functional of u while assuming the worst of v (that is, that the disturbance will tend to maximize the functional). This leads to *minimax* (rather that straight minimization) problems. Other problems such as controllability and observability must also take into account the (possibly deleterious) action of the disturbance. Sometimes, statistical assumptions are made on v. See Kurzhanski [1977] [1994] Kurzhanski-Khapalov [1986], Ahmed-Xiang [1994], Aubin-Frankowska [1990], Mordukhovich [1990], Mordukhovich-Zhang [1997] and other papers. When the parameter v is intelligently operated by an opponent bent on maximizing the functional we have a *differential game*. For references see Roubíček [1997].

We point out finally that some of the results on linear systems have counterparts in other theories, where the emphasis is more on the "input- output" approach of systems theory; see for instance Falb [1964] for a result on closedness of reachable sets, as well as for an example corresponding to our definition of rigid system. See also Knowles [1975] [1976] and other papers; many of the results have bearing on the theory of vector measures and are commented on in Diestel-Uhl [1977]. More on this can be found in the Miscellaneous Comments to Part 3.

The maximum principle for nonlinear systems in Banach spaces.   The earliest treatment is in the (somewhat enigmatic) papers by Yu. Egorov [1963:1] [1964], where the arguments follow the lines of the original approach by Pontryagin et al [1961]. Leaving aside the linear case, where development was faster, other nonlinear Banach space results are in Li-Yao [1981] [1985]. The maximum principle for a semilinear system in Banach space was proved in Li-Yong [1991] using Ekeland's variational principle, but accounting for irregularity of the norm in a different

way than that used in this work and in the author [1993:1]. For many other contributions of these authors and others of the "Fudan School" to the study of the
maximum principle for semilinear and quasilinear equations see the references in
Li-Yong [1995][1]; two of Li and Yao's fundamental insights are important in our
treatment, namely that finite codimensionality of reachable and/or target sets is of
the essence in infinite dimensional spaces, and the use of patch instead of spike
perturbations. Patch perturbations are now in general use in treatments of state
constrained problems. In their extension from one to several dimensions by Casas
[1994], Casas-Yong [1995] they have been applied by Casas, Yong and others (*vide
infra*) to both evolution and steady-state problems.

During the last decade, Ekeland's variational principle has become a tool of
choice in the treatment of necessary conditions for optimal problems governed by
partial differential equations. By its intercession, versions of the maximum principle have been discovered in situations far afield from Pontryagin's original formulation. One instance is that of steady-state problems governed by elliptic equations; see Bonnans-Casas [1989] [1992], Casas [1992] [1993], Casas-Fernández
[1993:1] [1993:2] and the references in these papers. For other related works
see Ahmed-Teo [1984], Altman [1990], Basile-Mininni [1990], Casas-Fernández-
Yong [1994], Casas-Yong [1995], Fursikov [1992], He [1987], Hu-Yong [1995],
Staib-Dobrowolski [1995]. In some of these papers, tools other than Ekeland's
principle are used, for instance the Dubovitski-Milyutin or Ioffe-Tijomirov theories of nonlinear programming. Optimal problems for partial differential equations
have also been approached along lines that stress direct partial differential equations methods, as in some of the papers above: see for instance Sakawa [1964]
[1966], Wang [1964:2], [1966], A. I. Egorov [1967], Wiberg [1967], Butkovski
[1969], Gregory-Lin [1992], Barbu [1976] [1984] [1987] [1993] [1997] [1998],
Barbu-Pavel [1993], Ginzburg-Ioffe [1996], Raymond [1996], Alibert-Raymond
[1997], Raymond-Zidani [i.p:1] [i.p:2], Tiba [1990] and references there. Some
of these methods also apply to partial differential inequations such as variational
inequalities.

It was observed by Frankowska that the general Banach space version of the
nonlinear programming theory (as presented in Frankowska [1990:2] or in Chapter
7 of this work) would apply handily to problems with state constraints. An outline
of this program was carried out in the author-Frankowska [1991:2]. A version of
(part of) this theory with full proofs and some improvements is given in the author
[1996] and reproduced in Chapter 10 of this work. The material on parabolic
equations in Chapter 11 is a further elaboration on this theme that attempts to
accomodate wilder nonlinearities; it appears here for the first time.

There exist a vast literature on the nonlinear maximum principle; for a short sample see Alt-Mackenroth [1989], Goldberg-Tröltzsch [1993] Mackenroth [1980]

---

[1] Including many in Chinese.

[1981] [1986:1], [1986:2], Roubíček [1987], Tröltzsch [1982] [1984], White [1983].
See also Li-Yong [1995] for many other references. The treatment of the minimal
surface problem is taken from McShane [1978]. Theorems 11.6.1 and 11.6.4 are
generalizations of results in the author [1995].

The treatment of the Navier-Stokes equations follows that in the author-Sritharan
[i.p.]. Earlier treatments under different hypotheses are in the author-Sritharan
[1992] [1994], Sritharan [1991], the latter two papers including the Hamilton-
Jacobi approach (*vide infra*). For other approaches that stress direct partial differential methods over operator theory see Abergel-Temam [1990], Barbu [1997] [1998]
and Casas [1994]. It may not be exaggerated to say that control theory of fluid flow
(aero and hydrodynamics) did not exist fifteen years ago, except for scattered and
almost purely experimental results, most of them on attenuation or elimination of
turbulence and drag reduction. Since then, a true explosion of research has taken
place. For works with a more computational orientation see Brutyan-Krapivski
[1984] Hou-Sbovodny [1991], Gunzburger-Hou-Sbovodny [1992] and references
in these papers. See also Sritharan [1994] for a survey of recent developments in
the field.

The distance $d(u, v) = |\{t; u(t) \neq v(t)\}|$ given to the control space to apply
Ekeland's principle requires boundedness of the control set for completeness. Thus,
unbounded controls must be handled differently. A treatment lacking in details is
outlined in the author-Sritharan [1994:1] for the Navier-Stokes equations. Versions
of the maximum principle for unbounded controls were discovered by Raymond-
Zidani [i.p:1] [i.p:2] and independently by the author [1997:1] [i.p.]. They work
in somewhat different settings.

Invariance of the Hamiltonian.   As pointed out in Remark 6.6.6, Pontryagin's
original criterion to distinguish free arrival time optimal control problems from
their fixed arrival time counterparts—vanishing of the Hamiltonian—is not applicable in general to infinite dimensional problems; the Hamiltonian may not even
be defined, since it includes a term $\langle z, Ay \rangle$ where, in general, neither $y \in D(A)$
nor $z \in D(A^*)$. This can be circumvented (and Pontryagin's criterion extended)
for certain parabolic equations: see the author [1990:1] [1994:3] for results under
different hypotheses. See also Raymond-Zidani [i.p:2]. We don't know of results
for equations without smoothing properties.

Convergence of suboptimal controls.   Examples 11.10.4 and 11.10.6 are slight
extensions of a result in [1987:5]. The other results in **11.9** and **11.10** are taken
from the author [1997:3] and generalize results in the author-Frankowska [1990:1]
[1990:3], where state constraints are absent. Finite dimensional ancestors can be
found in Frankowska-Olech [1982:1] [1982:2].

Ekeland's variational principle deals with *approximate* solutions of optimal
control problems, thus it automatically yields results on suboptimal controls. This

is obvious from the statement of the various theorems of Kuhn-Tucker type in this book, and is exploited in the author [1987:2] [1987:3] [1990:2]. There exists a vast literature on perturbations of diverse types of abstract minimization problems; in some of these papers one tries for more precise continuous dependence results (say, differentiability with respect to the parameters under perturbation) under stronger hypotheses. For a small sample see Roubíček [1990], Colonius-Kunisch [1993], Ioffe [1994], Bonnans-Cominetti [1996], Lasiecka-Sokolowski [1991], Zălinescu [1984], Malanowski [1984:1] [1984:2] [1987:1], Malanowski-Sokolowski [1986], Sokolowski [1985] [1987], Mordukhovich [1994]. In other results, connected with numerical analysis, "perturbation" is discretization; see Tiba-Tröltzsch [1996]. For more information and numerous references see Malanowski [1987:2], Roubíček [1997].

Boundary control.   Early work of the author [1968:2] attempts to describe control on the boundary of a domain by means of an operator model. In this formulation, the Green or Neumann function of a boundary value problem appears as the composition of a "boundary operator" (transforming a space of functions on the boundary into a space of functions in the domain) and the semigroup generated by the operator with homogeneous boundary conditions. Since the objective in [1968:2] is to study approximate controllability, no metric properties of this composition are established. Operator bounds are indispensable for the study of optimal problems and were first studied by Washburn [1979] elaborating on an example of Balakrishnan for a rectangle. Depending on what functions spaces are used, this approach produces bounds on the Green and Neumann kernels, some of them not immediately amenable to classical methods. Operator bounds on the kernels also produce integral equations for the solutions of certain nonlinear equations with nonlinear boundary conditions, which allow application of methods similar to those used in Chapters 7 and 11 of this book. These integral equations are frequently combined with direct partial differential equation methods. See von Wolfersdorf [1976/7], Glashoff-Weck [1976], Sachs [1978:1], Lasiecka [1980], Tröltzsch [1984] [1993] Schmidt [1989], Goldberg-Tröltzsch [1993], the author-Murphy [1994:1] [1994:2] [1995], Raymond-Zidani [i.p:1], [i.p:2], Desch-Lasiecka-Schappacher [1985], Lasiecka-Triggiani [1991], Mordukhovich-Zhang [1997]. Other approaches (such as that initiated by Lions [1969]) stress direct methods.

Much sharper results exist for linear boundary control systems. We mention the author [1976], where the subject is the heat equation under boundary control, and the result is the maximum principle for the time optimal problem with a point target. (these targets, for the moment, are inaccessible in the nonlinear case). For a different optimal problem for a parabolic equation see Glashoff-Weck [1976]. In these papers, bang-bang theorems following from the maximum principle are established under analyticity conditions on the coefficients of the equation and the boundary of the domain (*vide infra*). Analyticity is relaxed to infinite differentiability in Schmidt-Weck [1978] and Schmidt [1980].

Evanescence of the costate.   Interest in nodal sets of solutions of elliptic equations antedates control theory by a more than a century and was probably sparked by vibration phenomena, where the solutions under study are eigenfunctions of the Laplacian. In contrast, nodal sets of parabolic equations seem to have received scant attention and it may have been through control theory that they first attracted attention more than twenty years ago.

Although there is partial information on Problem 11.6.7, to the best of our information a complete (positive or negative) answer is not yet available. Under analyticity (rather than infinite differentiability) Theorem 11.6.8 is essentially equivalent to a result by the author in [1976]. Its main trick (to reduce the parabolic to the elliptic case via eigenfunction expansions) can be used to obtain results of the same type under weaker smoothness assumptions. The downward step from analyticity to infinite differentiability was taken by Schmidt-Weck [1978] and Schmidt [1980] in a somewhat different setting (*vide infra*) and then extended to minimal smoothness by Lin [1990] via various recent advances on nodal sets of solutions of elliptic equations (Garofalo-Lin [1987], Jerison-Kenig [1985], Robbiano [1988], Sogge [1990]). In these, papers, however, what is proved is that solutions $z(x)$ of elliptic equations having a *zero of infinite order* x_0 (roughly, $z(x) = o(|x - x_0|^k)$ for all $k = 1, 2, \ldots$) must vanish identically, and this is the statement that is extended by Lin [1990] to parabolic equations with time-independent coefficients. The time-dependent case has been recently dealt with by by Poon [1996]. If $z(x)$ is infinitely differentiable and vanishes in a set of positive measure it must have a zero of infinite order somewhere, but in the absence of smoothness the nodal sets must be studied directly. This has been carried out in Lin [1991] and Han-Lin [1994:1] [1994:2]; however, the Hölder continuity assumptions on the lower order coefficients make the results apparently inapplicable to equations like (11.3.5) and (11.5.52). Also, further restrictions must be placed on the solutions.

The one dimensional case deserves mention. In fact, when the domain is an interval the nodal set of $z(t, x)$ for t fixed is *finite* for each t if $z(t, x)$ is not identically zero and its number of elements (the *lap number*) is nonincreasing in t. (see Matano [1982], Angenent [1988]). However, as in the higher dimensional case, the assumptions on the lower order coefficients exceed what is available for (11.3.5) and (11.5.52).

Evanescence of the costate for boundary control.   Versions of the maximum principle for boundary control systems (see **Boundary control** above) lead to the following companion of Problem 11.6.7, which we state somewhat short on generality.

Problem Let Ω be a domain in $\mathbb{R}^m$, $z(t, x)$ a solution of

$$\frac{\partial z(t, x)}{\partial t} = \Delta y(t, x) \quad (0 \le t \le \bar{t}, \ x \in \Omega)$$

satisfying the Dirichlet boundary condition or a variational boundary condition β on the boundary Γ. What can be said about the set

$$e(\bar{t}, z) = \left\{ (t, x) \in [0, \bar{t}\,] \times \Gamma; \ z(t, x) = \frac{\partial z(t, x)}{\partial v} = 0 \right\} ?$$

That e must have measure zero was proved in the author [1976] for the Dirichlet boundary condition and Glashoff-Weck [1976] (for more general differential operators and boundary conditions) assuming analyticity of the domain and the coefficients. The assumptions were then downgraded to infinite differentiability by Schmidt-Weck [1978] and Schmidt [1980], who use deep continuation arguments due to Aronszajn and Mizohata. The results available at the moment seem to require too much smoothness to be applicable to adjoint variational equations of nonlinear systems, and we don't know of any results under more permissive smoothness assumptions.

Functional differential equations.   For various versions of the maximum principle and related results see Jaratishvili [1961], Friedman [1964], Halanay [1968], Banks-Jacobs [1970], Banks-Kent [1972], Colonius-Hinrichsen [1978], Colonius [1982], Seierstad [1996]. For the linear-quadratic problem see Delfour [1977] [1986], Gibson [1983], Kappel [1991]. For an detailed treatment of controlled functional differential systems see Bensoussan-Da Prato-Delfour-Mitter [1992/3]; this work includes also references on functional differential systems where instantaneous values are infinite dimensional, for instance, parabolic or hyperbolic equations with delays. For earlier papers in this subject see Wang [1964:1] [1975]. Systems with hysteresis have been examined in Brokate-Friedman [1989], Brokate-Sprekels [1989].

The Hamilton-Jacobi approach.   The same tools that facilitated the treatment in finite dimensions—viscosity solutions and nonsmooth analysis among them—have paved the way for a rapid expansion of this branch of control theory of infinite dimensional systems starting in the eighties. The bibliography is now substantial. For a (very small) sample, see Barbu-Da Prato [1983], Barbu-Barron-Jensen [1988], Cannarsa-DaPrato [1990], Cannarsa-Frankowska [1992], Cannarsa-Gozzi-Mete Soner [1993], Cannarsa-Tessitore [1994], Frankowska [1993], Tataru [1992], the author-Sritharan [1994]. For a survey of the Hamilton-Jacobi method in flow control see Sritharan [1994]. Numerical solution of the Hamilton-Jacobi equations in infinite dimensions (or even in high finite dimensions) is mostly beyond the scope of present day computers but advances are being made. See Falcone [1994], Camilli-Falcone [1996] and references there.

Shape optimization.   In these problems, the control parameters affect the shape of a domain rather than (sometimes as well as) the coefficients of an equation; among industrial applications is the optimal design of shell structures (say, with minimum weight). For information and references see Banichuk [1983], Haslinger-Neittanmäki [1988] and Sokolowski-Zolésio [1992].

Parameter estimation and identification.   A typical problem in this area is that of obtaining information on the coefficients of an equation from input-output observations; a specific example is that of determining $a(x)$ in a given class of functions in such a way that the solution of $\nabla(a(x)\nabla y(x)) = f(x)$ in a domain (suitable boundary conditions) matches, or approximates as closely a possible, a "target" $y_0(x)$. As in other *inverse problems,* identification problems are usually ill-posed and their numerical solution is delicate. See the monograph of Banks and Kunisch [1989]. For other references, see Hoffmann-Sprekels [1986], Zolésio [1990], Avdonin-Seidman [1995], Barbu-Kunisch [1996].

Part III

Relaxed Controls

12

Spaces of Relaxed Controls. Topology and Measure Theory

12.0. Weak Topologies in Linear Spaces

Consider the unit ball $B^* = \{y^* \in X^*; \|y^*\|_{X^*} \leq 1\}$ of the dual X^* of a Banach space X. Let $\eta(t)$ be a continuous increasing function in $t \geq 0$ with $\eta(0) = 0$, $\eta(s + t) \leq \eta(s) + \eta(t), \eta(t) \leq 1$ (for instance, $\eta(t) = t/(1 + t)$). Then, if $\{\alpha_m\}$ is a convergent sequence of positive numbers with $\Sigma \alpha_m < \infty$ (say, $\alpha_m = 2^{-m}$) and $\{y_m\}$ is a sequence dense in X,

$$\rho(y^*, z^*) = \sum_{m=1}^{\infty} \alpha_m \eta(|\langle y^* - z^*, y_m \rangle|) \tag{12.0.1}$$

is easily seen to be a distance in B^*. Symmetry and the triangle inequality are obvious; on the other hand, if $\rho(y^*, z^*) = 0$ then $\langle y^* - z^*, y_m \rangle = 0$ for all m, which implies $y^* = z^*$ by denseness of $\{y_m\}$. It is obvious that, if $\{y_n^*\}$ is a sequence in X^*, then $\langle y_n^*, y_m \rangle \to \langle y^*, y_m \rangle$ for all m if and only if $\rho(y_n^*, y^*) \to 0$; on the other hand, writing $\langle y_n^*, y \rangle = \langle y_n^*, y_m \rangle + \langle y_n^*, y - y_m \rangle$, we see that $\langle y_n^*, y_m \rangle \to \langle y^*, y_m \rangle$ for all m implies $\langle y_n^*, y \rangle \to \langle y^*, y \rangle$ for $y \in X$. Hence, we have

Lemma 12.0.1. *Let X be a separable Banach space, $\{y_m\}$, a countable dense set in X, $\{y_n^*\}$ a sequence in B^*. Then $y_n^* \to y^* \in X^*$ in the distance (12.0.1) if and only if*

$$\langle y_n^*, y \rangle \to \langle y^*, y \rangle \tag{12.0.2}$$

as $n \to \infty$ for all $y \in X$.

We recall (**2.0**) that if V is a metric space and $K \subseteq V$, then K is compact if every sequence in K has a subsequence convergent to an element of K. In the next result, the definition is applied with $K = V$.

Theorem 12.0.2. B^*, *equipped with the metric* (12.0.1), *is compact*.

The proof is the same as that for separable Hilbert spaces (Theorem 2.0.11). If $\{y_n^*\}$ is a sequence in B^* we use Cantor's diagonal sequence trick to obtain a subsequence (equally named $\{y_n^*\}$) such that $\langle y_n^*, y \rangle$ is convergent for $y = y_m$, thus for all $y \in X$. The functional $\phi(y) = \lim_{n \to \infty} \langle y_n^*, y \rangle$ is linear and bounded by 1, thus it defines an element of B^* satisfying (12.0.2).   ■

When X is not separable, it is not possible to define a metric in B^* with convergence equivalent to (12.0.2). We must then use topological spaces, where open and closed sets are *postulated* rather than defined in terms of other notions.

Let U be an arbitrary set. A **topology** in U is a collection Υ of subsets of U satisfying the following postulates:

(*a*) The empty set $\emptyset$ and the entire set U belong to Υ.

(*b*) The union of an arbitrary subfamily of Υ belongs to Υ.

(*c*) The intersection of a finite subfamily of Υ belongs to Υ.

The space U equipped with Υ is called a **topological space** and denoted by (U, Υ). The sets in Υ are called the **open sets** of U; their complements are the **closed sets** of U. A **Hausdorff topological space** is one where

(*d*) given two distinct points of U there exist two disjoint open sets each containing one of the points.

Since all spaces from now on will be Hausdorff, we simply say "topological space" for "Hausdorff topological space"; also, "topology" means "Hausdorff topology."

In a metric space, a set W is open if and only if, for every $u \in W$, there exists an open ball $B_0(u, \varepsilon)$ with $B_0(u, \varepsilon) \subseteq W$. We check that the collection of all open sets in a metric space satisfies (*a*), (*b*), (*c*), and (*d*) so that every metric space is a topological space. Keeping this particular case in mind, we will have no trouble in translating metric space statements to topological spaces. For instance, if V is an arbitrary set in U, a point $u \in V$ is **interior** if there exists an open set W with $u \in W \subseteq V$; a point $u \in V$ is a **boundary point** if it is not interior. A map G from a topological space (U_1, Υ_1) to a topological space (U_2, Υ_2) is **continuous** at $u \in U_1$ if, for every open set $W_2 \in \Upsilon_2$ with $W_2 \supseteq G(u)$, there exists an open set $W_1 \in \Upsilon_1$, $W_1 \supseteq u$ such that $G(W_1) \subseteq W_2$. A sequence $\{u_n\}$ in a topological space U **converges to** $u \in U$ if for every $W \in \Upsilon$ with $u \in W$ there exists n_0 such that $u_n \in W$ for $n \geq n_0$. Sequences are not enough in topological spaces; there are spaces such that a point u is an accumulation point of a set V (that is, every open set W that contains u contains an infinite number of elements of V), but there is no sequence in V converging to u. (Kelley [1955, Example B, p. 76]). Sequences must be replaced by *generalized sequences*, defined below.

A relation "$\leq$" in an arbitrary set K is an **order** if (i) $\iota \leq \iota$ for all $\iota \in K$ (ii) $\iota \leq \kappa$ and $\kappa \leq \lambda$ imply $\iota \leq \lambda$ for all $\iota, \kappa, \lambda \in K$. The set K equipped with $\leq$ is an **ordered set** and denoted $(K, \leq)$. An ordered set $(K, \leq)$ is **directed** if, given $\iota, \kappa \in K$ there exists $\lambda \in K$ such that $\iota \leq \lambda, \kappa \leq \lambda$. Examples of directed sets are the positive integers (or any subset of the reals) with the usual order relation, or all the subsets of a set containing a fixed set V with $\subseteq$ or $\supseteq$ as order relation. On the other hand, the set of all subsets contained either in V or in W with $V \cap W = \emptyset$, although ordered with $\subseteq$ is not directed.

A **net** or **generalized sequence** $\{u_\kappa\} = \{u_\kappa; \kappa \in K\}$ in a set U consists of a directed set K and a function $\kappa \to u_\kappa \in U$. The set K is the **index set** of $\{u_\kappa\}$. A generalized sequence $\{u_\kappa\}$ is **convergent** to $u \in U$ (in symbols $u_\kappa \to u$ or $u = \lim_\kappa u_\kappa$) if, given an arbitrary open set V with $u \in V$ there exists $\kappa_0 \in K$ such that $u_\kappa \in V$ for $\kappa \geq \kappa_0$. Obviously, an ordinary sequence is a generalized sequence (integers with usual order), and the notion of convergence generalizes the ordinary definition.

A **subnet** (or **generalized subsequence**) of a generalized sequence $\{u_\kappa; \kappa \in K\}$ is a pair $(\{v_\lambda; \lambda \in \Lambda\}, \phi)$, where $\{v_\lambda; \lambda \in \Lambda\}$ is another generalized sequence and the map $\phi : \Lambda \to K$ satisfies (i) for every $\kappa \in K$ there exists $\lambda \in \Lambda$ such that $\phi(\mu) \geq \kappa$ if $\mu \geq \lambda$ (ii) $v_\lambda = u_{\phi(\lambda)}$. A particular type of generalized subsequence is that where Λ is a cofinal subset of K equipped with the order inherited from K (Λ is **cofinal** in K if and only if for every $\kappa \in K$ there exists $\lambda \in \Lambda$ with $\kappa \leq \lambda$), and ϕ is the identity map, that is $v_\lambda = u_\lambda$. This type of generalized subsequence is called **strict**. Strict generalized subsequences are the obvious generalization of subsequences and it would be nice to use only them. However, the general definition of generalized subsequence cannot be ignored (Kelley [1955, Example E, p. 77]). We refer the reader to Kelley [1955, Chapter 2] for a complete treatment of generalized sequences. A more streamlined but less intuitive description of convergence in general topological spaces in terms of *filters* can be found in Bourbaki [1966].

Example 12.0.3. Let (U, Υ) be a topological space. Then (a) a set $V \subseteq U$ is closed if and only if it contains the limit of every generalized sequence contained in V. (b) Let V be an arbitrary set. The **closure** of V (denoted $Cl(V)$) is the intersection of all closed sets containing V. Every $u \in Cl(V)$ is the limit of a generalized sequence in V.

Example 12.0.4. Let (U_1, Υ_1), (U_2, Υ_2) be topological spaces, G a map from U_1 into U_2. G is continuous at u if only if $G(u_\kappa) \to G(u)$ for every generalized sequence $\{u_\kappa\}$ such that $u_\kappa \to u$.

Let E be a vector space over the real or complex numbers. A function $y \to |y| \in \mathbb{R}$ is called a **semi-norm** if

$$|y| \geq 0, \qquad |\lambda y| = |\lambda||y|, \qquad |y + z| \leq |y| + |z| \qquad (12.0.3)$$

where $y, z \in E$ and λ is a scalar. Semi-norms are the same as norms except that we do not require that $|y| = 0$ imply $y = 0$. A family $\mathfrak{S}$ of semi-norms is **total** if

$$|y| = 0 \text{ for all } |\cdot| \in \mathfrak{S} \quad \text{implies } y = 0. \tag{12.0.4}$$

A space E equipped with a total family of semi-norms $\mathfrak{S}$ is a **locally convex space**. (A normed space is a locally convex space; the family of semi-norms contains just one element, the norm of E.)

In a locally convex space, we can define a topology Υ as follows: $V \in \Upsilon$ (that is, V is open) if and only if, for every $y \in V$ there exists a finite subset $\{|\cdot|_1, |\cdot|_2, \ldots, |\cdot|_n\} \subseteq \mathfrak{S}$ and $\delta > 0$ such that

$$\{z; |z - y|_1 < \delta, \ldots, |z - y|_n < \delta\} \subseteq V. \tag{12.0.5}$$

We leave to the reader the proof that Υ is a topology. We check easily that a generalized sequence $\{y_\kappa\} \subseteq E$ is convergent to $y \in E$ in this topology if and only if

$$|y_\kappa - y| \to 0 \quad \text{for all } |\cdot| \in \mathfrak{S}. \tag{12.0.6}$$

Example 12.0.5. In a locally convex space, the operations $x, y \to x + y$ and $\alpha, y \to \lambda y$ are continuous. Continuity of $x + y$ means: for every open set V with $x + y \in V$ there exist open sets W, Z with $x \in V, y \in Z$ and $W + Z \subseteq V$. A corresponding definition characterizes continuity of λy.

If E is a locally convex space, we denote by E^* (the **dual** of E) the space of all continuous linear functionals in E. It is a linear space with the natural (pointwise) operations. It is not *a priori* obvious that a locally convex space E will have nontrivial continuous linear functionals, that is, that $E^* \neq \{0\}$. That this is true is a very particular case of the separation theorems later in this section.

We assume from now on that the scalars are the real numbers; the modifications needed to handle the complex case are left to the reader.

Let $E = \{y, \ldots\}$ be a linear space, $F = \{y^*, \ldots\}$ a linear space of linear functionals in E; we say that E and F (or the pair E, F) are **in duality** (no topology is *a priori* given to any of the two spaces). Application of $y^* \in F$ to $y \in E$ is written $\langle y^*, y \rangle$. F is **total** if

$$\langle y^*, y \rangle = 0 \quad \text{for all } y^* \in F \text{ implies } y = 0. \tag{12.0.7}$$

If F is total, then every element of E gives rise to a (unique) linear functional on F, so that (F, E) are in duality as well. The F-**weak topology** of E is the topology defined by the family $\mathfrak{S}$ of semi-norms

$$|y| = |\langle y^*, y \rangle| \quad (y^* \in F). \tag{12.0.8}$$

The family $\mathfrak{S}$ is total if F is total. E is now a locally convex space, and, as such, it has a dual space E^*.

Lemma 12.0.6. $E^* = F$; in other words, every linear functional ϕ on E continuous in the F-weak topology of E is given by

$$\phi(y) = \langle y^*, y \rangle \tag{12.0.9}$$

for some $y^* \in F$.

Proof. Obviously, $F \subseteq E^*$. Conversely, let ϕ be a linear functional in E. If ϕ is continuous in the F-weak topology of E then, given $\varepsilon > 0$ there exist $y_1^*, \ldots, y_m^*$ in F and $\delta > 0$ such that $|\phi(y)| < \varepsilon$ if $|\langle y_1^*, y \rangle| < \delta, \ldots, |\langle y_m^*, y \rangle| < \delta$. This means in particular that

$$\langle y_1^*, y \rangle = \cdots = \langle y_m^*, y \rangle = 0 \quad \text{implies } \phi(y) = 0. \tag{12.0.10}$$

Define a linear map $\Phi : E \to \mathbb{R}^m$ by $\Phi(y) = (\langle y_1^*, y \rangle, \ldots, \langle y_m^*, y \rangle)$ and a linear functional on the subspace $N = \Phi(E) \subseteq \mathbb{R}^m$ by

$$\psi((\langle y_1^*, y \rangle, \ldots, \langle y_m^*, y \rangle)) = \phi(y).$$

This functional is well defined because of (12.0.10). Extend ψ to $\mathbb{R}^m$. It must have the form $\psi((\eta_1, \ldots, \eta_m)) = \Sigma a_j \eta_j$ so that

$$\phi(y) = \Sigma a_j \langle y_j^*, y \rangle = \langle \Sigma a_j y_j^*, y \rangle.$$

Hence (12.0.9) holds as claimed. ∎

Let E be a locally convex space. Given two sets $A, B \subseteq E$ we say that $y^* \in E^*$ **separates** A and B if $y^* \neq 0$ and

$$\langle y^*, y \rangle \leq \langle y^*, z \rangle \quad (y \in A, z \in B). \tag{12.0.11}$$

We say that y^* **separates strictly** A and B if $y^* \neq 0$ and there exists $\varepsilon > 0$ such that

$$\langle y^*, y \rangle + \varepsilon \leq \langle y^*, z \rangle \quad (y \in A, z \in B). \tag{12.0.12}$$

(These definitions extend the Banach space versions in **6.7**.)

Theorem 12.0.7. Let E be a locally convex space, $A \subseteq E$ convex with interior points, y_0 not an interior point of A. Then there exists a continuous linear functional y^* on E separating A and y_0, that is, there exists $y^* \neq 0$ such that

$$\langle y^*, y \rangle \leq \langle y^*, y_0 \rangle \quad (y \in A).$$

Proof. A functional y^* separates A and y_0 if and only if separates $A - \{y\}$ and $y_0 - y$, thus we may assume that 0 is an interior point of A. This means there exist $|\cdot|_1, |\cdot|_2, \ldots, |\cdot|_n \in \mathfrak{S}$ and $\delta > 0$ such that

$$|y|_1, |y|_2, \ldots, |y|_n < \delta \quad \text{implies } y \in A.$$

It follows that the support function of A,

$$p(y) = \inf\{\lambda; \lambda^{-1}y \in A\} \tag{12.0.13}$$

is finite for all $y \in E$. Obviously,

$$p(\lambda y) = \lambda p(y) \quad \text{for } \lambda \geq 0. \tag{12.0.14}$$

Moreover, if $\lambda, z \in E$ then there exist λ, μ with $\lambda \leq p(y) + \varepsilon$, $\mu \leq p(z) + \varepsilon$ and $\lambda^{-1}y, \mu^{-1}z \in A$. By convexity, $\lambda(\lambda + \mu)^{-1}(\lambda^{-1}y) + \mu(\lambda + \mu)^{-1}(\mu^{-1}z) = (\lambda + \mu)^{-1}(y + z) \in A$ so that we have $p(y + z) \leq \lambda + \mu$. Since, on the other hand $\lambda + \mu \leq p(y) + p(z) + 2\varepsilon$ and $\varepsilon > 0$ is arbitrary,

$$p(y + z) \leq p(y) + p(z) \quad (y, z \in E). \tag{12.0.15}$$

Since y_0 is not an interior point of A we have $p(y_0) \geq 1$. Define a linear functional in the one-dimensional subspace generated by y_0 setting $\phi(\lambda y_0) = \lambda$; then

$$\phi(\lambda y_0) \leq p(\lambda y_0).$$

This is obvious for $\lambda < 0$; for $\lambda \geq 0$, $\phi(\lambda y_0) = \lambda \leq \lambda p(y_0) = p_0(\lambda y_0)$. We then apply the Hahn-Banach theorem (Theorem 5.0.2) and extend ϕ to a linear functional on E satisfying

$$\phi(y) \leq p(y) \quad (y \in E). \tag{12.0.16}$$

It is clear that ϕ does the separation job, and it only remains to show that it is continuous. This follows from the fact that if $|y|_1, |y|_2, \ldots, |y|_n \leq \delta$ then $p(y) \leq 1$ and from (12.0.16).   ■

Corollary 12.0.8. *Let A, B be disjoint convex sets in E, and assume A has an interior point. Then there exists a nonzero continuous linear functional $y^* \in E^*$ that separates A and B.*

Proof. The convex set $A - B$ does not contain 0 and has an interior point, thus $A - B$ and $y_0 = 0$ can be separated by Theorem 12.0.7. If y^* is the linear functional there, we have

$$\langle y^*, y - z \rangle \leq 0 \quad (y \in A, z \in B)$$

which is (12.0.11).   ■

Corollary 12.0.9. *Let A be a nonempty convex closed set and $y_0 \notin A$. Then there exists $y^* \in E^*$ that strictly separates A and y_0, that is, $y^* \neq 0$ and*

$$\langle y^*, y \rangle + \varepsilon \leq \langle y^*, y_0 \rangle \quad (y \in A)$$

for some $\varepsilon > 0$.

Proof. Since A is closed its complement A^c is open, thus there exist $|\cdot|_1, |\cdot|_2, \ldots, |\cdot|_n \in \mathfrak{S}$ and $\delta > 0$ such that the convex open set

$$B = \{y; |y - y_0|_1 < \delta, \ldots, |y - y_0|_n < \delta\}$$

is disjoint from A. Corollary 12.0.8 produces a functional $y^* \in E^*$, $y^* \neq 0$ separating A and B, thus separating A and y_0. Since $y^* \neq 0$ there exists $\bar{y} \in E$ such that $\langle y^*, \bar{y} \rangle = 1$; for $\varepsilon > 0$ sufficiently small $y_0 - \varepsilon \bar{y} \in B$, so that $\langle y^*, y_0 \rangle - \varepsilon \geq \langle y^*, y \rangle$ $(y \in A)$. ∎

Corollary 12.0.9 obviously implies the existence of numerous linear functionals in locally convex spaces; in particular, the elements of E^* **distinguish points** in the sense that, given $y, z \in E$ with $y \neq z$ there exists $y^* \in E^*$ with $\langle y^*, y \rangle \neq \langle y^*, z \rangle$.

We shall use weak topologies in a normed space E and its dual E^*, where E^* denotes the space of all linear functionals continuous in the norm. Obviously, E and E^* are in duality and E^* is total. The E^*-weak topology of E is the one defined by this duality, and it follows from Lemma 12.0.6 that the dual space in this topology coincides with E^*. Convergence of a generalized sequence $\{y_\kappa\}$ to y in this topology means

$$\langle y^*, y_\kappa \rangle \rightarrow \langle y^*, y \rangle \quad \text{for all } y^* \in E^*. \tag{12.0.17}$$

Likewise, E^* and E are in duality with E total, and the E-weak topology of E^* is the one defined by this duality. Since we use it later in this chapter, we isolate this particular case of Lemma 12.0.6.

Corollary 12.0.10. *Let E^{w*} denote the dual of E^* with respect to the E-weak topology. Then*

$$E^{w*} = E. \tag{12.0.18}$$

Note that E^* and E^{**} are in duality as well with E^{**} total, thus E^* can also be equipped with the E^{**}-weak topology, with respect to which topology the dual space coincides with E^{**}. Convergence of a generalized sequence $\{y_\kappa^*\} \subseteq E^*$ to $y^* \in E^*$ in the E-weak topology means

$$\langle y, y_\kappa^* \rangle \rightarrow \langle y, y_\kappa^* \rangle \quad \text{for all } y \in E, \tag{12.0.19}$$

while convergence in the E^{**}-weak topology is

$$\langle y^{**}, y^*_\kappa \rangle \to \langle y^{**}, y^*_\kappa \rangle \quad \text{for all } y^{**} \in E^{**}. \qquad (12.0.20)$$

Obviously convergence in norm in E or E^* implies convergence in any of the weak topologies. Convergence in the E^{**}-topology of E^* implies convergence in the E-topology. Open (resp. closed) sets in any of the weak topologies are also open (resp. closed) in the sense of the norm. The result below is a converse for convex sets.

Theorem 12.0.11. *Let* $A \subseteq E$ *be convex and closed in the norm topology. Then it is closed in the* E^*-*weak topology.*

Proof. Let $\{y_\kappa\}$ be a generalized sequence in A converging to $y_0 \in E$ in the E^*-weak topology. If $y_0 \notin A$, using Corollary 12.0.8, get a functional $y^* \in E^*$ such that $\langle y^*, y \rangle \leq \langle y^*, y_0 \rangle - \varepsilon$ for $y \in A$. Contradiction: $\{y_\kappa\}$ cannot converge to y_0 E^*-weakly. ∎

Theorem 12.0.11 is of course true in E^* with the E^{**}-weak topology, but not with the E-weak topology.

We take a look below at compact subsets K of a topological space (U, Υ). We shall assume that $K = U$; if $K \neq U$, all considerations can be applied to the topological space (K, Υ_K), where the sets of Υ_K are the intersections of the sets of Υ with K.

We say that U is **compact** if every generalized sequence $\{u_\kappa\}$ in U has a convergent generalized subsequence $\{v_\lambda\}$.

A family $\Xi = \{V, \ldots\}$ of open sets in U is a **cover** of U if

$$U \subseteq \bigcup_{V \in \Xi} V.$$

We say that U is **cover compact** if, given an arbitrary cover Ξ of U there exists a finite subcover, that is, a finite subfamily $\Xi_f \subseteq \Xi$ such that Ξ_f is a cover of U.

A family Δ of closed sets has the **finite intersection property** if the intersection of every finite subfamily is nonempty. It is easy to see (taking complements) that U is cover compact if and only if every family $\Delta = \{D, \ldots\}$ of closed sets in U having the finite intersection property has nonempty intersection, that is

$$\bigcap_{D \in \Delta} D \neq \emptyset.$$

Theorem 12.0.12. *U is cover compact if and only if it is compact.*

Proof. Assume U is cover compact, and let $\{u_\kappa; \kappa \in K\}$ be a generalized sequence in U. Since K is directed, the family of sets $\{\{u_\iota; \iota \geq \kappa\}; \kappa \in K\}$ has the finite intersection property; hence, the family $\{Cl(\{u_\iota; \iota \geq \kappa\}); \kappa \in K\}$ has the same property. Accordingly,

$$D = \bigcap_{\kappa \in K} Cl(\{u_\iota; \iota \geq \kappa\}) \neq \emptyset.$$

Let $u \in D$, V an open set containing u. Then $\{u_\iota; \iota \geq \kappa\} \cap V \neq \emptyset$ for all $\kappa \in K$ (otherwise, $\{u_\iota; \iota \geq \kappa\} \subseteq V^c$, *a fortiori* $u \in Cl(\{u_\iota; \iota \geq \kappa\}) \subseteq V^c$, a contradiction). In other words, given an open set V containing $u \in D$ and $\kappa \in K$, there exists $\iota \geq \kappa$ such that $u_\iota \in V$.

Below, we pick $u \in D$ fixed and construct a generalized subsequence of $\{u_\kappa; \kappa \in K\}$ converging to u. Let Λ be the set of pairs (κ, V) with $\kappa \in K$, V an open set containing u, and $u_\kappa \in V$. Order this set thus: $(\kappa, V) \leq (\lambda, W)$ if $\kappa \leq \lambda$, $V \supseteq W$. The arguments above show that Λ is directed. Define $\phi(\kappa, V) = \kappa$, $v_{(\kappa, V)} = u_{\phi(\kappa, V)} = u_\kappa$. Then $(\{v_{(\kappa, V)}; (\kappa, V) \in \Lambda\}, \phi)$ is a generalized subsequence of $\{u_\kappa; \kappa \in K\}$, and $\{v_{(\kappa, V)}; (\kappa, V) \in \Lambda\}$ converges to u.

Conversely, assume that every generalized sequence $\{u_\kappa\}$ in U has a generalized subsequence $\{v_\lambda\}$ converging to a point of U. Let Δ be a collection of closed sets having the finite intersection property, and let Δ_i be the family of all finite intersections of elements of Δ. Equip Δ_i with the order $\supseteq$ and let $u_F \in F$ ($F \in \Delta_i$). Then the generalized sequence $\{u_F; F \in \Delta_i\}$ has a generalized subsequence $(\{v_\lambda; \lambda \in \Lambda\}, \phi)$ converging to some $u \in U$. It is enough to show that

$$u \in D \quad \text{for all } D \in \Delta.$$

If this is not the case, there exist $D \in \Delta$ and $V \ni u$ open with $D \cap V = \emptyset$. However, there must exist $F \in \Delta_i$ with $F \subseteq D$ and $u_F \subseteq V$; since, by definition $u_F \subseteq F$, a contradiction ensues. ∎

If U is a metric space, we have at play two definitions of compactness. One is the sequential one in **2.0** and the beginning of this section; the other is the definition by covers. As seen below, both coincide.

Theorem 12.0.13. *Let U be a metric space. Then U is cover compact if and only if every sequence has a convergent subsequence.*

Proof. Cover compactness is equivalent to: every family of closed sets having the finite intersection property has nonempty intersection. Let $\{u_n\}$ be a sequence in U. The family $\{Cl(\{u_k; k \geq n\}), n = 1, 2, \ldots\}$ has the finite intersection property, thus, it has nonempty intersection F. If $u \in F$ it is plain that we may select a subsequence $\{u_{n_k}\}$ such that $u_{n_k} \to u$ as $k \to \infty$.

Conversely, assume that every sequence has a convergent subsequence. Then, given $\varepsilon > 0$ there exists a finite sequence $\{u_n(\varepsilon); n = 1, 2, \ldots, n(\varepsilon)\}$ such that

$$U \subseteq B_0(u_1(\varepsilon), \varepsilon) \cup B_0(u_2(\varepsilon), \varepsilon) \cup \ldots \cup B_0(u_n(\varepsilon), \varepsilon) \qquad (12.0.21)$$

where $B_0(u, \varepsilon)$ is the open ball of center u and radius ε. In fact, if this is not true, we can construct a countable sequence $u_1, u_2, \ldots$ such that $d(u_n, u_k) \geq \varepsilon$ for $k < n$; this sequence cannot have a convergent subsequence. We use (12.0.21) in an obvious way with a sequence $\{\varepsilon_n\}$, $\varepsilon_n \to 0$ to show that U is *separable*, that is, that there exists a sequence $\{u_n\}$ dense in U.

Let Ξ be a cover of U. Denote by B the (countable) set of all open balls $B_0(u_n, \varepsilon_m)$ with ε_m rational, and by $B_\Xi \subseteq B$ the subset of all these balls entirely contained in some set $V \subseteq \Xi$. Denseness of $\{u_n\}$ is easily seen to imply that every set $V \subseteq \Xi$ is the union of a subset of the balls in B_Ξ (so that B_Ξ is a cover of U) and it is obviously sufficient to show that a finite number of balls in B_Ξ cover U, since each one is contained in a set of the family Ξ. Assume then that U is not contained in the union of a finite number of balls in B_Ξ and let v_n be a point not in the union of the first n elements of B_Ξ. Select a convergent subsequence $\{v_{n_k}\}$ such that $v_{n_k} \to v$. Since the complement of each ball is closed, $v \notin B_0(u_n, \varepsilon_n)$ for all balls in B_Ξ, a contradiction. ∎

Let $\{(U_\beta, \Upsilon_\beta); \beta \in B\}$ be a family of topological spaces. We look for a topology Υ in the Cartesian product $\mathbf{U} = \Pi_{\beta \in B} U_\beta$ having the property that for every generalized sequence $\{\mathbf{u}_\kappa\} \subseteq \mathbf{U}$, $\lim \mathbf{u}_\kappa = \mathbf{u}$ is equivalent to

$$\lim P_\beta \mathbf{u}_\kappa = P_\beta \mathbf{u} \qquad (12.0.22)$$

in the topology of U_β, where P_β is the projection of $\Pi_{\beta \in B} U_\beta$ into U_β. This topology, called the **product topology**, is constructed as follows. Let $\mathbf{V}$ be the family of all finite intersections of sets of the form $P_\beta^{-1}(V_\beta)$, where $V_\beta \in \Upsilon_\beta$. Then $\mathbf{V}$ satisfies (a) and (c) of the definition of topology, and it is easy to see that the family Υ of all unions of sets in $\mathbf{V}$ is a topology. We check just as easily that convergence in this topology is characterized by (12.0.22).

Example 12.0.14. Let $\{U_m, \rho_m\}$ be a sequence of metric spaces. Then the product topology Υ of $\mathbf{U} = \Pi_m U_m$, as defined above, coincides with the topology produced by the distance

$$\rho(\mathbf{u}, \mathbf{v}) = \sum_{m=1}^{\infty} \alpha_m \eta(\rho_m(u_m, v_m)), \qquad (12.0.23)$$

where $\mathbf{u} = \{u_m\}$, $\mathbf{v} = \{v_m\}$, $\eta(t)$ is a continuous increasing function in $t \geq 0$ with $\eta(0) = 0$, $\eta(s + t) \leq \eta(s) + \eta(t)$, $\eta(t) \leq 1$ and $\{\alpha_m\}$ is a convergent sequence of positive numbers.

The proof is the same as that of Lemma 12.0.1.

Theorem 12.0.15 (Tijonov). [1] *Assume each U_β is compact. Then $\mathbf{U}$ is compact.*

---

[1] Pronounce the "j" in Spanish.

Proof. Consider an arbitrary family Δ of closed subsets of $\mathbf{U}$ having the finite intersection property and the family $\Im(\Delta)$ of all families of subsets (closed or not) of $\mathbf{U}$ having the finite intersection property and containing Δ. Order $\Im(\Delta)$ with $\subseteq$. Then, if L is a linearly ordered subfamily of $\Im(\Delta)$, its union belongs to $\Im(\Delta)$ thus, by Zorn's Lemma (Lemma 5.0.4), $\Im(\Delta)$, it has a maximal element $\Delta_{\max}$. We shall use $\Delta_{\max}$ to show that

$$\bigcap_{\mathbf{D} \in \Delta} \mathbf{D} \neq 0. \tag{12.0.24}$$

Maximality of $\Delta_{\max}$ implies that (*i*) each set that contains a set of $\Delta_{\max}$ belongs to $\Delta_{\max}$, (*ii*) the intersection of a finite number of members of $\Delta_{\max}$ belongs to $\Delta_{\max}$, (*iii*) each set that intersects each member of $\Delta_{\max}$ intersects every finite intersection (by (*ii*)), thus it belongs to $\Delta_{\max}$.

If $\mathbf{V}, \mathbf{W}$ are sets in $\mathbf{U}$ we have

$$P_\beta(\mathbf{V} \cap \mathbf{W}) \subseteq P_\beta(\mathbf{V}) \cap P_\beta(\mathbf{W})$$

so, the fact that $\Delta_{\max}$ possesses the finite intersection property implies that each family $P_\beta(\Delta_{\max}) = \{P_\beta(\mathbf{D}); \mathbf{D} \in \Delta_{\max}\}$ has the finite intersection property in U_β. Since U_β is compact, for each β there exists u_β such that $u_\beta \in Cl(P_\beta(\mathbf{D}))$ for all $\mathbf{D} \in \Delta_{\max}$.

Let now $\mathbf{u} = \{u_\beta; \beta \in B\}$. We aim to show that

$$\mathbf{u} \in \bigcap_{\mathbf{D} \in \Delta} \mathbf{D} \tag{12.0.25}$$

thus proving (12.0.24). Let $u_\beta \in V_\beta \in \Upsilon_\beta$. Then V_β intersects every $P_\beta(\mathbf{D}), \mathbf{D} \in \Delta_{\max}$ so that $P_\beta^{-1}(V_\beta)$ intersects every $\mathbf{D} \in \Delta_{\max}$, thus it belongs to $\Delta_{\max}$ by (*iii*). By (*ii*), $\bigcap_{\beta \in B'} P_\beta^{-1}(V_\beta) \in \Delta_{\max}$, where B' is any finite subset of B. It follows that every open set in $\mathbf{U}$ containing $\mathbf{u}$ intersects every $\mathbf{D} \in \Delta_{\max}$, in particular every $\mathbf{D}$ in the original family Δ; since members of Δ are closed, $\mathbf{u} \in \mathbf{D}$ for all $\mathbf{D} \in \Delta$ as claimed. ∎

Example 12.0.16. Show Tijonov's theorem in the particular case in Example 12.0.14 (the proof is the same as that of Theorem 12.0.2).

Theorem 12.0.17 (Alaoglu). *Let E be an arbitrary Banach space. Then the closed unit sphere $B^* = \{y^*; \|y^*\|_{E^*} \leq 1\}$ of E^* is compact in the E-weak topology.*

Proof. Let B be the closed unit ball in E, ϕ an arbitrary element of E^*. Then $\Phi = \{\phi(y); y \in B\}$ is an element of the product space $\Pi_{y \in B} I_y$, where for each $y \in B$ I_y is a copy of the interval $[-1, 1]$ equipped with its usual topology. This identification is one-to-one; moreover, a generalized sequence $\{\phi_\kappa\} \subseteq E^*$

converges to zero in the E-weak topology of E^* if and only if its counterpart $\{\Phi_\kappa\} \subseteq \Pi_{y \in B} I_y$ converges to zero in the product topology of $\Pi_{y \in B} I_y$. Finally, we check easily that the set $\{\Phi; \phi \in B^*\}$ is closed in $\Pi_{y \in B} I_y$ (note that we only have to check linearity $\phi(\lambda y + \mu z) = \lambda \phi(y) + \mu \phi(z)$ for small $|\lambda|, |\mu|, \|y\|, \|z\|$.) By Tijonov's theorem $\Pi_{y \in B} I_y$ is compact, then so is $\{\Phi; \phi \in B^*\}$ and *a fortiori* B^* is compact. ∎

A topological space U is called **normal** (Kelley [1955, p. 112], Dunford-Schwartz [1958, p. 15]) if, for every disjoint pair C, D of closed sets, there exists a disjoint pair of open sets V, W such that $C \subseteq V, D \subseteq W$. Normal topological spaces have a variety of real valued continuous functions, as the following result shows (Dunford-Schwartz [1958, p. 15], Kelley [1955, p. 115]).

Lemma 12.0.18 (Urysohn). *Let C and D be disjoint closed sets in a normal topological space U. Then there exists a real continuous function f defined on U such that $0 \le f(u) \le 1$, $f(u) = 0$ $(u \in C)$, $f(u) = 1$ $(u \in D)$.*

Metric spaces are easily seen to be normal, but Urysohn's Lemma is elementary in these spaces. If d is the distance, the functions $\text{dist}(u, C) = \inf_{v \in C} d(u, v)$ and $\text{dist}(u, D)$ are continuous functions of u and we may take

$$f(u) = \frac{\text{dist}(u, C)}{\text{dist}(u, C) + \text{dist}(u, D)}.$$

12.1. Existence Theory of Optimal Control Problems: Measure-Valued Controls

We take a second look at the examples in **2.2**. We begin with Example 2.2.9,

$$\begin{aligned}
y'(t) &= u(t), & y(0) &= 0, \\
U &= \{-1\} \cup \{1\}, & Y &= \mathbb{R}, & \bar{t} = 1,
\end{aligned} \tag{12.1.1}$$

with cost functional

$$y_0(t, u) = \int_0^t y(\tau)^2 d\tau. \tag{12.1.2}$$

As seen in **2.2**, there exist minimizing sequences $\{u^n(\cdot)\}$, for instance,

$$u^n(t) = \begin{cases} 1 & (2k/n \le t < (2k+1)/2n) \\ -1 & (2k+1)/2n \le t < (2k+2)/2n \end{cases} \tag{12.1.3}$$

However there is no optimal control since the minimum of the functional (12.1.2) is produced by the control $u(t) \equiv 0$, which does not belong to the control space $C_{\text{ad}}(0, \bar{t}; U)$ of all measurable U-valued functions.

There is a way out. Instead of looking upon this system as if driven by functions $u(t)$ taking values in the control set U, we consider it under "controls" $\mu(t)$ whose values are probability measures in U. Since $U = \{-1\} \cup \{1\}$, these measures are $\mu(t) = \alpha(t)\delta(u - 1) + (1 - \alpha(t)) \delta(u + 1), 0 \le \alpha(t) \le 1$. With the new control space, the dynamics becomes

$$y'(t) = \int_U u\mu(t, du), \qquad y(0) = 0. \tag{12.1.4}$$

The cost functional does not need changing since it does not depend directly on the control. The new control space contains the original one; a control $u(\cdot) \in C_{\mathrm{ad}}(0, \bar{t}; U)$ can be replicated by one of the new controls setting $\alpha(t) = 1$ (resp. $\alpha(t) = 0$) where $u(t) = 1$ (resp. where $u(t) = -1$). We do have an optimal control in the larger class, namely

$$\mu(t) = \frac{1}{2}\delta(u - 1) + \frac{1}{2}\delta(u + 1). \tag{12.1.5}$$

It may be objected that we have transformed the original optimal control problem into another one. An engineer may be stuck for design reasons with (12.1.1), the control space indicating for instance an actuator with only two positions. On the other hand, an optimal control problem where solutions do not exist is defective, and conclusions on optimal controls are necessarily senseless. For instance, let's try to characterize an optimal control $\bar{u}(t)$ for (12.1.1)–(12.1.2) by the maximum principle. In the notation of the examples in Chapters 3 and 4, the Hamiltonian is

$$H(y, p, u) = \mu y^2 + pu$$

and the canonical equation for p is $p'(t) = -2\mu y(t)$. If $\mu = 0$, $p(t) \equiv c \ne 0$ since $(\mu, p(1))$ cannot be zero; the maximum principle then gives $\bar{u}(t) \equiv \operatorname{sign} c$, which is admissible for any c but is not an optimal control. If $\mu < 0$, we cannot have $p(t) \equiv 0$ since this would imply $y(t) \equiv 0$, corresponding to the nonadmissible control $\bar{u}(t) \equiv 0$. It follows then that $p(t) \ne 0$ in some subinterval, where fraudulent information on the optimal control is again obtained.

The process whereby (12.1.1)-(12.1.2) becomes (12.1.4)-(12.1.2) is called **relaxation**; the second system is the **relaxed system**, and its controls are **relaxed controls**. For the system in Example 2.2.10,

$$y'(t) = u(t), \qquad y(0) = 0,$$
$$U = [-1, 1], \qquad Y = \mathbb{R}, \qquad \bar{t} = 1, \tag{12.1.6}$$

$$y_0(t, u) = \int_0^t \{(y(\tau))^2 + (u(\tau)^2 - 1)^2\}d\tau, \tag{12.1.7}$$

relaxed controls are probability measures in $U = [-1, 1]$. The relaxed dynamics is again given by (12.1.4). This time, the cost functional itself has to be relaxed:

(12.1.7) becomes

$$y_0(t, \mu) = \int_0^t \int_U \{y(\tau)^2 + (u^2 - 1)^2\} \mu(\tau, du) d\tau. \tag{12.1.8}$$

We check easily that the relaxed control (12.1.5) is optimal and that the maximum principle applied in the class $C_{ad}(0, \bar{t}; U)$ produces counterfeit information on optimal controls.

These two examples suggest a general way to "relax" a general ordinary differential control system

$$y'(t, u) = f(t, y(t, u), u(t)), \qquad y(0) = \zeta \tag{12.1.9}$$

with cost functional

$$y_0(t, u) = \int_0^t f_0(\tau, y(\tau, u), u(\tau)) d\tau + g_0(t, y(t, u)). \tag{12.1.10}$$

The space $C_{ad}(0, \bar{t}; U)$ is replaced by a space $R(0, \bar{t}; U)$ of functions whose values are probability measures $\mu(t)$ on the control set U. The new control system and cost functional will then be

$$y'(t, \mu) = \int_U f(t, y(t, \mu), u) \mu(t, du), \qquad y(0) = \zeta, \tag{12.1.11}$$

$$y_0(t, \mu) = \int_0^t \int_U f_0(\tau, y(\tau, \mu), u) \mu(\tau, du) d\tau + g_0(t, y(t, \mu)). \tag{12.1.12}$$

Conditions on t-dependence of $\mu(t)$ are needed to insure that everything makes sense. For U compact, it is enough that for every function $\phi(u)$ continuous in U the function, $t \to \int_U \phi(u) \mu(t, du)$ should be measurable and bounded in $0 \le t \le \bar{t}$. Convergence of sequences $\{\mu_n(\cdot)\}$ of relaxed controls is understood weakly: $\mu_n(\cdot) \to \mu(\cdot)$ if, for every $\phi(u)$ continuous in U, $\int_U \phi(u) \mu_n(t, du) \to \int_U \phi(u) \mu(t, du)$. Ordinary controls in the space $C_{ad}(0, \bar{t}; U)$ are special cases of relaxed controls:

$$\mu(t) = \delta(\cdot - u(t)). \tag{12.1.13}$$

That relaxed controls will provide existence is easy to guess. Consider for simplicity the fixed time problem and let $\{u^n(\cdot)\} \subseteq C_{ad}(0, \bar{t}; U)$ be a minimizing sequence. Let $\{\mu^n(\cdot)\}$ be the minimizing sequence realized by (12.1.13) as a sequence of relaxed controls. Write the trajectory $y(t, \mu)$ using (12.1.11) and the cost functional $y_0(t, \mu)$ using (12.1.12). Assume we can show, if necessary passing to a subsequence, that

(a) the sequence $\{\mu^n(\cdot)\}$ is weakly convergent to $\mu(\cdot) \in R(0, \bar{t}; U)$

(*b*) the sequence of trajectories $y(t, \mu^n)$ is uniformly convergent to $y(t)$ in $0 \le t \le \bar{t}$.

Under suitable assumptions on $f(t, y, u)$ and $f_0(t, y, u)$, writing (12.1.11) in integral form

$$y(t, \mu) = \zeta + \int_0^t \int_U f(\tau, y(\tau, \mu^n), u) \mu^n(\tau, du) d\tau, \qquad (12.1.14)$$

and taking limits we show that $y(t, \mu)$ is a solution of the optimal control problem. The reason we can take limits is that *now the controls appear linearly*; if we try to take limits in the original equation,

$$y(t, u^n) = \zeta + \int_0^t f(\tau, y(\tau, u^n), u^n(\tau)) d\tau, \qquad (12.1.15)$$

weak convergence may be defeated by nonlinearity.

Still, the fact that the system has been changed has to be justified. One indication that the relaxed system is closely related to the original system is: for every solution or trajectory $y(t, \mu)$ of (12.1.11) and every $\varepsilon > 0$ there exists a trajectory $y(t, u)$ of the original system (12.1.9) with

$$\|y(t, \mu) - y(t, u)\| \le \varepsilon. \qquad (12.1.16)$$

This is known as a **relaxation theorem** and will be proved in Chapter 13. One example is the zero trajectory for the system (12.1.4) corresponding to the relaxed control (12.1.5). It can be uniformly approximated by the trajectories $y(t, u^n)$ corresponding to the controls (12.1.3). Typically, in a relaxed system certain trajectories $y(t, \mu)$ can only be approximated by ordinary trajectories $y(t, u^n)$ where the (ordinary) controls u^n become more and more oscillatory, with $u^n \to \mu$ weakly. This motivates the name **chattering controls** sometimes given to relaxed controls.[1]

To implement relaxed controls changes a control system. If we are stuck with a system like (12.1.1)–(12.1.2), relaxation plays a role not unlike, say, convergence results for Newton's method as we implement it on a computer. The computer works with floating point numbers of a certain length. Convergence is to a real number, possibly not even rational, thus outside computer scope. However, convergence information is not useless; it ensures that approximations will "eventually stabilize," that is, that large oscillations or overflow are not to be expected. In a similar way, relaxed control theory establishes that, if convergence is correctly

---

[1] Chatter *of a tool*: to vibrate rapidly in cutting; to operate with an irregularity that causes rapid intermittent noise or vibration (Webster's Dictionary).

understood, oscillating sequences such as (12.1.3) are convergent to an optimal relaxed control. Sometimes we go the opposite way; assuming we can figure out the optimal relaxed control, an ordinary control approximating it can be constructed.

Fortunately, we don't need a new chapter (or book) where everything we know for ordinary controls (say the maximum principle) has to be proved anew. The payoff for having done control theory in sufficient generality is that proofs automatically apply to relaxed controls.

12.2. Spaces of Vector Valued Functions and Their Duals, I

Let X be an arbitrary Banach space. We recall **(5.0)** that the space $L^1(0, T; X)$ consists of all (equivalence classes of) strongly measurable X-valued functions $f(\cdot)$ defined in $0 \le t \le T$ such that $\| \cdot \|_1 = \int \| f(t) \| dt < \infty$. Equipped with $\| \cdot \|_1$, the space $L^1(0, T; X)$ is a Banach space whose dual will be characterized below using a result of Dunford and Pettis (Theorem 12.2.4).

Given a Banach space X, we say that an X^*-valued function $g(\cdot)$ is X-**weakly measurable** if

$$t \to \langle g(t), y \rangle \tag{12.2.1}$$

is a measurable function for each $y \in X$.

We recall that there is another notion of weak measurability for X^*-valued functions, namely that of X^{**}-weak measurability, where y in (12.2.1) is taken in X^{**}. These concepts are not the same.

We denote by $L_w^\infty(0, T; X^*)$ the linear space of all X^*-valued X-weakly measurable functions $g(\cdot)$ such that there exists $C \ge 0$ with

$$|\langle g(t), y \rangle| \le C \| y \|_X \quad \text{a.e. in } 0 \le t \le T \tag{12.2.2}$$

for each $y \in X$ (the null set where (12.2.2) fails to hold may depend on y). Two functions $g(\cdot), h(\cdot)$ are equivalent in $L_w^\infty(0, T; X^*)$ (in symbols, $g \approx h$) if $\langle g(t), y \rangle = \langle f(t), y \rangle$ a.e. in $0 \le t \le T$ for each $y \in X$. The infimum of all constants C such that (12.2.2) holds for all $y \in X$ is named $\| g \|_{L_w^\infty(0,T;X^*)}$; we have $|\langle g(t), y \rangle| \le \| g \|_{L_w^\infty(0,T;X^*)} \| y \|_X$ a.e. in $0 \le t \le T$. We check easily that $\| \cdot \|$ is a norm in (the equivalence classes of) $L_w^\infty(0, T; X^*)$. The equivalence relation in $L_w^\infty(0, T; X^*)$ does not coincide with the traditional one in L^p spaces, which is: $g(\cdot)$ is equivalent to $h(\cdot)$ if and only if $g(t)$ and $h(t)$ coincide almost everywhere (see **5.0**). The following example illustrates the discrepancy.

Example 12.2.1. Let X consist of all families $x = \{x_\alpha; 0 \le \alpha \le 1\}$ of real numbers with

$$\| x \|^2 = \sum_{0 \le \alpha \le 1} |x_\alpha|^2 < \infty$$

(this requires $x_\alpha = 0$ except for a finite or countable number of indices). Equipped with pointwise operations and the norm $\|\cdot\|$, X is a Hilbert space and we show easily that $L_w^\infty(0, T; X^*) = L_w^\infty(0, T; X)$ consists of all functions $t \to g(t) = \{g_\alpha(t)\}$ such that each $g_\alpha(\cdot)$ is measurable and essentially bounded, with ess.sup g_α bounded independently of α. Define a function $\delta(t) = \{\delta_\alpha(t)\}$ by $\delta_\alpha(t) = 1$ if $\alpha = t$, $\delta_\alpha(t) = 0$ if $\alpha \neq t$. Then $\delta(\cdot) \approx 0$ in $L_w^\infty(0, T; X)$, but $\|\delta(t)\| = 1$ for all t. This example shows that a function in the equivalence class of zero may be nonzero everywhere. Multiplying $\delta(t)$ by a scalar function $\eta(t)$, we obtain another element in the same class with norm $\eta(t)$. This shows that $t \to \|g(t)\|$ need not be essentially bounded or even measurable for an element $g(\cdot) \in L_w^\infty(0, T; X^*)$.

The result below shows these pathologies disappear when X is separable (see also the comments before Lemma 7.7.13).

Lemma 12.2.2. *Let X be separable. Then if $g(\cdot) \in L_w^\infty(0, T; X^*)$ the function $t \to \|g(t)\|_{X^*}$ is measurable and essentially bounded and*

$$\|g\|_{L_w^\infty(0,T;X^*)} = \underset{0 \leq t \leq T}{\text{ess. sup.}} \|g(t)\|_{X^*}. \tag{12.2.3}$$

Proof. Let $\{y_n\}$ be a countable dense set in the unit ball of X. We have

$$\|g(t)\|_{X^*} = \sup |\langle g(t), y_n \rangle|$$

so that $t \to \|g(t)\|_{X^*}$ is measurable and $\|g\|_{L^\infty(0,T;X^*)} \leq \|g\|_{L_w^\infty(0,T;X^*)}$. On the other hand, $|\langle g(t), y \rangle| \leq \|g\|_{L^\infty(0,T;X^*)} \|y\|_X$ a.e. in $0 \leq t \leq T$, thus we have $\|g\|_{L^\infty(0,T;X^*)} \geq \|g\|_{L_w^\infty(0,T;X^*)}$. ∎

Theorem 12.2.4 is based on the following magical result, which eliminates the "wild null set" of all functions in L^∞ at once.

Lifting Theorem 12.2.3 (Von Neumann). *Let $B[0, T]$ be the space of all bounded functions in $0 \leq t \leq T$ endowed with the supremum norm. Then there exists a* **linear lifting** $\mathcal{L}$ *in* $L^\infty(0, T)$, *that is, a linear operator*

$$\mathcal{L} : L^\infty(0, T) \to B(0, T)$$

such that, for every $f(\cdot) \in L^\infty(0, T)$ (a) $(\mathcal{L}f)(\cdot)$ belongs to the equivalence class of $f(\cdot)$ in $L^\infty(0, T)$, (b) $\mathcal{L}(1) = 1$ (1 the function identically 1), (c) $\mathcal{L}(f)(t) \geq 0$ in $0 \leq t \leq T$ if $f(t) \geq 0$ a.e. in $0 \leq t \leq T$.

Proof. Let $\text{LIM}_{n\to\infty}$ be a Banach limit (**5.0**). Define

$$(\mathcal{L}f)(t) = \underset{n\to\infty}{\text{LIM}} f_n(t) = \underset{n\to\infty}{\text{LIM}} \frac{1}{a_n} \int_{I_n} f(\tau)d\tau$$

where $I_n = (0, T) \cap (t - 1/n, t + 1/n)$ and a_n is the length of I_n. Using the properties of $\text{LIM}_{n \to \infty}$ we check that (b) and (c) hold. By virtue of Theorem 2.0.6 $\lim_{n \to \infty} f_n(t) = f(t)$ (thus $\text{LIM}_{n \to \infty} f_n(t) = f(t)$) at every Lebesgue point t of $f(\cdot)$, hence (a) holds as well. ∎

Theorem 12.2.4 (Dunford–Pettis Theorem Part 1). *Let X be a Banach space, $B : L^1(0, T) \to X^*$ a bounded linear operator. Then there exists a unique $g(\cdot) \in L_w^\infty(0, T; X^*)$ (up to equivalence) such that*

$$\|g\|_{L_w^\infty(0,T;X^*)} = \|B\|_{(L^1(0,T),X^*)} \tag{12.2.4}$$

$$\langle Bf, y \rangle = \int_0^T f(t)\langle g(t), y \rangle dt \quad (f \in L^1(0, T), y \in X). \tag{12.2.5}$$

Moreover, the map

$$B \to g \tag{12.2.6}$$

from the space $(L^1(0, T), X^*)$ (of all linear bounded operators from $L^1(0, T)$ into X^*) into $L_w^\infty(0, T; X^*)$ is linear. Finally, (12.2.6) is onto, that is, an arbitrary function $g(\cdot) \in L_w^\infty(0, T; X^*)$ defines an operator $B : L^1(0, T) \to X^*$ by (12.2.5), with norm (12.2.4).

Proof. Fix $y \in X$. Then $\phi_y(f) = \langle Bf, y \rangle$ is a linear functional in $L^1(0, T)$. Since $|\phi_y f| \leq \|y\|_X \|Bf\|_{X^*} \leq \|B\|_{(L^1(0,T),X^*)} \|y\|_X \|f\|_{L^1(0,T)}$, this linear functional is bounded (with norm $\leq \|B\|_{(L^1(0,T),X^*)} \|y\|_X$) and there exists a function $g_y(\cdot) \in L^\infty(0, T)$ with

$$\langle Bf, y \rangle = \int_0^T f(t)g_y(t)dt, \tag{12.2.7}$$

$$\|g_y\|_{L^\infty(0,T)} \leq \|B\|_{(L^1(0,T),X^*)} \|y\|_X. \tag{12.2.8}$$

The function $g_y(\cdot)$ is uniquely determined by y, and the map

$$y \to g_y(\cdot) \in L^\infty(0, T) \tag{12.2.9}$$

is obviously linear; in view of (12.2.8), it is bounded with norm $\leq \|B\|(L^1(0, T), X^*)$. If $\mathcal{L}$ is one of the linear liftings constructed in Theorem 12.2.3, and t is a point in the interval $0 \leq t \leq T$, the map $y \to (\mathcal{L}g_y)(t)$ is a linear functional in X with norm $\leq \|B\|_{(L^1(0,T),X^*)}$, thus there exists $g(t) \in X^*$ with $\|g(t)\|_{X^*} \leq \|B\|_{(L^1(0,T),X^*)}$ and

$$\langle g(t), y \rangle = (\mathcal{L}g_y)(t) \quad (y \in X). \tag{12.2.10}$$

As a function of t, $g(\cdot) \in L_w^\infty(0, T; X^*)$ and

$$\|g(t)\|_{X^*} \leq \|B\|_{(L^1(0,T),X^*)} \quad (0 \leq t \leq T) \tag{12.2.11}$$

so that $\|g\|_{L_w^\infty(0,T;X^*)} \leq \|B\|_{(L^1(0,T),X^*)}$. Formulas (12.2.7) and (12.2.10) yield (12.2.5), and this formula implies $\|B\|_{(L^1(0,T),X^*)} \leq \|g\|_{L_w^\infty(0,T;X^*)}$, so that (12.2.4) holds. The map (12.2.6) is clearly linear.

Conversely, let $g(\cdot) \in L_w^\infty(0, T; X^*)$. Then (12.2.5) defines a unique $Bf \in X^*$ for each $f(\cdot) \in L^1(0, T)$; moreover, $f \to Bf$ is linear and $\|B\|_{(L^1(0,T),X^*)} \leq \|g\|_{L_w^\infty(0,T;X^*)}$. Applying the arguments in the first half of the proof to B we obtain (12.2.5) for another function $g_0(\cdot)$ in $L_w^\infty(0, T; X^*)$ with norm $\|g_0\|_{L_w^\infty(0,T;X^*)} \leq \|B\|_{(L^1(0,T),X^*)}$. Since $g(\cdot) - g_0(\cdot)$ produces the zero operator through (12.2.5), it follows that $g(\cdot) \approx g_0(\cdot)$; in fact, if $\langle g(t), y \rangle \neq \langle g_0(t), y \rangle$ in a set of positive measure for some $y \in X$ then $f(t) = \mathrm{sign}(\langle g(t), y \rangle - \langle g_0(t), y \rangle)$ makes the integral (12.2.5) for $g(\cdot) - g_0(\cdot)$ nonzero. ∎

Given $B : L^1(0, T) \to X^*$ we call the function $g(\cdot)$ constructed in Theorem 12.2.4 the **representing function** of B. We have proved more than announced in Theorem 12.2.4; in fact, not only (12.2.4) holds but also

$$\sup_{0 \leq t \leq T} \|g(t)\|_{X^*} = \|B\|_{(L^1(0,T),X^*)} \tag{12.2.12}$$

in view of (12.2.11). We exploit this in the following result.

Corollary 12.2.5. *There exists a linear bounded operator*

$$\mathcal{L} : L_w^\infty(0, T; X^*) \to L_w^\infty(0, T; X^*)$$

such that (a) $\mathcal{L}g \approx g$ *in* $L_w^\infty(0, T; X^*)$, *(b)*

$$\sup_{0 \leq t \leq T} \|(\mathcal{L}g)(t)\|_{X^*} = \|g\|_{L_w^\infty(0,T;X^*)}. \tag{12.2.13}$$

In fact, let $g(\cdot) \in L_w^\infty(0, T; X^*)$. Define $B : L^1(0, T) \to X^*$ by (12.2.5). Then apply Theorem 12.2.4 and obtain an equivalent representing function $g_0(\cdot) \in L_w^\infty(0, T; X^*)$ satisfying (12.2.12). The map $\mathcal{L}g = g_0$ is linear. ∎

We shall call $\mathcal{L}$ a **lifting** in $L_w^\infty(0, T; X^*)$. Obviously, this operator depends on the lifting in $L^\infty(0, T)$ constructed in Theorem 12.2.3 (this is why we give it the same name).

When X is separable, it follows from Lemma 12.2.2 that the representing function $g(\cdot)$ of B is uniquely defined almost everywhere with measurable norm and

$$\|g\|_{L^\infty(0,T;X^*)} = \|B\|_{(L^1(0,T);X^*)}. \tag{12.2.14}$$

Example 12.2.1 dynamites all hopes of extending this to the nonseparable case. The result below is a sort of substitute.

Theorem 12.2.6 (Dunford–Pettis Theorem Part 2).[1] *Let $\mathcal{L}$ be the lifting operator in* Corollary 12.2.5. *Then* $t \to \|(\mathcal{L}g)(t)\|_{X^*}$ *is measurable for every* $g(\cdot) \in L_w^\infty(0, T; X^*)$.

For the proof, we need an auxiliary result. Note that if X is a Banach space and Y is a subspace of X (equipped with the same norm), every $y^* \in X^*$ defines a continuous linear functional in Y (the restriction of y^* to Y), and

$$\|y^*\|_{Y^*} \leq \|y^*\|_{X^*} \tag{12.2.15}$$

with possibly strict inequality (for instance, a nonzero $y^* \in X^*$ may vanish on Y).

Lemma 12.2.7. *Let Z be a separable subspace of X^*. Then there exists a closed separable subspace Y of X such that Z is metrically and algebraically isomorphic to a subspace of Y^* (that is, we have equality in (12.2.15) for every $y^* \in Z$).*[2]

Proof. Let $\{y_n^*\}$ be a countable dense set in Z. Pick a double sequence $\{y_{mn}\}$ in X such that

$$\|y_{mn}\| = 1, \qquad |\langle y_n^*, y_{mn} \rangle| \geq \left(1 - \frac{1}{m}\right)\|y_n^*\|. \tag{12.2.16}$$

The closed subspace Y generated by the $\{y_{mn}\}$ in X is separable. Equality in (12.2.15) is obvious. We note that Y can also be taken to be the closed subspace of X generated by $\{y_{mn}\}$ and an arbitrary set $U \subseteq X$, since adding to Y does not decrease the norm $\|y^*\|_{Y^*}$. ∎

End of Proof of Theorem 12.2.6. Since $L^1(0, T)$ is separable, the range $Z = B(L^1(0, T)) \subseteq X^*$ is separable. Applying Lemma 12.2.7, we construct a separable subspace Y of X such that $Z \subseteq Y^*$. The spaces and maps can be visualized as follows,

$$Y_{(\text{separable})} \xrightarrow{i} X, \quad L^1(0, T)_{(\text{separable})} \xrightarrow{B} B(L^1(0, T)) = Z_{(\text{separable})} \begin{matrix} \overset{i}{\nearrow} Y^* \\ \\ \underset{i}{\searrow} X^* \end{matrix}$$

the arrows $\xrightarrow{i}$ indicating isometric mappings. We now have two bounded operators in $L^1(0, T)$:

$$B : L^1(0, T) \to X^*, \qquad B : L^1(0, T) \to Y^*,$$

---

[1] This result is not used in the sequel.
[2] The norm of Z is that of X^*; the norm of Y is that of X.

and, since the X^*-norm and the y^*-norm coincide in Z, both operators have the same norm. We name provisionally B_0 the second operator and apply Theorem 12.2.4 to both B and B_0. If $y \in Y$ then, since $Bf \in Z \subseteq Y^*$, the function $g_y(\cdot)$ defined by (12.2.7) and the corresponding function $g_{0,y}(\cdot)$ corresponding to B_0 coincide. The representing function $g_0(t)$ of B_0 is defined by

$$\langle g_0(t), y \rangle_{Y^* \times Y} = (\mathcal{L}g_y)(t) \quad (y \in Y) \tag{12.2.17}$$

while $g(t)$, the representing function of B is defined by

$$\langle g(t), y \rangle_{X^* \times X} = (\mathcal{L}g_y)(t) \quad (y \in X), \tag{12.2.18}$$

where the subindices indicate duality pairings. If $y \in Y$ the definition (12.2.7) of the function $g_y(\cdot)$ is obviously the same for both operators B and B_0, thus we obtain from (12.2.10) that

$$\langle g_0(t), y \rangle_{Y^* \times Y} = \langle g(t), y \rangle_{X^* \times X} \quad (y \in Y) \tag{12.2.19}$$

and it follows that

$$\|g_0(t)\|_{Y^*} = \sup_{y \in Y, \|y\|_X \leq 1} \langle g_0(t), y \rangle_{Y^* \times Y} \leq \sup_{y \in X, \|y\|_X \leq 1} \langle g(t), y \rangle_{X^* \times X} = \|g\|_{X^*}$$

(recall that Y inherits its norm from X). If we can show that

$$\|g(t)\|_{X^*} \leq \|g_0(t)\|_{Y^*} \quad \text{a.e. in } 0 \leq t \leq T,$$

then $\|g(t)\|_{X^*} = \|g(t)\|_{Y^*}$ a.e., thus $t \to \|g(t)\|_{X^*}$ will be measurable as well. It is plain that the inequality above will follow from

$$\langle g(t), y \rangle_{X^* \times X} = (\mathcal{L}g_y)(t) \leq \mathcal{L}(\|g_0(\cdot)\|_{Y^*})(t)\|y\|_X \quad (0 \leq t \leq T) \tag{12.2.20}$$

for every $y \in X$, with $\mathcal{L}$ the lifting operator in Theorem 12.2.3. It is enough to prove the weaker inequality

$$\langle g(t), y \rangle_{X^* \times X} \leq \|g_0(t)\|_{Y^*}\|y\|_X \quad \text{a.e. in } 0 \leq t \leq T \tag{12.2.21}$$

for each $y \in X$, "a.e." depending on y. In fact, assume this inequality holds. Taking into account that the function on the left side and $g_y(t)$ belong to the same equivalence class in $L^\infty(0, T)$ we have $g_y(t) \leq \|g_0(t)\|_{Y^*}\|y\|_X$ a.e. in $0 \leq t \leq T$ and we obtain (12.2.20) applying $\mathcal{L}$ to both sides.

Assume then that for some $y \in X$ with $\|y\|_X = 1$ (12.2.21) fails in a set of positive measure. We can then find $\varepsilon > 0$ and another set e of positive measure such that

$$\langle g(t), y \rangle_{X^* \times X} \geq \|g_0(t)\|_{Y^*}\|y\|_X + \varepsilon \quad (t \in e).$$

We apply (12.2.5) to $f = \chi_e$ = characteristic function of e and $\|y\| = 1$:

$$\|B\chi_e\|_{X^*} \geq \langle B\chi_e, y\rangle_{X^* \times X} = \int_0^T \chi_e(t)\langle g(t), y\rangle_{X^* \times X}\, dt$$

$$= \int_e \langle g(t), y\rangle_{X^* \times X} dt \geq \int_e \|g_0(t)\|_{Y^*} dt + \varepsilon|e| \qquad (12.2.22)$$

On the other hand, $B\chi_e \in Z \subseteq X^* \cap Y^*$ so that there exists $y_\varepsilon \in Y$, $\|y_\varepsilon\| = 1$ with $\langle B\chi_e, y_\varepsilon\rangle_{X^* \times X} = \langle B\chi_e, y_\varepsilon\rangle_{Y^* \times Y} \geq \|B\chi_e\|_{Y^*} - \varepsilon|e|/2 = \|B\chi_e\|_{X^*} - \varepsilon|e|/2$. Combining with (12.2.5) and (12.2.22),

$$\int_e \langle g_0(t), y_\varepsilon\rangle_{Y^* \times Y} dt = \int_0^T \chi_e(t)\langle g_0(t), y_\varepsilon\rangle_{Y^* \times Y} dt$$

$$= \langle B\chi_e, y_\varepsilon\rangle_{Y^* \times Y} \geq \|B\chi_e\|_{X^*} - \frac{\varepsilon}{2}|e| \geq \int_e \|g_0(t)\|_{Y^*} dt + \frac{\varepsilon}{2}|e|, \qquad (12.2.23)$$

an obvious contradiction. This ends the proof of Theorem 12.2.6.   ∎

Example 12.2.8. In general, we have $L_w^\infty(0, T; X^*) \neq L^\infty(0, T; X^*)$ even if X is separable. In fact, let $X = C[0, 1]$ be the space of all continuous functions in $0 \leq t \leq 1$; X^* is the space $\Sigma[0, 1]$ of all finite regular Borel measures equipped with the total variation norm. Let $g(t) = \delta(\cdot - t)$ be the Dirac delta at t; $g(\cdot) \in L_w^\infty(0, T; X^*)$ but is not strongly measurable, nor equivalent to any strongly measurable function. To see this note that it follows from separability of X and Lemma 12.2.2 that if $g(\cdot) \approx h(\cdot)$ then $g(t) = h(t)$ a.e. Accordingly, $g(\cdot)$ itself is strongly measurable and there exists a countably valued function $g_1(\cdot)$ with $\|g(t) - g_1(t)\|_{X^*} < 1/2$. Let e be a set a of positive measure where $g_1(\cdot)$ is constant. Then

$$\|g(t') - g(t)\|_{X^*} \leq \|g(t') - g_1(t')\|_{X^*} + \|g_1(t') - g_1(t)\|_{X^*} + \|g_1(t) - g(t)\|_{X^*} < 1$$

for $t, t' \in e$, which is a contradiction since $\|g(t') - g(t)\|_{X^*} = 1$ for $t' \neq t$.

Remark 12.2.9. Theorem 12.2.6 implies that every equivalence class of the space $L_w^\infty(0, T; X^*)$ contains at least an element $g(\cdot)$ such that $t \to \|g(t)\|_{X^*}$ is measurable (and whose supremum norm coincides with the norm of the class in $L_w^\infty(0, T; X^*)$). There may be many other functions with $\|g(\cdot)\|_{X^*}$ measurable. An extravagant case is that in Example 12.2.1, where the equivalence class of zero contains functions with $\|g(t)\|_{X^*} = \eta(t)$ a totally arbitrary function.

Example 12.2.10. Note that the representing function $g(t)$ of an operator $B : L^1(0, T) \to X^*$ may not take values in the (strong) closure of the range $Z = B(L^1(0, T))$, even if X is separable. In fact, let again $X = C[0, 1]$, $X^* = \Sigma[0, 1]$. The operator B is the "identity operator" from $L^1(0, 1)$ into $\Sigma[0, 1]$;

precisely, $Bf = \mu \in \Sigma[0, 1]$, with μ defined by $\mu(e) = \int_e f(t)dt$. The operator B is an isometry from $L^1(0, 1)$ into $\Sigma[0, 1]$ and the range Z is the (closed) subspace $\Sigma_c[0, 1]$ of all absolutely continuous measures $\mu(d\sigma) = f(\sigma)d\sigma$ $(f(\cdot) \in L^1(0, 1))$. Let $g(t) = g(t, d\sigma) \in \Sigma[0, T]$ be the representing function of B. Then, by (12.2.5) we must have

$$\langle Bf, y \rangle = \int_0^1 f(t)y(t)dt = \int_0^1 \int_0^1 f(t)g(t, d\sigma)y(\sigma)dt$$

for every $f(\cdot) \in L^1(0, T)$ and $y(\cdot) \in C[0, 1]$. Accordingly,

$$\int_0^1 g(t, d\sigma)y(\sigma) = y(t)$$

for all $y(\cdot) \in C[0, t]$, thus $g(t) = \delta(\cdot - t)$ a.e. in t and $g(t) \notin Z$.

Theorem 12.2.11. *Let X be an arbitrary Banach space. The dual $L^1(0, T; X)^*$ is isometrically isomorphic to $L_w^\infty(0, T; X^*)$, the duality pairing between both spaces given by*

$$\langle g, f \rangle = \int_0^T \langle g(t), f(t) \rangle dt. \tag{12.2.24}$$

Proof. If $g(\cdot) \in L_w^\infty(0, T; X^*)$, the integral (12.2.24) makes sense for every $f(\cdot) \in L^1(0, T; X)$ and defines a bounded linear functional in $L^1(0, T; X)$. In fact, if $f(\cdot)$ is countably valued, the function $t \to \langle g(t), f(t) \rangle$ is measurable in each set where $f(t)$ is constant, thus it is measurable, and

$$|\langle g(t), f(t) \rangle| \leq \|g\|_{L_w^\infty(0,T;X^*)} \|f(t)\|_X \quad \text{a.e. in } 0 \leq t \leq T. \tag{12.2.25}$$

That $t \to \langle g(t), f(t) \rangle$ is measurable for any $f(\cdot) \in L^1(0, T; X)$ and that (12.2.25) holds follows approximating $f(\cdot)$ uniformly a.e. by countably valued functions. Inequality (12.2.25) also shows that, if $\|g\|_{L^1(0,T;X)^*}$ is the norm of $g(\cdot)$ as a linear functional in $L^1(0, T; X)$ then $\|g\|_{L^1(0,T;X)^*} \leq \|g\|_{L_w^\infty(0,T;X^*)}$. To prove the opposite inequality, let $\varepsilon > 0$. By definition of $\|g\|_{L_w^\infty(0,T;X^*)}$ there exists $y \in X$, $\|y\| = 1$ and a set e of positive measure in $0 \leq t \leq T$ such that $\langle g(t), y \rangle \geq \|g\|_{L_w^\infty(0,T;X^*)} - \varepsilon$ $(t \in e)$. Hence, if $\chi(\cdot)$ is the characteristic function of e we have $\langle g, \chi y \rangle \geq |e|(\|g\|_{L^1(0,T;X^*)} - \varepsilon)$ so that $\|g\|_{L^1(0,T;X)^*} \geq \|g\|_{L_w^\infty(0,T;X^*)} - \varepsilon$.

We show now that every bounded linear functional in $L^1(0, T; X)$ admits the representation (12.2.24). Let ϕ be such a functional. Given $f(\cdot) \in L^1(0, T)$, the map $y \to \phi(f(\cdot)y)$ is a linear functional in X. Since

$$|\phi(f(\cdot)y)| \leq \|\phi\|_{L^1(0,T;X)^*} \|f\|_{L^1(0,T)} \|y\|_X, \tag{12.2.26}$$

this functional is bounded (with norm $\leq \|\phi\|_{L^1(0,T;X)^*}\|f\|_{L^1(0,T)}$), thus there exists $y_f^* \in X^*$ with $\phi(f(\cdot)y) = \langle y_f^*, y \rangle$ and $\|y_f^*\|_{X^*} \leq \|\phi\|_{L^1(0,T;X)^*}\|f\|_{L^1(0,T)}$. Consider the map

$$Bf = y_f^* \tag{12.2.27}$$

from $L^1(0, T)$ into X^*. Obviously, B is a linear bounded operator from $L^1(0, T)$ into X^* with norm $\|B\|_{(L^1(0,T),X^*)} \leq \|\phi\|_{L^1(0,T;X)^*}$. By Theorem 12.2.4 (Dunford-Pettis Part 1), there exists a representing function $g(\cdot) \in L_w^\infty(0, T; X^*)$ satisfying (12.2.4) and such that

$$\phi(f(\cdot)y) = \langle y_f^*, y \rangle = \int_0^T f(t)\langle g(t), y \rangle dt \tag{12.2.28}$$

for $f(\cdot) \in L^1(0, T)$ and $y \in X$. The argument after (12.2.25) shows that $\|g\|_{L_w^\infty(0,T;X^*)} = \|\phi\|_{L^1(0,T;X)^*}$. Moreover, taking $f_1(\cdot), \ldots, f_n(\cdot) \in L^1(0, T)$ and $y_1, \ldots, y_n \in X$ we have

$$\phi(\Sigma f_k(\cdot)y_k) = \sum \int_0^T f_k(t)\langle g(t), y_k \rangle dt = \int_0^T \langle g(t), \Sigma f_k(t)y_k \rangle dt.$$

The proof of Theorem 12.2.11 ends observing that elements of the form $\Sigma f_k(\cdot)y_k$ with $f_k(\cdot) \in L^1(0, T)$ and $y_k \in X$ are dense in $L^1(0, T)$ (Theorem 5.0.27). ∎

Remark 12.2.12. It results from the proof of Theorem 12.2.11 that the function $g(\cdot) \in L_w^\infty(0, T; X^*)$ representing a bounded linear functional ϕ in $L^1(0, T; X)$ can be chosen in such a way that

$$\sup_{0 \leq t \leq T} \|g(t)\|_{X^*} = \|\phi\|_{L^1(0,T;X)^*}. \tag{12.2.29}$$

This is a consequence of Theorem 12.2.4. Theorem 12.2.6 insures that $t \to \|g(t)\|_{X^*}$ is measurable; for X separable, this is automatic (Lemma 12.2.2).

Example 12.2.13. If X^* is separable, strong and weak measurability coincide (Example 5.0.39) so that

$$L_w^\infty(0, T; X^*) = L^\infty(0, T; X^*) \tag{12.2.30}$$

with equality of norms. The equality also holds in reflexive spaces or, more generally, when X^* is a Gelfand space, that is, a Banach space where absolutely continuous functions have a derivative a.e. (see Theorem 5.0.14). In fact, let $g(\cdot)$ be an X^*-valued X-weakly measurable function defined in $a \leq t \leq b$ and satisfying $|\langle g(t), y \rangle| \leq \|y\|_X \alpha(t)$ a.e. there for every $y \in X$ ("a.e." depending on y) and $\alpha(\cdot) \in L^1(0, T)$. Define an X^*-valued function $G(t)$ by

$$\langle G(t), y \rangle = \int_0^t \langle g(\tau), y \rangle d\tau.$$

This function is everywhere defined and bounded. Let $\{(a_j, b_j)\}$ be a collection of n pairwise disjoint intervals in $[0, T]$. Select $y_j \in X$ such that $\|y_j\|_X = 1$, $\langle G(b_j) - G(a_j), y_j \rangle \geq \|G(b_j) - G(a_j)\|_{X^*} - \varepsilon/n$. Then

$$\sum \|G(b_j) - G(a_j)\|_{X^*} - \varepsilon \leq \sum \langle G(b_j) - G(a_j), y_j \rangle$$

$$= \sum \int_{a_j}^{b_j} \langle g(\tau), y_j \rangle d\tau \leq \sum \int_{a_j}^{b_j} \alpha(\tau) d\tau$$

so that $G(\cdot)$ is absolutely continuous. Since X^* is a Gelfand space $G'(t)$ exists a.e.; $G'(t)$ is the a.e. limit of the (continuous) functions $h^{-1}\{G(t + h) - G(t)\}$ as $h \to 0$, thus it is strongly measurable. Finally, we have $\langle G'(t), y \rangle = \langle g(t), y \rangle$ a.e. for every $y \in X$.

Note that we have *not* shown that every X-weakly measurable X^*-valued function is strongly measurable, just that for every such function $g(\cdot)$ there is a strongly measurable function $h(\cdot)$ with $h(\cdot) \approx g(\cdot)$ in the sense that $\langle h(t), y \rangle = \langle g(t), y \rangle$ a.e. ("a.e." depending on y). If this construction is applied to $g(\cdot) \in L_w^\infty(0, T; X^*)$, we have

$$\|g\|_{L_w^\infty(0,T;X^*)} = \|h\|_{L^\infty(0,T;X^*)}. \tag{12.2.31}$$

To see this, note that $\|g\|_{L_w^\infty(0,T;X^*)} = \|h\|_{L_w^\infty(0,T;X^*)}$. The inequality

$$\|h\|_{L_w^\infty(0,T;X^*)} \leq \|h\|_{L^\infty(0,T;X^*)} \tag{12.2.32}$$

is obvious from the definitions. If it is strict, we may approximate $h(\cdot)$ by a countably valued function $h_0(\cdot)$ so closely that the inequality is strict for $h_0(\cdot)$; we deduce the existence of a measurable set e of positive measure such that $h_0(t) = h_0$ is constant in e and $\|h_0\|_{X^*} \geq \|h\|_{L_w^\infty(0,T;X^*)} + \varepsilon$. A contradiction is then obtained integrating $\langle h_0(t), \chi(t)y \rangle$, χ the characteristic function of e, $y \in X$ such that $\|y\|_X = 1$, $\langle h_0, y \rangle \geq \|h_0\|_{X^*} - \varepsilon/2$.

The preceding argument works in an arbitrary Banach space X if $h(\cdot)$ is a strongly measurable member of the equivalence class of $g(\cdot)$. Note, however, that the space X in Example 12.2.8 illustrates the possibility that there may be equivalence classes in $L_w^\infty(0, T; X^*)$ without any strongly measurable dweller, such as that of the function $t \to \delta(\cdot - t)$ (although all elements of $L_w^\infty(0, T; X^*)$ have measurable norm).

Incidentally, note that Example 12.2.13 "cleans up" the pathologies associated with the space in Example 12.2.1. Since X is a Hilbert space it is reflexive, thus although many elements of $L_w^\infty(0, T; X^*)$ are rather unruly, each class of equivalence contains a strongly measurable representative $h(\cdot)$ satisfying (12.2.31) for every other element $g(\cdot)$ of the class.

Example 12.2.14. In a general Banach space X, the norm in $L_w^\infty(0, T; X^*)$ is given by

$$\|g(\cdot)\|_{L_w^\infty(0,T;X^*)} = \inf_{h\in[g]} \|h\|_{L^\infty(0,T;X^*)} = \inf_{h\in[g]} \operatorname*{ess\,sup}_{0\le t\le T} \|h(t)\|_{X^*}, \qquad (12.2.33)$$

the infimum taken over all members $h(\cdot)$ of the equivalence class $[g(\cdot)]$ of $h(\cdot)$ with measurable norm (by Theorem 12.2.6 one such member is $h = \mathcal{L}g$). Indeed, this results from inequality (12.2.32) and the argument that follows it.

12.3. Finitely Additive Measures: Integration

In this section U is an arbitrary set, Φ is a field of subsets of U and $\mu, \nu, \dots$ are finitely additive measures (see **2.0** for definitions). We assume the measures bounded; this means $|\mu|(U) < \infty$, where $|\mu|$ is the total variation of μ. We exhibit below examples of finitely additive measures which are not countably additive.

Example 12.3.1. Let $U = [a, b]$, and let Φ be the set of all finite disjoint unions $U_j I_j$ of subintervals $I_j = [c, d], [c, d), (c, d], (c, d)$ of $[a, b], a \le c \le d \le b$. Then Φ is a field. Define a measure μ in Φ by the formula $\mu(U_j I_j) = \Sigma \mu(I_j)$, where $\mu([c, d]) = \mu([c, d)) = \mu((c, d]) = \mu((c, d)) = f(d) - f(c)$, $f(\cdot)$ a finite function defined in $a \le t \le b$. A little work is needed to show that μ is well defined. This measure is finitely additive. However, if f is not continuous in $a \le t \le b$, the measure μ is not countably additive. (Hint: Let $\bar{t} \in (a, b]$ be a point where f is not left continuous, and let $\{t_k\}$ be a sequence with $a \le t_1 < t_2 < \cdots < \bar{t}$, $\lim t_k = \bar{t}$, but $\lim f(t_k) \ne f(\bar{t})$.) Then countable additivity

$$\mu\left(\bigcup e_k\right) = \sum \mu(e_k) \qquad (12.3.1)$$

fails with $e_k = [t_k, t_{k+1})$, $\cup_k e_k = [t_1, \bar{t})$. A similar argument exploits right discontinuity.

Example 12.3.2. Let $U = [0, 1)$ and $\text{LIM}_{n\to\infty}$ a Banach limit. (**5.0**). Pick an arbitrary sequence $\{u_n\}$ in $[0, 1)$ with $u_n \to 1$. Define

$$\mu(e) = \operatorname*{LIM}_{n\to\infty} \chi_e(u_n) \qquad (12.3.2)$$

where $\chi_e(\cdot)$ is the characteristic function of the set e. Obviously, μ is well defined for every $e \subseteq [0, 1)$, thus we take Φ as the field of all subsets of $[0, 1)$; moreover, we deduce from linearity of $\text{LIM}_{n\to\infty}$ that μ is a finitely additive measure. It is plain that $\mu(e) = 0$ for every set $e \subseteq [0, 1)$ bounded away from 1, whereas $\mu([0, 1)) = 1$. Accordingly, (12.3.1) fails for $e_k = [t_k, t_{k+1})$, where $\{t_k\}$ is a strictly increasing sequence in $[0, 1)$ tending to 1.

We sketch below an integration theory with respect to finitely additive measures due to N. Dunford and J. T. Schwartz (for full details see Dunford-Schwartz [1958, III.2.1]). We note that the definition of Lebesgue integral in **2.0** (extended to vector-valued functions in **5.0**) and based on approximation of measurable functions by countably valued functions is unworkable here. In fact, there are countably valued functions that coincide everywhere but which would have different integrals if defined as in **2.0**, as shown below.

Example 12.3.3. Let U, μ, $\{t_k\}$ be as in Example 12.3.2. Consider the following countably valued functions: $\chi(\cdot)$ (the characteristic function of the interval $[0, 1)$) and $\Sigma \chi_j(\cdot)$, $\chi_j(\cdot)$ the characteristic function of the interval $e_j = [t_j, t_{j+1})$. Although these functions coincide everywhere, the integral of the first as defined in **2.0** is 1; the integral of the second is zero.

We only consider below bounded measures; the functions under integration take values in a Banach space.

Given a positive finitely additive measure ν defined in a field Φ, we define a set function ν^* by

$$\nu^*(e) = \inf_{d \in \Phi, d \supseteq e} \nu(d) \tag{12.3.3}$$

for arbitrary $e \subseteq U$. The set function ν^* is called the **outer measure** associated with ν and satisfies $\nu^*(e) = \nu(e)$ for $e \in \Phi$, $\nu^*(c) \leq \nu^*(e)$ for arbitrary $c, e \subseteq U$ with $c \subseteq e$ and

$$\nu^*\left(\bigcup e_k\right) \leq \sum \nu^*(e_k) \tag{12.3.4}$$

for arbitrary subsets $e_1, e_2, \ldots, e_n$ of U. (This definition was already used to construct the outer Lebesgue measure; see **2.0**.)

If μ is an arbitrary bounded, finitely additive measure, we define a set $e \subseteq U$ to be a **null set with respect to** μ or simply a μ-**null set** if and only if

$$|\mu|^*(e) = 0, \tag{12.3.5}$$

$|\mu|$ the total variation of μ. Once null sets have been defined, "almost everywhere" definitions and expressions make sense. For instance, two functions, f, g defined in U coincide almost everywhere if and only if the set $\{u \in U; f(u) \neq g(u)\}$ is a null set. The reader should remember, however, that countable unions of null sets may no longer be null sets.

Example 12.3.4. Let μ be the measure in Example 12.3.1 with $f(t) = 0$ $(0 \leq t < \bar{t})$, $f(t) = 1$ $(\bar{t} \leq t \leq b)$, where $\bar{t} \in (a, b)$. This measure is positive, so that $|\mu| = \mu$. The sets e_k in Example 12.3.1 are μ-null sets but their union is not; $\mu^*(\cup_k e_k) = 1$.

Example 12.3.5. If U, μ are as in Example 12.3.2, the measure μ is positive so that $|\mu| = \mu$. On the other hand, μ is defined for every subset $e \subseteq U$ so that $|\mu|^* = \mu^* = \mu$. The sets $e_j = \{u_j\}$ are μ-null sets; their union is not.

Let E be a Banach space. We denote by $\mathcal{F}(U; E)$ the linear space (pointwise operations) of all E-valued functions defined in U. Given $\{f_n(\cdot)\} \subseteq \mathcal{F}(U; E)$ and $f(\cdot) \in \mathcal{F}(U; E)$, we say that $\{f_n(\cdot)\}$ **converges to** $f(\cdot)$ **in** μ-**measure** if and only if for every $\varepsilon > 0$

$$\lim_{n \to \infty} |\mu|^*(\{u \in U; \|f_n(u) - f(u)\| \geq \varepsilon\}) = 0. \qquad (12.3.6)$$

The space $\mathcal{F}(U, E)$ can be given a metric whose convergence is precisely convergence in μ-measure. Define

$$|f|_\mu = \inf_{\delta > 0}(\delta + |\mu|^*(\{u \in U; \|f(u)\| > \delta\})) \qquad (12.3.7)$$

for $f(\cdot) \in \mathcal{F}(U; E)$. We obviously have $|f|_\mu \geq 0$ and $|f|_\mu = |-f|_\mu$. If $|\lambda| \leq 1$, then $\{u \in U; \|\lambda f(u)\| > \delta\} \subseteq \{u \in U; \|f(u)\| > \delta\}$, thus

$$|\lambda f|_\mu = \inf_{\delta > 0}(\delta + |\mu|^*(\{u \in U; \|\lambda f(u)\| > \delta\}))$$

$$\leq \inf_{\delta > 0}(\delta + |\mu|^*(\{u \in U; \|f(u)\| > \delta\})) = |f|_\mu \qquad (12.3.8)$$

Finally,

$$\{u \in U; \|f(u) + g(u)\| > \delta + \varepsilon\}$$

$$\subseteq \{u \in U; \|f(u)\| > \delta\} \cup \{u \in U; \|g(u)\| > \varepsilon\},$$

hence

$$|f + g|_\mu \leq \delta + \varepsilon + |\mu|^*(\{u \in U; \|f(u) + g(u)\| > \delta + \varepsilon\})$$

$$\leq \delta + |\mu|^*(\{u \in U; \|f(u)\| > \delta\}) + \varepsilon + |\mu|^*(\{u \in U; \|h(u)\| > \varepsilon\}).$$

Taking infimum in the two terms on the right, we obtain

$$|f + g|_\mu \leq |f|_\mu + |g|_\mu. \qquad (12.3.9)$$

It follows that

$$d_\mu(f, g) = |f - g|_\mu \qquad (12.3.10)$$

has all properties of a metric except $d_\mu(f, g) = 0$ does not imply $f \equiv g$; equivalently, $|h|_\mu = 0$ does not imply $h(u) = 0$ for all $u \in U$. Functions with $|h|_\mu = 0$ are identified below.

Lemma 12.3.6. *Let* $h(\cdot) \in \mathcal{F}(U; E)$. *Then* $|h|_\mu = 0$ *if and only if*

$$|\mu|^*(\{u \in U; \|h(u)\| > \varepsilon\}) = 0 \quad \text{for all } \varepsilon > 0. \qquad (12.3.11)$$

Proof. It is obvious that if (12.3.11) holds then $|h|_\mu = 0$. Conversely, assume that $|h|_\mu = 0$. If $\delta \le \varepsilon$, we have $\{u \in U; \|h(u)\| > \varepsilon\} \subseteq \{u \in U; \|h(u)\| > \delta\}$, so that

$$|\mu|^*(\{u \in U; \|h(u)\| > \varepsilon\}) \le \delta + |\mu|^*(\{u \in U; \|h(u)\| > \delta\})$$

Since $|h|_\mu = 0$ the infimum in (12.3.7) must be approached as $\delta \to 0$ and we deduce that $|\mu|^*(\{u \in U; \|h(u)\| > \varepsilon\}) = 0$ as claimed. ∎

We denote by $\mathcal{F}(U, E; \mu)$ the space $\mathcal{F}(U, E)$ modulo the equivalence relation $f \approx g$ if $|f - g|_\mu = 0$; equivalently, $f \approx g$ if

$$|\mu|^*(\{u \in U; \|f(u) - g(u)\| > \varepsilon\}) = 0 \quad \text{for all } \varepsilon > 0.$$

Equipped with the distance (12.3.10) $\mathcal{F}(U, E; \mu)$ is a metric space. It is also a linear space, since the natural (pointwise) linear operations of $\mathcal{F}(U, E)$ are compatible with the equivalence relation. This is obvious for the sum from (12.3.9). For the product by scalars, we use (12.3.8) if $|\lambda| \le 1$; if $|\lambda| > 1$, we write $\lambda f = (\lambda/n)f + \cdots + (\lambda/n)f$ (n times) for sufficiently large n. Finally, convergence with respect to the metric d_μ in $\mathcal{F}(U, E; \mu)$ is convergence in μ-measure. A function h satisfying (12.3.11) (that is, with $|h|_\mu = 0$) will be called a μ-**null function**.

Example 12.3.7. A μ-null function is *not* necessarily zero almost everywhere. In fact, consider the measure in Example 12.3.2 and the function $h(u_n) = 1/n$ ($n = 1, 2, \ldots$), $h(u) = 0$ elsewhere. Then $|h|_\mu = 0$, but the set $\{u_n\}$ where $h(u) \ne 0$ has measure 1.

A function $g(\cdot) \in \mathcal{F}(U, E)$ is **simple** if

$$g(u) = \sum \chi_j(u) y_j, \tag{12.3.12}$$

where the χ_j are characteristic functions of a finite family of pairwise disjoint sets $e_1, e_2, \ldots, e_n$ in Φ. The function $g(\cdot)$ is μ-**simple** if it differs from a simple function by a μ-null function. In other words, μ-simple functions are elements of $\mathcal{F}(U, E; \mu)$ equivalent to a function of the form (12.3.12). The class of all μ-simple functions is denoted by $\mathcal{S}(U, E; \mu)$, and the class $\mathcal{M}(U, E; \mu) \subseteq \mathcal{F}(U, E; \mu)$ of all μ-**measurable** functions is defined as the closure of $\mathcal{S}(U, E; \mu)$ in $\mathcal{F}(U, E; \mu)$.

The **integral** of a μ-simple function is

$$\int_U f(u)\mu(du) = \sum \mu(e_j)y_j, \tag{12.3.13}$$

and a little work is needed to show that the definition does not depend on the particular $g(\cdot)$ used in the equivalence class of $f(\cdot)$ (Dunford-Schwartz [1958, p. 109]).

Let $f(\cdot) \in \mathcal{F}(U, E; \mu)$ and denote by $\|f\|(\cdot)$ the function $\|f\|(u) = \|f(u)\|$. The function $\|f\|$ belongs to $\mathcal{F}(U, \mathbb{R}; |\mu|)$, and the following result is a simple consequence of the definitions.

Lemma 12.3.8. (a) *Let f be μ-simple. Then $\|f\|$ is $|\mu|$-simple and*

$$\left\| \int_U f(u)\mu(du) \right\| \leq \int_U \|f(u)\| |\mu|(du). \tag{12.3.14}$$

(b) *Let f be μ-measurable. Then $\|f\|$ is $|\mu|$-measurable.*

A function $f(\cdot) \in \mathcal{F}(U, E; \mu)$ is μ-**integrable** (in symbols, $f(\cdot) \in \mathcal{I}(U, E; \mu)$) if there exists a sequence $\{f_n(\cdot)\}$ of μ-simple functions converging to $f(\cdot)$ in μ-measure (that is, in the metric of the space $\mathcal{F}(U, E; \mu)$) and such that

$$\lim_{m,n\to\infty} \int_U \|f_n(u) - f_m(u)\| |\mu|(du) = 0. \tag{12.3.15}$$

The integral is defined by

$$\int_U f(u)\mu(du) = \lim_{n\to\infty} \int_U f_n(u)\mu(du), \tag{12.3.16}$$

and again some work is needed to show that the value of the integral does not depend on the particular sequence $\{f_n\}$ used (Dunford-Schwartz [1958, p. 113]). Once we have insured this, the proof of the following result follows directly from the definition.

Theorem 12.3.9. *Let f be μ-integrable. Then $\|f\|$ is $|\mu|$-integrable, and (12.3.14) holds.*

We direct the reader to Dunford-Schwartz [1958] for much additional information on measurability and integration, limiting ourselves to the following result.

Theorem 12.3.10. (a) *The class $\mathcal{M}(U, E; \mu)$ of measurable functions is a linear subspace of $\mathcal{F}(U, E; \mu)$. (b) The class $\mathcal{I}(U, E; \mu)$ of integrable functions is a linear subspace of $\mathcal{F}(U, E; \mu)$ and*

$$\int_U (\alpha f(u) + \beta g(u))\mu(du) = \alpha \int_U f(u)\mu(du) + \beta \int_U g(u)\mu(du)$$

for $f, g \in \mathcal{I}(U, E; \mu)$ and arbitrary scalars α, β. (c) If $f \in \mathcal{I}(U, E; \mu)$ then $\|f\| \in \mathcal{I}(U, \mathbb{R}; |\mu|)$ and (12.3.14) holds. (d) If μ is positive and $f \in \mathcal{I}(U, \mathbb{R}; \mu)$,

$$\int_U f(u)\mu(du) \geq 0 \quad \text{if } f(u) \geq 0.$$

To prove (a), note that $\mathcal{S}(U, E; \mu)$ is a linear space. Let $f, g \in \mathcal{M}(U, E; \mu)$ and let $\{f_n\}, \{g_n\}$ be sequences in $\mathcal{S}(U, E; \mu)$ such that $|f_n - f|_\mu \to 0, |g_n - g|_\mu \to 0$. We have $|f_n + g_n - (f + g)|_\mu \le |f_n - f|_\mu + |g_n - g|_\mu$; accordingly $f + g \in \mathcal{M}(U, E; \mu)$. If λ is a scalar with $|\lambda| \le 1$ then $|\lambda f_n - \lambda f|_\mu \le |f_n - f|_\mu$, hence $\lambda f \in \mathcal{M}(U, E; \mu)$. If $|\lambda| > 1$, write $\lambda f = (\lambda/m)f + \cdots + (\lambda/m)f$ for sufficiently large m. The proof of (b) and (c) follows directly from the definition, and we leave that of (d) to the reader. ∎

When measures are not countably additive, almost everything that can go wrong does, as shown in the examples below.

Example 12.3.11. *Failure of the dominated convergence theorem.* Take U and μ as in Example 12.3.2. Let $\chi_n(\cdot)$ be the characteristic function of $(1 - 1/n, 1)$. Then $\chi_n(u) \to 0$ for all $u \in U$. However, each χ_n is a simple function with integral $\mu((1 - 1/n, 1)) = 1$.

Example 12.3.12. *The true dominated convergence theorem* (Dunford-Schwartz [1958, p. 124]). Let $g(\cdot) \in \mathcal{I}(U, \mathbb{R}; |\mu|), g(u) \ge 0$ and let $\{f_\kappa(\cdot)\}$ be a generalized sequence in $\mathcal{I}(U, E; \mu)$ such that $f_\kappa(\cdot) \to f(\cdot) \in \mathcal{M}(U, E; \mu)$ in μ-measure and

$$\|f_\kappa(u)\| \le g(u) \quad \text{a.e. in } U \text{ for all } \kappa.$$

Then $f(\cdot) \in \mathcal{I}(U, E; \mu)$ and

$$\int_U f_\kappa(u)\mu(du) \to \int_U f(u)\mu(du).$$

Example 12.3.13. *Supports cannot be reasonably defined.* Let μ be a finitely additive measure defined in a field Φ. Call a set d the **support** of μ if (i) $\mu(e) = 0$ if $e \cap d = \emptyset$ and (ii) d is minimal with respect to property (i), that is, if d' satisfies (i) then $d' \supseteq e$. Under this definition, the measure μ in Example 12.3.2 does not have a support. In fact, assume d has properties (i) and (ii) above. Then d must contain a subsequence of the sequence $\{u_n\}$ (otherwise we would have $\mu(d) = 0$, thus $\mu = 0$). Call d' the set obtained discarding from d a finite number of elements of the subsequence. Then d' satisfies (i) but $d' \subset d$ strictly, a contradiction.

Example 12.3.14. *Vector valued continuous functions may not be integrable.* In the realm of σ-additive measures, continuous vector valued functions (and many others) are integrable. Things are different for finitely additive measures. For instance, if μ is the measure in Example 12.3.2, a vector valued function $f(\cdot)$ continuous and bounded in U may not even be μ-measurable. To see this, assume the u_n are all different and let E be an infinite dimensional Banach space. Then there exists $\delta > 0$ and an infinite sequence $\{y_n\} \subseteq E$ such that $\|y_n\| = 1$, $\|y_n - y_m\| \ge \rho > 0$ for $n \ne m$ (Riesz-Nagy [1955, p. 218]). Let $f(\cdot)$ be a continuous bounded E-valued function in U with $f(u_n) = y_n$ (such a function can be constructed in the form $\sum \psi_n(u) y_n$, $\psi_n(u)$ a triangular spike of summit 1 at $u = u_n$ and sufficiently small base). If $g_n(\cdot)$ is a simple function then the set $\{u \in U; \|g_n(u) - f(u)\| < \rho/2\}$ can contain only a finite number of the u_n, thus its complement has measure 1.

On the other hand, the integral $\int_U f(u)\nu(du)$ exists if ν is a countably additive bounded measure in U (since f is bounded and continuous in U it can be uniformly approximated by functions constant in a countable number of subintervals).

Example 12.3.15. *Failure of Fubini's Theorem.* We consider one of the measures μ in $U = [0, 1)$ in Example 12.3.2 and the following function $f(u, v)$ in the product space $U \times U = [0, 1) \times [0, 1)$:

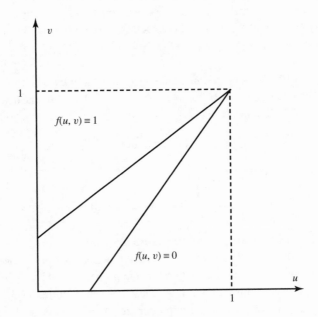

Figure 12.1.

(between the triangles $f = 0$ and $f = 1$ f is defined in such a way as to be continuous in $[0, 1) \times [0, 1)$.) Then we have

$$\int_{[0,1)} f(u, v)\mu(du) = 0 \quad (0 \le v < 1)$$

$$\int_{[0,1)} f(u, v)\mu(dv) = 1 \quad (0 \le u < 1). \quad (1)$$

---

[1] Strictly speaking, the counterexample should include construction of the *product measure* $\nu = \mu \otimes \mu$ in $U \times U$. See Dunford-Schwartz [1957, p. 184].

Example 12.3.16. *An impersonation of the delta.* Let $U = [0, 1]$, $\{u_n\}$ a sequence in U convergent to $\bar{u} \in U$, $u_n \neq \bar{u}$, μ defined as in Example 12.3.2 (that is, $\mu(e) = \mathrm{LIM}_{n \to \infty} \chi_e(u_n)$). Then $\mu(\{\bar{u}\}) = 0$. However, if $f(u)$ is continuous, $\int_U f(u)\mu(du) = f(u)$.

For other interesting pathologies of finitely additive measures, see Yosida-Hewitt [1952].

12.4. Measures and Linear Functionals in Function Spaces

Finitely additive measures are continuous linear functionals in certain function spaces, as Theorem 12.4.1 shows. This is the basis of all the other duality results; roughly, the more conditions are put on the space the better are the measures. The reader going after finite dimensional systems needs only Theorem 12.4.9.

Let U be an arbitrary set. We denote by $B(U)$ the (linear) space of all real valued bounded functions defined in U. Endowed with the supremum norm, $B(U)$ is a Banach space. If Φ is a field of subsets of U, $B(U, \Phi)$ denotes the closed subspace of $B(U)$ generated by all characteristic functions of sets in Φ. Finally, $\Sigma_{ba}(U, \Phi)$ is the space of all finitely additive measures λ in U with $|\lambda|(U) < \infty$. Linear operations in $\Sigma_{ba}(U, \Phi)$ are defined in the obvious way: $(\lambda + \mu)(e) = \lambda(e) + \mu(e)$, $(\alpha\mu)(e) = \alpha\mu(e)$, and make $\Sigma_{ba}(U, \Phi)$ a linear space. Finally,

$$\|\lambda\| = \|\lambda\|_{\Sigma_{ba}(U,\Phi)} = |\lambda|(U) \tag{12.4.1}$$

is a norm in $\Sigma_{ba}(U, \Phi)$. The fact that $\Sigma_{ba}(U, \Phi)$ is a Banach space under $\|\cdot\|$ (that is, that $\Sigma_{ba}(U, \Phi)$ is complete) is a byproduct of

Theorem 12.4.1. *The dual space $B(U, \Phi)^*$ can be identified algebraically and metrically with $\Sigma_{ba}(U, \Phi)$. The duality pairing of $B(U, \Phi)$ and $\Sigma_{ba}(U, \Phi)$ is*

$$\langle f, \lambda \rangle = \int_U f(u)\lambda(du). \tag{12.4.2}$$

Proof. We show first that a measure $\lambda \in \Sigma_{ba}(U, \Phi)$ creates a linear functional in $B(U, \Phi)$ through (12.4.2). Let $f \in B(U, \Phi)$. Then f is the limit in the norm of $B(U)$ of a sequence $\{f_n\}$,

$$f_n(u) = \sum \alpha_{nj} \chi_{nj}(u), \tag{12.4.3}$$

where the χ_{nj} are characteristic functions of sets in Φ, and the sum is finite for each n. Since convergence in the norm of $B(U)$ is stronger than convergence in λ-measure, we only need to check (12.3.15), which follows from (12.3.14) and the triangle inequality $\|f_n - f_m\|_{B(U)} \leq \|f_n - f\|_{B(U)} + \|f_m - f\|_{B(U)}$. Hence $f(\cdot)$

is λ-integrable. That the integral (12.4.2) defines a bounded linear functional is a consequence of Theorems 12.3.9 and 12.3.10.

We check next that the norm (12.4.1) of λ coincides with its norm $\|\lambda\|^*$ as a linear functional. We have

$$|\langle f, \lambda \rangle| \leq \int_U |f(u)||\lambda|(du) \leq \|f\|_{B(U)} |\lambda|(U) = \|f\|_{B(U)} \|\lambda\|_{\Sigma_{ba}(U,\Phi)},$$

so that $\|\lambda\|^* \leq \|\lambda\|_{\Sigma_{ba}(U,\Phi)}$. To show the opposite inequality select $\varepsilon > 0$ and a finite collection of sets in Φ such that $\Sigma|\lambda(e_j)| \geq \|\lambda\|_{\Sigma_{ba}(U,\Phi)} - \varepsilon$. Define $f(u) = \Sigma(\text{sign}\lambda(e_j))\chi_j(u)$, χ_j the characteristic function of e_j. Then we have $\langle f, \lambda \rangle = \Sigma|\lambda(e_j)|$, so that $\|\lambda\|^* \geq \|\lambda\|_{\Sigma_{ba}(U,\Phi)} - \varepsilon$.

It only remains to show that every bounded linear functional ψ in $B(U, \Phi)$ can be represented by (12.4.2) with $\lambda \in \Sigma_{ba}(U, \Phi)$. The measure λ is defined by

$$\lambda(e) = \psi(\chi_e(\cdot)), \tag{12.4.4}$$

$\chi_e(\cdot)$ the characteristic function of e. Linearity of ψ implies that λ is a finitely additive measure; boundedness of ψ implies that $|\lambda|(U) < \infty$. If f is a linear combination of characteristic functions of sets in Φ it is clear that (12.4.2) will hold. Since, by definition, these functions are dense in $B(U, \Phi)$, (12.4.2) holds in the entire space; we take limits on the integral on the right using (12.3.14). This completes the proof. ∎

Remark 12.4.2. Theorem 12.4.1 can be applied to the entire space $B(U)$. It suffices to take Φ as the field of *all* subsets of U and to show that $B(U) = B(U, \Phi)$, that is, that every function in $B(U)$ can be approximated in the supremum norm by a linear combination of characteristic functions. To see this define $h = 2\|f\|_{B(U)}/n$, $f_{nj} = -\|f\|_{B(U)} + jh$ $(j = 0, 2, \dots, n)$ and divide the interval $[-\|f\|_{B(U)}, \|f\|_{B(U)}]$ into the subintervals

$$I_{nj} = [f_{n,j-1}, f_{nj}) \quad (j = 1, 2, \dots, n-1), \qquad I_{nn} = [f_{n,n-1}, f_{nn}]. \tag{12.4.5}$$

If $\chi_{nj}(\cdot)$ is the characteristic function of the set $e_{nj} = f^{-1}(I_{nj})$ and

$$f_n(u) = \sum_{j=1}^{n} f_{nj}\chi_{nj}(u), \tag{12.4.6}$$

then we have $|f_n(u) - f(u)| \leq 2\|f\|_{B(U)}/n$ $(u \in U)$.

We use the shorthand $\Sigma_{ba}(U)$ for $\Sigma_{ba}(U, \Phi)$ when Φ is the field of all subsets of U.

Let U be a topological space. We denote by $BC(U)$ the space of all bounded continuous functions in U. Endowed with the supremum norm, $BC(U)$ is a Banach space (subspace of the space $B(U)$). Theorem 12.4.6 below identifies the dual $BC(U)^*$ as a space of measures when U is a normal topological space.

Let G be an arbitrary collection of subsets of U. The field $\Phi(G)$ *generated* by G is the intersection of the family of all fields that contain all elements of G (see **2.0**). If U is a topological space, we pay attention to the field $\Phi(C)$ generated by the collection C of all closed sets of U (which is the same as the field $\Phi(\Upsilon)$ generated by the collection Υ of all open sets of U).

A definition that relates measures on a topological space and its topology is the following. A bounded finitely additive measure $\mu \in \Sigma_{ba}(U, \Phi(C))$ is **regular** (Dunford-Schwartz [1958, p. 137]) if, for each $e \in \Phi(C)$ and each $\varepsilon > 0$ we have an open set $d \supseteq e$ and a closed set $c \subseteq e$ such that

$$|\mu|(d \setminus c) \leq \varepsilon, \qquad (12.4.7)$$

$|\mu|$ the total variation of μ. The set $\Sigma_{rba}(U, \Phi(C))$ of all regular measures is a subspace of $\Sigma_{ba}(U, \Phi(C))$. If μ is a positive measure, μ is regular if one of the two conditions

$$\mu(e) = \inf_{d \text{ open, } d \supseteq e} \mu(d) \quad \text{or} \quad \mu(e) = \sup_{c \text{ closed, } c \subseteq e} \mu(c) \qquad (12.4.8)$$

is satisfied. Note first that both are equivalent, that is, both hold if one does; for instance, to prove the second from the first, we apply it to the complement e^c of e. To obtain (12.4.7) from (12.4.8) we construct $d \supseteq e$ open and $c \subseteq e$ closed such that $\mu(d) \leq \mu(e) + \varepsilon/2$, $\mu(c) \geq \mu(e) - \varepsilon/2$. That (12.4.7) implies (12.4.8) is obvious.

In the following three examples the intervals $[a, b]$, $U = [0, 1)$ and $U = [0, 1]$ are appointed with their usual topology.

Example 12.4.3. Consider one of the measures μ in Example 12.3.1. If $f(\cdot)$ is discontinuous anywhere in $[a, b]$, then μ is not regular; in fact, if it were, it would be countably additive (Alexandrov's Theorem 12.4.10 below).

Example 12.4.4. The measure μ in Example 12.3.2 is regular. To see this, let $e \in U = [0, 1)$ be an arbitrary set. Take $c = \{u_n\} \cap e$, $d = [0, 1) \setminus (\{u_n\} \setminus e)$. Obviously we have $c \subseteq e \subseteq d$ and $\mu(c) = \mu(e) = \mu(d)$, since $\{u_n\} \cap c = \{u_n\} \cap e = \{u_n\} \cap d$. (the intersections may be empty). That c is closed and d is open follow from the fact that an arbitrary (finite or infinite) subset $\{u_n\}'$ of $\{u_n\}$ must be closed (the only possible limit point outside of $\{u_n\}'$ is 1, not a point of U).

Example 12.4.5. The pseudo delta in Example 12.3.16 is not regular. In fact, $\mu(\{\bar{u}\}) = 0$, while $\mu(d) = 1$ if d is an open set containing $\{\bar{u}\}$. We can also apply Alexandrov's theorem here; if regular, μ would be countably additive.

Theorem 12.4.6. *Let U be a normal topological space. Then the dual space $BC(U)^*$ can be identified algebraically and metrically with $\Sigma_{rba}(U, \Phi(C))$ equipped with the norm (12.4.1) inherited from $\Sigma_{ba}(U, \Phi(C))$. The duality pairing of $BC(U)$ and $\Sigma_{rba}(U, \Phi(C))$ is given by (12.4.2).*

Proof. We begin by proving that the integral (12.4.2) exists for $f \in BC(U)$ and $\mu \in \Sigma_{rba}(U, \Phi(C))$. We use again the approximation (12.4.5)-(12.4.6), but we must show that $f^{-1}(I_{nj}) \in \Phi(C)$. This is obvious for the closed interval I_{nn} in (12.4.4), since $f^{-1}(I_{nn})$ is closed; for the other intervals, $f^{-1}(I_{nj}) = f^{-1}([f_{n,j-1}, f_{nj})) = f^{-1}(\{f_{n,j-1}\}) \cup f^{-1}((f_{n,j-1}, f_{nj}))$, the first set in the union closed, the second open. ∎

We show that the norm $\|\mu\|^*$ of a measure μ as a linear functional coincides with its norm in $\Sigma_{rba}(U, \Phi(C))$. The proof that $\|\mu\|^* \le \|\mu\|_{\Sigma_{rba}(U,\Phi(C))}$ is the same as that in Theorem 12.4.1. To show the opposite inequality, let $\varepsilon > 0$ and $e_1, e_2, \dots, e_n$ disjoint sets in $\Phi(C)$ such that $\Sigma|\mu(e_j)| \ge \|\mu\|_{\Sigma_{rba}(U,\Phi(C))} - \varepsilon$. Taking advantage of the regularity of μ select closed sets $c_1, \dots, c_n$ such that $c_j \subseteq e_j$ and

$$|\mu|(e_j \setminus c_j) \le \varepsilon/n.$$

Since U is a normal topological space, there exist disjoint open sets $d_1, \dots, d_n$ such that $c_j \subseteq d_j$; using, again regularity of μ, these open sets can be chosen in such a way that

$$\|\mu\|(d_j \setminus c_j) \le \varepsilon/n.$$

By Urysohn's Lemma 12.0.18, we can construct continuous functions $f_j(\cdot)$ such that $0 \le f(u) \le 1$, $f_j(u) = 1$ $(u \in c_j)$, $f_j(x) = 0$ $(u \in d_j^c = \text{complement of } d_j)$. Then $f(u) = \Sigma(\text{sign } \mu(c_j)) f_j(u)$ is a function in $BC(U)$ with $\|f\|_{BC(U)} = 1$. We have

$$\langle f, \mu \rangle = \int_U f(u)\mu(du) = \sum |\mu(c_j)| + \sum \int_{d_j \setminus c_j} f(\mu)\mu(du)$$

$$\ge \sum |\mu(c_j)| - \varepsilon \ge \sum |\mu(e_j)| - 2\varepsilon \ge \|\mu\|_{\Sigma_{rba}(U,\Phi(C))} - 3\varepsilon.$$

This shows that $\|\mu\|^* \ge \|\mu\|_{\Sigma_{rba}(U,\Phi(C))}$.

The fact that every bounded linear functional ψ in $BC(U)$ admits the representation (12.4.2) with $\mu \in \Sigma_{rba}(U, \Phi(C))$ is a consequence of

Theorem 12.4.7. *Let $\lambda \in \Sigma_{ba}(U)$. Then there exists $\mu \in \Sigma_{rba}(U, \Phi(C))$ such that $\|\mu\|_{\Sigma_{rba}(U,\Phi(C))} = \|\lambda\|_{\Sigma_{ba}(U)}$ and*

$$\int_U f(u)\lambda(du) = \int_U f(u)\mu(du) \quad (f(\cdot) \in BC(U)). \tag{12.4.9}$$

In fact, assume Theorem 12.4.7 has been proved. Then it suffices to extend the functional ψ from $BC(U)$ to $B(U)$ (with the same norm) by the Hahn–Banach theorem and combine (12.4.2) with (12.4.9).

Proof of Theorem 12.4.7. It is enough to do it for a positive measure λ. In fact, if this is valid, we have $\int_U f(u)|\lambda|(du) = \int_U f(u)\mu_0(du)$ for $f \in BC(U)$ and some $\mu_0 \in \Sigma_{rba}(U, \Phi(C))$. We apply again the particular case, this time to the positive measure $\lambda + |\lambda|$, obtaining (12.4.9) after some rearrangements. That $\|\mu\|_{\Sigma_{rba}(U, \Phi(C))} = \|\lambda\|_{\Sigma_{ba}(U)}$ follows from Theorem 12.4.1 and what has been proved of Theorem 12.4.6.

Assume then that $\lambda \geq 0$. The idea of the proof is intuitive enough; we define two nonnegative set functions μ_1, μ_2 on arbitrary subsets of U by

$$\mu_1(e) = \inf_{d \text{ open}, d \supseteq e} \lambda(d), \quad \mu_2(e) = \sup_{c \text{ closed}, c \subseteq e} \mu_1(c) \tag{12.4.10}$$

and hope that μ_2, restricted to some field, will be the right μ. Note that both μ_1, μ_2 satisfy $\mu_i(e_1) \leq \mu_i(e_2)$ if $e_1 \subseteq e_2$ and that $\mu_1(\emptyset) = \mu_2(\emptyset) = 0$.

Let c_1, c_2 be disjoint closed sets in U. Since U is a normal topological space, there exist disjoint open sets d_1, d_2 with $d_1 \supseteq c_1, d_2 \supseteq c_2$. If $d \supseteq c_1 \cup c_2$ is open, then $\lambda(d) \geq \lambda(d \cap d_1) + \lambda(d \cap d_2) \geq \mu_1(c_1) + \mu_2(c_2)$, which implies

$$\mu_1(c_1 \cup c_2) \geq \mu_1(c_1) + \mu_1(c_2).$$

If e_1, e_2 are arbitrary disjoint subsets of U we take $c_1 \subseteq e_1, c_2 \subseteq e_2$ closed and apply the previous inequality; we obtain $\mu_2(e_1 \cup e_2) \geq \mu_1(c_1 \cup c_2) \geq \mu_1(c_1) + \mu_1(c_2)$, so that

$$\mu_2(e_1 \cup e_2) \geq \mu_2(e_1) + \mu_2(e_2). \tag{12.4.11}$$

Let c_1 be closed, d_1 open. Select d open with $d \supseteq c_1 \cap d_1^c$. Then $d_1 \cup d \supseteq c_1$; since λ is positive, $\lambda(c_1) \leq \lambda(d_1 \cup d) \leq \lambda(d_1) + \lambda(d)$, so that $\mu_1(c_1) \leq \lambda(d_1) + \lambda(d)$. Taking the infimum over all open $d \supseteq c_1 \cap d_1^c$,

$$\mu_1(c_1) \leq \lambda(d_1) + \mu_1(c_1 \cap d_1^c) = \lambda(d_1) + \mu_2(c_1 \cap d_1^c), \tag{12.4.12}$$

the last equality because $c_1 \cap d_1^c$ is closed.

We now take another closed set c and d_1 open with $d_1 \supseteq c_1 \cap c$. Since $c_1 \cap d_1^c \subseteq c_1 \cap (c \cap c_1)^c \subseteq c_1 \cap c^c$, we obtain from (12.4.12) that $\mu_1(c_1) \leq \lambda(d_1) + \mu_2(c_1 \cap c^c)$; taking into account that $d_1 \supseteq c_1 \cap c$ is arbitrary,

$$\mu_1(c_1) \leq \mu_1(c_1 \cap c) + \mu_2(c_1 \cap c^c) = \mu_2(c_1 \cap c) + \mu_2(c_1 \cap c^c)$$

where we have used that $\mu_1(c_1 \cap c) = \mu_2(c_1 \cap c)$ since $c_1 \cap c$ is closed. Finally, let e be an arbitrary set, $c_1 \subseteq e$ closed. Replacing c_1 by e on the right side and then taking the supremum in $c_1 \subseteq e$ we obtain

$$\mu_2(e) \leq \mu_2(e \cap c) + \mu_2(e \cap c^c).$$

Combining with (12.4.11) for $e_1 = e \cap c$, $e_2 = e \cap c^c$, we conclude that if e is an arbitrary subset of U and c is closed,

$$\mu_2(e) = \mu_2(e \cap c) + \mu_2(e \cap c^c). \tag{12.4.13}$$

We apply now Carathéodory's Theorem 2.0.1 to the set function μ_2 (note that the only requirement there is $\mu_2(\emptyset) = 0$) and deduce that the class of all the sets that satisfy (12.4.13) for arbitrary sets e (that is, the class of all μ_2-sets in the terminology of **2.0**) is a field Φ which contains all closed sets (thus $\Phi \supseteq \Phi(C)$) and that $\mu = \mu_2$ is a finitely additive measure in Φ, thus, in particular, in $\Phi(C)$. It follows from the second step in the definition (12.4.10) that $\mu_1(c) = \mu_2(c) = \mu(c)$ for c closed, then the second condition (12.4.8) is satisfied, and it is a consequence of the definition of μ_2 that μ is regular.

It results also from (12.4.10) that if c is closed then $\mu(c) = \mu_2(c) = \mu_1(c) \geq \lambda(c)$, hence

$$\lambda(c) \leq \mu(c) \quad (c \text{ closed}), \qquad \lambda(d) \geq \mu(d) \quad (d \text{ open}); \tag{12.4.14}$$

the second inequality results from the first taking complements.

It only remains to show (12.4.9), and we do not lose generality by assuming that $0 \leq f(u) \leq 1$ or that $\mu(U) \leq 1$. Given $\varepsilon > 0$ we choose n such that $1/n \leq \varepsilon$ and use the approximation (12.4.5)-(12.4.6). Then:

(i) Using regularity of μ, we select $c_{nj} \subseteq e_{nj}$ closed with

$$\mu(c_{nj}) \geq \mu(e_{nj}) - \varepsilon/n.$$

(ii) For every $u \in c_{nj}$ we use continuity of $f(u)$ to select an open set containing u and such that $f(u) \geq f_{n,j-1} - \varepsilon/n$. Then we take d_{nj} as the union of these sets for all $u \in c_{nj}$ so that

$$d_{nj} \supseteq c_{nj}, \qquad f(u) \geq f_{n,j-1} - \varepsilon/n \quad (u \in d_{nj}).$$

(iii) Using normality of the topological space U, we construct a family of n pairwise disjoint open sets $b_{nj} \supseteq c_{nj}$; replacing the d_{nj} in (iii) by $d_{nj} \cap b_{nj}$ we may add to (ii) the condition that the d_{nj} are disjoint.

All of this done, we estimate:

$$\int_U f(u)\mu(du) \leq \sum_{j=1}^n (f_{n,j-1} + 1/n)\mu(e_{nj}) \leq \sum_{j=1}^n f_{n,j-1}\mu(e_{nj}) + \varepsilon$$

$$\leq \sum_{j=1}^n f_{n,j-1}\mu(c_{nj}) + 2\varepsilon \leq \sum_{j=1}^n f_{n,j-1}\lambda(d_{nj}) + 2\varepsilon$$

$$\leq \sum_{j=1}^n \int_{d_{nj}} f_n(u)\lambda(du) + 2\varepsilon \leq \int_U f(u)\lambda(du) + 2\varepsilon,$$

where we have used the fact that $\mu(c_{nj}) \leq \mu(d_{nj})$ and the second inequality (12.4.14) for each d_{nj}. This shows that $\int_U f(u)\mu(du) \leq \int_U f(u)\lambda(du)$. Arguing in the same way with $1 - f(u)$ we obtain the opposite inequality. This ends the proof of Theorem 12.4.7, thus that of Theorem 12.4.6.   ∎

Remark 12.4.8. Theorem 12.4.7 does *not* say "given $\lambda \in \Sigma_{ba}(U)$ there exists $\mu \in \Sigma_{rba}(U, \Phi(C))$ with $\mu(e) = \lambda(e)$ for $e \in \Phi(C)$." In fact, if λ is the bogus delta in Example 12.3.5 μ is the Dirac delta at $\bar{u}$, so that $\lambda(\{\bar{u}\}) = 0$, $\mu(\{\bar{u}\}) = 1$.

We denote by $\Sigma_{rca}(U, \Phi(C))$ the space of all regular bounded countably additive measures defined in $\Phi(C)$. Obviously $\Sigma_{rca}(U, \Phi(C)) \subseteq \Sigma_{rba}(U, \Phi(C)) \subseteq \Sigma_{ba}(U)$, and we give $\Sigma_{rca}(U, \Phi(C))$ the norm inherited from $\Sigma_{ba}(U)$. Again, that $\Sigma_{rca}(U, \Phi(C))$ is closed is a byproduct of its identification as a dual in Theorem 12.4.9 below. In it, U is a compact topological space and $C(U)$ is the space of all continuous functions in U equipped with the supremum norm. Recall that in a compact space U every continuous function is bounded, so that $C(U) = BC(U)$.

Theorem 12.4.9 (F. Riesz). *Let U be a compact topological space. Then the dual space $C(U)^*$ can be identified algebraically and metrically with $\Sigma_{rca}(U, \Phi(C))$, equipped with the norm (12.4.1) inherited from $\Sigma_{ba}(U)$. The duality pairing of $C(U)$ and $\Sigma_{rba}(U, \Phi(C))$ is given by (12.4.2).*

Theorem 12.4.9 follows from

Theorem 12.4.10 (Alexandrov). *Let U be a compact topological space, Φ a field of subsets of U containing the closed sets, $\mu \in \Sigma_{rba}(U, \Phi)$. Then μ is countably additive.*

Proof. Arguing as in the proof of Theorem 12.4.7 we limit ourselves to positive measures. Let $\{e_j\}$ be a countable family of pairwise disjoint sets in Φ with $e = \cup e_j \in \Phi$. Finite additivity implies $\mu(e) \geq \mu(e_1) + \mu(e_2) + \cdots + \mu(e_n)$ for all n, so that

$$\mu(e) \geq \sum_{j=1}^{\infty} \mu(e_j). \tag{12.4.15}$$

To estimate the other way, use regularity of μ and select a closed set $c \subseteq e$ and open sets $d_j \supseteq e_j$ such that $\mu(c) \geq \mu(e) - \varepsilon$, $\mu(d_j) \leq \mu(e_j) + \varepsilon/2^n$. Since the space is compact and $c \subseteq \cup d_j$, by Theorem 12.0.12, we can select a finite subfamily such that $c \subseteq d_1 \cup \cdots \cup d_n$. This implies $\mu(e) \leq \mu(c) + \varepsilon \leq \mu(d_1) + \cdots + \mu(d_n) + \varepsilon \leq \mu(e_1) + \cdots + \mu(e_n) + 2\varepsilon$, so that the inequality opposite to (12.4.15) results.   ∎

Although to have μ defined in $\Phi(C)$ is all we need to integrate continuous functions, it is unnatural to use a countably additive measure in a field which is not a σ-field. This can be easily fixed.

Theorem 12.4.11. *Let U be a compact topological space, λ a regular countably additive measure defined in $\Phi(C)$. Then there exists a regular countably additive extension of λ to the σ-field $\Phi_b(C)$ generated by the closed sets of U (the Borel field of U).*

No new work is necessary; the proof can be read off the proof of Theorem 12.4.7. We define μ_1, μ_2 as in (12.4.10) (note that λ need only be defined in the open sets of U). Since λ is regular, $\mu = \mu_1(e) = \mu_2(e) = \lambda(C)$ for $e \in \Phi(C)$. On the other hand, $\mu_2 = \lambda$ is σ-additive, thus by Carathéodory's Theorem 2.0.2 the field Φ of all μ_2-sets is a σ-field (so that $\Phi \supseteq \Phi_b(C)$) and the extension μ_2 is σ-additive in Φ.   ∎

We use from now on the shorthand notations $\Sigma_{rba}(U)$ for $\Sigma_{rba}(U, \Phi(C))$ and $\Sigma_{rca}(U)$ for $\Sigma_{rca}(U, \Phi(C))$.

12.5. Spaces of Relaxed Controls

We introduce three different classes of relaxed controls under increasingly lenient requirements on the control set U.

Class 1: U is a compact metric space.
Class 2: U is a normal topological space.
Class 3: U is an arbitrary set.

The three definitions are similar and based on the characterization $L^1(0, T; X)^* = L_w^\infty(0, T; X^*)$ of the dual of $L^1(0, T; X)$ for an arbitrary Banach space X in Theorem 12.2.11, and on the characterizations $B(U)^* = \Sigma_{ba}(U)$, $BC(U)^* = \Sigma_{rba}(U)$, $C(U)^* = \Sigma_{rca}(U)$ of the duals of $B(U)$, $BC(U)$ and $C(U)$ in **12.4**. Only Definition 1 is (usually) used in finite dimensional systems.

Definition 1. The space $R_c(0, T; U)$ of relaxed controls consists of all $\mu(\cdot)$ in $L_w^\infty(0, T; \Sigma_{rca}(U)) = L^1(0, T; C(U))^*$ that satisfy

(i) if $f(\cdot, \cdot) \in L^1(0, T; C(U))$ is such that $f(t, u) \geq 0$ for $u \in U$ a.e. in $0 \leq t \leq T$ then

$$\int_0^T \int_U f(t, u)\mu(t, du) \, dt \geq 0. \tag{12.5.1}$$

(*ii*) if $\chi(t)$ is the characteristic function of a measurable set $e \subseteq [0, T]$ and $\mathbf{1} \in C(U)$ is the function $\mathbf{1}(u) = 1$, then

$$\int_0^T \int_U (\chi(t) \otimes \mathbf{1}(u))\mu(t, du) \, dt = |e| \qquad (12.5.2)$$

(note that $\chi(\cdot) \otimes \mathbf{1}(\cdot) \in L^1(0, T; C(U))$).

Definition 2. The space $R_{bc}(0, T; U)$ of relaxed controls consists of all $\mu(\cdot)$ in $L_w^\infty(0, T; \Sigma_{rba}(U)) = L^1(0, T; BC(U))^*$ that satisfy (*i*) and (*ii*).

Definition 3. The space $R_b(0, T; U)$ of relaxed controls consists of all $\mu(\cdot)$ in $L_w^\infty(0, T; \Sigma_{ba}(U))^* = L^1(0, T; B(U))^*$ that satisfy (*i*) and (*ii*).

We note that (*ii*) can be generalized to

$$\int_0^T \int_U (\phi(t) \otimes \mathbf{1}(u))\mu(t, du)dt = \int_0^T \phi(t)dt \qquad (12.5.3)$$

for any $\phi(\cdot) \in L^1(0, T)$; it suffices to approximate $\phi(t)$ in the L^1 norm by a sequence of countably valued functions and apply (12.5.2).

The reader will recognize in (*i*) and (*ii*) a disguised way to require that "the values $\mu(t)$ of $\mu(\cdot)$ are probability measures." In fact, (*i*) is "$\mu(t) \geq 0$," while (*ii*) is "$\mu(t, U) = 1$." As we shall see later, (*i*) and (*ii*) imply that $\mu(t)$ is an actual probability measure almost everywhere for any $\mu(\cdot) \in R_c(0, T; U)$ (Theorem 12.5.6). For the classes $R_{bc}(0, T; U)$ and $R_b(0, T; U)$ we can only say that $\mu(t)$ is a probability measure almost everywhere (in fact everywhere) for a certain member of the equivalence class of $\mu(\cdot)$ (Theorem 12.5.7).

Lemma 12.5.1. *Let $\mu(\cdot) \in R_c(0, T; U)$ (resp. $R_{bc}(0, T; U)$, $R_b(0, T; (U))$). Then*

$$\|\mu(\cdot)\| \leq 1^{(1)} \qquad (12.5.4)$$

in $L_w^\infty(0, T; \Sigma_{rca}(U))$ (resp. in $L_w^\infty(0, T; \Sigma_{rba}(U))$, $L_w^\infty(0, T; \Sigma_{ba}(U))$).

Proof. To prove (12.5.4), it is enough to show that

$$\int_0^T \langle f(t), \mu(t) \rangle dt \leq 1$$

for every $f(\cdot, \cdot) \in L^1(0, T; C(U))$ with $\|f\|_{L^1(0, T; C(U))} \leq 1$, and we may limit ourselves to $C(U)$-valued countably valued functions $t \to f(t, \cdot)$ in $L^1(0, T; C(U))$.

---

[1] In this section and the next $\|\mu(\cdot)\|$ denotes the norm of $\mu(\cdot)$ in the corresponding L_w^∞ space, not the function $t \to \|\mu(t)\|$.

Let $f(\cdot) \in C(U)$. Define $f^+(u) = \max(f(u), 0)$. Then $f^+(\cdot) \in C(U)$ and

$$\|f^+\|_{C(U)} \le \|f\|_{C(U)} \quad (f \in C(U)). \tag{12.5.5}$$

Moreover,

$$\|f^+ - g^+\|_{C(U)} \le \|f - g\|_{C(U)} \quad (f, g \in C(U)). \tag{12.5.6}$$

In fact, (12.5.6) follows from the inequality $|f^+(u) - g^+(u)| \le |f(u) - g(u)|$. This inequality is obvious when $f(u), g(u) \ge 0$ or $f(u), g(u) < 0$. If $f(u) \ge 0$ and $g(u) < 0$ then $|f^+(u) - g^+(u)| = f(u), |f(u) - g(u)| = f(u) - g(u)$, thus (12.5.6) holds as well; the symmetric case $f(u) < 0, g(u) \ge 0$ is similar. If we set $f^-(u) = (-f(u))^+$, we have

$$f = f^+ - f^-. \tag{12.5.7}$$

Let $f(\cdot, \cdot)$ be a countably valued function in $L^1(0, T; C(U))$,

$$f(t, u) = \sum \chi_j(t) f_j(u),$$

$\chi_j(\cdot)$ the characteristic function of e_j, $\{e_j\}$ a sequence of pairwise disjoint sets covering $[0, T]$. The function $f^+(t, u) = \Sigma \chi_j(t) f_j^+(u)$ is countably valued and $\|f^+(t)\|_{C(U)} \le \|f(t)\|_{C(U)}$, thus it belongs to $L^1(0, T; C(U))$ as well; in view of (12.5.7), so does $f^-(t, u)$. Using (12.5.1), (12.5.3) and the fact that $\mu(\cdot)$ is an element of the dual of $L^1(0, T; C(U))$ we obtain

$$\int_0^T \int_U f(t, u) \mu(t, du) dt$$

$$= \int_0^T \int_U f^+(t, u) \mu(t, du) dt - \int_0^T \int_U f^-(t, u) \mu(t, du) dt$$

$$\le \int_0^T \int_U \sum \chi_j(t) f_j^+(u) \mu(t, du) dt$$

$$= \sum \int_0^T \int_U \chi_j(t) f_j^+(u) \mu(t, du) dt$$

$$\le \sum \int_0^T \int_U \chi_j(t) \|f_j^+\|_{C(U)} \mu(t, du) dt$$

$$= \sum |e_j| \|f_j^+\|_{C(U)} = \|f^+\|_{L^1(0,T;C(U))} \le \|f\|_{L^1(0,T;C(U))}.$$

This ends the proof of Lemma 12.5.1 for the case $R_c(0, T; U)$. The proof is exactly the same in the other two cases ($R_{bc}(0, T; U)$, $R_b(0, T; U)$). ∎

As seen in **12.2**, the spaces $L_w^\infty(0, T; X^*)$ enjoy special properties when X is separable. This is significant for $R_c(0, T; U)$.

Theorem 12.5.2. *If U is a compact metric space, $C(U)$ is separable.*

We show Theorem 12.5.2 using the Stone–Weierstrass theorem below. A set $\mathcal{F} \subseteq C(U)$ **distinguishes points** in U if and only if, for arbitrary $u, v \in U, u \neq v$, there exists a function $f(\cdot) \in \mathcal{F}$ such that $f(u) \neq f(v)$. A linear subspace $\mathcal{S}$ of $C(U)$ (not necessarily closed) is called a **subalgebra** of $C(U)$ if it contains the product of any two of its elements.

Theorem 12.5.3 (Stone-Weierstrass). *Let U be a compact topological space and let $\mathcal{F}$ be a closed subalgebra of $C(U)$ that contains the function $\mathbf{1}(u) \equiv 1$. Then $\mathcal{F} = C(U)$ if and only if $\mathcal{F}$ distinguishes points in U.*

For the proof of Theorem 12.5.3, see Dunford-Schwartz [1958, p. 272]; we only show that Theorem 12.5.3 implies Theorem 12.5.2. As shown in the proof of Theorem 12.0.13, a compact metric space is separable, that is there exists a countable dense set $\{u_j\}$ in U. Define $f_j(u) = d(u, u_j)$ $(j = 1, 2, \ldots)$ and consider the set $\mathcal{F}$ of all finite linear combinations $\lambda_1 \Pi_1(u) + \lambda_2 \Pi_2(u) + \cdots + \lambda_n \Pi_n(u)$ with real coefficients λ_j, where each $\Pi_k(u)$ denotes either the identity function or a product of a finite number of elements of the sequence $\{f_j(u)\}$. Denote by $\mathcal{F}_r$ the subset of $\mathcal{F}$ corresponding to rational coefficients λ_j. Obviously, $\mathcal{F}$ is a subalgebra of $C(U)$, and $\mathcal{F}_r$ is a countable and dense subset of $\mathcal{F}$ (in the norm of $C(U)$). (Note that $\mathcal{F}_r$ is not a subalgebra, in fact, not even a subspace of $C(U)$.) We take the closure $Cl(\mathcal{F})$ in $C(U)$; $Cl(\mathcal{F}))$ is a closed subalgebra of $C(U)$ and the countable set $\mathcal{F}_r$ is still dense in $Cl(\mathcal{F})$. Obviously, the algebra $Cl(\mathcal{F})$ separates points in U. ∎

Hopes to extend Theorem 12.5.2 to $BC(U)$ collapse in view of

Example 12.5.4. Let U be a normal topological space. Then $BC(U)$ is separable if and only if U is a compact metric space.

The complete proof (of a more general result) is in Dunford-Schwartz [1958, Exercise 16, p. 340]. We limit ourselves to show that if U is a noncompact metric space, then $BC(U)$ cannot be separable. If U is not compact, then there is some $\varepsilon > 0$ such that there exists an infinite sequence $\{u_n\} \subseteq U$ with $d(u_n, u_m) \geq \varepsilon > 0$. Define $f_j(u) = (3/\varepsilon)d(u, B_0(u_n, \varepsilon/3)^c)$ $(B_0(u, r)$ the open ball of center u and radius r). Consider the (uncountable) set of functions of the form

$$f(u) = \sum_{j \in N} f_j(u), \tag{12.5.8}$$

where N is an arbitrary (finite or infinite) subset of the integers $\{1, 2, \ldots\}$. Two distinct elements of this set lie at distance 1, thus there cannot be a countable dense sequence in $BC(U)$.

Example 12.5.5. The space $B(U)$ is separable if and only if U is finite.

In fact, if U is finite, $B(U)$ is $\mathbb{R}^m$ with the supremum norm. Conversely, if U contains an infinite sequence $\{u_j\}$, consider the (uncountable) set A of elements of $B(U)$ defined by (12.5.8) with, say, $f_j(u)$ the characteristic function of $\{u_j\}$.

Theorem 12.5.6. *Let* $\mu(\cdot) \in R_c(0, T; U)$. *Then*

$$\mu(t) \geq 0, \qquad \mu(t, U) = 1 \quad a.e. \ in \ 0 \leq t \leq T. \tag{12.5.9}$$

In particular,

$$\|\mu(t)\| = 1 \quad a.e. \ in \ 0 \leq t \leq T \tag{12.5.10}$$

(norm in $\Sigma_{rca}(U)$*).*

Proof. Since $\mu(t, U) = \langle \mu(t), \mathbf{1} \rangle$, the function $\mu \to \mu(t, U)$ is measurable. If $\mu(t, U) < 1$ in a set of positive measure, we may find $\varepsilon > 0$ and a set e of positive measure such that

$$\mu(t, U) \leq 1 - \varepsilon \quad (t \in e).$$

If $\chi(\cdot)$ is the characteristic function of e, then

$$\int_0^T \int_U (\chi(t) \otimes \mathbf{1}(u))\mu(t, du)dt \leq |e|(1 - \varepsilon),$$

which contradicts *(ii)*. We show in the same way that we cannot have $\mu(t, U) > 1$ in a set of positive measure, thus $\mu(t, U) = 1$ a.e. By Lemma 12.5.1 the norm $\|\mu(\cdot)\|$ of $\mu(\cdot)$ in the space $L_w^\infty(0, T; \Sigma_{rba}(U))$ is ≤ 1, thus by Lemma 12.2.2 $\|\mu(t)\|_{\Sigma_{rca}(U)} \leq 1$ a.e. If t is such that $\mu(t, U) = 1$ and $\|\mu(t)\|_{\Sigma_{rca}(U)} \leq 1$, and there exists[2] $V \in \Phi(C)$ such that $\mu(t, V) < 0$, let W be the complement of V. Then $1 = \mu(t, U) = \mu(t, V) + \mu(t, W)$ whereas $\|\mu(t)\|_{\Sigma_{rca}(U)} = |\mu|(t, U) \geq |\mu(t, V)| + |\mu(t, W)| > 1$, a contradiction. Accordingly, the first condition (12.5.9) holds as well. Obviously, (12.5.10) is a consequence of (12.5.9). ∎

The result for the other two relaxed control spaces is

Theorem 12.5.7. *Let* $\mu(\cdot) \in R_{bc}(0, T; U)$, $R_b(0, T; U)$ *be such that*

$$\|\mu(t)\| \leq 1 \quad a.e. \ in \ 0 \leq t \leq T. \tag{12.5.11}$$

Then (12.5.9) *and* (12.5.10) *hold, the latter in the norms of* $\Sigma_{rba}(U)$, $\Sigma_{ba}(U)$.

---

[2] The measure $\mu(t)$ is defined in a σ-field Φ containing $\Phi(C)$ (see Theorem 12.4.11); we can take $V \in \Phi$ here.

The proof is the same.

The difference between Theorem 12.5.6 and Theorem 12.5.7 is: in the latter we must *assume* (12.5.11); in contrast, this inequality is automatically satisfied by controls in $R_c(0, T; U)$. On the other hand (Remark 12.2.13) every equivalence class in $R_{bc}(0, T; U)$, $R_b(0, T; U)$ contains an element $\mu(\cdot)$ satisfying

$$\sup_{0 \le t \le T} \|\mu(t)\| = \|\mu(\cdot)\| \tag{12.5.12}$$

where $\|\mu(\cdot)\|$ is the norm of $\mu(\cdot)$ in $L_w^\infty(0, T; \Sigma_{rba}(U))$, $L_w^\infty(0, T; \Sigma_{ba}(U))$. It then follows that $\|\mu(t)\| \le 1$ in $0 \le t \le T$ and we have

Theorem 12.5.8. *Let* $\mu(\cdot) \in R_{bc}(0, T; U)$ *(resp.* $R_b(0, T; U)$*). Then there exists* $v(\cdot)$ *in the equivalence class of* $\mu(\cdot)$ *in* $L_w^\infty(0, T; \Sigma_{rba}(U))$ *(resp. in* $L_w^\infty(0, T; \Sigma_{ba}(U))$*) such that*

$$\mu(t) \ge 0, \qquad \mu(t, U) = 1 \quad (0 \le t \le T). \tag{12.5.13}$$

In particular,[3]

$$\|\mu(t)\| = 1 \quad (0 \le t \le T). \tag{12.5.14}$$

The basis of all existence results for relaxed optimal controls is

Theorem 12.5.9. *(a) Let* U *be a compact metric space,* $\{\mu_n(\cdot)\}$ *a sequence in* $R_c(0, T; U)$*. Then there exists a subsequence* $L^1(0, T; C(U))$*-weakly convergent in* $L_w^\infty(0, T; \Sigma_{rca}(U))$ *to* $\mu(\cdot) \in R_c(0, T; U)$*. (b) Let* $\{\mu_\kappa(\cdot)\}$ *be a sequence or generalized sequence in* $R_{bc}(0, T; U)$*,* U *a normal topological space (resp. in* $R_b(0, T; U)$*,* U *an arbitrary set). Then there exists a generalized subsequence* $L^1(0, T; BC(U))$*-weakly convergent in* $L_w^\infty(0, T; \Sigma_{rba}(U))$ *to an element of* $R_{bc}(0, T; U)$ *(resp.* $L^1(0, T; B(U))$*-weakly convergent in* $L_w^\infty(0, T; \Sigma_{ba}(U))$ *to an element of* $R_b(0, T; U)$*).*

Proof. All the convergence statements follow from the fact that all three relaxed control spaces are contained in the unit ball of the dual of a Banach space. In case (a) the space is $L^1(0, T; C(U))$. Since $C(U)$ is separable, so is $L^1(0, T; C(U))$ (Corollary 5.0.28) and the unit ball of $L^1(0, T; C(U))^* = L_w^\infty(0, T; \Sigma_{rca}(U))$ equipped with the $L^1(0, T; C(U))$-weak topology is a compact metric space; Theorem 12.0.2 insures this. To show that the limit of the subsequence belongs to $R_c(0, T; U)$ we must show that $R_c(0, T; U)$ is $L^1(0, T; C(U))$-weakly closed

---

[3] Theorem 12.2.6 (the second part of the Dunford–Pettis Theorem) plays no role here. The only elements of $L_w^\infty(0, T; \Sigma_{rba}(U))$ or of $L_w^\infty(0, T; \Sigma_{ba}(U))$ we use satisfy $\|\mu(t)\| = 1$ (or have an equivalent $v(\cdot)$ that does). Thus, measurability of $t \to \|\mu(t)\|$ is moot.

in $L_w^\infty(0, T; \Sigma_{rca}(U))$. This is obvious, since conditions (12.5.1) and (12.5.2) are preserved through weak convergence. Case (*b*) is similar, but the spaces $L^1(0, T; BC(U))$ and $L^1(0, T; B(U))$ (except for trivial cases) are not separable; the unit ball of the duals, endowed with the weak topology, is a compact topological space (by Alaoglu's Theorem 12.0.17) but is not metric. Thus, we cannot do justice to compactness using only sequences. ∎

12.6. Approximation in Spaces of Measures and Spaces of Relaxed Controls

To avoid restating hypotheses, we agree that every time $C(U), \Sigma_{rca}(U)$, $R_c(0, T; U)$ (resp. $BC(U), \Sigma_{rba}(U), R_{bc}(0, T; U)$) are brought into play U is a compact metric space (resp. a normal topological space). Nothing is required of U if we use $B(U), \Sigma_{ba}(U), R_b(0, T; U)$. Measures in $\Sigma_{rca}(U)$ and $\Sigma_{rba}(U)$ are defined in the field $\Phi(C)$ generated by the closed (or open) sets in U;[1] measures in $\Sigma_{ba}(U)$ are defined in the field Φ of all subsets of U. We denote by $\Pi_{rca}(U)$ (resp. $\Pi_{rba}(U), \Pi_{ba}(U)$) the set of all probability measures μ in $\Sigma_{rca}(U)$ (resp. in $\Sigma_{rba}(U), \Sigma_{ba}(U)$), that is, the set of all measures $\mu \geq 0$ with $\mu(U) = 1$. In terms of these sets, Theorem 12.5.6 shows that every $\mu(\cdot) \in R_c(0, T; U)$ takes values in $\Pi_{rca}(U)$ almost everywhere. The corresponding statement for the classes $R_{bc}(0, T; U), R_b(0, T; U)$ is Theorem 12.5.8, which ensures the existence of a representative in each equivalence class taking values in $\Pi_{rba}(U), \Pi_{ba}(U)$.

We denote the Dirac measure with mass at u by the functional notation $\delta(\cdot - u)$ or by δ_u. The set $D = \{\delta_u; u \in U\}$ of all Dirac measures is a subset of $\Pi_{rca}(U)$, $\Pi_{rba}(U)$ and $\Pi_{ba}(U)$.

Lemma 12.6.1. *The $C(U)$-weak topology of $\Pi_{rca}(U)$ is defined by a metric.*

The proof follows from Lemma 12.0.1 and from the fact that $C(U)$ is separable (Theorem 12.5.2), which implies that the $C(U)$-weak topology of the unit ball of $\Sigma_{rca}(U)$ is defined by a metric. This applies to $\Pi_{rca}(U)$ which is a subset of the unit ball.   ∎

Lemma 12.6.2. $\Pi_{rca}(U)$ *(resp. $\Pi_{rba}(U), \Pi_{ba}(U)$) is $C(U)$-weakly closed in* $\Sigma_{rca}(U)$ *(resp. $\Sigma_{rba}(U), \Sigma_{ba}(U)$).*

Let $\{\mu_n\}$ be a sequence in $\Pi_{rca}(U)$ such that $\mu_n \to \mu$ $C(U)$-weakly. We have $\mu(U) = \langle \mu, \mathbf{1} \rangle = \lim \langle \mu_n, \mathbf{1} \rangle = \lim \mu_n(U) = 1$. Since the unit ball of $\Sigma_{rca}(U)$ is weakly closed, $\|\mu\| \leq 1$. If $\mu(V) < 0$ for some $V \in \Phi(C)$ and $W = V^c$ then $1 = \mu(U) = \mu(V) + \mu(W)$, hence, $\|\mu\| = |\mu|(U) = |\mu|(V) + |\mu|(W) > 1$, a

---

[1] In the case of $\Sigma_{rca}(U)$, each $\mu(t)$ is defined in a σ-field $\Phi(t)$ containing $\Phi(C)$, thus containing the *Borel field* $\Phi_b(C)$ (the intersection of all σ-fields containing $\Phi(C)$). Nothing but continuous functions will be integrated against $\mu(t)$, thus the exact field of definition is irrelevant.

contradiction. The proofs for $\Pi_{rba}(U)$ and Π_{ba} are the same, but we use a generalized sequence $\{\mu_\kappa\}$ instead of a sequence. ∎

Corollary 12.6.3. $\Pi_{rca}(U)$ *(resp. $\Pi_{rba}(U)$, $\Pi_{ba}(U)$) is $C(U)$-weakly compact in $\Sigma_{rca}(U)$ (resp. $BC(U)$-weakly compact in $\Sigma_{rba}(U)$, $B(U)$-weakly compact in $\Sigma_{ba}(U)$).*

In fact, $\Pi_{rca}(U)$ is a closed subset of the unit sphere of $\Sigma_{rca}(U)$; compactness then follows from Theorem 12.0.2. For the other two spaces we use Alaoglu's Theorem 12.0.17. ∎

Lemma 12.6.4. *Let $\overline{\text{conv}}$ denote closed convex hull (closure taken in the weak $C(U)$-topology). Then*

$$\Pi_{rca}(U) = \overline{\text{conv}}(D). \tag{12.6.1}$$

Similar statements for $\Pi_{rba}(U)$, $\Pi_{ba}(U)$.

Proof. $\Pi_{rca}(U)$ is convex and (by Lemma 12.6.2) $C(U)$-weakly closed so that $\overline{\text{conv}}(D) \subseteq \Pi_{rca}(U)$. If we view $\Sigma_{rca}(U)$ as a locally convex space equipped with its weak $C(U)$-topology, then the dual $\Sigma_{rca}(U)^*$ coincides with $C(U)$ as a linear space (Corollary 12.0.10). Let $\mu \notin \overline{\text{conv}}(D)$. Since $\overline{\text{conv}}(D)$ is closed in $\Sigma_{rca}(U)$, applying the strict separation theorem (Corollary 12.0.9) to the sets $\{\mu\}$ and $\overline{\text{conv}}(D)$ we obtain a $f(\cdot) \in B(U)$ such that

$$\langle \mu, f \rangle \geq c, \qquad \langle \delta_u, f(\cdot) \rangle = f(u) \leq c - \varepsilon \quad (u \in U). \tag{12.6.2}$$

Now, if $f^+(u) = \max(f(u), 0)$ then it follows from the second inequality (12.6.2) that $\|f^+(u)\|_{C(U)} \leq c - \varepsilon$. On the other hand, since $\mu \geq 0$ it follows from the first that $\langle \mu, f^+ \rangle \geq \langle \mu, f \rangle \geq c$, so that $\|\mu\| > 1$, a contradiction. The proof for $\Pi_{rba}(U)$, $\Pi_{ba}(U)$ is the same. ∎

Remark 12.6.5. For the spaces $\Pi_{rba}(U)$ and $\Pi_{ba}(U)$, Lemma 12.6.4 says that every $\mu \in \Pi_{rba}(U)$ (resp. $\mu \in \Pi_{ba}(U)$) can be approximated in the $BC(U)$-weak topology (resp. in the $B(U)$-weak topology) by a generalized sequence $\{\mu_\kappa\} \subseteq \text{conv}(D)$. The same holds for $\Pi_{rca}(U)$, but this space is metric in the $C(U)$-weak topology; we can then use an ordinary sequence $\{\mu_n\}$.

The next result will not be used but throws additional light on the deltas as probability measures. If B is a convex subset of a linear space X, an element $x \in B$ is **extremal** if it is not an interior point of any segment contained in B. In other words, if $x = \alpha y + (1 - \alpha)z$ with $y, z \in B, 0 < \alpha < 1$, then $x = y = z$.

Lemma 12.6.6. *Elements of D are extremal points of $\Pi_{rca}(U)$. The same is true for $\Pi_{rba}(U)$, $\Pi_{ba}(U)$.*

Proof. Assume that δ_u is not an extremal. Then there exist $\mu, \nu \in \Pi_{rca}(U)$ such that $\delta_u = \alpha\mu + (1 - \alpha)\nu$ with $0 < \alpha < 1$. Since both μ, ν are positive measures, it is clear that $\mu(U \setminus \{u\}) = \nu(U \setminus \{u\}) = 0$, thus $\mu = \nu = \delta_u$. ∎

It will be shown below (Example 12.6.12) that D is the set of *all* the extremals of $\Pi_{rca}(U)$. This result does not extend to the other spaces (Example 12.6.13); all we can say is that D is dense among the extremals of $\Pi_{rba}(U)$, $\Pi_{ba}(U)$.

At this point, additional generality is helpful. Let X be a Banach space, $U \subseteq X^*$ convex, bounded and X-weakly closed. A set $D \subseteq U$ is **total** in U if $\overline{conv}(D) = U$, where $\overline{conv}$ is closed convex hull in the X-weak topology of X^*. This means that for every $u \in U$ there exists a generalized sequence $\{u_\kappa\} \subseteq conv(D)$ with $u_\kappa \to u$ X-weakly. If X is separable the X-weak topology of the unit ball of X^* (thus of any bounded subset) is given by a metric, thus we may replace the generalized sequence $\{u_\kappa\}$ by a true sequence $\{u_n\}$. We denote by $R(0, T; U)$ the set of all functions $u(\cdot) \in L_w^\infty(0, T; X^*)$ such that there exists $v(\cdot)$ in the equivalence class of $u(\cdot)$ in $L_w^\infty(0, T; X^*)$ such that

$$v(t) \in U \quad \text{a.e. in } 0 \le t \le T. \tag{12.6.3}$$

If X is separable the equivalence relation in $L_w^\infty(0, T; X^*)$ is equality almost everywhere, thus $R(0, T; U)$ consists of all $v(\cdot) \in L_w^\infty(0, T; X^*)$ satisfying (12.6.3). We call $PC(0, T; U)$ the set of all U-valued functions defined in $0 \le t \le T$ and constant in intervals $t_{j-1} \le t < t_j$ where $0 = t_0 < t_1 < \cdots < t_{n-1} < t_n = T$. The space $PC(0, T; D)$ is similarly defined, but the functions are D-valued. In what follows, if $V \subseteq W$ are subsets of a topological space, V is **sequentially dense** in W if every element of W is the limit of a sequence of elements of V.

Theorem 12.6.7. *Let U be convex, bounded and X-weakly closed, $D \subseteq U$ total in U. Then $PC(0, T; D)$ is $L^1(0, T; X)$-weakly dense in $R(0, T; U)$. If X is separable, $PC(0, T; D)$ is $L^1(0, T; X)$-weakly sequentially dense in $R(0, T; U)$.*

The proof has several steps. In the first we denote by $C(0, T; U)$ the space of all (strongly) continuous, U-valued functions $u(\cdot)$ defined in $0 \le t \le T$. In all steps we assume that a function in $R(0, T; U)$ satisfies (12.6.3). This involves, at most, replacing $u(\cdot)$ by another function in the same equivalence class.

Lemma 12.6.8. *The space $C(0, T; U)$ is $L^1(0, T; X)$-sequentially dense in $R(0, T; U)$.*

Proof. Let ϕ be a nonnegative, infinitely differentiable function $\phi(t)$ with support in $|t| \le 1$ and $\int \phi(t)dt = 1$, $\{\phi_n\}$ the mollifier sequence $\phi_n(t) = n\phi(nt)$. Pick

$u(\cdot) \in R(0, T; \mathbf{U})$, extend it to $t < 0$ and $t > 0$ setting $u(t) = u$ there (u a fixed element of $\mathbf{U}$) and for each t define $u_n(t)$ by

$$\langle u_n(t), f \rangle = \int_{-\infty}^{\infty} \phi_n(t - \tau) \langle u(\tau), f \rangle d\tau \quad (f \in X). \tag{12.6.4}$$

Obviously, the functional $f \to \langle u_n(t), f \rangle$ is linear and, since $\mathbf{U}$ is bounded we have $|\langle u_n(t), f \rangle| \leq C\|f\|_X$, hence $u_n(t) \in X^*$. We check next that

$$u_n(t) \in \mathbf{U} \quad (0 \leq t \leq T). \tag{12.6.5}$$

To do this, we recall (Corollary 12.0.10) that X^*, endowed with its X-weak topology is a locally convex space whose dual is X. If (12.6.5) is false for some t we apply the strict separation theorem (Corollary 12.0.9) and obtain an $f \in X$ with

$$\langle u_n(t), f \rangle \geq c, \qquad \langle u(\tau), f \rangle \leq c - \varepsilon \quad \text{a.e. in } 0 \leq \tau \leq T.$$

Multiplying by $\phi_n(t - \tau)$, integrating with respect to τ in $(-\infty, \infty)$, and noting that $\int \phi_n(\tau) d\tau = 1$, an obvious contradiction ensues.

That each $u_n(\cdot)$ is norm continuous in X^* for each n follows from the estimate

$$|\langle u_n(t') - u_n(t), f \rangle| \leq \int_{-\infty}^{\infty} |\phi_n(t' - \tau) - \phi_n(t - \tau)| \|u(\tau)\|_{X^*} \|f\|_X d\tau$$

since $\|u(\tau)\|_{X^*}$ is bounded and $\int |\phi_n(t' - \tau) - \phi_n(t - \tau)| d\tau \to 0$ as $t' - t \to 0$. Finally, the convergence statement is

$$\int_0^T |\langle u_n(t) - u(t), f(t) \rangle| dt \to 0 \quad \text{as } n \to \infty \tag{12.6.6}$$

in $L^1(0, T)$ for every $f(\cdot) \in L^1(0, T; X)$. Since the sequence $\{u_n(\cdot)\}$ is bounded in $L_w^\infty(0, T; X^*)$ we only have to show (12.6.6) for a dense set in $L^1(0, T; X)$, for instance, for $f(\cdot) \in PC(0, T; X)$. Let $[t_{j-1}, t_j)$ be one of the intervals where $f(\cdot) = f$ is constant, and let t be a Lebesgue point of $\langle u(\cdot), f \rangle$ in (t_{j-1}, t_j). For sufficiently large n the support of $\phi_n(t - \cdot)$ is contained in (t_{j-1}, t_j), so that we have

$$|\langle u_n(t) - u(t), f(t) \rangle| = \int_{t-1/n}^{t+1/n} \phi_n(t - \tau) |\langle u_n(\tau) - u(t), f \rangle| d\tau$$

$$\leq n \int_{t-1/n}^{t+1/n} |\langle u_n(\tau) - u(t), f \rangle| d\tau \to 0 \quad \text{as } n \to \infty.$$

We obtain then (12.6.6) from the fact that almost every t is a Lebesgue point of $\langle u(\cdot), f \rangle$ and from the dominated convergence theorem.   ∎

Corollary 12.6.9. *The space $PC(0, T; \mathbf{U})$ is sequentially dense in $R(0, T; \mathbf{U})$.*

The proof is obvious: do the approximation with a sequence $u_n(\cdot)$ in $C(0, T; \mathbf{U})$ and then, using uniform continuity of $u_n(t)$ in $0 \le t \le T$, find $v_n(\cdot) \in PC(0, T; U)$ with $\|u_n(t) - v_n(t)\|_{X^*} \le 1/n$ in $0 \le t \le T$. ∎

Proof of Theorem 12.6.7. In view of Corollary 12.6.9, we only have to show

Lemma 12.6.10. *Let $\mathbf{U}$ be convex, bounded and X-weakly closed, $D \subseteq \mathbf{U}$ total in $\mathbf{U}$. Then $PC(0, T; D)$ is $L^1(0, T; X)$-weakly dense in $PC(0, T; \mathbf{U})$. If X is separable, then $PC(0, T; D)$ is $L^1(0, T; X)$-weakly sequentially dense in $PC(0, T; \mathbf{U})$.*

Proof. Every element $u(\cdot)$ of $PC(0, T; \mathbf{U})$ is constant in each of a finite number of intervals. We can construct the approximations in each interval and piece them together, thus we may assume that $u(\cdot) \equiv u$ is constant. Let $\{u_\kappa\}$ be a bounded generalized sequence in X^* such that $u_\kappa \to u$ X-weakly. Define $u_\kappa(t) \equiv u_\kappa$; then if $f(t) \in L^1(0, T; X)$ we have

$$\langle u_\kappa(\cdot), f(\cdot)\rangle = \int_0^T \langle u_\kappa, f(t)\rangle dt \to \int_0^T \langle u, f(t)\rangle dt \qquad (12.6.7)$$

by the dominated convergence theorem. Accordingly, since D is total in $\mathbf{U}$ we only have to prove the approximation for a function $u(\cdot)$ of the form

$$u(t) \equiv u = \sum_{k=1}^m \alpha_k u^k \quad (\alpha_k \ge 0, \Sigma\alpha_k = 1, u^k \in D). \qquad (12.6.8)$$

Take a partition $\{t_j\}, t_j = jh = jT/n$ ($j = 0, 1, \ldots, n$) of the interval $[0, T]$ and divide each interval $[t_{j-1}, t_j]$ into m subintervals I_{jk} of length $|I_{jk}| = \alpha_k h$. Define $u_n(\cdot) \in PC(0, T; D)$ as follows: $u_n(t) = u^k$ ($t \in I_{jk}, 1 \le j \le n, 1 \le k \le m$). Then, if $f(\cdot) \in C(0, T; X)$,

$$\langle u_n(\cdot), f(\cdot)\rangle = \sum_{j=1}^n \sum_{k=1}^m \int_{I_{jk}} \langle u^k, f(t)\rangle dt =$$

$$\to \sum_{k=1}^m \alpha_k \int_0^T \langle u^k, f(t)\rangle dt = \langle u(\cdot), f(\cdot)\rangle \quad \text{as } n \to \infty.$$

Since $C(0, T; X)$ is dense in $L^1(0, T; X)$ and the $u_n(\cdot)$ are bounded in $L_w^\infty(0, T; X)$ we deduce that $u_n(\cdot) \to u$ $L^1(0, T; X)$-weakly in $L_w^\infty(0, T; X^*)$. This ends the proof of Lemma 12.6.10 and thus that of Theorem 12.6.7. Note that all approximations are sequential except (12.6.7), where $u \in \mathbf{U}$ and $u_\kappa \in \text{conv}D$; this one is sequential only if X is separable. ∎

We can apply Theorem 12.6.7 to:

(a) $X = C(U), X^* = \Sigma_{rca}(U), \mathbf{U} = \Pi_{rca}(U), D = \{\text{deltas}\}$ (U a compact metric space); here $R(0, T; \mathbf{U}) = R_c(0, T; U)$.

(b) $X = BC(U), X^* = \Sigma_{rba}(U), \mathbf{U} = \Pi_{rba}(U), D = \{\text{deltas}\}$ (U a normal topological space); here $R(0, T; \mathbf{U}) = R_{bc}(0, T; \mathbf{U})$.

(c) $X = B(U), X^* = \Sigma_{ba}(U), \mathbf{U} = \Pi_{ba}(U), D = \{\text{deltas}\}$ (U an arbitrary set); here $R(0, T; \mathbf{U}) = R_b(0, T; U)$.

In the first case, $C(U)$ is separable, thus the denseness in Lemma 12.6.10 and Theorem 12.6.7 is sequential.

Example 12.6.11. Let Q be a compact set in a locally convex space E such that $\overline{\text{conv}}(Q)$ is compact. Then every extremal point of $\overline{\text{conv}}(Q)$ belongs to Q (Dunford-Schwartz [1958, p. 440]).

Example 12.6.12. Elements of D are *exactly* the extremal points of $\Pi_{rca}(U)$. In fact, in view of Lemma 12.6.6 it is enough to show that any possible extremal of $\Pi_{rca}(U)$ belongs to D, and this results from Example 12.6.11 (applied to $\Sigma_{rca}(U)$ equipped with the $C(U)$-weak topology) if we can show that D is $C(U)$-weakly compact. This is obvious: if $\{\delta_{u_n}\}$ is a sequence in D then, passing to a subsequence we may assume that $u_n \to u \in U$, which implies $\delta_{u_n} \to \delta_u$ $C(U)$-weakly. This observation applies equally well to a compact (not necessarily metric) space U; we use generalized sequences instead of sequences.

The argument in Example 12.6.12 is inapplicable in $\Sigma_{rba}(U)$ if U is not compact or in $\Sigma_{ba}(U)$ if U is not finite, thus one may suspect D does not exhaust the set of extremals. Example 12.6.13 confirms this.

Example 12.6.13. Let U be a topological space. A compact topological space V is said to be a **compactification** of U if there exists a 1–1 continuous map $i : U \to V$ such that (a) $i(U)$ is dense in V, (b) $i^{-1} : i(U) \to U$ is continuous (intuitively, "U can be made a dense subset of a compact space and inherits its topology.") If U is a normal topological space there a exists a compactification V of U called the **Stone-Čech compactification** (Kelley [1955, p. 153]) that satisfies (c) every continuous bounded function $f : U \to R$ can be extended to a function $g : V \to R$ with the same supremum (since U is dense in V, this extension is unique). In these conditions it is clear that

$$BC(U) \approx C(V) \tag{12.6.9}$$

with equality of norms, so that

$$\Sigma_{rba}(U) = BC(U)^* \approx C(V)^* = \Sigma_{rca}(V) \tag{12.6.10}$$

also with equality of norms. It follows from Lemma 12.6.6 that δ_v is extremal
in the unit ball of $\Sigma_{rca}(V) = \Sigma_{rba}(U)$ for arbitrary $v \in V$. If U is not compact
then $U \neq V$, and if $v \notin U$ we cannot have $\delta_v = \delta_u$ for some $u \in U$ since V, being
compact, is normal (Kelley [1955, p. 141]) and by Urysohn's lemma we may
construct a continuous function in V taking different values in u and v.

To apply a similar argument to $\Pi_{ba}(U)$ we appoint U with the topology where
all subsets of U are open. Under this topology all functions are continuous, so that
$BC(U) = B(U)$. The space U is trivially normal, so that all of the above applies
if U is not compact (this means U is infinite).

As a byproduct, we obtain that D is dense in the extremals of $\Pi_{rba}(U)$ (resp. of
$\Pi_{ba}(U)$). This can also be proved directly using Example 12.6.11 with $Q = \overline{D}$.

12.7. Topology and Measure Theory

A topological space U has the **measurable image property** if for every continuous
function $f : U \to \mathbb{R}$ the image $f(U)$ is Lebesgue measurable. Compact spaces
have this property; if U is compact $f(U)$ is compact, hence closed, hence measurable. It does not seem out of line to conjecture that Lebesgue measurable sets in
$\mathbb{R}$ (with the topology inherited from $\mathbb{R}$) have the measurable image property. This
turns out to be false even for Borel sets (Bourbaki [1966, Chapter 9, Historical
Note, p. 271]). This apparent anomaly was one of the motivations for development
of the surprisingly rich theory of spaces having the measurable image property.
We limit ourselves to the following result:

Measurable Image Theorem 12.7.1. *Separable complete metric spaces have the*
measurable image property.

Below, we consider diverse metric spaces (U, d), (V, ρ), $\ldots$, sometimes indicated simply as $U, V, \ldots$. Each U is a topological space (with the topology
generated by the metric) and some of the statements impinge only on the topology
of U, not on the metric. For instance, the statement "$f : U \to V$ is a continuous
function from the space U into the space V" is topological, since continuity of f
only depends on the open sets of U and V or, equivalently, on their convergent
sequences. On the other hand, the statement "U is complete" is not topological,
since the notion of Cauchy sequence depends on the metric.

Example 12.7.2. Let $U = [0, 1)$ equipped with the ordinary distance $d(u, v) = |u - v|$. Then U is not complete (the sequence $\{1 - 1/n\}$ is Cauchy but not con-
vergent in U). On the other hand, the same space endowed with the distance

$$\rho(u, v) = \left| \frac{1}{1-u} - \frac{1}{1-v} \right|$$

has the same convergent sequences (thus is topologically equivalent to (U, d)) but is complete (a sequence $\{u_n\}$ with $u_n \to 1$ cannot be a Cauchy sequence).

Let $I = [0, 1]$. We denote by I^∞ the topological product of I with itself a countable number of times, that is the space of all sequences $\mathbf{r} = \{r_1, r_2, \ldots\}$, $\mathbf{s} = \{s_1, s_2, \ldots\}, r_j, s_j \in I$, equipped with the distance

$$\rho(\mathbf{r}, \mathbf{s}) = \sum_{m=1}^{\infty} 2^{-m} |r_m - s_m|.$$

We check easily that I^∞ is a complete metric space; convergence of a sequence $\{\mathbf{r}_n\} = \{r_{mn}\}$ to $\mathbf{r} = \{r_m\}$ as $n \to \infty$ in this space is equivalent to convergence of $r_{mn} \to r_n$ as $n \to \infty$ for all m.

Lemma 12.7.3. I^∞ *is compact and separable.*

Proof. Compactness follows from Tijonov's Theorem 12.0.15, but a simpler proof along the lines of Theorem 12.0.2 is possible. Let $\{\mathbf{r}_n\} = \{r_{mn}\}$ be a sequence in I^∞. By compactness of the interval I, we may select a subsequence such that $\{r_{1n}\}$ is convergent, then a subsequence such that $\{r_{2n}\}$ is convergent, ... and so on. Using then the Cantor diagonal sequence trick we end up with a subsequence (denoted with the same symbol) such that $r_{mn} \to r_m \in I$ as $n \to \infty$. This implies that $\mathbf{r}_n \to \mathbf{r} = \{r_m\}$ as $n \to \infty$.

To show separability, note that the (countable) subset of all sequences $\{r_m\}$ with r_m rational, $r_m = 0$ except a finite number of indices, is dense in I^∞. ∎

A **homeomorphism** between two topological spaces U and V is a continuous 1–1 function ϕ from U onto V such that the inverse ϕ^{-1} is as well continuous.

Theorem 12.7.4. *Let (U, d) be a separable metric space. Then there exists a homeomorphism ϕ from U onto a subspace V of I^∞.*

Proof. We assume that the distance $d(u, v)$ in U satisfies $d(u, v) \le 1$; if not so, we replace $d(u, v)$ by the distance $d'(u, v) = d(u, v)/(1 + d(u, v))$ which is topologically equivalent to d (that is, generates the same topology). Let $\{u_m\}$ be a countable dense set in U. Define a map $\phi : U \to I^\infty$ by

$$\phi(u) = \{d(u, u_1), d(u, u_2), \ldots\}.$$

Since $|d(u, u_m) - d(v, u_m)| \le d(u, v)$ we have

$$\rho(\phi(u), \phi(v)) = \sum_{m=1}^{\infty} 2^{-m} |d(u, u_m) - d(v, u_m)| \le \sum_{m=1}^{\infty} 2^{-m} d(u, v) \le d(u, v);$$

accordingly ϕ is Lipschitz continuous. It is also one-to-one; in fact, if $\phi(u) = \phi(v)$ we have $d(u, u_m) = d(v, u_m)$ for all m and, in view of the denseness of $\{u_m\}$, $u = v$. It only remains to prove that ϕ^{-1} is continuous from $V = \phi(U)$ into U. To do this, fix $u \in U$ and let $v \in U$ be such that $d(u, v) \geq 3\varepsilon > 0$. Take u_n with $d(u, u_n) \leq \varepsilon$ (so that $d(v, u_n) \geq 2\varepsilon$). Then $d(v, u_n) - d(u, u_n) \geq \varepsilon$, and

$$\rho(\phi(u), \phi(v)) \geq 2^{-n}|d(v, u_n) - d(u, u_n)| \geq 2^{-n}\varepsilon.$$

Equivalently, $\rho(\phi(u), \phi(v)) < 2^{-n}\varepsilon$ implies $d(u, v) < 3\varepsilon$. Since n depends on u but not on v, the proof is finished.   ∎

Theorem 12.7.5. *Let U be a separable complete metric space, ϕ the map in Theorem 12.7.4. Then, if $V = \phi(U)$ there exists a sequence $\{V_n\}$ of subsets of I^∞ such that (a) each V_n is open, (b) each V_n is a countable union of compact sets in I^∞, (c) $V_1 \supseteq V_2 \supseteq \cdots \supseteq V$ and*

$$V = \bigcap_{n=1}^{\infty} V_n. \tag{12.7.1}$$

Proof. Let $Cl(V)$ be the closure of V in the topology of I^∞. Denote by $V_n \subseteq Cl(V)$ the set of all $\mathbf{r} \in Cl(V)$ such that there exists $\varepsilon = \varepsilon(\mathbf{r})$ such that if $B_0(\mathbf{r}, \varepsilon)$ is the open ball of center $\mathbf{r}$ and radius ε in I^∞ the d-diameter of $\phi^{-1}(B_0(\mathbf{r}, \varepsilon) \cap V)$ in U is $< 1/n$: this means

$$\rho(\phi(u), \mathbf{r}) < \varepsilon, \quad \rho(\phi(v), \mathbf{r}) < \varepsilon \qquad \text{implies } d(u, v) < 1/n. \tag{12.7.2}$$

That $V_1 \supseteq V_2 \supseteq \ldots$ is obvious. Clearly, each V_n is open in I^∞. If $\mathbf{r} = \phi(w)$, it follows from continuity of ϕ^{-1} at $\phi(w)$ that (12.7.2) will hold for sufficiently small ε, thus $V_n \supseteq V$ for all n and we have the inclusion $\subseteq$ in (12.7.1).

To show the opposite inclusion, let $\mathbf{r} \in \bigcap V_n$. Then there exists a sequence of open balls $B_0(\mathbf{r}, \varepsilon_n)$ in I^∞ such that diam $\phi^{-1}(B_0(\mathbf{r}, \varepsilon) \cap V) \leq 1/n$, and we may clearly assume that $\varepsilon_n \to 0$. Select a sequence $\{u_n\} \subseteq U$ such that $\phi(u_n) \in B_0(\mathbf{r}, \varepsilon_n)$ (that is, $u_n \in \phi^{-1}(B_0(\mathbf{r}, \varepsilon_n) \cap V)$; note that, since $\mathbf{r} \in Cl(V)$, $B_0(\mathbf{r}, \varepsilon_n) \cap V$ is nonempty, so that $\phi^{-1}(B_0(\mathbf{r}, \varepsilon_n) \cap V)$ is nonempty). Then $\{u_n\}$ is a Cauchy sequence, hence it is convergent: $u_n \to u \in U$. Accordingly, $\phi(u_n) \to \phi(u)$, which, shows that $\mathbf{r} = \phi(u) \in V$ and establishes (12.7.1).

It only remains to show that each V_n is a countable union of compact sets. In fact, any open set $W \subseteq I^\infty$ has this property. To see this, let $\{\mathbf{r}_n\}$ be a countable dense subset of W. Then every $\mathbf{r} \in W$ belongs to a closed ball $B(\mathbf{r}_n, \varepsilon_n) \subseteq W$ with rational radius: take the union of these balls, each of which is compact since I^∞ is compact.   ∎

Corollary 12.7.6. *Let U be a separable complete metric space, Y a compact metric space, $f : U \to Y$ a continuous function. Then there exists a compact metric space*

K, a continuous function $g : K \to Y$ and a sequence $\{F_n\}$ of subsets of K such that (a) each F_n is a countable union of compact sets in K, (b) $F_1 \supseteq F_2 \supseteq \ldots$ and

$$f(U) = g(F) \quad \left(F = \bigcap_n F_n \right). \tag{12.7.3}$$

Proof. Using Theorem 12.7.4, it is enough to do the proof for $V = \phi(U)$ endowed with the topology inherited from I^∞ and for the function $f \circ \phi^{-1}$; we may then assume that $U = V$. Let $g : I^\infty \times Y \to Y$ be the projection of $I^\infty \times Y$ into Y, $F = \{(u, f(u)); u \in U\} \subseteq U \times Y \subseteq I^\infty \times Y$ the graph of f, $K = Cl(F) \subseteq I^\infty \times Y$. (note that U may not be closed in I^∞, so that F may not be closed in $I^\infty \times Y$). The spaces and maps can be visualized this way:

$$U \xrightarrow{f} f(U) \subseteq Y \supseteq g(F) \xleftarrow{g} F \subseteq K = Cl(F) \subseteq I^\infty \times Y.$$

We obviously have $g(F) = f(U)$, thus it only remains to show that

$$F = \bigcap_n F_n \tag{12.7.4}$$

with each F_n a countable union of compact sets. By Theorem 12.7.5, U is the intersection of a decreasing sequence V_n of open subsets of I^∞, each of them a countable union of compact sets. If we take

$$F_n = Cl(F) \cap (V_n \times Y),$$

then (12.7.4) holds. In fact, the inclusion $\subseteq$ is obvious. For the opposite inclusion, note that if $(u, y) \in Cl(F)$ then $u = \lim u_n$ in U, $y = \lim f(u_n)$ in Y. If, in addition, u belongs to all V_n then $u \in U$ and continuity of the function f insures that $y = f(u)$, so that $(u, y) = (u, f(u)) \in F$.

Write $V_n = \bigcup_k C_{nk}$, where each C_{nk} is compact. Then $Cl(F) \cap (V_n \times Y) = \bigcup_k (Cl(F) \cap (C_{nk} \times Y))$ with each $Cl(F) \cap (C_{nk} \times Y)$ compact. This ends the proof. $\blacksquare$

We note a few facts on images of unions and intersections. If $\{Q_\kappa\}$ is an arbitrary family of subsets of a set X and $f : X \to Y$ then

$$f\left(\bigcup_\kappa Q_\kappa \right) = \bigcup_\kappa f(Q_\kappa), \tag{12.7.5}$$

$$f\left(\bigcap_\kappa Q_\kappa \right) \subseteq \bigcap_\kappa f(Q_\kappa), \tag{12.7.6}$$

with possibility of strict inclusion in (12.7.6) (take $\{Q_\kappa\} = \{Q_1, Q_2\}$ with Q_1, Q_2 nonempty and disjoint, $f \equiv$ constant). However, if X and Y are metric spaces,

f is continuous and $\{C_k\}$ is a sequence of compact subsets of X with $C_1 \supseteq C_2 \supseteq \ldots$ we have

$$f\left(\bigcap_k C_k\right) = \bigcap_k f(C_k). \tag{12.7.7}$$

In fact, let $y = f(x_k)$ for all k, where $x_k \in C_k$. By compactness of C_1, $\{x_k\}$ must have a convergent subsequence (denoted also $\{x_k\}$) which, except for a finite number of terms belongs to every C_n. It follows that $x_k \to x \in \bigcap C_n$ so that $y = f(x)$ belongs to the left-hand side of (12.7.7).

Proof of Theorem 12.7.1. Let U be a separable complete metric space, $f : U \to \mathbb{R}$ a continuous function. The objective is to show that $f(U) \subseteq \mathbb{R}$ is a measurable set. We note first that it is enough to show the result for f *bounded*. In fact, if $I_n = f(U) \cap [n, n+1]$, the set $U_n = f^{-1}(I_n)$ is closed in U, thus a separable complete metric space in its own right and f is continuous and bounded in U_n with $f(U_n) = I_n$. Accordingly, if each I_n is measurable so is $f(U)$.

Assuming then that f is bounded, let Y be the compact metric space $Cl(f(U)) \subseteq \mathbb{R}$. Apply Corollary 12.7.6 to express $f(U)$ in the form (12.7.3), where $g : K \to Y$ is continuous, K is a compact metric space and $\{F_n\}$ is a sequence of subsets of K satisfying the provisos there. Then it is clear that we may forget f entirely and just show

Theorem 12.7.7. *Let K be a compact metric space, $g : K \to \mathbb{R}$ a continuous function and $\{F_n\}$ a sequence of subsets of K, each a countable union of compact sets and such that $F_1 \supseteq F_2 \supseteq \ldots$. Let*

$$F = \bigcap_n F_n. \tag{12.7.8}$$

Then $g(F)$ is Lebesgue measurable.

Proof. We have

$$F_n = C_{n1} \cup C_{n2} \cup \ldots \tag{12.7.9}$$

with C_{nj} compact; for notational simplicity we assume all sequences $\{F_n\}$ and $\{C_{nj}\}$ are infinite. Using (12.7.5) we have $g(F_n) = g(C_{n1}) \cup g(C_{n2}) \cup \ldots$ with $g(C_{nj})$ compact, thus each $g(F_n)$ is a Borel set in $\mathbb{R}$. On the other hand, (12.7.6) implies

$$g(F) \subseteq G = \bigcap_{n=1}^{\infty} g(F_n)$$

where the set G is Borel, hence measurable; since $g(F_1) \supseteq g(F_2) \supseteq \ldots$ we have

$$|G| = \lim |g(F_n)|. \tag{12.7.10}$$

Measurability of $g(F)$ will follow if, for every $\varepsilon > 0$ we can find a set $C \subseteq F$ such that $g(C) \subseteq g(F)$ is measurable and

$$|g(C)| \geq |G| - \varepsilon. \tag{12.7.11}$$

In fact, if (12.7.11) holds for all ε then $G = g(F) \cup \{\text{null set}\}$, so that $g(F)$ is measurable.

We manufacture C as follows. Assume we can construct a sequence $\{D_n\}$ of compact subsets of K with

$$D_n \subseteq F_n, \quad |g(F_n)\backslash g(D_n)| \leq 2^{-n}\varepsilon \quad (n = 1, 2, \ldots) \tag{12.7.12}$$

and let $C_n = D_1 \cap D_2 \cap \cdots \cap D_n$. Then $C_n \subseteq F_n$ and

$$F_n \subseteq C_n \cup (F_n\backslash D_1) \cup (F_n\backslash D_2) \cup (F_n\backslash D_n)$$
$$\subseteq C_n \cup (F_1\backslash D_1) \cup (F_2\backslash D_2) \cup (F_n\backslash D_n).$$

Applying g to both sides, using (12.7.5) and taking measures,

$$|g(F_n)| \leq |g(C_n)| + \varepsilon. \tag{12.7.13}$$

Let now $C = \bigcap_n C_n$. Then, since $C_n \subseteq F_n$ we have $C \subseteq F$. Moreover, each C_n is compact and $C_n \supseteq C_{n+1}$, thus we have

$$g(C) = \bigcap_{n=1}^{\infty} g(C_n).$$

by (12.7.7), and, since $g(C_n) \supseteq g(C_{n+1})$,

$$|g(C)| = \lim |g(C_n)|. \tag{12.7.14}$$

We take limits in (12.7.13), using (12.7.10) on the left and (12.7.14) on the right. The result is (12.7.11).

It only remains to show that a sequence $\{D_n\}$ satisfying (12.7.12) can be constructed. For this it suffices to note that

$$\bigcup_{j=1}^{\infty} g(C_{nj}) = g\left(\bigcup_{j=1}^{\infty} C_{nj}\right) = g(F_n)$$

so that we may take as D_n the finite union of sufficiently many of the C_{nj}. ∎

12.8. The Filippov Implicit Function Theorem

A **set valued function** (not to be confused with a *set function*) assigns subsets $U(t)$ of a set U to each value of the variable t in a set e. A set valued function

$t \to U(t) \subseteq U =$ metric space, defined in a measurable set e is **measurable** if for every closed subset C of U the set

$$\{t \in e; U(t) \cap C \neq \emptyset\} \tag{12.8.1}$$

is measurable. This implies measurability of the set $\{t \in e; U(t) \cap D \neq \emptyset\}$ for every open set D as well. In fact, if D is open, let D^c be its complement and $C_k = \{u \in U; \mathrm{dist}(u, D^c) \geq 1/k\}$. Then each C_k is closed, and we have

$$\{t \in e; U(t) \cap D \neq \emptyset\} = \bigcup_{n=1}^{\infty} \bigcap_{k=n}^{\infty} \{t \in e; U(t) \cap C_k \neq \emptyset\}. \tag{12.8.2}$$

A function $u(t)$ defined in a measurable set e with values in a metric space U is **countably valued** if it is constant in each of the members of a pairwise disjoint countable family $\{e_j\}$ of measurable sets covering e. A function $u(\cdot)$ is **strongly measurable** if there exist a null set $e_0 \subseteq e$ and a sequence $u_n(\cdot)$ of countably valued functions such that $u_n(t) \to u(t)$ uniformly in $e \setminus e_0$. These definitions are literal extensions of those for Banach space valued functions (**5.0**).

A strongly measurable U-valued function $t \to u(t)$ is a **measurable selection** of the set valued function $t \to U(t)$ if

$$u(t) \in U(t) \quad \text{a.e. in } e. \tag{12.8.3}$$

Selection Theorem 12.8.1. *Assume that the metric space U is separable and complete, that each $U(t)$ is nonempty and closed and that $t \to U(t)$ is measurable. Then there exists a measurable selection of the set valued function $U(\cdot)$.*

Proof. Denote by $\rho(u, v)$ the distance in U and by $\mathrm{dist}(u, V) = \rho(u, V) = \inf_{v \in V} d(u, v)$ the distance from the point $u \in U$ to the set $V \subseteq U$. The function $t \to \rho(u, U(t))$ is measurable; in fact, if $B_0(u, \varepsilon)$ is the open ball of center u and radius ε,

$$\{t \in e; \rho(u, U(t)) < \varepsilon\} = \{t \in e; U(t) \cap B_0(u, \varepsilon) \neq \emptyset\}.$$

Let $\{u_j; j = 1, 2, \ldots\}$ be a countable dense set in U. Define a function $u_0(t) : e \to U$ by

$$u_0(t) = \text{the first } u_j \text{ such that } \rho(u_j, U(t)) \leq 1.$$

The function $u_0(t)$ is a countably valued function; in fact, it assumes countably many values, and the set where it assumes each value is measurable, since

$$\{t \in e; u_0(t) = u_j\}$$
$$= \{t \in e; \rho(u_j, U(t)) \leq 1\} \cap \bigcap_{k=1}^{j-1} \{t \in e; \rho(u_k, U(t)) > 1\}.$$

The function $t \to \rho(u_0(t), U(t))$ is measurable since it coincides with the measurable function $\rho(u_j, U(t))$ in the (measurable) set where $u_0(t) = u_j$; in particular, the set

$$e_0 = \{t \in e; u_0(t) \in U(t)\} = \{t \in e; \rho(u_0(t), U(t)) = 0\}$$

is measurable. Finally, it is plain that

$$\rho(u_0(t), U(t)) \leq 1 \quad (t \in e). \tag{12.8.4}$$

Let $1/2 < \beta < 1$. We define $u_1(t)$ as follows. If $t \in e_0$, we set $u_1(t) = u_0(t)$. Otherwise,

$$u_1(t) = \text{ the first } u_j \text{ with}$$

$$\rho(u_j, u_0(t)) \leq \beta \rho(u_0(t), U(t)), \qquad \rho(u_j, U(t)) \leq \beta \rho(u_0(t), U(t)).$$

(note that, on account of (12.8.4) and of the fact that $\beta > 1/2$ the open sets $\{u; \rho(u, u_0(t)) < \beta \rho(u_0(t), U(t))\}$ and $\{u; \rho(u, U(t)) < \beta \rho(u_0(t), U(t))\}$ have nonempty intersection). We show in the same way that $u_1(\cdot)$ is a countably valued function, and that

$$\rho(u_1(t), U(t)) \leq \beta \rho(u_0(u), U(t)) \leq \beta.$$

We construct $u_2(t)$ from $u_1(t)$ exactly in the same way $u_1(t)$ was constructed from $u_0(t)$ and continue the process recursively. At the end, we have a sequence $\{u_n(\cdot)\}$ of countably valued functions with

$$\rho(u_n(t), U(t)) \leq \beta \rho(u_{n-1}(\cdot), U(t)) \leq \cdots \leq \beta^n \rho(u_0(t), U(t)) \leq \beta^n, \tag{12.8.5}$$

$$\rho(u_{n+1}(t), u_n(t)) \leq \beta \rho(u_n(t), U(t)) \leq \cdots \leq \beta^{n+1} \rho(u_0(t), U(t)) \leq \beta^{n+1}. \tag{12.8.6}$$

The second chain of inequalities implies that $\{u_n(t)\}$ is a Cauchy sequence in U independently of t, that is, for each $\varepsilon > 0$ there exists n_0 such that

$$\rho(u_m(t), u_n(t)) \leq \varepsilon \quad (t \in e) \quad \text{for } m, n \geq n_0. \tag{12.8.7}$$

Due to completeness of U, $\{u_n(t)\}$ is convergent to a function $u(t)$ pointwise; letting $n \to \infty$ in (12.8.7) we see that convergence is uniform in e. Thus, by definition, $u(t)$ is a strongly measurable function. Inequalities (12.8.5) imply in turn that $u(t) \in U(t)$ as claimed. This ends the proof. ∎

A function $f(\cdot)$ defined in a metric space U with values in another metric space X is **locally uniformly continuous** if, for every $u \in U$, $f(\cdot)$ is uniformly continuous in some ball $B(u, \delta), \delta > 0$.

Implicit Function Theorem 12.8.2. *Let U be a complete separable metric space, X a metric space, e a bounded measurable set in $\mathbb{R}$, $\phi : e \times U \to X$ a function such that*

(a) $u \to \phi(t, u)$ *is locally uniformly continuous in u for each t fixed*

(b) $t \to \phi(t, u)$ *is strongly measurable in t for every $u \in U$ fixed.*

Let $g(\cdot)$ be a strongly measurable X-valued function such that

$$g(t) \in \phi(t, U) \quad a.e. \text{ in } e. \tag{12.8.8}$$

Then there exists a strongly measurable U-valued function $u(\cdot)$ such that

$$\phi(t, u(t)) = g(t) \quad a.e. \text{ in } e. \tag{12.8.9}$$

The proof requires a train of auxiliary results.

Lemma 12.8.3. *Let $f(t)$, $g(t)$ be strongly measurable X-valued functions defined in e. Then $\rho(f(t), g(t))$ is measurable in e.*

Proof. Assume first that $f(t)$, $g(t)$ are countably valued. Then there exists a partition of e in (countably many) measurable sets where both $f(t)$, $g(t)$ are constant; accordingly $d(f(t), g(t))$ is a countably real valued function. In the general case, let $\{f_n(\cdot)\}$, $\{g_n(\cdot)\}$ be the sequences of countably valued functions in the definition of strong measurability; we have $\rho(f_n(t), g_n(t)) \to \rho(f(t), g(t))$ uniformly outside a set of measure zero, thus the result follows. ∎

The next three results improve a given assumption in a bounded measurable set e to something better in a subset $c \subseteq e$ "close to e" in the sense that $|e \setminus c| \le \varepsilon$ or, equivalently,

$$|c| \ge |e| - \varepsilon. \tag{12.8.10}$$

In Egorov's theorem, almost everywhere convergence is upgraded to uniform convergence; in Lusin's and Scorza Dragoni's theorems, measurability becomes continuity. In all results, c is either a countable intersection $c = \bigcap_n c_n$ or a subset of a countable intersection $d = \bigcap_n d_n$ with $c_n, d_n \subseteq e$. Note that if we prove that $|c_n| \ge |e| - \varepsilon/2^n$ ($n = 1, 2, \ldots$) then $|e \setminus c_n| \le \varepsilon/2^n$ and $e \setminus c = \bigcup_n (e \setminus c_n)$ so that $|e \setminus c| \le \varepsilon$, which is (12.8.10). Likewise, if we prove that $|d_n| \ge |e| - \varepsilon/2^{n+1}$ then $|d| \ge |e| - \varepsilon/2$, thus (12.8.10) results taking $c \subseteq d$ with $|c| \ge |d| - \varepsilon/2$.

Theorem 12.8.4 (D. F. Egorov). *Let $\{g_n(\cdot)\}$ be a sequence of strongly measurable functions in a bounded measurable set e with values in a metric space X such that $g_n(t) \to g(t)$ almost everywhere. Then, given $\varepsilon > 0$, there exists a closed set $c \subseteq e$ such that (12.8.10) holds and $g_n(t) \to g(t)$ uniformly in c. The function $g(\cdot)$ is strongly measurable.*

Proof. Modifying $g(\cdot)$, $g_n(\cdot)$ in the null set where there is no convergence we may assume that $g_n(t) \to g(t)$ for all $t \in e$. This ensures that for each $m = 1, 2, \ldots$

$$\bigcup_{n=1}^{\infty} \bigcap_{j,k=n}^{\infty} \{t \in e; \rho(g_j(t), g_k(t)) \leq 1/m\} = \bigcup_{n=1}^{\infty} d_n(m) = e,$$

where each $d_n(m)$ is measurable by Lemma 12.8.3. Since $d_1(m) \subseteq d_2(m) \subseteq \ldots$, for each $m = 1, 2, \ldots$, there exists n_m such that $|d_{n_m}(m)| \geq |e| - \varepsilon/2^{m+1}$. Then

$$d = \bigcap_{m=1}^{\infty} d_{n_m}(m)$$

satisfies $|d| \geq |e| - \varepsilon/2$; moreover, if $t \in d$ and m is arbitrary we have

$$\rho(g_j(t), g_k(t)) \leq 1/m \quad (j, k \geq n_m),$$

which implies uniform convergence in d. The set c is any closed subset of d with $|c| \geq |d| - \varepsilon/2$. For strong measurability of $g(\cdot)$ see Example 12.8.8. ∎

Theorem 12.8.5 (Lusin). *Let X be a metric space, e a bounded measurable set, $g : e \to X$ a strongly measurable function. Then, given $\varepsilon > 0$ there exists a closed set $c \subseteq e$ such that (12.8.10) holds and $g : c \to X$ is continuous.*

Proof. Let $g_n(t)$ ($n = 1, 2, \ldots$) be a sequence of countably valued functions converging uniformly to $g(t)$ outside of a null set e_0; since we can modify $g(t)$ and the $g_n(t)$ in a null set we assume that $e_0 = \emptyset$. For each n, let $\{e_{nj}\}$ be the finite or countable family of pairwise disjoint measurable sets where $g_n(t)$ is constant. Select m_n so large that

$$\sum_{j=m_n+1}^{\infty} |e_{nj}| \leq \varepsilon/2^{n+1}.$$

This implies

$$\left| \bigcup_{j=1}^{m_n} e_{nj} \right| \geq \left| \bigcup_{j=1}^{\infty} e_{nj} \right| - \varepsilon/2^{n+1} \geq |e| - \varepsilon/2^{n+1}. \tag{12.8.11}$$

Pick closed sets $c_{nj} \subseteq e_{nj}$ ($j = 1, 2, \ldots, m_n$) with

$$|c_{nj}| \geq |e_{nj}| - \varepsilon/2^{n+j+1}, \tag{12.8.12}$$

and set

$$c_n = \bigcup_{j=1}^{m_n} c_{nj}.$$

It follows from (12.8.11) and (12.8.12) that

$$|c_n| \geq |e| - \varepsilon/2^{n+1} - \varepsilon/2^{n+1} = |e| - \varepsilon/2^n.$$

Since the e_{nj} are pairwise disjoint, so are the c_{nj} and $g_n(t)$ is continuous in each closed set c_{nj}, thus in c_n. Finally, let

$$c = \bigcap_{n=1}^{\infty} c_n.$$

Then c satisfies (12.8.10). Each $g_n(\cdot)$ is continuous in c_n, hence in the closed subset $c \subseteq c_n$. In view of the uniform convergence of $\{g_n(t)\}$ in c, $g(t)$ is as well continuous in c.   ∎

Theorem 12.8.6 (Scorza Dragoni). *Let* (U, ρ) *be a separable metric space,* (X, σ) *a metric space, e a bounded measurable set, $f : e \times U \to X$ such that*

(a) $u \to \phi(t, u)$ *is locally uniformly continuous in u for each t fixed*

(b) $t \to \phi(t, u)$ *is strongly measurable in t for every u fixed.*

Then for every $\varepsilon > 0$ there exists a closed set $c \subseteq e$ satisfying (12.8.10) such that $f(t, u)$ *is continuous in $c \times U$.*

Proof. We assume first that $f(t, u)$ is uniformly continuous in u for every $t \in e$. Given $\delta > 0$, define a function $\omega(t, \delta)$ in e by

$$\omega(t, \delta) = \sup_{\rho(u,v) \leq \delta} \sigma(f(t, u), f(t, v)). \qquad (12.8.13)$$

On account of continuity of $f(t, u)$ for t fixed and of separability of U the sup in (12.8.13) can be restricted to those u, v lying in a countable dense set, thus, by Lemma 12.8.3 $\omega(t, \delta)$ is measurable in e. On the other hand, uniform continuity of $f(t, u)$ for t fixed implies that $\omega(t, \delta) \to 0$ as $\delta \to 0$ for t fixed. By Egorov's theorem applied to the sequence $\{\omega(t, 1/n)\}$ there exists a closed set c_0 with $|c_0| > |e| - \varepsilon/2$ and

$$\omega(t, 1/n) \to 0 \quad \text{as } n \to \infty, \text{ uniformly in } c_0. \qquad (12.8.14)$$

Let $\{u_n\}$ be a countable dense set in U. By Lusin's Theorem, for each $n = 1, 2, \ldots$ there exists a closed set c_n with $|c_n| \geq |e| - \varepsilon/2^{n+1}$ and such that $t \to f(t, u_n)$ is continuous in c_n. Define

$$c = c_0 \cap \bigcap_{n=1}^{\infty} c_n.$$

Then $|c| > |e| - \varepsilon$, and it only remains to prove that $f(t, u)$ is continuous in $c \times U$. To do this, take $(t, u), (t', u') \in c \times U$. We have

$$\sigma(f(t', u'), f(t, u)) \le \sigma(f(t', u), f(t', u_k)) + \sigma(f(t', u_k), f(t, u_k))$$

$$+ \sigma(f(t, u_k), f(t, u)) \qquad (12.8.15)$$

where u_k is one of elements of $\{u_n\}$. Assuming that $\rho(u, u') \le \delta$, u_k can be chosen in the intersection of the balls $B(u, 2\delta)$ and $B(u', 2\delta)$ so that $\rho(u, u_k) \le 2\delta$, $\rho(u', u_k) \le 2\delta$. Using (12.8.14), we then take δ so small that the first and third term in (12.8.15) are less than $\varepsilon/3$ independently of t, t'. Once this is done and u_k has been chosen, we use the fact that $t \to f(t, u_k)$ is continuous in c and squeeze the middle term below $\varepsilon/3$ by taking t, t' close enough.

The general case where $f(t, u)$ is merely locally uniformly continuous is handled working not in the entire space U but in each ball $B(u_n, \delta)$. The corresponding set c_n such that $f(t, u)$ is continuous in $c_n \times B(u_n, \delta)$ is chosen in such a way that $|c_n| \ge |e| - \varepsilon/2^n$. If $c = \bigcap c_n$, then $|c| \ge |e| - \varepsilon$ and $f(t, u)$ is continuous in $c \times U$. ∎

Proof of Theorem 12.8.2. We claim that it is enough to dispose of the following particular case.

Lemma 12.8.7. *Let U be a complete metric space, X a metric space, e a bounded closed set in $\mathbb{R}$, $\phi : e \times U \to X$ a continuous function, $g : e \to X$ a continuous function such that*

$$g(t) \in \phi(t, U) \quad \text{a.e. in } t \in c. \qquad (12.8.16)$$

Then there exists a strongly measurable U-valued function $u(\cdot)$ such that

$$\phi(t, u(t)) = g(t) \quad \text{a.e. in } t \in c. \qquad (12.8.17)$$

In fact, assume that Lemma 12.8.7 has been proved. To show Theorem 12.8.2, we do as follows. Applying Lusin's theorem in e and then Scorza Dragoni's theorem in the set c produced by Lusin's theorem, we construct a closed set $c_1 \subseteq e$ such that $g(t)$ is continuous in c_1 and $\phi(t, u)$ is continuous in $c_1 \times U$. We then repeat this step in $e \setminus c_1$ and construct a closed set $c_2 \subseteq e \setminus c_1$ with $g(t)$ continuous in c_2 and $\phi(t, u)$ continuous in $c_2 \times U$; in the third step we construct a closed set $c_3 \subseteq e \setminus (c_1 \cup c_2)$ with $g(t)$ continuous in c_3 and $\phi(t, u)$ continuous in $c_3 \times U, \ldots$ and so on. The c_n so produced are pairwise disjoint, and we may select them in such a way that $|e \setminus (c_1 \cup \cdots \cup c_n)| \to 0$ as $n \to \infty$. We then apply Lemma 12.8.7 in each c_n and construct strongly measurable functions $u_n(t)$ such that (12.8.9) holds in c_n. Finally, we define

$$u(t) = u_n(t) \quad (t \in c_n), \; n = 1, 2, \ldots.$$

Then $u(t)$ is defined in $e' = \bigcup_n c_n \subseteq e$, $|e \setminus e'| = 0$ and satisfies (12.8.9) there.

Proof of Lemma 12.8.7. Consider the set valued function

$$t \to U(t) = \{u; \phi(t, u) = g(t)\} \tag{12.8.18}$$

in c. By hypothesis each $U(t)$ is nonempty, and u-continuity of ϕ implies that each $U(t)$ is closed. Thus, if we show that (12.8.18) is measurable, that is, that

$$\{t \in e; U(t) \cap C \neq \emptyset\} \tag{12.8.19}$$

is measurable for each closed $C \subseteq U$ we may construct $u(t)$ applying Theorem 12.8.1. Measurability of $U(\cdot)$ is proved as follows. Consider the complete separable metric space $e \times U$ endowed with the product topology. Continuity of $\phi(t, u)$ and $g(t)$ guarantees that the *graph*

$$G(U(\cdot)) = \{(t, u); u \in U(t), t \in e\} = \{(t, u); \phi(t, u) = g(t), t \in e\}$$

is closed in $e \times U$. Hence, the intersection

$$G(U(\cdot)) \cap \{(t, u); u \in C\}$$

of $G(U(\cdot))$ with the cylinder $\{(t, u); u \in C\}$ is closed as well, thus it is a separable complete metric space with the metric handed down by $e \times U$. Now, $(t, u) \in G(U(\cdot)) \cap \{(t, u); u \in C\}$ if and only if $u \in U(t) \cap C$, thus if $\Pi : e \times U \to e$ is the projection on the first coordinate,

$$\{t; B(t) \cap C \neq \emptyset\} = \Pi(G(U(\cdot)) \cap \{(t, u); u \in C\}).$$

By the Measurable Image Theorem 12.7.1, the set (12.8.19) is measurable. This ends the proof of Lemma 12.8.7 and thus that of Theorem 12.8.2.   ∎

Example 12.8.8. Let $\{g_n(\cdot)\}$ be a sequence of strongly measurable functions in a measurable set e with values in a metric space X such that $g_n(t) \to g(t)$ almost everywhere. Then the function $g(\cdot)$ is strongly measurable.

This result is a generalization of Example 5.0.35. We begin by observing that, approximating each $g_n(\cdot)$ uniformly by countably valued functions we may assume that the $g_n(\cdot)$ *are* countably valued functions; also, excising the null set where there is no convergence, we may assume that we have convergence everywhere. Finally, $g(t)$ takes values in the separable subspace of X consisting of the closure of the set of all values of all the $g_n(t)$, thus we may assume X separable. The rest of the proof is essentially the same as that of Example 5.0.35.

12.9. Spaces of Vector Valued Functions and Their Duals, II

We complete here the work in **12.2** identifying the dual of the space $L^p(0, T; X)$, $1 \le p < \infty$, for an arbitrary Banach space X.

Description of the dual requires "integrating" possibly nonmeasurable functions. The **Lebesgue upper integral** $°\!\int f(t)dt$ of an arbitrary nonnegative function $f(t)$ defined in $0 \leq t \leq T$ is

$$\int_0^{°T} f(t)dt = \inf \int_0^T \eta(t)dt, \qquad (12.9.1)$$

the infimum taken over all Lebesgue measurable functions $\eta(\cdot)$ with $\eta(t) \geq f(t)$ (if for all such η we have $\int \eta(t)dt = +\infty$ then $°\!\int f(t)dt = +\infty$ by definition). Obviously, $°\!\int f(t)dt = \int f(t)dt$ if $f(\cdot)$ is Lebesgue integrable.

Lemma 12.9.1.

$$\int_0^{°T} \lambda f(t)dt = \lambda \int_0^{°T} f(t)dt \quad (\lambda \geq 0), \qquad (12.9.2)$$

$$\int_0^{°T} f(t)dt \leq \sum \int_0^{°T} f_n(t)dt \quad \text{if } f(x) \leq \sum f_n(x) \qquad (12.9.3)$$

(*finite or countable sum*),

$$\text{if } \int_0^{°T} f(t)dt = 0 \quad \text{then } f(t) = 0 \quad a.e. \qquad (12.9.4)$$

Proof. (12.9.2) and (12.9.3) are obvious. To show (12.9.4), we note that if $°\!\int f(t)dt = 0$ there exists a sequence $\{\eta_n(\cdot)\}$ of measurable functions with $\eta_n(t) \geq f(t)$ and $\int \eta_n(t)dt \to 0$ as $n \to \infty$. Replacing this sequence by $\eta_1, \inf(\eta_1, \eta_2), \inf(\eta_1, \eta_2, \eta_3) \ldots$, we may assume that $\{\eta_n\}$ is nonincreasing, thus it converges to a measurable function $\eta(\cdot)$ with $\eta(t) \geq f(t)$ and $\int \eta(t)dt = 0$ by the dominated convergence theorem. Hence $\eta(t) = 0$ a.e. and *a fortiori* $f(t) = 0$ a.e. ∎

As in **12.2**, two X^*-valued X-weakly measurable functions $g(\cdot), h(\cdot)$ are *equivalent* (in symbols, $g \approx h$) if and only if $\langle g(t), y \rangle = \langle h(t), y \rangle$ a.e. in $0 \leq t \leq T$ for each $y \in X$, the set where equality fails depending on y. The equivalence class of a function $g(\cdot)$ is denoted by $[g]$. For $1 \leq q < \infty$ the space $L_w^q(0, T; X^*)$ consists of all X^*-valued X-weakly measurable functions $g(\cdot)$ such that there exists a function $h \in [g]$ with $°\!\int \|h(t)\|_{X^*}^q dt < \infty$. The norm is

$$\|g\|_{L_w^q(0,T;X^*)} = \inf_{h \in [g]} \left(\int_0^{°T} \|h(t)\|_{X^*}^q dt \right)^{1/q}. \qquad (12.9.5)$$

The space $L_w^q(0, T; X^*)$ equipped with this norm is a Banach space. There is no need to check this since $L_w^q(0, T; X^*)$ will be identified below as a dual space and (12.9.5) will be shown to coincide with the dual norm.

Theorem 12.9.2. *Let $1 < p < \infty$, X an arbitrary Banach space. Then the dual space $L^p(0, T; X)^*$ is algebraically and metrically isomorphic to $L_w^q(0, T; X^*)$, $1/p + 1/q = 1$, the duality pairing between both spaces given by*

$$\langle g, f \rangle = \int_0^T \langle g(t), f(t) \rangle dt. \tag{12.9.6}$$

For each equivalence class $[g] \in L_w^q(0, T; X^*)$, there exists $g_0 \in [g]$ such that $t \to \|g_0(t)\|$ is measurable in $0 \le t \le T$ and

$$\left(\int_0^T \|g_0(t)\|_{X^*}^q dt \right)^{1/q} = \|g\|_{L_w^q(0,T;X^*)}. \tag{12.9.7}$$

We prove first three auxiliary results.

Theorem 12.9.3 (Beppo Levi). *Let $\{\phi_n(\cdot)\}$ be a sequence of measurable nonnegative functions in $0 \le t \le T$ such that $\phi_1(t) \le \phi_2(t) \le \cdots$ and*

$$\int_0^T \phi_n(t)dt \le C \quad (n = 1, 2, \ldots). \tag{12.9.8}$$

Then $\phi(t) = \sup \phi_n(t) = \lim_{n\to\infty} \phi_n(t)$ exists a.e., is finite, integrable and

$$\int_0^T \phi(t)dt = \lim_{n\to\infty} \int_0^T \phi_n(t)dt. \tag{12.9.9}$$

Proof. The function $\phi(\cdot)$ is measurable, and the result would be obvious (from the dominated convergence theorem) if $\phi(\cdot)$ were known to be integrable. Noting that $\psi_m(t) = \min(\phi(t), m) = \lim_{n\to\infty} \phi_{nm}(t)$, where $\phi_{nm}(t) = \min(\phi_n(t), m)$ and applying the particular case above, we deduce that

$$\int_0^T \psi_m(t)dt = \lim_{n\to\infty} \int_0^T \phi_{nm}(t)dt \le \lim_{n\to\infty} \int_0^T \phi_n(t)dt \tag{12.9.10}$$

for all $m = 1, 2, \ldots$. It follows from (12.9.10) that

$$\sum_{m=1}^\infty \int_{e_m} \phi(t)dt < \infty$$

for $e_m = \{t; m \le \phi(t) < m + 1\}$. This implies that $\Sigma m |e_m| < \infty$. Accordingly, $\Sigma(m + 1)|e_m| < \infty$ as well and $\phi(t)$ is bounded by the integrable function $\Sigma(m + 1)\chi_m(t)$, χ_m the characteristic function of $[m, m + 1)$. This ends the proof. Note that the result would also follow from Fatou's Theorem (Example 8.1.7) if it were known in advance that $\phi(t)$ is a.e. finite. ∎

A set Φ of nonnegative functions is **directed** if, given ϕ, $\psi \in \Phi$ there exists $\eta \in \Phi$ with $\eta(t) \geq \max(\phi(t), \psi(t))$ a.e.

Corollary 12.9.4. *Let* $1 \leq q < \infty$, $\Phi \subseteq L^q(0, T)$ *a directed set with*

$$\|\phi\|_{L^q(0,T)} \leq C \quad (\phi \in \Phi). \tag{12.9.11}$$

Then there exists $\phi_{\max}(\cdot) \in L^q(0, T)$ *such that*

$$\phi(t) \leq \phi_{\max}(t) \quad a.e. \ in \ 0 \leq t \leq T \quad (\phi \in \Phi) \tag{12.9.12}$$

for all $\phi \in \Phi$ *("a.e." depending on* ϕ*) and*

$$\|\phi_{\max}\|_{L^q(0,T)} = M = \sup_{\phi \in \Phi} \|\phi\|_{L^q(0,T)}. \tag{12.9.13}$$

Proof. Call a sequence $\{\phi_n\} \subseteq \Phi$ **maximizing** if it is nondecreasing ($\phi_n(t) \leq \phi_{n+1}(t)$ a.e.) and $\|\phi_n\|_{L^q(0,T)} \to M$ as $n \to \infty$; since Φ is directed, maximizing sequences exist. If $\{\phi_n\}$ is maximizing, using Beppo Levi's Theorem 12.9.3 we deduce that $\phi_{\max}(t) = \sup \phi_n(t) = \lim_{n\to\infty} \phi_n(t)$ belongs to $L^q(0, T)$ with $\|\phi_{\max}\|_{L^q(0,T)} = M$. If $\{\psi_n(\cdot)\}$ is another maximizing sequence and $\psi_{\max}(t) = \lim_{n\to\infty} \psi_n(t)$, then $\psi_{\max}(\cdot) \in L^q(0, T)$ as well with $\|\psi_{\max}\|_{L^q(0,T)} = M$. Select a third maximizing sequence $\{\eta_n\} \subseteq \Phi$ with $\eta_n(t) \geq \max(\phi_n(t), \psi_n(t))$. Then $\eta_{\max}(\cdot) = \lim_{n\to\infty} \eta_n(\cdot) \in L^q(0, T)$ with $\eta_{\max}(t) \geq \phi_{\max}(t)$, $\eta_{\max}(t) \geq \psi_{\max}(t)$ a.e. and $\|\eta_{\max}\|_{L^q(0,T)} = M = \|\phi_{\max}\|_{L^q(0,T)} = \|\psi_{\max}\|_{L^q(0,T)}$, so that $\eta_{\max}(t) = \phi_{\max}(t) = \psi_{\max}(t)$ a.e. We have at this point shown that if $\phi_{\max}(t) = \lim_{n\to\infty} \phi_n(t)$ for a maximizing sequence $\{\phi_n(\cdot)\}$ and $\psi_{\max}(t) = \lim_{n\to\infty} \psi_n(t)$ for any other maximizing sequence, then

$$\phi_{\max}(t) = \psi_{\max}(t) \quad a.e. \ in \ 0 \leq t \leq T. \tag{12.9.14}$$

Since *any* element $\phi(\cdot)$ of Φ can be given membership in a maximizing sequence, (12.9.12) follows. ∎

Theorem 12.9.5 (Dinculeanu-Foias). *Let* X *be an arbitrary Banach space,* $1 < p < \infty$, Ψ *a bounded linear functional in* $L^p(0, T; X)$. *Then there exists a function* $\rho(\cdot) \in L^q(0, T)$ $(p^{-1} + q^{-1} = 1)$ *such that* $\rho(t) \geq 0$, $\|\rho\|_{L^q(0,T)} \leq \|\Psi\|_{L^p(0,T;X)^*}$ *and*

$$|\Psi(f)| \leq \int_0^T \|f(t)\|_X \rho(t) dt \quad (f \in L^p(0, T; X)). \tag{12.9.15}$$

Proof. Let $h(\cdot) \in L^\infty(0, T; X)$ with $\|h\|_{L^\infty(0,T;X)} \leq 1$. The map $\eta(\cdot) \to \Psi(\eta(\cdot)h(\cdot))$ is a linear functional in $L^p(0, T)$ with norm $\leq \|\Psi\|_{L^p(0,T;X)^*}$, thus

(Theorem 5.0.19) there exists a unique element $\phi_h(\cdot) \in L^q(0, T)$ with

$$\|\phi_h\|_{L^q(0,T)} \leq \|\Psi\|_{L^p(0,T;X)^*}, \tag{12.9.16}$$

$$\Psi(\eta(\cdot)h(\cdot)) = \int_0^T \eta(t)\phi_h(t)dt. \tag{12.9.17}$$

We call Φ the set of all such ϕ_h. If $\phi_h(t)$ is the function corresponding to a given $h(\cdot)$ in the unit ball of $L^\infty(0, T; X)$, then $|\phi_h(t)| = (\text{sign } \phi_h(t))\phi_h(t)$ is the function corresponding to $(\text{sign}\phi_h(t))h(t)$, hence

$$|\phi_h(\cdot)| \in \Phi. \tag{12.9.18}$$

Note next that if $h_1(t) = h_2(t)$ in a measurable set e, then $\phi_{h_1}(t) = \phi_{h_2}(t)$ in e (take $\eta \in L^p$ with support in e in (12.9.17)). Given $h_1(\cdot), h_2(\cdot)$ in the unit ball of $L^\infty(0, T; X)$, let $d = \{t; \phi_{h_1}(t) \leq \phi_{h_2}(t)\}$, $e = d^c$ the complement of d. Define $h(\cdot) \in L^\infty(0, T; X)$ as follows:

$$h(t) = h_1(t) \quad (t \in e), \qquad h(t) = h_2(t) \quad (t \in d).$$

The function $\phi_h \in L^q(0, T)$ then satisfies $\phi_h(t) = \phi_{h_2}(t) \geq \phi_{h_1}(t)$ in d, $\phi_h(t) = \phi_{h_1}(t) \geq \phi_{h_2}(t)$ in e, so that

$$\phi_h(t) \geq \phi_{h_1}(t), \phi_{h_2}(t) \quad (0 \leq t \leq T).$$

This means that the set Φ is directed, and in view of (12.9.18), the set F of absolute values $|\phi_h(\cdot)|$ is as well directed. By (12.9.16) and Corollary 12.9.4 there exists $\rho(\cdot) \in L^q(0, T)$ with $\|\rho\|_{L^q(0,T)} \leq \|\Psi\|_{L^p(0,T;X)^*}$ and $|\phi_h(t)| \leq \rho(t)$ a.e., thus (12.9.17) implies

$$|\Psi(\eta(\cdot)h(\cdot))| \leq \int_0^T |\eta(t)|\rho(t)dt. \tag{12.9.19}$$

We apply (12.9.19) to functions of $L^p(0, T; X)$ of the form $f(t) = \eta(t)h(t) = \eta(t)\Sigma\chi_j(t)y_j$ ($\chi_1, \ldots \chi_n$ characteristic functions of pairwise disjoint measurable sets $e_1, \ldots e_n$ covering $[0, T]$, $y_j \in X$, $\|y_j\| = 1$, $\eta(\cdot) \in L^p(0, T)$). The resulting inequality is (12.9.15); since these functions are dense in $L^p(0, T; X)$ (Theorem 5.0.27) the proof of Theorem 12.9.5 is complete. ∎

Proof of Theorem 12.9.2. We show first that (12.9.6) exists for every $f \in L^p(0, T; X)$, $g(\cdot) \in L_w^q(0, T; X^*)$ and satisfies

$$|\langle g, f \rangle| \leq \int_0^T |\langle g(t), f(t) \rangle|dt$$

$$\leq \left(\int_0^T \|f(t)\|_X^p dt\right)^{1/p} \left(\int_0^T \|g(t)\|_{X^*}^q dt\right)^{1/q}. \tag{12.9.20}$$

For this, consider

$$f(t) = \sum \chi_j(t) y_j \tag{12.9.21}$$

where the $\chi_j(\cdot)$ are the characteristic functions of pairwise disjoint measurable sets $e_1, \ldots, e_n$ and $y_1, \ldots, y_n \in X$. Now, if $g(\cdot)$ is an arbitrary function in $L_w^q(0, T; X^*)$ then $\langle g(t), f(t) \rangle = \Sigma \chi_j(t) \langle g(t), y_j \rangle$ is measurable. On the other hand, functions of the form (12.9.21) are dense in $L^p(0, T; X)$ so that $\langle g(\cdot), f(\cdot) \rangle$ is measurable for every $f(\cdot) \in L^p(0, T; X)$. Let now $\varepsilon > 0$, $\eta_\varepsilon(\cdot)$, a measurable function such that $\|g(t)\|_{X^*}^q \leq \eta_\varepsilon(t)$ a.e. and

$$\int_0^T \eta_\varepsilon(t) dt \leq \int_0^{\circ T} \|g(t)\|_{X^*}^q dt + \varepsilon.$$

We have $|\langle g(t), f(t) \rangle| \leq \|g(t)\|_{X^*} \|f(t)\|_X \leq \eta(t)_\varepsilon^{1/q} \|f(t)\|_X$. Thus, by Hölder's inequality, $|\langle g(\cdot), f(\cdot) \rangle|$ is integrable with

$$\int_0^T |\langle g(t), f(t) \rangle| dt \leq \left(\int_0^T \|f(t)\|_X^p dt \right)^{1/p} \left(\int_0^T \eta_\varepsilon(t) dt \right)^{1/q}.$$

Letting $\varepsilon \to 0$, (12.9.20) follows.

Let Ψ be a continuous linear functional in $L^p(0, T; X)$ and let $\rho(t)$ be the function built in Theorem 12.9.5 (inequality (12.9.5)). For $n = 1, 2, \ldots$, let $e_n = \{t \in [0, T]; \ \rho(t) \leq n\}$, $\chi_n(\cdot)$ the characteristic function of e_n. Define

$$\Psi_n(f(\cdot)) = \Psi(f(\cdot) \chi_n(\cdot)) \tag{12.9.22}$$

for $f(\cdot) \in L^p(0, T; X)$. Then, due to (12.9.15), Ψ_n is continuous in the $L^1(0, T; X)$-norm, thus (since $L^p(0, T; X)$ is dense in $L^1(0, T; X)$ for any $p < \infty$) it can be uniquely extended to a bounded linear functional in $L^1(0, T; X)$. By Theorem 12.2.11, there exists $g_n(\cdot) \in L_w^\infty(0, T; X^*)$ such that

$$\Psi_n(f) = \int_0^T \langle g_n(t), f(t) \rangle dt. \tag{12.9.23}$$

Moreover, $g_n(t) \approx g_{n+1}(t)$ in e_n, that is, $\langle g_n(t), y \rangle = \langle g_{n+1}(t), y \rangle$ a.e. in e_n for all $y \in X$. We may then define $g(t) = g_n(t)$ $(t \in e_n)$ and thus produce an X^*-valued X-weakly measurable function defined in the whole interval $0 \leq t \leq T$.

We show that $g(\cdot) \in L_w^q(0, T; X^*)$. Put $f(t) = \eta(t) y$, $\eta(t)$ a nonnegative function in $L^\infty(0, T)$, $y \in X$, and apply Theorem 12.9.5:

$$\int_0^T \eta(t) \langle g_n(t), y \rangle dt \leq \|y\|_X \int_0^T \eta(t) \rho(t) dt.$$

Using this inequality for arbitrary n and $\eta(\cdot)$ we obtain

$$\langle g(t), y \rangle \le \|y\|_X \rho(t) \quad \text{a.e.} \tag{12.9.24}$$

However, since "a.e." depends on y we cannot conclude that $\|g(t)\|_X \le \rho(t)$ right away; we must replace g by an equivalent function g_0. Note first that if we change the definition of g by setting $g(t) = 0$ in $e_0 = \{t \in [0, T]; \rho(t) = 0\}$, we obtain an equivalent function, thus we may and will assume that $g(t) = 0$ in e_0. Define next

$$h(t) = \begin{cases} 0 & (t \in e_0) \\ \rho(t)^{-1} g(t) & (t \in e_+), \end{cases}$$

where $e_+ = \{t \in [0, T]; \rho(t) > 0\}$. Then we have $h(\cdot) \in L_w^\infty(0, T; X^*)$ with norm $\|h\|_{L_w^\infty(0,T;X^*)} \le 1$. Let $\mathcal{L}$ be the lifting operator in Corollary 12.2.5: define

$$g_0(t) = \rho(t)(\mathcal{L}h)(t).$$

We have $\mathcal{L}h \approx h$, that is $\langle h(t), y \rangle = \langle (\mathcal{L}h)(t), y \rangle$ for every $y \in X$. In particular, $\langle (\mathcal{L}h)(t), y \rangle = 0$ a.e. in e_0, which implies $\langle g_0(t), y \rangle = \langle g(t), y \rangle$ a.e. in e_0. On the other hand, $\langle g_0(t), y \rangle = \rho(t)\langle (\mathcal{L}h)(t), y \rangle = \rho(t)\langle \rho(t)^{-1} g(t), y \rangle$ a.e. in e_+, so that $g_0 \approx g$. Finally, since $\|h\|_{L_w^\infty(0,T;X^*)} \le 1$, we have

$$\|(\mathcal{L}h)(t)\|_{X^*} \le 1 \quad \text{in } 0 \le t \le T$$

so that

$$\|g_0(t)\|_{X^*} \le \rho(t) \quad \text{a.e. in } 0 \le t \le T. \tag{12.9.25}$$

Using Theorem 12.9.5 once again, we obtain

$$\left(\int_0^T \|g_0(t)\|_{X^*}^q dt \right)^{1/q} \le \left(\int_0^T \rho(t)^q dt \right)^{1/q} \le \|\Psi\|_{L^p(0,T;X)^*} \tag{12.9.26}$$

keeping in mind that the norm $\|g_0(\cdot)\|$ is measurable by Theorem 12.2.6.

We go back now to (12.9.22) and (12.9.23). These relations guarantee the representation (12.9.6) for functions of the form $f(\cdot)\chi_n(\cdot)$. But (due to the dominated convergence theorem) $f(\cdot)\chi_n(\cdot) \to f(\cdot)$ in $L^p(0, T; X)$, thus the fact that (12.9.6) holds for f follows taking limits; convergence of the right side is insured by (12.9.20) applied to $f - f\chi_n$.

We note finally that (12.9.20) for g_0, combined with (12.9.26) produces

$$\left(\int_0^T \|g_0(t)\|_{X^*}^q dt \right)^{1/q} = \|\Psi\|_{L^p(0,T;X)^*},$$

which is (12.9.7). Using again (12.9.20) we obtain

$$\|g\|_{L^\infty_w(0,T;X^*)} = \inf_{h\in[g]} \left(\int_0^{\circ T} \|h(t)\|^q_{X^*} dt \right)^{1/q} = \left(\int_0^T \|g_0(t)\|^q_{X^*} dt \right)^{1/q}.$$

(12.9.27)

The proof of Theorem 12.9.2 is then complete. ∎

Example 12.9.6. If X is separable, every $g(\cdot) \in L^q_w(0, T; X^*)$ has measurable norm (that is, $t \to \|g(t)\|_X$ is measurable). Equivalence in $L^q_w(0, T; X^*)$ concides with a.e. equivalence: $g(\cdot) \approx h(\cdot)$ if and only if $g(t) = h(t)$ a.e. The norm in $L^q_w(0, T; X^*)$ is

$$\|g\|_{L^q_w(0,T;X^*)} = \left(\int_0^T \|g(t)\|^q_{X^*} \right)^{1/q}.$$

(12.9.28)

If X^* is separable, then every X^*-valued X-weakly measurable function is strongly measurable, thus

$$L^p(0, T; X)^* = L^q_w(0, T; X^*) = L^q(0, T; X^*).$$

(12.9.29)

The same equality obtains if X^* is a Gelfand space, in particular if X is reflexive (see Example 12.2.13). Note, however, that the equivalence classes of $L^q_w(0, T; X^*)$ may contain elements that are not strongly measurable (Example 12.2.1), thus the second equality (12.9.29) must be interpreted as in Example 12.2.13.

13

Relaxed Controls in Finite Dimensional Systems. Existence Theory

13.1. Installation of Relaxed Controls in Finite Dimensional Systems

We go back to the ordinary differential control system (2.1.1),

$$y'(t) = f(t, y(t), u(t)), \qquad y(0) = \zeta. \tag{13.1.1}$$

This time, the control set U is a compact metric space. Depending on the result, the system will be assumed to satisfy either Hypothesis I(C) or Hypothesis II(C) below on the function $f : [0, T] \times E \times U \to E = \mathbb{R}^m$.

Hypothesis I(C). $f(t, y, u)$ is measurable in t for y, u fixed and continuous in y, u for t fixed. For every $c > 0$ there exists $K(\cdot, c) \in L^1(0, T)$ such that

$$\|f(t, y, u)\| \le K(t, c) \quad (0 \le t \le T, \|y\| \le c, u \in U). \tag{13.1.2}$$

Hypothesis II(C). $f(t, y, u)$ is measurable in t for y, u fixed and continuous in y, u for t fixed. For every $c > 0$ there exists $K(\cdot, c), L(\cdot, c) \in L^1(0, T)$ such that (13.1.2) holds and

$$\|f(t, y', u) - f(t, y, u)\| \le L(t, c)\|y' - y\|$$
$$(0 \le t \le T, \|y\|, \|y'\| \le c, u \in U). \tag{13.1.3}$$

We call (13.1.1) the **original** control system. The **relaxed** control system is

$$y'(t) = F(t, y(t))\mu(t), \qquad y(0) = \zeta. \tag{13.1.4}$$

The control set is $\mathbf{U} = \Pi_{rca}(U) = \{\text{probability measures in } \Sigma_{rca}(U)\}$ and the admissible control space is $R_c(0, T; U)$ (see **12.5** for its construction). The function $F : [0, T] \times E \times \Sigma_{rca}(U) \to E$ is defined by

$$F(t, y)\mu = \int_U f(t, y, u)\mu(du), \tag{13.1.5}$$

the integral interpreted coordinatewise; for each t, y, $F(t, y)$ is a linear operator from $\Sigma_{rca}(U)$ into E. An independent study of (13.1.4) is not necessary, since, as proven in Lemma 13.1.2 below, this system fits the assumptions in **2.1**.

Lemma 13.1.1. *Let U be a compact metric space and $\phi : [0, T] \times U \to \mathbb{R}$ a function such that*

(a) $u \to \phi(t, u)$ *is continuous in u for each t fixed,*

(b) $t \to \phi(t, u)$ *is measurable in t for each u fixed.*

Then the function

$$t \to \phi(t, \cdot) \tag{13.1.6}$$

is a strongly measurable $C(U)$-valued function.

Proof. Since $C(U)$ is separable (Theorem 12.5.2), to show that (13.1.6) is strongly measurable as a $C(U)$-valued function it is enough to show that it is $C(U)^*$-weakly measurable, that is, that for every $\mu \in \Sigma_{rca}(U) = C(U)^*$ the function

$$\phi(t) = \int_U \phi(t, u)\mu(du) \tag{13.1.7}$$

is measurable. Since $\mu = (\mu + |\mu|) - |\mu|$, we may limit ourselves to the case $\mu \geq 0$, or, after multiplication by a constant, to the case $\mu \in \Pi_{rca}(U)$. Finally, using Lemma 12.6.4 and Remark 12.6.5, we may construct a sequence $\{\mu_n\} \subseteq \text{conv}(D)$ with $\mu_n \to \mu$ $C(U)$-weakly, where D denotes the set of all Dirac deltas in $\Pi_{rca}(U)$. Denote by $\phi_n(t)$ the function defined from μ_n by (13.1.7). Then $\phi_n(t) \to \phi(t)$ in $0 \leq t \leq T$, thus we only have to show that $\phi_n(t)$ is measurable. This is obvious, since each $\phi_n(t)$ is a convex combination of functions $\phi(t, u_n)$. ∎

Lemma 13.1.2. *Assume that the system* (13.1.1) *satisfies Hypothesis I(C) (resp. Hypothesis II(C)) in $0 \leq t \leq T$. Then the system* (13.1.4) *satisfies Hypothesis I (resp. Hypothesis II) in* **2.1** *for every $\mu(\cdot) \in R_c(0, T; U)$, with $K(\cdot, c)$, $L(\cdot, c)$ independent of $\mu(\cdot)$.*

Proof. Assume first that (13.1.1) satisfies Hypothesis I(C). Since the bound (13.1.2) is independent of u,

$$\|F(t, y)\mu\| \leq K(t, c)\|\mu\|_{\Sigma_{rca}(U)} \quad (0 \leq t \leq T, \|y\| \leq c, \mu \in \Sigma_{rca}(U)). \tag{13.1.8}$$

This shows that $F(t, y)$ is a linear bounded operator from $\Sigma_{rca}(U)$ into E. If $\mu(\cdot) \in R_c(0, T; U)$ we may assume that $\|\mu(t)\| \equiv 1$ (Theorem 12.5.6), thus the independence statement follows.

Since $f(t, y, u)$ is continuous in y, u for t fixed and U is compact, $f(t, y, u)$ is uniformly continuous for $u \in U$ and y bounded; in particular,

$$\|f(t, y', \cdot) - f(t, y, \cdot)\|_{C(U;\mathbb{R}^m)} \to 0 \quad \text{as } y' \to y. \tag{13.1.9}$$

We have

$$\|F(t, y')\mu - F(t, y)\mu\| \le \int_U \|f(t, y', u) - f(t, y, u)\|\mu(du) \tag{13.1.10}$$

so that that $F(t, y)\mu(t)$ is continuous in y for t fixed. To investigate t-dependence of $F(t, y)\mu(t)$ write $f = (f_1, \ldots, f_m)$ and take y fixed. By Lemma 13.1.1 each $t \to f_j(t, y, \cdot)$ is a strongly measurable $C(U)$-valued function and, in view of (13.1.2),

$$t \to f_j(t, y, \cdot) \quad \text{belongs to } L^1(0, T; C(U)).$$

On the other hand,

$$t \to \mu(t) \quad \text{belongs to } L_w^\infty(0, T; \Sigma_{rca}(U)).$$

By definition of the duality pairing between $L^1(0, T; C(U))$ and its dual $L_w^\infty(0, T; \Sigma_{rca}(U))$ (Theorem 12.2.11) the function

$$t \to \langle f_j(t, y, \cdot), \mu(t) \rangle = \int_U f_j(t, y, u)\mu(t, du)$$

belongs to $L^1(0, T)$, in particular is measurable. This takes care of Hypothesis I for the relaxed system (13.1.4).

Assume (13.1.1) satisfies Hypothesis II(C). To show that (13.1.4) satisfies Hypothesis II, we only have to prove the Lipschitz continuity property (2.1.4) for $F(t, y)\mu$. It follows integrating (13.1.3) against $\mu(du)$:

$$\|F(t, y')\mu - F(t, y)\mu\| \le L(t, c)\|y' - y\|\|\mu\|_{\Sigma_{rca}(U)}$$
$$(0 \le t \le T, \|y\|, \|y'\| \le c, \mu \in \Sigma_{rca}(U)), \tag{13.1.11}$$

and it clearly holds independently of $\mu(\cdot) \in R_c(0, T; U)$. This ends the proof of Lemma 13.1.2. We note that the arguments are not restricted to relaxed controls; (13.1.1) can be equipped equally well with arbitrary controls $\mu(\cdot) \in L_w^\infty(0, T; \Sigma_{rca}(U))$. ∎

All the work done, any of the results in Part I (in particular those in **2.2** on existence under Hypothesis I and on uniqueness and maximal intervals of existence under Hypothesis II) can be applied to the relaxed system (13.1.4). We denote by $y(t, \mu)$ the solution of (13.1.4) corresponding to $\mu(\cdot) \in R_c(0, T; U)$.

So far, we have not chosen an admissible control space $C_{ad}(0, T; U)$ for the original system (13.1.1). We do it now. As defined in **12.8**, a function $u(\cdot)$ defined in $0 \leq t \leq T$ with values in a metric space U is *countably valued* if it is constant in each of the members of a pairwise disjoint countable family $\{e_j\}$ of measurable sets covering $[0, T]$. A function $u(\cdot)$ is *strongly measurable* if there exists a null set $e \subseteq [0, T]$ and a sequence $u_n(\cdot)$ of countably valued functions such that $u_n(t) \to u(t)$ uniformly in the complement of e.

Lemma 13.1.3. *Let $C_{ad}(0, T; U)$ be the set of all strongly measurable U-valued functions defined in $0 \leq t \leq T$. Then, if the original system (13.1.1) satisfies Hypothesis I(C) (resp. Hypothesis II(C)) it satisfies Hypothesis I (resp. Hypothesis II) for every $u(\cdot) \in C_{ad}(0, T; U)$ with $K(\cdot, c), L(\cdot, c)$ independent of $u(\cdot)$.*

Lemma 13.1.3 can be proved directly. It also follows from Lemma 13.1.2 using the (independently interesting) result that a control $u(\cdot) \in C_{ad}(0, T; U)$ can be replicated by means of the relaxed control

$$\mu(t) = \delta(\cdot - u(t)). \tag{13.1.12}$$

Lemma 13.1.4. *The relaxed control $\mu(\cdot)$ defined by (13.1.12) belongs to $R_c(0, T; U)$.*

Proof. If $f(\cdot) \in C(U)$, then

$$\langle \mu(t), f \rangle = f(u(t)). \tag{13.1.13}$$

If $\{u_n(\cdot)\}$ is a sequence of countably valued functions such that $u_n(t) \to u(t)$ a.e. then $\{f(u_n(\cdot))\}$ is as well a sequence of countably valued functions such that $f(u_n(t)) \to f(u(t))$ a.e., so that (13.1.13) is measurable. Obviously, each $\mu(t)$ is a probability measure. ∎

In some results in this chapter, the use of the full space $C_{ad}(0, T; U)$ is unnecessary; we make do with the subspace $PC(0, T; U)$ of piecewise constant U-valued functions (that is, functions constant in intervals $t_{j-1} \leq t < t_j$, where $0 = t_0 < t_1 < \cdots < t_{n-1} < t_n = T$).

13.2. Approximation of Relaxed Trajectories by Ordinary Trajectories

Relaxation Theorem 13.2.1. *Let the original system (13.1.1) satisfy Hypothesis II(C), and let $\mu(\cdot) \in R_c(0, \bar{t}; U)$ be such that the solution $y(t, \mu)$ of the relaxed system (13.1.4) exists in $0 \leq t \leq \bar{t}$. Then there exists $u(\cdot) \in PC(0, \bar{t}; U)$ such that the solution $y(t, u)$ of the original system (13.1.1) exists in $0 \leq t \leq \bar{t}$ and*

$$\|y(t, u) - y(t, \mu)\| \leq \varepsilon \quad (0 \leq t \leq \bar{t}). \tag{13.2.1}$$

Theorem 13.2.1 follows from

Theorem 13.2.2. *Let the original system* (13.1.1) *satisfy Hypothesis II(C), and let* $\{\mu_n(\cdot)\}$ *be a sequence in* $L_w^\infty(0, \bar{t}; \Sigma_{rca}(U))$ *with* $\mu_n(\cdot) \to \mu(\cdot)$ $L^1(0, \bar{t}; C(U))$-*weakly. Assume that* $y(t, \mu)$ *exists in* $0 \leq t \leq \bar{t}$. *Then there exists* n_0 *such that if* $n \geq n_0$ *the solution* $y(t, \mu_n)$ *exists in* $0 \leq t \leq \bar{t}$ *and*

$$\|y(t, \mu_n) - y(t, \mu)\| \to 0 \qquad (13.2.2)$$

uniformly in $0 \leq t \leq \bar{t}$.

In fact, assume that Theorem 13.2.2 is proved. Using Corollary 12.6.9 we may construct a sequence $\{\mu_n(\cdot)\}$ in $PC(0, \bar{t}; D)$ (that is, a sequence $\mu_n(t) = \delta(\cdot - u_n(t))$, $u_n(\cdot) \in PC(0, \bar{t}; U)$) with $\mu_n(\cdot) \to \mu(\cdot)$ $L^1(0, \bar{t}; C(U))$-weakly in $L_w^\infty(0, \bar{t}; \Sigma_{rca}(U))$. Accordingly, Theorem 13.2.1 results from the fact that $y(t, \mu_n) = y(t, u_n)$.

Proof of Theorem 13.2.2. Let $[0, t_n]$ be the maximal subinterval of $[0, \bar{t}]$ such that $y(t, \mu_n)$ exists and satisfies

$$\|y(t, \mu_n) - y(t, \mu)\| \leq 1 \quad (0 \leq t \leq t_n). \qquad (13.2.3)$$

We have

$$y(t, \mu_n) - y(t, \mu) = \int_0^t \{F(\tau, y(\tau, \mu_n))\mu_n(\tau) - F(\tau, y(\tau, \mu))\mu_n(\tau)\}d\tau$$

$$+ \int_0^t F(\tau, y(\tau, \mu))(\mu_n(\tau) - \mu(\tau))d\tau.$$

We estimate using the Lipschitz condition (13.1.11) in the first integral and call $\rho_n(t)$ the second integral:

$$\|y(t, \mu_n) - y(t, \mu)\| \leq C \int_0^t L(\tau, c + 1)\|y(\tau, \mu_n) - y(\tau, \mu)\|d\tau + \|\rho_n(t)\|,$$

where C is a bound for $\|\mu_n\|$ and c is a bound for $\|y(t, \mu)\|$ in $0 \leq t \leq \bar{t}$. By Corollary 2.1.3 we have

$$\|y(t, \mu_n) - y(t, \mu)\| \leq C'\|\rho_n(t)\|. \qquad (13.2.4)$$

Thus, if we show that $\rho_n(t) \to 0$ uniformly in $0 \leq t \leq \bar{t}$ it will follow that $t_n = \bar{t}$ for n large enough and that (13.2.2) holds true. If this is not so, there exists a sequence $\{t_n\} \subset [0, \bar{t}]$ such that

$$\|\rho_n(t_n)\| \geq \varepsilon > 0, \qquad (13.2.5)$$

and we may assume that $t_n \to t \in [0, \bar{t}]$. If $\chi_t(\cdot)$ is the characteristic function of $[0, t]$ we have

$$\rho_n(t_n) = \int_0^{\bar{t}} \chi_{t_n}(\tau) F(\tau, y(\tau, \mu))(\mu_n(\tau) - \mu(\tau))d\tau.$$

However, each component of $\chi_{t_n}(\tau) F(\tau, y(\tau, \mu))$ converges to the corresponding component of $\chi_t(\tau) F(\tau, y(\tau, \mu))$ in $L^1(0, \bar{t}; C(U))$ while $\mu_n(\cdot) \to \mu(\cdot)$ $L^1(0, \bar{t}; C(U))$-weakly in $L_w^\infty(0, \bar{t}; \Sigma_{rca}(U))$, thus $\rho_n(t_n) \to 0$ and we have a contradiction with (13.2.5). This ends the proof.   ■

13.3. The Filippov Implicit Function Theorem in the Compact Case

Filippov's implicit function theorem (Theorem 12.8.2) was proved in Chapter 12 with a considerable investment in measure theory and topology. We show here that the particular case where U is a compact metric space can be dispatched by elementary means.

Recall (**10.8**) that a set valued function $t \to U(t) \subseteq U =$ metric space is *measurable* in $0 \le t \le T$ if for every closed $C \subseteq U$ the set

$$\{t \in [0, T]; U(t) \cap C \ne \emptyset\} \tag{13.3.1}$$

is measurable, and that a strongly measurable U-valued function $t \to u(t)$ is a *measurable selection* of the set valued function $t \to U(t)$ if

$$u(t) \in U(t) \quad \text{a.e. in } 0 \le t \le T. \tag{13.3.2}$$

We refer the reader to **12.8** for a proof of the *selection theorem*: if the metric space U is separable and complete, each $U(t)$ is nonempty and closed and $t \to U(t)$ is measurable, there exists a measurable selection of the set valued function $U(\cdot)$.

Implicit Function Theorem 13.3.1. *Let U be a compact metric space and $\phi : [0, T] \times U \to \mathbb{R}^m$ a function such that*

(a) $u \to \phi(t, u)$ is continuous in u for each t fixed,

(b) $t \to \phi(t, u)$ is measurable in t for each u fixed.

Let $g(\cdot)$ be a measurable $\mathbb{R}^m$-valued function such that

$$g(t) \in \phi(t, U) \quad \text{a.e. in } 0 \le t \le T. \tag{13.3.3}$$

Then there exists a strongly measurable U-valued function $u(\cdot)$ such that

$$\phi(t, u(t)) = g(t) \quad \text{a.e. in } 0 \le t \le T. \tag{13.3.4}$$

Proof. Consider the set valued function

$$t \rightarrow U(t) = \{u \in U; \phi(t, U) = g(t)\}. \qquad (13.3.5)$$

By hypothesis each $U(t)$ is nonempty, and u-continuity of ϕ implies that each $U(t)$ is closed. Moreover U, as a compact metric space, is complete and separable, thus if we show that (13.3.5) is measurable, the selection theorem produces the claimed measurable selection. Measurability of $U(\cdot)$ will be shown by means of an approximation trick based on Lemma 13.1.1. According to this result, each component of $t \rightarrow \phi(t, \cdot)$ is a strongly measurable $C(U)$-valued function, thus $t \rightarrow \phi(t, \cdot)$ itself is a strongly measurable $C(U; \mathbb{R}^m)$-valued function. Then for each $n = 1, 2, \ldots$ we may select a countably $C(U; \mathbb{R}^m)$-valued function $t \rightarrow \phi_n(t, \cdot)$ and (by measurability of each component of $g(\cdot)$) a countably $\mathbb{R}^m$-valued function $g_n(t)$ such that

$$\|\phi_n(t, \cdot) - \phi(t, \cdot)\|_{C(U;\mathbb{R}^m)} \le 1/n, \qquad \|g_n(t) - g(t)\| \le 1/n \qquad (13.3.6)$$

outside of a set e_n of measure zero. Doctoring the functions involved in null sets we may assume that both inequalities (13.3.6) actually take place in the whole interval $0 \le t \le T$. We have

$$U(t) = \{u \in U; \phi(t, u) = g(t)\}$$
$$= \bigcap_{n=1}^{\infty} \bigcap_{k=1}^{n} \{u \in U; \|\phi_k(t, u) - g_k(t)\| \le 2/k\} = \bigcap_{n=1}^{\infty} U_n(t). \qquad (13.3.7)$$

Now, each of the set functions $t \rightarrow U_n(t)$ is measurable since the set $U_n(t)$ is constant in any set where $\phi_k(t, \cdot)$ and $g_k(t)$ are constant for $k \le n$, thus $U_n(t)$ is constant in a countable family of pairwise disjoint measurable sets covering $[0, T]$. Finally, (by u-continuity of $\phi_k(t, u)$) each $U_n(t)$ is closed. We show next that

$$\{t; U(t) \cap C \ne \emptyset\} = \bigcap_{n=1}^{\infty} \{t; U_n(t) \cap C \ne \emptyset\}. \qquad (13.3.8)$$

To see this, note that

$$U_1(t) \supseteq U_2(t) \supseteq \cdots \supseteq U(t) \qquad (13.3.9)$$

so that if $U(t) \cap C \ne \emptyset$ then $U_n(t) \cap C \ne \emptyset$ and the inclusion $\subseteq$ in (13.3.8) holds. On the other hand, assume that $U_n(t) \cap C \ne \emptyset$ for all n. Choose $u_n \in U_n(t) \cap C$. Using compactness, we may assume that $u_n \rightarrow u \in U$. In view of (13.3.9) the sequence $\{u_k\}$ (except for a finite number of terms) is contained in each $U_n(t)$, so that the limit u belongs to every $U_n(t) \cap C$. Using (13.3.6) and (13.3.7) we deduce that $\|\phi(t, u) - g(t)\|_{\mathbb{R}^m} \le 4/n$ for all n, so that $u \in U(t) \cap C$ and we have $U(t) \cap C \ne \emptyset$. This ends the proof. ∎

The following result is the "relaxed" version of Filippov's implicit function theorem (see Young [1969, (34.7), p. 297]).

Lemma 13.3.2. *Let U, ϕ satisfy the assumptions in Theorem 13.3.1. Define $g(t) = \Phi(t)\mu(t)$ by*

$$\Phi(t)\mu(t) = \int_U \phi(t, u)\mu(t, du) \qquad (13.3.10)$$

for $\mu(\cdot) \in R_c(0, T; U)$. Then $g(\cdot)$ is measurable and

$$g(t) \in \operatorname{conv}(\phi(t, U)) \quad \text{a.e. in } 0 \le t \le T. \qquad (13.3.11)$$

Conversely, let $g(\cdot)$ be a measurable function such that (13.3.11) holds. Then there exists $\mu(\cdot) \in R_c(0, T; U)$ such that

$$\Phi(t)\mu(t) = g(t) \quad \text{a.e. in } 0 \le t \le T. \qquad (13.3.12)$$

Proof. The assumptions imply that $\Phi(t)\mu(t) \in \mathbb{R}^m$ is well defined for every $\mu(\cdot) \in L_w^\infty(0, T; \Sigma_{rca}(U))$ and

$$\|\Phi(t)\mu(t)\|_{\mathbb{R}^m} \le \|\phi(t)\|_{C(U;\mathbb{R}^m)} \|\mu\|_{L_w^\infty(0,T;\Sigma_{rca}(U))}. \qquad (13.3.13)$$

We have shown in Lemma 13.1.1 that each component of $t \to \phi(t, \cdot)$ is a strongly measurable $C(U)$-valued function, thus the $C(U; \mathbb{R}^m)$-valued function $t \to \phi(t, \cdot)$ is strongly measurable, *a fortiori* $t \to \|\phi(t)\|_{C(U;\mathbb{R}^m)}$ is measurable. Accordingly, the characteristic function χ_n of the set e_n where $n \le \|\phi(t)\| < n+1$ is measurable, and it is clearly enough to show the direct and converse parts of Lemma 13.3.2 for $\phi_n(t, u) = \chi_n(t)\phi_n(t, u)$. We may then assume from the start that $\|\phi(t)\|_{C(U;\mathbb{R}^m)}$ is bounded, although all we need is that $t \to \|\phi(t)\|_{C(U;\mathbb{R}^m)}$ belong to $L^1(0, T)$.

By u-continuity of ϕ and compactness of U the set $\phi(t, U) \subseteq \mathbb{R}^m$, *a fortiori* the set $\operatorname{conv}(\phi(t, U)) \subseteq \mathbb{R}^m$ is compact (Example 13.3.4) thus it is closed. Let $g(\cdot)$ be a measurable function such that (13.3.11) holds a.e. Since $g(\cdot)$ is measurable, if necessary after modifying $g(\cdot)$ in a null set there exists a countably valued function $g_\varepsilon(\cdot)$ with

$$\|g(t) - g_\varepsilon(t)\| \le \varepsilon \quad (0 \le t \le T). \qquad (13.3.14)$$

In particular,

$$g_\varepsilon(t) \in \{\operatorname{conv}(\phi(t, U))\}_\varepsilon \quad (0 \le t \le T), \qquad (13.3.15)$$

where $A_\varepsilon = \{y; \operatorname{dist}(y, A) \le \varepsilon\}$ for any set $A \subseteq E$.

Select a countably $C(U; \mathbb{R}^m)$-valued function $t \to \phi_\varepsilon(t, \cdot)$ such that, after modification in a null set we have $\|\phi_\varepsilon(t, \cdot) - \phi(t, \cdot)\|_{C(U;\mathbb{R}^m)} \le \varepsilon$, or

$$\|\phi_\varepsilon(t, u) - \phi(t, u)\| \le \varepsilon \quad (0 \le t \le T, u \in U). \qquad (13.3.16)$$

Obviously, there exists a countable family Ξ_ε of measurable sets such that both $\phi_\varepsilon(t, \cdot)$ and $g_\varepsilon(\cdot)$ are constant in each set. Let $e \in \Xi_\varepsilon$ be one of these sets and pick $t_0 \in e$. Using (13.3.15) for $t = t_0$, select a finite set of real numbers $\{\alpha_j(e)\}$ with $\alpha_j(e) \ge 0$, $\Sigma\alpha_j(e) = 1$ and a corresponding finite set $\{u_j(e)\} \subseteq U$ such that

$$\|g_\varepsilon(t_0) - \Sigma\alpha_j(e)\phi(t_0, u_j(e))\| \le \varepsilon.$$

It follows that

$$\|g_\varepsilon(t) - \Sigma\alpha_j(e)\phi_\varepsilon(t, u_j(e))\| \le 2\varepsilon \qquad (13.3.17)$$

for $t = t_0$, hence for all $t \in e$ since the left side of (13.3.17) is constant there. Finally, using again (13.3.14) and (13.3.16),

$$\|g(t) - \Sigma\alpha_j(e)\phi(t, u_j(e))\| \le 4\varepsilon \quad (t \in e).$$

Define a measure $\mu(t) \in \Sigma_{rca}(U)$ for all $t \in e$ by $\mu(t) = \Sigma\alpha_j(e)\delta(\cdot - u_j(e))$; we obviously have $\Phi(t)\mu(t) = \Sigma\alpha_j(e)\phi(t, u_j(e))$. Let $\mu_\varepsilon(\cdot)$ be the element of $L_w^\infty(0, T; \Sigma_{rca}(U))$ obtained piecing together these measures; then

$$\|g(t) - \Phi(t)\mu_\varepsilon(t)\| \le 4\varepsilon \quad (0 \le t \le T). \qquad (13.3.18)$$

It is clear that $\mu_\varepsilon(\cdot) \in R_c(0, T; U)$. Picking $\varepsilon_n \to 0$ and using Theorem 12.5.9 (a) select a $L^1(0, T; U)$-weakly convergent subsequence of the $\{\mu_{\varepsilon_n}(\cdot)\}$. If $\mu(\cdot)$ is the limit and $\chi_t(\cdot)$ is the characteristic function of the interval $0 \le \tau \le t$ we obtain from weak convergence and from (13.3.18) that

$$\int_0^{\bar{t}} \chi_t(\tau)g(\tau)d\tau = \int_0^{\bar{t}} \int_U \chi_t(\tau)\phi(\tau, u)\mu(\tau, du)d\tau = \int_0^{\bar{t}} \chi_t(\tau)\Phi(\tau)\mu(\tau)d\tau,$$

which implies (13.3.12). ∎

The fact that (13.3.11) will hold a.e. for arbitrary $\mu(\cdot) \in R_c(0, T; U)$ is a consequence of the following,

Lemma 13.3.3. *Let U be a compact metric space, $\phi : U \to \mathbb{R}^m$ a continuous function. Define $\Phi : \Pi_{rca}(U) \to \mathbb{R}^m$ by*

$$\Phi\mu = \int_U \phi(u)\mu(du), \qquad (13.3.19)$$

the integral understood coordinatewise. Then, for any $\mu \in \Pi_{rca}(U)$,

$$\Phi\mu \in \text{conv}(\phi(U)). \qquad (13.3.20)$$

Proof. We recall (Example 13.3.4) that $\text{conv}(\phi(U))$ is compact, hence closed. Let $\{\mu_n\}$ be a sequence in $\text{conv}(D)$ (that is, a sequence whose elements are finite convex combinations $\Sigma\alpha_j\delta(\cdot - u_j)$ of Dirac measures) with $\mu_n \to \mu$ (Lemma 12.6.4). We have $\Phi\mu_n = \Sigma\alpha_j\phi(u_j) \in \text{conv}(\phi(U))$. Replacing in (13.3.19) and taking limits, we see that $\Phi\mu$ itself satisfies (13.3.20). ∎

Example 13.3.4. Let B be a compact subset of $\mathbb{R}^m$. Then $\text{conv}(B)$ is compact. This result follows from the the fact that, if y is a finite convex combination of points $y_1, \ldots, y_k \in \mathbb{R}^m$ then y is a convex combination of a finite subset of

$\{y_1, \ldots, y_k\}$ containing $\leq m + 1$ elements. Accordingly, an arbitrary sequence $\{y_n\} \subseteq \operatorname{conv}(B)$ can be written in the form

$$y_n = \sum_{j=1}^{m+1} \alpha_{nj} y_{nj}$$

$(\alpha_{nj} \geq 0, \Sigma \alpha_{nj} = 1, y_{nj} \in B)$. Using compactness we may assume, passing to a subsequence, that $\{y_{nj}\}$ is convergent to $y_j \in B$ and that $\alpha_{nj} \to \alpha_j$, thus $y_n \to \Sigma \alpha_j y_j \in \operatorname{conv}(B)$.

13.4. Differential Inclusions

A **differential inclusion** is an inequation of the form

$$y'(t) \in \mathcal{F}(t, y(t)), \qquad y(0) = \zeta \tag{13.4.1}$$

where $(t, y) \to \mathcal{F}(t, y) \subseteq E = \mathbb{R}^m$ is a set valued function defined in $[0, T] \times E$. A function $y(\cdot)$ is a **solution** of (13.4.1) in $0 \leq t \leq \bar{t}$ if and only if $y(\cdot)$ is absolutely continuous there and (13.4.1) is satisfied almost everywhere; the initial condition is understood literally. A similar definition applies to other intervals.

A control problem for the differential equation

$$y'(t) = f(t, y(t), u(t)), \qquad y(0) = \zeta \tag{13.4.2}$$

with control constraint $u(t) \in U = $ compact metric space can be written as a control problem for the differential inclusion

$$y'(t) \in f(t, y(t), U), \qquad y(0) = \zeta. \tag{13.4.3}$$

It is clear that a solution of (13.4.2) will be a solution of (13.4.3), but the opposite implication is not obvious and will be proved below. Related with (13.4.3) we consider the **relaxed** differential inclusion

$$y'(t) \in \operatorname{conv}(f(t, y(t), U)), \qquad y(0) = \zeta \tag{13.4.4}$$

which, as we show below, is equivalent to the relaxed control system

$$y'(t) = F(t, y(t))\mu(t), \qquad y(0) = \zeta \tag{13.4.5}$$

under suitable assumptions. As a particular case of this proof, we shall obtain the equivalence of (13.4.2) and (13.4.3). For these purposes, even Hypothesis I(C) in **13.1** is too much: we weaken it to

Hypothesis $\emptyset(C)$. For every $y(\cdot) \in C(0, T; U)$ the function $t \to f(t, y(t), u)$ is measurable in t for u fixed and continuous in u for t fixed. For every $c > 0$ there exists $K(\cdot, c) \in L^1(0, T)$ such that

$$\|f(t, y, u)\| \leq K(t, c) \quad (0 \leq t \leq T, \|y\| \leq c, u \in U). \tag{13.4.6}$$

Under Hypothesis $\emptyset(C)$ the function $F(t, y)\mu$ in (13.1.5) is well defined; moreover, $t \to f(t, y(t), \cdot)$ belongs to $L^1(0, T; C(U))$ by Lemma 13.1.1 and (13.4.6), thus $F(t, y(t))\mu(t)$ is measurable for any $\mu(\cdot) \in L_w^\infty(0, T; \Sigma_{rca}(U))$ and (13.1.8) holds, that is,

$$\|F(t, y)\mu\| \leq K(t, c)\|\mu\|_{\Sigma_{rca}(U)} \quad (0 \leq t \leq T, \|y\| \leq c, \mu \in \Sigma_{rca}(U)).$$
$$(13.4.7)$$

These properties are enough to prove that an absolutely continuous E-valued solution $y(t)$ of the relaxed system (13.4.5) is a solution of the integral equation

$$y(t) = \zeta + \int_0^t F(\tau, y(\tau))\mu(\tau)d\tau \qquad (13.4.8)$$

and vice versa. The admissible control space $C_{ad}(0, T; U)$ for (13.1.1) consists of all strongly measurable functions $u(\cdot)$ in $0 \leq t \leq T$ (see definitions in **13.1**). Obviously, Hypothesis $\emptyset(C)$ implies that $t \to f(t, y(t), u(t))$ is measurable for every $y(\cdot) \in C(0, T; E)$ and every countably valued $u(t)$, thus using u-continuity and taking limits, $t \to f(t, y(t), u(t))$ is measurable for every $u(\cdot) \in C_{ad}(0, T; U)$. On the other hand (13.4.6) implies that

$$\|f(t, y, u(t))\| \leq K(t, c) \quad (0 \leq t \leq T, \|y\| \leq c), \qquad (13.4.9)$$

which guarantees that an absolutely continuous E-valued function $y(t)$ satisfies (13.4.2) if and only if it satisfies the integral equation

$$y(t) = \zeta + \int_0^t f(\tau, y(\tau), u(\tau))d\tau. \qquad (13.4.10)$$

In the result below $[0, \bar{t}]$ is a fixed subinterval of $[0, T]$ and "solution" means "solution in $0 \leq t \leq \bar{t}$."

Equivalence Theorem 13.4.1. *Let Hypothesis $\emptyset(C)$ hold for the original system (13.1.1). (a) A function $y(\cdot)$ is a solution of (13.4.2) with $u(\cdot) \in C_{ad}(0, \bar{t}; U)$ if and only if it is a solution of (13.4.3). (b) A function $y(\cdot)$ is a solution of (13.4.5) with $\mu(\cdot) \in R_c(0, \bar{t}; U)$ if and only if it is a solution of (13.4.4).*

Proof. We begin with *(b)*. Let $y(\cdot)$ be a solution of (13.4.5). To show that $y(\cdot)$ solves (13.4.4) it is enough to show that for arbitrary $\mu(\cdot) \in R_c(0, T; U)$ we have

$$F(t, y(t))\mu(t) \in \text{conv}(f(t, y(t), U)) \quad \text{a.e. in } 0 \leq t \leq \bar{t} \qquad (13.4.11)$$

for $y(\cdot) \in C(0, T; E)$, which is a consequence of Lemma 13.3.2. Conversely, let $y(\cdot)$ be a solution of the differential inclusion (13.4.3). Apply Lemma 13.3.2 to the measurable function $g(t) = y'(t)$ and to $\phi(t, u) = f(t, y(t), u)$. Then we obtain $\mu(\cdot) \in R_c(0, \bar{t}; U)$ such that $y'(t) = F(t, y(t))\mu(t)$.

We prove (a). That a solution $y(\cdot)$ of (13.4.2) has to be a solution of (13.4.3) is obvious. Conversely, let $y(\cdot)$ be a solution of (13.4.3). Applying Theorem 13.3.1 to the function $\phi(t, u) = f(t, y(t), u)$ and to $g(t) = y'(t)$ we obtain $u(\cdot) \in C_{ad}(0, T; U)$ with $y'(t) = f(t, y(t), u(t))$. ∎

13.5. Existence Theorems for Relaxed Optimal Control Problems

Continuing the work in **2.2**, we consider the existence problem for the general system

$$y'(t) = f(t, y(t), u(t)), \qquad y(0) = \zeta. \tag{13.5.1}$$

In general, optimal controls will be relaxed. However, with the help of Filippov's theorem, we also obtain results on existence of ordinary optimal controls in next section.

We assume (13.5.1) satisfies Hypothesis II(C) in **13.1** so that the relaxed system

$$y'(t) = F(t, y(t))\mu(t), \qquad y(0) = \zeta, \tag{13.5.2}$$

satisfies Hypothesis II with $K(\cdot, c)$, $L(\cdot, c)$ independent of $\mu(\cdot)$ (Lemma 13.1.2). The problem includes state constraints and a target condition, respectively,

$$y(t, \mu) \in M(t), \quad (0 \leq t \leq \bar{t}) \qquad y(\bar{t}, \mu) \in Y. \tag{13.5.3}$$

We assume the cost functional $y_0(t, u)$ can be defined for relaxed controls $\mu(\cdot) \in R_c(0, T; U)$. The cost functional is $L^1(0, T; C(U))$-**weakly lower semi-continuous** if for every sequence $\{\mu^n(\cdot)\} \subseteq R_c(0, \bar{t}; U)$ such that (a) $\mu^n(\cdot) \to \bar{\mu}(\cdot) \in R_c(0, \bar{t}; U)$ $L^1(0, \bar{t}; C(U))$-weakly, (b) $y(t, \mu_n)$ exists in $0 \leq t \leq \bar{t}$ and $y(\cdot, u^n) \to y(\cdot)$ in $C(0, \bar{t}; E)$ we have

$$y_0(\bar{t}, \bar{\mu}) \leq \limsup_{n \to \infty} y_0(\bar{t}, \mu^n). \tag{13.5.4}$$

where we can replace lim sup by lim inf taking subsequences. This definition is a companion of the corresponding one in **2.2**; so is the next one.

Let $\{\mu^n(\cdot)\}$ be a sequence with $\mu^n \in R_c(0, t_n; U)$, $t_n \leq T$. Assume that $t_n \to \bar{t}$. Extend $\mu^n(\cdot)$ (resp. $y(t, \mu^n)$) to $[0, T]$ setting $\mu^n(t) = \mu \in \Pi_{rca}(U)$ (resp. $y(t, \mu^n) = y(t_n, u^n)$) in $t \geq t_n$, keeping the same name for the extensions. The cost functional $y_0(t, \mu)$ is **equicontinuous with respect to** t if for every $\{\mu^n(\cdot)\}$ as above and such that (a) $\mu^n(\cdot) \to \bar{\mu}(\cdot) \in R_c(0, T; U)$ $L^1(0, T; C(U))$-weakly, (b) $y(t, \mu^n)$ exists in $0 \leq t \leq t_n$ and $y(\cdot, \mu^n) \to y(\cdot)$ in $C(0, T; E)$, we have

$$|y_0(t_n, \mu^n) \to y_0(\bar{t}, \mu^n)| \to 0 \quad \text{as } n \to \infty. \tag{13.5.5}$$

We look at the end of this section at specific cost functionals having these properties.

The **original** optimal control problem is that of minimizing $y_0(\bar{t}, u)$ among all $u \in C_{ad}(0, \bar{t}; U)$ whose corresponding trajectories $y(t, u)$ satisfy both conditions (13.5.3) (the time $\bar{t}$ may be free or fixed). Let

$$m = \inf y_0(\bar{t}, u), \tag{13.5.6}$$

the infimum taken over all $\mu \in C_{ad}(0, \bar{t}; U)$ whose trajectories $y(\cdot, u)$ satisfy (13.5.3), and

$$\mathbf{m} = \inf y_0(\bar{t}, \mu), \tag{13.5.7}$$

the infimum taken over all $\mu \in R_c(0, t; U)$ whose trajectories $y(t, \mu)$ satisfy (13.5.3) (in both definitions, $\bar{t}$ varies for the free arrival time problem). It is clear that

$$\mathbf{m} \leq m \tag{13.5.8}$$

(there are more relaxed than ordinary controls). In principle, strict inequality is possible, including situations where $\mathbf{m} < \infty$, $m = \infty$ (the target may be attained by a trajectory $y(\cdot, \mu)$ of the relaxed system but not by a trajectory $y(\cdot, u)$ of the original system.)

Minimizing sequences of relaxed controls are defined in the same way as for ordinary controls: $\{\mu^n(\cdot)\}$, $\mu^n(\cdot) \in R_c(0, t_n; U)$ is a **minimizing sequence** if $y(t, \mu^n)$ exists in $0 \leq t \leq t_n$ and

$$\lim_{n \to \infty} \text{dist}(y(t, \mu^n), M(t)) = 0 \quad (0 \leq t < t_n), \tag{13.5.9}$$

$$\lim_{n \to \infty} \text{dist}(y(t_n, \mu^n), Y) = 0, \tag{13.5.10}$$

$$\limsup_{n \to \infty} y_0(t_n, \mu^n) \leq \mathbf{m}. \tag{13.5.11}$$

Existence Theorem 13.5.1. *Assume that $f(t, y, u)$ satisfies Assumption II(C) in* **13.1**[1] *and that the cost functional $y_0(t, \mu)$ is weakly lower semicontinuous and equicontinuous with respect to t. Further, assume that (a) $-\infty < \mathbf{m} < \infty$, (b) the state constraint sets $M(t)$ are closed, (c) the target set Y is closed. Let $\{\mu^n(\cdot)\}$, $\mu^n(\cdot) \in R_c(0, t_n; U)$ be a minimizing sequence of relaxed controls such that $\{t_n\}$ is bounded and $\{y(\cdot, \mu^n)\}$ is uniformly bounded. Then there exists a relaxed optimal control $\bar{\mu}(\cdot)$ which is the $L^1(0, T; C(U))$-weak limit in $L_w^\infty(0, T; \Sigma_{rca}(U))$ of a subsequence of $\{\mu^n(\cdot)\}$.*

Proof. Essentially the same as that of Theorem 2.2.6. We may assume that $t_n \to \bar{t}$. Let $t_n < T$. Extend $\mu^n(\cdot)$ to $0 \leq t \leq T$ setting $\mu^n(t) = \mu = $ fixed element of $\Sigma_{rca}(U)$, and use Theorem 12.5.9 (a) to ensure $L^1(0, T; C(U))$-weak convergence of a subsequence of $\{\mu^n(\cdot)\}$ to an element $\bar{\mu}(\cdot) \in R(0, T; U)$. Then invoke the

---

[1] Modulo some modifications. Theorem 13.5.1 can be proved under Assumption I(C). See Remark 2.2.17.

Arzelá-Ascoli Theorem to obtain a subsequence such that $\{y(t, \mu^n)\}$ (extended to $t_n \leq t \leq T$ by $y(t, \mu^n) = y(t_n, \mu^n)$) is uniformly convergent in $0 \leq t \leq T$. The theorem applies since uniform boundedness is part of the hypotheses and equicontinuity follows from (13.1.8):

$$\|y(t', \mu^n) - y(t, \mu^n)\|$$
$$\leq \int_t^{t'} \|F(\tau, y(\tau, \mu^n))\mu^n(\tau)\| d\tau \leq \int_t^{t'} K(\tau, c) d\tau, \quad (13.5.12)$$

where c is a bound for $\|y(t, \mu^n)\|$, $n = 1, 2, \ldots$. In order to show that $y(t) = \lim_{n \to \infty} y(t, \mu^n) = y(t, \bar{\mu})$ we take limits for $t < \bar{t}$:

$$y(t, \mu^n) = \zeta + \int_0^t F(\tau, y(\tau, \mu^n))\mu^n(\tau) d\tau$$
$$= \zeta + \int_0^{\bar{t}} \int_U \chi_t(\tau) f(\tau, y(\tau, \mu^n), u)\mu^n(\tau, du) d\tau$$
$$\to \zeta + \int_0^{\bar{t}} \int_U \chi_t(\tau) f(\tau, y(\tau), u)\bar{\mu}(\tau, du) d\tau$$
$$= \zeta + \int_0^t F(\tau, y(\tau))\bar{\mu}(\tau) d\tau \quad (13.5.13)$$

where χ_t is the characteristic function of the interval $[0, t]$. To justify passage to the limit under the integral sign we use (13.1.9), uniform convergence of $y(t, \mu^n)$, (13.1.2) and the dominated convergence theorem to show that

$$\int_0^{\bar{t}} \|\chi_t(\tau) f(\tau, y(\tau, \mu^n), \cdot) - \chi_t(\tau) f(\tau, y(\tau), \cdot)\|_{C(U; \mathbb{R}^m)} d\tau \to 0.$$

Since $\mu^n(\cdot) \to \mu(\cdot)$ $L^1(0, \bar{t}; C(U))$-weakly, equality (13.5.13) is legitimate. That $y_0(t, \mu) \leq \mathbf{m}$ follows directly from weak lower semicontinuity and t-equicontinuity of y_0. Finally, that the state constraint and the target condition are satisfied is obvious. This ends the proof. ∎

Remark 13.5.2. We may also apply Theorem 13.5.1 to a minimizing sequence $\{u^n(\cdot)\}$ of *ordinary* controls. In this way, we obtain a limit *relaxed* control $\bar{\mu}(t)$ satisfying the state constraints and the target condition and such that

$$y_0(\bar{t}, \bar{\mu}) \leq m.$$

This control is also optimal among relaxed controls if $\mathbf{m} = m$.

We look at cost functionals of the form

$$y_0(t, u) = \int_0^t f_0(\tau, y(\tau, u), u(\tau)) d\tau + g_0(t, y(t, u)) \quad (13.5.14)$$

with $f_0 : [0, T] \times E \times U \to \mathbb{R}$, $g_0 : [0, T] \times E \to \mathbb{R}$. Under Hypothesis $I^0(C)$ below (the counterpart of Hypothesis $I(C)$ for f) $y_0(t, u)$ can be extended to a weakly lower semicontinuous, t-equicontinuous cost functional $y_0(t, \mu)$ defined for $\mu(\cdot) \in R_c(0, T; U)$.

Hypothesis $I^0(C)$. $f_0(t, y, u)$ is measurable in t for y, u fixed and continuous in y, u for t fixed. For every $c > 0$ there exists a function $K_0(\cdot, c) \in L^1(0, T)$ such that

$$|f_0(t, y, u)| \leq K_0(t, c) \quad (0 \leq t \leq T, \|y\| \leq c, u \in U). \tag{13.5.15}$$

We show in the same way as we did for $f(t, y, u)$ in **13.1** that $t \to f_0(t, y, \cdot)$ is a strongly measurable $C(U)$-valued function and that

$$\|f_0(t, y', \cdot) - f_0(t, y, \cdot)\|_{C(U)} \to 0 \quad \text{as } y' \to y. \tag{13.5.16}$$

The **relaxed cost functional** $y_0(t, \mu)$ is

$$y_0(t, \mu) = \int_0^t F_0(\tau, y(\tau, \mu))\mu(\tau)d\tau + g_0(t, y(t, \mu)), \tag{13.5.17}$$

where

$$F_0(t, y, \mu) = \int_U f_0(t, y, u)\mu(du). \tag{13.5.18}$$

Lemma 13.5.3. *Let $f_0(t, y, u)$ satisfy Hypothesis $I^0(C)$ and let $g(t, y)$ be continuous. Then the relaxed cost functional $y_0(t, \mu)$ is weakly lower semicontinuous and t-equicontinuous.*

Proof. That the part of the functional depending on $g(t, y)$ has all the claimed properties is obvious. Let $\{\mu^n(\cdot)\} \subseteq R_c(0, T; U)$ be such that $\mu^n(\cdot) \to \bar{\mu}(\cdot)$ $L^1(0, \bar{t}; C(U))$-weakly in $L_w^\infty(0, T; \Sigma_{rca}(U))$ and $y(t, \mu^n) \to y(t)$ uniformly in $0 \leq t \leq \bar{t}$. We then show using (13.5.16) that $f_0(t, y(t, \mu^n), \cdot)$ is convergent to $f_0(t, y(t), \cdot)$ in $L^1(0, \bar{t}; C(U))$ and take limits using weak convergence of the μ^n; the result is

$$y_0(\bar{t}, \bar{\mu}) = \lim_{n \to \infty} y_0(\bar{t}, \mu^n). \tag{13.5.19}$$

Equicontinuity follows from the estimation

$$|y_0(t_n, \mu^n) - y_0(\bar{t}, \mu^n)| \leq \left|\int_{\bar{t}}^{t_n} K_0(\tau, c)d\tau\right|, \tag{13.5.20}$$

where c is a bound for $\|y(t, \mu^n)\|$. ∎

Example 13.5.4. We apply the combination of Theorem 13.5.1, Remark 13.5.2, and Lemma 13.5.3 to the minimum drag problem in **1.7**. Although ordinary solutions were characterized at length in **4.9**, existence of solutions of any type was never proved. The problem is to minimize the cost functional

$$\int_0^{\bar{t}} \frac{x(\tau)u(\tau)^3}{u(\tau)^2 + v(\tau)^2} d\tau$$

among the trajectories of the system

$$x'(t) = u(t), \qquad y'(t) = -v(t), \qquad x(0) = 0, \qquad y(0) = h.$$

There is no state constraint; the target condition is

$$(x(\bar{t}), y(\bar{t})) = (r, 0),$$

and the control set is $U = [0, \infty) \times [0, \infty)$. However, it has been argued in **4.9** that we may assume all curves in competition for the minimum are parametrized by arclength, so that we may reduce the control set to the quarter arc

$$U = \{(u, v); u, v \geq 0; u^2 + v^2 = 1\}.$$

We have $f(t, x, y, u, v) = (u, -v)$ and $f_0(t, x, y, u, v) = xu^3/(u^2 + v^2)$, so that Hypotheses II($C$) and I^0($C$) are satisfied. Existence of the minimizing sequence required in Theorem 13.5.1 is automatic; in fact, the family of curves in competition for the minimum is nonempty and the functional is nonnegative, which means that we may always construct a minimizing sequence $\{(u^n(\cdot), v^n(\cdot))\}$. That the arrival time t_n to the target must be bounded follows from

$$t_n = \int_0^{t_n} \sqrt{u^n(\tau)^2 + v^n(\tau)^2} d\tau \leq \int_0^{t_n} (u^n(\tau) + v^n(\tau)) d\tau = r + h \quad (13.5.21)$$

assuming that $\sqrt{u^n(\tau)^2 + v^n(\tau)^2} = 1$, that is, that each curve in the minimizing sequence is parametrized by arclength. Theorem 13.5.1 then produces an element $\bar{\mu}(\cdot) \in R_c(0, \bar{t}; U)$ with

$$\int_0^{\bar{t}} x(t) \int_U u^3 \bar{\mu}(t, dudv) dt \leq m, \qquad (13.5.22)$$

where $x(t)$ and $y(t)$ are given by

$$x(t) = \int_0^t \int_U u\bar{\mu}(t, dudv), \qquad y(t) = h - \int_0^t \int_U v\bar{\mu}(t, dudv), \qquad (13.5.23)$$

and $\bar{\mu}$ satisfies

$$\int_0^{\bar{t}} \int_U u\bar{\mu}(t, dudv) = r \qquad \int_0^{\bar{t}} \int_U v\bar{\mu}(t, dudv) = h. \qquad (13.5.24)$$

The minimum drag nose shape problem is one for which relaxed solutions are problematic, to say the least: admitting the "relaxed curves" (13.5.23) into competition requires a complete (and unphysical) resetting of the problem. However, we shall see in **13.7** that the relaxed solution found is actually an ordinary solution.

13.6. Existence Theorems for Ordinary Optimal Control Problems

Theorem 13.6.2 below gives conditions for existence of ordinary optimal controls that are more general than those in **2.2**; in particular, linearity in the control is not needed. Lemma 13.6.1 is an auxiliary result whose proof can be simplified using Filippov's theorem in full force (Theorem 12.8.2), but we stick to the cut-rate version (Theorem 13.3.1).

Lemma 13.6.1. *Let U be a compact metric space, $\phi : [0, T] \times U \to \mathbb{R}^m$ and $\phi_0 : [0, T] \times U \to \mathbb{R}$ such that*

(a) $u \to \phi(t, u)$, $u \to \phi_0(t, u)$ are continuous in u for each t fixed,

(b) $t \to \phi(t, u)$, $t \to \phi_0(t, u)$ are measurable in t for each u fixed,

(c) for every t the set

$$\mathcal{Z}(t; \phi_0, \phi) = \{(y_0, \phi(t, u)) \in \mathbb{R} \times \mathbb{R}^m; y_0 \geq \phi_0(t, u); u \in U\}$$
$$= \bigcup_{u \in U} ([\phi_0(t, u), \infty) \times \{\phi(t, u)\}) \tag{13.6.1}$$

is convex.

    Then, given $\mu(\cdot) \in R_c(0, T; U)$ there exists a strongly measurable U-valued function $u(t)$ such that

$$\phi(t, u(t)) = \int_U \phi(t, u)\mu(t, du) \quad \text{a.e. in } 0 \leq t \leq T, \tag{13.6.2}$$

$$\phi_0(t, u(t)) \leq \int_U \phi_0(t, u)\mu(t, du) \quad \text{a.e. in } 0 \leq t \leq T. \tag{13.6.3}$$

Proof. Let $n = 1, 2, \ldots$. Denote by e_n the set of all $t \in [0, T]$ such that

$$n \leq \|\phi_0(t, \cdot)\|_{C(U)} + \|\phi(t, \cdot)\|_{C(U; \mathbb{R}^m)} < n + 1.$$

It was proved in Lemma 13.1.1 that $t \to \phi_0(t, \cdot)$ (resp. $t \to \phi(t, \cdot)$) is a strongly measurable $C(U)$-valued function (resp. a strongly measurable $C(U; \mathbb{R}^m)$-valued function), thus the (disjoint) sets e_n are measurable and $[0, T] = \cup e_n$. We only need to construct $u(t)$ in each e_n; in other words we may assume from the beginning that $\phi(t, u)$ is bounded and that $-C \leq \phi_0(t, u) \leq C$. Rig the space $\mathcal{U} = [0, 2C] \times U$ with the product distance and let $\psi : [0, T] \times \mathcal{U} \to \mathbb{R} \times \mathbb{R}^m$ be the function defined by

$$\psi(t, (u_0, u)) = (\min(u_0 + \phi_0(t, u), C), \phi(t, u)) \quad ((u_0, u) \in \mathcal{U} = [0, 2C] \times U).$$

The function $\psi(t, (u_0, u))$ is continuous in $\mathcal{U}$ for each t, thus $\psi(t, \mathcal{U})$ is compact in $\mathbb{R} \times \mathbb{R}^m$. On the other hand if $u \in U$, we have $([0, 2C] + \phi_0(t, u)) \cap [-C, C] =$

$([0, \infty) + \phi_0(t, u)) \cap [-C, C]$ so that

$$\psi(t, \mathcal{U}) = \mathcal{Z}(t, \phi_0, \phi) \cap ([-C, C] \times \mathbb{R}^m)$$

and we gather that $\psi(t, \mathcal{U})$ is convex. Given $\mu(\cdot) \in R_c(0, T; U)$ define a function $g(t)$ with

$$g(t) = (\Phi_0(t)\mu(t), \Phi(t)\mu(t)),$$

where

$$\Phi_0(t)\mu(t) = \int_U \phi_0(t, u)\mu(t, du), \quad \Phi(t)\mu(t) = \int_U \phi(t, u)\mu(t, du).$$

It follows from Lemma 13.3.3 that

$$g(t) \in \mathrm{conv}(\psi(t, \mathcal{U})) = \psi(t, \mathcal{U}) \quad \text{a.e.} \tag{13.6.4}$$

We note finally that, since $\phi_0(t, u)$ is bounded, $t \to \phi_0(t, \cdot)$ belongs to $L^1(0, T; C(U))$; then, since $\mu(\cdot) \in L_w^\infty(0, T; \Sigma_{rca}(U))$, $\Phi_0(t)\mu(t)$ is measurable. A similar observation applies to $\Phi(t)\mu(t)$, so that $g(t)$ is measurable. Accordingly, we can invoke Filippov's Theorem 13.3.1 if we show that $(u_0, u) \to \psi(t, (u_0, u))$ is continuous in (u_0, u) for each t fixed and that $t \to \psi(t, (u_0, u))$ is measurable in t for each u fixed. Both conditions follow immediately from the corresponding conditions for $\phi(t, u)$ and $\phi_0(t, u)$. If $(u_0(t), u(t)) \in \psi(t, \mathcal{U})$ is the function provided by Filippov's theorem, then

$$\min(u_0(t) + \phi_0(t, u(t)), C) = \int_U \phi_0(t, u)\mu(t, du),$$

$$\phi(t, u(t)) = \int_U \phi(t, u)\mu(t, du).$$

The second equality is (13.6.2); the first is (13.6.3) since $\phi_0(t, u(t)) \leq \min(u_0(t) + \phi_0(t, u(t)), C)$. ∎

Existence Theorem 13.6.2. *Let the original system*

$$y'(t) = f(t, y(t), u(t)) \qquad y(0) = \zeta \tag{13.6.5}$$

satisfy Hypothesis II(C) in **13.1**[(1)]. *Assume the cost functional $y_0(t, u)$ is defined by*

$$y_0(t, u) = \int_0^t f_0(\tau, y(\tau, u), u(\tau))d\tau + g_0(t, y(t, u)) \tag{13.6.6}$$

[(1)] With some modifications, Hypothesis I(C) is enough.

where f_0 satisfies Hypothesis $I^0(C)$ in **13.5** *and $g(t, y)$ is continuous. Further, let the set*

$$\mathcal{Z}(t, y; f, f_0) = \{(y_0, f(t, y, u)) \in \mathbb{R} \times \mathbb{R}^m; y_0 \geq f_0(t, y, u), u \in U\}$$
$$= \bigcup_{u \in U} ([f_0(t, y, u), \infty) \times \{f(t, y, u)\}) \tag{13.6.7}$$

be convex for every t, y. Finally, assume that the state constraint sets $M(t)$ and the target set Y are closed and that there exists a minimizing sequence $\{\mu_n(\cdot)\}$, $\mu^n(\cdot) \in R_c(0, t_n; U)$ of relaxed controls such that (a) $\{t_n\}$ is bounded, (b) $\{y(\cdot, \mu^n)\}$ is uniformly bounded. Then there exists an ordinary optimal control $u(\cdot) \in C_{ad}(0, T; U)$.

Proof. We use Theorem 13.5.1 and obtain an optimal relaxed control $\bar{\mu}(\cdot)$. Then we apply Lemma 13.6.1 and obtain a control $u(\cdot) \in C_{ad}(0, T; U)$ such that

$$f(t, y(t, \mu), u(t)) = \int_U f(t, y(t, \mu), u)\mu(t, du)$$
$$= F(t, y(t, \mu))\mu(t) \quad \text{a.e. in } 0 \leq t \leq \bar{t} \tag{13.6.8}$$
$$f_0(t, y(t, \mu), u(t)) \leq \int_U f_0(t, y(t, \mu), u)\mu(t, du)$$
$$= F_0(t, y(t, \mu))\mu(t) \quad \text{a.e. in } 0 \leq t \leq \bar{t}. \tag{13.6.9}$$

Equality (13.6.8) and the integral equations (13.4.8) and (13.4.10) show that $y(t, \mu) = y(t, u)$, while inequality (13.6.9) implies $y_0(t, u) \leq y_0(t, \mu)$. ∎

Remark 13.6.3. Under the assumptions of Theorem 13.6.2, (13.5.8) can be obviously improved to

$$\mathbf{m} = m. \tag{13.6.10}$$

Remark 13.6.4. Theorem 13.6.2 confirms our hunch that existence of solutions of the control problem in Example 2.2.12 was related to convexity of the set $f(t, y, U) = \{1 + u - u^2; -1 \leq u \leq 1\} = [-1, 5/4]$. In fact, since $f_0(t, y, u) = y^2$ the set $\mathcal{Z}(t, y; f, f_0)$ reduces to $[y^2, \infty] \times f(t, y, U)$, thus is convex as well and Theorem 13.6.2 applies.

13.7. The Minimum Principle for Relaxed Optimal Control Problems

Our only job here is to pile up hypotheses on the original system

$$y'(t) = f(t, y(t), u(t)), \qquad y(0) = \zeta \tag{13.7.1}$$

that guarantee that the relaxed system

$$y'(t) = F(t, y(t))\mu(t), \qquad y(0) = \zeta \qquad (13.7.2)$$

satisfies the assumptions needed for the proof of the maximum principle in Part I. The rest is a direct application of the results in **3.7**, **4.2**, and **4.6**.

As in the previous sections, the control set U is a compact metric space. We assume that $f(t, y, u)$ has partial derivatives $\partial_k f_j(t, y, u)$ in $[0, T] \times E \times U$ satisfying

Hypothesis III(C). $f(t, y, u)$ and each $\partial_k f_j(t, y, u) = \partial f_j(t, y, u)/\partial y_k$ are continuous in y, u for t fixed and measurable in t for y, u fixed. For every $c > 0$ there exist $K(\cdot, c), L(\cdot, c) \in L^1(0, T)$ such that

$$\|f(t, y, u)\| \le K(t, c) \quad (0 \le t \le T, \|y\| \le c, u \in U), \qquad (13.7.3)$$
$$\|\partial_y f(t, y, u)\| \le L(t, c) \quad (0 \le t \le T, \|y\| \le c, u \in U), \qquad (13.7.4)$$

$\partial_y f(t, y, u)$ the Jacobian matrix.

Lemma 13.7.1. *Assume that* $f(t, y, u)$ *in* (13.7.1) *satisfies Hypothesis III(C). Then* $F(t, y)\mu$ *in* (13.7.2) *satisfies Hypothesis III for every* $\mu(\cdot) \in R_c(0, T; U)$ *with* $K(t, c), L(t, c)$ *independent of* $\mu(\cdot)$.

Proof. We have already shown in Lemma 13.1.2 that $F(t, y)\mu(t)$ is continuous in y and that $t \to F(t, y)\mu(t)$ is measurable; moreover,

$$\|F(t, y)\mu\| \le K(t, c)\|\mu\|_{\Sigma_{rca}(U)} \quad (0 \le t \le T, \|y\| \le c, \mu \in \Sigma_{rca}(U)). \qquad (13.7.5)$$

If we show that the partial derivatives $\partial_k F_j(t, y)\mu$ exist and are given by

$$\partial_k F_j(t, y)\mu = \int_U \partial_k f_j(t, y, u)\mu(du) \qquad (13.7.6)$$

then, arguing as in Lemma 13.1.2 we prove that $\partial_k F_j(t, y)\mu(t)$ is continuous in y and that $t \to \partial F(t, y)\mu(t)$ is measurable, with

$$\|\partial F(t, y)\mu\| \le L(t, c)\|\mu\|_{\Sigma_{rca}(U)} \quad (0 \le t \le T, \|y\| \le c, \mu \in \Sigma_{rca}(U)). \qquad (13.7.7)$$

We then show (13.7.6). Let $h_k = (0, \ldots, h, \ldots, 0)$ (h in the kth place). For every t fixed, $h^{-1}\{f_j(t, y + h_k, u) - f_j(t, y, u)\} = \partial_k f_j(t, y + \theta_k(t, u)h_k, u)$ with $0 \le \theta_k(t, u) \le 1$. Since $\partial_k f_j(t, y, u)$ must be uniformly continuous for y bounded and $u \in U$, it follows that

$$h^{-1}\{f_j(t, y + h_k, \cdot) - f_j(t, y, \cdot)\} \to \partial_k f_j(t, y, \cdot)$$

in $C(U; \mathbb{R}^m)$. Accordingly, the limit

$$\partial_k F_j(t, y)\mu = h^{-1}\{F_j(t, y + h)\mu - F_k(t, y)\mu\}$$

exists and equals (13.7.6). This ends the proof.   ∎

To apply the theory in **3.7**, **4.2**, and **4.6**, we install in the spaces $R_c(0, \bar{t}; U)$ the distance

$$d(\mu(\cdot), \nu(\cdot)) = |\{t \in [0, \bar{t}]; \mu(t) \neq \nu(t)\}| \qquad (13.7.8)$$

where $| \cdot |$ denotes Lebesgue measure; measurability of the set between bars follows from the fact that, by separability of $C(U)$ the norm of any element of $L_w^\infty(0, T; \Sigma_{rca}(U))$ is measurable (**12.5**) and from the equality $\{t; \mu(t) \neq \nu(t)\} = \{t; \|\mu(t) - \nu(t)\| > 0\}$.

We do the minimum principle without state constraints on the basis of spike perturbations as in Part I. We leave it to the reader interested in state constraints to follow the route beginning with patch perturbations in Chapter 6 and ending with the minimum principle in Chapter 10.

The spike perturbations are

$$\mu_{s,h,\nu}(t) = \begin{cases} \nu & (s - h < t \leq s) \\ \mu(t) & \text{elsewhere} \end{cases}$$

where $\nu \in \Pi_{rca}(U)$.

Lemma 13.7.2. $R_c(0, T; U)$ *is spike complete and saturated* (*thus complete with respect to the distance* (13.7.8)).

Proof. Spike completeness is obvious. If $\{\mu_n(\cdot)\}$ is a stationary sequence in $R_c(0, \bar{t}; U)$ we have $\mu_n(t) = \mu_{n+1}(t) = \cdots$ outside of a null set, where n depends on t. Obviously, $\mu(t) = \lim_{n \to \infty} \mu_n(t)$ belongs to $R_c(0, \bar{t}; U)$ (it is the $L^1(0, T; C(U))$-weak limit of μ_n in $L_w^\infty(0, T; \Sigma_{rca}(U))$).   ∎

We assume below that the target set Y is closed. The minimum principle for the time optimal problem is

Theorem 13.7.3. *Let $\bar{\mu}(\cdot)$ be a relaxed time optimal control. Then there exists $z \in N_Y(y(\bar{t}, \bar{\mu}))$, $\|z\| = 1$ such that if $\bar{z}(s)$ solves*

$$\bar{z}'(s) = -\{\partial_y F(s, y(s, \bar{\mu}))\bar{\mu}(s)\}^*\bar{z}(s), \qquad z'(\bar{t}) = z \qquad (13.7.9)$$

then

$$\langle \bar{z}(s), F(s, y(s, \bar{\mu}))\bar{\mu}(s)) \rangle = \min_{\nu \in \Pi_{rca}(U)} \langle \bar{z}(s), F(s, y(s, \bar{\mu}))\nu \rangle \qquad (13.7.10)$$

almost everywhere in $0 \leq s \leq \bar{t}$.

With the aid of the reduced relaxed Hamiltonian $H(s, y, z, \mu) = \langle z, F(s, y)\mu \rangle$ $= z_1 F_1(t, y)\mu + \cdots + z_m F_m(t, y)\mu$ $(y, z \in E)$ we can write (13.7.1) and (13.7.9) as the canonical equations

$$y'(t, \bar{\mu}) = \nabla_z H(t, y(t, \bar{\mu}), \bar{z}(t), \bar{\mu}(t)) \tag{13.7.11}$$

$$\bar{z}'(s) = -\nabla_y H(s, y(s, \bar{\mu}), \bar{z}(s), \bar{\mu}(s)), \tag{13.7.12}$$

and the minimum principle (13.7.10) becomes

$$H(s, y(s, \bar{\mu}), \bar{z}(s), \bar{\mu}(s)) = \min_{\nu \in \Pi_{rca}(U)} H(s, y(s, \bar{\mu}), \bar{z}(s), \nu) \tag{13.7.13}$$

almost everywhere in $0 \leq s \leq \bar{t}$.

To prove the minimum principle for the general cost functional

$$y_0(t, u) = \int_0^t f_0(\tau, y(\tau, u), u(\tau))d\tau + g_0(t, y(t, u)), \tag{13.7.14}$$

we require

Hypothesis III^0(C). $f_0(t, y, u)$ and each $\partial_k f_0(t, y, u)$ are continuous in y, u for t fixed and measurable in t for y, u fixed. For every $c > 0$ there exist $K_0(\cdot, c)$, $L_0(\cdot, c) \in L^1(0, T)$ such that

$$|f_0(t, y, u)| \leq K_0(t, c) \quad (0 \leq t \leq T, \|y\| \leq c, u \in U) \tag{13.7.15}$$

$$\|\partial f_0(t, y, u)\| \leq L_0(t, c) \quad (0 \leq t \leq T, \|y\| \leq c, u \in U), \tag{13.7.16}$$

with $\partial f_0(t, y, u)$ the gradient vector.

Lemma 13.7.4. *Assume that $f_0(t, y, u)$ satisfies Hypothesis III^0(C). Then $F_0(t, y)\mu$ in the relaxed cost functional*

$$y_0(t, \mu) = \int_0^t F_0(\tau, y(\tau, \mu))\mu(\tau)d\tau + g_0(t, y(t, \mu)), \tag{13.7.17}$$

satisfies Hypothesis III^0 for every $\mu(\cdot) \in R_c(0, T; U)$ with functions $K_0(t, c)$, $L_0(t, c)$ independent of $\mu(\cdot)$.

The proof is similar to that of Lemma 13.7.1.

In the minimum principle below, we assume that $g(t, y)$ is differentiable with respect to y.

Theorem 13.7.5. *Let $\bar{\mu}(\cdot)$ be a relaxed optimal control. Then there exists $(z_0, z) \in \mathbb{R} \times E$, $z_0^2 + \|z\|^2 = 1$, $z_0 \geq 0$, $z \in N_Y(y(\bar{t}, \bar{\mu}))$ such that if $\bar{z}(s)$ is the solution of*

$$\bar{z}'(s) = -\{\partial_y F(s, y(s, \bar{\mu}))\bar{\mu}(s)\}^* \bar{z}(s) - z_0 \partial_y F_0(s, y(s, \bar{\mu}))\bar{\mu}(s),$$
$$\bar{z}(\bar{t}) = z + z_0 \partial_y g_0(\bar{t}, y(\bar{t}, \bar{\mu})) \tag{13.7.18}$$

then

$$z_0 F_0(s, y(s, \bar{\mu}))\bar{\mu}(s) + \langle \bar{z}(s), F(s, y(s, \bar{\mu}))\bar{\mu}(s) \rangle$$
$$= \min_{v \in \Pi_{rca}(U)} \{z_0 F_0(s, y(s, \bar{\mu}))v + \langle \bar{z}(s), F(s, y(s, \bar{\mu}))v \rangle\} \quad (13.7.19)$$

almost everywhere in $0 \le s \le \bar{t}$.

Bringing into play the relaxed Hamiltonian $H(t, \mathbf{y}, \mathbf{z}, \mu) = \langle \mathbf{z}, \mathbf{F}(t, y)\mu \rangle = z_0 F_0(t, y)\mu + z_1 F_1(t, y)\mu + \cdots + z_m F_m(t, y)\mu$, where $\mathbf{y} = (y_0, y_1, \dots, y_m) = (y_0, y)$, $\mathbf{z} = (z_0, z_1, \dots, z_m) = (z_0, z) \in \mathbb{R} \times E$ and $\mathbf{F} = (F_0, F_1, \dots, F_m)$ we combine the equation (13.7.1) (and the cost functional) and the adjoint variational equation (13.7.18) into the canonical equations

$$\mathbf{y}'(t, \bar{u}) = \nabla_{\mathbf{z}} H(t, \mathbf{y}(t, \bar{u}), \bar{\mathbf{z}}(t), \bar{\mu}(t)), \quad (13.7.20)$$

$$\bar{\mathbf{z}}'(s) = -\nabla_{\mathbf{y}} H(s, \mathbf{y}(s, \bar{u}), \bar{\mathbf{z}}(s), \bar{\mu}(s)), \quad (13.7.21)$$

and the minimum principle is

$$H(s, \mathbf{y}(s, \bar{\mu}), \bar{\mathbf{z}}(s), \bar{\mu}(s)) = \min_{v \in \Pi_{rca}(U)} H(s, \mathbf{y}(s, \bar{\mu}), \bar{\mathbf{z}}(s), v) \quad (13.7.22)$$

a.e. in $0 \le s \le \bar{t}$.

Application of Theorem 4.6.1 on the vanishing of the Hamiltonian for problems with free endtime requires verification that $R_c(0, T; U)$ has the reparametrization property (see **4.6**). Example 13.7.10 is an invitation to do this job.

Example 13.7.6. We try the relaxed minimum principle on the optimal control problem (12.1.6)-(12.1.7) in its relaxed version (12.1.4)-(12.1.8). We already know that the original problem has no solution, thus the conditions of Theorem 13.6.2 cannot be satisfied. This is confirmed noting that $\mathcal{Z}(t, y; f, f_0)$ is the set above the curve of equation $u \to y^2 + (u^2 - 1)^2$ in the interval $-1 \le u \le 1$. On the other hand, the functions $f(t, y, u) = u$ and $f_0(t, y, u) = y^2 + (u^2 - 1)^2$ fit comfortable into Theorem 13.5.1, thus a relaxed optimal control $\bar{\mu}(t, du)$ exists. The minimizing sequence (12.1.3) warns that the minimum is zero, thus the optimal relaxed trajectory $\bar{y}(t) = y(t, \bar{\mu})$ must be identically zero.

We use the notation $(z_0, z) = (\lambda, p)$ for the costate and change the minimum into a maximum, so that $\lambda \le 0$. The relaxed Hamiltonian is

$$H(y_0, y, \lambda, p, \mu) = \lambda \int_{-1}^{1} \{y^2 + (u^2 - 1)^2\}\mu(du) + p \int_{-1}^{1} u\mu(du),$$

and the canonical equation for $p(s)$,

$$p'(s) = -2\lambda \int_{-1}^{1} \bar{y}(s)\bar{\mu}(s, du) = -2\lambda \bar{y}(s). \quad (13.7.23)$$

Eliminating terms not depending on μ, the maximum principle for a candidate $\bar{\mu}(t, du)$ to relaxed optimal control becomes

$$\lambda \int_{-1}^{1} (u^2 - 1)^2 \bar{\mu}(s, du) + p(s) \int_{-1}^{1} u \bar{\mu}(s, du)$$

$$= \max_{v \in \Pi_{rca}(U)} \left(\lambda \int_{-1}^{1} (u^2 - 1)^2 v(du) + p(s) \int_{-1}^{1} uv(du) \right). \quad (13.7.24)$$

We have $Y = \mathbb{R}$ so that $N_Y(y(\bar{t}, \bar{\mu})) = \{0\}$. This implies that $p(\bar{t}) = 0$, hence $p(s) \equiv 0$. Since $(\lambda, p(\bar{t})) \neq 0$, $\lambda < 0$ and $\bar{\mu}(s, du)$ satisfies

$$\int_{-1}^{1} (u^2 - 1)^2 \bar{\mu}(s, du) = \min_{v \in \Pi_{rca}(U)} \int_{-1}^{1} (u^2 - 1)^2 v(du) \quad (13.7.25)$$

which means that, for every s, $\bar{\mu}(s)$ must be a convex combination of $\delta(\cdot - 1)$ and $\delta(\cdot + 1)$. On account of the fact that $\bar{y}(t) \equiv 0$ we must have

$$\bar{\mu}(t) = \frac{1}{2}\delta(t - 1) + \frac{1}{2}\delta(t + 1). \quad (13.7.26)$$

Example 13.7.7. A less jejune application of Theorem 13.7.5 is to the minimum drag problem, which makes here its next-to-last appearance. We use the control set and existence result in Example 13.5.4. The relaxed Hamiltonian is

$$H(x, y, \lambda, p, q, \mu) = \int_U \left(\lambda x \frac{u^3}{u^2 + v^2} + pu - qv \right) \mu(dudv)$$

so that the maximum principle is

$$\int_U \left(\lambda \bar{x}(s) \frac{u^3}{u^2 + v^2} + p(s)u - qv \right) \bar{\mu}(s, dudv)$$

$$= \max_{v \in \Pi_{rca}(U)} \int_U \left(\lambda \bar{x}(s) \frac{u^3}{u^2 + v^2} + p(s)u - qv \right) v(dudv) = 0 \quad (13.7.27)$$

where $\bar{x}(s) = x(s, \bar{\mu})$ is the first coordinate of the optimal solution. Since

$$U = \{(u, v); u, v \geq 0, u^2 + v^2 = 1\}, \quad (13.7.28)$$

the denominator can be omitted, but is convenient not to do so. The fact that the maximum is zero is due to the fact that there are no restrictions on the arrival time. The canonical equations for the costate are

$$p'(s) = -\lambda \bar{x}(s) \int \frac{u^3}{u^2 + v^2} \bar{\mu}(s, dudv), \qquad q'(s) = 0, \quad (13.7.29)$$

the functions $\bar{x}(s)$, $\bar{y}(s)$ given by (13.5.23). It follows from (13.7.27) and the fact that v is arbitrary that the function between parentheses in the first integral is ≤ 0 in U; since the function is homogeneous of degree 1, we must have

$$\lambda \bar{x}(s) \frac{u^3}{u^2 + v^2} + p(s)u - qv \leq 0 \qquad (u, v \geq 0) \quad (13.7.30)$$

for $0 \le s \le \bar{t}$. We do now an analysis of this function roughly in the style of **4.9**; the final objective is to show that

$$\bar{\mu}(s, dudv) = \delta(\cdot - (\bar{u}(s), \bar{v}(s))) \qquad (13.7.31)$$

with $\bar{u}(\cdot)$, $\bar{v}(\cdot)$ measurable, so that the optimal relaxed control constructed in Example 13.5.4 is actually an *ordinary* control.

If $\lambda = 0$, then $p(s) \equiv p$. Since $(\lambda, p(\bar{t}), q(\bar{t})) \ne 0$, we must have $(p, q) \ne 0$. Assuming $q = 0$ the integrand in (13.7.27) is pu, hence $\bar{\mu}(t) \equiv \delta_{(1,0)}$, a Dirac delta with mass in $(1, 0) \in U$. This relaxed control corresponds to the ordinary control $u(t) \equiv 1$, $v \equiv 0$, which cannot accommodate all conditions except in the trivial case $h = 0$. A similar argument disposes of the possibility $q = 0$. Note finally that (13.7.30) implies $p \le 0$, $q \ge 0$, thus if $p \ne 0$, $q \ne 0$ the integrand of (13.7.27) is everywhere negative and $\bar{\mu}(s) \equiv 0$, absurd.

The possibility that $\lambda = 0$ discarded, we may take for granted that $\lambda = -1$. We may also assume that $\bar{x}(s) > 0$ for $s > 0$. In fact $\bar{x}(t)$ is nondecreasing, thus if s is the largest point with $\bar{x}(s) = 0$ we have $x(t) \equiv 0$ $(0 \le t \le s)$, thus the control is ordinary in $[0, s]$ $(u(t) \equiv 0, y(t) \equiv 1)$ and we can restrict our attention to the interval $[s, \bar{t}]$.

Finally, we may also count on the condition $q > 0$. In fact, if $q = 0$ the left side of (13.7.30) is $-\bar{x}(s)u^3 + p(s)u$; since $\bar{x}(s) > 0$ and $p(s) \le 0$, the maximum occurs at $u = 0$ which means, as argued above, that the control is ordinary. In conclusion,

$$\lambda = -1, \qquad \bar{x}(s) > 0, \qquad q > 0. \qquad (13.7.32)$$

At this point, we are in the same situation as in **4.9** after (4.9.21), and we use the analysis of the Hamiltonian there. In particular, there exists exactly one maximum $(\bar{u}(s), \bar{v}(s))$ in U, hence the relaxed control $\bar{\mu}(s, dudv)$ satisfying (13.7.27) is a travelling delta of the form (13.7.31), thus an ordinary control $(\bar{u}(s), \bar{v}(s))$. This control is completely identified in **4.9**, and we have finally completed our analysis of the minimum drag nose shape problem and confirmed Newton's 1686 solution. A final crack at the problem will be taken in next section, but only for illustration of another approach.

The arguments above may in principle be applied to a general control problem (13.7.1) with a cost functional (13.7.17); the maximum principle for a relaxed optimal control $\bar{\mu}(s, du)$ (existence guaranteed by Theorem 13.5.1) is

$$\int_U \{z_0 f_0(s, y(s, \bar{\mu}), u) + \langle \bar{z}(s), f(s, y(s, \bar{\mu}), u)\rangle\}\bar{\mu}(s, du)$$

$$= \max_{v \in \Pi_{rca}(U)} \int_U \{z_0 f_0(s, y(s, \bar{\mu}), u) + \langle \bar{z}(s), f(s, y(s, \bar{\mu}), u)\rangle\}v(du).$$

If we can show that the function of u between curly brackets attains its maximum in u at exactly one point $\bar{u}(s)$, then we have $\bar{\mu}(s) = \delta(\cdot - \bar{u}(s))$ so the optimal control is actually an ordinary control. In general, the support of the measure $\bar{\mu}(s)$ must be contained in the set of points where $\{\cdot\}$ attains its maximum.

Example 13.7.8. The space $R_c(0, T; U)$ has the reparametrization property; if $\varepsilon < 1$, $t(\tau)$ is an absolutely continuous function with $|t'(\tau) - 1| \le \varepsilon$ and $\mu(\cdot)$ is an element of $R_c(0, \bar{t}; U)$ then $\mu(t(\cdot)) \in R_c(0, t(\bar{t}); U)$ and vice versa. That $\mu(t(\cdot)) \in L_w^\infty(0, T; \Sigma_{rca}(U))$ follows from changing variables in the integral defining $\langle \mu(t(\tau)), f(\tau) \rangle$ for $f(\cdot) \in L^1(0, \bar{t}; C(U))$.

13.8. Noncompact Control Sets

The assumption that the control set U is a compact metric space excludes unbounded control sets U in $\mathbb{R}^k$. We sketch how to deal with a closed, unbounded control set.

We denote by $\Sigma_{rca}(U)$ the space of all bounded, regular, countably additive measures defined the field $\Phi_b(C)$ of all Borel sets of U and equipped with the total variation norm. $C_0(U)$ is the (Banach) space of all continuous functions tending to zero at infinity equipped with the supremum norm.

Theorem 13.8.1. *Let* $U = \mathbb{R}^k$. *The dual space* $C_0(U)^*$ *can be identified algebraically and metrically with* $\Sigma_{rca}(U)$, *the duality pairing given by*

$$\langle f, \lambda \rangle = \int_U f(u)\lambda(du). \tag{13.8.1}$$

Proof. Let Φ be the stereographic projection of $\mathbb{R}^k$ into the surface S of its unit ball minus the north pole p. Then Φ and its inverse are continuous, so that the distance

$$\rho(u, v) = \|\Phi(u) - \Phi(v)\|$$

is equivalent to the Euclidean distance $\|u - v\|$, that is, it defines the same topology or, equivalently, the same convergent sequences than the Euclidean norm. Adjoin to $\mathbb{R}^k$ the point at infinity ∞; the distance from any $u \in \mathbb{R}^k$ to ∞ is $\rho(\infty, u) = \|\Phi(u) - p\|$. Then $\mathbb{R}^k \cup \{\infty\}$ is a compact metric space. Setting $f(\infty) = 0$, we see that the space $C_0(\mathbb{R}^k)$ is the same as the subspace $C_\infty(\mathbb{R}^k \cup \{\infty\})$ of $C(\mathbb{R}^k \cup \{\infty\})$ consisting of all continuous functions in $\mathbb{R}^k \cup \{\infty\}$ that vanish at ∞. Its dual can be identified with the subspace $\Sigma_{rca,\infty}(\mathbb{R}^k \cup \{\infty\})$ of $\Sigma_{rca}(\mathbb{R}^k \cup \{\infty\})$ defined by $\mu(\{\infty\}) = 0$. This subspace is $\Sigma_{rca}(\mathbb{R}^k)$.   ■

Corollary 13.8.2. *Let* $U \subseteq \mathbb{R}^k$ *be an arbitrary closed set. Then the dual space* $C_0(U)^*$ *can be identified algebraically and metrically with* $\Sigma_{rca}(U)$, *the duality pairing given by* (13.8.1).

Proof. Obviously, (13.8.1) defines a continuous linear functional on $C_0(U)$. Conversely, every continuous linear functional on $C_0(U)$ is a continuous linear functional on $C_0(\mathbb{R}^k)$ (we apply it to restrictions of functions to U) and is thus given by (13.8.1). We only have to show that μ is supported by U, that is, that $\mu(d) = 0$ for every open set d with $d \cap U = \emptyset$. This can be seen as follows. Let $d_\varepsilon = \{u \in d; \operatorname{dist}(u, d^c) \geq \varepsilon\}$. If $f_\varepsilon(u) = \operatorname{dist}(u, d^c)/(\operatorname{dist}(u, d^c) + \operatorname{dist}(u, d_\varepsilon))$ then $f_\varepsilon(u) \equiv 0$ in U so that $\int f_\varepsilon(u)\mu(du) = 0$. Since $f_\varepsilon(u) \to \chi(u)$ everywhere, χ the characteristic function of d, it follows from the dominated convergence theorem that $\mu(d) = 0$. ∎

This result and Theorem 12.2.11 imply the following.

Corollary 13.8.3. $L^1(0, T; C_0(U))^* = L_w^\infty(0, T; \Sigma_{rca}(U))$.

The definition of the space $R_c(0, T; U) \subseteq L_w^\infty(0, T; \Sigma_{rca}(U))$ is exactly the same as that for compact U; we require

$$\int_0^T \int_U f(t, u)\mu(t, du)dt \geq 0. \tag{13.8.2}$$

for $f(\cdot, \cdot) \in L^1(0, T; C_0(U))$ with $f(t, u) \geq 0$ ($u \in U$) a.e. in t and[1]

$$\int_0^T \int_U (\chi(t) \otimes \mathbf{1})\mu(t, du)dt = |e| \tag{13.8.3}$$

if $\chi(t)$ is the characteristic function of a measurable set $e \subseteq [0, T]$ and $\mathbf{1}(u) \equiv 1$, However, the function $\mathbf{1}$ does not belong to $C_0(U)$, thus we cannot ensure any more that $R_c(0, T; U)$ is weakly closed. For instance, if $\{u_n\} \subseteq U$ with $u_n \to \infty$ the sequence $\{\delta(\cdot - u_n)\}$ belongs to $R_c(0, T; U)$ but we have $\delta(\cdot - u_n) \to 0$ $L^1(0, T; C_0(U))$-weakly in $L_w^\infty(0, T; \Sigma_{rca}(U))$. A smaller space and a stronger notion of weak convergence are needed to prevent "escape of mass to ∞."

Let $\kappa(u)$ be a real-valued continuous function defined in U and such that

$$\kappa(u) \geq 1, \qquad \lim_{u \in U, \|u\| \to \infty} \kappa(u) = \infty, \tag{13.8.4}$$

and consider the space $R_{c,\kappa}(0, T; U) \subseteq R_c(0, T; U)$ consisting of all $\mu(\cdot)$ with

$$\|\mu(\cdot)\|_\kappa \in L^2(0, T),^{(2)} \tag{13.8.5}$$

where

$$\|\mu\|_\kappa^2 = \int_U \kappa(u)^2 \mu(du). \tag{13.8.6}$$

---

[1] Recall that each $\mu(t, du)$ is a bounded measure in U so that, the integral makes sense.
[2] Here $\|\mu(\cdot)\|_\kappa$ denotes the function $t \to \|\mu(t)\|_\kappa$.

To give sense to the definition, we must show that if $\mu(\cdot) \in L_w^\infty(0, T; \Sigma_{rca}(U))$ then the function $t \to \|\mu(t)\|_\kappa$ is measurable if finite a.e. We show a little more with a view to future use.

Lemma 13.8.4. *Let $\phi : [0, T] \times U \to \mathbb{R}$ be continuous in u for each t fixed, measurable in t for each u fixed, and let $\mu \in R_c(0, T; U)$. Then the function $\Phi(t) = \int_U \phi(t, u)\mu(t, du)$ is measurable if it is finite a.e.*

Proof. Let $\eta_n(u)$ be a continuous function in $\mathbb{R}^k$ with $0 \le \eta_n(u) \le 1$,

$$\eta_n(u) = \begin{cases} 1 & (\|u\| \le n), \\ 0 & (\|u\| \ge n + 1). \end{cases} \tag{13.8.7}$$

Then, $\eta_n(\cdot)\phi(t, \cdot) \in C_0(U)$, so that $t \to \Phi(t) = \int_U \eta_n(u)\phi(t, u)\mu(t, du)$ is measurable (the proof is the same as that of Lemma 13.1.1 since $C_0(U)$ is separable). The result follows from the fact that $\Phi_n(t) \to \Phi(t)$ where $\Phi(t) < \infty$, a consequence of the dominated convergence theorem. ∎

Theorem 13.8.5. *Let $\{\mu_n(\cdot)\}$ be a sequence in $R_{c,\kappa}(0, T; U)$ such that (a) $\mu_n(\cdot) \to \bar{\mu}(\cdot)$ $L^1(0, T; C_0(U))$-weakly in $L_w^\infty(0, T; \Sigma_{rca}(U))$, (b) $\|\mu_n(\cdot)\|_\kappa$ is bounded in $L^2(0, T)$. Then $\bar{\mu}(\cdot) \in R_{c,\kappa}(0, T; U)$.*

Proof. Let $n = 1, 2, \ldots$. Let $U_n = \{u \in U; \|u\| \le n\}$, U_n^c its complement, M_n the maximum of $\kappa(u)^{-2}$ in U_n^c. Then $M_n \to 0$ and

$$\int_{U_n^c} \mu_n(t, du) \le M_n \int_{U_n^c} \kappa(u)^2 \mu_n(t, du) \le M_n \|\mu_n(t)\|_\kappa^2$$

for all t. If $\chi(t)$ is the characteristic function of a measurable set $e \subseteq [0, T]$ we have

$$\int_0^T \int_{U_n} (\chi(t) \otimes 1)\mu_n(t, du)dt \ge \int_e \int_{U_n} \mu_n(t, du)dt$$

$$= \int_e \left(1 - \int_{U_n^c} \mu_n(t, du)dt\right) \ge |e|(1 - M_n\|\mu_n(\cdot)\|_\kappa^2)$$

so that (13.8.3) is satisfied in the limit, and it only remains to show (13.8.5). Let

$$\phi_m(t) = \int_U \eta_m(u)\kappa(u)^2\mu(t, du),$$

$$\phi_{mn}(t) = \int_U \eta_m(u)\kappa(u)^2\mu_n(t, du) \quad (0 \le t \le T).$$

We have $0 \le \phi_{mn}(t) \le \|\mu_n(t)\|_\kappa^2$, so that $\|\phi_{mn}\|_{L^1(0,T)} \le C$. On the other hand, if $\psi(\cdot) \in L^1(0, T)$ the function $t \to \psi(t)\eta_m(\cdot)\kappa(\cdot)^2$ belongs to $L^1(0, T; C_0(U))$, hence

$$\int_0^T \psi(t)\phi_{mn}(t)dt = \int_0^T \psi(t) \int_U \eta_m(u)\kappa(u)^2 \mu_n(t, du)dt$$
$$\to \int_0^T \psi(t) \int_U \eta_m(u)\kappa(u)^2 \mu(t, du)dt = \int_0^T \psi(t)\phi_m(t)dt$$

as $n \to \infty$. In particular, taking $\psi(\cdot) \in L^\infty(0, T)$ we deduce that $\phi_{mn}(\cdot) \to \phi_m(\cdot)$ L^∞-weakly in $L^1(0, T)$, which implies $\|\phi_m(\cdot)\|_{L^1(0,T)} \leq C$ for all m. On the other hand, we have $\phi_1(t) \leq \phi_2(t) \leq \ldots$, hence by Beppo Levi's Theorem 12.9.3 $\phi_m(u) \to \phi(t)$ a.e. and $\|\phi\|_{L^1(0,T)} \leq C$. It remains to justify that

$$\phi(t) = \int_U \kappa(u)^2 \mu(t, du)$$

and, since $\eta_1(u)\kappa(u)^2 \leq \eta_2(u)\kappa(u)^2 \leq \ldots$ and $\eta_m(u)\kappa(u)^2 \to \kappa(u)^2$ as $m \to \infty$, this follows from Fatou's Theorem (Example 8.1.7; the proof there is for Lebesgue measure in a set of finite measure but extends with the same proof to a finite Borel measure. Note that we have everywhere convergence thus we don't need to get involved with the null sets of $\mu(t, du)$). ∎

The assumptions on

$$y'(t) = f(t, y(t), u(t)), \qquad y(0) = \zeta \qquad (13.8.8)$$

are adaptations of those in **13.1**.

Hypothesis I(C, κ). $f(t, y, u)$ is measurable in t for y, u fixed and continuous in y, u for t fixed. For every $c > 0$ there exists $K(\cdot, c) \in L^2(0, T)$ such that

$$\|f(t, y, u)\| \leq K(t, c)\kappa(u) \qquad (0 \leq t \leq T, \|y\| \leq c, u \in U). \qquad (13.8.9)$$

The "natural" admissible control space $C_{ad}(0, \bar{t}; U)$ consists of all strongly measurable U-valued functions such that

$$\kappa(u(t)) \in L^2(0, T).$$

The relaxed control system is

$$y'(t) = F(t, y(t))\mu(t), \qquad y(0) = \zeta \qquad (13.8.10)$$

with control set $\mathbf{U} = \Pi_{rca}(U, \kappa) \subseteq \Pi_{rca}(U)$ consisting of all $\mu \in \Pi_{rca}(U)$ with $\|\mu\|_\kappa^2 < \infty$ ($\|\mu\|_\kappa^2$ defined in (13.8.6)), and $F : [0, T] \times E \times \Sigma_{rca}(U) \to E$ defined by

$$F(t, y)\mu = \int_U f(t, y, u)\mu(du), \qquad (13.8.11)$$

the integral interpreted coordinatewise. That (13.8.11) exists follows from the Schwarz inequality applied to $\kappa(u)$, 1:

$$\|F(t, y)\mu\| \leq \int_U \|f(t, y, u)\|\mu(du) \leq K(t, c) \int_U \kappa(u)\mu(du)$$
$$\leq K(t, c)\left(\int_U \kappa(u)^2\mu(du)\right)^{1/2}\left(\int_U \mu(du)\right)^{1/2}$$
$$\leq K(t, c)\left(\int_U \kappa(u)^2\mu(du)\right)^{1/2} = K(t, c)\|\mu\|_\kappa. \quad (13.8.12)$$

The relaxed control space is the subspace $R_{c,\kappa}(0, T; U) \subseteq R_c(0, T; U)$ whose elements satisfy $\|\mu(\cdot)\|_\kappa \in L^2(0, T)$ or, equivalently,

$$\|\mu(\cdot)\|_\kappa^2 = \int_U \kappa(u)^2\mu(\cdot, du) \in L^1(0, T). \quad (13.8.13)$$

Under Hypothesis $I(C, \kappa)$ for $f(t, y, u)$ the function $F(t, y)\mu$ satisfies Hypothesis I. In fact, (13.8.12) shows that $\|F(t, y)\mu(t)\| \leq K(t, c)\|\mu(t)\|_\kappa$, where $K(\cdot, c)\|\mu(\cdot)\|_\kappa \in L^1(0, T)$, and Lemma 13.8.4 implies that $t \to F(t, y)\mu(t)$ is measurable. To prove y-continuity, we note that by the Schwarz inequality,

$$\|F(t, y')\mu - F(t, y)\mu\|^2$$
$$\leq \int_U (\|f(t, y', u) - f(t, y, u)\|\kappa(u)^{-1})^2\mu(du)\int_U \kappa(u)^2\mu(du),$$

where the first integral tends to zero as $y' \to y$ by the dominated convergence theorem.

Hypothesis II(C, κ). $f(t, y, u)$ is measurable in t for y, u fixed and continuous in y, u for t fixed. For every $c > 0$ there exist $K(\cdot, c), L(\cdot, c) \in L^2(0, T)$ such that (13.8.9) holds and

$$\|f(t, y', u) - f(t, y, u)\| \leq L(t, c)\kappa(u)\|y' - y\|$$
$$(0 \leq t \leq T, \|y\|, \|y'\| \leq c, u \in U). \quad (13.8.14)$$

Under Hypothesis II(C, κ) the function $F(t, y)\mu$ satisfies Hypothesis II. We just have to notice that (13.8.14) and the Schwarz inequality imply

$$\|F(t, y')\mu(t) - F(t, y)\mu(t)\| \leq L(t, c)\|y' - y\|\int_U \kappa(u)^2\mu(t, du)$$
$$\leq L(t, c)\|\mu(t)\|_\kappa\|y' - y\|. \quad (13.8.15)$$

After an auxiliary result below, we prove an existence theorem under Hypothesis II(C, κ): for simplicity, we take fixed terminal time. The problem includes state constraints and a target condition

$$y(t, \mu) \in M(t), \quad (0 \leq t \leq \bar{t}) \qquad y(\bar{t}, \mu) \in Y. \quad (13.8.16)$$

Given $\mu \in \mathbf{U}$, define a measure λ by

$$\lambda(e) = \int_e \kappa(u)\mu(du) \qquad (13.8.17)$$

for e a Borel subset of U (informally, $\lambda(du) = \kappa(u)\mu(du)$). Then $\lambda \in \Sigma_{rca}(U)$ and

$$\|\lambda\| = \int_U \kappa(u)\mu(du) \leq \left(\int_U \kappa(u)^2 \mu(du) \right)^{1/2} = \|\mu\|_\kappa. \qquad (13.8.18)$$

In the next result, $BC(U)$ is the space of bounded continuous functions in U equipped with the supremum norm and $\Sigma_{rba}(U) = BC(U)^*$ is the space of all regular bounded finitely additive measures in U. See **12.9** for the definition of the space $L^2_w(0, T; BC(U)^*)$.

Lemma 13.8.6. *Let* $\mu(\cdot) \in R_{c,\kappa}(0, T; U)$ *and* $\lambda(\cdot)$ *the measure-valued function constructed from* $\mu(\cdot)$ *in* (13.8.17). *Then*

$$\lambda(\cdot) \in L^2_w(0, T; \Sigma_{rba}(U)) = L^2_w(0, T; BC(U)^*).$$

Proof. Obviously $\Sigma_{rca}(U) \subseteq \Sigma_{rba}(U)$ and, by (13.5.18), $\|\lambda(\cdot)\|_{\Sigma_{rca}(U)} = \|\lambda(\cdot)\|_{\Sigma_{rba}(U)} \leq \|\mu(\cdot)\|_\kappa$. The function $t \to \|\lambda(t)\|$ is measurable by Lemma 13.8.4, and it follows from (13.8.18) that it belongs to $L^2(0, T)$. We then only have to show that the real-valued function $t \to \langle f, \lambda(t) \rangle$ is measurable for every element $f(\cdot)$ of the space $BC(U)$. This follows from Lemma 13.8.4 since $\langle f, \lambda(t) \rangle = \int_U f(u)\lambda(t, du) = \int_U f(u)\kappa(u)\mu(t, du)$, the integral a.e. finite on account of (13.8.18). This ends the proof. ∎

Existence Theorem 13.8.7. *Let* $y_0(t, \mu)$ *be weakly lower semicontinuous. Assume that* $(a) -\infty < \mathbf{m} < \infty$, (b) *the state constraint sets* $M(t)$ *are closed,* (c) *the target set* Y *is closed. Let* $\{\mu^n(\cdot)\} \subseteq R_c(0, \bar{t}; U)$ *be a minimizing sequence of relaxed controls such that* $\{\|\mu^n(\cdot)\|_\kappa\}$ *is bounded in* $L^2(0, T)$ *and* $\{y(t, \mu^n)\}$ *is uniformly bounded in* $0 \leq t \leq \bar{t}$. *Then there exists a relaxed optimal control* $\bar{\mu}(\cdot)$ *which is the* $L^1(0, T; C_0(U))$-*weak limit in* $L^\infty_w(0, \bar{t}; \Sigma_{rca}(U))$ *of a subsequence of* $\{\mu^n(\cdot)\}$.

The proof is similar to that of Theorem 13.5.1. We select a (generalized) subsequence of $\{\mu^n(\cdot)\}$ (denoted with the same name) such that

$$\mu^n(\cdot) \to \bar{\mu}(\cdot) \in R_{c,\kappa}(0, \bar{t}; U) \qquad (13.8.19)$$

$L^1(0, \bar{t}; C_0(U))$-weakly in $L^\infty_w(0, \bar{t}; \Sigma_{rca}(U))$ (Theorem 13.8.5),

$$\lambda^n(\cdot) \to \bar{\lambda}(\cdot) \in L^2_w(0, \bar{t}; \Sigma_{rba}(U)) \qquad (13.8.20)$$

$L^2(0, \bar{t}; BC(U))$-weakly in $L^2_w(0, \bar{t}; \Sigma_{rba}(U))$ (Theorem 12.9.2 on the dual of $L^2(0, \bar{t}; BC(U))$ and

$$y(\cdot, \mu^n) \to y(\cdot) \in C(0, \bar{t}; E) \tag{13.8.21}$$

(the Arzelà-Ascoli theorem; $\lambda^n(t)$ is defined from $\mu^n(t)$ as in (13.8.17)). We have

$$
\begin{aligned}
y(t, \mu^n) &= \zeta + \int_0^t F(\tau, y(\tau, \mu^n))\mu^n(\tau)d\tau \\
&= \zeta + \int_0^{\bar{t}} \int_U \chi_t(\tau) f(\tau, y(\tau, \mu^n), u)\kappa(u)^{-1}\kappa(u)\mu^n(\tau, du)d\tau \\
&\to \zeta + \int_0^{\bar{t}} \int_U \chi_t(\tau) f(\tau, y(\tau), u)\kappa(u)^{-1}\bar{\lambda}(\tau, du)d\tau, \tag{13.8.22}
\end{aligned}
$$

thus we only have to show that $\bar{\lambda}(\tau, du) = \kappa(u)\bar{\mu}(\tau, du)$ as elements of $L^2_w(0, \bar{t}; BC(U)^*)$. Taking into account the equivalence relation in this space, this means to show that

$$\int_U f(u)\bar{\lambda}(t, du) = \int_U f(u)\kappa(u)\bar{\mu}(t, du) \tag{13.8.23}$$

for every $f(\cdot) \in BC(U)$. For this we need

Lemma 13.8.8. *Let* $\lambda(du) \in \Sigma_{rba}(U)$. *Then* $\mu(du) = \kappa^{-1}(u)\lambda(du)$ *belongs to* $\Sigma_{rca}(U)$.

Proof. By Alexandrov's Theorem 12.4.10, λ is countably additive (over the field $\Phi(C)$ generated by the closed sets) in any compact subset of U. Thus, by Theorem 12.4.11, it can be extended to a regular, bounded countably additive measure on the Borel σ-field of each compact subset of U.

Let now $\{e_j\}$ be a sequence of pairwise disjoint Borel sets in U, $e = \cup_j e_j$. Choose N so large that $\kappa(u)^{-1} \le \varepsilon$ for $\|u\| \ge N$. Then $\mu(e \cap B(0, N)^c) \le \varepsilon\lambda(e) \le \varepsilon\lambda(U)$. Using countable additivity on compacts, choose n so large that $\mu(e_1 \cap B(0, N)) + \cdots + \mu(e_n \cap B(0, N)) \ge \mu(e \cap B(0, N)) - \varepsilon$. We have

$$
\begin{aligned}
\mu(e_1) &+ \mu(e_2) + \cdots + \mu(e_n) \\
&\ge \mu(e_1 \cap B(0, N)) + \cdots + \mu(e_n \cap B(0, N)) \\
&\ge \mu(e \cap B(0, N)) - \varepsilon \ge \mu(e) - \mu(e \cap B(0, N)^c) - \varepsilon \\
&\ge \mu(e) - (1 + \lambda(U))\varepsilon \tag{13.8.24}
\end{aligned}
$$

so that $\sum_{j=1}^\infty \mu(e_j) \ge \sum_{j=1}^n \mu(e_j) \ge \mu(e)$; the opposite inequality is a consequence of finite additivity. ∎

End of Proof of Theorem 13.8.7. Define $\tilde{\mu}(t, du) = \kappa^{-1}(u)\bar{\lambda}(t, du)$. If η_m is the function in (13.8.7) and $f(\cdot) \in BC(U)$ we have

$$\int_0^{\bar{t}} \int_U \chi_t(\tau) f(u)\eta_m(u)\kappa(u)\tilde{\mu}(t, du) = \int_0^{\bar{t}} \int_U \chi_t(\tau) f(u)\eta_m(u)\kappa(u)\bar{\mu}(t, du)$$

(χ_t the characteristic function of $[0, t]$) as we see writing the equality for $\lambda^n(t, du)$, $\mu^n(t, du)$ and taking limits. Now, this implies

$$\int_U f(u)\eta_m(u)\kappa(u)\tilde{\mu}(t, du) = \int_U f(u)\eta_m(u)\kappa(u)\bar{\mu}(t, du) \qquad (13.8.25)$$

a.e. independently of m, where $f(\cdot)\kappa(\cdot)$ is integrable with respect to both countably additive measures $\tilde{\mu}(t, du)$ and $\bar{\mu}(t, du)$. Hence, (13.8.23) results from the dominated convergence theorem.   ∎

Example 13.8.9. (A relaxation theorem for unbounded controls). Let the original system (13.8.8) satisfy Hypothesis II(C, κ) and let $\mu(\cdot) \in R_{c,\kappa}(0, T; U)$ be such that the solution of the relaxed system (13.8.10) exists in $0 \le t \le \bar{t}$. Then there exists $u(\cdot) \in PC(0, T; U)$ such that the solution $y(t, u)$ of the original system (13.8.7) exists in $0 \le t \le \bar{t}$ and

$$\|y(t, u) - y(t, \mu)\| \le \varepsilon \qquad (0 \le t \le T). \qquad (13.8.26)$$

In fact, let μ_n be the truncation $\mu_n(du) = \chi_n(u)\mu(du)$, χ the characteristic function of $B(0, n)$. We have

$$y(t, \mu_n) - y(t, \mu) = \int_0^t \{F(\tau, y(\tau, \mu_n))\mu_n(\tau) - F(\tau, y(\tau, \mu))\mu_n(\tau)\}d\tau$$

$$+ \int_0^t F(\tau, y(\tau, \mu))(\mu_n(\tau) - \mu(\tau))d\tau.$$

By the dominated convergence theorem, the last integral tends to zero uniformly in $0 \le t \le \bar{t}$. Estimating the first with (13.8.14) and using Gronwall's inequality as in Theorem 13.2.2, we show that $y(t, \mu_n)$ exists in $0 \le t \le \bar{t}$. Then we apply Relaxation Theorem 13.2.1 with control set $U \cap B(0, n)$.

Example 13.8.10. The "fix" of requiring (13.8.5) from admissible controls is not necessarily germane to all problems. An (admittedly artificial) illustration is Example 2.2.4,

$$y'(t) = u(t), \qquad y(0) = 0,$$
$$U = [0, \infty), \qquad Y = \mathbb{R}, \qquad \bar{t} = 1,$$
$$y_0(t) = \int_0^{\bar{t}} \frac{1}{1 + u(\tau)}d\tau,$$

If we designate $L^2(0, \bar{t})$ as admissible control space (we may also take the space of all measurable finite functions in $0 \le t \le 1$ without summability conditions), then the sequence $\{u_n(t) \equiv n\}$ is a minimizing sequence, but no control gives the minimum value (zero) of the cost functional. If we fit this system with the space $R_{c,\kappa}(0, \bar{t}; U)$ for any function $\kappa(u)$ satisfying (13.8.4), there is no control $\mu(t, du)$ such that

$$y_0(t, \mu) = \int_0^t \frac{1}{1+u} \mu(\tau, du) d\tau = 0;$$

in fact, $\int (1 + u)^{-1} \mu(du) = 0$ implies $\mu = 0$ for any $\mu \in \Pi_{rca}(U)$. Theorem 13.8.7 cannot be applied to the minimizing sequence $\mu^n(t) \equiv \delta(\cdot - n)$, since $\|\mu^n(\cdot)\|_\kappa$ is unbounded. The "true" relaxation of this system will be seen in **14.1**.

To establish the minimum principle for optimal control problems, we just need to formulate the analogue of Hypothesis III(C) in **13.7** and generalize the arguments therein.

Hypothesis III(C, κ). $f(t, y, u)$ and each $\partial_k f_j(t, y, u) = \partial f_j(t, y, u)/\partial y_k$ are continuous in y, u for t fixed and measurable in t for y, u fixed. For every $c > 0$ there exist $K(\cdot, c), L(\cdot, c) \in L^2(0, T)$ such that

$$\|f(t, y, u)\| \le K(t, c)\kappa(u) \quad (0 \le t \le T, \|y\| \le c, u \in U), \quad (13.8.27)$$
$$\|\partial_y f(t, y, u)\| \le L(t, c)\kappa(u) \quad (0 \le t \le T, \|y\| \le c, u \in U), \quad (13.8.28)$$

$\partial_y f(t, y, u)$, the Jacobian matrix.

Example 13.8.11. Under Assumption III(C, κ), the function $F(t, y)\mu$ satisfies Assumption III with respect to the control space $R_{c,\kappa}(0, T; U)$.

The proof is essentially the same as that of Lemma 13.7.1.

After Example 13.8.11, we argue as in **13.7** and obtain the maximum principle; statements and proofs are the same as those of Theorems 13.7.3 and 13.7.5. The accoutrements regarding free terminal time are of course valid.

By way of applications, we limit ourselves to the example below, where (once again) the minimum drag problem illuminates a corner of optimal control theory.

Example 13.8.12. We approach optimal nose shape (last seen in Example 13.7.7) from scratch, that is, without using the reparametrization arguments in **4.9**. In these conditions,

$$U = [0, \infty) \times [0, \infty). \quad (13.8.29)$$

We have $f(t, x, y, u, v) = (u, -v)$, $f_0(t, x, y, u, v) = xu^3/(u^2 + v^2)$, so that

$$\|f(t, x, y, u, v)\|, |f_0(t, x, y, u, v)| \le C(|u| + |v|)$$

for $|x|$, $|y|$ bounded, thus one natural choice is $\kappa(u, v) = 1 + |u| + |v|$. Assuming that the minimizing sequence in Theorem 13.8.7 can be constructed, the optimal relaxed control $\bar{\mu}(\cdot)$ satisfies the relaxed maximum principle with free arrival time, thus we obtain again (13.7.27):

$$
\int_U \left(\lambda \bar{x}(s) \frac{u^3}{u^2 + v^2} + p(s)u - qv \right) \bar{\mu}(s, dudv)
$$

$$
= \max_{v \in \Pi_{rca}(U)} \int_U \left(\lambda \bar{x}(s) \frac{u^3}{u^2 + v^2} + p(s)u - qv \right) v(dudv) = 0, \quad (13.8.30)
$$

this time in a different control set. The analysis of (13.8.30) is exactly the same as that of (13.7.27) and we go from here to the complete solution of the problem as in Example 13.7.7. However, that we have avoided reparametrization arguments is an illusion; they are needed to construct the minimizing sequence in Theorem 13.8.7.

14

Relaxed Controls in Infinite Dimensional Systems. Existence Theory

14.1. Control Systems: Limits of Trajectories

The arguments used in **10.1** to justify introducing relaxed controls apply as well to infinite dimensional systems. We try here a simpler motivation. Consider a control system

$$y'(t) = A(t)y(t) + f(t, y(t), u(t)) \tag{14.1.1}$$

in a Banach space E, with a space $C_{\text{ad}}(0, T; U)$ of admissible controls. We say that (14.1.1) is **trajectory complete** if for every sequence $\{u_n(\cdot)\} \subseteq C_{\text{ad}}(0, \bar{t}; U)$ such that

$$y(t, u_n) \to y(t) \tag{14.1.2}$$

uniformly in $0 \le t \le \bar{t}$ there exists $u(\cdot) \in C_{\text{ad}}(0, \bar{t}; U)$ such that

$$y(t) = y(t, u). \tag{14.1.3}$$

Trajectory completeness is an essential ingredient of existence theorems for optimal problems. We have seen in **12.1** examples of finite dimensional systems that are not trajectory complete, the simplest one being

$$y'(t) = u(t), \qquad y(0) = 0,$$
$$U = \{-1\} \cup \{1\}, \qquad \bar{t} = 1. \tag{14.1.4}$$

As seen in **2.6**, a sequence of controls that "chatter" with increasing frequency between 1 and -1 (the sequence (2.2.18)) produces a sequence of trajectories such that $y(\cdot, u_n) \to 0$ uniformly in $0 \le t \le \bar{t}$; however, there is no control in $C_{\text{ad}}(0, \bar{t}; U)$ driving the zero trajectory. Systems such as this justify looking into the

Completion problem: Given a control system with control space $C_{\text{ad}}(0, \bar{t}; U)$, find a control space $R(0, T; U) \supseteq C_{\text{ad}}(0, \bar{t}; U)$ such that the system, equipped with $R(0, T; U)$, is trajectory complete.

A solution of the completion problem for finite dimensional systems was given in Chapter 13 using spaces $R(0, T; U)$ of controls that take values in the space of probability measures μ on the control set U. For the system (14.1.4), these controls are $\mu(t) = \alpha(t)\delta(t - 1) + (1 - \alpha(t))\delta(t + 1)$, where $0 \leq \alpha(t) \leq 1$. The dynamics of the expanded system is

$$y'(t) = \int_U u\mu(t, du), \qquad y(0) = 0,$$

and we do have a control in the larger class that produces the zero trajectory, namely

$$\mu(t) = \frac{1}{2}\delta(t - 1) + \frac{1}{2}\delta(t + 1).$$

Compactness of U was essential in all results in Chapter 13. What may happen without compactness can be seen in the system below (which is closely related to that in Example 13.8.10).

Example 14.1.1.

$$y'(t) = \frac{1}{1 + u(t)}, \qquad y(0) = 0,$$

$$U = [0, \infty), \qquad \bar{t} = 1.$$

The admissible control space $C_{ad}(0, \bar{t}; U)$ consists of all nonnegative measurable functions defined in $t \geq 0$. Here, the sequence $u_n(t) \equiv n$ produces a sequence of trajectories converging uniformly to zero, whereas the zero trajectory answers to no admissible control. We may relax this system in the style of Chapter 13,

$$y'(t) = \int_0^\infty \frac{1}{1 + u}\mu(t, du), \tag{14.1.5}$$

with $\mu(t, du)$ a countably additive probability measure in $[0, \infty)$, but this does no good; no such control produces the zero trajectory. However, if we allow $\mu(t, du)$ to be a *finitely additive* measure (precisely, an element of $\Sigma_{rba}([0, \infty)) = BC([0, \infty))^*$, we may easily produce a measure valued control whose trajectory is identically zero. In fact, let

$$\phi(f) = \operatorname*{LIM}_{n \to \infty} f(u_n)$$

where $\{u_n\}$ is a sequence in $[0, \infty)$ with $u_n \to \infty$ and $\operatorname{LIM}_{n \to \infty}$ is a Banach limit (5.0). Then ϕ is a bounded linear functional in $BC([0, \infty))$. Hence,

$$\phi(f) = \int_0^\infty f(u)\mu(du)$$

for some $\mu \in \Sigma_{rba}(U)$. The control $\mu(t) \equiv \mu$ produces the zero trajectory, since (14.1.5) is zero. Of course, simpler ad-hoc solutions to this particular problem are available. For instance, we may *compactify* the control set U by adjoining the point at infinity ∞; the new control set is $U_c = U \cup \{\infty\}$, which is compact under the distance

$$d(u, v) = \left| \frac{1}{1+u} - \frac{1}{1+v} \right|,$$

if we define $1/\infty = 0$ (the topology generated by this distance in U coincides with the original topology). In the new control set, the (ordinary) control producing the zero trajectory is $u(t) \equiv \infty$. This is in fact a particular case of a general way to approach the completion problem (see Miscellaneous Comments to Part III).

We show in this chapter that the completion problem can be solved, even with no conditions whatsoever on the control set U, if we give up countably additive measures in favor of finitely additive ones. In view of our glimpse into these measures in **12.3**, this solution is no doubt esoteric, and there are many situations where one can (and should) use smaller and more tractable spaces $R(0, T; U)$. One instance is that of the linear diffusion system in Example 9.2.13:

Example 14.1.2. Consider the system

$$y'(t) = A_1(\beta)y(t) + u(t) \quad y(0) = \zeta \tag{14.1.6}$$

in the space $L^1(\Omega)$ ($A_1(\beta)$ the operator in $L^1(\Omega)$ determined by a uniformly elliptic differential operator A and a boundary condition β; see **9.2**, especially (9.2.23)). In this system, the instantaneous values $u(t)$ of admissible controls $u(\cdot)$ belong to the unit ball of $L^1(\Omega)$. To make this system trajectory complete, it is enough to expand the control space to the set of all controls $\mu(\cdot)$ whose instantaneous values $\mu(t)$ are regular bounded Borel measures in $\overline{\Omega}$ with total variation norm $\|\mu(t)\| \leq 1$ a.e., that is, to the unit ball of $L_w^\infty(0, T; \Sigma(\overline{\Omega}))$. The proof that this makes (14.1.6) path complete is essentially the same as that of Lemma 9.1.5.

Examples like this indicate that, in infinite dimensional systems, one should not only consider measure-valued relaxed controls but try to theorize in such a way to include "intermediate" spaces like the one above.

14.2. Semilinear Systems Linear in the Control. Approximation by Extremal Trajectories

The systems are the infinite dimensional counterparts of those in **2.2**:

$$y'(t) = Ay(t) + f(t, y(t)) + F(t, y(t))u(t), \qquad y(0) = \zeta, \tag{14.2.1}$$

with A the infinitesimal generator of a strongly continuous semigroup $S(\cdot)$. The functions f, F map as follows:

$$f : [0, T] \to E, \qquad F : [0, T] \times E \to (X^*, E),$$

X^* the dual of an arbitrary Banach space X. The control set $\mathbf{U}$ is a subset of X^*, and the admissible control space is

$$C_{\text{ad}}(0, T; \mathbf{U}) = \{u(\cdot) \in L_w^\infty(0, T; X^*); u(t) \in \mathbf{U}\}, \tag{14.2.2}$$

where "$u(t) \in \mathbf{U}$" means "there exists $\tilde{u}(\cdot)$ in the equivalence class of $u(\cdot)$ in $L_w^\infty(0, T; X^*)$ such that $\tilde{u}(t) \in \mathbf{U}$ in $0 \le t \le T$." For details on the spaces L_w^∞ and on the meaning of this requirement, see **12.2**.

The assumptions on f, F are designed in such a way that the nonlinear term

$$f(t, y) + F(t, y)u(t) \tag{14.2.3}$$

satisfies Hypotheses I or II in **5.5**, so that the existence-uniqueness results there can be used.

Hypothesis I(f, F). (*a*) $f(t, y)$ is strongly measurable in t for y fixed and continuous in y for t fixed. For every $c > 0$ there exists $K_1(\cdot, c) \in L^1(0, T)$ such that

$$\|f(t, y)\|_E \le K_1(t, c) \quad (0 \le t \le T, \|y\| \le c). \tag{14.2.4}$$

(*b*) $F(t, y)^* E^* \subseteq X$ and for each $y \in E$ and $y^* \in E^*$ fixed, $F(t, y)^* y^*$ is strongly measurable in X; $F(t, y)$ is continuous in the norm of (X^*, E) for t fixed. For every $c > 0$ there exists $K_2(\cdot, c) \in L^1(0, T)$ such that

$$\|F(t, y)\|_{(X^*, E)} \le K_2(t, c) \quad (0 \le t \le T, \|y\| \le c). \tag{14.2.5}$$

Hypothesis II(f, F). Hypothesis I(f, F) is satisfied and in addition, for every $c > 0$ there exists $L_1(\cdot, c), L_2(\cdot, c) \in L^1(0, T)$ such that

$$\|f(t, y') - f(t, y)\|_E \le L_1(t, c)\|y' - y\|_E$$

$$(0 \le t \le T, \|y\|, \|y'\| \le c) \tag{14.2.6}$$

$$\|F(t, y') - F(t, y)\|_{(X^*, E)} \le L_2(t, c)\|y' - y\|_E$$

$$(0 \le t \le T, \|y\|, \|y'\| \le c). \tag{14.2.7}$$

Lemma 14.2.1. *Assume that f, F satisfy Assumption I(f, F) (resp. Assumption II(f, F)) and that E is separable. Then (14.2.3) satisfies Assumption I (resp. II) for every $u(\cdot) \in L_w^\infty(0, T; X^*)$.*

The proof is obvious except perhaps that $t \to F(t, y)u(t)$ is strongly measurable in E for $u(\cdot) \in C_{ad}(0, T; \mathbf{U})$. To see this, note that $t \to F(t, y(t))^*y^*$ is strongly measurable in X for any $y^* \in E^*$ so that $t \to \langle y^*, F(t, y)u(t) \rangle = \langle F(t, y(t))^*y^*, u(t) \rangle$ is measurable. It follows then that $t \to F(t, y(t))u(t)$ is E^*-weakly measurable in E, hence strongly measurable by separability of E.

Solutions $y(t, u)$ of (14.2.1) in $0 \le t \le \bar{t}$ are solutions of the integral equation

$$y(t, u) = S(t)\zeta + \int_0^t S(t - \tau)f(\tau, y(\tau, u))d\tau$$

$$+ \int_0^t S(t - \tau)F(\tau, y(\tau, u))u(\tau)d\tau \quad (0 \le t \le \bar{t}) \quad (14.2.8)$$

to which the results in Chapter 6 apply.

Lemma 14.2.2. *Assume that E is separable, that $S(t)$ is compact for $t > 0$ and that Hypothesis II(f, F) holds. Let $u(\cdot) \in L_w^\infty(0, T; X^*)$ be such that the solution $y(\cdot, u)$ of (14.2.1) exists in $0 \le t \le \bar{t}$, and let $\{u_\kappa(\cdot); \kappa \in K\}$ be a bounded generalized sequence in $L_w^\infty(0, \bar{t}; X^*)$ such that $u_\kappa(\cdot) \to u(\cdot)$ $L^1(0, \bar{t}; X)$-weakly. Then there exists $\kappa_0 \in K$ such that $y(\cdot, u_\kappa)$ exists in $0 \le t \le \bar{t}$ if $\kappa \ge \kappa_0$ ("$\ge$" the ordering in K) and*

$$y(t, u_\kappa) \to y(t, u)$$

uniformly in $0 \le t \le \bar{t}$.

The pattern of the proof is familiar. Let $[0, \bar{t}_\kappa] \subseteq [0, \bar{t}]$ be the maximal interval where $y(t, u_\kappa)$ exists and satisfies $\|y(t, u_\kappa) - y(t, u)\| \le 1$. For any κ we have

$$y(t, u_\kappa) - y(t, u)$$

$$= \int_0^t S(t - \tau)\{f(\tau, y(\tau, u_\kappa)) - f(\tau, y(\tau, u))\}d\tau$$

$$+ \int_0^t S(t - \tau)\{F(\tau, y(\tau, u_\kappa)) - F(\tau, y(\tau, u))\}u_\kappa(\tau)d\tau$$

$$+ \int_0^t S(t - \tau)F(\tau, y(\tau, u))(u_\kappa(\tau) - u(\tau))d\tau \quad (14.2.9)$$

in $0 \le t \le \bar{t}_\kappa$. Estimating and using Gronwall's inequality,

$$\|y(t, u_\kappa) - y(t, u)\| \le C\|\rho_\kappa(t)\| \quad (0 \le t \le t_\kappa) \quad (14.2.10)$$

with $\rho_\kappa(t) = \Lambda(F(\cdot, y(\cdot, u))(u_\kappa(\cdot)) - u(\cdot)))$, Λ the operator

$$\Lambda g(t) = \int_0^t S(t - \tau)g(\tau)d\tau \quad (14.2.11)$$

in Theorem 5.5.19. In view of (14.2.5),

$$\|F(t, y(t, u))(u_\kappa(t)) - u(t))\| \le K_2(t, c)\|u_\kappa(t) - u(t)\|,$$

thus, by boundedness of $\{u_\kappa\}$, the members of the generalized sequence

$$\{g_\kappa(\cdot)\} = \{F(\cdot, y(\cdot, u))(u_\kappa(\cdot) - u(\cdot))\} \subseteq L^1(0, \bar{t}; E) \qquad (14.2.12)$$

have equicontinuous integrals in $0 \le t \le \bar{t}$. By Theorem 5.5.19 there exists a generalized subsequence (denoted with the same symbol) with $\rho_\kappa(t) \to \rho(t)$ $\in C(0, \bar{t}; U)$ in the norm of $C(0, \bar{t}; U)$. Strictly speaking, this has only been proved in Theorem 5.5.19 for sequences. However, the statement for sequences expresses compactness of a metric space (namely, the closure in $C(0, \bar{t}; U)$ of the set $\{\Lambda g_\kappa\}$) and, in metric spaces, compactness is equivalent to sequential compactness (Theorem 12.0.13).

The proof of Lemma 14.2.2 will be complete if we prove that $\rho(t) = 0$; in fact, if this is true we have $\rho_\kappa(t) \to 0$ uniformly and (14.2.10) will imply that $t_\kappa = \bar{t}$ and the uniform convergence relation. To see that $\rho(t) = 0$ note that

$$\langle y^*, \rho_\kappa(t) \rangle = \int_0^{\bar{t}} \langle \chi_t(\tau) F(\tau, y(\tau, u))^* S(t - \tau)^* y^*, u_\kappa(\tau) - u(\tau) \rangle d\tau,$$

χ_t the characteristic function of $0 \le \tau \le t$. Now, since $S(t)$ is compact it is continuous in the (E, E)-norm for $t > 0$ (Lemma 5.5.18) so that $S(t)^*$ is as well continuous in the (E^*, E^*)-norm for $t > 0$. Using Assumption II(f, F)) and approximating $y(\cdot, u)$, $S(t - \cdot)^*$ by piecewise constant functions we deduce that $\chi_t(\cdot) F(\cdot, y(\cdot, u))^* S(t - \cdot) y^* \in L^1(0, \bar{t}; X)$. On the other hand, $u_\kappa(\cdot) \to u(\cdot)$ $L^1(0, \bar{t}; X)$-weakly in $L_w^\infty(0, \bar{t}; X^*)$. Accordingly, $\langle y^*, \rho(t) \rangle = \lim_\kappa \langle y^*, \rho_\kappa(t) \rangle$ $= 0$, thus $\rho(t) = 0$. This ends the proof.   ∎

Recall **(12.6)** that if $\mathbf{U} \subseteq X^*$ is convex, a subset $D \subseteq \mathbf{U}$ is *total* in $\mathbf{U}$ if $\overline{\text{conv}}(D) = \mathbf{U}$, where $\overline{\text{conv}}$ means closed convex hull in the X-weak topology of X^*. $PC(0, \bar{t}; D)$ denotes the space of all piecewise constant D-valued functions defined in $0 \le t \le \bar{t}$.

Theorem 14.2.3 (First relaxation theorem). *Assume that $\mathbf{U} \subseteq X^*$ is convex, bounded and X-weakly compact, that D is total in $\mathbf{U}$, that E is separable, $S(t)$ is compact for $t > 0$ and that Hypothesis II(f, F) holds. Let $u(\cdot) \in C_{\text{ad}}(0, \bar{t}; \mathbf{U})$ be such that the solution $y(\cdot, u)$ of (14.2.1) exists in $0 \le t \le \bar{t}$ and $\varepsilon > 0$. Then there exists $v(\cdot) \in PC(0, \bar{t}; D)$ such that $y(t, v)$ exists in $0 \le t \le \bar{t}$ and*

$$\|y(t, u) - y(t, v)\| \le \varepsilon \quad (0 \le t \le \bar{t}). \qquad (14.2.13)$$

Proof. Combine Theorem 12.6.7 with Lemma 14.2.2.   ∎

Remark 14.2.4. If U satisfies the assumptions in Theorem 14.2.3, the Krein-Milman theorem (Dunford-Schwartz [1958, p. 440]) shows that if D is the set of **extremal** points of U (see the definition in **12.6**), then D is total in U. In this case, it makes sense to call **extremal** those controls taking values in D (in particular, controls in $PC(0, \bar{t}; D)$) and to call their trajectories **extremal** as well. Abusing this name a lot, we may call extremal any control with values in D, where D is an arbitrary set total in U, hence the heading of this section. The word *relaxation* used in Theorem 14.2.3 and other results below comes from relaxed controls and is used loosely here.

The second approach to relaxation does not require compactness of the semigroup. Given $V \subseteq X^*$ the space $CV(0, T; V)$ consists of all countably valued functions $u(\cdot)$ such that $u(t) \in V$ a.e. (the definition includes the requirement that values are taken in measurable sets). The control set U is assumed convex.

Lemma 14.2.5. *Let E be reflexive, $U \subseteq X^*$ a convex X-weakly compact set and $F \in (X^*, E)$ an operator such that $F^* : E^* \to X$. Then $F(U) = \{Fu; u \in U\}$ is convex and closed in E. If $D \in U$ is total in U we have $\overline{\text{conv}}(F(D)) = F(U)$ (closure taken in the strong topology).*

Proof. Obviously, $F(U)$ is convex. Let $\{y_n\} \subseteq F(U)$. Then $y_n = Fu_n$, $\{u_n\} \subseteq U$. Select a generalized subsequence of $\{u_n\}$ (same name) X-weakly convergent to $u \in U$, and let $y^* \in E^*$. Then $\langle y^*, y_n \rangle = \langle y^*, Fu_n \rangle = \langle F^*y^*, u_n \rangle \to \langle F^*y^*, u \rangle = \langle y^*, Fu \rangle$, so that $Fu_n \to Fu$ E^*-weakly in E. It follows that $F(U)$ is convex and E^*-weakly compact in E, hence weakly closed; since E is reflexive, $F(U)$ is as well strongly closed (Theorem 12.0.11).

A similar argument shows that, if $u \in U$ and $\{u_\kappa\} \subseteq \text{conv}(D)$ is a generalized sequence with $u_\kappa \to u$ X-weakly then $Fu_\kappa \to Fu$ E-weakly. It follows that the weak closure of conv $(F(D))$ equals $F(U)$. Since conv $(F(D))$ is convex, its strong closure equals its weak closure. ∎

Throughout the rest of the section we assume that E is reflexive and that U is convex, bounded and X-weakly compact.

Lemma 14.2.6. *Let E be reflexive, $t \to g(t) \in E$ strongly measurable and $t \to F(t) \in (X^*, E)$ strongly measurable in the norm of (X^*, E). Assume that D is total in U and*

$$g(t) \in F(t)(U) \quad a.e. \text{ in } 0 \le t \le T.$$

Then, given $\varepsilon > 0$ there exists $w(\cdot) \in CV(0, T; \text{conv}(D))$ such that

$$\|F(t)w(t) - g(t)\| \le \varepsilon \quad (0 \le t \le T). \tag{14.2.14}$$

Proof. Choose a countably (X^*, E)-valued function $F_\varepsilon(t)$ and a countably E-valued function $g_\varepsilon(t)$ such that

$$\|F_\varepsilon(t) - F(t)\| \leq \varepsilon, \qquad \|g_\varepsilon(t) - g(t)\| \leq \varepsilon$$

a.e. in $0 \leq t \leq T$. Let Ξ_ε be a countable collection of pairwise disjoint measurable sets where $F_\varepsilon(t) = F_\varepsilon$ and $g_\varepsilon(t) = g_\varepsilon$ are constant. Since $g_\varepsilon \in F(\mathbf{U})_\varepsilon = \{y; \mathrm{dist}(y, F(\mathbf{U})) \leq \varepsilon\}$, for every $e \in \Xi_\varepsilon$ there exists $w(e) \in \mathrm{conv}(D)$ with $\|F_\varepsilon w(e) - g_\varepsilon\| \leq 2\varepsilon$. Defining $w(t) = w(e)$ in each e in the collection Ξ_ε we have $\|F(t)w(t) - g(t)\| \leq (3 + M)\varepsilon$ a.e. in $0 \leq t \leq T$, M a bound for $\mathbf{U}$. ∎

Lemma 14.2.7. *Let E be reflexive and separable, $t \to F(t) \in (X^*, E)$ strongly measurable in $0 \leq t \leq T$ in the norm of (X^*, E) with $\|F(\cdot)\| \in L^1(0, T)$ and $F(t)^* : E^* \to X$. Let $t \to S(t) \in (E, Y)$ (Y a Banach space) be strongly continuous in $0 \leq t \leq T$. Assume that D is total in $\mathbf{U}$. Then, if $u(\cdot) \in C_{\mathrm{ad}}(0, T; \mathbf{U})$ there exists $v(\cdot) \in CV(0, T; D)$ such that*

$$\left\| \int_0^T S(t)\{F(t)v(t) - F(t)u(t)\}dt \right\|_Y \leq \varepsilon. \tag{14.2.15}$$

Proof. We show first that $g(\cdot) = F(\cdot)u(\cdot)$ is strongly measurable in E. Since E is separable, E^*-weak measurability is enough, and follows from the equality $\langle y^*, F(t)u(t)\rangle = \langle F(t)^*y^*, u(t)\rangle$; in fact, the hypotheses imply that $t \to F(t)^*$ is strongly measurable as a (E^*, X)-valued function.

We use next Lemma 14.2.6 to select $w(\cdot) \in CV(0, T; \mathrm{conv}(D))$ such that $\|F(t)w(t) - F(t)u(t)\| \leq \varepsilon$. Then we pick a countably valued (X^*, E)-valued function $F_\varepsilon(\cdot)$ such that $\|F_\varepsilon(t) - F(t)\| \leq \varepsilon$ a.e., so that

$$\|F_\varepsilon(t)w(t) - F_\varepsilon(t)u(t)\| \leq \|F(t)w(t) - F(t)u(t)\|$$
$$+ \|(F_\varepsilon(t) - F(t))w(t)\| + \|(F_\varepsilon(t) - F(t))u(t)\| \leq (1 + 2M)\varepsilon,$$

M a bound for $\mathbf{U}$. Let Ξ_ε be a collection of pairwise disjoint measurable sets covering $[0, T]$ and such that both $F_\varepsilon(\cdot) = F_\varepsilon$ and $w(\cdot) = w \in \mathrm{conv}(D)$ are constant in each $e \in \Xi_\varepsilon$. Pick $e \in \Xi_\varepsilon$ and let

$$w = \Sigma\alpha_j(e)v_j(e) \quad (v_j(e) \in D)$$

be the value of $w(\cdot)$ in e. Denote by $X^*(e)$ the subspace of X^* generated by $v_1(e), \ldots, v_n(e)$. The operator function $S(\cdot)F_\varepsilon$ maps X^* into Y and is strongly continuous in e. Accordingly, its restriction to $X^*(e)$ (equally named) is strongly continuous, and, since $X^*(e)$ is finite dimensional, it is continuous in the $(X^*(e), Y)$-norm. Thus there exists a piecewise constant function $S_\varepsilon(\cdot) : E \to Y$ defined in e and such that

$$\|S_\varepsilon(t)F_\varepsilon - S(t)F_\varepsilon\|_{(X^*(e);Y)} \leq \varepsilon \quad (t \in e).$$

Let, finally $d \subseteq e$ be a set where $S_\varepsilon(t) = S_\varepsilon$ is constant. Divide d into n disjoint measurable sets with $|d_j| = \alpha_j(e)|d|$ ($|\cdot|$ indicates Lebesgue measure), and define $v(t) = v_j(e)$ for $t \in d_j$. Then

$$\int_d S_\varepsilon F_\varepsilon v(t) dt = \int_d S_\varepsilon F_\varepsilon w \, dt.$$

The control $v(\cdot) \in CV(0, T; D)$ obtained piecing together all the $v(\cdot)$ constructed in each $d \subseteq e$ for all $e \in \Xi_\varepsilon$ satisfies (14.2.15) with a different ε. This ends the proof. ∎

Lemma 14.2.8. *Same assumptions as in Lemma* 14.2.7. *Let the (E, E)-valued function $S(t, s)$ be defined and strongly continuous in $0 \le s \le t \le T$. Then, if $u(\cdot) \in C_{ad}(0, T; U)$ and $\varepsilon > 0$ there exists $v(\cdot) \in CV(0, T; D)$ such that*

$$\left\| \int_0^t S(t, \tau)\{F(\tau)v(\tau) - F(\tau)u(\tau)\}d\tau \right\|_E \le \varepsilon \quad (0 \le t \le T). \tag{14.2.16}$$

Proof. Extend $S(t, s)$ to the square $0 \le s, t \le T$ setting $S(t, s) = 0$ for $s > t$. Let C be the maximum of $\|S(t, s)\|$ in the square. Then construct a strongly continuous operator-valued function $S_\varepsilon(t, s)$ with the same maximum C in the square and such that $S_\varepsilon(t, s) = S(t, s)$ in $s \le t$ and $s \ge t + \varepsilon$; for instance, we may use

$$S_\varepsilon(t, s) = \frac{t + \varepsilon - s}{\varepsilon} S(t, t) \quad (t \le s \le \min(t + \varepsilon, T), 0 \le t \le T).$$

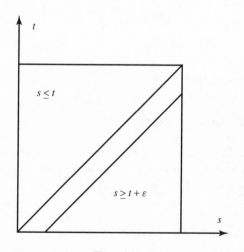

Figure 14.1.

Since $S_\varepsilon(t, s)$ is strongly continuous, the function $\tau \to S_\varepsilon(\cdot, \tau) \in Y = C(0, T; E)$ is strongly continuous in the norm of $C(0, T; E)$ in $0 \le t \le T$. Applying

Lemma 14.4.7 to this function, (14.2.16) results for $S_\varepsilon(t, \tau)$, with the integral over $0 \le \tau \le T$. That the inequality will also hold for the extended $S(t, \tau)$ follows from

$$\int_0^T \|(S_\varepsilon(t, \tau) - S(t, \tau))F(\tau)u(\tau)\|_E \, d\tau \le C \int_t^{\min(t+\varepsilon, T)} \|F(\tau)u(\tau)\|_E \, d\tau$$

and equicontinuity of the integral of $\|F(\cdot)u(\cdot)\|_E$. ∎

Theorem 14.2.9 (Second relaxation theorem). *Let the space E be reflexive and separable, **U** convex, bounded and X-weakly compact in X^*, and D total in **U**. Let the control system* (14.2.1) *satisfy Hypothesis II(f, F); further, assume that for every y fixed the function $t \to F(t, y)$ is strongly measurable in the norm of (X^*, E). Let $u(\cdot) \in C_{ad}(0, T; \mathbf{U})$ be such that $y(t, u)$ exists in $0 \le t \le \bar{t}$ and $\varepsilon > 0$. Then there exists $v(\cdot) \in CV(0, T; D)$ such that $y(t, v)$ exists in $0 \le t \le \bar{t}$ and*

$$\|y(t, u) - y(t, v)\| \le \varepsilon \quad (0 \le t \le \bar{t}).$$

Proof. Let $v(\cdot) \in CV(0, T; D)$ be the control that performs (14.2.16) for $S(t, s) = S(t - s)$, $F(t) = F(t, y(t, u))$. We have

$$y(t, v) - y(t, u) = \int_0^t S(t - \tau)\{f(\tau, y(\tau, v)) - f(\tau, y(\tau, u))\}d\tau$$

$$+ \int_0^t S(t - \tau)\{F(\tau, y(\tau, v))v(\tau) - F(\tau, y(\tau, u))v(\tau)\}d\tau$$

$$+ \int_0^t S(t - \tau)\{F(\tau, y(\tau, u))v(\tau) - F(\tau, y(\tau, u))u(\tau)\}d\tau$$

in the maximal interval $0 \le t \le t_v$ where $y(t, v)$ exists and satisfies $\|y(t, v) - y(t, \mu)\| \le 1$. Estimating and using Gronwall's inequality,

$$\|y(t, v) - y(t, u)\| \le C\varepsilon.$$

The proof ends like that of Theorem 14.2.3. ∎

Remark 14.2.10. Theorem 14.2.9 obviously admits an extension to the time dependent case $A = A(t)$ under the usual assumption that the Cauchy problem for $y' = A(t)y$ is well posed; $S(t - \tau)$ is replaced by the solution operator $S(t, \tau)$.

One of the uses of relaxation results like Theorems 14.2.3 and 14.2.9 is in systems outfitted with relaxed controls. However, there are many interesting applications that involve only ordinary controls, such as the one below involving Theorem 14.2.9. Theorem 14.2.3 applies to parabolic control systems, but it becomes more interesting in the context of the $^\odot$-theory. The corresponding version and applications will be seen in **14.7**.

Example 14.2.11. We do the semilinear wave equation

$$y_{tt}(t, x) = \sum_{j=1}^{m} \sum_{k=1}^{m} \partial^j (a_{jk}(x) \partial^k y(t, x)) + \sum_{j=1}^{m} b_j(x) \partial^j y(t, x) + c(x) y(t, x)$$
$$- \phi(t, x, y(t, y)) + \psi(t, x, y(t, x)) u(t, x) \tag{14.2.17}$$

modelled as the abstract differential equation

$$\mathbf{y}'(t) = \mathbf{A} y(t) + \mathbf{f}(t, \mathbf{y}(t)) + \mathbf{F}(t, \mathbf{y}(t)) u(t) \tag{14.2.18}$$

in $\mathbf{E} = H_0^1(\Omega) \times L^2(\Omega)$ or $H^1(\Omega) \times L^2(\Omega)$ according to the boundary condition. The operators are

$$\mathbf{f}(t, \mathbf{y})(x) = \mathbf{f}(t, (y, y_1))(x) = \begin{bmatrix} 0 \\ -\phi(t, x, y(x)) \end{bmatrix}, \tag{14.2.19}$$

$$(\mathbf{F}(t, \mathbf{y}) u)(x) = (\mathbf{F}(t, (y, y_1)) u)(x) = \begin{bmatrix} 0 \\ \psi(t, x, y(x)) u(x) \end{bmatrix}, \tag{14.2.20}$$

and the admissible control space $C_{\mathrm{ad}}(0, \bar{t}; \mathbf{U})$ the unit ball of $L^\infty((0, \bar{t}) \times \Omega) = L_w^\infty(0, T; L^\infty(\Omega))$. We take $X = L^1(\Omega)$ and $\mathbf{U} = $ unit ball of $X^* = L^\infty(\Omega)$, thus our control space is of the form prescribed in this section. The linear part of the operator satisfies the requirements in Chapter 5 (that is, the coefficients are measurable and bounded) and the domain Ω is of class $C^{(1)}$, thus the semilinear theory in **5.7** can be applied. The functions $\phi(t, x, y)$, $\psi(t, x, y)$ satisfy the assumptions there; this means that both are measurable in t, x for y fixed and they satisfy (5.7.5) and (5.7.6) when $m \geq 2$, (5.7.7) and (5.7.8) when $m = 1$. Under these requirements it has been proved that the operator $\Phi(t, y)(x) = \phi(t, x, y(x))$ maps $H^1(\Omega)$ into $L^2(\Omega)$, is strongly measurable for $y(\cdot)$ fixed and

$$\|\Phi(t, y)\|_{L^2(\Omega)} \leq K(t, c) \quad (0 \leq t \leq T, \|y\|_{H^1(\Omega)} \leq c), \tag{14.2.21}$$

$$\|\Phi(t, y') - \Phi(t, y)\|_{L^2(\Omega)} \leq L(t, c) \|y' - y\|_{H^1(\Omega)}$$
$$(0 \leq t \leq T, \|y\|_{H^1(\Omega)}, \|y'\|_{H^1(\Omega)} \leq c), \tag{14.2.22}$$

where $L(\cdot, c) \in L^1(0, T)$. The operator $\Psi(t, y)(x) = \psi(t, x, y(x))$ enjoys exactly the same privileges; it maps $H^1(\Omega)$ into $L^2(\Omega)$, it is strongly measurable for y fixed and satisfies the estimates (14.2.21) and (14.2.22). (For both operators, the estimates are actually much more precise if $m \geq 2$; see (5.7.10) and (5.7.11)).

The properties of $\Phi(t, y)$ guarantee that $\mathbf{f}(t, \mathbf{y})$ satisfies its share of Assumption II($\mathbf{f}, \mathbf{F}$). Inequality (14.2.21) for $\Psi(t, y)$ implies that $\mathbf{F}(t, \mathbf{y})$ is a bounded operator from $L^\infty(\Omega)$ into $\mathbf{E}$ and insures (14.2.5); on the other hand, (14.2.7) is a consequence of (14.2.22) for $\Psi(t, y)$. Finally, the adjoint of $\mathbf{F}(t, \mathbf{y})$ is given by

$$\mathbf{F}(t, \mathbf{y})^* \mathbf{z} = \mathbf{F}(t, (y, y_1))(z, z_1)(x) = \psi(t, x, y(x)) z_1(x) \tag{14.2.23}$$

thus is a bounded operator from $\mathbf{E}$ into $L^1(\Omega) = X$; that $t \to \Psi(t, y)$ is strongly measurable in $L^2(\Omega)$ for each $y \in H^1(\Omega)$ implies that $t \to \mathbf{F}(t, \mathbf{y})^*\mathbf{z}$ is strongly measurable in $\mathbf{E}$ for each $\mathbf{y}, \mathbf{z}$.

At this point we have completely verified Hypothesis II($\mathbf{f}, \mathbf{F}$). However, our objective is to apply Theorem 14.2.9, which requires $t \to \mathbf{F}(t, \mathbf{y})$ to be strongly measurable as a $(L^\infty(\Omega), \mathbf{E})$-valued function. To see this, note that $t \to \Psi(t, y)$ is strongly measurable in $L^2(\Omega)$ for each $y \in H^1(\Omega)$ so that, given $\varepsilon > 0$ there exists a countably $L^2(\Omega)$-valued function $f(t)$ with $\|\Psi(t, y) - f_\varepsilon(t)\|_{L^2(\Omega)} \leq \varepsilon$. The operator function $\mathbf{F}_\varepsilon(t) : L^\infty(\Omega) \to \mathbf{E}$ defined from (14.2.20) with $f_\varepsilon(t, x)$ instead of $\psi(t, x, y(x))$ is countably valued and

$$\|\mathbf{F}(t, \mathbf{y}) - \mathbf{F}_\varepsilon(t)\|_{(L^\infty, \mathbf{E})} \leq \varepsilon \quad (0 \leq t \leq \bar{t}).$$

All assumptions checked, we are poised for application of Theorem 14.2.9. After Remark 14.2.4 we can take D as the set of all extremal points of the unit ball of $L^\infty(\Omega)$. These are the functions $u(\cdot)$ that satisfy $|u(x)| = 1$ a.e., thus the result for the equation (14.2.17) reads: given $\varepsilon > 0$ and an arbitrary control $u(t, x) \in L^\infty((0, \bar{t}) \times \Omega)$ with $|u(t, x)| \leq 1$ a.e. and such that the solution $y(t, x, u)$ exists in $0 \leq t \leq \bar{t}$ there exists a control $v(t, x)$ such that

$$|v(t, x)| = 1 \quad \text{a.e. in } (0, \bar{t}) \times \Omega,$$

and such that the solution $y(t, x, v)$ of (14.2.17) with $v(t, x)$ and the same initial conditions exists in $0 \leq t \leq \bar{t}$ and satisfies

$$\|y(t, \cdot, u) - y(t, \cdot, v)\|^2_{H^1(\Omega)} + \|y_t(t, \cdot, u) - y_t(t, \cdot, v)\|^2_{L^2(\Omega)} \leq \varepsilon^2 \quad (0 \leq t \leq \bar{t}).$$
$$(14.2.24)$$

Similar conclusions can be obtained with other control spaces. For instance, let $X = X^* = \mathbb{R}, U = [-1, 1]$, corresponding to the control term $\psi(x, y(t, x))u(t)$. The operator $\mathbf{F} : \mathbf{E} \to (\mathbb{R}, \mathbf{E})$ is defined by

$$(\mathbf{F}(t, \mathbf{y})u)(x) = (\mathbf{F}(t, (y, y_1))u)(x) = \begin{bmatrix} 0 \\ \psi(t, x, y(x))u \end{bmatrix}, \qquad (14.2.25)$$

and the adjoint $\mathbf{F}(t, \mathbf{y})^* : \mathbf{E} \to \mathbb{R}$ is

$$\mathbf{F}(t, \mathbf{y})^*\mathbf{z} = \mathbf{F}(t, (y, y_1))^*(z, z_1) = \int_\Omega \psi(t, x, y(x))z_1(x)dx. \qquad (14.2.26)$$

The conclusion is the same; this time the control $u(\cdot)$ is a measurable function with $|u(t)| \leq 1$ and $v(\cdot)$ satisfies $|v(t)| = 1$ a.e.

14.3. Installation of Relaxed Controls in Infinite Dimensional Systems

We use the three relaxed control spaces $R(0, T; U) = R_c(0, T; U), R_{bc}(0, T; U)$, $R_b(0, T; U)$ introduced in **10.5**, the first if U is a compact metric space, the second if U is a normal topological space, the third for a general control set U. In all cases, the **original** control system is

$$y'(t) = Ay(t) + f(t, y(t), u(t)), \qquad y(0) = \zeta \qquad (14.3.1)$$

and the **relaxed** system is

$$y'(t) = Ay(t) + F(t, y(t))\mu(t), \qquad y(0) = \zeta \qquad (14.3.2)$$

equipped with the admissible control space $R(0, T; U)$. The function $F(t, y)$ is defined by

$$\langle y^*, F(t, y)\mu \rangle = \int_U \langle y^*, f(t, y, u) \rangle \mu(du) \qquad (14.3.3)$$

for $y^* \in E^*$. Note (Example 12.3.14) that, due to mere finite additivity of μ, the "obvious" definition of $F(t, y)$ by E-valued integration may fail except for $R_c(0, T; U)$; this forces the "weak" definition (14.3.3) and places $F(t, y)\mu$ in E^{**} rather than in E, which demands restrictions on the space.

The first (routine) task is to impose conditions on $f(t, y, u)$ that give sense to (14.3.3) and put the equation (14.3.2) under the setup in **14.2**, that is, conditions that insure that (14.3.2) satisfies Hypotheses I(f, F) or II(f, F). This is achieved along the lines of **13.1**. We do in detail the space $R(0, T; U) = R_{bc}(0, T; U)$ and then point out the modifications pertaining to the other spaces. Note first that the space $R_{bc}(0, T; U)$ can be identified with the space $C_{ad}(0, T; \mathbf{U}) \subseteq L_w^\infty(0, T; \Sigma_{rba}(U)) = L_w^\infty(0, T; BC(U)^*)$, where $\mathbf{U}$ is the set $\Pi_{rba}(U)$ of all probability measures in $\Sigma_{rba}(U)$. Below, $B(U; E)$ (resp. $BC(U; E)$) denotes the space of all bounded (resp. bounded and continuous) E-valued functions defined in U, both spaces equipped with the supremum norm; if U is an arbitrary set, only $B(U; E)$ makes sense.

Hypothesis I(BC). $f(t, y, \cdot) \in B(U; E)$ for t, y fixed; $f(t, y, u)$ is continuous in y for t fixed uniformly with respect to u.[1] For every $y^* \in E^*$ the function $t \to \langle y^*, f(t, y, \cdot) \rangle$ is a strongly measurable $BC(U)$-valued function. For every $c > 0$ there exists $K(\cdot, c) \in L^1(0, T)$ such that

$$\|f(t, y, u)\| \le K(t, c) \quad (0 \le t \le T, \|y\| \le c, u \in U). \qquad (14.3.4)$$

Hypothesis II(BC). $f(t, y, \cdot) \in B(U; E)$ for t, y fixed and, for every $y^* \in E^*$ the function $t \to \langle y^*, f(t, y, \cdot) \rangle$ is a strongly measurable $BC(U)$-valued function. For every $c > 0$ there exist $K(\cdot, c), L(\cdot, c) \in L^1(0, T)$ such that (14.3.4) holds and

$$\|f(t, y', u) - f(t, y, u)\| \le L(t, c)\|y' - y\|$$
$$(0 \le t \le T, \|y\|, \|y'\| \le c, u \in U). \qquad (14.3.5)$$

---

[1] That is, $\|f(t, y', u) - f(t, y, u)\| \le \varepsilon$ for $\|y' - y\| \le \delta$ independently of $u \in U$.

Lemma 14.3.1. *Let E be reflexive and separable. If Hypothesis $I(BC)$ (resp. Hypothesis $II(BC)$) holds, then* (14.3.3) *defines an operator function $t \to F(t, y) \in (\Sigma_{rba}(U), E) = (BC(U)^*, E)$, satisfying Hypothesis $I(f, F)$ (resp. Hypothesis $II(f, F)$) in* **14.2** *with $X = BC(U)$ and $K(\cdot, c), L(\cdot, c)$ independent of $\mu(\cdot) \in R_{bc}(0, T; U)$.*

Proof. Clearly, under Hypothesis $I(BC)$, (14.3.3) defines a unique element $F(t, y)\mu$ of $E^{**} = E$ for any $\mu \in \Sigma_{rba}(U)$. By virtue of (14.3.4) we have

$$\|F(t, y)\mu\|_E \le K(t, c)\|\mu\|_{\Sigma_{rba}} \quad (0 \le t \le T, \|y\| \le c, \mu \in \Sigma_{rba}(U)). \quad (14.3.6)$$

The continuity assumptions on $f(t, y, u)$ imply that $y \to F(t, y)$ is continuous in the $(BC(U)^*, E)$-norm. The adjoint operator $F(t, y)^*$ is given by $F(t, y)^* y^* = \langle y^*, f(t, y, \cdot) \rangle \in BC(U)$, thus maps E^* into $X = BC(U)$. By hypothesis, $t \to \langle y^*, f(t, y, \cdot) \rangle$ is a strongly measurable function of t. If (14.3.5) holds, then

$$\|F(t, y')\mu - F(t, y)\mu\|_E \le L(t, c)\|y' - y\|\|\mu\|_{\Sigma_{rba}}$$
$$(0 \le t \le T, \|y\|, \|y'\| \le c, \mu \in \Sigma_{rba}(U)). \quad (14.3.7)$$

This ends the proof. ∎

Installation of relaxed controls in $R_b(0, T; U)$ follows along the same lines. Hypotheses $I(BC)$ and $II(BC)$ are replaced by Hypotheses $I(B)$ and $II(B)$, where the space $BC(U)$ is replaced by $B(U)$ and Lemma 14.3.1 has an immediate translation.

The case where U is a compact metric space is handled via Hypotheses $I(C)$ and $II(C)$, where the space is $C(U)$ instead of $BC(U)$ (actually, the two spaces coincide). If we upgrade weak u-continuity to strong continuity, however, we may give up reflexiveness of the space. The resulting assumptions are the literal translation of Hypotheses $I(C)$ and $II(C)$ in **13.1**; here, they are called $I(C, s)$ and $II(C, s)$ (s means "strong").

Hypothesis I(C, s). $f(t, y, u)$ is continuous in u for t, y fixed and continuous in y uniformly with respect to u for t fixed. The function $t \to \langle y^*, f(t, y, u) \rangle$ is measurable in t for y^*, y, u fixed. For every $c > 0$ there exists $K(\cdot, c) \in L^1(0, T)$ such that (14.3.4) holds.

Hypothesis II(C, s). $f(t, y, u)$ is continuous in u for t, y fixed and the function $t \to \langle y^*, f(t, y, u) \rangle$ is measurable in t for y^*, y, u fixed. For every $c > 0$ there exist $K(\cdot, c), L(\cdot, c) \in L^1(0, T)$ such that (14.3.4) and (14.3.5) hold.

Lemma 14.3.2. *Let E be separable. If Hypothesis $I(C, s)$ (resp. Hypothesis $II(C, s)$) holds, then* (14.3.3) *defines a function $t \to F(t, y) \in (\Sigma_{rca}(U), E) = (C(U)^*, E)$,* *satisfying Hypothesis I(f, F) (resp. Hypothesis II(f, F)) with $K(\cdot, c)$, $L(\cdot, c)$ independent of $\mu(\cdot) \in R_c(0, T; U)$.*

The operator $F(t, y)$ can be directly constructed with the vector integral

$$F(t, y)\mu = \int_U f(t, y, u)\mu(du), \tag{14.3.8}$$

which we define as in **5.0** since μ is countably additive. Given that the space E is separable, it is enough to show that $t \to \langle y^*, F(t, y)\mu \rangle$ is measurable for all $y^* \in E^*$, which follows from strong measurability of $t \to \langle y^*, f(t, y, \cdot) \rangle$ as a $C(U)$-valued function; this, in turn, is a consequence of Lemma 13.1.1. For both hypotheses, y-continuity of $F(t, y)$ in the norm of $(\Sigma_{rca}(U), E)$ follows from

$$\|F(t, y')\mu - F(t, y)\mu\| \le \int_U \|f(t, y', u) - f(t, y, u)\| \|\mu\|(du). \qquad \blacksquare$$

Back in the general case, the admissible control space $C_{ad}(0, T; U)$ for the original system (14.3.1) consists of all U-valued strongly measurable functions. These functions are defined as pointwise a.e. limits of countably valued functions (the definition as uniform limit may not make sense, as uniform convergence is not necessarily defined in arbitrary topological spaces.[2] If $u(\cdot)$ is a strongly measurable function, then

$$\mu(t) = \delta(\cdot - u(t)) \tag{14.3.9}$$

belongs to $R_{bc}(0, T; U)$, since $\langle \mu(t), f \rangle = f(u(t)) = \lim f(u_n(t))$ with $u_n(\cdot)$ countably valued.

For a control set U with no topological structure, we simply take as $C_{ad}(0, T; U)$ the set of all countably U-valued functions. In all cases we have, assuming that E is reflexive and separable,

Lemma 14.3.3. *Assume that U is a normal topological space and that the original system (14.3.1) satisfies Hypothesis I(BC) (resp. II(BC)). Then (14.3.1), equipped with $C_{ad}(0, T; U)$ satisfies Hypothesis I (resp. Hypothesis II) with $K(\cdot, c)$, $L(\cdot, c)$ independent of $u(\cdot)$. Same statement for a compact metric space U and Hypotheses I(C), II(C), or for an arbitrary set U and Hypotheses I(B), II(B).*

Once it has been established that (14.3.2) satisfies Assumption I(f, F) or II(f, F), all results in **14.2** and **14.3** can be applied, in particular, the relaxation theorems. For the record, we write below the corresponding version of both relaxation theorems for a normal topological control space U. The assumption that **U** is

---

[2] Uniform convergence can be defined in certain topological spaces, namely A. Weil's *uniform spaces* (see Kelley [1955, Ch. 6]).

X-weakly compact in X^* is automatically satisfied, since $\mathbf{U} = \Pi_{rba}(U)$ is $BC(U)$-weakly compact in $\Sigma_{rba}(U)$, with corresponding statements for $\Pi_{ba}(U)$, $\Pi_{rca}(U)$ (Corollary 12.6.3); moreover, the set D of all Dirac deltas is total in $\Pi_{rba}(U)$, $\Pi_{ba}(U)$, $\Pi_{rca}(U)$ (Lemma 12.6.4).

Theorem 14.3.4. *Assume that $S(t)$ is compact and that Hypothesis II(BC) holds. Let $\mu(\cdot) \in R_{bc}(0, T; U)$ be such that the solution $y(\cdot, \mu)$ of (14.3.2) exists in $0 \le t \le \bar{t}$ and $\varepsilon > 0$. Then there exists $v(\cdot) \in PC(0, \bar{t}; U)$ such that the solution $y(t, v)$ of (14.3.1) exists in $0 \le t \le \bar{t}$ and*

$$\|y(t, u) - y(t, v)\| \le \varepsilon \quad (0 \le t \le \bar{t}). \tag{14.3.10}$$

Theorem 14.2.9 needs strong measurability of $t \to F(t, y)$ in the norm of $(\Sigma_{rba}(U), E)$, which in turn requires stronger conditions on $f(t, y, u)$.

Lemma 14.3.5. *Let U be a normal topological space, $\phi : [0, T] \times U \to E$ a function such that $\phi(t, \cdot) \in BC(U; E)$ and that the function $t \to \phi(t, \cdot)$ belongs to $L^1(0, T; BC(U; E))$. Then the function $t \to \Phi(t) \in (\Sigma_{rba}(U); E)$ defined by*

$$\langle y^*, \Phi(t)\mu \rangle = \int_U \langle y^*, \phi(t, u) \rangle \mu(du), \tag{14.3.11}$$

is strongly measurable in the norm of $(\Sigma_{rba}(U); E)$.

Proof. Since $t \to \phi(t, \cdot)$ is strongly measurable, given $\varepsilon > 0$ there exists a countably valued $BC(U; E)$-valued function $t \to \phi_\varepsilon(t, \cdot)$ with

$$\|\phi(t, \cdot) - \phi_\varepsilon(t, \cdot)\|_{BC(U;E)} \le \varepsilon \quad \text{a.e. in } 0 \le t \le T,$$

which implies

$$\|\Phi(t)\mu - \Phi_\varepsilon(t)\mu\|_E \le \varepsilon \|\mu\|_{\Sigma_{rba}(U)} \quad \text{a.e. in } 0 \le t \le T,$$

where $\Phi_\varepsilon(t)$ is the function defined from $\phi_\varepsilon(t, u)$ by (14.3.11); since each $\Phi_\varepsilon(\cdot)$ is a countably valued $(\Sigma_{rba}(U), E)$-valued function, $\Phi(\cdot)$ is a strongly measurable $(\Sigma_{rba}(U), E)$-valued function as claimed. ∎

Theorem 14.3.6. *Let E be reflexive and separable and let the control system (14.1.1) satisfy Assumption II(BC); further, assume that for every y fixed, the function $t \to f(t, y, \cdot)$ belongs to $L^1(0, T; BC(U; E))$. Let $\mu(\cdot) \in R_{bc}(0, T; U)$ be such that the solution $y(t, \mu)$ of (14.3.2) exists in $0 \le t \le \bar{t}$ and $\varepsilon > 0$. Then there exists $v(\cdot) \in CV(0, T; D)$ such that the solution $y(t, v)$ of (14.3.1) exists in $0 \le t \le \bar{t}$ and (14.3.10) holds.*

The results for the spaces $B(U)$ and $C(U)$ (U compact and metric) are the same and we omit them.

14.4. Differential Inclusions

Let $t \to \mathcal{F}(t, y) \subseteq E$ be a set valued function defined in $0 \le t \le T$. According to the definition in Frankowska [1990 : 1], the E-valued function $y(t)$ is a **solution of** the differential inclusion

$$y'(t) \in Ay(t) + \mathcal{F}(t, y(t)), \qquad y(0) = \zeta \qquad (14.4.1)$$

in $0 \le t \le T$ if there exists a function $g(\cdot) \in L^1(0, T; E)$ such that

$$g(t) \in \mathcal{F}(t, y(t)) \quad \text{a.e. in } 0 \le t \le T, \qquad (14.4.2)$$

$$y(t) = S(t)\zeta + \int_0^t S(t - \tau)g(\tau)d\tau \quad (0 \le t \le T). \qquad (14.4.3)$$

We call the pair $(y(\cdot), g(\cdot))$ a **trajectory** of (14.4.1).

The first result is on equivalence of the control system

$$y'(t) = Ay(t) + f(t, y(t), u(t)), \qquad y(0) = \zeta \qquad (14.4.4)$$

and the differential inclusion

$$y'(t) \in Ay(t) + f(t, y(t), U), \qquad y(0) = \zeta. \qquad (14.4.5)$$

No special assumptions are required from the space E. As befits Filippov's theorem, U is required to be a complete separable metric space. We equip (14.4.4) with the admissible control space $C_{ad}(0, T; U)$ of all strongly measurable U-valued functions and assume that for every $y(\cdot) \in C(0, T; U)$ and every $u(\cdot) \in C_{ad}(0, T; U)$ the function $t \to f(t, y(t), u(t))$ belongs to $L^1(0, T; E)$, so that the solution of (14.4.4) can be defined in the usual way as the solution of the integral equation

$$y'(t) = S(t)\zeta + \int_0^t S(t - \tau)f(\tau, y(\tau), u(\tau))d\tau. \qquad (14.4.6)$$

Equivalence Theorem 14.4.1. *Assume that the function $f(t, y, u)$ is locally uniformly continuous in u for each t, y fixed. Then $y(\cdot)$ satisfies (14.4.4) with $u(\cdot) \in C_{ad}(0, T; U)$ if and only if it satisfies the differential inclusion (14.4.5).*

Proof. The "only if" part is obvious. Conversely, assume that $(y(\cdot), g(\cdot))$ is a trajectory of the differential inclusion (14.4.5). Then

$$g(t) \in f(t, y(t), U) \quad \text{a.e. in } 0 \le t \le T. \qquad (14.4.7)$$

The assumptions imply that the function $\phi(t, u) = f(t, y(t), u)$ is locally uniformly continuous in u for each t fixed, and that $t \to \phi(t, u)$ is strongly measurable in t for every $u \in U$ fixed. Accordingly, by Filippov's Theorem 12.8.2, there exists a strongly measurable U-valued function $u(\cdot)$ (that is, an element of $C_{ad}(0, T; U)$) such that

$$g(t) = f(t, y(t), u(t)) \quad \text{a.e. in } 0 \leq t \leq T \tag{14.4.8}$$

so that $y(\cdot)$ satisfies (14.4.6). This completes the proof. ∎

The results below are a counterpart of the finite dimensional theory in **13.4** and are proved in similar ways. We do first the space $R_{bc}(0, T; U)$ with U a normal topological space. The basic assumption for the original system (14.4.4) is

Hypothesis $\emptyset(BC)$. For every $y(\cdot) \in C(0, T; U)$ and every $y^* \in E^*$, $t \to \langle y^*, f(t, y(t), \cdot) \rangle$ is a strongly measurable $BC(U)$-valued function. There exists $K(\cdot, c) \in L^1(0, T)$ such that

$$\|f(t, y, u)\| \leq K(t, c) \quad (0 \leq t \leq T, \|y\| \leq c, u \in U). \tag{14.4.9}$$

Hypothesis $\emptyset(BC)$ and the fact that E is reflexive imply that the function $F(t, y)$ in (14.3.3),

$$\langle y^*, F(t, y)\mu \rangle = \int_U \langle y^*, f(t, y, u) \rangle \mu(du) \quad (y^* \in E^*), \tag{14.4.10}$$

is well defined; moreover,

$$\|F(t, y)\mu\| \leq K(t, c)\|\mu\|_{\Sigma_{rba}(U)}$$
$$(0 \leq t \leq T, \|y\| \leq c, \mu \in \Sigma_{rba}(U)). \tag{14.4.11}$$

In general, $t \to F(t, y(t))\mu(t)$ is weakly measurable; if E is separable, $t \to F(t, y(t))\mu(t)$ is strongly measurable, thus it belongs to $L^1(0, T; E)$ for every $\mu(\cdot) \in L_w^\infty(0, T; \Sigma_{rba}(U))$. Accordingly, the solution of the relaxed system

$$y'(t) = Ay(t) + F(t, y(t))\mu(t), \qquad y(0) = \zeta \tag{14.4.12}$$

can be defined by means of the integral equation

$$y'(t) = S(t)\zeta + \int_0^t S(t - \tau)F(\tau, y(\tau))\mu(\tau)d\tau. \tag{14.4.13}$$

Equivalence Theorem 14.4.2. *Assume the space E is reflexive and separable and that Hypothesis $\emptyset(BC)$ holds. Then $y(\cdot) \in C(0, T; E)$ is a solution of the relaxed system* (14.4.12) *with $\mu(\cdot) \in R_{bc}(0, T; U)$ if and only if it is a solution of the differential inclusion*

$$y'(t) \in Ay(t) + \overline{\text{conv}} f(t, y(t), U). \tag{14.4.14}$$

The proof follows from several auxiliary results. The second, Lemma 14.4.4, is the "relaxed" version of Filippov's Theorem; see **13.3** for the finite dimensional counterpart.

Lemma 14.4.3. *Let E be a reflexive separable Banach space, U a normal topological space, $\phi(\cdot) \in B(U; E)$ a function such that $\langle y^*, \phi(\cdot)\rangle \in BC(U)$ for all $y^* \in E^*$. Define $\Phi : \Sigma_{rba}(U) \to E$ by*

$$\langle y^*, \Phi\mu\rangle = \int_U \langle y^*, \phi(u)\rangle\mu(du) \tag{14.4.15}$$

for $y^* \in E^*$. Then, for any $\mu \in \Pi_{rba}(U)$,

$$\Phi\mu \in \overline{\text{conv}}\,\phi(U). \tag{14.4.16}$$

Proof. Note first that the assumptions imply that $\Phi\mu \in E$ is well defined for any $\mu \in \Sigma_{rba}(U)$ with

$$\|\Phi\mu\|_E \leq \|\phi\|_{B(U;E)}\|\mu\|_{\Sigma_{rba}(U)}.$$

The set $\overline{\text{conv}}\,\phi(U)$ is an intersection of half-spaces

$$\{y; \langle y^*, y\rangle \leq a\} \tag{14.4.17}$$

over a family $\{y^*\}$ of elements of E^* and a family $\{a\}$ of real numbers. Making use of Lemma 12.6.4, construct a generalized sequence $\{\mu_\kappa\}$ in conv(D) with $\mu_\kappa \to \mu$ $BC(U)$-weakly. Each μ_κ is a finite convex combination $\Sigma\alpha_j\delta(\cdot - u_j)$ of Dirac measures, so that $\Phi\mu_\kappa = \Sigma\alpha_j\phi(u_j) \in \text{conv}\phi(U)$. Replacing in (14.4.17), $\langle y^*, \Phi\mu_\kappa\rangle \leq a$; taking limits, $\langle y^*, \Phi\mu\rangle \leq a$. ■

Lemma 14.4.4. *Let E and U be as in Lemma 14.4.3. Let $\phi : [0, T] \times U \to E$ be a function such that $\phi(t, \cdot) \in B(U; E)$ and such that, for every $y^* \in E^*$, $\langle y^*, \phi(t, \cdot)\rangle \in BC(U)$ in $0 \leq t \leq T$ and $t \to \langle y^*, \phi(t, \cdot)\rangle$ is a strongly measurable $BC(U)$-valued function. Let $\mu(\cdot) \in R_{bc}(0, T; U)$. Define $g(t) = \Phi(t)\mu(t)$, where*

$$\langle y^*, \Phi(t)\mu(t)\rangle = \int_U \langle y^*, \phi(t, u)\rangle\mu(t, du). \tag{14.4.18}$$

Then $g(\cdot)$ is strongly measurable and

$$g(t) \in \overline{\text{conv}}\,\phi(t, U) \quad a.e. \text{ in } 0 \leq t \leq T. \tag{14.4.19}$$

Conversely, let $g(\cdot)$ be a strongly measurable function such that (14.4.19) holds. Then there exists $\mu(\cdot) \in R_{bc}(0, T; U)$ such that

$$\Phi(t)\mu(t) = g(t) \quad a.e. \text{ in } 0 \leq t \leq T. \tag{14.4.20}$$

Proof. The assumptions imply that $\Phi(t)\mu(t)$ is well defined for every $\mu(\cdot) \in L_w^\infty(0, T; \Sigma_{rba}(U))$ and

$$\|\Phi(t)\mu(t)\|_E \leq \|\phi(t)\|_{B(U;E)}\|\mu\|_{L_w^\infty(0,T;\Sigma_{rba}(U))}. \tag{14.4.21}$$

Let $\{z_k^*\}$ be a countable dense set in the unit ball of E^*. If $0 \leq t \leq T$ and $\varepsilon > 0$ we can pick $u = u(t)$ such that $\|\phi(t, u(t))\| \geq \|\phi(t)\|_{B(U;E)} - \varepsilon$ and then $z_k^* = z_k^*(t)$ such that $\langle z_k^*(t), \phi(t, u(t)) \rangle \geq \|\phi(t, u(t))\| - \varepsilon$. Accordingly, we have $\langle z_k^*(t), \phi(t, u(t)) \rangle \geq \|\phi(t)\|_{B(U;E)} - 2\varepsilon$, which means

$$\|\phi(t)\|_{B(U;E)} = \sup_k \sup_{u \in U} |\langle z_k^*, \phi(t, u) \rangle| = \sup_k \|\langle z_k^*, \phi(t) \rangle\|_{BC(U)}$$

and it follows that

$$t \to \|\phi(t)\|_{B(U;E)}$$

is measurable. Accordingly, the characteristic function χ_n of the set e_n where $n \leq \|\phi(t)\|_{B(U;E)} < n + 1$ is measurable and it is clearly enough to show the direct and converse parts of Lemma 14.4.4 for $\phi_n(t, u) = \chi_n(t)\phi_n(t, u)$, so that we may assume from the start that $\|\phi(t)\|_{B(U;E)}$ is bounded; actually all we need is that $t \to \langle y^*, \phi(t, \cdot) \rangle$ belong to $L^1(0, T; BC(U))$ for all $y^* \in E^*$.

If $g(t) = \Phi(t)\mu(t)$ and $y^* \in E^*$ then it follows from (14.4.18) that $g(\cdot)$ is weakly measurable, thus strongly measurable by separability of E. That (14.4.19) holds has been proved in Lemma 14.4.3.

Conversely, let a strongly measurable $g(\cdot)$ satisfy (14.4.19). Given $\varepsilon > 0$ select a countably valued function $g_\varepsilon(\cdot)$ with

$$\|g(t) - g_\varepsilon(t)\| \leq \varepsilon \quad (0 \leq t \leq T) \tag{14.4.22}$$

outside a null set. In view of (14.4.19),

$$g_\varepsilon(t) \in \{\overline{\text{conv}}\, \phi(t, U)\}_\varepsilon \quad (0 \leq t \leq T), \tag{14.4.23}$$

where $A_\varepsilon = \{y; \text{dist}(y, A) \leq \varepsilon\}$ for any set A.

Let $Y = \{y_k^*; k = 1, 2, \ldots, m\}$ be a finite set in E^*. For each k, the $BC(U)$-valued function $t \to \langle y_k^*, \phi(t, \cdot) \rangle$ is strongly measurable, thus there exists a finite set $\{\phi_{\varepsilon,k}(t, \cdot)\}$ of countably $BC(U)$-valued functions such that

$$|\langle y_k^*, \phi(t, u) \rangle - \phi_{\varepsilon,k}(t, u)\rangle| \leq \varepsilon \quad (u \in U, k = 1, 2, \ldots, m) \tag{14.4.24}$$

outside of a null set $e(Y)$. We can divide $[0, T]$ in a countable family $\Xi_{\varepsilon,Y}$ of pairwise disjoint measurable sets such that each $\phi_{\varepsilon,k}(t, \cdot)$ $(k = 1, 2, \ldots, m)$ and $g_\varepsilon(\cdot)$ are constant in each set $e \in \Xi_{\varepsilon,Y}$. Pick a fixed $e \in \Xi_{\varepsilon,Y}$ and let $t_0 \in e$. Using (14.4.23) for $t = t_0$, select a finite set of real numbers $\{\alpha_j(e)\}, \alpha_j(e) \geq 0$, $\Sigma\alpha_j(e) = 1$, and a corresponding finite set $\{u_j(e)\} \subset U$ such that

$$\|g_\varepsilon(t_0) - \Sigma\alpha_j(e)\phi(t_0, u_j(e))\| \leq 2\varepsilon.$$

Then $|\langle y_k^*, g_\varepsilon(t_0)\rangle - \Sigma\alpha_j(e)\langle y_k^*, \phi(t_0, u_j(e))\rangle| \le 2\varepsilon\|y_k^*\|$ and, by (14.4.24),

$$|\langle y_k^*, g_\varepsilon(t)\rangle - \Sigma\alpha_j(e)\phi_{\varepsilon,k}(t, u_j(e))| \le 3\varepsilon\|y_k^*\| \quad (k = 1, \ldots, m)$$

for $t = t_0$, hence for all $t \in e$ since both functions on the left side are constant in e. Finally, using again (14.4.22) and (14.4.24),

$$|\langle y_k^*, g(t)\rangle - \langle y_k^*, \Sigma\alpha_j(e)\phi(t, u_j(e))\rangle| \le 5\varepsilon\|y_k^*\|$$
$$(t \in e, k = 1, \ldots, m). \quad (14.4.25)$$

Define $\mu_{\varepsilon,Y}(t) \in \Sigma_{rba}(U)$ for each $t \in [0, T]$ thus: if $t \in e \in \Xi_{\varepsilon,Y}$ then $\mu_{\varepsilon,Y}(t)$ $= \Sigma\alpha_j(e)\delta(\cdot - u_j(e))$. Obviously, $\mu_{\varepsilon,Y}(\cdot) \in R_{bc}(0, T; U)$ and, if $e \in \Xi_{\varepsilon,Y}$ we have $\Phi(t)\mu_{\varepsilon,Y}(t) = \Sigma\alpha_j(e)\phi(t, u_j(e))$ for $t \in e$. We can then rewrite (14.4.25) in the form

$$|\langle y_k^*, g(t) - \Phi(t)\mu_{\varepsilon,Y}(t)\rangle| \le 5\varepsilon\|y_k^*\| \quad (0 \le t \le T, k = 1, \ldots, m) \quad (14.4.26)$$

outside of a null set.

Select a countable dense set $\{y_k^*\}$ in E^* and apply inductively this construction as follows.

Step 1: $\varepsilon = 1$, $Y = \{y_1^*\}$; call $\mu_{\varepsilon,Y}(\cdot) = \mu_1(\cdot)$.

Step 2: $\varepsilon = 1/2$, $Y = \{y_1^*, y_2^*\}$; call $\mu_{\varepsilon,Y}(\cdot) = \mu_2(\cdot)$.

Step 3: $\varepsilon = 1/3$, $Y = \{y_1^*, y_2^*, y_3^*\}$; call $\mu_{\varepsilon,Y}(\cdot) = \mu_3(\cdot)$,

... and so on. We obtain in this way a sequence $\{\mu_n(\cdot)\} \subset R_{bc}(0, T; U)$ having the property that

$$|\langle y_k^*, g(t)\rangle - \langle y_k^*, \Phi(t)\mu_n(t)\rangle| \to 0 \quad (14.4.27)$$

uniformly in $0 \le t \le T$ (outside of a null set) for each $k = 1, 2, \ldots$.

We use now Theorem 12.5.9 (b) to select a generalized subsequence $\{\mu_\iota(\cdot)\}$ of $\{\mu_n(\cdot)\}$, $L^1(0, T; BC(U))$-weakly convergent in $L_w^\infty(0, T; \Sigma_{rba}(U))$ to $\mu(\cdot) \in R_{bc}(0, T; U)$. This implies

$$\int_0^T f(\tau)\langle y_k^*, \Phi(\tau)\mu_\iota(\tau)\rangle d\tau = \int_0^T \int_U f(\tau)\langle y_k^*, \phi(\tau, u)\rangle\mu_\iota(\tau, du)d\tau$$

$$\to \int_0^T \int_U f(\tau)\langle y_k^*, \phi(\tau, u)\rangle\mu(\tau, du)d\tau$$

$$= \int_0^T f(\tau)\langle y_k^*, \Phi(\tau)\mu(\tau)\rangle d\tau \quad (14.4.28)$$

for every $f(\cdot) \in L^\infty(0, T)$. We use (14.4.28) with $f(\tau) = \chi_t(\tau)$ the characteristic function of $[0, t]$; on the left-hand side we employ (14.4.27) and the dominated convergence theorem. The result is

$$\int_0^t \langle y_k^*, g(\tau)\rangle d\tau = \int_0^t \langle y_k^*, \Phi(\tau)\mu(\tau)\rangle d\tau$$

for all $t \in [0, T]$. Differentiating, $\langle y_k^*, g(\tau)\rangle = \langle y_k^*, \Phi(\tau)\mu(\tau)\rangle$ except in a null set d_k; using denseness of $\{y_k^*\}$, (14.4.20) results outside of $d = \cup d_k$. This ends the proof of Lemma 14.4.4. ∎

Proof of Theorem 14.4.2. Hypothesis $\varnothing(BC)$ guarantees that the function

$$\phi(t, u) = f(t, y(t), u) \tag{14.4.29}$$

satisfies the requirements of Lemma 14.4.3 and Lemma 14.4.4. If $y(\cdot)$ is a solution of (14.4.12) for some $\mu(\cdot) \in R_{bc}(0, T; U)$ then by Lemma 14.4.3 we have

$$F(t, y(t))\mu(t) = g(t) \in \overline{\text{conv}} f(t, y(t), U)$$

so that $y(\cdot)$ is a solution of the differential inclusion (14.4.14). Conversely, let $(y(\cdot), g(\cdot))$ be a trajectory of (14.4.14). Then, by definition of solution, we have $g(t) \in \overline{\text{conv}} f(t, y(t), U)$. We apply Lemma 14.4.4 to the function (14.4.29) and deduce that there exists $\mu(\cdot) \in R_{bc}(0, T; U)$ with

$$g(t) = F(t, y(t))\mu(t) \quad \text{a.e.}$$

so that $y(\cdot)$ is a solution of (14.4.12) for this $\mu(\cdot)$. ∎

To handle an arbitrary control set we simply change the space $BC(U)$ by the space $B(U)$; in this case, the admissible control space $C_{\text{ad}}(0, T; U)$ is the set of all countable U-valued functions. For the rest of the results the corresponding companion of Hypothesis $\varnothing(B)$ is Hypothesis $\varnothing(B)$, formulated in the same way but exchanging the space $BC(U)$ by the space $B(U)$.

The case where U is a compact metric space deserves a few comments. Theorem 14.4.1 is the same under the same assumptions for $f(t, y, u)$, although local uniform continuity (in fact, uniform continuity) is automatic. We only need the cheap version of Filippov's Theorem (Theorem 13.3.1), which can be extended from $\mathbb{R}^m$ to an arbitrary Banach space with the same proof. For the rest of the results, Hypothesis $\varnothing(C)$ is formulated the same as $\varnothing(BC)$ interchanging $BC(U)$ by $C(U)$.

14.5. Existence Theorems for Optimal Control Problems, I

Theorem 14.5.1 below applies to the system

$$y'(t) = Ay(t) + f(t, y(t)) + F(t, y(t))u(t) \tag{14.5.1}$$

under Assumption II(f, F) in **14.2**, and is an infinite dimensional counterpart of Theorem 2.2.6. It includes state constraints and a target condition, respectively,

$$y(t) \in M(t) \quad (0 \le t \le \bar{t}), \qquad y(\bar{t}) \in Y, \qquad (14.5.2)$$

in a fixed or variable time interval $0 \le t \le \bar{t}$, and a weakly lower semicontinuous cost functional $y_0(t, u)$ equicontinuous with respect to t. The definitions are the same as those in **2.2**, only that they involve *generalized* sequences rather than sequences. We say that the cost functional $y_0(t, u)$ is **weakly lower semicontinuous** if, for every generalized sequence $\{u_\kappa(\cdot)\} \subseteq C_{\mathrm{ad}}(0, \bar{t}; U)$ such that

$$u(\cdot) \to \bar{u}(\cdot) \in C_{\mathrm{ad}}(0, \bar{t}; U) \quad L^1(0, \bar{t}; X)\text{-weakly in } L_w^\infty(0, \bar{t}; X^*) \qquad (14.5.3)$$

and such that $y(t, u_\kappa)$ exists in $0 \le t \le \bar{t}$ and

$$y(\cdot, u_\kappa) \to y(\cdot) \text{ in } C(0, \bar{t}; E), \qquad (14.5.4)$$

we have

$$y_0(\bar{t}, \bar{u}) \le \limsup y_0(\bar{t}, u_\kappa). \qquad (14.5.5)$$

Let $\{u_\kappa(\cdot)\}$, $u_\kappa(\cdot) \in C_{\mathrm{ad}}(0, t_\kappa; U)$ be a generalized sequence $(t_\kappa \le T)$ such that $y(t, u_\kappa)$ exists in $0 \le t \le t_\kappa$. Extend $u_\kappa(\cdot)$ and $y(t, u_\kappa)$ to $[0, T]$ setting $u_\kappa(t) = u =$ fixed element of U and $y(t, u_\kappa) = y(t_\kappa, u_\kappa)$ in $t \ge t_\kappa$; the extensions are named in the same way. The cost functional $y_0(t, u)$ is **equicontinuous with respect to** t if, for every generalized sequence as above and such that $t_\kappa \to \bar{t}$ and that (14.5.3) and (14.5.4) hold we have

$$|y_0(t_\kappa, u_\kappa) \to y_0(\bar{t}, u_\kappa)| \to 0 \quad \text{as } n \to \infty. \qquad (14.5.6)$$

Theorem 14.5.1. *Let E be reflexive and separable, $S(t)$ compact for $t > 0$, and let Assumption II(f, F) be satisfied. Assume the cost functional $y_0(t, u)$ is weakly lower semicontinuous and equicontinuous with respect to t, that $M(t)$ and Y are closed and that $C_{\mathrm{ad}}(0, T; \mathbf{U})$ is $L^1(0, T; X)$-weakly compact in $L_w^\infty(0, T; X^*)$. Then, if there exists a minimizing sequence $\{u^n(\cdot)\}$, $u^n(\cdot) \in C_{\mathrm{ad}}(0, t_n; U)$ with $\{t_n\}$ bounded and $\{y(t, u^n)\}$ uniformly bounded there exists an optimal control $\bar{u}(\cdot)$, $L^1(0, \bar{t}; X)$-weak limit of a generalized subsequence of $\{u^n(\cdot)\}$.*

The proof is essentially the same as that of Theorem 2.2.6. We assume (if necessary selecting a subsequence) that $t_n \to \bar{t}$ and write the integral equation (14.2.8) for each $y(t, u^n)$,

$$y(t, u^n) = S(t)\zeta + \int_0^t S(t - \tau) f(\tau, y(\tau, u^n)) d\tau$$

$$+ \int_0^t S(t - \tau) F(\tau, y(\tau, u^n)) u^n(\tau) d\tau \quad (0 \le t \le t_n). \qquad (14.5.7)$$

Then we use Theorem 12.5.9, the boundedness assumptions on the trajectories and Theorem 5.5.19 to show that (a generalized subsequence of) $\{u^n(\cdot)\}$ is $L^1(0, \bar{t}; X)$-weakly convergent in $L_w^\infty(0, T; X^*)$ and the corresponding sequence of trajectories $\{y(t, u^n)\}$ is uniformly convergent to $\bar{y}(\cdot) \in C(0, \bar{t}; E)$ in $0 \le t \le \bar{t}$. To show that $\bar{y}(t) = y(t, \bar{u})$ we apply an arbitrary linear functional $y^* \in E^*$ and take limits in the expression

$$\langle y^*, y(t, u^n)\rangle = \langle y^*, S(t)\zeta\rangle + \int_0^t \langle S(t-\tau)^* y^*, f(\tau, y(\tau, u^n))\rangle d\tau$$

$$+ \int_0^t \langle F(\tau, y(\tau, u^n))^* S(t-\tau)^* y^*, u^n(\tau)\rangle d\tau \quad (0 \le t \le t_n). \quad (14.5.8)$$

∎

Theorem 14.5.1 applies to all spaces $R_{bc}(0, T; U)$, $R_b(0, T; U)$, $R_c(0, T; U)$ of relaxed controls, with $X = BC(U)$, $B(U)$, $C(U)$ respectively; in each case, $C_{ad}(0, \bar{t}; U)$ is $L^1(0, T; X)$-weakly compact in $L_w^\infty(0, T; X^*)$. The main difference between this result and finite dimensional analogues such as Theorem 2.2.6 is that in the latter, uniform convergence of the trajectories is a consequence of the Arzelà-Ascoli theorem; in the infinite dimensional setting, it has to be brought about by compactness of the semigroup.

The following example shows that, even in the linear case, compactness cannot be completely given up.

Example 14.5.2. Consider the linear control system

$$y'(t) = Ay(t) + u(t), \qquad y(0) = 0 \quad (14.5.9)$$

in the space $E = L^2(0, 2\pi)$. The semigroup is

$$S(t)y(x) = y(x + t), \quad (14.5.10)$$

($y(\cdot)$ continued 2π-periodically outside of $(0, 2\pi)$). The infinitesimal generator is

$$Ay(x) = y'(x) \quad (14.5.11)$$

with domain consisting of all $y(\cdot) \in E$ absolutely continuous, with square integrable derivative and $y(0) = y(2\pi)$. The operator A is skew-adjoint and $S(t)$ is a unitary group: $S(t)^* = S(-t)$. In particular, $\|S(t)y\| = \|y\|$ for all $y \in E$. We take $X = X^* = E$ and the unit ball of of E as control set U, so that the assumptions regarding the control space in Theorem 14.5.1 are satisfied. The (fixed) control interval is $0 \le t \le \pi$. There are no state constraints or target condition, and the cost functional is

$$y_0(\pi, u) = \int_0^\pi \{(S(-t)\eta, y(t, u))^2 + (t^2 - \|y(t, u)\|^2)^2\}dt, \quad (14.5.12)$$

where η is a fixed element of E. We check immediately that this functional is weakly lower semicontinuous; it is also equicontinuous with respect to t, but the latter is a moot point since the control interval is fixed.

We construct a minimizing sequence. Let

$$u^n(t) = S(t)y_n, \qquad y_n(x) = \frac{1}{\sqrt{\pi}} \cos nx. \tag{14.5.13}$$

Trajectories of the system are

$$y(t, u) = \int_0^t S(t - \tau)u(\tau)d\tau, \tag{14.5.14}$$

so that

$$y(t, u^n) = t S(t)y_n \tag{14.5.15}$$

and, since $y_n \to 0$ in E weakly, the same is true of $y(t, u^n)$ for all t. On the other hand, $\|y(t, u^n)\| = t$ so that the integrand of $y_0(\pi, u^n)$ tends to zero a.e., thus $y_0(\pi, u^n) \to 0$ by the dominated convergence theorem. Since the functional is nonnegative, the minimum m is zero and the sequence $\{u^n(\cdot)\}$ satisfies all the assumptions of Theorem 14.5.1. However, we show below that if

$$\eta(x) = \sum_{n=0}^{\infty} e^{-n^2} \cos nx, \tag{14.5.16}$$

there is no optimal control.

Lemma 14.5.3. *Let E be a Hilbert space and $f(\tau)$ a strongly measurable E-valued function defined in a measurable set e, with $\|f(\tau)\| \leq 1$. Assume that*

$$\left\| \int_e f(\tau)d\tau \right\| = |e| \tag{14.5.17}$$

Then there exists a one-dimensional subspace E_0 of E such that $f(t) \in E_0$ a.e.

Proof. Assume our claim is bogus; then, if $y \neq 0$ is arbitrary, we can write

$$f(\tau) = f_0(\tau)y + g(\tau) \tag{14.5.18}$$

where $\|y\|\,|f_0(\tau)| \leq 1$ and $g(\tau)$ belongs to the orthogonal complement of the subspace generated by y. We apply the decomposition (14.5.18) to $y = \int_e f(\tau)d\tau$. Taking the scalar product with y we obtain $(y, f(\tau)) = (y, y)f_0(\tau)$; integrating in e, $|e|^2 = (y, y) = (y, y) \int_e f_0(\tau)d\tau = |e| \int_e \|y\| f_0(\tau)d\tau$. We finally obtain that $\|y\|\,|f_0(\tau)| = 1$ a.e., hence $g(\tau) = 0$ a.e. in (14.5.18). ∎

Assume there is an optimal control $\bar{u}(\tau)$. Then we must have $\|y(t, \bar{u})\| = t$, in particular $\|y(\pi, \bar{u})\| = \pi$. Since $\|S(t - \tau)\bar{u}(\tau)\| \le 1$, we may apply Lemma 14.5.3 in $e = [0, 2\pi]$ and obtain that $S(\pi - \tau)\bar{u}(\tau) = r(\tau)y$, where $\|y\| = 1$ and $|r(\tau)| = 1$ a.e.; *a fortiori*, $\bar{u}(\tau) = S(\tau - \pi)r(\tau)y$. We then have

$$y(t, \bar{u}) = S(t - \pi)y \int_0^t r(\tau)d\tau.$$

Since $\|y(t, \bar{u})\| = t$ we must have $r(\tau) \equiv 1$ or $r(\tau) \equiv -1$, hence

$$y(t, \bar{u}) = t S(t)z \tag{14.5.19}$$

with $\|z\| = \|\pm S(-\pi)y\| = 1$. Replacing in the cost functional,

$$0 = y_0(\pi, \bar{u}) = \int_0^\pi t^2 (S(-t)\eta, S(t)z)^2 dt = \int_0^\pi t^2 (S(-2t)\eta, z)^2 dt$$

so that

$$(S(-2t)\eta, z) = 0 \quad (0 \le t \le \pi). \tag{14.5.20}$$

We have $S(-2t)(\cos nx) = \cos n(x - 2t) = \cos nx \cos 2nt + \sin nx \sin 2nt$, thus (14.5.20) is

$$\phi(t) = \sum_{n=0}^\infty e^{-n^2} \left(\cos 2nt \int_0^{2\pi} z(x) \cos nx \, dx - \sin 2nt \int_0^{2\pi} z(x) \sin nx \, dx \right) = 0$$

in $0 \le t \le \pi$. The function $\phi(t/2)$ is then identically zero in $0 \le t \le 2\pi$ and has a uniformly convergent Fourier series. Accordingly, all of its Fourier coefficients are zero and we have

$$\int_0^{2\pi} z(x) \cos nx \, dx = \int_0^{2\pi} z(x) \sin nx \, dx = 0$$

for $n = 1, 2, \ldots$, which implies that $z(x) = 0$ a.e., a contradiction since $\|z\| = 1$.

The fact that Theorem 14.5.1 "fails" in this example is not due to the functional, which is weakly lower semicontinuous as required. It is because the sequence of trajectories (14.5.14) corresponding to the minimizing sequence is not convergent in $C(0, \bar{t}; E)$, nor does it have any subsequence with this property, which means (14.5.4) is not satisfied.

We assume below a cost functional of the form $y_0(t, y, u)$ defined for $u(\cdot) \in C_{\mathrm{ad}}(0, T; U)$ and $y(\cdot) \in C(0, T; E)$. What makes it different from the cost functionals at play until now is that $y(\cdot)$ is not necessarily the trajectory $y(\cdot, u)$. We call $y_0(t, y, u)$ **weakly-weakly lower semicontinuous** if, for every generalized sequence $\{u_\kappa(\cdot)\} \subseteq C_{ad}(0, \bar{t}; U)$ such that (14.5.3) holds, $y(t, u_\kappa)$ exists in $0 \leq t \leq \bar{t}$, the $y(t, u_\kappa)$ are uniformly bounded and

$$\langle y^*, y(t, u_\kappa)\rangle \to \langle y^*, \bar{y}(t)\rangle \quad (0 \leq t \leq \bar{t}, y^* \in E^*) \tag{14.5.21}$$

for some $\bar{y}(\cdot) \in C(0, \bar{t}; U)$, we have

$$y_0(\bar{t}, \bar{y}, \bar{u}) \leq \limsup y(\bar{t}, y(u_\kappa), u_\kappa) \tag{14.5.22}$$

(the full generality of not requiring $\bar{y}(\cdot)$ to be a trajectory will only be used in **14.10**). Under this new definition, an existence theorem for linear systems

$$y'(t) = Ay(t) + Bu(t) \tag{14.5.23}$$

holds under the only assumption that A generates a strongly continuous semigroup $S(\cdot)$ in an arbitrary Banach space E. However, we shall assume E reflexive and X^* separable, so that $L_w^\infty(0, T; X^*) = L^\infty(0, T; X^*)$ and controls are strongly measurable; this insures that the trajectories

$$y(t, u) = S(t)\zeta + \int_0^t S(t - \tau)Bu(\tau)d\tau$$

are well defined. We assume B bounded with

$$B : X^* \to E, \qquad B^* : E^* \to X. \tag{14.5.24}$$

The result runs along rather predictable lines. Since we won't be able to extract more than weak convergence of (a subsequence of) the trajectories corresponding to a minimizing sequence, the definition of t-equicontinuity of the cost functional must be modified; we only require uniform boundedness of the trajectories and the weak convergence relation (14.5.21). For the same reason, the state constraint sets $M(t)$ and the target set Y will be requested to be E^*-weakly closed. To simplify, we retain the strong convergence relations (2.2.3) and (2.2.4) for minimizing sequences, but passing to the limit in (2.2.4) without strong convergence of the trajectories requires to show that $y(t_n, u^n) - y(\bar{t}, \bar{u}) \to 0$ E^*-weakly.

Theorem 14.5.4. *Assume $M(t)$ and Y are E^*-weakly closed, that $C_{ad}(0, T; U)$ is $L^1(0, T; X)$-weakly compact in $L_w^\infty(0, T; X^*)$ and that $y_0(\bar{t}, u)$ is weakly-weakly lower semicontinuous and equicontinuous with respect to t. Then, if there exists a minimizing sequence $\{u^n(\cdot)\}, (u^n(\cdot) \in C_{ad}(0, t_n; U))$ with $\{t_n\}$ bounded there exists an optimal control $\bar{u}(\cdot)$, which is the $L^1(0, \bar{t}; X)$-weak limit of a generalized subsequence of $\{u^n(\cdot)\}$.*

Proof. The assumptions imply that the set $\mathbf{U}$ must be bounded, so that the trajectories $y(t, u^n)$ are uniformly bounded. If $y^* \in E^*$ we have

$$\langle y^*, y(t, u^n) \rangle = \langle y^*, S(t)\zeta \rangle + \int_0^t \langle B^* S(t - \tau)^* y^*, u^n(\tau) \rangle d\tau. \quad (14.5.25)$$

We extend the $u^n(\cdot)$ as in Theorem 14.5.1 and then select a subsequence of $\{u^n(\cdot)\}$ (denoted with the same name) such that $t_n \to \bar{t}$ and $u^n(\cdot) \to \bar{u}(\cdot) L^1(0, \bar{t}; X)$-weakly. It then follows from (14.5.25) that

$$\langle y^*, y(t, u^n) \rangle \to \langle y^*, y(t, \bar{u}) \rangle \quad (0 \le t \le \bar{t}, y^* \in E^*) \quad (14.5.26)$$

so that the passage to the lim sup in the cost functional is justified; it also follows from (14.5.26) and E^*-weak closedness of $M(t)$ that $y(t, \bar{u})$ will satisfy the state constraint. Similarly, to prove that $y(t, \bar{u})$ satisfies the target condition we need to show that

$$\langle y^*, y(t_n, u^n) \rangle \to \langle y^*, y(\bar{t}, u) \rangle \quad (y^* \in E^*).$$

This follows from (14.5.25) for $t = t_n$, written in the form

$$\langle y^*, y(t_n, u^n) \rangle = \langle y^*, S(t_n)\zeta \rangle + \int_0^T \chi_n(\tau) \langle B^* S(t_n - \tau)^* y^*, u^n(\tau) \rangle d\tau, \quad (14.5.27)$$

χ_n the characteristic function of $[0, t_n]$, and the fact that

$$\chi_n(\cdot) B^* S(t_n - \cdot)^* y^* \to \chi(\cdot) B^* S(\bar{t} - \cdot)^* y^*$$

in $L^1(0, \bar{t}; X)$, χ the characteristic function of $[0, \bar{t}]$.   ∎

The following example shows that Theorem 14.5.4 does not admit an extension to nonlinear systems.

Example 14.5.5. Consider the control system

$$y'(t) = Ay(t) + \phi(t^2 - \|y(t)\|^2)y + u(t), \quad y(0) = 0 \quad (14.5.28)$$

with the same space E, operator A and admissible control space $C_{\text{ad}}(0, \bar{t}; \mathbf{U})$ as in Example 14.5.2. The function $\phi(s)$ is infinitely differentiable, bounded, positive for $s \ne 0$ and $\phi(0) = 0$ (for instance, $\phi(s) = s^2/(1 + s^2)$), and y is an arbitrary nonzero element of E. Obviously, Hypothesis II is satisfied, and solutions exist globally (Corollary 5.5.5). We consider the optimal control problem in the fixed interval $0 \le t \le 2\pi$ with cost functional

$$y_0(2\pi, u) = \int_0^{2\pi} \{(S(-t)\eta, y(t, u))^2 + (S(t)\eta, y(t, u))^2 + (S(t)\eta, u(t))^2\}dt, \quad (14.5.29)$$

and no target condition or state constraints: $\eta \in E$ is again given by (14.5.16). The functional $y_0(2\pi, u)$ is weakly-weakly lower semicontinuous. That this is true for the first two terms is a consequence of the dominated convergence theorem. For the third term, we note that $\{u^n(\cdot)\}$ is a sequence in $L^\infty(0, \bar{t}; E)$ such that $u^n(\cdot) \to \bar{u}(\cdot)$ $L^1(0, \bar{t}; E)$-weakly in $L^\infty(0, \bar{t}; E)$, then $(S(\cdot)\eta, u^n(\cdot)) \to (S(\cdot)\eta, \bar{u}(\cdot))$ $L^1(0, \bar{t})$-weakly in $L^\infty(0, \bar{t})$. A fortiori, $(S(\cdot)\eta, u^n(\cdot)) \to (S(\cdot)\eta, \bar{u}(\cdot))$ $L^2(0, \bar{t})$-weakly in $L^2(0, \bar{t})$, hence

$$\int_0^{\bar{t}} (S(\tau)\eta, \bar{u}(\tau))^2 d\tau \le \limsup \int_0^{\bar{t}} (S(\tau)\eta, u^n(\tau))^2 d\tau.$$

The sequence (14.5.13) is a minimizing sequence also for this problem, that is,

$$y_0(2\pi, u^n) \to 0 \quad \text{as } n \to \infty. \tag{14.5.30}$$

To see this note that $(S(-t)\eta, y(t, u^n)) = (S(-t)\eta, tS(t)y_n) = t(S(-2t)\eta, y_n)$, $(S(t)\eta, y(t, u^n)) = (S(t)\eta, tS(t)y_n) = t(\eta, y_n)$, $(S(t)\eta, u^n(t)) = (S(t)\eta, S(t)y_n)$ $= (\eta, y_n)$, thus (14.5.30) follows from the fact that $y_n \to 0$ weakly and from the dominated convergence theorem.

Again, there is no optimal control. In fact, assume $\bar{u}(\cdot)$ is optimal. Then we must have

$$(S(-t)\eta, y(t, \bar{u})) = (S(t)\eta, y(t, \bar{u}))$$

$$= (S(t)\eta, \bar{u}(t)) = 0 \quad (0 \le t \le 2\pi), \tag{14.5.31}$$

thus

$$\begin{aligned}
0 &= (S(t)\eta, y(t, \bar{u})) \\
&= \left(S(t)\eta, \int_0^t S(t-\tau)\{\phi(\tau^2 - \|y(\tau, \bar{u})\|^2)y + \bar{u}(\tau)\}d\tau \right) \\
&= \int_0^t (S(t)\eta, S(t-\tau)y)\phi(\tau^2 - \|y(\tau, \bar{u})\|^2)d\tau \\
&\quad + \int_0^t (S(t)\eta, S(t-\tau)\bar{u}(\tau))d\tau \\
&= \int_0^t (S(\tau)\eta, y)\phi(\tau^2 - \|y(\tau, \bar{u})\|^2)d\tau + \int_0^t (S(\tau)z, \bar{u}(\tau))d\tau \\
&= \int_0^t (S(\tau)\eta, y)\phi(\tau^2 - \|y(\tau, \bar{u})\|^2)d\tau \quad (0 \le t \le 2\pi), \tag{14.5.32}
\end{aligned}$$

where the second integral drops out in view of the third equality (14.5.31). We deduce that

$$(S(t)\eta, y)\phi(t^2 - \|y(t, \bar{u})\|^2) = 0 \quad (0 \le t \le 2\pi). \tag{14.5.33}$$

Assume that

$$(S(t)\eta, y) = 0 \quad (0 \le t \le 2\pi). \tag{14.5.34}$$

Then the argument following (14.5.20) reveals that $y = 0$, a contradiction, and (14.5.34) has to be ruled out. Since $t \to (S(t)\eta, y)$ is analytic we must have $(S(\tau)\eta, y) \ne 0$ a.e., hence (14.5.33) and the fact that $\phi(s) > 0$ for $s \ne 0$ imply that $\|y(t, \bar{u})\| = t$ a.e. By continuity, $\|y(t, \bar{u})\| = t$ in $0 \le t \le 2\pi$. It follows that for the optimal control $\bar{u}(\cdot)$ the equation (14.5.28) reduces to the linear equation

$$y'(t) = Ay(t) + \bar{u}(t), \tag{14.5.35}$$

and the arguments after Lemma 14.5.3 show that (14.5.19) holds, that is, that $y(t, \bar{u}) = t S(t)z$ with $\|z\| = 1$. The first equality (14.5.31) yields $(S(-2t)\eta, z) = 0$ $(0 \le t \le 2\pi)$. We can then apply once again the argument after (14.5.20) to conclude that $z = 0$, again a contradiction.

Nonexistence of optimal controls in this example is caused by the fact that the weak limit of the sequence $\{y(\cdot, u_n)\}$ is not the solution $y(\cdot, \bar{u})$ of the equation (14.5.28), a phenomenon typical in nonlinear equations.

14.6. Existence Theorems for Optimal Control Problems, II

Lemma 14.6.1. *Let U be a separable complete metric space, E a separable reflexive Banach space and $\phi : [0, T] \times U \to E$, $\phi_0 : [0, T] \times U \to \mathbb{R}$ functions such that*

(a) $\phi(t, \cdot)$, $\phi_0(t, \cdot)$ are bounded and locally uniformly continuous in u for each t fixed,

(b) $t \to \phi_0(t, \cdot)$ and $t \to \langle y^*, \phi(t, \cdot) \rangle$ are strongly measurable $BC(U)$-valued functions for every $y^* \in E^*$,

(c) for every t, the set

$$\mathcal{Z}(t; \phi_0, \phi) = \{(y_0, \phi(t, u)) \in \mathbb{R} \times E; y_0 \ge \phi_0(t, u); u \in U\}$$
$$= \bigcup_{u \in U} ([\phi_0(t, u), \infty) \times \{\phi(t, u)\}) \tag{14.6.1}$$

is convex and closed.

Then, given $\mu(\cdot) \in R_{bc}(0, T; U)$ there exists a strongly measurable U-valued function $u(t)$ such that

$$\phi(t, u(t)) = \int_U \phi(t, u)\mu(t, du) \quad \text{a.e. in } 0 \le t \le T \tag{14.6.2}$$

$$\phi_0(t, u(t)) \le \int_U \phi_0(t, u)\mu(t, du) \quad \text{a.e. in } 0 \le t \le T. \tag{14.6.3}$$

Proof. Outfit the space $\mathcal{U} = \mathbb{R}_+ \times U$ with the product distance and consider the function $\psi : [0, T] \times \mathcal{U} \to \mathbb{R} \times E$ defined by

$$\psi(t, (u_0, u)) = (u_0 + \phi_0(t, u), \phi(t, u)).$$

This function is locally uniformly continuous in $\mathcal{U}$ for every t, and

$$\psi(t, \mathcal{U}) = \mathcal{Z}(t; \phi_0, \phi)$$

is convex in $\mathbb{R} \times E$. Given $\mu(\cdot) \in R_{bc}(0, T; U)$, define a function $g(t)$ with values in $\mathbb{R} \times E$ by

$$g(t) = (\Phi_0(t)\mu(t), \Phi(t)\mu(t)),$$

the second integral below defined as in (14.4.18):

$$\Phi_0(t)\mu(t) = \int_U \phi_0(t, u)\mu(t, du), \quad \Phi(t)\mu(t) = \int_U \phi(t, u)\mu(t, du).$$

By virtue of Lemma 14.4.4, $g(t)$ is strongly measurable and

$$g(t) \in \overline{\mathrm{conv}}\, \psi(t, \mathcal{U}) = \psi(t, \mathcal{U}) \quad \text{a.e.} \tag{14.6.4}$$

The proof ends by applying Filippov's Theorem 12.8.2 and obtaining a strongly measurable $(\mathbb{R}_+ \times U)$-valued function $(u_0(t), u(t))$ such that

$$g(t) = \psi(t, u_0(t), u(t)) \quad \text{a.e. in } 0 \le t \le T,$$

which implies (14.6.2) and (14.6.3). However, for application of Theorem 12.8.2 we must check that:

(*i*) $t \to \phi_0(t, u)$ is measurable for u fixed, and

(*ii*) $t \to \phi(t, u)$ is strongly measurable for u fixed.

For (*i*) we approximate the function $t \to \phi_0(t, \cdot) \in BC(U)$ uniformly with a sequence of countably valued $BC(U)$-valued functions $\phi_{0n}(t, \cdot)$ and note that, if $P_u : BC(U) \to U$ is the projector $P_u \phi = \phi(u)$ then $P_u \phi_{0n}(t)$ is countably valued and approximates uniformly $\phi(t, u) = P_u \phi(t, \cdot)$. For (*ii*) we use the same argument to show that $t \to \langle y^*, \phi(t, u) \rangle$ is measurable for all y^*; since E is separable $t \to \phi(t, u)$ is strongly measurable. ∎

The existence results below (Theorems 14.6.3 and 14.6.4) are for the system

$$y'(t) = Ay(t) + f(t, y(t), u(t)), \qquad y(0) = \zeta \tag{14.6.5}$$

and its relaxed companion

$$y'(t) = Ay(t) + F(t, y(t))\mu(t), \qquad y(0) = \zeta \tag{14.6.6}$$

under the assumptions in **14.3**. The space E is reflexive and separable and $f(t, y, u)$ satisfies Assumption II(BC) there. Until Theorem 14.6.4 we only need U to be a normal topological space. The cost functional is

$$y_0(t, u) = \int_0^t f_0(\tau, y(\tau, u), u(\tau))d\tau + g_0(t, y(t, u)) \qquad (14.6.7)$$

with $f_0 : [0, T] \times E \times U \to \mathbb{R}$, $g_0 : [0, T] \times E \to \mathbb{R}$, under

Hypothesis I^0(BC). $f_0(t, y, \cdot)$ is bounded and continuous in u for t, y fixed and continuous in y for t fixed, uniformly with respect to u.[1] The function $t \to f_0(t, y, \cdot)$ is a strongly measurable $BC(U)$-valued function and for every $c > 0$ there exists $K_0(\cdot, c) \in L^1(0, T)$ such that

$$|f_0(t, y, u)| \le K_0(t, c) \quad (\|y\| \le c, u \in U). \qquad (14.6.8)$$

The **relaxed cost functional** $y_0(t, \mu)$ is

$$y_0(t, \mu) = \int_0^t F_0(\tau, y(\tau, \mu))\mu(\tau)d\tau + g_0(t, y(t, \mu)), \qquad (14.6.9)$$

where

$$F_0(t, y, \mu) = \int_U f_0(t, y, u)\mu(du). \qquad (14.6.10)$$

Lemma 14.6.2. *Let $f_0(t, y, u)$ satisfy Hypothesis I^0(BC) and let $g_0(t, y)$ be continuous. Then the relaxed cost functional $y_0(t, \mu)$ is weakly lower semicontinuous and t-equicontinuous.*

Proof. We look first at the part of the functional depending on g_0. Weak lower semicontinuity is obvious, and equicontinuity follows from the estimate

$$|g_0(t_\kappa, y(t_\kappa, \mu_\kappa)) - g_0(\bar{t}, y(\bar{t}, \mu_\kappa))|$$
$$\le |g_0(t_\kappa, y(t_\kappa, \mu_\kappa)) - g_0(\bar{t}, y(t_\kappa, \mu_\kappa))| + |g_0(\bar{t}, y(t_\kappa, \mu_\kappa)) - g_0(\bar{t}, y(\bar{t}, \mu_\kappa))|$$

and the fact that the sequence $\{y(\cdot, \mu_\kappa)\}$, being convergent, is equicontinuous.

As for the other part, assume $g_0 = 0$. Let $\{\mu_\kappa(\cdot)\} \subseteq R_{bc}(0, T; U)$ be a generalized sequence such that $\mu_\kappa(\cdot) \to \bar{\mu}(\cdot)$ $L^1(0, \bar{t}; BC(U))$-weakly in $L_w^\infty(0, \bar{t}; \Sigma_{rba}(U))$ and $y(t, \mu_\kappa) \to \bar{y}(t)$ uniformly in $0 \le t \le \bar{t}$. The hypotheses imply that $f_0(t, y(t, \mu_\kappa), \cdot)$ converges to $f_0(t, y(t), \cdot)$ in $L^1(0, \bar{t}; BC(\cdot))$, so we can take limits using weak convergence of the μ_κ; the result is

$$y_0(\bar{t}, \bar{\mu}) = \lim y_0(\bar{t}, \mu_\kappa). \qquad (14.6.11)$$

---

[1] That is, $|f_0(t, y', u) - f_0(t, y, u)| \le \varepsilon$ for $\|y' - y\| \le \delta$ independently of $u \in U$.

Equicontinuity follows from the estimate

$$|y_0(t_\kappa, \mu_\kappa) - y_0(\bar{t}, \mu_\kappa)| \leq \left| \int_{\bar{t}}^{t_\kappa} K_0(\tau, c) d\tau \right|, \tag{14.6.12}$$

where c is a bound for $\|y(t, \mu_\kappa)\|$.   ∎

In the state constraints and target condition below, $M(t)$ and Y are closed:

$$y(t) \in M(t) \quad (0 \leq t \leq \bar{t}), \qquad y(\bar{t}) \in Y. \tag{14.6.13}$$

Existence Theorem 14.6.3. *Let $S(t)$ be compact for $t > 0$. Assume that there exists a minimizing sequence $\{\mu^n(\cdot)\}$, $\mu^n(\cdot) \in R_{bc}(0, t_n; U)$ of relaxed controls such that (a) $\{t_n\}$ is bounded (b) $\{y(t, \mu^n)\}$ is uniformly bounded. Then there exists a relaxed optimal control $\bar{\mu}(\cdot) \in R_{bc}(0, \bar{t}; U)$ which is the $L^1(0, \bar{t}; BC(U))$-weak limit of (a generalized subsequence) of $\{\mu^n(\cdot)\}$ in the space $L_w^\infty(0, T; BC(U)^*) = L_w^\infty(0, T; \Sigma_{rba}(U))$.*

Theorem 14.6.3 is a straight consequence of Theorem 14.5.1. That Assumption II(f, F) is satisfied for the relaxed system (14.6.6) was shown in Lemma 14.3.1, and that $C_{ad}(0, T; \mathbf{U}) = R_{bc}(0, T; U)$ is $L^1(0, T; BC(U))$-weakly compact in $L_w^\infty(0, T; \Sigma_{rba}(U))$ is proved in Theorem 12.5.9.

The other existence result is for the original system (14.6.5). We upgrade U to a separable complete metric space and take as the space $C_{ad}(0, T; U)$ of ordinary admissible controls all strongly measurable U-valued functions. All the other hypotheses above are in force.

Existence Theorem 14.6.4. *Let U be a separable complete metric space. Assume that $f(t, y, u)$ and $f_0(t, y, u)$ are locally uniformly continuous in u for t, y fixed and that the set*

$$\mathcal{Z}(t, y; f_0, f) = \{(y_0, f(t, y, u)) \in \mathbb{R} \times E; y_0 \geq f_0(t, y, u), u \in U\}$$

$$= \bigcup_{u \in U}([f_0(t; y, u), \infty) \times \{f(t, y, u)\}) \tag{14.6.14}$$

is convex and closed for every t, y. Finally, assume that there exists a minimizing sequence $\{\mu^n(\cdot)\}$, $\mu^n(\cdot) \in R_{bc}(0, t_n; U)$ of relaxed controls such that (a) $\{t_n\}$ is bounded and (b) $\{y(t, \mu_n)\}$ is uniformly bounded. Then there exists an ordinary optimal control $\bar{u}(\cdot) \in C_{ad}(0, T; U)$.

Proof. We use Theorem 14.6.3 and obtain an optimal relaxed control $\bar{\mu}(\cdot)$. Then we apply Lemma 14.6.1 and obtain a control $\bar{u}(\cdot) \in C_{ad}(0, T; U)$ (that is, a strongly measurable U-valued function) such that

$$f(t, y(t, \bar{\mu}), \bar{u}(t)) = \int_U f(t, y(t, \bar{\mu}), u)\bar{\mu}(t, du)$$

$$= F(t, y(t, \bar{\mu}))\bar{\mu}(t) \qquad (14.6.15)$$

$$f_0(t, y(t, \bar{\mu}), \bar{u}(t)) \le \int_U f_0(t, y(t, \bar{\mu}), u)\bar{\mu}(t, du)$$

$$= F_0(t, y(t, \bar{\mu}))\bar{\mu}(t). \qquad (14.6.16)$$

That the functions $\phi(t, u) = f(t, y(t), u), \phi_0(t, u) = f_0(t, y(t), u)$ satisfy the assumptions in Lemma 14.6.1 is a consequence of Hypotheses $I(BC)$ and $I^0(BC)$. Equality (14.6.15) shows that $y(t, \bar{\mu}) = y(t, \bar{u})$, while inequality (14.6.16) (after integration in $0 \le t \le \bar{t}$) expresses that $y_0(t, \bar{u}) \le y_0(t, \bar{\mu})$ for all t. This ends the proof. ∎

The case of U arbitrary and control space $R_b(0, T; U)$ is disposed of with the customary change of $BC(U)$ by $B(U)$. When U is a compact metric space, $BC(U)$ is replaced by $C(U)$ and the space need not be reflexive. Moreover, the sets $\mathcal{Z}(t; \phi_0, \phi)$ and $\mathcal{Z}(t, y; f_0, f)$ are automatically closed as in the finite dimensional case.

14.7. Abstract Parabolic Equations, I

This and the next section are on extensions of the results in the chapter to the setup in **7.7**; the semigroup $S(t)$ is analytic, the space E is $\odot$-reflexive and $E, E^\odot$ are separable. We also assume that $\|S(t)\| \le Ce^{-ct}$ $(t \ge 0)$ for $c > 0$, so that fractional powers $(-A)^\alpha$ and the spaces $E_\alpha, (E^*)_\alpha$ can be defined (see **7.4**). The equation is the companion of (14.2.1),

$$y'(t) = Ay(t) + f(t, y(t)) + F(t, y(t))u(t), \qquad (14.7.1)$$

and the admissible control space is defined exactly as in **14.2**. The functions f, F act as follows:

$$f : [0, T] \times E_\alpha \to ((E^\odot)^*)_{-\rho}, \quad F : [0, T] \times E_\alpha \to (X^*, ((E^\odot)^*)_{-\rho}),$$

where $\alpha, \rho \ge 0, \alpha + \rho < 1$. We say that f, F satisfy Hypothesis $I_{\alpha,\rho}(f, F)$ (resp. $II_{\alpha,\rho}(f, F)$) if the functions

$$g(t, \eta) = ((-A^\odot)^{-\rho})^* f(t, (-A)^{-\alpha}\eta) \qquad (14.7.2)$$

$$G(t, \eta) = ((-A^\odot)^{-\rho})^* F(t, (-A)^{-\alpha}\eta) \qquad (14.7.3)$$

satisfy Hypothesis $I(g, G)$ (resp. $II(g, G)$) with $K_j(t, c) = K_j(c), L_j(t, c) = L_j(c)$ independent of t.

Lemma 14.7.1. *Assume* f, F *satisfy Hypothesis* $I_{\alpha,\rho}(f, F)$ *(resp.* $II_{\alpha,\rho}(f, F)$*). Then the function*

$$f(t, y) + F(t, y)u(t) \tag{14.7.4}$$

satisfies Hypothesis $I_{\alpha,\rho}$ *(resp.* $II_{\alpha,\rho}$*) in* **7.7** *for every* $u(\cdot) \in L_w^\infty(0, T; X^*)$.

Lemma 14.7.1 is a counterpart of Lemma 14.2.1 and is proved in the same way. The only thing we need to check is strong measurability of $t \to G(t, \eta)u(t)$ for $\eta \in E$ and $u(\cdot) \in L_w^\infty(0, T; X^*)$. To see this note that if $\eta^* \in E^*$ then the function

$$t \to \langle y^*, G(t, \eta)u(t) \rangle = \langle G(t, \eta)^*\eta^*, u(t) \rangle \tag{14.7.5}$$

is measurable, hence $t \to G(t, \eta)u(t)$ is E^*-weakly measurable, thus measurable by separability of E.

For applications, it is convenient to weaken Hypotheses $I_{\alpha,\rho}(f, F)$ and $II_{\alpha,\rho}(f, F)$ in the style of **7.7**. Hypothesis $I_{\alpha,\rho}(f, F)(wm)$ is Hypothesis $I_{\alpha,\rho}(f, F)$ with the difference that the functions map as follows:

$$g : [0, T] \times E \to (E^\odot)^*, G : [0, T] \times E \to (X^*, (E^\odot)^*), G(t, \eta)^* E^\odot \subseteq X,$$

and that $t \to g(t, y)$ is merely $E^\odot$-weakly measurable. The function $t \to G(t, \eta)^*z$ is strongly measurable for $y \in E$ and $z \in E^\odot$, and $G(t, \eta)$ is continuous in η in the norm of $(X^*, (E^\odot)^*)$. Inequalities (14.2.4), (14.2.5), (14.2.6) and (14.2.7) are the same, but $K_j(t, c) = K_j(c)$, $L_j(t, c) = L_j(c)$ are independent of t and in (12.2.4), (12.2.6) (resp. (12.2.5), (12.2.7)) the norm on the left is the $(E^\odot)^*$-norm (resp. the $(X^*, (E^\odot)^*)$-norm).[2]

Lemma 14.7.2. *(a) Assume* f, F *satisfy Hypothesis* $I_{\alpha,\rho}(f, F)(wm)$ *(resp.* $II_{\alpha,\rho}(f, F)(wm)$*). Then the function* (14.7.4) *satisfies Hypothesis* $I_{\alpha,\rho}(wm)$ *(resp.* $II_{\alpha,\rho}(wm)$*) in* **7.7** *for every* $u(\cdot) \in L_w^\infty(0, T; X^*)$. *(b) If* $\rho' > \rho$ *then* (14.7.2) *satisfies Hypothesis* $I_{\alpha,\rho'}$ *(resp.* $II_{\alpha,\rho'}$*).*

Proof. For (a) it is only necessary to check that the $(E^\odot)^*$-valued function $t \to G(t, \eta)u(t)$ is $E^\odot$-weakly measurable. This follows from (14.7.5) and the assumptions. To show (b) we use Corollary 7.7.12. ∎

The integral equation for $\eta(t) = (-A)^\alpha y(t)$, $y(\cdot)$ the solution of (14.7.1) is

$$\eta(t) = (-A)^\alpha S(t)\zeta$$
$$+ \int_0^t (-A)^{\alpha+\rho} S^\odot(t - \tau)^*((-A^\odot)^{-\rho})^* f(\tau, (-A)^{-\alpha}\eta(\tau))d\tau$$
$$+ \int_0^t (-A)^{\alpha+\rho} S^\odot(t - \tau)^*((-A^\odot)^{-\rho})^* F(\tau, (-A)^{-\alpha}\eta(\tau))u(\tau)d\tau.$$

$$\tag{14.7.6}$$

---

[2] The difference between Hypotheses $I_{\alpha,\rho}(f, F)$, $II_{\alpha,\rho}(f, F)$ and their (wm) counterparts appears only when $\rho = 0$. See Lemma 7.7.10, Lemma 7.7.11 and Corollary 7.7.12.

Versions of the results in the previous sections are given below with sketches of proofs. The first two are the analogues of Lemma 14.2.2 and Theorem 14.2.3.

Lemma 14.7.3. *Assume that $S(t)$ is compact for $t > 0$ and that Hypothesis $II_{\alpha,\rho}(f, F)(wm)$ holds with $\alpha + \rho < 1$. Let $u(\cdot) \in L_w^\infty(0, T; X^*)$ be such that the solution $y(\cdot, u)$ of (14.7.1) exists in $0 \le t \le \bar{t}$ and let $\{u_\kappa(\cdot); \kappa \in K\}$ be a bounded generalized sequence in $L_w^\infty(0, \bar{t}; X^*)$ such that $u_\kappa(\cdot) \to u(\cdot)$ $L^1(0, \bar{t}; X)$-weakly. Then there exists $\kappa_0 \in K$ such that $y(\cdot, u_\kappa)$ exists in $0 \le t \le \bar{t}$ if $\kappa \ge \kappa_0$ ("$\ge$" the ordering in K) and*

$$y(t, u_\kappa) \to y(t, u) \quad \text{in } E_\alpha$$

uniformly in $0 \le t \le \bar{t}$.

The proof is essentially that of Lemma 14.2.2. Equality (14.2.9) becomes

$$
\begin{aligned}
&\eta(t, u_\kappa) - \eta(t, u) \\
&= \int_0^t (-A)^{\alpha+\rho} S^\circ(t - \tau)^* \\
&\quad \times ((-A^\circ)^{-\rho})^* \{(f(\tau, (-A)^{-\alpha}\eta(\tau, u_\kappa)) - f(\tau, (-A)^{-\alpha}\eta(\tau, u))\}d\tau \\
&\quad + \int_0^t (-A)^{\alpha+\rho} S^\circ(t - \tau)^* \\
&\quad \times ((-A^\circ)^{-\rho})^* \{(F(\tau, (-A)^{-\alpha}\eta(\tau, u_\kappa)) - F(\tau, (-A)^{-\alpha}\eta(\tau, u))\}u_\kappa(\tau)d\tau \\
&\quad + \int_0^t (-A)^{\alpha+\rho} S^\circ(t - \tau)^* \\
&\quad \times ((-A^\circ)^{-\rho})^* F(\tau, (-A)^{-\alpha}\eta(\tau, u))(u_\kappa(\tau) - u(\tau))d\tau
\end{aligned}
\tag{14.7.7}
$$

in $0 \le t \le \bar{t}_\kappa$, $[0, \bar{t}_\kappa]$ the maximal interval where $y(\cdot, u_\kappa)$ exists and satisfies $\|y(t, u_\kappa) - y(t, \bar{u})\| \le 1$. The subsequent estimation uses the generalized Gronwall inequality (Lemma 7.7.6), and produces

$$\|\eta(t, u_\kappa) - \eta(t, u)\| \le C\|\rho_\kappa(t)\| \quad (0 \le t \le t_\kappa) \tag{14.7.8}$$

where

$$\rho_\kappa(t) = \Lambda_{\alpha+\rho}\big(((-A^\circ)^{-\rho})^* F(\cdot, y(\cdot, u))(u_\kappa(\cdot) - u(\cdot))\big), \tag{14.7.9}$$

$\Lambda_{\alpha+\rho}$ the operator in (7.7.28); this operator is compact from $L_w^\infty(0, \bar{t}; X^*)$ into $C(0, \bar{t}; E)$ (Theorem 7.7.16). The proof ends as that of Lemma 14.2.2.

Theorem 14.7.4 (Relaxation Theorem). *Assume that the set $\mathbf{U} \subseteq X^*$ is convex, bounded and X-weakly compact, that D is total in $\mathbf{U}$, that $S(t)$ is compact and that Hypothesis $II_{\alpha,\rho}(f, F)(wm)$ holds, $\alpha + \rho < 1$. Let $u(\cdot) \in C_{ad}(0, \bar{t}; U)$ be* such that the solution $y(t, u)$ of (14.1.1) exists in $0 \leq t \leq \bar{t}$. Then there exists $v(\cdot) \in PC(0, T; D)$ such that $y(t, v)$ exists in $0 \leq t \leq \bar{t}$ and

$$\|y(t, u) - y(t, v)\|_\alpha \leq \varepsilon \quad (0 \leq t \leq \bar{t}).$$

The proof is the same as that of Theorem 14.2.3.

Example 14.7.5. We consider a particular case of the system in **8.4**,

$$y_t(t, x) = A_c(\beta)y(t, x) - \phi(t, x, y(t, x), \nabla y(t, x))$$
$$+ \psi(t, x, y(t, x), \nabla y(t, x))u(t, x), \quad (14.7.10)$$

assuming both functions $\phi(t, x, y, \mathbf{y})$, $\psi(t, x, y, \mathbf{y})$ satisfy the hypotheses in Theorem 8.4.8. The control set is the unit ball of $L^\infty(\Omega)$, so that $X = L^1(\Omega)$, $X^* = L^\infty(\Omega)$, and $C_{ad}(0, T; U)$ is the unit ball of $L^\infty_w(0, \bar{t}; L^\infty(\Omega)) = L^\infty((0, \bar{t}) \times \Omega)$. The space is $E = C(\overline{\Omega})$ or $E = C_0(\overline{\Omega})$ according to the boundary condition. We have $E^\odot = L^1(\Omega)$, $(E^\odot)^* = L^1(\Omega)^* = L^\infty(\Omega)$. The function $f(t, y)$ and operator function $F(t, y)$ involved in the application of Theorem 14.7.4 are

$$f(t, y)(x) = -\phi(t, x, y(x), \nabla y(x)) \quad (14.7.11)$$

$$(F(t, y)u)(x) = \psi(t, x, y(x), \nabla y(x))u(x) \quad (14.7.12)$$

It follows from Theorem 8.4.8 that if $\alpha > 1/2$ the function $f(t, y)$ satisfies its share of Hypothesis $II_{\alpha,0}(f, F)(wm)$. It also follows that $y(\cdot) \to \psi(t, \cdot, y(\cdot), \nabla y(\cdot))$ is a continuous map from E_α into $L^\infty(\Omega)$, so that $F(t, y) \in (L^\infty(\Omega), L^\infty(\Omega))$ for $y \in E_\alpha$. The necessary inequalities for the function $g(t, \eta)$ and the operator function $G(t, \eta)$ are a consequence of (8.4.24) and (8.4.25). Finally, $F(t, y)^*$ restricted to $E^\odot = L^1(\Omega)$ is given by the same multiplicative formula (14.7.12). As proved in Theorem 8.4.8, if $y(\cdot) \in E_\alpha$ then the function $t \to \psi(t, \cdot, y(\cdot), \nabla y(\cdot)) \in L^\infty(\Omega)$ is $L^1(\Omega)$-weakly measurable; this is easily seen to imply that if $z(\cdot) \in L^1(\Omega)$ then the function

$$t \to \psi(t, \cdot, y(\cdot), \nabla y(\cdot))z(\cdot)$$

is L^∞-weakly measurable in $L^1(\Omega)$, hence strongly measurable by separability. All assumptions verified, we recall that the extremals of the unit ball of $L^\infty(\Omega)$ are characterized by $|v(x)| = 1$ a.e. and apply Theorem 14.7.4. We deduce that, if $y(t, \cdot, u)$ is a solution of (14.7.10) corresponding to a control satisfying $|u(t, x)| \leq 1$ and existing for $0 \leq t \leq \bar{t}$ there exists a control $v(t, x)$ whose solution $y(t, \cdot, v)$ exists in the same interval and satisfies

$$|v(t, x)| = 1 \quad \text{a.e. in } (0, \bar{t}) \times \Omega, \quad (14.7.13)$$

$$\|y(t, \cdot, u) - y(t, \cdot, v)\|_\alpha \leq \varepsilon \quad (0 \leq t \leq \bar{t}), \quad (14.7.14)$$

Since $\alpha > 1/2$, $E_\alpha \hookrightarrow C^{(1)}(\overline{\Omega})$, hence (14.7.14) produces approximation in the norm of $C^{(1)}(\overline{\Omega})$. Similar results are obtained with other control spaces, for instance $X = \mathbb{R}$, $X^* = \mathbb{R}$, $U = [-1, 1]$; here $u(\cdot)$ is a measurable scalar function in $0 \leq t \leq \bar{t}$ with $|u(t)| \leq 1$ a.e. and $|v(x)| = 1$ a.e.

The next result is companion of Theorem 14.5.1. Consider the optimal control problem for (14.7.1) under Assumption $I_{\alpha,\rho}(f, F)(wm)$, including state constraints and a target condition

$$y(t) \in M(t) \subseteq E_\alpha \quad (0 \leq t \leq \bar{t}), \quad y(\bar{t}) \in Y, \tag{14.7.15}$$

in a variable time interval $0 \leq t \leq \bar{t}$ and a weakly lower semicontinuous cost functional $y_0(t, u)$ equicontinuous with respect to t. The definitions are the same as those in **14.5**, only that in (14.5.4) convergence is understood in $C(0, \bar{t}; E_\alpha)$.

Theorem 14.7.6. *Let $S(t)$ be compact and let Assumption $II_{\alpha,\rho}(f, F)$ be satisfied. Assume the cost functional $y_0(t, u)$ is weakly lower semicontinuous and equicontinuous with respect to t, that $M(t)$ and Y are closed and that $C_{ad}(0, \bar{t}; U)$ is $L^1(0, T; X)$-weakly compact in $L_w^\infty(0, T; X^*)$. Then, if there exists a minimizing sequence $\{u^n(\cdot)\}$, $u^n(\cdot) \in C_{ad}(0, t_n; U)$ with $\{t_n\}$ bounded and $\{y(t, u^n)\}$ uniformly bounded in E_α there exists an optimal control $\bar{u}(\cdot)$, which is the $L^1(0, \bar{t}; X)$-weak limit of a generalized subsequence of $\{u^n(\cdot)\}$.*

The proof is essentially the same as that of Theorem 14.5.1. We assume that $t_n \to \bar{t}$ and write the integral equation (14.7.7) for the trajectories $y(t, u^n)$; then, using Theorem 7.7.16 and passing if necessary to a generalized subsequence, we may assume that $y(\cdot, u^n)$ is convergent in $C(0, \bar{t}; E_\alpha)$. Verification that the limit $\bar{y}(\cdot)$ satisfies $\bar{y}(t) = y(t, \bar{u})$ is effected applying a linear functional $y^* \in E^\odot$:

$$\langle y^*, \eta(t, u^n) \rangle = \langle y^*, (-A)^\alpha S(t)\zeta \rangle$$

$$+ \int_0^t \langle (-A^\odot)^{\alpha+\rho} S^\odot(t - \tau)y^*, ((-A^\odot)^{-\rho})^* f(\tau, (-A)^{-\alpha}\eta(\tau, u^n)) \rangle d\tau$$

$$+ \int_0^t \langle (-A^\odot)^{\alpha+\rho} S^\odot(t - \tau)y^*, ((-A^\odot)^{-\rho})^* F(\tau, (-A)^{-\alpha}\eta(\tau, u^n))u^n(\tau) \rangle d\tau \tag{14.7.16}$$

and taking limits.

The relaxation and existence results above can be used for systems equipped with relaxed controls. We use the three relaxed control spaces $R(0, T; U) = R_c(0, T; U)$, $R_{bc}(0, T; U)$, $R_b(0, T; U)$ in **14.4**, and, as usual, we do in some detail the case $R_{bc}(0, T; U)$, U a normal topological space. The control system is

$$y'(t) = Ay(t) + f(t, y(t), u(t)), \quad y(0) = \zeta. \tag{14.7.17}$$

We assume that

$$f : [0, T] \times E_\alpha \to ((E^\odot)^*)_{-\rho}.$$

and the two basic hypotheses involve the function

$$g(t, \eta, u) = ((-A^{\odot})^{-\rho})^* f(t, (-A)^{-\alpha} \eta, u). \qquad (14.7.18)$$

Hypothesis I$_{\alpha,\rho}$(BC). $g(t, \eta, \cdot) \in B(U; (E^{\odot})^*)$ for t, y fixed; $g(t, \eta, u)$ is continuous in y for t fixed uniformly with respect to u. For every $\eta^* \in E^{\odot}$ the function $t \to \langle \eta^*, \phi(t, \eta, \cdot) \rangle$ is a strongly measurable $BC(U)$-valued function. For every $c > 0$ there exists $K(c)$ such that

$$\|g(t, \eta, u)\|_{(E^{\odot})^*} \leq K(c) \quad (0 \leq t \leq T, \|\eta\| \leq c, u \in U). \qquad (14.7.19)$$

Hypothesis II$_{\alpha,\rho}$(BC). $g(t, y, \cdot) \in B(U; (E^{\odot})^*)$ for t, y fixed and, for every $\eta^* \in E^{\odot}$ the function $t \to \langle y^*, g(t, y, \cdot) \rangle$ is a strongly measurable $BC(U)$-valued function. For every $c > 0$ there exist $K(c)$, $L(c)$ such that (14.7.19) holds and

$$\|g(t, \eta', u) - g(t, \eta, u)\|_{(E^{\odot})^*} \leq L(c)\|\eta' - \eta\|$$
$$(0 \leq t \leq T, \|\eta\|, \|\eta'\| \leq c, u \in U). \qquad (14.7.20)$$

The construction of the relaxed system

$$y'(t) = Ay(t) + F(t, y(t))\mu(t), \qquad y(0) = \zeta \qquad (14.7.21)$$

is similar to that in **14.3**. However, instead of defining $F(t, y)$ as in (14.3.3) we define $G(t, \eta) = \text{``}((-A)^{\odot})^{-\rho})^* F(t, (-A)^{-\alpha}\eta)\text{''}$ directly. Given $\mu \in \Sigma_{rba}(U)$ and $\eta \in E$, $G(t, \eta)\mu$ is the unique element of $(E^{\odot})^* \supseteq E$ satisfying

$$\langle y^*, G(t, \eta)\mu \rangle = \int_U \langle y^*, g(t, \eta, u) \rangle \mu(du) \qquad (14.7.22)$$

for $y^* \in E^{\odot}$. It follows directly from the definition that $G(t, \eta) \in (\Sigma_{rba}(U), (E^{\odot})^*)$. The adjoint $G(t, \eta)^*$ restricted to $E^{\odot}$ is

$$G(t, \eta)y^* = \langle y^*, g(t, \eta, \cdot) \rangle \in BC(U). \qquad (14.7.23)$$

Although $F : [0, T] \times E_\alpha \to (\Sigma_{rba}(U), ((E^{\odot})^*)_{-\rho})$ will not be used directly, we may define

$$F(t, y) = ((-A^{\odot})^{\rho})^* G(t, (-A)^{\alpha} y), \qquad (14.7.24)$$

with $((-A^{\odot})^{\rho})^*$ extended as in **7.4** (or we may look at the right side of (14.7.24) merely as a notation). We have

Lemma 14.7.7. *Assume the function $f(t, y, u)$ satisfies Hypothesis I$_{\alpha,\rho}$(BC) (resp. Hypothesis II$_{\alpha,\rho}$(BC)). Then the operator function $F(t, y)$ satisfies Hypothesis I$_{\alpha,\rho}$(0, F)(wm) (resp. II$_{\alpha,\rho}$(0, F)(wm)) with $X = BC(U)$.*

The proof only uses $G(t, \eta) = ((-A^{\odot})^{-\rho})^* F(t, (-A)^{-\alpha}\eta)$. We have already seen that $G(t, \eta) \in (\Sigma_{rba}(U), (E^{\odot})^*)$. Continuity in η follows from the assumptions, as do the estimates corresponding to both assumptions. The operator $G(t, \eta)^*$ is given by

$$G(t, \eta)^* y^* = \langle y^*, f(t, (-A)^{-\alpha}\eta, \cdot) \rangle \quad (y^* \in E^{\odot})$$

and thus performs to specifications.   ∎

As an application of Theorem 14.7.4, we obtain the following relaxation result for (14.7.21), corresponding to Theorem 14.3.4.

Theorem 14.7.8. *Let $S(t)$ be compact and let $\mu(\cdot) \in R_{bc}(0, T; U)$ be such that $y(t, \mu)$ exists in $0 \leq t \leq \bar{t}$ and $\varepsilon > 0$. Then there exists $v(\cdot) \in PC(0, T; U)$ such that $y(t, v)$ exists in $0 \leq t \leq \bar{t}$ and*

$$\|y(t, \mu) - y(t, v)\|_{\alpha} \leq \varepsilon \quad (0 \leq t \leq \bar{t}).$$

Remark 14.7.9. If $\rho > 0$ we have $((-A^{\odot})^{-\rho})^*((E^{\odot})^*)_{-\rho} \subseteq E$, (Lemma 7.7.10), thus the norm on the left sides of (14.7.19) and (14.7.20) is that of E. It does not seem to follow from (14.7.22) that $G(t, \eta)\Sigma_{rba}(U) \subseteq E$. However, this can be insured raising ρ in Hypotheses I(BC) or II(BC). In fact, subindexing G to indicate which ρ is used in the definition, we have $G_{\rho'}(t, \eta) = ((-A^{\odot})^{-(\rho'-\rho)})^* G_{\rho}(t, \eta)$.

14.8. Abstract Parabolic Equations, II

Let $t \to \mathcal{F}(t, y) \subseteq ((E^{\odot})^*)_{-\rho}$ be a set valued function defined in $[0, T] \times E_{\alpha}$. A E_{α}-valued function $y(\cdot)$ is a **solution** of the differential inclusion

$$y'(t) \in Ay(t) + \mathcal{F}(t, y(t)), \qquad y(0) = \zeta \qquad (14.8.1)$$

in $0 \leq t \leq T$ if there exists a function $g(\cdot) \in L_w^{\infty}(0, T; (E^{\odot})^*)$ such that[1]

$$g(t) \in ((-A^{\odot})^{-\rho})^* \mathcal{F}(t, y(t)) \quad \text{a.e. in } 0 \leq t \leq T, \qquad (14.8.2)$$

and $\eta(t) = (-A)^{\alpha} y(t)$ satisfies

$$\eta(t) = (-A)^{\alpha} S(t)\zeta + \int_0^t (-A)^{\alpha+\rho} S^{\odot}(t - \tau)^* g(\tau)d\tau \quad (0 \leq t \leq T). \quad (14.8.3)$$

We call the pair $(y(\cdot), g(\cdot))$ a **trajectory** of (14.8.1).

---

[1] $L_w^p(0, T; (E^{\odot})^*)$ for p large enough is sufficient. Same in other places.

The first result is a companion of Theorem 14.4.1 on the equivalence of the original system

$$y'(t) = Ay(t) + f(t, y(t), u(t)), \qquad y(0) = \zeta \qquad (14.8.4)$$

and the differential inclusion

$$y'(t) \in Ay(t) + f(t, y(t), U), \qquad y(0) = \zeta, \qquad (14.8.5)$$

with U a complete separable metric space. The system (14.8.4) is equipped with the admissible control space $C_{ad}(0, T; U)$ of all strongly measurable U-valued functions and we only assume that for each $y(\cdot) \in C(0, T; E_\alpha)$ and every $u(\cdot) \in C_{ad}(0, T; U)$ the function $t \to ((-A^\odot)^{-\rho})^* f(t, y(t), u(t))$ belongs to $L_w^\infty(0, T; (E^\odot)^*)$ so that solutions can be defined by the customary integral equation.

Equivalence Theorem 14.8.1. *Assume that the function $((-A^\odot)^{-\rho})^* f(t, y, u)$ is locally uniformly continuous in u for each $t, y \in E_\alpha$ fixed. Then $y(\cdot)$ satisfies (14.8.4) with $u(\cdot) \in C_{ad}(0, T; U)$ if and only if it satisfies (14.8.5).*

The "only if" part is obvious. Conversely, if $(y(\cdot), g(\cdot))$ is a trajectory of (14.8.5) we have $g(t) \in ((-A^\odot)^{-\rho})^* f(t, y(t), U)$ a.e. so that, if $\varepsilon > 0$,

$$((-A^\odot)^{-\varepsilon})^* g(t) \in ((-A^\odot)^{-\rho-\varepsilon})^* f(t, y(t), U).$$

By virtue of Lemma 7.7.11 the function on the left is a strongly measurable E-valued function; the same Lemma shows that $t \to ((-A^\odot)^{-\rho-\varepsilon})^* f(t, y(t), u)$ is a strongly measurable E-valued function for u fixed. By Filippov's Theorem 12.8.2 there exists a strongly measurable U-valued function $u(\cdot)$ (that is, an element of $C_{ad}(0, T; U)$) such that

$$((-A^\odot)^{-\varepsilon})^* g(t) = ((-A^\odot)^{-\rho-\varepsilon})^* f(t, y(t), u(t)),$$

thus, since $((-A^\odot)^{-\varepsilon})^*$ is one-to-one,

$$g(t) = ((-A^\odot)^{-\rho})^* f(t, y(t), u(t)) \quad \text{a.e.,}$$

so that $y(\cdot)$ satisfies (14.8.4).

We extend to the present setup the other results in **14.4** and **14.5** under the assumption below on the function

$$g(t, \eta, u) = ((-A^\odot)^{-\rho})^* f(t, (-A)^{-\alpha}\eta, u). \qquad (14.8.6)$$

Hypothesis $\emptyset_{\alpha,\rho}$(BC). For every $\eta(\cdot) \in C(0, T; E)$ and for every $y^* \in E^\odot$, the function $t \to \langle y^*, g(t, \eta(t), \cdot) \rangle$ is a strongly measurable $BC(U)$-valued function. There exists $K(c)$ such that

$$\|g(t, \eta, u)\| \le K(c) \quad (0 \le t \le T, \|\eta\| \le c, u \in U). \qquad (14.8.7)$$

This assumption is enough to ensure that the operator functions $G(t, \eta)$ and $F(t, y)$ in (14.7.22) and (14.7.24) are well defined. Moreover, it implies that for every $y(\cdot) \in C(0, T; E_\alpha)$ the function $t \to G(t, (-A)^\alpha y(t))\mu(t) = ((-A^\odot)^{-\rho})^* F(t, y(t))\mu(t)$ belongs to $L_w^\infty(0, T; (E^\odot)^*)$, so that solutions of the relaxed system

$$y'(t) = Ay(t) + F(t, y(t))\mu(t), \qquad y(0) = \zeta \qquad (14.8.8)$$

can be defined by the associated integral equation.

Equivalence Theorem 14.8.2. *Under Hypothesis $\emptyset_{\alpha,\rho}(BC)$ the relaxed system* (14.8.8) *is equivalent to the differential inclusion*

$$y'(t) \in Ay(t) + \overline{\text{conv}} \, f(t, y(t), U), \qquad y(0) = \zeta. \qquad (14.8.9)$$

In this inclusion, $\overline{\text{conv}}$ is used in a purely symbolic way; an E_α-valued function $y(\cdot)$ is a solution of (14.8.9) if there exists $g(\cdot) \in L_w^\infty(0, T; (E^\odot)^*)$ such that (14.8.3) holds and

$$g(t) \in \overline{\text{conv}} \, ((-A^\odot)^{-\rho})^* f(t, y(t), U) \quad \text{a.e. in } 0 \le t \le T. \qquad (14.8.10)$$

The proof of Theorem 14.8.2 requires the two results below, counterparts of Lemma 14.4.3 and 14.4.4. In them, $\overline{\text{conv}}$ means closed convex hull in the $E^\odot$-weak topology of $(E^\odot)^*$.

Lemma 14.8.3. *Let $\phi(\cdot) \in B(U, (E^\odot)^*)$ be a function such that $\langle y^*, \phi(\cdot)\rangle \in BC(U)$ for all $y^* \in E^\odot$. Define $\Phi : \Sigma_{rba}(U) \to (E^\odot)^*$ by*

$$\langle y^*, \Phi\mu \rangle = \int_U \langle y^*, \phi(u)\rangle \mu(du) \qquad (14.8.11)$$

for $y^* \in E^\odot$, so that $\Phi\mu \in (E^\odot)^*$. Then, for any $\mu \in \Pi_{rba}(U)$,

$$\Phi\mu \in \overline{\text{conv}} \, \phi(U). \qquad (14.8.12)$$

Proof. The assumptions imply that $\Phi\mu \in (E^\odot)^*$ is well defined by (14.8.11) and

$$\|\Phi\mu\|_{(E^\odot)^*} \le \|\phi\|_{B(U;(E^\odot)^*)} \|\mu\|_{\Sigma_{rba}(U)}. \qquad (14.8.13)$$

If we give to $(E^\odot)^*$ the $E^\odot$-weak topology, then $(E^\odot)^*$ becomes a linear topological space whose dual coincides with $E^\odot$ (Corollary 12.0.10). Accordingly, the separation theorem (Corollary 12.0.9) implies that the set $\overline{\text{conv}} \, \phi(U)$ is an intersection of halfspaces $\{y; \langle y^*, y\rangle \le a\}$ over a family $\{y^*\}$ of elements of $E^\odot$ and a family $\{a\}$ of real numbers. The passage to the limit is the same as in the proof of Lemma 14.4.3. ∎

Lemma 14.8.4. *Let* $\phi : [0, T] \times U \to (E^\odot)^*$ *be a function such that* $\phi(t, \cdot) \in$ $B(U; (E^\odot)^*)$ *and such that for every* $y^* \in E^\odot$, $\langle y^*, \phi(t, \cdot) \rangle \in BC(U)$ *in* $0 \le t \le T$ *and* $t \to \langle y^*, \phi(t, \cdot) \rangle$ *is a strongly measurable* $BC(U)$*-valued function. Let* $\mu(\cdot) \in R_{bc}(0, T; U)$. *Define* $g(t) = \Phi(t)\mu(t)$, *where*

$$\langle y^*, \Phi(t)\mu(t) \rangle = \int_U \langle y^*, \phi(t, u) \rangle \mu(t, du) \tag{14.8.14}$$

for $y^* \in E^\odot$. *Then* $g(\cdot)$ *is a* $E^\odot$*-weakly measurable* $(E^\odot)^*$*-valued function and*

$$g(t) \in \overline{\text{conv}}\, \phi(t, U) \quad \textit{a.e. in } 0 \le t \le T. \tag{14.8.15}$$

Conversely, let $g(\cdot)$ *be a* $E^\odot$*-weakly measurable* $(E^\odot)^*$*-valued function such that* (14.8.15) *holds. Then there exists* $\mu(\cdot) \in R_{bc}(0, T; U)$ *such that*

$$\Phi(t)\mu(t) = g(t) \quad \textit{a.e. in } 0 \le t \le T. \tag{14.8.16}$$

This result is proved much as Lemma 14.4.4, but there are some technical complications. Inequality (14.4.21) extends verbatim,

$$\|\Phi(t)\mu(t)\|_{(E^\odot)^*} \le \|\phi(t)\|_{B(U;(E^\odot)^*)} \|\mu\|_{L^\infty_w(0,T;\Sigma_{rba}(U))}, \tag{14.8.17}$$

and we show as in **14.4** that $t \to \|\phi(t)\|_{(E^\odot)^*}$ is measurable, this time using a countable dense set $\{z^*_\kappa\}$ in the unit ball of $E^\odot$. Accordingly, we may limit ourselves to the case where $\|\phi(t)\|_{(E^\odot)^*}$ is bounded (see the proof of Lemma 14.4.4).

If $g(t) = \Phi(t)\mu(t)$ then it follows directly from (14.8.14) that $g(\cdot)$ is $E^\odot$-weakly measurable; that (14.8.15) holds has been proved in Lemma 14.8.3.

Conversely, let $g(\cdot)$ be an $E^\odot$-weakly measurable function such that (14.8.15) holds. The proof of (14.4.20) in Lemma 14.4.4 needs modification since the space is not reflexive and $g(\cdot)$ is merely $E^\odot$-weakly measurable, so that we cannot approximate as in (14.4.22). The construction in the proof is modified as follows. Let $Y = \{y^*_k\}$ be a countable dense set in $E^\odot$. Since, for each k the function $\langle y^*_k, g(\cdot) \rangle$ is measurable, for each $\varepsilon > 0$ there exists a countably valued scalar valued $g_{\varepsilon,k}(t)$ such that

$$|\langle y^*_k, g(t) \rangle - g_{\varepsilon,k}(t)| \le \varepsilon \tag{14.8.18}$$

outside of a null set $e(Y)$. Using this inequality and (14.4.24) (established as in Lemma 14.4.4) we show (14.4.25) in each of the sets e in a countable family of pairwise disjoint, measurable sets and construct $\mu_{\varepsilon,Y}(\cdot)$ satisfying (14.4.26) for each y^*_k. The proof concludes in the same way. ∎

End of proof of Theorem 14.8.2. Let $y(\cdot) \in C(0, \bar{t}; E_\alpha)$ be a solution of the relaxed system (14.8.8). Hypothesis $\text{II}_{\alpha,\rho}(BC)$ guarantees that the function

$\phi(t, u) = ((-A^{\odot})^{-\rho})^* f(t, y(t), u)$ fits into the assumptions of Lemma 14.8.4, so that we can manufacture $\mu(\cdot) \in R_{bc}(0, T; U)$ with

$$\Phi(t, y(t))\mu(t) = g(t) \in \overline{\mathrm{conv}}\,\phi(t, U) = \overline{\mathrm{conv}}\,((-A^{\odot})^{-\rho})^* f(t, y(t), U).$$

∎

We extend the existence results in **14.6**, beginning with Theorem 14.6.3, under Hypothesis $I_{\alpha,\rho}(BC)$. The state constraints and target conditions are

$$y(t, u) \in M(t) \quad (0 \le t \le \bar{t}), \qquad y(\bar{t}, u) \in Y, \tag{14.8.19}$$

where $M(t) \subseteq E_\alpha$ and $Y \subseteq E_\alpha$ are closed in E_α. The cost functional is

$$y_0(t, u) = \int_0^t f_0(\tau, y(\tau, u), u(\tau))d\tau + g_0(t, y(t, u)), \tag{14.8.20}$$

with $f_0 : [0, T] \times E_\alpha \times U \to \mathbb{R}$, $g_0 : [0, T] \times E_\alpha \times \mathbb{R}$. We assume

Hypothesis $I_\alpha^0(BC)$. $f_0(t, y, \cdot)$ is bounded and continuous in u for t, $y \in E_\alpha$ fixed and continuous in $y \in E_\alpha$ for t fixed, uniformly with respect to u. The function $t \to f_0(t, y, \cdot)$ is a strongly measurable $BC(U)$-valued function and for every $c > 0$ there exists $K_0(\cdot, c) \in L^1(0, T)$ such that

$$|f_0(t, y, u)| \le K_0(t, c) \quad (\|y\|_\alpha \le c, u \in U). \tag{14.8.21}$$

The **relaxed cost functional** $y_0(t, \mu)$ is

$$y_0(t, \mu) = \int_0^t F_0(\tau, y(\tau, \mu))\mu(\tau)d\tau + g_0(t, y(t, \mu)), \tag{14.8.22}$$

with

$$F_0(t, y, \mu) = \int_U f_0(t, y, u)\mu(du). \tag{14.8.23}$$

We prove exactly as in Lemma 14.6.2 that if $g_0(t, y)$ is continuous in $[0, T] \times E_\alpha$, the relaxed cost functional is weakly lower semicontinuous and t-equi-continuous.

Existence Theorem 14.8.5. *Let* $S(t)$ *be compact. Assume that there exists a minimizing sequence* $\{\mu^n(\cdot)\}$, $\mu^n(\cdot) \in R_{bc}(0, t_n; U)$ *of relaxed controls such that* (a) $\{t_n\}$ *is bounded,* (b) $\{y(t, \mu^n)\}$ *is uniformly bounded in* E_α. *Then there exists a relaxed optimal control* $\bar{\mu}(\cdot) \in R_{bc}(0, \bar{t}; U)$, *which is the* $L^1(0, \bar{t}; BC(U))$-*weak limit of (a generalized subsequence) of* $\{\mu^n(\cdot)\}$ *in* $L_w^\infty(0, \bar{t}; BC(U)^*) = L_w^\infty(0, \bar{t}; \Sigma_{rba}(U))$.

Theorem 14.8.5 leads to an existence theorem for ordinary controls in the same way that Theorem 14.6.3 produces Theorem 14.6.4. The assumptions are the same as those of Theorem 14.8.5, plus: (*i*) U is a separable complete metric space, (*ii*) the functions $f_0(t, y, u)$ and $((-A^\odot)^{-\rho})^* f(t, y, u)$ are locally uniformly continuous in u for t, y fixed, (*iii*)

$$\mathcal{Z}(t, y; f_0, f)$$
$$= \left\{ \left(y_0, ((-A^\odot)^{-\rho})^* f(t, y, u)\right) \in \mathbb{R} \times (E^\odot)^*; \ y_0 \geq f_0(t, y, u), u \in U \right\}$$
$$= \bigcup_{u \in U} \left([f_0(t, y, u), \infty) \times \{((-A^\odot)^{-\rho})^* f(t, y, u)\}\right) \subseteq \mathbb{R} \times (E^\odot)^* \quad (14.8.24)$$

is convex and closed for every $t, y \in E_\alpha$. The result follows from

Lemma 14.8.6. *Let U be a separable complete metric space and let the functions $\phi : [0, T] \times U \to (E^\odot)^*$ and $\phi_0 : [0, T] \times U \to \mathbb{R}$ be such that*

(a) $\phi(t, \cdot)$, $\phi_0(t, \cdot)$ are bounded and locally uniformly continuous in u for each t fixed,

(b) $t \to \phi_0(t, \cdot)$ and $t \to \langle y^*, \phi(t, \cdot)\rangle$ are strongly measurable $BC(U)$-valued functions for every $y^* \in E^\odot$,

(c) for every t, the set

$$\mathcal{Z}(t; \phi_0, \phi) = \{(y_0, \phi(t, u)) \in \mathbb{R} \times E; \ y_0 \geq \phi_0(t, u); u \in U\}$$
$$= \bigcup_{u \in U} \left([\phi_0(t, u), \infty) \times \{\phi(t, u)\}\right) \subseteq \mathbb{R} \times (E^\odot)^* \quad (14.8.25)$$

is convex and $(\mathbb{R} \times E^\odot)$-weakly closed in $\mathbb{R} \times (E^\odot)^*$.

Then, given $\mu(\cdot) \in R_{bc}(0, T; U)$ there exists a strongly measurable U-valued function $u(t)$ such that

$$\phi(t, u(t)) = \int_U \phi(t, u)\mu(t, du) \quad a.e. \ in \ 0 \leq t \leq T \quad (14.8.26)$$

$$\phi_0(t, u(t)) \leq \int_U \phi_0(t, u)\mu(t, du) \quad a.e. \ in \ 0 \leq t \leq T. \quad (14.8.27)$$

The proof is essentially the same as that of Lemma 14.6.1. We set $\mathbf{U} = \mathbb{R}_+ \times \mathcal{U}$ equipped with the product distance and define $\psi : [0, T] \times \mathcal{U} \to \mathbb{R} \times E$ by $\psi(t, (u_0, u)) = (u_0 + \phi_0(t, u), \phi(t, u))$, which is locally uniformly continuous in $\mathcal{U}$ for every t. By assumption, the set $\psi(t, \mathcal{U}) = \mathcal{Z}(t; \phi_0, \phi)$ is convex and closed in $\mathbb{R} \times U$. Given $\mu(\cdot) \in R_{bc}(0, T; U)$, define a $(\mathbb{R} \times E^\odot)$-weakly measurable $(\mathbb{R} \times (E^\odot)^*)$-valued function $g(t)$ by $g(t) = (\phi_0(t)\mu(t), \Phi_0(t)\mu(t))$. That $g(t) \in \psi(t, \mathcal{U})$ a.e. follows from Lemma 14.8.3, so that we have

$$((-A^\odot)^{-\varepsilon})^* g(t) \in ((-A^\odot)^{-\varepsilon})^* \psi(t, \mathcal{U}),$$

where both the function on the left and the set function on the right satisfy the assumptions in Filippov's Theorem 12.8.2 (the argument uses Lemma 7.7.11 and the considerations in Lemma 14.6.1). Accordingly, we may construct a strongly measurable $(\mathbb{R} \times U)$-valued function $(u_0(t), u(t))$ such that

$$((-A^{\odot})^{-\varepsilon})^* g(t) = ((-A^{\odot})^{-\varepsilon})^* \psi(t, u_0(t), u(t)) \quad \text{a.e. in } 0 \le t \le T.$$

"Crossing out" $((-A^{\odot})^{-\varepsilon})^*$ from both sides, (14.8.26) and (14.8.27) result.

14.9. Existence Under Compactness of the Nonlinear Term

The main result in this section applies to the semilinear wave equation

$$y_{tt}(t, x) = \sum_{j=1}^{m} \sum_{k=1}^{m} \partial^j (a_{jk}(x) \partial^k y(t, x)) + \sum_{j=1}^{m} b_j(x) \partial^j y(t, x) + c(x) y(t, x)$$
$$- \phi(t, x, y(t, x)) + u(t, x) \tag{14.9.1}$$

with a boundary condition β; $u(t, \cdot) \in U \subseteq L^2(\Omega)$ (U possibly unbounded). As customary, we handle (14.9.1) as the abstract differential equation

$$\mathbf{y}'(t) = \mathbf{A}(\beta)\mathbf{y}(t) + \mathbf{f}(t, \mathbf{y}(t)) + \mathbf{B}u(t) \tag{14.9.2}$$

in $\mathbf{E} = H_0^1(\Omega) \times L^2(\Omega)$ or $H^1(\Omega) \times L^2(\Omega)$ depending on β, with

$$\mathbf{A}(\beta) = \begin{bmatrix} 0 & I \\ A(\beta) & 0 \end{bmatrix}, \quad \mathbf{f}(t, \mathbf{y})(x) = \begin{bmatrix} 0 \\ -\phi(t, x, y(x)) \end{bmatrix}, \quad \mathbf{B} = \begin{bmatrix} 0 \\ I \end{bmatrix}. \tag{14.9.3}$$

The section title does not mean that we require $\mathbf{y} \to \mathbf{f}(t, \mathbf{y})$ to be compact for each t, an exacting standard that would exclude the nonlinearity in (14.9.1); rather, compactness refers to the map $\mathbf{E} \ni \mathbf{y} \to \mathbf{f}(\cdot, \mathbf{y}) \in L^2(0, T; \mathbf{E})$.

The assumptions on the linear part and on ϕ are those in **5.7** or in Example 14.2.11 with $\phi(t, x, y) \equiv 1$; more general control terms can be handled in the same way (see Example 14.9.9). Some of the results below are not restricted to (14.9.2), thus are proved for the model

$$y'(t) = Ay(t) + f(t, y(t)) + Bu(t), \qquad y(0) = \zeta \tag{14.9.4}$$

in a Banach space E, where A generates a strongly continuous semigroup $S(t)$ and f satisfies Hypothesis II$(f, 0)$. Controls $u(\cdot)$ take values in a second Banach space F and $B \in (F, E)$. The first result is a "weak equicontinuity" statement.

Lemma 14.9.1. *Let E be reflexive, $\mathcal{B}$ a bounded subset of $L^p(0, T; F)$, $p > 1$. Assume that the trajectories $y(t, u)$ $(u(\cdot) \in \mathcal{B})$ exist in $0 \le t \le T$ and are uniformly bounded. Then, for each $y^* \in E^*$ we have*

$$\lim_{|t'-t| \to 0} \langle y^*, y(t', u) - y(t, u) \rangle = 0 \quad (0 \le t, t' \le T) \tag{14.9.5}$$

uniformly with respect to $u(\cdot) \in \mathcal{B}$.

Proof. Since E is reflexive, the adjoint semigroup $S^*(t) = S(t)^*$ is strongly continuous. Given $t \le t'$ we have

$$\langle y^*, y(t', u) - y(t, u) \rangle = \langle y^*, (S(t') - S(t))y \rangle$$
$$+ \int_0^t \langle (S^*(t' - \tau) - S^*(t - \tau))y^*, f(\tau, y(\tau, u)) + Bu(\tau) \rangle d\tau$$
$$+ \int_t^{t'} \langle y^*, S(t' - \tau)\{f(\tau, y(\tau, u)) + Bu(\tau)\} \rangle d\tau$$
$$= \langle y^*, (S(t') - S(t))y \rangle + I_1(t, t') + I_2(t, t').$$

Obviously, the first term performs to specifications. To estimate the second we bound the first element inside the angled brackets by the maximum of its norm and note that if c is a bound for all the trajectories $y(t, u)$ ($u \in \mathcal{B}$) then we have $\| f(\tau, y(\tau, u)) \| \le K_1(\tau, c)$, $K_1(\cdot, c)$ the function in (14.2.4). In the integral involving $Bu(\tau)$ we use Hölder's inequality with first function $= 1$ and second function $= \|Bu(\tau)\|_E \le C\|u(\tau)\|_F$. The result is

$$\| I_1(t, t') \| \le \max_{0 \le \tau \le T} \|S^*(t' - \tau)y^* - S^*(t - \tau)y^*\|_{E^*}$$
$$\times \{\|K_1(c)\|_{L^1(0,T)} + CT^{1-1/p}\|u\|_{L^p(0,T;F)}\}.$$

The right side tends to zero uniformly in $u(\cdot)$ by uniform continuity of $S^*(\tau)y^*$ in $0 \le \tau \le T$. On the other hand, again by Hölder's inequality,

$$\| I_2(t, t') \| \le C \int_t^{t'} K(\tau, c)d\tau + C(t' - t)^{1-1/p}\|u\|_{L^p(0,T;F)},$$

and this is the only place where we need $p > 1$; obviously, if we take $p = 1$ we must request equicontinuity of the integrals of the $u(\cdot)$. If $t' < t$ the roles of t and t' are reversed. ∎

Remark 14.9.2. Lemma 14.9.1 admits the following variant. Under the same hypotheses, let $\{t_n\} \subseteq [0, T]$ be a sequence with $t_n \to \bar{t}$ and $\{u_n(\cdot)\}$, $u_n(\cdot) \in C_{\text{ad}}(0, t_n; U)$ a sequence such that $y(t, u_n)$ exists in $0 \le t \le t_n$ and such that $\|u\|_{L^p(0,t_n;F)}$, $\max_{0 \le t \le t_n} \|y(t, u_n)\|$ are uniformly bounded. Then

$$\lim_{n \to \infty} \langle y^*, y(t_n, u_n) - y(\bar{t}, u_n) \rangle = 0. \tag{14.9.6}$$

The proof is exactly the same.

In all existence results so far, the required convergence properties of trajectories corresponding to minimizing sequences are a consequence of compactness of the semigroup. In the present situation, they follow from the Rellich-Kondrachev theorem below (Adams [1975, p. 144]) that insures that in a bounded domain, all except one of the imbeddings in Sobolev's Theorem 5.7.1 are compact.

Theorem 14.9.3. *Let $\Omega \subseteq \mathbb{R}^m$ be a bounded domain having the cone property, $kp < m$. Then, if $1 \le p \le q < mp/(m - kp)$ the imbedding $W^{k,p}(\Omega) \hookrightarrow L^q(\Omega)$ is compact. If $kp = m$, the imbedding is compact for all $1 \le p \le q < \infty$.*

The only imbedding in Theorem 5.7.1 missed by Theorem 14.9.3 is $kp < m$, $q = mp/(m - kp)$.

Lemma 14.9.4 (Lions). *Let $Q \subseteq \mathbb{R}^m$ be measurable, $1 < p < \infty$, $\{f_n(\cdot)\}$ a sequence in $L^p(Q)$ such that $f_n(x) \to f(x)$ a.e. in Q and $\|f_n\|_{L^p(Q)} \le C$. Then we have (a) $f(\cdot) \in L^p(Q)$, (b) $f_n(\cdot) \to f(\cdot)$ $L^q(Q)$-weakly in $L^p(Q)$, $1/p + 1/q = 1$.*

Proof. To prove (a) apply Fatou's theorem to $\{|f_n(x)|^p\}$. To show (b), let d_n be the set where $|f_j(x) - f(x)| \le (1 + |x|)^{-(m+1)}$ $(j \ge n)$. Then $d_1 \subseteq d_2 \subseteq \dots$ and $\cup d_n$ has full measure in Q, thus if $e_n = d_n \setminus d_{n-1}$, $\cup e_n$ has full measure in Q as well and

$$\|g\|_{L^p(Q)}^p = \Sigma_n \|g\|_{L^p(e_n)}^q$$

for every $g(\cdot) \in L^q(\Omega)$ due to the dominated convergence theorem. This implies that the set $\mathcal{G}$ of functions $g(\cdot) \in L^q(Q)$ such that the support of g is contained in a finite union of the e_j is dense in $L^q(Q)$. That $\langle g, f_n - f \rangle \to 0$ for $g \in \mathcal{G}$ follows from the dominated convergence theorem since

$$|f_j(x)| \le |f(x)| + (1 + |x|)^{-(m+1)} \quad (j \ge n, x \in e_n).$$

∎

Theorem 14.9.5 below is specific to the system (14.9.2). To gain some generality on the cheap we assume controls take values in the dual X^* of a reflexive separable Banach space, so that $\mathbf{B}$ is given by

$$\mathbf{B} = \begin{bmatrix} 0 \\ B \end{bmatrix}, \tag{14.9.7}$$

$B : X^* \to L^2(\Omega)$ a bounded operator. The control space is $C_{ad}(0, T; U) \subseteq L^2(0, T; X^*)$ defined by $u(t) \in U = $ control set $\subseteq X^*$ a.e. and we require that $C_{ad}(0, T; U)$ be $L^2(0, T; X)$-weakly compact in $L^2(0, T; X^*)$. The cost functional $y_0(t, y, u)$ is weakly-weakly lower semicontinuous but, due to the use of controls in L^2 rather than in L^∞ we modify the definition by requiring $L^2(0, \bar{t}; X)$-weak convergence in (14.5.3). The definition of equicontinuity with respect to t is modified in the same way. The state constraint sets $\mathbf{M}(\bar{t}) \in \mathbf{E}$ and the target set $\mathbf{Y} \in \mathbf{E}$ are required to be weakly closed.

Theorem 14.9.5. *Let $\{u^n(\cdot)\}$ be a minimizing sequence such that $\{t_n\}$ is bounded and the sequence of trajectories $\{\mathbf{y}(t, u^n)\} = \{(y(t, u^n), y_t(t, u^n))\}$ is uniformly* *bounded in* $\mathbf{E}$ *for* $0 \le t \le t_n$. *Then (a subsequence of)* $\{u^n(\cdot)\}$ *converges weakly in* $L^2(0, \bar{t}; F)$ *to an optimal control* $\bar{u}(\cdot)$.

Proof. We may assume that $t_n \to \bar{t}$. Take $T > $ all t_n and extend each $u^n(\cdot)$ to $[0, T]$ in the time-honored way setting $u^n(t) = u = $ fixed element of U. Since $C_{ad}(0, T; U)$ is weakly compact in $L^2(0, T; F)$ we may select a subsequence of $\{u^n(\cdot)\}$ such that

$$u^n(\cdot) \to \bar{u}(\cdot) \in C_{ad}(0, T; U) \subseteq L^2(0, T; X^*) \quad L^2(0, T; X)\text{-weakly.} \quad (14.9.8)$$

Extend the trajectories to $t > t_n$ by $\mathbf{y}(t, u^n) = \mathbf{y}(t_n, u^n)$. Uniform boundedness implies boundedness in $L^2(0, t_n; \mathbf{E}) \approx H^1((0, t_n) \times \Omega)$, thus we may also take for granted that

$$y(\cdot, \cdot, u^n) \to \bar{y}(\cdot, \cdot) \in H^1((0, t_n) \times \Omega) \quad \text{weakly in } H^1((0, t_n) \times \Omega). \quad (14.9.9)$$

We apply Theorem 14.9.3 in the domain $(0, T) \times \Omega$ (which has the cone property if Ω does), and deduce that the imbedding $H^1((0, T) \times \Omega) \hookrightarrow L^p((0, T) \times \Omega)$ is compact in the range $p < 2(m + 1)/((m + 1) - 2) = 2(m + 1)/(m - 1)$ (in particular, if $p = 2$), so that we may also assume that

$$y(\cdot, \cdot, u^n) \to \bar{y}(\cdot, \cdot) \quad \text{strongly in } L^2((0, t_n) \times \Omega). \quad (14.9.10)$$

Thinning out again the subsequence, we may secure that

$$y(t, x, u^n) \to \bar{y}(t, x) \quad (14.9.11)$$

a.e. in $(0, T) \times \Omega$. Using then y-continuity of $\phi(t, x, y)$, the estimate (5.7.10) and Lemma 14.9.4 we deduce that

$$\phi(\cdot, \cdot, y_n(\cdot, \cdot)) \to \phi(\cdot, \cdot, \bar{y}(\cdot, \cdot)) \quad \text{weakly in } L^2((0, T) \times \Omega). \quad (14.9.12)$$

Hence

$$\mathbf{f}(\cdot, \mathbf{y}(\cdot, u^n)) + \mathbf{B}u^n(\cdot) \to \mathbf{f}(\cdot, \bar{\mathbf{y}}(\cdot)) + \mathbf{B}\bar{u}(\cdot) \quad \text{weakly in } L^2(0, T; \mathbf{E}), \quad (14.9.13)$$

where $\bar{\mathbf{y}}(t)(x) = (\bar{y}(t, x), \bar{y}_t(t, x))$ (note that the left sides of (14.9.12) and (14.9.8) express the second coordinate of the limit relation (14.9.12), the first coordinate being zero on both sides).

Let $\mathbf{S}(\cdot)$ be the group generated by the operator $\mathbf{A}(\beta)$ in (14.9.3) and let $\mathbf{z} \in \mathbf{E}$. It follows from the integral equation satisfied by $\mathbf{y}(t, u^n)$ and from the fact that $t_n \to \bar{t}$ that if $t < \bar{t}$ then

$$(\mathbf{y}(t, u^n), \mathbf{z}) = (\mathbf{S}(t)\zeta, \mathbf{z}) + \int_0^t (\mathbf{f}(\tau, \mathbf{y}(\tau, u^n)) + \mathbf{B}u^n(\tau), \mathbf{S}(t - \tau)^*\mathbf{z})d\tau \quad (14.9.14)$$

for n large enough, the scalar product that of $\mathbf{E}$. Direct taking of limits in the left side of (14.9.14) is indecorous since we have no information on weak convergence of $\mathbf{y}(t, u^n)$ for fixed t. Instead, we integrate both sides in $0 \leq t \leq \bar{t}$ against a test function $\eta(t)$ with support in $(-\infty, \bar{t})$ obtaining

$$\int_0^{\bar{t}} \eta(t)(\mathbf{y}(t, u^n), \mathbf{z})dt = \int_0^{\bar{t}} \eta(t)(\mathbf{S}(t)\zeta, \mathbf{z})dt$$
$$+ \int_0^{\bar{t}} \left(\mathbf{f}(\tau, \mathbf{y}(\tau, u^n)) + \mathbf{B}u^n(\tau), \int_\tau^{\bar{t}} \eta(t)\mathbf{S}(t - \tau)^* \mathbf{z}\, dt \right) d\tau.$$

Letting $n \to \infty$ and using (14.9.13) we obtain

$$\int_0^{\bar{t}} \eta(t)(\bar{\mathbf{y}}(t), \mathbf{z})dt = \int_0^{\bar{t}} \eta(t)(\mathbf{S}(t)\zeta, \mathbf{z})dt$$
$$+ \int_0^{\bar{t}} \left(\mathbf{f}(\tau, \bar{\mathbf{y}}(\tau)) + \mathbf{B}\bar{u}(\tau), \int_\tau^{\bar{t}} \eta(t)\mathbf{S}(t - \tau)^* \mathbf{z}\, dt \right) d\tau. \quad (14.9.15)$$

Switching back orders of integration and keeping in mind our free choice of $\eta(\cdot)$,

$$(\bar{\mathbf{y}}(t), \mathbf{z}) = (\mathbf{S}(t)\zeta, \mathbf{z})$$
$$+ \int_0^t \left(\mathbf{f}(\tau, \bar{\mathbf{y}}(\tau)) + \mathbf{B}\bar{u}(\tau), \mathbf{S}^*(t - \tau)\mathbf{z}^* \right) d\tau \quad (0 \leq t < \bar{t}) \quad (14.9.16)$$

for arbitrary $\mathbf{z}$, hence

$$\bar{\mathbf{y}}(t) = \mathbf{S}(t)\zeta + \int_0^t \mathbf{S}(t - \tau)\{\mathbf{f}(\tau, \bar{\mathbf{y}}(\tau)) + \mathbf{B}\bar{u}(\tau)\}d\tau \quad (0 \leq t < \bar{t}) \quad (14.9.17)$$

which shows that $\bar{\mathbf{y}}(t) = \mathbf{y}(t, \bar{u})$. Using (14.9.13) to take limits on the right side of (15.9.14) we obtain the right side of (14.9.16), so that

$$\mathbf{y}(t, u^n) \to \bar{\mathbf{y}}(t) \quad \text{weakly in } \mathbf{E} \quad (0 \leq t < \bar{t}) \quad (14.9.18)$$

and we conclude that $\mathbf{y}(t) \in \mathbf{M}(t)$ for all t. That the target condition is satisfied follows writing

$$\mathbf{y}(\bar{t}, \bar{u}) = \mathbf{y}(t_n, u_n) + (\mathbf{y}(\bar{t}, u_n) - \mathbf{y}(t_n, u_n))$$

and using Remark 14.9.2. Finally, the fact that $y_0(\bar{t}, \bar{u})$ attains the minimum m results from (14.9.8), uniform boundedness of the trajectories, the weak convergence relation (14.9.18) and weak-weak lower semicontinuity and t-equicontinuity of the functional.

Remark 14.9.6. Theorem 14.9.5 handles nonintegral control constrains as well; for instance, for the constraint

$$|u(t, x)| \leq 1 \quad (0 \leq t \leq \bar{t}, x \in \Omega)$$

we take $X^* = L^2(\Omega)$, $B = I$: the control set $U \subseteq L^2(\Omega)$ is the unit ball of $L^\infty(\Omega)$.

Remark 14.9.7. Theorem 14.9.5 extends to unbounded domains. In fact, the only part of our argument where boundedness of $Q = (0, T) \times \Omega$ has been used is in securing (14.9.10) by Theorem 14.9.3; however, we only need the pointwise convergence relation (14.9.11). It is shown in Adams [1975, p. 145] that if Q has the cone property we may write $Q = \cup_k Q_k$, $Q_1 \subseteq Q_2 \subseteq \ldots$, where each Q_k is bounded and has the cone property. Hence, we may apply Theorem 14.9.3 in each Q_k and obtain subsequences $\{y(t, x, u^{k,n})\}, k = 1, 2, \ldots$ (each a subsequence of the former) with $y(t, x, u^{k,n}) \to \bar{y}(t, x)$ a.e. in Q_k. Using Cantor's diagonal sequence trick, we may extend a.e. convergence to all of Q.

Example 14.9.8. Some cost functionals of interest for (14.9.1) (with control set $U \subseteq L^2(\Omega)$) are of the form

$$y_0(t, y, u) = \int_{(0,t) \times \Omega} f_0(\tau, x, y(\tau, x, u), y_t(\tau, x, u), \nabla y(\tau, x, u), u(\tau, x)) dx d\tau.$$

$$(14.9.19)$$

A trivial example is $f_0 \equiv 1$ (the time optimal problem), where all requirements check. Another is the quadratic cost functional

$$f_0(\tau, x, y, y_t, \mathbf{y}, u) = \alpha + |y|^2 + |y_t|^2 + \|\mathbf{y}\|^2 + \gamma |u|^2, \qquad (14.9.20)$$

$(\mathbf{y} = (y_1, \ldots y_m), \|\mathbf{y}\|^2 = \Sigma |y_k|^2)$ related to minimizing energy while keeping down time (if $\alpha > 0$) and control costs (if $\gamma > 0$). If $U \in L^2(\Omega)$ is bounded then the functional is equicontinuous. If U is unbounded this is not necessarily true, as weak convergence of the u^n does not prevent concentration of mass between t_n and $\bar{t}$. On the other hand, the functional is weak-weak lower semicontinuous for any U. To see this note that (14.9.8) is equivalent to $u_n \to \bar{u}$ weakly in $L^2((0, T) \times \Omega)$; we then use (14.9.9) and the fact that $\|\text{weak lim } v_n\| \leq \lim \sup \|v_n\|$ in Hilbert spaces.

Example 14.9.9. The arguments is this section apply equally well to a control term of the form $\psi(t, x, y(x))u(t, x)$. The assumptions are the same as those in Example 14.2.11 and $U \subseteq L^\infty(\Omega)$.

14.10. Existence Without Compactness

Control systems such as those in Example 14.5.2 and 14.5.5 are linear in the control thus nonexistence cannot be fixed by relaxation of the controls (note also that the set (14.6.14) is convex and closed in both examples). At this juncture, we may concede that problems like these have no reasonable solution and leave them well enough alone, or doggedly insist in extending the definitions and bring solutions out of the blue. We take the die-hard's way and outline below two such extended definitions. The first is a generalization of Gamkrelidze's [1962] "sliding optimal states." Given the control system

$$y'(t) = Ay(t) + f(t, y(t), u(t)), \tag{14.10.1}$$

a **sliding trajectory** is any continuous E-valued function $y(t)$ such that

$$y(t) = \lim_{n \to \infty} y(t, u_n) \tag{14.10.2}$$

E^*-weakly in $0 \le t \le \bar{t}$ for a sequence $\{u^n\} \subseteq C_{ad}(0, \bar{t}; U)$; we require that each $y(t, u^n)$ be defined in $0 \le t \le \bar{t}$ and that $\{y(t, u^n)\}$ be uniformly bounded. If $y_0(\bar{t}, u^n)$ approaches the minimum value of the functional, we have a **sliding optimal trajectory**, not necessarily corresponding to any admissible control (as is the case in Examples 14.5.2 and 14.5.5). To give interest to this definition, (i) one should prove that sliding optimal regimes fill the gap left by nonexistent optimal trajectories, (ii) one should have some version of Pontryagin's maximum principle for optimal sliding regimes or for elements u^n of minimizing sequences. We do (i) for the system

$$y'(t) = Ay(t) + f(t, y(t)) + u(t) \tag{14.10.3}$$

in Theorem 14.10.1 below and a particular case of (ii) in Example 14.10.2.

We assume in the rest of the section that A generates a strongly continuous semigroup $S(t)$ in a reflexive separable Banach space E. The control space $C_{ad}(0, T; U)$ is the subspace of $L^\infty(0, T; U)$ determined by the condition $u(t) \in U \subseteq E$, and we request it to be $L^1(0, T; E^*)$-weakly compact in $L^\infty(0, T; E)$. In the first result, $f(t, y)$ satisfies Assumption II$(f, 0)$ with $K_1(\cdot, c) = K(c)$ $(K(\cdot, c) \in L^p(0, T),$ $p > 1$ is enough) and the cost functional $y_0(t, y, u)$ is weakly-weakly lower semicontinuous; for simplicity, we take fixed terminal time. Predictably, the state constraint sets $M(t)$ and the target set Y must be E^*-weakly closed.

Theorem 14.10.1. (a) *Let $\{u^n(\cdot)\}$ be an arbitrary sequence in $C_{ad}(0, \bar{t}; U)$ with $\{y(t, u^n)\}$ uniformly bounded in $0 \le t \le \bar{t}$. Then, if necessary passing to a subsequence,*

$$y(t, u^n) \to \bar{y}(t) \quad E^*\text{-weakly in } 0 \le t \le \bar{t} \tag{14.10.4}$$

where $\bar{y}(t)$ is a sliding trajectory. (b) If $\{u^n(\cdot)\} \subseteq C_{ad}(0, \bar{t}; U)$ is in addition a minimizing sequence there exists $\bar{u}(\cdot) \in C_{ad}(0, \bar{t}; U)$ such that $\bar{u}(\cdot)$ is the $L^1(0, \bar{t}; E^*)$-weak limit of (a subsequence of) $\{u^n(\cdot)\}$, (14.10.4) holds, $\bar{y}(t)$ satisfies the state constraints and the target condition and

$$y_0(\bar{t}, \bar{y}, \bar{u}) = m. \tag{14.10.5}$$

Proof. Selecting a subsequence we may assume that $\{u^n(\cdot)\}$ is $L^1(0, \bar{t}; E^*)$-weakly convergent to $\bar{u}(\cdot) \in C_{ad}(0, \bar{t}; U)$ and that $\{f(\cdot, y(\cdot, u^n))\}$ is $L^1(0, \bar{t}; E^*)$-weakly convergent to $g(\cdot) \in L^\infty(0, \bar{t}; E)$. Applying a functional $y^* \in E^*$ to both sides of the integral equation defining $y(t, u^n)$ we obtain

$$\langle y^*, y(t, u^n) \rangle = \langle y^*, S(t)\zeta \rangle + \int_0^t \langle S^*(t - \tau)y^*, f(\tau, y(\tau, u^n)) \rangle d\tau.$$

We deduce that $y(t, u^n) \to \bar{y}(t)$ E^*-weakly for $0 \le t \le \bar{t}$, $\bar{y}(t)$ given by

$$\bar{y}(t) = S(t)\zeta + \int_0^t S(t - \tau)g(\tau)d\tau \qquad (14.10.6)$$

and thus continuous. That (14.10.5) holds when the sequence is minimizing is obvious from the definitions. ∎

Example 14.10.2. We consider the time optimal problem for (10.4.3) with A the infinitesimal generator of a group $S(\cdot)$ in a separable Hilbert space H, with $U = $ unit ball of E. There are no state constraints and the target condition is

$$y(\bar{t}) \in Y$$

with $Y \subseteq H$ closed. We assume the existence of a minimizing sequence $\{u^n(\cdot)\}$ such that

$$\text{dist}(y(t_n, u^n), Y) \le \varepsilon_n \to 0 \quad \text{as} \quad n \to \infty$$

with $t_n < \bar{t} = $ optimal time, and

$$\|y(t, u^n)\| \le C \quad (0 \le t \le t_n). \qquad (14.10.7)$$

Using Theorem 3.3.2 we obtain a sequence $\{\tilde{u}^n\}$, $\tilde{u}^n \in C_{ad}(0, t_n; U)$ with

$$d_n(u^n, \tilde{u}^n) = |\{t \in [0, t_n]; u^n(t) \ne \tilde{u}^n(t)\}| \le \sqrt{\varepsilon_n} \qquad (14.10.8)$$

and sequences $\{\tilde{y}^n\} \subseteq Y, \{z_n\} \subseteq E$ such that

$$\|z_n\| = 1, \qquad (z_n, \xi(\bar{t}, s, \tilde{u}^n, v) - w^n) \le \sqrt{\varepsilon_n}(1 + \|w^n\|) \qquad (14.10.9)$$

for w^n in the contingent cone $K_Y(\tilde{y}^n)$, s in a set of full measure in $[0, \bar{t}]$ and $v \in U$. Going to a subsequence we may assume that $\{z_n\}$ is weakly convergent, and using the comments after Theorem 6.7.1 we deduce that

$$z = \text{weak} \lim_{n \to \infty} z_n \ne 0.$$

Assumption (14.10.7) on global boundedness of $y(t, u^n)$, (14.10.8) and a simple application of Gronwall's lemma imply that the trajectories $y(t, \tilde{u}^n)$ exist globally and are uniformly bounded:

$$\|y(t, \tilde{u}^n)\| \le C \quad (0 \le t \le t_n). \qquad (14.10.10)$$

Using (3.7.6) and setting $w^n = 0$ in (14.10.9) we obtain

$$(S(t_n, s; \tilde{u}^n)^* z_n, v - \tilde{u}^n(s)) \le \sqrt{\varepsilon_n} \quad (\|v\| \le 1), \qquad (14.10.11)$$

a.e. in $0 \le t \le \bar{t}$, $S(t, s; u)$ the solution operator of the variational equation $z'(t) = \{A + \partial_y f(t, y(t, u))\}z(t)$. This inequality is exploited using

Lemma 14.10.3. *Let* $\|z\| = 1, \|u\| \le 1, 0 < \delta < 1$,

$$(z, v - u) \le \delta \quad (\|v\| \le 1) \tag{14.10.12}$$

Then

$$\|u - z\| \le \sqrt{2\delta} \tag{14.10.13}$$

Proof. (14.10.12) implies $(z, z - u) \le \delta$, while we obviously have $(z, z - u) \ge 0$, so that the proof follows from examination of Figure 14.2 below and the equality $1 - (1 - \delta)^2 + \delta^2 = 2\delta$. ∎

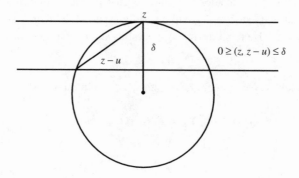

Figure 14.2.

Since $S(t)$ is a group, the solution operator $S(t, s; u)$ is defined for $s \le t$ as well as for $s \ge t$. Moreover (14.10.10) implies that

$$\|S(t, s; \tilde{u}^n)\| \le C \quad (0 \le s, t \le t_n).$$

We have $S(t, s; \tilde{u}^n)^{-1} = S(s, t; \tilde{u}^n)$, thus the inverses are uniformly bounded as well. The same statements apply to the adjoint operator $S(t, s; \tilde{u}^n)^*$, so that $\|S(t, s; \tilde{u}^n)^* z\| \ge \rho \|z\|$ $(0 \le s, t \le t_n)$ with $\rho > 0$; accordingly,

$$\|S(t_n, s; \tilde{u}^n)^* z_n\| \ge \rho > 0 \tag{14.10.14}$$

for n large enough.

We deduce from (14.10.11) divided by $\|S(t_n, s; \tilde{u}^n)^* z_n\|$, from (14.10.14) and from Lemma 14.10.3 that, if we define

$$\bar{u}^n(s) = \frac{S(t_n, s; \tilde{u}^n)^* z_n}{\|S(t_n, s; \tilde{u}^n)^* z_n\|} \quad (0 \le t \le t_n) \tag{14.10.15}$$

then

$$\|\tilde{u}^n(t) - \bar{u}^n(t)\| \to 0 \tag{14.10.16}$$

uniformly in $0 \le t \le t_n$ so that, due to continuous dependence of the solution of (14.10.3) on the control, $\bar{u}^n(\cdot)$ is a minimizing sequence as well. It finally results that although the time optimal control problem for (14.10.3) may not have a solution, there exists a sequence $\{z_n\}$ with $\|z_n\| = 1$, weak $\lim z_n \neq 0$ and such that (14.10.15) is a minimizing sequence; moreover, combining (14.10.11) with (14.10.16) we obtain that

$$\|u^n - \bar{u}^n\|_{L^p(0,t_n)} \to 0$$

for $p < \infty$. This may be considered "almost as good" as the maximum principle. Note, however that the existence of a minimizing sequence with $t_n <$ optimal time is a serious restriction.

Results for cost functionals other than time can be obtained using the corresponding "approximate maximum principle" (Theorem 4.1.1).

The second fix for nonexistence of optimal controls involves relaxation of the trajectories themselves. Let $B(0, c)$ be the closed ball of center 0 and radius c in a Banach space E, $BC(B(0, c))$ the space of bounded continuous functions in E equipped with the supremum norm; according with the theory in Chapter 12 we have $BC(B(0, c))^* = \Sigma_{rba}(B(0, c), \Phi)$ (Theorem 12.4.6) and

$$L^1(0, T; BC(B(0, c))^* = L_w^\infty(0, T; \Sigma_{rba}((0, c), \Phi)), \tag{14.10.17}$$

(Theorem 12.2.11); here, $\Phi = \Phi(C)$ indicates the field generated by the closed sets of $B(0, c)$. As customary, we denote by $\Pi_{rba}(B(0, c), \Phi)$ the set of all probability measures η in $\Sigma_{rba}(B(0, c), \Phi)$ ($\eta \geq 0, \eta(B(0, c)) = 1$), and by $L_w^\infty(0, T; \Pi_{rba}(B(0, c), \Phi))$ the subspace of $L_w^\infty(0, T; \Sigma_{rba}(B(0, c), \Phi))$ consisting of all $\eta(\cdot)$ such that there exists an element $\mu(\cdot)$ in the equivalence class of $\eta(\cdot)$ in $L_w^\infty(0, T; \Sigma_{rba}(B(0, c), \Phi))$ with $\mu(t) \in \Pi_{rba}(B(0, c), \Phi)$ a.e.; equivalently, $L_w^\infty(0, T; \Pi_{rba}(B(0, c), \Phi))$ can be defined, just as relaxed controls, by conditions corresponding to (*i*) and (*ii*) in **12.5**.

To install the relaxed trajectories in (14.10.3) we assume: for every $c > 0$, $f(t, \cdot) \in B(B(0, c); E)$, and $y^* \in E^*, t \to \langle y^*, f(t, \cdot) \rangle \in BC(B(0, c))$ is strongly measurable, and there exists $K(\cdot, c) \in L^1(0, \bar{t})$ with

$$\|f(t, y)\| \leq K(t, c) \quad (0 \leq t \leq \bar{t}, \|y\| \leq c). \tag{14.10.18}$$

Call $\eta(\cdot, d\xi, u) \in L_w^\infty(0, \bar{t}; \Pi_{rba}(B(0, c), \Phi))$ a **measure solution** of equation (10.4.3) in $0 \leq t \leq \bar{t}$ if

$$\int_{B(0,c)} \langle y^*, \xi \rangle \eta(t, d\xi, u) = \langle y^*, S(t)\zeta \rangle$$

$$+ \int_0^t \int_{B(0,c)} \langle S(t-\tau)^* y^*, f(\tau, \xi) \rangle \eta(\tau, d\xi, u) d\tau$$

$$+ \int_0^t \langle S(t-\tau)^* y^*, u(\tau) \rangle d\tau \tag{14.10.19}$$

for $y^* \in E^*, 0 \leq t \leq \bar{t}$. Since $y \to \langle y^*, \cdot \rangle$ belongs to $L^1(0, \bar{t}; BC(B(0, c)))$, the only thing we need for this definition to make sense is that the function $\tau \to \langle S(t - \tau)^* y^*, f(\tau, \cdot) \rangle$ belong to $L^1(0, \bar{t}; BC(B(0, c)))$; this follows from the hypotheses, strong continuity of the adjoint semigroup and a simple approximation argument. The notation $\eta(t, d\xi, u)$ (often abbreviated to $\eta(u)$ below) does not imply that there exists a unique measure solution satisfying (14.10.19) (in general, there are many); it only indicates the association of η and u in (14.10.19). The **relaxed trajectory** $y(t, \eta(u))$ corresponding to the measure solution $\eta(t, d\xi, u)$ is the E-valued function $y(t, \eta(u))$ defined by

$$\langle y^*, y(t, \eta(u)) \rangle = \int_{B(0,c)} \langle y^*, \xi \rangle \eta(t, d\xi, u) \tag{14.10.20}$$

for $y^* \in E^*, 0 \leq t \leq \bar{t}$. Since $y = \langle y^*, \cdot \rangle$ belongs to $L^1(0, \bar{t}; BC(B(0, c)))$, $y(t, \eta(u))$ is E^*-weakly measurable, thus strongly measurable. We also have $|\langle y^*, y(t, \eta(u)) \rangle| \leq c \|y^*\|$ (η is a probability measure) so that

$$\|y(t, \eta(u))\| \leq c \qquad (0 \leq t \leq \bar{t}). \tag{14.10.21}$$

A usual solution $y(t, u)$ of (14.10.3) with $\|y(t, u)\| \leq c$ corresponds to a measure solution

$$\eta(t, d\xi, u) = \delta_{y(t,u)}(d\xi), \tag{14.10.22}$$

δ_y the Dirac delta with mass at y. We generalize this below showing that sliding trajectories are (the relaxed trajectories corresponding to) certain measure solutions. The assumptions are the intersection of those needed in the definition of sliding trajectory and in that of measure solution.

Theorem 14.10.4. *Let $\bar{y}(t) = \lim y(t, u^n)$ be a sliding trajectory. Then there exists a control $\bar{u}(\cdot)$ and a measure solution $\eta(\cdot, \bar{u})$ such that*

$$\bar{y}(t) = y(t, \eta(\bar{u})).$$

Proof. Choose c such that $\|y(t, u^n)\| \leq c$. Let $\eta(t, d\xi, u^n) = \delta_{y(t,u^n)}(d\xi)$. Using Alaoglu's Theorem, select a (generalized) subsequence $\{u^\kappa(\cdot), \eta(\cdot, u^\kappa)\}$ such that

$$u^\kappa(\cdot) \to \bar{u}(\cdot) \quad L^1(0, \bar{t}; E^*)\text{-weakly in } L^\infty(0, \bar{t}; E) \tag{14.10.23}$$

$$\eta(\cdot, u^\kappa) \to \bar{\eta}(\cdot) \quad L^1(0, \bar{t}; BC(B(0, c)))\text{-weakly in } L^\infty_w(0, \bar{t}; \Sigma_{rba}(B(0, c), \Phi)) \tag{14.10.24}$$

where $\bar{\eta} \in L^\infty_w(0, \bar{t}; \Sigma_{rba}(B(0, c), \Phi))$. Write (10.4.19) for $\eta(\cdot, u^\kappa), u^\kappa(\cdot)$; on the right side replace integration in $[0, t]$ by integration in $[0, \bar{t}]$ inserting the characteristic function χ_t of $[0, t]$ under the integral. Then take advantage of the two convergence relations (14.10.23) and (14.10.24); we deduce that $y(t, u^\kappa) = y(t, \eta(u^\kappa))$

on the left side of (14.10.19) is E^*-weakly convergent. Since the entire sequence $y(t, u^n)$ is E^*-weakly convergent to the sliding trajectory $\bar{y}(t)$, the limit must be $\bar{y}(t)$. Taking limits on both sides we achieve the proof.   ∎

Remark 14.10.5. Let $\eta(t, d\xi, u)$ be a measure solution. Let

$$F(\tau, u) = \int_{B(0,c)} f(\tau, \xi)\eta(\tau, d\xi, u), \qquad (14.10.25)$$

the integral defined as in (14.4.18). Lemma 14.4.4 implies that $F(\cdot)$ is strongly measurable and

$$F(\tau, u) \in \overline{\mathrm{conv}} f(\tau, B(0, c)) \qquad (14.10.26)$$

a.e. in $0 \le t \le \bar{t}$; moreover, it follows from (14.10.18) that $\|F(\tau, u)\| \le K(\tau, c)$. This means that that y^* can be "simplified from" the integral equation (14.10.19) defining measure solutions; in other words, the equation can be written

$$y(t, \eta(u)) = S(t)\zeta + \int_0^t S(t - \tau)\{F(\tau, u) + u(\tau)\}d\tau \qquad (14.10.27)$$

which shows that relaxed trajectories are actually continuous, not just strongly measurable. The inclusion relation (14.10.26) indicates that the role of the measure $\eta(\tau, d\xi, u)$ in the integral equation (14.10.19) is that of providing a "moving average" of the values of $f(\tau, y)$, a kind of situation typical in relaxation arguments.

Remark 14.10.6. It is apparently unknown whether the relaxed trajectory corresponding to a measure solution is a sliding trajectory, that is, it can be approximated weakly for all t by an uniformly bounded sequence of ordinary trajectories. A result of this type would be a "trajectory analogue" of the relaxation theorems in this chapter.

Miscellaneous Comments for Part III

Relaxed controls for finite dimensional systems and Young measures. It was (essentially) stated by Hilbert [1904] that "every problem of the calculus of variations has a solution, provided the word 'solution' is suitable understood"[1]. This dictum is illustrated in [1904] with a construction of weak solutions of the Dirichlet problem, a forerunner of much twentieth century mathematics (generalized derivatives, Sobolev spaces, theory of distributions). Hilbert's apothegm was put in practice systematically for variational problems by Young [1937], who showed that certain functionals not having minima due to lack of lower semicontinuity could be "given" a minimum by extending the class of curves in competition. This involves averaging the functional with probability measures, nowadays called *Young measures,* and an active subject of research in calculus of variations and other areas of analysis. See Young [1969] for more on this subject. The counterparts of Young's measures in the field of control problems are the *relaxed controls* (controls whose instantaneous values are probability measures) introduced by Warga [1962:1] [1962:2]; our treatment of finite dimensional relaxed controls follows essentially these works and others of the same author. In particular, the approximation theorem in **13.2** and the existence result in **13.5** are due to Warga [1967] [1970]. See also Warga [1972] and Gamkrelidze [1978] for conditions that guarantee that the relaxed maximum **m** and the ordinary maximum *m* coincide, and for additional information on relaxed controls.

A different approach to relaxation that uses differential inclusions rather than measure-valued controls was initiated by Filippov [1959]. Later, Gamkrelidze [1962] defined relaxed trajectories (in his terminology, "sliding regimes") as uniform limits of ordinary trajectories; under suitable assumptions, this approach is equivalent to those of Warga and Filippov.

The material in **13.8** is related to Ball [1990], where the problem of avoiding "escape of mass to ∞" is handled in a different way. The present treatment is a particular case of the author-Sritharan [1995:2], where the control set is much more general and the setting is infinite dimensional.

The way the minimum principle is handled in this work confirms the observation (Young [1969, §39]) that if the control set is allowed sufficient generality no special version for relaxed controls needs to be proved. For other approaches to the relaxed maximum principle see Warga [1962:1] [1967] [1972]. For a direct approach to optimal control theory of differential inclusions see Blagodatski [1984], Blagodatski-Filippov [1985], Vinter-Loewen [1987] and references in these papers.

Example 13.5.4 (as well as the rest of the material on Newton's problem) is taken from McShane [1978]. It illustrates the curious fact that, although ordinary solutions exist, relaxed controls are needed to find them, since the problem misses

---

[1] In the (very free) translation of L. C. Young. See Young [1969] for a discussion of the word "regular," added to the statement in Hilbert's collected papers.

the convexity criteria needed to apply results of the type of Theorem 13.6.2. A more compelling case for relaxed controls would be hard to imagine.

Functional analysis and measure theory.   The proof of the Dunford-Pettis theorem follows the original and, more closely, Dunford-Schwartz [1958, VI.8.6] with some modifications. Theorem 12.2.11 is outlined in Dieudonné [1947/8] without reference to Dunford-Pettis [1940]. The proof of the complete result, as well as that of Theorem 12.9.2 is that of A. and C. Ionescu Tulcea [1969, Chapter 7]. The results in **12.6** on approximation in spaces of measures and relaxed controls are due to the author [1991:2] [1994:2], and generalize results corresponding to compact U due to Warga [1967] [1970]. That the Dirac deltas do not exhaust the extremal points of Π_{rba} was pointed out to the author by T. Roubíček, to whom the argument in Example 12.6.13 is due. Roubíček also observes that one may avoid the use of finite additive measures by replacing the control set U by its Stone-Čech compactification. For this and other questions see Roubíček [1997].

The few facts on the measurable image property in **12.7** follow Bourbaki's English translation [1966] (the French original covers only some of them). In its present version, the implicit function Theorem 12.8.2 (which, according to common lore we have called Filippov's Theorem) is due to Himmelberg, Jacobs and Van Vleck [1969] and is a far reaching generalization, in contents and method of proof, of the original result. See also Jacobs [1969] and Datko [1970]. The original Filippov Theorem is in Filippov [1959] with U a compact set in Euclidean space (possibly depending on t).

Relaxed controls for infinite dimensional systems.   In the context of evolution equations in Banach spaces these were introduced in Ahmed [1983]; see also Papageorgiou [1989] and other papers of these authors, where the emphasis is on obtaining relaxed controls whose values are countably additive measures. The approach followed in Chapter 14 is different and is based on the author [1991:2] [1993:2] [1993:3] [1994:1] and [1994:2], where very lenient assumptions are placed on the control set at the cost of working with finitely additive measures. Most results on relaxed controls in Chapter 14 are (sometimes generalized) versions of results in these papers. Some have ancestors in different settings, for instance the approximation theorems in **14.2**; see Ahmed [1983], Papageorgiou [1989] and other works of these authors. Theorem 14.6.3 is Theorem 2.12 in Yong [1992] under somewhat different hypotheses. The nonexistence examples in **14.5** are taken from the author [1997:1], as is the patching up of nonexistence by means of sliding regimes and measure valued solutions. It may of course be objected that these "solutions" are too tenuously connected with the equation (they all but ignore the nonlinearity), but existence of optimal controls is perhaps not an overriding concern in science and technology; all that we need are controls that are "sufficiently suboptimal," and some distinguishing feature that may simplify their computation, such as formula (14.10.15) for the suboptimal controls constructed in Example 14.10.2. The idea of proving "approximate" versions of the maximum principle that one can apply to sequences of increasingly suboptimal controls has been employed earlier in the context of partial differential equations. For references and further information see Roubíček [1997]. See also the Miscellaneous Notes to Part II for approximate minimum principles obtained by direct application of Ekeland's principle.

We mention in passing that measure-valued solutions of partial differential equations have been at play for some time in other contexts; among many other works see DiPerna [1985] for hyperbolic equations, Slemrod [1991] for parabolic equations and Foias-Temam [1980], Vishik-Fursikov [1988] for the Navier-Stokes equations. These measures are countably additive, in the parabolic and Navier-Stokes cases due to compactness properties of the equations.

In view of the vagaries associated with finitely additive measures (these vagaries come with the territory when one takes duals of nonseparable spaces) it might be justifiable objected that spaces of relaxed controls like $R_{bc}(0, T; U)$ and $R_b(0, T; U)$, although providing a simple solution to the completion problem, are too large to be useful. On the other hand, playing the devil's advocate one might argue that we rarely if ever implement relaxed controls in practice, just approximations; besides, relaxation results or the minimum principle became more inclusive as the class of relaxed controls becomes larger, and "stand proved" for smaller subspaces.

There is a large literature on relaxed controls for problems governed by partial differential equations; in most of these papers, equations are treated directly (rather than converted into abstract evolution equations) and there are also results for steady state problems. The emphasis is not on "off the rack" loosely fitting relaxed control spaces good for any equation but on tailor-made spaces custom cut for the equation and the subsidiary conditions. As an example, see Hoffmann-Roubíček [1995] and, for additional information and references, Roubíček [1997]. In the opposite direction of increased abstraction and generality see Roubíček [1986] [1989] [1990] and [1997] for a relaxation theory of constrained minimization problems in topological spaces.

Approximation of trajectories by extremal trajectories. The finite dimensional forerunner of the relaxation theorems in **14.2** is Hermes [1964], which studies approximation of trajectories of a nonlinear ordinary differential system by extremal trajectories. The principal result in this paper is the common finite dimensional ancestor of Theorems 14.2.3 and 14.2.9. See also Hermes-LaSalle [1969, p. 112]. In the context of relaxed controls, Warga [1967] does the same job for relaxed trajectories.

Approximation by extremal trajectories for a linear infinite dimensional system was studied in the author [1968:1]; the main result is the linear version of Theorem 14.2.9. In their present form, Theorems 14.2.3 and 14.2.9 are in the author [1993:2]. These results generalize an earlier result of Seidman, proved under stronger hypotheses. A variant can be found in the author-Sritharan [1995:2] for the Navier-Stokes equations; the estimations are based on a further generalization of the generalized Gronwall lemma.

For a general study of relaxation theorems (denseness of extremal trajectories in the set of all trajectories of differential inclusions) see Tolstonogov [1991], Papageorgiou [1993] and references in these papers, where some of the assumptions are weaker and the setting more general.

Some finite dimensional approximation results such as Hermes [1964] are based on *Lyapunov's theorem* below. In it, U is an arbitrary set, Φ is a σ-field of subsets of U and μ is a bounded $\mathbb{R}^m$-valued countably additive measure. A set $e \subseteq \Phi$ is an *atom* of μ if $\mu(e) \neq 0$ and, for every $d \in \Phi$, $d \subseteq e$ we have $\mu(d) = 0$ or $\mu(d) = \mu(e)$; μ is *nonatomic* if it has no atoms. The *range* $\mathcal{R}(\mu)$ of μ is

$$\mathcal{R}(\mu) = \{\mu(e); e \subseteq \dot{\Phi}\} \subseteq \mathbb{R}^m.$$

Theorem 1. (A. A. Lyapunov)[2] *Assume μ is nonatomic. Then $\mathcal{R}(\mu)$ is compact and convex.*

For a proof, see A. A. Lyapunov [1940] or Diestel-Uhl [1977, p. 264]. The control application in mind concerns the linear finite dimensional system

$$y'(t) = A(t)y(t) + b(t)u(t), \quad y(0) = \zeta \tag{1}$$

in $0 \leq t \leq \bar{t}$, with $A(\cdot) \in L^1(0, \bar{t}; \mathbb{R}^{m \times m})$, $b(\cdot) \in L^1(0, \bar{t}; \mathbb{R}^m)$ and control set $U = [-1, 1]$. We denote by $R(\bar{t}, \zeta; U)$ (resp. $R_0(\bar{t}, \zeta; U)$) the subset of $\mathbb{R}^m$ attainable in time $\bar{t}$ by trajectories of (1) (resp. by trajectories driven by *extremal* controls satisfying $|u(t)| = 1$ a.e.)

Corollary 2. (LaSalle) $R(\bar{t}, \zeta; U) = R_0(\bar{t}, \zeta; U)$ *for any $\bar{t} > 0$.*

For the proof, consider the $\mathbb{R}^m$-valued measure $\mu(e) = \int_e S(\bar{t}, \tau)b(\tau)d\tau$ defined in the field Φ of all Lebesgue measurable subsets of $[0, \bar{t}]$, $S(t, \tau)$ the solution operator of (1). Obviously, $R_0(\bar{t}, \zeta; U)$ is a translation of conv$\mathcal{R}(\mu) = \mathcal{R}(\mu)$, the equality a consequence of Lyapunov's theorem, which also implies that $R_0(\bar{t}, \zeta; U)$ is closed. Since, by Lemma 14.2.7, $R_0(\bar{t}, \zeta; U)$ is dense in $R(\bar{t}, \zeta; U)$, the result follows.

---

[2] A. A. Lyapunov $\neq$ A. M. Lyapunov.

Applying the same argument in each coordinate, it can be easily extended to multidimensional control sets, for instance $U = [-1, 1] \times \ldots \times [-1, 1]$, which is the setting in LaSalle [1960].

LaSalle's Corollary is strictly finite dimensional; a quick infinite dimensional counterexample can be produced using a rigid system $y'(t) = Ay(t) + bu(t)$ with one-dimensional control. It is also strictly linear; it fails for nonlinear ordinary differential systems, even if the control appears linearly. For a counterexample see Hermes-LaSalle [1969, p. 124]. In infinite dimensional spaces one can only prove denseness, but, as a compensation, the whole trajectory can be uniformly approximated as in the author [1968:1]. For infinite dimensional restricted variants of the Corollary as well as of Liapunov's theorem see Datko [1970]. Many other versions of Lyapunov's theorem for infinite dimensional spaces exist, some not particularly germane to control theory; see Diestel-Uhl [1977].

Existence of solutions of optimal control problems. In the finite dimensional realm, an approach to existence results based on differential inclusions and convex-ification was developed in Filippov [1959]; see also Roxin [1962] and Ważewski [1962:1] [1962:2]. In this setting, as well as in that of relaxed controls, finite dimen-sionality makes it possible to apply the Arzelà-Ascoli theorem. For *linear* infinite dimensional systems weak convergence of the trajectories is not only enough but also cheaply available, as the result (for the time optimal problem) in the author [1964] shows. This result is a direct ancestor of 14.5.4 (the author [1997:1]). The-orems for linear partial differential equations exploiting similar weak convergence schemes appeared at about the same time: see Russell [1966].

The first existence results for nonlinear partial differential equations are in Lions [1966] and Schmaedeke [1967], the first for parabolic equations, the second for hyperbolic equations in one space variable. These results make plain that the compactness needed to insure convergence of the trajectories has to be supplied by the equation or by the special form of the nonlinearity. This basic paradigm has been implemented many times since, for partial differential equations and for abstract evolution equations, and appears in various places in this book. See also Slemrod [1974:2], Papageorgiou [1991], Ahmed [1986], Yong [1992]. The hyperbolic case is harder as the equation does not contribute compactness; a complete treatment in the n-dimensional case is due to Yong [1992] (see also below). Except for some additional generality in the functional, the material in **14.9** follows this paper and Li-Yong [1995, **3.5**]. For other early results, in the spirit of classical calculus of variations, see Cesari [1969] [1970] and, for numerous theorems in this vein, Cesari [1983]. It should be mentioned that, in the setting of partial differential equations, many existence theorems for control problems are closely related in statement and method of proof to existing theorems in calculus of variations.

One way of disposing of the bane of existence results—controls appearing nonlinearly—is: use relaxed controls and then Filippov type theorems. Another way is: employ convexity conditions directly, as in Roxin [1962] or, in the infinite dimensional case, Yong [1992]. For yet another, unusual way to tame a control nonlinearity see Seidman [1994].

While existence of optimal controls may be sometimes be surmised on "physical grounds," there are usually no expectations for uniqueness and,—except in the linear case—uniqueness seems to be a scarcely studied problem. For a nonlinear result, see Seidman-Zhou [1982].

REFERENCES

The Russian transliterations are obvious except perhaps for "x = j", the latter pronounced in Spanish.

F. Abergel and R. Temam [1990] On some control problems in fluid mechanics, *Theor. Comp. Fluid Dynamics* **1** (1990) 303–325.

R. A. Adams [1975] *Sobolev Spaces,* Academic Press, New York, 1975.

N. U. Ahmed

[1982] Sufficient conditions for controllability of a class of distributed parameter systems, *Systems & Control Letters* **2** (1982) 237–242.

[1983] Properties of relaxed trajectories for a class of nonlinear evolution equations in a Banach space, *SIAM J. Control & Optimization* **21** (1983) 953–967.

[1985:1] Finite-time null controllability for a class of linear evolution equations on a Banach space with control constraints, *J. Optimization Theory & Appl.* **47** (1985) 129–158.

[1985:2] A note on the maximum principle for time optimal controls for a class of distributed-boundary control problems, *J. Optimization Theory & Appl.* **45** (1985) 147–157.

[1986] Existence of optimal controls for a class of systems governed by differential inclusions in a Banach space, *J. Optimization Theory & Appl.* **50** (1986) 213–237.

N. U. Ahmed and K. Teo [1984] *Optimal Control of Distributed Parameter Systems,* North-Holland, Amsterdam 1981.

N. Ahmed and X. Xiang [1994] Optimal control of infinite-dimensional uncertain systems *J. Optimization Theory & Appl.* **80** (1994) 261–272.

J.-J. Alibert and J.-P. Raymond [1997] Boundary control of semilinear elliptic equations with discontinuous leading coefficients and unbounded controls, *Numer. Func. Anal. & Optimization* **18** (1997) 235–249.

W. Alt and U. Mackenroth [1989] Convergence of finite element approximations to state constrained convex parabolic boundary control problems, *SIAM J. Control & Optimization* **27** (1989) 718–736.

M. Altman [1990] *A Theory of Optimization and Optimal Control for Nonlinear Evolution and Singular Equations; Applications to Nonlinear Partial Differential Equations,* World Scientific Publishing, River Edge, NJ 1990.

H. Amman, M. Hieber and G. Simonett [1994] Bounded H^∞ calculus for elliptic operators, *Differential & Integral Equations* **7** (1994) 1–32.

S. Angenent [1998] The zero set of a solution of a parabolic equation, *J. Reine Angew. Math.* **390** (1988) 79–96.

H. A. Antosiewicz [1963] Linear control systems, *Archive Rat. Mech. Analysis* **12** (1963) 313–324.

K. E. Atkinson [1989] *An Introduction to Numerical Analysis,* Wiley, New York 1989.

J.-P. Aubin and I. Ekeland [1984] *Applied Nonlinear Analysis,* Wiley-Interscience, New York 1984.

J.-P. Aubin and H. Frankowska [1990] Controllability and observability of control systems with uncertainty, *Ann. Pol. Math.* **51** (1990) 37–76.

S. A. Avdonin and S. A. Ivanov [1995] *Families of Exponentials,* Cambridge University Press, New York 1995.

S. A. Avdonin and T. Seidman [1995] Identification of $q(x)$ in $u_t = \Delta u - qu$ from boundary observations, *SIAM J. Control & Optimization* **33** (1995) 1247–1255.

A. V. Balakrishnan

[1965]   Optimal control problems in Banach spaces, *SIAM J. Control* **3** (1965) 152–180.

[1968]   On a new computing technique in optimal control, *SIAM J. Control* **6** (1968) 149–173.

[1978]   Boundary control of parabolic equations; LQR theory, *Abh. Akad. Wiss. DDR Abt. Math. Naturwiss. Tech.* Berlin (1978) 11–23.

J. M. Ball [1990] A version of the fundamental theorem for Young measures, Springer Lecture Notes in Physics, vol. 344 (1990) 207–215.

J. M. Ball, J. Marsden and M. Slemrod [1982] Controllability of distributed bilinear systems, *SIAM J. Control & Optimization* **33** (1982) 575–597.

S. Banach [1932] *Théorie des Opérations Linéaires,* Monografje Matematyczne I, Warszawa 1932.

N. V. Banichuk [1983] *Problems and Methods of Optimal Structural Design,* Plenum Press, New York 1983.

H. T. Banks and M. Q. Jacobs [1970] The optimization of trajectories of linear functional differential equations, *SIAM J. Control* **8** (1970) 461–488.

H. T. Banks and G. A. Kent [1972] Control of functional differential equations of retarded and neutral type to target sets in function space, *SIAM J. Control* **10** (1972) 567–591.

H. T. Banks and K. Kunisch

[1984]   The linear regulator problem for parabolic systems, *SIAM J. Control & Optimization* **22** (1984) 684–699.

[1989]   *Estimation Techniques for Distributed Parameter Systems,* Systems and Control: Foundations & Applications I, Birkhäuser, Boston 1989.

V. Barbu

[1976]   Constrained control problems with convex cost in Hilbert space, *J. Math. Anal. Appl.* **56** (1976) 502–528.

[1984]   The time optimal control problem for parabolic variational inequalities, *Appl. Math. & Optimization* **11** (1984) 1–22.

[1987]   The time optimal problem for a class of nonlinear distributed systems, Springer Lecture Notes in Control and Information Sciences, vol. 97 (1987) 16–39.

[1993]   *Analysis and Control of Nonlinear Infinite Dimensional Systems,* Academic Press, San Diego 1993.

[1997]   The time optimal control of the Navier-Stokes equations, *Systems and Control Letters* **30** (1997) 93–100.

[1998]   Optimal control of Navier-Stokes equations with periodic inputs, *Nonlinear Analysis* **31** (1998) 15–31.

V. Barbu, E. N. Barron and R. Jensen [1988] The necessary conditions for optimal control in Hilbert spaces, *J. Math. Anal. Appl.* **133** (1988) 151–162.

V. Barbu and G. Da Prato [1983] *Hamilton-Jacobi equations in Hilbert Spaces,* Pitman Research Notes in Mathematics Series, vol. 86, Longman, Harlow 1983.

V. Barbu and K. Kunisch [1996] Identification of nonlinear elliptic equations, *Appl. Math. & Optimization* **33** (1996) 139–167.

V. Barbu and N. H. Pavel [1993] Optimal control problems with two-point boundary conditions, *J. Optimization Theory & Appl.* **77** (1993) 51–78.

V. Barbu and T. Precupanu [1986] *Convexity and Optimization,* Editura Academiei, Bucureşti, D. Reidel, Boston, 1986.

C. Bardos, C. G. Lebeau and J. Rauch [1992] Sharp sufficient conditions for the observation, control and stabilization of waves from the boundary, *SIAM J. Control & Optimization* **30** (1990) 1024–1065.

S. Barnett and R. G. Cameron [1984] *Introduction to Mathematical Control Theory,* The Clarendon Press, Oxford 1984.

R. G. Bartle [1956] A general bilinear vector integral, *Studia Math.* **15** (1956) 337–352.

N. Basile and M. Mininni [1990] An extension of the maximum principle for a class of optimal control problems in infinite-dimensional spaces, *SIAM J. Control & Optimization* **28** (1990) 1113–1135.

R. E. Bellman [1957] *Dynamic Programming,* Princeton University Press, Princeton 1957.

R. E. Bellman, I. Glicksberg and O. A. Gross [1956] On the "bang-bang" control problem, *Quart. Appl. Math.* **14** (1956) 11–18.

R. E. Bellman and R. Kalaba [1964] *Selected Papers on Control Theory,* Dover, New York 1964.

A. Bensoussan, G. Da Prato, M. C. Delfour and S. K. Mitter [1992/3] *Representation and Control of Infinite Dimensional Systems,* vol. I (1992), vol. II (1993) Birkhäuser, Basel.

L. D. Berkowitz [1974] *Optimal Control Theory,* Springer, Berlin 1974.

V. Bernstein [1933] *Leçons sur les Progrés Récents de la Théorie des Series de Dirichlet,* Gauthier-Villars, Paris 1933.

V. I. Blagodatski [1984] The maximum principle for differential inclusions, *Proc. Steklov Inst. Mat.* **166** (1984) 23–43.

V. I. Blagodatski and A. F. Filippov [1985] Differential inclusions and optimal control, *Proc. Steklov Inst. Mat.* **169** (1985) 194–252.

G. A. Bliss
[1925] *Calculus of Variations,* Carus Math. Monographs, vol. 1, Chicago 1925.
[1946] *Lectures on the Calculus of Variations,* Univ. Chicago Press Chicago, 1946.

R. P. Boas [1954] *Entire Functions,* Academic Press, New York 1954.

V. G. Boltyanski [1966] Sufficient conditions for optimality and the justification of the dynamic programming principle, *SIAM J. Control* **4** (1966) 326–361.

V. G. Boltyanski, R. V. Gamkrelidze and L. S. Pontryagin [1956] On the theory of optimal processes, *Dokl. Akad. Nauk. SSSR* **110** (1956) 7–10.

O. Bolza [1914] Über Variationsprobleme mit Ungleichungen als Nebenbedingungen, *Math. Abhandlungen* **23** (1914) 1–18.

J. F. Bonnans and E. Casas
[1989] Optimal control of semilinear multistate systems with state constraints, *SIAM J. Control & Optimization* **27** (1989) 446–455.
[1992] A boundary Pontryagin's principle for the optimal control of state constrained elliptic systems, International Series of Numerical Mathematics, vol. 107, Birkhäuser (1992) 241–249.

J. F. Bonnans and R. Cominetti [1996] Perturbed optimization in Banach spaces I, II, *SIAM J. Control & Optimization* **34** (1996) 1151–1171, 1172–1189.

N. Bourbaki [1966] *General Topology* (part 2), Addison-Wesley, Reading, 1966.

J. Bourgain [1983] Some remarks on Banach spaces in which martingale difference sequences are unconditional, *Ark. Mat.* **21** (1983) 163–168.

R. Brockett [1973] Lie algebras and Lie groups in control theory, *Geometric Methods in Control Theory,* Reidel, Boston (1973) 43–82.

M. Brokate and A. Friedman [1989] Optimal design for heat conduction problems with hysteresis, *SIAM J. Control & Optimization* **27** (1989) 697–717.

M. Brokate and J. Sprekels [1989] Existence and optimal control of mechanical processes with hysteresis in viscous solids, *IMA J. Applied Math.* **43** (1989) 219–229.

M. A. Brutyan and P. L. Krapivski [1984] On the optimal control of incompressible fluid flow, *Prikl. Math. Mekh.* **48** (1984) 678–682.

A. E. Bryson and Y.-C. Ho [1969] *Applied Optimal Control,* Blaisdell, Waltham 1969.

D. L. Burkholder [1983] A geometric condition that implies the existence of a certain singular integral of Banach space valued functions, *Wadsworth Math. Series,* Wadsworth (1983) 270–286.

A. G. Butkovski [1969] *Theory of Optimal Control of Distributed Parameter Systems,* Izd. Nauka, Moscow 1965. English translation: Elsevier, New York 1969.

A. G. Butkovski and L. M. Pustylnikov [1980] *The Theory of Moving Controls for Systems with Distributed Parameters,* Nauka, Moscow 1980.

P. Butzer and H. Berens [1967] *Semi-Groups of Operators and Approximation,* Springer, Berlin-Heidelberg, 1967.

A. P. Calderón [1964] Intermediate spaces and interpolation, the complex method, *Studia Math.* **26** (1964) 133–190.

F. Camilli and M. Falcone [1996] Approximation of optimal control problems with state constraints; estimates and applications, *Nonsmooth Analysis and Geometric Methods in Deterministic Optimal Control,* IMA Volumes in Mathematics and its Applications, vol. 78, Springer (1996) 23–57.

P. Cannarsa and G. Da Prato [1990] Some results on nonlinear optimal control problems and Hamilton-Jacobi equations in infinite dimensions, *J. Functional Analysis* **90** (1990) 27–47.

P. Cannarsa and H. Frankowska

[1992]   Value function and optimality conditions for semilinear control problems, *Appl. Math. & Optimization* **26** (1992) 139–169.

[1996]   Value function and optimality conditions for semilinear control problems, II: parabolic case, *Appl. Math. & Optimization* **33** (1996) 1–33.

P. Cannarsa, F. Gozzi and H. Mete Soner [1993] A dynamic programming approach to nonlinear boundary control problems of parabolic type, *J. Functional Analysis* **117** (1993) 25–61.

P. Cannarsa and M. E. Tessitore [1994] Optimality conditions for boundary control problems of parabolic type, International Series of Numerical Mathematics, vol. 118, Birkhäuser (1994) 79–96.

C. Carathéodory [1927] *Vorlesungen Über Reelle Funktionen,* Teubner, Leipzig 1927.

O. Cârjă

[1979]   Local controllability of nonlinear evolution equations in Banach spaces, *An. Stiintifice ale Univ. "Al. I. Cuza" Iasi* **25** (1979) 117–125.

[1984]   The time optimal problem for boundary-distributed control systems, *Bull. Unione Mat. Italiana* **3-B** (1984) 563–581.

[1988]   On constraint controllability of linear systems in Banach spaces, *J. Optimization Theory & Applications* **56** (1988) 215–225.

[1989]   Range inclusion for convex processes on Banach spaces; applications in controllability *Proc. Amer. Math. Soc.* **105** (1989) 185–191.

E. Casas

[1992]   Optimal control in coefficients of elliptic equations with state constraints, *Appl. Math. & Optimization* **26** (1992) 21–37.

[1993]   Boundary control of semilinear elliptic equations with pointwise state constraints, *SIAM J. Control & Optimization* **31** (1993) 993–1006.

[1994]   Pontryagin's principle for optimal control problems governed by semilinear elliptic equations, International Series of Numerical Mathematics, vol. 118, Birkhäuser (1994) 97–114.

[1996]   Boundary control problems for quasi-linear elliptic equations: a Pontryagin's principle, *Appl. Math. Optimization* **33** (1996) 265–291.

[1994]   Optimality conditions for some control problems of turbulent flows, *Flow Control,* IMA Volumes in Mathematics and its Applications, vol. 68, Springer (1994) 127–147.

E. Casas and L. A. Fernández

[1993:1] Optimal control of semilinear elliptic equations with pointwise constraints on the gradient of the state, *Appl. Math. & Optimization* **27** (1993) 35–56. Corrigendum, *Appl. Math. & Optimization* **28** (1993) 337–339.

[1993:2] Distributed control of systems governed by a general class of quasilinear elliptic equations, *J. Diff. Equations* **104** (1993) 20–47.

E. Casas, L. A. Fernández and J. Yong [1994] Optimal control of quasilinear parabolic equations, Springer Lecture Notes in Control and Information Sciences, vol. 197 (1994) 23–49.

E. Casas and J. Yong [1995] Maximum principle for state-constrained optimal control problems governed by quasilinear elliptic equations, *Differential & Integral Equations* **8** (1995) 1–18.

C. Castaing and M. Valadier [1977] *Convex Analysis and Measurable Multifunctions,* Springer Lecture Notes in Mathematics, vol. 580, Berlin 1977.

L. Cesari

[1969]   Optimization with partial differential equations and conjugate problems, *Archive Rat. Mech. Analysis* **33** (1969) 339–357.

[1970]   Existence theory for abstract multidimensional control problems, *J. Optimization Theory & Appl.* **6** (1970) 210–236.

[1983]   *Optimization Theory and Applications,* Springer, New York, 1983.

G. Chen [1979] Control and stabilization for the wave equation in a bounded domain, *SIAM J. Control & Optimization* **17** (1979) 66–81.

G. Chen and J. Zhou [1993] *Vibration and Damping in Distributed Systems,* Vols. I, II, CRC Press, Boca Raton 1993.

W. Chewning

[1976:1] Controllability of the nonlinear wave equation in several space variables, *SIAM J. Control & Optimization* **14** (1976) 19–25.

[1976:2] Control of nonlinear parabolic equations to meet finitely many terminal conditions, *Jour. Math. Anal. Appl.* **56** (1976) 185–194.

M. A. Cirinà [1969] Boundary controllability of nonlinear hyperbolic systems, *SIAM J. Control* **7** (1969) 198–212.

F. Clarke

[1976:1] The maximum principle under minimum hypotheses, *SIAM J. Control & Optimization* **14** (1976) 1078–1091.

[1976:2] Necessary conditions for a general control problem, *Calculus of Variations and Optimal Control Theory* (D. L. Russell, Ed.) Academic Press, New York 1976.

[1983] *Optimization and Nonsmooth Analysis,* Wiley-Interscience, New York 1983.

Ph. Clément, O. Diekmann, M. Gyllenberg, H. J. A. M. Heijmans and H. R. Thieme

[1987] Perturbation theory for dual semigroups I. The sun-reflexive case, *Math. Ann.* **277** (1987) 709–725.

[1988] Perturbation theory for dual semigroups II. Time-dependent perturbations in the sun-reflexive case, *Proc. Royal Soc. Edinburgh* **109A** (1988) 145–172.

[1989] Perturbation theory for dual semigroups III. Nonlinear Lipschitz continuous perturbations in the sun-reflexive case, Pitman Research Notes in Mathematics Series, vol. 190, Longmans (1989) 67–89.

E. A. Coddington and N. Levinson [1983] *Theory of Ordinary Differential Equations,* McGraw-Hill, New York 1983.

F. Colonius

[1982] The maximum principle for relaxed hereditary differential systems with function space end condition, *SIAM J. Control & Optimization* **20** (1982) 695–712.

[1988] *Optimal Periodic Control,* Springer Lecture Notes in Mathematics, vol. 1313, Berlin 1988.

F. Colonius and D. Hinrichsen [1978] Optimal control of functional differential systems, *SIAM J. Control & Optimization* **16** (1978) 861–879.

F. Colonius and K. Kunisch [1993] Sensitivity analysis for optimization problems in Hilbert spaces with bilateral constraints, *J. Math. Systems, Estimation & Control* **3** (1993) 265–299.

R. Conti

[1968] Time-optimal solution of a linear evolution equation in Banach spaces, *J. Optimization Theory & Appl.* **2** (1968) 277–284.

[1982] *Infinite Dimensional Linear Autonomous Controllability,* University of Minnesota Mathematics Report 82-127, 1982.

R. F. Curtain and A. J. Pritchard

[1974] The infinite dimensional Riccati equation, *J. Math. Anal. Appl.* **47** (1974) 43–56.

[1976] The infinite-dimensional Riccati equation for systems defined by evolution operators, *SIAM J. Control & Optimization* **14** (1976) 951–983.

[1977] An abstract theory for unbounded control action for distributed parameter systems, *SIAM J. Control & Optimization* **15** (1977) 566–611.

[1978] *Infinite Dimensional Linear Systems Theory,* Springer Lecture Notes in Control and Information Sciences, vol. 8, Heidelberg 1978.

R. Curtain and H. Zwart [1995] *An introduction to Infinite-Dimensional Linear Systems Theory,* Springer, New York 1995.

G. Da Prato and A. Ichikawa [1985] Riccati equations with unbounded coefficients, *Ann. Mat. Pura Appl.* **140** (1985) 209–221.

G. Da Prato, I. Lasiecka and R. Triggiani [1986] A direct study of the Riccati equation arising in hyperbolic boundary control problems, *J. Differential Equations* **64** (1986) 26–47.

R. Datko

[1970]   Measurability properties of set-valued mappings in a Banach space, *SIAM J. Control* **8** (1970) 226–238.

[1971]   A linear control problem in an abstract Hilbert space, *J. Diff. Equations* **9** (1971) 346–359.

[1972]   Uniform asymptotic stability of evolutionary processes in a Banach space, *SIAM J. Math. Anal.* **3** (1972) 428–445.

[1974]   Unconstrained control problems with quadratic cost, *SIAM J. Control & Optimization* 11 (1973) 32–52.

[1988]   Not all feedback stabilized hyperbolic systems are robust with respect to small time delays in their feedback, *SIAM J. Control & Optimization* **26** (1988) 697–713.

R. Datko and Y. You [1991] Some second-order vibrating systems cannot tolerate small time delays in their damping, *J. Optimization Theory & Appl.* **70** (1991) 521–537.

M. Delfour

[1977]   The linear quadratic optimal control problem for hereditary differential systems: theory and numerical solution, *Applied Math. & Optimization* **3** (1977) 101–162.

[1986]   The linear quadratic optimal control problem with delays in state and control variables, *SIAM J. Control & Optimization* **24** (1986) 835–883.

W. Desch, I. Lasiecka and W. Schappacher [1985] Feedback boundary control problems for linear semigroups, *Israel Jour. Math.* **51** (1985) 177–201.

W. Desch and R. L. Wheeler [1989] Destabilization due to delay in one dimensional feedback, International Series of Numerical Mathematics, vol. 91, Birkhäuser (1989) 61–83.

L. De Simon [1964] Un'applicazione della teoria degli integrali singolari allo studio delle equazioni differenziali astratte del primo ordine, *Rend Sem. Mat. Univ. Padova* **34** (1964) 205–223.

J. Diestel and J. J. Uhl [1977] *Vector Measures,* Amer. Math. Soc., Providence 1977.

J. Dieudonné.

[1947/8] Sur le théoreme de Lebesgue-Nikodym III, *Ann. Université Grenoble* **23** (1947/8) 25–53.

[1960]   *Foundations of Modern Analysis,* Academic Press, New York 1960.

R. J. DiPerna [1985] Measure-valued solutions to conservation laws, *Archive Rat. Mech. Analysis* **88** (1985) 223–270.

G. Dore and A. Venni [1987] On the closedness of the sum of two closed operators, *Math. Zeitschrift* **196** (1987) 189–201.

N. Dunford and B. J. Pettis [1940] Linear operations on summable functions, *Trans. Amer. Math. Soc.* **47** (1940) 323–392.

N. Dunford and J. T. Schwartz

[1958]   *Linear Operators,* part I, Interscience, New York 1958.

[1963]   *Linear Operators,* part II, Interscience, New York 1963.

N. V. Efimov and E. R. Rozendorf [1975] *Linear Algebra and Multi-Dimensional Geometry,* Mir Publishers, Moscow 1975.

A. I. Egorov [1967] Necessary optimality conditions for distributed parameter systems, *SIAM J. Control* **5** (1967) 352–408.

Yu. B. Egorov

[1962]   Certain problems in the theory of optimal control, *Dokl. Akad. Nauk SSSR* **145** (1962) 122–125.

[1963:1] Optimal control in Banach spaces *Dokl. Akad. Nauk SSSR* **150** (1963) 241–244.

[1963:2] Certain problems in the theory of optimal control, *Z. Vycisl. Mat. Fiz.* **5** (1963) 887–904.

[1964]   Some necessary conditions for optimality in Banach spaces, *Mat. Sbornik* **64** (1964) 79–101.

I. Ekeland

[1974]   On the variational principle, *J. Math. Anal. Appl.* **47** (1974) 324–353.

[1979]   Nonconvex minimization problems, *Bull. Amer. Math. Soc.* **1** (NS) (1979) 443–474.

L. Elsgolts [1970] *Differential Equations and the Calculus of Variations,* Mir Publishers, Moscow 1970.

P. Falb [1964] Infinite dimensional control problems I: On the closure of the set of attainable states for linear systems, *J. Math. Anal. Appl.* **9** (1964) 12–22.

M. Falcone [1994] Discrete time high-order schemes for viscosity solutions of the Hamilton-Jacobi equation, *Numer. Math.* **67** (1994) 315–344.

H. O. Fattorini

[1964] Time-optimal control of solutions of operational differential equations, *SIAM J. Control* **2** (1964) 54–59.

[1966:1] Control in finite time of differential equations in Banach space, *Comm. Pure Appl. Math.* **19** (1966) 17–34.

[1966:2] Some remarks on complete controllability, *SIAM J. Control* **4** (1966) 686–694.

[1966/7] On Jordan operators and rigidity of linear control systems, *Rev. Un. Mat. Argentina* **23** (1966/7) 67–75.

[1967:2] On complete controllability of linear systems, *J. Differential Equations* **3** (1967) 391–402.

[1967:3] Controllability of higher order linear systems, Mathematical Theory of Control (A. V. Balakrishnan and L. W. Neustadt, Eds.) Academic Press (1967) 301–311.

[1968:1] A remark on the "bang-bang" principle for linear control systems in infinite dimensional spaces, *SIAM J. Control* **6** (1968) 109–113.

[1968:2] Boundary control systems, *SIAM J. Control* **6** (1968) 349–385.

[1968:3] An observation on a paper of A. Friedman, *J. Math. Anal. Appl.* **22** (1968) 382–384.

[1969] Control with bounded inputs, Springer Lecture Notes in Economics and Mathematical Systems, vol. 14 (1969) 92–100.

[1974/5] The time optimal control problem in Banach spaces, *Appl. Math. & Optimization* **1** (1974/5) 163–188.

[1975:1] Boundary control of temperature distributions in a parallelepipedon, *SIAM J. Control* **13** (1975) 1–13.

[1975:2] Local controllability of a nonlinear wave equation, *Math. Systems Theory* **9** (1975) 30–45.

[1975:3] Exact controllability of linear systems in infinite dimensional spaces, Springer Lecture Notes in Mathematics, vol. 446 (1975) 166–183.

[1976] The time-optimal problem for boundary control of the heat equation, *Calculus of Variations and Control Theory* (D. L. Russell, Ed.) Academic Press (1976) 305–320.

[1977:1] The time optimal problem for distributed control of systems described by the wave equation, *Control Theory of Systems Governed by Partial Differential Equations* (A. K. Aziz, J. W. Wingate, M. J. Balas, Eds.) Academic Press (1977) 151–175.

[1977:2] Estimates for sequences biorthogonal to certain complex exponentials and boundary control of the wave equation, Springer Lecture Notes in Control and Information Sciences, vol. 2 (1977) 111–124.

[1978] Reachable states in boundary control of the heat equation are independent of time, *Proc. Royal Soc. Edinburgh* **81A** (1978) 71–77.

[1983] *The Cauchy Problem*, Encyclopedia of Mathematics and its Applications vol. 18, Addison-Wesley, Reading, 1983.

[1985] *Second Order Linear Differential Equations in Banach Spaces,* North-Holland Mathematical Studies vol. 108 (Notas de Matematica vol. 99), Elsevier-North Holland, Amsterdam 1985.

[1986] The maximum principle for nonlinear nonconvex systems in infinite dimensional spaces, Springer Lecture Notes in Control and Information Sciences, vol. 75 (1986) 162–178.

[1987:1] A unified theory of necessary conditions for nonlinear nonconvex control systems, *Appl. Math. Optimization* **15** (1987) 141–185.

[1987:2] Convergence of suboptimal controls for point targets, International Series of Numerical Mathematics, vol. 78, Birkhäuser (1987) 91–107.

[1987:3] Optimal control of nonlinear systems: convergence of suboptimal controls I, Lecture Notes in Applied Mathematics, vol. 108, Marcel Dekker (1987) 159–199.

[1987:4] Optimal control of nonlinear systems: convergence of suboptimal controls II, Springer Lecture Notes in Control and Information Sciences, vol. 97 (1987) 230–246.

[1987:5] Some remarks on convergence of suboptimal controls, *A. V. Balakrishnan 60th. Anniversary Volume*, Software Optimization (1987) 359–363.

[1989] Convergence of suboptimal elements in infinite dimensional nonlinear programming problems, Springer Lecture Notes in Control and Information Sciences, vol. 114 (1989) 23–34.

[1990:1] Constancy of the Hamiltonian in infinite dimensional control problems, International Series of Numerical Mathematics, vol. 91, Birkhäuser (1990) 123–133.

[1990:2] Convergence of suboptimal controls: the point target case, *SIAM J. Control & Optim.* **28** (1990) 320–341.

[1990:3] Some remarks on Pontryagin's maximum principle for infinite dimensional control problems, *Perspectives on Control Theory* (B. Jakubczyk, K. Malanowski, W. Respondek, Eds.) Progress in Systems and Control, Birkhäuser (1990) 12–25.

[1991:1] Optimal control problems for distributed parameter systems governed by semilinear parabolic equations in L^1 and L^∞ spaces, Springer Lecture Notes in Control and Information Sciences, vol. 149 (1991) 68–80.

[1991:2] Relaxed controls in infinite dimensional control systems, International Series of Numerical Mathematics, vol. 100, Birkhäuser (1991) 115–128.

[1993:1] Optimal control problems for distributed parameter systems in Banach spaces, *Applied Math. & Optimization* **28** (1993) 225–257.

[1993:2] Relaxation in infinite dimensional control systems, *L. Markus Festschrift Volume* (K. D. Elworthy, W. N. Everitt, E. B. Lee, Eds.) Lecture Notes in Pure and Applied Mathematics, vol. 152, Marcel Dekker (1993) 505–522.

[1993:3] Relaxed controls, differential inclusions, existence theorems and the maximum principle in nonlinear infinite dimensional control theory, Lecture Notes in Pure and Applied Mathematics, vol. 152, Marcel Dekker (1993) 177–195.

[1994:1] Existence theory and the maximum principle for relaxed infinite dimensional optimal control problems, *SIAM J. Control & Optimization* **32** (1994) 311–331.

[1994:2] Relaxation theorems, differential inclusions and Filippov's theorem for relaxed controls in semilinear infinite dimensional systems, *Jour. Diff. Equations* **112** (1994) 131–153.

[1994:3] Invariance of the Hamiltonian in control problems for semilinear parabolic distributed parameter systems, International Series of Numerical Mathematics vol. 118, Birkhäuser (1994) 115–130.

[1995] The maximum principle for linear infinite dimensional control systems with state constraints, *Discrete & Continuous Dynamical Systems* **1** (1995) 77–101.

[1996] Optimal control problems with state constraints for semilinear distributed parameter systems, *Jour. Optimization Theory & Appl.* **88** (1996) 25–59.

[1997:1] A remark on existence of solutions of infinite dimensional noncompact optimal control problems, *SIAM J. Control & Optimization* **35** (1997) 1422–1433.

[1997:2] Nonlinear infinite dimensional problems with state constraints and unbounded control sets, *Rend. Ist. Mat. Univ. Trieste Suppl.* **28** (1997) 127–146.

[1997:3] Robustness and convergence of suboptimal controls in distributed parameter systems, *Proc. Royal Soc. Edinburgh.* **127A** 1153–1179.

[i.p.] Control problems for parabolic equations with state constraints and unbounded control sets, to appear in International Series of Numerical Mathematics, Birkhäuser.

H. O. Fattorini and H. Frankowska

[1988] Necessary conditions for infinite dimensional control problems, Springer Lecture Notes in Control and Information Sciences, vol. 111 (1988) 381–392.

[1989] Explicit estimates for convergence of suboptimal controls, Lecture Notes in Pure and Applied Mathematics, vol. 119, Marcel Dekker (1989) 71–87.

[1990:1] Explicit convergence estimates for suboptimal controls I, *Problems in Control & Information Theory* **19** (1990) 3–29.

[1990:2] Explicit convergence estimates for suboptimal controls II, *Problems in Control & Information Theory* **19** (1990) 69–93.

[1991:1] Necessary conditions for infinite dimensional control problems, *Math. Control Signals & Systems* **4** (1991) 41–67.

[1991:2] Infinite dimensional control problems with state constraints, Springer Lecture Notes in Control and Information Sciences, vol. 154 (1991) 52–62.

H. O. Fattorini and T. Murphy

[1994:1] Optimal control problems for nonlinear parabolic boundary control systems: the Dirichlet boundary condition, *Differential & Integral Equations* **7** (1994) 1367–1388.

[1994:2] Optimal boundary control of nonlinear parabolic equations, Lecture Notes in Pure and Applied Mathematics, vol. 160, Marcel Dekker (1994) 91–109.

[1994:3] Optimal problems for nonlinear parabolic boundary control systems, *SIAM J. Control & Optimization* **32** (1994) 1578–1596.

H. O. Fattorini and D. L. Russell

[1971] Exact controllability theorems for linear parabolic equations in one space dimension, *Arch. Rat. Mech. Analysis* **43** (1971) 272–292.

[1974/5] Uniform bounds on bi-orthogonal functions for real exponentials with an application to the control theory of parabolic equations, *Quart. Appl. Math.* **32** (1974/5) 45–69.

H. O. Fattorini and S. S. Sritharan

[1992] Existence of optimal controls for viscous flow problems, *Proc. Royal Soc. London* **439A** (1992) 81–102.

[1994] Necessary and sufficient conditions for optimal control in viscous flow problems, *Proc. Royal Soc. Edinburgh* **124A** (1994) 211–251.

[1995:1] Optimal chattering control for fluid flow, *Nonlinear Analysis* **25** (1995) 763–797.

[1995:2] Relaxation in semilinear infinite dimensional systems modelling fluid flow control problems, *Control and Optimal Design of Distributed Parameter Systems,* IMA Volumes in Mathematics and its Applications, vol. 70, Springer (1995) 93–111.

[i.p.] Optimal control problems with state constraints in fluid mechanics and combustion, to appear in *Appl. Math. Optimization.*

A. F. Filippov [1959] On certain questions in the theory of optimal control, *Vestnik Moskov. Univ. Ser. Mat. Mech. Astronom.* **2** (1959) 25–32. English translation: *SIAM J. Control* **1** (1962) 76–84.

W. E. Fleming and R. W. Rishel [1975] *Deterministic and Stochastic Optimal Control,* Springer, Berlin 1975.

C. Foias and R. Temam [1980] Homogeneous statistical solutions of the Navier-Stokes equations, *Indiana Univ. Math. J.* **29** (1980) 913–957.

H. Frankowska

[1989] On the linearization of nonlinear control systems and exact reachability, Springer Lecture Notes in Control and Information Science, vol. 114 (1989) 132–143.

[1990:1] A priori estimates for operational differential inclusions, *Jour. Diff. Equations* **84** (1990) 100–128.

[1990:2] Some inverse mapping theorems, *Ann. Inst. Henri Poincaré* **7** (1990) 183–234.

[1993] Set-valued approach to the Hamilton-Jacobi equations, Progress in Systems and Control Theory, vol. 16, Birkhäuser (1993) 105–118.

H. Frankowska and Cz. Olech

[1982:1] *R*-convexity of the integral of set-valued functions, *Amer. Jour. Math.* (1982) 117–129.

[1982:2] Boundary solutions of differential inclusions, *Jour. Differential Equations* **44** (1982) 243–260.

A. Friedman

[1964] Optimal control for hereditary processes, *J. Math. Anal. Appl.* **15** (1964) 396–414.

[1967] Optimal control in Banach spaces, *J. Math. Anal. Appl.* **19** (1967) 35–55.

[1968] Optimal control in Banach spaces with fixed end-points, *J. Math. Anal. Appl.* **24** (1968) 161–181.

P. A. Fuhrmann [1981] *Linear Systems and Operators in Hilbert Space,* McGraw-Hill, New York, 1981.

N. Fujii and Y. Sakawa [1974] Controllability for nonlinear differential equations in Banach space, *Aut. Control Theory & Appl.* **2** (1974) 44–46.

H. Fujita and T. Kato

[1962] On the nonstationary Navier-Stokes system, *Rend. Sem. Mat. Univ. Padova* **32** (1962) 243–260.

[1964] On the Navier-Stokes initial value problem, I, *Arch. Rat. Mech. Anal.* **16** (1964) 269–315.

D. Fujiwara and H. Morimoto [1977] An L_r-theorem of the Helmholtz decomposition of vector fields, *J. Fac. Science Univ. Tokyo Sec.* I **24** (1977) 685–700.

A. T. Fuller [1963] Study of an optimum non-linear control system, *J. Electronics & Control* **15** (1963) 63–71.

A. V. Fursikov

[1992] Lagrange principle for problems of optimal control of ill-posed or singular distributed systems, *J. Math. Pures Appl.* **139** (1992) 139–195.

[1995]   Exact boundary zero controllability of the three dimensional Navier-Stokes equations, *J. Dynamics & Control Systems* **1** (1995) 325–350.

A. V. Fursikov and O. Yu. Imanuvilov

[1994]   On exact boundary zero-controllability of the two-dimensional Navier-Stokes equations, *Acta Appl. Math.* **37** (1994) 67–76.

[1996]   *Controllability of Evolution Equations,* Research Institute of Mathematics, Global Analysis Research Center, Seoul National University, 1996.

L. I. Galchuk [1968] Optimal control of systems described by parabolic equations, *Vest. Mosk. Univ. Mat. Mekh.* **3** (1968) 21–23. English translation: *SIAM J. Control* **7** (1969) 546–558.

R. Gamkrelidze

[1968]   On sliding optimal states, *Dokl. Akad. Nauk SSSR* **143** (1962) 1243–1245.

[1978]   *Principles of Optimal Control Theory,* Plenum Press, London 1978.

R. V. Gamkrelidze and G. L. Jaratishvili [1967] Extremal problems in linear topological spaces I, *Math. Systems Theory* **3** (1967) 229–256.

F. Gantmajer

[1954]   *Matrix Theory,* Gostejizdat, Moscow 1954. English translation (of some chapters): *Applications of the Theory of Matrices,* Interscience, New York 1959.

[1970]   *Lectures in Analytical Mechanics,* Mir Publishers, Moscow 1970.

P. Garabedian [1964] *Partial Differential Equations,* Wiley, New York 1964.

L. Gårding [1953] On the asymptotic distribution of the eigenvalues of elliptic differential operators, *Math. Scand.* **1** (1953) 237–255.

N. Garofalo and F.-H. Lin [1987] Unique continuation for elliptic operators: a geometric-variational approach, *Comm. Pure Appl. Math.* **40** (1987) 347–366.

I. M. Gelfand and S. V. Fomin [1963] *Calculus of Variations,* Prentice-Hall, Englewood Cliffs, 1963.

J. S. Gibson

[1979]   The Riccati integral equations for optimal control problems in Hilbert spaces, *SIAM J. Control & Optimization* **17** (1979) 537–565.

[1980]   A note on stabilization of infinite dimensional linear oscillators by compact linear feedback, *SIAM J. Control & Optimization* **18** (1980) 311–316.

[1983]   Linear-quadratic optimal control of hereditary differential systems: infinite dimensional Riccati equations and numerical approximations, *SIAM J. Control & Optimization* **21** (1983) 95–139.

Y. Giga

[1981]   Analyticity of the semigroup generated by the Stokes operator in L_r spaces, *Math. Zeitschrift* **178** (1981) 297–328.

[1985]   Domains of fractional powers of the Stokes operator in L_r, *Arch. Rat. Mech. Analysis* **89** (1985) 251–265.

Y. Giga and T. Miyakawa [1985] Solutions in L^r of the Navier-Stokes initial value problem, *Archive Rat. Mech. Analysis* **89** (1985) 267–281.

B. Ginzburg and A. Ioffe [1993] The maximum principle in optimal control of systems governed by semilinear equations, *Nonsmooth Analysis and Geometric Methods in Deterministic Optimal Control,* IMA Volumes in Mathematics and its Applications, vol. 78, Springer (1995) 93–111.

E. Giusti [1967] Funzioni coseno periodiche, *Bull. Un. Mat. Italiana* **22** (1967) 478–485.

K. Glashoff and N. Weck [1976] Boundary control of parabolic differential equations in arbitrary dimensions: supremum-norm problems, *SIAM J. Control & Optimization* **14** (1976) 662–681.

A. N. Godunov [1974] Peano's theorem in infinite-dimensional Hilbert space is false, even in a weakened formulation, *Math. Zametki* **15** (1974) 467–477.

C. Goffman and G. Pedrick [1983] *First Course in Functional Analysis,* Chelsea, New York 1983.

H. Goldberg and F. Tröltzsch [1993] Second-order sufficient optimality conditions for a class of nonlinear parabolic boundary control problems, *SIAM J. Control & Optimization* **31** (1993) 1007–1025.

H. H. Goldstine [1980] *A History of the Calculus of Variations,* Springer, New York 1980.

E. Goursat [1942] *Course d'Analyse Mathematique,* Gauthier-Villars, Paris 1942.

J. Gregory and C. Lin [1992] *Constrained optimization in the calculus of variations and optimal control theory*, Van Nostrand-Reinhold, New York 1992.

E. Güichal [1985] A lower bound on the norm of the control operator for the heat equation, *Jour. Math. Anal. Appl.* **110** (1985) 519–527.

M. Gunzburger, L. Hou and T. Sbovodny [1992] Boundary velocity control of incompressible flow with an application to viscous drag reduction, *SIAM J. Control & Optimization* **30** (1992) 167–181.

R. Haberman [1977] *Mathematical Models*, Prentice-Hall, Englewood Cliffs 1977.

A. Halanay [1968] Optimal control systems for systems with time lag, *SIAM J. Control* **6** (1968) 215–234.

H. Halkin [1966] Optimal control as programming in infinite dimensional spaces, Centro Internazionale Matematico Estivo (CIME), Bressanone (1966) 23–37.

P. R. Halmos [1957] *Introduction to Hilbert Space and the Theory of Spectral Multiplicity*, Chelsea, New York, 1957.

Q. Han and F.-H. Lin
[1991:1] Nodal sets of solutions of elliptic and parabolic equations II, *Comm. Pure App. Math.* **47** (1994) 1219–1238.
[1991:2] On the geometric measure of nodal sets, *Jour. Partial Differential Equations* **7** (1994) 111–131.

J. Haslinger and P. Neittanmäki [1988] *Finite Element Approximation for Optimal Shape Design*, Wiley, Chichester, 1988.

Z.-X. He [1987] State constrained control problems governed by variational inequalities, *SIAM J. Control & Optimization* **25** (1987) 1119–1144.

D. Henry [1981] *Geometric Theory of Semilinear Parabolic Equations*, Springer, Berlin 1981.

H. Hermes [1964] A note on the range of a vector measure: application to the theory of optimal control, *J. Math. Anal. Appl.* **8** (1964) 78–83.

H. Hermes and J. La Salle [1969] *Functional Analysis and Time-Optimal Control*, Academic Press, New York 1969.

M. Hestenes
[1950] A general problem in the calculus of variations with applications to paths of least time, Rand Corporation RM-100 (1950) Astia Document AD 112382.
[1966] *Calculus of Variations and Optimal Control Theory*, Wiley, New York, 1966.

D. Hilbert [1904] Über das Dirichletsche Prinzip, *Math. Ann.* **59** (1904) 161–186.

E. Hille and R. S. Phillips [1957] *Functional Analysis and Semi-Groups*, Amer. Math. Soc., Providence, 1957.

C. J. Himmelberg, M. Q. Jacobs and F. S. Van Vleck [1969] Measurable multifunctions, selectors, and Filippov's implicit function lemma, *J. Math. Anal. Appl.* **25** (1969) 276–284.

L. M. Hocking [1991] *Optimal Control: an Introduction to the Theory with Applications*, The Clarendon Press, Oxford 1991.

K. Hoffman [1962] *Banach Spaces of Analytic Functions*, Prentice-Hall, Englewood Cliffs, 1962.

K.- H. Hoffmann and T. Roubíček [1995] About the concept of measure-valued solutions to distributed parameter systems, *Math. Methods Appl. Science* **18** (1995) 671–685.

K.-H. Hoffmann and J. Sprekels [1986] On the identification of parameters in general variational inequalities by asymptotic regularization, *SIAM J. Math. Anal.* **17** (1986) 1198–1217.

L. Hou and T. Sbovodny [1991] Optimization problems for the Navier-Stokes equations with regular boundary controls, *J. Math. Anal. Appl.* **177** (1991) 342–367.

B. Hu and J. Yong [1995] Pontryagin maximum principle for semilinear and quasilinear parabolic equations with pointwise state constraints, *SIAM J. Control & Optimization* **33** (1995) 1857–1880.

A. Hurwitz [1895] Über die Bedingungen unter Welchen eine gleichung nur Wurzeln mit negativen reelen Teilen besitzt, *Math. Ann.* **46** (1895) 273–384.

A. Ioffe
[1984] Necessary conditions in nonsmooth optimization, *Math. Operations Research* **9** (1984) 159–188.
[1994] On sensitivity analysis of nonlinear programs in Banach spaces; the approach via composite unconstrained optimization, *SIAM J. Control & Optimization* **4** (1994) 1–43.

A. D. Ioffe and V. M. Tijomirov [1979] *Theory of Extremal Problems*, North-Holland, Amsterdam 1979.

A. Ionescu Tulcea and C. Ionescu Tulcea [1969] *Topics in the Theory of Lifting*, Springer, Heidelberg 1969.

M. Jacobs [1967] Remarks on some recent extensions of Filippov's implicit function lemma, *SIAM J. Control* 5 (1967) 622–627.

G. L. Jaratishvili [1961] The maximum principle in the theory of optimal processes with a delay, *Dokl. Akad. Nauk SSSR* 136 (1961) 39–42.

D. Jerison and C. E. Kenig [1985] Unique continuation and absence of positive eigenvalues for Schrödinger operators, *Annals of Math.* 121 (1985) 463–488.

D. W. Jordan and P. Smith [1987] *Nonlinear Ordinary Differential Equations*, The Clarendon Press, Oxford 1987.

V. Jurdjevic and H. Sussmann [1972] Control systems on Lie groups, *J. Diff. Equations* (1972) 470–486.

S. Kaczmarz and H. Steinhaus [1935] *Theorie der Orthogonalreihen*, Monografje Matematyczne VI, Warszawa-Lwów 1935.

R. E. Kalman [1960] New results in linear filtering theory, *Bol. Soc. Mat. Mexicana* 12 (1960) 102–119.

R. E. Kalman and R. S. Bucy [1961] New results in linear prediction and filtering theory, *J. Basic. Engineering (Trans ASME, Ser. D)* 83 (1961) 189–213.

R. E. Kalman, Y.-C. Ho and K. S. Narendra [1963] Controllability of linear dynamical systems, *Contributions to Differential Equations* 1 (1963) 189–213.

L. V. Kantorovich and G. P. Akilov [1964] *Functional Analysis in Normed Spaces*, Fizmatgiz, Moscow, 1959. English translation: Pergamon Press, New York, 1964.

F. Kappel [1991] Approximation of *LQR* problems for delay systems: a survey, Progress in Systems and Control Theory, vol. 11, Birkhäuser (1991) 187–224.

T. Kato [1995] *Perturbation Theory of Linear Operators*, Springer, New York 1995.

H. Keller [1968] *Numerical Methods for Two-point Boundary Value Problems*, Blaisdell, Boston, 1968.

J. L. Kelley [1955] *General Topology*, Van Nostrand, New York 1955.

A. Yu. Khapalov

[1995:1] Controllability of the wave equation with moving point control, *Appl. Math. & Optimization* 31 (1995) 156–175.

[1995:2] Exact observability of the time-varying hyperbolic equation with finitely many moving internal observations, *SIAM J. Control & Optimization* 33 (1995) 1256–1269.

J. Klamka [1993] Constrained controllability of a linear retarded dynamical system, *Appl. Math. & Computational Science* 7 (1993) 647–672.

G. Knowles

[1975] Lyapunov vector measures, *SIAM J. Control & Optimization* 13 (1975) 294–303.

[1976] Time optimal control of infinite dimensional systems, *SIAM J. Control & Optimization* 14 (1976) 919–933.

T. Kobayashi [1978] Some remarks on controllability of distributed parameter systems, *SIAM J. Control & Optimization* 16 (1978) 733–742.

A. N. Kolmogorov and S. V. Fomin [1970] *Introduction to Real Analysis*, Dover, New York 1970.

A. S. Kompaneyets [1978] *A Course of Theoretical Physics*, vols. 1, 2, MIR Publishers, Moscow 1978.

W. Krabs

[1985] On time-minimal distributed control of vibrating systems governed by an abstract wave equation, *Applied Math. Optimization* 3 (1985) 137–149.

[1989] On time-minimal distributed control of vibrations, *Applied Math. Optimization* 19 (1989) 65–73.

[1992] *On Moment Theory and Controllability of One-Dimensional Vibrating Systems and Heating Processes*, Springer Lecture Notes in Control and Information Sciences, vol. 173, Heidelberg 1992.

W. Krabs and E. J. P. G. Schmidt [1980] Time-minimal controllability of linear systems, *Math. Methods Operations Research*, Sofia (1980) 65–80.

S. G. Krein, Ju. I. Petunin and E. M. Semenov [1982] *Interpolation of Linear Operators,* Amer. Math. Soc., Providence 1982.

K. Kunisch [1982] Approximation schemes for the linear-quadratic optimal control problem associated with delay equations, *SIAM J. Control & Optimization* **20** (1982) 506–540.

L. M. Kuperman and J. M. Repin [1971] On controllability in infinite dimensional spaces, *Dokl. Akad. Nauk SSSR* **100** (1971) 767–769.

I. A. K. Kupka [1990] Fuller's phenomenon, Progress in Systems and Control Theory, vol. 2, Birkhäuser (1990) 129–142.

A. B. Kurzhanski

[1977]   *Control and Observation Under Conditions of Uncertainty,* Nauka, Moscow 1977.

[1994]   On the stabilization of uncertain differential systems, Lecture Notes in Pure and Applied Mathematics, vol. 162, Marcel Dekker (1994) 217–225.

A. B. Kurzhanski and A. Yu. Khapalov [1986] On the state estimation problem for distributed systems, Springer Lecture Notes in Control and Information Sciences, vol. 83 (1986) 102–113.

E. Landau [1927] *Vorlesungen über Zahlentheorie,* vol. I, Hirzel, Leipzig 1927.

J. P. LaSalle [1960] The time optimal control problem, *Contributions to the Theory of Nonlinear Oscillations* **5**, Princeton University Press (1960) 1–24.

I. Lasiecka [1980] Unified theory for abstract parabolic boundary problems — a semigroup approach, *Appl. Math. & Optimization* **6** (1980) 287–333.

I. Lasiecka and J. Sokolowski [1991] Sensitivity analysis of optimal control problems for wave equations, *SIAM J. Control & Optimization* **29** (1991) 1128–1149.

I. Lasiecka and R. Triggiani

[1987:1] The regulator problem for parabolic equations with Dirichlet boundary control, Part I: Riccati's feedback synthesis, *Appl. Math. & Optimization* **16** (1987) 147–168.

[1987:2] The regulator problem for parabolic equations with Dirichlet boundary control, Part II: Galerkin approximation, *Appl. Math. & Optimization* **16** (1987) 187–216.

[1991]   *Differential and Algebraic Riccati Equations with Applications to Boundary/Point Control Problems: Continuous Theory and Approximation Theory,* Springer Lecture Notes in Control and Information Sciences, vol. 164, Heidelberg 1991.

G. Lebeau and L. Robbiano [1995] Contrôle exact de l'équation de la chaleur, *Comm. Partial Differential Equations* **20** (1995) 335–356.

G. Lebeau and E. Zuazua [i.p.] Null controllability of a system of linear thermoelasticity.

U. Ledzewicz and A. Novakowski [1994] Necessary and sufficient conditions for optimality for nonlinear control problems in Banach spaces, Lecture Notes in Pure and Applied Mathematics, vol. 160, Marcel Dekker (1994) 195–216.

E. B. Lee and L. Markus

[1961]   Optimal control for nonlinear processes, *Arch. Rat. Mech. Anal.* **8** (1961) 36–58.

[1967]   *Foundations of Optimal Control Theory,* Wiley, New York 1967.

G. Leugering

[1987:1] Optimal controllability in viscoelasticity of rate type, *Math. Methods Applied. Science* **8** (1986) 368–386.

[1987:2] Time optimal boundary controllability of a simple linear viscoelastic liquid, *Math. Methods Applied Science* **9** (1987) 413–430.

[1987:3] Exact boundary controllability of an integro-differential equation, *Applied Math. & Optimization* **15** (1987) 223–250.

X. Li and J. Yong

[1991]   Necessary conditions of optimal control for distributed parameter systems, *SIAM J. Control & Optimization* **29** (1991) 895–908.

[1995]   *Optimal Control Theory for Infinite Dimensional Systems,* Birkhäuser, Boston 1995.

X. Li and Y. Yao

[1981]   On optimal control for distributed parameter systems, Proc. IFAC 8th. Triennial World Congress, Kyoto (1981) 207–212.

[1985]   Maximum principle of distributed parameter systems with time lags, Springer Lecture Notes in Control and Information Sciences, vol. 75 (1985) 410–427.

F.-H. Lin
[1990]   A uniqueness theorem for parabolic equations, *Comm. Pure Appl. Math.* **43** (1990) 127–136.
[1991]   Nodal sets of solutions of elliptic and parabolic equations, *Comm. Pure Appl. Math.* **44** (1991) 287–308.
J. L. Lions
[1966]   Optimisation pour certaines classes d'équations d'évolution non linéaires, *Annali Mat. Pura Appl.* **82** (1966) 275–294.
[1968]   *Contrôle Optimal des Systèmes Gouvernés par des Équations aux Derivées Partielles,* Dunod/Gauthier-Villars, Paris, 1968.
[1969]   *Quelques Méthodes de Résolution des Problémes aux Limites non Linéaires,* Dunod/Gauthier-Villars, Paris, 1969.
[1990]   *Controllabilité Exacte, Perturbations et Stabilisation des Systèmes Distribués,* vols. I, II, Masson, Paris, 1990.
D. G. Luenberger [1984]   *Linear and Nonlinear Programming,* Addison-Wesley, Reading, 1984.
D. A. Lukes and D. L. Russell [1969]   Quadratic criterion for distributed systems, *SIAM J. Control* **7** (1969) 101–121.
A. A. Lyapunov [1940]   Sur les fonctions-vecteurs complètement additives, *Izv. Akad. Nauk SSSR, Ser. Mat.* **4** (1940) 465–478.
A. M. Lyapunov [1882]   On the general problem of stability of motion, Dissertation, Jarkov 1882. 1907 French translation reproduced by Princeton University Press, 1947.
L. A. Lyusternik [1934]   Conditional extrema of functionals, *Mat. Sbornik* **41** (1934) 390–401.
U. Mackenroth
[1980]   Some general considerations on optimality conditions for state constrained parabolic control problems, Springer Lecture Notes in Control and Information Sciences, vol. 22 (1980) 402–411.
[1981]   Time-optimal parabolic boundary control problems with state constraints, *Numer. Funct. Analysis & Optimization* **3** (1981) 285–300.
[1986:1] On some elliptic control problems with state constraints, *Optimization* **17** (1986) 595–607.
[1986:2] Convex parabolic control problems with pointwise state constraints, *J.Math. Anal. Appl.* **48** (1986) 102–116.
K. Malanowski
[1969]   On optimal control of the vibrating string, *SIAM J. Control* **7** (1969) 260–271.
[1984:1] Differential stability of solutions to convex, control constrained optimal control problems, *Appl. Math. Optimization* **12** (1984) 1–14.
[1984:2] On differentiability with respect to the parameter of solutions to convex optimal control problems subject to state space constraints, *Appl. Math. Optimization* **12** (1984) 231–245.
[1987:1] Stability and sensitivity of solutions to optimal control problems for systems with control appearing linearly, *Appl. Math. Optimization* **16** (1987) 73–91.
[1987:2] *Stability of Solutions to Convex Problems of Optimization,* Springer Lecture Notes in Control and Information Sciences, vol. 93, Heidelberg 1987.
[1992]   Second order conditions and constraint qualifications in stability and sensitivity analysis of solutions to optimal problems in Hilbert space, *Appl. Math. & Optimization* **25** (1992) 51–79.
K. Malanowski and J. Sokolowski [1986]   Sensitivity of solutions to convex, control constrained optimal control problems for distributed parameter systems, *J. Math. Anal. Appl.* **120** (1986) 240–263.
L. Markus [1994]   A brief history of control, Lecture Notes in Pure and Applied Mathematics, vol. 152, Marcel Dekker (1994) xxv-xl.
K. Masuda [1975]   On the stability of incompressible viscous fluid motions past objects, *J. Math. Soc. Japan* **27** (1975) 294–327.
H. Matano [1982]   Nonincrease of the lap-number of a solution for a one-dimensional semilinear parabolic equation *J. Fac. Sci. Tokyo Sect. IA Math.* **29** (1982) 401–441.
J. C. Maxwell [1867/8]   On governors, *Proc. Royal Soc. London* **16** (1867/8) 270–283.
R. C. McCamy, V. J. Mizel and T. I. Seidman [1968]   Approximate boundary controllability of the heat equation, *Jour. Math. Anal. Appl.* **23** (1968) 699–703.

E. J. McShane

[1939] On multipliers for Lagrange problems, *Amer. Jour. Math.* **61** (1939) 151–171.

[1978] The calculus of variations from the beginning through optimal control theory, *Optimal Control and Differential Equations* (A. B. Schwartzkopf, W. B. Kelley, S. B. Eliason, eds.) Academic Press (1978) 3–51. Reprinted (abridged, with changes) in *SIAM J. Control & Optimization* **27** (1989) 446–455.

M. Megan and V. Hiriş [1978] On the space of linear controllable systems in Hilbert space, *Glasnik Mat.* **10** (1975) 161–167.

T. Miyakawa [1981] On the initial value problem for the Navier-Stokes equation in L^p spaces, *Hiroshima Math. J.* **11** (1981) 9–20.

V. J. Mizel and T. I. Seidman [1969] Observation and prediction for the heat equation, *Jour. Math. Anal. Appl.* **28** (1969) 303–312.

B. Mordukhovich

[1990] Minimax design for a class of distributed control systems, *Automatics & Remote Control* **50** (1990) 1333–1340.

[1994] Sensitivity analysis for constraint and variational systems by means of set-valued differentiation, *Optimization* **31** (1994) 13–45.

B. Mordukhovich and B. S. Zhang [1997] Minimax control of parabolic systems with Dirichlet boundary conditions and state constraints, *Applied Math. & Optimization* **36** (1997) 323–360.

K. Naito

[1987] Controllability of semilinear control systems dominated by the linear part, *SIAM J. Control & Optimization* **25** (1987) 715–722.

[1989] Approximate controllability for trajectories of semilinear control systems, *J. Optimization Theory & Appl.* **60** (1989) 57–65.

K. Naito and T. I. Seidman [1991] Invariance of the approximately reachable set under nonlinear perturbations, *SIAM J. Control & Optimization* **29** (1991) 731–750.

T. Nambu

[1979] Remarks on approximate boundary controllability for distributed parameter systems of parabolic type: supremum norm problem, *J. Math. Anal. Appl.* **69** (1979) 194–204.

[1984] On the stabilization of diffusion equations: boundary observation and feedback, *J. Differential Equations* **52** (1984) 204–233.

K. Narukawa

[1981] Admissible null controllability and optimal control, *Hiroshima Math. J.* **11** (1981) 533–551.

[1982] Admissible controllability of vibrating systems with constrained controls, *SIAM J. Control & Optimization* **20** (1982) 770–782.

[1984] Complete controllability of one-dimensional vibrating systems with bang-bang controls, *SIAM J. Control & Optimization* **22** (1984) 788–804.

K. Narukawa and T. Suzuki [1986] Nonharmonic Fourier series and their applications, *Appl. Math. & Optimization* **14** (1996) 249–264.

I. P. Natanson [1957] *Theory of Functions of a Real Variable,* 2nd ed., Gostejizdat, Moscow, 1957. English translation [1955] Vol. I (Chapters 1–9 of 1st ed.) Ungar, New York 1955, [1960] Vol. II (Chapter 10–16, 18 of 2nd ed.), Ungar, New York 1960.

J. von Neumann [1931] Algebraische Repräsentanten der Funktionen "bis auf eine Menge von Masse Null", *J. de Crelle* **165** (1931) 109–115.

L. W. Neustadt [1966] An abstract variational theory applied to a broad class of optimization problems I, *SIAM J. Control* **4** (1966) 505–527, II, *SIAM J. Control* **5** (1967) 90–137.

H. Nyquist [1932] Regeneration theory, *Bell Systems Tech. Jour.* **11** (1932) 126–147.

Z. Páles and V. Zeidan [1994] First and second-order necessary conditions for control problems with constraints, *Trans. Amer. Math. Soc.* **346** (1994) 421–453.

R. E. A. C. Paley and N. Wiener [1934] *Fourier Transforms in the Complex Domain,* Amer. Math. Soc., New York 1934.

N. S. Papageorgiou

[1989] Properties of the relaxed trajectories of evolution equations and optimal control, *SIAM J. Control & Optimization* **27** (1989) 267–288.

[1991]   Existence of optimal controls for a class of nonlinear distributed parameter systems, *Bull. Austral. Math. Soc.* **43** (1991) 211–224.

[1993]   On the "bang-bang" principle for nonlinear evolution equations, *Aeq. Math.* **45** (1993) 267–280.

N. H. Pavel [1976/7] Invariant sets for a class of semi-linear equations of evolution, *Nonlinear Analysis, Theory & Appl.* **1** (1976/7) 187–196.

A. Pazy

[1975]   A class of semi-linear equations of evolution, *Israel J. Math.* **20** (1975) 23–36.

[1983]   *Semigroups of Linear Operators and Applications to Partial Differential Equations,* Springer, New York 1983.

G. Peichl and W. Schappacher [1986] Constrained controllability in Banach spaces *SIAM J. Control & Optimization* **24** (1986) 1261–1275.

V. I. Plotnikov and M. I. Sumin [1982] The construction of minimizing sequences in problems of control systems with distributed parameters, *Z. Vichysl. Mat. i Mat. Fiz.* **22** (1982) 49–56.

L. S. Pontryagin [1961] *Ordinary Differential Equations,* Gostejizdat, Moscow 1961. English translation: Addison-Wesley, New York 1962.

L. S. Pontryagin, V. G. Boltyanski, R. V. Gamkrelidze and E. F. Mischenko [1961] *The Mathematical Theory of Optimal Processes,* Gostejizdat, Moscow 1961. English translation: Wiley, New York, 1962.

C.-C. Poon [1996] Unique continuation for parabolic equations, *Comm. Partial Differential Equations* **21** (1996) 521–539.

A. J. Pritchard [1969] Stability and control of distributed parameter systems, *Proc. IEEE* **116** (1969) 1433–1438.

A. J. Pritchard and D. Salamon [1987] The linear-quadratic control problem for infinite dimensional systems with unbounded input and output operators, *SIAM J. Control & Optimization* **25** (1987) 121–144.

A. J. Pritchard and J. Zabczyk [1981] Stability and stabilizability of infinite dimensional systems, *SIAM Review* **23** (1981) 25–52.

J. Prüss and H. Sohr [1993] Imaginary powers of elliptic second order differential operators in L^p spaces, *Hiroshima Math. J.* **23** (1993) 161–192.

J.-P. Raymond [1996] Optimal control problems for semilinear parabolic equations with pointwise state constraints, *Modelling and Optimization of Distributed Parameter Systems: Applications to Engineering,* Chapman & Hall (1996) 216–222.

J.-P. Raymond and H. Zidani

[i.p:1]   Pontryagin's principle for state-constrained control problems governed by parabolic equations with unbounded controls.

[i.p:2]   Hamiltonian Pontryagin's principles for control problems governed by semilinear parabolic equations.

M. Reghis and V. Hiriş [1973] Controllabilité complète des équations différentielles linéaires dans des espaces de Hilbert, *Rev. Roum. Math. Pures Appl.* **7** (1973) 1091–1098.

F. Riesz and B. Sz.-Nagy [1955] *Functional Analysis,* Ungar, New York 1955.

L. Robbiano

[1988]   Dimensions des zéros d'une solution faible d'un opérateur elliptique, *J. Math. Pures Appl.* **67** (1988) 339–357.

[1991]   Théorème d'unicité adapté au contrôle des solutions des problèmes hyperboliques, *Comm. Partial Diff. Equations* **16** (1991) 789–800.

R. T. Rockafellar [1970] *Convex Analysis,* Princeton University Press, Princeton 1970.

H. H. Rosenbrock [1970] *State Space and Multivariable Theory,* Wiley, New York 1970.

T. Roubíček

[1986]   Generalized solutions of constrained optimization problems, *SIAM J. Control & Optimization* **24** (1986) 951–960.

[1987]   Optimal control of a Stefan problem with state-space constraints, *Numer. Math.* **50** (1987) 724–744.

[1989]   Stable extensions of constrained optimization problems, *Jour. Math. Anal. Appl.* **141** (1989) 120–135.

[1990] Constrained optimization: a general tolerance approach, *Aplikace Mat.* **35** (1990) 99–128.

[1997] *Relaxation in Optimization Theory and Variational Calculus*, De Gruyter, Berlin 1997.

E. J. Routh [1877] *Stability of a Given State of Motion,* London, 1877.

E. Roxin [1962] On the existence of optimal controls, *Michigan Math. J.* **9** (1962) 109–112.

W. Rudin [1966] *Real and Complex Analysis,* McGraw-Hill, New York 1966.

D. L. Russell

[1966] Optimal regulation of linear symmetric hyperbolic systems with finite dimensional controls, *SIAM J. Control* **4** (1966) 276–294.

[1967:1] Nonharmonic Fourier series in the control theory of distributed parameter systems, *J. Math. Anal. Appl.* **18** (1967) 542–560.

[1967:2] On boundary value controllability of linear symmetric hyperbolic systems, *Mathematical Theory of Control* (A. V. Balakrishnan and L. W. Neustadt, Eds.) Academic Press (1967) 312–321.

[1969] Linear stabilization of the linear oscillator in Hilbert space, *J. Math. Anal. Appl.* **25** (1969) 663–675.

[1971] Boundary value control theory of the higher-dimensional wave equation, *SIAM J. Control* **9** (1971) 29–42, II, *SIAM J. Control* **9** (1971) 401–419.

[1973:1] A unified boundary controllability theory for hyperbolic and parabolic partial differential equations, *Studies in Applied Math.* **52** (1973) 189–211.

[1973:2] Exact boundary controllability theorems for wave and heat processes in star-complemented regions, Lecture Notes in Pure and Applied Mathematics, vol. 10, Marcel Dekker (1974) 115–142.

[1978] Controllability and stabilizability theorems for linear partial differential equations: recent progress and open questions, *SIAM Review* **20** (1978) 639–739.

E. Sachs

[1974] Controllability for partial differential equations of parabolic type, *SIAM J. Control* **12** (1974) 389–399.

[1975] Observability and related problems for partial differential equations of parabolic type, *SIAM J. Control & Optimization* **13** (1975) 14–27.

[1978:1] A parabolic control problem with a boundary condition of the Stefan-Boltzmann type, *Z. Angew. Math. Mech.* **58** (1978) 443–449.

[1978:2] Differentiability in optimization theory, *Math. Operationsforsch. Statist., Ser. Optimization* **9** (1978) 497–513.

E. Sachs and E. H. P. G. Schmidt [1989] On reachable states in boundary control for the heat equation and an associated moment problem, *Appl. Math. & Optimization* **7** (1989) 225–232.

Y. Sakawa

[1964] Solution of an optimal control problem in a distributed-parameter system, *IEEE Trans. Automatic Control* **AC-9** (1964) 420–426.

[1966] On a solution of an optimization problem in linear systems with quadratic performance index, *SIAM J. Control* **4** (1966) 382–395.

W. Schmaedeke [1967] Mathematical theory of optimal control for semilinear hyperbolic systems in 2 independent variables, *SIAM J. Control* **5** (1967) 138–152.

E. J. P. G. Schmidt

[1980] Boundary control of the heat equation with steady state targets, *SIAM J. Control & Optimization* **18** (1980) 145–154.

[1986] Even more states reachable by boundary control for the heat equation, *SIAM J. Control & Optimization* **24** (1986) 1319–1322.

[1989] Boundary control of the heat equation with steady state targets, *J. Differential Equations* **78** (1989) 89–121.

E. J. P. G. Schmidt and R. Stern [1980] Invariance theory for infinite dimensional linear control systems, *Appl. Math. Optimization* **6** (1980) 113–122.

E. J. P. G. Schmidt and N. Weck [1978] On the boundary behavior of solutions to elliptic and parabolic equations—with applications to boundary control of parabolic equations, *SIAM J. Control & Optimization* **16** (1978) 593–598.

J. T. Schwartz [1969] *Nonlinear Functional Analysis,* Gordon & Breach, New York 1969.

L. Schwartz [1969] *Études des Sommes d'Exponentielles,* Hermann, Paris 1959.

I. Segal [1963]   Non-linear semigroups, *Ann. of Math.* **78** (1963) 339–364.

T. Seidman

[1976]   Boundary observation and control for the heat equation, *Calculus of Variations and Control Theory* (D. L. Russell, Ed.) Academic Press (1976) 321–351.

[1977]   Observation and prediction for the heat equation IV: Patch observability and controllability, *SIAM J. Control & Optimization* **15** (1977) 412–427.

[1978]   Exact boundary control for some evolution equations, *SIAM J. Control & Optimization* **16** (1978) 979–999.

[1979]   Time-invariance of the reachable set for linear control problems, *Jour. Math. Anal. Appl.* **72** (1979) 17–20.

[1984]   Two results on exact boundary control of parabolic equations, *Applied Math. & Optimization* **11** (1984) 145–152.

[1987]   Invariance of the reachable set under nonlinear perturbations *SIAM J. Control & Optimization* **25** (1987) 1173–1191.

[1988]   How violent are fast controls? *Math. Control Signals & Systems* **1** (1988) 89–95.

[1994]   Existence of optimal controls for some nonlinear systems, Lecture Notes in Pure and Applied Mathematics, vol. 165, Marcel Dekker (1994) 179–188.

T. Seidman and H.-X. Zhou [1982] Existence and uniqueness of optimal controls for a quasilinear parabolic equation, *SIAM J. Control & Optimization* **20** (1982) 747–762.

A. Seierstad [1996] Constrained retarded nonlinear optimal control problems in Banach state space, *Appl. Math. Optimization* **34** (1996) 191–230.

G. E. Silov [1961] *Mathematical Analysis,* Gostejizdat, Moscow 1961.

M. Slemrod

[1972]   The linear stabilization problem in Hilbert space, *J. Functional Analysis* **11** (1972) 334–345.

[1974:1] A note on complete controllability and stabilizability for linear control systems in Hilbert space, *SIAM J. Control* **12** (1974) 500–508.

[1974:2] Existence of optimal controls for control systems governed by nonlinear partial differential equations, *Ann. Scuola Normale Superiore Pisa Serie* IV **I** (1974) 229–246.

[1974:3] An application of maximal dissipative sets, *J. Math. Anal. Appl.* **46** (1974) 369–387.

[1976]   Stabilization of boundary control systems, *J. Diff. Equations* **22** (1976) 402–415.

[1991]   Dynamic of measure-valued solutions to a backward-forward heat equation, *J. Dynamics Diff. Equations* **3** (1991) 1–28.

P. E. Sobolevski

[1959]   On the non-stationary equations of hydrodynamics for viscous fluids, *Dokl. Akad. Nauk SSSR* **128** (1959) 45–48.

[1960]   On the smoothness of generalized solutions of the Navier-Stokes equations, *Dokl. Akad. Nauk SSSR* **131** (1960) 758–760.

[1964]   The use of fractional powers of operators in studying the Navier-Stokes equations, *Dokl. Akad. Nauk. SSSR* **155** (Russian) (1964) 57–60.

C. Sogge

[1990]   Strong uniqueness theorems for second order elliptic differential equations, *Amer. J. Math.* **112** (1990) 943–984.

[1993]   *Fourier integrals in classical analysis,* Cambridge University Press, New York, 1993.

J. Sokolowski

[1985]   Differential stability of solutions to constrained optimization problems, *Applied Math. Optimization* **13** (1985) 97–115.

[1987]   Sensitivity analysis of control constrained optimal control problems for distributed parameter systems, *SIAM J. Control & Optimization* **25** (1987) 1542–1556.

J. Sokolowski and J.-P. Zolésio [1992] *Introduction to Shape Optimization; Shape Sensitivity Analysis,* Springer, Berlin 1992.

V. A. Solonnikov [1977] Estimates for solutions of nonstationary Navier-Stokes equations, *J. Soviet Math.* **8** (1977) 467–529.

S. S. Sritharan

[1991]   Dynamic programming of the Navier-Stokes equations, *Systems & Control Letters* **16** (1991) 1212–1244.

[1994]   Optimal feedback control of hydrodynamics, *Flow Control,* IMA Volumes in Mathematics and its Applications, vol. 68, Springer (1994) 1–18.

T. Staib and M. Dobrowolski [1995] Optimality conditions for state constrained nonlinear control problems, *Math. Methods Oper. Research* **42** (1995) 107–116.

E. M. Stein and G. Weiss [1971] *Introduction to Fourier Analysis in Euclidean Spaces,* Princeton University Press, Princeton 1971.

H. B. Stewart [1974] Generation of analytic semigroups by strongly elliptic operators, *Trans. Amer. Math. Soc.* **199** (1974) 141–162.

D. Tataru

[1992]   Boundary value problems for first order Hamilton-Jacobi equations, *Nonlinear Analysis* **19** (1992) 1091–1110.

[1994]   A priori estimates of Carleman's type in domains with boundary, *J. Math. Pures Appl.* **73** (1994) 355–387.

D. Tiba [1990] *Optimal Control of Nonsmooth Distributed Parameter Systems,* Springer Lecture Notes in Mathematics, vol. 149, Heidelberg 1990.

D. Tiba and F. Tröltzsch [1996] Error estimates for the discretization of state constrained convex control problems, *Num. Functional Analysis & Optimization* **17** (1996) 1005–1028.

A. Tolstonogov [1991] Extreme continuous selectors of multivalued maps and the bang-bang principle for evolution inclusions, *Dokl. Akad. Nauk SSSR* **317** (1991) 589–594.

L. Tonelli [1921] *Fondamenti del Calcolo delle Variazioni,* vol. I, Zanichelli, Bologna 1921.

P. Topuzu and E. Topuzu [1979] Asupra controlabilității sistemelor liniare in spații Banach, *Bul. Ştiinţific sj Technic Inst. Pol. "Traian Vuia", Timişoara* **24** (1979) 40–44.

J. S. Treiman [1983] Characterization of Clarke's tangent and normal cones in infinite dimensions, *Nonlinear Analysis* **7** (1983) 771–783.

H. Triebel [1978] *Interpolation Theory, Function Spaces, Differential Operators,* North-Holland, Amsterdam 1978.

R. Triggiani

[1975:1]   Controllability and observability in Banach space with bounded operators, *SIAM J. Control* **13** (1975) 462–491.

[1975:2]   On the lack of exact controllability for mild solutions in Banach spaces, *J. Math. Anal. Appl.* **50** (1975) 438–446.

[1975:3]   Pathological asymptotic behavior of control systems in Banach space, *Jour. Math. Anal. Appl.* **49** (1975) 411–429.

[1976]   Extensions of rank conditions for controllability and observability to Banach spaces and unbounded operators, *SIAM J. Control & Optimization* **14** (1976) 313–338.

[1977]   A note on the lack of exact controllability for mild solutions in Banach spaces, *SIAM J. Control & Optimization* **15** (1977) 407–411. Addendum: *SIAM J. Control & Optimization* **18** (1980) 98–99.

[1979]   On Nambu's boundary stabilizability problem for diffusion processes, *J. Differential Equations* **33** (1979) 189–200.

[1997]   Carleman estimates and exact boundary controllability for a system of coupled non-conservative Schrödinger equations, *Rend. Ist. Mat. Univ. Trieste Suppl.* **28** (1997) 453–504.

F. Tröltzsch

[1982]   On some parabolic boundary control problems with constraints on the control and functional controls on the state, *Z. Anal. Anwendungen* **1** (1982) 1–13.

[1984]   *Optimality Conditions for Parabolic Control Problems and Applications,* Teubner, Leipzig 1984.

[1989]   On the semigroup approach for the optimal control of semilinear parabolic equations including distributed and boundary control, *Z. Anal. Anwendungen* 8 (1989) 431–443.

K. Tsujioka [1970] Remarks on controllability of second order evolution equations in Hilbert spaces, *SIAM J. Control* **8** (1970) 90–99.

F. Valentine [1937] The problem of Lagrange with differential inequalities as added side conditions, Contributions to the Calculus of Variations 1933–37, University of Chicago Press (1937) 407–448.

R. B. Vinter and T. L. Johnson [1977] Optimal control of nonsymmetric hyperbolic systems in n variables in the half-space, *SIAM J. Control & Optimization* 17 (1977) 129–143.

R. B. Vinter and P. D. Loewen [1987] Pontryagin-type necessary conditions for differential inclusion problems, *Systems & Control Letters* 9 (1987) 263–265.

M. J. Vishik and A. V. Fursikov [1988] *Mathematical Problems in Statistical Hydromechanics*, Kluwer, Boston 1988.

J. Vyshnegradski [1876] Sur la théorie générale des régulateurs, *Compt. Rend. Acad. Sci. Paris* 833 (1876) 318–321.

W. von Wahl

[1980]   Regularity questions for the Navier-Stokes equations, Springer Lecture Notes in Mathematics, vol. 771 (1980) 538–542.

[1985]   *The Equations of Navier-Stokes and Abstract Parabolic Equations,* Vieweg & Sohn, Braunschsweig 1985.

P. K. C. Wang

[1964:1] Optimum control of distributed parameter systems with time delays, *IEEE Trans. Automatic Control* **AC-9** (1964) 13–22.

[1964:2] Control of distributed parameter systems, *Advances in Control Systems,* Academic Press (1964) 75–172.

[1966]   On the feedback control of distributed parameter systems, *Int. Jour. Control* 3 (1966) 255–273.

[1975]   Time optimal control of time-lag systems with time-lag controls, *J. Math. Anal. Appl.* **52** (1975) 366–378.

[1986]   On a class of nonlinear distributed systems with linear finite-dimensional feedback, *J. Math. Anal. Appl.* **114** (1986) 433–449.

J. Warga

[1962:1] Relaxed variational problems, *J. Math. Anal. Appl.* **4** (1962) 11–128.

[1962:2] Necessary conditions for minimum in relaxed variational problems, *J. Math. Anal. Appl.* **4** (1962) 129–145.

[1967]   Functions of relaxed controls, *SIAM J. Control* **5** (1967) 628–641.

[1970]   Control problems with functional restrictions, *SIAM J. Control* **8** (1970) 372–382.

[1972]   *Optimal Control of Differential and Functional Equations,* Academic Press, New York 1972.

D. Washburn [1979] A bound on the boundary input map for parabolic equations with applications to time optimal control, *SIAM J. Control & Optimization* 17 (1979) 652–671.

T. Ważewski

[1962:1] Sur une géneralisation de la notion des solutions d'une équation au contingent, *Bull. Acad. Sci. Polon. Ser. Sci. Math. Astronom. Phys.* 10 (1962) 11–15.

[1962:2] Sur les systèmes de commande non linéaires dont le contredomaine de commande n'est pas forcemment convexe, *Bull. Acad. Sci. Polon. Ser. Sci. Math. Astronom. Phys.* 10 (1962) 17–21.

N. Weck [1984] More states reachable by boundary control for the heat equation, *SIAM J. Control & Optimization* 22 (1984) 699–710.

L. W. White [1983] Control of a hyperbolic problem with pointwise state constraints, *J. Optimization Theory & Appl.* 41 (1983) 359–369.

D. M. Wiberg [1967] Feedback control of linear distributed parameter systems, *Trans ASME, J. Basic Engineering* 89 (1967) 379–384.

L. von Wolfersdorf [1976/7] Optimal control processes governed by mildly nonlinear differential equations of parabolic type I, II, *Z. Angew. Math. Mech.* 56 (1976) 531–538, 57 (1977) 11–17.

W. M. Wonham

[1967]   On pole assignment in multi-input controllable linear systems, *IEEE Trans. AC* 12 (1967) 660–665.

[1979]   *Linear Multivariable Control: a Geometric Approach,* Springer, Berlin 1979.

M. Yamamoto [1991] Feedback stabilization for nondissipative hyperbolic equations, International Series of Numerical Mathematics, vol. 100, Birkhäuser (1991) 379–389.

Y. Yao [1985] The regulator pointwise control for parabolic systems, *Proc. 9th. IFAC Triennial World Congress,* Budapest (1985) 46–49.

Y. Yao and Y. You [1985] Closed-loop optimal solutions for some parabolic systems with pointwise control *J. Appl. Science* **3** (1985) 223–232.

J. Yong [1991] Time-optimal control for semilinear distributed parameter systems — existence theory and necessary conditions, *Kodai Math. J.* **14** (1991) 239–253.

J. Yong [1992] Existence theory of optimal control for distributed parameter systems, *Kodai Math. J.* **15** (1992) 193–220.

K. Yosida [1978] *Functional Analysis,* Springer, Berlin 1978.

K. Yosida and E. Hewitt [1952] Finitely additive measures, *Trans. Amer. Math. Soc.* **72** (1952) 46–66.

Y. You

[1991]   Boundary stabilization of two-dimensional Petrovski equation: vibrating plate, *Diff. & Integral Equations* **4** (1991) 617–638.

[1992]   Pointwise boundary stabilization of hyperbolic evolution equations: two-dimensional hybrid elastic structures, *J. Math. Anal. Appl.* **165** (1992) 239–265.

L. C. Young

[1937]   Generalized curves and the existence of an attained absolute minimum in the calculus of variations, *C. R. Sci. Lettres Varsovie C* III **30** (1937) 212–234.

[1969]   *Lectures on the Calculus of Variations and Optimal Control Theory,* W. B. Saunders, Philadelphia, 1969.

C. Zălinescu [1984] Continuous dependence on data in abstract control problems, *J. Optimization Theory & Appl.* **43** (1984) 277–305.

H.-X. Zhou

[1982]   A note on approximate controllability for semilinear one-dimensional heat equation, *Applied Math. & Optimization* **8** (1982) 275–285.

[1983]   Approximate controllability for a class of semilinear abstract equations, *SIAM J. Control & Optimization* **21** (1983) 551–565.

[1984]   Controllability properties of linear and semilinear abstract control systems, *SIAM J. Control & Optimization* **22** (1984) 405–422.

J.-P. Zolésio [1990] Identification of coefficients with bounded variation in the wave equation, Springer Lecture Notes in Control and Information Sciences, vol. 147 (1990) 248–254.

J. Zowe and S. Kurcyusz [1979] Regularity and stability for the mathematical programming problem in Banach spaces, *Appl. Math. & Optimization* **5** (1979) 49–62.

NOTATION AND SUBJECT INDEX